2nd International Conference on Science (ICOS 2017)

Journal of Physics: Conference Series Volume 979

Makassar, Indonesia
2 - 3 November 2017

Editor:

Dahlang Tahir

ISBN: 978-1-5108-5989-0
ISSN: 1742-6588

Printed from e-media with permission by:

Curran Associates, Inc.
57 Morehouse Lane
Red Hook, NY 12571

Some format issues inherent in the e-media version may also appear in this print version.

Printed by Curran Associates, Inc. (2018)

For permission requests, please contact the Institute of Physics
at the address below.

Institute of Physics
Dirac House, Temple Back
Bristol BS1 6BE UK

Phone: 44 1 17 929 7481
Fax: 44 1 17 920 0979

techtracking@iop.org

Additional copies of this publication are available from:

Curran Associates, Inc.
57 Morehouse Lane
Red Hook, NY 12571 USA
Phone: 845-758-0400
Fax: 845-758-2633
Email: curran@proceedings.com
Web: www.proceedings.com

Table of contents

Volume 979

The 2nd International Conference on Science (ICOS)
2–3 November 2017, Makassar, Indonesia

Accepted papers received: 20 February 2018
Published online: 13 March 2018

Preface

Papers

Bioscience and Bioinformatics

Chemistry

Computational Science

Mathematics

Physics

IOP Conf. Series: Journal of Physics: Conf. Series **979** (2018) 011001 doi:10.1088/1742-6596/979/1/011001

Preface

Following the successful of the 1st International Conference on Science (ICOS), held by Faculty of Mathematics and Natural Sciences, Hasanuddin University, we have organised the 2nd International Conference on Science, ICOS 2017 in November 2017.

The 2nd International Conference on Science (ICOS) is a conference organized by Faculty of Mathematics and Natural Sciences, Hasanuddin University, Indonesia in cooperation with Indonesian Mathematical Society Sulawesi region (IndoMS), Indonesian Physical Society (HFI), Indonesian Association of Geophysicists (HAGI) and Indonesian Biology Consortium (KOBI). It is a three-year conference activity of the Faculty of Mathematics and Natural Science, started three years back on 19th – 20th of November 2014.

The objectives are to bring together scholars, researchers and experts from diverse backgrounds and applications areas. Special emphasis is placed on promoting interaction between the theoretical, experimental, and case studies, so that a high level exchange in new and emerging areas within Physics, bioscience, computational science and mathematics in science and technology.

The keynote presentations are provided especially to show the contribution of the various fields of sciences, with the main theme "Science for Sustainable Development and Better Quality of Life". We have eight keynote speakers coming from The University of Melbourne, Australia, Prof Alexander Babanin, Directorate Research and Community Services, Ministry of Research and Technology, Higher Education, Indonesia, Prof Ocky Karna Rajadsa, United Nations, Pulse Lab Jakarta, Dr Jong Gun Lee, The University of Tokyo, Jepang, Prof Koji Inoue, La Trobe University, Australia, Assoc Prof Agus Salim, Universiti Kebangsaan Malaysia, Malaysia, Prof Mohammad B Kassim, Geophysical Adviser, PT Pertamina Indonesia, Dr Alvius Dwi Guntara and Universitas Riau, Indonesia, Erman Taer, Ph.D.

Sri Astuti Thamrin (Chairman) and Dahlang Tahir (Editor in Chief)*
The 2nd International Conference on Science (ICOS 2017) Publication
Faculty of Mathematics and Natural Sciences
Hasanuddin University
Tamalanrea, Makassar, 90245, Indonesia
*E-mail: dtahir@fmipa.unhas.ac.id

IOP Conf. Series: Journal of Physics: Conf. Series **979** (2018) 011002 doi:10.1088/1742-6596/979/1/011002

The Committees

The Second International Conference on Science (ICOS 2017)

Organizing Committee

Sri Astuti Thamrin	Chairman
Fredryk Mandey	Co-Chair
Naimah Aris	Treasurer
Sulfahri	Secretary
Kasbawati	Secretariat Member
Andi Masniawati	Secretariat Member
Agustinus Ribal	Secretariat Member

Advisory Committee

Amiruddin	Dean of Faculty of Mathematics and Natural Sciences
Moh Ivan Azis	Vice-Dean for Academic Affairs
Muhammad Zakir	Vice-Dean for General Administration, Finance, and Human Resources
Andi Ilham Latunra	Vice-Dean for Student Affairs and Alumni Relations

Steering Committee

Alexander Babanin (the University of Melbourne, Australia)
Koji Inoue (The University of Tokyo, Japan)
Mohammad . B. Kassim (Universiti Kebangsaan Malaysia, Malaysia)
Jong Gun Lee (United Nations, Pulse Lab Jakarta, Indonesia)
Agus Salim (La Trobe University, Australia)
Erman Taer (University of Riau, Indonesia)
Alfian Noor (Hasanuddin University, Indonesia)
Moh Ivan Azis (Hasanuddin University, Indonesia)
Dadang Suriamihardja (Hasanuddin University, Indonesia)
Alpius Dwi Guntara (Geophysical PT Pertamina)
Ocky Karna Radjasa (Ministry of Research, Technology and Higher Education, Indonesia)

Scientific and Reviewer Committee

Dahlang Tahir (Hasanuddin University)
Isnaeni (Indonesian Institute of Sciences)
Halmar Halide (Hasanuddin University)
Taufik Sutanto (Syarif Hidayatullah State Islamic University)
Paulina Taba (Hasanuddin University)

Misita Anwar (Monash University, Australia)
Utami Dyah Syafitri (Bogor Agricultural University)
Junaidi (Tadulako University)
Sri Astuti Thamrin (Hasanuddin University)
Achmad Effendi (Brawijaya University)
Mawardi Bahri (Hasanuddin University)

Sponsor or funding acknowledgements

We would like to thank our main sponsor, who has supported the 2nd International Conference on Science (ICOS 2017).

Conference Photographs

The Second International Conference on Science (ICOS 2017)

Peer review statement

All papers published in this volume of *Journal of Physics: Conference Series* have been peer reviewed through processes administered by the proceedings Editors. Reviews were conducted by expert referees to the professional and scientific standards expected of a proceedings journal published by IOP Publishing.

IOP Conf. Series: Journal of Physics: Conf. Series **979** (2018) 012001 doi:10.1088/1742-6596/979/1/012001

Thermostable α-Amylase Activity from Thermophilic Bacteria Isolated from Bora Hot Spring, Central Sulawesi

F M Gazali and I N Suwastika

Department of Biology, Faculty of Sciences, Tadulako University, Jl. Soekarno-Hatta Km. 9, Palu 94118, Indonesia

E-mail: farismuhammadgazali@gmail.com

Abstract. α-Amylase is one of the most important enzyme in biotechnology field, especially in industrial application. Thermostability of α-Amylase produced by thermophilic bacteria improves industrial process of starch degradation in starch industry. The present study were concerned to the characterization of α-Amylase activity from indigenous thermophilic bacteria isolated from Bora hot spring, Central Sulawesi. There were 18 isolates which had successfully isolated from 90°C sediment samples of Bora hot spring and 13 of them showed amylolytic activity. The α-Amylase activity was measured qualitatively at starch agar and quantitatively based on DNS (3,5-Dinitrosalicylic acid) methods, using maltose as standard solution. Two isolates (out of 13 amylolytic bacteria), BR 002 and BR 015 showed amylolytic index of 0.8 mm and 0.5 mm respectively, after being incubated at 55°C in the 0.002% Starch Agar Medium. The α-Amylase activity was further characterized quantitatively which includes the optimum condition of pH and temperature of α-Amylase crude enzyme from each isolate. To our knowledge, this is the first report on isolation and characterization of a thermostable α-Amylase from thermophilic bacteria isolated from Central Sulawesi particularly from Bora hot spring.

1. Introduction

α-Amylase (E.C. 3.2.2.1.) is a hydrolase enzymes belongs to family 13 (GH-13) of glycoside hydrolase, which catalyzes the hydrolysis of internal α-1,4-glycosidic bonds in starch to yield products like glucose, maltose, and other oligosaccharides units [1,2]. These enzymes can be produced by plants, animals, and microbes [3]. The most widely used α-Amylase in industrial application is α-Amylase coming from microbes [4]. It possesses some advantages such as cost effectiveness, large productivity, consistency, less time and space needs for production, and easy to modify and optimize the process of production [1].

α-Amylase plays an important role in biotechnology field, particularly for industrial needs [5]. It constitutes a class of industrial enzymes having approximately 25-30% of the world enzyme market [6]. Various industry had been known to use α-Amylase in their industrial process ranging from starch, detergent, fuel alcohol, food, textile to paper industry [1].

The most widespread applications of α-Amylase are in the starch industry [1]. Starch industry needs a thermostable α-Amylase in the starch processing as the process of gelatinization, liquefaction, and saccharification are performed in high temperature [7]. Thermostable α-Amylase could be isolated from mesophilic microbes, thermophilic microbes as well as hyperthermophilic microbes. These microbes possess unique properties of α-Amylase, including specificity, thermostability, and pH response which are a critical properties for industrial use [8]. Screening of microbes with high α-Amylase activity, high thermostability, proper optimum condition of pH and temperature could therefore facilitate the discovery of newly suitable α-Amylase for industrial application [9].

Hot spring is one of the habitat of thermophilic bacteria which offers great potential for isolating a thermostable α-Amylase [7,10]. Bora hot spring is one of the hot spring located at Central Sulawesi with water temperature up to 90,1°C [11]. Therefore, it is mostly possible to isolate the bacteria which could produce highly thermostable α-Amylase from that hot spring.

Little is known about the properties of the α-Amylase produced by indigenous thermophilic bacteria from Indonesia. As far as our concern, this is the first report regarding the isolation and characterization of α-Amylase activity produced by indigenous thermophilic bacteria isolated form Bora hot spring, Central Sulawesi. Hence, it is necessary to know character of these local α-Amylase and its potential to be implemented in industrial sector.

2. Methods

2.1. Sample collection and Isolation of thermophilic bacteria
Sediment samples were collected from Bora hot spring, Central Sulawesi, Indonesia. Sediment samples were homogenized by continuously mixing 10 g of sediment samples with 90 ml of 0.9% NaCl solution in 150 rpm of orbital shaker at 55°C for 30 minutes. A serial dilution technique was performed up to 10^{-9} in dilluting the samples. Luria Bertani Agar was used as an isolation and cultivation media of thermophilic bacteria isolates which incubated at 55°C for 24 hours to obtain the colonies.

2.2. Screening of amylase producing thermophilic bacteria
Starch Agar plate was used as a media for screening of amylase producing thermophilic bacteria. The pure isolate colonies were inoculated and streaked on 0.002% starch agar plates and incubated at 55°C for 24 hours. After incubation, the plates were flooded by Gram's Iodine (2% I_2 and 0.2% KI) to produce starch-iodine complex which is visualized by deep blue color. No color would be appeared at starch-free zone which is known to be the zone of clearance. The appearance of clear zone at starch agar plate was used as the main indicator of the activity of starch degrading enzyme produced by the bacterial strain. The bacterial strains which produced clear zone at starch agar plate were further characterized.

2.3. Amylolytic index value determination
Amylase producing bacteria colonies were sub-cultured at Luria Bertani Broth media and incubated at 55°C for 24 hours. After incubation, 10 µl of bacterial culture was inoculated to filter paper at 0.002% starch agar plate. The agar plates were incubated at 55°C for 48 hours. After incubation, the plates were flooded by Gram's Iodine (2% I_2 and 0.2% KI) and the amylolytic index value of each isolate was determined based on the ratio of clear zone diameter (mm) formed by the bacterial colony and its diameter of colony (mm) [12]. These were performed in triplicates and the standard error were determined based on the deviation number obtained. Bacterial isolates in high amylolytic index value was further characterized.

2.4. Crude enzyme production
Two isolate were initially sub-cultured at basal medium containing 1% starch according to Asgher *et al.* (2007) [9] and incubated at 55°C for 48 hours without shaking. After incubation, 5 ml of the cultures was transferred to 45 ml sterile basal medium and incubated at 55°C for 48 hours on 130 rpm of orbital shaker. The crude enzyme of each isolate was obtained after centrifugation process of 48 hours cultures. The cultures were centrifuged at 8,000 rpm on 4°C to sediment the cell. Cells free supernatant was used as the crude enzyme for enzyme assay. The enzyme assays were performed immediately after centrifugation process to prevent the lost activity of enzyme.

2.5. Enzyme assay

α -Amylase activity was determined by measuring the increase of reducing sugar as maltose formed by the hydrolysis of starch. The quantity of the releasing reducing sugar was determined by 3,5-dinitrosalicylic acid method according to Bernfeld (1955) [13]. A 1.0% (w/v) solution of starch in 50 mM Phosphate Buffer (pH 7.0) was used as a substrate. The reaction mixture (100 μl), containing 50 μl crude enzymes and 50 μl substrate was incubated at 55°C for 15 minutes. After incubation, 50 μl of 3,5-dinitrosalicylic acid was added to stop reaction. This reaction mixture was boiled for 15 minutes and diluted 4 times in dH_2O. The blank was made by adding 3,5-dinitrosalicylic acid before the crude enzyme. Also, the control was made by adding 3,5-dinitrosalicylic acid before the substrate. The absorbance was measured at 540 nm, where one unit of α-Amylase enzyme corresponded to the formation of 1 μmol reducing sugar as maltose per minute in 50 mM Phosphate Buffer (pH 7.0) with 1% soluble starch at 55°C. These were performed in duplicates and the standard error were determined based on the deviation number obtained.

2.6. Determination of optimum temperature and pH of the crude enzyme

Effect of temperature on α-Amylase activity was determined by performing enzyme assay at a various temperature ranging on 32°C - 80°C. Effect of pH on α-Amylase activity was determined by performing enzyme assay at various pH ranging on 5 - 8.6 using 50 mM citrate buffer, 50 mM phosphate buffer, and 50 mM tris-HCl buffer. The optimum condition (of both temperature and pH) was determined by polynomial regression equation with derivative = 0 using software Microsof Excel 2016 where the peak of the curve showed the maximum point of both temperature and pH.

3. Result and Discussion

Eighteen thermophilic bacteria were successfully isolated from sediment samples, collected from Bora hot spring, Central Sulawesi, Indonesia. Out of 18 isolates, 13 isolates showed amylolytic activity at 0.002% starch agar plates. Two isolate, BR 002 and BR 015 were further described throughout this paper, due to their consistence in showing high amylolytic activity on early selection using starch agar plates (figure 1). The amylolytic indexes (ratio of clear zone diameter to colony diameter) of BR 002 and BR 015 were 0.8 mm and 0.5 mm respectively (table 1).

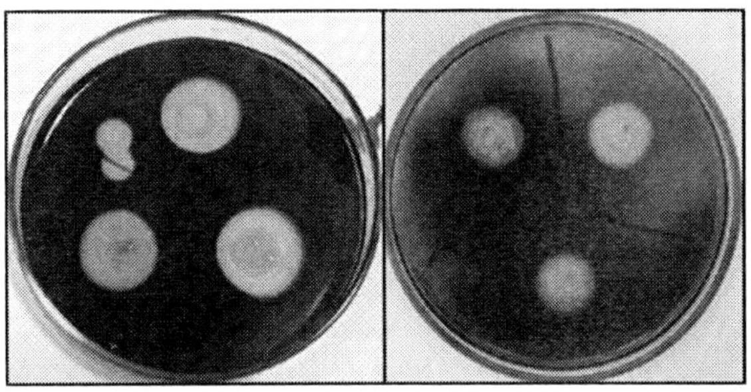

Figure 1. Zone of clearance on starch agar plates from BR 002 isolates (*left*) and BR 015 isolates (*right*).

The enzyme activity was further characterized quantitatively using 3.5-dinitrosalicylic acid method. α-Amylase activity from both isolates, BR 002 and BR 015 were around 2 U/ml and 3.7 U/ml, respectively (table 1). The data showed that α-Amylase activity from both isolates didn't correspond to the value of amylolytic index. The same works were carried out by Goyari *et al.* (2014) [14] which described the relative enzyme index on plate agar was not corresponded to the enzyme activity.

Table 1. Comparative analysis of amylolytic activities of both thermophilic bacteria isolates.

No.	Bacterial Isolates	Amylolytic Index (mm) ± SE[a]	Enzyme Activity (U/ml) ± SE
1	BR 002	0.8 ± 0.05	2.0 ± 0.3
2	BR 015	0.5 ± 0.03	3.7 ± 0.2

[a]Standard Error

Determination effect of temperature on α-Amylase activity was shown in figure 2. The α-Amylase activity of BR 002 was found to increase at temperature ranging on 32°C to 52°C and start to significantly decrease at 55°C to 80°C where no activity of α-Amylase was observed. The optimum temperature of the α-Amylase activity of BR 002 was around 50°C with enzyme activity around 2 U/ml. Meanwhile, the α-Amylase activity of BR 015 was found to increase at temperature ranging on 32°C to 69°C and start to decrease after 69°C. The optimum temperature of the α-Amylase activity of BR 015 is around 70°C with enzyme activity around 6.4 U/ml. Hence, it was explained that the α-Amylase activity from BR 015 isolates was found to have broader range of active temperature as well as higher optimum temperature compare with BR 002 isolates. However, the optimum temperature exceeding 50°C from both isolates confirm the suitable character needed by most of the industrial sector as had been mentioned by Goyal *et al* (2005) [15].

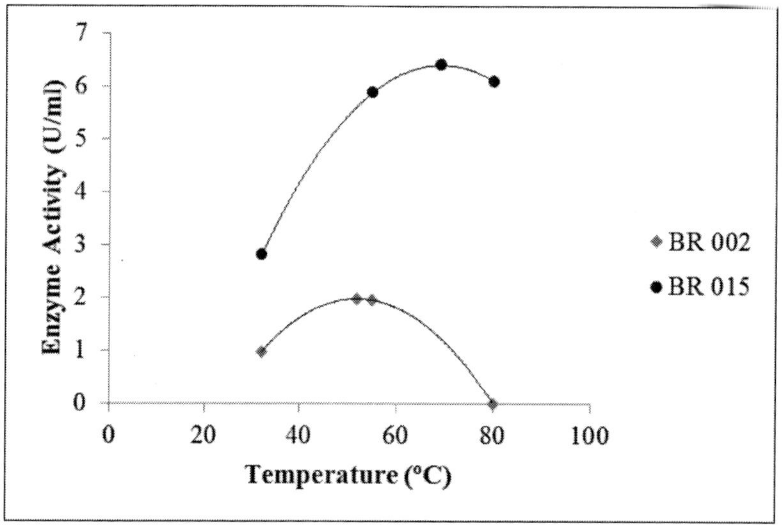

Figure 2. Effect of temperature on α-Amylase activity.

The effect of pH on α-Amylase activity was shown in Figure 3. The effect of pH on α-Amylase activity of BR 002 was found to be similar with BR 015. No enzyme activity to a very slight enzyme activity was shown at pH 5 in α-Amylase from both isolates. The enzyme activity from both isolates starts to significantly increase until it reaches pH around 8.0 and begin to decrease afterwards. Hence, the optimum pH of both BR 002 and BR 015 isolates was around 8.0 with enzyme activity 2.2 U/ml and 6.5 U/ml respectively. It was suggested that, both isolates had optimum pH at slightly basic. The same works were performed by Sen *et al.* (2016) [16] which had characterized optimum pH of α-Amylase at slightly basic produced by thermophilic bacteria isolated from hot spring.

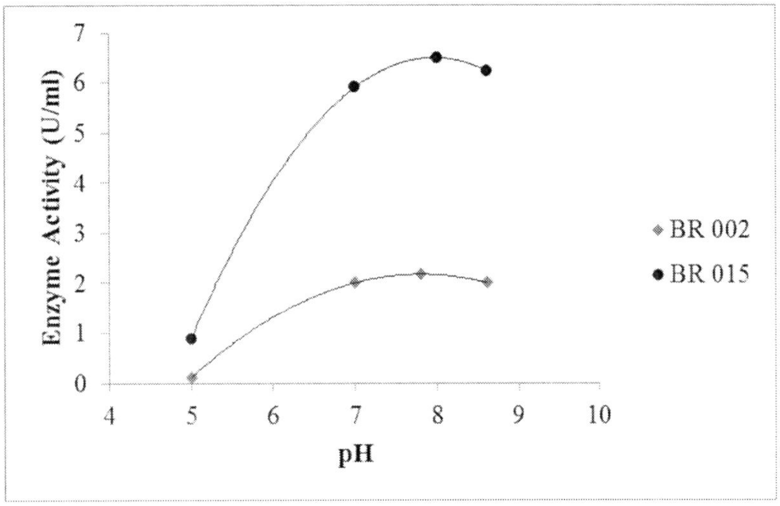

Figure 3. Effect of pH on α-Amylase activity.

The α-Amylase activity of both isolates, particularly when it was in the optimum condition, showed good potential for the application in industrial sector. However, it was still far from the real implementation regarding the α-Amylase activity which was relatively low compare with the α-Amylase activity of *Bacillus* sp. (up to 57 U/ml) [17] or *Bacillus subtilis* (up to 100 U/ml) which commonly had already been implemented in industrial sector [9]. Hence, the α-Amylase activity of both isolates should reach at least more than 57 U/ml (α-Amylase activity of *Bacillus* sp.) when it was planned to be implemented in industrial sector. However, this limitation could be solved by performing further characterization such as enzyme purification, determination of enzyme stability against temperature and pH, effect of cofactor, effect of inhibitor, effect of substrate concentration, and optimum enzyme production as those properties would affect the activity of the enzyme. By knowing the exact properties of the enzyme, the optimization process of enzyme including the improvement in enzyme activity could obviously be performed. More than that, here we show the potential genetic material from the indigenous bacteria which could be improved on its enzyme activity by such as advance genetic engineering.

4. Conclusion

The crude α-Amylase from indigenous thermophilic bacteria isolate has high optimum temperature and pH where the other enzyme would denature. However, these characteristics were not enough to use the enzyme in industrial application because of the low of enzyme activity. Furthermore, further characterization should be performed in order to know the exact properties of the enzyme.

Acknowledgements

The authors would like to thank Biology Department, Faculty of Sciences, Tadulako University for facilitating this work. Also, the author would like to thank Meryany Ananda for the valuable suggestion during the research. The authors also wish to thank Ms. Sami Bukang and all laboratory staffs at Biology Department, Faculty of Sciences, Tadulako University.

References

[1] Souza P M de and Magalhães P de O e 2010 *Brazilian J. Microbiol.* **41** 850–61
[2] Sivaramakrishnan S, Gangadharan D, Madhavan K, Soccol C R, Pandey A, Nampoothiri K M, Soccol C R and Pandey A 2006 *Food Technol. Biotechnol.* **44** 173–84
[3] Sundarram A and Murthy T P K 2014 *J. Appl. Environ. Microbiol.* **2** 166–75
[4] Reddy N S, Nimmagadda A and Rao K R S S 2003 *African J. Biotechnol.* **2** 645–8
[5] Pandey A, Nigam P, Soccol C R, Soccol V T, Singh D and Mohan R 2000 *Biotechnol. Appl.*

IOP Conf. Series: Journal of Physics: Conf. Series **979** (2018) 012001 doi:10.1088/1742-6596/979/1/012001

Biochem. **31** 135
[6] Deb P, Talukdar S A, Mohsina K, Sarker P K and Sayem S A 2013 *Springerplus* **2** 154
[7] Mehta D and Satyanarayana T 2016 *Front. Microbiol.* **7** 1–21
[8] Vieille C, Zeikus G J and Vieille C 2001 *Microbiol. Mol. Biol. Rev.* **65** 1–43
[9] Asgher M, Asad M J, Rahman S U and Legge R L 2007 *J. Food Eng.* **79** 950–5
[10] Madigan M T and Marrs B L 1997 *Sci. Am.* **276** 82–7
[11] Idral A and Mansoer W R 2015 *Proc. World Geothermal Congress* 19-25 April 2015 Melbourne Australia pp 1–5
[12] Jamilah I, Meryandini A, Rusmana I, Suwanto A and Rachmania Mubarik N 2009 *Microbiol. Indones.* **3** 67–71
[13] Bernfeld P 1955 *Methods Enzymol.* **1** 149–58
[14] Goyari S, Devi S S, Kalita M C and Talukdar N C 2014 *Springerplus* **3** 700
[15] Goyal N, Gupta J K and Soni S K 2005 *Enzyme Microb. Technol.* **37** 723–34
[16] Sen S K, Jana A, Bandyopadhyay P, Das Mohapatra P K and Raut S 2016 *Sustain. Chem. Pharm.* **3** 59–68
[17] Cordeiro C A M, Martins M L L and Luciano A B 2002 *Brazilian J. Microbiol.* **33** 57–61

The potential for Probiotic Bacteria from milkfish intestine in reducing mercury metals in skimmed milk media

Zaraswati Dwyana, D Priosambodo, N Haedar, A E Erviani, A K Djabura and R Sukma

Department of Biology, Faculty of Mathematics and Natural Sciences, Hasanuddin University, South Sulawesi 90245, Indonesia

E-mail: zaraswatidwyana@unhas.ac.id

Abstract. Mercury (Hg) is one of the heavy metals that is harmful to humans. The accumulation of mercury in the body is generally derived from food. Several types of bacteria from intestine of milkfish are known to reduce mercury concentration. People can take advantage of this bacterial ability by eating it through probiotic foods. This research conducted to figure out the potential for probiotic bacteria from milkfish intestine in reducing mercury. Isolation from probiotic bacteria from milkfish intestine conducted with grown the isolates in MRSA medium with addition of 1% $CaCO_3$. Twelve isolate were obtained from milkfish intestine. Mercury resistance tested was performed by measuring cell density using a spectrophotometer at concentrations of 10, 15 and 20 ppm respectively in skim milk media. Probiotic tests (gastric acid, bile salts and antimicrobial activity) for MRSB media was also conducted. Results showed that seven isolate were resistant to mercury in all concentrations and potential as probiotics. All resistant isolate then tested for skim milk media with addition of 5, 10, 20 ppm mercury acetate respectively. Result showed that only one isolated was able to reduce the concentration of mercury (Hg) in all variations on concentration and potential as mercury reducer probiotic bacteria.

1. Introduction

One of the most dangerous heavy metal pollutant in the level of toxicity is mercury (Hg). Although all forms into mercury are toxic chemicals, mostly concerns focus on methyl mercury. The main routes of exposure to these toxic elements are through fish or biota from contaminated waters. Mercury mostly present in methyl form as result of bioaccumulation and biomagnification of CH_3Hg in aquatic food chain [1].

An important step in tackling heavy metal pollution, especially mercury is by using potential microbes that can transform mercury (Hg) in the environment (bioconversion). Intestinal microbial from Milkfish used in heavy metal detoxification because of its ability in heavy metal binding. Fish intestinal bacteria are expected to be potent as probiotics and also resistant to heavy metals. It is important to note that isolates from mercury-resistant bacteria especially from probiotic groups have ability to detoxify mercury. Heavy metals from aquatic environment were accumulated mainly in the cell walls of bacteria when marine biota i.e. milkfish directly feeds contaminated food mercury in a pond.

2. Material and Methods

Preparation and isolation of probiotics candidate bacteria protocols were conducted base on Djide and Wahyuddin [2].The milkfish intestines are cleaned and crushed with mortar. The scraped banding of milkfish intestine is then weighed as much as 1 g and fed into a sterile physiologic NaCl solution and diluted with dilated dilution (10^{-1} - 10^{-6}). A total of 1 ml of a solution of 10^{-6} dilution was inoculated on MRSA medium (Man Ragosa Sharpe Agar) which added $CaCO_3$ 1%, then Incubated for 24-48 hours at 37°C.

Selection of probiotics resistant bacterial candidates (Hg) and resistance test of probiotics candidate bacteria in mercury concentration (Hg) were conducted according to Badjoeri and Hafidh (2010). A total of 0.5 ml of suspension of each isolate of probiotic bacteria were inoculated on an MRSB (Man Ragosa Sharpe Broth) medium added with Mercury Acetate $Hg(CH_3COO)_2$, as much as 5 ppm (5 μgml^{-1}). The isolate of probiotic candidate bacteria was inoculated on MRSB medium which added some concentration of mercury acetate $Hg(CH_3COO)_2$, i.e 10 ppm (10 μgml^{-1}), 15 ppm (15 μgml^{-1}) and 20 ppm (20 μgml^{-1}). Then incubated for 72 hours. Growth was observed by looking at the value of bacterial cell density grown in culture using a uv-vis spectrophotometer with a wavelength of 580 nm. The highest optical density values are considered to be the most resistant.

3.1 Characterization of Probiotic Mercury-resistant Bacterial Isolates

3.1.1. Observation of Cell Morphology. On morphological observations every single colony formed after purification was then observed. Observations include colony shape, colony color, colony edge, and colony surface. Observation of cell morphology was done by gram staining technique.

3.1.2. Probiotic Test. Probiotic test conducted following Djide and Wahyuddin [2].

3.1.3. Test of Resilience to Gastric acidity (pH). The acid resistance test was performed using an MRSB medium added with 0.1 N HCl to obtain a pH of 2.5-3 (according to gastric pH). One ose of each bacterial isolate was taken from culture stock and then inoculated on MRSB-HCl medium. Then incubated for 2x24 hours at 370C. Positive result indicated if bacterial growth occurs on MRSB-HCl medium and negative result showed if no bacterial growth occurs on MRSB-HCl medium.

3.1.4. Resistance Test against Gall bladder. The MRSB medium is added with synthetic bile salts (ox bite), with a concentration of 5%. One ose, each bacterial isolate extracted from culture stock was inoculated on MRSB-bile salt, then incubated for 2-3 x 24 hours at 37°C.

3.2. Fermentation of Probiotic Bacteria IPB 9 on Scaly Milk Media with Addition of Mercury

A total of 10% starter was introduced into 1200 ml of skimmed milk medium then homogenized. Skimmed milk media was then put into a sterile glass container each of 100 ml then treated with the addition of Mercury Acetate ($Hg(CH_3COO)_2$) as follows:
P1: Milk Media Skim + mercury concentration 5 ppm
P2: Milk Media Skim + mercury concentration 10 ppm
P3: Milk Media Skim + mercury concentration 20 ppm
P4: Skimmed Milk Media without mercury addition (control)
All subsequent media at close the meeting then incubated at room temperature for 3 days. In incubation day 1, day 2 and day 3, mercury, acidity (pH) and calculation of bacterial count were measured.

3.3. Calculation of Number of Bacteria

Calculation of bacterial count is done by Standard Plate Count (SPC) method. A total of 1 ml of sample, diluted using physiological NaCl (0.9%) with dilated dilution, to dilution 10^{-9}. A total of 1 ml of dilution result was grown on MRSA + $CaCO_3$ 1% medium by pouring method. Then incubated for 1x24 hours. Furthermore, the colonies number of bacteria that grow in the plate were calculated.

3.4. Mercury Analysis on Skim Milk Media
After fermentation, each sample of milk media that has been added with mercury acetate (Hg (CH3COO) 2) was tested to analyze mercury levels contained in the media after fermentation process. Mercury analysis was conducted at the Center of Makassar Health Laboratory using AAS (Atomic Absorbtion Spectrophotometry) method. At first, the sample was filtered with filter paper 0.045 m. The pH is set to about 3.5-4 by adding concentrated HNO_3 and then added 1 ml of concentrated HNO_3 solution and 5 ml of acetate buffer abrasive mixture. Then 5 ml of Ammonium Pyrrolidine Diphocarbonate (APDC) was added and shaken for about 5 min. An organic solvent, 10 ml Methyl Isobutyl Ketone (MIBK), were added to the solution, shake about 3 minutes and leave it to two separate phases.

3.5. Accommodated water phase
This aqueous phase is used for the preparation of laboratory and standard blank solutions. Addition of 10 ml distilled water in solution is required. The solution then will shake about 5 seconds until both phases apart.

3.6. Discard the water phase
Addition of 5 ml of HCl concentration dropped to the solution. The solution then shaken briefly and left for about 15 minutes. Addition of 10 ml of Cl3La7H2O then gave to the solution that will shake about 2 minutes. After this process, sample are ready to be measured with AAS using dry ice method with wavelength 358 nm.

3.7. Data Analysis
Data obtained from mercury level analysis, acidity change (pH), number of microorganisms, discussed descriptively, and processed in the form of tables and graphs.

4. Results and Discussion
An incubation of 2 x 24 h from 10^{-4} dilution was obtained by 12 isolates showing clear zone as probiotic bacterial candidate. Medium MRSA is a medium often used to grow lactic acid bacteria. The result of bacterial selection of probiotic candidate on mercury showed from 12 isolates, only 7 isolates showed growth in MRSB media added mercury i.e: IPB 1, IPB 2, IPB 3, IPB 8, IPB 9, IPB 11, and IPB 12.

The absence of growth in some isolates because it is suspected that mercury is toxic to bacterial cells. Mercury kill the bacterial cell while other isolates that grow on mercury-containing MRSB medium have their own way in overcoming the toxic environment. The bacteria adapted to this extreme condition through the mechanism of resistance either the mechanism of metal ion biosorbtion with the compilation of complex cell walls or induce mer operon systems in plasmids containing mercury-resistant genes. Mercury acetate $Hg(CH_3COO)_2$ that used in this study has the same characteristic with methyl mercury (CH_3Hg). Both compound is the most common form of organic mercury that accumulated in the body of aquatic organisms and has known for its very high toxicity.

In relation to mercury resistance levels, Zulaika [3] explained that isolates which is able to grow on synthetic media containing mercury (Hg) $\geq$ 5 mg/l categorized as high resistance to mercury. Whereas those isolates that only grow at 1 mg/l mercury has weak resistance.

The results of the isolate resistance test on some mercury concentrations can be seen from the value of OD (optical density). This value determine whether or not there is growth on mercury-added media. According to [4], this growth can be measured by cell density (dry weight of cells of cell cultures).

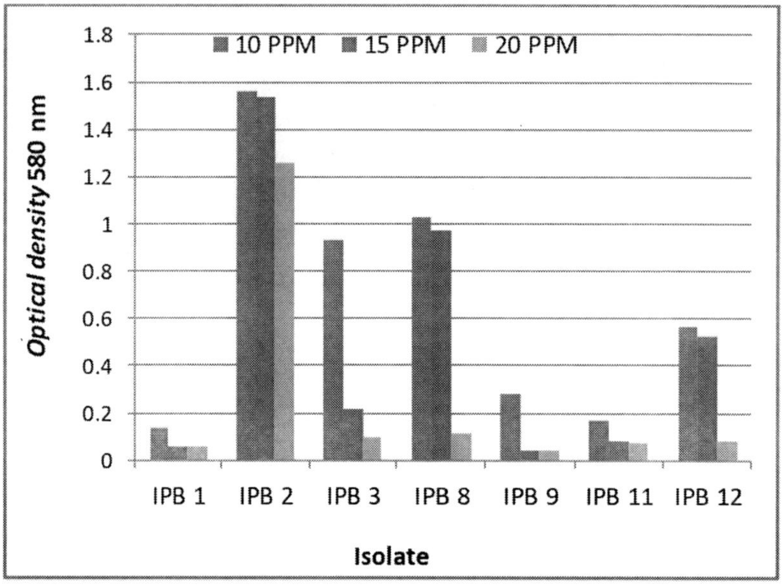

Figure 1. Isolated Probiotic Bacteria (IPB) on different Mercury concentration after 3 days incubation.

Turbidity level analysis showed that in 10 ppm of mercury concentration, IPB 2, IPB 3, IPB 8 and IPB 12 have much turbidity and cell precipitation compared with other isolates in accordance with the results of cell density measurements showing the highest value (figure 1). This is because the isolates are able to tolerate the presence of mercury in the media so that it is still experiencing growth. At concentrations of 15 ppm, IPB 2 and IPB 8 still showed high growth, while the low growth occurred in IPB 1, IPB 9 and IPB 11 ranged from 0.04 - 0.08 OD. At the concentration of 20 ppm only IPB 2 showed the highest growth (1.26 OD), while the other isolates ranged from 0.04 - 0.1 OD which indicates the growth is still low. Bacterial resistance to mercury can be through biosorbtion and bioaccumulation mechanisms. The biosorbtion is a passive mechanism, so the metal does not poisoned the bacterial cells. On the contrary, bioaccumulation mechanism is an active process in which heavy metals can poisoned the bacterial cells [5].

4.1. Characterization of Probiotic Bacterial Isolates

The result of pH test after incubation 2 x 24 hours showed that isolate IPB 2 had very high turbidity and sediment level, IPB 1 had moderate growth, while in other isolates had less precipitate but media did not have turbidity. Lactic acid bacteria are capable of maintaining an alkali-pH of intracellular rather than extracellular pH, but intracellular pH decrease persists with decreasing extracellular pH that supports tolerance to acid [6].

Table 1. Characterization of Probiotic Isolate from Milkfish Intestine

Type of Characterization	Isolate						
	IPB 1	IPB 2	IPB 3	IPB 8	IPB 9	IPB 11	IPB 12
Cell Morphology	coccus	bacil	bacil	coccus	bacil	coccus	coccus
Bile Salt	-	+	+	+	+	-	-
Probiotic Test:							
pH 2.5 Resistance	+	+	+	+	+	+	+
Bile Salt Resistance 5%	+	+	+	+	+	+	+

The bile salt resistance test showed excellent growth for all isolates characterized by turbidity in the media and there were many deposits at the bottom of the tube, but the most common sediment was found in IPB isolate 2. The growth occurring in bile salts for all isolates indicated that all isolates of ALB (Acid Lactid Bacteria) were able to withstand high bile salt conditions. Bile salts affect the permeability of bacterial cells. Bacteria that are not resistant to bile salts are suspected of permeability of cell membranes resulting in leakage of large intracellular material and causing cell lysis [7]. The ALB group in general can withstand with high bile salt stress conditions.

4.2. Growth of Probiotic Bacteria IPB 9 on Skimmed Milk Media with Addition of Mercury

Probiotic bacterial fermentation is done by inserting a starter of 10% into a skim milk production medium. Then added mercury with various concentrations. The absorption of heavy metals by other probiotic bacteria has now been developed. Some members of the intestinal microbiota, such as Lactobacillus used in food applications, are potentially additive to reduce the toxicity of metals in humans. This is because they have an effective resistance mechanism in preventing damage to their cells and can bind and absorb heavy metals onto the cell surface, then later the cells will come out through the feces. Heavy metal resistance genes and antibiotics are often encoded together on the same plasmid, so selective pressure can be suppressed in the intestinal tract [8].

4.3. Dynamics of Growth of Bacteria IPB 9 on Milk Media Skim with Various Concentrations of Mercury.

Calculation of the number of colonies of probiotic bacteria L. plantarum starts from calculating T0 treatment to T4 treatment with 1 day interval. The dilution results of each dilution were inoculated on the selective medium of lactic acid bacteria i.e., MRSA (Man Ragosa Sharpe Agar) which added 1% $CaCO_3$ using the casting method.

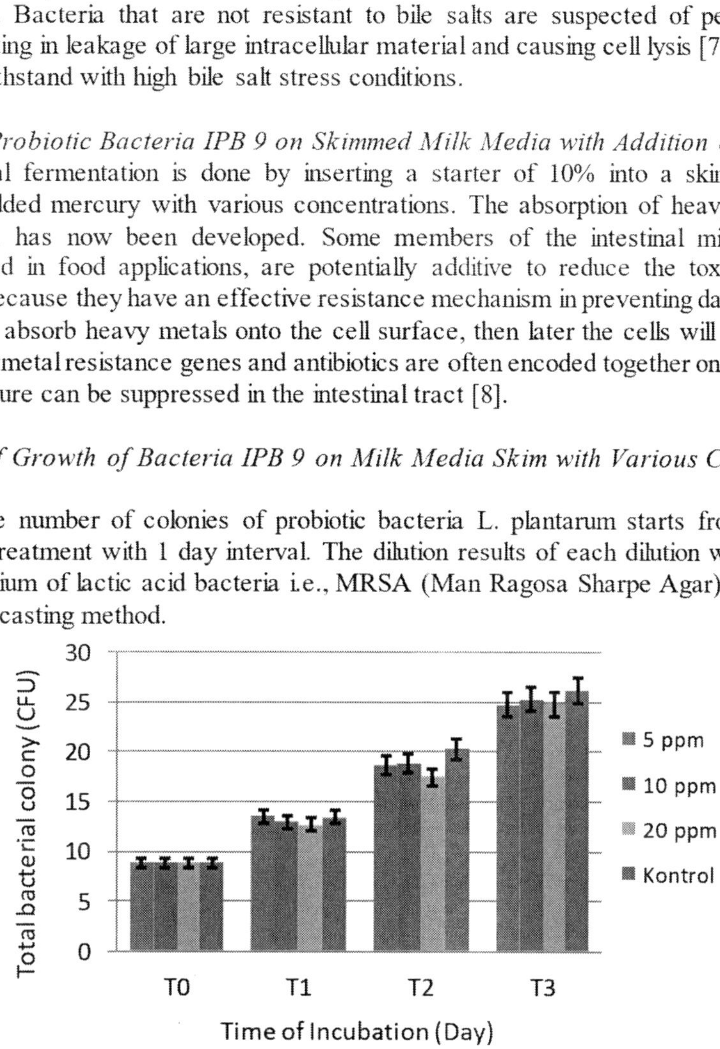

Figure 2. Diagram of isolated probiotic bacteria IPB 9 grown on different mercury concentration.

The addition of mercury concentration did not significantly affect the amount of bacteria produced. Even in the addition of high mercury concentrations, the bacteria are still able to grow and multiply because in addition to fermenting lactose in skim milk, the bacteria also bind and absorb heavy metals onto the surface of their cells.

4.4. Changes in pH in Skimmed Milk Media during Fermentation Process

Day 1 incubation (T1 treatment) pH control value was 6.0, pH value of mercury concentration 5 ppm was 6.2, pH value of 10 ppm concentration was 6.2 seta pH value of 20 ppm was 6.0. As for incubation day 2 (T2 treatment) pH control value is 5,9, pH value of 5 ppm is 5,9, pH value of 10 ppm is 6,0 and pH value of 20 ppm is 5,9. While for 3 days incubation (T3 treatment) the pH value of control was 5.9,

IOP Conf. Series: Journal of Physics: Conf. Series **979** (2018) 012002 doi:10.1088/1742-6596/979/1/012002

the pH value of 5 ppm was 5.9, the pH concentration of 10 ppm was 6.0 and the pH value of 20 ppm was 6.0.

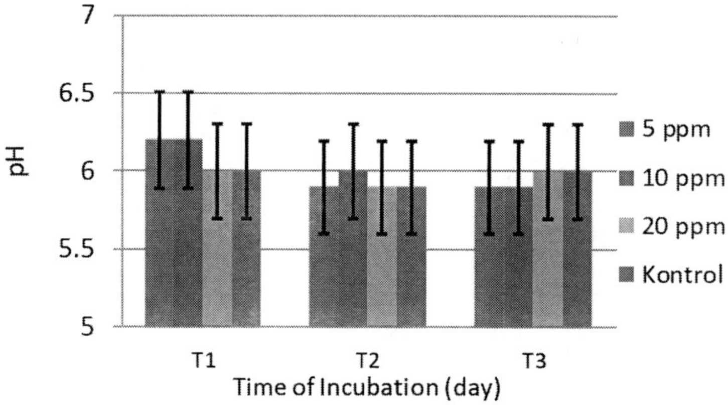

Figure 3. Graph of pH change on skimmed milk medium during fermentation process.

The pH of skim milk ranges from 6.5 to 6.7, IPB 9 isolates can decrease the pH of milk during fermentation. The increase of pH along with the destruction of fermented milk is characterized by the rupture of solids in fermented milk, the texture turns roughly, the increase of the liquid separates from the solid and the incidence of unpleasant odors.

4.5. Analysis of Mercury Levels Using AAS (Atomic Absorbtion Spectrophotometry) with Dry Method (Dry Ice)

Mercury-resistant bacteria and a mechanism for dealing with mercury stress in the presence of mercury-resistant genes called mer-operon genes comprising the metaregulator (merR) genes, mercury transport genes (merT, merP, merC), mercury reductase (merA) and gene organomerkuri liase (merB) (Chein, et al., 2010). The merB gene will catalyze the bonding of the Me-Hg yielding the organic compound and the Hg^{2+} ion, the Hg^{2+} ion will be bound to merC or merT. While Hg^{2+} outside the bacterial environment will enter the periplasm with cysteine merp residue pairs. Then merP transfer Hg^{2+} to cysteine residues merT or merC. Furthermore, Hg^{2+} ions cross the cytoplasmic membrane through a ligand exchange reaction process toward NADPH dependent on mercury reductase (the merA gene). Mercury reductase gives its 2 electrons to NADPH so that Hg^{2+} turns into a volatile Hg 0 without generating energy for the bacteria, then Hg0 is removed from the cell [9].

Operon Mer function is to give the ability of bacteria as absorbent agent molecule Hg (II). Combination of specific Mer genes as transport functions (MerTP) with metallothionein belonging to eukaryotes located in the cytoplasm of the Escherichia coli strain. This system is very specific. MerR has an affinity feature for reducing Hg (II) by sorbent by fusion with polypeptide (ELPs) [10].

Based on the results of mercury (Hg) analysis, results obtained that probiotic bacteria IPB 9 is able to reduce various mercury concentrations. At T1 the mercury concentration of 5 ppm was reduced to 3.56 ppm, the concentration of 10 ppm was reduced to 8.71 ppm, the concentration of 20 ppm was reduced to 11.93 ppm and on the undetectable control of mercury (Hg). While at T2 the mercury concentration of 5 ppm was reduced to 3.45 ppm, the concentration of 10 ppm was reduced to 7.41 ppm, the concentration of 20 ppm was reduced to 10.02 ppm and the control of undetectable mercury metal (Hg) was not detected. While at T3 the mercury concentration of 5 ppm was reduced to 3.02 ppm, the concentration of 10 ppm was reduced to 6,69 ppm, the concentration of 20 ppm was reduced to 9,00 ppm and on the undetectable control of mercury (Hg).

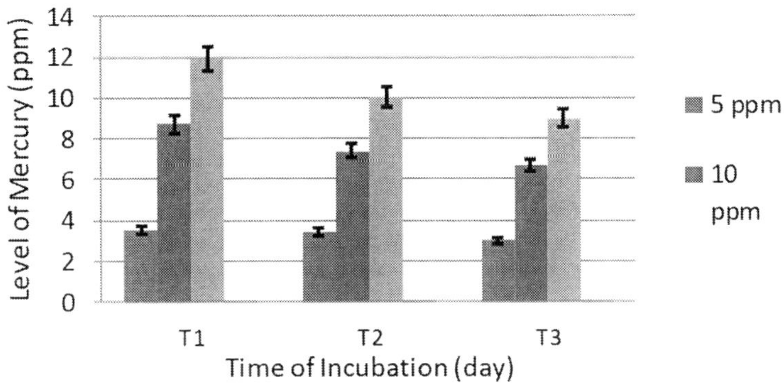

Figure 4. Graphic decrease of mercury (Hg) level by probiotic bacteria IPB 9 which analyzed by AAS (Atomic Absorbtion Spectrophotometry) with dry freeze method (Dry Ice).

5. Conclusion

Isolation of milkfish intestine obtained 7 isolates of probiotic bacteria that had growth in MRSB media added mercury. Probiotics IPB 9 able to grow and can reduce mercury metal (Hg) at concentrations of 5 ppm, 10 ppm and 20 ppm on skim milk media. The number of *Lactobacillus plantarum* bacteria during the fermentation process from T0 treatment to T3 treatment increased.

References

[1] Barkay, T., and W. D. Irene, 2005. Microbial Transformations of Mercury: Potentials, Challenges, and Achievements in Controlling Mercury Toxicity in The EnvironmentDepartment of Biochemistry and Microbiology. New Jersey. Advances In Applied Microbiology. Vol **57**

[2] Djide, M. N., and E., Wahyudin, 2008. Isolation of Lactic Acid Bacteria from Mother's Milk, and Its Potential in Lowering Cholesterol In Vitro. Pharmacy and Pharmacology Magazine. Vol.**12** (3)

[3] Zulaika, E., A. Widiyanti, and M. Shovitri, 2011. Endogenic Mercury Endogenous Bacteria Downstream Kalimas Surabaya. Journal. Biology Study Program, FMIPA ITS. Surabaya

[4] Khodijah, S., B. J. Tuasical, I. Sugoro and Yusneti, 2006. Growth of Streptococcus agalactiae As Bacteria Cause Subclinical Mastitis in Dairy Cattle. National Seminar on Livestock Technology. UIN Syarif. Jakarta

[5] Chojnacka, K., 2010. Biosorption and Bioaccumulation, The Prospects for Practical Applications. Environment International. **36**:299 – 307

[6] Siegumfeldt, H., B. K. Rechninger and M. Jacobsen, 2000. Dynamic Changes of Intracellular pH in Individual Lactid Acid Bacterium Celss in Response to a Rapid Drop in Extracellula pH. Appl. Environ Microbiol. **66**:2330 - 2335

[7] Kusumawati, N., L. J. Bettysri, S. Siswa, Ratihdewanti and Hariadi, 2003. Selection of Lactic Acid Bacteria Indigenous as Probiotics Galur with Cholesterol Lowering Ability. Journal of Microbiology Indonesia.Vol. **8** (2): 39-43

[8] Monachese, M., J.P. Burton. and G. Reid, 2012. Bioremediation and Tolerance of Humans to Heavy Metals through Microbial Processes: a Potential Role for Probiotics?. Applied and Environmental Microbiology.**78** (18). P. 6397-6404

[9] Brown, N., Y. Shih, C. Leang, K. Glendinning, J. Hobman, and J. Wilson, 2002. Mercury Transport and Resistance. Biometals, International Biometals Symposium

[10] Kostal, J., A. Mulchandani, K. E. Gropp and W. Chen, 2003. A Temperature Responsive Biopolymer for Mercury Remediation. Environ. Sci. Technol. **37**.4457 4462

Marine tunicates from Sangkarang Archipelago Indonesia: recent finding and bio-prospecting

Magdalena Litaay

Department of Biology, Faculty of Mathematics and Natural Sciences Hasanuddin University, Makassar, South Sulawesi, Indonesia

E-mail: mlitaay@fmipa.unhas.ac.id

Abstract. Tunicate belongs to urochordata that inhabit benthic area of coral reefs. This paper attempts to compile result of several studies on diversity of marine tunicates from Sangkarang Archipelago of South Sulawesi Indonesia. Method for tunicates sample collection was line transect method that applied at two 3 and 7 m depth. A 50 m line transect was applied parallel to a shore line in each depth and was done in duplicate. A 2.5 m plot was places side by side of transect, in which all tunicate inside plot was counted, identified, and photographed, respectively. Tunicates identification was based on morphological characteristics. The latest finding shows that eighteen species of tunicates were recorded in Samalona waters as also the same number of species in Barrang Caddi waters. The result from this finding was compared to previous studies done at other part of Sangkarang area. This study concluded that this archipelago is rich in tunicates and these resources can be utilized for various purposes. Bio-prospect of marine tunicates is also discus in the present study.

1. Introduction

Coral reefs are known as tropical marine natural resources, hence habitat for various marine biotas. An annual review by expert on tropical natural resourced argued that this unique ecosystem is source for bioactive compound origin from its associated biotas such as sponge, acsidians, molluscs, bryozoans, cnidarian [1]. Marine invertebrates as major group living at coral reefs of Indo-Pacific region rich for secondary metabolite and are targeted for studying lead compound as marine drug discovery [2].

Amongst marine invertebrates, sponge is better known for their secondary metabolite and extensively studied [3-11]. Other invertebrate that may have potential in producing active compound is far less studied compared to sponge. Biodiversity of marine biotas has pushed discovery marine natural material that can be developed as a therapeutics candidate [11]. Some examples of marine natural product, tarabine, Ara-C and vidarabine Ara-A have been used to cure human diseases [12].

In order to explore bio-prospecting of marine biota, a basic research to know species diversity and distribution is needed. Sangkarang archipelago which is also known as Spermonde archipelago is located in South-west of Makassar, capital city of South Sulawesi province of Indonesia. In comparison with previous funding in marine biodiversity in Sangkarang area [13-19], diversity of marine tunicates in the Sangkarang region is less studied [20-23]. Information on biodiversity and distribution of tunicate in this region will provide useful baseline data to support sustainable use of marine resources. This article aims to update information on marine tunicates from this area based on a recent finding and literature review from several sources.

2. Materials and Methods

2.1 Data Sources

Data was obtained from the result from different studies in Sangkarang Archipelago (figure 1). A similar method was applied for sample collection in all studied. SCUBA was using to collect sample and LIT method combination with plot was applied at 3 m and 7 m depth at three stations (figure 2). In each station, a 50 line transect was placed parallel to shore line and plot size 2.5 m x 2.5 m was applied side by side of transects (figure 2). All tunicates inside plot was noted, identified, counted and photographed. Sub-sample was taken and brought to laboratory for further study. Environmental parameters were measured in situ. Underwater camera was used to obtain images of species and habitat [24,25]. LIT method combination with plot was similar with one applied by [20-21] and [22-23]. Identification of tunicates was based on a main morphological character according to [26-28].

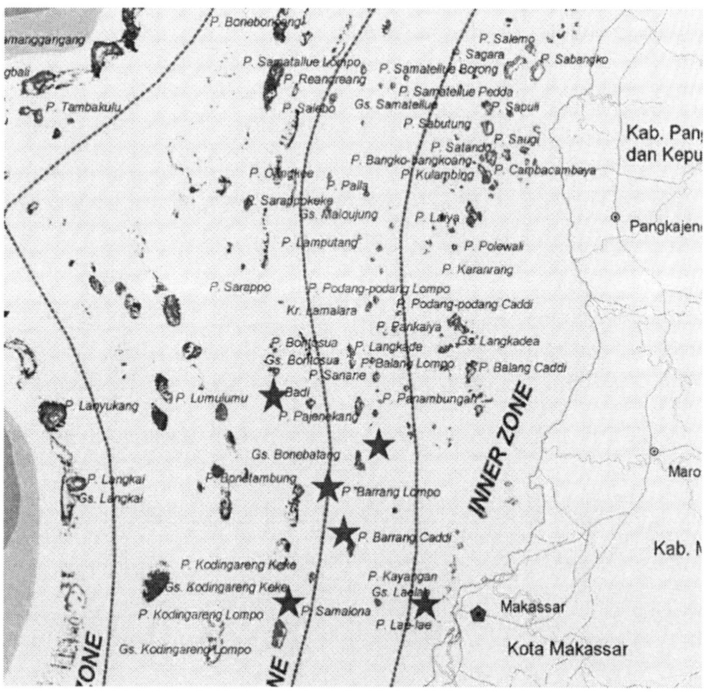

Figure 1. Sangkarang Archipelago South Sulawesi Indonesia (★location of data sources)

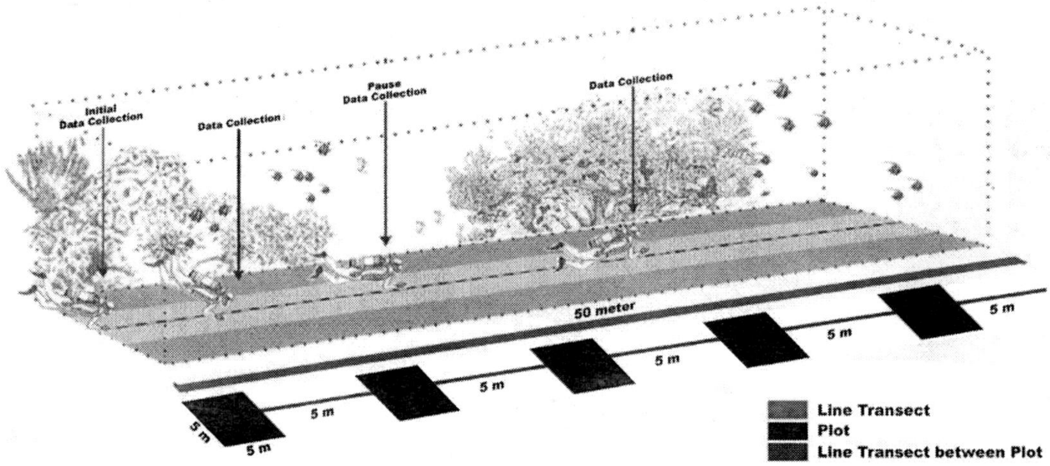

Figure 2. Line intercept transect method used for data collection

3. Result and Discussion

3.1 Distribution of marine tunicate

The latest finding on tunicate species composition in Sangkarang Archipelago indicates that number of species found as follows Samalona (18) [23], Barrang Caddi (18) [22]; Barranglompo (23) [21]; Lae-lae (7), Bone Batang (9), and Badi (10) [20] (table 1). In comparison with neighbor region, [29] has reported the number of ascidian records for Singapore to a total of 50. They also reported that 155 species were recorded from the South China Sea. [30] has documented close to 3000 species of ascidian worldwide. Furthermore, [31] have estimated there may be as many as 1500-2000 ascidian species worldwide still to be discovered and named. In the present study, as described in table 1, even though Sangkarang area is rich in marine tunicates, more studied still need to be carried out.

Table 1. List of Tunicates from different location in Sangkarang Archipelago Indonesia.

No.	Tunicate	Samalona a)	Barrang Caddi b)	Barrang Lompo c)	Lae-lae d)	Bone-Batang e)	Badi f)
1	*Clavelina robusta* Kott 1990 (white spot)	√	-	√	-	-	-
2	*Clavelina lepadiformis* Muller, 1776	√	√	-	-	-	-
3	*Clavelina sp.*	-	√	√	-	√	√
4	*Clavelina arafuensis*	-	-	√	-	-	-
5	*Clavelina moluccensis*	-	-	√	-	-	-
6	*Eudistoma sp.*	-	-	√	-	-	-
7	*Eudistoma gilboviride*	-	-	√	-	-	-
8	*Eudistoma reginum*	-	-	√	-	-	-
9	*Oxycornia fascicularis* Drasche, 1882	√	√	-	-	-	-
10	*Nephtheis fasicularis*	-	-	√	-	-	-
11	*Didemnid sp.* (brown white)	-	-	√	-	-	-
12	*Didemnid sp.* (purple)	-	-	√	-	-	-
13	*Didemnum molle* Hermann, 1986	√	√	√	√	√	√
14	*Didemnid sp.*	-	-	√	-	-	-
15	*Didemnid sp.* (white)	-	-	√	-	-	-
16	*Didemnum sp.*	-	-	√	-	-	-
17	*Trididemnum della* Ritter & Forsyth, 1917	√	-	-	-	-	-
18	*Trididemnum sp.*	-	√	-	-	-	-
19	*Lissoclinum patella* Gottschaldt 1989	√	-	√	-	-	-
20	*Diplosoma simile*	-	-	√	-	-	-

21	*Diplosoma sp.*	-	-	√	-	-	-
22	*Polycarpa aurata* Quoys & Gaimard, 1834	√	√	√	√	√	√
23	*Polycarpa captiosa*	-	-	√	-	-	-
24	*Polycarpa contecta*	-	-	√	-	-	-
25	*Polycarpa sp.*	-	-	√	-	-	-
26	*Polycarpa sp.* (brown)	-	-	√	-	-	-
27	*Polycarpa sp.* (white dots)	-	-	√	-	-	-
28	*Polycarpa papillata* Sluiter, 1886	√	√	-	-	√	√
29	*Polycarpa nigricans* Heller, 1878	√	√	-	-	-	-
30	*Polycarpa spongiabilis* Traustedt 1898	√		-	-	-	-
31	*Ascidia sp.* (brown)	-	-	√	-	-	-
32	*Ascidia kreagran*	-	-	√	-	-	-
33	*Ascidia sp.* (blue)	-	-	√	√	√	√
34	*Ascidia sp.*	-	√	-	-	-	√
35	*Ascidia ornate*	-	√	√	-	-	-
36	*Ascidia sydneinsis* Monniot F. 1898	√	-	-	-	-	-
37	*Diazona sp.*	-	-	√	-	-	-
38	*Nephtheis fasicularis*	-	-	√	-	-	-
39	*Rhopalaea sp.* (blue)	-	-	√	-	-	-
40	*Rhopalaea sp.* (yellow spot)	-	-	√	-	-	-
41	*Rhopalaea sp.*	-	√	-	√	-	√
42	*Rhopalaea crassa*	√	-	-	√	√	√
43	*Rhopalaea abdominalis*	-	-	-	√	-	-
44	*Herdmania momus*	-	-	√	-	√	√
45	*Pyura molina* Blainville, 1824	√	-	-	-	-	-
46	*Pyura sp.*	-	√	√	-	-	-
47	*Halocynthia verril* Dumosa Simpson, 1885	√	-	-	-	-	-
48	*Halocynthia sp.*	-	√	-	-	-	-
49	*Microcosmus juinii* Drasche 1884	√	-	-	-	-	-
50	*Plurella sp.*	-	-	√	-	-	-
51	*Atriolum robustum*	-	-	-	-	√	√
52	*Botryllus sp.*	-	√	-	√	√	-
53	*Pherophora sp.* (soft blue)	√	-	-	-	-	-
54	*Pherophora sp.* (orange)	√	-	-	-	-	-
55	*Perophora annectens* Ritter, 1893	√	-	-	-	-	-

Note: a) [23]; b) [22]; c) [21]; d) [20].
√ = present - = absent

3.2 Bioprospecting of marine tunicates

Sangkarang Archipelago is located South West off Makassar, consisting more than hundred islands, hence having abundant marine natural resources. This area hold importing role as habitat for various marine biotas including tunicates. Previous studies indicates that marine tunicates were potency for inoculum source for endo-symbiontic bacteria that can produce anti-bacterial and anti-fungi [7,32,33]; [34-36]. Other studies has also indicated that tunicates are also had a potency as antiviral [37], anticancer [38], inhibit and induces apoptosis of breast (MCF-7; MDA-MB) cancer cells and also used for phase II cancer treatment [39-41], as inhibits breast cancer cells by JNK dependent apoptosis [42], breast and prostate cancer [43]. Yondelin one of bioactive compound produced by tunicate is use to cure refractory soft-tissue sarcomas [44].

References

[1] Radjasa OK, Vaske YM, Navarro G, Vervoort HC, Tenney K, Linington RG and Crews P 2011 Highlights of marine invertebrate-derived biosynthetic products: Their biomedical potential and possible production by microbial associants. *Bioorganic & Medicinal Chemistry* **19** 6658–74

[2] Sabdono A and Radjasa OK. 2008. Microbial symbionts in marine sponges: Marine natural product factory. *J Coast Dev.* **11** 57–61
[3] Bugni TS and Ireland CM 2004 Marine derived fungi: a chemically and biologically diverse group of microorganisms. *Nat. Prod. Rep.* **21** 143–63
[4] Thakur NL and Müller MEG 2004 Biotechnological potential of marine sponges. *Curr. Sci.* **86** 1506–1512
[5] Wang G. 2006. Diversity and biotechnological potential of the sponge-associated microbial consortia *J. Ind. Microbiol. Biotechnol.* **33** 545–51
[6] Jackson, K 2009 The Halichondrins and E7389 *Chem. Rev.* **109** 3044–79
[7] Karthikeyan MM, Ananthan G and Balasubramanian T 2009 Antimicrobial Activity of Crude Extracts of Some Ascidians (Urochordata: Ascidiacea), from Palk Strait, (Southeast Coast of India). *World J. of Fish and Mar. Sci.* **1** (4) 262-7
[8] Kennedy JP, Baker C, Piper PD, Cotter M, Walsh MJ, Mooij MB, Bourke MC, Rea, O'Connor PM, Ross RP, Hill C, O'Gara F,. Marchesi JR and Dobson ADW 2009 Isolation and analysis of bacteria with antimicrobial activities from the marine sponge *Haliclona simulans* collected from Irish waters. *Mar. Biotechnol.* **11** 384–96
[9] Kuznetsov G 2009 Tubulin-based antimitotic mechanism of E7974, a novel analogue of the marine sponge natural product hemiasterlin. *Mol. Cancer Ther* **8** 2852–60
[10] Li, Z. 2009. Advances in marine microbial symbionts in the China Sea and related pharmaceutical metabolites. Mar. Drugs, 7: 113–129
[11] Thomas TRA, Kavlekar DP and LokaBharathi PA 2010 Marine Drugs from Sponge-Microbe Association—A Review *Mar. Drugs* **8** 1417-1468; doi:10.3390/md8041417
[12] Mayer AMS, Glaser KB, Cuevas C, Jacobs RS, Kem, W, Little R, Mcintosh JM, Newman DJ, Potts BC and Shuster DE 2010 The odyssey of marine pharmaceuticals: a current pipeline perspective. *Trends in Pharmaceutical Sci.* **31** 255-65
[13] Moll H 1983 Zonation and diversity of scleractinian on reefs of South Sulawesi Indonesia. [Thesis]. Netherland: Leiden University
[14] Verheij E 1993 Marine plants on the reefs of the Sangkarang Archipelago, SW Sulawesi, Indonesia: aspect of taxonomy, floristics and ecology. [Dissertation] Leiden: Rijksherbarium/Hortus Botanicus
[15] Massin 1999 Reef Dwelling Holothuroidea (Echinodermata) of The Sangkarang Archipelago (South-West Sulawesi, Indonesia). *Zoologische Verhandelingen* 329. National Natuurhistorisch Museum. Leiden
[16] Renema WT and Simon R 2001 Larger foraminifera distribution on a mesotrophic carbonate shelf in SW Sulawesi (Indonesia). *Palaeogeography, Palaeoclimatology, Palaeoecology* **175** 125-46
[17] de Voogd NJ, Cleary DFR, Hoeksema BW, Noor A and van Soest RWM 2006 Sponge beta diversity in the Sangkarang Archipelago, SW Sulawesi, Indonesia. *Mar. Ecol. Prog. Ser.* **309** 131–42
[18] Pogoreutz C, Kneer D, Asmus H, Anelt H and Litaay M 2012 The influence of canopy structure and tidal level on fish assemblages in tropical Southeast Asian seagrass meadows. *Eustarine Coastal and Shelf Science* **107** (2) 58-68
[19] Priosambodo D, Kneer D, Asmus H, Zamani NP, von Juterzenka K, Litaay M, and Soekendarsi E 2014 Community analysis of burrower shrimp in Bone Batang seagrass bed South Sulawesi. Proceeding of The First International Conference on Science (ICOS-1) 2014 ISBN 978-602-72198-0-9: 209 -219
[20] Fikruddin MBAbdH 2013 Distribution and diversity of Tunicate (Ascidiacea) at different habitat at waters of Badi, Bonebatang and Laelae Island. BSc Hon Thesis FIKP Hasanuddin University [Indonesian]
[21] Mawaleda R 2014 Distribution and habitat preference Urochordata class Ascidiacea at coral reefs area of Barrang Lompo island of Makassar. BSc Hon Thesis B.ScHon FIKP

Hasanuddin University [Indonesian]
[22] Litaay M, S Johannes E, Santosa S, Moka W, Tanjung J D and Annas B. 2017. Diversity of tunicate from Barrang Caddi waters. Presented at the 4th International Marine and Fisheries Symposium, Makassar 18 May 2017
[23] Litaay M, Santosa S, Johannes E, Agus R, Moka W and Tanjung JS. 2018. Biodiversity of marine tunicate from Samalona waters off Indonesia. Spermonde J Mar. Sci. **3** (1). In press
[24] English S, Wilkinson C and Baker V 1997 Survey Manual for Tropical Marine Resources. Australian Institute of Marine Science. Townsville
[25] Brower JE, ZR JH and Von Ende CN 1998 Field and Laboratory Methods for General Ecology, Mc Graw Hill Company
[26] Kott P 2005 Catalogue of Tunicata in Australian Waters. Queensland Museum, Brisbane, Australia. ISBN 0 642 56842 1.301. 301 p
[27] Page M and Kelly M 2013 Awesome Ascidians, a Guide to the Sea Squirts of New Zealand. TC Media Ltd.53 p
[28] WoRMS. 2017. World Register of Marine Species http://www.marinespecies.org/ aphia.php?p=taxdetails&id=146420
[29] Lee S SC, Chan J YH, Teo S LM and Lambert G 2016 State of knowledge of ascidian diversity in South China Sea and new records for Singapore *Raffles Bul. of Zoology Suppl.* **34** 718-43
[30] Shenkar N and Swalla BJ 2011 Global diversity of Ascidiacea. *PloS ONE* **6**(6):e20697. doi.10.1371/journal.pone0020657
[31] Appeltans W, Ahyong ST, Anderson G, Angel MV, Artois T, Bailly N, Bamber R, Barber A, Bartsch I, Berta A, BłażewiczPaszkowycz M, Bock P, Boxshall G, Boyko CB, Brandão SN, Bray RA, Bruce NL, Cairns SD, Chan TY, Cheng L, Collins AG, Cribb T, Curini-Galletti M, Dahdouh-Guebas F, Davie PJ, Dawson MN, De Clerck O, Decock W, De Grave S, de Voogd NJ, Domning DP, Emig CC, Erséus C, Eschmeyer W, Fauchald K, Fautin DG, Feist SW, Fransen CH, Furuya H, Garcia-Alvarez O, Gerken S, Gibson D, Gittenberger A, Gofas S, Gómez-Daglio L, Gordon DP, Guiry MD, Hernandez F, Hoeksema BW, Hopcroft RR, Jaume D, Kirk P, Koedam N, Koenemann S, Kolb JB, Kristensen RM, Kroh A, Lambert G, Lazarus DB, Lemaitre R, Longshaw M, Lowry J, Macpherson E, Madin LP, Mah C, Mapstone G, McLaughlin PA, Mees J, Meland K, Messing CG, Mills CE, Molodtsova TN, Mooi R, Neuhaus B, Ng PK, Nielsen C, Norenburg J, Opresko DM, Osawa M, Paulay G, Perrin W, Pilger JF, Poore GC, Pugh P, Read GB, Reimer JD, Rius M, Rocha RM, Saiz-Salinas JI, Scarabino V, Schierwater B, Schmidt-Rhaesa A, Schnabel KE, Schotte M, Schuchert P, Schwabe E, Segers H, Self-Sullivan C, Shenkar N, Siegel V, Sterrer W, Stöhr S, Swalla B, Tasker ML, Thuesen EV, Timm T, Todaro MA, Turon X, Tyler S, Uetz P, van der Land J, Vanhoorne B, van Ofwegen LP, van Soest RW, Vanaverbeke J, Walker-Smith G, Walter TC, Warren A, Williams GC, Wilson SP and Costello MJ 2012 The magnitude of global marine species diversity. *Current Biology* **22** 1–14
[32] Litaay M, Christine G, Gobel RB and Dwyana Z 2015 Bioactivity of endo-symbiont bacteria of tunicate *Polycarpa aurata* as antimicrobial. In: Ohee (eds) *Proceeding of The 23 National Seminar of Indonesia Biology Society*. Jayapura, 18 September 2015. [Indonesian]
[33] Christine G, Litaay M, Gobel RB and Dwyana Z 2015 Potency of tunicate *Polycarapa aurata* as inoculum source for endosymbiotic bacteria; characterization of isolates. In: Tahir (eds). Proceeding National Seminar on Marine and Fisheries. FIKP Unhas. Makassar, May 2015. [Indonesian]
[34] Nurfadillah A, Litaay M, Gobel RB and Haedar N 2015 Potency of tunicate *Polycarpa aurata* as inoculum source of sebagai sumber endosimbyotic fungsi that produce antimicrobe. J. Alam dan Lingkungan **6** (12) 10-6 [Indonesian]
[35] Sardiani N, Litaay M, Gobel RB, Dwyana Z 2015 Potency of tunicate *Rhopalaea sp* as source of bacterial inoculum that produce antibacterial; isolates characterization. J. Alam dan Lingkungan **6** 11 [Indonesian]

[36] Tahir E, Litaay M, Gobel RB, Haedar N, Prisambodo D and Syahribulan. 2016 Potency of tunicate *Rhopalaea crassa* as inoculum source of endosymbiont fungi that produce antimirobe. *J Mar. Sci. Sangkarang* **2** 2 33-7 [Indonesian]
[37] Murti Y and Agrawal T 2010 Marine derived pharmaceuticals development of natural health products from marine biodiversity. *Int. J of Chem. Tech. Res.* **2** (4) 2198-217
[38] Shaala LA and Diaa Youssef TA 2015 Identification and Bioactivity of Compounds from the Fungus Penicillium sp. CYE-87 Isolated from a Marine Tunicate. *Mar. Drugs* **13** 1698-709; doi:10.3390/md13041698
[39] Zelek L, Yovine A, Brain E, Turpin F, Taamma A, Riofrio M, Spielmann M, Jimeno J and Misset J 2006 A phase II study of Yondelis®(trabectedin, ET 743) as a 24-h continuous intravenous infusion in pretreated advanced breast cancer. *Brit. J Cancer* **94**
[40] Michaelson M, Bellmunt J, Hudes G, Goel S, Lee R, Kantoff P, Stein C, Lardelli P, Pardos I and Kahatt C 2012 Multicenter phase II study of trabectedin in patients with metastatic castrationresistant prostate cancer. *Annals of Oncology* **23** 1234-40
[41] Atmaca H, Bozkurt E, Uzunoglu S, Uslu R and Karaca B 2013 A diverse induction of apoptosis by trabectedin in MCF-7 (HER2- /ER+) and MDA-MB-453 (HER2+/ER-) breast cancer cells. *Toxicology Letters* **221** 128-36
[42] Gonzalez-Santiago L, Suarez Y, Zarich N, Munoz-Alonso M, Cuadrado A, Martinez T, Goya, Iradi A, Saez-Tormo G and Maier J 2006 Aplidin® induces JNK-dependent apoptosis in human breast cancer cells via alteration of glutathione homeostasis, Rac1 GTPase activation, and MKP-1 phosphatase downregulation. *Cell Death & Differentiation* **13** 1968-81
[43] Kalimuthu S, Venkatesan J and Se-Kwon k 2014 Marine Derived Bioactive Compounds for Breast and Prostate Cancer Treatment: A Review. *Current Bioactive Compounds* **10** 62-74
[44] Sinko J, Rajchard J, Balounova Z and Fikotova L 2012 Biologically active substances from water invertebrates: a review. *Veterinarni Medicina* **57** (4) 177–84

Preliminary Study on Testicular Germ Cell Transplantation of Endemic Species *Oryzias celebensis*

I Andriani[1], F Agustiani[2], M Hassan[2], A Parenrengi[3], K Inoue[4]

[1]Dept. of Biology, Hasanuddin University, Indonesia
[2]Dept. of Biotechnology, Bogor Agricultural Institute, Indonesia
[3]Maros Brackishwater Aquaculture Research Centre, Indonesia
[4]Atmosphere and Ocean Research Institute, The University of Tokyo, Japan

E-mail: irmaandriani@unhas.ac.id

Abstract. The research has been conducted to study some technical steps for male germ-plasm from endemic fish species such as some species of *Oryzias* fish in Indonesia to preserve and propagate through germ cell transplantation technology. For preliminary research, the study was started with germ cell characterization of testes, cryopreservation of TGC and the transplantation of *Oryzias celebensis* as candidates for surrogate broodstock of *Oryzias* fish male germ plasm. The data analized included the potential number of TGC as donor, the viability of cryopreserved TGC in two types of cryoprotectans and the survival rate of *O.celebensis* larvae as recipient after transplantation. The result showed that the average amount of TGC yielded after dissociation was 131000 ± 31349 with 74.2 % viability of TGC each. Cryoprotectan10% DMSO +glucose yielded higher viable of TGC. More than 80 % of *O.celebensis* larvae survived after transplantation. In conclusion, these preliminary data of *O.celebensis* as surrogate broodstock candidate will support the application of TGC transplantation technology in *Oryzias* endemic species.

1. Intoduction

Genus *Oryzias* is one of the endemic fishes in South Sulawesi which some of them have been included in the red list of endangered fish. Overfishing so far has not been accompanied by efforts to the cultivation lead to extinction. Therefore, *Oryzias* fish conservation efforts as part of the diversity or biodiversity of fish, especially in South Sulawesi needs to be done. Recent studies have indicated that germ cell transplantation and cryopreservation technology have become technical breakthrough and prospect approach to solve the production of seed for commercially valuable fish, as well as threatened spesies.

Surrogate broodstock production through germ cell transplantation technology may eventually answer problems to the production of valuable commercially fish with long generation time. This technology has been applied in some fishes mainly in salmonids [1, 2] successfully produced surrogate broodstock and in tilapia resulted donor-derived spermatogenesis [3]. Male germ cell contain spermatogonia, the only cell responsible for generating offspring [4]. Spermatogonia donor cell transplanted into the peritoneal cavity of newly hatched embryo using microinjection technique. Transplanted spermatogonia wil be migrated toward and colonized in the genital ridge of the recipient followed by proliferation and differentiation of donor derived germ cell into mature gametes in the recipient. If spermatogonia of *Oryzias* fish could be successfully transplanted into short generation time fish then gametes of *Oryzias* fish could be easily and rapidly be produced. Therefore, germ cell transplantation technique is a potential breakthrough technique to preserve germ plasm of *Oryzias* fish.

Some basic techniques need to be established for the application of male germ cell transplantation in *Oryzias* fish including the preparation of male germ cell enriched in spermatogonia. To make the technique effective for transplantation, preparation of single germ cell suspension from a solid tissue is of critical importance. In general, it is important to maximize the number of viable cell, disperse clump of cells and minimize the reaggregation [5]. There are six variable conditions should be well defined to get suitable results, those are dissociation medium, type of enzyme(s), enzyme preparation, concentration(s) of enzyme(s), temperature and incubation times [6].

Cryopreservation is a technique of storage or preservation of animal cells, plants or other genetic material (including semen and oocytes) in a frozen state by reduction of metabolic activity without affecting the organelles inside cells that the function of physiology, biology and morphology remain. In germ cell transplantation technology, cryopreservatioan is basic technique that should be done to synchronize availability of recipient and donor cell. The successfull of cryopreservation technique is usually determined by several factors, including the type of extender solution used. Currently there are sperms of ± 200 species of fish in the world successfully preserved using the techniques of cryopreservation, including carp (*Cyprinus carpio*) and Atlantic salmon (*Salmo salar*). However, none of them is found for reference of the cryopreservation done on celebensis medaka fish neither on cryopreservation of spermatogonial . Based on this, the research on cell cryopreservation of medaka fish germ cell needs to be done and as a treatment in this study will be observed the influence of some kind of solution extenders.

2. Materials And Methods

2.1 Collection of testes

Testes were collected from adult *Oryzias* fish (4 cm) that were obtained from Bantimurung river, Maros, South Sulawesi. Testes tissues for histologycal analysis (n=4) were fixed immediately after collection. For the dissociation of germ cell, collected testes were washed twice with Phosphate Buffer Saline (PBS) pH=8.2 (Gibco). All connective tissue at posterior and anterior ends, blood capillary and fat attached to gonad were removed from the collected testes and then testes were cut into pieces of 20-30 mg each for dissociation analysis .

2.2 Germ cell morphological analysis

Four testicular samples were fixed in Bouin fixative for 24 hours, dehydrated along an ascending series of ethanol according to [7], embedded in paraffin, and sectioned (5µm thickness). The block of tissue were sliced serially, and to avoids counting the same cell twice, 6 sections of 5 µm thickness per fish were taken every 35 µm sections. The sections were stained with Hematoxilin-Eosin for microscopic observation under the light microscope. Photographs were taken with an automatic Nikon camera. Germ cell staging was refers to [8]. For quantification of diameter and morphological character of germ cell type , three fields of view per section were selected at random from the middle portion of the testis as much 30 germ cell each type. Additionally, six cysts per type of spermatogonia were analyzed to obtain the average number of each type of spermatogonia per cyst.

2.3 Dissociation of germ cell

Testes were minced with scissors and incubated in dissociation medium at room temperature. For the purpose of defining the effect of dissociaton medium against the number and viability of spermatogonia germ cell, testes were incubated in dissociation medium. Dissociation medium contain PBS (Gibco) containing 0.5% trypsin and 30 µl of 10 IU/µl DNase (Sigma) in 1 ml PBS supplemented with fetal bovine serum (Sigma), Ca2+ (1mM) and 25 mM HEPES (Sigma). To obtain optimal yielded high number of viable cell, testes were incubated in dissociation medium for two hours. During incubation, the cell suspension was pipetting gently every 30 minutes to dissociate germ cell tissue. Then, cell suspension was filtered through a 35 µm pore-size nylon screen to eliminate non dissociated cell clumps and washed twice with PBS. The number and viability of spermatogonia were analyzed using CX 10 microscope and haemocytometer. Viability of the cells was analysed using the trypan blue dye exclusion test (1:1). Blue cell indicated damaged or death cell, while normal cell was transparant.

2.4 Cryopreservation of testicular cells
Cryopreservation method used in this study refers to the method. The cell suspension was diluted gametes dissociation results using extender solution with a ratio of 1: 9. A total of 10 mL of testicular cells added to a solution of 90 mL extender. In this study, we used two different extender solution in 10% DMSO cryoprotectan : 5% glucose and Ringer solution. Before the suspension was added cryoprotectants testicular cells, testicular cells drops pipette and cell concentration is calculated in advance. Furthermore testicular cell suspension in cryoprotectants immediately put in a 10 mL size straw and put in cryotube. Cryotube left not closed tightly and put in bicell biofreezing vessel. Before freezing, gamete cells in Biocell containers are stored beforehand in the refrigerator -80°C for 20 hours. After that, cryotube containing testicular cells be closed tightly and put in a tube canister and stored in liquid nitrogen. Straw was stored in liquid N2 containers for more than one week. To analyze testicular cell viability after cryopreservation, thawing process is done. Thawing carried out by dipping the straw into the water with temperature 30-40°C for 10-15 minutes. The testicular cells then ready to be observed using trypan blue vital staining.

2.5 Data analysis
Testes samples from 10 *Oryzias* fish were used for tissue dissociation and four testes for histological examination. Characterization of germ cell morphology was presented descriptively. Parameters for the testes dissociation were number and viability of spermatogonia and presented as mean.

3. Results and Discussion

3.1 Result

3.1.1 Morphological characteristic of Celebes Oryzias fish male germ cell. The testes of *Oryzias* fish form bilobed organ, with the right lobe slightly skinnier and thinner than the left one (figure 1A). Cross section of seminiferous tubules at the middle part of *Oryzias* fish testes was depicted in Figure 1B. Histological analysis showed that all testes examined were producing germ cell at all the stages of development. Testes of *Oryzias* fish were unrestricted tubular in which of spermatogonia were distributed along the lengths of *Oryzias* fish testis tubules (figure 1B).

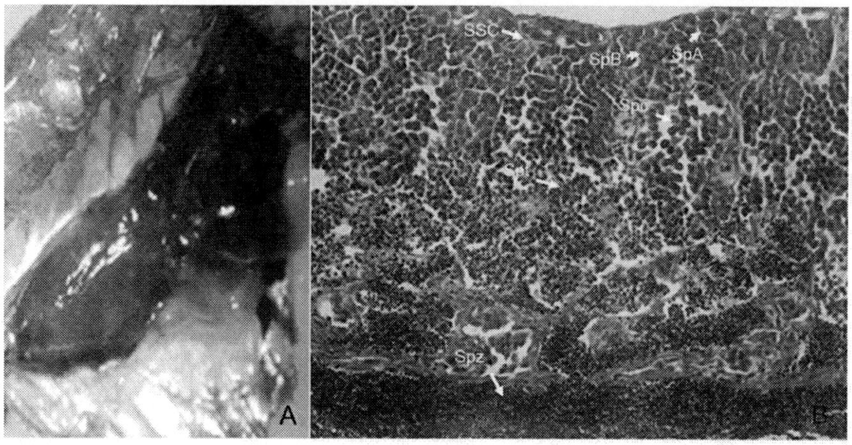

Figure 1 A. *Oryzias celebensis* fish testes. B. Histological structure of male germ cell of *Oryzias celebensis*. SSC = spermatogonial stem cell, SpA = Type A spermatogonia, SpB = Type B= spermatogonia, Spc = spermatocyte, Spd = spermatid, Spz ; spermatozoa.

Morphological characterization of germ cells were determined based on some criteria such as cell and nuclear size, nuclear morphology, topological organization of germ cell and cell number per cyst. Germ cell staging identified in *Oryzias* fish was as follows : Primary A spermatogonium (stem cell spermatogonium), A spermatogonia, transitional spermatogonium, B spermatogonium, primary spermatocyte, secondary spermatocyte, spermatid and spermatozoon. All proliferation stages between late A spermatogonia and early B spermatogonia were categorized as transitional spermatogonia. Thus, these morphological analysis allowed us to determine that there were three type of spermatogonia in immature 4 mm TL Of *Oryzias* fish. Spermatogonial stem cells were the largest single spermatogonia mostly located close to the basement membrane of the germinal epithelium. The nuclear membrane had an irregular outline, some of which remained unclear membrane as well. On the other hand, A spermatogonia were smaller in size, surrounded by Sertoli cells, forming 2-4 germ cell in a cyst located close to the lumina of the tubules. Morphologically, spermatogonial stem cell and some of type A spermatogonia were ellipsoid. The type B spermatogonia were smaller than A spermatogonia with the number of germ cells inside the cyst increases (18-48 germ cells within cyst).

3.1.2 Cell viability of testicular germ cell before and after cryopreservation. Observations were made in cells from three different treatment. Sample 1 was added to a solution of the extender in the form of glucose and DMSO as cryoprotectants, while sample 2 was added to a solution of the extender ringer and DMSO, and samples to 3 as a control is given only cryoprotectant DMSO alone without extenders , The results obtained from 3. The treatment of post-cryopreservation can be seen in Table 1. The average yield testing I and II showed that glucose extender is a solution that is more effective than the use of Ringer's solution and DMSO alone. In Table 1 are shown that the average yield of cell viability were using DMSO + extenders glucose after cryopreservation higher than the results of two other treatments, namely 1.964% on the replay I and 2,248 in replicates II, while the average viability of cells using extenders ringer and without extender solution are in the same position on the replay I is 0% and the second replay respectively 0.449% and 0.899%.

Table 1. Number of spermatogonia before and after cryopreservation with different extender.

	Sample	Extender	Viability before cryopreservation (%)	Viability after cryopreservation (%)
I	1	Ringer	*79.28*	*0*
	2	Glukosa 5%	*79.83*	*1.96*
	3	DMSO	*79.83*	*0*
II	1	Ringer	*64.39*	*0.45*
	2	Glukosa 5%	*64.39*	*2.25*
	3	DMSO	*64.39*	*0.89*

3.2 Discussion

Spermatogonia are stem cells that can restock their own population and simultaneously producing sperm on a renewable basic. This fact has become a benefit for technical application of assisted reproduction based on spermatogonia such as transplantation, cell culture or preservation of endangered and genetically valuable animals. This research is a first step prior to the effort to make surrogate broodstock using *Oryzias* fish as donor and transplantation technique assay. One of major limitation to the transplant procedure is that the donor germ cell mixture contained an unknown number of spermatogonia [9].

To increase the efficiency of transplantation, the percentage of spermatogonia with stem characteristic (Undifferentiated spermatogonia including spermatogonia stem cell and A spermatogonia) should be increased [10]. One of first successful studies increased the relative number of spermatogonia colonized in recipient by using testicular cell that were selected based on the ability of the cell to interact with the basement membrane of the seminiferous tubules [11]. In this research we just observed the type of testicular cells including type of spermatogonia morphologically using histological approach. Moreover, for viability assay, we just used trypan blue vital staining day to characterized normal viable cells. Although those technique approach had major limited information, but it helped us to prepare mixed germ cell enriched high number and viable spermatogonia for *Oryzias* fish transplantation assays.

Prior to dissociation of testes, we studied the morphological structure of testes and male germ cell type in *Oryzias* fish. Testicular parenchyma of *Oryzias* fish show typical organization as most of fish with two major component, the tubular and intertubular compartment. Similar to vertebrate, clearly it has been described that the tubular compartment organized in testes comprising the seminiferous epithelium, where the Sertoli/germ cells units called cysts are located, the tunica propria composed of basal lamina and peritubular myoid cells, and the tubular lumen filled with fluid secreted by Sertoli cells. While the intertubular compartment containing Leydig cells, connective tissue cells, and blood vessels [12-14]. The seminiferous tubules contain spermatogenic cysts, a group of synchronously developing germ cells that enveloped by sertoli cells, which are derived in a clonal manner from one stem cell or A spermatogonia in which interconnected by cytoplasmic bridge [3, 15].

The morphological characteristic of germ cell types observed for *Oryzias* fish was in general similar to most species of teleost fish that has been described previously [16, 17]. With histological approach, we have analyzed morphological features of *Oryzias* fish germ cell type based on nuclear and cytoplasm morphology and the number of germ cell per cyste. Only three generations of spermatogonia could be classified in this research descriptively as 5 μm histological sections were disable to describe the nuclear morphology in clear performance. A spermatogonia in *Oryzias* fish were found at random sites along the seminiferous tubule throughout the testis. This type of germ cell organization has been classified as unrestricted type, a primitive character that appears to exist in a number of teleost fish orders such as cypriniformes, perciformes and others [18, 19].

Since each species has own anatomical characteristics including topological posistion of germ cell type, the relative amount of connective tissue between tubules, lobulation, the ease of removing tunica, need to be isolated then the protocol for the preparation of donor cell are species specific [20]. For these reasons, we compared two protocols, those of modified protocol from [21] and modified protocol from [3, 22], the latter of which proved more efficient in *Oryzias* fish. However, 2 mechanical steps were required before testes exposure to digestive enzymes,: first, removal of testicular blood vessels to eliminate the contamination of blood cell. Second, minced process of testes and gently pipetting of cell suspension provide an effective and fast enzimatic digestion, however being carefull not to cause damage by squeezing the testes or by roughly pipetting. According to our observation, these mechanical steps could become critical point to get high number of spermatogonia yielded.

In conclusion, the dissociation of *Oryzias celebensis* testes yielded sufficient number of spermatogonia as donor cells for germ cell transplantation and cryopreservation still given a small number of donor cells avalaible for germ cell transplantation.

Acknowledgements
We thank Prof Yusuke Naruse of The National Institute For Basic Biology, Okazaki, Japan for the assitance during training of 8th NIBB Practical Course for laboratory animals in his laboratory. We thank to Hasanuddin University for research grant. We also thank Brackish Water Fish Research Centre of Maros for some equipment and support of *Oryzias* fish dissociation which was performed in the laboratorium. Thank to Pharmacology Laboratorium of Pharmacy Faculty, UNHAS for allowing using liquid N2 containers.

References

[1] Takeuchi Y, Takeuchi T and Yoshizaki G 2004 Surrogate broodstock produces salmonids *Nature* **430** : 630-639
[2] Okutsu T, Suzuki K, Takeuchi Y, Takeuchi T and Yoshizaki G 2005 Testicular germ cells can colonize sexually undifferentiated embryonic gonad and produce functional eggs in fish *Proc Natl Acad Sci* **103 (8)** 2725–2729
[3] Lacerda S M S N, Batlouni S R, Assis L H, Resende F M, Campos-Silva S M, Campos-Silva R, Segatelli T M and França L R 2008 Germ cell transplantation in tilapia (Oreochromis niloticus) *Cybium* **32**(2) 115-118
[4] Brinster R L and Avarbock M R 1994 Germline transmission of donor haplotype following spermatogonial transplantation *Proc Natl Acad Sci* **91** 11303–11307
[5] Freshney R I 2005 Culture of animal cells : a manual of basic technique (5th edition) (John Wiley and Sons Inc) 580 p
[6] Worthington 2009 Tissue dissociation guide (Worthington Biochemical Corporation)
[7] Kiernan J A 1990 Histological and histochemical methods : theory and practice (Toronto: Pergamon Press) 480p
[8] Takashima F and Hibiya F 1995 An atlas of fish histology: normal and pathological features 195p
[9] Griswold M D, McLean D and Russell L 2001 Promise and limitations of germ cell transplantation in the testis *Journal of Andrology* **22** 5
[10] Shinohara T and Brinster R L 2000 Enrichment and transplantation of spermatogonial stem cells. *Int J Androl* **23** 89–91
[11] Shinohara T, Avarbock M R and Brinster R L 1999 Beta1- and alpha6-integrin are surface markers on mouse spermatogonial stem cells *Proc Natl Acad Sci* **96** 5504–5509
[12] Grier H J 1993 Comparative organization of sertoli cells including the sertoli cell barrier *In Russell LD, Griswold MD (eds) The Sertoli cell* (Cache River Press) 704–739 p
[13] Pudney J 1993 Comparative cytology of the non-mammalian vertebrate sertoli cell *In Russell LD and Griswold MD (eds). The sertoli cell. Clearwater, FL* (Cache River Press) 657p.
[14] Nobreg R H and Quagio-Grassiotto I 2007 Morphofunctional changes in Leydig cells throughout the continuous spermatogenesis of the freshwater teleost fish, Serrasalmus spilopleura (Characiformes, Characidae): an ultrastructural and enzyme study. *Cell Tissue*

Res **329(2)** 339–349

[15] Loir M, Sourdaine P, Mendis-Handagama S M L C and Jégou B 1995 Cell-cell interactions in the testis of teleost and elasmobranchs *Micr Res Tech* **32** 533-552

[16] Vilela D A R, Silva S G B, Peixoto M T D, Godinho H P and França L R 2003 Spermatogenesis inteleost: insights from the Nile tilapia (Oreochromis niloticus) model *Fish Physiol Biochem* **28** 187-190

[17] Schulz R W, Menting S, Bogerd J, França L R, Vilela D A R and Godinho H P 2005 Sertoli cell proliferation in the adult testis–evidence from two fish species belonging to different orders *Biology of Reproduction* **73** 891-898

[18] Parenti L R and Grier H J 2004 Evolution and phylogeny of gonad morphology in bony fishes *Integ Comp Biol* **44** 333-348

[19] Leal M C, Cardoso E R, Nóbrega R H, Batlouni S R, Bogerd J, França L R and Schulza R W 2009 Histological and stereological evaluation of zebrafish (Danio rerio) spermatogenesis with an emphasis on spermatogonial generations *Biology of Reproduction* Papers in Press

[20] Kim Y, Selvaraj V, Dobrinski I, Lee H, Mcentee M C and Travis A J 2006 Recipient preparation and mixed germ cell isolation for spermatogonial stem cell transplantation in domestic cats *Journal of Andrology* **27** 2

[21] Takeuchi Y, Yoshizaki G, Kobayashi T and Takeuchi T 2002 Mass isolation of primordial germ cells from transgenic rainbow trout carrying the green fluorescent protein gene driven by the vasa gene promoter *Biology of Reproduction* **67** 1087-1092

[22] Takeuchi Y, Higuchi K, Yatabe T, Miwa M and Yoshizaki G 2009 Spermatogonial cell transplantation in marine teleost

Characterization of rice physicochemical properties local rice germplasm from Tana Toraja regency of South Sulawesi

A Masniawati, Nur Al Marwah Asrul, E Johannes and M Asnady

Department of Biology, Faculty of Mathematics and Natural Sciences, Hasanuddin University, Makassar, South Sulawesi, Indonesia

E-mail: nuralmarwahasrul@gmail.com

Abstract. The research about the characterization of physicochemical properties from local rice germplasm of Tana Toraja's Regency, South Sulawesi aims to determine the physicochemical properties of rice as a parameter to indicate the quality of cooking. Local varieties categorized as germplasm that needs to be protected for future varietal improvement.In this research, the researchers used seven varieties of local rice. The parameters analyzed including physicochemical properties of amylose content, protein content, gel consistency, and gelatinization temperature. Percentage of amylose content ranged from 2 to 18 %. Pare Bumbungan and Pare Lalodo are categorized as waxy rice and Pare Ambo, Pare Bau, Pare Kobo, Pare Rogon and Pare Tallang are categorized as low amylose content. The percentage of protein content ranged from 7.3 to 9.5 %. Gelatinization temperature of rice showed high gelatinization temperature. Pare Bumbungan, Pare Kobo, Pare Lalodo, and Pare Rogon are categorized as soft gel consistency (>50 mm). Pare Ambo, Pare Bau and Pare Tallang are categorized as medium gel consistency (36-50m). Pare Rogon and Pare Kobo are two kinds of rice varieties according to the quality of cooking criteria for consumers in Indonesia.

1. Introduction

Rice (*Oryza sativa* L.) is the carbohydrate sources consumed by humans. Increased productivity of rice plants is needed to fulfill these needs. Agricultural productivity is related to the potential of genetic varieties. Assembling of new varieties requires the availability of germplasm with a wide genetic diversity to be used as a crossing parent. Local varieties, in this case, local rice is categorized as germplasm that needs to be protected for future varietal improvement. The disappearance of existing local rice varieties is considered to threaten its genetic diversity. In fact, localized seed erosion local varieties of local rice varieties will disappear. Identification and description of the existence of local rice are necessary to prevent the extinction of local rice varieties [1].

In rice breeding, the identification of local rice quality is one of the indicators to assemble the new varieties. The good quality of local rice can be determined by the characteristic of cooking quality. It is divided into three kinds including sticky, medium, and fluffy rice. The cooking quality is determined by genetic factors than post-harvest treatment. [2].

Based on the explanation above, the researchers conducted the research of characterization of rice physicochemical properties from local rice germplasm in Tana Toraja, South Sulawesi. In this research, physicochemical properties are used as parameters in identifying the cooking quality of rice. It is analyzed by testing amylose content, gelatinization temperature, gel consistency, and protein content of seven local rice germplasms.

2. Materials and Methods

2.1. Materials

This research was conducted at the Chemical Laboratory, Center of Health Laboratory, Makassar, and Chemical Laboratory, Cereals Plant Research Institute, Maros. The materials in this research were 7 local rice germplasms (Pare Ambo, Pare Bau, Pare Bumbungan, Pare Lalodo, Pare Kobo, Pare Rogon, Pare Tallang) accession from Tana Toraja, South Sulawesi.

2.2. Methods

2.2.1. Analysis of Amylose Level. Measurement of amylose level was conducted according to the Rice Systems Evaluation Standard [3]. At the Standard Curve preparation stage, 40 mg pure amylose was fed into a 100 ml volumetric flask and added 1 ml of 95% ethanol and 9 ml of NaOH 1 N. The standard solution were sterilized for 24 hours and added aquatic until the mark on volumetric flask. Furthermore, the solution was dropped as much as 1, 2, 3, 4, and 5 ml and then put into a 100 ml volumetric flask. 2 ml Iodine solution was added in each volumetric flask. The solution also added of 0.5 N acetic acid each amount of 0.2; 0.4; 0.6; 0.8; and 1 ml. Add aquatic until the mark on volumetric flask. The solution was shaken and stand for 20 minutes. The solution was measured by the intensity of the color formed by the spectrophotometer at a wavelength of 620 nm. The standard curve describes the link between amylose concentration and absorbance.

At the measurement of the sample stage, a total of 200 mg of non-fat samples was introduced into the test tube and 1 ml of ethanol and 9 ml of NaOH 1 N were added. The sample solution was heated for 10 minutes then cool and transferred to a 100 ml 100ml flask, then adjusted until the mark with aquades. The solution was dropped as much as 5 ml, then put into 100 ml volumetric flask and added 2 ml of iodine solution and 1 ml of 0.5 N acetic acids. The subsequent solution and aquades were added until the mark, then the solution shaken, stuck for 20 min, and measured the color intensity which is formed by a spectrophotometer at a wavelength of 620 nm. Amylose content was calculated by equation 1:

$$Amylose\ Content\ (\%) = \frac{A}{S} \times \frac{FP}{W} \times 100\% \quad (1)$$

A : Sample absorbance at a wavelenght of 620 nm
S : slope on the standard curve
FP: dilution factor, 20
W : sample weight (gram)

Based on amylose content levels, rice is grouped into five groups[4], glutinous rice (0-2 %), very low amylose content (2-9%), low amylose content (10-19%), medium amylose content (20 – 25 %), high amylose content (>25%).

2.2.2. Analysis of Protein Levels, Micro Kjeldahl Method. The protein measurements based on the Micro-Kjeldhal Method were performed in 3 stages: destruction, distillation, and titration. A sample of 150 mg was fed into a volumetric flask and added 1.94 g of selenium reagent mixture catalyst, and 10 ml of concentrated H2SO4. Samples were boiled for 20 minutes until the liquid clear. The samples were boiled slowly, then cool again. Aquades was added until the mark and 3 drops of PP indicator. Then additional NaOH until the color of liquid changed into purple while cool using water the samples were distilled. Erlenmeyer contains 20 ml of H3BO3 solution placed under the condenser. The condenser tube shall be submerged under the H3BO3 solution. The distilled solution was added 3 drops of BCG

indicator. Then it was titrated with standardized HCl 0.02N. Titration was stopped until the colour change to the purple. The blanking was done using the same method as the sample.
The protein content was calculated by using the following equation:

$$N\ Content(\%) = \frac{(v_s - v_b) \times c \times 14.007}{W} \times 100\% \quad (2)$$

$$Protein\ Content\ (\%\ wet\ basis) = \ \%N \times \ Rice\ conversion\ factor \quad (3)$$

N : Nitrogen
v_s : HCL volume of titration sample (ml)
v_b : volume of blanko titration (ml)
C : HCL concentration (N)
W : sample weight (mg)
Rice Convertion Factor: 5.95

2.2.3. Gelatinization Temperature (GT). The gelatinization temperature was indicated by alkali spreading to test. The samples were evaluated by the degree of spreading of individual milled rice kernel in a weak alkali solution (1.7% KOH) at room temperature (30°C). 6 intact milled grains were placed in a petri dish to which 15 ml of 1.7% KOH was added. The grains were carefully separated from each other and incubated at ambient temperature for 23 hrs to allow spreading of the grains.

2.2.4. Gel Consistency Gel Consistency was determined by 100 mg rice flour. The flour was put into the tubes. 0.2 ml alcohol 95% (contain 0.025% thymol blue) and 2 ml KOH 0.2 N was added and then shaken tubes by a vortex mixer. The tubes were heated using a water bath to a temperature of 90°C for 15 minutes, lifted and then stand for 5 minutes, then cooled using ice water for 20 minutes. Tubes were laid horizontally for 1 hrs on a table lined with millimeter graph paper and the total length of the gel measured in millimeters.

Category of gel consistency [5] , high gel consistency <36 mm, medium gel consistency 36-50 mm, low gel consistency >50 mm.

3. Result and discussion

The measurement of amylose content based on International Rice Research Institute (IRRI) method [3]. The amylose content of samples was showed around 2.517- 18.295 %. The categories based on Dela and Khush (2000)[4], rice with an amylose content of 0-2% can be categorized as sticky rice. Rice with amylose content of 10 - 19% can be categorized as rice with low amylose content. The local rice from Tana Toraja categorized as glutinous rice were Pare Bumbungan (2.676 %) and Pare Lalodo (2.517%). Another local rices categorized as rice with low amylose content were Pare Ambo (18.295%), Pare Bau (17.570%), Pare Kobo (15.916%), Pare Rogon (17.797%), Pare Tallang (17.797%).

Table 1. Categorized local rice based on amylose content

Local Rice	Amylose Content (%)	Category
Pare Ambo	18.295	Low- Amylose Content
Pare Bau	17.540	Low- Amylose Content
Pare Bumbungan	2.676	Glutinous Rice
Pare Kobo	15.916	Low- Amylose Content
Pare Lalodo	2.517	Glutinous Rice
Pare Rogon	17.797	Low- Amylose Content
Pare Tallang	17.797	Low- Amylose Content

Pare Bumbungan and Pare Lalodo are categorized as glutinous rice. It contains no amylose (whole starch consisting of amylopectin), when it is cooked, the rice condition is very sticky, soft, moist, solid, less water-absorbing, and less expanding. While Pare Ambo, Pare Bau, Pare Kobo, Pare Rogon, dan Pare Tallang are categorized as rice with low amylose content. The low-amylose content of rice has a sticky rice characteristic, not too wet or dry. There is no high amylose rice in this study. Rice with the high amylose content has a fluffy rice characteristic, hard, and dry.

Differences in amylose content of rice are influenced by varieties, air temperatures of planting sites, and N levels in the soil. The study by Juliano [6] showed that the rice with the same variety but planted in the regions which have different nitrogen content in the soil and the air temperature of different planting locations will produce the rice with different amylose content. In line with this, the research by Rusmiati [7] also showed the glutinous rice and low-amylose content rice in Enrekang, South Sulawesi. Pare Pallan (2.033%), Pare Pulu Lotong (2.122%) and Pare Pinjan (2.033%) were categorized as glutinous rice, while Pare Solo (18.595%), Pare Salle (16.193 %) dan Pare Lambau (15.538 %) were categorized as low-amylose content rice. Enrekang and Toraja have the same natural form. Most of the rice was grown in the highlands.

Indonesia has many varieties of rice with various levels of amylose. Bengawan Solo, Tukad Petanu, Sentani, Membrano, Cilosari, and Cisadane are rice local varieties that categorized as low-amylose content. Bondoyudo, Pandan wangi, Rojolele, IR-64, Cibodas, Maros, and Way Opo Buru are categorized as medium-amylose content. IR-68, Batang Anai, Digul Indrasari Ratih, and IR-36 are categorized as high-amylose content [8].

Micro Kjeldahl Method was used to analyze the protein levels. The results of protein contents in 7 samples local rice from Tana Toraja showed amount of 7.3-8.5 %. The highest protein level was Pare Ambo (8.5%), and the lowest protein level was Pare Bumbungan (7.3%).

Table 2. Protein levels of local rice by micro kjeldahl method.

Local Rice	Protein Levels (%)
Pare Ambo	8.5
Pare Bau	8.2
Pare Bumbungan	7.3
Pare Kobo	8.4
Pare Lalodo	7.4
Pare Rogon	7.6
Pare Tallang	8

According to Juliano [6], the protein content of rice was in the range of 7%. The protein content of milled rice was strongly influenced by rice milling degree and the soil conditions in which the rice was grown. Rice grown on soils rich in N elements tends to have high levels of protein

Overall, the local rice from Tana Toraja has a protein content less than or equal to 8.5%. This shows the tendency of sticky rice texture [9]. Proteins have the dominant fraction of water-insoluble glutelin. The lower protein content can form a sticky rice texture. Sticky rice texture in accordance with amylose content analysis that indicates that local rice has a sticky rice texture.

Based on gelatinization temperature, rice can be classified into three types: low gelatinized (55-69 °C), medium temperature gelatinization (70-74 °C) and high gelatinization temperature (> 74 °C) [4].

All samples of local rice from Tana Toraja classified as high gelatinization temperature. For Pare Ambo, Pare Kobo, and Pare Lalodo were black rice, the color fades but did not affect rice texture changes. Pare Bau, Pare Bumbungan, Pare Rogon, and Pare Tallang did not change the texture of rice. Overall, it indicates a low spread of alkali in rice that can lead to low corrosion of the granules.

Table 3. Categorized local rice based on gelatinization temperature.

Local Rice	Rice Condition	Alkali Test	Gelatinization Temperature
Pare Ambo	Not affected, whiten	Low	High
Pare Bau	Not affected	Low	High
Pare Bumbungan	Not affected	Low	High
Pare Kobo	Not affected, whiten	Low	High
Pare Lalodo	Not affected, whiten	Low	High
Pare Rogon	Not affected	Low	High
Pare Tallang	Not affected	Low	High

According to Aryunis [9] gelatinization temperature associated with the time of cooking rice showed the increase of gelatinization temperatures will lengthen the cooking time of rice. Rice with low gelatinization temperature will absorb water and expands at a lower temperature than rice that has high gelatinization temperature [9].Gelatinization temperature influenced by granule characteristic and that presence protein, fat, and sugar components. Rice that has high protein content will require more water and more time to cook. This is related to the structure of the seed. Starch granules that wrapped by proteins and granular protein fill the spaces between starch granules in the endosperm.

The gel consistency value can be used as the index of the softness of the rice. The consistency of rice gel consistency from local rice from Tana Toraja which ranges from 36-50 mm is Pare Ambo, Pare Bau, and Pare Tallang which shows the index of moderate softness. While Pare Kobo, Pare Rogon, Bumbungan and Pare Lalodo exceed 50 mm including low gel consistency category which shows the index of high softness.

Table 4. Categorized Local Rice based on Gel Consistency.

Local Rice	Scale (mm)	Type of Gel Consistency
Pare Ambo	43	Medium
Pare Bau	48	Medium
Pare Bumbungan	108	Low
Pare Kobo	58	Low
Pare Lalodo	122	Low
Pare Rogon	59	Low
Pare Tallang	45	Medium

According to Aryunis [9] in his research, the consistency of gel associated with the texture of rice, the higher the consistency of the gel the more texture produced. The rice with the hardest consistency gel is Pare Bau, while the rice with the softest consistency gel is Pare Lalodo.

Pare Ambo is the rice with the lowest consistency of the gel scale (43 mm) which shows the gel consistency is harder than the others. This is in accordance with amylose content analysis, Pare Ambo is the highest amylose content compared to the others. While Pare Bumbungan and Pare Lalodo showed the consistency of the gel above 100 mm. Both rice is categorized as very low amylose content. Amylose levels affect gel consistency.

In the research conducted by Damardjati and Purwani [2], it showed that several rice varieties in Indonesia have a variety of gel consistency. Rice varieties of japonica (Padi Bulu from Central Java) include Rojolele (79.3 mm), Cendrawati (66.5 mm), and Hawarabatu (74.0 mm) categorized in low gel consistency (> 50 mm) with a high softness index. Indica rice varieties (Padi Cere from West Java) such as Angkong (63.0 mm), Jambe (70 mm), Ciamis (50 mm) are also included in low gel consistency category (> 50 mm) with high soft index.

4. Conclusion

Based on this research, it can be concluded that the local rice from Tana Toraja which categorized as sticky rice are Pare Bumbungan (2.676%) and Pare Lalodo (2.517%), while rice with low amylose content includes Pare Ambo (18.295 %), Pare Bau (17.570%), Pare Kobo (15.916%), Pare Rogon (17.797%), Pare Tallang (17.797%). The whole local rice contains protein content ranging from 7.3-8.5%. Overall, rice has a high gelatinization temperature. The consistency of rice gel from local rice from Tana Toraja regency ranging from 36-50 mm are Pare Ambo, Pare Bau and Pare Tallang, while Pare Kobo, Pare Rogon, Pare Bumbungan and Pare Lalodo are more than 50 mm including low gel consistency category.

Acknowledgement
Authors extend their appreciation to Dr. Eddy Soekandarsi, M.Sc as the head of department of biology who gave us support and technical guidance within the framework of the implementation of this study.

References

[1] Irawan, B and K Purbayanti. 2008. *Karakterisasi dan Kekerabatan Kultivar Padi Lokal di Desa Rancakalong, Kecamatan Rancakalong, Kabupaten Sumedang.* Thesis. Department of Biology. Faculty of Mathematics and Natural Sciences. Padjadjaran University. Bandung
[2] Damardjati, D S and E Y Purwani, 1991. *Mutu Beras. Di dalam : Padi Buku-3.* Balai Penelitian dan Pengembangan Tanaman Pangan, Bogor, pp.875-914
[3] IRRI, 1996. *Standard Evaluation System For Rice.* INGER Genetic Resources Center, International Rice Research Institute, Manila, Philippines
[4] Dela, N C and G S Khush, 2000. *Rice Grain Quality Evaluation Procedures.* In Singh, R.K., U.S. Singh, G.S. Khush. 2000. *Aromatic Rices.* Oxford & IBH Publishing Co. Pvt. Ltd., New Delhi, India, pp. 16-28
[5] Suprihatno *et al.* 2009. *Deskripsi Vaietas Padi.* Balai Besar penelitian tanaman padi. Departemen pertanian. Subang
[6] Juliano. 1972. *The Rice Caryopsis and It's Composition*, in: Houston.D.F. (Ed) *Rice Chemistry and Technology*. P. 16-74. Amer. Assoc. Cereal Chem. Inc, St. Paul Minn
[7] Rusmiati. 2012. *Karakterisasi Sifat Fisikokimia Beras Dari Plasma Nutfah Padi Lokal Di Kabupaten Enrekang Sulawesi Selatan.* Thesis. Department of Biology. Faculty of Mathematics and Natural Sciences. Hasanuddin University. Makassar
[8] Deliani, L. 2004. *Mempelajari Pengaruh Penyimpanan Beras Terhadap Beberapa Varietas Beras yang Berbeda Tingkat Kepulenannya.* Thesis. Department of Food Technology and Nutrition. Faculty of Agricultural Technology. Bogor Agricultural University (IPB). Bogor
[9] Aryunis, 2010. *Karakterisasi dan Identifikasi Mutu Beras dari Padi Lokal Asal Kabupaten Tanjung Jabung Barat.* Journal of Indonesian Center of Rice Research. Subang

IOP Conf. Series: Journal of Physics: Conf. Series **979** (2018) 012006 doi:10.1088/1742-6596/979/1/012006

Physical structure of artificial seagrass affects macrozoobenthic community recruitment

R Ambo-Rappe and C Rani
Marine Science Department, Hasanuddin University, Makassar 90245, Indonesia

E-mail: rohani.amborappe@gmail.com

Abstract. Seagrass ecosystems are important in supporting marine biodiversity. However, the worldwide decline in seagrass areas due to anthropogenic factors leads to a decrease in the marine biodiversity they can support. There is growing awareness of the need for concepts to conserve and/or rehabilitate seagrass ecosystems. One option is to create artificial seagrass to provide a physical structure for the marine organisms to colonize. The objective of this research was to analyze the effect of some artificial seagrasses and seagrass transplants on marine biodiversity, with a focus on the macrozoobenthic community. The experimental design compared two types of artificial seagrass (polypropylene ribbons and shrub-shaped plastic leaves), and seagrass transplants from nearby seagrass meadows. The experimental plots were 4 x 4 m^2 with 3 replicates. Macrozoobenthic communities were sampled fortnightly for 3.5 months. At the end of the experiment, makrozoobenthos were also sampled from a natural seagrass bed nearby. Of 116 macrozoobenthic species in the artificial seagrass plots, 91 were gastropods. The density of the macrobenthic fauna increased from the beginning to the end of the study in all treatments, but the increase was only significant for the artificial seagrass treatment (i.e. shrub-like plastic leaves). There was a distinct separation between the macrozoobenthic community structure found in the restoration plots (artificial seagrass and transplanted seagrass) compared to natural seagrass beds.

1. Introduction

Seagrass ecosystems are a shallow-water coastal marine ecosystem. In Indonesia, seagrass beds are often found adjacent to or between mangrove and/or coral reef ecosystems, and the importance of the multiple ecological interactions among these ecosystems is now universally recognised, if not yet fully understood. As the community structure and physical properties of the three ecosystems are mutually supportive, if one is disturbed, the other ecosystems will be affected [1].

On the one hand, seagrass beds provide valuable ecosystem services such as nursery grounds, beach protection, and water quality maintenance. On the other hand, industrial development and the growth of human populations in coastal areas have placed severe pressures on seagrass ecosystems, resulting in the degradation or even loss of seagrass ecosystems in many locations [2]. A decline in marine biodiversity is likely as a result of the partial or total loss of ecological functions following damage to or loss of seagrass beds [3]. One measure to mitigate this impact is to plant artificial seagrass in areas where seagrasses have been severely damaged or eradicated.

Artificial seagrass has been shown to serve as habitat for some marine organisms; artificial seagrass units could therefore be used as a refuge by some marine organisms, especially epifauna, when natural seagrasses are in critical condition [4]. The results of this pervious study inspired a more in-depth study on the effects of different artificial seagrass types on the macrozoobenthic community.

The objectives of the research were: 1) to analyze the effect of two types of artificial seagrass on marine biodiversity, with a focus on the macrozoobenthic community; 2) to evaluate the ecological function of artificial seagrasses by comparing macrozoobenthic communities found in artificial and natural seagrass habitat.

2. Materials and Methods

2.1. Study site
The experiment was conducted in the waters surrounding Barrang Lompo Island, Makassar, Indonesia (5°03'S, 119°20'E). Artificial seagrasses were deployed in the sub-tidal zone (± 1.5 meters at low tide) on the west side of the island, in areas not covered by natural seagrass, but with seagrass *Enhalus acoroides* stands nearby.

2.2. Treatments
Three treatments were used in this study. The first artificial seagrass treatment (A) used ribbon-like leaves made from plastic rope (width 1.6 cm; length 50 cm). The second artificial seagrass treatment (B) comprised plastic floral decorations used to make shrub leaves. The third treatment (C) was transplants of natural seagrass *E. acoroides*. The artificial seagrasses and transplant were tied to or planted in 4 m x 4 m iron frames.

All treatments were randomly deployed on the substrate parallel to the shoreline. The three replicates of each treatment were tagged with symbols A1, A2, A3 (Treatment A); B1, B2, B3 (Treatment B); C1, C2, and C3 (Treatment C). The distance between the treatment levels was 50 m.

2.3. Sampling of macrozoobenthos
Macrozoobenthos samples were collected using a grab sampler with an area of 20 x 20 cm^2. The first sampling was one week after deployment, followed by sampling every fortnight for three months. During the final data collection, macrozoobenthos samples were also collected from a natural seagrass bed nearby to enable comparison between the restoration plots and a natural seagrass ecosystem. The samples were sieved using a mesh size of 1 mm to separate the benthic fauna from sediments. The benthic fauna found were then preserved in 70% alcohol, stored in small jars and labelled. The macrozoobenthic organisms were identified based on [5,6,7].

2.4. Data Analysis
The macrozoobenthos community structure parameters or indices calculated were species diversity, population density, diversity index, evenness index, and dominance index based on [8].

Macrozoobenthos density was calculated using the formula:

$$Y = \frac{100 \times a}{b} \tag{1}$$

where: Y = Density (ind/m^2)
a = Number of individuals (ind)
b = Area sampled by the grab sampler (cm^2)

Diversity Index was calculated as follows:

$$H' = -\sum pi \ln pi\,;\ \ pi = ni/N \tag{2}$$

where: H'= Diversity Index
ni = Number of individual for each species
N = Total number of individuals

Evenness Index was calculated as follows:

$$E = \frac{H'}{\ln S} \tag{3}$$

where: E = Evenness Index
H'= Diversity Index
S = Number of species

Index of Dominance was calculated using the formula:

$$C = \sum \left(\frac{ni}{N}\right)^2 \quad (4)$$

where: C = Index of Dominance
ni = Number of individuals for the ith species (i = 1 to S)
N = Total number of individuals

Macrozoobenthic community structure parameters/indices were calculated and grouped according to treatment (3 treatments) and then analyzed using One-Way Analysis of Variance (ANOVA) for each observation period. To evaluate the effect of the presence of artificial seagrass and natural seagrass transplants on the macrozoobenthic community structure, data from before and after the deployment of each treatment were compared using a t-student test. In addition, similarities in macrozoobenthic community structure were also analyzed between treatments and natural seagrass using Cluster Analysis (CA) [9].

3. Results and Discussion

Macrobenthic fauna found in the artificial seagrass treatment plots comprised 116 genera. Of these, the majority were molluscs, with 91 gastropod genera (21825 individuals) and 12 bivalve genera (1250 individuals). The remainder comprised crustaceans (2 genera, 63 individuals), asteroids (2 genera, 625 individuals), polychaetes (2 genera, 88 individuals), sipunculids (one genus, 100 individuals), scapopods (3 genera, 800 individuals), and echinoids (3 genera, 200 individuals). The macrozoobenthic community structure indices (diversity, evenness, and dominance) for each treatment are shown in Figure 1.

Macrozoobenthic diversity index values ranged from 2.03 to 2.47. Overall, the macrozoobenthos diversity in all treatments was in the high category ($2.0 < H' \leq 3.0$). The Evenness Index was in the range of 0.83 to 0.96. This value indicates that the individuals of each genus or species were evenly distributed ($0.75 < E \leq 1.00$) [8]. This result was further supported by the Dominance Index of around 0.09, showing no signs of dominance of certain species of macrozoobenthos under any treatment.The fortnightly sampling data showed no significant differences among treatments in macrozoobenthic species number or density during any sampling period (Fig. 2).

Macrobenthic fauna density tended to increase during the study. The highest density was obtained from the final sampling in week 14 (571 ind/m^2), and was approximately double the baseline (before planting) density (267 ind/m^2), as shown in Fig. 3. This indicates that artificial seagrass and seagrass transplants can both attract some macrozoobenthos to inhabit the area and use it as a refuge.

The increase over time in the number of species and the macrozoobenthos density present in the two types of artificial seagrass and transplanted natural seagrass plots is in line with previous findings, which indicated that the number of species and density of epifaunal organisms inhabiting artificial seagrass is more strongly correlated with the length of deployment period of the artificial seagrass than with the type of material used to construct it [4]. This suggests that artificial seagrass may possibly be used as a foraging site as well for protection; epiphytes are a food source for many types of macrozoobenthic organisms, and the biomass of epiphytes attached to the structure would seems to be determined by the time elapsed since the artificial seagrasses were deployed.

Macrozoobenthos density increased from the beginning to the end of the study in all treatments, but the increase was only significant for treatment B (shrub-like plastic leaves) (Fig. 3). Artificial seagrass with shrub-like leaves may provide a higher protection because its leaf structure is more complex [10]. However, the increase in macrozoobenthic species was not significant under any of the treatments.

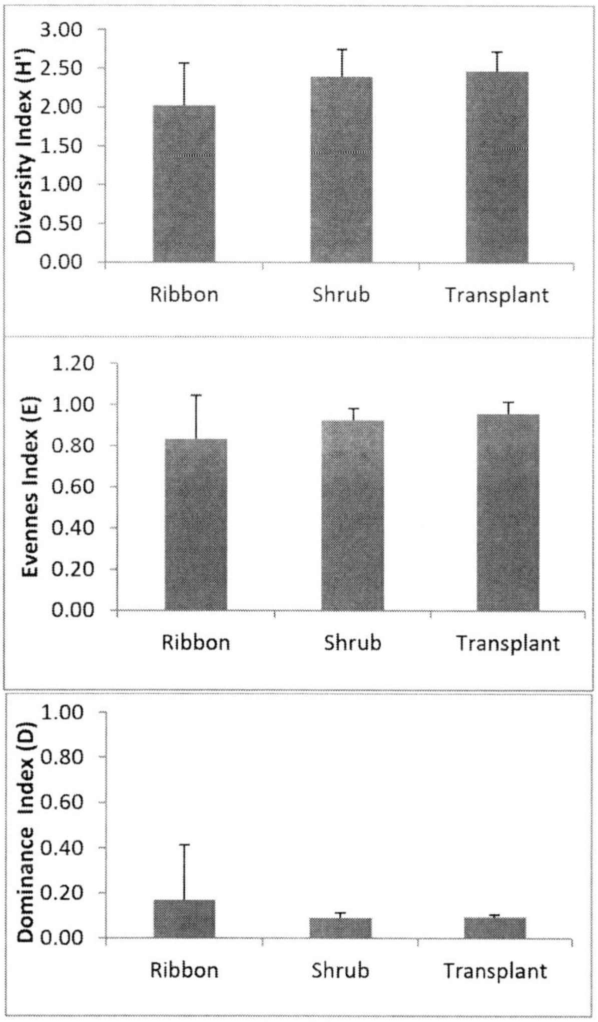

Figure 1. Macrozoobenthic community structure indices (diversity, evenness, and dominance) for each treatment (mean values over time and between replicates).

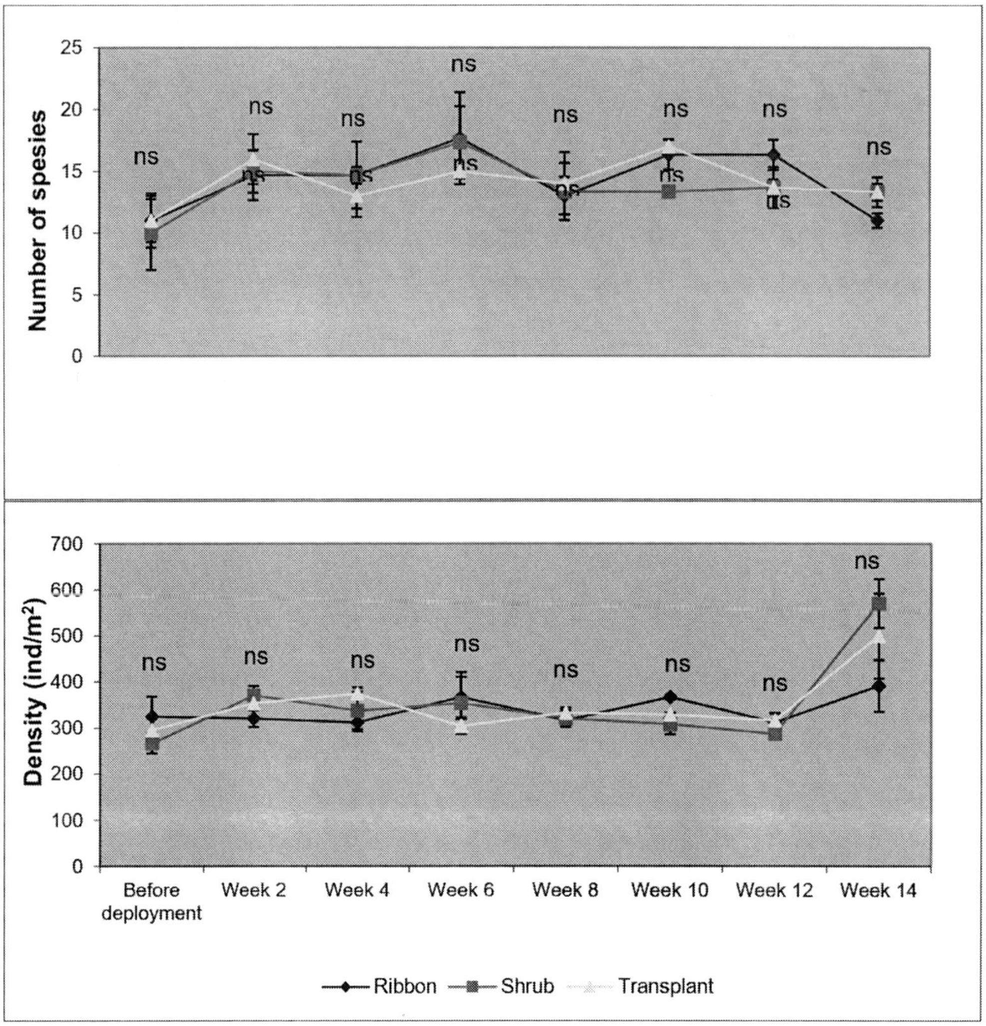

Figure 2. Mean number of species and density of macrozoobenthos during the observation period for each treatment.

The minimal separation in the cluster analysis (Fig. 4) shows a high degree of similarity in community structure between the three treatments. However there was a clear separation between the macrozoobenthic community structure of the three treatments and that of the nearby natural seagrass bed. One reason for this difference is the number of species found; whereas the treatment plots contained 116 species, there were 118 species in the natural seagrass area. The two distinct species found only in natural seagrass areas were *Nereis pelagica* and *Asterias astrubs*. Starfish of the class Asteroidea (*A. astrubs*) require seagrass habitats that contain a large amount of litter as food. While *N. pelagica* (class Polychaeta) require seagrass roots as a place to attach.

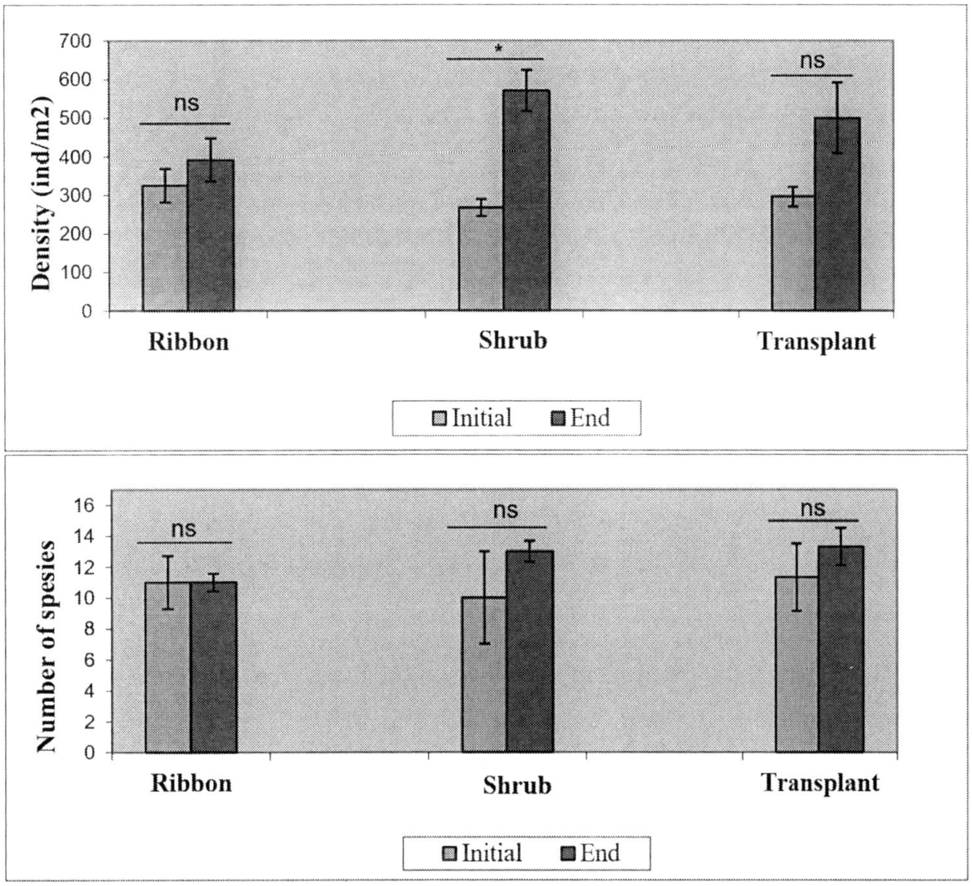

Figure 3. Comparison of initial and final mean macrozoobenthos density and species number by treatment.

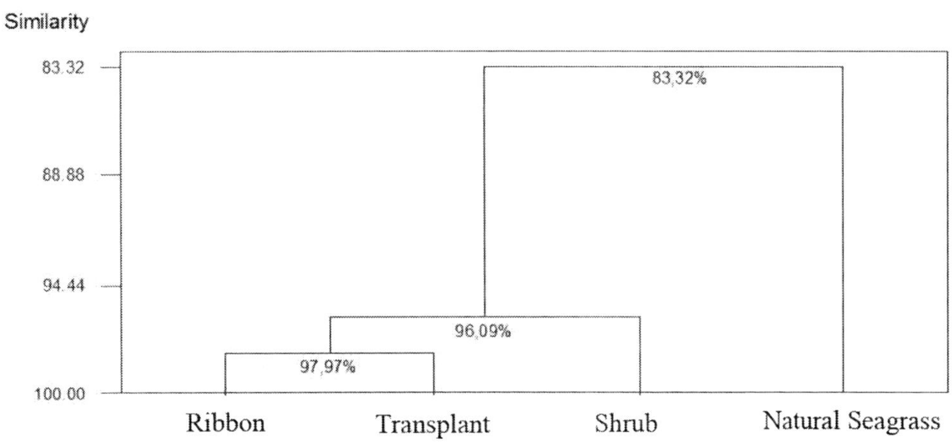

Figure 4. Similarity of macrozoobenthic community structure of the three treatments and natural seagrass bed.

References

[1] Unsworth R F K, De Leon P S, Garrard S L, Jompa J, Smith D J and Bell J J 2008 High connectivity of Indo-Pacific seagrass fish assemblages with mangrove and coral reef habitats *Mar Ecol Prog Ser* **353** 213-224

[2] Waycott M, Duarte C M, Carruthers T J B, Orth R J, Dennison W C, Olyarnik S, Calladine A, Fourqurean J W, Heck Jr KL, Hughes A R, Kendrick G A, Kenworthy W J, Short F T and Williams S L 2009 Accelerating loss of seagrasses across the globe threatens coastal ecosystems *Proc Natl Acad Sci USA* **106** 12377-12381

[3] Hughes A R, Williams S L, Duarte C M, Heck Jr K L and Waycott M 2009 Associations of concern: declining seagrasses and threatened dependent species *Fron Ecol Environ* **7** 242-246

[4] Ambo-Rappe R 2008 Study on the effect of seagrass fragmentation on associated mobile epifauna *Jurnal TORANI* **18** 418-426.

[5] Fauchal K 1997 *The Polychaeta Worms: Definitions and Keys to The Orders, Families and Genera* (Los Angeles: Natural Museum)

[6] Carpenter K E and Niem P H 1998 *The Living Marine Resources of the Western Central Pacific. FAO Species Identification Guide for Fishery Purposes* (Rome: Food and Agriculture Organization of the United Nations)

[7] Morton J 1990 *The Shore Ecology of The Tropical Pacific* (Jakarta: UNESCO Regional Office for Science and Technology for South-East Asia)

[8] Odum E P 1983 *Basic Ecology* (New York: Saunders College Publishing)

[9] Ter Braak C J F 1986 Canonical correspondence analysis: a new eigen vector technique for multivariate direct gradient analysis *Ecology* **67** 1167-1179

[10] Ambo-Rappe R 2016 Differences in richness and abundance of species assemblages in tropical seagrass beds of different structural complexity *J Environ Sci Tech* **9** 246-256

IOP Conf. Series: Journal of Physics: Conf. Series **979** (2018) 012007 doi:10.1088/1742-6596/979/1/012007

Using site-selection model to identify suitable sites for seagrass transplantation in the west coast of South Sulawesi

Mahatma Lanuru, S Mashoreng and K Amri

Department of Marine Science, Faculty of Marine Science and Fisheries, Hasanuddin University, Makassar 90425, Indonesia

E-mail: mahat70@gmail.com

Abstract. The success of seagrass transplantation is very much depending on the site selection and suitable transplantation methods. The main objective of this study is to develop and use a site-selection model to identify the suitability of sites for seagrass (*Enhalus acoroides*) transplantation. Model development was based on the physical and biological characteristics of the transplantation site. The site-selection process is divided into 3 phases: Phase I identifies potential seagrass habitat using available knowledge, removes unnecessary sites before the transplantation test is performed. Phase II involves field assessment and transplantation test of the best scoring areas identified in Phase I. Phase III is the final calculation of the TSI (Transplant Suitability Index), based on results from Phases I and II. The model was used to identify the suitability of sites for seagrass transplantation in the West coast of South Sulawesi (3 sites at Labakkang Coast, 3 sites at Awerange Bay, and 3 sites at Lale-Lae Island). Of the 9 sites, two sites were predicted by the site-selection model to be the most suitable sites for seagrass transplantation: Site II at Labakkang Coast and Site III at Lale-Lae Island.

1. Introduction

Seagrass bed is a valuable marine resource both ecologically and economically in Indonesia. The ecological role of this vegetation in the marine environment has been the subject of many studies. Increased human activities in coastal areas such as dredging or catchment run-off, transportation, cultivation activities, reclamation, overfishing and continuous domestic waste, can damage the seagrass habitat and degrade the conditions of the vegetation [1].

To overcome the destruction of seagrass beds, it is necessary to do restoration efforts one of which is by transplanting the seagrass plants on a substrate or a suitable habitat. Seagrass transplant success varies considerably. Of the possible factors that can directly influence the survival of transplanted seagrass, poor site-selection has been identified as the major limitation. Thus, site selection is an important step in any seagrass transplantation attempt [2-3].

Several methods to select suitable site for seagrass transplantation have been reported on the tropical and subtropical area. In the Thai Coastal water, [4] simply chosen a site where current and wave action are reduced for the seagrass tropical seagrass *Enhalus acoroides* (L.) transplantation experiments. Whereas, [5] selected the optimal site for restoration of the eelgrass *Zostera marina* in the Taehwa River estuary (Korea) based on the density, morphology, and growth rate of the transplanted seagrasses. In the Bangka Island water (Indonesia), [6] used several parameters including similarity environmental condition of donor site to the transplantation site, water depth, bed morphology, and wave exposure of the transplantation site as determined parameter in selecting suitable site for seagrass transplantation.

In the United States, a site-selection model for seagrass transplantation has been recently developed by [3]. The model synthesizes available historic and literature-based information, reference data and simple field measurements to identify and prioritise locations for large-scale seagrass *Z. marina* restoration. The model has been tested on several seagrass transplantation projects on the Northeastern coast of the United States with a high success rate (survival exceeds 62%). The site selection model introduced by [3] used parameters that commonly require for seagrass growth such as sediment type, wave exposure, depth, and water quality so that the model can be applied widely. To our knowledge, a site-selection model to identify suitable sites for tropical seagrass transplantation has not been attempted in Indonesia. Therefore, the main objective of this study is to develop and use a site-selection model to select suitable sites for tropical seagrass *Enhalus acoroides* transplantation in the West coast of South Sulawesi.

2. Material and Methods

2.1. Study site

This study was conducted on June - November 2009 at nine sites in the West coast of South Sulawesi (Indonesia), which are three sites in the Lae-Lae Island (Makassar City), Awerange Bay (Barru Regency) and coastal area of Labakkang (Pangkep Regency) (figure 1).

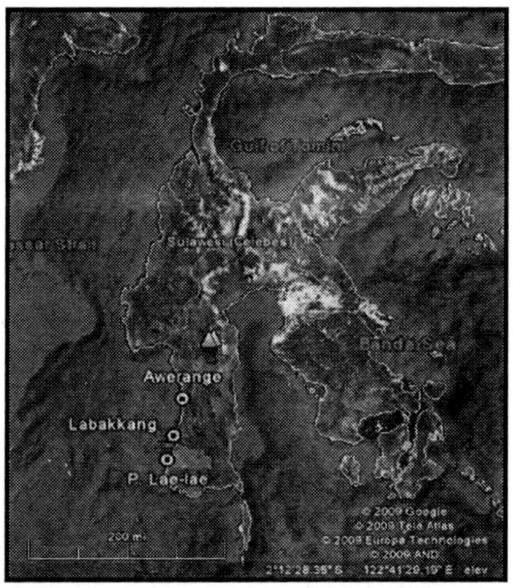

Figure 1. Map showing of the study sites in the West coast of South Sulawesi.

2.2. Methods

2.2.1. Development of Site-selection Model. The site-selection model was developed from the model introduced by [3] by making several modifications that suitable for Indonesia's tropical waters. The model is developed based on the physical and biological characteristics of transplantation sites such as the history of seagrass existence, seagrass distribution, sediment type, wave exposure, water depth, water quality, light intensity in the waters, and the existence of bioturbating marine organisms. The physical and biological characteristic data of the transplant sites is obtained from the results of field measurements. The physical and biological characteristic data of the transplant site was then compiled to formulate a quantitative site-selection model.

2.2.2. Application of Site-selection Model. The site selection model consists of three phases: Phase I identifies potential seagrass habitat using available knowledge, removes unnecessary sites before the transplantation test is performed. Phase II includes field assessment or transplantation tests at potential sites identified in Phase I. Phase III is the last stage in which a TSI (Transplant Suitability Index) score was calculated based on the results of Phase I and Phase II. Details of the activities of each stage of the site-selection model are presented as follows:
Phase I: Identify potential habitat for seagrass and PTSI Value Calculation. Phase I involves a review of available information to select and prioritize potential transplant sites. In this phase, information on characteristics of vegetated areas and areas without vegetation, sediment type, wave exposure, depth, and water quality are compiled to identify possible sites for seagrass transplantation, especially potential seagrass habitats. These characteristics contribute to the Preliminary Transplant Suitability Index (PTSI) calculation, which is used as initial screening in site selection (table 1) [3]. Due to information for the PTSI were not available for the study sites in published and unpublished literature, basic field measurements of the various PTSI parameters were performed.

Table 1. Data used during Phase I in the preliminary transplant suitability index (PTSI) for identification of potential seagrass habitat.

Parameter	PTSI rating	Reference
Historical seagrass distribution	1 for previously unvegetated 2 for previously vegetated	[7]
Current seagrass distribution	0 for currently vegetated 1 for currently unvegetated	
Proximity to natural seagrass bed	0 for < 100 m 1 for ≥ 100 m	[8]
Sediment	0 for rock or cobble 1 for > 70% silt/clay 2 for cobble free with <70% silt/clay	[9]
Wave exposure	0 for > mean + 2 SD of measurement result relative exposure index at each location 1 for ≤ mean + 2 SD of measurement result relative exposure index at each location	[10]
Water depth	0 for > 2 m during low tide 1 for < 0,5m or 1,5 – 2m during low tide 2 for 0,5 m – 1,5 m during low tide	Modified from [11]
Water quality (total suspended solid/TSS concentration)	1 for > 20 mg l^{-1} 2 for < 20 mg l^{-1}	[12]

Each of the parameters presented in table 1 receives a rating. The PTSI score is calculated by multiplying the ratings of each parameter in Table 1. PTSI scores that may appear are 0, 1, 2, 4, 8, and 16. Site that has any parameter with 0 (zero) rating will cause the score PTSI becomes 0, and the site will be excluded and no longer considered in the next site selection stage. Whereas a site with a higher

PTSI score has greater likelihood as a successful area for seagrass planting and the site is ranked for further evaluation to obtain a TSI (Transplant Suitability Index) score.

Phase II: field assessments and test-transplanting. Phase II is a continuation of Phase I where at this stage a field assessment and transplantation test at a site with a high PTSI score identified at Phase I. At this stage, site-specific light intensity data was collected, assessment of bioturbation activity and monitoring the rate of survival of transplanted seagrasses were conducted.

The light conditions at each site were measured with a lux meter to determine the intensity of light arriving at the bottom of the water. The existence of fauna that potentially disrupts seagrasses (bioturbators) such as crabs, shrimp diggers, worms, stingrays, green urchin, and swans was observed through an underwater visual survey with the help of video or divers. At each selected site (with a high PTSI score), seagrass (*E. acoroides*) transplantation tests were conducted using Sprig method [13, 16]. 50 planting units were planted in each 1 x 2 m plot with intervals of 0.2 m. Planting unit survival was monitored every month for two months.

Phase III: calculating the TSI (Transplant Suitability Index) score. In Phase III, the TSI Score was calculated for each test-transplanting site (table 2). TSI was determined by combining the results of PTSI in Phase I and measurement results of light intensity, bioturbator activity and transplantation test (seagrass survival rate) obtained in Phase II.

Each site is assigned a rating for each parameter in Table 2. First, the PTSI rating is assigned: 0 if the PTSI score is ≤1; 1 if the PTSI is 2 or 4; 2 if the PTSI is ≥8. TSI ratings for light are assigned as follows: 0 for light <20% surface irradiance; 1 for light ≥20% surface irradiance; 2 for light greater than light measured in the reference seagrass beds. The TSI rating for bioturbation is based on the number of bioturbating organisms observed at a site; 0 if abundant bioturbators were observed, 1 for the observed presence of bioturbators; 2 if no bioturbating organisms were found. After assessment of the test transplants, 0 is assigned for <20% survival, 1 for 20 to 40% survival, and 2 for >40% survival of the transplanted seagrass. The TSI score is finally calculated as follows: TSI = PTSI x light conditions x bioturbator activity x seagrass survival rate [3]. Site that has a zero TSI score will not be considered for transplantation. Similarly, sites that have a TSI score <8 were also rejected for transplantation. Sites with the highest TSI score at the end of Phase III will be selected as the most suitable site for seagrass transplantation.

Table 2. Data from Phase II for calculation of the transplant suitability index (TSI) to select the most suitable seagrass transplant sites.

Parameter	TSI Rating	Reference
PTSI	0 for PTSI = 0 - 1 1 for PTSI = 2 – 4 2 for PTSI = 8 – 16	
Light (field data)	0 for <20% surface irradiance 1 for ≥20% surface irradiance 2 for irradiance > reference seagrass	[11, 12]
Bioturbation (field data)	0 for abundant 1 for present 2 for not present	[13]
Seagrass survival (test-transplants)	0 for < 20 % 1 for 20 – 40 % 2 for > 40 %	[3]

3. Results and Discussion

3.1. Phase I: identify potential habitat for seagrass and PTSI Value calculation

The seagrass history distribution parameters are recommended by [7]) as one of the important considerations in choosing a transplant location. The seagrass existence history at the study sites obtained from interviews with local communities showed that all site was previously occupied/vegetated by seagrass except at the Site I of Awerange Bay and Site II in the Lae-Lae Island. Previously vegetated sites receive a PTSI rating of 2 and previously unvegetated sites receive a rating of 1 (table 1).

Current seagrass distribution is important knowledge relative to transplanting activities because restoration should not be done at sites where seagrass already exists. The sites were assigned a PTSI rating of 0 if the sites currently vegetated with seagrass and a rating of 1 when the sites are currently unvegetated. The results of field measurements and based on the available seagrass distribution map indicate that all current sites were not overgrown with seagrass and hence all sites receive a PTSI rating of 1 for this parameter.

The distance of a possible transplant site from natural seagrass was included to ensure that transplanting is taking place outside an area that could be naturally revegetated by seed. If a site is less than 100 m from a natural seagrass bed, it is considered within the range of natural revegetation, based on [8], and the sites receive a PTSI rating of 0. If a site is over 100 m from a natural seagrass bed, it is rated 1. The results of field measurements and based on the available seagrass distribution map showed that the distance of all sites from natural seagrass beds was more than 100m and therefore all sites achieving a rating of 1.

The sediment particle size is an important parameter that has a major influence on the growth and spread of seagrass [9]. The results of sediment grain size analysis at each study site indicated that the sediment on the site was all sandy (free of pebbles) with mud content <70% except at Site I of Awerange Bay and Site II of Lae-Lae Island where seabed sediment of the two sites was rocky (corals). Sites that composed of sandy sediment with mud content <70% received a PTSI rating of 2 whereas sites with rocky (coral) substrate received a PTSI rating of 0 (not suitable for seagrass growth).

The wave exposure was incorporated as one of the parameters in PTSI because it has been shown by previous studies that wave energy can damage the leaves and remove plants that can obviously interfere with transplanted seagrasses [7]. The exposure of the transplant sites to the waves was determined by calculating the relative exposure index following the method introduced by [10] based on wind speed and wind direction data and effective fetch length. The relative exposure index calculation results showed that the relative exposure index value at each location was less than mean + 2 SD at each measurement location, and therefore all locations achieving a PTSI rating of 2.

In the site selection model developed by [3], very shallow waters (<0.5 m at low tide) or deeper water (> 2 m at low tide) are given a PTSI value of 0 because they are not suitable for *Z. marina* (eelgrass) growth. Seagrass *E. acoroides* that we tested in this study can still grow well at depths of less than 0.5 m or even in the intertidal zone exposed for several hours at the lowest tide [17] Therefore, the parameters of depth parameters in the model that developed in this study were given PTSI 1 for the location where the depth is less than 0.5 m and 0 for the depth > 2 m because the light arriving at the bottom of the waters is limited to depth > 2m. While the optimal depth for transplantation is 0.5 – 1.5 m at low tide and given a PTSI value of 2. The result of depth measurement at each study site indicated that almost all site had a depth between 0.5 m and 1.5 m except at Site I of Labakkang coastal area and sites II and III of Awerange Bay where water depth was less than 0.5 m.

Similar to water depth, water quality is a location-specific parameter, with various types of data suitable to rate water quality for the PTSI, depending on what is available [3]. In this study, the total suspended solids (TSS) were used as water quality parameter. According to [12], seagrass plants can grow well if water TSS concentration less than 20 mg l^{-1}. Therefore sites where the TSS concentration < 20 mg l^{-1} are rated 2 while sites with the TSS concentration > 20 mg l^{-1} are rated 1 (see table 1). The measurement of TSS concentrations at each study site showed that all sites had a concentration of TSS > 20 mg l^{-1}, and therefore all sites received a PTSI rating of 1.

PTSI rating of each physical and biological parameter explained above is summarized in table 3. Ratings for each parameter in the table 3 are multiplied to determine PTSI score for each site. Results of the PTSI Score calculation are also presented in table 3. As explained earlier, a site with a higher PTSI score has greater likelihood as a successful area for seagrass planting and the site is ranked for further evaluation to obtain a TSI score. Based on the PTSI scores presented in table 3, sites II and III of Labakkang Coast, Sites II and III of Awerange Bay, and Sites I and III of Lae-Lae Island were sites that have a greater likelihood of being successful areas for seagrass planting so that these sites were evaluated further to get a TSI score. The three other sites were no longer considered in the next site selection stage because of their small PTSI score.

Table 3. Parameter ratings of the PTSI and the final PTSI score for the 9 transplant sites. Ratings for each parameter are multiplied to determine PTSI score for each site.

Site name	Historical seagrass	Seagrass distribution	Proximity	Water depth	Water quality (TSS)	Sediment	Wave exposure	PTSI Score
Site I (Labakkang)	2	1	1	1	1	2	1	4
Site II (Labakkang)	2	1	1	2	1	2	1	8
Site III (Labakkang)	2	1	1	2	1	2	1	8
Site I (Awerange)	1	1	1	2	1	0	1	0
Site II (Awerange)	2	1	1	1	1	2	1	4
Site III (Awerange)	2	1	1	1	1	2	1	4
Site I (Lae-lae)	2	1	1	2	1	2	1	8
Site II (Lae-lae)	1	1	1	2	1	0	1	0
Site III (Lae-lae)	2	1	1	2	1	2	1	8

3.2. *Phase II: field assessments and transplantation test*

At this phase, site-specific light intensity data was collected, assessment of bioturbation activity and monitoring survival rate of the transplanted seagrasses were conducted. The result of measurement of light intensity and macrofauna condition at each study site and the survival rate of seagrass transplants are presented in table 4.

Table 4. The light intensity, macrofauna condition (bioturbator), and survival rate of seagrass transplants.

Site name	Light intensity (%)	Bioturbation activity	Survival rate of transplanted seagrass (%)
Site II (Labakkang Coast)	75	not present	86
Site III (Labakkang Coast)	69	not present	72
Site II (Awerange Bay)	68	present	62
Site III (Awerange Bay)	58	present	40
Site I (Lae-Lae Island)	73	not present	84
Site III (Lae-Lae Island)	69	not present	90

Light is a critical factor of the survival and growth of seagrass [14]. Reduced light conditions that get to the bottom have been implicated in limiting seagrass metabolic activity which can eventually cause seagrass death [15]. The results of the measurement of light intensity on the surface and the bottom of the waters at each study site by using lux meters showed that the intensity of light reaching the bottom was greater than 20% in all sites, and therefore all sites received a TSI rating of 1.

Bioturbation has been noted as one of the important factors affecting natural seagrasses and transplanted seagrasses. The visual observations of epifauna combined with infauna observations in the sediments showed that almost all transplantation sites were free of fauna that potentially interferes seagrass transplants except in two sites at the Awerange Bay. At the two sites, several gastropods of *Cerithium sp* were observed inside the transplantation plots and hence these the two sites of Awerange Bay received a TSI rating of 1. While sites in the Labakkang coast and Lae-Lae Island were rated 2 because bioturbators that potentially disrupt the transplants were not found at these locations.

A final and critical step in assessing a site for seagrass transplantation is to conduct a test-transplant, in which the transplants respond to the prevailing environmental conditions and integrate the combination of site-specific factors which might limit their survival and growth [3]. The result of seagrass transplantation tests showed that the transplants survival rates varied from 40% to 90%. Site III in the Awerange Bay received a TSI rating of 1 because the seagrass survival rate of the site was only 40%, and other sites were rated 2 because the survival rates of the seagrass were > 40%. Low survival rates observed in the sites of Awerange Bay were most likely due to the presence of epibenthic bioturbating organism, i.e. gastropods *Cerithium sp* that damage seagrass transplants. In addition, observed lower light intensity in the sites of Awerange Bay compared to those in the Labakkang coast and Lae-Lae Island may be responsible for lower seagrass survival rate in Awerange Bay. As explained earlier that reduced light conditions that get to the bottom has been implicated in limiting seagrass metabolic activity which can eventually cause seagrass death.

3.3. Phase III: calculating the TSI (Transplant Suitability Index) score

In Phase III, the TSI Score was calculated for each test-transplanting site. TSI was determined by combining the results of PTSI in Phase I and measurement results of light intensity, bioturbator activity and transplantation test (seagrass survival rate) obtained in Phase II. Each site is assigned a rating for PTSI, light conditions, bioturbator activity, and seagrass survival rate (table 5). Ratings for each parameter in the table 5 were multiplied to determine TSI score for each site. Site that has a zero TSI score will not be considered for transplantation. Similarly, sites that have a TSI value < 8 were rejected for transplantation. Sites with the highest TSI score at the end of Phase III will be selected as the most suitable site for seagrass transplantation.

Table 5. Parameter ratings of the TSI and TSI Score for 6 seagrass tansplantation testing sites. Ratings for each parameter are multiplied to determine TSI score for each site.

Site name	PTSI	Light intensity	Bioturbator activity	Survival rate of transplanted seagrass	TSI Score
Site II (Labakkang Coast)	2	1	2	2	8
Site III (Labakkang Coast)	2	1	2	2	8
Site II (Awerange Bay)	1	1	1	2	2
Site III (Awerange Bay)	1	1	1	1	1
Site I (Lae-Lae Island)	2	1	2	2	8
Site III (Lae-Lae Island)	2	1	2	2	8

Based on the TSI score presented in table 5, the most suitable sites for the seagrass transplantation were Site II at Labakkang coast and Site III at Lae-Lae Island because in addition to having a high TSI score, the transplantation test results also indicated high seagrass survival rates at these two sites. The sites in the Aweange Bay were rejected to be a suitable seagrass transplant site because they had a small TSI score (< 8). The site selection model was developed and tested in the West coast of South Sulawesi. For future applications beyond this region, the model requires additional testing and evaluation sites to proof that our site selection is working in the other tropical waters.

4. Conclusion

Application of site-selection model on nine sites in the West coast of South Sulawesi indicated that the most suitable sites for seagrass *E. acoroides* transplantation were Site II in the Labakkang Coast and

Site III in the Lae-Lae Island. These two sites were selected as the most suitable sites for seagrass transplantation because in addition to having high TSI (Transplant Suitability Index) scores, the test-transplanting showed high seagrass survival rates at the two sites.

References

[1] Tomascik T, Mah A J, Nontji A and Moosa M K 1997 *The Ecology of The Indonesian Seas: Part Two* (Singapore: Periplus Edition)

[2] van Katwijk, M M, Bos A R, de Jonge V N, Hanssen L S A M, Hermus D C R. and de Jong D J 2009 Guidelines for seagrass restoration: Importance of habitat selection and donor population, spreading of risks, and ecosystem engineering effect *Mar. Pollut. Bull.* **58**: 179–188

[3] Short F T, Davis R C, Kopp B S, Short C A, Burdick D M 2002 Site-selection model for optimal transplantation of eelgrass Zostera marina in the northeastern US *Mar. Ecol. Prog. Ser.* **227** 253–267

[4] Vichkovitten T, Intarachart A, Khaodon K and Rermdumri S 2016 Transplantation of Tropical Seagrass Enhalus acoroides (L.) in Thai Coastal Water: Implication for Habitat Restoration *GMSARN International Journal* **10** 113 – 120

[5] Park J I, Kim J B, Lee K S and Son M H 2013 An Experimental Transplantation to Select the Optimal Site for Restoration of the Eelgrass *Zostera marina* in the Taehwa River Estuary *Ocean Sci. J.* **48** 311-318

[6] Seprianti R, Karlina I and Irawan H. Laju Pertumbuhan Jenis Lamun Thalassia hemprichii dengan Teknik Transplantasi Sprig Anchor dan Polybag pada Jumlah Tegakan yang Berbeda dalam Rimpang di Perairan Kabupaten Bintan *Intek Akuakultur* **1** 56 – 70

[7] Fonseca M S, Kenworthy W J, Thayer G W 1998 *Guidelines for conservation and restoration of seagrass in the United States and adjacent waters*. NOAA/NMFS Coastal Ocean Program Decision Analysis Series, No. 12. NOAA Coastal Ocean Office, Silver Spring, MD

[8] Orth R J, Luckenbach M and Moore K A 1994 Seed dispersal in a marine macrophyte: implications for colonization and restoration *Ecology* **75** 1297–1939

[9] Short F T 1987 Effects of sediment nutrients on seagrasses, literature review and mesocosm experiment *Aquat. Bot* **27** 41–57

[10] Murphey P L and Fonseca M S 1995 Role of high and low energy seagrass beds as nursery areas for *Penaeus duorarum* in North Carolina *Mar. Ecol. Prog. Ser.* **121** 91–98

[11] Davis R C 1999 The effects of physical and biological site characteristics on the survival and expansion of transplanted eelgrass (*Zostera marina* L.). PhD dissertation. University of New Hampshire, Durham

[12] Republik Indonesia 2004 Keputusan Menteri Negara Lingkungan Hidup No. 51 Tahun 2004 Tentang Baku Mutu Air Laut. Sekretariat Negara. Jakarta.

[13] Calumpong H P and Fonseca M S 2001 Seagrass transplantation and other seagrass restoration methods *Global Seagrass Research Methods* ed F T Short and R G Coles (Amsterdam: Elsevier) pp 425–443

[14] Dennison W C 1987 Effects of light on seagrass photosynthesis, growth and depth distribution *Aquat. Bot* **27** 15–26

[15] Zimmerman R C, Reguzzoni J L, Alberte R S 1995 Eelgrass (*Zostera marina* L.) transplants in San Francisco Bay: role of light availability on metabolism, growth and survival *Aquat. Bot* **51** 67–86

[16] Davis R C and Short F T 1997 Restoring eelgrass, Zostera marina L., habitat using a new transplanting technique: the horizontal rhizome method *Aquat. Bot* **59** 1–15

[17] Stapell J, Manuntun R and Hemminga H A 1997 Biomass loss and nutrient redistribution in an Indonesian *Thalassia hemprichii* seagrass bed following seasonal low tide exposure during daylight *Mar. Ecol. Prog. Ser.* **148** 251-262

Gene distribution of ABO blood type system on the Dengue Hemorrhagic Fever (DHF) patients in the working area of Puskesmas Bonto Bangun, District of Rilau Ale, Bulukumba

Sjafaraenan, D N Alvionita, R Agus and A Sabran

Department of Biology, Faculty of Mathematics and Natural Sciences, Hasanuddin University, Makassar, South Sulawesi, Indonesia

E-mail: enan.gininnur@gmail.com

Abstract. This research is about gene distribution of ABO blood type system on the Dengue Hemorrhagic Fever (DHF) patients in the working area of Puskesmas Bonto Bangun, District of Rilau Ale, Bulukumba. This research aimed to determine the blood type which is most affected by DHF using ABO blood type system. In this research, there are 104 samples, 8 of them were attacked by DF and 96 were attacked by DHF. From the 96 patients of DHF, there were 38 patients with A-blood type, 17 patients with B-blood type, 36 patients with O-blood type and 5 patients of AB-blood type. The data were tested using genotype frequency test and the results showed that the percentage of A-homozygous blood type (I^AI^A) is 0:09%; A heterozygous blood type (I^AI^O) is 0:36%; B-homozygous blood type (I^BI^B) is 0.01%; B heterozygous blood type (I^B I^O) is 0.12%; AB blood type (I^AI^B) is 0.06% and O blood type (I^OI^O) is 12:36%. So the biggest frequency of genotype are I^AI^O (0.36%) and I^OI^O (0.36%). The results showed that O blood type gene is the most affected by DHF. Then continued by the regression test between blood type and DHF, it is obtained that the correlation value is 1 which indicated that there is a strong relationship.

1. Introduction

Blood is a liquid tissue consisting of two parts, namely blood plasma and blood cells. Blood cells consist of three types: erythrocytes, leukocytes and platelets. The overall blood volume is one-twelfth of the body weight or about five liters. About 55% are blood plasma, while the remaining 45% consists of blood cells. The main function of blood in the circulatory system is as a medium of transport, regulation of temperature, maintenance of fluid balance, and balance of erythrocyte base during his life remain in the body. The red blood cells are capable of transporting effectively without abandoning its function in the tissues, while its presence in the blood, passes only [1].

According to Land-Steiner in Suryo [2] ABO blood type system can be inherited from to their children. ABO blood type system is divided into four groups namely A, B, AB and O. Blood classification is caused by the type of antigen contained by erythrocytes (red blood cells). The antigens can be carbohydrates, proteins, glycoproteins, or glycolipids and also some of these antigens are in cells of various muscle tissues. Human blood type is hereditary and highly dependent on the blood group of both humans concerned [3].

Dengue Hemorrhagic Fever (DHF) has become an important concern of the health community in the world. Indonesia occupied the highest ranking in ASEAN in the incidence of Dengue Infection. In

2014, until mid-December recorded dengue cases in 34 provinces in Indonesia as many as 71,668 people, and 641 of them died. The number is lower than the previous year, namely in 2013 with the number of people as many as 112,511 people and died as many as 871 patients [4]. Meanwhile data from Disease Control and Environmental Health, Department of Health of South Sulawesi, released data of DHF patients during January 2016 as many as 528 cases. Head of Disease Control and Environmental Health said the data is based on reports from the Regional Public Hospital in 24 districts in South Sulawesi [5]. Meanwhile, data obtained from Puskesmas Bonto Bangun is the number of DHF patients from January to March as many as 142 cases.

Based on the epidemiology paradigm, the outline of a DHF sufferer is influenced by host, agent and environment interaction [6]. Therefore, in addition to the need to deepen the problem of the agent and the handling of environment that support the transmission of dengue is very important also studied more in the role of host (human) in DHF disease. Manifestations of dengue virus infection in humans are varied due to the pathogenesis of dengue disease is very complex. Based on existing research, host factors that may affect the manifestation of dengue infection are ethnicity, heredity, gender, age, nutritional status, secondary infection and immune response [7].

The immune response factor in the host body is different because the immune response is controlled by the natural genes of the human body. Several studies have shown that genes that are suspected to be associated with manifestations of dengue hemorrhagic clinical symptoms include genes that encode the Human Leucosyte Antigen (HLA) molecule, Human Platelet Antigen (HPA), ABO blood group system, IgG receptor cells, Vitamin D receptor and Mannose Binding Lectin.

Based on research conducted Kalayanarooj, et al., [8] one that may be associated with severe dengue virus infection is a factor of blood type. one that may be related to the severity of Dengue virus infection is a factor of blood type. This blood type factor is based on the ABO blood type system, because this blood type system is commonly used. This study suggests that blood type factors play a role in viral elimination. However, due to limited samples there was no significant difference in each blood group. In addition, according to Rahayu, et al., [9] also obtained a description of blood type AB is a risk factor for the occurrence of Dengue Shock Syndrome (DSS) in children with DHF. This phenomenon is thought to be caused by cross-reactions between antigens and antibodies owned by DHF patients.

Another opinion expressed by Hartanto [10] in his research on blood type O relationship with shock occurrence in DHF patients found no significant association between blood type O and shock occurrence in DHF patients but there were differences about clinical manifestations of significant bleeding among blood group O and non blood group O. This can lead to worsening of shock in DHF. Further research conducted by Kurniawan [11] in his research on the relationship of blood type ABO against dengue virus infection, from the study concluded that blood type O has a relationship to dengue virus infection which is the factor of occurrence of DHF. Based on the above background then conducted a study on DHF patients in the work area of Puskesmas Bonto Bangun to know the blood type most affected by DHF based on the ABO blood type system.

In this study, we have analyzed serological parameters of DHF carriage in a group of 104 adults. The 8 participants were confirmed positive for Dengue Fever (DF) and 96 participants were confirmed positive for Dengue Fever (DF) at Puskesmas Bonto Bangun, Carriers were tested for identified bood type and analysed it by gene frequency analysis with Weinberg and regression analysis test.

2. Materials and Methods

Materials used in this study were blood samples of dengue patients, 70% alcohol, serum anti-A type, serum anti-B type, and cotton. For blood sampling, we inserted the lanset into the autoclick and set it at the appropriate depth of the number. Then, apply 70% alcohol on the surface of the respondent's fingers. Next, press the autoclick over the fingers that have been smeared with alcohol. Then press the stabbed hand to get the blood out. The blood is placed on the glass of preparations. Then, drop the serum anti-A and serum anti-B. Observe the changes that occur on the glass that has been dropped blood, serum anti-A and serum anti-B.

The population in this study were all people suspected of having Dengue Hemorrhagic Fever who checked himself into Puskesmas Bonto Bangun. From the preliminary data taken from puskesmas there are about 140 population.

The sample is an affordable part of the population that can be used as a research subject by sampling. Sample size can be calculated by this formula:

$$n = \frac{N}{1 + N(d^2)} \quad (1)$$

n = sample size
N = population size
d = level of significance (e)

Based on that formula, sample that needed in this research are 104 people.

Data analysis was done in 2 ways, ie gene frequency analysis with Weinberg and regression analysis test. Gene frequency analysis was performed using the Hardy-Weinberg equilibrium law. The equations of the Hardy-Weinberg equilibrium law are:

$$p + q + r = 1 \quad (2)$$

p = Person with blood type A
q = People with blood type B
r = People with blood type O

Regression analysis test is used to determine whether there is influence of blood group with DHF. Regression analysis test was done by using SPSS 22 tools.

3. Result and discussion

Based on data from research that has been done in the work area of Puskesmas Bonto Bangun, Rilau Ale, Bulukumba District in September to December with the number of samples of 104 people. In this study other than blood type, also researched the sex and the type of infection in the respondent and obtained the following results:

Table 1. Data based on the respondent sex.

Sex	Total	Percentage (%)
Male	57	54.8
Female	47	45.2

From the research data, there are 104 respondents, it is known that the number of dengue male respondents is 57 people (54.8%) and female (47.2%).

Table 2. Data based on infection type.

Infection Type	Total	Percentage (%)
DF	8	7.7
DHF	96	92.3

Based on the type of dengue infection, the data obtained as many as 8 people (7.7%) of DF and as many as 96 people (92.3%) attacked by DHF.

Table 3. Data based on blood type.

Blood Type	Total	Percentage (%)
A	42	40.3
B	18	17.3
AB	5	4.9
O	39	37.5

Based on blood type, it was found that A blood type result were 42 people (40,3%), B blood type were 18 people (17.3%), AB blood type were 5 people (4.9%) and O blood type were 39 people (37.5%).

Based on research conducted, DF patients as many as 8 people. Patients with A blood type as many as 4 people (50%), people with B blood type as many as 1 person (12.5%), people with O blood type as many as 3 people (37.5%) and AB blood type sufferers.

Table 4. Patients with DF based on blood type.

Blood Type	Total	Percentage (%)
A	4	50
B	1	12.5
AB	0	0
O	3	37.5

Table 5. DHF based on Blood Type.

Blood Type	Total	Percentage (%)
A	38	37.59
B	17	17.70
AB	5	5.20
O	36	37.5

Based on the research conducted on 96 DHF patients, there were 38 people (37,59%) with DHF, DHF patients with B blood type as many as 17 people (17.70%), DHF patients with O blood type as many as 36 people (37.5%), DHF patients with AB blood type as many as 5 people (5.20%).

Table 6. Allele Frequency of blood type on DHF patients.

Allele Frequency	Frequency (%)
I^A (p)	0.3
I^B (q)	0.1
I^O (r)	0.6

Based on the research conducted on 96 DHF patients known Frequency allele for I^A is 0.3, the allele frequency for I^B is 0.1 and the frequency of allele I for I^O is 0.6.

From the results of genotypes frequency calculation results obtained for A-homozygous blood type (I^AI^A) of 0.09%, A-heterozygous blood type (I^AI^O) of 0.36%, B-homozygous blood type (B^BI^B) of 0.01%, B-heterozygous blood type (I^BI^O) of 0.12%, AB blood type (I^AI^O) of 0.06% and O blood type (I^OI^O) of 0.36%.

Table 7. Genotype frequency of blood type on DHF patients.

Genotype Frequency	Frequency (%)
$I^A I^A$	0.09
$I^A I^O$	0.36
$I^B I^B$	0.01
$I^B I^O$	0.12
$I^A I^B$	0.06
$I^O I^O$	0.36

Research on the gene distribution of ABO blood type in patients with DHF in Puskesmas Bonto Banguns, Rilau Ale, Bulukumba District. This study aims to determine the blood type most affected by DHF. From the research that has been done in September - December with the number of samples as many as 104 people were obtained as many as 57 people (54.8%) men and 47 people (45.2%) women. This shows no significant difference between men and women. No susceptibility difference for DHF is associated with sex differences. In the Philippines it was reported that the sex difference was 1:1. Similarly in Thailand there was no reported difference in susceptibility to dengue fever between male and female.

In this study found two kinds of dengue infection caused by *Aedes aegypty* mosquito bites DF and DHF. The number of DF patients in this study were 8 people and the number of DHF patients was 96 people. The main difference between DF and DHF is found in the presence of leakage of blood fluid in the blood vessels, which causes the blood to thicken. The presence of leakage of this blood fluid is indicated by an increased haematocrit value. In this study more specifically discusses the blood type in DHF patients.

After the blood group test on 96 DHF patients, it was found that the data of DHF were 38 people (37,59%), DHF patients with B blood type as many as 17 people (17.70%), DHF patients with O blood type as many as 36 people (37.5%), DHF patients with AB blood type as many as 5 people (5.20%). From these data it can be seen that the percentage of blood type most affected by DHF is blood type A and O. This is in accordance with research that has been done by Hartanto[4] namely that blood type O affects dengue virus infection. Further research conducted by Kurniawan[6] in his research on the correlation of blood type ABO against dengue virus infection, from the study concluded that blood type O has a relationship to dengue virus infection which is the factor of occurrence of DHF.

In some studies showed that von Willebrand factor (vWf) and factor in each ABO blood type system is very different, especially in O blood type which has lower levels than non O blood type. The von Willebrand factor is a glycoprotein consisting of several oligosaccharides that are supplemented by the ABO blood type system.

The von Willebrand factor is encoded by chromosome 12 and is made by endothelial cells and megakaryocytes. Its presence in the blood is required by the liver to make factor VIII. If there is deficiency or defect vWf, then production factor VIII decreases. The vWf function will stimulate the liver to release factor VIII and platelet function to form clumps and attach to sub endothelial tissue. VWf serves to help attach platelets to sub endothelium. Sub endothelial is the main source of vWf plasma.

Capillary endothelial cells are assumed to play a role in the pathogenesis of DHF and begin to be widely investigated in vitro. This is due to vascular leakage and thrombocytopenia in DHF is pathogenesis and related to the integrity of the impaired capillary endothelium. In essence, haemostatic is a very complex physiological mechanism that will keep the blood in the circulation. A very diverse path can initiate the formation to prevent those following a trauma. When the tissue is injured, the blood vessels will be disturbed and bleeding will occur through wounds in the blood vessels.

From the data of each blood group in DHF patients then the frequency of allele is calculated. Based on the results of research that has been done allele frequency for I^A is 0.3, allele frequency for I^B is 0.1 and allele frequency I for I^O is 0.6. After the allele frequency is known then the next calculation for each genotype frequency. Gene frequency is the frequency of the presence of a gene in a population in relation to the frequency of all alleles.

Based on the data from the research, the results obtained for A-homozygous blood type (I^AI^A) of 0.09%, A-heterozygous blood type (I^AI^O) of 0.36%, B-homozygous blood type (0.01%), B-heterozygotes blood type (I^BI^O) of 0.12%, AB blood type AB (I^AI^O) of 0.06 and O blood type (I^OI^O) of 0.36%. Based on the calculation of genotype frequency it can be seen that A-heterozygous blood type (I^AI^O) has the same frequency with O blood type (I^OI^O). It shows that most people with DHF are O blood type because A heterozygous blood type still have recessive O blood type gene.

After done Test Regression analysis got result of correlation value of blood type with DHF equal to 1. This suggests that the relationship between blood type and DHF is very strong. After testing, the Anova table of probability value (F-count) obtained by 0.000 <0.05 thus the regression equation model based on the research data is significant, meaning that the linear regression model meets the linearity criteria, and in the table coefficient obtained the probability value t-count 0.000 <0.05 indicates rejected and H_1 accepted.

4. Conclusion
Genotype of O and A blood type O which is the blood type most affected by DHF.

References
[1] Pearce. 2006. *Anatomi dan Fisiologi untuk Paramedis.* Gramedia Pustaka Utama. Jakarta.
[2] Suryo. 2005. *Genetika Manusia.* Gajah Mada University Press. Yogyakarta
[3] Siegal, F. 1999. *The Nature of the Principal Type 1 Interferon-Producing Cell in Human Blood. Science Magazine*, 284 (5421), pp. 1835-1837
[4] Departemen Kesehatan. 2016. *Buletin DemamBerdarah.*www.depkes.go.id. (Accessed at 9 Maret 2016 pukul 15.00 WITA, Makassar)
[5] Dinas Kesehatan Sulawesi Selatan. 2016. *Laporan Kasus Demam Berdarah Provinsi Sulawesi Selatan.* Dinas Kesehatan Provinsi Sulawesi Selatan
[6] Thimmerck, T. 2004. *Epidemiologi Suatu Pengantar.* Edisi 2. EGC. Jakarta
[7] Hadinegoro, S. R. H., S. Soegjanto, dan S. Muryadi. 1999. *DSS: Clinical Manifestation,Management and Outcome. A hospital based study in Jakarta, Indonesia.* Dengue Bull. ;23:12
[8] Kalayanarooj, S., D. W. Vaughn, S. Nimmannitya, S. Green, S. Suntayakorn, N. Kunentrasai, W. Viramitrachai, S. Ratanachu-eke, S. Kiatpolpoj, B. L. Innis, A. L. Rothman, A. Nisalak, and F. A. Ennis. 2007. *Blood Group AB is Associated with Increased Risk for Severe Dengue Disease in Secondary Infection. JID.; 195:1014-7*
[9] Rahayu, D. Hilmanto dan D. Setiabudi. 2008. *Golongan Darah AB sebagai Faktor Risiko Sindrom Syok Dengue Pada Anak. Majalah Kedokteran Indonesia* Vol 58 (10). (Accessed at 27 Maret 2016 pukul 18.00 WITA, Makassar)
[10] Hartanto, F. 2005. *Hubungan Golongan Darah O dengan Kejadian Syok Pada Penderita DBD.* http.repository.undip. ac.id. (Accessed at 27 Maret 2016 pukul 21.00 WITA, Makassar)
[11] Kurniawan, A. A. 2013. *Hubungan Golongan Darah ABO Terhadap Infeksi Virus Dengue. Skripsi.* Universitas Syiah Kuala, Banda Aceh. (Accessed at 20 Mei 2016, pukul 20.00 WITA, Makassar)

Chlorella sp : Extraction of fatty acid by using avocado oil as solvent and its application as an anti-aging cream

T W Putri[1], I Raya[2], H Natsir[2] and E Mayasari[2]

[1] Magister Program Study, Chemistry Departement of Hasanuddin University, Makassar, Sulawesi Selatan, Indonesia
[2] Department of Chemistry Hasanuddin University, Makassar, Sulawesi Selatan, Indonesia

E-mail : indahraya05@gmail.com

Abstract. The study aimed to analyze the fatty acid content of *Chlorella sp* crude extract by using avocado oil solvent and determining the effectiveness of fatty acids *Chlorella sp* as the anti-aging cream. The extraction of fatty acids from *Chlorella sp* using avocado oil as a solvent with three ratios were 1:10, 1:20 and 1:25 w/V. The highest lipid content was obtained at 1:20 w/V (gram microalgae: mL avocado oil) yielding 52.73%. Crude extracted were analysis by GC-MS and FTIR, and skin condition was determined by skin analyzer. The effectiveness test of Chlorella sp cream was applicated on the face of the panelists aged 20-60 years. From 10 panelists, the applied of *Chlorella sp* cream was 90% increased on the facial skin yielded moisture and oil content, 70% repair the skin structure. The composition of fatty acids *Chlorella sp* extract was palmitic acid, linoleic, oleic and stearate. Fatty acids crude extract of *Chlorella sp* can improve the effectiveness of anti-aging cream. The cream from *Chlorella sp* was more effective than the cream without containing microalgae. This is very promising because it is alternative to organic solvents i.e. green chemistry.

1. Introduction

Human skin will dry up when aged 25 years and over. This is due to the dense activity, UV exposure, pollution, and cosmetic usage, resulting in the decrease of oil gland and sweat production on the skin, with the result that the water within the skin becomes more volatile [1] Moreover, the factor of increasing age will reduce the thickness of the dermis layer. This condition occurs which does elastin and collagen fibres is loss in the skin [2]

Many research studies have found that ingredient in anti-aging cream is a moisturizing. Moisturizers are a product designed to restore and maintain optimal hydration at the stratum corneum. Moisturizers contain lipids such as triglycerides and mineral oils [3].

Currently, the main producers of fatty acids are marine microorganisms that are microalgae such as phytoplankton. The phytoplankton synthesizes omega-3 and omega-6 fatty acids which include EFA (Essential Fatty Acid) and DHA (Docosahexaenoic acid). EFA and DHA serve as an energy source, maintain the body heat loss, prevent skin from drying and peeling and protect tissues and other organs [4].

Phytoplankton contained fatty acid were needed by human skin such as Palmitic Acid, Stearate acid, Oleic acid, and Linoleic acid. Phytoplankton also possesses several advantages that are abundant as natural resources, not dependent on climate and weather, fast-growing time that can be harvested not in the long-term period. In addition, phytoplankton can be produced continuously, causing no adverse impact on the environment, its production might be controlled according to the need and desire, and safety for health [5].

One of the phytoplankton that has an active component is *Chlorella sp* which possesses a fatty acid content approximately 31% - 68% [6]. *Chlorella sp* is the type of phytoplankton that has commercial potential that had been used to produce fats, fatty acid, proteins, and pigments [7].

The common solvent employed in the production of anti-aging cream is a toxic organic solvent so that the need for a safe alternative solvent used for the skin. Human skin contains fatty acid palmitoleic, palmitate, stearate, oleate, linolenic and linoleic which is also contained in avocado oil [8][9]. So that avocado oil can be used as a natural solvent that can be used to extract fatty acids from phytoplankton by sonication method [10] with a heating temperature of 50-60°C [11].

Based on the description of the background, it will be conducted the research on "Extraction Fatty Acid of *Chlorella sp* with Avocado Oil and Its Application as an Anti-Aging Cream".

2. Material and Methods

The materials used in this research work include *chlorella sp* powder (Production of Laboratorium Pakan Hidup Laboratory, Balai Besar Perikanan Budidaya Air Payau Jepara), avocado oil (Production Healthy Organic Store) methanol pa, stearic acid ($CH_3(CH_2)_{16}COOH$), glycerol, triethanolamine ($(HOC_2H_4)_3N$), nipagin, Nipasol, propylene glycol, DPPH, vitamin C pa (Production of PT Brataco obtained from CV Sumber Rejeki), aquadest, aquabidest (Production Science Building FMIPA UNHAS), filter paper, label paper, aluminum foil and tissue roll.

The apparatus used in this research work included glass tools which are generally used in the laboratory, Centrifuge, digital scales, mortar and pestle, plastic pot, Erlenmeyer, pipette, Brookfield viscometer, pH meter, pycnometer, Ultrasonic Cleaner Krisbow 42 kHz, spectrophotometer, UV-Vis US-110PC, *Skin analyzer Luvance*, FTIR (*Fourier Transform Infra Red*) Prestige-21 Shimadzu, and GC-MS QP2010S SHIMADZU.

2.1. Exraction Lipid of Chlorella sp

The powder of *Chlorella sp* was added to Erlenmeyer 50 mL, then added avocado oil with various ratios (g microalgae: mL of avocado oil) 1:10 w/V (weight/Volume), 1:20 w/V, 1:25 w/V. The mixture sonicated for 6 hours with temperature range 50-55°C. The mixture was added aquadest and shaken. The oil and aquadest were separated by centrifugation. The obtained lipids analyzed using FTIR (*Fourier Transform Infra Red*) Prestige-21 Shimadzu, and GC-MS QP2010S SHIMADZU to know the fatty acid content.

2.2. Production of Chlorella sp Cream

Preparation of creams was separated into two phase are oil phase (avocado oil, and microalgae extract, stearic acid, triethanolamine) and aqueous phase (aquadest, nipagin, nipasol).

2.2.1 *Chlorella sp* crude extract is dissolved with stearic acid at 60-70 °C (oil phase) and added Triethanolamine.
2.2.2 Nipagin and nipasol were weighed and dissolved in propylene glycol (water phase).
2.2.3 Oil phase was moved to a hot mortar and added gradually aquadest while stirring until creamy.

Chlorella sp cream then tested according to SNI 16-4399-1996 is the organoleptic test, pH, density, viscosity and microbial contamination. *Chlorella sp* cream effectiveness test was carried out randomly to 10 panelist aged 20-60 years, and one panelist as a control. Skin Analyzer was used to analyze the texture facial skin of panelist before and after the use of *Chlorella sp* cream and control cream.

2.3. Antioxidant Activity Test of Chlorella sp Cream and Avocado oil Cream (Control Cream)
25 mg *Chlorella sp* was dissolved in methanol pro analysis. Standard series in various concentrations 20, 40, 80, 160 and 320 ppm. As much as 0.01; 0.02; 0.03; 0.04 and 0.05 mL put in volumetric flask and added 1 mL of 0.4 mM DPPH solution in methanol pro analysis. The solution was shaken and stored at room temperature for 30 min, after that, the absorption was measured at a maximum wavelength. Vitamin C was utilized for a positive control and for a reference. Antioxidant activity is expressed by percent of DPPH immersion.

3. Result and Discussion

3.1. Extraction Lipid of Chlorella sp
The most lipid result obtained from microalgae crude extract of *Chlorella sp* with avocado oil was at ratio 1: 20 with the percentage to 52,73% .

Table 1. Result of Isolation Lipid *Chlorella sp.*

Sample			Extract	
Chlorella sp (gram)	Avocado oil (mL)	Dry Residue Sample (gram)	Extracted Sample (gram)	Lipid Sample (%b/b)
1,15	10	0,89	0,26	22,61
1,10	20	0,52	0,58	**52,73**
1,12	25	0,76	0,36	32,14

3.2. Characteristics of Avocado Oil and Lipid of Chlorella sp with Gas Chromatography - Mass Spectrometry (GC-MS)
The content of avocado oil and lipid of *Chlorella sp* crude extract based on the dominant peak was shown in Table 2. The dominant number of fatty acids identified of avocado oil were 7, 8, 9, and lipid of *Chlorella sp* crude extract were 1, 3, 5. The result of the dominant peak was compared to database WILLEY229.LIB.

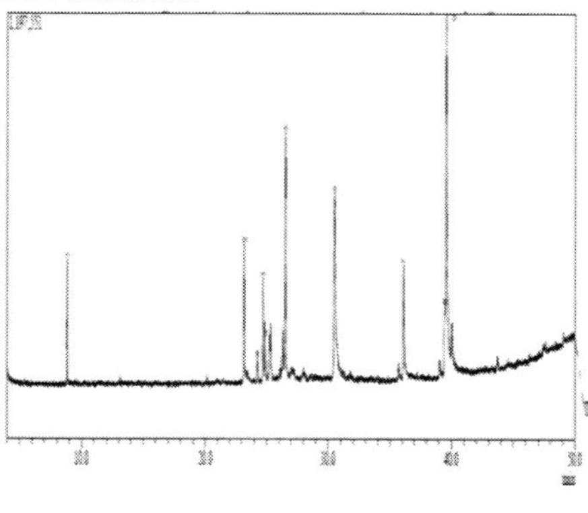

Figure 1. Avocado Oil Chromatogram.

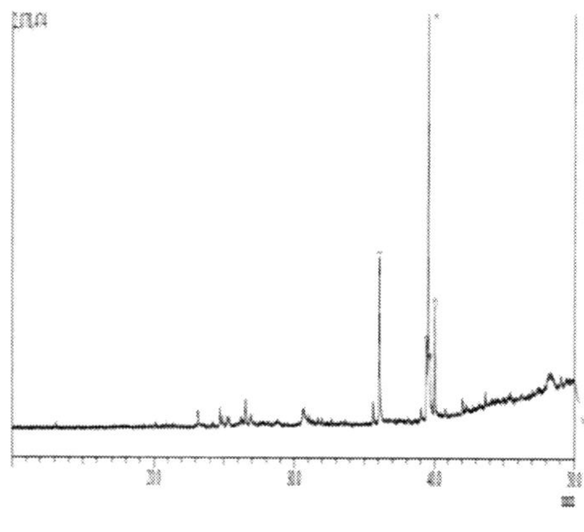

Figure 2. Lipid extract of *Chlorella sp* with avocado oil.

Table 2. Composition of Avocado Oil and lipid of *Chlorella sp* crude extract with avocado oil

Fatty Acid	Avocado Oil (%)	Lipid of *Chlorella sp* crude extract (%)	Fatty Acid Decrease (%)	Fatty Acid Increase (%)
Palmitit acid	6,64	23,07	-	16,43
Linoleic Acid	3,57	8,04	-	4,47
Oleic Acid	32,73	56,73	-	24
Stearate Acid	-	12,16	-	12,16

The GC-MS analysis of avocado oil showed the presence of 9 peaks, with the highest abundance existed at the 9th peak with a retention time of 39.567 minutes. The lipid of *Chlorella sp* crude extract exhibited 5 peaks with the highest abundance also at the 3rd peak with retention time 39,553. Based on WILLEY229.LIB databases, peak 9 and 6 referred to 9-octadecanoic acid or known as oleic acid. This compound had MW at m / z 296. Oleic acid content on avocado oil was 32,573% and lipid of *Chlorella sp* crude extract was an enhancement to 56,73%.

Oleic acid is the most dominant compound in the composition of the avocado oil. Oleic acid also has a function as an antioxidant [12].

Peak number 7 with a retention time of 36.093 minutes for avocado oil and peak number 1 for Lipid of *Chlorella sp* crude extract with a retention time of 36.060 referred to a hexadecanoic acid compound known as palmitic acid. This compound had MW at m / z 270. This compound in avocado oil before sonication levels was 6.64% and after sonication reached 23.07%.

The third dominant peak with a retention time of 39.458 minutes for avocado oil was the number 8 peak referring to the 9: 12-octadecanoic acid compound or known as linoleic acid. This compound possessed molecular weight at m / z 294 with a compound content reached 3.57%. And peak number 5 for Lipid of *Chlorella sp* crude extract with a retention time of 39.967 referred to Octadecanoic acid compounds or known as stearic acid. This compound had molecular weight at m/z 298 with a compound content reached 12.16%

The increase and decrease of fatty acid concentration in each comparison are due to the time of sonication process the percentage of the oleic acid fraction was more detached in triglycerides than palmitic acid, linoleic acid, stearate acid, laurate acid and palmitoleate acid, so the percentage of fatty acid was reduced.

3.3. Characteristics of Avocado Oil and Lipid of Chlorella sp with Fourier Transform Infra Red (FTIR)

The analysis of fatty acid using by FTIR aimed to determine the functional groups presented in the sample. Figure 3 and figure 4 illustrated the IR spectrum of avocado oil and lipid of *Chlorella sp* crude extract.

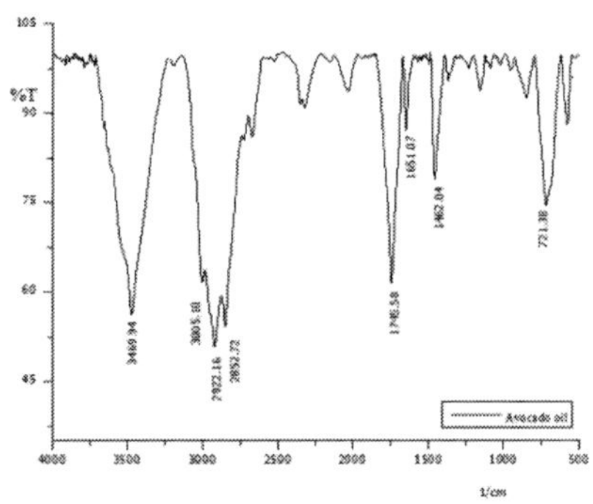

Figure 3. Spectra Avocado oil.

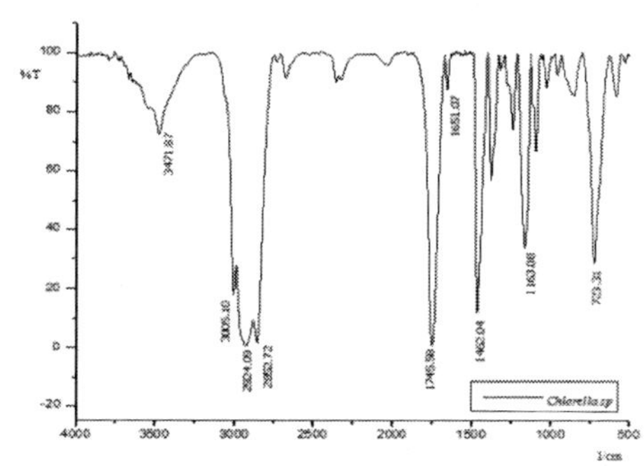

Figure 4. Lipid extract of *Chlorella sp* with avocado oil

The FT-IR spectrum of avocado oil and lipid of *Chlorella sp* crude extract exhibited an absorption range in 3469,94-3471,87 cm^{-1} indicating the presence of hydroxyl group (-OH). The wave number at 1651.07 cm^{-1} was a stretch that arose as a result of -C = C-. This was supported by the absorption peak at 3005.10 cm-1 which showed a stretching bond of = C-H. Furthermore, strong absorption also occurred in the wave number 2852,72-2924,09 cm^{-1} which were the stretching of the saturated carbon chain. The typical absorption of triglycerides appeared at the wave number 1745.58 cm^{-1} which indicated the absorption of the carbonyl group -C=O and supported by C-O stretching at wavenumber 1163,08cm^{-1}, while absorption at wave number 1462,04 cm^{-1} was -C-H bending and wave number of 721,38 - 723,31 cm^{-1} was CH_2.

Table 3. Results of FTIR Analysis Lipid Avocado Oil

Number	Wave Numbers of Avocado Oil (cm^{-1})	Wave Numbers of Lipid Chlorella sp Extract with Avocado Oil (cm^{-1})	Information
1	3469,94	3471,87	-OH
2	2852,72 2922,16 3005,10	2852,72 2924,09 3005,10	strecthing bond C-H
3	1745,58	1745,58	-C=O
4	1651,07	1651,07	-C=C- Streching
5	1462,04	1462,04	–C-H bending
6	721,38	723,31	CH_2
7	-	1163,08	C-O

3.4. Antioxidant Activity Test from Chlorella sp Cream

The result of antioxidant activity test of *Chlorella sp* cream (sample A), and Avocado Oil as a control cream (sample B) can be seen in table 4 below:

Table 4. Results of Antioxidant Activity Test from *Chlorella sp* cream and Avocado Oil cream

NO	Cream Sample	Value of IC_{50} (µg/mL)
1	A	565,330
3	B	1.103,711
4	Ascorbit acid	2,896

Based on the table 4 samples of cream preparations A and B possessed weaker antioxidant activity than ascorbic acid which had an extremely strong antioxidant activity. The smaller the value of IC_{50} then the larger the antioxidant activity of a compound. This was because the compounds extracted from the microalgae were triglycerides that had no antioxidant activity.

The IC_{50} value of sample A was lower than that of sample B indicating that sample A underwent an increase in antioxidant activity than sample B. Sample A was *Chlorella sp* cream and sample B was avocado oil cream as a control cream. This suggested an increase of % inhibition on sample A which was probably due to a reaction between fatty acid compounds from *Chlorella sp* extract and DPPH thus raising their antioxidant activity.

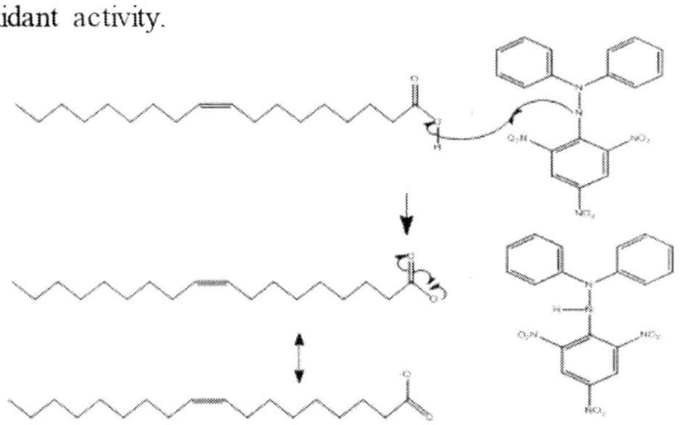

Figure 5. Prediction reaction of Fatty acid with DPPH.

3.5. *Profile Anti Aging Cream*

There are two cream, the first is cream was made from the fatty acid extract of *Chlorella sp* has pH 5,5, density is 0,967 g/cm^3 and viscosity is 15,066,667 Cps. The second is cream made from avocado oil without adding the phytoplankton extract has pH 6.5, density is 0,951 g/cm^3 and viscosity is 12,400 Cps. Both of creams showed the homogeneous result. Both of cream has phytoplankton smell. The use of cream does not irritate or does not cause itchy skin, does not make skin red and not rough or scaly. The results of *Chlorella sp* cream and cream control tests were still in accordance with SNI 16-4399-1996 quality standard.

3.6. *Effectiveness of Anti Aging Cream Evaluation*

The effectiveness test of cream was undertaken to 10 panelists with age range around 20-60 years. The results apply of *Chlorella sp* microalgae extract cream to human skin exhibited a smoother skin texture.

Control cream and cream containing *Chlorella sp* microalga were applied to the 43-year-old panelist with cream usage for one month. Figure 4 illustrated the skin texture of the use of control creams and *Chlorella sp*.

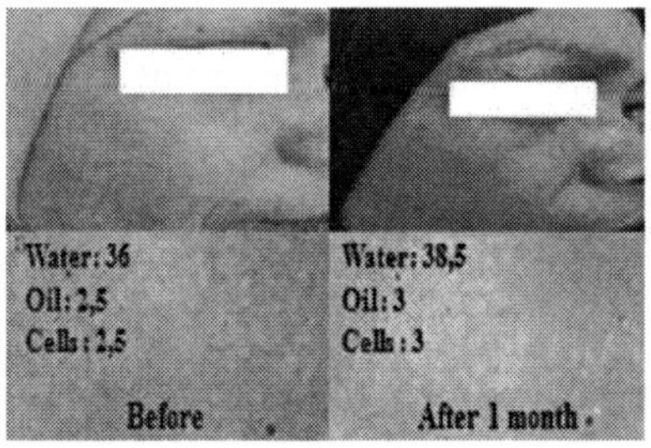

Figure 6. Control cream.

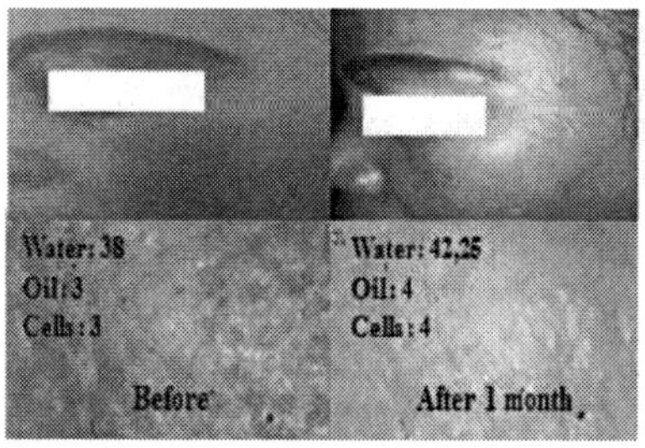

Figure 7. *Chlorella sp* cream.

The use both of cream for one month period showed result better improvement of facial skin textures. The results indicated that creams containing microalgae (figure 7) produced a better effect than no microalgae or control cream (figure 6).

The skin-analyzer tested results represented the used of creams that contained microalgae were able to enhance the water content and oil on the panelist skin. The epidermal layer especially the stratum corneum is the outer layer surface that has a balance between water and lipids to keep the skin elastic and smooth. Lipids role to maintain the natural moisture factor remains in the cell so that no excessive evaporation of water, and water contents determine the skin's moisture level [13].

Figure 8 and Table 5 illustrated a change in the facial skin of the panelists in which 90% of panelist using cream, moisture, and oil content on the face increased, and 70% of panelists experienced an improvement on skin texture.

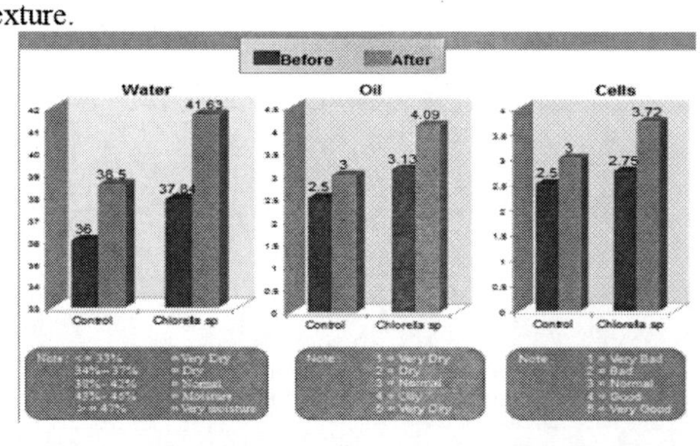

Figure 8. Graph of Facial Skin Changes Panelist.

Table 5. Percentage of Changes in Facial Skin of Panelists

Observation	Up (%)	Down (%)	Constantly (%)
Water	90	10	-
Oil	90	-	10
Cells	70	-	30

4. Conclusions

Based on these results, it can be concluded that lipid extract of *Chlorella sp* content are oleic acid, linoleic acid, palmitate acid, and stearate acid. The fatty acids of *Chlorella sp* increasing of anti-aging activity.

References

[1] Jusuf N K 2005 Kulit Menua *Majalah Kedokteran Nusantara* **2** p 184-188
[2] Sharma B, Sharma A 2012 Future Prospetcs of Nanotechnology in Development of Anti-Ageing Formulations Review Artikel *International Journal of Pharmacy and Pharmaceutical Sciences* **4**
[3] Leyden J J 2002 *Skin Moisturization* Marcel Dekker Inc New York Basel ISBN0824706439
[4] Perez-Garcia O , Escalante F M E, Luz de-Bashan, Bashan Y 2011 Heterotrophic Cultures of Microalgae: Metabolism and Potential Products *Water Research* **45** p 11-36
[5] Barr T and Pearson M 2004 *Method Of Creating Antigens Using Emus, Patent* WO2004026339A1
[6] Widianingsih R, Hartati H, Endrawati J, Mamuaja 2013 Fatty Acid Composition Of Marine Microalgae in Indonesia, *Journal of Tropical Biology and Conservation* **10** p 75-82
[7] Raya I, Anshar A M, Mayasari E, Dwiyana Z and Asdar M 2016 Chorella vulgaris and Spirulina Platensis : Concentration of Protein, Docosahexaenoic Acid Chorella (DHA), Eicosapentaenoic Acid (EPA) and Variation Concentration of Maltodextrin via Microencapsulation Method, *International Journal of Applied Chemistry* **12** p 539-548
[8] Indrawati T dan Sari H Y 2011 Pengaruh Buffer Alpukat (*Avocado Butter*) Terhadap Stabilitas Krim Pelembut Avocado *Majalah Farmasi Indonesia* **22** p 98-105
[9] Ferreyra R G, Sellés G, Saavedra J, Ortiz J, Zúñiga C, Troncosa S A, Rivera M, González-Agüero and B G Defilippi 2016 Identification of pre-harvest factors that affect fatty acid profiles of avocado fruit (Persea americana Mill) cv. 'Hass' at harvest South African *Journal of Botany* **104** p 15–20
[10] Supardan M D 2011 Penggunaan Ultrasonik untuk Transesterifikasi Minyak Goreng Bekas *Jurnal Rekayasa Kimia dan Lingkungan* **8** p 11-16
[11] Kwangdinata R, Raya I, Zakir M 2014 Production of Biodiesel from Lipid of Phytoplankton Chaetoceros calcitrans through Ultrasonic Method *The Scientific World Journal* Article ID 231361
[12] Zulharmita, Afrina R, Wahyuni R 2013 Ekstraksi Asam Lemak Dari Daging Buah Alpukat *Jurnal Farmasi Higea* **5**
[13] Baumann L dan Saghari S 2009 Photoaging dalam Baumann L, Saghari S, Weisberg E Eds *Cosmetic Dermatology: Principles and Practice, 2nd Ed.* McGraw-Hill Professional New York p 34–41

IOP Conf. Series: Journal of Physics: Conf. Series **979** (2018) 012010 doi:10.1088/1742-6596/979/1/012010

Synthesis, structurale elucidation and antioxidant study of *Ortho*-substituted *N,N'*-bis(benzamidothiocarbonyl)hydrazine derivatives

Syadza Firdausiah[1], S A Hasbullah[2] and B M Yamin[2]

[1] Chemistry Department, Faculty of Mathematics and Natural Sciences, Hasanuddin University, Jl. Perintis Kemerdekaan 10 Tamalanrea Makassar, Sulawesi Selatan, 90245, Indonesia

[2] School of Chemical Science and Food Technology, Faculty Science and Technology, National University of Malaysia, Bangi, Selangor, 43600, Malaysia

E-mail: syadza.f@gmail.com

Abstract. Some bis(thiourea) compounds have been reported to posses excellent performance in pharmaceutical and environmental fields because of their ability to form chelating complexes with various anions and metal ions. Structurally for carbonyl thiourea derivatives, to become a chelating agent, it must adopt *cis*-configuration. In the present study, four new bis(thiourea) derivatives namely *N,N'*-bis(*o*-fluorobenzamidothiocarbonyl)hydrazine **(1)**, *N,N'*-bis(*o*-chloro-benzamidothiocarbonyl)hydrazine **(2)**, *N,N'*-bis(*o*-nitrobenzamidothiocarbonyl)-hydrazine **(3)**, and *N,N'*-bis(*o*-methylbenzamidothiocarbonyl)hydrazine **(4)** were successfully synthesized and characterized by CHNS microelemental analysis, FTIR, UV-Vis, and ^{1}H and ^{13}C NMR spectroscopy. However chemical crystallography study showed that both thiourea moieties in compound **(2)** and **(3)** adopt *trans* geometry. Therefore they are potential monodentate ligand with two active moieties. DPPH radical scavenging experiment showed that compound **(1)**, **(2)**, and **(4)** exhibited higher antioxidant activity than ascorbic acid (Vitamin C).

1. Introduction

Thiourea derivatives have diverse potential applications in many fields. In pharmaceutical field, some thiourea derivatives showed the potential application as inhibitor of urease [1], antimicrobial agents [2–5] and in human cancer treatment [6]. Thiourea derivatives also applicable as potentiometric sensor due to the presence of C=S and amine groups which are easy to form complexes with metal ions [7]. Studies on the structure-activity relationship revealed that thiourea moiety played important role in their antioxidant study [8].

The modifications on the N side of thiourea have been reported by many researchers. Some *N,N'*-disubstituted thiourea compounds showed good performance as HIV binding inhibitors [9]. Thiourea compound bearing chitosan moiety was also reported as good antibacterial and antifungal agent [3]. Other thiourea derivatives bearing allyl and triazole moieties were also active as antiviral and antituberculosis agent [10]. Carbonyl thioureas also showed potential in industrial [11] and agrochemical fields [12].

The metal complexes of some thiourea derivatives have been found as new classes of antifungal agent [13,14] and showed moderate toxicity against human cancer cell lines [15]. The other thiourea-metal complexes have also been used for metal ion removal from water environment and as component of chemosensors [16,17].

Bis(thiourea) compounds with hydrazine spacer have been widely used as building blocks in synthesis of many heterocyclic compounds. For example, bis(thiourea)hydrazines which substituents are alkyl groups such as methyl, ethyl, propyl and buthyl have been used for synthesis of triazolidine compounds [18]. Those bis(thiourea)hydrazine compounds have also been reported as building blocks for synthesis of novel thiazolone azines which are ideal candidates for electrochemical absorption and redox indicators due to their capability to form stable radical cations [19]. The synthesis of biologically active compounds, bis-oxathiaaza[3.3.3]propellanes, have been carried out using bis(thiourea)hydrazines containing aromatic groups such as phenyl, benzyl and allyl [20].

The simplest bis(thiourea)hydrazine has been used for building of supramolecular networks by interaction with crown ether via N-H···O or N-H···S hydrogen bonds [21]. This bis(thiourea) compound has also been reported as chelating resin for removal some metal ions from a wastewater with high efficiency [22]. Bis(thiourea)hydrazine bearing pyridine as substituent group also showed good performance as chelating agent for separation of toxic metal ions from aqueous solutions [23]. Other aromatic bis(thiourea)hydrazine compounds were also potential in pharmaceutical field as anticonvulsant agent and selective agent for grand mal epilepsy [24].

There are less structure of bis(thiourea) have been reported so far compared to monothiourea compounds. To the best of our knowledge, there are only five crystal structures of bis(thiourea) compounds utilizing hydrazine linker have been reported in CCDC (The Chambridge Crystallographyc Data Centre). They are dithiobiurea [25], bis(*N*-phenylthiourea) [26], 1-(2,3,4,6-tetra-*O*-β-*D*-glucopyranosyl)-3-thioureidothiourea monohydrate [27], *N,N'*-bis(benzamidothiocarbonyl)hydrazine [28] and 1,6-bis-phosphorylated-2,5-dithiobiurea [29].

In this research, bis(carbonylthiourea) with the shortest linker N-N (N-C_n-N; n=0), namely bis(aroylthiourea)hydrazine (Figure 1), is designed and synthesized, which X is *o*-fluoro, *o*-chloro, *o*-nitro and *o*-methyl. Figure 1 shows predicted structure of synthesized compounds. If all C=S and C=O bonds are in *cis*-configuration about the C-N and N-N bonds, the compound can act as tetradentate ligand. However, if both C=S bonds are in *trans*-cofiguration about the N-N bond but C=S and C=O bonds are in *cis*-configuration about their C-N bonds, the compound can provide two bidentate active sites metal chelations. The characterizaion and molecular structure of the compounds including their antioxidant properties are presented.

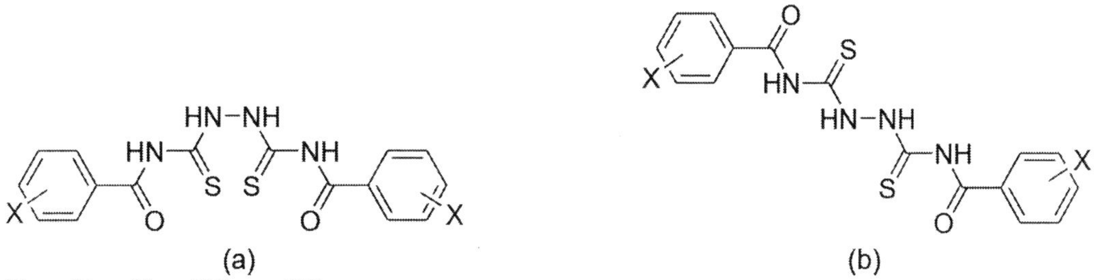

X= *o*-F, *o*-Cl, *o*-NO_2, *o*-CH_3

Figure 1. The predicted structure of synthesized compounds. (a) Tetradentate ligand; (b) bidentate ligand

2. Experimental

2.1. Material and methods
All reagents were used without further purification. The solvent was treated with molecular sieve before used. C, H, N and S analyses were performed using a ThermoFinnigan Instrument. FTIR spectra were recorded using KBr pellet on Perkin Elmer 100 Spectrophotometer in the region of 400 – 4000 cm^{-1}. ^{1}H and ^{13}C NMR spectra were obtained using Bruker Avance III 400 MHz. Electronic spectra were recorded on UV-1800 240V Shimadzu Spectrophotometer. Crystal data was recorded on Bruker SMART APEX CCD Diffractometer. The structure was solved using SHELXS97 [30] and was refined by full matrix least squares on F^2 [30].

2.2. Synthesis
A solution of aroyl chloride (4 mmol) was added to 0.304 g of ammonium thiocyanate (4 mmol) in acetone. The mixture was stirred at room temperature for 15 minutes and the precipitate was filtered out. Hydrazine hydrate as many as 0.1 g (2 mmol) was added to the filtrate and the mixture was refluxed for 3 hours. Then, the residue was filtered out. The precipitate was washed with dichloromethane and ethanol/H_2O 1:1.

N,N'-bis(*o*-fluorobenzamidothiocarbonyl)hydrazine (**1**): Yield 57%; colourless solid, m.p 271.2-272.0 °C. IR (KBR pellet, cm^{-1}): υ(N-H)3424; υ(C=O)1671; δ(N-H)1612; υ(C-N)1229; υ(C=S)1079; υ(C_{Ar}-F)1150. ^{1}H-NMR (DMSO-d_6, 400 MHz): δ 7.35-7.74 (m, 4H); 12.35 (s, 1H). ^{13}C-NMR (DMSO-d_6, 100 MHz): δ 116.8; 125.3; 131.1; 134.9; 158.5; 161.1; 166.1; 171,6. Anal. Calc. For $C_{16}H_{12}N_4O_2S_2F_2$: C,48.72; H,3.07; N,14.20; S,16.26. Found: C,49.12; H,3.18; N,14.83; S,16.73. λ_{max} 335 nm, ε 15,578.75 $L.mol^{-1}.cm^{-1}$.

N,N'-bis(*o*-chlorobenzamidothiocarbonyl)hydrazine (**2**): Yield 88%; colourless solid, m.p °C. IR (KBR pellet, cm^{-1}): υ(N-H)3237, υ(C=O)1686, δ(N-H)1591, υ(C-N)1225, υ(C=S)1100, υ(C_{Ar}-Cl)1040. ^{1}H-NMR (DMSO-d_6, 400 MHz): δ 7.45-7.68 (m, 4H); 12.57 (s, 1H); 13.94 (s, 1H). ^{13}C-NMR (DMSO-d_6, 100 MHz): δ 127.4; 130.0; 130.2; 130.6; 132.9; 134.3; 168.3; 171.6. Anal. Calc. For $C_{16}H_{12}N_4O_2S_2Cl_2$: C,44.97; H,2.83; N,13.11; S,15.01. Found: C,45.22; H,2.98; N,13.83; S,15.68. λ_{max} 334 nm, ε 14,580.25 $L.mol^{-1}.cm^{-1}$.

N,N'-bis(*o*-nitrobenzamidothiocarbonyl)hydrazine (**3**): Yield 52%; yellowish solid, m.p °C. IR (KBR pellet, cm^{-1}): υ(N-H)3271; υ(C=O)1693; δ(N-H)1614; υ(C-N)1250; υ(C=S)1067; υ(C_{Ar}-NO_2) Asym. 1518, sym. 1349. ^{1}H-NMR (DMSO-d_6, 400 MHz): δ 7.81-8.27 (m, 4H); 12.67 (s, 1H); 13.86 (s, 1H). ^{13}C-NMR (DMSO-d_6, 100 MHz): δ 124.8; 130.1; 130.4; 132.5; 135.2; 145.9; 168.0; 171.29. Anal. Calc. For $C_{16}H_{12}N_6O_6S_2$: C,42.85; H,2.70; N,18.74; S,14.30. Found: C,43.23; H,3.24; N,19.40; S,14.69. λ_{max} 331.5 nm, ε 6,501.46 $L.mol^{-1}.cm^{-1}$.

N,N'-bis(*o*-methylbenzamidothiocarbonyl)hydrazine (**4**): Yield 50%; colourless solid, m.p °C. IR (KBR pellet, cm^{-1}): υ(N-H)3246, υ(C=O)1677, δ(N-H)1600, υ(C-N)1226, υ(C=S)1061. ^{1}H-NMR (DMSO-d_6, 400 MHz): δ 2.41 (3H, s); 7.29-7.54 (m, 4H); 12.26 (s, 1H). ^{13}C-NMR (DMSO-d_6, 100 MHz): δ 19.9; 126.1; 128.8; 131.2; 131.7; 133.8; 136.6; 171.0; 171.8. Anal. Calc. For $C_{18}H_{18}N_4O_2S_2$: C,55.94; H,4.69; N,14.50; S,16.59. Found: C,56.26; H,5.44; N,15.12; S,17.27. λ_{max} 332.5 nm, ε 14,392.52 $L.mol^{-1}.cm^{-1}$.

2.3. Single crystal X-Ray determination
Single crystal data were collected using Bruker SMART APEX CCD Diffractometer with graphite monochromatic Mo Kα radiation source. A suitable crystal with size less than 0.5 mm was selected under microscope and mounted on the end of the glass fibre. The fibre was then mounted on a goniometer head before the X-ray experiment started. Raw data was collected by using SMART system (Siemens, 1997) and data reduction was done by using SAINT software (Siemens, 1997). Structure was solved directly by using SHELXTL [30].

2.4. Antioxidant Screening

The antioxidant activity of the thiourea was screened by using DPPH radical scavenging method. 300 μM each solution of DPPH (1.5 mL) in 96% ethanol was added to sample solutions (each 1.5 mL) of different concentrations (100, 200, 300, 400, 500, and 600 μM) in DMSO. The mixtures were allowed to react in the dark place at room temperature. After 30 minutes, the absorbance values were measured at 517 nm. DMSO was used as negative control. The scavenging activity of each sample was calculated as the percentage reduction of DPPH absorbance (Q) using the following formula:

$$Q = 100(A_c - A_s)/A_c\,\% \qquad (1)$$

Where A_s is the absorbance of the sample and A_c is the absorbance of negative control. The results were compared to ascorbic acid activity as positive control. The EC_{50} values of the sample and ascorbic acid were determined by using Four Parameter Logistics in SigmaPlot 12.0.

3. Results and discussion

3.1. Synthesis

The procedure and reaction schemes are outlined in Figure 2. The synthesis of bis(aroylthiourea)hydrazine (**1-4**) was carried out by two steps of reaction. The first step was the nucleophilic substitution of chlorine atom to form the isothiocyanate compound. The next step was the nucleophilic addition of hydrazine nitrogen on the isothiocyanate carbon atom to produce the thiourea compounds.

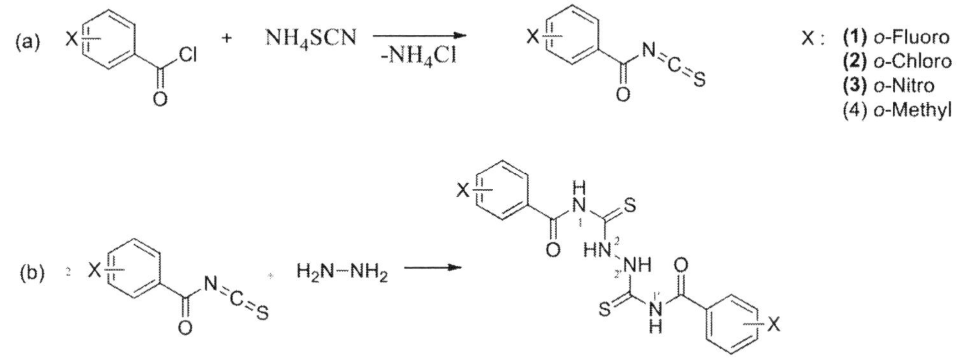

Figure 2. Synthesis of bis(aroylthiourea)hydrazine consist of (a) Reaction of aroyl chloride with ammonium thiocyanate at room temperature for 15 minutes, (b) Reaction of aroyl isothiocyanate with hydrazine at reflux condition for 3 hours

3.2. Chemistry

All compounds show their UV-vis spectra with only one maximum absorption peak between 331,5-335 nm due to n→π* transition of C=O bonds. The expected two maximum absorption of n→π* and π→π* transitions of C=S bond were not observed due to the cut point of DMSO solvent occured at 268 nm and the π→π* transition normally occurs at shorter wavelength (<250 nm).

The compound which has amide group contains both π systems and unshared electron pairs and will show two absorption peaks due to n→π* transition at longer wavelength (>300 nm) and a π→π* transition at shorter wavelength (<250 nm) [31]. Refat & El-Metwaly reported that the thiosemicarbazide compound, (*E*)-*N*-(5-bromo-2-hydroxy-benzylidene)-1,4-dioxo-3,4-dihydrophthalazine-2(*1H*)-carbo-thioamide, showed two absorption peak at 350 and 270 nm for n→π* and π→π* transitions, respectively[32]. In 2013, Huang *et al.* revealed that the benzoyl thiourea compound in acetonitrile has two absorption peaks at 211 and 288 nm which assigned to E_2 and B bands, respectively [33]. Osman reported that each thiourea compounds showed an absorption of n→π* transition of C=S bond at 272-309 nm as shoulder peak [34].

IOP Conf. Series: Journal of Physics: Conf. Series **979** (2018) 012010 doi:10.1088/1742-6596/979/1/012010

The infrared spectra of the products are almost similar. The thione C=S stretching bands appeared in the range of 1100-1061 cm^{-1} indicating the presence of thioketo functional groups. The sharp and strong bands about 1680 cm^{-1} are the carbonyl C=O amide stretching frequencies. The C-N stretching frequencies are between 1250 and 1225 cm^{-1}. Compound **3** has higher C=O stretching frequencies than the other compounds due to the electron withdrawing characteristic of nitro group. The strong bands between 3424 and 3237 cm^{-1} are due to N-H stretching frequencies. The bending N-H frequencies can be observed around 1616-1572 cm^{-1}.

Due to the low solubility of the products in many organic solvents, less than 5 mg of each compound were able to dissolve in 560 µl of DMSO-d_6, they are insoluble in other deuterated solvents. Thus, in some compounds, the signals of all NH protons were not observable. It is due to exchange of proton with deuterium of the solvent. The exchangeable protons like –OH and –NH give broad signals in their NMR spectra, and it completely disappear if the concentration of the sample is low [31]. ^{1}H NMR spectrum of all compounds exhibited broad singlet signals at 12.26 – 12.67 ppm, which were assigned to the NH(1) and NH(1') protons.

The thioamide proton NH(2) and NH(2') appear at lower field (13.94-13.86 ppm) than amide proton NH(1) and NH(1'). The protons are very deshielded due to the interaction with d_6-DMSO and intramolecular hydrogen bondings [35]. The integration value of thioamide and amide protons are one, indicating the symmetry properties of the molecule at N-N bond. The multiplet signals between 7.29 and 8.27 ppm are assigned to protons of aromatic ring. But due to the low concentration of the samples, the multiplicity and the coupling constant in their NMR spectrum were not observable. Their ^{13}C NMR spectra showed signals at 166.1-171.0 and 171.29-171.8 ppm for C=O (amide) and C=S (thioamide), respectively. The signal of aromatic carbons was observed at 116.80-161.1 ppm.

3.3. X-ray Crystallography

In the present study, crystal of compound **2** and **3** were obtained from DMSO. The crystals are not stable in the air. They became white powder after about 3 hours upon exposure to air. Therefore, the crystals were coated with special oil for the X-ray crystallography analysis.

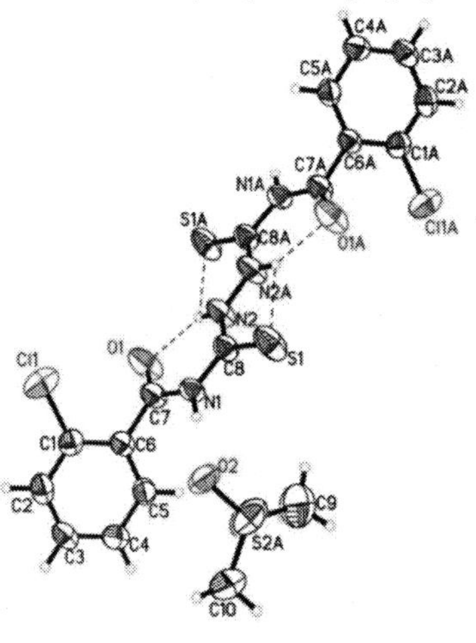

Figure 3. The molecular structure of **2** drawn at 50% probability displacement ellipsoid. The dashes lines indicated intramolecular hydrogen bonds.

N,N'-bis(2-chlorobenzamidothiocarbonyl) hydrazine **(2)** crystallized in monoclinic system with space group of C2/c. The crystal system and refinement data are given in Table 1. The crystal is centrosymmetric about the mid-point of the N2-N2a bond (Figure 3). The sulfur S2 atom is disordered and was treated by parting the atom into two different occupancies.

The benzene (C1-C6 and C1A-C6A) and the bis(thiourea) fragment, O1/N1/N2/C6/C7/C8/O1A/N1A/N2A/C6A/C7A/C8A, are planar with maximum deviation of 0.030(3)Å for C8 and N2 atoms from the least square planes. The dihedral angle between these two planes is 67.73(11)°. The thiourea moieties have *trans* geometry. This configuration is contributed by intramolecular N2-H2···S1A and N2-H2···O1 hydrogen bonds.

There are four intramolecular hydrogen bonds of **2** (Table 2). Both N2-H2···O1 and N2-H2···S1A of the centro-symmetric molecule form two pseudo six-membered rings {(C7/N1/C8/N2/H2···O1) and (C7A/N1A/C8A/N2A/H2A···O1A)} and two five-membered rings {(C8/N2/N2A/H2A···S1) and (C8/N2/N2A/H2A···S1A)}.

Table 1. Crystal data and structure refinement of **2**.

Parameter	Data
Empirical formula	$C_{16}H_{12}N_4O_2S_2Cl_2$, $2(C_2H_6OS)$
Formula weight	583.57
Temperature	301(2) K
Wavelength	0.71073 Å
Crystal system, space group	Monoclinic, C 2/c
Unit cell dimensions	a = 29.7268(11) Å α= 90°
	b = 5.1077(2) β=
	103.0055(12)°
Volume, Z	c = 18.3219(7) γ= 90°
Density (calculated)	2710.56(18) $Å^3$, 4
Absorption coefficient	1.430 Mg/m^3
F(000)	0.581 mm^{-1}
Crystal size	1208
Theta range for data collection	0.500 x 0.220 x 0.110 mm^3
Index ranges	2.938 to 28.359°.
Reflections collected	-38<=h<=39, -6<=k<=6, -
Independent reflections	24<=l<=24
Completeness to theta = 25.242°	32985
Refinement method	3372 [R(int) = 0.0448]
Data / restraints / parameters	99.8 %
Goodness-of-fit on F^2	
Final R indices [*I*>2sigma(*I*)]	Full-matrix least-squares on F^2
R indices (all data)	3372 / 1 / 166
Largest diff. peak and hole	1.109
	R1 = 0.0596, wR2 = 0.1572
	R1 = 0.0887, wR2 = 0.1760
	0.777 and -0.508 e.$Å^{-3}$

IOP Conf. Series: Journal of Physics: Conf. Series **979** (2018) 012010 doi:10.1088/1742-6596/979/1/012010

Table 2. Hydrogen bonds geometry of **2** (Å and °)

D-H···A	d(D-H)	d(H···A)	d(D···A)	<(DHA)
N1-H1···O2#2	0.86	1.99	2.845(3)	178
N2-H2A···O1	0.86	1.91	2.580(4)	134
N2-H2A···S1	0.86	2.55	2.932(3)	108
C10-H2A2···O1#4	0.96	2.54	3.437(6)	156
C4-H4···O2#5	0.93	2.54	3.230(4)	132

Symmetry transformations used to generate equivalent atoms:
#1 -x,-y+1,-z #2 x,y,z #3 x+1/2,y+3/2,z #4 -x+1/2,y+5/2,-z+1/2 #5 x,y+1,z

The molecular packing of **2** showed that the molecules are linked by (N1-H1···O2), (C10-H2A2···O1) and (C4-H4···O2) intermolecular hydrogen bonds, where the oxygen atom of the DMSO is connected to two thiourea molecules to form 2-dimensional network (Figure 4).

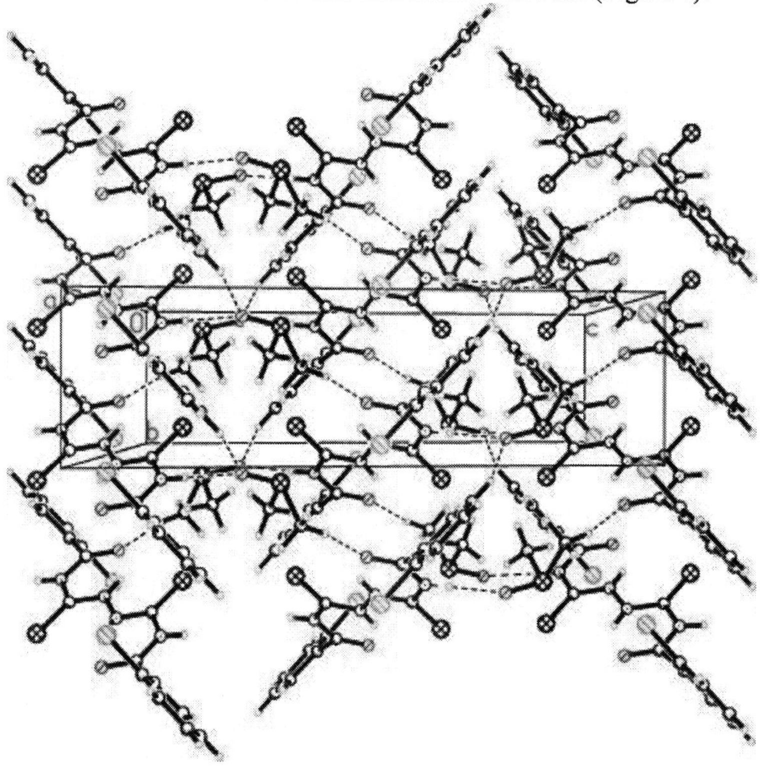

Figure 4. Molecular packing of compound **2** viewed down a axis. The dashed lines indicate intermolecular hydrogen bonds.

The selected bond lengths and angles of **2** is shown in Table 3. All bond lengths and angles are in normal ranges [36]. The bond length of N2-C8 is shorter than N1-C7. This double bond character is due to the resonance effect with C=S thiocarbonyl groups.

Table 3. Selected bond lengths and angles of **2**

Bond Length		Bond Angle	
S1 – C8	1.655(3)	O1 – C7 – C6	121.7(3)
N2 – N2a	1.372(3)	C6 – C7 – N1	114.7(2)
N2 – C8	1.331(3)	O1 – C7 – N1	123.6(3)
N1 – C8	1.391(4)	N1 – C8 – N2	115.0(2)
N1 – C7	1.362(4)	N2 – C8 – S1	123.9(2)
O1 – C7	1.220(3)	C8 – N2 – N2a	119.4(2)
C – Cl	1.733(3)		

N,N'-bis(2-nitrobenzamidothiocarbonyl)hydrazine (**3**) crystallized in triclinic crystal system with space group of P$\bar{1}$. The unit cells of the crystal are a=8.6373(12) Å, b=8.7962(12) Å, c=13.5691(18) Å, α=94.440(4)°, β=93.877(4)° and γ=118.458(4)°, V=897.2(2) Å^3 and Z=1. Crystal system and refinement parameters are shown in Table 4.

The molecule has the centre of inversion at N3-N3A bond (Figure 5). Both thiourea moieties have *trans* geometry. The nitrobenzoyl group is *trans* to the thiono group about the C-N bond. The bis(thiourea) fragment (S1/O3/N2/N3/C7/C8/S1A/O3A/N2A/N3A/C7A/C8A) and benzene {(C1-C6) and (C1A-C6A)} groups are planar with maximum deviation 0.046(4) Å for N2 atom. The bis(thiourea) fragment is perpendicular to the benzene fragment by dihedral angle of 87.9(3)°. The large dihedral angle is due to the bulky nature of the nitro group that cause the steric hindrance between the benzene and the thiourea fragment.

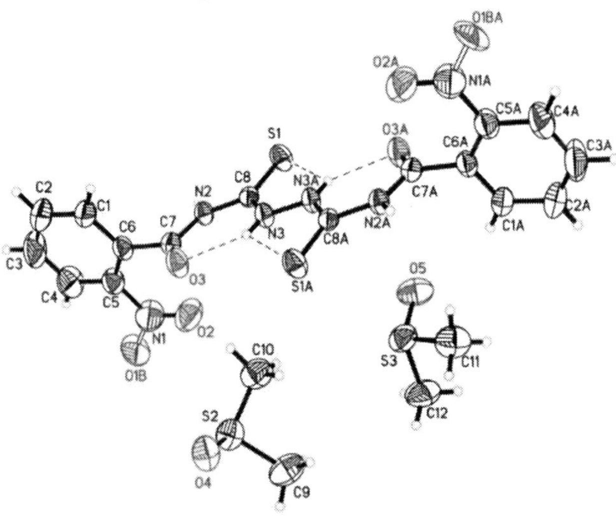

Figure 5. The molecular structure of **3** drawn at 50% probability displacement ellipsoid. The dashes lines indicated intramolecular hydrogen bonds.

Table 4. Crystal data and structure refinement of **3**.

Parameter	Data
Empirical formula	$C_{16}H_{12}N_6O_6S_2$, 4(C_2H_6OS)
Formula weight	761.01
Temperature	301(2) K
Wavelength	0.71073 Å
Crystal system, space group	Triclinic, $P\bar{1}$
Unit cell dimensions	a = 8.6373(12) Å α= 94.440(4)°
	b = 8.7962(12) Å β= 93.877(4)°
	c = 13.5691(18) Å γ= 118.458(4)°
Volume, Z	897.2(2) $Å^3$, 1
Density (calculated)	1.464 Mg/m^3
Absorption coefficient	0.551 mm^{-1}
F(000)	412
Crystal size	0.490 x 0.440 x 0.310 mm^3
Theta range for data collection	2.888 to 26.497°.
Index ranges	-10<=h<=10, -11<=k<=11, -12<=l<=12
Reflections collected	45496
Independent reflections	3674 [R(int) = 0.0588]
Completeness to theta = 25.242°	99.5 %
Refinement method	Full-matrix least-squares on F^2
Data / restraints / parameters	3674 / 2 / 230
Goodness-of-fit on F^2	1.157
Final R indices [I>2sigma(I)]	R1 = 0.0684, wR2 = 0.1928
R indices (all data)	R1 = 0.0856, wR2 = 0.2093
Largest diff. peak and hole	0.627 and -0.488 e.$Å^{-3}$

There are two pseudo six-membered rings, {(C7/N2/C8/N3/H3A···O3) and (C7A/N2A/C8A/N3A/H3AA···O3A)} and two five-membered rings, {(C8/N3/N3A/H3AA···S1) and (C8A/N3A/N3/H3A···S1A)} in the molecule due to N3-H3···O3A and N3-H3A···S1 intramolecular hydrogen bonds (table 5). In the crystal structure, the molecule is linked to two DMSO solvent by N2-H2A···O4 intermolecular hydrogen bonds. In addition, the linked DMSO molecule is also connected to another DMSO molecule by C10-H10B···O5 intermolecular hydrogen bonds which form short polymeric chain consist of five molecules and arrange along ab face (Figure 6).

Table 5. Hydrogen bonds geometry of **3** (Å and °)

D-H···A	d(D-H)	d(H···A)	d(D···A)	<(DHA)
N2-H2A···O4#2	0.86(4)	1.92(4)	2.759(5)	165(4)
N3-H3A···O3	0.85(4)	2.00(4)	2.609(4)	127(4)
N3-H3A···S1	0.85(4)	2.47(4)	2.922(5)	114(3)
C10-H10B···O5#4	0.96	2.45	3.336(8)	153

Symmetry transformations used to generate equivalent atoms:
#1 -x+1,-y,-z+1 #2 x+1,y,z #3 x,y,z #4 -x+1,-y+1,-z+1

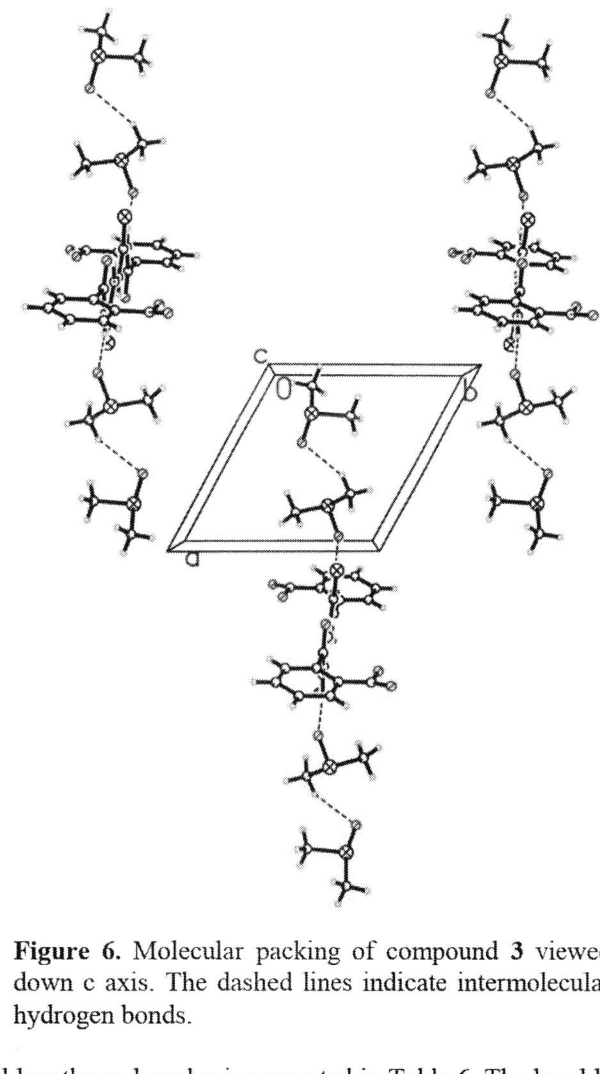

Figure 6. Molecular packing of compound **3** viewed down c axis. The dashed lines indicate intermolecular hydrogen bonds.

The important bond lengths and angles is presented in Table 6. The bond lengths and angles are in normal ranges (Table 6). All bond lengths and angles are in normal ranges [36].

Table 6. Selected bond lengths and angles of **3**.

Bond Length (Å)		Bond Angle (°)	
S1 – C8	1.663(4)	O3 – C7 – C6	121.6(4)
N3 – N3a	1.371(6)	C6 – C7 – N2	113.4(4)
N3 – C8	1.322(5)	O3 – C7 – N2	124.7(4)
N2 – C8	1.387(5)	N2 – C8 – N3	115.8(3)
N2 – C7	1.365(5)	N3 – C8 – S1	123.7(3)
O3 – C7	1.210(5)	C8 – N3 – N3a	119.4(4)
C5 – N1	1.458(7)		

3.4. Antioxidant Screening

DPPH is a stable free radical molecule that has maximum absorption at 515-518 nm, thus it has purple colour. When DPPH accepts an electron or H radical from other molecule, it is converted into a non-radical molecule. The purple colour generally disappears and it can be measured from the changes in absorbance. Based on this property, DPPH was extensively used to evaluate the antioxidant activity of many compounds.

Table 7. The EC_{50} values of the products and ascorbic acid

Product	EC_{50} (μM)
1	123.09
2	374.89
3	611.89
4	215.60
Ascorbic Acid	561.36

The antioxidant activity of the synthesized compounds on the DPPH radical was evaluated according to the methods of Molyneux with some modifications [37]. Various research groups have used widely different experimental procedures which differed in the concentration of DPPH, incubation time and reaction solvent. It caused the EC_{50} values for even the standard ascorbic acid significantly different. For examples, by using the DPPH concentration of 50 μM in buffered methanol and 30 minutes of incubations, the EC_{50} of ascorbic acid is 11.5 μM [38]. But when using the DPPH concentration of 100 μM in DMSO and 60 minutes of incubation, the EC_{50} value of ascorbic acid is 68 μM [8]. Therefore, it is not possible to compare the results of the similar compounds and ascorbic acid of different experimental protocols [37].

The antioxidant activity of all compounds and ascorbic acid is presented in table 7. It can be seen that the EC_{50} values of all compounds by DPPH free radical method are higher than the antioxidant activity of ascorbic acid, except for compound **3**.

The proposed reaction mechanism of DPPH scavenging by synthesized compounds is shown in figure 6. The thiourea compounds stabilized the DPPH radical by two reaction mechanism involving hydrogen donor and electron transfer [8,39]. The thione C=S group transferred an electron to form a new bond with the DPPH radical. To stabilize the radical formed, the H atom from N-H thioamide group was donated to the other DPPH radical. The presence of two thiourea moieties, a bis(thiourea) molecule should be able to stabilize 4 DPPH radicals much effectively.

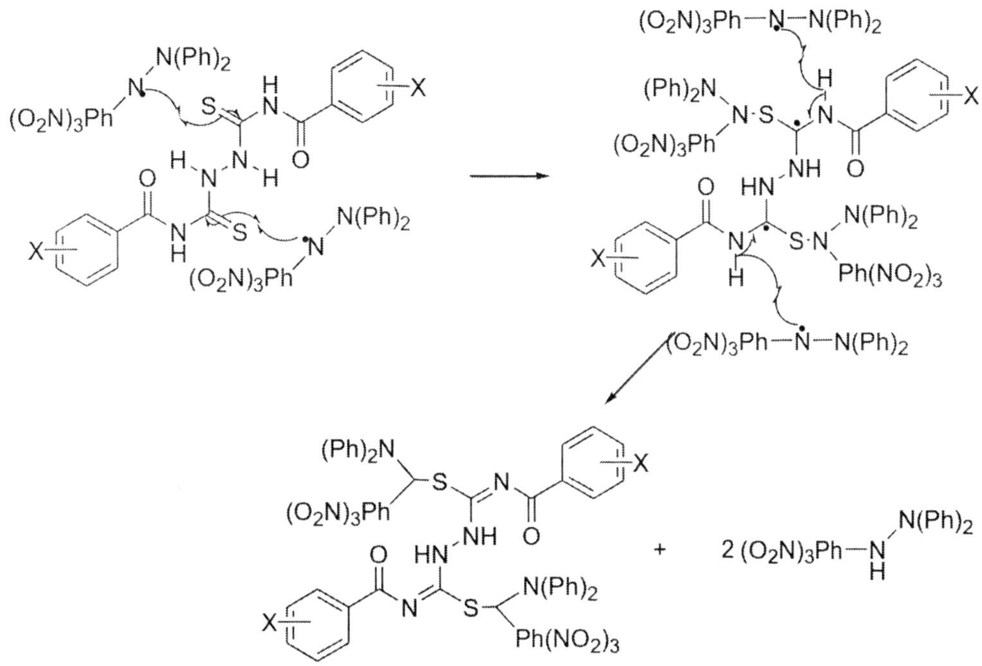

Figure 7. Proposed reaction mechanism of DPPH scavenging by *N,N'*-bis(benzamidocarbonyl)hydrazine derivatives

Generally, the presence of electron withdrawing groups can enhance the antioxidant activity because they could stabilize the radical formed [40]. The low antioxidant activity of compound **3** may be due to the steric hindrance to the DPPH radicals to attack the (NH)1 groups. Further study about the structure activity relationship is needed to reveal the influence of the structures of the synthesized compounds to their antioxidant activity.

4. Conclusion

Four new *N,N'*-bis(benzamidothiocarbonyl)hydrazine (X-PhC(O)NHC(S)NH), where X = *o*-fluoro, *o*-chloro, *o*-nitro and *o*-methyl groups have been successfully synthesized and characterized by spectroscopic methods. Chemical crystallographic study of *N,N'*-bis(*o*-fluorobenzamidothiocarbonyl)hydrazine and *N,N'*-bis(*o*-chlorobenzamidothiocarbonyl)hydrazine showed that both thiourea moieties adopt *trans* geometry. Thus,monodentate coordination with metal is most likely to occur in a complexation reaction.

All the compound showed higher antioxidant activity against DPPH free radical than ascorbic acid (EC_{50} of 561.36 μM), except compound **3** (EC_{50} of 611.89 μM). Compound **1** has the highest antioxidant activity with EC_{50} of 123.09 μM, due to the presence of fluoro atom in *ortho* position. The computational docking study is also needed to identify the potential important interaction between the products and the radical molecule for further antioxidant study.

Acknowledgement

We thank all staff of School of Chemical Science and Food Technology, FST UKM and The Centre for Research and Instrumentation (CRIM) UKM for providing all the facilities. We also acknowledged LPDP (Indonesia Endowment Fund for Education) of Indonesian Ministry of Finance for the scholarship.

References

[1] Khan K M, Naz F, Taha M, Khan A, Perveen S, Choudhary M I and Voelter W 2014 Synthesis and in vitro urease inhibitory activity of N,N'-disubstituted thioureas. *Eur. J. Med. Chem.* **74** 314–23
[2] Karipcin F, Atis M, Sariboga B, Celik H and Tas M 2013 Structural , spectral , optical and antimicrobial properties of synthesized *J. Mol. Struct.* **1048** 69–77
[3] Mohamed N a and Abd El-Ghany N a 2012 Preparation and antimicrobial activity of some carboxymethyl chitosan acyl thiourea derivatives. *Int. J. Biol. Macromol.* **50** 1280–5
[4] Saeed S, Rashid N, Jones P G, Ali M and Hussain R 2010 Synthesis, characterization and biological evaluation of some thiourea derivatives bearing benzothiazole moiety as potential antimicrobial and anticancer agents. *Eur. J. Med. Chem.* **45** 1323–31
[5] Vega-Pérez J M, Periñán I, Argandoña M, Vega-Holm M, Palo-Nieto C, Burgos-Morón E, López-Lázaro M, Vargas C, Nieto J J and Iglesias-Guerra F 2012 Isoprenyl-thiourea and urea derivatives as new farnesyl diphosphate analogues: synthesis and in vitro antimicrobial and cytotoxic activities. *Eur. J. Med. Chem.* **58** 591–612
[6] Koca İ, Özgür A, Coşkun K A and Tutar Y 2013 Synthesis and anticancer activity of acyl thioureas bearing pyrazole moiety. *Bioorg. Med. Chem.* **21** 3859–65
[7] Jumal J, Yamin B M, Ahmad M and Heng L Y 2012 Mercury Ion-Selective Electrode With Self-plasticizing Poly(n–buthylacrylate) Membrane Based On 1,2-Bis-(N' benzoylthioureido)cyclohexane As Ionophore *APCBEE Procedia* **3** 116–23
[8] Arif A, Abdul N, Yehye W A, Alhadi A A and Kadir F A 2014 European Journal of Medicinal Chemistry PASS-assisted design , synthesis and antioxidant evaluation of new butylated hydroxytoluene derivatives **87** 564–77
[9] Sivan S K, Vangala R and Manga V 2013 Molecular docking guided structure based design of symmetrical N,N'-disubstituted urea/thiourea as HIV-1 gp120-CD4 binding inhibitors. *Bioorg. Med. Chem.* **21** 4591–9
[10] Tatar E, Rollas S and Clercq E De 2008 Synthesis of some novel thiourea derivatives obtained from 5- [(4- thiones and evaluation as antiviral / anti-HIV and anti-tuberculosis agents **43**
[11] Alkan C, Tek Y and Kahraman D 2011 Preparation and characterization of a series of thiourea derivatives as phase change materials for thermal energy *Turk J Chem* **35** 769–77
[12] Xu X, Qian X, Li Z, Huang Q and Chen G 2003 Synthesis and insecticidal activity of new substituted N-aryl-N'-benzoylthiourea compounds *J. Fluor. Chem.* **121** 51–54
[13] Criado J J, Gheorghe R, Gonz F J, Hermosa M R, Sanz F, Manzano J L, Monte E and Rodr E 2004 N -benzoyl- N 0 -alkylthioureas and their complexes with Ni (II), Co (III) and Pt (II) – crystal structure of 3-benzoyl-1-butyl-1-methyl-thiourea : activity against fungi and yeast **98** 1307–14
[14] Rodríguez-Fernández E, Manzano J L, Benito J J, Hermosa R, Monte E and Criado J J 2005 Thiourea, triazole and thiadiazine compounds and their metal complexes as antifungal agents. *J. Inorg. Biochem.* **99** 1558–72
[15] Rauf M K, Badshah A, Gielen M, Ebihara M, Vos D De and Ahmed S 2009 Synthesis , structural characterization and in vitro cytotoxicity and anti-bacterial activity of some copper (I) complexes with N , N 0 -disubstituted thioureas *J. Inorg. Biochem.* **103** 1135–44
[16] Vinithra G, Suganya S and Velmathi S 2013 Naked eye sensing of anions using thiourea based chemosensors with real time application *Tetrahedron Lett.* **54** 5612–5
[17] Fan L, Luo C, Lv Z, Lu F and Qiu H 2011 Removal of Ag+ from water environment using a novel magnetic thiourea-chitosan imprinted Ag+. *J. Hazard. Mater.* **194** 193–201
[18] Taylor P, Jing X, Xu F, Zhu Q, Ren X, Yan C and Wang J 2005 Novel and Efficient Cyclization Procedure for the Synthesis of 2,5-Disubstituted-1,3,4-thiadiazoles Without Using Any Ring-Closing Reagents *Synth. Commun.* **42** 3251–60
[19] Hammami A, C. Kosheiry H and Armand M 2004 A New Family of 2(3H)-Thiazolone Azines as Precursors to Air-Stable Radical Cations *Bull. Chem. Soc. Jpn.* **77** 165–7

[20] Hassan A A, Mohamed N K, Makhlouf M M, Brase S and Nieger M 2015 Synthesis of Oxa-aza- and Bis-oxathiaaza[3.3.3]propellanes from Dicyanomethylene-1,3-indanedione and 2,5-Dithiobiureas *Synthesis (Stuttg).* **47** 3036–42
[21] Simonov Y A, Fonari M S, Zaworotko M J, Abourahma H, Lipkowski J, Ganin V and Yavolovskii A A 2003 From 1D strands to extended molecular assemblies in the binary compounds of dithiooxamide and dithiobiurea with crown ethers *Org. Biomol. Chem* **1** 2922–9
[22] Donia A M, Atia A A and Heniesh A M 2008 Efficient removal of Hg (II) using magnetic chelating resin derived from copolymerization of bisthiourea / thiourea / glutaraldehyde *Sep. Purif. Technol.* **60** 46–53
[23] El S, Ghazy S, Mohamed G, El A, Ahmed O, Gammal A and Yousef T 2010 Flotation-separation of toxic metal ions from aqueous solutions using thiosemicarbazide derivatives as chelating agents and oleic acid as a surfactant *Eur. J. Chem.* **1** 76–83
[24] Nevagi R J, Dhake A S, Narkhede H I and Kaur P 2014 Bioorganic Chemistry Design , synthesis and biological evaluation of novel thiosemicarbazide analogues as potent anticonvulsant agents *Bioorg. Chem.* **54** 68–72
[25] Pignedoli A, Peyronel G and Antolini L 1975 The Crystal and Molecular Structure od Dithiobiurea, N,N'-bis(thiocarbamoyl)hydrazine (NH2.CS.NH)2 *Acta Crystallogr. Sect. B Struct. Crystallogr. Cryst. Chem.* **B31** 1903–6
[26] Akinchan N T, Drozdzewski P M and Battaglia L P 2002 Crystal structure and spectroscopic characterization of bis (N-phenylthiourea) *J. Chem. Crystallogr.* **32** 91
[27] Weidong S, Jin Y, Lifei B and Wang X 2009 1-(2,3,4,6-tetra-O-β-D-glucopyranosyl)-3-thioureidothiourea monohydrate *Acta Crystallogr. Sect. E Struct. Reports Online* **E65** o242
[28] Yusof M S M, Yamin B M and Shamsuddin M 2003 N-(N-benzoylhydrazinocarbothioyl)benzamide *Acta Crystallogr. Sect. E Struct. Reports Online* **59** o810–1
[29] Safin D a., Babashkina M G, Bolte M and Klein A 2009 The influence of the spacer Z on N-phosphorylated bis-thioureas and 2,5-dithiobiurea Z[C(S)NHP(O)(OiPr)2]2 (Z=NHCH2CH2NH, NHC6H4-2-NH, NHNH) crystal design *Polyhedron* **28** 1403–8
[30] Sheldrick G M 2008 A short history of SHELX. *Acta Crystallogr. A.* **64** 112–22
[31] Pavia D L, Lampman G M and Kriz G S 2001 *Introduction to Spectroscopy* (Thomson Learning, Inc.)
[32] Refat M S and El-Metwaly N M 2012 Spectral, thermal and biological studies of Mn(II) and Cu(II) complexes with two thiosemicarbazide derivatives. *Spectrochim. Acta. A. Mol. Biomol. Spectrosc.* **92** 336–46
[33] Huang Y, Dong X, Zhang L, Chai W and Chang J 2013 Structure – property correlation of benzoyl thiourea derivatives as organogelators *J. Mol. Struct.* **1031** 43–8
[34] Osman U M and Yamin B M 2012 Synthesis and Structural Study of Bis-Thiourea Moieties with Aromatic Linkers 1535–8
[35] Wilson D, Arada M D L A, Alegret S and del Valle M 2010 Lead(II) ion selective electrodes with PVC membranes based on two bis-thioureas as ionophores: 1,3-bis(N'-benzoylthioureido)benzene and 1,3-bis(N'-furoylthioureido)benzene. *J. Hazard. Mater.* **181** 140–6
[36] Allen F H, Kennard O, Watson D G, Brammer L and Orpen A G 2002 Tables of Bond Lengths determined by X-Ray and Neutron Diffraction. Part 1–19
[37] Molyneux P 2004 The use of the stable free radical diphenylpicryl- hydrazyl (DPPH) for estimating antioxidant activity *Songklanakarin J. Sci. Technol.* **26** 211–9
[38] Sharma O P and Bhat T K 2009 DPPH antioxidant assay revisited *Food Chem.* **113** 1202–5
[39] Nguyen D T, Le T H and Bui T T T 2013 Antioxidant activities of thiosemicarbazones from substituted benzaldehydes and N-(tetra-O-acetyl-β-D-galactopyranosyl)thiosemicarbazide. *Eur. J. Med. Chem.* **60** 199–207

[40] Khan I, Ali S, Hameed S, Rama N H, Hussain M T, Wadood A, Uddin R, Ul-Haq Z, Khan A, Ali S and Choudhary M I 2010 Synthesis, antioxidant activities and urease inhibition of some new 1,2,4-triazole and 1,3,4-thiadiazole derivatives *Eur. J. Med. Chem.* **45** 5200–7

Heavy metal concentrations in cocoa beans *(Theobroma cacao* L.*)* originating from East *Luwu,* South Sulawesi, Indonesia

A Assa[1,2], A Noor[2], M R Yunus[1], Misnawi[3] and M N Djide[4]

[1] Center for Plantation Based Industry, Agency for Research and Development of Industry, Ministry of Industry, Makassar, South Sulawesi, Indonesia
[2] Radiation Chemistry Laboratory, Department of Chemistry, Hasanuddin University, Makassar, South Sulawesi, Indonesia
[3] Indonesian Coffee and Cocoa Research Institute, Jember, East Java, Indonesia
[4]Department of Pharmacy, Hasanuddin University, Makassar, South Sulawesi, Indonesia

E-mail: *emma_bbihp@yahoo.com*

Abstract. Concentrations of some heavy metals (Pb, Cu, Cd, As and Hg) were assessed for cocoa beans *(Theobroma cacao L)* originating from East *Luwu*, South Sulawesi, Indonesia after five-day fermentation. Consisting of PB 123, BR 25, and MCC 02 cocoa clones, the spectrophotometric analysis showed that concentrations of Pb, Cd, As and Hg in the cocoa beans over the three clones was below the detection limits of 0.100; 0.050, 0.010 and 0.005 mg/kg. For Cu, they were 19.343; 10.391, and 18.594 mg/kg respectively, but still below the maximum critical levels, established by the European Food Safety Authority (EFSA). Concentrations of those five heavy metals in the bean shells were found to be parallel to those in the cocoa beans, except for Pb.

1. Introduction

Cocoa (*Theobroma cacao* L.) is cultivated mainly in West Africa, Central and South America and Asia. Based on the ICCO 2017 data, eight largest cocoa producing countries today are Ivory Coast, Ghana, Indonesia, Nigeria, Brazil, Equador, and Malaysia. The total world cocoa production in 2015/2016, is estimated 3.972 thousand tonnes, in which Indonesia contributed for about 8.0 % of the total world production [1].

Studies on cocoa chemical composition and properties have proven that a moderate consumption of cocoa may be beneficial for human health, mainly due to cocoa contains high polyphenolic compounds, particular flavonoids as antioxidants [2], which are responsible for the overall health of the humans. Antioxidant capacity of cocoa has been reported higher than that of green tea and red wine [3].

The beneficial effects of cocoa range from its free radical scavenging capacity [4], reduction of the risk of cardiovascular diseases [4,5], prevention of autoimmune diseases and hepatopathy [6,4,7], anti inflammatory [8], reduction of anxiety and depressive symptoms [9], to anti cancer effect [10,4]. Moreover, extract prepared from cocoa beans or cocoa powder shows to exhibit an effect on syrtocotozin induced diabetes mellitus [11] and antibacterial effect against Streptococcus mutants formulated in mouth washes [12].

Despite the health benefits of cocoa, recently, there has been a high food safety concern regarding the presence and levels of heavy metals, such as nickel (Ni), cadmium (Cd), chromium (Cr), and lead (Pb), both in cocoa beans and cocoa products i.e. cocoa butter, cocoa liquor, cocoa powder and chocolates. Rankin *et al.* [13] reported a relatively high level contaminants of lead in chocolates and other cocoa products, produced in Nigeria. The average lead level for 23 chocolates and cocoa products, measured by WCAS was 32.5 ng/g. The samples were taken in 2002. Dahiya *et al.* [14] have examined Cd, Ni and Pb levels of 69 chocolate brands sold locally in the suburban areas of Mumbai, India. They found that the Cd levels ranged from 0.001 to 2.730 g/kg. The Ni levels ranged from 0.041 to 8.290 g/kg, while the Pb levels ranged from 0.049 to 8.040 g/kg. If using the Polish National Standard for Pb, established by the Poland Regulatory Agency [15] as a reference, 42 chocolates out of 69 contained Pb, exceeding the authorized limits of 0.30 g/kg. In the case of Cd, 19 chocolates exceeded the authorized limits of 0.05 g/kg.

Cocoa beans, as a key raw material in the cocoa derived products, could also contain those heavy metals, obtained from the growing soils and/ or during the production process. High levels of Cd, Pb, Cu, and As in the cocoa beans are of the interest as these trace elements are generally considered as toxic to human [16]. Chaves *et al.* [17] in investigating Cd contamination in cocoa plants and the growing soil in Southern Ecuador, found that the accumulation of Cd in the surface soil of the cocoa farms was most probably caused by the anthropogenic activities, such as irrigation of Cd rich water used. The contamination caused accumulation of Cd in the cocoa beans, in which 12 out of 19 sampling sites had a bean Cd level exceeding 0.6 mg/kg, the maximum critical level for Cd, established by European Union (EU). Yanus *et al.* [9] reported significant levels of Bi, Pb, and As in the cocoa bean shells, but not in the cocoa beans. Samples of the cocoa beans were received from the chocolate brand importers in Israel. This indicated that the metal contaminants occurred likely after the beans are harvested, sun-dried, and shipped.

Due to the toxicity of those heavy metals, there have been several agitations to reduce the concentrations to the safe levels by setting up stringent standards by international or state legislative bodies, and by chocolate manufacturers as well, in order to protect the health of consumers.

This study assessed the heavy metal concentrations in the cocoa beans originating from East *Luwu*, South Sulawesi, Indonesia, and to ascertain their compliance with standards. Those heavy metals included lead (Pb), copper (Cu), cadmium (Cd), arsen (As), and mercury (Hg). As an additional information, East *Luwu* is one of the main producing areas of cocoa in Indonesia, located in South Sulawesi Province.

2. Materials and Methods

2.1. Preparation of the cocoa beans

The cocoa beans for assessment were prepared as dried fermented cocoa beans. They consisted of three cocoa clones, namely PBC 123 (Sulawesi 1), BR 25 (Sulawesi 2), and MCC 02 (M 45).

Cocoa pods were procured from Sahar Cocoa Village (SCV) farm, a plasma cocoa farm of Mars Symbioscience Indonesia Ltd., located in *Maliowowo* Village, District of Angkona, East *Luwu,* South Sulawesi, Indonesia, while fermentation and sun-drying of cocoa beans took place in Mars Cocoa Research Station, about 12 km from the SCV location.

The mature cocoa pods were stored for six days, before the beans were removed from the pod husk. The cocoa beans were then separated from its placenta. Fermentation took place for five days, as recommended in the Sime-Cadbury method, but using fermentation boxes made of *styrofoam,* instead of wood materials. The capacity of each box is 7 kg of wet beans. After fermentation, the cocoa beans were sun-dried for 5 days, until reaching 6 to 7% moisture contents.

2.2. Preparation of the samples and analysis

Preparation and analysis of the cocoa bean samples were conducted at Laboratory of Chemical and Microbiologycal Testing, Center for Plantation Based Industry, in Makassar, South Sulawesi, Indonesia.

The cocoa nibs were separated from their bean shells and then ground with ceramic mortar. One gram of each sample was weighed into a teflon polytetrafluoroethylene (PTFE) vessel and digested with 5 mL HNO_3 and 3 mL H_2O_2. The digestion was carried out in a microwave for 50 minutes using the digestion protocol. Upon cooling to room temperature, the digested sample was transferred into 50 mL polypropylene tubes and made up to 25 mL with deionized water. The digested solution was filtered through a Whatman No. 42 filter paper, and diluted prior to analysis. The concentration of heavy metals was determined using Atomic Absorption Spectrophotometer (Shimadzu, AA-7000). Heavy metals of Pb, Cu, Cd, As, and Hg concentrations were determined at a wavelength of 248.3 nm, 324.8 nm, 228.8 nm, 193.7 nm, and 253.7 nm respectively, using acetylene as the fuel and air as the support gas [18].

The heavy metal concentration levels were then ascertained for their compliance with standards of European Food Safety Authority for cocoa beans.

3. Result and Discussion

The spectrophotometric analysis of Pb, Cu, Cd, As, and Hg concentrations in the cocoa beans are presented in Table 1. The cocoa beans were prepared as dried five-day fermented cocoa beans, consisting of PBC 123 (Sulawesi 1), BR 25 (Sulawesi 2), and MCC 02 (M 45) cocoa clones. Table 2 shows the same analysis results, but for the cocoa bean shells.

Table 1. Concentration levels of Pb, Cu, Cd. As, and Hg in the cocoa beans over the three clones.

Heavy metals	Cocoa clones	Concentration level (mg/kg)	Maximum critical level (mg/kg)
Lead (Pb)	PBC 123	< 0.10	1.0 mg/kg*
	BR 25	< 0.10	
	MCC 02	< 0.10	
Copper (Cu)	PBC 123	19.343	50 mg/kg*
	BR 25	10.391	
	MCC 02	18.594	
Cadmium (Cd)	PBC 123	< 0.05	0.10 mg/kg**
	BR 25	< 0.05	
	MCC 02	< 0.05	
Arsenic (As)	PBC 123	< 0.01	
	BR 25	< 0.01	
	MCC 02	< 0.01	
Mercury (Hg)	PBC 123	< 0.005	
	BR 25	< 0.005	
	MCC 02	< 0.005	

* European Commission No.1881/2006
** European Union No. 488/2014

Concentrations of Pb in the cocoa beans over the three clones, shown in Table 1, were below detection limits of 0.10 mg/kg, thus far below the maximum critical level of 1.0 mg/kg, established by EFSA [19]. On the other hand, the concentrations of Pb in the cocoa bean shells shows quite high levels; those are 9.724 mg/kg for the PBC 123 clone, 11.146 mg/kg for the BR 25 clone, and 5.810 for the MCC 02 clone. All the Pb intrusions were retained in the bean shells and did not pass into the nibs. Regarding this phenomenon, Rankin *et al.* [13] found that the cocoa bean shell can be regarded as an excellent protective shield against intrusions of lead into the nib from external sources before the cocoa beans are harvested. In fact, Meunier *et al.* [20] in their study have reported that the cocoa bean shells were particularly efficient in the removal of lead from very acidic solution (q_{max} = 6.2 mg/g of Pb). The presence of other metals and cations in the solution did not seem to affect the recovery of lead.

Table 2. Concentration levels of Pb, Cu, Cd, As, and Hg in the cocoa bean shells over the three clones.

Heavy metals	Cocoa clones	Concentration level (mg/kg)
Lead (Pb)	PBC 123	9.724
	BR 25	11.146
	MCC 02	5.810
Copper (Cu)	PBC 123	12.861
	BR 25	11.089
	MCC 02	10.516
Cadmium (Cd)	PBC 123	< 0.10
	BR 25	< 0.10
	MCC 02	< 0.10
Arsenic (As)	PBC 123	< 0.02
	BR 25	< 0.02
	MCC 02	< 0.02
Mercury (Hg)	PBC 123	< 0.005
	BR 25	< 0.005
	MCC 02	< 0.005

Related to lead contamination in the cocoa beans, especially in some African countries, Rankin *et al.* [13] and Manton [21] hypothesized to derive from the atmospheric emission of leaded gasoline, (which is still used in the countries) during fermentation and drying of the cocoa beans.

As stated by Ellen [22], lead can be very harmful even at low concentrations if taken for a long period. After absorption, lead is initially distributed in soft tissue throughout the body via blood and then deposited in bone. It may cause damage to kidneys, cardiovascular, central nervous, and reproductive system [10].

Concentrations of Cu in the cocoa beans were 19.343 mg/kg for the PBC 123 clone, 10.391 mg/kg for the BR 25 clone, and 18.594 mg/kg for the MCC 02 clone, as shown in Table 1. All these concentrations were below the maximum critical level of 50 mg/kg, regulated by the EFSA (2006). Moreover, concentrations of Cu in the bean shells were found to be 12.861 mg/kg for the PBC 123 clone, 11.089 mg/kg for the BR 25 clone, and 10.516 mg/kg for the MCC 02 clone, as shown in Table 2. This indicated that some of the Cu intrusions were retained in the bean shells, and some others passed into the nibs. Compared to the removing lead from an acidic solution, the bean shells have less capacity to the removal of copper [20].

Some studies have proposed that high levels of copper in blood could increase the risk of artherosclerosis by promoting the oxidation of low-density lipoprotein (LDL) [23]. Increase in serum copper levels lead to increase in cardiovascular risk [23,24].

The concentration of Cd, As, and Hg in both the cocoa beans and the bean shells over the three clones, as shown Tables 1 and 2, were virtually below the detection limit of 0.050; 0.010 and 0.005

mg/kg Cd, As and Hg, respectively. The maximum critical level of Cd itself is 0.10 mg/kg, as established by the EFSA [25], while for As and Hg, so far, have not been regulated for the cocoa beans.

Some evidence suggest that many cocoa farmers now apply pesticides, insecticides, and fertilizers to control pests and insects and to enhance maximum cocoa bean yields. These activities, if not controlled and managed well, could increase contamination of the heavy and trace metals like Cd, Fe, Pb, Cu, Mn, and Zn into the soil which eventually will end up in the edible parts of the cocoa through phytoextraction by the crops and translocation throughout the plant system [26,27].

High level of cadmium can lead to disturbance in calcium metabolism and formation of kidney stones, softening of the bones [28] and synthesis of pancreatic DNA [29]. Arsenic, in inorganic form is special of concern because of its cancer causing properties. Arsenic has been classified by the International Agency for Research into Cancer (IARC), as a human carcinogen. Mercury is often associated with a wide spectrum of adverse health effects, especially damage to the central nervous system and kidney [15].

Regarding the presence of Pb in the bean shells in our study, in which its levels were quite high, there was no available data whether the cocoa has uptaken Pb dominantly from atmospheric emission of leaded gasoline during the fermentation and sun-drying of the cocoa beans or from the soil. It was similar for the presence of Cu in both the cocoa beans and the bean shells, whether it comes from Cu contaminated soil due to the fertilizer and other agrochemical uses and/or from Cu rich sedimentary rocks. Alloway [30] reported his study that concentrations of Pb in the cocoa beans related to the presence of Fe, Zn and Cu in the soil. This can be attributed to the interactions of Fe, Zn, and Cu with Pb in competing for absorption by the plant. On the other hand, concentrations of Cu in the cocoa beans are mainly related to the concentrations of Pb and Mn in the soil. To examine the sources of these Pb and Cu contaminants in the cocoa beans, requires further studies.

Conclusion

Of the five heavy metals assessed, Pb, Cd, As and Hg were not detected in the cocoa beans over the PBC 123, BR 25 and MCC 02 cocoa clones. Cu was detected, but their concentrations were still below the maximum critical levels, established by the European Food Safety Authority (2006). Moreover, concentrations of those heavy metals in the bean shells seemed to be parallel with those in the cocoa beans (nibs), except for Pb.

Acknowledgments

This study was made possible through "*Program Rintisan Gelar S3*" Grant, from the Agency for Research and Development of Industry, Ministry of Industry, The Republic of Indonesia. Cocoa pods and other facilities were provided by Sahar Cocoa Village (SCV) Farming and Mars Symbioscience Indonesia Ltd., particularly in the preparation of the cocoa beans.

References

[1] ICCO 2017 *ICCO Quarterly Bulletin of Cocoa Statistics* Vol. XLIII, No.2

[2] Lamuela R M, A I R Perez, C A Lacueva and A Tornero 2005 Review: health effect of cocoa flavonoids. Food Sci. and Techn. Int. **11** 3 p 159-176

[3] Hii C L, C L Law, S Suzannah, Misnawi, and M Cloke 2009 Polyphenols in cocoa (Theobroma cacao L). *As. J. Food Ag-Ind.* Vol. **2** 04 p 702-722

[4] Bertoldi D, A Barbero, F Camin, A Caligiani, and R Larcher 2016 Multielemental fingerprinting and geographic traceability of Theobroma cacao beans and cocoa product *Food Control* **65** 46-53

[5] Keen C L, Holt R R, Oteiza P I, Fraga C G, and Schmitz H H 2005 Cocoa antioxidants and cardiovascular health. *American J. of Clinical Nutrition* **81** 298S -303S

[6] Arlorio M, C Bottini, F Travaglia, M Locatelli, M Bordiga M, J D Coisson, A Martelli and L Tessitore 2009 Protective activity of Theobroma cacao L. phenolic extract on AML12 and

MLP29 liver cells by preventing apoptosis and inducing autophagy *J. of Agric. and Food Chem.* **57** 10612-10618

[7] Sanbongi C, N Suzuki and T Sakane 1997 Polyphenols in chocolate, which have antioxidant activity, modulate immune functions in humans in vitro *Cell Immunology* **177** 129-136

[8] Selmi C, Cocchi C A, Lanfredini M, and Keen C L 2008 Chocolate at heart: The anti inflammatory impact of cocoa flavanols *Mol. Nutr. Food Res.* **52** 1340-1348

[9] Yanus R L, H Sela, E J C Borojovich, Y Zakon, M Saphier, A Nikolski, E Gutflais, S Lorber, and Z Karpas 2014 Trace elements in cocoa solids and chocolate: An ICPMS study *Talanta* **119** 1-4

[10] Ames N M K, Sigenaga and T M Hagen 1993 Oxidant, antioxidant, and the degenerative diseases of aging *Proc. of the National Academic of Science*, USA 90 p 7915-7922

[11] Scalbert A, I T Johnsons, M Saltmarsh 20015 Polyphenols: antioxidants and beyond 1'2'3'. *Am. J. Clin. Nutr.* Vol **81** 1 p 2158-2175

[12] Loppies J E, Rosniati, M Yumas, and M R Yunus 2013 In-vitro antibacterial activity of mouth washes supplemented with cocoa pod husk extract against Streptococcus mutans. *J. Ind. Hasil Perkebunan,* Vol. **8** No 2 p 59-65

[13] Rankin C W, J O Nriagu, J K Aggarwal, T A Arowolo, K Adebayo and A R Flegal 2005 Lead contamination in cocoa and cocoa products: Isotopic evidence of global contamination. *Environmental Health Prospective* Vol.**113** 10 p 1344-1348

[14] Dahiya S, R Karpe, A Hegde and R M Sharma 2005 Lead, cadmium, and nickel in chocolates and candies from suburban areas of Mumbai India, *J. Food Composition and Analysis*, Vol. 18, Issue 6 p 517-522

[15] FAO/WHO 2001 Draft standards for chocolate and chocolate products *Report of the 19th session of the Codex Committee on Cocoa Products and Chocolate* Fri-bourg 3-5 October 2001 ALINORM 03/14

[16] Vitola V and I Ciprovica 2016 The effect of cocoa benas heavy and trace elements on safety and stability of confectionary products *Rural Sustainability Research* 35 330

[17] Chaves E, Z L He, P J Stoffella, R D Mylavarapu, Y C Li, B Moyanio and V C Baligar 2015 Concentration of cadmium in cacao beans and its relationship with soil cadmium in Southern Equador. *Sci. of the Total Environment* **533** 205-214

[18] AOAC 2005 Official Methods iof Analysis 999.10

[19] European Commission Regulation (EC) No 1881/2006 of 19 December 2006 Setting maximum levels for certain contaminants in foodstuffs *Official J. of European Union* Brussels L364/5

[20] Meunier N, Laroulandie J, Blais J F, Tyagi R D 2003 Cocoa shells for heavy metal removal from acidic solutions *Bioresour Technol.* **90** 255-263

[21] Manton, W I. 2010 Determination of the provenance of cocoa by soil protolith ages and assessment of lead contamination of anthropogenic by Pb/Nd and lead isotop ratios. *J. Agric. and Food Chem.* **58** 713-728

[22] Ellen G, Loon J W and Tolsma K 1990 Heavy metals in vegetables grown in the Netherlands and in domestic and imported fruits *Zeitschrift fur Lebensmittel-Untersuchungund Forschung* **190** 34-39

[23] Higdon J, Drake J V, and Ho E 2007 Micronutrient research for optimum health *Linus Pauling Institute* Oregon State University

[24] Ford E S 2000 Serum copper concentration and coronary heart disease among US adults *Am. J. Epidemiol* **151** 12 p 1182-1188

[25] European Commission Regulation (EU) N 488/2014 of 12 May 2014 amending Regulation (EC) No. 1881/2006 as regards maximum levels of cadmium in foodstuffs. *Official Journal of the European Union*, L 138/75-79

[26] Aikpopodion P E, A. Odule, O C Osobamiro, T Oduwole, and Ademola S M 2013 A survey of copper, lead, cadmium and zinc residues in cocoa beans obtained from selected plantations in Nigeria *J. Chem. Pharm. Res.* **56** 88-98

[27] Thomas, E.Y. 2015. Asessment of heavy metal concentrations and fractination on selected dump oils within Ibadan Metropolis, Nigeria. *J. Agric and Ecology Research Intern.* Vol. **4** No.3 p 117-127
[28] Jarup 2003 Hazards of heavy metal contamination *British Medical Bulletin* 68 p 167-182
Keen C L, Holt R R, Oteiza P I, Fraga C G, and Schmitz H H 2005 Cocoa antioxidants and cardiovascular health. *American J. of Clinical Nutrition* **81** 298S -303S
[29] Schwartz and Reis 2000 Trace elements as contaminants and nutrien: https://books.google.co.id/books?isbn=0470355093 accessed September 27, 2017
[30] Alloway B J 2013 Heavy Metals in Soils: Trace Metals and Metallloids in Soils and their Bioavailability. 3rd ed. *Environmental Pollution* Vol 22 Springer

IOP Conf. Series: Journal of Physics: Conf. Series **979** (2018) 012012 doi:10.1088/1742-6596/979/1/012012

Effect of Fe^{2+} and Mn^{2+} addition on growth and β-carotene production of *Dunaliella salina*

E Mayasari, I Raya and H Natsir

Department of Chemistry, Natural Science and Mathematics Faculty, Hasanuddin University, Perintis Kemerdekaan Km 10, Makassar South Sulawesi, Indonesia

E-mail: emaysa13@gmail.com

Abstract. *Dunaliella salina* is a unicellular chlorophyte, which could grow in a wide range of harsh environmental condition. Under growth-limitation conditions, *D.salina* was known as famous for a high β-carotene producer with up to 10% of its dry cell weights, therefore it could be a β-carotene supplement. The research aimed to find out the effect of Iron (Fe^{2+}) and Manganese (Mn^{2+}) towards β-carotene productivity as result of oxidative stress from the second photosystem (PS II). The analysis method was carried out by the ultrasonic extraction for short and cheap lyses of phytoplankton biomasses; infra-red (IR) to find out the interaction metal ion, and UV/VIS spectrophotometer to determine the β-carotene concentration from phytoplankton the crude extract. The result showed that the interaction occurred between metal ions and M-N, -O-M and M←OH-C groups in the amino acid of phytoplankton. The highest impact was indicated on 0.3 ppm Fe^{2+} to *D. Salina*. The highest β-carotene concentration was 13.08 μg/g DW for 0.3 ppm Fe^{2+} and 8.08 μg/g DW for 0.8 ppm Mn^{2+}. The dry weight concentrations of β-carotene indicated that *D. salina* with 0.6 ppm Fe^{2+} addition had more potential as a β-carotene supplement.

1. Introduction

Carotenoid was a natural product with antioxidant activity. Phytoplankton was known widely had many antioxidants. Species of phytoplankton had been used in many countries in the world [1]. Oxidative stress plays an important role in the pathogenesis of various diseases such as atherosclerosis, alcoholic liver cirrhosis, cancer, etc. Oxidative stress was initiated by free radicals, especially reactive oxygen species (ROS). Oxidative stress occurs when the generation of free radicals or reactive oxygen species (ROS) exceeds the antioxidant capacity of a biological [2,3,4,5].

Dunaliella, *Chlorella,* and *Spirulina* are three types of phytoplankton that had commercial potential that had been used to produce fats, proteins, and pigments [6,7].

Several studies had shown that the use of metal ions play an active role in the process of photosynthesis and oxidative stress. As known oxidative processes were intimately associated with the production of antioxidant compounds such as carotenoids. Research had been done on production astaxanthin [8], using *Haematococcus pluvialis* with the addition of Fe^{2+}-EDTA and Fe^{3+}-EDTA. The addition of Fe^{2+}-EDTA was known to stimulate the production of more effective astaxanthin, whereas Fe^{3+}-EDTA was more effective in improving cell growth. The Fe^{2+} ion is a catalyst that can cause decomposition of H_2O_2 in the Haber-Weiss reaction and Fenton oxidative reactions. Both of these reactions produce hydroxyl radical (HO●), which could stimulate production of astaxanthin on microalgae *Haematococcus pluvialis*. Other studies have also shown that induction of oxidative stress occurs with the addition of high concentrations of Co^{2+} and Mn^{2+} on microalgae *Pavlopa viridis*. In this case, the addition of metal Co^{2+} and Mn^{2+} were known to stimulate the activity of catalase (CAT) and glutathione (GSH) [9].

Antioxidant activity had shown in the methanol crude extract from *Isochrysis aff galbana*, *Chlorella vulgaris*, *Nannochloropsis oculata*, *Tetraselmis* and *Chaetoceros calcitrans* [10]. Cell damage caused by free radicals was believed due to the aging process and degenerative nature of the disease. β-carotene supplementation was one of the preferred sources of antioxidants that could be consumed. However, β-carotene supplements contain only synthetic product isomer trans-β-carotene (all-trans-β-carotene) a pro vitamin A [11,12] that would be converted into vitamin A. While the 9-cis isomer-β-carotene (9-cis-β-carotene), which had a higher antioxidant activity than the trans-β-carotene (all-trans-β-carotene) in natural products only. So that should be pursued natural product which is maximal a 9-cis-β-carotene produced.

Dunaliella bardawil and *D. salina* showed potential as a source of β-carotene [7, 12-17] (Pisal and Lele, 2005; Oren, 2005; Tafreshi and Shariati, 2009; Abusara, 2011). El-Baky et al obtained β-carotene levels of 54.08% of the total carotenoid, at a condition of 16% NaCl [6]. Interclinical Laboratories data provide information that β-carotene of Dunaliella salina is higher than that of carrots. The amount of β-carotene of *D. salina* is 1,100-2,100 mg/100 g while in carrots it was only 5.8 mg/100 g [12].

This study has investigated that the addition of metal Fe^{2+} and Mn^{2+} can determine their effect on the productivity of β-carotene in *D. salina*. The addition of metal ions Fe^{2+} is expected to stimulate oxidative stress that could induce productivity of β-carotene which was a natural product predominantly consists of the isomer trans-β-carotene (all-trans-β-carotene) and 9-cis-β-carotene (9-cis-β-carotene). Infrared spectrophotometer (IR) was used to determine the initial data of functional groups that interact with metal ions Fe^{2+} on the mechanism of formation of β-carotene productivity by these species.

2. Material and Method

Materials used were pure culture of *D. salina* (the phytoplankton cultivation of Jepara, Yogyakarta, Indonesia), acetone pa, $FeSO_4 \cdot 7H_2O$, $MnSO_4.H_2O$; Stock A medium ($FeCl_3.6H_2O$; $MnCl_2.4H_2O$; H_3BO_3; Na_2EDTA; $NaH_2PO_4.2H_2O$; $NaNO_3$); Stock B ($ZnCl_2$; $CoCl_2.6H_2O$; $(NH_4)_6Mo_7O_{24}.4H_2O$; $CuSO_4.5H_2O$); Stock C (Vitamin B_{12} dan Vitamin B_1); β-carotene powder and distilled water.

Equipment used was commonly used glassware, microscopes, haemocytometer (Neubauer-improved), Hand counter, Salinometer, SPNISOSFD oven, digital balance Ohauss NO AP 110, FT-IR Shimadzu 820 1PC, centrifugation, Ultrasonic, and UV-VIS spectrophotometer.

2.1 Procedures

2.1.1 Preparation of Medium Conwy. Medium conwy were three stocks, namely A, B and C. The A nutrient solution was 1000 mL which were added 2 mL to the B stock. The mixture was added to a solution of Conwy in sterile sea water in the ratio of 2 mL per 1000 ml of sea water that has been sterilized. Furthermore, a mixture of stocks A and B added 1 drop of stock C solution of vitamin.

2.1.2. Phytoplankton growth. Determination phytoplankton growth pattern had been done by counting the number of cells per milliliter of medium every day. Sample was taken with a Pasteur pipette, dropped about 1-2 mL on a haemocytometer, and then observed through a microscope. The cell density was calculated using equation (1).

$$\frac{cell}{mL} = \frac{number\ of\ cells\ in\ 4\ blocks}{number\ of\ blocks\ (=4)}\ x\ 10^4 \qquad (1)$$

2.1.3. Identification by FT IR. Identification of functional groups in phytoplankton that could be potential to bind Fe^2. Phytoplanktons were harvested, dried, crushed and screened. Biomass without and with exposure to metal ions were analysed using an infrared spectrophotometer (IR).

2.1.4. β-carotene extraction by sonication method. Working performed in a room with low intensity lighting and at a cold room. Phytoplankton biomass obtained was put into a centrifuge tube and mixed with acetone as much as 3 mL, with a sealed tube, the mixture was sonicated for 50 min at 40°C. Samples were centrifuged for 20 min at 5000 rpm and the supernatant was taken for analysis of β-carotene.

3. Result and Discussion

3.1. Phytoplankton Cell Growth of D. salina
This research was used phytoplankton species *D. salina*. The pattern of growth without the addition of metal ions is shown in figure 1.

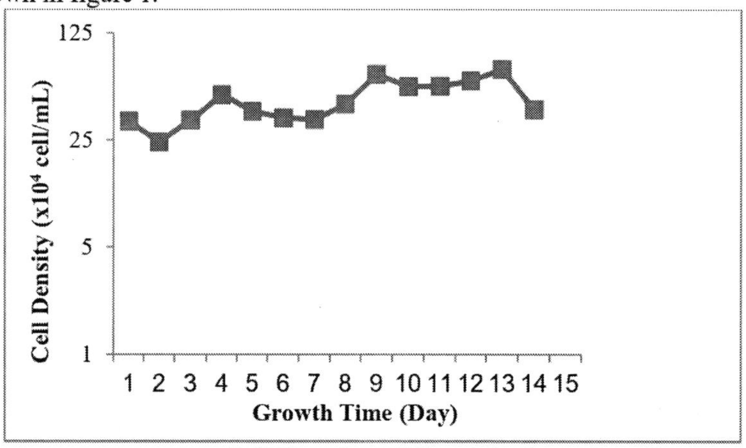

Figure 1. Cell Density of Phytoplankton *D. salina.*

It is clear that *D. salina* requires two days for adaptation to the culture medium Conwy. This was relatively short period. Significant increase observes on the third day, which means optimal cell division begins. This figure also shows the growth pattern of the phytoplankton. The nutrient requirement was highly correlated with the cell size and the cell movement rate, where the cell size was 9-11 µm [18]. It had a flagellum using nutrients in addition to the process of cell growth and division, nutrients were also required for movement. The end of cultivation was stopped at fourteenth days.

3.2 Determination of Fe^{2+} and Mn^{2+}concentration to Tolerable by Phytoplankton D. salina
The growth pattern of *D. salina* with and without the addition of Fe^{2+} and Mn^{+2} is given in figure 2. The concentration of Fe^{2+} and Mn^{2+} ion having the highest and the lowest cell density can be seen in table 1.

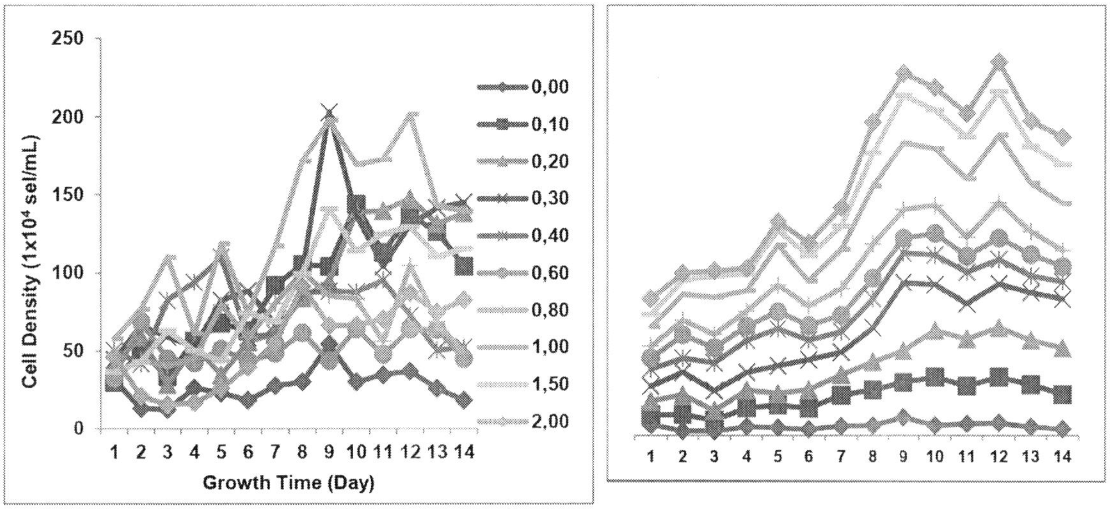

(a)

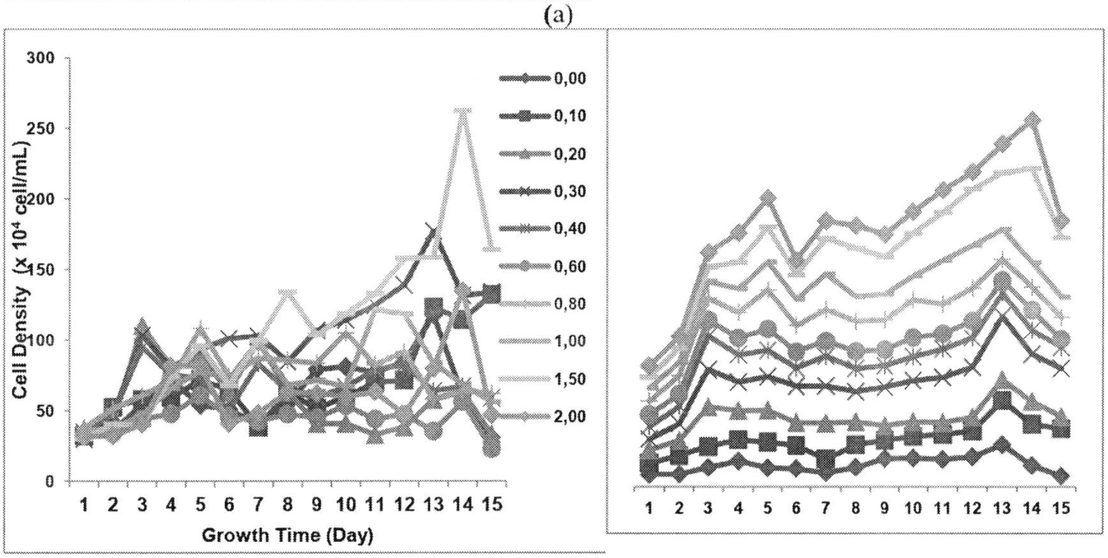

(b)

Figure 2. Growth Pattern of *D. salina* (a) without and with the addition of Fe^{2+} and (b) without and with the addition of Mn^{2+} at various concentrations.

Table 1. The Concentration of Fe^{2+} and Mn^{2+} with the highest and the lowest cell density

Density	$[Fe^{2+}]$ (ppm)	Cell Density ($x10^4$ sel/mL)	$[Mn^{2+}]$ (ppm)	Cell Density ($x10^4$ sel/mL)
Highest Growth Density	0,3	203	1,5	158,5
Lowest Growth Density	0,6	43,5	0,6	35,5

It is clear that the concentration of Fe^{2+} and Mn^{2+} with the highest cell density is different. This occurred because there was differences in morphological properties (cell wall, cell size, movement of phytoplankton) and metal ion functions in metabolic processes. In microalgae, Fe metal ions played a very important role in the regulation of cell metabolism as an essential element, in addition to nutrients and in the process of photosynthesis [19-21].

High growth is shown in the high concentration of Mn^{2+} to the phytoplankton of *D. salina* (table 1). While low growth was in a low concentration of the ion, this suggests that photosynthesis requires more Mn metal ions to produce O_2 located in the process of photosystem II (PS II). The important role of the Mn metal ion in the photosynthesis process lies in the water-solving system of Photosystem II, which prepares electrons for the electron transfer of photosynthesis. Groups of four Mn associated with the oxygen complex (OEC¬_Oxygen Evolving Complex) were bound to the photosystem II (PSII) protein [22].

The table also showed that flagged phytoplankton require in a higher concentrations of Mn^{2+} for cell division and development. It was suitable with [23] and [21] that Manganese ions had functions in photosynthesis process and cell structure.

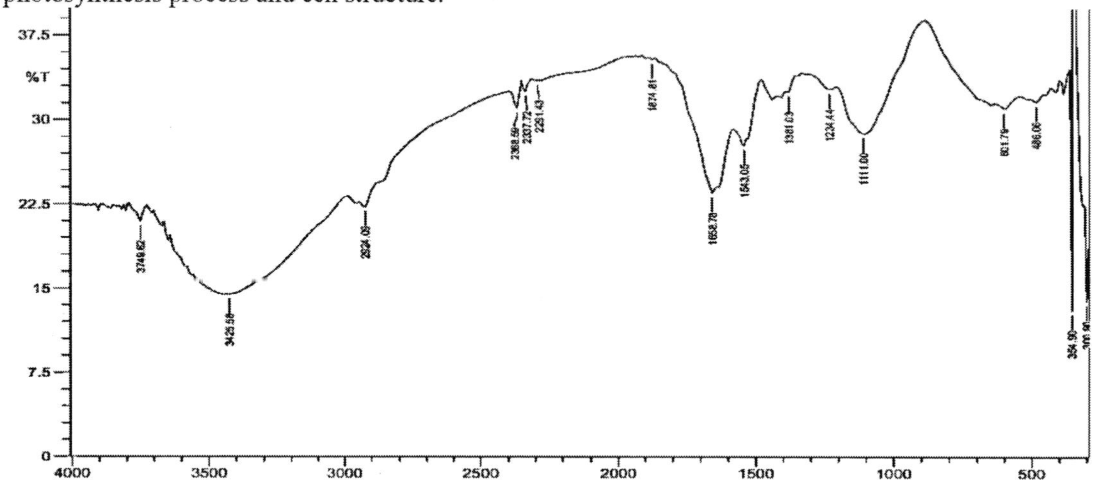

(a)

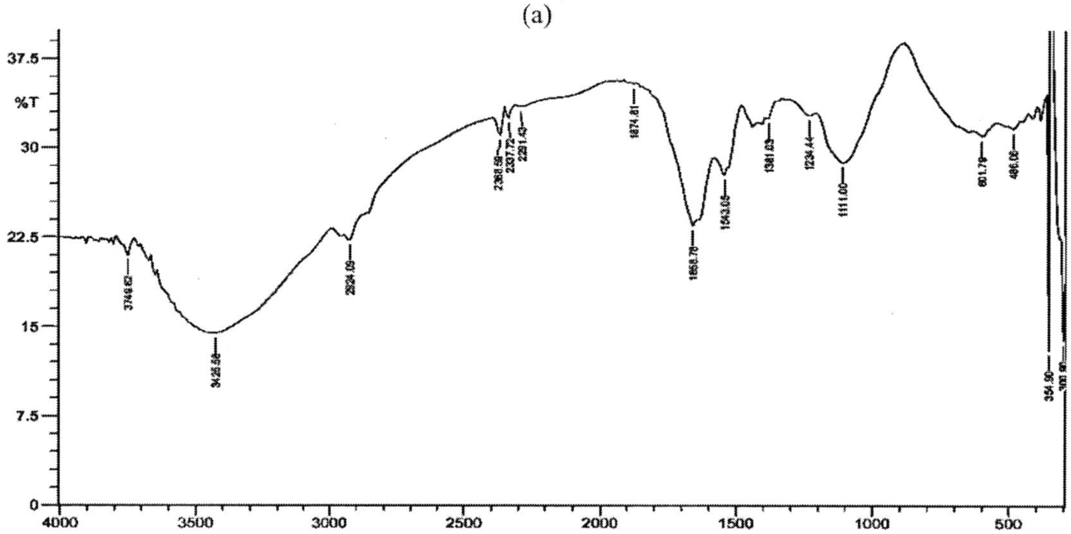

(b)

Figure 3. IR Spectrum IR of *D. salina* (DS), (a) DS ; (b) DS + Fe^{2+}.

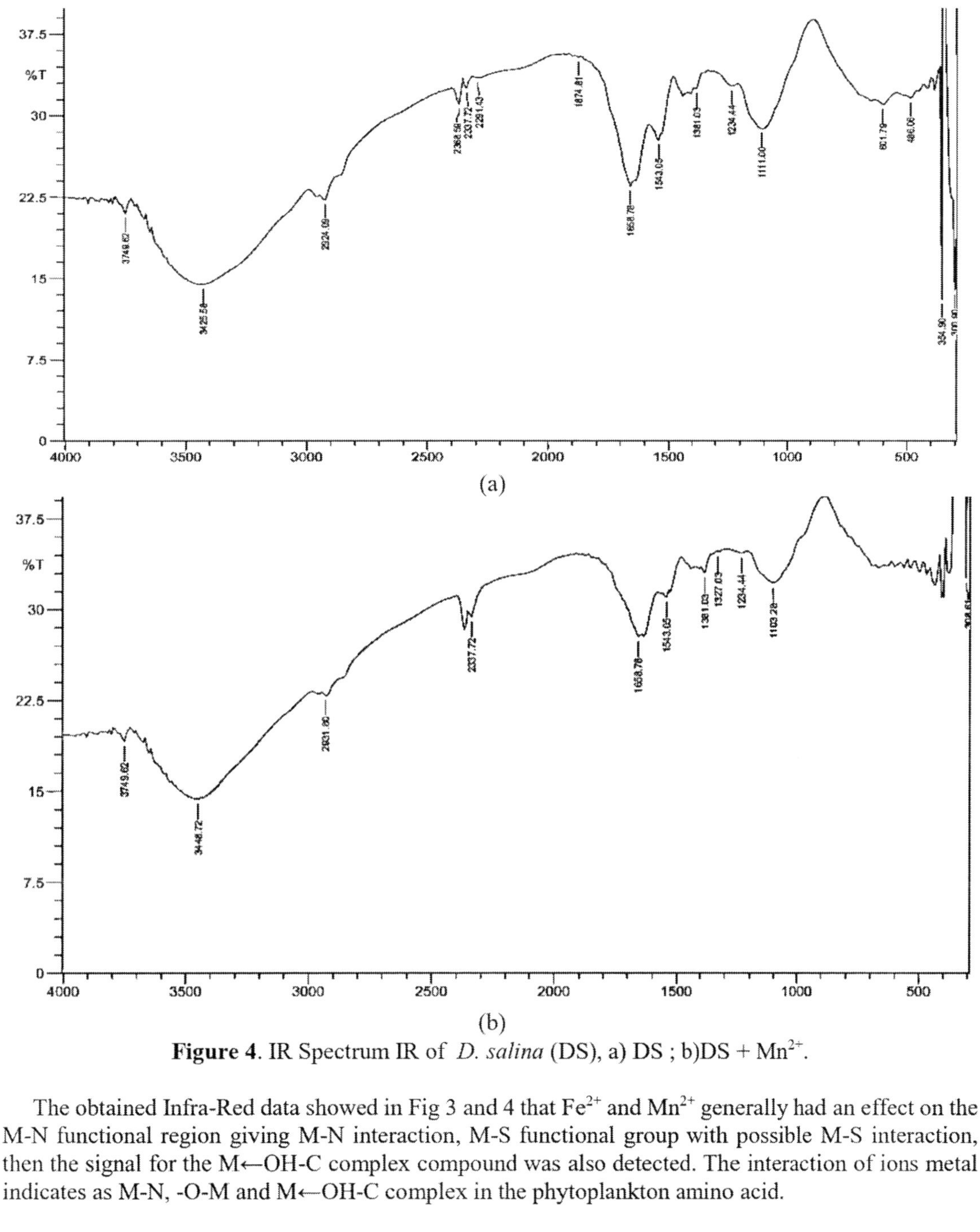

Figure 4. IR Spectrum IR of *D. salina* (DS), a) DS ; b)DS + Mn^{2+}.

The obtained Infra-Red data showed in Fig 3 and 4 that Fe^{2+} and Mn^{2+} generally had an effect on the M-N functional region giving M-N interaction, M-S functional group with possible M-S interaction, then the signal for the M←OH-C complex compound was also detected. The interaction of ions metal indicates as M-N, -O-M and M←OH-C complex in the phytoplankton amino acid.

3.3 Data of β-Carotene Analysis on D. salina, Fe^{2+} and Mn^{2+}

Biomass of each species was extracted, either by the addition of Fe^{2+} and Mn^{2+} at various concentrations. The extraction process was done by a sonication method to streamline the time and get more extract and use less solvent volume. The method utilized ultrasonic waves that can destroy the cell to accelerate the process of mass transfer of compounds from the cell into the solvent. Figure 3 shows β-Carotene analysis on *D. salina* after the addition of Fe^{2+} and Mn^{2+} ions.

The highest amount of β-carotene is obtained after the addition of Fe^{2+} with the concentration of 0.6 ppm. The amount of β-carotene produced after the addition of 0.6 ppm Fe^{2+} is 13.08 μg/g dried weight which is three times higher than produced by the *D. salina* control. This indicates that the addition of 0.6 ppm Fe^{2+} is highly potential to produce β-carotene. From table 1, it is shown that the addition of Fe^{2+} with the same concentration gives low growth of *D. salina*. A low growth in carotenoid productivity was also expressed by Hadi *et al.* [23].

The low growth followed by a decline in the amount of *D.salina* biomasswas also reported on *Haemotococcus pluvialis* with Fe^{2+}-EDTA [8]. This was possibly due to the action of Fe^{2+} as an effective catalyst in the breakdown of H_2O_2 [24,25], which was produced from superoxide anion radicals by the action of the enzyme superoxide dismutase, the catalase enzyme that broke hydrogen peroxide to produce O_2 and H_2O. However, the presence of Fe^{2+} as the catalyst will produce hydroxyl radicals (OH •), given the Haber-Weiss reaction [24] as follows:

$$O_2^{\bullet-} + H_2O_2 + Fe^{2+} \rightarrow Fe^{3+} + O_2 + OH^- + HO^{\bullet},$$

and the Fenton reaction:

$$H_2O_2 + Fe^{2+} \rightarrow Fe^{3+} + OH^- + HO^{\bullet}$$

Therefore, the increase in the productivity of β-carotene as a type of antioxidant in the protection of cells from radical compounds will be observed.

The dry weight concentrations of β-carotene was generally not significantly affected by the addition of Mn^{2+}. It is possible because the Mn metal ion has a small reduction potential compared to the Fe metal ion. In addition, Mn^{2+} has a dominant role as a cofactor of catalase, peroxidase and superoxidase (SOD) that play a protective role against radical pressure [24].

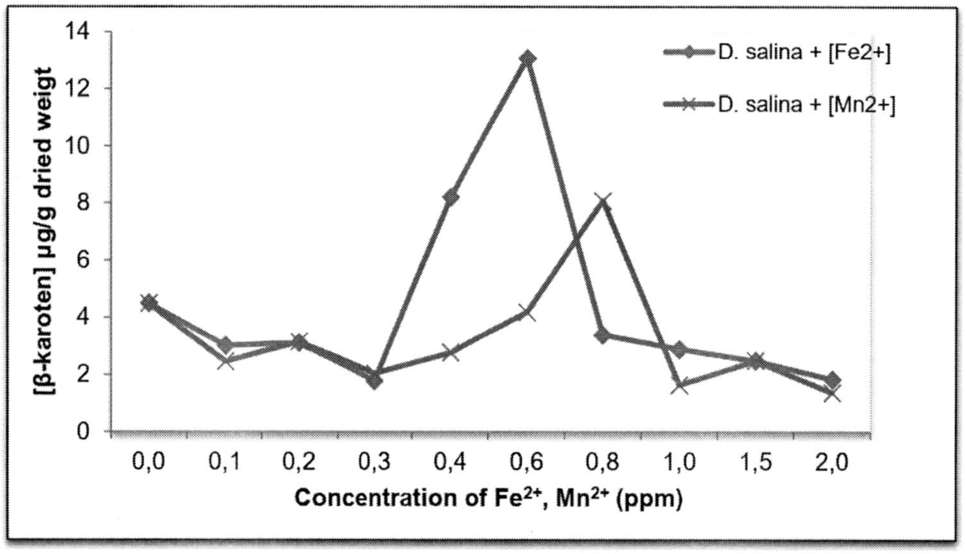

Figure 5. Analysis of β-carotene in *D. salina* after the addition of Fe^{2+} or Mn^{2+} at various concentrations.

4. Conclusion

Based on the results presented above, it can be summarized that the addition of Fe^{2+} in general has a greater influence on the growth of *D. salina*. The highest amount of β-carotene, *i.e.* 13.08 and 8.08 µg/g dried weight, was obtained after the addition of 0.6 ppm Fe^{2} and 0.3 ppm Mn^{2+}, respectively. The interaction occurred between ions metal indicates and M-N, -O-M and M←OH-C complex of amino acid in the phytoplankton. *D. salina* with the addition of 0.6 ppm Fe^{2+} addition was potential to be used as the β-carotene supplement.

References

[1] Guedes, A. C., Amaro, H. M. dan Malcata, F. X., 2011, Microalgae As Source of High Added-Value Compound-Brief Riview of Recent Work, *Biotechnol.Prog*, **27 (3)**: 597-613. DOI: 10.1002/btpr.575

[2] Farasat, M., Khavari-Nejad, R., Nabavi, S. M. B. dan Namjooyan, F., 2013, Antioxidant Properties of Some Filamentous Green Algae (*Chaetomorpha* Genus), Braz. arch. biol. technol, 56 (6), 921-927. ISSN 1516-8913 Printed in Barazil

[3] Rahman., M., Hossain, S., Rahaman, A., Fatima, N., Nahar, T., Uddin, B., Basunai, M. A., 2013, Antioxidant Activity of *Centella asiatica* (Linn.) Urban: Impact of Extraction Solvent Polarity, *phytojournal*, ISSN 2278-4236

[4] Schaich, K. M., 2005, *Lipid Oxidation: Theoretical Aspects*, Bailey's Industrial Oil and Fat Product, Sixth Edition, Six Volume Set. John Wiley & Sons

[5] Shebis, Y., Kinel-Tahan, Y., Dubinsky, Z., Yehoshua, Y., 2013, Natural Antioxidant: Function and Sources, *Food and Nutrition Sciences*, **4**, 643-649. http://dx.doi.org/10.4236/fns.2013.46083

[6] El-Baky, H. H. A., El-Baz, F. K. dan El-Baroty, G. S., 2004, Production of Antioxidant by The Green Alga *Dunaliella salina*, Int. J. Agric. Biol , 6 (1): 49-57. DOI: 10.1.1.317.7556

[7] Ramos, A. A., Polle, J., Tran, D., Cushman, J. C., Jin, E. S., Varela, J.C., 2011, The Unicelluler Green Alga *Dunaliella salina* Teod. as a Model for Abiotic Stress Tolerance: Genetic Advanced and Future Perspective, *The Korean Society of Phycology*, **26 (1)**: 3-20. OI: 10.4490/algae.2011.26.1.003

[8] Cai, M., Zhe, L. dan Anxiang, Q., 2009, Effect of Iron Electrovalence and Spesies on Growth and Astaxanthin Production of *Haemotococcus pluvialis*, *Chin. J. Oceanol. Limnol*, **27 (2)** : 370-375

[9] Li, M., Ahu, Q., Hu, C. W., Chen, L., Liu, Z. L. dan Kong, Z.M., 2007, Abstrak: Cobalt and Manganese Stress in the Microalga Pavlova viridis (*Prymnesiophyceae*): Effect on Lipid Peroxidation and Antioxidant enzymes, *J. Environ Sci*, **19 (11)**: 1330-1335

[10] Gouveia, L., Batista, A. P., Sousa, I., Raymundo, A. dan Bandarra, N. M., 2008, Microalgae in Novel Food Product, Food Chemistry Research Developments, Nova Science Publishers, Inc: 1-37

[11] Hess, S. Y., Thurnham, D. I. dan Hurrell, R. F., 2005, Technical Monograph Series: *Influence of Provitamin A Carotenoids of Iron, Zinc and Vitamin A Status*, Harvest Plus-UK: 4-28

[12] *Interclinical Laboratories*, 2010, Dunaliella salina-Marine Phytoplankton, Botanical Monograph, *Interclinical Laboratories* Pty Limited, Australia

[13] Pisal, D. P. dan Lele, S. S., 2005, Carotenoid Production from Microalga, *Dunaliella salina*, *Indian J Biotechnol*, **4**: 476-483

[14] Oren, A., 2005, Review: A Hundred Years odf *Dunaliella* Research: 1905-2005, BioMed Central, *Saline System*, **1 (2)**: 1-14

[15] Tafreshi, A. H. dan Shariati, M., 2009, *Dunaliella* Biotechnology: Methods and Applications, *J. Appl. Microbiol*, **107**: 14-35

[16] Abusara, N. F., Emeish, S. dan Sallal, A. K. J., 2011, The Effect of Certain Environmental Factors on Growth and β-carotene Production by *Dunaliella sp.* Isolated from Dead Sea, *Jordan J. Bio.Sci.*, **4 (1)**: 29-36

[17] Trun, D., Vo, Trung, 2014, Carotene and Antioxidant Capacity of *Dunaliella salina* Strains, *J.NutrHealth*, 2 **(2)**, 21-23. DOI: 10.12691/jnh-2-2-2
[18] Garofalo, R., 2010, *Aqua Fuels*: *Algae and aquatic biomass for a sustainable production of 2nd generation biofuels*, seven world programme, UK
[19] Matsuzawa, T., Sera, K., Futatsugawa, S. dan Shiraiwa, Y., 2004, Abstrack: Change in element composition of microalgal cells under iron-deficient conditions: basic study on homeostasis of element utilization, *NMCC Annual Report*, **12**: 103
[20] Pessarakli, M., 2005b, *Handbook of Photosynthesis-Chapter 44: Heavy Metal Toxicity Induced Alteration in Photosynthetic Metabolims in Plants*, 2nd edition, CRC Press, USA: 2
[21] Millaleo, R., Reyes-Diaz, M., Ivanov, A. G., Mora, M. L. dan Alberdi, M., 2010, Manganese as Essential and Toxic Element for Plants: Transport, Accumulation and Resistance and Mechanisms, *J. Soil Sci.Plant Nutr*, **10 (4)**: 476-494
[22] Allen, J. F., de Paula, B. M. W/. Puthyiyaveetil, S., dan Nield, J., 2011, Review: A structural Phylogenetic Map for Chloroplast Photosynthesis, *Trends Plant Sci.*, **16 (12)**: 645-655
[23] Hadi, M. R., Shariati, M. dan Afsharzadeh, S., 2008, Microalgal Biotechnology: Carotenoid and Glycerol Production by The Green Algae *Dunaliella* Isolated from the Gave-Khooni Salt Marsh, Iran, *Biotechnol. Bioprocess Eng.*, **13**: 540-544
[24] Manahan, S., 2002, *Toicological Chemistry and Biochemistry*, CRC Press, USA: 401
[25] Koolman dan Roehm, J. dan Roehm, K. H., 2005, *Color Atlas of Biochemistry: revised and enlarged*, 2nd edition, Thieme Stuttgart, New York: 24

IOP Conf. Series: Journal of Physics: Conf. Series **979** (2018) 012013 doi:10.1088/1742-6596/979/1/012013

Contribution of Co^{2+} in increasing chlorophyll *a* concentration of *Nannochloropsis salina* in controlled Conwy medium

Y Hala[1], P Taba[1], E Suryati[2], P Kasih[1] and N F Firman[1]

[1] Faculty of Mathematic and Natural Sciences Hasanuddin University, Makassar Jl. Perintis Kemerdekaan Km. 10, Kampus Tamalanrea Makassar 90245, Indonesia
[2] Research institute for Coastal Aquaculture, Maros, Jl. Makmur Dg. Sitakka, Maros 90511, Indonesia

E-mail: yusafirhala@unhas.ac.id

Abstract. A research in determining the contribution of Co^{2+} on the increase of chlorophyll *a* concentration of *Nannochloropsis salina* has been caried out. The cultivation of *N. salina* was conducted in the Conwy medium with a salinity of 5‰ and 25‰ and various Co^{2+} concentration (2, 4, and 8 ppm). In this research, Co^{2+} was exposed early in the cultivation of *N. salina*. The growth of *N. salina* was observed daily by counting the number of populations using a haemocytometer while the chlorophyll *a* concentration was determined by a Uv-Vis spectrophotometer. The results showed that the growth of *N. salina* in the control was higher than that in the medium containing Co^{2+}. The optimum growth time was achieved on 15th days (5‰) and 8th days (25‰). In the cultivation medium with a salinity of 5‰, Co^{2+} with a concentration of 2 ppm increased the chlorophyll *a* level while Co^{2+} with concentrations of 4 and 8 ppm decreased it. In the medium of cultivation with a salinity of 25‰, the increase in chlorophyll *a* level was observed at Co^{2+} concentrations of 2 and 4 ppm whereas the decrease in chlorophyl *a* level was given at a concentration of 8 ppm. It can be concluded that at low concentrations, Co^{2+} increased the concentration of chlorophyll *a* in *N. salina*.

1. Introduction

Biosorption of bivalent metal ions of Zn^{2+} and Cd^{2+} by *N. salina* has been implemented[1]. This process is very much affected by chemical components of the cell surface and the structure of cell walls. Various polysaccharide compounds such as cellulose, chitin, alginate, can be found in the cell wall of microalgae, as well as various proteins. Some of the functional groups, especially carboxyl -and other functional groups which contain -O, –N, –S, and –P have the potential to actively involved in bonding between heavy metals and biomass [2]. Chu and Hashim [3] reported that hydroxyl, carboxyl, amino, esther, carbonyl, phosphate, and sulphohydryl functional groups have the potential to bind metal ion by biomass.

Chlorophyll is a pigment produced by microalgae which plays an important role in photosynthesis. It utilizes energy from sun rays to turn carbon dioxide and water into glucose while releasing oxygen to the environment. One of the factors affecting the concentration of chlorophyll and microalgae growth is salinity [4]. *Nannochloropsis salina* (Eustigmatophyceae) is one of the microalgae containing chlorophyll *a* in its cell, other than violaxanthin, and esther vaucherioxanthin. The other minor pigments found are β-karoten, zeaxanthin, and some other unknown chlorophyll *a* derivatives [5]. The formation of chlorophyll in plants requires an essential element.

Cobalt is heavy metal which is naturally found in various forms in the environment. It is also an essential element that affects plant or animals growth. However, in high concentration, Co^{2+} can be very toxic [6]. Cobalt can displace Fe, Mn, Zn, and Cu from physiologically important binding sites and thus might decrease uptake and translocation of essential micronutrient [7].

This research was conducted to investigate the role and contribution of Co^{2+} in increasing chlorophyll *a* concentration in *N. salina* cultivated in the medium with a salinity of 5‰ and 25‰.

2. Material and Methods

2.1. Materials

Materials used in this research were sterile seawater, sterile freshwater, the unialgal strain of *N. Salina*, and Conwy cultivated medium [1] was obtained from The Research Institute for Coastal Aquaculture, Maros, Indonesia; HNO_3 (p.a) solution; double distilled water. Stock solutions of Co^{2+} (1000 ppm) were prepared by dissolving 0.4941 g of $Co(NO_3)_2.6H_2O$ with a few of HNO_3 (p.a) in a 100 mL volumetric flask diluted with double distilled water to 100 mL. The stock solutions were then diluted to the certain concentration.

2.2. Apparatus

Apparatus used in this research included haemocytometer Marienfeld LOT-No 4551, hand counter, aerator Amara, autoclave All American No. 1925X, phase contrast microscope Olympus IX71 magnification 40 times, stirrer Hettich Mikro 22R, and oven Memmert, cellulose nitrate membrane filter Millipore (0.45 μm), Uv-Vis spectrophotometer Shimadzu model Prestige-21.

2.3. Procedures

2.3.1. Optimum time of N. salina growth. Sterile freshwater with a salinity of 5‰ or sterile seawater with a salinity of 25‰ was put into a 1 L container, added with 2 mL of Conwy medium and unialgal strain of *N. salina* with the initial populations of about 30 x 10^4 cells/mL. The volume of the mixture was adjusted with sterile water to be 1 L. The solution was mixed, connected to an aerator with the following culture conditions: continuous irradiation (≈ 4.000 lux), aeration and a room temperature of 20 °C [1]. The growth of *N. salina* was observed with a haemocytometer every day until the maximum growth was obtained. The population number based on the 4 fields observation A, B, C, and D on the haemocytometer was calculated using equation (1).

$$\sum cell = \frac{A + B + C + D}{4} \times 10^4 \text{ cells/mL} \quad (1)$$

2.3.2. Exposure of metal ions into the culture of N. salina. Sterile freshwater or sterile seawater was put into 4 containers of 1 L, solutions of Co^{2+} with a concentration of 2, 4, and 8 ppm were separately put into 3 containers. The other container was used as a control. A Conwy medium solution (2 mL) and the unialgal strain of *N. salina* (4 mL) with the initial population of 30 x 10^4 cells/mL was added into each container, and the volume was adjusted into 1 L using sterile freshwater or sterile seawater. The solution was aerated and the growth of *N. salina* was observed every day.

2.3.3. Determination of chlorophyll a. 10 mL of sample was filtered using an aspirator and a vacuum pump, equipped with cellulose paper. Cellulose filter paper was prepared and folded and then put into a testing jar containing 7.5 mL of 90% acetone solution. The mixture was then kept in the freezer for 24 hours, centrifuged at a speed of 3500 rpm for 10 minutes and left for 2 hours. The clear liquid was poured into a cuvette and its absorbance was measured using a visible spectrophotometer at wavelengths of 750; 664; 647; and 630 nm. Acetone was used as a blank. Absorbance was recorded and chlorophyll *a* concentration calculated using Eq. (2), where V_s is the sample volume, V_1 is the volume of acetone extract. The OD664 (absorbance 664 minus absorbance 750), OD647 (absorbance 647 minus absorbance 750), OD630 (absorbance 630 minus absorbance 750).

$$\text{Chlorophyll} - \text{a}\left(\frac{mg}{m^3}\right) = \frac{(911.85 \times \text{OD664}) - (1.54 \times \text{OD647}) - (0.08 \times \text{OD630})}{V_s} \times V_1 \qquad (2)$$

3. Results and Discussion

3.1. Growth Rate of N. salina

The growth rate of *N. salina* in the control and Co^{2+} exposed sample in freshwater with a salinity of 5‰ is shown in Figure 1.

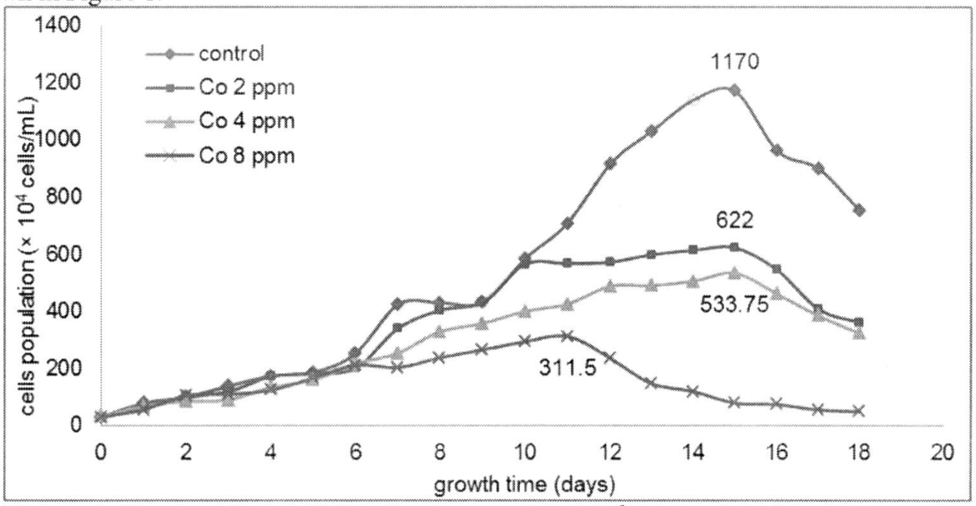

Figure 1. The growth rate of *N. salina* in control and Co^{2+} exposed medium with a salinity of 5‰.

It is clear that the optimum growing time of *N. salina* in the control medium was on day-15, with the cell population of 1170 x10^4 cells/mL. It also occurrs in the medium containing Co^{2+} with concentrations of 2 and 4 ppm. The optimum growing time was on day-15 with the population of 622 and 533.75 x10^4 cells/mL, respectively. On the other hand, the optimum growing time of *N. salina* sample with the Co^{2+} exposure of 8 ppm is on day-11 with the population of 311.25 x10^4 cells/mL. *N. salina* in the control medium was observed to have slow growth rate until day-6 as it needs time to adapt to new environment. This stage was called adaptation phase. In this stage, *N. salina* is more sensitive to nutrients, temperature, and other parameters which differ from its former condition [8]. *N. salina* will be at the growing phase after that, where it experiences rapid growth until day-15. *N. salina* constantly carries out cell division, and the increase of population is well-balanced with the nutrients.

After day-16 to day-25, it is observed that the population of *N. Salina* decreases. At this stage, *N. Salina* experienced death phase. The decrease in the population was caused by limited nutrients in the medium, corresponding with the increase of population. At this stage, death cells were counted more than the live ones. The dead cells of microalgae which sedimented to the bottom of the medium becoming a new competitor to the live *N. salina*. The nutrients were getting more limited, and the deposition of toxic organic compounds increased [1].

N. salina in the Co^{2+}exposed medium with a concentration of 2 and 4 ppm also experiences adaptation phase until day-6. Then the growth increases rapidly until day-15. Day-16 to day-20 is the death phase for this variance. In contrast with the other variances, the optimum growing time for *N. salina* with Co^{2+}exposure of 8 ppm occurs on day-11, while day-12 to day-18 is the death phase. It is considerably earlier than the variance of control, 2 ppm exposure and 4 ppm exposure. This was because of the high extent of heavy metal exposed to the cells. Cobalts is one of the essential heavy metals needed in trace amount. However, at excessive concentration, cobalts can be very toxic, due to its ability to inhibit chlorophyll biosynthesis caused by magnesium insersion to the protoporphyrin ring [9]. The population decrement of *N. salina* in the environment was affected by the metals concentration in the medium. The higher the concentration of metals in the medium, the higher the decrement in population *N. salina*. This shows the reverse correlation between metals concentration in the medium to the growth of *N. salina*. Bark [10] reported that *N. salina* grows very well in the low salinity environment (2.5‰), with the optimum growing time reached on day 14-15, while Fakhri et. al. [4] reported that the optimum growth and chlorophyll production in *Nannochloropsis* sp. occurred in a salinity of 15‰.

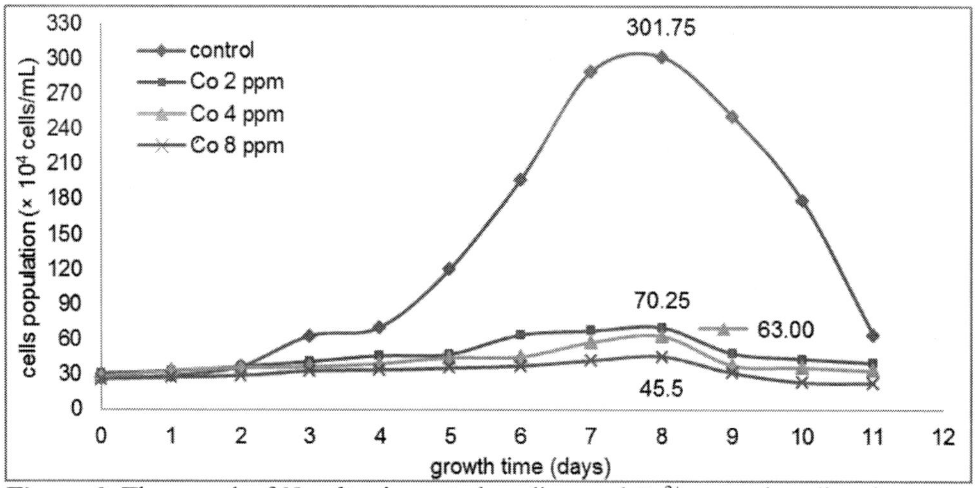

Figure 2. The growth of *N. salina* in control medium and Co^{2+}exposed medium in a salinity of 25‰.

In a salinity 25‰, the growth of *N. salina* in the control and Co^{2+}exposed medium is shown in Figure 2. The highest population of *N. salina* in the control medium is observed on day-8 with the population of 301.75×10^4 cells/mL. The optimum growth in this variance was obtained earlier than that in a salinity of 5‰ with the population of 4 times lower than that in a salinity of 5‰. The same phenomena is also observed in the growth of *N. salina* in the Co^{2+}exposed medium with a concentration of 2, 4 and 8 ppm. The optimum growth time iss on day-8 with the population of 70.25; 63.00 dan 45.5 (×10^4 cells/mL), respectively. The population is 8 times less than the population in 5‰ salinity. On day-9 to day-11 the population of *N. salina* in both control and Co^{2+}exposed medium experienced decrement.

The population decrement was also caused by the accumulation of organic compound from dead biomass which sedimented at the bottom of the medium. This biomass will be the competitor for living *N. salina* to the use of dissolved oxygen in culture medium [11]. There is a gap between the population of *N. salina* in control sample and Co^{2+}exposed sample. The higher the concentration of Co^{2+} the lower the population of *N. salina*. The growth inhibition was caused by Co^{2+} metals which is essential but relatively toxic.

3.2 Effect of Co^{2+} to the chlorophyll a concentration in N. Salina

The Uv-Vis spectrophotometer determination to detect changes in chlorophyll *a* concentration afer exposure of Co^{2+} compared to the control is shown in Table 1. In a salinity of 5‰, the concentration of chlorophyll *a* in the control was 446.37 mg/m^3 while its concentration in the Co^{2+}exposed medium with a concentration of 2, 4 and 8 ppm was 507.40; 432.15; and 400.13 mg/m^3 respectively. The exposure of 2 ppm Co^{2+} increases the concentration of chlorophyll *a* in *N. salina* around 61.03 mg/m^3. This was because, in a low concentration of Co^{2+} exposure, the biosynthesis rate of chlorophyll is sped up. Cobalt is needed to synthesize vitamin B_{12} as cobalamin, where this vitamin is one of the micronutrients needed by microalgae [12]. On the other hand, the exposure of Co^{2+} with a concentration of 4 and 8 ppm decrease the chlorophyll *a* concentration in *N. salina* of 14.22 mg/m^3 dan 46.24 mg/m^3 respectively. It was because the high concentration of cobalt inhibit the biosynthesis of chlorophyll *a* and the inhibition effect was stronger in the higher concentration [13].

Table 1. Chlorophyll *a* concentration in *N. salina* in control and Co^{2+} exposed medium at a salinity of 5‰ and 25‰

Salinity (‰)	Chlorophyll *a* conc. (mg/m^3)				Δ Chlorophyll *a* conc. with control (mg/m^3)			Δ Chlorophyll *a* conc. with control (%)		
	Control	Co^{2+} 2 ppm	Co^{2+} 4 ppm	Co^{2+} 8 ppm	Co^{2+} 2 ppm	Co^{2+} 4 ppm	Co^{2+} 8 ppm	Co^{2+} 2 ppm	Co^{2+} 4 ppm	Co^{2+} 8 ppm
5	446.37	507.4	432.15	400.13	61.03	-14.22	-46.24	13.67	-2.80	-10.70
25	176.12	198.45	189.46	171.91	22.33	13.34	-4.21	12.68	6.72	-2.22

Δ = different of sample concentration with exposure of Co^{2+} with control

At a salinity of 25‰, the concentration of chlorophyll *a* of *N. salina* in the control is 176.12 mg/m^3, while in the Co^{2+}exposed medium with a concentration of 2, 4 and 8 ppm is 198.45; 189.46 and 171.91 mg/m^3 respectively. The chlorophyll *a* concentration in the control and 2 ppm Co^{2+}exposed medium is 3 times lower than in the sample at 5‰ salinity, while the 4 ppm and 8 ppm Co^{2+}exposed medium are 2 times lower than that at 5‰ salinity. The consequence of increasing the concentration of Co^{2+} is an increment of chlorophyll *a* concentration in *N. salina*. It was also proven in the data that the addition of metal ions Co^{2+} increase the chlorophyll *a* concentration produced by *N. salina*, which is shown in 2 ppm and 4 ppm variances. However, at higher concentration of 8 ppm Co^{2+}, the decrement of chlorophyll *a* concentration was observed far less than the control. The similar result was also reported by El-Sheekh et al. [12], that at a lower concentration of Co^{2+} (0.5 and 1.5 ppm), the chlorophyll and carotenoid concentration increase significantly in *N. perminuta*. On the other hand, at higher concentration (2.5; 3.5; and 5 ppm) Co^{2+} caused the decrease in pigment concentration proggresively. It was also observed that at 3.5 and 5 ppm, the inhibition effect to chlorophyll *a* was much stronger than to carotenoid. This inhibition occurred as the result of magnesium displacement mechanism in chlorophyll molecules, causing the accumulation of protoporphyrin and blockage of chlorophyll synthesis.

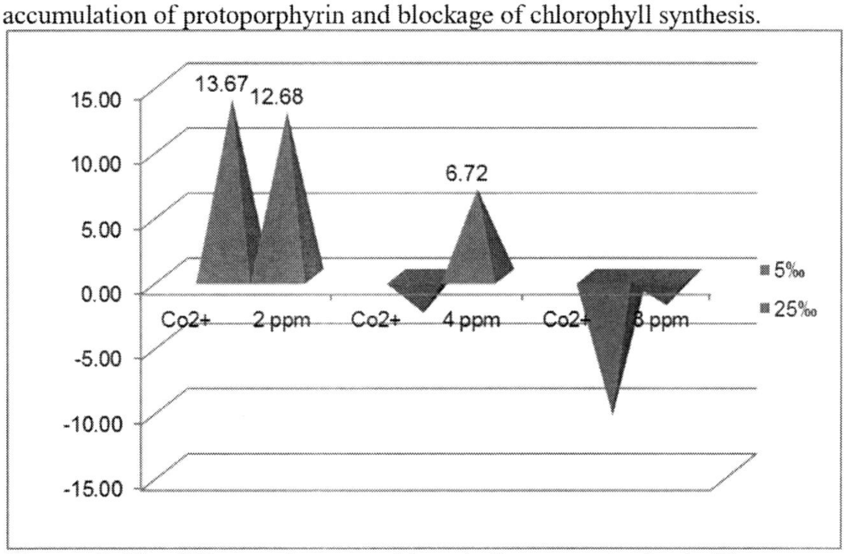

Figure 3. The difference between chlorophyll *a* concentration in *N. salina* in the control and Co^{2+}exposed medium.

It is obvious that the chlorophyll *a* concentration of *N. salina* in freshwater with a salinity of 5‰ is relatively higher than that of *N. salina* in seawater with a salinity of 25‰, contrary to Ak et al. [14] report that chlorophyll *a* content of *Dunaliella viridis* increased with increasing salinity.

4. Conclusions

The growth of *N. salina* and the concentration of chlorophyll *a* in cultivating medium with a salinity of 5‰ is better than in the medium with a salinity of 25‰. The optimum growing time was obtained on day-15 for a medium with a salinity of 5‰ and day-8 for a medium with a salinity of 25‰. The contribution of Co^{2+} to increase chlorophyll *a* concentration in *N. salina* in both cultivating medium occurred at a concentration of Co^{2+} 2 ppm, while at 8 ppm inhibition was observed. For a medium with a salinity of 5‰, Co^{2+} with a concentration of 4 ppm was also observed causing increment to the concentration of chlorophyll *a*. It can be concluded that at low concentrations, Co^{2+} has an increasing effect of chlorophyll *a* on *N. salina*.

References

[1] Hala, Y., Syahrul, M., Suryati, E., Taba, P., Soekamto, N. H. (2013). *Eur. Chem. Bull.*, **2**(5), 238-241
[2] Wang, J., dan Chen, C., 2009, *Biotechnol. adv.*, **27**, 195-226
[3] Chu, K. H and Hashim M. A. 2007. *J. Environ. Sci.* **19**, 928-932
[4] Fakhri, M, Arifin, N. B, Yuniarti, A., and Hariati, A. M., 2017, *Nature Environtment and Pollution Technology*, **16**(1), 209-212
[5] Suda, S., Atsumi, M., and Miyashita, H. (2002). *Phycologia*, **41**(3), 273-279
[6] Jayakumar, K., Jaleel, C. A., Vijayarengan, P., 2009, *Bot. Res. Int.*, **2**(3), 153-156
[7] Chatterjee, J.,Chatterjee, C. 2000. *Environ. Pollut.* **109**, 69-74
[8] Fogg, G. E., 1975, *Algae Culture and Phytoplankton Ecology*, The University of Wisconsin Press
[9] Csatorday, K., Gombos, Z., Szalontai, 1984, *Proc. Nat. Acad. Sci. USA*, 81, 476-478
[10] Bark, M., 2012, *Cultivation of Eleven Different Species of Freshwater Microalgae Using Simulated Flue Gas Mimicking Effluents from Paper Mills as Carbon Source*, Published Thesis, Departement of Chemical and Biological Engineering, Chalmers University of Technology
[11] Sjahrul, M., 2013, *Res. J. Sci. IT. Manag.*, **3**(1), 1-6
[12] El-Sheek, M. M., El-Naggar, A. H., Osman, M. E. H., dan El-Mazaly, E., 2003, Braz. *J. Plant Physiol*, **15**(3), 159-166
[13] Al-Hejuje, M. M., 2008, *Marsh Bulletin*, **3**(2), 136-146
[14] Ak, I., Cirik, S. and Goksan, T. 2008. *J. Biol. Sci.*, **8**(8), 1356-1359

Utilization of Diamine Oxidase Enzyme from Mung Bean Sprouts (*Vigna radiata* L) for Histamine biosensors

Abdul Karim, A W Wahab, I Raya, H Natsir and A R Arif

Department of Chemistry, Mathematics and Natural Sciences Faculty, Hasanuddin University, Jl. Perintis Kemerdekaan Km. 10 Makassar 90245, Indonesia

E-mail: karimkimia@yahoo.com

Abstract. This research is aimed to utilize the diamine oxidase enzyme (DAO) which isolated from mung bean sprouts (*Vigna radiata* L) to develop histamine biosensors based on electode enzyme with the amperometric method (cyclic voltammetry).The DAO enzyme is trapped inside the membrane of chitin-cellulose acetate 2:1 and glutaraldehyde which super imposed on a Pt electrode. Histamine will be oxidized by DAO enzyme to produce aldehydes and H_2O_2 that acting as electron transfer mediators.The performance of biosensors will be measured at various concentrations of glutaraldehyde, temperature changes and different range of pH. Recently, it has been found that the optimal conditions obtained from the paramaters as follows; at 25% of glutaraldehyde, temperature of 37°C and pH of 7.4. Eventually, the results provided an expectation for applying histamine biosensors in determining the freshness and safety of fish specifically skombroidae families.

1. Introduction

Histamine is extremely decisive for the quality and freshness of fish specifically the skombroidae family such as tuna and skipjack. Histamine is formed from oxidation processof essential amino acid, histidin, either by bacterial or enzymatic activity that can cause poisoning and allergies. According to Rudi of Research and Development Product Processing and Biotechnology of Marine and Fisheries (BBP4BKP), the content of histamine in fresh fish is below 100 ppm and the levels of 500-1000 ppm is generally considered as indangerous extentand can cause poisoning in people who are sensitive [1].Therefore, it is required a fast and accurate histamine analyses on fish and its derivatives.

Research has been undergone in order to discover the best method to dectect and determine the levels of histamine in fish particularly on skombridae family. Currently histamine analyses can be applied on a variety of ways and methods. According to the Indonesian National Standard, histamine analysis was conducted using spectrofluorometry and HPLC [1]. Both of these methods require specialized personnel trained and skilled because its initial preparation of samples is quite complicated with a long time, besides its equipment is relatively expensive. Thus, to reduce the time required in the analysis and to offera rapid screening method to test the quality of the food industry products, some enzymatic method and amount of enzyme sensors have been described by several researchers [2]. Aziz [3] using the biosensors constructed of DAO enzyme and immobilized on PVA membrane. Keow, et al [4], DAO enzyme was immobilized on a poly-HEMA membrane to detect histamine quickly using SPE-carbon paste electrodes. Research Telsnig, *et al* [5] using the DAO enzyme of pea seeds amine oxidase (PSAO), with MnO_2 electrode modified with carbon paste. Biosensor applications generally show various advantages such as enabling faster analysis with preparation/processing samples less even not needed at all [6].

DAO enzyme is widely found in plants Leguminaceae (pulses), such as *Cicer arietinum, Lathyrus sativus* and *Vigna radiata* (mung bean) [7]. DAO enzyme was first purified from pea sprouts. This enzyme oxidizes diamine, such as histamine, and also some primary monoamines, with little or no activity against secondary and tertiary amine [8]. Currently DAO is commercially available which obtained from pig kidney. DAO has a low specific activity even when it has been purified, in contrast to the enzyme from plants that possessing a higher activity. Likewise the DAO enzyme which isolated from green beans has a relatively high activity [9].

Analysis of biosensor is expected to replace the procurement of modern chemistry instruments which is relatively much more expensive [10]. An advantageous of biosensors analysis because it is easily taken to the field, relatively low acquisition costs and sensitive enough to identify a particular substance or chemical compound. Biosensors analysis can be performed with the use of transducers electrochemical (amperometric, potentiometric). Transducer design with amperometric or potentiometric method can be done with the development of enzyme electrodes.

Enzyme electrodes that is effective, sensitive, and accurate is highly determined by membrane composition [11]. One component of the membrane that play a role in binding the DAO enzyme is glutaraldehyde. Enzyme electrode which will be developed in this study is the DAO enzyme electrode using glutaraldehyde at various concentrations, temperatures and pH for the analysis of histamine contained in certain materials, for example in the fish from skombroidae family such as tuna and skipjack. Levels of histamine in fish and food derivatives will determine the degree of freshness and safety of food stuffs.

In this research, enzyme biosensors based histamine amperometric will be developed, namely the biosensors to detect and determine the content of histamine in certain materials. The use of histamine biosensors with DAO enzyme electrodes which immobilized in the membrane of chitin-cellulose acetate and glutaraldehyde can be done. Research objective is to develop histamine biosensors based on electode enzyme with the amperometric method (cyclic voltammetry).

2. Material and Methods

2.1 Materials

Ammonium sulfate, wire Pt, wire Cu, wire Ag, distilled water, HCl, NaOH, tin, glutaraldehyde, acetone, histamine, mung beans, cellophane, $Na_2HPO_4.2H_2O$, $NaH_2PO_4.H_2O$, sephadex G-75.

2.2 Equipment

Cyclic voltammetry–eDAQ ED410-159, pH meter/ potentiometer Orion 710 A, an enzyme electrode, magnetic stirrer, spectrophotometer IR Shimadzu Prestige-21, spectrophotometer UV/Vis, oven, analytical balance, solder, and glasswaren used in the laboratory.

2.3 Procedure

2.3.1 Preparation of phosphate buffer solution

a. Preparation of $NaH_2PO_4.H_2O$ solution 0.2 M.

A number of 2.7598 gram $NaH_2PO_4.H_2O$ was diluted with distilled water and put in a 100 mL volumetric flask. After wards it was diluted until reach the boundary mark (solution A).

b. Preparation of $Na_2HPO_4.2H_2O$ solution 0.2 M

A number of 1.7799 gram $Na_2HPO_4.H_2O$ was diluted with distilled water and put in a 50 mL.Then it was diluted until the boundary mark (solution B).

The creation of phosphate buffer solution was conducted by mixing both solution A and solution B in accordance with a ratio of each to desired pH buffer.

2.3.2 Isolation of DAO Enzyme of Mung Beans (Vigna radiata L.)

a. Crude extract of DAO Enzyme.

A total of 100 grams of mung beans were washed and soaked in distilled water overnight. Furthermore, drained and leveled above porous containers are lined with filter paper, stored at room temperature humid. To help the process of growing sprouts in order to more quickly and evenly, the container can be placed in the pot (lid slightly open) or covered with a napkin and moisture is maintained by sprinkling water on a regular basis. For the isolation and purification of the DAO enzyme of mung bean sprouts, sprouts aged 3 days were collected and washed with distilled water and the outer skin removed. Mung bean sprouts that have clean homogenized with an equal amount (w / v) of 0.05 M phosphate buffer pH 7.0 cold by using a blender. Crude extract obtained by squeezing homogenates through four layers of gauze and centrifuged at 10,000 rpm for 20 min at 4°C. The obtained result (solution) is called DAO enzyme crude extract [9].

b. Ammonium Sulfate Fractionation

DAO enzyme crude extract was fractionated with ammonium sulfate at saturation degree of 0-80 % with an interval of 20 units which added gradually while stirring with a magnetic stirrer until completely dissolved and stored over night at 4°C. After that, it was centrifuged at 4°C for 20 minutes. The attained supernatant was proceeded for further fractionation, whereas the precipitate was dissolved in a small volume of 0.2 M phosphate buffer pH 7.0 and its activity was measured. Fractions that have the highest activity continued to a process which called dialysis [9].

c. Dialysis.

Ammonium sulphate precipitate fraction with high activity was dissolved in 5 mL of 0.2 M phosphate bufferpH 7.0. The solution was put in a cellophane bag and then dialyzed with 0.05 M phosphate buffer pH 7.0 and stirred with a magnetic stirrer for 12 hours at 4°C. The buffer is changed every 3 hours [9].

d. Gel Filtration.

The active fraction which has undergone dialysis process in corporated into the matrix column with sephadex G-75 (column 35 cm long and 1 cm in diameter), previously equilibrated with 0.05 M phosphate buffer pH 7.0 for one night. Subsequently it was eluted with the same buffer. The volume of each fraction as much as 3.0 mL and each fraction is measured for its activity and protein content. Fractions that have high activity will be saved and used as the DAO enzyme for further research [9].

2.3.3 Design and characterisation of histamine biosensors

Plastic-coated copper wire with a size of 5 cm, 1 mm diameter which connected to a Pt wire of1.5 cm with 0.4 mm diameter by soldering used lead wire. Furthermore, the tip of the Pt wire is curved back into the solder point (Pt electrode, Ep). Ep Pt electrode wire section was immersed in a homogeneous solution of chitin-cellulose acetate 2:1 in acetone [10]. As soon as the membrane layer is formed, the electrode is dipped into the water repeatedly. Then, on the wire that has been coated with chitin-cellulose acetate membrane is soaked in a solution of glutaraldehyde (E1=15, E2=20, E3=25%) for 6 hours. After the membrane layer containing gluteraldehyde formed, the electrode was washed with phosphate buffer pH 7.0 (Em). DAO enzyme electrode was made by soaking Em in phosphate buffer pH 7.0 containing DAO enzyme for 16 hours. Further DAO enzyme electrode design results were characterized by cyclic voltammetry at various temperatures and pH. Characterization is conducted with a standard solution of histamine: 1, 10, 100, 1000, 10000 ppm. Each of these enzymes is done by scanning

IOP Conf. Series: Journal of Physics: Conf. Series **979** (2018) 012014 doi:10.1088/1742-6596/979/1/012014

electrode potential from -1.5 to +2.0 V at a rate of 100 mV/ sec from a low concentration to a higher concentration. The formed cyclic voltammogram results are analyzed to obtain a peak flow changes caused by alteration in the concentration of the solution used.

3. Results and Discussion

3.1 DAO Enzyme Isolation

The yield of isolation and fractionation DAO enzyme protein with $(NH_4)_2SO_4$ can be seen in Table 1. It is shown that 40-60% of saturated fraction was the fraction with the highest DAO enzyme activity (3326.00 mU/mL). Therefore 40-60% of saturated fraction was followed by dialysis and gelfiltration process in order to obtain morepure DAO enzyme quantity.

After having done the dialysis and gel filtration processes on sephadex G-75 column, it was gained the fraction 14 which had protein level at 0.63 mg/mL, 116.00 mU/mL of activity and its specific activity was 185.01 mU/mg protein. The specific activity of DAO enzyme from results isolation of mung bean sprouts was 185.01 mU/mg protein which was much larger than DAO enzyme from pig kidney in the market (50 mU/mg protein). So the DAO enzyme isolated from mung bean sprouts can be further used in the manufacture of electrodes for the DAO enzyme for histamine biosensor.

Table 1. The isolation and fractionation results of DAO enzyme protein

No	Protein Fraction	[Protein] (mg/mL)	Activity (mU/mL)	Specific Activity (mU/mg protein)
1	Crude extract	25.20	1226.00	48.65
2	0 - 20 %	20.49	1336.00	65.20
3	20 - 40 %	24.34	2042.67	83.92
4	40 - 60 %	29.03	3326.00	114.57
5	60 - 80 %	24.65	1329.33	53.93
6	Fraction 14	0.63	116.00	185.01

3.2 DAO enzyme electrode characterization with cyclic voltammetry

3.2.1 The influence of the concentration of glutaraldehyde

Results of measurement (cyclic voltammogram) various concentrations of histamine in various glutaraldehyde concentrations are shown in figure 1.

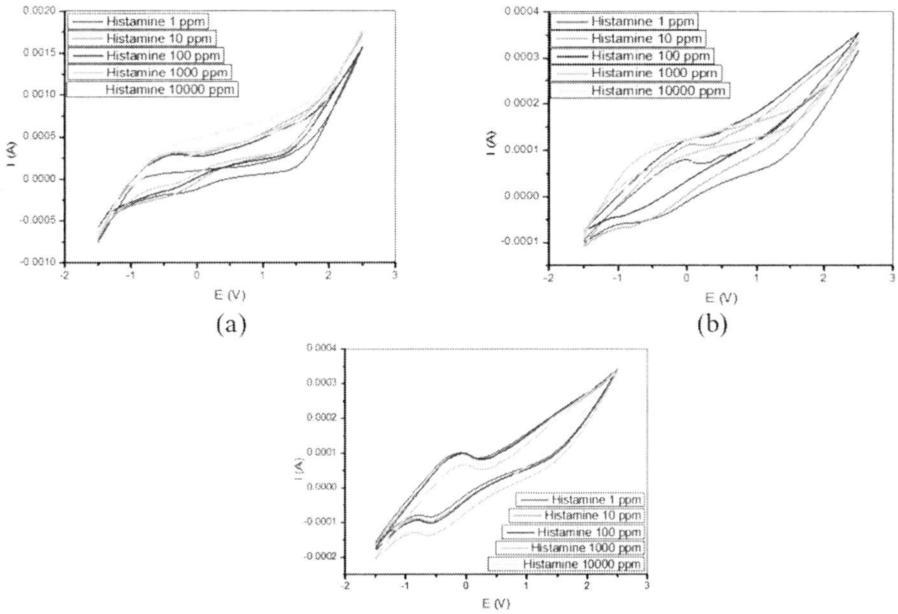

Figure 1. Cyclic voltammograms of various concentrations of histamine at a glutaraldehyde concentration of; 15% (a), 20% (b) and 25% (c).

At Figure 1 it was appeared that the type of electrode E3, glutaraldehyde concentration of 25% (1c) showed a clearer peak current and ocurred to the histamine concentration of 1 ppm up to 10,000 ppm compared with the type of electrode E2 and E1 namely glutaraldehyde concentration of 20% (1b) and a 15% glutaraldehyde (1a). On the membrane with glutaraldehyde concentration of 20%, the peak current is clearly visible on the histamine concentration of 1 ppm to 100 ppm. At a concentration of 15% glutaraldehyde, peak currents only occur at a concentration of 100 ppm histamine. Based on these results it can be said that the type of electrode with glutaraldehyde concentration of 25% is better than 20% glutaraldehyde and 15% on the membrane of chitin-cellulose acetate 2:1. It means that the higher the concentration of glutaraldehyde, the better the DAO enzyme electrode is produced [12]. This happens because the higher the concentration of glutaraldehyde,the easier the glutaraldehyde under going the polymerization process inter molecules and binding to DAO enzyme. Thus the membrane electrode will contain more DAO enzymes that will eventually oxidize more histamine.

3.2.2 The Influence of pH. Cyclic voltammograms of various concentrations of histamine at various pH measurements can be seen in figure 2.

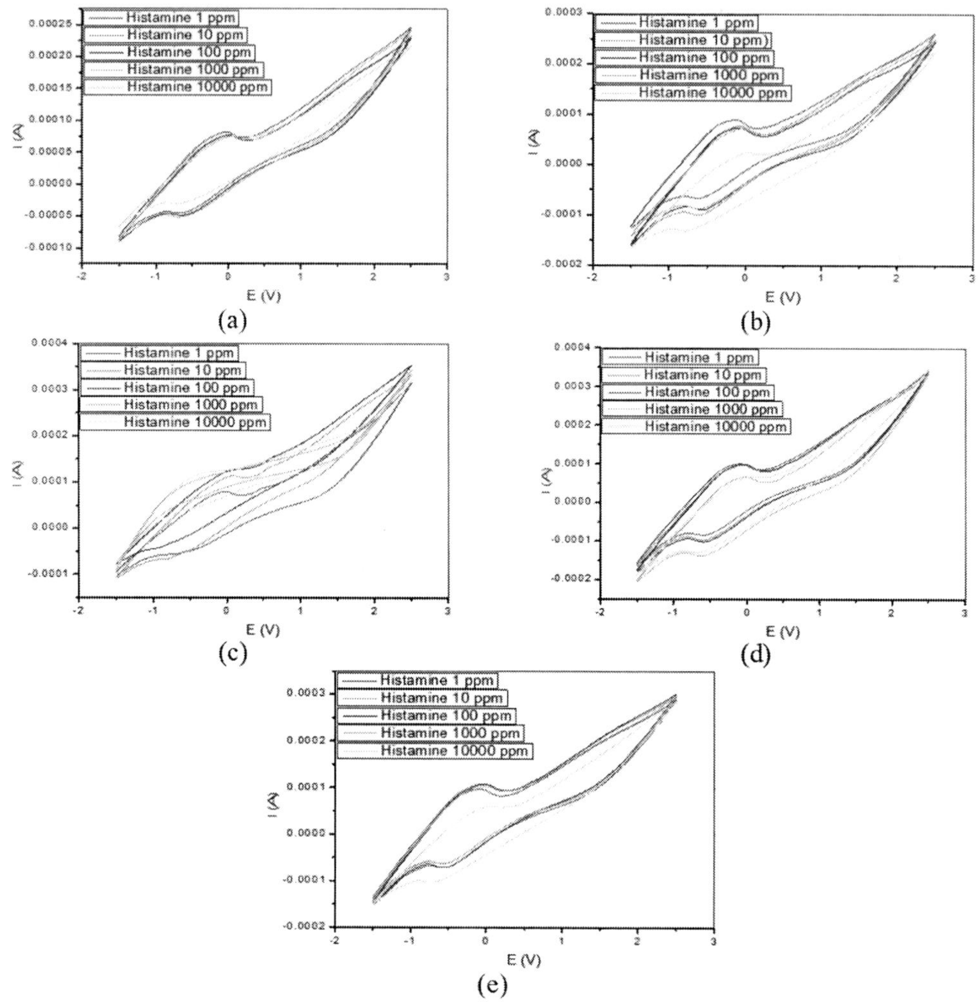

Figure 2. Voltammogram cyclic various concentrations of histamine at pH; 6.2 (a), 6.6 (b), 7.0 (c), 7.4 (d) and 7.8 (e).

It was illustrated that at pH 6.2 the peak current (though not too sharp) occurs in histamine concentration of 1 ppm to 100 ppm. Whereas at pH 6.4 to 7.8 peak flow occurs at all concentrations of histamine. This indicates that the DAO enzyme electrode works at pH ranges are quite wide and of course this is a profitable thing [13]. However, to see the most optimum pH, it can be noted regarding the linearity of the relationship between the peak current (anodic or cathodic) with concentrations of histamine. The Linearity can be seen via coefficient linearity (R) as shown in Table 2. Table 2 hows that the linearity flow of the highest peak of histamine concentrations of histamine shown in cathodic current at pH 7.4 (R = 0.9935), whereas the highest linearity on the anodic current at pH 7.8 (R = 0.9884). R value is calculated without involving concentration of 10000 ppm because at this concentration peak currents generated is far deviating and it is also shown that the measurement does not reach the maximum limit of 10000 ppm.

Table 2.The relationship between the concentration of histamine with anodic and cathodic peak currents at various pH.

pH	[Histamine] (ppm)	I anodic (mA)	R anodic	I cathodic (mA)	R cathodic
6.2	1 - 10000	0.063 - 0.090	0.9072	-0.010 - -0.059	0.9089
6.6	1 - 10000	0.083 - 0.101	0.9518	-0.052 - -0.084	0.9576
7.0	1 - 10000	0.065 - 0.100	0.9560	-0.022 - -0.086	0.9760
7.4	1 - 10000	0.079 - 0.100	0.9685	-0.041 - -0.079	0.9935
7.8	1 - 10000	0.080 - 0.097	0.9884	-0.051 - -0.088	0.9824

3.2.3 The Influence of temperature. Cyclic voltammograms on various concentrations of histamine at diverse temperatures can be seen in Figure 3.

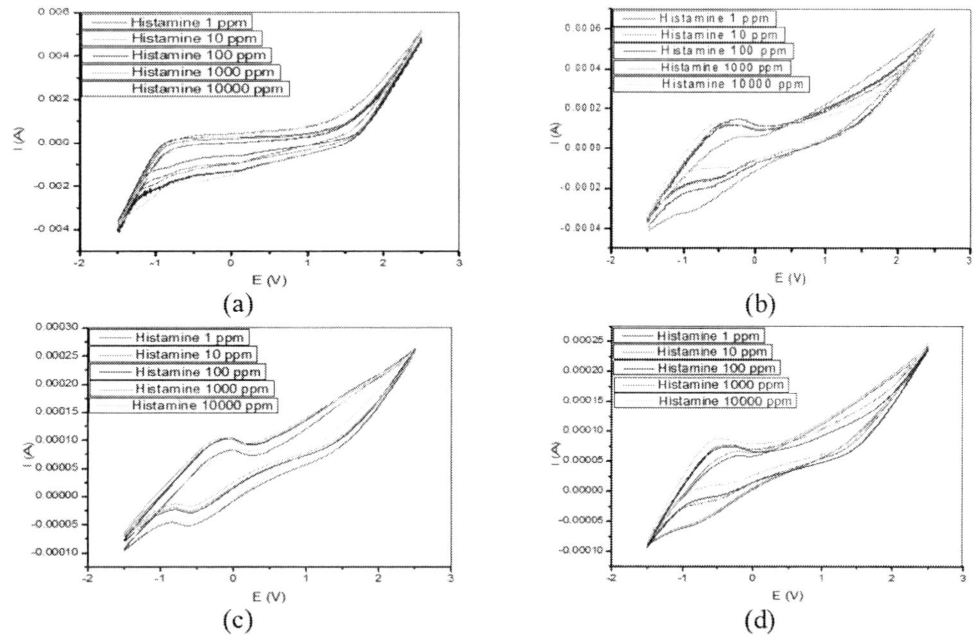

Figure 3. Voltammoggram in a variety of histamine concentrations at a temperature of; 31°C (a), 34°C (b), 37°C (c) and 40°C (d).

It was described that the temperature of 31°C (3a) does not give a peak current for the entire concentration of histamine, it means that the DAO enzyme electrode can not work and detect histamine at that temperature. Both temperature of 34°C and 40°C (3b and 3d) provided anodic peak currents were quite good but the peak cathodicwas unclear. At 37°C (3c) seen that for the whole concentration of histamine giving anodic and cathodic peak currents were clear and orderly. It can be concluded that the DAO enzyme electrode can work optimally at 37°C. This temperature was the same as the optimum temperature of the DAO enzyme isolated from mung bean sprouts [14].

4 Conclusions

Some principal findings are: DAO enzyme isolation of mung bean sprouts obtained specific activity of 185.01 mU/ mg of protein and can be used in the make of DAO enzyme electrodes for histamine biosensors. Comparison of chitin-cellulose acetate 2:1 with 25 % of glutaraldehyde levels which is the best membrane composition on making the DAO enzyme electrode for histamine biosensors. The DAO enzyme electrode obtained will work optimally at a temperature of 37°C and pH 7.4.

References

[1] Rudi, R.J., *Histakit Detector Histamine Fishery products*, Aquamina, 88th IV: 2014, 24

[2] Wimmerova. M. and Macholan, L., Sensitif Amperometric Biosensor for the Determination of Biogenic and Synthetic Amines Using Pea Seedlings Amine Oxidase: A Novel Approach for Enzyme Immobilization. *Biosensor and Bioelectronics*. 1999,14: 695-702

[3] Aziz, A. A., 2007, *Development of A Biosensor Based On Amine Oxidase From Cicerarietinum For The Detection Of Biogenic Amines*, Faculty of Chemical and Natural Resources Engineering, Universiti Teknologi Malaysia

[4] Keow, C.M., Bakar, F.A., Salleh, A.B., Heng, L.Y., and Wagiran R., Screen-Printed Histamine Biosensor Fabricated from the Entrapment of Diamine Oxidase in a Photocured Poly(HEMA) Film, *Int. J. Electrochem. Schi.*,2012, **7**:4702-4715

[5] Telsnig, D., Terzic, A., Krenn, T., Kassarnig, V., Kalcher, K., and Ortner A., Development of a Voltammetric Amine Oxidase-Modified Biosensor for the Determination of Biogenic Amines in Food, *Int. J. Electrochem. Schi*., 2012, **7**: 6893-6903

[6] Lange, J. and Wittmann, C., 2001, Enzyme Sensor Array for the Determination of Biogenic Amines in Food Samples, *Anal. Bioanal. Chem,* 372.2 (2002): 276-283

[7] Choundary, A., Singh, R.P., Cadmium-Induced Changes in Diamine Oxidase Activity and Polyamine Levels in Vigna radiate Wilczek Seedings, *Journal of Plant Physiology*, 2013, 156:704-710

[8] Chang, A., Scheer, M., Grote, A., Schomburg, I., Schomburg, D., Brenda, Amenda and Frenda the Enzyme Information System: New Content and Tools in 2009. *Nucleid Acids Res*. 2008, **37**: 588-592

[9] Karim, A., Patong, A. R., Wahab, A. W., and Raya, I., Isolation and Characterization of Diamine Oxidase Enzyme from Mung Bean Sprouts (*Vignaradiata*L), *AJBLS*, 2015, **3**(1): 7-11

[10] Ralph, L.S , Mathias, W, Steffen, M, Kristine, Iain, A. W, Colin, J. M, Edwin, C. C, Axel, F, Christian, S, and Michel, C., Implementing Silicon Nanoribbon Field-Effect Transistors as Arrays for Multiple Ion Detection, *Biosensors*, 2016, **6**, (21), 1-10

[11] Xuesong, Y., Timothy, F. , Pavan, K. K. , Maggie, T. , Cuong, M. N. and Jung-C. C., Lactate Sensors on Flexible Substrates, *Biosensors*, 2016, **6**, (48), 1-15

[12] Tarasov, A.; Wipf, M.; Stoop, R.L.; Bedner, K.; Fu, W.; Guzenko, V.A.; Knopfmacher, O.; Calame, M.; Schönenberger, C. Understanding the Electrolyte Background for Biochemical Sensing with Ion-Sensitive Field-Effect Transistors. *ACS Nano*, 2012, **6**, 9291–9298

[13] Bedner, K.; Guzenko, V.A.; Tarasov, A.; Wipf, M.; Stoop, R.L.; Just, D.; Rigante, S.; Fu, W.; Minamisawa, R.A.; David, C., pH Response of Silicon Nanowire Sensors: Impact of Nanowire Width and Gate Oxide. *Sens. Mater*. 2013, **25**, 567–576

[14] Karim, A., Patong, A. R., Wahab, A. W., and Raya, I., Utilization of Chitin As Membrane Electrode Materials Diamine Oxidase Enzyme for Histamine Biosensor, *Al Kimia*, 2014, **2** (2): 13-23

Effect of pH fermentation on production bioethanol from jackfruit seeds (Artocarpus heterophyllus) through separate fermentation hydrolysis method

A R Arif [1,2], H Natsir[1,2], H Rohani[2] and A Karim[1,2]

[1] Departement of Chemistry, Faculty of Mathematics and Natural Sciences, Hasanuddin University, Makassar, South Sulawesi, 90245, Indonesia
[2] Laboratory of Biochemistry, Chemistry Department,, Faculty of Mathematics and Natural Sciences, Hasanuddin University, Km 10 Tamalanrea, Makassar, South Sulawesi, 90245, Indonesia

E-mail: arrahman.arif@unhas.ac.id

Abstract. Bioethanol is one of the alternative energy sourced from natural products containing carbohydrates through hydrolysis and fermentation process. Jackfruit seeds is one of the feedstock that contain high carbohydrate content but less utilized. The aims of this study to determine the effect of pH hydrolysis in the process of production bioethanol from jackfruit seeds (Artocarpus heterophyllus) through separate fermentation hydrolysis (SHF) method. The hydrolysis process uses H_2SO_4 as a hydrolyzing agent. The fermentation process used Saccharomyces cereviceae as a fermentor with a variation of pH 2,3 4 and 5 for 70 hours. The results showed that glucose content of 75% and pH 3 was the optimum pH of fermentation with the content of bioethanol 57.94%. The fermentation stage has an important role in increasing the levels of glucose and bioethanol in linear. The content of glucose and bioethanol of jackfruit seeds showed a great potential for development as the feedstock in bioethanol production.

1. Introduction

Bioethanol is the product of the sugar fermentation process from carbohydrate sources (starch) such as cassava, sugar cane, sap, sorghum and sweet potato. However, the use of carbohydrate sources is also used as food sources cause a new problem that is the occurrence of competition with human needs. The solution of the problem is to search for raw materials that are underutilized but rich in carbohydrates [1].

Jackfruit (Artocarpus heterophyllus) is a multifunctional tropical fruit plant that can be grown in tropical areas less than 1,000 meters above sea level from South India. In Indonesia jackfruit is widely used as food and basic ingredients in making traditional cakes. However, the underutilized of jackfruit is in the seeds. Jackfruit seeds contain 36.7% of carbohydrates. High carbohydrate content in jackfruit seeds is very potential to be used as a raw material in the production of bioethanol by fermentation [2].

The fermentation process is one of the stages in bioethanol synthesis. The fermentation process is used a bread yeast (Saccharomyces cerivisiae), because this yeast highly produce and stable in alcohol with high content [3]. In addition, the pH factor of fermentation solution determines the process of sugar reaction to bioethanol. Fermentation solutions that are very acidic or alkaline can inhibit the fermentation process, thus reducing the amount of bioethanol produced [4]. This study aims to examine the effect of fermentation pH in bioethanol production from jackfruit seeds through a separate fermentation hydrolysis method.

2. Material and Methods

2.1 Materials
The materials used in this research are jackfruitseeds, Saccharomyces cerevisiae, aquadest, Whatman No. 42 filter paper, H_2SO_4 0.3 M, NaOH, NPK fertilizer, urea, Nelson-Somogyi reagents.

2.2 Preparation of Jackfruit
Jackfruit seeds are cleaned and then dried under the sun. The dried jackfruit seeds are crushed using a flour mill. Jackfruit seed flour is continued at the hydrolysis stage.

2.3 Chemical Hydrolysis
Jackfruit seed powder is hydrolyzed with H_2SO_4 0.3 M at a ratio of 1:10, then refluxed at 100 ° C for 1 hour. The sample is cooled and then filtered. Glucose levels of filtrate were analyzed by Nelson-Somogyi method [5]

2.4 Fermentation

2.4.1 Making a Stater. Yeast bread weighed as much as 9 grams and then put in a solution of hydrolysis result as much as 9% of total volume. The pH of the starter solution was varied at pH 2, 3, 4 and 5. Urea and NPK were added to a solution of 3 g/L. The mixed solution is then shaken for 30 minutes [5].

2.4.2 Fermentation Process. The hydrolysis solution is added with NaOH to adjust the pH. Starter Saccharomyces cerevisiae as much as 9% of the volume of the solution is mixed into the hydrolysis solution. The mixed solution is introduced into the fermentation reactor. The fermentation process is carried out anaerobically for 1 - 5 days at a temperature of 30 ℃. The same process was performed for pH 3, 4 and pH 5 [6].

2.5 The Process of Distillation and Dehydration of Bioethanol. The fermented product is filtered and then fed into the distillation flask, the temperature is kept at 78-88 ℃. Distillation is carried out for 6 hours until bioethanol stops dripping. The bioethanol fraction is dried by CaO. Bioethanol content was analyzed using Gas Cromatography [7].

2.6 Gas Chromatography Analysis. Measurement of bioethanol content using gas chromatography. Bioethanol concentration was obtained from the calculation of area area ratios in which the area of the sample bioethanol was divided by the area of pure ethanol which was the comparison [8].

$$ethanol\ content = \frac{area\ of\ ethanol\ sample}{standard\ area\ of\ ethanol} \times 100\% \quad (1)$$

3. Results and Discussion

3.1 Preparation sample
In the process of making jackfruit seeds powder as much as 4.5 kg washed to remove dirt. The jackfruit seeds have been cleaned and dried for 12 days in the light sun. Dry jackfruit seeds are crushed into jackfruit seed powder. Jackfruit seed flour obtained as much as 2.5 kg of turbid white as in figure 1.

Figure 1. (a). Jackfruit seed clearance, (b) Seeds of jackfruit, (c) Jackfruit seeds powder.

3.2 Chemical Hydrolysis Process
The hydrolysis process is carried out by using sulfuric acid with a ratio of 1:10 (w / v). Samples of 30 grams were dissolved with 0.3 M sulfuric acid in 300 mL and heated at 100 ℃ for 1 hour. Sulfuric acid is used as a catalyst because it is effective for breaking the glycoside bond on starch to glucose. The reaction mechanism can be seen in figure 2.

$$(C_6H_{10}O_6)n + nH_2O \xrightarrow{H_2SO_4} nC_6H_{12}O_6$$

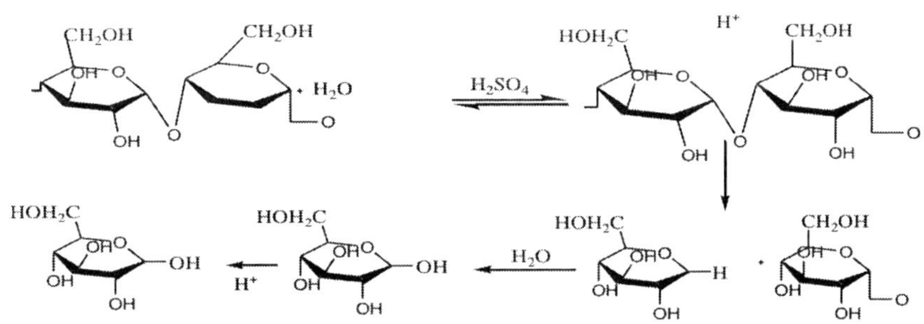

Figure 2. Mechanism of reaction of carbohydrate hydrolysis with sulfuric acid.

In the process of hydrolysis occurs a physical change of the solution from a turbid white to a dark reddish brown. The changes indicate the compound in the jackfruit seed flour has decomposed. The result of the hydrolysis process was cooled and then filtered using filter paper whatman no 42. The filtrate obtained was 298 mL. The same treatment is used for all pH. Glucose level of the filtrate of hydrolysis result determined by the Nelson Somogy method. Based on the measurement obtained glucose conten of 75%.

3.3 Making a Starter
Starter media serves to activate bacteria before the fermentation process. The addition of NaOH serves to increase the pH. the pH of the initial solution is at pH 1 so that NaOH is added until it reaches pH 2. The pH setting to optimize the activation process because bacteria can not work optimally in a solution that is moreover acidic. After reaching pH 2, nutrient NPK and Urea were added 3 g/L. The addition of nutrients

IOP Conf. Series: Journal of Physics: Conf. Series **979** (2018) 012015 doi:10.1088/1742-6596/979/1/012015

serves as a source of food for yeast Saccharomyces cerevisiae. Saccharomyces cerevisiae as much as 9 grams was added in the solution and stirred until homogeneous and then allowed to stand for 30 minutes. The same treatment is applied for pH 3, 4 and 5.

3.4 Fermentation Process

The fermentation process lasts for 1-5 days. At pH 2 incubated only 1 day because it is still too acidic solution, so the bacteria Saccharomyces cerivisiae can not survive while pH 3, 4, and pH 5 incubated for 2, 4 and 5 days. After the fermentation is complete, the volume of the solution is reduced to 280 mL. This happens because the formation of residues in the form of CO_2 gas in the process of fermentation [9]. After the fermentation process is complete, the color of the sample solution changes from brownish yellow to brownish red and forms 2 layers. According prasetyo [5] the formation of these two layers resulting from the process of separation of bioethanol by impurities.

The sample distillation was carried out at a temperature range of 78-88 ° C for 6 hours. The temperature range used is due to the difficulty of obtaining bioethanol distillate if used the boiling point of ethanol. Each pH produces a different volume of the distillate, the result can be seen in table 1.

Table 1. Volume of distillate in various pH

No	pH	Volume of distillate
1.	2	3 mL
2.	3	3 mL
3.	4	4 mL
4.	5	6 mL

Distillate bioethanol is added CaO to bind water. Distillates were filtered and analyzed qualitatively using gas chromatography (GC). The bioethanol distillate chromatogram can be seen in figure 3.

The retention time for pH 2 was 1.902, pH 3 of 1.910, pH 4 and pH 5 with retention times of 1.894 and 1.907. The standard ethanol retention time difference between the four samples can be affected by the purity rate of the bioethanol content. According to Brexo, [10] Uncultible samples can occur due to impurities in the form of by products formed in the form of acetic acid or other organic acids formed from bioethanol that undergo further reactions or the dehydration process of bioethanol is not good so that it affects the separation process of the components contained in the distillate.

The sample content of bioethanol can be determined by using chromatograms from pure ethanol as standard. The determination of the sample content was determined by comparing the area between the standard ethanol and the sample bioethanol. The results showed the standard ethanol area (Rt: 1,875) = 66587191, area of bioethanol pH 2 (Rt: 1,902) = 10004822, area of bioethanol pH 3 (Rt: 1,910) = 38657956, area of bioethanol pH 4 (Rt: 1,894) = 34386838, area of bioethanol pH 5 (Rt: 1,907) = 31874859. Bioethanol content of each sample can be seen in table 2.

Table 2. Bioethanol content based on GC analysis results

No	pH	Bioethanol Content (%)
1	2	14, 99
2	3	57,94
3	4	51.53
4	5	47.77

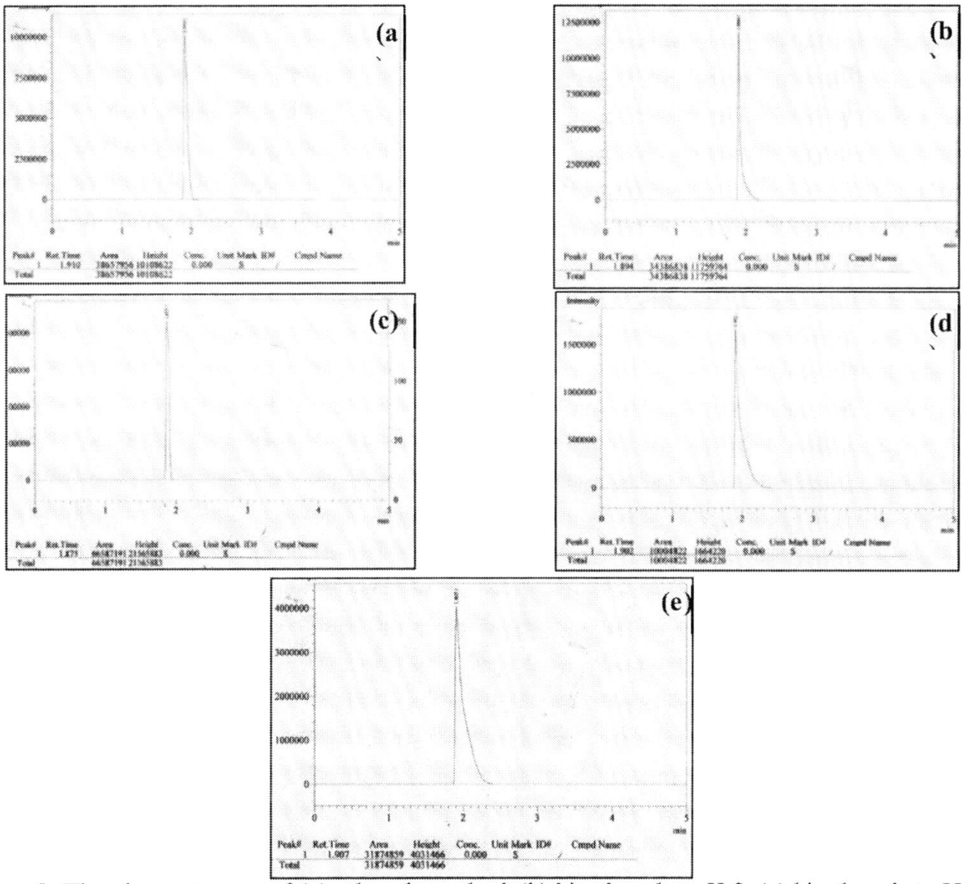

Figure 3. The chromatogram of (a) ethanol standard, (b) bioethanol at pH 2, (c) bioethanol at pH 3, (d) bioethanol at pH 4 (e) bioethanol at pH 5.

Figure 4 showed that the difference in pH affects the glucose and bioethanol levels produced. The optimum pH was obtained at pH 3 with bioethanol content of 57.94%. The smallest content of bioethanol was obtained at pH 2 with a value of 14.99%. The condition of the solution that is still too acid cause yeast does not last long the process of transformation of glucose into bioethanol is not optimal.

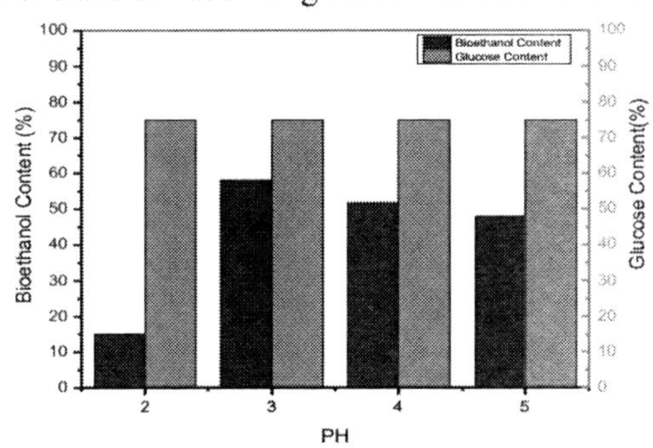

Figure 4. Glucose and bioethanol content during various pH fermentations.

Jackfruit seeds as the feedstock of bioethanol is very potential to be developed refers to the content of glucose content of 75% and bioethanol content obtained by 57.94%. Several other studies on bioethanol exploration from several sources such as seaweed glacilaria sp yielded 5,50% bioethanol content [11] and woodflour acid hydrolyzate of 6.77% [12]. When compared to the amount of bioethanol content produced from the two studies, the potential of jackfruit seeds as the basic ingredients of bioethanol production is very prospective to be developed in the future.

4. Conclusions

Bioethanol production process from jackfruit seed flour can be done by hydrolysis of H2SO4 and anaerobic fermentation. Total carbohydrate content in hydrolysis result is 75% while based on gas chromatography analysis obtained high bioethanol content at pH 3 equal to 57,94% while at pH 2, 4 and pH 5 bioethanol content obtained equal to 14,99%, 51,53 % and 47.77%. The results showed that high carbohydrate content in jackfruit seeds could potentially be developed as the feedstock of bioethanol production on a large scale.

References

[1] Andayana, Y., and Ramli, L., 2010., Bioethanol Production Of Rice Straw With Hydrolysis And Fermentation Process, Chem. Eng, Faculty of Industrial Technology, UPN Vetran, East Java

[2] Zulkifli, E.R., 2012, Making Of Bioethanol From Durian Seeds Through Enzymatic Hydrolysis And Fermentation Using Saccharomyces Cerevisiae, Chemistry Department, Faculty Of Mathematics And Natural Sciences, Andalas University, Padang

[3] Andini, C., Mahajoeno, E., and Setyaningsih, R., 2013, Production Of Bioethanol Citrus Fruit (Citrus sp) Waste By Acid Hydrolysis And Fermentation Using Saccharomyces Cerevisiae, National Seminar X Educational Biology FKIP UNS, Sebelas Maret University, Surakarta

[4] Hapsari, MA, and Pramashinta, A., 2013, Manufacture of Bioethanol From Rubber Sengkong (Manihot glaziovii), For Household Stove Fuel For Efforts to Accelerate Kerosene Conversion to Biofuels, *Journal Chem. and Industrial Tech.*, 2 (2): 240 -245

[5] Zabed, H., Sahu, J. N., & Suely, A. 2017. Bioethanol Production from Lignocellulosic Biomass: An Overview of Pretreatment, Hydrolysis, and Fermentation. *Sustainable Utilization of Natural Resources*, 145

[6] Nikolić, S., Lazić, V., Veljović, Đ., & Mojović, L. (2017). Production of bioethanol from pre-treated cotton fabrics and waste cotton materials. *Carbohydrate polymers*, *164*, 136-144

[7] Retno, E., Kriswiyanti, E., Nur, A. 2009. Bioethanol Fuel Grade from Talas (Colocasia Esculenta). *Ekuilibrium* V 8. (1). Chemical Engineering Department Sebelas Maret University

[8] Osvaldo, Z.S., Panca, P.S., and Faizal, M., 2012, Effect of Acid and Time Concentrations on Hydrolysis and Fermentation Process of Bioethanol Fertilization from Alang-Alang, *Jou. Chem. Eng.*, 2 (18), Sriwijaya University, Palembang

[9] Rahman, R.A., Sumarni, N.K., and Rahmat, B., 2012, Optimization of Sucrose Hydrolysis Using Cation Resin Sulfonate Type, *Jou. Nat. Sci*, 1 (1), Tadulako University, Palu

[10] Brexó, R. P., and Sant'Ana, A. S. 2017. Impact and significance of microbial contamination during fermentation for bioethanol production. *Renewable and Sustainable Energy Reviews*, *73*, 423-434

[11] Saniha A., Endang K, and Anto B., 2015, Production of Bioethanol From Seaweed and Waste Agar Gracilaria sp. with Different Methods of Sacrification, *Bioma*, Vol. 16, No. 2, Hal. 65 - 75

[12] I Wayan A. , Dwi S., Nur R., 2015, Production of Bioethanol From Hydrolysis of Wood Across of Two Wood Mixed Culture Trichoderma viride and Saccharomyces cerevisiae, *Agritech*, Vol. 33, No. 2

Phenethyl ester and amide of Ferulic Acids: Synthesis and bioactivity against P388 Leukemia Murine Cells

Firdaus [1], N H Soekamto [1], Seniwati [1], M F Islam [2] and Sultan [3]

[1] Chemistry Department, Hasanuddin University, Makassar, Indonesia
[2] Student of Chemistry Master Program, Hasanuddin University, 90245 Makassar, Indonesia
[3] Student of Chemistry Bachelor Program, Hasanuddin University, 90245 Makassar, Indonesia

E-mail: firdaus@unhas.ac.id

Abstract. Bioactivity of a compound is closely related to the molecular structure of the compound concerned, its strength being the quantitative relation of the strength of the activity of the group it possesses. The combining of moieties of the active compounds will produce more active compounds. Most phenolic compounds as well as compounds containing moiety phenethyl groups have potential activity as anticancer. Combining phenolic groups and phenethyl groups in a compound will result in compounds having strong anticancer bioactivity. This study aims to combine the feruloyl and phenethyl groups to form esters and amides by synthesize of phenethyl *trans*-3-(4-hydroxy-3-methoxyphenyl)acrylate (**5**) and *trans*-3-(4-hydroxy-3-methoxyphenyl)-*N*-phenethylacrylamide (**6**) from ferulic acid with phenethyl alcohol and phenethylamine, and to study their bioactivity as anticancer. The synthesis of both compounds was conducted via indirect reaction, including acetylation, chlorination, esterfication/amidation, and deacetylation. Structures of products were characterized by FTIR and NMR data, and their bioactivity assay of the compounds against P388 Leukemia Murine Cells was conducted by an MTT method. Results showed that the compound **5** was obtained as a yellow gel with the IC_{50} of 10.79 μg/mL (36.21 μM), and the compound **6** was a yellowish solid with a melting point of 118-120°C and the IC_{50} of 29.14 μg/mL (97.79 μM). These compounds were more active than the analog compounds.

1. Introduction

Ferulic acid or *trans*-3-(4-hydroxy-3-methoxyphenyl) acrylic acid is a derivative compound of hydroxycinnamic acid having various interesting bioactivity [1]. The structure of these compounds consists of a phenolic group which has been associated with anticancer [2]. Ferulic acid and its derivatives have a therapeutic effect that is potentially useful in the treatment of cancer, diabetes, lung and cardiovascular diseases, as well as hepatic, neuro and photoprotective effects, and has antimicrobial and anti-inflammatory activities [3]. Hexacosyl-(*E*)-ferulate and leucosceptoside displayed significant antioxidant activity in the DPPH assay with the RC_{50} of 0.0976 mg/mL and 0.0148 mg/mL, respectively [4].

The same phenomenon, 2-phenyl ethanol exhibits a wide spectrum of activity as it is highly active against *S. aureus*, *E. faecalis*, *E. coli* and *P. aeruginosa* [5]. Several ester and amide compounds containing phenyl group exhibit biological activity such as antitumor [6], antioxidant [7], anti-inflammatory [8,9], immune regulation [10], antimicrobial [11], and antidiabetic [12]. These facts indicated that phenethyl group played an important role on bioactivity of compounds. Therefore, the merge of phenethyl and ferulic groups in an ester or amide compounds like **5** or **6** (figure 1) can be expected to be more active as anticancer.

Attempts to synthesize ester and amide derivatives of *p*-hydroxycinnamic acid directly by using a catalyst have been conducted. However, the method was failed. This is thought to be due to the reaction of condensation between hydroxycinnamic compounds involving the phenolic group with the active carbonyl group. Based on these assumptions, the *p*-hydroxyic acid amide has been successfully synthesized after the phenolic group is protected with an acetyl group [13]. Therefore, the conversion of *trans*-3-(4-hydroxy-3-methoxyphenyl)acrylic acid to ester **5** and amide **6** have been conducted via protection, chlorination, esterification/amidation, and deprotection reactions (figure 1). Their activities were assayed against P388 leukemia murine cells using the MTT method.

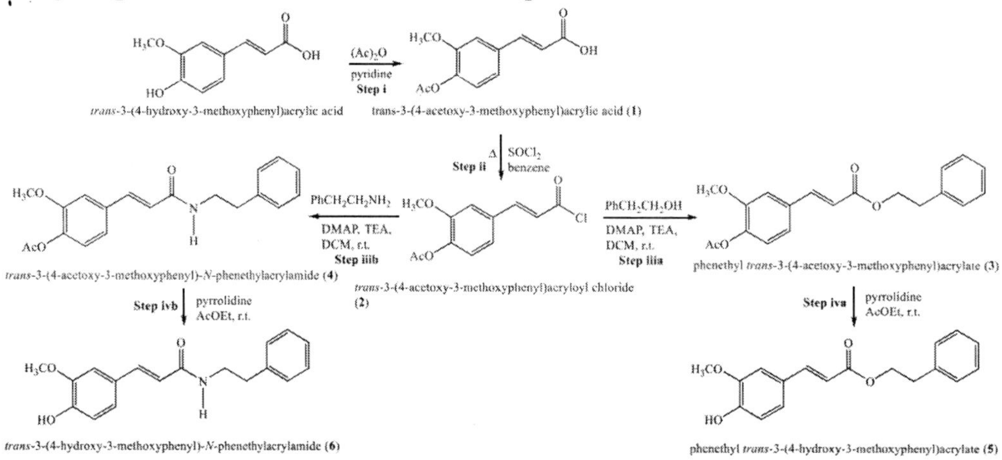

Figure 1. The pathway of synthesis reaction of compounds **5** and **6**.

2. Methodology

2.1 Materials

Trans-Ferulic acid, acetic anhydride (Ac_2O), phenethylalcohol, phenethylamine, ethyl acetate (EtOAc), pyridine, thionyl chloride, dichloromethane (DCM), benzene, methanol, pyrrolidine, triethylamine (TEA), *n*-hexane, 4-(dimethylamino)pyridine (DMAP), NH_4Cl, HCl, and anhydrous Na_2SO_4, were pro synthesis and pro analysis grades, used without purification.

2.2 Equipment

The melting points were taken using electro-thermal apparatus, the NMR spectra were recorded on a Spectrometer Agilent operating at 500 MHz for ^1H and 125 MHz for ^{13}C nuclei, and the FT-IR were recorded on a Spectrophotometer Shimadzu Prestige 21, and the activity against P388 Leukemia murine cells were analyzed using a MTT method in the Natural Product Laboratory, Institute Technology of Bandung (ITB).

2.3 procedures

2.3.1 Synthesis of compound 1. The mixture of *trans*-3-(4-hydroxy-3-methoxyphenyl)acrylic acid (10.0 mmol) and acetate anhydride (28.0 mmol) was dissolved into pyridine and stirred at room temperature for 30 minutes. Cold water was added into the reaction product while stirred until the formation of white precipitate has been completed. The precipitate was filtered, washed with water, dried, and recrystallized from hot methanol to obtain the compound **1** as white solids with a melting point of 194-196°C (56.42% yields). The pure compound **1** was analyzed with FT-IR spectrometer. IR (KBr): υ (cm^{-1}) = 2500-3200 (O-H, carboxyl), 1761.01 (C=O, acetyl ester), 1687.71 (C=O, conj. carboxyl), 1631.78 (C=C, olefin), 3010.88 (unsat. C-H), 1600.99 & 1506.41 (C=C, Ar), 2976.16, 2943.37 & 2841.15 (sat. C-H), 1485.90 &1371.39 (CH_3), 985.64 (*trans*-olefin), 856.39 & 837.11 (1,2,4-trisubst. Ar).

2.3.2 Synthesis of compound 2. The compound **1** (2.56 mmol) was dissolved in 20 mL benzene and added with thionyl chloride (14.4 mmol). After atmosphere of reaction system has been changed with nitrogen, the reaction mixture was refluxed for 4 hours. The reaction product was cooled until room temperature, and the solvent was removed by evaporation to give compound **2** as an orange solids. This solid was used in further procedure without purification.

2.3.3 Synthesis of compound 3. The compound **2** was dissolved in 20 mL of dichlomethane and added with 2-phenylethanol (4.17 mmol), pyridine (4.0 mmol), and triethylamine (4.0 mmol). The solution was stirred at room temperature for 4 hours. The reaction product was diluted with 20 mL of dichloromethane and washed with saturated ammonium chloride solution and water respectively, dried with anhydrous sodium nitrate, filtered, and evaporated to give a solid. Purification the product was conducted by gravity column chromatography and recrystallized to give compound **3** as a white crystal with a melting point of 79-80°C (41.58%). The pure compound **3** was analyzed with FT-IR spectrometer. IR (KBr): υ (cm^{-1}) = 1761.01 (C=O, acetyl ester), 1707.00 (C=O, conj. ester), 1641.42 (C=C, olefin), 3030.17 & 3070.88 (unsat. C-H), 1595.13 & 1510.26 (C=C, Ar), 2968.45, 2943.37 & 2843.07 (sat. C-H), 1452.40 &1371.39 (CH_3), 989.48 (*trans*-olefin), 852.54 & 829.39 (1,2,4-trisubst. Ar), 754.17 & 702.09 (monosubst. Ar).

2.3.4 Synthesis of compound 5. The compound **3** (0.75 mmol) was dissolved in 1 mL pyrrolidine while stirred, and added with 50 mL of ethyl acetate. Stirring was continued for 1 hour. The reaction product was washed with unsaturated solution of ammonium chloride and water sequently, dried with anhydrous sodium sulfate, and the solvent was removed by evaporation to give compound **4** as a yellow gel (51.96% yields). The pure compound **4** was analyzed with FT-IR, and NMR spectrometer. IR (KBr): υ (cm^{-1}) = 3412.08 (O-H fenolic), 1705.07 (C=O ferulate ester), 1631.78 (C=C olefin), 3062.96 & 3028.31 (unsat. C-H), 1597.08 & 1514.12 (C=C Ar), 2956.87, 2927.94 & 2852.72 (sat. C-H), 1458.18 &1384.89 (CH_3), 981.77 (*trans*-olefin), 848.68 & 817.62 (1,2,4-trisubst. Ar), 748.38 & 702.09 (monosubst. Ar); ^{13}C-NMR ($CDCl_3$): δ (ppm) = 36.37, 56.08, 65.02, 109.40, 114.83, 115.51, 123.27, 126.69, 127.09, 128.65, 129.06, 138.07, 145.09, 146.87, 148.09, and 167.34; ^{1}H-NMR ($CDCl_3$): δ (ppm) = 3.02 (*t*, 2H, *J* = 7.05 Hz, CH_2), 3.93 (*s*, 3H, OCH_3), 4.42 (*t*, 2H, *J* = 7.05 Hz, CH_2), 5.90 (*s*, 1H, O-H), 6.29 (*d*, 1H, *J* = 15.95 Hz and 7.62 (*d*, 1H, *J* = 15.90 Hz, *trans* olefin), 6.93 (*d*, 1H, *J* = 8.15 Hz, H-Ar), 7.07 (*dd*, 1H, *J* = 1.75 & 8.15 Hz, H-Ar), 7.24 (*m*, 3H, H-Ar), 7.27 (*m*, 2H, H-Ar), 7.32 (*dd*, 2H, *J* = 7.2 & 0.95 Hz, H-Ar).

2.3.5 Synthesis of compound 4. The compound **2** was added with phenethylamine (1.1 eq. mol) and dissolved into 30 mL of dichloromethane, added with pyridine and triethylamine (1:1) and stirred at room temprature for 4 hours. The product of this reaction was washed with 20 mL of HCl 1 N, the organic layer was washed again with 20 mL of saturated NH_4Cl, dried with anhydrous Na_2SO_4, and evaporated to give a white solid. The solid was purified with gravity chromatography column to obtaine a white solid with a melting point of 118-120°C (63.47% yields). The pure compound **5** was analyzed with FT-IR, and NMR spectrometer. IR (KBr): υ (cm^{-1}) = 3435.22 (sec. N-H), 1753.29 (C=O, ferulate ester), 1651.57 (C=O, amide), 3070.68 (C-H unsat.), 1612.99 (C=C olefin), 1583.56 & 1496.26 (C=C Ar), 2962.66 & 2926.01 (C-H sat.), 1456.26 &1365.60 (CH_3), 977.91 (*trans*-olefin), 910.40 & 833.25 (1,2,4-trisubst. Ar), 756.02 & 709.80 (monosubst. Ar).

2.3.6 Synthesis of compound 6. The compound **4** was dissolved in 1 mL of pyrrolidine and added with 50 mL of ethyl acetate, and stirred at room temperature for 4 hours. The product of this reaction was washed with 20 mL of 1 M H_2SO_4 the organic layer was washed again with 20 mL of saturated NH_4Cl, dried with anhydrous Na_2SO_4, and evaporated to give a white solid. The solid was dissolved into ethyl acetate, added with *n*-hexane to give turbid solution, and stored at room temperature until the crystal was completely formed. The crystal was filtered and dried to obtaine white crystalline with a melting point of 94-96°C (63.47% yields). The pure compound **6** was analyzed with FT-IR, and

NMR spectrometer. IR (KBr): υ (cm^{-1}) = 3352.26 (N-H, amide), 3288.63 (O-H, phenolic), 3062.96 & 3024.36 (unsat. C-H), 1595.13 & 1514.12 (C=C, Ar), 2926.01, & 2852.72 (C-H sat.), 1454.33 &1368.46 (CH_3), 977.91 (*trans*-olefin), 904.61 & 817.82 (1,2,4-trisubst., Ar), 748.36 & 700.16 (monosubst. Ar); ^{13}C-NMR ($CDCl_3$): δ (ppm) = 35.76, 40.91, 55.97, 109.83, 114.90, 118.13, 122.17, 126.61, 127.31, 128.75, 129.90, 138.99, 141.24, 146.91, 147.60, and 166.49; ^{1}H-NMR ($CDCl_3$): δ (ppm) = 2.88 (*t*, 2H, *J* = 6.95 & 13.90 Hz, CH_2), 3.64 (*dd*, 2H, *J* = 6.75 & 13.20 Hz, CH_2), 3.85 (*s*, 3H, OCH_3), 5.85 (*t*, 1H, *J* = 5.15 Hz, N-H), 6.21 (*d*, 1H, *J* = 15.50 Hz & 7.53 (*d*, 1H, *J* = 15.90 Hz, *trans* olefin), 6.87 (*d*, 1H, *J* = 8.20 Hz, H-Ar), 6.94 (d, 1H, J = 1.60 Hz, H-Ar) 7.00 (*dd*, 1H, *J* = 1.55 & 8.15 Hz, H-Ar), 7.20 (*m*, 1H, H-Ar), 7.22 (*m*, 2H, H-Ar), 7.30 (*m*, 2H, H-Ar).

3. Discussion

Approximately a quarter registered medicines have molecules containing amide groups and, concequently, amidation is the most commonly executed conversion to obtain a particular medicine. [14]. However, the common methods for conversion of carboxylic acids to amides suffer from inherent drawbacks. This case is much more pronounced in the conversion of *p*-hydroxycinnamic acids to their amides, due to the presence of phenolic groups capable of performing dimerization reactions with activated carbonyl groups [15]. Therefore, the conversion of *p*-coumaric acid to its amide was successful when the phenolic group was protected by an acetyl group [13]. This method has been adopted and successfully used in converting ferulic acid compounds into ester and amide, i.e compound **5** and compound **6**.

The activity both compounds against P388 murine leukemia cells has been evaluated by a MTT method, and gave the IC_{50} of 10.79 μg/mL for compound **5** and 29.14 μg/mL for compound **6.** Pursuant to these values, the compound **5** can be thought gave strong activity and the compound **6** was moderate [16]. This fact indicated that the activity of the ester compound were stronger than the amide compound. Structurally, this phenomenon is a proper case. When the compounds released hydrogen radicals leave phenolic radicals, the phenolic radical of ester was easily to delocalize its single electron to its carbonyl group; but in the phenolic radical of amide, the delocalization was interfered by the resonance of lone pair electron of nitrogen to its carbonyl group.

Although the activity of ester compounds was higher than amide compounds, the amide derivatives were the most widely used in the medicinal field. It is due to the amide compounds which are neutral, stable, and have hydrogen-bond receptor accepting and donating properties [7]. Moreover, some ester compounds are toxic to the cells [17], so the ester compounds should pass the toxicity assay against the cell before it becomes a drug.

Comparing the result of the previous research about bioactivity of analog amide compounds against P388 murine leukemia cells, i.e. *N*-feruloylpiperidine with IC_{50} = 46.67 μg/mL and *N*-feruloylmorpholine IC_{50} = 57.10 μg/mL [13], the activity of compound **6** was higher than both compounds. This fact indicated that beside the *p*-hydroxycinnamyl group, the residue amine groups also influenced the bioactivity of *p*-hydroxycinnamide compounds.

4. Conclusion

The compounds **5** and **6** can be synthesized from ferulic acid with 2-phenethylalcohol and phenethylamine, respectively via indirect conversion giving compound **5** as a yellow gel, and compound **6** as a yellowish solid with a melting point of 118-120°C. Both compounds were active against P388 murine leukemia cells with the IC_{50} of 10.79 and 29.14 μg/mL, respectively. These compounds were more active than the analog compounds synthesized in the previous research [13].

Acknowledgments

The authors are thankful to the Rector of Hasanuddin University to provide funds for PUPT project, and the NMR Laboratory staff and the Natural Product Laboratory of Institute Technology of Bandung for providing the NMR spectra and for the evaluating bioactivity of both compounds against P388 murine leukemia cells, respectively, as well as the Integrated Laboratory staff of Chemistry Department, Hasanuddin University for providing FTIR spectra.

References

[1] Boz H 2015 *rev. Czech J. Food Sci.* **33** 1
[2] Cai Y, Luo Q, Sun M, and Corke H 2004 *Life Sci.* **74** 2157
[3] de Paiva B L, Goldbeck R, dos Santos W D, and Squina F M 2013 *Braz. J. Pharm. Sci.* **49** 395
[4] Delazar A, Shoeb M, Kumarasamy Y, Byres M, Nahar L, Modarresi M, Sarker S D 2004 *DARU* **12** 49
[5] Saharan R., Verma P S, Sharma I K, and Joshi J 2014 *Res. J. Pharm. Bio. Chem Sci.* **5** 738
[6] Cavell B E, Alwi S S S, Donlevy A M, Proud C G,and Packham G 2012 *J. Nat. Prod.* **75** 1051
[7] Aswathanarayanappa C, Bheemappa E, Bodke Y D, Ningegowda R, and Peethambar S K 2014 *Del Phar. Chemica* **4** 399
[8] Ozturk G, Giniz G, Akyol S, Erden G, Gurel A, Akyol O 2012 *Eur. Rev.* for *Med. & Pharm. Sci.* **16** 2064
[9] Kuo Y Yu, Jim W Tim, Su L Cheng, Chung C Jun, Lin C Yu, Huo C, Tseng J Chih, Huang S Han, Lai C Jen, Chen B Chih, Wang B Juan, Chan T Min, Lin H Ping, Chang W W Shaing, Chang C Rung, and Chuu C Pin 2015 *Int. J. Mol. Sci.* **16** 10748
[10] Zhang P, Tang Y, Li N Guang, Zhu Y, and Duan J Ao 2014 *Molecules* **19** 16458
[11] Kim M G and Lee H S 2009 *J. Food Sci.* **74** M467
[12] Ho Y Jin, Chen W Pin, Chi T Cherng, Chien C C Chia, Lee A Seng, Chiu H Lin, Kuo Y Hsiung, and Su M Jai 2013 *Cardiovas. Diabet.* **12** 99
[13] Firdaus, Husain D R, Naid T, Seniwati, Soekamto, Sumarna S, Islam M F 2017 *Int. J. Chemtech Res.* **10** 27
[14] McPherson C G, Caldwell N, Jamieson C, Simpson I, and Watson A J B 2012 *Org. Biomol. Chem.* **15** 3507
[15] Firdaus, Seniwati, Husain D, Rasyid H, Sukarti 2014 *Proceeding at International Conference on Science (ICOS)* **I** 339
[16] Cao S Geng, Sung V H L, Wu X Hua, Sim K Yeow, Tan B H K, Fereira J T, Goh S H 1998 *Tetrahedron* **54** 10915
[17] Yang J, Marriner G A, Wang X, Bowman P D, Kerwin S M, Stavchansky S 2010 *Bioorg. Med Chem.* **18** 5032

Dengue antiviral activity of polar extract from *Melochia umbellata* (Houtt) Stapf var. Visenia

Nunuk Hariani Soekamto[1], S Liong[1], S Fauziah[1], I Wahid[2], Firdaus[1], P Taba[1] and F Ahmad[3]

[1] Department of Chemistry, Hasanuddin University, Jl. PerintisKemerdekaan, Km 10, Tamalanrea, Makassar 90245, Indonesia
[2] Department of Parasitology, Faculty of Medicine Hasanuddin University
[3] Postgraduate student in Department of Chemistry Hasanuddin University, Jl. Perintis Kemerdekaan, Km 10, Tamalanrea, Makassar 90245, Indonesia

E-mail: nunukhariani@unhas.ac.id

Abstract. A research on the dengue antiviral activity test on the polar bark extract of *M. umbelatta* (Houtt.) Stapf var. Vicenia have been done to determine the relation to its activity againts brine shrimp *Artemia salina*. The bark was extracted by maceration with n-hexane, chloroform, and ethylacetate. The activity of the ethyl acetate extract was then tested against *A. salina* and *dengue virus*. It was found that the ethyl acetate extract was active to *A. salina* with the LC_{50} value of 101.66 μg/mL and also very active to *dengue virus* with the IC_{50} value of 1.67μg /mL. It is clear that the toxicity to brine shrimp *A. salina* has a positive correlation with the *dengue* anti virus.

1. Introduction

One of the diseases caused by the virus is dengue hemorrhagic fever (DHF). DHF is caused by dengue virus infection (DENV) transmitted by *Aedes aegypti* and *Aedes albopictus* female infected by DENV. The World Health Organization (WHO) reports that 2.5 billion (or more than 70%) of the human population is at risk for dengue fever, including in Southeast Asia and the Western Pacific [1]. An effective antiviral drug or patent vaccine against dengue virus is not currently available. One alternative approach that may be used for the prevention or treatment of dengue fever is to use natural products derived from plants.

Some species of the Malvaceae family have been tested for their efficiency as antiviral and antitumor agents, partly based on information on plants traditionally used as medicines to treat various human diseases. [2-5]. In addition, studies of the genus *Melochia* (Malvaceae) have also been performed, for example the roots of *M. chamaedrys* are used to treat various diseases such as cancer and hypertension [6]. Leaves *M. corchorifolia* as anti-cancer [7], *M. odorata* Inhibits HIV P24 [8], root wood and stem wood *M. umbellata* (Houtt.) Stapf var. Degrabrata K. has an activity as antitumor of murine leukemia P-388 cell [9,10].

Plants with the same family or genus will produce similar bioactivity. it can thus be assumed that the bark of M. umbellata (Houtt) Stapf var. Visenia has an antiviral bioactivity, especially dengue antiviral.

2. Experimental

2.1. Materials
The materials used in this research are plant stem bark powder *M. umbellata* (Houtt) Stapf var. Visenia, some organic solvents such as technical methanol, technical n-hexane, chloroform pa, ethyl acetate, technical acetone, Liebermann-Buchard phytochemical reagent, Dragendroff, Wagner, iron (III) chloride medium MEM, Fetal Bovin Serum (FBS), MEM medium, Vero cell ((derived from the kidney of an African green monkey), cell line C6/36, DENV-2, Phosphate-Buffered Saline (PBS), 1% L-glutamine, 1.5% CMC, tetramethylbenzidine (TMB), $NaHCO_3$, H_2SO_4 0 ,1 M.

2.2. Instruments
Instrument used in this research were tools of glass, blender, funnel, separating funnel, Buchner funnel, rotary evaporator, digital scales, Vigreux distillation device, concave tube, T25 flask, CO_2 incubator, incubator without CO2, centrifuge 4°C, freezer, hemositometer, microwell plate, inverted microscope, ELISA reader (Benchmark).

2.3. Procedure

2.3.1. *Extraction.* Dried powder of *M. umbellata* stem of 50 g is macerated with n-hexane for 1 x 24 hours 4 times. The macerate collection obtained is then concentrated to obtain a concentrated n-hexane extract. The residue was macerated with chloroform solvent 4 times, then the residue was again macerated with ethyl acetate 3 times. The obtained macerate was then concentrated to obtain concentrated ethyl acetate extract as much as 85,1 mg of green color. The residue of the next partition was given a notation as methanol extract of 47.2 mg. The water extract was obtained from boiling the dried sample of the plant with water solvent and obtained the result of 6.21 g.

2.3.2. Bioassay. The extract obtained was tested toxicity to A. salina using Meyer method [11] and dengue Antivirus test is done by using Focus Formation Unit Reduction Assay method (FFURA)

2.3.3.Dengue Antivirus test. Dengue antivirus test is done by using Focus Formation Unit Reduction Assay (FFURA) method to know IC_{50} (Inhibition Consentration 50%) on extract obtained and Enzyme-Linked Immunosorbent Assay (ELISA reader) method. Stages of bioactivity testing extracts of dengue virus among other:

2.3.4.Cell culture Vero Line. Cell line was added to DMEM (Gibco) medium which added nutrient in the form of 10% Fetal Bovine Serum (FBS), L-glutamine 1%, incubated at 37 °C under 5% CO_2 condition. Vero cells are taken from storage in the freezer. Then diluted at room temperature for 2-3 minutes. The cells in the cryovial were transferred into a sterile conical tube containing 5 mL MEM free, then centrifuged 800 rpm for 5-10 min and the supernatant was discarded. The pellet was added 5 mL Culture medium, resuspended, then transferred to T25 flask, incubated in CO_2 incubator at 37 °C and 5% CO_2. The medium is replaced after 72 hours and the cells are allowed to grow back until confluent. Confluent cells are harvested, the medium in the flask removed and the cell washed with 1 X PBS 2 times by rinsing it slowly. The cells in the flask were added 0.25% aqueous solution of 0.5 mL, incubated in CO_2 incubator at 37 °C and 5% and centrifuged 800 rpm for 5-10 min.

2.3.5 Cell Culture line C6/36. Cell line vero was added to the MEM (Gibco) medium which added nutrients in the form of 10% FBS, 1% glutamine, incubated at 37 °C under 5% CO_2. Cell vero is taken from storage in the freezer. Then diluted at room temperature for 2-3 minutes. The cryovial cell was transferred to a sterile conical tube containing 5 mL MEM free, then centrifuged 800 rpm for 5-10 min and the supernatant was discarded. The pellets were added 5 mL of culture medium, slowly suspended, then the cells transferred to T25 flask, incubated in CO_2 incubator at 37 °C and 5% CO_2. The medium is replaced after 72 hours and the cells are allowed to grow back until confluent (evenly filling the flask). Confluent cells are harvested, by means of medium in flask removed and cell washed with PBS 2 times by rinsing them slowly. The cells in the flask were added 0.5% aqueous trypsin solution 0.25%, incubated in CO_2 incubator at 37 °C and 5% CO_2 then centrifuged 800 rpm for 5-10 min.

2.3.6. Preparation of virus stock in cell line C6/36. Cell line C6/36 is used to multiply the DENV-2 virus (stock preparation) by 400 μL of the virus added from the previous stock diluted with MEM medium (containing 2% FBS, 1% glutamine 1 to 1 mL) into the flask which contains C6/36 cells then shaken, then incubated at 28 °C. Every 15-20 minutes (for 1 hour) gently beaten, after 1 hour added MEM medium (with nutrient FBS 2% 5 mL) then incubated until the cytopathic effect (morphological cell changes) slowly, then centrifuged 4000 rpm at 4 oC for 10 minutes, then the virus supernatant (stock virus) is stored at -80 °C

2.3.7. Virus stock creation on cell line Vero. Virus infected by C6/36 infection was then infected again in vero cells in flask by 50 μL viral infected with diluted MEM 1 mL (containing 2% FBS, 1% glutamine L), then incubated in CO_2 incubator at 37 °C and 5% CO_2. Every 15-20 minutes (for 1 hour) the mixture is shaken gently, after 1 hour added MEM medium (with 2% FBS 2%) 5 mL then incubated in CO_2 incubator at 37 °C and 5% CO_2 until cytopathic effect (observed for 7 days after being infected). The supernatant was harvested by slowly transferring, then centrifuged 4000 rpm at 4 ° C for 10 min, then the viral supernatant (stock virus) was stored at -80 °C.

2.3.8. Focus Formation Unit Reduction Assay (FFURA). The antiviral activity of the extract was determined by measuring the result of the reduction of virus amount through Foci Forming Unit (FFU). The supernatant of the vero cell is placed in 24 microplate holes, during the viral transmission process in cells added CMC 1.5% and 2% FBS into the MEM medium. Reduced virus was observed using standard peroxidation based on discoloration for 4 days after transmission. Furthermore, the amount of DENV-2 was calculated using a stereomicroscope and the viral titer was calculated through FFU.

2.3.9. Enzyme Linked Immunosorbent Assay (ELISA reader). The ELISA plate was coated by an antigen with optimum dilution (1: 1000) using a $NaHCO_3$ buffer (pH 9,6), incubated at room temperature overnight at 4 °C or 2 h at 37 °C. The platter is then washed with PBS 0.05% 3 times then add blockers, BSA 200 μL. The plates were incubated at room temperature for 2 hours and washed again with PBS 3 times. At incubation, positive and negative control serum was diluted in 1% PBS, and 100 μL was included in the washed plate. The plates were incubated at 37 °C for 2 hours and washed again with PBS 3 times. Subsequently, 100 μL of tetramethylbenzidine (TMB) substrate was added, then incubated in dark space for 10 min and the cessation of reaction was carried out with 100 μL of 0.1 M sulfuric acid (H_2SO_4) solution. Inspection results are read on ELISA reader with wavelength 450 nm. For comparison, an ELISA kit was obtained from the Virology Laboratory of the Elizabeth Macarthur Agriculture Institute (EMAI), Woodbridge Rd. Menangle, NSW, Australia.

3. Results and discussion

Infectious diseases caused by dengue virus (Dengue Fever) is a dangerous disease that strikes in almost all tropical regions and subtropics. Until now there has been no drug or special therapy used for the treatment of the disease. For that we need to find an alternative drug that is safe and cheap, so that people can easily. This can help reduce the number of people with Dengue Fever (DHF).

One of the dengue virus that generally attacks some areas is dengue virus type 2 (DENV-2). DENV-2 anti-viral assay was performed on ethyl acetate extract and methanol. The antiviral test is followed by cytotoxicity test against Vero cells in order to provide more accurate information about the activity of the extract against DENV-2.

Before performing these anti-viral tests, screening begins by testing the toxicity of the extract on Brine shrimp *Artemia salina*. It was done as a preliminary test to determine whether the extract has high toxicity to *A. salina*. Previous studies have suggested that high toxicity of compounds/extracts against *A. salina* is positively correlated as antitumor/antivirus [11]. For example, walterion C compound has isolated from wood stem *Melochia umbellata* var. Degrabrata K. was very active against *A. salina* (LC50 0.29 µg / mL). Waltherione C has been reported as having anti-HIV activity (EC_{50} 0.3 µg/mL) and toxicity against CEM-TART cells (LC_{50} 3.8 µg/mL) (Jadulco et al., 2014). In our hands, waltherione C significant cytotoxicity against P-388 murine leukemia cells with IC_{50} values of 0.26 µg/mL [12]. Other compounds Moracin M which is a 2-arylbenzofuran derivative also has a high toxicity to *A. salina* (LC_{50} = 61.7 µg/mL), is very active against murine leukemia P-388 (IC_{50} 2.4 µg/mL [13,14].

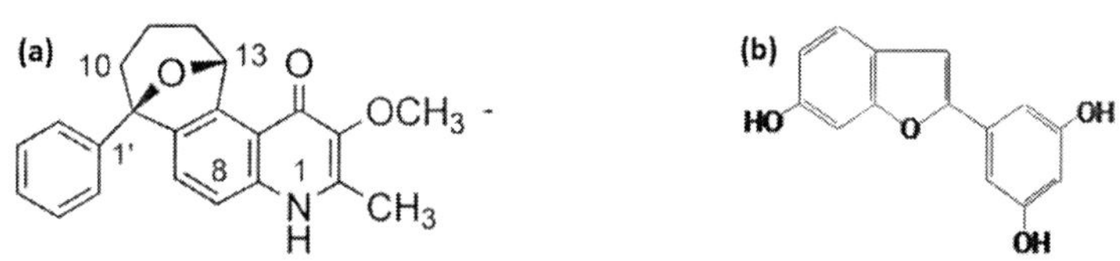

Figure 1. (a) Waltherione C and (b) Moracin M

Toxicity test results on *A. salina*, cytotoxicity to DENV-2 and Vero Cell of Ethyl acetic, methanol and water extract can be seen in table 1.

Table 1. Bioassay test of polar extract (ethyl acetic, methanol and water).

No	Extract	*A. salina* (µg/mL)	DENV-2 (µg/mL)	Vero Cell (µg/mL)
1.	Ethyl acetic	101.66	2.81	12.92
2.	Methanol	30.22	1.67	15.59
3.	Water	82.90	2.61	11.30

Based on the data in Table 1, it shows that the three of extracts toxicity (ethyl acetic, methanol and water) against *A. salina* are positively correlated with their cytotoxicity. While the cytotoxicity of Vero cells is significant to show that the three extracts are very active against DENV-2.

4. Conclusion

The toxicity to brine shrimp *Artemia salina* has a positive correlation with the dengue anti virus. The cytotoxic data of ethyl acetic, methanol, and water extract are showing active on vero cells proved that the three extracts had strong potential against dengue virus type 2 (DENV-2).

Acknowledgments

We thank to Hasanuddin University for financial support and my student undergraduate and Magister Program.

References

[1] WHO, 2002, *Dengue and Dengue Hemorhagic Fever*, Geneva: World Health Organisation, Fact Sheet No. 17

[2] Kapadia, G. J., Shukla Y. N., Morton, J. F., and Lloyd, A., 1977, New Cyclopetide Alkaloids from Melochia tomentosa, *Phytochemistry*, **16(9)**: 1431-1433

[3] Ananil, K., Hudson, J.B., Souzal. C., Akpaganal K, Tower 3, G.H.N.J.T., Amason, J.T. and Gbeassor, M. 2000. Investigation Of Medicinal Plants Of Togo For Antiviral and Antimicrobial Activities. *Pharmaceutical Biology* **38 (1):** 40-45

[4] Saravanakumar, A., Venkateshwaran, K., Vanitha, J., Ganesh, M., Vasudevan, M. and Sivakumar T. 2009. Evaluation of antibacterial activity, phenol and flavonoid contents of *Thespesia populnea* flower extracts. Pak. *Journal Pharmachy Science* **22 (3):** 282-286

[5] Arullappan, S., Muhamad, S., and Zakaria, Z., 2013. Cytotoxic Activity of the Leaf and Stem Extracts of *Hibiscus rosa sinensis* (Malvaceae) against Leukaemic Cell Line (K-562). *Tropical Journal of Pharmaceutical Research* **12 (5):** 743-746

[6] Dias, G.C.D., Gressler, V., Hoenzel, S.C.S.M., Silva, U.F., Dalcol, I.I., and Morel, A.F., 2007, Constituents of the Roots of *Melochia chamaedrys*, *Phytochemistry*, **68**, 668-672

[7] Palaksha, M.N., Ravishankar, K., and Sastry, V. G., 2015. Evaluation of in-vitro Anticancer Activity and Quantitative Estimation of Phenolics and Flavonoids of *Melochia corchorifolia* and *Saccharum officinarum* Leaf Extracts. *European Journal of Biomedical Pharmaceutical Sciences* **2 (4):** 1410-1420

[8] Jadulco, C.J., Pond, C.D., Wagoner, R.M.V., Koch, M., Gideon, O.G., Matainaho, T.K., Piskaut, P., and Barrows, L.R., 2014, 4-Quinolone Alkaloids from *Melochia odorata*. *Journal Natural Product* **77 (1):** 183-187

[9] Erwin, Noor, A., Soekamto, N.H., and Harlim, T. 2010. 6,6'-dimethoxy-4,4'-dihydroxy-3',2'-Furano-Isoflavane, A New Compound From *Melochia umbellata* (Houtt.) Stapf var. Degrabrata K. (Paliasa). *Indonesia Journal Chemistry* **10 (2)**: 222 – 225

[10] Erwin, Noor., A., Soekamto , N.H., Altena, I.V., dan Syah, Y.N., 2014, Waltherione C and cleomiscosin from *Melochia umbellata* var. Degrabrata K. (Malvaceae), Biosynthetic and Chemotaxonomic, Significance. *Biochemical Systematics and Ecology* **55**: 358-361

[11] Meyer, B.N., Ferrigni, N.R., Putnam, J.E., Jacobson, L.B., Nichols, D.E., and McLaughlin, J.L. 1982. Brine shrimp: A convenient general bioassay for active plant constituents. *Plant Medical* **45**: 31-34

[12] Alley, M.C., D.A. Scudiero, A. Monks, M.L. Hursey, M.J. Czerwinski, D.L. Fine, B.J. Abbott, J.G. Mayo, R.H. Shoemaker, M.R. Boyd (1988), *Cancer Res.,* **48**, 589-601

[13] Soekamto, N. H., Achmad, S.A., Ghisalberti, E. L., Aimi, N., Hakim, E. H., and Syah, Y. M. 2003 Beberapa Senyawa Fenol dari Tumbuhan *Morus macroura* Miq. *Jurnal Matematika dan Sains* Vol. 8 No. 1

[14] Soekamto, N.H., S. A. Achmad, Ghisalberti, E. L., Hakim, E. H., and Syah, Y. M. (2000), Stilben and arylbenzofuran from the heartwood of *Morus macroura*, Proceeding of International Seminar on the role of Chemistry in indutry and environment, *Andalas University*, Padang Indonesia, 30-31 Agustus, 190

IOP Conf. Series: Journal of Physics: Conf. Series **979** (2018) 012018 doi:10.1088/1742-6596/979/1/012018

Utilization of α-amylase enzyme from *Bacillus stearothermophilus* RSAII$_{1B}$ for maltodextrin production from sago starch

R A Arfah[1], A Ahmad[1], S Dali[1], M N Djide[2], Mahdalia[1] and A R Arif[1]

[1] Laboratory of Biochemisrty, Department of Chemistry, Faculty of Mathematics and Natural Sciences, Hasanuddin University, Jl. Perintis Kemerdekaan KM 10, Makassar 90245, Indonesia

[2] Laboratory of Microbiology, Pharmacy Faculty, Hasanuddin University, Jl. Perintis Kemerdekaan KM 10, Makassar 90245, Indonesia

E-mail: rugaiyaha.arfah@gmail.com

Abstract. The dried sago flour derived from Palopo contains 28.80% amylose and 91.23% total carbohydrate. Based on the data, sago starch has the potential to become an alternative raw material for themaltodextrin production. Maltodextrin is one of the starch derivative products produced by hydrolysis process using the α-amylase enzyme with amaximum DE (dextrose equivalent) value of 20. The use of maltodextrin for food and pharmaceutical industries is increasing because of maltodextrin is widely used as thickener filler, surfactant and sugar substitute in milk powder. The aims of this study are to optimize the addition of enzyme concentration and hydrolysis time of α-amylase enzyme to obtain high quality ofmaltodextrin This study also aimed to characterization the obtained maltodextrin.The first step was isolation and purificationα-amylase from the isolate of *Bacillus stearothermophilus* RSAII$_{1B}$, followed by determination of the α-amylase concentration (0.05%, 0.07% and 0.09%) in 2.0% starch substrate, and the hydrolysis time ofα-amilase (60, 90, 120, 240 minutes). Maltodextrin characters observed were dextrose equivalent (DE), reducing sugar, moisture content, pH changes, color, solubility, viscosity, and total plate count (TPC). The results showed that the value of DE was 12.31, reducing sugar was11.4%; water content was 10.92%; pH was 4.85; The color of maltodextrin powder was white bone color; solubility was 153.2 g/L; Viscositywas 210-240 cps, TPCwas 380 cfu/g. Maltodextrins produced from sago starch using the α-amylase enzyme from *B.stearothermophillus* RSAII$_{1B}$ met the quality requirements of SNI 7599: 2010.

1.Introduction

Bacillus stearothermophillus RSAII$_{1B}$ is a thermophilic bacteria derived from the hot spring of Lejja South Sulawesi that can produce the thermostable α-amylase enzyme. The α-amylaseenzymederived from *B.stearothermophillus* RSAII$_{1B}$ has the characteristic that the optimum pH and the stability pH of 5.0; the optimum temperature of 50-60°C and thetemperature stabilityof 60°C for 3 hours, Mr range 47.32 KDa [1]. The α-amylase enzyme can be used to partially hydrolyze starch to maltodextrin. The production process of maltodextrin using high temperature was 95°C for wheat starch [2], 75°C for cassava starch [3], and75°C for banana starch [4]. Based on these data, it appears that the process of hydrolysis of starch to maltodextrin requires high temperatures so that the enzyme used has to bea thermostable enzyme which remains active and stable at high temperatures.

Maltodextrin is a mixture of glucose, maltose, oligosaccharide, and dextrin produced from partial hydrolysis of starch using enzymes, acids or irradiated with a maximum DE value of 20. The DE value of 2-5 can be used as a substitute for milk fat in desserts mouth, yogurt, bakery products and ice cream [5]. Maltodextrin with the DE value of 15-20 is used as a high-calorie food product [6] and [7]. Maltodextrin from wheat starch with the DE values of 3-9 values can be used as a source of carbohydrates in isotonic sports drinks [2]. Most of the needs of maltodextrin in Indonesia are still filled with imported products, it is estimated that the import value of modified starch to Indonesia each year reaches the US $ 150 million. Indonesia has the great natural potential to meet these needs with domestic products because Indonesia has many plants that are the source of starch. One potential source of starch to modify into maltodextrin is sago starch.

Sago (*Metroxylon sagu Rottb*) is native to Southeast Asia. In Papua, the area of sago is more than 1 million ha, while the utilization of sago is still low *i.e.* 0.1% of the total national sago area [8]. Sago that is managed well can give a dried starch of 25 tons/ha/year [9]. Due to the highproduction of sago and the low price of wet sago (6000 IDR/kg), it is necessary to convert sago into high-value products such as maltodextrin.

The dried starch of sago from Palopo, South Sulawesi contained a total carbohydrate of 91.23% with the water content of 7.42% and the calcium content of 408.39 ppm [1]. Purnawani [10] reported that the dried starch of sago from Palopo contained amylose of 26.80% with the water content of 8.30% and the gelatinization temperature of 70.5°C. Based on the characteristics, sago starch from Palopo can be used as raw material in making maltodextrinusing the thermostable α- amylaseenzyme.

The objectives of the study were to optimize the addition of α-amylase enzyme concentration and time used for hydrolizing the α-amylaseenzyme to fit the quality requirement of maltodextrin.

2.Research Methods

2.1 Isolation and Purification of the α-amylase enzyme from B.stearothermophillus RSAII$_{1B}$

2.1.1 Inoculum Preparation. The culture stock of *B,stearothermophilus* RSAII$_{1B}$ was grown in Luria agar (LA) medium with 1.0% soluble star in order to obtain fresh isolates. Inoculum preparation was then conducted in the production medium with the inoculum medium used by Arfah *et al.* [11]consisting of 1.5% soluble starch, 0.4% bacto-yeast extract, 1.0% bacto-pepton, 0.005% NaCl, 0.05% $MgSO_4.7H_2O$ and 0.08% $CaCl_2$.The medium pH was 7.0,incubatedat 55 °C for 24 h with anagitation speed of 200 rpm.

2.1.2 Production of α-amylase Enzymes. The composition of the production medium was equal to the composition of the inoculum medium. Theactive inoculum (10%)was incorporated into an Erlenmeyer containing the production medium, incubated in a shaker incubator with an agitation speed of 200 rpm at 55°C for 33 hours [11]. The centrifugation was then conductedat 3500 rpm for 30 minutes. The supernatant obtainedwas a crude extract α-amylase enzyme. Furthermore, the crude extract enzyme obtained was purified by using ammonium sulfate with saturation of 40-60% and followed by dialysis in the cellophane pouch.. dialysis of theα-amylase enzymewas used to hydrolyze the sago starch partially into maltodextrin.

2.2 Optimization of maltodextrin production

2.2.1.Effect of α-amylase enzyme concentration from B.stearothermophillusRSAII$_{1B}$ on the production of maltodextrin. Sago starch wasdissolved in a phosphate buffer pH 5 solution containing 10 ppm $CaCl_2$. Subsequently, the starch solution was heated at 70-80 ° C for 2 hours on a hotplate stirrer for the gelatinization process. After that, the temperature was lowered to 60 °C, then α-amylasewith a concentration of 0.05%; 0.07%; and 0.09%was added. The mixture was heatedat 60-65 °C for 60 min.

The temperature of the mixture was then adjusted to30 °C. The solution pH was adjusted to 3.7-3.9 with 0.1N HCl. Furthermore,the solution was heated in boiling water to stop the enzyme activity. The resulting liquid (maltodextrin) produced was dried in an oven at 50 °C for 3 day and crushed [12]. The obtained maltodextrin powder was analyzed for thetotal sugar content by aphenol-H_2SO_4 method, thereducing sugar content by a DNS method.

2.2.2 Effect of hydrolysis of the α-amylase enzyme from B.stearothermophillus RSAII$_{1B}$ on the production of maltodextrin. The procedure for studying the effect of hydrolysis time of the α-amylase enzyme on the resulting value of DE maltodextrin was similar to the procedure in point 2.2.1 but the concentration of the α-amylase enzyme used was 0.09% and the time used was 60, 120, 180, 240 minutes.

2.3 Maltodextrin production using α-amylase enzyme from B. stearothermophillusRSAII$_{1B}$.
The procedure for maltodextrin production was similar to the procedure in point 2.2.1 but the concentration of the α-amylase enzyme used was 0.09% and the hydrolysis time was 120 min.

2.4 Characterization of maltodextrin resulting from partial sago starch hydrolysis using the α-amylase enzyme from B. stearothermophillusRSAII$_{1B}$

2.4.1 Determination of equivalent Dextrose value (DE). The determination of the equivalent Dextrose value begins with the determination of reducing sugar and the determination of total reducing sugar in maltodextrin products, as follows:

2.4.1.1 Determination of reducing sugar by DNS method [13]
Maltodextrin (1000 ppm) was taken as much as 1.5 mL and then put into a test tube containing 1.5 mL of the DNS reagent. After that, the solution was shaken with a vortex for 10 seconds and heated in boiling water for 10 minutes. Theequivalent Dextrose was calculated by using equation (1).

$$DE = \frac{Reduced\ sugar\ content}{Sugar\ (carbohydrate)\ total} \times 100 \quad (1)$$

2.4.1.2 Determination of sugar (carbohydrate) total by phenol-sulfuric acid method
Determination of sugar total using the same maltodextrin solution with the determination of reducing sugar. A 1.5 mL maltodextrin solution was added with 0.75 mL of 5% phenol solution and 3.75 mL concentrated H_2SO_4 slowly through the tube wall. After that, the solution was allowed to stand for10 min and shaken in a vortex for 10 sec. After that, the solution was allowed to stand for 20 min at a room temperature until the color change occured. The standard solution usedwas glucose solution at various concentrations of 0.04-0.12 mg/mL and distilled water was utilized as blank. Measurement of absorbance was conductedat 490 nm wavelength with a Spectrophotometer. The calculation of carbohydrate total was donebysubstituting the absorbance of the solution obtained on the determination of glucose concentration into the regression equation of the standard glucose solution of the phenol-sulfate method.

2.4.2 Determination of type and concentration of reducing sugar in maltodextrin by HPLC method.
The resulting maltodextrin was analyzed by high-performance liquid chromatography (HPLC) to determine the type and concentration of reducing sugar of maltodextrin. Determination of the type of reducing sugar in maltodextrin product by HPLC method was by comparingthe sample retention time with the standard retention time. HPLC was performed at a flow rate of 2 mL/min, thetemperatureof 40°C, acetonitrile:double distilled water phase (80:20), carbohydrate column (4.6 mm x 250 mm), particle size 4 μm, 20 μL injection volume and Detector Refractive Index Detector. The sugar content can be calculated by using equation (2).

IOP Conf. Series: Journal of Physics: Conf. Series **979** (2018) 012018 doi:10.1088/1742-6596/979/1/012018

$$Sugar\ content\ (\%) = \frac{(peak\ area - intercept) \times factor}{slope \times weight\ of\ sample} \quad (2)$$

2.4.3 Determination of water content. A total of 1 g of maltodextrin sample was weighed and put into a container. The sample was dried in an oven at 105 °C. After that, it was cooled and weighed. The work was repeated until achieved a constant weight. The water content was calculated by using equation (3).

$$Water\ content = \frac{A - B}{C} \times 100\% \quad (3)$$

A = weight of container + sample before draining (g)
B = weight of container + sample after drying (g)
C = sample weight (g)

2.4.4 Determination of pH. The determination of pH was performed by dipping the electrode into the maltodextrin suspension placed on a hotplate stirrer so that the suspension is always homogeneous.

2.4.5 Density Measurement. The density was measured by comparing the weight of maltodextrin from a certain amount of powders of the same volume. The maltodextrin powder was put into a 10 mL measuring bottleand weighed. The density maltodextrin was calculated by equation (4).

$$Density = \frac{mass}{volume} \quad (4)$$

2.4.6 Total Plate Count Test (TPC). TPC testing aims to determine the number of microbial colonies that grow on a material. In the TPC test, Plate Count Agar (PCA) was weighed as many as 2.82 gramadded with distilled water until the volume up to 120 mL in an Erlenmeyer. The solution was then heated until dissolved and sterilized in an autoclave at a temperature of 121 °C and a pressure of 2 atmospherefor 15 minutes [14].

3 Result and Discussion

3.1 Effect of α-amylase enzyme concentration from B.stearothermophillus RSAII1B on DE value of maltodextrin

Figure 1 shows the value of DE as a function of the concentration of α-amylase enzyme. It is clear that the DE value increases with increasing the concentration of the α-amylase enzyme. This is due to the increasing concentration of enzyme that can increase the reaction speed. This happens because the more the α-amylase enzymecontacts with the substrate, the higger the enzyme-substrate complex is formed, this contact occurs on the enzyme's active side.

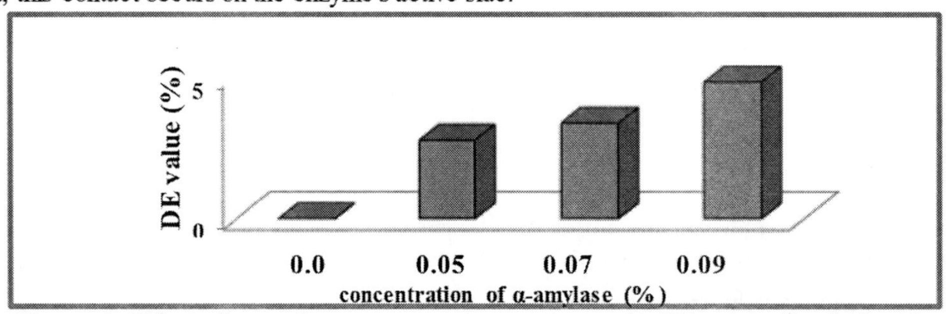

Figure 1. Effect of α-amylase enzyme concentration on the DE value of maltodextrin.

The data in figure 1 also show that without the use of the α-amylase enzyme there is no reducing sugar (DE = 0), although it was heated above its gelatinization temperature at 75 ° C for 2 hours, followed by hydrolysis for 60 min, at 60-65 ° C. This phenomenon proves that enzymes play an important role in the process of hydrolysis of starch substrate into reducing sugar products. In addition,the enzyme concentrationof 0.09% produced the maximum DE value (4.78%) with the time of hydrolysisof 60 min.

3.2 Effect of hydrolysis of the α-amylase enzyme from B.stearothermophillus RSAII$_{1B}$ on the value of dextrose Equvalent (DE) maltodextrin

The effect of the hydrolysis time on the DE value of maltodextrin is shown in figure 2.There is an increase in the value of DE as the hydrolysis time increases. The more enzymes bind to the substrate the more the product of reducing sugars produced, it follows the theory of hydrolysis, that the longer the hydrolysis time the more hydrolyzed the material.The time of hydrolysis used for the α-amylase enzyme to produce maltodextrin with the DE value of about 5-10% is 60-120 minutes. Ni'maturohmah *et al.* [15] reported that dextrin produced from sago starch using the β-amylase enzyme had a DE value of 2.1% required the hydrolysis time of 15 hours with the enzyme concentration of 0.1%, the hydrolysis temperature of 60°C, and the starch concentration of 30%. This indicated that the α-amylase enzyme from bacterium thermophile isolate of *B. stearothermophillus* RSAII1B required relatively faster hydrolysis time to hydrolyze 1,4-glycosidic bond compared with the β-amylase enzyme. It can be explained that the α-amylase enzyme works randomly to break the α-1,4-glycosidic bond on the starch, especially on long chains, to produce maltotriose and maltose. Due to its nature, which can break the glycosidic bond randomly, the enzyme works faster than the β-amylase enzyme.

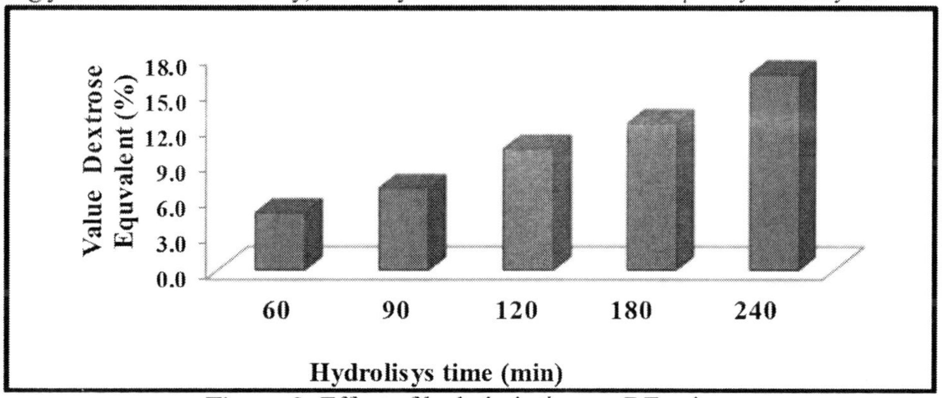

Figure 2. Effect of hydrolysis time on DE values.

The β-amylase enzyme acts to hydrolyze starch from the non-reduction portion and hydrolyze the 1.4-glycosidic bond in the second stage of starch hydrolysis to form maltose and glucose [16, 17]. The equivalent dextrose value of 5-12% in this study requires a hydrolysis time of about 60-180 minutes. It is expected that the resulting maltodextrins have properties and characters as margarine additives, and as a substitute for milk fat in desserts, yogurt, bakery products, and ice cream.

3.3 Characteristics of maltodextrin produced from hydrolysis of sago starch by the α-amylase enzyme

The characteristics of maltodextrin obtained in this study are shown in table 1. It is clear that the characteristics of maltodextrin is in parallel with the Indonesian National Standard (SNI) requirement in terms of the DE value, sugar reduction, the moisture content, pH and the color.

Table 1. Characteristics of maltodextrin produced from hydrolysis of sago starch by the α-amylase enzyme.

Characteristics	maltodextrin of sago	Syarat mutu SNI 7599:2010 (18)
Dextrose Equivalent (DE)	12.31	Maximum 20
Sugar reduction (%)	11.14	11 - 14
Water content (%)	10.92	Maximum 11
pH	4.85	Maximum 5
Color	White beige	White yellowish
Solubility (g/L)	153.2	-
Density (g/mL)	0.3	-
Viscosity of maltodextrin 10% (cps)	240	-
Glucose (%)	2.82	-
Maltose (%)	5.54	-

4. Conclusion

The results showed that the DEvalue was 12.31, reducing sugar was 11.4%; water content was 10.92%; pH was 4.85; the color of maltodextrin powder was white bone color; solubility was 153.2 g/L; viscosity was 210-240 cps, and the total plate count (TPC) was 380 cfu/g. Maltodextrins produced from sago starch using the α-amylase enzyme from *B.stearothermophillus* $RSAII_{1B}$ met the quality requirements of SNI 7599: 2010.

Acknowledgments

The authors thankthe Ministry of Research and Higher Education Indonesia, for the Doctoral Dissertation Grant in 2016 to the head of laboratory Biochemistry,Hasanuddin Universityfor the facilitation given.

References

[1] Arfah, R A., 2016. *Isolation, Purification and Characterization of the α-Amylase Enzyme from Termofil BacteriaSour Lejja Hot Spring South Sulawesi.* Unpublished dissertation, MIPA-UNHAS Graduate School, Makassar

[2] Hidayat, B., Ahza, A B., and Sugiyono, 2003. Characterization of Maltodextrin DP 3-9 and Study of Potential of Use as Source of Carbohydrate at Sports Drink. *Journal. Teknol. And Food Industry,* XIV (1): 51-58

[3] Suryani R and Nisa F.Ch., 2015, Modification of Cassava Starch (Manihot esculenta) with Enzyme α-Amylase as Agent of Fertilizer and Its Application on Marshmallow Making Process, *Journal of Food and Agroindustry,* **3** (2): 723-33

[4] Yusriani E., Hariyadi P., Kuswanda F., 2007. Characterization of Maltodextrin Production Process from banana starch (Moses sp) enzymatically with α-amylase. *Graduate Forum,* **30** (2); 159-68

[5] Kuntz, A.L., 1997, *Making the Most of Maltodextrins.* (online) (www.foodproductdesign.com)

[6] Anwar E., Joshita D., Yanuar A., Baktiar A., 2009, Utilization of Wheat Flour Maltodextrin as Essipien in Tablet and Niosomian Tablet Formulations. *Pharmaceutical Science Magazine* **1**(1): 34-36

[7] Takelti, C.Y., Kiecbusch, T.G., and Collares-Queiroz, F.P., 2010. Morphological and Physicochemical Characterization of Commercial Maltodextrins With Different Degrees of Equivalent Dextrose. *Journal of Food Properties,* (13): 411-25

[8] Hermiati, E., 2012. *Maltodextrin production from Sago Uses Microwave Irradiation.* PKKP. Ristek Lipi

[9] Indonesian Plantation Biotechnology Research Institute, 2007. Sago Plant as Alternative Energy Source. *Journal of Warta Agricultural research and development.* 29(4) (online) (http: // library, R & D, Publications Department Go.id)

[10] Purwani, E.Y., Setiawaty, Y., Setiawanti and Widaningrum. 2006. Characteristics and Case Study of Reception of Sago Community Sago Noodle. *Agritech*, **26** (1): 24-33

[11] Arfah, RA, Ahmad, A., Djide, MN, Anis, M., Zakir, M., 2015. Production Optimization and Characterization of Amylase Enzyme Isolated from Termofil Bacteria Bacillus sp.RSAII-1b from Lejja Hot Spring South Sulawesi. *AJBLS*, **3** (6): 115-119

[12] Jufri, M., Anwar E., Djajadisastra J., 2004. Making Niosom-Based Maltodextrin ED 5-10 from Pati Singkong (Manihot Utllissima). *Pharmaceutical Science Magazine*, **1** (1): 10- 20

[13] Miller, G.L., 1959. Use of Dinitrosalicylic Acid Reagent for Determination of Reducing Sugar. *Anal Chem.*, **31** (3): 426-28

[14] Djide, M.N, and Sartini, 2002. *Analysis of Pharmaceutical Microbiology*. Second printing. Laboratory of Pharmaceutical Microbiology, Faculty of Pharmacy, Hasanuddin University, Makassar

[15] Ni'maturohmah, E., and Yunianta, 2015. Sago Starch Hydrolysis (Metroxylon sagu Rottb) by Enzyme β-Amylase for Preparation of Dextrin. *Journal of Food and Agroindustry*, **3** (1): 292-302

[16] Aiyer, P.V. 2005. Amylases and their applications. African Journal of Biotechnology. **4** (13): 1525-1529

[17] Swargiari, BN. 2015. Microbial α-amylase: Structur Function and Aplication-A Revew. *Int. J. Pharm Bio Sci.*, **6** (4): 1133-1140

[18] SNI, 2010. *Maximum Limit of Microbial Contamination in Food*. National Standard of BSN

Inhibitory activities of *Moringa oleifera* leaf extract against α-glucosidase enzyme in vitro

H Natsir [1,3], A W Wahab [2,3], A Laga [3] and A R Arif [1,3]

[1] Laboratory of Biochemistry, Chemistry Department, Faculty of Mathematics and Natural Sciences, Hasanuddin University, Tamalanrea, Makassar, South Sulawesi, 90245, Indonesia
[2] Laboratory of Analytical Chemistry, Faculty of Mathematics and Natural Sciences, Hasanuddin University, Tamalanrea, Makassar, South Sulawesi, 90245, Indonesia
[3] Departement of Food and Science Technology, Faculty of Agriculture, Hasanuddin University, Tamalanrea, Makassar, South Sulawesi, 90245, Indonesia

E-mail: hasnahnatsir@unhas.ac.id

Abstract. Alpha-glucosidase is a key enzyme in the final process of breaking carbohydrates into glucose. Inhibition of α-glucosidase affected more absorption of glucose, so it can reduce hyperglycemia condition. The aims of this study is to determine the effectiveness of inhibition wet and dried *Moringa oleifera* leaf extract through α-glucosidase activity in vitro. The effectiveness study of inhibition on the activity of α-glucosidase enzyme obtained from white glutinous rice (*Oryza sativa* glutinosa) was carried out using wet and dried kelor leaf extract of 13% (w/v) with 10 mM α-D-glucopyranoside (PNPG) substrate. A positive control used 1% acarbose and substrate without addition of extract was a negative control. Inhibitory activity was measured using spectrophotometers at a wavelength of 400 nm. The result showed that the inhibition activity against α-glucosidase enzyme of dried leaf extract, wet leaf extract and acarbose was 81,39%, 83,94%, and 95,4%, respectively on pH 7,0. The effectiveness inhibition of the wet Moringa leaf extract was greater than the dried leaf extract. The findings suggest that *M. oleifera* leaf has the potential to be developed as an alternative food therapy for diabetics.

1. Introduction

Glucosidase enzyme is an enzyme that plays a role in the process of complex polysaccharide hydrolysis into glucose that will be distributed into the blood circulation [1]. Reduction in the absorption of carbohydrates from food by the intestines is a therapeutic approach for postprandial hyperglycemia [2,3]. One of the agents that can be used to minimize the increased contents of sugar in the blood is an inhibitor of α-glucosidase (alpha glucosidase inhibitor, AGI). AGI is one of the antidiabetic agents that works in inhibiting α-glucosidase enzymes [4]. During this time, synthetic inhibitors for α-glucosidase such as acarbose have been widely used for the treatment of type II diabetics but reportedly this drug causes various side effects [5,6]. Therefore, a lot of efforts have been made to find the AGI from a natural source to treat diabetes. According to Shibano *et al.* [7], combination of AGI and antioxidant will be more effective in the prophylaxis of type II diabetics.

Moringa oleifera is a plant of the Moringaceae, a family of angiosperms consisting of 14 species [8]. In Indonesia, this plant is easy to cultivate because the climate suitability, so it can grow easily. Currently, Moringa leaves become interested subject of study for researchers because the nutritional content is very high. In composition, the leaves of *Moringa oleifera* contains a varied chemical compound and very complete. Several studies report that leaves M. oleifera contains of vitamins A and E, complete amino acids such as hydroxyproline, aspartic acid, threonine, serine, glutamic acid, proline, glycine, alanine, valine, cysteine, methionine, isoleucine, leucine, tyrosine, phenylalanine, hydroxylysine, ornithine, lysine, arginine, tryptophan, histidine [9-11]. A study on phytochemical analysis by HPLC analysis on M. oleifera showed that there were polyphenols such as quercetin glycosides, kaempferol glycoside, and chlorogenic acid in the powder leaf [12,13]. An ethanol extract of M. oleifera leaf contained flavonoids, tannins, anthraquinones, cardiac glycosides, alkaloids, saponins, triterpenoid, and reducing sugars [12,14].

Various studies on the application of moringa plants have been conducted, among others, Sashidhara *et al.* [15] reported antiinflammatory and anti nociceptive effects of the root part of Moringa plant. Moringa was also reported to have activity as hepatoprotective and antibiotic. The ethanol extract of Moringa leaf combined with 5-fluorouracil was used in chemotherapy of colon cancer cells. Moringa was also reported to have chemopreventive activity, antiinflammatory, antispasmodic, antidiuretic, cholesterol-lowering, antioxidant and antifungal [14-18]. The potential of Moringa leaf as AGI has a very big chance to be developed, referring to the content of antioxidant compounds, amino acids and some other secondary metabolite compounds contained therein. This study was conducted to test the inhibitory activity of moringa leaf extract on the activity of the α-glucosidase enzyme in vitro.

2. Materials and Methods

2.1 Materials
Materials used were dried and wet *M. oleifera* leaves from Soppeng district, a α-glucosidase enzyme from white glutinous rice, p-nitrophenyl α-D-glucopyranoside (PNPG), 1% acarbose, phosphate buffer, distilled water, Whatman 42 filter paper.

2.2 Preparation of M oleifera leaves
M. oleivera leaves were selected from young and fresh leaves from the plantation area. Samples were processed by leaf sorting (fresh young leaf separated from the dirt). The sorted samples were divided into two parts of the wet M. oleivera leaf (WML) and dried M. oleivera leaf (DML).

2.3 M. oleivera extraction
Dried and wet M. oleivera leaves were mashed with a blender, weighed as many as 15 g and then added with 100 mL of phosphate buffer. The mixed solution stirred for 1 hour then the extract of M. oleivera was filtered with a filter paper. Extracts will be used in an inhibitory test against α-glucosidase enzyme activity.

2.4 α-Glucosidase inhibition assay
The α-Glucosidase inhibitory activity was measured using a modification method and adaptation of the method described in Shim *et al.* [19]; Subramanian, Asmawi, and Sadikun [20]. 0.2 mL of enzyme solution was added with a 5 mL buffer of pH 7 into the test tube, and 0.2 mL of M. oleivera leaf extract was added with concentrations of 1%, 5%, 10% and 15% (b/v) on a different tube, 0.5 mL of PNPG was added to a concentration of 10 mM, incubated at 37 °C for 20 min and 8 mL of 100 mM Na_2CO_3 was

added to the mixture. The absorbance was recorded at 400 nm using a spectrophotometer. The optimum concentration was used in the effect study of pH against α-glucosidase inhibition. The negative control used an enzyme solution without the addition of *M.oleifera* leaf extract and positive control was 1% acarbose. The enzyme activity was calculated by equation 1:

$$\text{Enzyme activity}\left(\frac{\text{U}}{\text{mL}}\right) = \frac{\text{Abs sample} - \text{Abs blanko x Total volume analyzed x Volume (enzyme + substrat + buffer)}}{\text{Molar extinction coefficients x Incubation time x Mixtured volume x Enzyme volume}} \quad (1)$$

Abs = absorption

3. Result and discussion

The process of extracting wet and dried moringa leaves using buffer pospat pH 7 to prevent decomposition. The inhibitory test against α-glucosidase enzyme is done by variation of concentration of extract of wet and dried moringa leaves. The concentration variations used were 1%, 5%, 10% and 15%. This is intended to determine the effect of wet and dried leaves on the inhibition of α-glucosidase enzyme activity. Table 1 shows the activity percentage and the inhibition percentage of M. oleifera on α-glucosidase.

Table 1. Value of α-glucosidase activity, percentage of activity and percentage of inhibition on α-glucosidase enzyme.

Sample	α-glucosidase activity (mU/mL)	% Activity	% Inhibition
Negative Control (NC)	158.07	100	0
Acarbose 1% (PC)	6.85	4.64	95.36
DML 1%	157.17	99.43	0.57
DML 5%	141.05	89.23	10.77
DML 10%	94.71	59.91	40.09
DML 15%	29.42	18.61	81.39
WML 1%	142.66	90.25	9.75
WML 5%	108.81	68.84	31.16
WML 10%	72.94	46.15	53.85
WML 15%	25.39	16.06	83.94

The effects of the dried and wet M. oleifera extract on the inhibition of the α-glucosidase enzyme is given in figure 1.

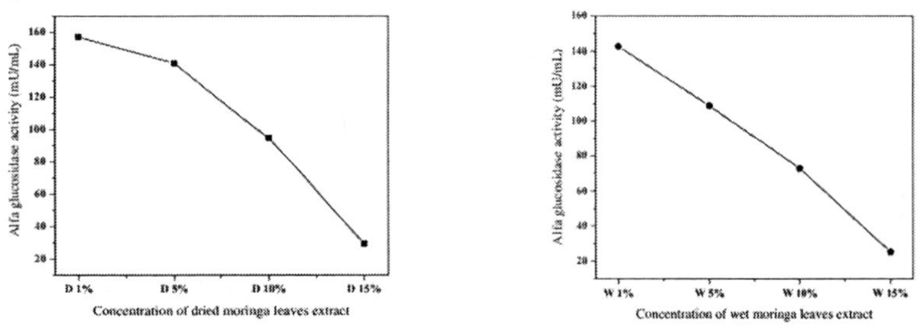

Figure 1. The effect of the concentration of dried and wet M.oleivera leaves extract on the inhibition effectiveness of the α-glucosidase enzyme.

It is clear that the higher the concentration of the extract the higher the activity of inhibition enzyme will be. The extract of DML has the lowest activity inhibition i.e. 0.57 % with the activity of 157.17 mU/mL and the activity percentage of 99.43% at the concentration of 1%. For WML, the lowest activity inhibition is 9.7% with the activity of 142.66 mU/mL and the activity percentage of 90.25% at the leaf concentration of 1%. The optimum activity of dried and wet M. oleifera leaves on the α-glucosidase enzyme was 29.42 mU/mL (18.61%) and 25.39 mU/mL (16.06%), respectively at the concentration of the extract of 15%. Based on the value of α-glucosidase inhibition obtained at a different percentage, it is clear that the M. oleivera leaf extract has the potential of anti-hyperglicemia activity by inhibiting the enzyme α-glucosidase in the brush border of intestinal smooth. Inhibition of the α-glucosidase enzyme decreased the rate of digestion of carbohydrates into monosaccharides that can be absorbed by the intestine.

The pH effect on the activation and inhibition of α-glucosidase by dried and wet M. oleivera leaves extract was studied using buffer phosphate solutions of pH 6.5 and 7.0. Table 2 shows the α-glucosidase activity, the activity percentage and the inhibition percentage on the α-glucosidase enzyme at the used pH.

Table 2. α-glucosidase activity, percentage of activity and percentage of inhibition of α-glucosidase enzyme at pH 6.5 and 7.0.

Sample	α-glucosidase activity (mU/mL)	% Activity	% Inhibition
Negative control (NC), 6.5	153.79	100	0
DML 15%, 6.5	45.94	29.87	70.13
WML 15%, 6.5	29.42	19.13	80.17
Acarbose 1% (PC), 6.5	7.25	7.89	92.11
Negative control (NC), 7.0	157.30	100	0
DML 15%, 7.0	29.47	18.70	81.30
WML 15%, 7.0	25.79	16.40	83.60
Acarbose 1% (PC), 7.0	1.61	5.48	94.52

The effectiveness of inhibition of wet M.oleifera leaf extract is higher than that of dried M.oleifera leaf extract. This is occured because the heat used for drying process of M.oleifera leaves caused oxidation or decomposition of active compounds contained in the sample [21][22]. According to Adisakwattana and Chanathong [23], the inhibition ability of M.oleifera leaf against α-glucosidase enzymes showed a good anti hyperglycemia activity, so that it is very potential to be developed. An intake of M. oleivera leaf may delay glucose absorption to blood circulation in pre-diabetic patients that help to prevent the increase of type 2 diabetes [23].

4. Conclusion

The optimum concentration of the dried and wet M.oleifera leaves extract in inhibiting the activity of the α-glucosidase enzyme was 15% with the inhibition percentage of 81.39% and 83.94%, respectively. The effective pH in the process of inhibition was 7.0. There was quantitatively the difference activity between wet and dried leaves. The degradation factor of the chemical compound composition due to sample drying has a significant effect on the decrease of inhibition activity on the α-glucosidase enzyme. Besides of the nutritional composition of the M. oleifera leaf, the effectiveness of inhibition on the α-glucosidase enzyme showed that M. oleifera is potential to be used as a natural α-glucosidase inhibitor (AGI).

Acknowledgments

Authors gratefully acknowledge Hasanuddin University for fully supported in this research through Unggulan UNHAS Research Grant 2017.

References

[1] Ben Gara, Amel et al. 2017. "Inhibitory Activities of *Cystoseira Crinita* Sulfated Polysaccharide on Key Enzymes Related to Diabetes and Hypertension: *In Vitro* and Animal Study." *Arch. Physiolog Biochem.* 123(1): 31–42. https://www.tandfonline.com/doi/full/10.1080/13813455.2016.1232737

[2] Mbikay, Majambu. 2012. "Therapeutic Potential of Moringa Oleifera Leaves in Chronic Hyperglycemia and Dyslipidemia: A Review." *Frontiers in Pharm.* 3 MAR(March): 1–12

[3] Safithri, Mega, and Puspita Sari. 2017. "Release Profile and Inhibition Test of The Nanoparticles A. Paniculata Extract as Inhibitor of α-Glucosidase in The Process of Carbohydrates Breakdown Into Glucose Diabetes Mellitus Release Profile and Inhibition Test of The Nanoparticles A . Paniculata." *IOP Conference Series: Materials Science and Engineering,* 214(1). http://iopscience.iop.org/article/10.1088/1757-899X/214/1/012026/ meta;jsessionid=2C387E6344A796737DD2770310FED12A.ip-10-40-1-105

[4] Sulistiyani, Mega Safithri, and Yoana Puspita Sari. 2016. "Inhibition of α-Glucosidase Activity by Ethanolic Extract of Melia Azedarach L. Leaves." *IOP Conference Series: Earth and Environmental Science* 31: 12025. http://stacks.iop.org/1755-1315/31/i=1/a=012025?key=crossref.94f5a610d827f4b9fde55405277e70d0

[5] Rengasamy, Kannan R.R. et al. 2017. "Cuparane Sesquiterpenes from Laurencia Natalensis Kylin as Inhibitors of Alpha-Glucosidase, Dipeptidyl Peptidase IV and Xanthine Oxidase." *Algal Res.* 25(May): 178–83

[6] Sun, Wenlong et al. 2017. "Baicalein Improves Insulin Resistance via Regulating SOCS3 and Enhances the Effect of Acarbose on Diabetes Prevention." *J. Funct. Foods* 37: 339–53. http://linkinghub.elsevier.com/retrieve/pii/S1756464617304693

[7] Shibano, M et al. 2008. "Antioxidant Constituents in the Dayflower (Commelina Communis L.) and Their Alpha-Glucosidase-Inhibitory Activity." *J Nat Med* 62: 691–701. http://journal.ipb.ac.id/index.php/jtip/article/view/7712

[8] Road, Sardar Patel. 2017. Nutritional Quality of Moringa Oleifera for Its Bioactivity and Antibacterial Properties. *Int. Res. Food J.* 24(April): 825–33.

[9] Ching, L S, and S Mohamed. 2001. "Alpha-Tocopherol Content in 62 Edible Tropical Plants." *J. Agric. Food Chem.* 49(6): 3101–5

[10] Al Juhaimi, Fahad et al. 2017. "The Biochemical Composition of the Leaves and Seeds Meals of Moringa Species as Non-Conventional Sources of Nutrients." *J. Food Biochem* 41(1): 1–7.

[11] Lalas, Stavros et al. 2017. "Nutritional Characterization of Leaves and Herbal Tea of *Moringa Oleifera* Cultivated in Greece." *J. Herbs, Spices & Med. Plants* 0(0): 1–14. https://www.tandfonline.com/doi/full/10.1080/10496475.2017.1334163

[12] Perumal Siddhuraju and Klaus Becker. 2003. "Antioxidant Properties of Various Solvent Extracts of Total Phenolic Constituents from Three Different Agroclimatic Origins of Drumstick Tree (Moringa Oleifera Lam.) Leaves",*J. Agric. Food Chem.*, 51 (8), Pp 2144–2,2003." *J. Agric. Food Chem* 51: 2144–55

[13] Valdez-Solana, Monica A. et al. 2015. "Nutritional Content and Elemental and Phytochemical Analyses of Moringa Oleifera Grown in Mexico." *J. Chem* 2015

[14] Alhakmani, Fatma, Sokindra Kumar, and Shah Alam Khan. 2013. "Estimation of Total Phenolic Content, in-Vitro Antioxidant and Anti-Inflammatory Activity of Flowers of Moringa Oleifera." *Asian Pacific. J. Tropical Biomed.* 3(8): 623–27

[15] Sashidhara, Koneni V. et al. 2009. "Rare Dipeptide and Urea Derivatives from Roots of Moringa Oleifera as Potential Anti-Inflammatory and Antinociceptive Agents." *European J. Med. Chem.* 44(1): 432–36. http://dx.doi.org/10.1016/j.ejmech.2007.12.018

[16] Farooq, Fozia et al. 2012. "Medicinal Properties of Moringa Oleifera: An Overview of Promising Healer." *J. Med. Plants. Res* 6(27): 4368–74. BMC Complementary and Alternative Medicine.

[17] Kumar, E, Kn Harsha, and Shabana Shaik. 2013. "Of In Vitro Antioxidant Activity and In Vivo Hepatoprotective Activity Of Moringa Oleifera Seeds Extract Against Ethanol Induced Liver

Damage In Wistar Rats." *Iosr J Pharmacy* 3(1): 10–15http://www.researchgate.net/publication/259173297_In_vitro_antioxidant_activity_and_in_vivo_hepatoprotective_activity_of_aqueous_extract_of_Allium_cepa_bulb_in_ethanol_induced_liver_damage_in_Wistar_rats/file/9c96052de85b71d1b6.pdf

[18] Pari, L, and N Ashok Kumar. 2002. "Hepatoprotective Activity of Moringa Oleifera on Antitubercular Drug-Induced Liver Damage in Rats." *J.Med. Food* 5(3): 171–77

[19] Shim, Young Jun et al. 2003. "Inhibitory Effect of Aqueous Extract from the Gall of Rhus Chinensis on Alpha-Glucosidase Activity and Postprandial Blood Glucose." *J. Ethnopharmacology* 85(2–3): 283–87

[20] Subramanian, Rammohan, M Zaini Asmawi, and Amirin Sadikun. 2008. "In Vitro Alpha-Glucosidase and Alpha-Amylase Enzyme Inhibitory Effects of Andrographis Paniculata Extract and Andrographolide." *Acta biochim Pol* 55(2): 391–98

[21] Blessy, M., Ruchi D. Patel, Prajesh N. Prajapati, and Y. K. Agrawal. 2014. "Development of Forced Degradation and Stability Indicating Studies of Drugs - A Review." *J. Pharm. Analysis* 4(3): 159–65

[22] Sheraz, Muhammad Ali et al. 2014. "Photo, Thermal and Chemical Degradation of Riboflavin." *Beilstein J. Org. Chem.* 10: 1999–2012

[23] Adisakwattana, S. and Chanathong, B., 2011. Alpha-glucosidase inhibitory activity and lipid-lowering mechanisms of Moringa oleifera leaf extract. *Eur Rev Med Pharmacol Sci*, *15*(7), pp.803-8

Study of characteristic of tsunami base on the coastal morphology in north Donggala, Central Sulawesi

W S N Rahmadaningsi[1], **A H Assegaf**[1], **W Setyonegoro**[2] **and Paharuddin**[1]

[1] Study Program of Geophysics, Faculty of Mathematics and Natural Sciences, Hasanuddin University
[2] Research and Development Center Meteorological Climatological and Geophysical Agency, Indonesia 10720

E-mail: ningsiataka@gmail.com

Abstract. The northern arm of Sulawesi potentials to generate earthquake and Tsunami due to the existence of subduction zone in sulawesi sea. It makes the North Donggala as an area with active seismicity. One of the earthquake and Tsunami events occurred is the earthquake and tsunami of Toli-Toli 1996 (M 7.9) causing 9 people are killed and severe damage in Tonggolobibi, Siboang, and Balukang. This earthquake induced tsunami runup of 3.4 m and inundated as far as 400 meters. The aims of this study is to predict runup and inundation area using numerical model and to find out the characteristics of Tsunami wave on straight, bay and cape shape coastal morphology and slopes of coastal. The data in this research consist of are the Etopo2 bathymetry data in data obtained from NOAA (National Oceanic and Atmospheric Administration) , Toli-toli's main earthquakes focal mechanism data 1st January1996 from GCMT (Global Centroid Moment Tensor) , the data gained from the SRTM (Shuttle Radar Topography Mission) data 30 m and land cover data in 1996 from Ministry of environment and forestry . Single fault model is used to predict the high of tsunami run-up and to inundation area along Donggala coastal area. Its reviewed by morphology of coastal area that higher run up shows occurs at coastline type like bay have higher run up compare to area with cape and straight coastline. The result shows that the slopes have negative or contras correlation with Tsunami runup and its inundation area.

1. Introduction

The coastal area is susceptible to inundation of natural disasters like tsunami, storm surges, etc. The propagation of tsunami is related to the morphological bathymetry. There are parameters for coastal morphology such as bays, Capes, straight shoreline, curved shoreline, gentle, stib coastal, etc.

The north arm is a region with highest seismicity especially northern coast, south and eastern part of the norh arm. There were occurred earthquakes and tsunami in 1968 and 1996 along north arm which is an area of seismic prone with highest activity. All phenomena of the shallow earthquakes generated by the motion of Palu-koro fault, pool of north arm Sulawesi and Spreading canter north Makassar strait [1]. In Toli-toli (1996) happened earthquake and tsunami (M 7.9), which suffered 9 victims, and physical defect in Tonggolobibi, Siboang, and Balukang [2]. The earthquake also generated tsunami with 3.4 meters level and inundation 400 meters [3]. When 1996 tsunami in toli-toli is caused by subduction zone in Sulawesi Sea based on previous study. The earthquake is caused by palu-koro fault but it's not true because palu-koro fault is moving horizontally.

Therefore, this research focus on numerical modelling of tsunami in west coast north Donggala, that generated by tectonic earthquakes in which can be mapping by considering the coastal morphological impact due to tsunami. This research is mean to the mitigation of tsunami in the coastal area because from the previous research because there were only view research about the correlation of tsunami due to coastal morphology and inundation area in Toli-toli's tsunami 1996. Finally this research trying to developed the previous research with to make runup and inundation using numerical modelling and characterize tsunami due to the coastal morphology, bay, cape, straight, slope, etc.

2. Method

This research located in -0.350 S-0.7040 N to 119.760-119.990 E, west coast northern part of Donggala. The data in this research consist of are the Etopo2 bathymetry data in data obtained from NOAA (National Oceanic and Atmospheric Administration) , Toli-toli's main earthquakes focal mechanism data 1st January1996 from GCMT (Global Centroid Moment Tensor) , the data gained from the SRTM (Shuttle Radar Topography Mission) data 30 m and land cover data in 1996 from Ministry of environment and forestry. There are three step in this research as follows:

2.1.Runup simulation using single fault method

In this step, the runup simulation calculation of fault dimension will be done by observe the distribution of main shock, foreshock, and aftershock [4]. Empirically, the fault dimension can be known by Papazachos et al, (2004) equation [5], as follows:

$$Log\ Lo = 0.55\ Mw - 2.19 \tag{1}$$

$$Log\ Wo = 0.31\ Mw - 0.63 \tag{2}$$

The calculation of Wide field fault using this equation below :

$$A = Lo\ x\ Wo \tag{3}$$

Where Lo is fault length (m), Wo is width of fault (m), and A is wide field fault (m^2).

Seismic moment is the amount of energy that released in earthquakes event and it can be determine using equation below [6] :

$$Log\ M0 = (6.08 + Mw) * 3/2 \tag{4}$$

Dislocation of water body define using equation below [9]

$$S = \frac{Mo}{\mu A} \tag{5}$$

Where Mo is seismic moment (Nm), μ is Rigidity (stiffness object, the harder the object is the energy required to move it greater, meaning greater seismic moment) (N/m2), S is Dislocation (m) and Mw is moment magnitude (Mw).

This calculation result inputted to L-2008 software by Mamoru Nakamura to illustrate vertical displacement and runup tsunami.

2.2. Simulation of tsunami Inundation. The simulation of tsunami using Arcgis 10.1 software with the Raster calculator functions. Hills, J. G. & Mader, C. L. 1997 [7], formulated an equation to calculated the runup tsunami as follows :

$$Xmax = (Hs)^{1.33} * n^{-2} * k \cos S \tag{6}$$

Where Xmax is maximum Inundasi (m), Hs is maximum tsunami run-up (m), k is constants corresponding to 0.06 for many tsunamis, n is Manning coefficient of land covers and S is slope of land.

3. Result and explaination

3.1 Runup simulation using single fault method
The calculation result from of the empirical approach (table 1) inputted in L-2008 software. Output of that process is visualization of vertical displacement.

Table 1. Empirical approaches with Scaling law.

Mw	L (km)	W (km)	A(km^2)	A (m^2)	Mo (Nm)	μ(N/m^2)	S(m)
7.9	142.8894	65.91739	9.42E+03	9.42E+09	9.33254E+20	3.00E+10	3.30E+00

In this study an earthquake with a large 7.9 Mw, source modeling of tsunami Toli-Toli use single fault model with fault parameter obtained by Donald L. Well & Kevin J. Coppersmith, 1994 scalling law. We get fault dimension with length 142.8894 km, width 65.91739 km, and 3.3 m for maximum displacement of Toli-Toli tsunami 1996.

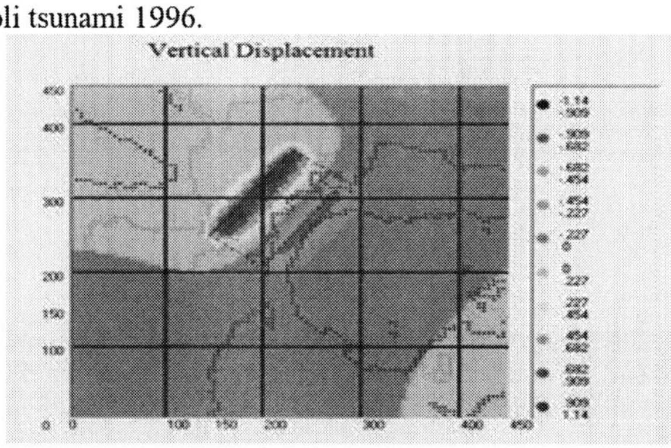

Figure 1. Vertical displacement of tsunami modelling.

Results from the model design (source modeling) of vertical displacement, which shows deformation occurring on the sea bottom. Variation of vertical displacement value for this event describes the variation of the value of the maximum vertical displacement for this event is 1.14 meters and the minimum vertical displacement value is -1.14 meters from maximum vertical displacement value (+) show the rising of bottom sea. In the other hand, the minimum vertical displacement values (-) show the dropping in of sea floor (figure 1) [8]. Strike related to the fault direction. When the strike lead to the area with shallow morphology cause higher level of tsunami. In the other hand, if strike direction lead to the area with deep of bottom sea morphology cause lower level of tsunami [9]. This is understandable, because the vertical movement of the sea bottom can cause changes in the mass of water over the sea bottom in motion. If the upper oceans rise (uplift) or descend rapidly in response to an earthquake, it will rise and fall sea water on a large scale, from sea bottom to surface. The value of strike has a relationship with the direction of the fault. When this Strike leads to a more morphological area shallow the tsunami run-up is higher whereas if the direction of this strike leads to the morphological area of the seafloor in the small tsunami height.

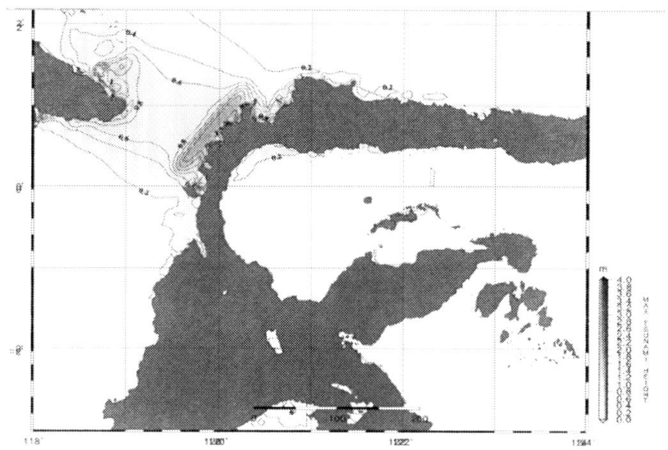

Figure 2. Tsunami propagation model.

In tsunami propagation model figure 2) until Maximum time that choosen is during 3600 second, this parameter useful to define the maximum time that used to see propagation time of the tsunami. The model output can describe tsunami propagation during maximum time which we choosen before. Its data can be saved in propagation wave form every minutes, so that we can know when tsunami wave arrived to continent [3].

Table 2. Runup distribution result in few point from tsunami source modelling.

Longitude	Latitude	H	Point
120.2374	0.895056	1.98	Simuntu
120.2449	0.820917	2.77	Dongko
120.2044	0.773473	2.1	Soni
120.0367	0.585327	2.7	Balukang
120.0494	0.520936	2.1	Siwalempu
120.0354	0.502564	3.52	Siboang
120.0032	0.482956	2.39	Taipah 3
119.9978	0.48295	3.31	Taipah 2
119.9976	0.482281	3.24	Taipah 1
119.9917	0.481658	2.71	Taipah (river Mouth)
119.9781	0.480986	3.51	Tonggolobibi 3
119.9759	0.475939	3.21	Tonggolobibi 2
119.9578	0.475631	2.74	Tonggolobibi 1
119.9478	0.474539	2.78	Limbosu
119.9311	0.479928	2.72	Pangalaseang 2
119.9071	0.480939	2.39	Pangalaseang 1
119.8851	0.423511	2.32787	Tonggoloibi 4
119.8869	0.390006	2.09888	Tonggoloibi 5
119.8636	0.368844	2.09402	lemba Mukti 1
119.8452	0.342183	2.25418	lemba Mukti 2

Runup value in the map of impact area above show in table tsunami affected areas map above (table 2).

IOP Conf. Series: Journal of Physics: Conf. Series **979** (2018) 012020 doi:10.1088/1742-6596/979/1/012020

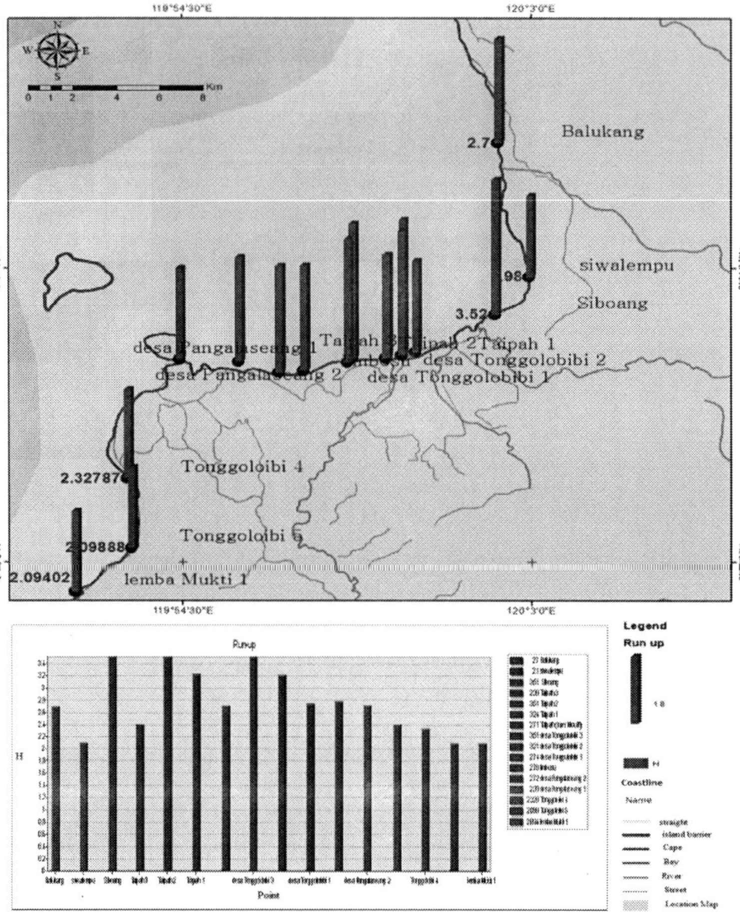

Figure 3. Map of run up tsunami distribution in for each village.

The simulated track record (figure 3).The simulation results show that if an area is near to the tsunami source, the region will experience a tsunami in the first wave, this is because the distance is less than 46 km so it can be categorized as near field or local a tsunami that is a tsunami occurring where the distance between the source and the tsunami generated by the coast is very near. It is likely that the area around the earthquake felt or even damaged the building, based on tsunami data of the area that had severe damage ie the villages of Tongolobibi, Balukang and Siboang. These areas are areas near to fault summers, so it is important to conduct disaster mitigation efforts in this area if a tsunami is likely to occur at any time in the region. The results show that the area of the bay-shaped coastline, based on the wave theory, states that the wave region will converge or overcome the tsunami or maturity of the bay due to a resonance process in which the incoming wave meets the reflected wave causing the area the bay becomes strong, the ocean waves reflect each other and each other or a large wave in a straight line with a straight coastline and a headland. This research also shows the highest run-up is located at the end of the bay around it there is a river mouth like the village Tonggolobibi.

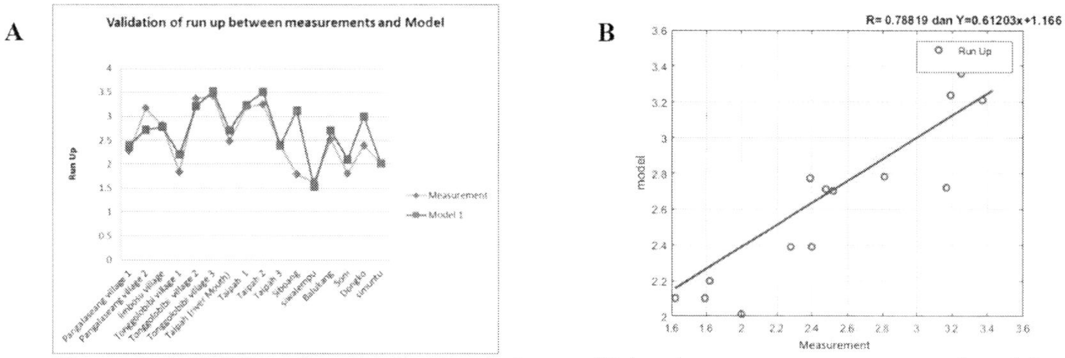

Figure 4. A.Validity of model. B. Correlation coefficient between measurements and model.

To measure the validity of model, the correlation coefficient calculated in matlab. From calculation show the correlation coefficient value between measurement and model about 0.78 (figure 4).

3.2 Inundation Simulation

Research for simulation of extents inundation Toli-toli's for tsunami have not been there before. Application of geographic information system to create tsunami inundation area of Toli-toli 1996 by utilizing. When tsunami arrive in land, it propagation determine by slopenes of coastal morphology, where in the steep coastal, the effect of tsunami will not too far reach the land because it stuck and reflect back the coastal cliff, while in the sloping coastal, effect of tsunami can be reach few kilometres to the land. Simulation result show that inundation area affected by land topography, higher land topography it cause lower inundation area and otherwise. The simulation show inundation area with runup value about 3.52 is 598 Ha (figure 5) and Inundation simulation analysis results show that tsunami inundation will travel far to the more gentle and will turn if it encounters high topography and low inundation area on terrestrial vegetation in the form of forest both primary forest, secondary forest and mangrove forest and the greater the extent of inundation if passing land with fine vegetation in this area including vacant land or dry land, water bodies and ponds.

Figure 5. inundation simulation with morphological condition.

4. Conclusion

Based on this research can be concluded that tsunami runup with 7.9 Mw will produce highest runup about 3.52 meters and inundation area about 286300 m^2. From simulation, describe the shoreline with bay shape have runup higher than straight and cape shape. Also, more sloping an area then the tsunami will higher, this case, have opposite relation with runup and inundation area or negative correlation with coastal characteristic.

Acknowledgments

I would like to express my thankfulness to supervisor Dr. Alimuddin Hamzah M.Eng, Wiko Setyonegoro, S.Si, M.Kom and Dr. Paharuddin M.Si. Special to Ikawati Basri and Akramunnisa for kindly helping me translate and Sulastri S.Si, M.Si and Another special to Dr. Mamoru Nakamura for that given me user manual of tsunami L-2008 software.

References

[1] Baeda Y A 2011 Seismic and Tsunami Hazard Potential in Sulawesi Island, Indonesia *Journal of International Development and Coorperation* vol 17 No. 1, pp 17-30

[2] Pelinovsky E, Yuliadi, D, Prasetya G, and H Rahman 1997 The January 1, 1996 Sulawesi Island Tsunami *Natural Hazards* **B16**

[3] S. Aswad, et al. 2013 Numerical Simulation Of The 1996 Toli-Toli Tsunami And The 2000 Peleng Tsunami Central Of Celebes With Single Fault Model, *Proceedings of the 7th International Conference on Asian and Pacific Coasts* (APAC 2013) Bali, Indonesia, September 24-26

[4] Suparto, Eka T P and Surono 2006 *Katalog gempabumi merusak di Indonesia tahun 1629-2006* No.3

[5] Papazachos B C, Scordilis E M, Panagiotopoulos D G, Papazachos CB, and Karakaisis G F 2004 Global Relations between Seismic Fault Parameters and Moment Magnitude of Earthquakes Bull. Geol. Soc. Greece, Vol. XXXVI. *Proceedings of the 10 International Congress*, Thessaloniki, April 2004Th

[6] Hanks, Thomas and Kanamori H 1979 Moment magnitude scale *Journal of Geophysical Research,* 84 (B5): 2348–2350. Retrieved 2007-10-06

[7] Hills J G & Mader C L 1997 Tsunami producted by the impacts of the small asteroids. *Annals of the New York Accademy of Sciences* 822

[8] Nakamura M 2006 Source fault model of the 1771 Yaeyama tsunami- Southern Ryukyu island Japan inferred from numerical simulation, *Pure Appl Geophys* no 163

[9] Setyonegoro W 2011 Tsunami Numerical Simulation Applied To Tsunami Early Warning System Along Sumatra Region *Jurnal Meteorologi dan Geofisika* Vol. 12, No. 1Smid, Op.cit., pp 3

Dislocation, crystallite size distribution and lattice strain of magnesium oxide nanoparticles

I W Sutapa[1], Abdul Wahid Wahab[2], P Taba[2] and N L Nafie[2]

[1] Department of Chemistry, Faculty Mathematic and Natural Science, University of Pattimura Jl. Ir. Putuhena Poka 97233, Ambon, Maluku, Indonesia
[2] Department of Chemistry, Faculty Mathematic and Natural Science, University of Hasanuddin, Jl. Perintis Kemerdekaan 90245, Makassar, Indonesia

E-mail: wahidwhb@yahoo.com

Abstract. The oxide of magnesium nanoparticles synthesized using sol-gel method and analysis of the structural properties was conducted. The functional groups of nanoparticles has been analysed by Fourier Transform Infrared Spectroscopy (FT-IR). Dislocations, average size of crystal, strain, stress, the energy density of crystal, crystallite size distribution and morphologies of the crystals were determined based on X-ray diffraction profile analysis. The morphological of the crystal was analysed based on the image resulted from SEM analysis. The crystallite size distribution was calculated with the contention that the particle size has a normal logarithmic form. The most orientations of crystal were determined based on the textural crystal from diffraction data of X-ray diffraction profile analysis. FT-IR results showed the stretching vibration mode of the Mg–O–Mg in the range of 400.11-525 cm^{-1} as a broad band. The average size crystal of nanoparticles resulted is 9.21 mm with dislocation value of crystal is 0.012 nm^{-2}. The strains, stress, the energy density of crystal are 1.5×10^{-4}; 37.31 MPa; 0.72 MPa respectively. The highest texture coefficient value of the crystal is 0.98. This result is supported by morphological analysis using SEM which shows most of the regular cubic-shaped crystals. The synthesis method is suitable for simple and cost-effective synthesis model of MgO nanoparticles.

1. Introduction

Oxide materials such as MgO nanoparticles have attracted some researchers for a variety of unique properties owned by this material. In general, the properties of the solid material can be determined using computational methods or conventional (laboratory method) [1–3]. Various methods can be used to produce these nanoparticles in accordance with the raw materials and processes used such as through the process of emulsion with assisted by microwave [4], using a hydrothermal process [5], through the process of co-precipitation [6], and through the formation of sol and followed by gel formation [7].

The results of nanoparticles synthesis obtained should be conducted the analysis of advanced characteristics. Characterization of further properties of nanoparticles is needed to ensure the success of the synthesis process as well as other characters possessed by these materials. The results of calculating characteristic values such as dislocations and microstructures of MgO nano crystals are important factors in determining the physical properties to ensure the success of the formation process of these nanomaterials [8]. Various types of instrument have been used to define the microstructure of solid nanomaterials and to understand the character of nanomaterials [9]. X-ray diffraction is one of the most effective instruments for estimating average crystal size, dislocations and microstructures. Estimated values are based on FWHM data (full width at half maximum), intensity obtained and result of integration analysis of X-ray diffraction profiles. The stress, strain, and energy density values possessed by the crystals were calculated using the Williamson-Hall analysis method [10,11].

The crystal size distribution parameter can be illustrated using the calculation of the log-normal function from the angle of X-ray diffraction results from nanocrystallite. In addition, to know the formation of functional groups contained in the resulting nanoparticles is required FT-IR (Fourier Transmittance Infra-Red) instrument that will provide information about the functional groups of the compounds contained. The best authors know, based on the literature study, there has been no report on the analysis of nanoparticles synthesis results using methanol for the sol-gel formation process and calcination process at high temperatures.

In this paper, the analysis and determination of microstructural properties, crystal value textures, the size of crystal, the distribution of size and the dislocation of structures, which occur during the calcination process to produce nanocrystalline MgO materials have been studied by X-ray diffraction. It is then supported by FTIR analysis to know the functional group changes contained in the resulting nanoparticles and SEM to analyse the morphology of the crystals.

2. Experimental

2.1 Materials
Oxalic acid dihydrate (Merck, >98 % purity), Magnesium acetate tetra hydrate (Merck, 99.5% purity), and methanol (Merck, 99.9% purity) were used in the synthesis of the MgO nanoparticles.

2.2 Synthesis of MgO nanoparticles
Firstly, magnesium acetate tetrahydrate is dissolved using methanol followed by constant stirring until a clear mixture is produced. The mixture is then adjusted to a pH value of 5 using 0.1 M oxalic acid and stirred vigorously until a white gel is formed. The resulting gel is left for 12 hours for further gelation process then dried in an oven at 200°C for 24 hours. Subsequently, the resulting solid is crushed and sieved to produce a magnesium oxalate complex powder[12]. Then, it was calcined at 950 °C for 6 hours in the furnace to produce MgO nanoparticle crystals. The crystals obtained are then characterized by FT-IR, X-ray diffraction and SEM.

3. Results and discussion

3.1 FTIR Analysis
The analysis of functional groups present on MgO nanoparticle materials was scanned in the range of 350 cm^{-1} to 4000 cm^{-1}. The results of the analysis show that the broad vibration band has seen in the range of wavenumber 3250-3550 cm^{-1} resulted by stretching vibration from the O-H group of water molecules absorbed into the solid and hydroxyl groups on the solid surface. This is due to the adsorption of water molecules in the air to the surface of MgO when the nanomaterial is in the free air.

In addition, the wave numbers appearing at 1018.24-1062.51 cm^{-1} and 1225-1512 cm^{-1} are generated by the bending vibration of the hydroxyl group (-OH) water molecules absorbed on the surface (-OH). While the functional groups of the Mg-O-Mg compound bonds were seen in the range 400.11-525 cm^{-1} as wide band. The absorption peaks of FT-IR seen at wavenumber 1417.68 cm^{-1} were assigned to asymmetric stretching of carbonate ion species (CO_3^{-2}).

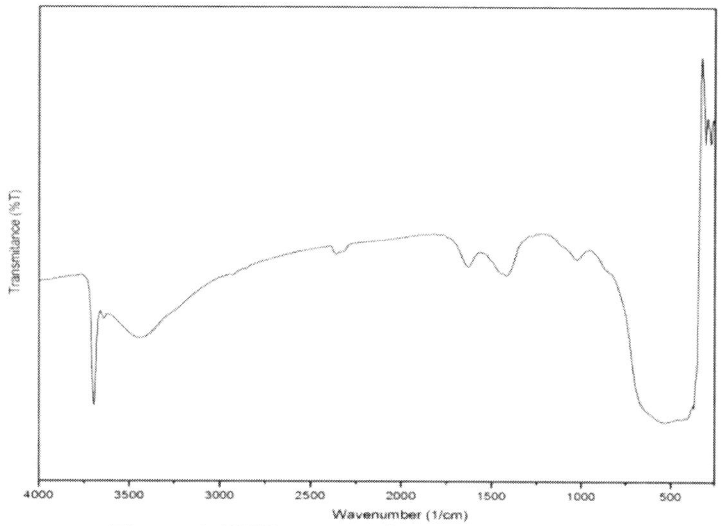

Figure 1. FTIR spectra of MgO nanoparticles.

3.2 XRD Analysis

Further nanoparticle analysis was performed using XRD. Results of XRD difractogram of calcined MgO nanoparticles at 950 °C for 6 hours are shown in Figure 2. Based on the difractrogram there are several peaks that characterize MgO nanoparticles. The presence of a strong and sharp diffraction peak lies in the values of 2θ 38.73°, 43.09°, 55.82°, 62.44°, 74.79° and 78.73° which have Miller Index (1 1 1), (2 0 0) , (2 2 0), (2 2 0), (3 1 1), and (2 2 2) are derived originally from MgO nanoparticles, whereas the diffraction peak lies in the value of 2θ 37.13° derived from $Mg(OH)_2$ which is a hydroxyl group absorbed in the solid material.

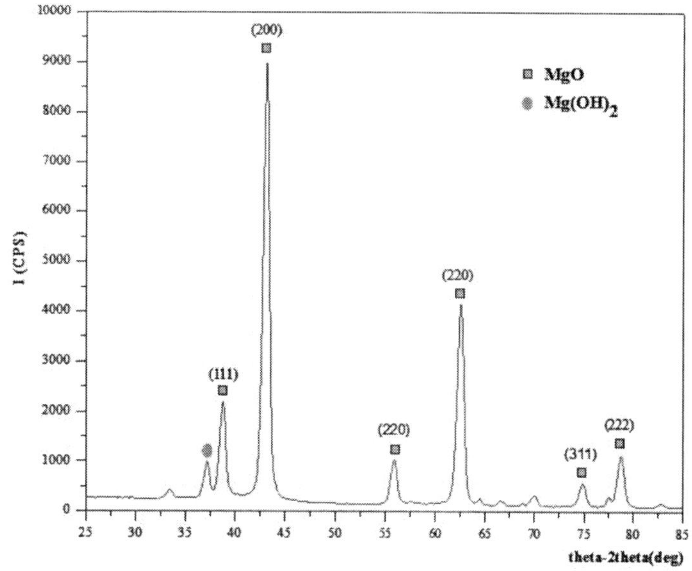

Figure 2. X-ray diffraction pattern of the MgO nanoparticle calcined at 950°C 6 hours.

3.3 Determination average the particle size and the dislocation density

To determine the average particle size and density of nanoparticle dislocations based on XRD analysis results was done using the following Debye-Scherer equation 1 [13].

$$D_{ave} = \frac{K\lambda}{\beta cos\theta} \quad (1)$$

D_{ave} is the average crystallite size of the particles, K is Debye – Scherer's constant (0.94), λ is the wavelength of the CuK_α radiation (0.154 nm), β is the full width half maximum (FWHM) of the peak, θ is the Bragg's angle samples. The average crystal size obtained for the synthesis of MgO nanoparticles at a temperature of 950 °C 6 h is 9.21 nm. Based on the results obtained shows that the synthesis process with sol-gel method and continued with the calcination process in this procedure can produce nanoparticles with average size lower than 10 nm.

Based on the average particle size values obtained from equation 1, the dislocation density value of the MgO nanoparticles can be determined. The dislocation density (δ) can be calculated using equation 2 [14]. The dislocation density value (δ) denotes the number of dislocation lines per unit volume of crystal, which is the size of the crystal defects possessed by a crystal. In other words, the dislocation density value will illustrate the degree of crystallinity of the nanoparticles profile.

$$\delta_{np} = 1/D^2 \quad (2)$$

The result of the dislocation density (δ_{np}) of the MgO nanoparticles is 0.012 $(nm)^{-2}$. Based on the results of this calculation is known that the dislocation density (δ_{np}) of the nanoparticles obtained in this study is quite small. Small dislocation density value is indicates that MgO nanoparticles have been produced had a high degree of crystallinity.

3.4 Crystallite Size Distribution

Crystallite Size Distribution (CZD) of a crystal is determined using the assumption that the crystal particles have a size distribution with a normal distribution in all planes. The CZD function is defined as equation 3 [11].

$$f(y) = \frac{1}{\sqrt{2\pi}\,\sigma}\frac{1}{y}exp\left\{-\frac{[ln(y/m)]^2}{2\sigma^2}\right\} \quad (3)$$

The y value is the size of a crystallite from the size distribution, σ *is* the variance and m is the median of the size distribution function. The crystallite size distribution function of the MgO nanoparticles was calcined at temperature 950 °C is plotted in Figure 3. The resulting CZD function can also be used to determine the area and the volume of the mean of crystallite size. The area, volume means crystallite size is obtained on equation 4 and 5[8].

$$\langle y \rangle_{area} = m\,\exp(2.5\,\sigma^2) \quad (4)$$

$$\langle y \rangle_{volume} = m\,\exp(3.5\,\sigma^2) \quad (5)$$

The MgO nanoparticles have particle size distributions lower than 12.5 nm. The median values and variance of the crystal size distribution based on CZD function calculations are 9.21 and 0.72 respectively. Using the median and the variant obtained the value of the area and the volumes of the crystal are 34.08 and 57.51 nm respectively. Referring to the value of crystal volume is 57.51 nm, it indicates that the amorphous portion of the specimen does not give impact to the overall material properties or nanocrystalline properties produced has a good crystal form. This is also supported by XRD results showing that the resulting diffractogram with a sharp peak and a flat background [8].

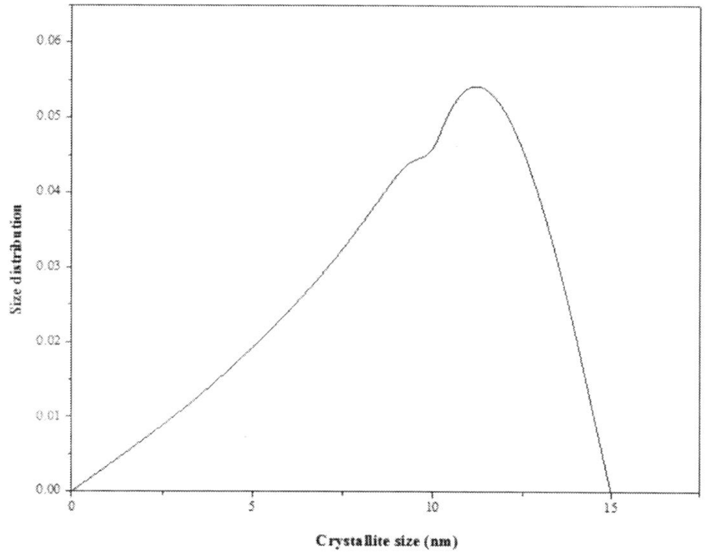

Figure 3. Crystallite size distribution functions of MgO nanoparticles.

3.5 Determination of the stress of crystal (σ)

The strain calculation of nanoparticle crystals resulted conducted using the Williamson-Hall equation is modified into a model of uniform voltage deformation (USDM). In USDM, the deformation of the lattice strain in all directions of the uniform crystal plane. In addition, the Hook Law equation used only maintains linear proportionality between stress and strain ($\sigma = \varepsilon Y$), σ is the crystal voltage and Y is the Young (or) modulus of elasticity, β_i is an instrumental FWHM.

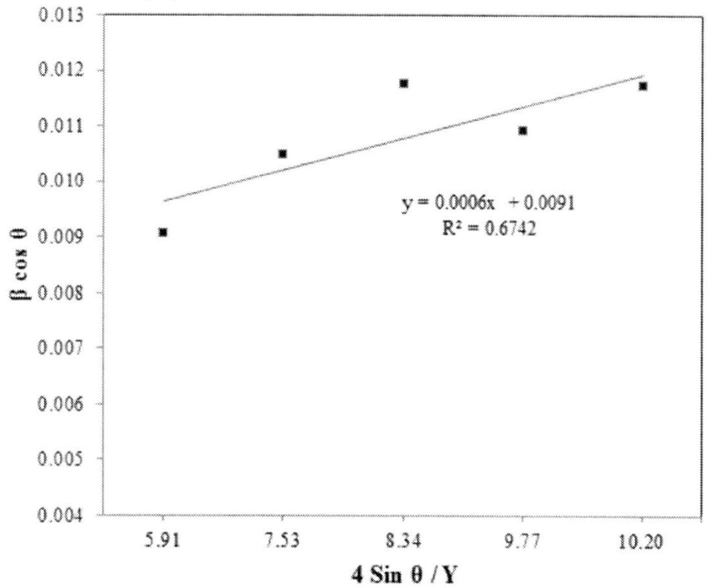

Figure 4. Plot $\beta_{hkl}\cos\theta$ versus $\frac{4sin\theta}{Y}$.

$$\beta_i \cos\theta = \frac{K\lambda}{D} + \frac{4\sigma\sin\theta}{Y} \quad (6)$$

The young's modulus (Y) for cubic MgO nanoparticle is 248.73 GPa. The USDM plots for MgO nanoparticle calcined at 950 °C are shown in Figure 4. The calculated result of the stress values calculated using slope of the straight line of the graph is 37.31 MPa.

3.6 Determination of the strain of crystal (ε)

The strain value of a crystal shows the alignment of a crystal that was produced. The crystalline strain was determined by using the diffraction pattern relationship of the line broadening that obtained in the measurement using XRD [10]. β_{hkl} is resulted by each the diffraction peak of MgO. The total peak broadening is the sum of the crystallite size and strain present in the material.

$$\beta_{hkl} = \left[\left(\beta_{hkl}^2\right)_{measured} - \left(\beta_{hkl}^2\right)_{standr} \right]^{1/2} \tag{7}$$

$$\beta_{hkl} = \beta_a + \beta_b \tag{8}$$

β_a is the contribution of crystallite size, β_b is strain-induced broadening and β_{hkl} is the width of the full wide half-maximum (FWHM) intensity of the instrumental corrected broadening.

$$\beta_{hkl} \cos\theta = \frac{K\lambda}{D} + 4\varepsilon \sin\theta \tag{9}$$

The solution of equation 8 uses the assumption that the crystalline strain is uniform in all directions of crystallography. The use of this assumption is known as a uniform deformation model (UDM). All crystals are regarded as having isotropic properties and have independent properties in all directions along their measurements [10].

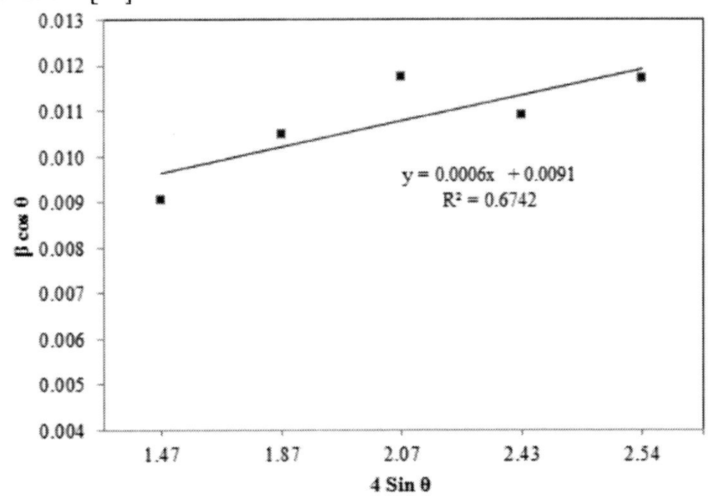

Figure 5. Plot $\beta_{hkl} \cos\theta$ versus $4sin\theta$.

The strain value of ε-crystal nanoparticles of MgO estimated from the slope (Figure 5) is 1.5 x 10^{-4}. The calculation result of strain value is very low and the slope of the line MgO calcined at 950 °C is R^2 = 0.6772. The value indicates that all the crystals are not far apart. In addition, this result also shows that the nanoparticle product is a lack of strain [15].

3.7 Determination of The Energy Density of crystal (U_{ed})

Based on the reference that the nanoparticle crystals are of the same nature and are isotropic, to determine the magnitude of the energy streams possessed by the crystals can be used a model of energy density having a uniform deformation (UDEDM. On the elastic system that follows Hooke's law, the energy density U_{ed} (energy per unit) can be calculated from the strain function $U_{ed} = (\varepsilon^2 Y)/2$. The equation (9) can be modified according to the energy and strain relation.

$$\beta_{hkl}\cos\theta = \frac{K\lambda}{D} + \left(4sin\theta\left(\frac{2U_{ed}}{Y}\right)^{1/2}\right) \quad (10)$$

Plots of $\beta_{hkl}\cos\theta$ versus $4sin\theta\left(\frac{2u_{ed}}{Y}\right)^{1/2}$ were constructed and the data fitted to lines (Fig. 6). The anisotropic energy density *Ued* was estimated from the slope of the linear fit is 0.712 MPa.

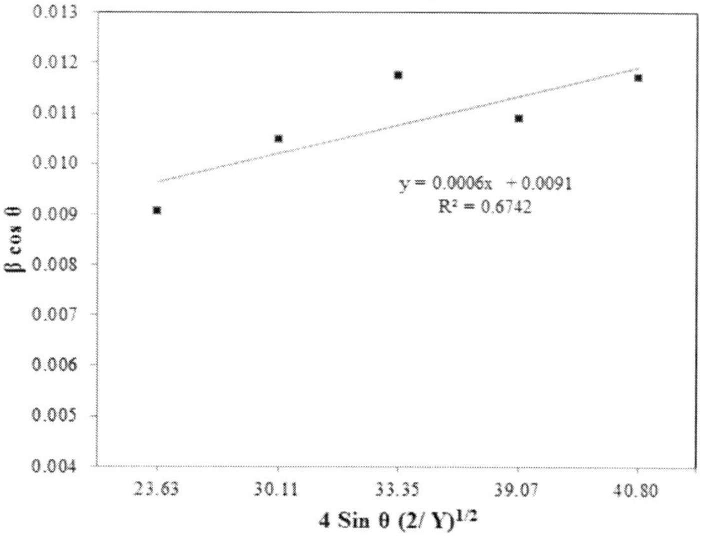

Figure 6. Plot $\beta_{hkl}\cos\theta$ versus $4sin(2/Y)^{1/2}$.

3.8 Analysis Texture coefficient $TC_{(hkl)}$

Quantitative information of crystal orientation preference can be analyzed using the texture coefficient calculation. The texture coefficient will give a percentage value for each X-ray diffraction of the MgO nanoparticles. The coefficient of texture can be calculated using equation 11 [16,17]:

$$TC_{(hkl)} = \frac{I_{(hkl)}/I_{o(hkl)}}{\sum_N I_{(hkl)}/I_{o(hkl)}} x\ 100\% \quad (11)$$

$I_{(hkl)}$, $I_{o(hkl)}$ and N are the intensity relative of diffraction peak, the intensity of standard measurement results and the number of diffraction peaks resulted respectively.

Table 1. $TC_{(hkl)}$ values of MgO nanoparticles product.

No.	*h k l*	$I_{hklmeasured}$	$I_{o(hkl)standard}$	TC_{hkl}
1	(2 0 0)	100	100	0.19
2	(2 2 0)	10	9	0.43
3	(2 2 0)	47	45	0.61
4	(3 1 1)	5	5	0.78
5	(2 2 2)	12	12	0.97

The TC (hkl) values calculated for MgO nanoparticles calcined at 950 °C for 6 hours show at Table 1. The calculation result showed that the highest of $TC_{(hkl)}$ value at (2 2 2) plane is 0,97. The results indicates that most of the crystal MgO nanoparticles resulted is preferred orientation of (2 2 2).

3.9 Morphological analysis

The SEM image of the MgO nanoparticles is shown in Figure 7. SEM analysis is a method for showing the surface morphology of the sample. Figure 7 shows that the results of MgO nanoparticles have a good agreement on the results that have been done by previous researchers with other methods [18,19].

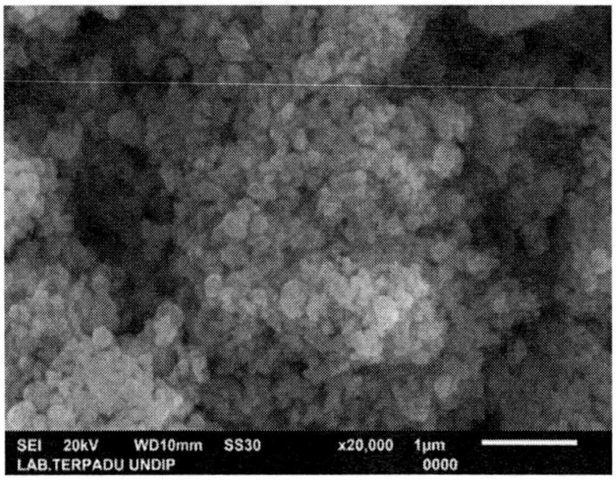

Figure 7. SEM of micrograph of MgO nanoparticles.

The results of SEM analysis showed that the crystal structure of MgO nanoparticles has a regular morphology. The shape of the crystal structure of the nanoparticles is dominated by crystals of the cubic shape. The MgO nanoparticles produced through these pathways are capable to produce the highest quality crystals with dominating cubic structures.

4. Conclusions

FTIR and XRD results show that MgO nanoparticles are formed through this synthesis process. This can be seen in the FT-IR wavelength range of 400.11-525 cm^{-1} as the broad band produced by the stretching vibration mode for the Mg-O-Mg compound. The average crystallite of MgO nanoparticles was 9.21 nm with a crystalline dislocation value is 0.012 nm^{-2} whereas the distribution of the MgO nanoparticle crystallite size showed a particle size of the distribution lower than 12.5 nm. Strain, stress, and density of crystal energy are 1.5×10^{-4}; 37.31 MPa; 0.72 MPa respectively. The highest texture coefficient value of the crystal is 0.98 in the crystal plane (2 2 2). This result is supported by morphological analysis using SEM which shows most of the regular cubic-shaped crystals. The synthesis method is suitable for a simple and cost-effective synthesis model of the MgO nanoparticles.

Acknowledgments

The research is funded by Ministry of Research and Higher Education Indonesia, for the Doctoral Grand Program on year 2017 and Scholarship for Postgraduate Education in the Country.

References

[1] Hadanu R, Idris S and Sutapa I W 2015 *Indones. J. Chem.* **15** 86–92
[2] Klabunde K J, Stark J, Koper O, Mohs C, Park D G, Decker S, Jiang Y, Lagadic I and Zhang D 1996 *J. Phys. Chem.* **100** 12142–53
[3] Sutapa I W, Armunanto R and Wijaya K 2010 *Indones. J. Chem.* **10** 184–8
[4] Svensson E E, Nassos S, Boutonnet M and Järås S G 2006 *Catal. Today* **117** 484–90
[5] Ding Y, Zhang G, Wu H, Hai B, Wang L and Qian Y 2001 *Chem. Mater.* **13** 435–40
[6] Das S and Srivasatava V C 2016 *Smart Sci.* **4** 190–5
[7] Athar T 2013*Mater. Focus* **2** 493–6
[8] Gubicza J, Ribárik G, Bakonyi I and Ungár T 2001 *J. Nanosci. Nanotechnol.* **1** 343–8
[9] Misriyani M, Kunarti E S and Yasuda M 2015 *Indones. J. Chem.* **15** 43–9

[10] Bindu P and Thomas S 2014 *J. Theor. Appl. Phys.* **8** 123–34
[11] Ungár T, Gubicza J, Ribárik G and Borbély A 2001 *J. Appl. Crystallogr.* **34** 298–310
[12] Mastuli M S, Kamarulzaman N, Nawawi M A, Mahat A M, Rusdi R and Kamarudin N 2014 *Nanoscale Res. Lett.* **9** 134
[13] Patterson A L 1939 *Phys. Rev.* **56** 978–82
[14] Khan Z R, Zulfequar M and Khan M S 2010 *Mater. Sci. Eng. B* **174** 145–9
[15] Zhang X, Zheng Y, Feng X, Han X, Bai Z and Zhang Z 2015 *RSC Adv.* **5** 86102–12
[16] Manifacier J C, Gasiot J and Fillard J P 1976 *J. Phys. [E]* **9** 1002
[17] Wang J, Yang F, Wei X, Zhang Y, Wei L, Zhang J, Tang Q, Guo B and Xu L 2014 *Phys. Chem. Chem. Phys.* **16** 16711–8
[18] Dianzeng J, Li Z, Yali C, Lang L and Dingquan X 2008 *J. Electroceramics* **21** 218–21
[19] Lakshmi Reddy S, Ravindra Reddy T, Siva Reddy G, Endo T and Frost R L 2014 *Spectrochim. Acta. A. Mol. Biomol. Spectrosc.* **123** 25–9

IOP Conf. Series: Journal of Physics: Conf. Series **979** (2018) 012022 doi:10.1088/1742-6596/979/1/012022

Benzene derivatives from the marine sponges *Cinachyrella*

Abraham[1,3], Y M Syah[2], H Natsir[1] and Nunuk Hariani Soekamto[1]

[1] Department Of Chemistry, Faculty of Mathematics and Natural Sciences, University of Hasanuddin, Jl. Perintis Kemerdekaan Km. 10 Tamalanrea Makassar, Indonesia
[2] Division of Organic Chemistry, Faculty of Mathematics and Natural Sciences, Institut Teknologi Bandung, Jl. Ganesa 10 Bandung, Indonesia
[3] Department of Education Chemistry, FKIP University of Halu Oleo (UHO), Kampus Bumi Tridharma Jl. HEA. Mokodompit Kendari, Indonesia

E-mail: nunukhariani@unhas.ac.id

Abstract. The benzene derivatives from the Spermonde Archipelago sponges *Cinachyrella* was isolated. 4-hydroxy-benzenepropanoic acid **(1)** from *Cinachyrella australiensis*, 4-methyl-benzoic acid **(2)** and 4-hydroxy-benzonitrile **(3)** from *Cinachyrella* sp. The structues of **1**, **2**, and **3** have been determined by ^{1}H-NMR, ^{13}C-NMR, and HR-MS.

1. Introduction

The benzene derivatives compounds are commonly referred to as aromatic compounds. Some very important compounds that we come across on a daily basis are manufactured from derivatives of benzene. One of the most obvious is polystyrene, manufactured through the polymerisation of styrene.

Some benzene derivatives are reported isolated from Australian marine sponge, *Ianthella* sp. are 4-hydroxybenzaldehyde, 4-hydroxybenzoic acid, 4-methoxybenzoic acid, and ethyl 4-hydroxybenzoate [1].

The GC/MS analysis revealed that 55 compounds were identified in the supernatant static extract and mycelial extract of the fungus *Aspergillus unguis* RSPG_204 isolated from the Sponge (*Agelas* sp.) among others 4-hydroxy-benzenepropanoic acid **(1)** identified in the supernatant static extract [2]

Isolation of the chemical compounds from marine sponges of *Cinachyrella* genus reported in this paper, found three known compounds of benzene derivatives; compound **1**, 4-methyl-benzoic acid **(2)**, and 4-hydroxy-benzonitrile **(3)**. Compounds **1**, **2**, and **3** are described for the first time from this genus.

2. Eksperiment

2.1. General

The NMR spectra were recorded on a Agilent 500 MHz. HR-MS spectra were performed with a Waters LCT Premier XE HRMS ESI-TOF.

The specimen of sponges were collected from Spermonde Archipelago, South Sulawesi, Indonesia, by SCUBA, in Sep-Oct 2015. A voucher specimen *Cinachyrella australiensis* (SVP 01/10/15) and *Cinachyrella* sp. (SVP 02/10/15) were deposited at the Research Center for Oceanography Indonesian Institute of Sciences in North Jakarta [3].

2.2. Extraction and Isolation
The sponges samples (10 kg, dry wt.), were immersed in methanol. The methanol extracts were collected and dried by rotary evaporator to obtain the crude extracts. The crude methanol extract was partitioned with n-hexane and followed by ethyl acetate. Each fraction was dried to get of n-hexane, ethyl acetate, and methanol extracts from sponges *C. autraliensis* and *Cinachyrella* sp. [3].

*2.2.1. Isolatian compound **1**.* The ethyl acetate extract sponge *C. autraliensis* was fractionated using vacuum-liquid column chromatograph (VLC), obtaining 20 fractions. The fractions 18-20 (567,1 mg) were united, then fractionated using Centrifugal Thin-Layer Chromatograph (Chromatotron), obtaining 13 fractions. The fractions 9-11 (27,5 mg) were united, then isolated using Chromatotron, obtaining 6 fractions. The fractions 2-4 (14,7 mg) were united, then purified, is obtained compound **1** (4,6 mg. as a yellowish green gel).

2.2.2. Isolatian compounds 2 and 3. The ethyl acetate extract sponge *Cinachyrella* sp. was fractionated using VLC, obtaining 12 fractions. Fraction 5 (53 mg) is isolated by using Chromatotron, obtaining 10 fraction. Fraction 9 (3,0 mg. as a brown gel) as compound **2** and fraction 7 (2,3 mg. as a brown gel) as compound **3**.

2.3. Spectral Data

2.3.1. Compound 1 : ^{1}H NMR ($CDCl_3$, 500 MHz) : δ_H 2.6639 (2H, t, J = 7.5 ; 15 Hz), 2.8876 (2H, t, J = 7.5 ; 15 Hz), 6.7612 (2H, d, J = 8.2 Hz), 7.0698 (2H, d, J = 8.2 Hz) ; ^{13}C NMR ($CDCl_3$, 125 MHz) : δ_C 35.68, 37.11, 115.38, 129.41, 132.23, 154.19, 176.47 ; HR-MS (negative ion mode) for $C_9H_9O_3$ [M-H] : Calc. Mass 165.0552, Mass 165.0547.

2.3.2. Compound 2 : ^{1}H NMR ($CDCl_3$, 500 MHz) : δ_H 2.5614 (3H, s), 6.8930 (2H, d, J = 8.8 Hz), 7.9048 (2H, d, J = 8.8 Hz) ; HR-MS (negative ion mode) for $C_8H_7O_2$ [M-H] : Calc. Mass 135.0446, Mass 135.0441.

2.3.3. Compound 3 : ^{1}H NMR ($CDCl_3$, 500 MHz) : δ_H 5.3659 [1H, bs], 6.9448 (2H, d, J = 8.4 Hz), 7.5356 (2H, d, J = 8.4 Hz) ; HR-MS (negative ion mode) for C_7H_4NO [M-H] : Calc. Mass 118.0293, Mass 118.0298.

3. Discussion

3.1. Compound 1
^{1}H-NMR, shown the doublet signal of the aromatic proton at 6.7612 and 7.0698 on the *ortho* coupling (J = 8.2 Hz), these data indicates the presence of symmetry in the benzene ring. ^{13}C NMR, shows that there are three aromatic carbon signals at 115.38, 129.41, 132.23, and one aromatic carbon bonded to oxygen at 154.19 ppm. The structure skeleton can be described as

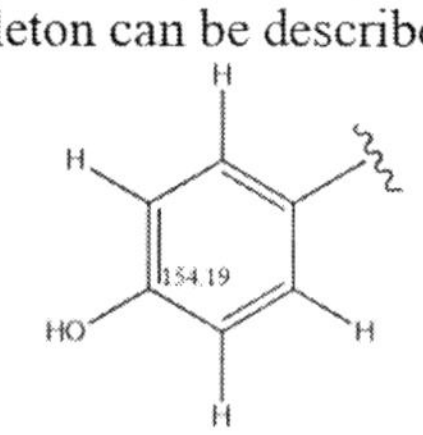

Figure 1. Compound 1 skeleton structure.

The triplet signals at 2.6639 and 2.8876 are the -CH_2 groups adjacent to the carbonyl (δ_C 176.47) and aromatic groups, aliphatic protons are bound to carbon 37.11 and 35.68 ppm. the presence of vicinal coupling (7.5 Hz) and geminal coupling (15 Hz) indicates that the protons do not rotate because they are flanked by two large groups (the carbonyl and aromatic). Based on NMR data analysis, supported by HR-MS data, the structure of compound **1** is :

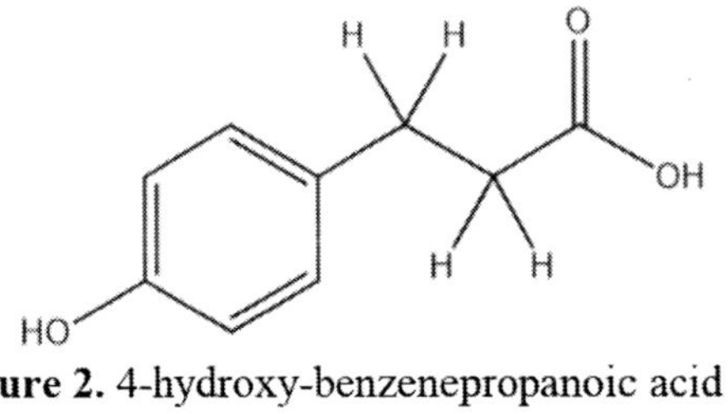

Figure 2. 4-hydroxy-benzenepropanoic acid **(1)**.

Confirmation of NMR data of compound **1** with previous research, indicates a conformity that reinforces the estimated structure of compound **1** as follows :

^{1}H NMR (CD_3OD, 300 MHz) δ: 7.02 (2H, d, *J*=8.4 Hz), 6.94 (2H, d, *J*=8.4 Hz), 2.80 (2H, t, *J*=7.5 Hz), 2.53 (2H, t, *J*=7.8 Hz); ^{13}C NMR (CD_3OD, 75 MHz) δ: 177.1, 156.3, 132.9, 130.2, 116.1, 37.0, 31.0 [4].

^{1}H NMR (400 MHz, $CDCl_3$) δ 7.07 (2H, d, 8.0 Hz, H-2 and H-6), 6.76 (2H, d, 8.0 Hz, H-3 and H-5), 2.89 (2H, t, 8.0 Hz, H-7), 2.64 (2H, t, 8.0 Hz, H-8); ^{13}C NMR (100 MHz, $CDCl_3$) δ 29.8 (C-7), 35.5 (C-8), 115.4 (C-3 and C-5), 129.4 (C-2 and C-6), 131.1 (C-1), 155.6 (C-4), 179.9 (C-9) [5].

^{1}H NMR ($CDCl_3$, 500.13 MHz), δ: 2.60 (2 H, t, CH_2COOH, *J* = 7.5 Hz); 3.86 (2 H, t, $ArCH_2$, *J* = 7.5 Hz); 6.67 (2 H, d, Ar, *J* = 8.5 Hz), 7.01 (2 H, d, Ar, *J* = 8.5 Hz) [6].

3.2. Compound 2

^{1}H NMR, shown the doublet signal of the aromatic proton at 6.8930 and 7.9048 on the *ortho* coupling (*J* = 8.8 Hz). These data indicate the presence of symmetry in the benzene ring. The singlet signal at 2.5614 indicates that -CH3 is bound to aromatic.

The HR-MS data shows the molecular formula $C_8H_7O_2$ [M-H] : Calc. Mass 135.0446, Mass 135.0441. Based on NMR and HR-MS data, the structure of compound **2** is :

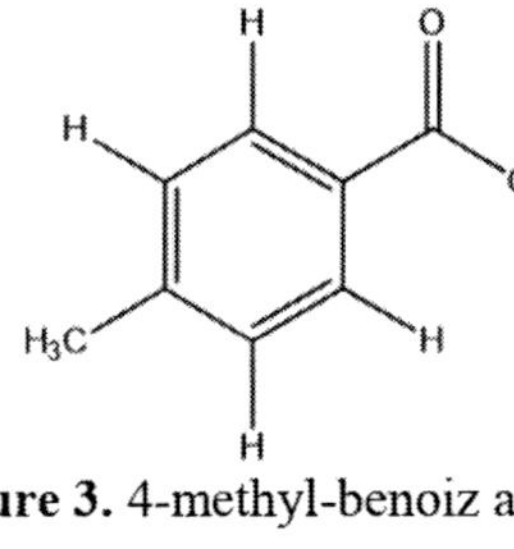

Figure 3. 4-methyl-benoiz acid **(2)**.

Confirmation of NMR data of compound **2** with previous research, indicates a conformity that reinforces the estimated structure of compound **2** as follow : ^{1}H NMR (400 MHz) δ 2.43 (3H, s), 7.28 (2H, d, *J* = 8.1 Hz), 8.01 (2H, d, *J* = 8.1 Hz) [7].

3.3. Compound 3

^{1}H NMR, shows that the doublet signal of the aromatic proton at 6.9448 and 7.5356 on the *ortho* coupling (J = 8.4 Hz), these data indicate the presence of symmetry in the benzene ring. The broad singlet signal at 5.3659 indicates -OH is bound to aromatic.

The HR-MS data shows the molecular formula C_7H_4NO [M-H] : Calc. Mass 118.0293, Mass 118.0298. Based on NMR and HR-MS data, the structure of compound **3** is :

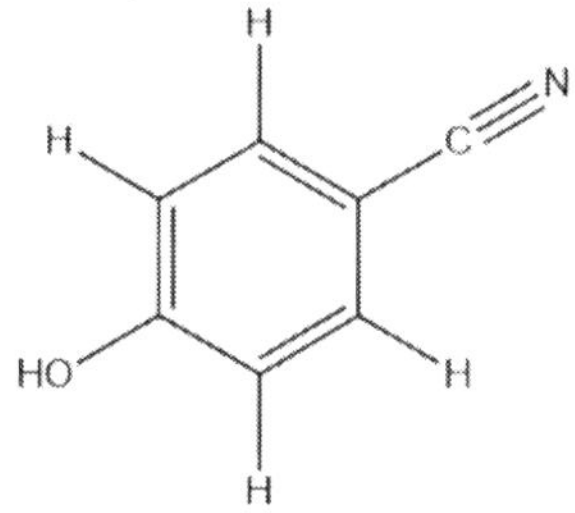

Figure 4. 4-hydroxy-benzonitrile **(3)**.

Confirmation of NMR data of compound **3** with previous research, indicates a conformity that reinforces the estimated structure of compound **3** as follow : ^{1}H NMR ($CDCl_3$, 300 MHz) δ: 7.56 (2H, d, J= 8.6 Hz), 6.94 (2H, d, J=8.6 Hz), 6.79 (1H, bs) [8].

4. Conclusion

Ethyl acetate extracts of sponge *C. australiensis* and *Chynachyrella* sp. from the Spermonde Archipelago contain the 4-hydroxy-benzenepropanoic acid **(1)**, 4-methyl-benzoic acid **(2)**, and 4-hydroxy-benzonitrile **(3)**.

Acknowledgements

We gratefully acknowledge the financial support from the ministry of research, technology and higher education, Republic of Indonesia (Doctor Dissertation Grand, 2017).

References

[1] Zhang, H., Conte, M.M., Huang, X.-C., Khalil, Z., and Capon, R.J. 2012 *Ianthella* sp. *Org. Biomol. Chem.* 10, 2656-2663
[2] Abd El-Hady, F.K., Abdel-Aziz, M.S., Shaker, K.H., El-Shahid, Z.A., Ibrahim, L.S. 2015. *Int. J. Pharm. Sci. Rev. Res.* 30(1), No. 48, 272-278
[3] Abraham, Syah, Y.M., Natsir, H., and Soekamto, N.H. 2017. *Res. J. Pharm. Biol. Chem. Sci.* 8(4), 833-837
[4] Li, Y., Yang, F., Xu, X., Pan, S., Wang, L., Xia, C. 2009. *Chin. J. Chem.* 27, 433—436
[5] Takahashi, T., Miyazawa, M. 2010. *Pharm.* 65, 913–918
[6] Prosenko, A.E., Skorobogatov, A.A., Dyubchenko, O.I., Pinko, P.I., Kandalintseva, N.V., Shakirov, M.M., and Pokrovsky, L.M. 2007. *Russi. Chem. Bull., Int. Ed.* 56(6), 1119—1124
[7] Nemoto, K., Yoshida, H., Egusa, N., Morohashi, N., and Hattori, T. 2010. *J. Org. Chem.*, 75, 7855–7862
[8] Chankeshwara, S.V., Chebolu, R., and Chakraborti, A.K. 2008. *J. Org. Chem.* 73, 8615–8618

Pilarization TiO_2 onto De-oiled spent bleaching clay using *Rarasaponin* as surfactant

N Hindryawati, Daniel, Erwin and N D Fadillah

Faculty of Mathematic and Natural Sciences, Mulawarman University, Gunung Kelua, 75113 Samarinda, East Kalimantan, Indonesia

E-mail: ienwati@yahoo.com

Abstract. Synthesis and characterization TiO^2 pillared deoiled spent bleaching clay (DSBC) with rarasaponin as surfactant had been done. Activation DSBC have been done with H_2SO_4 1N, followed by pillarization with TiO_2 using rarasaponin as surfactant. Characterization has done with Fourier transform infrared spectroscopy showed the rarasaponin as surfactant was successfully carried out in DSBC with the presence of absorption peak C=O stretching group in a sharp 1720.50 cm-1 wavelength range. As well as the $C-CH_2$ stretching uptake peak is represented on wave number 1462.04 cm^{-1} and 1033,85 cm^{-1} for aromatic functional group C=C stretching. After pillared by TiO_2, the XRD pattern on DSBC showed new peak appears on $2\theta =$ 27,4460°; 36,0850° and 55,3216° and the mineral contain on DSBC is rectorite with dioctahedral mica layer and dioctahedral smectite with ratio 2:1. This molecule have formula $Na.Al_4(Si, Al)_8.O_{20}.(OH)_4.\ H_2O$. Crystallinty of pillared clay showed 72,5014 % after calcination and there is some Ti suspected on the layer based on SEM.

1. Introduction

In palm oil processing there are several steps had been through such as oil removal, neutralization, tanning and odor removal. In bleaching process they are used a porous material such as clay [1]. Palm oil processing in Indonesia in 2016 touched 34 million with estimated residual clay (spent bleaching clay) are 34 thousand [2]. Spent bleaching clay (SBC) is classified as hazardous waste (B3) based on government rules no. 18 in 1999, because there is a 20-40 % oil content which is can aggravate the soil and air quality [3,4] and a lot of company dealing this problem by dumping them into landfills [2]. Bleaching clay is one of pore materials which is consist an aluminosilicate mineral as negative charge and balanced by Na, Ca, Mg or H as positive charge. Bleaching Clay can be regenerated by extraction to remove oil content and this material called deoiled spent bleaching clay (D-SBC) [1]. Since this material was used before, the active site of this material become inactive. There are several steps to activated this material and acid treatment is the best way [5]. Hydrogen ions that produced by adding some acid will charge the alluminosillicate in interlayer region. This charge will activated the active site and alters the structure, chemical composition and physical properties of the DSBC, afterwards this material called DSC-A [6]. The presence of ion hydrogen in interlayer of DSBC-A could be replaced by another ion with different size and similar charge [7,8]. To hence ions exchanged in interlayer DSBC-A there is needed support by another substance. Using polyoxypropylene (POP) as surfactant had been reported can hence ion exchange in interlayer area [9]. However, by using synthetic surfactant also impact environment because they are non biodegradable and toxic. Klerak fruit contain

IOP Conf. Series: Journal of Physics: Conf. Series **979** (2018) 012023 doi:10.1088/1742-6596/979/1/012023

rarasaponin as main substance has been reported can be used as natural surfactant, this natural surfactant also performed similar properties such as synthetic surfactant [10]

To overcome this phenomenon, in this work a novel approach to utilize DSBC as support material and pillaring with TiO_2. SBC needs some handling before those material can be used, first oil contained in the material need to remove with extraction step, SBC will turn to deoiled spent bleaching clay (DSBC) after these process. This work will perform synthesis and characterization of deoiled spent bleaching clay (DSBC) by addition rarasaponin as surfactant to increase the basal spacing of the material [10] which is expected these material can be used as photocatalyst material of liquid textile waste. In the process, the photocatalyst material will characterized using X-ray fluorescence, X-Ray diffraction, Fourier transform infrared spectra (FTIR) and scanning electron microscopy (SEM).

2. Method

2.1 Rarasaponin Extraction

Rarasaponin was extracted from Klerak fruit was find from Yogyakarta. Klerak fruits dried for 4 days, then crushed until 80 mesh. The ration of distillate water (mL): klerak frutis (g) is 2:1. The extraction was done with soxhlet extractor at 70 °C for 180 minutes. The results will separated by evaporator at 100 °C.

2.2 Oil extraction on SBC

Spent bleaching clay heated in an oven at 105 °C for 150 minutes. The ration between SBC and hexane used is 1:2 (g/g). The mixture was put into an ultrasonic cleaner (using a frequency of 42 kHz, a voltage of 235 W) at 55± 2 °C for 60 minutes. The results will be separated with centrifuged at 1000 rpm for 5 minutes. % oil content in SBC will calculated after this process and SBC will be named DSBC (deoiled spent bleaching clay). The oil content was determine using the following formula:

$$\%\ \text{oil content} = \left(\frac{M_1}{M_0}\right) \text{x } 100 \qquad (1)$$

M_1 and M_0 are the masses of the oil and SBC in g, respectively. The extraction of oil from SBC via ultrasound extraction yield about 21 wt% within 60 mins which was more efficient than that of soxhlet extraction. Furthermore, it was reported by Khaeng et al. [2] that extraction of oil using soxhlet extraction showed the yield about 21 wt% within 4 h. Soxhlet extraction is associated with high energy consumption, involves longer extraction time (3 h for extracted O-DC) and the use of large quantities of organic solvents. For these reasons, an improved or better extraction technique is desirable. Thus, in agreement with Maniam et al. [11], the ultrasound-assisted extraction is explored because it can enhance the extraction efficiency through acoustic cavitations and some mechanical effects.

2.3 Synthesis TiO_2 pillared DSBC

Deoiled spent bleaching clay need to activated first using H_2SO_4 1N with ration 2:1 (g/g) and then refluxed for 2 hours with temperature 96-98 °C, activated spent bleaching clay washed with distillate water up to pH 7 then dried into oven with temperature 105 °C for 1 hour afterward this is named DSBC-A. Subsequently DSBC-A need to add *rarasaponin* as surfactant with ration *rarasaponin*, aquades and DSBC-A is 1(g):50(g):10(g). the mixture as refluxed for 24 hours at 28 °C. The results will be separated and named DSBC-S. After add some surfactant DSBC-S will pillared by TiO_2 with mixed 04 g of TiO_2 into 100 mL on beaker glass filled with 1 g DSBC-S, added with 6 mL of 95% ethanol and stirring used magnetizer for 5 hours. The results will be separated by centrifugation followed by drying at 120 °C for 5 hours. Afterward proceed at calcination at 500 °C for 2 hours.

3. Results and discussion

3.1 XRF Analysis of DSBC and DSBC-A

Table 1 showed that there is an increase of Si metal and reduction of Ca and Mn metal. That is because the metal was dissolved during the activation process by H_2SO_4 solution. The activation of DSBC using acid treatment was aim to open active site of the DSBC by dissolving small ion [9].

Table 1. XRF result analysis of DSBC and DSBC-A

No	Sample Name	Compound									
		Fe_2O_3	CaO	K_2O	SiO_2	MnO	SO_3	ZnO	P_2O_5	Cr_2O_3	CuO
1.	DSBC	28,618	21,713	3,599	29,363	0,522	2,276	0,059	13,711	0,105	0,034
2.	DSBC-A	28,077	2,346	4,076	63,493	0,143	1,674	0,071	-	0,097	0,024

Other impurities such as SO_3, P_2O_5, Cr and Cu metals also decreased. This phenomenon suspected because the metal was dissolved in activation process resulted the material got bigger pore. DSBC-A will be further pillaritation.

3.2 Fourier Transform Infrared Analysis of DSBC-S and Ti-DSBC

Fourier transform infrared analysis (FTIR) is one of the important analyses to confirm formation of DSBC-A after adding of surfactant and pillarization of TiO_2 on DSBC. The FTIR spectrum of deoiled spent bleaching clay with *Rarasaponin* as surfactant and DSBC after pillarization with TiO_2 was shown in Figure 1.

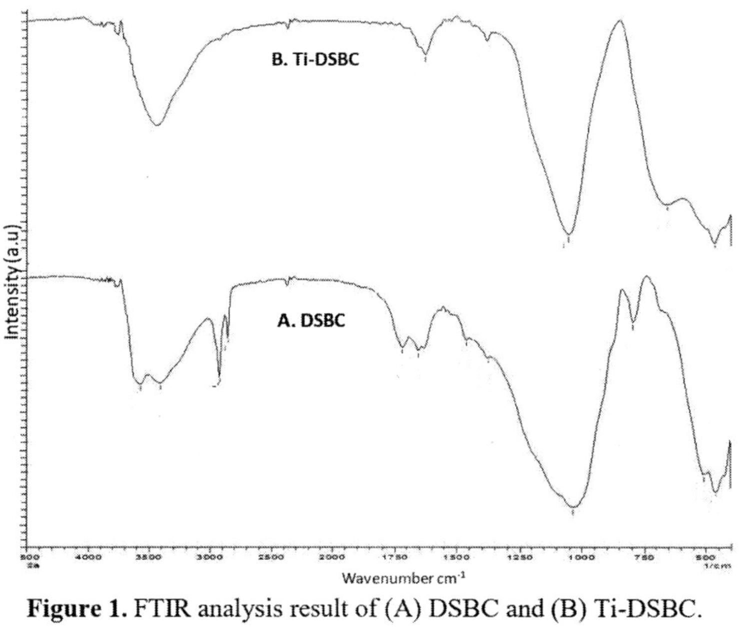

Figure 1. FTIR analysis result of (A) DSBC and (B) Ti-DSBC.

FTIR analysis showed functional group of material by interpreting absorption peaks of the infrared spectrum. *Rarasaponin* as surfactant was successfully carried out in the presence of carbonyl group absorption peak is C=O stretching group in a sharp 1720.50 cm^{-1} wavelength range. As well as the C-CH_2 stretching uptake peak is represented by spectrum with absorption peak of wave number 1462.04 cm^{-1} and 1033,85 cm^{-1} for aromatic functional group C=C stretching [10]. The structure of montmorrilonite also unchanged with the presence of spectra with absorption peak of 511.14 cm^{-1} and 462 cm^{-1} which is typical of Al-OH stretching and Si-O stretching. This information also reinforced by the presence of SiO_2 with absorption wavelengths 796.60 cm^{-1} [10]. The functional groups of C=O and

C-O esters are the main functional groups of *rarasaponin* and based on these information adding *rarasaponin* as surfactant had been successes. After Ti pillared to material DSBC-S this material losses some functional group like C=O marked with no absorption in 1720.50 cm^{-1}. There is some losses H_2O molecule with absorption wavelength only appeared in 3429.50 cm^{-1} and 1631.76 cm^{-1}.

3.3 X-ray Diffraction Analysis of DSBC-A and Ti-DSBC

Based on Figure 2 (A) showed XRD pattern on DSBC-A with peak on 20,8595°, 25,5835, 26,6393°, 35,1356, 36,5431° and 50,1375° with *hkl* value; [101], [012]; [104]; [110] and [112] indicated kind of this mineral contain on this material is rectorite. There is 2θ = 20,8595°, 26,6393°, 36,5431° with *hkl* value on [100], [110] and [101] indicated hexagonal crystal of SiO_2, furthermore 2θ = 25,5835 °, 35,1356 °, 50,1375° with *hkl* value [012], [104] and [112] indicated Al_2O_3 with rhombohedral crystal. Rectorite is one of Na-montmorrilonite with dioctahedral mica layer and dioctahedral smectite with ratio 2:1. This molecule have formula $Na.Al_4(Si , Al)_8.O_{20}.(OH)_4.H_2O$ or allevardite for common name [12]. Rectorite layer have thickness around 40 μm and wide around 5 μm. This size cause rectorite has bigger layer size than montmorrilonite. On Figure 2 we can conclude this material was ready for pillared because the amorphous value showed 55,8946 %. This material tends to amorphous so that possible to substitute another TiO_2 to this layer material.

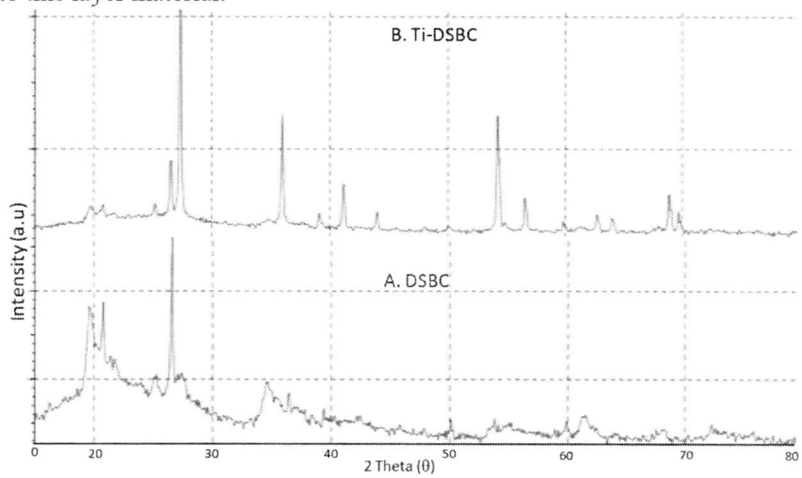

Figure 2. XRD results on (A) DSBC-A and (B) Ti-DSBC.

Based on Figure 2 (B) showed XRD Ti-DSBC result. JCPDS no. 88-1175 showed this material have rutile Ti with tetragonal geometric, this information can seen with new peak appears on 2θ = 27,4460°; 36,0850 ° and 55,3216 ° with *hkl* value [110], [101] and [211]. Otherwise, the main component of DSBC (rectorite) SiO_2 and Al_2O_3 appears on 2θ = 42,4490°, 68,3161 with *hkl* value [111] and [311] showed shifted of SiO_2 and 2θ = 25,5779° with *hkl* value [012] showed there is Al_2O_3. TiO_2 in this XRD pattern which is crystalline phase and perform as sustain for the metal oxide in the catalyst. After pillarization the crystallinity of Ti-DSBC was increased (72,50%). In line with Schneider et al. and Thamaphat et al.[7,13], the material also has different cristallinity before and after calcinated.

3.4 Scanning Electron Microscopy (SEM) of DSBC-A and Ti-DSBC

The micrograph of DSBC-A showed that DSBC-A has porous with irregular particle shape and small particle in the surface where this indicated its has specific surface area.

Figure 3. SEM image of DSBC-A 10.000 expand.

Figure 3 showed layer morphology of DSBC-A, there is have same size particle distribution. From this Figure also seen there is cavity for some metal into. This layer suspected as rectorite and has possibility to add TiO_2.

Figure 4. SEM image of Ti-DSBC 10.000 expand.

Figure 4 showed morphology Ti-DSBC had some difference and expected there is Ti with white colour stick on the layer DSBC-S. Form this Figure also seen the particle have uniform size if compared with DSBC-A, it happened because these material had calcination process.

4. Conclusion

XRD analysis showed crystallinity had increase to 72.5014 % which is only 50.0556% before treatment. Based on JCPDS no.88-1175 this material have rutile Ti various. This information reinforced by SEM analysis showed surface morphology on DSBC-A had equal particle size and presence of cavities in the material. However, in Ti-DSBC material the particle size become smaller and there are other white particles in Ti-DSBC pore suspected as Ti.

Acknowledgement

The authors acknowledge BOPTN Grant from the Faculty of Mathematic and Natural Sciences, Mulawarman University.

References

[1] Daous MM, Zahrani, Al A. *Reycycling Of Spent Bleaching Clay And Oil Recovery*. Trans Ichem. 2000;78(May).

[2] Kheang LS, Foon CS, May CY, Ngan MA. A *study of residual oils recovered from spent bleaching earth: their characteristics and applications*. Application Science. 2006;3(10):2063–7
[3] Debdoubi A, El Amarti A, Colacio E. *Production Of Fuel Briquettes From Esparto Partially Pyrolyzed*. Energy Convers Manag. 2005;46(11–12):1877–84
[4] Elyza F, Gofar N. *Identifikasi Dan Uji Potensi Bakteri Lipolitik Dari Limbah Sbe (Spent Bleaching Earth) Sebagai Agen Bioremediasi*. 2015;13(1):12–8
[5] Bieseki L, Bertell F, Treichel H. *Acid Treatments of Montmorrilonite-rich Clay for Fe REmoval Using a Factorial Design Method*. Material Science. 2013;16(5): 1122–1127
[6] K. Ojha, N. C. Pradhan, and A. N. Samanta, *Kinetics of batch alkylation of phenol with tert-butyl alcohol over a catalyst synthesized from coal fly ash*, Chemical Engeenering Journal 2005; 112(1–3):109–115
[7] Schneider J, Matsuoka M, Takeuchi M. *Understanding TiO_2 Photocatalyis: Mechanisms and Materials*. 2014;114: 9919–9986
[8] Aziz BK, Shareef FH. *Using Natural Clays and Spent Bleaching Clay as Cheap Adsorbent for the Removal of Phenol in Aqueous Media*. 2013;(2)
[9] Chen D, Zhu Q, Zhou F, Deng X, Li F. *Synthesis and photocatalytic performances of the TiO_2 pillared montmorillonite*. Journal Hazard Mater. 2012;235–236:186–93
[10] Kurniawan A, Sutiono H, Ju YH, Soetaredjo FE, Ayucitra A, Yudha A, et al. *Utilization of rarasaponin natural surfactant for organo-bentonite preparation: Application for methylene blue removal from aqueous effluent*. Microporous Mesoporous Mater 2011;142(1):184–93
[11] Maniam GP, Hindryawati N, Nurfitri I, Jose R, Rahim MHA, Dahalan FA, Yusoff MM. *Decanter cake as a feedstock for biodiesel production: A first report*. Energy Convers Manag. 2013; 76 (57)
[12] Wang Y, Zhang H, Wu Y, Yang J, Zhang L. *Preparation, Structure, And Properties Of A Novel Rectorite/Styrene-Butadiene Copolymer Nanocomposite*. J Application Polymer Science
[13] Thamaphat K, Limsuwan P, Ngotawornchai B. *Phase Characterization of TiO_2 Powder by XRD and TEM\n*. Natural Science. 2008;42:357–61

Determination of specific capacitance of modified candlenut shell based carbon as electrode material for supercapacitor

M Zakir[1], P Budi[1], I Raya[1], A Karim[1], R Wulandari[1] and A B J Sobrido[2]

[1] Department of Chemistry, Faculty of Mathematics and Natural Sciences, Hasanuddin University Jl. Perintis Kemerdekaan Km. 10 Makassar 90245, Indonesia
[2] Material Research Institute, School of Engineering and Material Sciences, Queen Mary University, Mile End Rd, E1 4NS London

E-mail: muh.zakir@science.unhas.ac.id

Abstract. Surface modification of candlenut shell carbon (CSC) using three chemicals: nitric acid (HNO_3), hydrogen peroxide (H_2O_2), and sulfuric acid (H_2SO_4) has been carried out. Activation of CSC was performed using H_3PO_4 solution with different ratio between CSC and activator. Carbon surface area was determined by methylene blue adsorption method. Surface characterization was performed using FTIR spectroscopy and Boehm titration method. Specific capacitance of electrode prepared from CSAC (candlenuts shell activated carbon) materials was quantified by Cyclic Voltammetry (CV) measurement. The surface area before and after activation are 105,127 m^2/g, 112,488 m^2/g, 124,190 m^2/g, and 135,167 m^2/g, respectively. Surface modification of CSAC showed the improvement in the chemical functionality of CSAC surface. Analyses using FTIR spectroscopy and Boehm titration showed that modifications with HNO_3, H_2SO_4 and H_2O_2 on the surface of the CSAC increased the number of oxygen functional groups. As a consequence, the specific capacitance of CSAC modified with 65% HNO_3 attained the highest value (127 μF/g). There is an incredible increase by a factor of 298% from electrode which was constructed with un-modified CSAC material. This increase correlates to the largest number of oxygen functional groups of CSAC modified with nitric acid (HNO_3).

1. Introduction

Capabus or Capacitor Bus that today are operating in developed countries, is one example of supercapacitor applications which are used for storing and generating energy to drive the bus. This example is an answer for the needs of alternative energy other than fuel [1,2].

Supercapacitor is utilizing the principle of EDL (electric double layer) formed between the electrode material and electrolyte interface [3,4,5]. Therefore, electrode materials and electrolyte types are important parts that determine the performance of supercapacitors [6,7]. Commercial carbons are widely used as a supercapacitor electrode active material, but due to the preparation of those materials which are quite expensive and environmentally unfriendly, biomass based carbon is an attractive alternative because it is cheap, plentiful and environmentally friendly [8].

Biomass wastes in Indonesia are in large quantities from plantations, households, agriculture and industry which can be a potential pollutant if they are not managed properly. During this time, the biomass wastes are usually dumped or burned away and not used optimally. In fact, biomass wastes such as candlenut shell are potentially producing chemicals and materials of economic value for fuel and electrode material in battery or supercapacitor [9].

The main problems encountered in the preparation of biomass based material for energy storage application are variations of pore size ranging from micropores (<2 nm), mesoporous (2-50 nm) and macropores (> 50 nm). Ion transport for electrolytes in biomass based material with such pore characteristics will be very slow, so it will limit the formation of electric double layer, which in turn will lower the capacitance of electrodes for application in supercapacitors [10,11]. This phenomenon is also found by other researchers in the utilization of biomass based activated carbon for energy storage application [12,13].

Surface chemistry modification of biomass based porous carbons might be a practical and promising way to improve their chemical properties toward many utilizations of these materials. A modified porous carbon comprising various functional groups could be used for many applications such as heavy metals extractions from aqueous and non-aqueous solutions, in catalysis, toxic and radiolytic effluents treatment from industrial processes, and energy storage [14,15]. It was found that chemical modification using nitric acid introduced various oxygen functional groups [16] which have a significant effect on the capacitive properties of carbon material [17].

In this study, the surface of CSAC was modified using oxidizing agents: HNO_3, H_2SO_4 and H_2O_2 to improve the chemical functionalities of CSAC surface and to apply these materials for supercapacitor electrode. The effect of modification conditions on the electrochemical properties of CSAC electrode-based supercapacitor was studied. The relationship between specific capacitance and the number of oxygen functional groups was discussed.

2. Materials and Methods

2.1. Materials

Candlenut shell, an agricultural based waste was collected from nearby farming as solid wastes. The collected materials were cleaned with distilled water for three times to pull out all the dirt particles. The washed materials were cut into smaller pieces (1-3 cm) and treated with sun-drying for 6 hours and then put into a hot air oven at 110°C for 1h. Carbonization was undertaken in furnace at 380°C for 1h. Carbonization products (CSC) were cooled, crushed, and sieved into 100mesh particles. All chemicals used are of pro analyst (p.a.) grade.

2.2. Methods

Candlenut Shell based Activated Carbon (CSAC) were prepared by immersing the precursor CSC with H_3PO_4 solution (85 vol.%) for 24 hours. The cooled activated carbon was sieved through Buchner funnel and washed with hot distilled water so as removing extra acid and to attain pH values = 6.5 in the washing solution, and finally dried at 110 °C for 1 hour and stored in a desiccator. Surface modification of CSAC was carried out with chemical agents, 98 wt.% H_2SO_4, 65 wt.% HNO_3, and 30 wt.% H_2O_2 solutions.

Determination of carbon surface area was carried out based on the ability of carbon in methylene blue adsorption. The amount of adsorbed methylene blue is proportional to the surface area of adsorbent as shown in the following formula:

$$S = \frac{X_m . N . a}{M} \quad (1)$$

where S is surface are (m^2/g), N is Avogadro's number (6.02×10^{23} mol^{-1}), Xm is fractional coverage (g/g), a is coverage by 1 molecule MB (197×10^{-20} m^2), and M is molecular weight of MB (320.5 g/mole).

Characterization of surface chemistry of CSAC was conducted with Boehm titration method (BTM, [18,19]) and FTIR (Fourier Transform Infra-Red) spectroscopy. The BTM can be illustrated as follows: 0.5 g of CSAC were put into a series of flask containing 50mL of 0.05N $NaHCO_3$, Na_2CO_3,

NaOH and HCl. The flasks were sealed and shaken for 24h. The solutions were, then, filtered and 10mL of each solution was placed to a flask. The solution was titrated with 0.05N NaOH or HCl, depending on the original solution employed. Calculations of Boehm titration results are based on the following assumptions: (i) $NaHCO_3$ neutralizes carboxyl groups, (ii) Na_2CO_3 neutralizes carboxyl groups and lactones, (iii) NaOH neutralizes carboxyl groups, lactones and phenol groups; and (iv) HCl neutralizes the total base groups. FTIR spectroscopy was performed by Shimadzu 8400S FTIR instrument in wavenumbers range of 4000-500 cm^{-1}.

The electrode body was prepared by connecting copper and platinum wires using a steam solder. The electrode body was then inserted into a tulip pipette and glued together using parafilm. CSAC sample was mixed with parafilm wax with a mass ratio of 1: 1 and stirred until homogeneous by using a spatula in a petri dish. The carbon paste was inserted into the electrode body by being pressed using a spatula for solidifying and flattening.

Specific capacitance (Cs) of the carbon paste electrode was measured using Cyclic Voltammetry (CV) technique. This measurement employed three electrodes system i.e. Pt electrode, Ag/AgCl electrode and carbon paste electrode. Electrode testing was performed with scan rate of 50mV/s using electrolyte solution of H_2SO_4 0.1M to obtain voltage-current voltammogram. Based on that data, specific capacitance (Cs) was calculated.

3. Results and Discussions

3.1. Carbonization of Candlenut Shell

Carbonization of candlenut shell was carried out in furnace at 380°C for 1h. This temperature is the optimum temperature for candlenut shell carbonization. Ash product was observed beyond this temperature. Figure (1-a) is the carbonization product of candlenut shell.

3.2. Activation of Candlenut Shell based Carbon (CSC)

Activation is the process of removing hydrocarbons layer from the carbon surface thereby increasing the carbon porosity. Figure (1-b) is activated carbon prepared from candlenut shell.

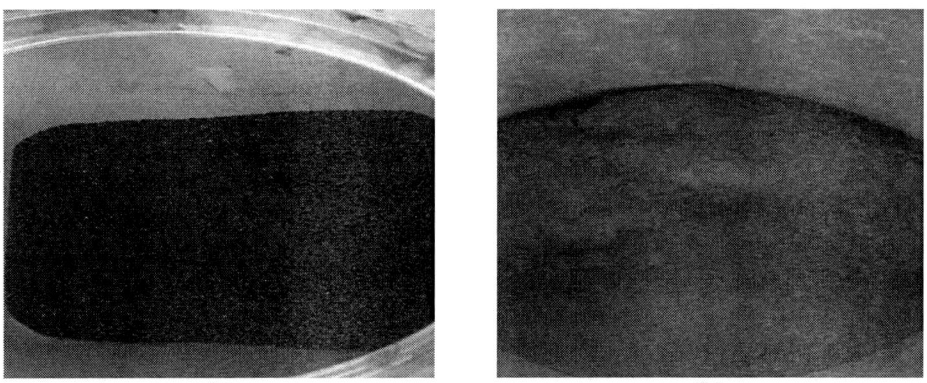

(a) (b)

Figure 1. (a) Carbonization product of candlenut shell, and (b) 100mesh size of Candlenut Shell based Activated Carbon (CSAC).

3.3. Determination of surface area of CSAC

Specific surface area of CSAC (Candlenut Shell Activated Carbon) is determined using Methylene Blue (MB) method. This method is a very simple method compared to commonly used BET method. In addition, the MB method is easier to carry out in the laboratory especially when no sophisticated equipment is available, such as N_2 adsorption instruments. The surface area is calculated based on the amount of MB absorbed by the CSAC as shown in Equation (1). It was found that the surface area increased by 30% after activation of candlenut shell based carbon.

3.4. Surface modification of CSAC

Acid treatment of activated carbon was generally used to oxidize the surface of porous carbon. The treatment will increase the acidic properties, pull off mineral contents and enhance the hydrophilic properties of surface of activated carbon. In addition, surface modification of CSAC will create new functional groups on the carbon surface which is reported to have an effect in increasing the value of specific capacitance of activated carbon as electrode material in supercapacitor [13, 14, 17, 20].

Surface modification of CSAC in this study was carried out using HNO_3, H_2SO_4, and H_2O_2 solutions, respectively. The addition of this solution will fabricate new functional groups, especially oxygen-containing functional groups (or surface oxides) that are widely known and the most common species formed on the surface of CSAC, which notably effects their performance in energy storage application [17, 20].

The functional groups generated from this modification are characterized using Boehm titration (back titration). The results of the Boehm titration were further compared with the results of functional group analysis using FTIR (Fourier Transform Infra-Red) spectroscopy.

Concentration of acid- and base-groups of CSAC determined by Boehm titration method is shown in the following figure.

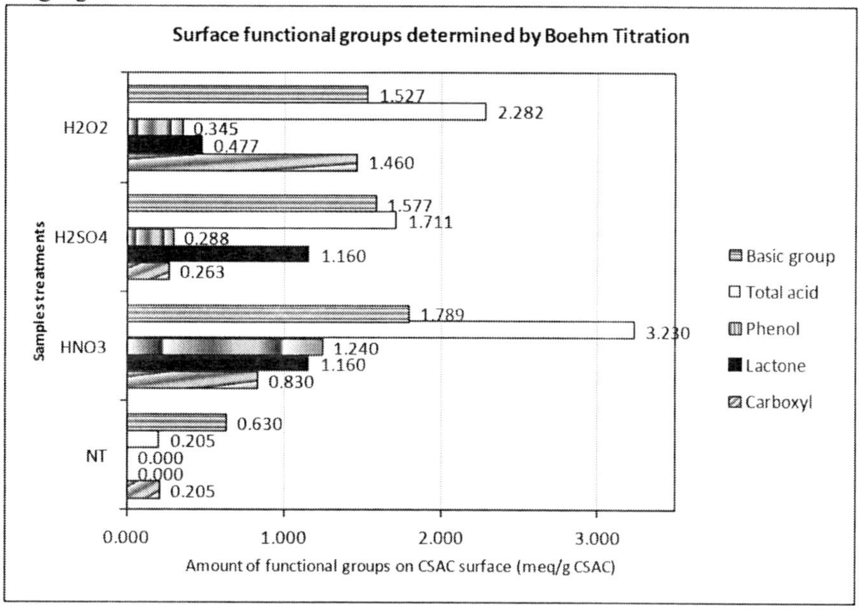

Figure 2. Comparison of functional groups concentration of CSAC before and after modification with HNO_3, H_2SO_4, and H_2O_2 solutions. Note NT = Non-Treatment (before modification).

FTIR results of candlenut shell based activated carbon showed absorption of functional groups on the surface which are identified as functional groups for -C-O, -C=O, and –OH. Those groups are oxygen-containing functional groups. Modification with nitric acid (HNO_3), hydrogen peroxide (H_2O_2), and sulfuric acid (H_2SO_4) all give OH-carboxylate, C-O carboxylate, and C=O either in lactone or quinone. This is consistent with the previous study results [14, 16, 17, 20]. Acidic oxygen-containing functional groups were incorporated into CSAC surface by acid oxidation. Carboxylic acid groups were the main species found on the surface. The oxidation process also presented quinone and lactone groups.

Peak analysis performed before modification showed that there is an -O-H group in the region of 3616 cm^{-1}. The regional absorption band at 1166 cm^{-1} indicates the existence of a tertiary C-O group. In addition, the area of 1691 cm^{-1} indicates the presence of a typical -C=O group of carboxylic acids. The absorption band in the area of 1591 cm^{-1} shows the presence of an -C=C bond of aromatic group.

FTIR results on HNO_3 modified activated carbon showed the presence of carbonyl groups in 1716 and 1606cm^{-1}. Absorption at 1716 cm^{-1} indicates the presence of -C=O groups of lactones, while 1606 cm-1 is usually identified as C=O groups of quinone, and an absorption band at 1529cm^{-1} indicates the presence of a C=C group. Absorption at 3415cm^{-1} indicates the presence of OH groups of carboxylates which is reinforced by C-O groups at 1217cm^{-1} wave numbers. FTIR results of H_2SO_4- and H_2O_2-modified CSAC almost showed the similar pattern with HNO_3-modified CSAC (Table 1).

Table 1. FTIR data of CSAC

CSAC treatment	Wave number (cm-1)	Functional groups	Notes
NT	3616	-OH	weakest
	1691	C=O carboxylic acid	weakest
	1591	C=C aromatic	
	1166	C-O tertier	
HNO_3	3415	-OH carboxylate	
	1217	-C-O carboxylate	
	1716	C=O lactone	
	1606	C=O quinone	
	1529	C=C	
H_2SO_4	3385	-OH carboxylate	
	1028	C-O carboxylate	
	1168	C-O tertier	
	1701	C=O lactone	
	1587	C=O quinone	
	1429	C=C	
H_2O_2	3435	-OH carboxylate	widest
	1278, 1037	-C-O carboxylate	
	1705	C=O lactone	
	1597	C=O quinone	

Both data, Boehm titration and FTIR spectroscopy, demonstrated that HNO_3-modified CSAC increases the oxygen-containing functional groups greater than other CSACs (H_2SO_4-, H_2O_2-modified CSAC and un-modified carbon).

3.5. Relationship between surface modification and specific capacitance of CSAC

We would like to discuss the effect of modification on specific capacitance changes of CSAC. Specific capacitance values were 40.2µF/g, 124µF/g and 127µF/g after modification with H_2O_2, H_2SO_4, and HNO_3, respectively. Surface oxidation with H_2O_2 only led to slightly increase in capacitance of CSAC from the unmodified CSAC (31.9µF/g). Incredible increase (±300%), however, was observed when using HNO_3 and H_2SO_4 as oxidizing agents. The order of these capacitance values is consistent with the changing of total number of basic functional groups after modification where $H_2O_2 < H_2SO_4 < HNO_3$ (Table 2). This fact showed that HNO_3 modification increased the number of oxygen oxygen-containing functional groups more than the other chemical agents and gave positive effects on the performance of CSAC-based electrodes [13, 17, 20].

On the contrary, the order of total number of acid functional groups is as follows $HNO_3 > H_2O_2 > H_2SO_4$. Even though sulfuric acid is a stronger acid than nitric acid, but nitrate ion is a stronger oxidant than sulphate ion. Nitric acid, for example, can oxidize Cl^- to Cl_2, but sulfuric acid cannot. In a nitrate ion, the molecular structure is inherently not very stable, as the nitrogen atom has to bear a positive charge to maintain the molecular structure. Besides, nitrogen itself is a very electronegative element, so that makes the nitrate ion even more unstable, hence its oxygen atom readily reacts with other atoms or molecules that oxygen can oxidize, including chloride ions [5]. On the other hand, hydrogen peroxide (H_2O_2) is a weak acid. H_2O_2 has an interesting chemistry because of its ability to function as oxidant as well as a reductant in both acid and basic solutions. On the whole, H_2O_2 is a very powerful oxidizing agent and a poor reducing agent. From this explanation, it might be understood that the order of total number of acid products will be HNO_3 > H_2O_2 > H_2SO_4, but quantitative order of the products chemical properties might be HNO_3> H_2SO_4 > H_2O_2 [11, 14]. This assumption is supported by the diversity of chemical functional groups observed by FTIR in Table 1, where H_2SO_4-modified CSAC produced 6 types of functional groups compared to H_2O_2-modified CSAC. It was reported that specific surface area (physical properties) is not the only factor which affects the value of specific capacitance. The chemical properties of activated carbon also have an important role in the performance of biomass based-activated carbon electrode. Of course, the formation of new chemical functional groups after modification also contributed to the increase of capacitance value of carbon electrodes. That is why the capacitance of H_2SO_4-modified CSAC is higher than H_2O_2-modified CSAC [4, 13, 14].

Table 2. The specific capacitances of CSAC-electrodes

Sample modification	Scan rate (mV/s)	Mass (g)	Ic (μA)	Ic (μA)	Cs (μF/g)
Unmodified	50	0.0189	3.63	3.33	31.9
HNO_3	50	0.0164	3.26	2.93	127
H_2SO_4	50	0.0075	3.31	2.83	124
H_2O_2	50	0.01755	3.44	2.34	40.2

The capacitance values obtained in this study are still in the order of milli-Farad and are much smaller when compared to the capacitance values obtained in the literature [20]. This may be due to the use of carbon paste electrodes in the determination of capacitance by cyclic voltammetry. In the future research, it will be examined the construction of electrodes as widely used in the literature [17, 20]. However, changes in capacitance values before and after CSAC modification suggested that, at least, biomass carbon modification is necessary to increase the capacitances value of biomass-based carbon for electrode material, especially biomass from agricultural wastes.

4. Conclusions

It has been shown that activation of candlenut shell based carbon increased the surface area of the carbon material. The increase of surface area is supposed by the formation of new pores after activation. The surface area of the candlenuts shell carbons before and after modification are 105,127 m^2/g, 112,488 m^2/g, 124,190 m^2/g, and 135,167 m^2/g, respectively. Surface modification of candlenuts shell based activated carbon with HNO_3, H_2SO_4, and H_2O_2 solutions has produced new functional groups on the carbon surface. The functional groups were mostly oxygen-containing functional groups as observed by Boehm titration and FTIR analyses. The formation of these functional groups is attributed to the increase of specific capacitance value of candlenut shell based carbon electrode. Specific capacitances of HNO_3-, H_2SO_4- and H_2O_2-modified candlenut shell activated carbon are 127; 124; and 40,2 μF/g, respectively. All modification treatment shows an increase compared to untreated candlenut shell activated carbon.

Acknowledgements

This work was supported by International Research Collaboration–World Class University Programme (Program "Word Class University–Kerjasama Riset Luar Negeri" (WCU-KLN)) 2017 of Hasanuddin University, Kampus Unhas Tamalanarea, Makassar, Indonesia.

References

[1] Gonzales A, Goikolea E, Barrena J A and Mysyk R 2016 Review on supercapacitors: technologies and materials *Renewable and Sustainable Energy Reviews* **58** p 1189
[2] Beguin F, Presser V, Balducci A and Frackowiak E 2014 Carbon and electrolytes for advanced supercapacitors *Adv. Mater.* 1-33. DOI: 10.1002/adma.201304137

[3] Winter M and Brodd R J 2004 What are batteries, fuel cells, and supercapacitors? *Chem. Rev.* **104** p 4245

[4] Frackowiak E and Beguin F 2001 Carbon materials for the electrochemical storage of energy in capacitors *Carbon* **39** p 937

[5] Simon P and Gogotsi Y 2008 Materials for electrochemical capacitors *Nature Mat.* **7** p 845

[6] Zhang Y, Xing H, Wu X, Wang L, Zhang A, Xia T C, Dong H C, Li X F and Zhong L S 2009 Progress of electrochemical capacitor electrode materials: a review *Int. J. Hydrogen Tech.* **34** p 4889

[7] Zhong C, Deng Y, Hu W, Qiao J, Zhang L and Zhang J 2015 A review of electrolyte materials and compositions for electrochemical supercapacitors *Chem. Soc. Rev.* **44** p 7484

[8] Ioannidou O and Zabaniotou A 2007 Agricultural residues as precursor for activated carbon production-a review *Renewable and Sustainable Energy Reviews* **11** p 1966

[9] Kwangdinata R, Raya R and Zakir M 2014 Production of biodiesel from lipid of phytoplankton Chaetores calcitrans trough ultrasonic method *The Scientific World J.* Article ID 23136, 5 pages

[10] Wang J, Xin H L and Wang D 2013 Recent Progress on Mesoporous Carbon Materials for Advanced Energy Conversion and Storage (Review) *Part. Part. Syst. Charact.* 1-25. DOI: 10.1002/ppsc.201300315

[11] Dutta S, Bhaumik A and Wu K C W. **2014** Hierarchichally porous carbon derived from polymers and biomass: effect of interconnected pores on energy application *Energy Environ. Sci.* **7** p 3574

[12] Chen Y, Zhu Y, Wang Z, Li Y, Wang L, Ding L, Gao X, Ma Y and Guo Y 2011 Application studies of activated carbon derived from rice husks produced by chemical-thermal process – a review *Adv. Coll. Inter. Sci.* **163** p 39

[13] Ma G, Yang Q, Sun K, Peng H, Ran F, Zhao X and Lei Z 2015 Nitrogen-doped porous carbon derived from biomass waste for high-performance supercapacitor *Bioresource Tech.* **197** **p** 137

[14] Shen W, Li Z and Liu Y 2008 Surface chemical functional groups modification of porous carbon *Recent Patents on Chem. Eng.* **1** p 27

[15] Sekine T and Zakir M 2008 Oxidative dissolution of Tc(IV)O2.nH2O colloid by sonolysis *Radiochim. Acta* **96** p 625

[16] Jia Y F and Thomas K M 2000 Adsorption of cadmium ions on oxygen surface sites in activated carbon *Langmuir* **16** p 1114

[17] Liu Y, Hu Z, Xu K, Zheng X and Gao Q 2008 Surface modification and performance of activated carbon electrode material *Acta Phys.-Chim.Sin.* **24** p 1143

[18] Boehm H P 1994 Some aspects of the surface chemistry of carbon blacks and other carbons *Carbon* **32** p 759

[19] Boehm H P 2002 Surface oxides on carbon and their analysis: a critical assessment *Carbon* **40** p 145

[20] Ismanto A E, Wang S, Soetaredjo F E and Ismadji S 2010 Preparation of capacitor's electrode from cassava peel waste *Bioresource Tech.* **101** p 3534

Application of MSSIP-2 nutrient in marine phytoplankton culture to support the production of biomass for biofuel industry

Paulina Taba, S Kasim and I Raya

Department of Chemistry, Hasanuddin University, Jl. Perintis Kemerdekaan, Km 10, Tamalanrea, Makassar 90245, Indonesia

E-mail: paulinataba@unhas.ac.id

Abstract. A research on the application of MSSIP-2 nutrient in marine phytoplankton culture has been conducted to produce biomass to be used as raw material for biofuel. Marine phytoplankton was cultivated using the nutrient media and the growth rates were studied by measuring the cell solidity at various growth times. Seven phytoplanktons; *Isochrysis aff galbana, Spirulina sp., Thalassiosira sp.*, and *Nitzchia sp., Chlorella vulgaris, Chaetoceros calcitrans and Isochrysis tahiti* were used in the research. The experimental temperature, salinity, and pH of the media were measured and the specific growth rates of phytoplanktons were determined using the first order rate equation. Results showed that the highest specific growth rate was given by *Chlorella vulgaris* (0.0322 cells/hour) and the lowest one was by *Thalassiosira sp.* (0.0277 cells/hour). The highest biomass weight was obtained from *Isochrysis aff galbana* (0.329 g), whereas the lowest one (0.27 g) was from *Nitzchia sp.* The carbohydrate content was various, the highest content was 34.07% found in *Isochrysis aff galbana* and the lowest was 28.16% in *Thalassiosira sp.*

1. Introduction

Fossil fuels (coal and oil) become a primary energy nowadays. Their contribution as a source of energy is around 80% where more than 50% of the total energy was for transport sector [1]. The fossil fuel is non-renewable resulting in the drastically decrease of its sources. This kind of fuel can produce harmful gases that have a negative impact. Furthermore, the high use of the fossil fuel can affect the global economic activities due to the increase in the prices of crude oil. The high-speed modern world travels by both industrialization and motorization and it is being the main cause for the unpredictable fuel demand [2]. Therefore, many efforts have conducted in order to find alternative energy sources which are renewable. Biofuels are alternative energy sources that are renewable and can be produced from biomass. Researchers are constantly trying to find the biofuel production from the sustainable biomass since it is being an efficient alternative to replace non-renewable fuels [3].

There are several advantages of using biofuels compared to petroleum fuels. It is easy to extract them from the biomass obtained from various sources, they have biodegradable property for the sustainability purpose, its combustion based on carbon-dioxide cycle, and they are more environment-friendly. The use of biofuel in automobile market will enhance rapidly for the next decade because of its environmental merits. Therefore, the strong growth in the agricultural sector for more production and associated by-products will occur [4,5].

Biofuels produced mainly from biomass can be in the form of solid, liquid and gaseous fuels. They are classified into first, second and third generations depending on the chemical composition and the natural complexity of the biomass. The first one, biodiesel and vegetable oils have been produced from the crop plants and the second, bioethanol and biohydrogen have been produced from agricultural by-products and energy plants which requires fertile lands for growth. The sources of the third generation which are attractive sources are marine algae [5,6].

It has been reported that biomass is one of the most important energy sources of the world [7]. It can also be considered as the natural and inexpensive source of energy and that energy could be utilized at any time [8,9]. The biomass was usually generated from major agricultural crops. However, the use of agricultural crops as resources of biofuels can compete with the need of food. Production of biofuels predominantly dependent on terrestrial plants, which have the limitation of cultivable land exploitation, marine biomass include seaweeds are new sources of biofuel feedstock produces the highest percentage of biomass productivity in lesser time and therefore such bioresources can be exploited for renewable biomass-based energy production. The various terrestrial and marine resources for sustainable and renewable resources for biofuel production can be used. The photosynthesis fuel has been initiated as another blooming field of biofuels. The plants and algae effectively utilize the atmospheric CO_2 and storing the energy as biomass which is converted into any form of energy in reverse. The photosynthetic microbe cyanobacteria have the ability to convert the CO_2 directly into ethylene a fuel chemical without production of biomass.

Biofuel (bioethanol dan biodiesel) become expectation sources that can open the new work field, help to decrease the number of poor people and to protect the damage of forestry as well as the lack of food material. Therefore, the effort to find biofuel sources have to be focused to the marine resources of Indonesia because 2/3 part of Indonesia is watered with the longest beach line, i.e. ± 80 791.42 km, having microscopic unit cell plants called phytoplankton about 35.000 species [10]. Utilization of phytoplankton for biofuel is environmentally friendly. Another advantage of the source is that phytoplankton can absorb CO_2 and convert it to oxygen gas. About 90% of phytoplankton's dried weight can absorb CO_2 that can decrease the gas to 1000 tonne/ha/year.

To produce biofuel, materials containing carbohydrate are required. Phytoplankton can be one of the sources because it can contain 29 – 30% carbohydrate that can be converted to bioethanol. In addition, it can also be a source of biodiesel because of its lipid content, *i.e.* 21% [10]. To produce 1 liter of bioethanol, it required about 6.5 kg of cassava. However, with the same amount of phytoplankton, bioethanol can be produced 100 times because harvesting can be conducted once in 7-10 days.

Besides the light, the phytoplankton growth also depends on the availability of nutrient for its growth. In this research, the excellent nutrient called MSSIP-2 was used to cultivate phytoplankton under investigation.

2. Experimental

2.1. Materials

Materials used was MSSIP-2 (Muh. Syahrul, Syahruddin Kasim, Indah Raya and Paulina Taba) consisting of sodium chloride (Merck), $MgSO_4.7H_2O$ (Merck), KNO_3 (Merck), KH_2PO_4 (Merck), $CaCl_2.6H_2O$ (Merck), H_3BO_3 (Merck), $ZnSO_4.7H_2O$ (Merck), $MnSO_4.4H_2O$ (Merck), $CuSO_4.5H_2O$ (Merck), $CoCl_2.6H_2O$ (Merck), $(NH_4)_6Mo_7O_{24}.4H_2O$, NaFeEDTA (Merck), $NaSiO_3.9H_2O$ (Merck), Mixture of vitamin, HCl (Merck), NaOH (Merck), Marine product waste, organic waste mixture, urea, ZA and KCl fertilizer, 1M Fe^{2+} ion, 1M Ni^{2+} ion, and distilled water.

2.2. Instruments

Instruments used were haemocytometer, High-Performance Liquid Chromatography (HPLC).

2.3. Procedure

2.3.1. Preparation of MSSIP-2 Nutrient. The MSSIP-2 nutrient was prepared using components listed in table 1.

Table 1. Composition of MSSIP-2.

Components	Concentrations (ppm)	Components	Concentrations (ppm)
NaCl	250	NaFeEDTA	25
$MgSO_4.7H_2O$	5	$NaSiO_3.9H_2O$	10
KNO_3	2 (3 mL)	Mixture of Vitamin	1
KH_2PO_4	5	HCl	50
$CaCl_2.6H_2O$	5 (10 mL)	NaOH	60
H_3BO_3	200	Marine product waste	3 mL
$ZnSO_4.7H_2O$	1	Mixture of organic waste	3 mL
$MnSO_4.4H_2O$	1	Urea, ZA, and KCl fertilizer	3 g each
$CuSO_4.5H_2O$	0,1	Fe^{2+} (1M)	3 mL each
$CoCl_2.6H_2O$	0,2	Ni^{2+} (1M)	1 mL each
$(NH_4)_6.Mo_7.O_{24}.4H_2O$	0,2		

All components were dissolved in distilled water up to a total volume of 1000 mL as nutrient stock for culture, its use was 1% - 20% depending on the culture volume.

2.3.2. Cultivation of marine phytoplankton and analysis of density. Phytoplanktons used in this study were *Isochrysis aff galbana*, *Spirulina sp.*, *Thalassiosira sp.*, and *Nitzchia sp.*, *Chlorella vulgaris*, *Chaetoceros calcitrans* and *Isochrysis tahiti* provided by the Laboratory of Plankton, Shrimp Seeds Centre, Maros. Phytoplankton was cultured in a medium with superior nutrient, then put into a 100 mL Erlenmeyer. After 4 days, the culture was transferred to a 500 mL bottle. During the cultivation, the physicochemical parameters were maintained including lighting of 40 watt with TL lamps continuously, CO_2 gas aeration, temperatures of between 25-27°C, sea medium pH of between 8.3-9.0; and medium salinity for the optimum condition of phytoplankton species. All equipment and materials used in culture were sterilized. To find out the growth pattern of phytoplankton, calculation of cell count per mL of medium was observed every 12 hours with a haemositometer using a microscope.

At this culture, the specific growth was calculated using equation (1):

$$\mu = \frac{\ln N_t - \ln N_0}{t} \quad (1)$$

N_t = cell population density at the time of t (cell mL^{-1})
No = cell population density at the early time (cell mL^{-1})
μ = fixed flow of specific growth (h^{-1})
t = time (h).

2.3.3. *Culturing biomass of chosen phytoplankton and determination of the biomass Weight.* Based on physical chemistry properties, the parameters affecting the growth of phytoplankton in the field were measured in situ, a great amount of phytoplankton chosen was cultivated in the 60 L aquarium. After the optimal density of phytoplankton obtained, 1 L of the culture was taken to determine its biomass by using gravimetric method through weighing with suitable filter paper. The light and aeration were stopped for about 3 days followed by separation of phytoplankton from water by sedimentation.

2.3.4. Analysis of the carbohydrate content in the chosen phytoplankton. The carbohydrate and monosaccharide content was analyzed using HPLC following a procedure developed by Mannheim,

1987. The total amount of 4 grams biomass of selected marine phytoplankton species was inserted into a 250 mL measuring flask, then added with 40 ml doubly distilled water + 2 mL of TCA 3M, and left for 10 minutes, neutralized with 1 M NaOH, added with 200 mL of doubly distilled water, filtered with Whatman no. 1 followed by syringe 0.45 μm. The clear filter solution was used for HPLC measurement. The sample solution was inserted into the venojek which was first sterilized, then sealed tightly and immediately analyzed by HPLC using an Aminex HPX-87H column. All the data obtained were then analyzed quantitatively by comparing the chromatogram results with the standard chromatogram.

3. Results and discussion

3.1. Physico-chemical parameter of various types of marine phytoplankton

Physico-chemical parameters in the research location were taken in-situ, then combined with culture parameters such as lighting, aeration rate blower for the CO_2 gas pump, and initial density of marine phytoplankton. The result is given in table 2.

Table 2. Physico-chemical parameters of marine phytoplankton cultures.

Phytoplanktons	Parameters of culture growth of marine phytoplankton					
	Salinity (‰)	pH	Tempe-rature (°C)	Lighting	Aeration	Initial density (cell/mL)
Isochrysis aff galbana	30	8 - 9	27-30	TL lamp, 40 watt	Air	30,000
Spirulina sp.	30	8 - 9	28-30	TL lamp, 40 watt	Air	30,000
Thalassiosira sp.	30	7 - 9	29-30	TL lamp, 40 watt	Air	30,000
Nitzchia sp.	30	8 - 9	27-30	TL lamp, 40 watt	Air	25,000
Chlorella vulgaris	30	7 - 9	28-30	TL lamp, 40 watt	Air	30,000
Chaetoceros calcitrans	33	8 - 9	29-30	TL lamp, 40 watt	Air	25,000
Isochrysisi tahiti	33	8 - 9	28-30	Sda	Air	30,000

3.2. Cell Density and the specific growth rate of Phytoplanktons

After cultivation using MSSIP-2 nutrient, the density of phytoplankton was determined using a haemacytometer. The results are given in Table 3.

The cell density of samples of marine phytoplankton as a function of the growth time is given in Figure 1.

It is clear that the cell density of marine phytoplankton generally increases with the increase of the cultivation time until achieving the maximum. All species of phytoplankton show similar growth patterns consisting of 3 stages, *i.e.* adaptation phase, exponential phase and stationary phase. The optimum density was achieved at 8 – 10 days. After the optimum growth of phytoplankton, the cell density decreases. The lowest cell density at the optimum growth time is given by Nitzia *sp.* (1.18 x 10^7 cells/mL) and the highest cell density is shown by Spirulina sp (2.029 x 10^7 cells/mL). The specific density rate of the marine phytoplankton is given in Table 4.

IOP Conf. Series: Journal of Physics: Conf. Series **979** (2018) 012025 doi:10.1088/1742-6596/979/1/012025

Table 3. Cell density of phytoplankton.

No.	Types of phytoplankton	Growth density (x 10^4 cell/mL)													
		1	2	3	4	5	6	7	8	9	10	11	12	13	14
1	*Isochrysis aff galbana*	3.0	37	94	217	587	904	1221	1305	1449	1415	1360	1206	1079	918
2	*Spirulina sp.*	3.0	39	96	207	453	881	1308	1565	1813	2029	1978	1619	1479	1315
3	*Thalassiosira sp.*	3.0	18	58	93	262	672	911	1019	1110	1186	1020	969	811	737
4	*Nitzchia sp.*	2.5	21	106	243	333	523	819	1045	1097	1180	1128	930	721	619
5	*Chlorella vulgaris*	3.0	23	104	317	782	994	1185	1370	1293	1102	1008	896	672	415
6	*Chaetoceros Calcitrans*	2.5	15	37	90	263	598	962	1209	1152	835	703	514	472	395
7	*Isochrysis tahiti*	3.0	18	89	237	562	803	1048	1196	1265	1149	1027	831	650	317

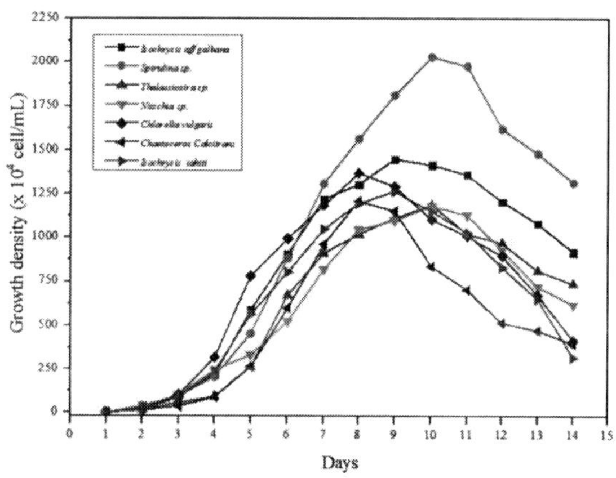

Figure 1. Growth density rate of 7 types of marine phytoplankton.

Table 4. The specific growth density rate of marine phytoplankton.

Species of marine phytoplankton	The optimum growth time (day)	Cell density		μ(hour^{-1})
		N_o	N_t	
Isochrysis aff galbana	9	3 x 10^4	1.448 x 10^7	0.0321
Spirulina sp.	10	3 x 10^4	2.029 x 10^7	0.0302
Thalassiosira sp.	10	3 x 10^4	1.186 x 10^7	0.0277
Nitzchia sp.	10	2.5 x 10^4	1.180 x 10^7	0,0284
Chlorella vulgaris	8	3 x 10^4	1.370 x 10^7	0.0322
Chaetoceros Calcitrans	8	2.5 x 10^4	1.209 x 10^7	0.0365
Isochrysis tahiti	9	3 x 10^4	1.265 x 10^7	0.0315

It is obvious that the lowest specific growth rate is given by *Thalassiosira sp,* whereas the highest one is by *C. vulgaris*.

3.3. Biomass of marine phytoplankton
Table 5 shows the dried weight of the phytoplankton biomass under investigation.

Table 5. The dry weight of phytoplankton biomass.

No.	Types of Phytoplankton	Biomass dry weight (mg L^{-1})
1.	*I. aff galbana*	0,329
2.	*Spirulina sp.*	0,325
3.	*Thalassiosira sp.*	0,315
4.	*Nitzchia sp.*	0,270
5.	*C. calcitrans*	0.285
6.	*C. vulgaris*	0.320
7.	*I. tahiti*	0.299

From the data in table 5, it can be concluded that the highest biomass obtained from *I. aff galbana,* whereas the lowest one is from *Nitzchia sp.*

3.4. Carbohydrate content of marine phytoplankton
The carbohydrate content of various types of marine phytoplankton is given in table 6.

Table 6. Carbohydrate content in phytoplankton.

No.	Types of Phytoplankton	Carbohydrate content (%w/w)
1.	*I. aff galbana*	33.98
2.	*Spirulina sp.*	32.85
3.	*Thalassiosira sp.*	28.16
4.	*Nitzchia sp.*	29.74
5.	*C. calcitrans*	32.63
6.	*C. vulgaris*	34.07
7.	*I. tahiti*	33.78

All types of phytoplankton cultivated in the MSSIP-2 nutrient contain high carbohydrate (28.16 – 33.98 %w/w). There are several factors affected the biochemical content including carbohydrate content [11] such as the intensity of light [12], the composition of the species [13] and growth steps [14]. However, the most factor affected the biochemical content were nutrient availability especially nitrogen source [15,16]. From the data of the carbohydrate content in this study, it can be said that MSSIP-2 is a good nutrient for growing phytoplankton which are potential to be used in producing biofuel.

4. Conclusion
The MSSIP-2 nutrient has good ability to grow all types of phytoplankton used in the study. The optimum growth time of all types of phytoplankton is in the range of 8 – 10 days. The highest biomass produced is given by *I. aff galbana* with the carbohydrate content of 33.98 %w/w. The lowest biomass is produced by Nitzia sp with the carbohydrate content of 29.74 %w/w.

Acknowledgments
Authors wishing to acknowledge The Minister of Research, Technology and High Education for the financial support to conduct this research and to Hasanuddin University.

References
[1] Escobar J C, Lora E S, Venturini O J, Yanez E E, Castillo E F, Almazan O 2009. *Renew. Sustain Energy Rev.* **13** 1275–87
[2] Agrawal A K 2007 Prog. *Energy Combust. Sci.* **33** 233–71
[3] Weldemichael Y and Assefa G 2016 *J. Clean Prod.* **112** 4257–64.

[4] Kim S and Dale B E 2005 *Biomass- Bioenergy* **29** 426–39
[5] Demirbas A 2008 *Energy Convers. Manag.* **49** 2106–16
[6] Kang Q, Appels L, Tan T and Dewil R 2014 *Sci. World J.* **4** 1–13 http://dx.doi.org/10.1155/2014/298153
[7] Acma H and Yaman S 2010 *Renew. Energy* *35* 288–92
[8] Balat H 2009 *Energy Educ. Sci. Technol. Part A* **24** 85-111
[9] Kan A 2009 *Energy Educ. Sci. Technol. Part A* **23** 55-69
[10] Kasim S and Sjahrul M 2012 *Eur. Chem. Bull.* **1**(6) 311-316
[11] Kanga J J, Jooa H T, Leea J H, Leea J H , Leea H W, Leea D, Kangb C K, Yuna M S and Heon Lee S 2017 *Deep Sea Research Part II: Topical Stud. Oceanogr.* **143** 82-90
[12] Smith A E and Morris I 1980 *Limnol. Oceanogr.* **25** 865–872
[13] Rivkin R B and Voytek M A 1987 *Limnol. Oceanogr.* **32** 249–259
[14] Smith R E H, Clement P, Cota G F and Li W K W 1987J. Phycol. **23** 251–263
[15] Granum E, Kirkvold S and Myklestad S M 2002 Mar. Ecol. Prog. Ser. **242** 83–94
[16] Kim B K, Lee J H, Joo H, Song H J, Yang E J, Lee S H and Lee S H 2016 *Deep-Sea Res. Part II* **123**, 42–49

Determination of the optimum concentration cellulose baggase in making film bioplastic

S Chadijah[1], W O Rustiah[2] and M I D Munir[1]

[1] Department of Chemistry, Faculty of Science and Technology, UIN Alauddin, Makassar, Indonesia
[2] Makassar Muhammadiyah Academy of Health Analyst. Jl. Dr. Ratulangi No. 101 Makassar, Indonesia

E-mail: chadijah19@yahoo.com

Abstract. The hoarding rubbish synthetic plastic caused pollution and demage in life circles, to cope it can be done with synthesizing the plastic from agriculture substance or called biopolymer (bioplastic). It was that potentially as bioplastic was biopolymer from agriculture substance baggase that contain cellulose 40 %. This research aimed to determine the optimum concentration cellulose baggase in making bioplastic film with adding chitosan and sorbitol plasticizer and also to know the result of characterization film bioplastic. The steps in this research were; the extraction of cellulose, making film bioplastic, tensile strenght test and used characterization spectrofotometer FTIR. In this research showed that optimum concentration cellulose baggase in making film bioplastic was 2% with adding chitosan and sorbitol plasticizer. The optimal result of tensile strenght test was 0,089 Kgf/cm^2 with elongation percent 15,90 %. The analyzing FTIR in all of variation that looked almost same with characterization with tapes -OH, -NH and C-O.

1. Introduction

The hoarding plastic waste in the world improved anytime [1], especially in indonesia exactly in urban areas including the largest plastic waste producer. According to the Ministry of Environment and Marine Affairs (KLHK) in 2016, each year approximately 1,3 billion tons of plastic waste is produced around the world. This excessive plastic consumption, putting Indonesia ranked second in the world of plastic waste to the Sea after China. Excessive plastic consumption was potentially a material that threaten the continuity of living things on earth, because it faced many environmental problem such as cannot be recycled and cannot be deciphered naturally by microbes in the soil. So there was a hoarding of plastic waste that caused pollution and damage to the environment [2].

The effort that can be done to reduce the consumption of plastic was to synthesize a polymer (the material of polymer or plastic manufacture) easily decomposed or was called biodegradable plastic (Bioplastic) [3]. In Indonesia the utilization of agricultural and industrial waste was still very less developed, while the waste generated very much and still contain benefits, one of which was a natural polymer found in waste bagasse.

Sugarcane waste was one of the wastes that contain natural polymers, especially cellulose [4], bagasse was found in many areas in Indonesia because it grows fertile and the main ingredient in the sugar industry, to get it was not so difficult too abundant. Sugarcane waste in Indonesia on average obtained 35-40% of each processed cane [5]. Cane production in Indonesia in 2007 amounted to 21 million tons, the potential for pulp produced about 6 million tons of pulp per year. The bagasse contained up to 44% cellulose, which cellulose content of more than 40% can already be used as raw material for making bioplastik [6]. Cellulose was the basic ingredient of plants that belong to the primary metabolite, forming the material structure of most plant cell walls. *and 10 mm after the title.*

2. *Experimental*

2.1 Materials
Chitosan was purchased from CV. Chi Multiguna, cellulose derived from bagasse from PT. PN XIV (Persero) Takalar Sugar Factory Pa'rappunganta Village Kec. Polobangkeng Utara.

2.2 Cellulose Extraction of Sugar Cane
The bagasse cleaned, washed and dried under direct sunlight. Then it mashed with a blender and sieved with a sieve of 40 mesh, then the sample was put into a maceration container and dissolved with CH_3OH to 7 days or the solution was not longer colored by replacing CH_3OH every 2 days. After the maceration process, the sample filtered with using whatman filter paper no. 42. The residue from the filtration process washed with H_2O and dried. Then it fed into a 1000 mL glass and 17.5% NaOH added until the sample submerged, then it heated for 60 min. After that, it filtered again and washed with H_2O. It hydrolyzed using HCl for 3 h, then washed with H_2O then dried in oven at 105 °C for 1 hour [7].

2.3 Making Bioplastic Films with Addition of Chitosan and Sorbitol
Chitosan of 0.8 g put into a 250 mL glass and dissolved with CH_3COOH 0.6 M, then it also put into ultrasonic for 8 min and sterilized until the temperature dropped to 50 °C. The results were added to 0.4, 0.6, 0.8, 1 and 1.2 g cellulose which had homogenized with sorbitol. The mixture stirred using a magnetic stirrer for 15 minutes, then it heated to 80 °C for 7 minutes. After that, it poured on a mold plate (glass plate). Repeat the procedure on cellulose by the addition of chitosan and cellulose with the addition of sorbitol [8].

3. Result and discussion

3.1 Cellulose bioplastic film
Cellulose extraction carried out on bagasse originating from Takalar District of South Sulawesi. The bagasse were macerated, delignified and hydrolyzed,it result in α-cellulose [7]. Based on research conducted in the manufacture of cellulosic bioplastic film with the addition of sorbitol and chitosan, the bioplastic film obtained a little stiff but flexible, yellowish brown, can be removed from the mold as a whole and the surface is rather slippery.

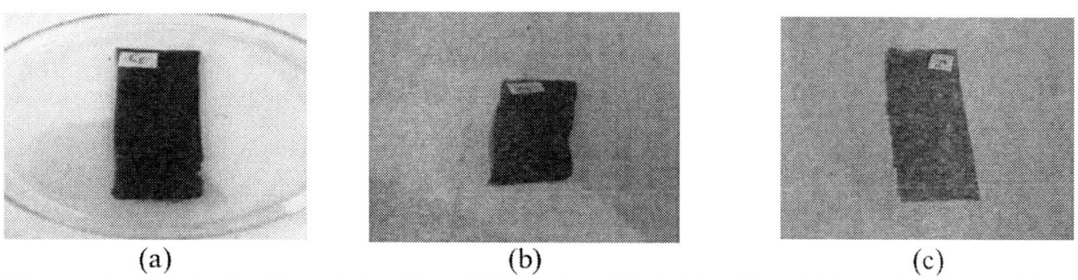

(a) (b) (c)

Figure 1. Bioplastic Films (a) with addition of sorbitol (b) with addition of chitosan (c) with addition of sorbitol and chitosan

IOP Conf. Series: Journal of Physics: Conf. Series **979** (2018) 012026 doi:10.1088/1742-6596/979/1/012026

The production of bioplastic film with the addition of sorbitol and chitosan aimed to increase the mechanical characteristic the bioplastics. Cellulose variations performed to increase tensile strength and percent elongation of the bioplastic film itself, where cellulose molecules form linear long chains and have a tendency to form intra- and inter-molecular hydrogen bonds. The hydrogen bond can be derived from chitosan and sorbitol and greatly affects tensile strength and percent elongation of bioplastic films. The result obtained when compared with research [6], have similarities that are flexible and can be released intact, but the color of the bioplastic film there was a difference because in the study added bleach so the film was white.

3.2 Tensile Strength and Elongation
Mechanical test also performed to measure tensile strenght at the time the three samples loaded [9]. The result of the tensile strenght of the three bioplastic showed in the figure below:

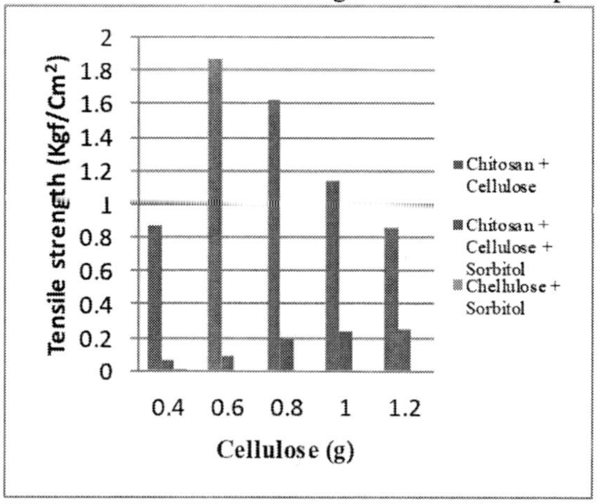

Figure 2. Tensile strength of cellulose bioplastic bagasse

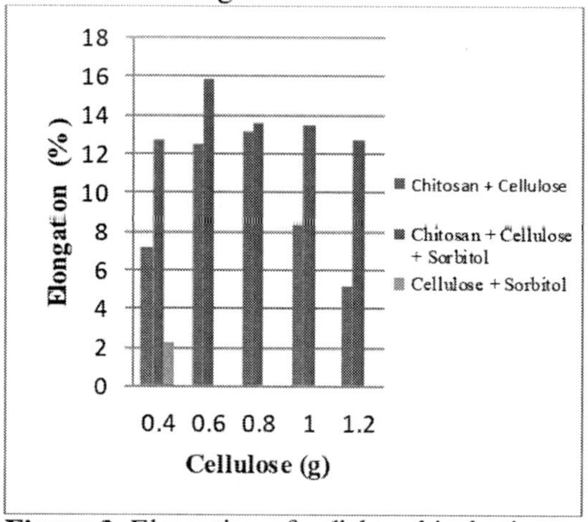

Figure 3. Elongation of cellulose bioplastic bagasse

The figure 2 performed to measure tensile strength at the time the three samples loaded [9]. The bioplastic from cellulose with the addition of chitosan and sorbitol greatly affect the value of tensile strength. The functional groups on the cellulose chain can interact with the -O and -N groups, forming hydrogen bonds where the hydrogen bond is longer than the covalent bond but the bond was weaker. The more hydrogen bonds that form the longer the chain [10]. Therefore, in the addition of chitosan and sorbitol interactions occur between the hydrogen bonds cellulose, chitosan and sorbitol.

In cellulose with the addition of chitosan and sorbitol, tensile strength of 0.4 g concentration of 0.071 Kgf/cm^2 and an increase of 0.6 g with a tensile strength of 0.089 Kgf/cm^2. The elongation tensile we improved 0.191 Kgf/cm^2 by 0.8 g of cellulose, 0.241 Kgf/cm^2 by 1.0 g of cellulose, 0.245 Kgf/cm^2 by 1.2 g of cellulose. The increasing in tensile strength test on these five concentrations was caused by the more free OH on the cellulose given with chitosan. In cellulose with the addition of chitosan and sorbitol, the resulting tensile strength increases. In addition to OH in cellulose there was also a free OH from sorbitol that will bind to cellulose and chitosan [11]. Sorbitol also had the ability to increase the tensile strength of bioplastic films.

The higest tensile strenght obtained was 1,868 Kgf/cm^2. This means that, this film when compared with research [12] still have much different of 2,5-36,7 MPa as well as in the research [11] that was equal to 9,12 MPa.

The figure 3 performed to measure the percent elongation of the three bioplastic samples until they disconnected [13]. On cellulose with the addition of chitosan and sorbitol elongation of 0,4 g concentration of 12,73% and an increase of 0,6 g with an elongation percentage of 15,90%. The increasing in percent elongation test at these three concentrations, were due to the more free OH on cellulose binding to sorbitol and vice versa [14]. Then decreased at 0,8 g cellulose concentration by 13,57% and continued to decrease at concentrations of 1,0 g and 1,2 g by 13,46% and 12,66%. The decreasing in elongation test caused by decreasing biopolymer flexibility due to decrease of hydrogen bond between chitosan and cellulose which can be decided by sorbitol. The optimum concentration obtained in bioplastic film was 15,90%, the results obtained and also have been resemblance to the results of the research [15], a percentage of elongation of 13,46%. This means, the percent of elongation produced with increasing considerably.

3.3 Characterization of functional groups using FTIR

The analyzis FTIR performed to prove that the results of the bioplastic film making process have distinctive bioplastic functional groups.

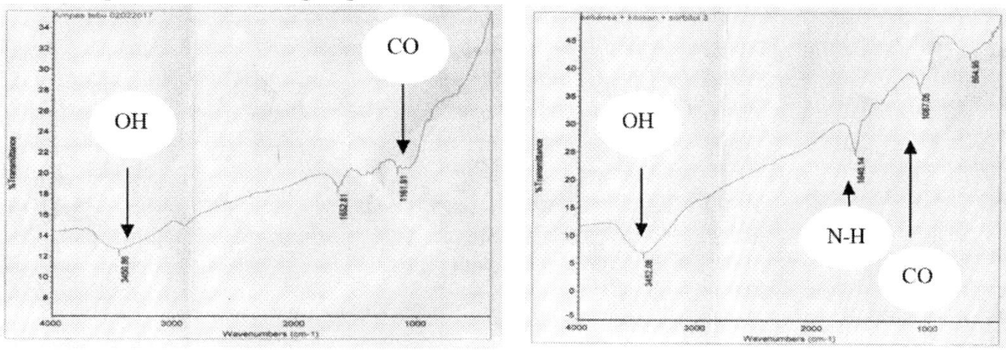

(a) (b)

Figure 4. Functional groups (a) The baggase cellulose (b) Cellulose bioplastic film with the addition of chitosan and sorbitol.

The figure 4, cellulose of bagasse had an absorption in the area of 3450.86 cm^{-1} showing the -OH group, reinforced with an area of 1161.91 cm^{-1} showing the C-O group. In the cellulose figure of bagasse with the addition of chitosan and sorbitol, the -OH group (alcoholic group) shown at the peak of 3452.08 cm^{-1} [16], this uptake was almost identical to the N-H group because it has a nearly equal absorption region. In addition, there was a C-O-C group at the peak of 1087.05 cm^{-1} or indicating a glycosidic bond in cellulose. That is, bioplastic film was a film with a physical mixing method, where in each mixing there was not new functional group formed, but the intensity of the uptake is different. This is clearly seen in the uptake of cellulose, cellulose with chitosan and sorbitol, in which the addition of chitosan resulted in increasing absorption intensity due to the hydrogen bond that was formed more and more and will decrease due to the addition of sorbitol, thus reducing the hydrogen bond. Moreover, the -OH group appearing in the uptake area of 3452.08 cm^{-1} proves that bioplastic film can be decomposed by microorganism activity in the soil [17]. The resulting FTIR characterization was similar to research [15], where O-H appeared at wavelengths of 3080 cm^{-1} and N-H at 1664 cm^{-1}.

4. Conclusions

The optimum concentration of cellulose of bagasse in the making of bioplastic film with the addition of chitosan and sorbitol plastisizer is at cellulose concentration 2% (w / v) with percent elongation of 15.90% and tensile strength of 0.089 Kgf/cm^2. The result of FTIR characterization show that there were functional groups -OH, -NH and C-O-C.

IOP Conf. Series: Journal of Physics: Conf. Series **979** (2018) 012026 doi:10.1088/1742-6596/979/1/012026

References

[1] Liew K C and Khor L 2013 *J. Of King Saud Univ. – Eng. Sci.* 137-41
[2] Schulze C, Juraschek M, Herrmann C and Thiede S 2017 *J. Proc. CIRP* **61** 600-5
[3] Mehdi E S, T. Onay T and Demirel B 2016 *J. Wast. Manag.* 1-11
[4] Brodin M, Vallejos M, Tanase O M, Cristina A M and Chinga-C G 2017 *J. Of Clea. Prod.* 1-36
[5] Mubarik S, Saeed A, Athar M M and Iqbal M 2015 *J. Of Indu. and Eng. Chem.* 1-7
[6] Sumartono N W, Handayani F, Desiriana R, Novitasari W and Hulfa D S 2015 *J. Peli.* **10** 13-25
[7] Lenihan P, Orozco A, O'Neill E, Ahmad M N M, Rooney D W and Walker G M 2010 *Chemical Engineering Journal* 395-403.
[8] Isroi, Cifriadi A, Panji T, A Wibowo Nendyo, Syamsu K, 2017, *Earth and Environmental J. Science* **65** 1-10
[9] Perez V, Felix M, Romero A and Guerrero A 2016 *J. Food and Biop. Proc.* 100-8
[10] González-G J, Partal P, García-M M and Gallegos C 2011 *J. Carb. Poly.* 308–15
[11] Maulida, Siagian M and Tarigan P 2016 *J. Of Phys. : Conf. Seri.* 1-8
[12] Sapei L, Padmawijaya K S, Sijayanti O and Wardhana P J 2017 *J. Mate. Scie. and Eng.* 1-8
[13] González-G J, Partal P, García-M M and Gallegos C 2011 *J. Carb. Poly.* 308–15
[14] Zárate-R L S, Romero A, Bengoechea C, Partal P and Guerrero A 2014 *J. Carb.e Poly.* 24–31
[15] Fajrin A, Sari L A, Rahmawati N, Saputra O A and Suryanti V 2017 *J. Mate. Scie. and Eng.* 1-7
[16] Silverstein R M, Webster F X and Kiemle D J 2005 *J. John Wiley and Sons Inc* 1-550
[17] Sato S, Saika A, Shinozaki Y, Watanabe T, Suzuki K, Sameshima-Y Y, Fukuoka T, Habe H and Morita T K H 2017 *J. Poly. Degr. and Stab.* 26-32

Voice Based City Panic Button System

Febriansyah, Zahir Zainuddin, M. Bachtiar Nappu

Department of Electrical Engineering, Faculty of Engineering Hasanuddin University.

E-mail: febriansyahey@gmail.com

Abstract. The development of voice activated panic button application aims to design faster early notification of hazardous condition in community to the nearest police by using speech as the detector where the current application still applies touch-combination on screen and use coordination of orders from control center then the early notification still takes longer time. The method used in this research was by using voice recognition as the user voice detection and haversine formula for the comparison of closest distance between the user and the police. This research was equipped with auto sms, which sent notification to the victim's relatives, that was also integrated with Google Maps application (GMaps) as the map to the victim's location. The results show that voice registration on the application reaches 100%, incident detection using speech recognition while the application is running is 94.67% in average, and the auto sms to the victim relatives reaches 100%.

1. Introduction

Criminality is a social disorder and a deviating behavior, which is breaking social norms and criminal law, against humanitarian morals, and harmful, thus opposed by society. In the urban area, criminality issues are difficult to overcome as it might happen anywhere instantaneously [1].

Criminality problems obviously contradict to the concept of comfortable and safe life in smart city. Smart city itself is an urban planning concept, which utilizes technological developments that would make life easier and healthier with high level of efficiency and effectiveness. One of the smart city classifications is the smart living, which requires comfortable and safe environment [2].

In anticipating such problems, several protections involving the police in crime-prone locations have already been done yet it is considered less efficient and effective as the crime might occur anywhere instantaneously in urban area while the police is far from the location.

Other anticipation is done by employing technology and smart system where one that has been successfully implemented is the panic button, which utilizes android platform of smart phones. This panic button works with several touched-combinations on smart phone then the danger notification and the user coordinate point would be received by the control quarters then forwarded to the police by the central controller admin. Nevertheless, this method still requires development using any faster and more efficient method.

One possible way to develop panic button is using speech recognition as user identification on the condition of danger. Some previous related researches are as follows: Research on Emergency Call and Location Tracking System with Enhanced Functionality for Android [3]. Review on Android Based SOS Emergency Alert Button [4]. The Framework of Navigation and Voice Recognition

System of Robot Guidance for Supermarket [5]. A Computer Remote Control System Based on Speech Recognition Technologies of Mobile Devices and Wireless Communication Technologies [6]. Zigbee Based Voice Controlled Wireless Smart Home System [7]. An integrated system for voice command recognition and emergency detection based on audio signals [8]. Android Based Emergency Alert Button [9]. A Survey on Speech Recognition [10]. Emergency Panic Button [11]. Voice Recognition using HMM with MFCC for Secure ATM [12].

On the basis of those previous researches, this study focuses on the development of panic button system by using speech as the danger identification of application user then the danger notification, including the incident location, would be sent to the police and victim relatives, whose mobile number already inputted, in the form of location map and SMS.

2. Research Method

This is a quantitative research that aims to design and implement the voice-activated panic button application by employing the voice recognition in android system. The data collection method used is accidental sampling. The built-application was tested by direct implementation to the respondents, which are citizen, in several locations. The data taken were the respondents, as victim, coordinate points and the police to measure the accuracy of the program.

The user voice was an input used as the detection when in danger conditions. The voice that was detected by the application would transmitted the coordinates point and auto SMS to the web server while the nearest police coordinates and victim coordinate were compared by using Haversine formula, as illustrated in Figure 1.

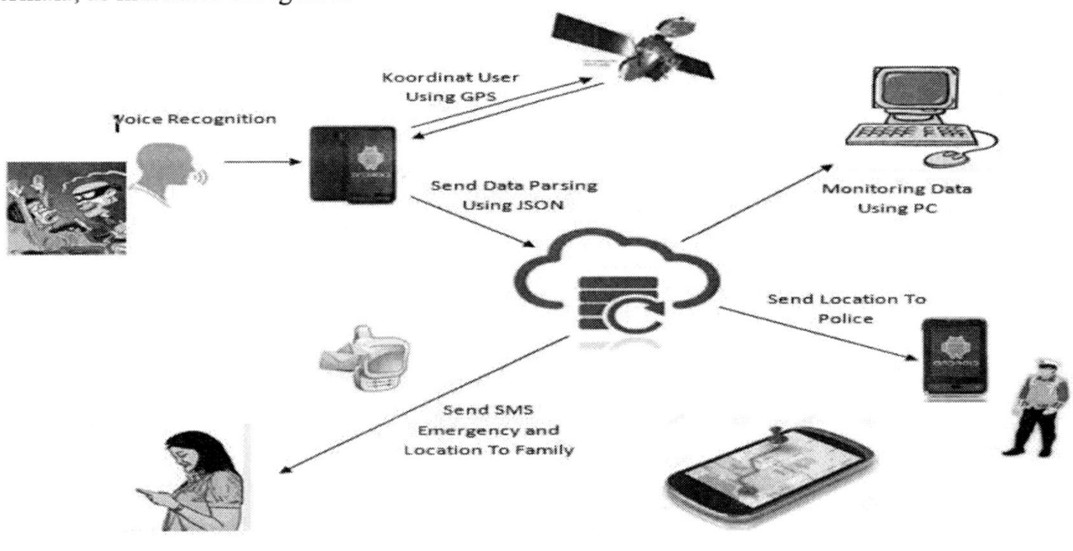

Figure 1. Application Architecture

2.1. GPS Tracking Via Java Programming

The location coordinate tracking of this panic button application is built with script access mobile device which arranged in java script android. GPS tracking script is useful to capture the location coordinates that are sent to the MySQL Server database.

Location coordinates are represented in the form of longitude and latitude - the height of coordinate system. Latitude is defined as 0-90 degrees north or south of the equator while longitude is defined as 0-180 degrees east or west of the prime meridian, which passes through Greenwich, England. Latitude is represented in meters above sea level.

2.2. Haversine Formula Application

This study shows that the application of Haversine formula in determining the closest distance of each coordinate point suggests a fairly high accuracy. The Haversine formula is a significant equation in navigation where it provides the distance between two points in the spherical sphere of any longitude and latitude [13].

The Haversine formula in this study is used in calculating the distance between two GPS coordinate points. Assuming that the earth is perfectly spherical with radius R 6,367.45 km and the location of the two spherical coordinates (latitude and longitude) are lon1, lat1 and lon2, lat2 respectively, then the Haversine formula is as follows:

Haversine formula

$$x = (lon2 - lon1) * cos\,((lat1 + lat2)/2) \tag{1}$$

$$y = (lat2 - lat1) \qquad d = sqrt(x * x + y * y) * R \tag{2}$$

where the formula x is the longitude, y is the latitude and d is the distance. In this research, the coordinate points that belong to the citizen user and the police user are the primary key for distance comparison in the closest distance determination.

3. Results and Testing

3.1. Result

The design of this voice-activated panic button has produced an android based application system that provides two levels of user in the application. Figure 2 shows the registration form of both user levels, namely citizen user and police user.

Figure 2. Registration Form

The citizen registration menu contains the input column of relatives number that will be used by the system as the destination number of notification which informs that the user is in danger. The notification is sent from victim mobile phone via auto SMS of the application, as shown in Figure 3 below:

Figure 3. Citizen Registration Form

Furthermore, the citizen registration menu requires voice input as an identification of the danger situation experienced by the user when the application is running. The user voice is inputted twice and the system would stop if the first and second voice inputs do not match each other therefore the registration process need to be repeated by input the identical sound twice. The first sound is recognized by using the voice recognition method thus the voice input in citizen registration menu must be done twice. The first voice input would be the data in template database while the second voice input is used for comparison and confirmation of the first data.

When the app is starting, the speaker phone will continually detect the sound. If a voice is detected, the system would automatically process the sound by comparing with the sound in database. When it is recognized as user voice, the data from GPS (Global Positioning System) coordinate point would be sent to the web server.

There would be a notification received by nearest police as shown in Figure 4 below. Anytime the web server receives reporting of coordinate points, it would compare them with the police coordinates that located nearest the victim. The search method is done by comparing the distance within a radius of five kilometers using Haversine formula. If nearest police is not found in first attempt, the search radius will be increased by the multiple of five kilometers in each. The point coordinates search radius is limited to 30 kilometers

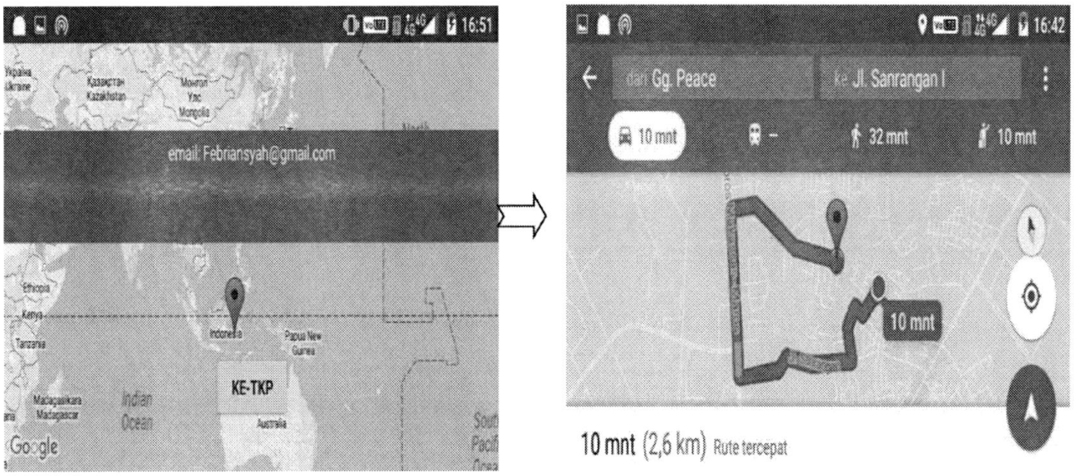

Figure 4. Police Notification

Furthermore, the victim relatives would receive a message from the victim phone number simultaneously with the notification received by the police as soon as the web server recognizes the victim coordinate point. The contents of message received by the relatives can be seen in Figure 5 below:

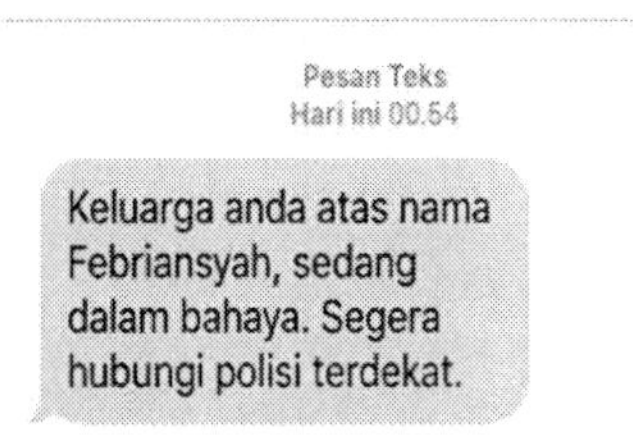

Figure 5. The Message Received by The Relatives

3.2. Testing

The testing stage was performed through direct implementation of the built system. The test involved 30 random respondents in different locations. Respondents were asked to test the application using variant keywords. This test used 7 test variables. The recapitulation of the system test is shown in table 1 below:

Table 1. Test Result Recapitulation

Num.	Test Variables	Presentation of Test Result (%)		
		Registration	Detection	Auto SMS
1	Registration Test Result	100	-	-
2	Incident Detection in Quiet Condition Test Result	-	100	100
3	Incident Detection in Crowded Condition Test Result	-	96,6666667	100
4	Incident Detection in Quiet Condition Test Result with Less than 1 M in Distance (0 to < 1 M)	-	96,6666667	100
5	Incident Detection in Quiet Condition Test Result with More than 1 M in Distance (1 to < 2 M)	-	96,6666667	100
6	Incident Detection in Crowded Condition Test Result with Less than 1 M in Distance (0 to < 1 M)	-	93,3333333	100
7	Incident Detection in Crowded Condition Test Result with More than 1 M in Distance (1 to < 2 M)	-	90	100
	Average	**100**	**94,6666667**	**100**

The recapitulation above shows significant result that the voice registration in application reaches 100%, the incident detection by using voice when the application is running is 94.67% in average, and auto sms to the victim relatives reaches 100%.

4. Conclusion

In accordance with the research results, the discussion and testing stage, it can be concluded that the design and implementation of voice-activated panic button application using voice recognition as user detector and haversine formula as closest data coordinate comparison tool with several additional features such as Gmaps and auto SMS are successfully done. It can be said that this research has contributed in providing alternative solution for criminality problem solving in the urban area to support the smart city program. To be more specific, this research results show that the use of voice registration on the application reaches 100%, the incident detection using voice when the application is running is 94.67% in average, and the use of auto SMS to the victim relatives reaches 100%.

Acknowledgments

The author would like to thank their colleagues for providing moral support in this research.

References

[1] Kartini K 2003 *Patalogi Sosial Jilid I. Cet. VIII* (Jakarta : PT RajaGrafindo Persada)
[2] Hendro Muliarto 2005 *Konsep Smart City : Smart Mobility* (Bandung : SAPPK-MPWK Institut Teknologi Bandung)
[3] Dinesh B, Raut and Pragati P 2015 *J. Advance Research in Computer Science and*

Management Studies **3** 2321-7782

[4] Rajani T, Thakare and Shahade 2014 *J. Research In Science & Engineering* **1** 2394-8299.

[5] Halim S and Budiharto W 2014 *J. Software Engineering and Its Applications* **8(10)** 143-152

[6] Hae-Duck J, Sang-Kug Y, Jiyoung L, Ilsun Y, and WooSeok H 2014 *J. Computer Science and Information Systems* **11(3)** 1001–1016.

[7] Thoraya O, Haliemah R, Ali A E N, Rehan M, Mussab M S, and Tarique M 2014 Zigbee *J. Wireless & Mobile Networks (IJWMN)* **6**

[8] Emanuele P, Stefano S, Roberto B, Giacomo F and Francesco P 2014 *J. Expert Systems with Applications.* **8**

[9] Dhrubajyoti G and Rupam K S 2013 *J. Innovative Technology and Exploring Engineering (IJITEE).* **2** 2278-3075.

[10] Harpreet S and Ashok K B 2013 *J. Advanced Research in Computer Engineering & Technology (IJARCET)* **2(6)**

[11] Priyanka S, Pranita T, Snehal T and Tejashree W 2012 *J. Scientific & Engineering Research.* **3(3)**

[12] Shumaila I, et al 2011 *J. Computer Science.* **6(3)**

[13] Chopde R N and Nichat K M 2013 *J. Innovative Research in Computer and Communication Enggineering.* **1**

Impact of Penetration Wind Turbines on Transient Stability in Sulbagsel Electrical Interconnection System

Andi Nurtrimarini Karim [1,2], Sri Mawar Said[2], Indar Chaerah Gunadin[2], Mustadir Darusman B [2,3]

[1]PT. PLN (Persero) Region Sulselrabar Area Bulukumba
Jl. Lanto Daeng Pasewang No.1 Bulukumba, South Sulawesi, 92517, Indonesia
[2]Department of Electrical Engineering Hasanuddin University
Jl. Poros Malino KM. 6 Borongloe Gowa, South Sulawesi, 92171, Indonesia
[3]Health Ministry of the Republic of Indonesia
Jl. Perintis Kemerdekaan KM. 11 Makassar, South Sulawesi, 90245, Indonesia

E-mail: rini07anika@gmail.com

Abstract. This paper presents a rotor angle analysis when transient disturbance occurs when wind turbines enter the southern Sulawesi electrical interconnection system (Sulbagsel) both without and with the addition of a Power Stabilizer (PSS) control device. Time domain simulation (TDS) method is used to analyze the rotor angle deviation (δ) and rotor angle velocity (ω). A total of 44 buses, 47 lines, 6 transformers, 15 generators and 34 loads were modeled for analysis after the inclusion of large-scale wind turbines in the Sidrap and Jeneponto areas. The simulation and computation results show the addition of PSS devices to the system when transient disturbance occurs when the winds turbine entering the Sulbagsel electrical system is able to dampen and improve the rotor angle deviation (δ) and the rotor angle velocity (ω) towards better thus helping the system to continue operation at a new equilibrium point.

1. Introduction

The electric power system generally consists of generating units connected to the line to serving the load [1]. Electrical power systems with multiple machines typically channel load power through interconnection lines [2, 3]. The main objective of the interconnection channel system is to maintain continuity and availability of electricity towards increasing need of load. The growing power system can lead to weak performance of the system when experiencing interference and greatly affect the stability of the system [4].

The stability itself is divided into three major parts namely steady state stability, dynamic stability and transient stability [5]. This paper will focus on the discussion of transient stability in southern Sulawesi electrical system (Sulbagsel). It needs to get a review because the system in Sulbagsel will enter to the Power Plant Bayu (PLTB / wind turbines) large capacity in two districts of Sidrap and Jeneponto which will certainly affect the tupologi in Sulbagsel electrical network. Changing network tupologi has impact on the flow of power that will affect too for the stability when a transient disturbance in the system.

Large and complex systems and their changes in instantaneous system conditions [6], usually happen to short circuit interruptions in the power system, and the sudden release or substantial burden of the large loads. As a result of changes in working conditions of this system and the state of the system will change from the old to the new state. A short period between the two states is called a parallel or transient period. Transient stability is based on a first swing stability condition with a period of investigation in the first second of the disturbance.

To analyzing the entry of wind turbine transient condition of the Sulbagsel interconnection system is used Power System Analysis Toolbox (PSAT). PSAT is a special software to analyze power system [7]. For the computation process results used Matlab software [8]. The system will also be tested using Power System Stabilizer (PSS) which will be placed on the system to see the damping process due to transient disturbance after the PLTB is integrated in Sulbagsel system.

2. Research Methodology

Analysis of the stability of a generator can be seen from the angle and rotor speed using Time based program. One time-based method is Time Domain Simulation (TDS) which is an indirect method that solves the transit stability problem by solving nonlinear differential equations from numerical integration with step-by-step techniques to calculate each machine of the swing curve (angle rotor against time). The modeled Sulbagsel system was then analyzed using the help of the PSAT toolbox. This toolbox is an open source software used to analyze and study the power system [9]. This TDS approach will simulate the system during periods during the faulted and postfault interruptions as well as the interruption time found between stable and unstable times (indirectly). The maximum simulation period depends on the characteristics of a good power system in terms of modeling. This usually does not exceed 15 seconds for complex system modeling. The rotor angle velocity is used to see the deviation at the entry of the wind turbine in the system after a short circuit 3 phase impedance [10]. All operations of computing can be assessed using Graphical User Interfaces (GUI).

2.1. Power System Modeling

In Sulbagsel system consists of 15 generators, 44 buses, 47 lines and 34 load centers [11] that scattered in the model where the transient voltage at q-axis is constant. The wind turbine model uses the doubly fed induction generator (DFIG) type which is connected to Sidrap and jeneponto buses. All generators are connected to the Autamitic Voltage Regulator (AVR). The researher used model 3 to facilitate the analysis process [12]. Model 3 presents hydro generator. In this model, some properties of the system are regarded like inelastic penstocks which water inertia is considered, as well as ideal turbines. For analysis implicity, static load models are also used.

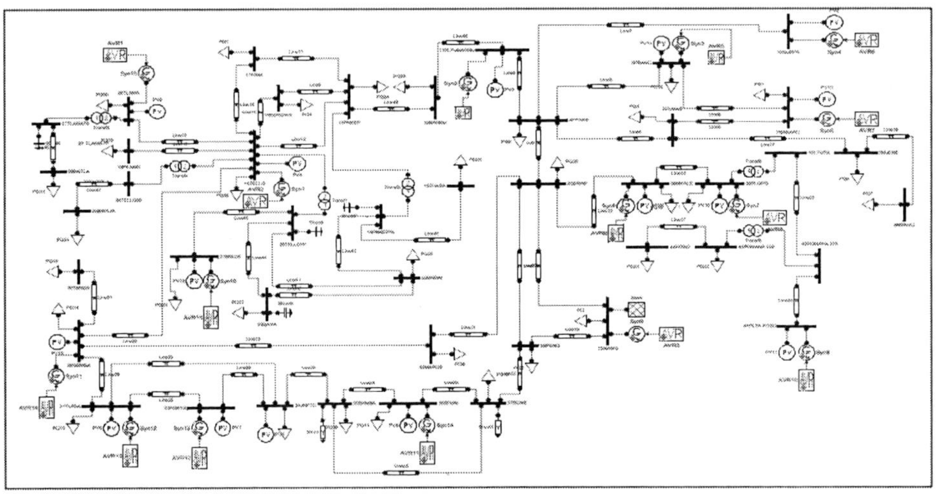

Figure 1. The Sulbagsel Interconection Power Grid Model Use PSAT

2.2. Double Feed Induction Generator (DFIG)

Double Feed Induction Generator (DFIG) is one type of wind turbine used in this simulation. The DIFG selection is considered suitable to represent the type of wind turbine that will enter in Sulbagsel interconnection system. DFIG power generated output depends on the reference power used. DFIG uses maximum power point tracking (MPPT). Figure 2 shows the DIFG system.

DFIG is an induction generator with two types of output (fed) used at variable speed PLTB [13]. Unlike the conventional, singly-fed induction generator, DFIG electrical power does not depend on speed. So, it is possible to realize variable speed wind generators by determining the mechanical speed at wind speed and improving turbine operation at an aerodynamic optimal point for a certain wind speed rating.

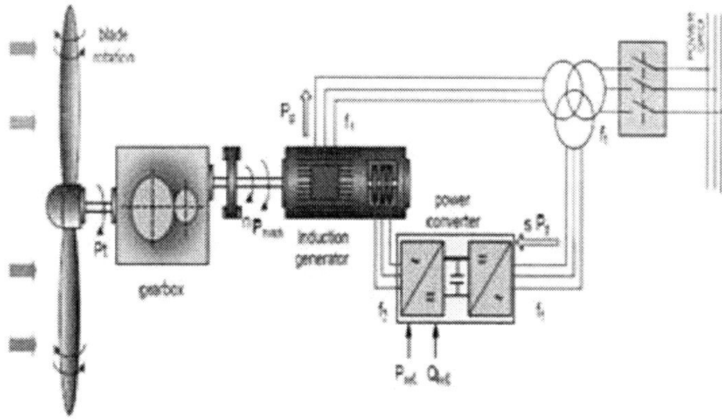

Figure 2. System Double Fed Induction Generator (DIFG)

2.3. Power System Stabilizer (PSS)

Inadequate response can cause oscillation of frequency in long period. This will result in reduced power transfer power that can be overcome using PSS. Implementation of a PSS in the power system connected to the stabilizer port that can be seen in Figure 3.

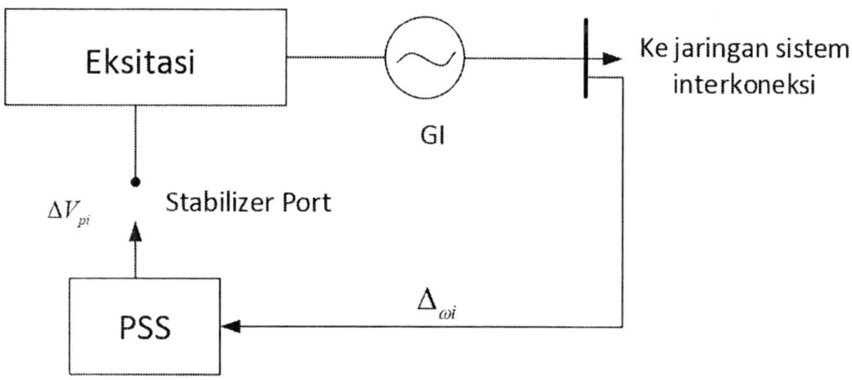

Figure 3. A PSS System in the i-Generator

PSAT provides of five PSS models [4]. The PSS model used in this study is type II model shown in Figure 4.

IOP Conf. Series: Journal of Physics: Conf. Series **979** (2018) 012028 doi:10.1088/1742-6596/979/1/012028

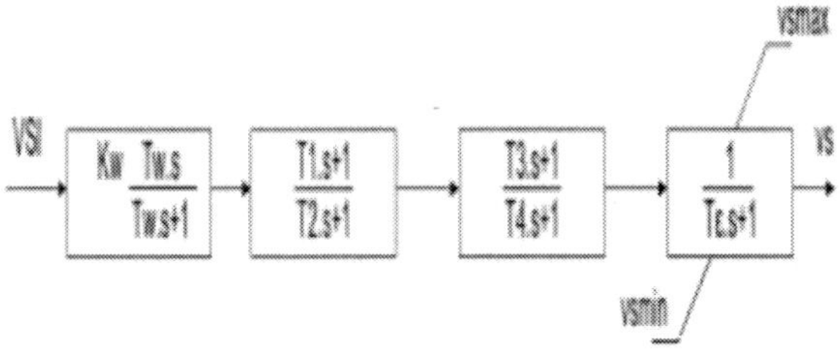

Figure 4. Second type of PSS

where :
K_w = Stabilizer gain
T_w = Wash-out time constant
T_1 = First stabilizer time constant
T_2 = Second stabilizer time constant
T_3 = Third stabilizer time constant
T_4 = Fourth stabilizer time constant
T_ε = Small time constant

3. Case Study

In this case study, the researher discuss about the study of transient stability after entry of wind turbine in Sulbagsel interconnection system. Figure 1 shows the modeling of the electrical interconnection system in PSAT Sulbagsel. The dynamic parameters of each plant are shown in Table 1. Power flow is used to see the system responses and time domain simulation for see the response of the generator during the disturbance [13]. The simulation will be done in two stages. The first stage modeled by system with the entry of wind turbine on Sidrap and Jeneponto buses and then given 3 phase disturbances to the ground. The second stage of adding PSS devices in the system to see the damping response of each generator rotor angle velocity (ω) in the event of a transient disturbance.

Table 1. Dynamic Parameter Dynamic Parameter Data

No	Pembangkit	Xd (pu)	X'd (pu)	X''d (pu)	Xq (pu)	X'q (pu)	X''q (pu)	ra (pu)	xl (pu)
1	PLTA Bakaru	0,924	0,268	0,27	0,553	0,276	0,27	0	0.12
2	PLTA Teppo (Pinrang)	2,08	0,385	0,261	1,12	0,274	0,261	0	0.186
3	PLTD Suppa	2,08	0,385	0,261	1,12	0,267	0,261	0	0,186
4	PLTU Barru	2,363	0,199	0,204	2,182	0,395	0,204	0	0.107
5	PLTU Tello	1,182	0,0995	0,102	1,091	0,1975	0,102	0	0.107
6	PLTD Agrekko (Tello Lama)	2,363	0,199	0,204	2,182	0,395	0,204	0	0.107
7	PLTD Sgmnsa	2,08	0,385	0,261	1,12	0,337	0,261	0	0.186
8	PLTU Arena (Jeneponto)	2,08	0,385	0,261	1,12	0,485	0,261	0	0.186
9	PLTA Tangka Manipi Sinjai	1,924	0,268	0,27	1,553	0,256	0,27	0	0.12
10	PLTGU Sengkang	2,31	0,2	0,12	0,553	0,6	0,12	0	0.6
11	PLTD Malea (Toraja)	2,08	0,385	0,261	1,12	0,337	0,261	0	0.186
12	PLTD Palopo	2,08	0,385	0,261	1,12	0,337	0,261	0	0.186
13	PLTA Bili-Bili	2,08	0,385	0,261	1,12	0,330	0,261	0	0.186
14	PLTA Poso	0,924	0,268	0,27	0,553	0,368	0,27	0	0.12
15	PLTD Tallasa	2,08	0,385	0,261	1,12	0,485	0,261	0	0.186

3 phase interruption is placed in the middle of lane Sengkang bus with fault time 0.05 second and fault clearing time 0.10 second. For the first interval breaker is 0.5 second and 1.0 for the second

intervention. There are 44 buses, 47 lines, 6 transformers, 15 generators and 34 loads that will be used in this research. The study will use a conventional thermal generating unit, which the Gevernor Turbine (TG) is used for Model 1, while the Hydroelectric unit uses the Model 3 TG applied to the entire system. As for wind turbines in the wind speed setting with nominal value around 15.00 (m/s).

4. Results and Discussions

The simulated results are rotor angle velocity (ω) and each generating unit power flow studies after the entry of the wind turbine in Sulbagsel interconnection system before and after the addition of the PSS device in the transient state. Time domain simulation is used to obtain angular rotor speed (ω).

Table 2 shows the power flow before and after the application of PSS on the system when a transient disturbance occurs. It can be seen that the stresses on each bus application of PSS decreased significantly below the tolerance limits permitted by the Regulatory Commission of the Energy Regulatory Commission in N-0 conditions (0.95-1.05 p.u.). Conversely, after the voltage of PSS application has improved or increased on almost all of buses when transient disturbance occurs in the system.

Table 2. Power Flow Before and after PSS Device Installation

No Bus	Name Bus	Without SPSS		With Spss		Load Flow 2 Conditions			
		V	Phase	V	Phase	Generator		Load	
		[p,u]	[rad]	[p,u]	[rad]	[p,u]	[p,u]	[p,u]	[p,u]
1	Sengkang	0.72	-20.01	0.99	-13.33	1.558	0.266	0.284	0.115
2	Sidrap	0.70	-20.72	0.90	-13.67	0.000	0.000	0.265	0.103
3	Soppeng	0.57	-20.32	0.94	-13.49	0.000	0.000	0.141	0.034
4	Parepare	0.80	-20.79	0.94	-13.73	0.000	0.000	0.187	0.047
5	Pinrang	0.84	-20.80	0.95	-13.73	0.143	-0.425	0.044	0.062
6	Polmas	0.85	-20.84	0.97	-13.73	0.000	0.000	0.171	0.041
7	Bakaru	0.92	-20.84	1.01	-13.77	0.163	0.886	0.150	0.021
8	Majene	0.83	-20.90	0.93	-13.83	0.000	0.000	0.233	0.037
9	Mamuju	0.82	-20.91	0.93	-13.84	0.000	0.000	0.096	0.048
10	Suppa	0.82	-20.78	0.95	-13.72	0.311	-0.097	0.000	0.000
11	PLTU Barru	0.77	-20.79	0.92	-13.75	0.604	-0.186	0.000	0.000
12	Barru	0.73	-20.84	0.93	-13.80	0.000	0.000	0.101	0.024
13	Pangkep	0.64	-20.95	0.95	-13.89	0.000	0.000	0.221	0.080
14	Pangkep70	0.65	-20.92	1,00	-14.05	0.000	0.000	0.000	-0.134
15	Tonasa	0.65	-20.93	0.99	-14.06	0.000	0.000	0.189	0.021
16	Bosowa	0.64	-20.97	0.95	-13.90	0.000	0.000	0.331	0.015
17	Kima	0.63	-20.97	0.96	-13.89	0.000	0.000	0.180	0.058
18	Tello	0.62	-20.99	0.98	-13.89	0.210	3.440	0.633	0.183
19	Panakukang	0.59	-21.04	0.93	-13.94	0.000	0.000	0.683	0.177
20	Tello70	0.70	-20.72	0.97	-14.20	0.000	0.000	0.000	-0.215
21	Borongloe	0.76	-20.59	0.92	-14.24	0.052	-0.280	0.114	0.000
22	Mandai	0.69	-20.75	0.96	-14.21	0.000	0.000	0243	0.026
23	Daya	0.70	-20.74	0.97	-14.21	0.000	0.000	0.245	-0.186
24	Tello30	0.62	-20.99	0.98	-13.89	0.000	0.000	0.000	0.000
25	Barawaja	0.62	-20.99	0.98	-13.89	0.000	0.000	0.000	0.000

26	T.Lama	0.62	-20.98	0.96	-13.90	0.226	-0.668	0.197	0.047
27	T.Lama70	0.63	-21.09	0.97	-14.01	0.000	0.000	0.000	-0,124
28	Bontoala	0.62	-21.11	0.96	-14.03	0.000	0.000	0.265	0.077
29	Sungguminasa	0.48	-21.00	0.94	-13.81	0.200	-1.104	0.157	0.036
30	Tanjung Bunga	0.47	-21.02	0.92	-13.84	0.000	0.000	0.552	0.167
31	Tallasa	0.00	-20.60	0.93	-13.57	0.790	0.501	0.206	0.047
32	Maros	0.65	-20.77	0.91	-13.71	0.000	0.000	0.186	0.005
33	Pagayya	0.00	-26.23	0.94	-13.47	1,961	-0.371	0.000	0.0000
34	Jeneponto	0.04	-81049	0.94	-13.48	0.625	0.001	0.174	0.034
35	Bulukumba	0.00	-18.06	0.95	-13.54	0.000	0.000	0.271	-0.001
36	Sinjai	0.30	-20.45	0.97	-13.54	0.040	0.070	0.009	0.006
37	Bone	0.37	-20.39	0.95	-13.54	0.000	0.000	0.321	0.016
38	Makale	0.69	-20.77	0.69	-13.44	0.008	0.274	0.119	0.015
39	Palopo	0.50	-20.65	0.48	-13.26	0.004	0.049	0.492	0.073
40	Latuppa	0.00	0.49	0.00	-12.31	0.000	0.000	0.000	0.000
41	PLTA Poso	0.39	32.17	0.34	-10.36	1.950	0.258	0.000	0.000
42	Pomana 275	0.33	32.18	0.29	-10.35	0.000	0.000	0.000	0.000
43	Pomana 150	0.33	32.17	0.29	-10;37	0.000	0.000	0.049	0.005
44	Poso	0.32	32.02	0.95	0.18	0.000	0.000	0.995	0.018

Figure 5 shows a graph of rotor angle velocity (ω) when a transient disturbance occurs prior to the application of PSS. It also shows the angle of the rotor coming out of its inertia moment due to a system interruption. Most of the generators have been off sync which resulted in a blackout on the system iterconnection in Sulbagsel.

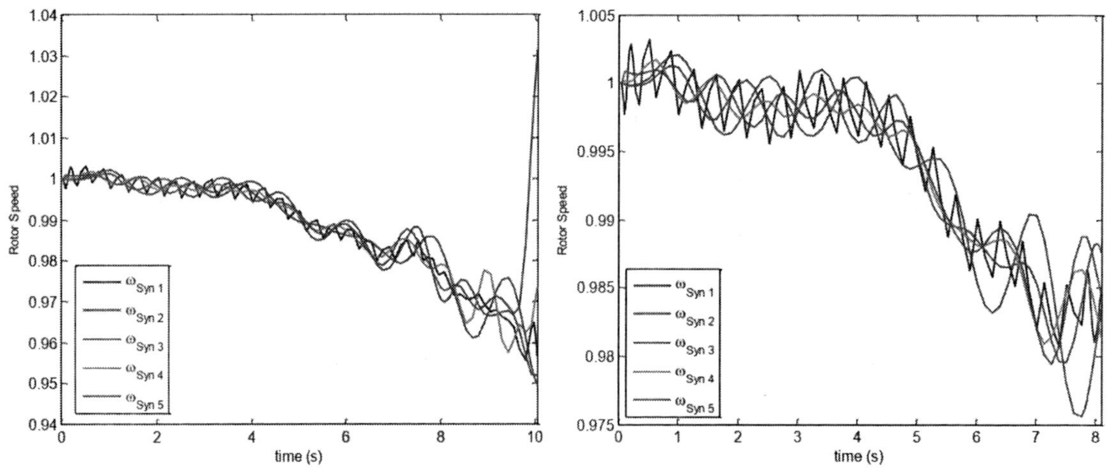

Figure 5. Graph of Rotor angle velocity for generator 1-5. (a) Without PSS, (b) With PSS

Figures 5a and 5b show the rotor angle velocity of generator 1-5. From the graph, it is seen that the rotor angle velocity (ω) is able to dampen and maintain the velocity condition at the limit of 0.98 (rad / s) in the appeal without using PSS during the application of transient disturbance PSS.

Figures 6a and 6b show the same thing in generator 6-10. Where the application of PSS is able to reduce transient noise and maintain the density at the limit of 0.98 (rad / s). Compared without using PSS, the rotor angle density continued to decline to below 0.96 (rad / s). This condition indicates that the influence of the entry of the PLTB which impact on the change of tupologi Sulbagsel network also influence system stability when transient disturbance occurs.

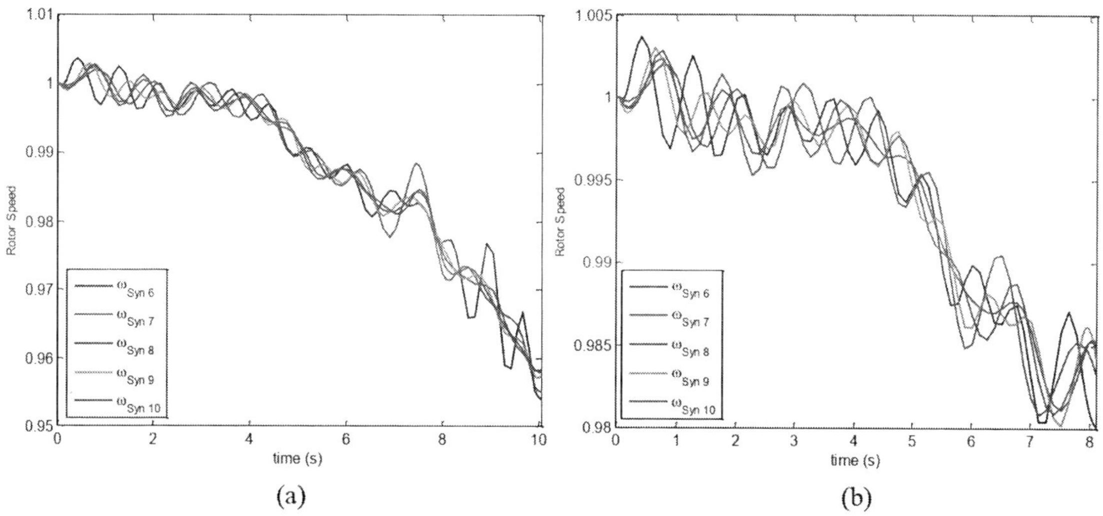

Figure 6. Graph of Rotor angle velocity for generator 6-10. (a) Without PSS, (b) With PSS

Figures 7a and 7b show slightly different things. The speed of the generator fluctuates with the density. Specifically for generator 14, the density increases 1.7 (rad/s) before PSS is installed in the system. After PSS is installed in the system, the rotor angle velocity of the generator 14 decreases below 1.1 (rad / s). Graph below is the presented of Figure 7a and Figure 7b.

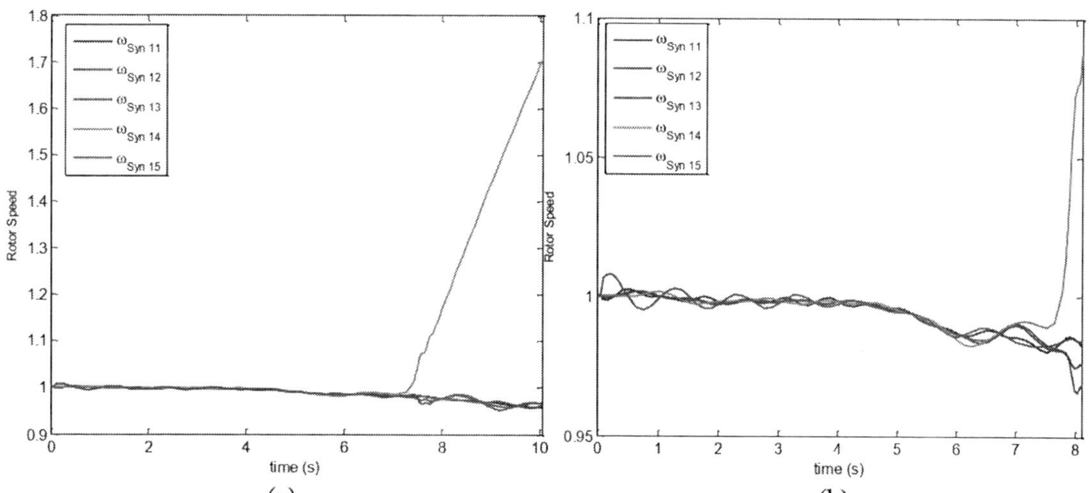

Figure 7. Graph of Rotor angle velocity for generator 11-15. (a) Without PSS, (b) With PSS

5. Conclusions

The TDS approach, which is an indirect method of solving three phase problems and analyzing the transient stability when the entry of wind turbine in the Sulbagsel interconnection system provides satisfactory results in the stability study of the system. From the results of research using TDS method it is found that the system voltage decreased quite extreme when the disturbance occurred. After the addition of PSS control devices mounted on the generator is able to reduce the deviation of the rotor angle and maintain the speed of the rotor angle to find a new equilibrium point. For further research can use a larger case study and control equipment such as STATCOM, UPFC or TCSC in analyzing the transient stability.

References

[1] Anju G P, Thomas P C, Sreerenjini K., Sarin B and Sasidhran S 2013 *J. Sci. Eng. Res* **4**
[2] Indar C G, Soeprijanto A and Penangsang O 2012 *International Review of Electrical Engineering* vol xx
[3] Indar C G, Soeprijanto A and Penangsang O 2012 *International Review of Electrical Engineering* vol xx
[4] Gamit M G., Jigar S S 2015 *Int. J. Res. Eng. Technol* **4(3)** 604–613
[5] Indar C G, Sri M S and Muhammad I 2016 *J. Theor. Appl. Inf. Technol* **90(1)** 161–167
[6] Anıl K N, Ramesh K 2007 *I* **2(3)** 38–45
[7] Federico M 2005 *IEEE Transaction On Power Systems* **20(3)** 1199–1206
[8] Rahul R R, Visakhan R, Sebin J, Aney S V 2015 *Int. J. Innov. Res. Sci. Eng. Technol* **4(11)** 11437–11446.
[9] Federico M, Luigi V and Juan C 2008 *IEEE Transactions on Education* **51(1)** 17–23
[10] Jeevajothi R and Devaraj D 2012 *Int. J. Comput. Electr. Eng.* **2(5)** 259–265.
[11] Ardiaty A and Muhammad B N 2016 *ICITACEE* p 169–172.
[12] Yuwa C, et al 2012 *IEEE* p 1–8
[13] Mustadir D, Ansar S, Indar C G 2017 *Analisis Stabilitas Small Signal Saat Masuknya PLTB Pada Sistem Interkoneksi Sulbagsel, Tesis* (Makassar: Hasanuddin University)

Classifying Physical Morphology of Cocoa Beans Digital Images using Multiclass Ensemble Least-Squares Support Vector Machine

Armin Lawi[1] and Yudhi Adhitya[2]

[1]Department of Computer Science, Hasanuddin University, Makassar
[2]Department of Informatics, Al-Asy'ariah Mandar University, Polewali Mandar

E-mail: yudhiadhitya@gmail.com

Abstract. The objective of this research is to determine the quality of cocoa beans through morphology of their digital images. Samples of cocoa beans were scattered on a bright white paper under a controlled lighting condition. A compact digital camera was used to capture the images. The images were then processed to extract their morphological parameters. Classification process begins with an analysis of cocoa beans image based on morphological feature extraction. Parameters for extraction of morphological or physical feature parameters, i.e., Area, Perimeter, Major Axis Length, Minor Axis Length, Aspect Ratio, Circularity, Roundness, Ferret Diameter. The cocoa beans are classified into 4 groups, i.e.: Normal Beans, Broken Beans, Fractured Beans, and Skin Damaged Beans. The model of classification used in this paper is the Multiclass Ensemble Least-Squares Support Vector Machine (MELS-SVM), a proposed improvement model of SVM using ensemble method in which the separate hyperplanes are obtained by least square approach and the multiclass procedure uses One-Against-All method. The result of our proposed model showed that the classification with morphological feature input parameters were accurately as 99.705% for the four classes, respectively.

1. Introduction

Cocoa plant (*Theobroma cacao L.*) is one of the important plantation commodities as a source of industrial raw materials trade commodities that can increase the country's foreign exchange and income of cocoa farmers. Indonesia is one of the producers of cocoa [1], most of Indonesia's cocoa production is exported to America, Singapore, Malaysia, Brazil, and China. Indonesia's cocoa beans production significantly increases, but the quality produced is very low and various of them are less fermented, not dry enough, beans size is evenly high skin content, high acidity, taste is very diverse and inconsistent. This is reflected in the relatively low price of Indonesian cocoa beans and discounted prices compared to similar products from other producer countries [2].

Indonesian cocoa farmers generally apply a variety of fermentation ways in terms of beans quantity, fermentation means and time. Fermentation is done in baskets, simple wooden crates or plastic bags. What farmers do is not real fermentation because most farmers keep the harvested beans in plastic bags for 1-2 days then dried by drying in direct sunlight on cement floors, mats or woven bamboo. The requirements or conditions used to determine the quality of cocoa beans in Indonesia are contained in the Indonesian national standard of cocoa beans SNI 2323-2008. The Indonesian national standard regulates the classification of the quality of dry cocoa beans as well as general requirements and in particular to maintain the consistency of the quality of the cocoa beans produced. At the exporter level, the separation is carried out by using machinery primarily for the classification of cocoa beans. The results of this cocoa bean sorting will be determined by taking samples of cocoa beans to be analyzed in the laboratory in accordance with the standards of cocoa beans quality classification [3].

Quality examination of cocoa beans is done using traditional and manual procedures; i.e., using the visual method on cocoa beans by choosing one-by-one. The human vision must accurately see the object on the surface of the cocoa beans. In plain view, a human without special knowledge can differentiate and classify cocoa beans. Usually, they only armed with experience and knowledge gained earlier. However, manual checks have limitations such as tired eyes and different analytical results of each examiner.

The development of automation technology and intelligent system is one form of technology created by humans to facilitate human activities and has made many changes in the production and processing of plantation commodities. Image processing technology provides an alternative to manual inspection. Digital technology has a variety of input devices including the camera. The output produced by the camera is an image (digital image). Imagery can be analyzed and processed to get useful information for the user.

This system has two important stages, i.e., image analysis and pattern recognition. Image analysis has standard techniques for identifying, measuring, and acquiring large quantities of quantitative data. Image processing techniques include image capture, pre-processing, interpretation, quantization and image classification. But unfortunately, the resulting image is still not in accordance with the results expected by the user. Therefore, the existence of a process that can process an image is needed by the user. The discipline that gave birth to the techniques to process the image is called Digital Image Processing [4, 5].

2. Related Works

Image processing techniques have been widely used in the field of agriculture such as determining the type of defects of coffee beans, edamame quality determination, quality inspection of RSS rubber, mango quality determination, and identification of maturity level of lemon and mangos teen, identification of defective cocoa beans, quality determination of cocoa beans.

I Wayan Astika, *et* al. use ANN structures to develop the relationship between input parameters, quality components of cocoa beans and outputs. ANN classified cocoa beans into 4 parameters namely: Normal Beans, Broken Beans, Fractured Beans, and Skin Damaged Beans. The classification of the size of the damage beans reaches an accuracy of 79.25%, which consists of accuracy: 84.38% for Normal Beans, 51.72% for Broken Beans, 98.29% for Fractured Beans, and 20% for Skin Damaged Beans [6].

S. Nurmuslimah, created a software system that starts with taking pictures of files to display on the system interface, image is processed using edge detection sobel to get numeric data value. Furthermore, these data are used as data input training Neural Network Back propagation. After training data is obtained, then the data is used for the testing process. From the testing process, the output of the created system is able to provide information about the quality of cocoa beans. Using Back propagation method with alpha = 0.6, hidden layer = 3, fault tolerance = 0.0001, target = 0.9, resulting in a system that has a level of accuracy of (76%) has an error rate (24%) in determining the quality of cocoa beans [7].

Data mining is a process uses statistical techniques, calculations, artificial intelligence and machine learning to extract and identify useful information and related knowledge from large databases [8]. SVM initially can only classify data in two classes [9, 10, 11]. However, further research SVM was developed so that it can classify data over two classes (multiclass) [9, 12, 13, 14]. Classifying M-classes means predicting the class labels $C_m, m = 1, \dots, M$ one way to solve the M-class problem by formulating it into binary L classification problems [11, 12].

SVM concept is simply described as trying to find the best hyper plane that provides as a separator of two classes in the input space. Pattern which is a member of two classes: +1 and -1 and share alternate field separators. The best dividing fields can not only separate the data but also have the largest margins. Margin is the distance between the fields of separator (hyper plane) with the closest pattern of each class.

Let $\{x_1, \ldots, x_n\}$ be the dataset and $y_i \in \{+1, -1\}$ is the class label of the x_i data. The two classes are separated by a pair of parallel bounding plane. The first delimiter field limits the first class while the second delimiter field limits the second class, so it is obtained [11, 12]:

$$\begin{aligned} \boldsymbol{x}_i.\boldsymbol{w} + b \geq +1 \text{ for } y_i = +1 \\ \boldsymbol{x}_i.\boldsymbol{w} + b \leq -1 \text{ for } y_i = -1 \end{aligned} \quad (1)$$

The best dividing fields with the largest margin values can be formulated into quadratic programming problems:

$$\min_{w,b,\xi} \frac{1}{2}\|\boldsymbol{w}\|^2 + C\sum_{i=1}^{l} \xi_i \quad (2)$$

with constraints: $y_i(\boldsymbol{x}_i.\boldsymbol{w} + b) \geq 1 - \xi_i$ and $\xi_i \geq 0$,

where $\mathrm{l} = 1, .., n$ is a slack slack variable that determines the level of misclassification of the data samples, whereas $C > 0$ is a parameter.

Method for classifying data that cannot be separated linearly is kernel method. The kernel method transforms the data into the feature space dimension so that it can be linearly separated on the feature space. The kernel method can be formulated:

$$K(\boldsymbol{x}_i.\boldsymbol{x}_j) = \varphi(\boldsymbol{x}_i).\varphi(\boldsymbol{x}_j) \quad (3)$$

Commonly used kernel functions are as follows:

a. The linear kernel: $K(\boldsymbol{x}_i, \boldsymbol{x}_j) = \boldsymbol{x}_i^T \boldsymbol{x}_j$
b. Kernel polynomial: $K(\boldsymbol{x}_i, \boldsymbol{x}_j) = (\gamma.\boldsymbol{x}_i^T \boldsymbol{x}_j + r)^p, \gamma \geq 2$
c. RBF Kernel (Radial Basis Function): $K(\boldsymbol{x}_i, \boldsymbol{x}_j) = exp\left(-\gamma\|\boldsymbol{x}_i - \boldsymbol{x}_j\|^2\right), \gamma > 0$

LS-SVM was first introduced by Suykens and Vandewalle in 1999. LS-SVM is one of the SVM modifications that solve linear equations [12, 13, 14, 15, 16, 17]. If the SVM separator field is given as in (3), then for LS-SVM is given as the following objective function.

$$\min_{w,b,\xi} \frac{1}{2}\|\boldsymbol{w}\|^2 + \frac{C}{2}\xi^T\xi \quad (4)$$

with constraint: $y_i(\boldsymbol{x}_i.\boldsymbol{w} + b) \geq 1 - \xi_i$

The above equation can be solved after forming Lagrangian:

$$L = \frac{1}{2}\|\boldsymbol{w}\|^2 + \frac{C}{2}\xi^T\xi - \sum_{i=1}^{l} \alpha_i (y_i(\varphi(\boldsymbol{x}_i).\boldsymbol{w} + b) - 1 + \xi_i) \quad (5)$$

where α_i is a Lagrangian multiplier whose value can be either positive or negative.

To optimize the conditions in (5), a decrease of w, b, ξ, and α is equal to zero. The results of the process are as follows.

$$\frac{\partial L}{\partial \boldsymbol{w}} = 0 \rightarrow \boldsymbol{w} = \sum_{i=1}^{l} \alpha_i y_i \varphi(\boldsymbol{x}_i) \quad (6)$$

$$\frac{\partial L}{\partial b} = 0 \rightarrow \sum_{i=1}^{l} \alpha_i y_i = 0 \quad (7)$$

$$\frac{\partial L}{\partial \xi} = 0 \rightarrow \alpha = \gamma\xi_i, \quad i = 1, \ldots, N \quad (8)$$

$$\frac{\partial L}{\partial \alpha} = 0 \rightarrow y_i(\varphi(\boldsymbol{x}_i).\boldsymbol{w} + b) - 1 + \xi_i = 0, i = 1, \ldots, N \quad (9)$$

Using the One-Against-All (OAA) method, a binary model is constructed k (k is the number of classes). Each i-class model is trained by using the entire data. For example, there is a classification problem with 3 classes. For training use 3 pieces of binary classification and thus its objective function is given as follows.

$$\min_{w_i, b_i, \xi_{t,i}} \frac{1}{2}\sum_{i}^{m}(w_i)^T w_i + \frac{C}{2}\sum_{i}^{m}\xi_{t,i}^2 \quad (10)$$

with constraint: $y_{t,i}(\varphi_i(x_t)(w_i)^T + b_i) \geq 1 - \xi_{t,i}$.

Confusion matrix is a table that states the amount of test data that is correctly classified or not. The table calculates the following parameters in order to evaluate accuracy, sensitivity and false discovery rate of the learner LS-SVM model from the tested data.

True Positive (TP) is the number of documents from class 1 is correctly classified as class 1.
True Negative (TN) is the number of documents from class 0 is are correctly classified as class 0.
False Positive (FP) is the number of documents from class 0 incorrectly classified as class 1.
False Negative (FN) is the number of documents from class 1 that are misclassified as class 0.

The calculation of accuracy is expressed by the equation:

$$Accuracy = \frac{TP + TN}{TP + FN + FP + TN} \times 100\% \quad (11)$$

The calculation of sensitivity is expressed by the equation:

$$Sensitivity = \frac{TP}{TP + FN} \times 100\% \quad (12)$$

The calculation of false discovery rate is expressed by the equation:

$$False\ Discovery\ Rate = \frac{FP}{FP + TP} \times 100\% \quad (13)$$

3. Experimental Details

The process of classifying cocoa beans is explained by the following activities:

1. Cocoa bean image capture
2. Extraction of morphological features using image processing.
3. Conducting the training process to obtain the classification model
4. Conducting the classification process using test data
5. From the classification process in obtaining the results of the classification of cocoa beans based on morphological features

3.1. Physical Morphological Features

Morphology is the geometric property of an image. In this case is the size and shape of the characteristics of cocoa beans, obtained from binary image analysis. The morphological features are extracted from binary imagery, and then displaying the following morphological features: Area, Perimeter, Major Axis Length, Minor Axis Length, Aspect Ratio, Circularity, Roundness, Ferret Diameter.

Cocoa bean image is processed using grayscale and thresholding and edge detection is done by using kernel sobel to detect the edge of the image. After that, the process of segmentation and then performed the process of morphological feature extraction.

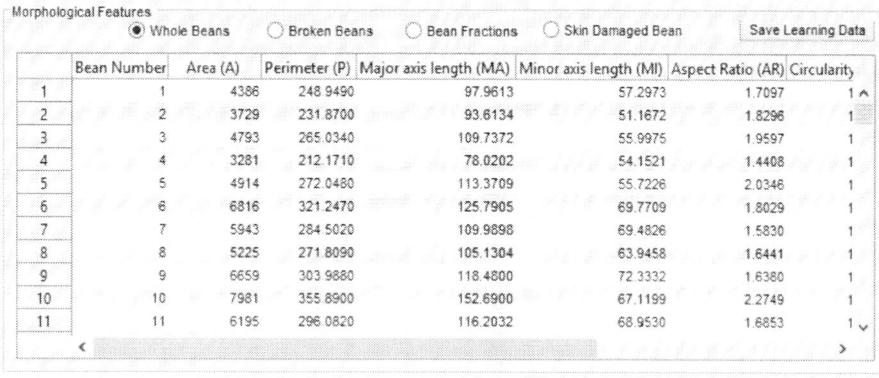

	Bean Number	Area (A)	Perimeter (P)	Major axis length (MA)	Minor axis length (MI)	Aspect Ratio (AR)	Circularity
1	1	4386	248.9490	97.9613	57.2973	1.7097	1
2	2	3729	231.8700	93.6134	51.1672	1.8296	1
3	3	4793	265.0340	109.7372	55.9975	1.9597	1
4	4	3281	212.1710	78.0202	54.1521	1.4408	1
5	5	4914	272.0480	113.3709	55.7226	2.0346	1
6	6	6816	321.2470	125.7905	69.7709	1.8029	1
7	7	5943	284.5020	109.9898	69.4826	1.5830	1
8	8	5225	271.8090	105.1304	63.9458	1.6441	1
9	9	6659	303.9880	118.4800	72.3332	1.6380	1
10	10	7981	355.8900	152.6900	67.1199	2.2749	1
11	11	6195	296.0820	116.2032	68.9530	1.6853	1

Figure 1. Results of the image analysis process for morphological features.

3.2. Dataset

Based on the morphological features there are four classes namely, first class is Normal Beans, second class is Broken Beans, third class is Fractured Beans and fourth class is Skin Damaged Beans. Attributes for dataset consist of 8 pieces; i.e., Area, Perimeter, Major Axis Length, Minor Axis Length, Aspect Ratio, Circularity, Roundness, Feret Diameter. The data sample for dataset 2 is 2,400 items.

Variables used in this research are:

Normal Beans, Broken Beans, Fractured Beans, and Skin Damaged Beans

y_i = class of dataset that is class 1, class 2, class 3 and class 4.

x_i = data features consist of 8 variables, i.e., Area, Perimeter, Major Axis Length, Minor Axis Length, Aspect Ratio, Circularity, Roundness, Feret Diameter.

α = Lagrange multiplier.

w = normal field to support vector.

b = distance of the bounding plane to the center point.

Values of α and b will be obtained after the training is completed. Value of α is then used to find the value of w.

3.3. Classification Process

The classification process is divided into two, training and testing process. The data used is the data of color feature extraction and morphological features as much as 100% data so it is expected to get the accuracy of multiclass classification on LS-SVM. Training process aims to build the MELS-SVM model by finding the parameters, ie values α, w, and b. After forming the MELS-SVM model, proceed with the testing process on the data to see the accuracy of the LS-SVM technique using the OAA method.

Based on the workflow, the classification process begins with the training process on the train data to obtain α and b values using the Matlab r2016a software and the additional toolbox LSSVMlabv which can be downloaded on the site http://www.esat.kuleuven.be/sista/lssvmlab/ toolbox.html. In the training process that generates α and b values, the RBF kernel is used. Furthermore, after the obtained values of α and b, the next step is to find the value of w then the value of w and b are used to arrange the separator function.

After getting the separation function model from the training post, proceed to the testing process. The testing process begins by inserting the item values (x_i) of the test data, then obtaining the prediction class. From this prediction class will be calculated the accuracy level of the method by finding the total class that is correctly predicted by the class of the test data.

3.4. *Multiclass Algorithm Method*

The analysis used in this multiclass classification is the OAA classification method. The multilingas method algorithm as follows:

1. Dataset input.
2. Identify the input dataset
   a. The values of the training data feature (x_i)
   b. Class of training data (yi)
   c. Values feature test data (xt_i)
   d. Class of test data (yt_i)
3. Initiate objects on LS-SVM before performing the training process with the initlssvm function
   a. Specify data of training data feature (x_i)
   b. Specifies the training data class (y_i),
   c. Choose a classifier to classify data
   d. Selects the *kernel* and its parameters to use
4. Selecting the multilingual method code used (*code_OneVsAll* for OAA)
5. Conduct training process with trainlssvm function
6. Calculating values w
7. Make predictions based on the model obtained and determine data feature test data (xt_i) with simlssvm function
8. Create a *confusion matrix*
9. Calculate the level of accuracy with the formula:

$$\lambda = \frac{C}{N} \times 100\%$$

where C is the correct total of predictions and N is the total of all data tested.

3.5. *Implementation Results on dataset*

The separator function for the one against all method with the RBF (Radial Basis Function) kernel using the parameter $\sigma = 0.5$ for dataset is as follows.

$$f_1(\boldsymbol{x}) = \boldsymbol{x}_i.\boldsymbol{w}_1 + b = \boldsymbol{x}_i.\begin{pmatrix} 474.060 \\ 620.798 \\ 676.654 \\ 1,036.838 \\ 57.949 \\ -1,039.173 \\ -270.937 \\ 1,086.816 \end{pmatrix} + (-0.1041), \quad (14)$$

$$f_2(\boldsymbol{x}) = \boldsymbol{x}_i.\boldsymbol{w}_2 + b = \boldsymbol{x}_i.\begin{pmatrix} -2,866.019 \\ -1,880.457 \\ -2,295.395 \\ -1,234.875 \\ -2,077.650 \\ -598.255 \\ 443.043 \\ -2,034.123 \end{pmatrix} + (-0,7476), \quad (15)$$

$$f_3(\boldsymbol{x}) = \boldsymbol{x}_i.\boldsymbol{w}_3 + b = \boldsymbol{x}_i.\begin{pmatrix} -342.200 \\ -393.094 \\ -379.771 \\ -476.487 \\ -160.096 \\ -152.709 \\ 925.336 \\ -483.535 \end{pmatrix} + (-0,4019). \quad (16)$$

$$f_4(\boldsymbol{x}) = \boldsymbol{x}_i.\boldsymbol{w}_4 + b = \boldsymbol{x}_i.\begin{pmatrix} -2,830.185 \\ -1,829.274 \\ -2,250.925 \\ -1,171.307 \\ -2,050.838 \\ -517.928 \\ 130.739 \\ -1,976.306 \end{pmatrix} + (-0,7463). \quad (17)$$

4. Results and Discussion

Values of w and b are based on RBF kernel and its parameters for the use of the One Against All method on dataset can be seen on table 1.

Table 1: w and b values

w value				*b* value			
w_1	w_2	w_3	w_4	b_1	b_2	b_3	b_4
474.060	−2,866.019	−342.200	−2,830.18				
620.798	−1,880.457	−393.094	−1,829.27				
676.654	−2,295.395	−379.771	−2,250.92				
1,036.838	−1,234.875	−476.487	−1,171.30				
57.949	−2,077.650	−160.096	−2,050.83	−0.1041	−0.7476	−0,4019	−0,7463
−1,039.173	−598.255	−152.709	−517.928				
−270.937	443.043	925.336	130.739				
1,086.816	1,086.816	−483.535	−1,976.30				

Based on table 1 and figure 2, the use of RBF kernel types and using the parameter $\sigma = 0.5$ has the highest accuracy on dataset has the highest accuracy (99.705). Accuracy for classification with the number of classes: 4 classes, i.e.: Normal Beans, Broken Beans, Fractured Beans, and Skin Damaged Beans.

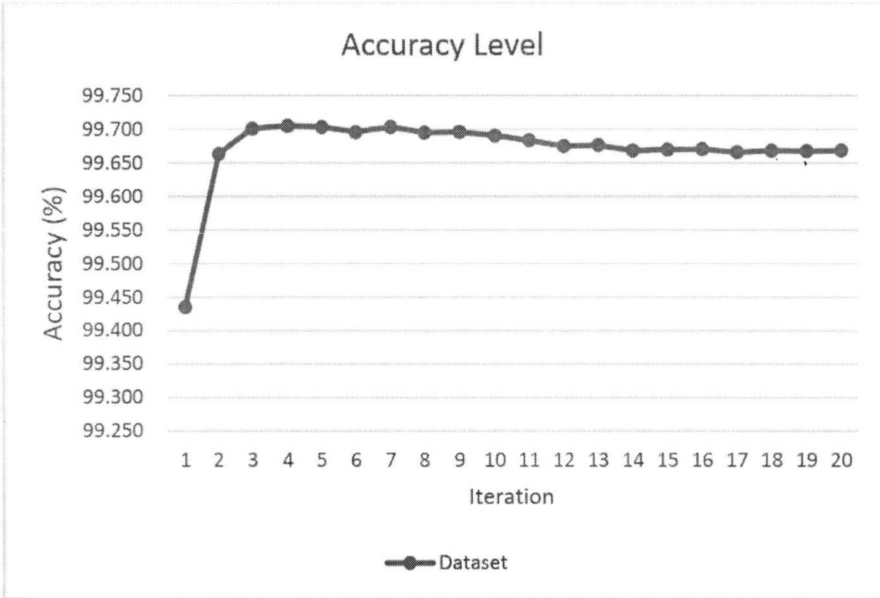

Figure 2. Accuracy level of OAA method for each kernel in dataset.

5. Conclusion

In this research, the cocoa beans image is processed using grayscale and thresholding, and then edge detection is done by using a sobel kernel to detect the edges of the image. After segmentation process which extracts feature that produces data based on the extraction of physical morphological features parameters i.e. Area, Perimeter, Major Axis Length, Minor Axis Length, Aspect Ratio, Circularity, Roundness, Ferret Diameter.

The accuracy level is derived from the number of data items categorized into the correct class by the MELS-SVM model. The level of accuracy using one against all method on datasets with RBF type (Radial Basis Function) using parameter $\sigma = 0.5$ resulting 99,705% with four class classification.

Acknowledgments

The authors would like to express heartfelt thanks to Department of Computer Science, Hasanuddin University and Department of Informatics, Al-Asy'ariah Mandar University, Polewali Mandar for the research support and assistances.

References

[1] ICCO Annual Report 2012/2013 International Cocoa Organization
[2] Haryadi M and Supriyanto 2001 *Pengolahan Kakao Menjadi Bahan Pangan, Pusat Antar Universitas Pangan dan Gizi* (Yogyakarta: Universitas Gajah Mada)
[3] Biji kakao [SNI] Standard Nasional Indonesia 2008 SNI; (SNI 2323:2008, ICS 67.140.30 Badan Standardisasi Nasional)
[4] Sutoyo T, Mulyanto E, Suhartono V, Nurhayati D, Oky W 2009 *Teori Pengolahan Citra Digital* (Yogyakarta: ANDI)
[5] Pratt, William K 2007 *Digital image processing, PIKS Scientific inside* (John Wiley) 4th Edition
[6] Astika I W, Solahudin M, Kurniawan A, Wulandari Y 2010 *Determination of Cocoa Bean Quality with Image Processing and Artificial Neural Network* (Indonesia: University Bogor)
[7] Nurmuslimah S 2016 *Implementasi Metode Backpropagation Untuk Mengidentifikasi Jenis Biji Kakao Yang Cacat Berdasarkan Bentuk Biji* (Indonesia: ITATS)

[8] Turban E, et al. 2005 *Decision Support Systems and Intelligent Systems* (Yogyakarta: Andi Offset)
[9] Sembiring K 2007 *Tutorial SVM Bahasa Indonesia* (Bandung: Institut Teknologi Bandung)
[10] Kurniawan D and Supriyanto C 2013 *Journal Teknologi Informasi*
[11] Vapnik V and Cortes C 1995 *Support-Vector Networks. Machine Learning* 273-297
[12] Gestel T, Suykens J, Lanckriet G, Lambrechts A, Moor B and Vandewalle J 2002 *Multiclass LS-SVMs: Moderated Outputs and Coding-Decoding Schemes. Neural Processing Letters* 45-58
[13] Jafar N, Thamrin S A, Lawi A 2016 *Proceeding of International Conference on Computational Intelligence and Cybernetic*
[14] Nurkamila J, Lawi A and Thamrin S A 2016. *Least Squares Support Vector Machine Untuk Klasifikasi Data Multikelas* (Makassar : Hasanuddin University)
[15] Marliani R R, Lawi A and Thamrin S A 2016 *Algoritma Adaboost untuk Optimasi Ensemble Least Squares Support Vector Machine* (Makassar : Hasanuddin University)
[16] Xu Y, Lv X, Wang Z and Wang L 2014 *Journal of Information Science and Engineering* 1773-1787
[17] Zhou L, Lai K K and Yu L 2010 *J. Expert Systems with Applications* 127-133

Combination of Rivest-Shamir-Adleman Algorithm and End of File Method for Data Security

Dian Rachmawati, Amalia Amalia, Elviwani

Departemen Ilmu Komputer, Fakultas Ilmu Komputer dan Teknologi Informasi, Universitas Sumatera Utara, Jl. Universitas No. 9-A, Medan 20155, Indonesia.

Email: dian.rachmawati@usu.ac.id

Abstract. Data security is one of the crucial issues in the delivery of information. One of the ways which used to secure the data is by encoding it into something else that is not comprehensible by human beings by using some crypto graphical techniques. The Rivest-Shamir-Adleman (RSA) cryptographic algorithm has been proven robust to secure messages. Since this algorithm uses two different keys (i.e., public key and private key) at the time of encryption and decryption, it is classified as asymmetric cryptography algorithm. Steganography is a method that is used to secure a message by inserting the bits of the message into a larger media such as an image. One of the known steganography methods is End of File (EoF). In this research, the cipher text resulted from the RSA algorithm is compiled into an array form and appended to the end of the image. The result of the EoF is the image which has a line with black gradations under it. This line contains the secret message. This combination of cryptography and steganography in securing the message is expected to increase the security of the message, since the message encryption technique (RSA) is mixed with the data hiding technique (EoF).

1. Introduction

Message security is urgently needed to safeguard the privacy of the owner of the message. There are many methods that can be used to protect the message. Cryptography is a way which is commonly used to provide data security. Cryptography, in general, is a science and art to keep the confidentiality of the information. Cryptography has encryption algorithm, some of them have been classified as symmetric algorithm and the others have been categorized as asymmetric algorithm [1]. Rivest-Shamir-Adleman (RSA) is an asymmetric cryptographic algorithm whose public key component is built with two distinct primes. In order to build an exact key, it requires some mathematical operations because the processed data is in numerical form, and will go through several stages of calculations until a secure key pair is obtained. The RSA algorithm consists of some discrete mathematics calculations, most importantly modulo exponentiation. [2]. The RSA has three stages: key generation, enciphering, and deciphering. The key generation stage begins by generating two random prime numbers (p, q), computing and publishing the public key pair (n, e), and computing the decryption key (d). The enciphering or encryption stage begins by receiving the public key pair and calculating the cipher text from the original message. The enciphering or decryption stage begins by getting the cipher text and calculating the value of the plaintext using the decryption key.

Delivering the cipher text by using RSA algorithm to the recipient may cause suspicion to the unwanted parties since the cipher text which appears random may cause them to think that that there is something hidden in it. In order to minimize this suspicion, the cipher text is embedded into a larger media without noticeable changes. The technique of hiding confidential data in digital media so that the existence of the confidential data is not known by the other person is called steganography. Steganography is done by hiding (or embedding) the confidential data into a media. The media is then sent to the recipient. In order to recover the data, the recipient extracts the hidden (or embedded) data from the media. Steganography has several methods, one of them is the End of File (EoF) method which appends the data to the end of the file. EoF method is adapted from the end of file marker method which used by the windows operating system. The combination of RSA algorithm and EoF is expected to improve data security and reduce the suspicion of third-party about the existence of confidential data. Cryptography and steganography are the fundamental methods which use to protect the data from unauthorized access [3].

2. Methods

2.1. Rivest Shamir Adleman Algorithm

RSA which was built by modular exponential function consists of three main processes, namely: key generation, encryption, and decryption [4].

Step 1. RSA Key Generation
Decryption must generate public and private keys in order to use RSA. Both of these keys require two large primes to make it difficult to be factored. The following RSA key generation algorithm is [5]:

$(p, q) \leftarrow$ large random primes
$n \leftarrow p \cdot q$
$\phi(n) \leftarrow (p-1) \cdot (q-1)$
$e \leftarrow$ the odd random numbers in range $1 < e < \phi(n)$, $\gcd(e, \phi(n)) = 1$
$d \leftarrow e^{-1} \bmod \phi(n)$
$K_{public} = (e, n)$
$K_{private} = d$

Step 2. RSA Encryption
The encryption process uses the RSA public key that has been raised before. RSA encryption algorithm uses a modular exponential function as follows:

$K_{public} = (e, n)$
P = plain text
$C = P^e \bmod n$

Step 3. RSA Decryption
To restore the cipher text into plain text, the RSA decryption also a form of a modular exponential function N by using the private key. The following algorithm RSA decryption are:

$K_{private} = d$
C = cipher text
$P \leftarrow C^d \bmod n$

2.2. End Of File

End-of-File (EoF) is a method or technique of steganography that hides a message by appending it to the end of the file horizontally. This technique can be used to insert the message into the needed size. The size of the files that have been inserted are equal with the size of the file before it is inserted into the message plus the size of the message that is inserted into the file. In this technique, the message is appended at the end of the file with the given identifier as the distinctive mark at the beginning of the message and the identifier of the end of the data.

In order not to cause any suspicions, it has to be arranged in some ways so that the inserted message does not change the size and shape of the image at the time is displayed, because the image-forming element is the pixel that has a value of Red, Green, and Blue [6]. The example of an image will be shown in Table 1, and already inserted by cipher text using steganography End of File method. Cipher text value is marked with a red number.

Table 1. The RGB value of an image before and after End of File process.

R : 23	R : 35	R : 16	R : 23	R : 35	R : 16
G : 17	G : 50	G : 23	G : 17	G : 50	G : 23
B : 87	B : 97	B : 56	B : 87	B : 97	B : 56
R : 70	R : 88	R : 48	R : 70	R : 88	R : 48
G : 42	G : 89	G : 72	G : 42	G : 89	G : 72
B : 48	B : 39	B : 35	B : 48	B : 39	B : 35
			R : 13	R : 15	R : 19
			G : 44	G : 80	G : 23
			B : 47	B : 18	B : 36

3. Results and Discussion

In some case, the stenography is still not secure enough to hide the data. Therefore the merger by cryptographic methods is needed [7]. For example End Of File is a steganography method which is easy to recognize because the secret massage was hidden in the bottom of the cover object. Therefore in this research, it uses a cryptography algorithm such as RSA to encode the message first, before it was embedded to the cover object. The combination of RSA method and the End of File method can be illustrated in the Figure 1.

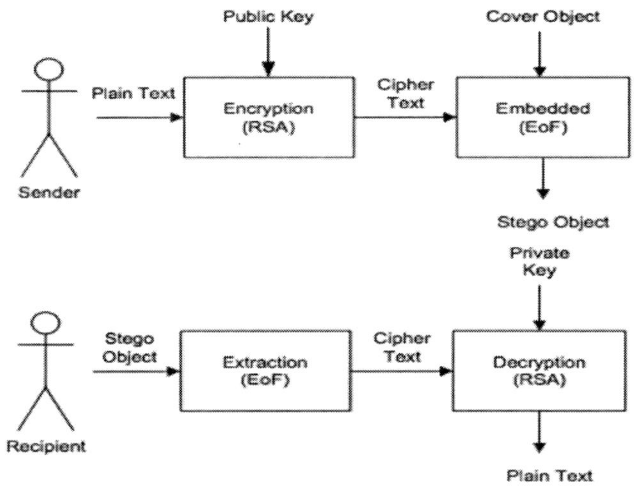

Figure 1. The combination of RSA algorithm and End of File in the security message.

In the image above are explained that the sender of the encrypted message with the RSA algorithm and perform the insertion with End of File algorithm, while the recipients doing the extraction process by using the End of File algorithm and decrypt the message by RSA algorithm.

The example of message security process is: if we have a message “Dian USU Medan”, then the first step is to change the message into the ASCII to 68 105 97 110 32 85 83 85 32 77 101 100 97 110. And then make a key generation process by using the RSA method. The first step is to take the two pieces of prime numbers, such as p = 11 and q = 13. Next step is to do a multiplication p and q so that we could retrieve n = 143 and ϕ (n) = 120 from the obtained formula of e and d are e=7 d=103. By using the encryption formula then it will gain the cipher text 29 118 59 33 98 123 8 123 98 77 62 100 59 33. The result of the cipher text then inserted into the cover image as illustrated in Table 2.

Table 2. Insertion Process in Cover Image.

R=13	R=10	R=18	R=42	R=58	R=97	R=56	R=50	R=80	R=61	R=49	R=56	R=56
G=72	G=54	G=96	G=18	G=16	G=120	G=87	G=86	G=111	G=93	G=77	G=87	G=87
B=18	B=27	B=56	B=78	B=72	B=134	B=107	B=110	B=132	B=116	B=98	B=107	B=107
R=73	R=50	R=48	R=61	R=80	R=97	R=52	R=52	R=56	R=80	R=52	R=50	R=80
G=102	G=86	G=82	G=93	G=111	G=120	G=89	G=89	G=87	G=111	G=89	G=86	G=111
B=116	B=110	B=107	B=116	B=132	B=134	B=118	B=118	B=107	B=132	B=118	B=110	B=132
R=69	R=49	R=46	R=97	R=54	R=75	R=58	R=52	R=75	R=58	R=50	R=69	R=54
G=65	G=77	G=78	G=120	G=98	G=111	G=86	G=89	G=111	G=86	G=86	G=65	G=98
B=32	B=98	B=101	B=134	B=125	B=133	B=107	B=118	B=133	B=107	B=110	B=32	B=125
R=29	*R=33*	*R=8*	*R=77*	*R=59*	*R=0*	*R=0*	*R=0*	*R=0*	*R=0*	*R=0*	*R=0*	*R=0*
G=118	*G=98*	*G=123*	*G=62*	*G=33*	*G=0*	*G=0*	*G=0*	*G=0*	*G=0*	*G=0*	*G=0*	*G=0*
B=59	*B=123*	*B=98*	*B=100*	*B=0*	*B=0*	*B=0*	*B=0*	*B=0*	*B=0*	*B=0*	*B=0*	*B=0*

After the insertion, the recipient do an extraction on stego image by taking the RGB value at the very bottom of the file. So the cipher text is obtained as follows: 29 118 59 33 98 123 8 123 98 77 62 100 59 33. By using the decryption process, the original message, i.e., Dian USU Medan, is then recovered. From the result of these calculations, it can be seen that the combination of RSA algorithm and End of File can restore message into the original form.

The message "Dian Amalia Elviwani Ilmu Komputer USU Medan" takes 6.43 seconds during an encryption, 0.56 second during the insertion, 0.41 seconds during the extraction and 7.1 seconds during the decryption. Meanwhile, the message "Ilmu Komputer USU" 4.23 seconds for the encryption, 0.54 second for the insertion, 0.39 second for the extraction, and 4.34 seconds for the decryption. In Table 3, it is shown that the usage of the value of pixel size for keeping the message.

Table 3. An Experiment by using various pixel size.

No.	Pixel	Cover Image	Stego Image
1	1800 x 1300	 7280 KB	 9561 KB
2	900 x 600	 1779 Kb	 2237 Kb

3	300 x 250	 204 Kb	 273 Kb

4. Conclusion

The use of RSA algorithms to encrypt messages can obfuscate the messages into the form that are not understood by third parties since the resulting cipher text appears random. The combination of RSA and EoF algorithms in this study is able to encode and hide messages into the image media by not significantly changing the quality of the cover image, so the inserted image does not look much different from the original image. In the EoF method, the one that distinguishes the original image from the inserted image is the line below the image. The larger the size of the cover image, the smaller the black line under the image appears to be. The extraction and decryption of data by this algorithm can return the data to the original size. The longer the message, the longer time it takes to perform the encryption and decryption of that message.

Acknowledgments

The authors gratefully acknowledge that the present research is supported by Fund Dissemination IPTEKS Research Results for Lecturers / Researchers University Sumatera Utara.

References

[1] George D and Bonifus P L 2013 *International Conference on Advanced Computing and Communication Systems* p 1-4. doi: 10.1109/ICACCS.2013.6938727

[2] Saxena and Kapoor B *2014 IEEE International Advance Computing Conference (IACC)* p 850-854. doi: 10.1109/IAdCC.2014.6779433

[3] Kothari L, Thakkar R and Khara S 2017 *International Conference on Computer, Communications and Electronics (Comptelix)* p 448-452. doi: 10.1109/COMPTELIX.2017.8004011

[4] Paar C and Pelzl J 2010 *Understanding Cryptography* (Berlin: Springer-Verlag)

[5] Mollin R A 2002 *RSA and Public-Key Cryptography.* 2nd Edition (Florida: Chapman & Hall/CRC)

[6] Rachmawati D et al 2017 *J. Phys.: Conf. Ser. 801 012059* doi : https://doi.org/10.1088/1742-6596/801/1/012059

[7] Duluta A, Mocanu S, Pietraru R, Merezeanu D and Saru D 2017 *21st International Conference on Control Systems and Computer Science (CSCS)* p 453-458. doi: 10.1109/CSCS.2017.70

Implementation of Canny and Isotropic Operator with Power Law Transformation to Identify Cervical Cancer

A Amalia[1], D Rachmawati[1], I A Lestari[1], C Mourisa[2]

[1]Departement of Computer Science, Faculty of Computer Science and Information Technology, Universitas Sumatera Utara, Medan, Indonesia.
[2]Departement of Pharmacology, Faculty of Medicine, University of Muhammadiyah Sumatera Utara

Email: amalia@usu.ac.id

Abstract. Colposcopy has been used primarily to diagnose pre-cancer and cancerous lesions because this procedure gives an exaggerated view of the tissues of the vagina and the cervix. But, the poor quality of colposcopy image sometimes makes physician challenging to recognize and analyze it. Generally, Implementation of image processing to identify cervical cancer have to implement a complex classification or clustering method. In this study, we wanted to prove that by only applying the identification of edge detection in the colposcopy image, identification of cervical cancer can be determined. In this study, we implement and comparing two edge detection operator which are isotropic and canny operator. Research methodology in this paper composed by image processing, training, and testing stages. In the image processing step, colposcopy image transformed by *n^{th} root power transformation* to get better detection result and continued with edge detection process. Training is a process of labelling all dataset image with cervical cancer stage. This process involved pathology doctor as an expert in diagnosing the colposcopy image as a reference. Testing is a process of deciding cancer stage classification by comparing the similarity image of colposcopy results in the testing stage with the image of the results of the training process. We used 30 images as a dataset. The result gets same accuracy which is 80% for both Canny or Isotropic operator. Average running time for Canny operator implementation is 0.3619206 ms while Isotropic get 1.49136262 ms. The result showed that Canny operator is better than isotropic operator because Canny operator generates a more precise edge with a fast time instead.

1. Introduction

Cervical cancer is the second most maglinancy among women [1]. Cervical cancer is an abnormal growth cell occurring in the uterine cervix that usually detected when cancer enters an advanced stage. Therefore women are recommended to undertake precancer examination such as colposcopy. Colposcopy is the examination to see the surface of the cervix by inserting a tool called colposcope into the vagina. The cervical surface image can be documented with a small camera on the colposcope. Colposcopy has been used primarily to diagnose pre-cancer and cancerous lesions because this procedure gives an exaggerated view of the tissues of the vagina and the cervix [2]. Furthermore, the colposcopy can directly visualize the quality and location of lesions of the cervix [3]. One of the challenge factors is difficulties in training the physicians to recognize the pathology of colposcopy image [4]. The poor quality of colposcopy image sometimes makes physician challenging to analyze it. Identification of cancer should be identified more accurate if the colposcopy images are transformed into edge detection forms. Edge detection is the process to characterize the intensity changes in the image [5]. This process aims to simplify the analysis of images by reducing the amount of data from the image but still preserving useful information about boundaries of the object [6].

Lesions are an area of abnormal tissue on the cervix. Every stage of cervical cancer has different broad and lesion location on cervix [7]. Every colposcopy with abnormal cell will generate different edge detection pattern because it corresponds to lesion boundaries. This model depends on wide and position of lesions on the cervix.

This work is experimental research to identify and classify cervical cancer stage based on colposcopy image processing. In this study, we implemented two edge detection algorithms which are Canny and Isotropic operator. The Canny algorithm is an optimal edge detection that already applied noise reduction in its step. On the other hand, Isotropic is one of gradient edge detection with more straightforward action. Classification of cancer stage identified by comparing edge detection result with edge detection of reference images. The parameters in this research are the accuracy of stadium cancer classification, quality of edge detection and the running time. The benefit of this study this method could be a choice to transform colposcopy image into edge detection form that easier to identify the cancer lesions.

2. Related Works

Previous research which implemented image processing in health field had been done by [1,4,7,8]. Research by [1] proposes to develop an automated diagnostic system for cervical pre-cancerous. The study by [4] describes a texture image analysis technique for characterizing and recognizing typical vascular patterns relating to cervical lesions from colposcopy image. The investigation by [8] has developed a prototype digital imaging colposcopy system for use in clinical, research and teaching environments. The previous research about edge detection algorithm had done by [5, 6, 9, 10]. From all the previous research declared that edge detection is a useful method to represent a clear image while still preserving useful information about object boundaries.

3. Methodology

The Methodology of this research can be described in Figure 1.

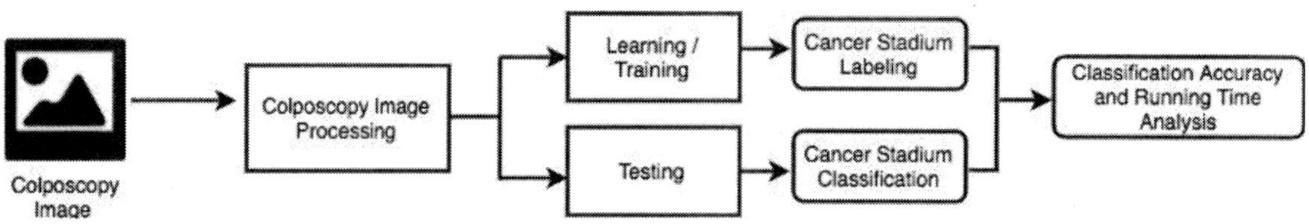

Figure 1. Methodology.

3.1 Colposcopy Image

Colposcopy Images are dataset for training and testing stage. The dataset is from a health official government sites at URL *www.pptm.depkes.go.id.* These colposcopy images are 30 png images with 200 x 200 pixels containing vascular pattern lesions from different stages of cervical cancer.

3.2 Colposcopy Image Processing

Colposcopy image processing is a stage to convert colposcopy image into edge detection form. The detail from this step can be seen in Figure 2.

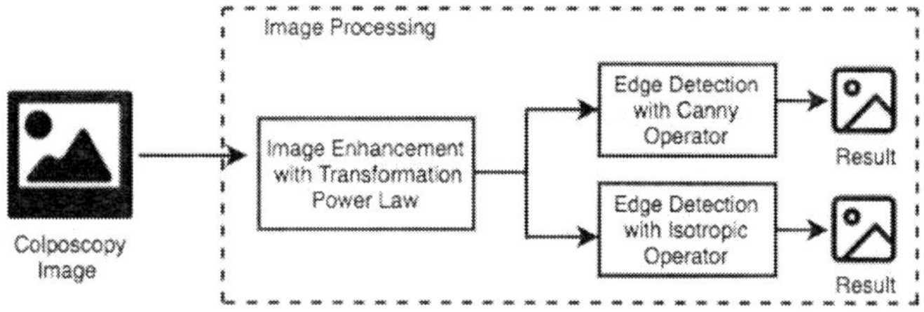

Figure 2. Colposcopy Image Processing.

3.2.1 Image Enhancement

Image enhancement is a process to convert an image into suitable form than the original image. In this research, we implemented n^{th} Root Power law transformation method. The principal objective of this process is to improve the detail of edge detection. This technique is a spatial transformation that operates directly on pixels. The advantage of spatial-based domain technique is easier to understand and easier to implement in a real-time implementation because the complexity of these methods is low [11].
The nth root power law is defined by equation 1.

$$G = c.\ F^{\frac{1}{\Upsilon}} \tag{1}$$

where:

G = Image Result, F = Original Image, c = Constanta, Υ = gamma value.

In this research we implemented c = 1 and gamma value = 0.53. The result of this transformation process becomes input for the next process.

3.2.2 Edge Detection with Canny Operator

There are several processes in determining edge detection by the canny method. The process includes a Gaussian filter, gradient magnitude calculation, non-maximum suppression and determination of thresholding and edge direction. Analysis of Gaussian filter process is to eliminate noise on the image before trying to locate and detect any edges. Gaussian filtering is convolution operation of multiplication of kernel matrix with an original image. This kernel matrix helpful for functions of Gaussian computation. In this study, we implemented 3 x 3 kernel matrix.

$$\text{Gx} = \begin{bmatrix} 1 & 0 & -1 \\ 1 & 0 & -1 \\ 1 & 0 & -1 \end{bmatrix} \tag{2}$$

The next step is to get the edge strength by taking the gradient of the image. The approximate of edge strength at each point can be calculated with convolution mask of 3 x 3 Sobel operator. One estimating the gradient in the x-direction (columns) and the other estimating the gradient in the y-direction (rows). They are shown below:

$$\text{Gx} = \begin{bmatrix} -1 & 0 & 1 \\ -2 & 0 & 2 \\ -1 & 0 & 1 \end{bmatrix} \qquad \text{Gy} = \begin{bmatrix} 1 & 2 & 1 \\ 0 & 0 & 0 \\ -1 & -2 & -1 \end{bmatrix} \tag{3}$$

The magnitude of the gradient is then approximated using the equation 4:

$$|G| = |G_x| + |G_y| \tag{4}$$

Edge direction can be calculated by inverse tangent of Gx / Gy.

The next process is Non-maximum suppression. This stage is used to trace along the edge in the edge direction. This process will eliminate the non-maximum edges into 0

3.2.3 Edge Detection With Isotropic
We implemented Isotropic as one of the gradient operators. Process stages in this operator contain isotropic mask determination, gradient magnitude calculation, and thresholding determination. Isotropic applies the nearer pixels to the corner kernel with value$\sqrt{2}$.

$$\text{Gx} = \begin{bmatrix} -1 & 0 & 1 \\ -\sqrt{2} & 0 & \sqrt{2} \\ -1 & 0 & 1 \end{bmatrix} \qquad \text{Gy} = \begin{bmatrix} 1 & \sqrt{2} & 1 \\ 0 & 0 & 0 \\ -1 & -\sqrt{2} & -1 \end{bmatrix} \tag{5}$$

The magnitude of the gradient can be calculated with equation 6:

$$\text{G} = \sqrt{{G_x}^2 + {G_y}^2} \tag{6}$$

3.3 Learning / Training and Cancer Stadium Labeling
Training or learning process is a process of labeling all dataset image with cervical cancer stage. This labeling process involved pathology doctor as an expert in diagnosing the colposcopy image as a reference. The Four stadiums are normal, stage 1, stage 2, stage 3 and stage 4. We provided 15 datasets to the training stage or 3 dataset image reference for each cancer stadium.

3.4 Testing Stage and Cancer Stadium Classification
Testing stage is a stage to determine the cancer stage classification of colposcopy image input. The process of deciding cancer stage classification is done by comparing the similarity image of colposcopy results in the testing stage with the image of the results of the training process. Classification of each colposcopy in a testing stage will follow the classification colposcopy in training stage with has the closest similarity value. In this study, the similarity distance is calculated by using Euclidean distance method. Euclidean Distance has a high accuracy so it is suitable to be used as a method to calculate the similarity or dissimilarity of some images. Euclidean distance from two vector x and y with n training data can be seen in equation 7.

$$\text{d(x,y)} = \sqrt{\sum_{i=1}^{n} (x_i - y_i)^2} \tag{7}$$

The smaller value of d (x, y) means objects are more similar conversely the bigger value of d (x, y) means objects are more dissimilar.

3.5 Classification Accuracy and Running Time Analysis
Classification Accuracy and running time analysis are processes to measure the accuracy of classification and running time process. In this research we implemented percentage of correct predictions with a formula:

$$\text{Percentage Accuracy} = \frac{\text{Number of Correct Classification}}{\text{Total number of samples}} \text{ x } 100\% \tag{8}$$

4. Results and Discussion

The result of stage cancer classification can be seen in Table 1.

Table 1. Stage Cancer Classification Result.

Stage Cancer Classification	**Original Image**	***n^{th} root power Law Transformation***	***Canny Operator***	**Isotropic Operator**
Normal Cervix	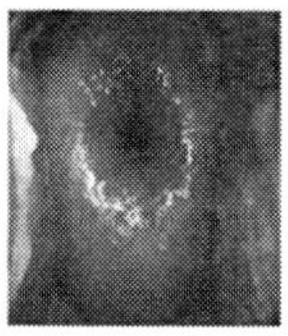	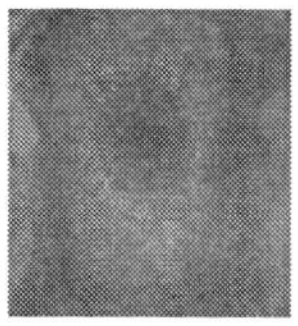		
Cervical Cancer Stage 1	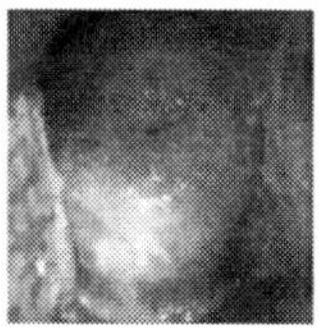	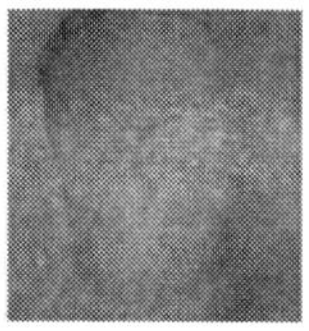		
Cervical Cancer Stage 2	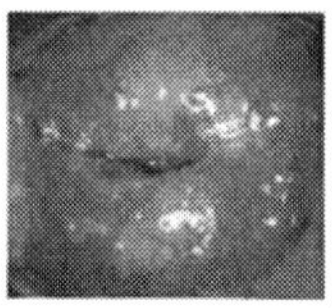			
Cervical Cancer Stage 3	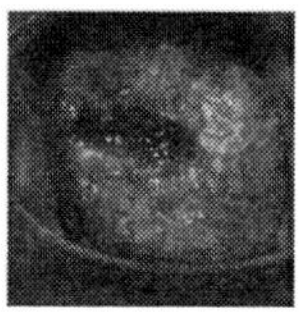	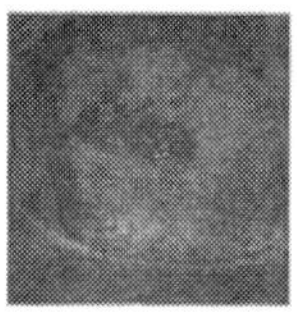	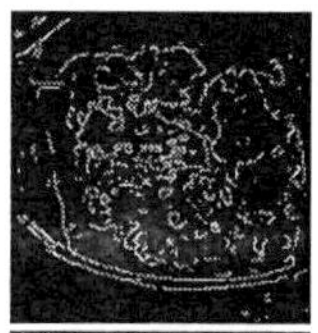	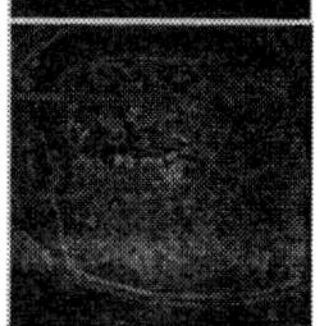
Cervical Cancer Stage 4	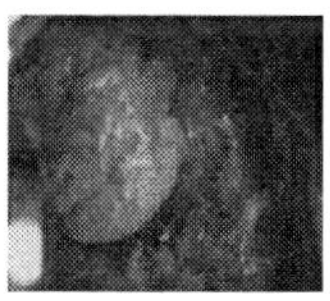		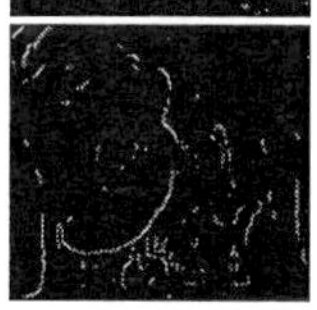	

Based on Table 1, the result of edge detection with Canny operator generates better result compared to Isotropic Operator. With Canny operator, the cancer lesions are brighter and sharper.

Running Time of Canny and Isotropic process time (in millisecond) can be seen in Table 2

Table 2. Running Time of Canny and Isotropic Process.

	Cervic Normal	Stage 1	Stage 2	Stage 3	Stage 4
Canny	0.3744006	0.3588006	0.3432006	0.3588006	0.3744006
Isotropic	1.2168022	1.2012021	1.2948022	2.0124035	1.7316031

According to running time process, the Canny operator has better time efficiency comparing to the Isotropic operator. The accuracy of stage cancer classification from this research can be seen in Table 3.

Table 3. Classification Stage Cancer Accuracy.

No	ID Colposcopy Image	Classification Accuracy
1	Image 1	Accurate
2	Image 2	Accurate
3	Image 3	Accurate
4	Image 4	Accurate
5	Image 5	Not Accurate
6	Image 6	Accurate
7	Image 7	Accurate
8	Image 8	Accurate
9	Image 9	Not Accurate
10	Image 10	Accurate
11	Image 11	Not Accurate
12	Image 12	Accurate
13	Image 13	Accurate
14	Image 14	Accurate
15	Image 15	Accurate

Based on this result, we can calculated percentage accuracy of this classification is 80 %.

5. Conclusion

In this study, we wanted to prove that by only implementing the identification of edge detection in the colposcopy image, identification of cervical cancer can be determined. In this research, we implemented and compared two edge detection algorithms which are Canny and Isotropic operator. The Canny algorithm is an optimal edge detection that already applied noise reduction in its step, in this study, we implement 3 x 3 kernel matrix. On the other hand, Isotropic is one of gradient edge detection with more straightforward action. The methodology of this research is image pre-processing, training and testing stage. The pre-processing step is image enhancement or the process to improve the detail of edge detection. In this research, we implemented n^{th} Root Power law transformation method with Constanta = 1 and gamma value = 0.53. Training or learning process is a process of labelling all dataset image with cervical cancer stage. In this study, this labelling process involved pathology doctor as an expert in diagnosing the colposcopy image as a reference. We provided 15 datasets to the training stage or 3 dataset image reference for each cancer stadium. Testing stage is a stage to determine the cancer stage classification of colposcopy image input. The process of deciding cancer stage classification is done by comparing the similarity image of colposcopy results in the testing stage with the image of the results of the training process. In this study, the similarity distance is calculated by using Euclidean distance method. We use 15 colposcopy images for testing step. The result of edge detection with Canny operator generates better result compared to Isotropic Operator. With Canny operator, the cancer lesions are clearer and sharper. Furthermore, the Canny operator has better time efficiency comparing to the Isotropic operator. Average running time for Canny operator implementation is 0.3619206 ms while Isotropic get 1.49136262 ms. The result showed that Canny operator is better than isotropic operator because Canny operator generates a clearer edge with a fast time instead. The percentage accuracy of cancer stage stadium with this experiment reach 80% for both methods.

References

[1] Mat-Isa N A, Mashor M Y and Othman N H 2008 *Artif. Intell. Med* **42(1)** 1–11
[2] Fernandes M P, Venkatesh S, Narayanan S and Prabitha V G 2015 *Int. J. of Tech. Enhancement and Emerging Eng. Research* **3(5)** 133–140
[3] Burghardt E, Winter R, Tamussino K, Pickel H, Lahousen M, Haas J, Girardi F, Ebner F, Hackl A and Pfister H 1994 *Crit. Rev. Oncol. Hematol* **17(3)** 181–231
[4] Ji Q, Engel J, and Craine E 2000 *IEEE Trans. Med. Imaging* **19(11)**1144–1149
[5] Torre V and Poggio T A 1986 *IEEE Trans on Pattern Anal. Mach. Intell* **(2)** 147–163
[6] Canny J 1986 *IEEE Trans. Pattern Anal. Mach. Intell* **PAMI-8(6)** 679–698
[7] *Signs, Symptoms of Cervical Cancer* – Treatment; Stages – Uterine Cervix, Pelvic | Columbia

Asia India." [Online]. Available: http://india.columbiaasia.com/health-articles/cervical-cancer-signs-symptoms-stages-treatment. [Accessed: 29-Sep-2017].
[8] Neil T, John R, Crame R, Eric R, Brian L, and Craine L 1988 *Med. Imaging II* **914** 505-511
[9] Fan J, Aref W G, Hacid M S, and Elmagarmid A K 2001 *Pattern Recognit. Lett.* **22(13)** 1419–1429
[10] Narendra V G and Hareesh K S 2011 S *Int. J. Agric. and Biol Eng* **4(2)**
[11] Singh G and Mittal A 2014 *Int. J. Innov. Sci. Res.* **10(2)** 267–274

Facial Expression Recognition using Multiclass Ensemble Least-Square Support Vector Machine

Armin Lawi[1], M Sya'Rani Machrizzandi[2]

[1]Department of Computer Science, Hasanuddin University, Makassar

[2]Department of Informatics, Al-Asy'ariah Mandar University, Polewali Mandar

E-mail: machrizandi@gmail.com

Abstract. Facial expression is one of behavior characteristics of human-being. The use of biometrics technology system with facial expression characteristics makes it possible to recognize a person's mood or emotion. The basic components of facial expression analysis system are face detection, face image extraction, facial classification and facial expressions recognition. This paper uses Principal Component Analysis (PCA) algorithm to extract facial features with expression parameters, i.e., happy, sad, neutral, angry, fear, and disgusted. Then Multiclass Ensemble Least-Squares Support Vector Machine (MELS-SVM) is used for the classification process of facial expression. The result of MELS-SVM model obtained from our 185 different expression images of 10 persons showed high accuracy level of 99.998% using RBF kernel.

1. Introduction

Facial expression in everyday human life is used as a form of natural human response that describes the feelings felt by a person in interacting with a certain thing. In the interaction between fellow human, expression is use as a part of communication. On the other hand, technological developments from time to time have great leaps when a new interaction is implemented such as the use of keys, monitor screen, GUI, touch screen, to voice commands. The key to technological development lies in the user, therefore computer interaction is an important part in technological development and understanding the user's response becomes very important. One approach taken to understand the response of the user is by a facial recognition system. By understanding the expression of the user, the computer can be made as if it can understand human feelings.

The research on facial expressions has been started since 1978 [1] and was extensively researched in the computer science area of the 90s and continues to the present day. an expression recognition system of two important stages, namely facial features representation and classification design. Facial feature representation is obtained by extracting input in the form of digital image by using certain method. The method of feature extraction used by researchers to build a facial expression recognition system that is Principal Component Analysis Algorithm (PCA). Principal Component Analysis (PCA) algorithm is one method that can be used to process the image of a person's face so that the system will automatically recognize a person's face through its main features such as eyes, nose, lips, and eyebrows as identity [2]. The identity of the person's face image by the system will be recognized through various training (training) stored in the database. The training phase is the result of extraction from different collections of different faces then collected and stored in a database. The results of facial images that have been extracted using PCA algorithm will be compared with the new face image as the face image to be tested and testing whether it has similarity or almost similar to recognizable by the system [3].

2. Related Works

Yubo Wang, *et* al., [4] have proposed a new method of recognition of facial expressions. The facial expressions are extracted from the human face by the expression classifier learned from the Haar feature based on the LUT Weak Classifiers. The expression recognition system consists of three modules, face detection, facial feature landmark extraction and facial expression recognition. This system is applied automatically can recognize seven expressions in real time that include anger, disgust, fear, pleasure, neutrality, sadness and surprise. Reda Shbib and Shikun Zhou in [5] studied the introduction of facial expressions using the Active Shape Model method by adopting AdaBoost classifier and Haar-Like feature to detect images. AdaBoost is a method to increase classifier accuracy gained from the component learner of the Support Vector Machine (SVM) method [6, 7, 8].

Data mining is a process that uses statistical techniques, calculations, artificial intelligence and machine learning to extract and identify useful information and related knowledge from large databases [9]. SVM initially can only classify data in two classes [8, 10, 11]. However, further research SVM was developed so that it can classify data over two classes (multiclass) [12, 13, 14]. Classifying M-classes means predicting the class labels $C_m, m = 1, \dots, M$ one way to solve the M-class problem by formulating it into binary L classification problems [15]. SVM concept is simply trying to find the best hyperplane that serves as a separator of two classes in the input space. Pattern that is a member of two classes into +1 and -1, and share alternate field separators. The best dividing fields can not only separate the data but also have the largest margins. Margin is the distance between the fields of separator (hyperplane) with the closest pattern of each class.

Let $\{x_1, \dots, x_n\}$ be the dataset and $y_i \in \{+1, -1\}$ is the class label of the x_i data. A pair of parallel bounding plane separates the two classes. The first delimiter field limits the first class while the second delimiter field limits the second class, it is obtained therefore as follows [16].

$$\begin{aligned} \boldsymbol{x}_i.\boldsymbol{w} + b \geq +1 \text{ for } y_i = +1 \\ \boldsymbol{x}_i.\boldsymbol{w} + b \leq -1 \text{ for } y_i = -1 \end{aligned} \tag{1}$$

The best separator field has the largest margin values that can be formulated into a quadratic programming problem as follows.

$$\min_{w,b,\xi} \frac{1}{2}\|\boldsymbol{w}\|^2 + C\sum_{i=1}^{l} \xi_i \tag{2}$$

with constraints: $y_i(\boldsymbol{x}_i.\boldsymbol{w} + b) \geq 1 - \xi_i$ and $\xi_i \geq 0$,

where $l = 1, .., n$ is a slack variable that determines the level of misclassification of the data samples. Whereas, $C > 0$ is a parameter.

Kernel method is used when the ordinary SVM classifier cannot separate the data linearly. The kernel method transforms the data into the features space dimension so that it can be linearly separated on the features space. The kernel method can be formulated as follows.

$$K(\boldsymbol{x}_i.\boldsymbol{x}_j) = \varphi(\boldsymbol{x}_i).\varphi(\boldsymbol{x}_j) \tag{3}$$

Commonly used kernel functions are as follows.

a. The linear kernel: $K(\boldsymbol{x}_i, \boldsymbol{x}_j) = \boldsymbol{x}_i^T \boldsymbol{x}_j$
b. Kernel polynomial: $K(\boldsymbol{x}_i, \boldsymbol{x}_j) = (\gamma.\boldsymbol{x}_i^T \boldsymbol{x}_j + r)^p, \gamma \geq 2$
c. RBF Kernel (Radial Basis Function): $K(\boldsymbol{x}_i, \boldsymbol{x}_j) = exp\left(-\gamma\|\boldsymbol{x}_i - \boldsymbol{x}_j\|^2\right), \gamma > 0$

LS-SVM is one of the SVM modifications that is solved linearly [7, 8, 11, 13, 14]. If the SVM separator field is given as in (3), then the LS-SVM is given as follows.

$$\min_{w,b,\xi} \frac{1}{2}\|\boldsymbol{w}\|^2 + \frac{C}{2}\xi^T\xi$$
$$\text{with constraint: } y_i(\boldsymbol{x_i}.\boldsymbol{w} + b) \geq 1 - \xi_i \quad (4)$$

The above equation can be solved with Lagrangian multiplier.

$$L = \frac{1}{2}\|\boldsymbol{w}\|^2 + \frac{C}{2}\xi^T\xi - \sum_{i=1}^{l}\alpha_i\,(y_i(\varphi(x_i).w + b) - 1 + \xi_i) \quad (5)$$

where the Lagrangian multiplier α_i can be either positive or negative.

To optimize the conditions in (5), we decrease of w, b, ξ, and α is equal to zero. The results of the process are as follows.

$$\frac{\partial L}{\partial w} = 0 \rightarrow \boldsymbol{w} = \sum_{i=1}^{l}\alpha_i y_i \varphi(\boldsymbol{x}_i) \quad (6)$$
$$\frac{\partial L}{\partial b} = 0 \rightarrow \sum_{i=1}^{l}\alpha_i y_i = 0 \quad (7)$$
$$\frac{\partial L}{\partial \xi} = 0 \rightarrow \alpha = \gamma\xi_i,\ i = 1, \ldots, N \quad (8)$$
$$\frac{\partial L}{\partial \alpha} = 0 \rightarrow y_i(\varphi(x_i).w + b) - 1 + \xi_i = 0, i = 1, \ldots, N \quad (9)$$

Using the One-Against-All (OAA) method, k binary model can be constructed (in which k is the number of classes). Each i-class model is trained by using the entire data. For example, there is a classification problem with 3 classes. Training step constructs 3 pieces of binary classification with the following objective function.

$$\min_{w_i, b_i, \xi_{t,i}} \frac{1}{2}\sum_{i}^{m}(\boldsymbol{w}_i)^T\boldsymbol{w}_i + \frac{C}{2}\sum_{i}^{m}\xi_{t,i}^2$$
$$\text{with constraint: } y_{t,i}(\varphi_i(\boldsymbol{x}_t)(\boldsymbol{w}_i)^T + b_i) \geq 1 - \xi_{t,i}. \quad (10)$$

Confusion matrix is a table that expresses the amount of test data that is correctly classified and misclassified. The table using the following parameters in order to calculate accuracy, sensitivity and false discovery rate. True Positive (TP) is the number of documents from class 1 is correctly classified as class 1. True Negative (TN) is the number of documents from class 0 is are correctly classified as class 0. False Positive (FP) is the number of documents from class 0 incorrectly classified as class 1. False Negative (FN) is the number of documents from class 1 that are misclassified as class 0.

Performance measurements of accuracy, sensitivity and false discovery rate can be calculated using the following formulas respectively.

$$Accuracy = \frac{TP + TN}{TP + FN + FP + TN} \times 100\% \quad (11)$$

$$Sensitivity = \frac{TP}{TP + FN} \times 100\% \quad (12)$$

$$False\ Discovery\ Rate = \frac{FP}{FP + TP} \times 100\% \quad (13)$$

3. Experimental Details

The process of facial classification is explained by the following activities:

1. Face image capture
2. Face image processing
3. PCA feature extraction
4. Perform Data Training Process
5. Pattern model Classification in get after doing the data training process.
6. Conducting Testing Process/Data Classification after obtained model pattern from result of training process data.

3.1. Principal Component Analysis (PCA)

The sequence of features extraction process using PCA is given as follows.

1. Read input data
2. Calculate the average of each column
   Input of all image of training data is input parameter on PCA function that is "ImageTraining" mean column image can be calculated using matlab program code:
   ```
   I = imread ('YM.AN1.61.tiff');
   Average = mean (I);
   ```
3. Calculate the zero mean
   To calculate the zero mean, the average matrix of training data must be doubled as much as the sample data, which is multiplied by 240 lines with the command:
   ```
   ZeroMean = bsxfun (@ minus, 256, Average);
   ```
4. Calculate the covariance
   To calculate the covariance matrix by multiplying the zero mean against its transpose, and the result is divided by "Amount Sample-1" with the command:
   ```
   CovMatric = (1 / (10-1)) * (ZeroMean * ZeroMean ');
   ```
5. Calculate eigen value and eigen vector
   The dimensions resulting from the covariance matrix have 240 row dimensions and 240 column dimensions. The dimension has been significantly reduced from 10304 to 240, so the dimension number has decreased by 10064 (10304-240 = 10064). The value of the covariance matrix is used further process to determine the value of eigen value and eigen vector with the command:
   ```
   [eigvector, eigvalue] = eig (CovMatric);
   ```
6. Sort the eigen value in descending order
   To perform a descending order of "eigen value", simply take the diagonal value followed by the sequencing process. The "eigen value" sorting result is stored in a variable, that is, the "Result" variable and the sorting index are stored in the "Index" variable with the command:
   ```
   eigvalue = diag (eigvalue);
   [Results, index] = sort (eigvalue, 'descend');
   eigvalue = eigvalue (index);
   ```
7. Sort the eigen vector column to change as the eigen value index changes
8. Check the value of eigen value close to 0
   The most non-dominant eigenvector values tend to have a correlation of the value of eigen value that is close to zero, therefore it is necessary to check the value of eigen value and the number of close to zero with the command:
   ```
   NotDominant = find (eigvalue <0.00000000001);
   Amount = length (Not Dominate);
   ```
9. Calculate the projection matrix
   The projection matrix can be calculated by multiplying the transpose data variable by "eigen vector". The code of the program is as follows:
   ```
   ProjectionMatric = ZeroMean '* eigvector;
   ```
   The result of the program code is used to form a diagonal matrix with the command:
   ```
   RootSumProjectionMatric = (1 ./ (sum (ProjectionMatric. ^ 2). ^ 0.5));
   ```
   New projection matrix can be obtained with the command:
   ```
   ProjectionMatric = (bsxfun (@ times, ProjectionMatric, RootSumProjectionMatric));
   ```
10. Throw away the eigen vector column whose position is equal to the eigen value position which has a value close to 0.
    If the value of eigen value is found to be close to zero, then the eigenvector column correspondingto the eigen value position need not be used for the next process with the command:
    ```
    eigvector = eigvector (:, 1: end-Amount);
    ```

3.2. *Dataset*

We conducted a simulation for the proposed method; i.e., face detection, feature extraction using PCA algorithm, and facial expression classification process using Ensemble Multiclass Least-Square Support Vector Machine (LS-SVM). The data used in this simulation is data about facial expressions from the Olivetty Research Laboratory (ORL) and Jaffe Images can be downloaded, respectively, on the following URL addresses.

1. http://www.cl.cam.ac.uk/research/dtg/attarchive/facedatabase.html [16], and
2. http://www.kasrl.org/jaffe.html [17] and

The type of expression is divided into six classes, i.e., angry, disgust, fear, happy, neutral, and sad. The sample data on this dataset amounts to 240 images of facial expressions.

3.3. *Classification Process*

In the process of expression classification, starting with image taking of the file, then image is processed using image processing software to get numerical data value which will be used as input data for learning process and validation. After the learning data in the next can be the result of data learning is used for the testing process.

3.4. *Multiclass Algorithm Method*

The analysis used in this multiclass classification is the One Against All classification method. The multilingas method algorithm as follows:

1. Dataset input.
2. Identify the input dataset
   a. The values of the training data feature (x_i)
   b. Class of training data (yi)
   c. Values feature test data (xt_i)
   d. Class of test data (yt_i)
3. Initiate objects on LS-SVM before performing the training process with the initlssvm function
   a. Specify data of training data feature (x_i)
   b. Specifies the training data class (y_i),
   c. Choose a classifier to classify data
   d. Selects the *kernel* and its parameters to use
4. Selecting the multilingual method code used (*code_OneVsAll* for *One Against All*)
5. Conduct training process with train lssvm function
6. Calculating values w
7. Make predictions based on the model obtained and determine data feature test data (xt_i) with simlssvm function
8. Create a *confusion matrix*
9. Calculate the level of accuracy with the formula:

$$\lambda = \frac{C}{N} \times 100\%$$

where C is the correct total of predictions and N is the total of all data tested.

3.5. *Implementation Results on dataset*

The separator function for the one against all method with the RBF (Radial Basis Function) kernel using the parameter $\sigma = 0.5$ for dataset is as follows.

$$f_1(x) = \boldsymbol{x}_i . \boldsymbol{w}_1 + b = \boldsymbol{x}_i . \begin{pmatrix} 10{,}5967 \\ 361{,}2859 \\ 31{,}8372 \\ 15{,}3328 \end{pmatrix} + (-0.8302), \quad (14)$$

$$f_2(\boldsymbol{x}) = \boldsymbol{x}_i.\boldsymbol{w}_2 + b = \boldsymbol{x}_i.\begin{pmatrix}149{,}1081\\285{,}1159\\35{,}5804\\148{,}3523\end{pmatrix} + (-0.5724), \quad (15)$$

$$f_3(\boldsymbol{x}) = \boldsymbol{x}_i.\boldsymbol{w}_3 + b = \boldsymbol{x}_i.\begin{pmatrix}60.8802\\352{,}8413\\71{,}1501\\64{,}0558\end{pmatrix} + (-0.6154). \quad (16)$$

$$f_4(\boldsymbol{x}) = \boldsymbol{x}_i.\boldsymbol{w}_1 + b = \boldsymbol{x}_i.\begin{pmatrix}-39{,}1510\\324{,}6996\\55{,}4214\\-38{,}4858\end{pmatrix} + (-0.4468), \quad (17)$$

$$f_5(\boldsymbol{x}) = \boldsymbol{x}_i.\boldsymbol{w}_2 + b = \boldsymbol{x}_i.\begin{pmatrix}4{,}7389\\352{,}3402\\81{,}7228\\3{,}0785\end{pmatrix} + (-0.8428), \quad (18)$$

$$f_6(\boldsymbol{x}) = \boldsymbol{x}_i.\boldsymbol{w}_3 + b = \boldsymbol{x}_i.\begin{pmatrix}-19{,}9091\\376{,}3541\\81{,}1863\\-19{,}6398\end{pmatrix} + (-0.6924). \quad (19)$$

4. Results and Discussion

The values of w and b based on the kernel type and their parameters for the use of the One Against All method on the expression dataset can be seen in table 1:

Table 1: *w* and *b* values.

kernel dan parameter		*w* value			*b* value		
		w_1	w_2	w_3	b_1	b_2	b_3
RBF	$\alpha = 0.5$	10,5967 361,2859 31,8372 15,3328	149,1081 285,1159 35,5804 148,3523	60.8802 352,8413 71,1501 64,0558	−0.8302	−0.5724	−0.6154

kernel dan parameter		*w* value			*b* value		
		w_4	w_5	w_6	b_4	b_5	b_6
RBF	$\alpha = 0.5$	−39,1510 324,6996 55,4214 −38,4858	4,7389 352,3402 81,7228 3,0785	−19,9091 376,3541 81,1863 −19,6398	−0.4468	−0.8428	−0.6924

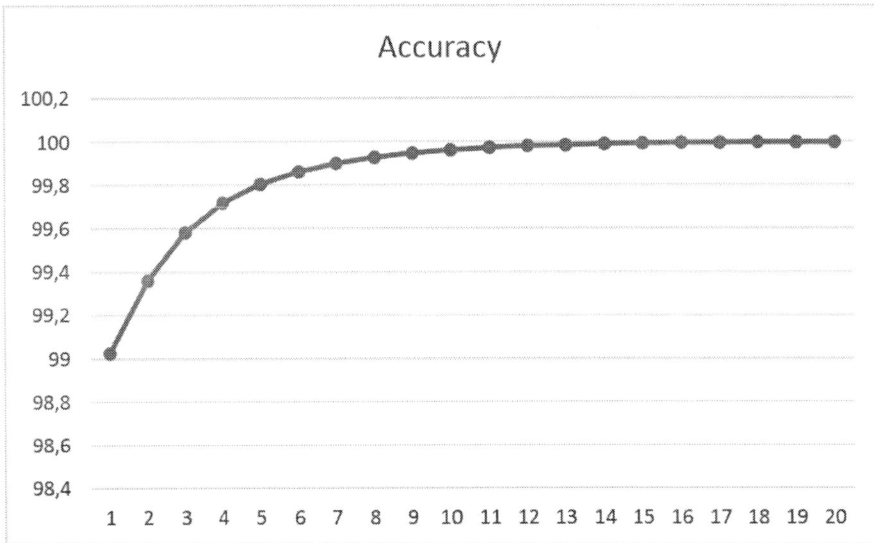

Figure 1. Accuracy level of the OAA for each kernel in dataset.

Based on table 1 and figure 1, the use of RBF kernel types and using the parameter $\sigma = 0.5$ has the highest accuracy on dataset has the highest accuracy of 99.99832%. Accuracy for classification with the 6 number of classes.

5. Conclusion

The facial expression recognition have been proposed using the Multiclass Ensemble Least-Squares Support Vector Machine (MELS-SVM) method. The image of the face is extracted using Principal Component Analysis (PCA) algorithm to take the value of Projection Matrix and Weight Matrix. The classification process uses the MELS-SVM to construct the classifier model from the training images. Finally, the testing images is used to measure accuracy using RBF kernel and the our result showed the accuracy level is 99.99832%.

Acknowledgments

The authors would like to express heartfelt thanks to Department of Computer Science, Hasanuddin University and the Department of Informatics, Al-Asy'ariah Mandar University, Polewali Mandar for the research support and assistances.

References

[1] Ojala T, Pietikäinen M and Harwood D 1994 *Proceedings of the 12th IAPR International Conference on Pattern Recognition (ICPR1994)* **1** p 582 – 585
[2] Pratt, William K 2007 *Digital image processing, PIKS Scientific inside* (John Wiley) 4th Edition
[3] Nugraheny D *Hasil Ekstraksi Algoritma Principal Component Analysis (PCA) Untuk Pengenalan Wajah* (Yogyakarta: STTA)
[4] Wang Y, Ai H, Wu B, Huang C 2004 *Real Time Facial Expression Recognition With Adaboost* (China : Tsinghua University)
[5] Shbib R, Zhou S *Facial Expression Analysis using Active Shape Model* (United Kingdom: School of Engineering, University of Portsmouth)
[6] Kurniawan D and Supriyanto C 2013 *Journal Teknologi Informasi*
[7] Marliani R R, Lawi A and Thamrin S A 2016 *Algoritma Adaboost untuk Optimasi Ensemble*

Least Squares Support Vector Machine (Makassar : Hasanuddin University)
[8] Zhou L, Lai K K and Yu L 2010 *J. Expert Systems with Applications* 127-133
[9] Turban E et.al 2005 *Decision Support Systems and Intelligent Systems* (Yogyakarta: Andi Offset) (Bahasa Indonesia translation version)
[10] Vapnik V and Cortes C 1995 *Support-Vector Networks. Machine Learning* 273-297
[11] Xu Y, Lv X, Wang Z and Wang L 2014 *Journal of Information Science and Engineering* 1773-1787
[12] Sembiring K 2007 *Tutorial SVM Bahasa Indonesia* (Bandung: Institut Teknologi Bandung)
[13] Jafar N, Thamrin S A, Lawi A 2016 *Proceeding of International Conference on Computational Intelligence and Cybernetic*
[14] Nurkamila J, Lawi A and Thamrin S A 2016. *Least Squares Support Vector Machine Untuk Klasifikasi Data Multikelas* (Makassar : Hasanuddin University)
[15] Gestel T, Suykens J, Lanckriet G, Lambrechts A, Moor B and Vandewalle J 2002 *Multiclass LS-SVMs: Moderated Outputs and Coding-Decoding Schemes. Neural Processing Letters* 45-58
[16] http://www.cl.cam.ac.uk/research/dtg/attarchive/facedatabase.html
[17] http://www.kasrl.org/jaffe.html

Super-Encryption Implementation Using Monoalphabetic Algorithm and XOR Algorithm for Data Security

Dian Rachmawati[1], Mohammad Andri Budiman[1], Indra Aulia[2]

[1] Departemen Ilmu Komputer, Fakultas Ilmu Komputer dan Teknologi Informasi, Universitas Sumatera Utara, Jl. Universitas No. 9-A, Medan 20155, Indonesia.
[2] Departemen Teknologi Informasi, Fakultas Ilmu Komputer dan Teknologi Informasi, Universitas Sumatera Utara, Jl. Universitas No. 9-A, Medan 20155, Indonesia.

Email: dian.rachmawati@usu.ac.id

Abstract. The exchange of data that occurs offline and online is very vulnerable to the threat of data theft. In general, cryptography is a science and art to maintain data secrecy. An encryption is a cryptography algorithm in which data is transformed into cipher text, which is something that is unreadable and meaningless so it cannot be read or understood by other parties. In super-encryption, two or more encryption algorithms are combined to make it more secure. In this work, Monoalphabetic algorithm and XOR algorithm are combined to form a super-encryption. Monoalphabetic algorithm works by changing a particular letter into a new letter based on existing keywords while the XOR algorithm works by using logic operation XOR Since Monoalphabetic algorithm is a classical cryptographic algorithm and XOR algorithm is a modern cryptographic algorithm, this scheme is expected to be both easy-to-implement and more secure. The combination of the two algorithms is capable of securing the data and restoring it back to its original form (plaintext), so the data integrity is still ensured.

1. Introduction

The technology of internet and multimedia industry rapidly develop, which makes people realize the importance of data security [1]. Protecting confidentiality and authenticity of some transactions by the use of cryptography method has become standard practice [2]. Cryptography is the art and science of using mathematical methods to secure messages from unauthorized parties [3]. To preserve security with cryptography, messages in analog forms or symbols (which are called "plaintext") are generally transformed using tables (for example, ASCII table) into numbers. A secret key or password is then generated by the sender and by using an encryption function (which usually includes some mathematical formulas), these numbers are transformed into meaningless information which are called "cipher text". The cipher text is then sent via a channel to the recipient. Using different (and preferably more secure) channel, the secret key is also sent to the recipient. Having got the cipher text and the key, the recipient can use the decryption function to transform the cipher text back to the original plaintext. Mathematically, the decryption function is simply the inverse of the encryption function.

Monoalphabetic algorithm is a classical cryptographic algorithm that works by switching each letter in the plaintext into another letter based on some secret keys or keywords. Since there are 26 letters in the alphabet, a cryptanalyst (a person who wants to convert the cipher text into the plaintext without knowing the secret keys) can try a brute-force attack for at most 26 times and it is guaranteed that he will be able to succeed. Therefore, the security of Monoalphabetic algorithm is weak and it should not be used for encrypting very sensitive and classified information such as those in banking and military.

In modern cryptography, XOR function plays an important role since it is very easy to perform [4]: to encrypt a message (or plaintext), one should transform it using an encoding table (such as ASCII) to become decimal numbers, and convert these numbers into binaries. Then, some random bits of the same length of those binaries are generated as the secret key. The binaries and the secret key are then XOR-ed together, and we get the cipher text. The cipher text is then sent to the recipient via a channel and the secret key is sent via another channel (which is preferably more secure than the previous channel). In order to convert the cipher-text back to the plaintext, the recipient just can just XOR the cipher-text with the secret key, convert the resulted binaries to decimal numbers, and by using the same encoding table, the decimal numbers can be converted back to the plaintext. This whole encryption and decryption system is called XOR algorithm.

In order to create a stronger cryptosystem, two or more cryptography algorithms can be combined together. This new cryptosystem is called super-encryption. The purpose of this research is to combine Monoalphabetic algorithm with XOR algorithm in a super-encryption scheme in order to make the whole cryptosystem more secure yet still easy-to-implement.

This super-encryption scheme is unique, since it conjoins a classical cryptographic algorithm (Monoalphabetic algorithm) and a modern cryptographic algorithm (XOR algorithm). The proposed method may be used to encrypt data or password that needs to be transmitted across a public network, to ensure its confidentiality and integrity against the attackers [5]. A Python source-code for the new crypto-system is provided. Python was designed to be a highly readable language. Python aims to be simple and consistent in the design of its syntax [6].

2. Methods

In this section, we explain and show an example of how Monoalphabetic algorithm works an example of how XOR algorithm works the model of the new crypto-system, and the Python source-code.

2.1 Monoalphabetic algorithm

Suppose we have a `plaintext = "KOMPUTER"` and a `secret key = "DIAN RACHMAWATI"`. The secret key should have no repeating letters and there should be nothing inside except letters. Thus, we omit the repeating letters and space, and we have the `keyword = "DIANRCHMWT"`.

The keyword is placed underneath the alphabet. The other letters that are not within the keyword is appended to the keyword, and the whole is called cipher alphabet.

```
alphabet       = ABCDEFGHIJKLMNOPQRSTUVWXYZ
cipheralphabet = DIANRCHMWTBEFGJKLOPQSUVXYZ
```

Through the encryption, key mapping can be done by searching the couple's alphabet to cipher alphabet so to plaintext = "KOMPUTER" cipher text obtained "BJFKSQRO" as follows:

```
alphabet       = ABCDEFGHIJKLMNOPQRSTUVWXYZ
cipheralphabet = DIANRCHMWTBEFGJKLOPQSUVXYZ
```

To restore the cipher text into the plain text then use the form of the mapping as follow:

```
cipheralphabet = DIANRCHMWTBEFGJKLOPQSUVXYZ
alphabet       = ABCDEFGHIJKLMNOPQRSTUVWXYZ
```

So the ciphertext `"BJFKSQRO"` back into `"KOMPUTER"`.

IOP Conf. Series: Journal of Physics: Conf. Series **979** (2018) 012033 doi:10.1088/1742-6596/979/1/012033

2.2 XOR
In cryptography, making a cipher (text encryption results) through the operation of XOR, which is an encryption of algorithm, are relatively simple. This technique operates in accordance with the principle of:
A XOR 0 = A,
A XOR A = 0,
(B XOR A) XOR A = B XOR 0 = B,

With this logic, a string text could be encrypted by applying XOR operation based of bit (binary digit), for each character using the specific key. To decrypt the cipher text is applied the operation XOR on the cipher text, so that plain text back to its original form. For example: if we have D message and the key is a so the encryption process is as follows:

D = 68 = 01000100
a = 97 = <u>01100001</u> XOR
00100101 → 37 = %

The resulting cipher text is %, as for the decryption process then performed the operation XOR between cipher text with a key so that it returns to the original message.

% = 37 = 00100101
a = 97 = <u>01100001</u> XOR
01000100 → 68 = D

2.3. The model of the new crypto-system
This research aims to improve security in the process of encoding a message by using a combination of Monoalphabetic and XOR algorithms, so this coded messages increasingly difficult to solve by any third party due to use of the concept Super-encryption. Super-encryption is a cryptographic concept that combines two or more symmetric encryption algorithms to ensure stronger confidentiality. The illustration of the super -encryption process in this research can be seen in Figure 1.

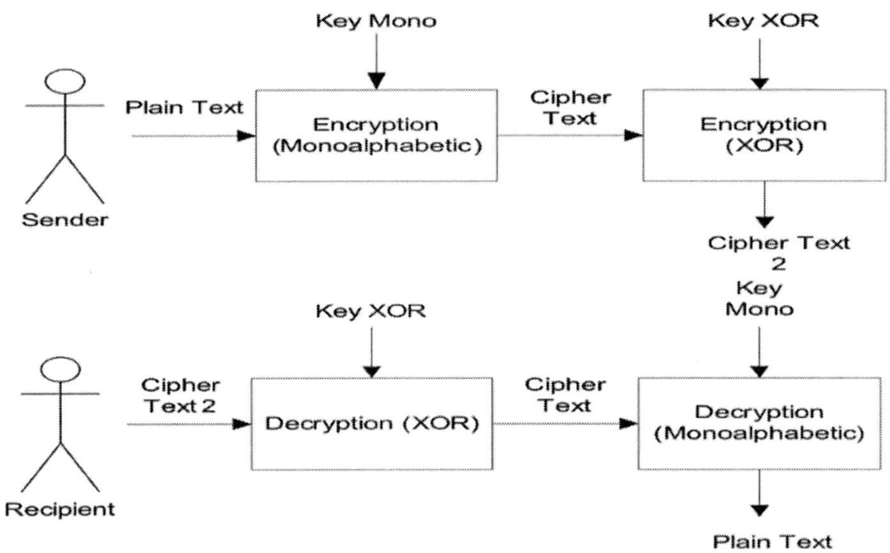

Figure 1. Super-encryption process with algorithm mono-alphabetic and XOR

The super-encryption in Figure 1 is explained as follows. Firstly, the sender encrypts the plaintext with Monoalphabetic encryption algorithm, resulting in the first cipher text. Secondly, the sender uses XOR encryption algorithm to encrypt the first cipher text into ciphertext2. The nature of the XOR algorithm that uses random bits as its key makes the ciphertext2 appears more random than the first cipher text. Thirdly, the ciphertext2 is sent to the recipient. Fourthly, the recipient decrypts the ciphertext2 into cipher text using XOR algorithm. Lastly, the recipient decrypts the cipher text with Monoalphabetic algorithm and gets the original plaintext.

2.4. The Python source code

The code below is used to perform the key generation based on user input string:

```
for c in key:
    if c not in alphabet:
        continue
    if c not in keyword:
        keyword += c

cipheralphabet = keyword

for c in alphabet:
    if c not in cipheralphabet:
        cipheralphabet += c
```

Source code below are used to encrypt and decrypt messages:

```
for p in plaintext:
    if p not in alphabet:
        continue
    i = alphabet.index(p)
    ciphertext1 += cipheralphabet[i]

decryptedtext = ""
for c in ciphertext1:
    i = cipheralphabet.index(c)
    decryptedtext += alphabet[i]
```

Source code below are used to encrypt and decrypt messages using XOR operation:

```
def XOR(bits1, bits2):
    #assert len(bits1) == len(bits2)
    result = ""
    for i in range(len(bits1)):
        if bits1[i] == bits2[i]:
            result += "0"
        else:
            result += "1"
    return result
```

3. Results and Discussions

By using super-encryption algorithm Monoalphabetic and XOR then obtained the following results:
First Encryption (Mono-alphabetic Cipher)

Key Initialization
key = DIAN RACHMAWATI
keyword = DIANRCHMWT

Table Initialization
alphabet = ABCDEFGHIJKLMNOPQRSTUVWXYZ

cipheralphabet = DIANRCHMWTBEFGJKLOPQSUVXYZ

Encryption
plaintext = KOMPUTER
ciphertext1 = BJFKSQRO

Second Encryption (XOR Cipher)
Intialization
ciphertext1 = BJFKSQRO
plainbit = 0100001001001010010001100100101101010011010100010101001001001111

Encryption
random key = 0010000000110100000001111010101001110110011110110100101001010100
plainbit = 0100001001001010010001100100101101010011010100010101001001001111
cipherbit = 0110001001010000010001011001111001101000011011001111011101100101
ciphertext = bPEž hl÷e

First Decryption (XOR Cipher)

Decryption
Ciphertext = bPEž hl÷e
Cipherbit = 0110001001010000010001011001111001101000011011001111011101100101
random key = 0010000000110100000001111010101001110110011110110100101001010100
decryptedbit = 0100001001001010010001100100101101010011010100010101001001001111
decryptedtext1 = BJFKSQRO

Second Decryption (Monoalphabetic Cipher)

Key Initialization
Key = DIAN RACHMAWATI
Keyword = DIANRCHMWT

Table Initialization
Alphabet = ABCDEFGHIJKLMNOPQRSTUVWXYZ
Cipheralphabet = DIANRCHMWTBEFGJKLOPQSUVXYZ

Decryption
decryptedtext1 = BJFKSQRO
decryptedtext2 = KOMPUTER

4. Conclusions

It can be seen that the obtained encryption scheme which uses two encryption algorithms (i.e., Monoalphabetic and XOR algorithms) is more secure than the scheme that only uses single algorithm (i.e., Monoalphabetic algorithm) alone. This is because the XOR algorithm converts the cipher text resulted by the Monoalphabetic algorithm into its corresponding ASCII values (in binary form) then performs operation on bit by XOR-ing it with randomly generated bits as the keyword. The cipher text resulted by the combination of these two algorithms appears random, thus it is completely dissimilar the original message. Therefore, without knowing the passwords of both algorithms, a hacker or a cryptanalyst will have a much longer computational time in order to derive the original message from the cipher text. It is also noticed that encrypted message can return to its original plaintext, and, therefore, the data integrity of this scheme is maintained.

Although it can be noticed that the resulting cipher text appears more random, the cryptanalyst attacks are beyond the scope of this study. Therefore, in order to get more information on how well this scheme performs against a considerable number of different attacks, more studies are required.

Since XOR algorithm requires random bits, which are as long as the bits of its input, it is expected that in the future works, these random bits are generated by cryptographically secure pseudo-random number generator algorithms to ensure that the bits are sufficiently random to be used as a secure key for the XOR algorithm. Moreover, it is also considered more secure and more appropriate if in the future works the XOR algorithm can be broken-down into two stages: key scheduling algorithm (KSA) and pseudo-random number generator algorithm (PRGA). By using the newest KSA and PRGA available, the XOR algorithm can be transformed into the state-of-the-art stream cipher.

Acknowledgments

The authors gratefully acknowledge that the present research is supported by Fund Dissemination IPTEKS Research Results for Lecturers / Researchers University Sumatera Utara.

References

[1] Zhang D, Zhang J, Wang Y, Xu S and Li Y 2010 *International Conference on Web Information Systems and Mining* (Sanya) p 71-73. doi: 10.1109/WISM.2010.74
[2] Baelde D, Delaune S, Gazeau I and Kremer S 2017 *IEEE 30th Computer Security Foundations Symposium (CSF)* (Santa Barbara: CA) p 234-248 doi: 10.1109/CSF.2017.22
[3] El-Aziz A E A A A and Kannan A 2012 *International Conference on Computer Communication and Informatics* (Coimbatore) p 1-3. doi: 10.1109/ICCCI.2012.6158815
[4] Paar C and Pelzl J 2010 *Understanding Cryptography* (Berlin: Springer-Verlag)
[5] Roy A and Nath A 2016 *8th International Conference on Computational Intelligence and Communication Networks (CICN)* (India: Tehri) p 390-393 doi: 10.1109/CICN.2016.83
[6] Python Software Foundation *PEP 20 - The Zen of Python 2004-08-23 Retrieved 2008-11-24.*

The Observation of Bahasa Indonesia Official Computer Terms Implementation in Scientific Publication

D Gunawan[1], A Amalia[2], M S Lydia[2], M I Muthaqin[1]

[1]Department of Information Technology, Universitas Sumatera Utara, Jl. dr. Mansur No. 9 Kampus USU Medan 20155

[2]Department of Computer Science, Universitas Sumatera Utara, Jl. dr. Mansur No. 9 Kampus USU Medan 20155

Email: danigunawan@usu.ac.id

Abstract. The government of the Republic of Indonesia had issued a regulation to substitute computer terms in foreign language that have been used earlier into official computer terms in Bahasa Indonesia. This regulation was stipulated in Presidential Decree No. 2 of 2001 concerning the introduction of official computer terms in Bahasa Indonesia (known as Senarai Padanan Istilah/SPI). After sixteen years, people of Indonesia, particularly for academics, should have implemented the official computer terms in their official publications. This observation is conducted to discover the implementation of official computer terms usage in scientific publications which are written in Bahasa Indonesia. The data source used in this observation are the publications by the academics, particularly in computer science field. The method used in the observation is divided into four stages. The first stage is metadata harvesting by using Open Archive Initiative – Protocol for Metadata Harvesting (OAI-PMH). Second, converting the harvested document (in pdf format) to plain text. The third stage is text-preprocessing as the preparation of string matching. Then the final stage is searching the official computer terms based on 629 SPI terms by using Boyer-Moore algorithm. We observed that there are 240,781 foreign computer terms in 1,156 scientific publications from six universities. This result shows that the foreign computer terms are still widely used by the academics.

1. Introduction

The era of the early Internet in Indonesia brought new computer terms in foreign language that had no suitable translation in Bahasa Indonesia. In 2001, the government of Republic of Indonesia responded to these computer terms by issued the Presidential Decree No. 2 of 2001 concerning the introduction of official computer terms in Bahasa Indonesia. The official computer terms are also known as Senarai Padanan Istilah (SPI). There are 629 terms to substitute the computer terms in foreign language. These terms were generated from the Computer Generation Specific Formatting Guidelines.

Sixteen years after being officially published by the government of Republic of Indonesia, the official computer terms in Bahasa Indonesia should have been implemented widely in scientific publications which are written in Bahasa Indonesia. A research that conducted in 2015 had analyzed the implementation of official computer terms in Bahasa Indonesia. The implementation of official computer terms in Bahasa Indonesia was analyzed by conducting questionnaire to the students. The research shows that the students only use as much as 33% of the official computer terms in Bahasa Indonesia [1].

Recent research in 2016 conducted the observation of the implementation of official computer terms in Bahasa Indonesia for the government websites [2]. The research observed 1,084 Indonesian government websites. It focused on the government websites' template. It means the research only observed the repeated part in the websites, not the content. This research found that there are 47% of Indonesian government websites which implement 88.4% official computer terms in Bahasa Indonesia compared to the foreign language. This result does not depict the condition of the whole websites because the observation was done for the template only, not included the content of the website.

This research will analyze the implementation of official computer terms in Bahasa Indonesia in scientific publications which are written by the academics, especially by the students who majoring computer science (or related to computer). We use the publications written by the higher education students instead of questionnaire to analyze the implementation of official computer terms in Bahasa Indonesia. The research stages include data harvesting by using Open Archive Initiative – Protocol for Metadata Harvesting (OAI-PMH), converting harvested documents (in pdf format) to plaintext, conducting text pre-processing and the last stage is comparing the computer terms by using a string matching algorithm. The OAI-PMH was used in previous research to harvest scholarly research journal's content [3]. The result of this research is proposed to be the starting point for the government to evaluate the implementation of official computer terms in Bahasa Indonesia.

The challenges of this research are the terms that have more than one meaning and the term with suffix. The terms that have more than one meaning, for example "address", might lead to the miscalculation of the number of computer terms. Because the term "address" can be used as personal address or computer address. The term that should be calculated is the one which has computer related meaning. The next challenge is the term with suffix. However, this can be covered by implementing word stemming or a dictionary. As an alternative, the dictionary can be replaced by lexical database. The recent research to build a lexical database for Bahasa Indonesia is still in design phase [4]. Another challenge in natural language processing is multiword expressions (MWE). Multiword expressions are the combination of two or more words that form a meaning. Current research is still determining the candidates of the multiword expressions for Bahasa Indonesia [5]. However, because of this research used the official computer terms which listed in the Presidential Decree No. 2 of 2001 (the list includes the terms with two or more words), then the multiword expressions handling is sufficiently done by implementing string matching. This research uses Boyer-Moore algorithm to implement string matching.

The organization of this paper is as follow: the first section is the introduction about the conducted research. Then, section two discusses the observation approach used in this research. Section three discusses the result of this research. The last section is conclusion.

2. Research Methodology

The research was done by conducting several steps. The first step is collecting the documents from several major universities in Indonesia. As we research the usage of official computer terms, we limit the documents to the fields which are related to computer, such as computer science, information technology and informatics. Second, we extracted the documents into tokens to prepare for the next process. According to the ordinance of writing in Bahasa Indonesia, the foreign language must be written in italics format. As the format has no effect on the calculation of the official computer terms, this research does not consider the format of the writing. Therefore, we implemented string matching algorithm to compare the extracted tokens with the official computer terms for Bahasa Indonesia in the third step. The last but not least, we observed the result and draw conclusion. The details of the research are as follow:

2.1 Harvesting the Documents

The aim of this research is to demonstrate the usage of official computer terms for Bahasa Indonesia in academics, especially in computer field. We harvested the thesis documents which are published by the universities. We utilize Open Archive Initiative – Protocol for Metadata Harvesting (OAI-PMH) to

IOP Conf. Series: Journal of Physics: Conf. Series **979** (2018) 012034 doi:10.1088/1742-6596/979/1/012034

harvest all the documents in each university. OAI-PMH provides services to publish metadata based on HyperText Transfer Protocol (HTTP) and eXtensible Markup Language (XML). To enable data interoperability, the harvested metadata should be available in agreed format. Usually, metadata is available in Dublin Core standard. The OAI-PMH provides data interoperability between two providers namely service providers and data providers. The service providers harvest the metadata published by the repositories. These repositories are called the data providers. Both service providers and data providers communicate using the same protocol called OAI-PMH.

There are six different types of protocol request and response in OAI-PMH, such as *GetRecord*, *Identify*, *ListRecords*, *ListIdentifier*, and *ListMetadataFormats*. *GetRecord* is a verb to retrieve single metadata record from a data provider (repository). *Identify* is a verb to retrieve information about a data provider (repository). Usually, part of the retrieved information is required for other request. *ListRecords* is used to obtain or harvest records from a repository. *ListIdentifiers* is similar to *ListRecords*. However, *ListRecords* only harvests headers instead of records. The last, *ListMetadataFormats* will response metadata formats available in a data provider (repository).

As mentioned previously, data interoperability in OAI-PMH will work if both data providers and service providers communicate with the same format. In this research, we use simplified Dublin Core standard because this format is common standard to create digital web catalog. It consists of 15 metadata elements, such as *title, creator, subject, description, publisher, contributor, date, type, format, identifier, source, language, relation, coverage,* and *rights*. These metadata elements describe the published document. Figure 1 shows the example of Dublin Core implementation in XML. All the Dublin Core elements are part of the dc scheme. This scheme is pointed to http://purl.org/dc/elements/1.1/ (by using xmlns attribute). Therefore, as shown in figure 1, each tag of Dublin Core elements should be followed by dc. For example, <dc:title>Revitalisasi Perpustakaan</dc:title> means the *title* elements belongs to the dc (Dublin Core) scheme and has the value: "Revitalisasi Perpustakaan". All the metadata do not have to be published or available. As shown in figure 1, the data provider only publishes 9 metadata elements.

```xml
<oai_dc:dc xmlns:dc="http://purl.org/dc/elements/1.1/" xmlns:doc="http://www.lyncode.com/xoai"
xmlns:oai_dc="http://www.openarchives.org/OAI/2.0/oai_dc/" xsi:schemaLocation="http://www.openarchives.org/OAI/2.0/oai_dc/
http://www.openarchives.org/OAI/2.0/oai_dc.xsd">
<dc:title>Revitalisasi Perpustakaan</dc:title>
<dc:creator>Siregar, A. Ridwan</dc:creator>
<dc:subject>revitalisasi</dc:subject>
<dc:subject>perpustakaan</dc:subject>
<dc:description>08E00533</dc:description>
<dc:date>2010-03-19T08:22:25Z</dc:date>
<dc:date>2010-03-19T08:22:25Z</dc:date>
<dc:date>2008-10-18T00:00:00Z</dc:date>
<dc:type>Lecture Papers</dc:type>
<dc:identifier>Administrator</dc:identifier>
<dc:identifier>http://repository.usu.ac.id/handle/123456789/1754</dc:identifier>
<dc:language>id</dc:language>
<dc:format>application/pdf</dc:format>
</oai_dc:dc>
```

Figure 1. Dublin Core implementation in XML.

Initially, we observed that there are 63 universities in Indonesia which have computer related study program. However, not all of the universities support OAI-PMH for their repositories. In addition, not all universities open the access to the documents. Some of them do not allowed guest access. This will prevent the service provider to harvest all the available documents. Therefore, this research only used 10 universities in Indonesia. The list of universities and their OAI-PMH addresses are shown in table 1.

The universities listed in table 1 provides open access to their documents. Those documents can be downloaded by using *GetRecord* protocol. In addition, this research limits the documents to PDF format and written in Bahasa Indonesia only.

Table 1. The list of universities

University	OAI-PMH Address
Universitas Airlangga	http://repository.unair.ac.id/cgi/oai2
Universitas Andalas	http://teknosi.fti.unand.ac.id/index.php/teknosi/oai
Universitas Diponegoro	http://jtsiskom.undip.ac.id/index.php/jtsiskom/oai
Universitas Sumatera Utara	http://repository.usu.ac.id/oai/request
Universitas Sultan Syarif Kasim	http://repository.uin-suska.ac.id/cgi/oai2
Universitas Sriwijaya	http://eprints.unsri.ac.id/cgi/oai2
Institut Teknologi Sepuluh November	http://iptek.its.ac.id/index.php/index/oai
Universitas Gadjah Mada	http://i-lib.ugm.ac.id/oai-ilib/oai21.php
Universitas Bengkulu	http://jamal.ub.ac.id/index.php/index/oai
Universitas Malikussaleh	http://ejurnal.tif.unimal.ac.id/index.php/index/oai

2.2 Text Pre-processing

This research requires single word (token) to be compared with official computer terms in Bahasa Indonesia. Therefore, after the metadata and document harvesting process, we extracted all the text in the documents and applied text pre-processing such as tokenizing, stop-word removal, case folding and stemming. Tokenizing is used to extract all the text into single word. Stop-word removal is intended to remove all the words that cannot be used as a single word. This research uses stop-word list provided by Tala [6]. Case folding will convert all the text to lowercase. This step will avoid case sensitive issue when comparing each token. The last text pre-processing is stemming. As the official computer terms are available in root word format, hence the token should be available in root word format also. Stemming process will extract the token to its root word. We used confix-stripping approach to provide words stemming [7]. In addition, this research does not consider the abbreviation of English terms. Therefore, the abbreviation of English terms is considered as the other terms and will not be calculated.

2.3 Comparing Token by Using String Matching Algorithm

The string matching algorithm has two types, namely: 1) exact string matching and 2) approximate string matching. This research uses extract string matching type. We utilize Boyer-Moore algorithm to match the token with the official computer terms for Bahasa Indonesia. This algorithm is known as the most famous and efficient algorithm in model matching [8].

2.4 The Result Observation

After comparing tokens with the official computer terms for Bahasa Indonesia, we observed the result for each university. Then we observed the result from the all available universities.

3. Results and Discussion

3.1 Harvesting Document Result

As mentioned earlier, the data are obtained by harvesting documents via OAI-PMH. This research limits to the documents that related to computer field only. To obtain only the specific documents, we should know the category structure in each university repository. The category structure can be observed via *ListSets* protocol. For example, we can observe the categories in Universitas Sumatera Utara repository by accessing http://repository.usu.ac.id/oai/request?verb=ListSets. This requires manual approach to observe each repository as there is no convention by name or code to classify specific document type. After determining the category which will be harvested for each repository, the metadata is ready to be harvested by specifying *SetSpec* attribute. All the documents' location can be extracted by using *ListRecords* protocol.

There are several issues in harvesting process. There are three universities that are registered as data provider but they do not provide required metadata. Therefore, we exclude those three universities from this research. They are Universitas Gadjah Mada, Universitas Bengkulu and Universitas Malikussaleh. In addition, we exclude Institut Teknologi Sepuluh Nopember from this research because it uses English instead of Bahasa Indonesia.

3.2 Data Extraction

Figure 2 shows the example of harvested XML from Universitas Sumatera Utara. We extract several fields such as published date (represented by <atom:published>), file name (represented by <atom:title>), and link to file download (represented by <atom: link>). For example, the extracted data are shown in figure 2 (a) (b) (c) respectively are published date, title and link. We notice that there is difference between Universitas Sumatera Utara and the other universities in providing link to file download. Universitas Sumatera Utara separates a document to several files according to its chapter. However, the other universities provide only one file for each document. These XMLs are extracted by using a DOM Parser.

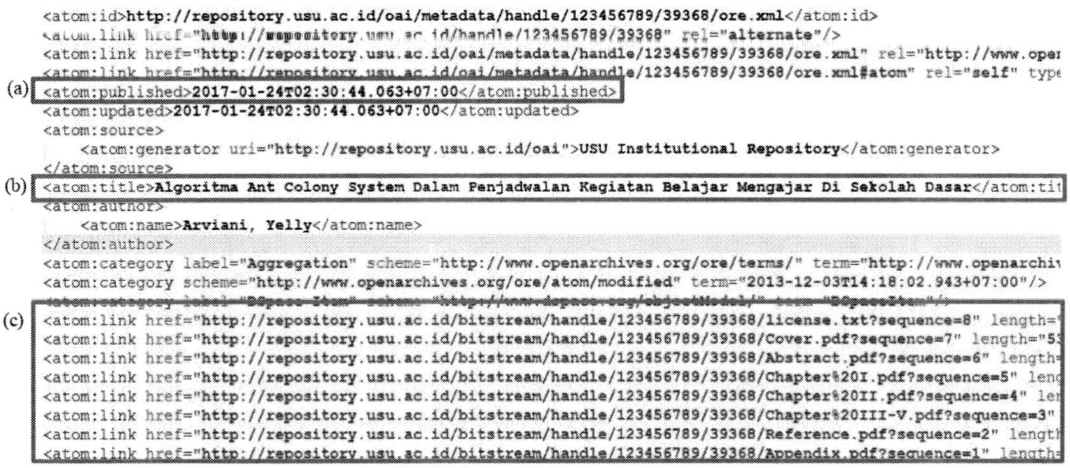

Figure 2. XML from Universitas Sumatera Utara.

Table 2. The downloaded files.

University	Number of documents from the XML	Number of downloaded files
Universitas Airlangga	130	123
Universitas Andalas	43	43
Universitas Diponegoro	186	185
Universitas Sumatera Utara	325	325
Universitas Sultan Syarif Kasim	654	395
Universitas Sriwijaya	99	85

3.3 Document Download

After extracting required field from the XML, the document will be downloaded to be processed for the next step. We found that some repositories do no provide file download links for the certain documents. Therefore, the number of downloaded files are fewer than the number of documents available in the XML data.

3.4 Results

Before implementing the algorithm, the downloaded files are required to be converted to plain text. Next, we implement text pre-processing such as tokenizing, stop-word removal, and case folding. To search the official computer terms in Bahasa Indonesia we utilize a string matching algorithm called Boyer-Moore. By utilizing the algorithm, we found ten most used computer terms in English such as user, file, input, database, server, output, text, image, error, and password.

The result of the searching official term for Bahasa Indonesia is shown in table 3. The column Year describes the year when the documents are published. This column is followed by UNAND, UA, UNDIP, UNSRI, UIN-SUSKA, USU which represent Universitas Andalas, Universitas Airlangga, Unversitas Diponegori, Universitas Sriwijaya, Universitas Islam Negeri – Sultan Syarif Kasim and Universitas Sumatera Utara respectively.

Table 3. Number of computer terms in English for each university.

Year	UNAND	UA	UNDIP	UNSRI	UIN-SUSKA	USU	Sub Total
2001	-	9	-	-	-	-	9
2002	-	0	-	-	-	-	0
2003	-	-	-	-	-	-	0
2004	-	21	-	-	-	-	21
2005	-	0	-	243	-	-	243
2006	-	0	-	291	-	-	291
2007	-	4	-	-	-	-	4
2008	-	0	-	-	-	-	0
2009	-	39	-	83	1,974	-	2,096
2010	-	85	-	334	3,925	-	4,344
2011	-	85	-	547	25,619	-	26,251
2012	-	28	-	1,218	7,751	-	8,997
2013	-	38	1,945	1,900	28,797	5,331	38,011
2014	-	45	2,802	5,893	3,437	32,634	44,811
2015	564	510	5,181	1,406	-	63,581	71,242
2016	3,309	1,477	4,458	-	1,094	25,044	35,382
Total	3,873	2,341	14,386	11,915	72,597	126,590	
Number of documents	43	123	185	85	325	395	

Most of the repositories do not have any documents from 2001 except Universitas Airlangga. However, we cannot convert the documents from UA which are published in 2005, 2006 and 2008 because the PDF files are scanned using image format. According to the table 3, the usage of the computer terms in foreign language is still very high. This result is proposed be the starting point for the government to do profound research to evaluate the implementation of official computer terms in Bahasa Indonesia.

4. Conclusion

Sixteen years after being published by the government of Republic of Indonesia, the official computer terms in Bahasa Indonesia should have been implemented widely, especially by the academics. This research observed the usage of official computer terms in scientific publications by the academics in computer related field. We observed 1,084 scientific publications which are published by six universities in Indonesia. We found that the usage of computer terms in foreign language is still widely used by the academics in their scientific publications.

References

[1] Sari C A 2015 *Tanggapan Mahasiswa di Kota Surakarta Terhadap Pengindonesiaan Istilah Asing Bidang Pengkomputeran (Kajian Sosiolinguistik)* (Universitas Sebelas Maret)
[2] Amalia A, Gunawan D, Lydia M S, and Charlie C 2017 *IOP Conf. Ser. Mater. Sci. Eng.* vol 180 p 12060
[3] Simek P, Jarolimek J, Vanek J, and Stoces M 2011 *Proc. of the 5th International Conference on Information and Communication Technologies for Sustainable Agri-production and Environment (HAICTA 2011)* p 827–834
[4] Gunawan D and Amalia A 2017 *IOP Conf. Ser. Mater. Sci. Eng.* vol 180 p 12052
[5] Gunawan D, Amalia A, and Charisma I 2016 *6th IEEE Intl. Conf. on Control System, Computing and Engineering (ICCSCE)* p 304–309
[6] Tala F Z 2003 A Study of Stemming Effects on Information Retrieval in Bahasa Indonesia (Universiteit van Amsterdam)
[7] Adriani M, Asian J, Nazief B, and Tahaghoghi S M M 2007 *ACM Trans. Asian Lang. Inf. Process.* vol 6 p 1–33
[8] Yuan L 2011 *Intl. Conf. on Computer Science and Service System (CSSS)* pp 182–184

Automated Robot Movement in the Mapped Area Using Fuzzy Logic for Wheel Chair Application

B Siregar[1], S Efendi[1], H Ramadhana[1], U Andayani[1], F Fahmi[2]
[1]Department of Information Technology, Faculty of Computer Science and Information Technology, University of Sumatera Utara, Medan 20155, Indonesia
[2]Department of Electrical Engineering, Faculty of Engineering, University of Sumatera Utara, Medan 20155, Indonesia

E-mail: fahmimn@usu.ac.id

Abstract. The difficulties of the disabled to move make them unable to live independently. People with disabilities need supporting device to move from place to place. For that, we proposed a solution that can help people with disabilities to move from one room to another automatically. This study aims to create a wheelchair prototype in the form of a wheeled robot as a means to learn the automatic mobilization. The fuzzy logic algorithm was used to determine motion direction based on initial position, ultrasonic sensors reading in avoiding obstacles, infrared sensors reading as a black line reader for the wheeled robot to move smooth and smartphone as a mobile controller. As a result, smartphones with the Android operating system can control the robot using Bluetooth. Here Bluetooth technology can be used to control the robot from a maximum distance of 15 meters. The proposed algorithm was able to work stable for automatic motion determination based on initial position, and also able to modernize the wheelchair movement from one room to another automatically.

1. Introduction

Currently, robotic technology is widely used in the industrial world to undertake the production process. The robot is also used to replace the human work that considered to be difficult and dangerous to humans.We only need to control the robot remotely, and so the jobs can be done more securely and efficiently [1,2].

The use of robots to support daily needs plays a role in the development of information technology. Various forms of the robot had been developed: mobile robot, network robot, robot manipulator (for hand), Humanoid Robot, Legged Robot, Flying Robot (Flying Robot), and Robot In Water (Under Water Robot). The robot has been widely applied in everyday life and used to replace human activities that are considered severe and dangerous[1].

The difficulties of the disabled to move, make them unable to live independently. People with disabilities need tools to move from place to place. For that, we need a solution that can help people with disabilities to move from one room to another automatically.

In 2010 a study was conducted to make micro-mouse robot by using depth-first search algorithm automatically, and the robot was paired with an infrared sensor to follow the black line. Here the robot did not have control and looking only for the path that will pass from one point to another [3].

In 2012 Febriani et al. conducted research to control the robot using a camera sensor that can recognize a predetermined object and follow wherever the object was move. Robot motion control using image-based random method was used [4]. Previews conducted research to control the robot using a smartphone with Bluetooth media. Robot received commands sent by smartphones such as: forward, backward, right, and left movement [5]. Moreover, furthermore, in 2015 Choudhury researched controlling robots using ultrasound and using the pointbug algorithm. Control of the robot was done using voice command from Google API so that that robot can receive voice message as well [6]. Some previous research also focused on specific location-based control algorithm for example in person locator [7] and crowd mobility simulation [8].

This study was designed to build wheelchair with robotic motion systems in the mapped area with the support of sensor technology and fuzzy logic algorithm.

2. Method

The starting point of the location for this robot first been determined. The input target of this app is the destination location. Sensors are needed as a tool to provide information about surrounding environment to facilitate the movement of robots; We used ultrasonic and infrared sensors that function to transmit information regarding objects that block the movement of robots and infrared sensors as black line readers. Furthermore, the command will be sent using Bluetooth media and processed into commands. The general architecture of the system can be seen in Figure 1.

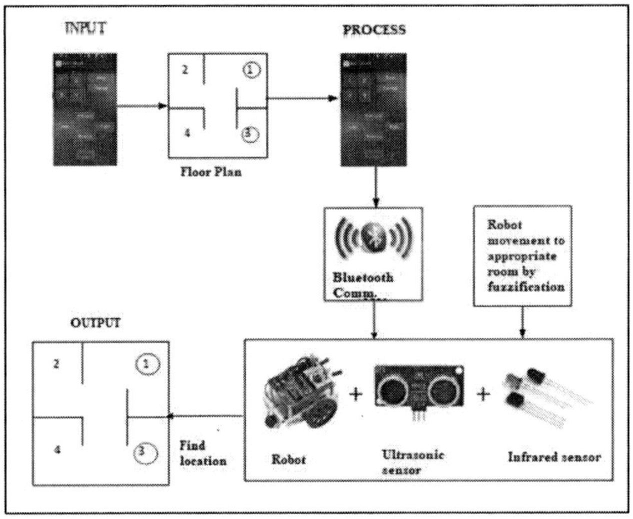

Figure 1. General Architecture.

The input of the system was the destination location marked with a blue circle, the starting point of the robot has been determined that be marked with a red circle. The movement of the robot to the location was controlled by using the smartphone so that the robot moves according to the desired destination set by the user. Smartphones connect with robots via Bluetooth media as the process of sending a destination command. The robot will move automatically to the location that has been determined by the user using a smartphone as a robot control system.

Wheelchair robot has embedded a program for determining the predetermined location automatically by using the fuzzy logic algorithm. Fuzzy logic specified one of some destination locations recursively. The robot was then also paired with ultrasonic sensors and infrared sensors. The ultrasonic transducer worked by transmitting a wave and calculating the reflection time of the stream. Ultrasonic sensors worked in a way so that the robot can avoid collisions with objects that are in front of the robot so that the movement to the location goes smoothly without any obstacles. The infrared sensor marked the black line for the robot to move above the black line.

The output expected was the movement of the robot to get to the location desired by the user. Users can control the flow of robots automatically or manually. It aims to modernize the wheelchair system

IOP Conf. Series: Journal of Physics: Conf. Series **979** (2018) 012035 doi:10.1088/1742-6596/979/1/012035

automatically from one room to another. Floor plan design for this research can be seen in Figure 2, while schematic diagram for the whole system in this study explained in Figure 3.

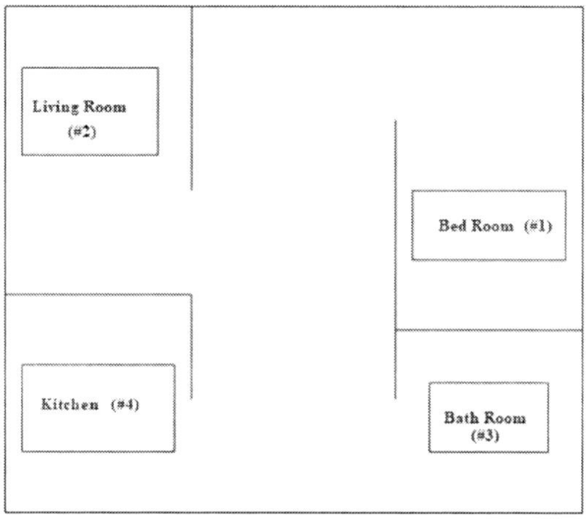

Figure 2. Floor Plan.

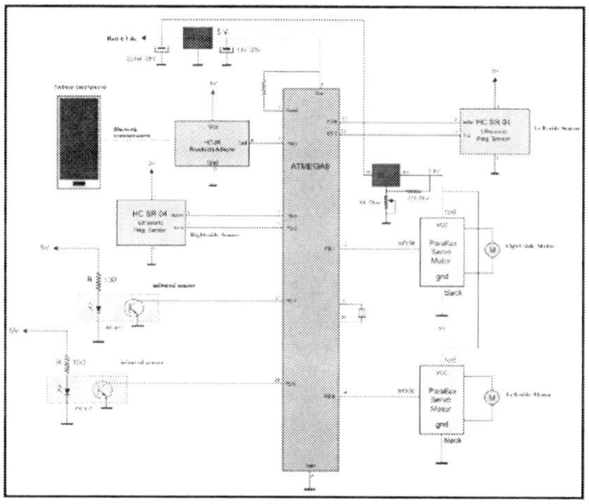

Figure 3. Schematic Diagram.

The design form of the system proposed in this study can be seen in Figure 4 which shows steps that users perform on the robot control application and the process done to move the robot to the destination location automatically based on the command given by the user.

Robot control process started when the robot and application on the smartphone have been connected. The connectivity between the two systems was done by utilizing Bluetooth technology as a media liaison between devices.

The control that can be done on the robot application of this location control was:

1. Automatic movement.

This process is done by giving the command to the robot to move automatically to the location that has been addressed and mapped.

2. Manually-controlled movement.

In this process, the application can perform the control manually. This process is done by giving orders to the robot to move manually with the command forward, backward, right, and left.

2.1. System Design Analysis

To analyze the system we designed, here we use process diagram which shows process chain in the application and robot control in both automatic and manual movement.

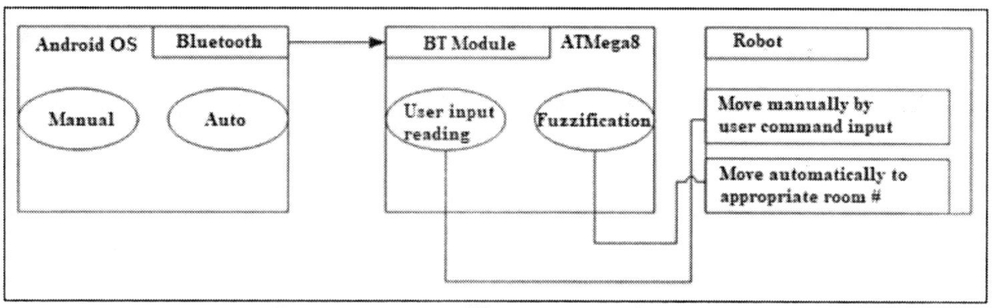

Figure 4. Controlling Process Diagram.

The robot control process to get the wheelchair robot to a location using this Android app was as follows:

1. Users pair-to-pair android with robot device using Bluetooth.

2. In this pair-to-pair process, the user inputs the Bluetooth code of the robot device on the smartphone.
3. The Bluetooth code sent by Android smartphone will be processed by ATmega8 to be customized and to connect to each other.
4. If the code sent users using the smartphone accordingly, then the smartphone and robot device will be connected.
5. Once connected, the user can control the robot to get to a location automatically or manually using the smartphone.
6. Users use the app to control the movement of the robot. The control performed by the user is the movement automatically and manually.
7. In this control, there are two types of utilization of signals used, namely, analog signals and digital signals. Analog signals are used in the process of pairing between Bluetooth from smartphones and robots, while the digital signal is used in the output produced by ATmega8.

2.2. System Flowchart

A system flowchart is a chart showing the workflow of what was being done by the system as a whole as well as the procedures used in its operation (Figure 5). The stages of the system on the robot according to the flowchart are:

1. When the system powered the device will detect all devices connected to it. Such as Bluetooth, infrared sensors, and ultrasonic sensors.
2. After all, devices were detected, the system will wait for the pairing command performed by the user using the smartphone.
3. If the pairing process were successful, then the robot device and smartphone would be connected.
4. If the pairing process failed, the system would prompt the user to enter the device code correctly one more time.

This process ensures the robot device with the smartphone to be connected.

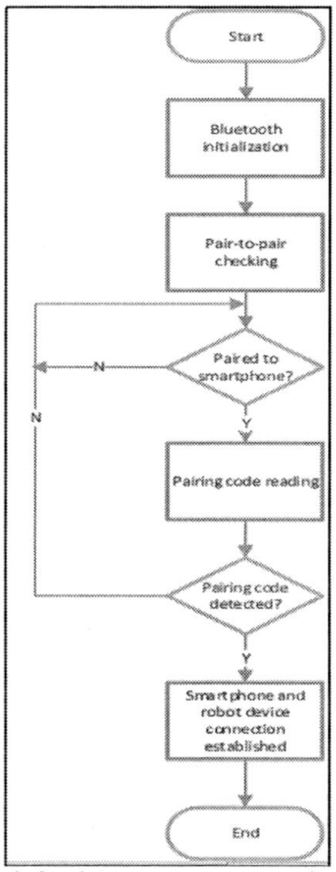

Figure 5. Flowchart connectivity between a smartphone and wheelchaired robot.

After smartphones and robot devices were connected, the user can control the robot to get to a location that has been mapped both manually and automatically using applications that have been installed on the smartphone.

The steps to control the robot explained in Figure 6 were as follows:

1. Enter the bot control app on the smartphone.
2. Perform a pair-to-pair device and smartphone.
3. Once the smartphone and devices are connected, the next process is to enter the robot control menu to get to the mapped location.
4. The control was done by the user is to control the robot automatically and manually to go to a location that has been mapped.
5. In the process of automatic movement, the system used a fuzzy logic algorithm to determine the direction of motion based on the initial position.
6. In the manual control process, commands from the smartphone, i.e., right, left, forward, backward were given.

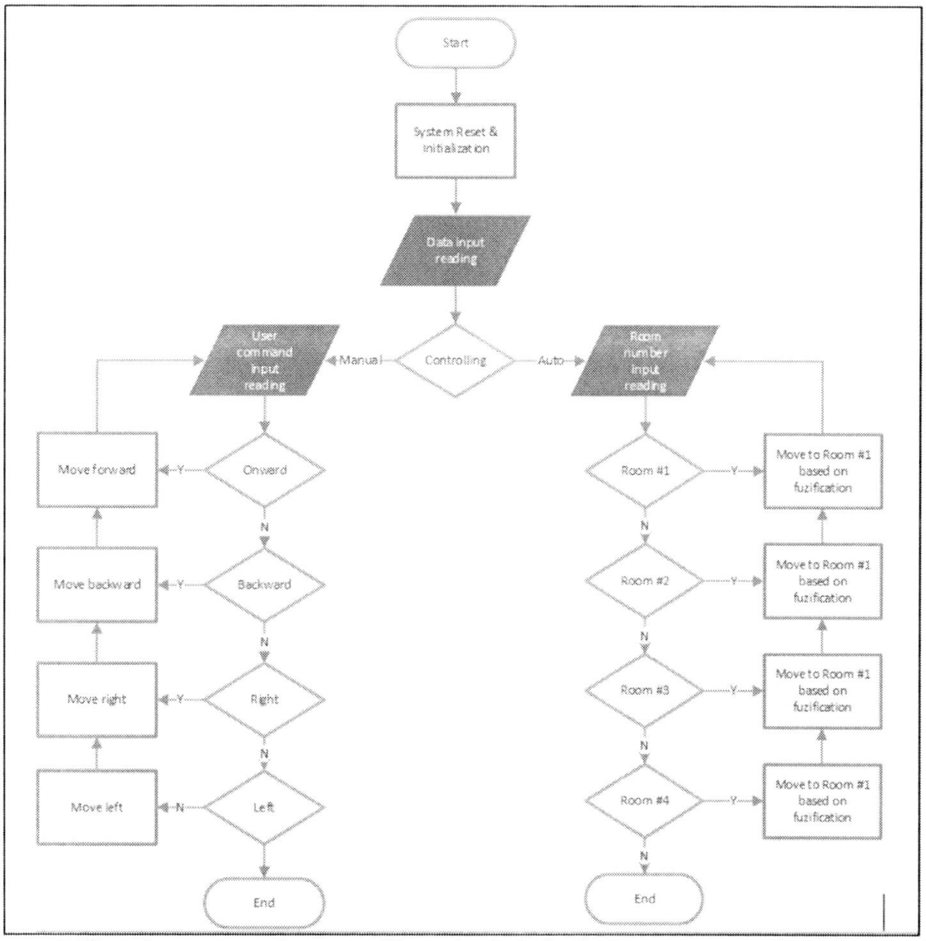

Figure 6. Flowchart system for controlling wheel chaired robot automatically and manually.

3. Results and Discussions

A set of system testing was done to find out the ability of the robot to move automatically or manually in the mapped area. In this test, we focused on the accuracy of the system in determining the space to be addressed. Also, testing also took into account the performance of devices attached to the robot. The plot was mounted on the robot as one of the most important parts because the attached device captured data that was processed by ATmega8 for the robot to move appropriately. In testing the performance of this system to analyze the transmission, voltage is divided into two variations of distance that is five voltages, (±) 5 voltages. The prototype was designed to test the performance of the system using the power coming from the installed battery.

The results of the processing done by the robot device and its effect on the robot system were shown in Table 1. To test the effectiveness of robotic movement, then the testing device was done based on location. Taking into account the position of the test to calculate a good position when the robot arrived and went to the specific location set by the user, as for the test results can be seen in Table 2.

Table 1. Testing Design.

No	Devices	Sent Voltage	Sent Voltage (±5)
1	Ultrasonic Sensor	Good	Not Good
2	Infrared Sensor	Good	Not Good
3	Bluetooth	Good	Not Good
4	Servo Motor	Good	Not Good

Table 2. Performance Testing.

No	Destination	End Position
1	Room #1	Good
2	Room #2	Good
3	Room #3	Good
4	Room #4	Good

4. Conclusions

Smartphones with the Android operating system can control the robot using Bluetooth. Here Bluetooth technology can be used to control the robot from a maximum distance of 15 meters. Controller ATmega8 must obtain current and voltage supply from the appropriate battery so that the sensors and wheels can work properly. The robot in several testing successfully arrived at a specified location in a favorable position as set by the user. Automation of robot movement for disables to a location hasbeen developed to modernize the wheelchair system.

References

[1] Siregar B, Purba HA, Efendi S and Fahmi F 2016 *IOP Conference Series: Materials Science and Engineering* **180(1)** 012106

[2] Siregar B, Gunawan D, Andayani U, Lubis SE and Fahmi F 2016 *Journal of Physics: Conference Series* **801(1)** 012038

[3] Anita NS 2010 *Robot Micromouse dengan Menggunakan Algoritma Deep-First Search* (Depok: Universitas Gunadarma)

[4] Febriani R and Suprijadi 2010 *Jurnal Otomasi Kontrol & Instrumentasi* **2(1)** 21

[5] Indriyanto C, Pratama MR and Sitompul D 2015 *Perancangan Sistem Kendali Remote Control Mobil Menggunakan Smartphone Android* (Medan: Universita Sumatera Utara)

[6] Choudhury N and Singh C T 2015 *International Journal of Computer Applications* **119** 13

[7] Fahmi F, Evelyn S, Siregar B, Gunawan D and Andayani U 2017 *Journal of Physics: IOP Conference Series* The 1st IConICT2017 Under Publication

[8] Siregar B, Silitonga AI, Nababan EB, Andayani U and Fahmi F 2017 *IEEE publisher* 2017 ICSIIT 2017 Under Publication

Content Abstract Classification Using Naive Bayes

Syukriyanto Latif , Untung Suwardoyo , Edwin Aldrin Wihelmus Sanadi

Electrical Engineering Department, Hasanuddin University, Makassar, Indonesia
E-mail: syukriyanto24@gmail.com

Abstract. This study aims to classify abstract content based on the use of the highest number of words in an abstract content of the English language journals. This research uses a system of text mining technology that extracts text data to search information from a set of documents. Abstract content of 120 data downloaded at www.computer.org. Data grouping consists of three categories: DM (Data Mining), ITS (Intelligent Transport System) and MM (Multimedia). Systems built using naive bayes algorithms to classify abstract journals and feature selection processes using term weighting to give weight to each word. Dimensional reduction techniques to reduce the dimensions of word counts rarely appear in each document based on dimensional reduction test parameters of 10% -90% of 5.344 words. The performance of the classification system is tested by using the Confusion Matrix based on comparative test data and test data. The results showed that the best classification results were obtained during the 75% training data test and 25% test data from the total data. Accuracy rates for categories of DM, ITS and MM were 100%, 100%, 86%. respectively with dimension reduction parameters of 30% and the value of learning rate between 0.1-0.5.

1. Introduction.

Technological developments have resulted in the increase of large amounts of data. Large data that has been stored is rarely used optimally because of humans often do not have sufficient time and ability to manage them. Data large volumes like text data, goes far beyond the ultimate human processing capacity limited.

The case in detail is the classification of abstract content in English. There are 3 areas of abstract content categories used as training data and test data, namely Intelligent Transport System, Data Mining and Multimedia. but abstract content labeling is currently still done manually, thus taking a long time in knowing which category of abstract content is. To solve it. Automatic content category system needs to be required, to minimize errors. Data mining approach has been widely used in manygopecto espesially evoluting data series[1-3]

As for some research on classification of documents that become reference source in this research is classification text with naive bayes classifier to classify news text and abstract academic by amir hamzah in year 2012, with method naive bayes can help user in classification with acquisition rate of 91% for text news and 85% for academic abstracts [4].

Further research conducted by Ashis Kumar Mandal et al in 2014, apply 4 algorithms DT (Decision Tree), KNN (K-Nearest Neighbors), NBC (Naive Bayes Classifier) and SVM (Support Vector Machine) in classifying web documents. The accuracy is 80.65%, KNN 74,24%, NB 85,22% and SVM 89,14%, research of G. Keerthika and D. Saravana Priya in 2015 apply NBC method by combining 2 feature selection said GR (Gain Ratio) and IG (information Gain) combined these 2 methods are able to produce accuracy of 92% [5].

In G. Keerthika and D. Saravana Priya in 2015, applying the NBC method by combining 2 features of the said GR (Gain Ratio) and IG (information Gain), combined these two methods are capable of achieving an accuracy of 92% [6]. Amalia Anjani, Djumardi et al in 2015 also apply NBC Algorithm (Naive Bayes Classifier) with Confix Stripping Stemmer as the classification of Indonesian news article. Based on the validation results using 10 Cross validation NBC achieved the best accuracy of 86.97% [7]. Fajar Rohman Hariri et al in 2015 also conducted a thesis abstract classification research using the method of learning vector quantization by dividing 3 categories of fields namely SI RPL (Software Engineering Information System), CAI (Computation - Artificial Intellegence) and Multimedia, in feature selection research used namely dimensional reduction so as to increase accuracy in the validation process. 100% accuracy obtained in SI RPL and CAI, 70% for Multimedia interest area, so that overall accuracy can be 90%, best condition get with dimension reduction parameter 20% [8]. I Gusti A Socrates et al in 2016 conducted a study using the NBC method by applying the Gain Ratio weighting feature selection in classifying Indonesian text, the accuracy of NBC reached 91% [9].

From some of the above research, the authors propose to use naive bayes method and Term Weighting algorithm also dimension reduction on the feature selection so that the process of classification accuracy gets better.

2. Methodology

This research is an experimental research because to get the performance best algorithm naive Bayes in classifying, conducted several times experiments with different parameters. The abstract content data can be on the www.computer.org page. From 3 categories of Intelligent Transport System, Data mining and Multimedia. There are 120 abstract content file data with pdf extension. For data analysis, from the abstract document obtained, will be experimented with several trial scenarios described in Table 1:

Table 1. Scenario test.

No	Amount of data train	Amount of data testing	Information
1	100%	100%	The entire abstract content becomes data learning as well as data testing
2	75%	25%	75% of the data will be used as data learning, and the remaining 25% is used as data testing.
3	50%	50%	50% of the data will be used as data learning, and the remaining 50% is used as data testing..
4	25%	75%	25% of the data will be used as data learning, and the remaining 75% is used as data testing.

It will also be analyzed by different dimension reduction with dimension reduction process, the amount of data learning and data testing that are different in expect to know the best accuracy result from Naive Bayes method in classifying abstract content. The design of the proposed system is shown in Figure 1.

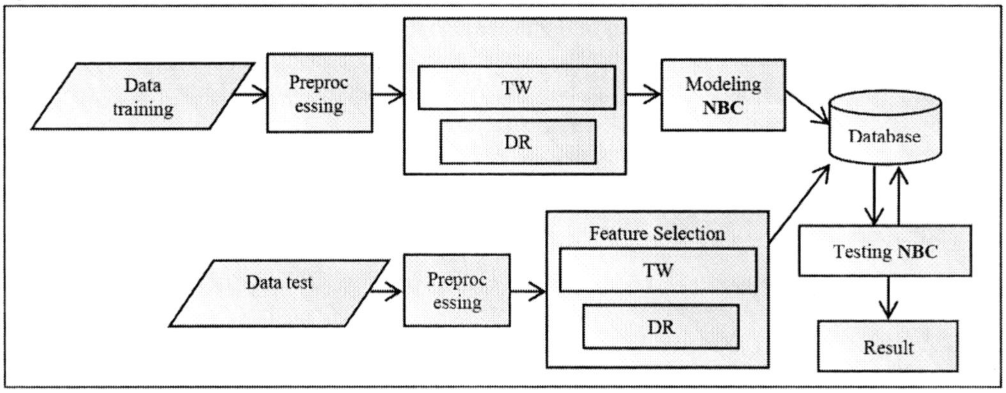

Figure 1. Proposed system design.

2.1 Preprocessing

There are several steps that need to be done in this process is case folding that change the letters in ducumen into lowercase, tokenizing as input string cutter based on each word that make up, filtering take important words but in this research use stoplist algorithm (throw less word important) and stemming is the stage of searching for root words (base word) of each word.

2.2 Feature Selection

In this stage there are 2 processes that will be done that is Term Weighting is giving the term weight that appears on each abstract content and dimension reduction by removing words that will only appear in one or a few on the abstract content. The following shows the implementation of the results of the Term Weighting word in Figure 2.

Document Training	Word And Term Frequency
A collision prevention mechanism for the multicast transport in IEEE 802.11 networks	ag => 1, based => 1, bs => 2, busy => 1, caused => 2, collision => 4, compliant=>1, content => 1, define => 1, deliver => 1, deployed => 1, excee => 1, factor => 1, failure => 1, fair => 1, now => 1, guarantee => 1, ieee => 2, interfere => 1, introduce => 1, loss => 2, lost => 1, measure => 1, mechan => 2, medium => 1, missing => 1, mode => 1, modification => 1, multicast => 4, network => 2, packet => 4, paper => 1, path => 1, perfective => 1, policy => 1, prevent => 1, rate => 1, receiver => 1, require => 1, retransmitted => 1, retry => 1, sharing => 1, show => 1, standard => 2, study => 1, symbol => 1, time => 1, transmission => 1, transmitted => 1, transport => 2, unicast => 1, wlan => 1
A composition toolkit for experimental wireless transport processes	algorithm => 1, architecture => 1, area => 1, compare => 1, compon => 1, conduct => 2, control => 1, decompose => 1, demonstrate => 1, describe => 1, develop => 2, developing => 1, differ => 1, easier => 1, effort => 1, enable => 1, evaluate => 1, expert => 2, experimental => 2, favorably => 1, greatly => 1, illustrate => 1, implementate => 1, improving => 1, kernel => 1, linux => 1, multiloop => 1, network => 2, opensource => 1, paper => 1, perform => 3, performed => 1, plugndisplay => 1, present => 1, previously => 1, protocol => 4, rapid => 1, race => 1, releasing => 1, reliable => 1, reported => 1, required => 1, research => 2, result => 1, show => 1, significantly => 1, simplifying => 1, smaller => 1, tcp => 1, toolkit => 2, transport => 4, userspace => 1, utility => 1, variance => 1, wifi => 5, wireless => 4, work => 1
A flash transportation layer for multimedia storages	accessed => 1, address => 1, application => 1, based => 1, block => 1, called => 1, collection => 1, data => 1, device => 3, offer => 1, drive => 1, experimental => 1, ffti => 3, file => 1, filter => 1, filtering => 1, flash => 2, frequently => 2, fti => 2, garbage => 1, idea => 1, improve => 1, index => 1, layer => 1, main => 1, manage => 2, manner => 1, mapping => 1, metadata => 2, multimedia => 2, nand => 1, overhead => 1, pagelevel => 1, paper => 1, perform => 1, propose => 1, proposed => 1, read => 1, reducing => 1, request => 2, result => 1, scheme => 5, separately => 1, sequential => 1, server => 1, show => 1, significantly => 1, soliditate => 1, ssd => 1, storage => 3, term => 1, tradition => 1, transition => 1, update => 1, updated => 2, user => 2, widely => 2

Figure 2. Term Weighting.

Dimensional reduction process is done to minimize the dimension of data so that computation time is needed less. However dimensional reduction processes must pay attention to data characteristics, since the missing dimensions may also eliminate the data characteristics [10]. Therefore in this study selected dimensional reduction by removing sparse part. Term that has a frequency that is mostly zero will be eliminated, since the term represents a very rare term in the data or in other words term may be ignored.

The amount of dimension dimension of the term in accordance with the parameters specified. Because it is feared that this process will eliminate the characteristics of data, then created value parameters that serve to limit dimensional reduction. So even if in a term most values are zero, the term will not be omitted if any of its frequency values exceed the specified parameter.

TF data that has been entered in the database has different values. The term that has a zero value TF is negligible, with no TF values exceeding the specified parameters. In other words dimensional reduction process is by eliminating the feature or term that the frequency of all / mostly worth 0. If there are some that are not worth 0 then the TF value must exceed certain parameters. This is so that data does not lose its characteristics. So the process can run well. The following dimensional reduction illustration is shown in Figure 3.

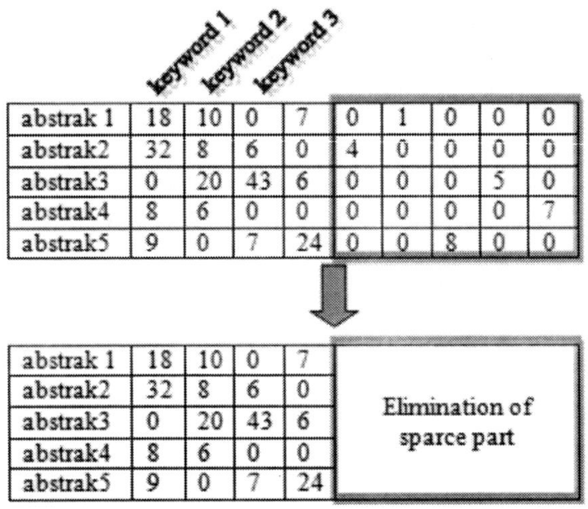

Figure 3. Illustration of dimensional reduction.

In Figure 3, the parameter is shown as 40%. Then dimensional reduction is done by omitting a column whose value is not 0 less than 40% of the data. The 5-9 columns will be removed because the column is not eligible.

Of the 120 abstract content data after being processed it produces 5,344 terms (words). Of the 5,344 words available, there are some words that appear only in one or two abstract content only, so the word can be ignored. Therefore, to eliminate these words, and also to speed up the calculation process, a dimensional reduction process is performed. The result of the number of terms produced after the dimensional reduction process is shown in Table 2 below:

Table 2. Dimension Reduction Results.

Dimension Reduction Value	Number of Term
10%	95
20%	70
30%	39
40%	25
50%	14
60%	8
70%	3
80%	3
90%	1

The % value above means the words taken are words that appear in more than how many% of documents. The fewer parameters, the less the term will be generated.

2.3 Classification

From the weighting result of Term Weighting and dimension reduction will then be used as training data and test data on Naive bayes classifier process. The naive bayes process includes two stages: learning stages and testing stages. The main characteristic of the naive bayes classifier is a very strong (naive) assumption of the independence of each condition or event. The Naive Bayes Classifier is a simplified model of a suitable bayes algorithm in the classification of texts or documents as in the following equations:

$$P(C_i \mid W_k) = \frac{P(Wk|Ci) x \, P(Ci)}{P(Wk)} \quad (1)$$

Where,

- $P(C_i \mid W_k)$ is the probability of occurrence of categories C_i with the word W_k
- $P(W_k)$ "Constant" for all categories so it is only formed $P(W_k \mid C_i)$x $P(Ci)$ which needs to be maximized
- C_i is the category available (*C1,C2,...Ci*)
- $P(Ci)$ is the probability of occurrence of categories *Ci*
- $P(W_k \mid C_i)$ is the probability of occurrence of words W_k in categories C_i

So the equation for solving this problem is as follows:

$$P(words|category) \frac{P(words \mid category)P(category)}{P(category)} \quad (2)$$

a. Training process

In the training phase, the process of analyzing the sample of documents in the form of vocabulary selection, which is a word that may appear in the collection of sample documents that can represent the documents as much as possible, as well as the formation of classes and as a reference how abstract content will be classified. to model the probability of the training process the following equations are used.

$$P(c_i) \frac{fd(ci)}{|D|}$$

$$P(w_k|c_i) = \frac{n_k+1}{n+|vocabulary|} \quad (3)$$

where.

- $P(c_i)$ is the amount of abstract content that has categories c_i
- $|D|$ is the sum of all abstract content training.
- n_k the value of word occurrence in each category c_i
- n is the total number of words in a category c_i
- $|$ *vocabulary* $|$ is the total number of words.

So the equation for modeling the probability of the training process is used by the following equation.

$$P(category) \frac{Abstract\ Content\ Amount\ in\ category}{The\ sum\ of\ all\ data\ is\ training\ abstract\ content} \quad (4)$$

$$P(words|category) \frac{The\ value\ of\ word\ occurrences\ in\ each\ category+1}{Number\ of\ all\ words\ in\ category + number\ of\ all\ data\ coaching\ words} \quad (5)$$

b. The testing process

This stage as the core of the naive bayes is to know the accuracy of the model built on the training process to predict the unknown class label

- Generate probabilities for each class according to equation (1) using the P (category) and P (category | categories) that have been obtained from the training.
- The maximum class probability value is the category of the selected class as the result of classification.

3. Results

After dimensional reduction process, the training process is done using Naive Bayes Classifier. From the final weights generated, then predicted by calculating the probability of the occurrence of the word testing on the NBC method. With 100% data used as data testing as well as training data and with different dimension reduction scenarios, it produces accuracy as in Table 3 below:

Table 3. The result of the dimension reduction scenario accuracy.

Dimension Reduction Value	Accuracy (%)
10%	68.56
20%	76.66
30%	88.48
40%	65.33
50%	58.45
60%	51.88
70%	51.88
80%	51.88
90%	51.88

From the above tables and graphs, the dimensional reduction of 60% -90% yields only 51.88% accuracy, all the abstract data are classified into the field of multimedia interest, and from table 3 it is known that the best accuracy is obtained when 30% dimension reduction with accuracy reached 88.48%. With the introduction details for each category are listed in table 4:

Table 4. Details of dimensional reduction dimension recognition accuracy 30%.

Category	The amount of content abs	The number of successful classified	accuracy
ITS	40	38	96%
DM	40	34	88%
MM	40	30	80%

From the above table it is known that the best NBC method in classifying abstracts for ITS category with accuracy of confusion matrix is 96%, 88% in DM category, and lowest accuracy in classifying abstract for Multimedia category which only succeeds in classifying 30 abstract content from 40 content abstract and produce an accuracy value of 80%.

After that, trial scenario is done by changing the data of learning and testing. Accuracy is generated as in Table 5.

Table 5. Results Accuracy Scenario Total Data Training.

Learning Rate	75% Training 25% testing	50% Training 50% testing	25% Training 75% testing
0.1	92%	86%	76,77%
0.2	92%	86%	75,55%
0.3	92%	86%	76,89%
0.4	92%	86%	76,56%
0.5	92%	86%	76,56%
0.6	72,33%	81,33%	75,33%
0.7	60%	81,33%	74,33%
0.8	55,53%	65.66%	71.66%
0.9	65,50%	44,66%	61,66%

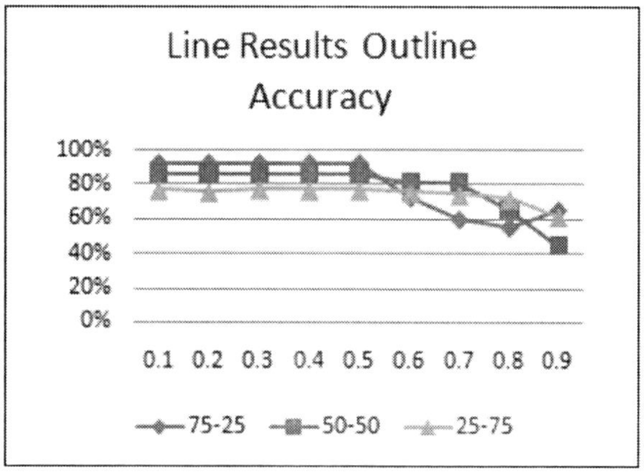

Figure 4. Line Diagram Results Accuracy.

From table 4 and figure 5, it is known that the best accuracy is 92%, with 75% of Data Training scenario - 25% data testing and with learning rate of 0.1-0.5. With details of classification results for each category in table 6:

Table 6. Details of 75-25 accuracy and 0.1-0.5 learning rate.

Category	The amount of content	The number of successful classified	Accuracy
ITS	40	40	100%
DM	40	40	100%
MM	40	33	86%

From the above table it is known that the NBC method is good in classifying abstract for ITS and DM category with 100% accuracy, and abstract classification for MM category only yields an accuracy value of 86%. From several test scenarios with dimensional dimension value variation, the amount of learning and test data produces different accuracy. In the scenario of the amount of data learning and testing, the more data learning the better the accuracy. The best results were obtained by dimensional reduction of 30%, using 75% data learning, and from 25% of test data, the NBC method successfully classified 95.33% of the abstract content correctly. With the best result using learning rate between 0.1 – 0.5, can classify 100% in category of ITS and DM, but MM abstract content is only 86%.

IOP Conf. Series: Journal of Physics: Conf. Series **979** (2018) 012036 doi:10.1088/1742-6596/979/1/012036

4. Conclusion

From the result of the research, it is concluded that the accuracy produced by NBC method in classifying abstract content reaches 95,33% for all data, by successfully classifying abstract content in ITS category, DM with 100% accuracy and MM with 86% accuracy. The best accuracy is found in dimension reduction condition with 30% parameter, data testing and training ratio is 75% data testing, 25% training data and learning rate between 0,1-0,5.

References

[1] Indrabayu, Harun N, Pallu M S and Achmad A 2013 *International Journal Of Engineering And Technology*
[2] Sespajayadi A, Indrabayu, Nurtanio I 2015 *International Seminar on Intelligent Technology and Its Application,. ISITIA Proceeding*
[3] Sidehabi S W, Indrabayu ,Tandungan S 2017 *Preceedings- CYBERNETICSCOM: International Conference on Computational Intelligenceand Cybernetics*
[4] Hamzah A 2012 *National Seminar on Science & Technology Applications)* Periode III (Yogyakarta) ISSN:1979-911X
[5] Manda A K and Sen R 2014 *International Journal of Artificial Intelligence & Applications* **5(5)**
[6] Keerthika G and Saravana P D 2015 *Journal of Network Communications and Emerging Technologies* 1(1)
[7] Amalia A A, Djunaidy A, Kusumawardani R P 2015 *Classification Articles News On The Net Based Naive Bayes Classifier Using Confix-Stripping Stemmer,(ITS-Undergraduate*)-38449-5210100106
[8] Fajar H R, Utami E and Amborowati A 2015*Citec Journal* **2(2)** ISSN: 2354-5771
[9] Guna I, Socrates A, Afrizal L, Akbar, and Akbar M S 2016 *J.Lontar komputer* **7(1)**
[10] Yang Y, Pedersen J O 1997 *Proceedings of the Fourteenth International Conference on Machine Learning* (Nashville, Tennessee, USA)

Effect of Fruits Waste in Biopore Infiltration Hole Toward The Effectiveness of Water Infiltration Rate on Baraya Campus Land of Hasanuddin University

Slamet Santosa

Department of Biology, Faculty of Mathematics and Natural Sciences, Hasanuddin University, Makassar 90245, South Sulawesi, Indonesia

E-mail: slamet_santosa@science.unhas.ac.id

Abstract. The infiltration of water into the soil decreases due to the transfer of soill function or the lack of soil biopores. This study aims to determine the effectiveness of the use of fruits waste toward the water infiltration rate. Observation of the water level decrease is done every 5 minutes interval. Observation of biopore water infiltration rate was done after fruits waste decomposed for 15 and 30 days. Result of standard water infiltration rate at the first of 5 minutes is 2.18 mm/min, then decreases at interval of 5 minutes on next time as the soil begins to saturate the water. Baraya campus soil observed in soil depths of 100cm has a dusty texture character, grayish brown color and clumping structure. Soil character indicates low porosity. While biopore water infiltration rate at the first of 5 minute interval is 6.61and 6.95 mm/min on banana waste; 5.55 and 6.61mm/min on papaya waste and 4.26 and 5.39 mm/min on mango waste. The effectiveness of water infiltration rate is 44.45% and 41.93% on banana; 44.61% and 30.09% on papaya and 15.79% and 28.36% on mango. Study concluded that banana waste causes the water infiltration rate most effective in biopore infiltration hole.

1. Introduction

The amount of rainwater that flows on baraya campus land of Hasanuddin University continues to increase due to reduced of water infiltration into the soil. The water infiltration rate decreases due to the transfer of soil function or the lack of soil biopores. The rainwater that flows on the soil surface will cause flooding in the rainy season; droughts in the dry season and the depletion of underground water reserves. The conversion of land use into a settlement causes the hydrological function of the land to be disturbed [1]. The part of open land also undergoes a process of compaction, and biopore is reduced due to the decrease of soil plants and fauna as the maker of biopore in the soil. The need for water continues to increase and the main source of water comes from rainfall, so there is an effort to absorb the effective rain water into the soil. The infiltration of water into the soil can be facilitated by the presence of biopore created by soil fauna and plant roots [2]. The cylindrical opening of the hole will be a depression store that can with stand the flow of the surface to allow seepage into the soil. The cylindrical hole wall provides additional water-absorption surfaces as wide as the walls of holes created. If the cylindrical hole is filled with organic waste, then the absorption surface will not be damaged or blocked because it is protected by organic waste.

Organic waste continues to grow as the population, market and food industries grow. Organic waste produced by households, including fruits waste. Fruits waste that is often produced by all layers of society that is banana , papaya and mango. Fruits waste is easily decayed and favored soil fauna such as ants, soil insects and worms. Therefore, this research uses banana, papaya and mango waste that is expected to be the driving force of the presence of soil fauna into the biopore infiltration hole.

Soil fauna such as earthworms will move to search for food sources and traces to form pores. Pores formed into a means of circulation of water and air in the soil.

2. Materials and Methods

This research was conducted on baraya campus land of Hasanuddin University, Makassar for 3 months. The tool used in this research is 'paralon' pipe of 100cm long and diameter of 10cm, drill, ground spoon, knife, ruler and water scoop. The materials used are papaya fruit waste, banana fruit waste, mango fruit waste and water. The implementation of the research consists of the following stages:

2.1. Measurement of water infiltration rate

Measurement of water infiltration rate begins with measuring the standard water infiltration rate then measuring the biopore water infiltration rate. The measurement of the water infiltration rate was carried out using a 'paralon' pipe of 100cm long and diameter of 10cm. Observation of the standard water infiltration rate is carried out on a drop in the water surface which has been filled in full water (mark) into the 'paralon' pipe. Observation of the water level decrease is done every five minutes interval. Observation of the biopore water infiltration rate is done after the fruits waste ei. papaya, banana and mango has been decomposed for 15 and 30 days in infiltration hole.

2.2. Analysis of water infiltration rate

Analysis of water infiltration rate using Horton method [3]. The water infiltration rate model used by Horton method with the following equation:

$$F = fc + (f0\text{-}fc)\, e^{-kt} \tag{1}$$

where: F = water infiltration rate (mm/min)
fc = constant water infiltration rate (mm/min)
f0 = standard water infiltration rate (mm/min)
e = 2.78
k = constantsta
t = time (min)

The effectiveness of the use of fruits waste toward the water infiltration rate was calculated based on the following formula [3] :

$$E = \frac{\text{Bwar} - \text{Swar}}{\text{Bwar}} \times 100\% \tag{2}$$

where: E = effectiveness of water infiltration rate
Bwar = biopore water infiltration rate
Swar = standard water infiltration rate

3. Results and Discussion

The results of standard water infiltration rate average for 30 minutes of measurement in the three infiltration holes is 1.43 mm/min (Figure 1). While the average of standard water infiltration rate at the first five minutes interval is 2.18 mm /min then decreases at interval five minutes next time as the soil begins to saturate the water (Table 1). The average of standard water infiltration rate in the four infiltration holes is 1.69 mm/min [3]. The water infiltration rate at the first five minutes interval is 2.74 mm/min on media of soil and sand mixture. The water infiltration rate is affected by the amount of soil pores. The observation of soil samples at 100cm depth has a dusty texture character, grayish brown color and clumping structure. Soils with dusty clay textures and clumping structures have low porosity [4]. Porosity is the ratio between the pore volume and the total volume of soil. The pore volume is the total number of cavities or open spaces formed between the grains of soil particles. This pore cavity will later function to absorb runoff water in the biopore hole mechanism.

The results of biopore water infiltration rate average for 30 minutes of measurement in the three infiltration holes is 2.65 and 2.48 mm/min on banana fruit waste; 2.59 and 2.06 mm/min on papaya fruit waste; 1.70 and 2.01 mm/min on mango fruit waste after decomposed for 15 and 30 days (Figure 2). While the average of biopore water infiltration rate at the first five minutes interval is 6.61 and 6.95 mm/min on banana fruit waste; 6.61 and 5.55 mm /min on papaya fruit waste; 4.26 and 5.39 mm /min on mango fruit wastes. The using of fruits waste causes a higher water infiltration rate than without fruits waste in the infiltration hole (Table 1). The biopore water infiltration rate is high due to the increasing amount of pores in the soil. The fruits waste attracts the presence of soil organisms. The banana, papaya and mango fruits waste that are inserted into the infiltration hole are still flavorful and sweet taste encouraging the presence of earthworms, ants, insects and soil microbes. Then the activity of the soil organism causes forms of new pores in the soil. Earthworms can make a path in the ground [5]. The activity of soil organisms causes soil fertility due to available circulation of water and air in the soil [6]. The number and size of biopore will continue to increase following the increase in population and soil fauna activity [7]. The biopore structure is a longitudinal hole branching so that it can smooth water absorption in all directions. Biopore is reinforced by organic compounds derived from soil organisms, so it is quite steady and not easily damaged. Thus, there will always be avenues for pervasive water and air into the soil. In the biopore there is enough organic material, water, oxygen, and nutrients to be suitable for the development of plant roots and soil organisms, including microorganisms that aid in waste generation. Fruits waste is larger in absorbing water than the leaf and vegetable waste [8]. The biopore water absorption rate using fruits waste reaches of 5.77 mm /min, whereas in leaf waste of 4.00 mm /min and vegetable waste of 2.63 mm /min. The interaction of earthworms causes the soil ecology system is very good because the earthworm makes the system of water and air circulation in the soil so that water and oxygen are available [9]. The biopore infiltration hole was developed on the basis of ecohydrological principles, namely by improving soil ecosystem conditions to improve the hydrological function of the ecosystem [1]. The effectiveness of rainwater infiltration needs to be done to reduce the flow of the flooded surface

Table 1. Biopore and standard water infiltration rate

Time of Interval (minutes)	Average of water infiltration rate (mm/min)						
	Standard	Biopore					
		Time of Decomposition (Day*)					
		Banana fruit waste		Papaya fruit waste		Mango fruit waste	
		(15*)	(30*)	(15*)	(30*)	(15*)	(30*)
5	2.18	6.61	6.95	6.61	5.55	4.26	5.39
10	1.87	4.23	3.93	4.29	2.90	2.37	3.05
15	1.37	2.21	1.52	1.97	1.27	1.13	1.28
20	1.19	0.98	0.92	1.07	1.02	0.88	0.82
25	1.02	0.94	0.81	0.86	0.88	0.81	0.77
30	0.94	0.90	0.76	0.75	0.72	0.80	0.75
	1.43	2.65	2.48	2.59	2.06	1.70	2.01

Figure 1. Standard water infiltration rate

Figure 2. Biopore water infiltration rate

Effectiveness of biopore water infiltration rate in infiltration holes is 44.45% and 41.93% on banana fruit waste; 44.61% and 30.09% on papaya fruit waste; 15.79% and 28.36% on mango fruit waste after decomposed for 15 and 30 days. The use of fruits waste in biopore infiltration holes is more effective than without fruit waste [3]. The effectiveness of the use of the fruit waste ranges from 21.30% - 58.58%. This research showed that the effectiveness of the use of banana and papaya fruits waste was not significantly different which was decomposed for 15 days but significantly different from mango fruit waste. Decomposition of fruit waste for 15 days was already provide a significant level of effectiveness enough to absorb water that is 44.45% on banana fruit waste and 44.61% on papaya fruit waste. (Figure 3). The most ideal composting age of organic waste ie 14 days in biopore infiltration hole [8]. The decomposition time on each waste is different because there is organic waste that is easy, medium and difficult degradation [10]. Banana fruit waste that has been mashed and mixed with tofu dregs can be feed ingredients on the cultivation of earthworms *Lumbriscus rubellus* [11]. Papaya fruit waste and crushed cabbage can be feed ingredients on the cultivation of earthworm *Lumbricus rubellus* [12].

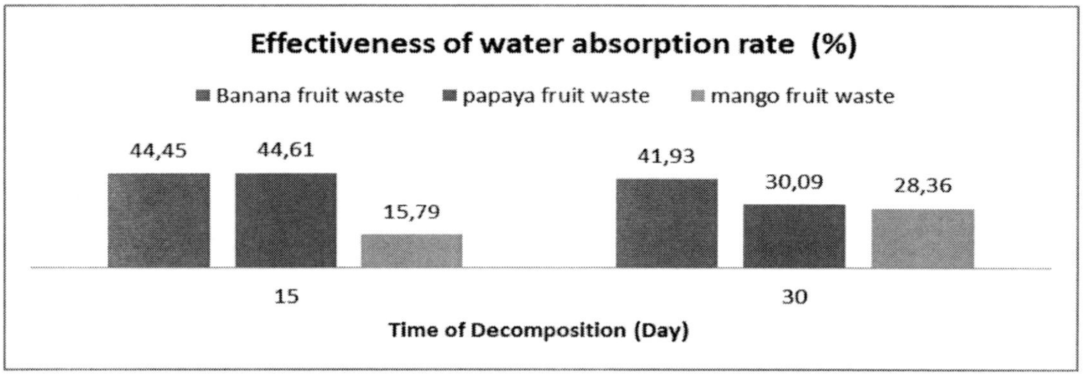

Figure 3. Effectiveness of water infiltration rate (%) in infiltration hole

However, the effectiveness of water infiltration rate on banana fruit waste decomposed for 30 days was different from papaya fruit waste and mango fruit waste (Figure 2). The banana fruit waste has one side of crumb and broad surface, easily destroyed and if stacked there is still air circulation. While the papaya and mango fruits waste have two sides of the slippery surface and if stacked clumps that cause less aeration. The decomposition factors such as particle size, aeration and humidity. The larger surface area will increase the contact between the microbe with the material and the decomposition process will run faster [13]. The rapid composting can occur in conditions of sufficient oxygen [14]. Any organism degrading organic materials requires different environmental conditions and materials [15]. If the conditions are appropriate, then the decomposer will work diligently to decompose organic waste. If the condition is less appropriate or inappropriate, then the organism will dormant, move to another place, or even die.

4. Conclusion

This study concludes that the use of banana fruit waste are more effective in absorbing water in biopore infiltration hole.

Acknowledgments

The researcher would like to thank Warjito and Riski Amalia Puji for financial aid and statisticall analysis.

References

[1] Budi B S 2013 *Wahana Teknik Sipil* **18(1)** 1-12
[2] Brata K R. and Nelistya A 2008 *Lubang resapan biopori* (Jakarta: Penebar Swadaya)
[3] Juliandri M, Azwa Nirmala and Erniyuniarti 2016 *J. Env. Tech* **2** 10-17
[4] Hardjowigeno S 2003 Ilmu Tanah (Jakarta: Maduataman Sarana Pratama)
[5] Smettem K R J 1992 *J. Soil Biol. Biochem* **24** 1539-1543
[6] Lavelle P 1988 *J. Biol. Fertil., Soil* **6** 237-251
[7] Brata K R 1990 *MSc Thesis The Effects of plant residue addition on the aggre-gation of a hardsetting western australia wheatbelts soil* (Australia: The University of Western)
[8] Sibarani R T and Didik B S 2016 *J. Env. Tech.* **2** 23-32
[9] Lee K E 1985 *Earthworm: Their ecology and relationships with soil and land use* (London: Academic Press)
[10] Samosir R D *2010 Studi pengaruh waktu pengomposan terhadap kandungan karbon dan nitrogen didalam kompos Hydrilla sp* (Medan: Universitas Sumatera Utara)
[11] Junaidin, Zohra H dan Santosa S 2016 *Pemanfaatan limbah kulit pisang dan ampas tahu sebagai bahan pakan cacing tanah Lumbricus rubellus* (Makassar: Hasanuddin University)
[12] Robin S, Zohra H and Santosa S 2016 *J. Bioma.* **1(1)** 8-15
[13] Rynk R 1992 *On Farm Composting Handbook. Northeast regional agricultural engineering service pub* (N.Y: Ithaca)
[14] Hartono R D 2012 *Decomposition of fruits waste in biopore infiltration hole on various land use* (Bogor: Bogor Agriculture Institute)
[15] Mulyadi A 2008 *Karakteristik kompos dari bahan tanaman kaliandra jerami dan sampah Sayuran* (Bogor: Bogor Agriculture Institute)

IOP Conf. Series: Journal of Physics: Conf. Series **979** (2018) 012038 doi:10.1088/1742-6596/979/1/012038

Analysis of Chemical and Physical Properties of Biochar from Rice Husk Biomass

Bidayatul Armynah[1,2], Atika[2], Zuryati Djafar [1], Wahyu H Piarah[1], Dahlang Tahir[2]

[1]Department of Mechanical Engineering, Hasanuddin University, Bontomarannu Gowa 92171 Indonesia
[2]Department of Physics, Hasanuddin University, Tamalanrea Makassar 90245 Indonesia

E-mail : wahyupiarah@unhas.ac.id

Abstract: Chemical and physical properties of Rice Husk as a potential energy resource were analyzed by means Fourier transform infrared (FTIR), x-ray diffraction (XRD), scanning electron microscope (SEM), and energy disperse spectroscopy (EDS). Rice husk is heated with varied temperature of 250°C, 350°C, 450°C and 30, 60, 90 minutes respectively combine with time variation. The results show that the calorific value decreases whenever the temperature and time increase. The heating time of 30 minutes at 250°C of temperature gives calorific value of 10.4 MJ/Kg. While at the 450°C of temperature, the calorific value decrease to 4.7 MJ/Kg. The EDS shows that the time of heating is an important parameter where carbon and nitrogen were decreasing with the increment of the heating time while the oxygen increase when the heating time increase. The XRD shows that the broad (002) reflections between 20° and 30° indicate carbon disordered with small domains of coherent and parallel stacking of the graphene sheets, which consists of surface morphology from SEM. FTIR shows that the O-H stretching pronounced at around 3452 cm^{-1} and 3412 cm^{-1} and pronounced clearly at the highest temperature. The aromatic group from lignin gives rise to C=C asymmetric stretching at cm^{-1} as a G band corresponds to the sp2-hybradization bonding of carbon atoms and C-H bending modes at 2927 at 796 cm^{-1}. This results of the characteristic of chemical and physical properties of the rice husk examination provide the prominent source of useful energy that can eventually replace the fossil fuel.

1. Introduction

Biomass is the third largest primary energy resource in the world, after coal and oil. Biomass is a major source of energy in developing countries, where it provides 35% of all the energy requirements. The use of biomass is importance as concerns global warming since biomass combustion has the potential to be CO2 neutral [1]. Biomass materials agricultural from residues such as straw, bagasse, and groundnut shell, coffee husks and rice husks as well as residues from forest-related activities such as wood chips, sawdust and bark having high energy potential. Residues from forest-related activities account for 65% of the biomass energy potential whereas 33% comes from residues of agricultural crops [2].

Among the proven combustion technologies (grate-fired, suspension-fired and fluidized bed systems), the fluidized bed technology is reported to be the most efficient and suitable for converting agricultural and wood residues into energy [1,3]. Extensive experimental investigation has been carried out to date on the feasibility and performance of the fluidized bed combustion of different alternative fuels. CO and NOx (generally, as NO) are also the major harmful pollutants emitted from biomass combustion in fluidized bed systems [4]. For a selected fuel, CO emission (strongly affecting the combustion efficiency) is a function of operating variables, such as excess of combustion air as well as combustor load, and can be effectively controlled by the air supply [5].

Today, various methods have been tried to obtain energy from biomass. Typically, all of those methods can be grouped into two categories, thermochemical process (e.g. combustion, gasification and pyrolysis) and biochemical process (e.g. fermentation and anaerobic digestion), which have been carefully summarized in previous report [6] .This research of Rice Husk from South Sulawesi Indonesia conduct to carry out experiments as function of heating temperature and time.

This research analyze carbon, oxygen, and Hydrogen content from the EDS the results while the Chemical bonding and structural properties of biochar analyzed by using Fourier transform infra-red (FTIR) spectroscopy and X-ray diffraction spectroscopy (XRD), respectively.

2. Research Methods

This research uses the carbonization method to examine the rice husk. The carbonization processes examine the burned biomass in the absence of oxygen. But due to the insufficient equipment conditions, this research is not capable to remove the oxygen in the feeding chamber. This research observes the temperature and time parameter which conducts with the variations temperature of 250°C, 350°C, and 450°C with the time variations of 30 minutes, 60 minutes, and 90 minutes for each temperature.

Infrared spectroscopy was carried out on IR Prestige-21 FT-IR spectrometer (Shimadzu Corp) equipped with a bright ceramic light source, a KBr beamsplitter, and a deuterated triglycine sulfate doped with L-alanine (DLATGS) detector. The measurements of the sample were collected over the range of 4000-600 cm^{-1} and 16 co-added scans. All samples were ground into powders prior to the spectral acquisition. All spectra were in Transmittance units.

The X-ray diffraction (XRD) pattern was collected on an X-ray diffraction (XRD) spectroscopy (Shimadzu 7000) with Cu Kα radiation (λ = 1.5405 Å) over the angular range $15° \leq 2\theta \leq 80°$, operating at 30 kV and 10 mA. It was performed to examine the structural properties of the samples.

Morphological and chemical characterization (composition of carbon, oxygen, nitrogen and another element in the samples) of particles has been performed by a Scanning electron microscope-energy-dispersive X-ray spectrometry (SEM-EDS) (JEOL JSM-IT- 300) with acceleration voltage 10 kV, beam current 7.475 nA, and the lowest vacuum is 50 pa.

3. Results and Discussion

Formation Charcoal Rice Husk with temperature and time varied

Table 1. Chemical composition for Rice Husk as a function of temperature and time of heating were measured by using energy disperses spectroscopy (EDS) from Experiment

Element	250°C /Time (Min.)			350°C /Time (Min.)			450°C /Time (Min.)		
(%)	30	60	90	30	60	90	30	60	90
C	18.62	16.91	22.33	14.27	14.77	7.72	18.29	5.68	6.15
N									
O	43.11	43.95	41.22	45.44	45.05	48.79	42.02	49.82	49.04
Na	0.01		0.04		0.01	0.05			
Mg	0.05	0.05	0.04	0.07	0.1	0.14	0.12	0.15	0.23
Al	0.39	2	0.41	0.17	0.96	0.25	5.67	0.87	0.47
Si	37.43	36.95	35.8	39.61	38.65	42.4	32.24	42.84	42.14
K	0.33	0.07	0.15	0.31	0.43	0.54	0.79	0.64	1.65
Ca	0.06	0.06	0.01	0.13	0.02	0.11			0.33

Table 1 shows EDS from experiment biomass from Rice Husk with the heating temperature from 250°C, 350°C, 450°C and time varied for 30, 60, and 90 minutes respectively. The ash composition shows that the temperatures of 350°C and 450°C increase when the heating time increment as the release of CO_2 and CH_4 gasses from hemicelluloses, cellulose, and lignin decomposition [6]. The temperature of 450°C for the heated time of 90 minutes shows a discoloration in the samples from black to gray which indicates the high content of ash. Some study which stated that fixed carbon increase with the increment of heating temperature, is not found in this research since the mass density of Rice Husk is lower than another source of biomass[6,7,8,]. As showed in Table 1, the element composition also show N, Na, Mg, Si, AL, K, and Ca with dominant content are C, O, and S for bio-chart from Rice Husk.

Based on the X-ray fluorescence spectroscopy (XRF), this research shows that Si, Ca, K, and P are the most abundant ash element in Rice Husk biomass with composition of 87%, 2%, 9%, and 0.9 %, respectively.

Table 2. Chemical composition for Rice Husk as a function of temperature and time of heating were measured by using ultimate analysis

NO	ELEMENT (%)				REF.
	C	N	O	H	
1	37.65	1.63	55.5	5.13	[9]
2	34.6	0.47	31.7	4.23	[10]
3	38.30	0.30		5.60	[11]
4	37.9	0	57.2	4.9	[12]
5	35	0.32	37	5	[13]

The detailed of elemental detected by ultimate analysis from biomass reference from Rice Husk are showed in Table 2. Table 2 shows that the element of C, N, O, and H with dominant content for bio-chart from Rice Husk is C, O, and H.

The heating time affects the produced calories. The results show that the calories decrease when the heating time increase. The heating time of 30 minutes for every temperature provides the highest calories as shown in Figure 1.

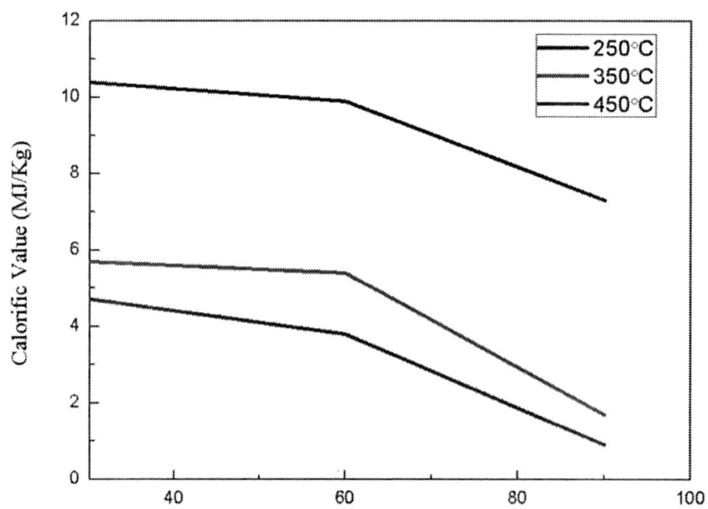

Figure 1. Calorific value determined of biochar from Rice husk by bomb calorimeter

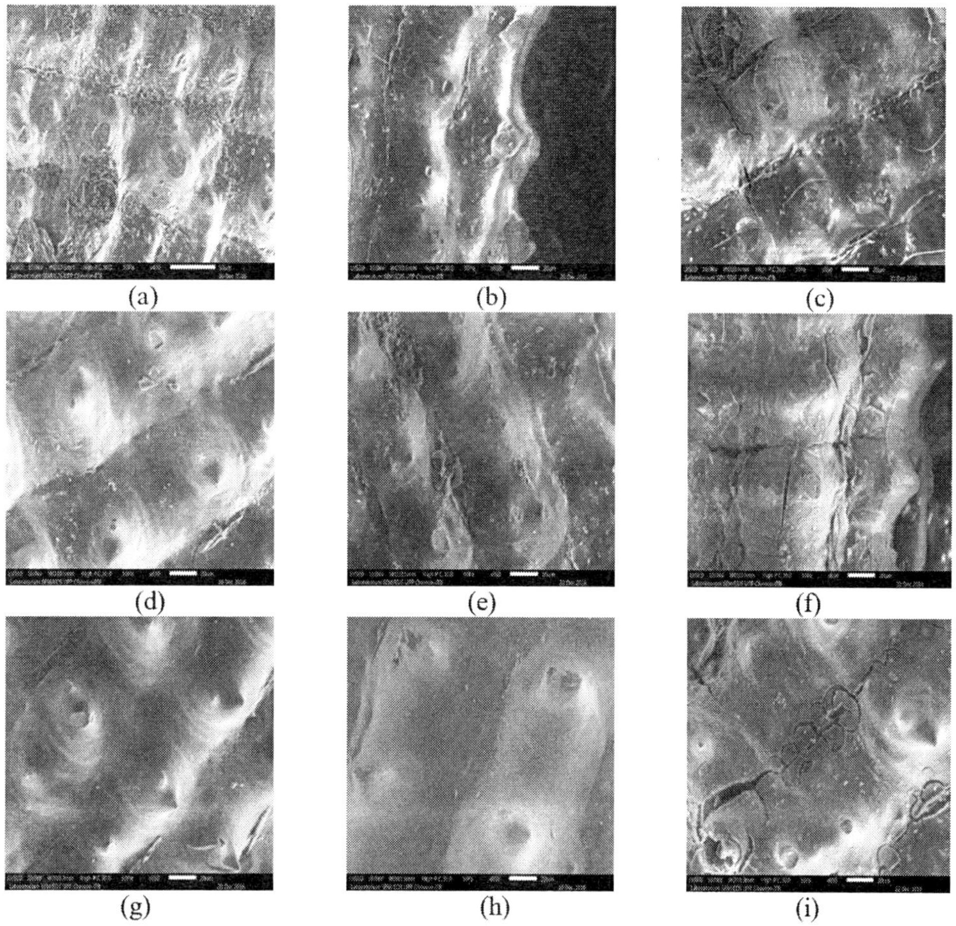

Figure 2. Scanning electron microscopy (SEM) image as a function of temperature and time of heating in free oxygen environments. From the left to right, rows 1, 250°C, row 2, 350°C, and row 3, 450°C for [(a), (d), (g)] 30 minutes, [(b), (e), (h)] 60 minutes, [(c), (f), (i)] 90 minutes.

Figure 2 shows the SEM image as a function of temperature and time of heating in free oxygen environments. From the left to right, rows 1, 250°C, row 2, 350°C, and row 3, 450°C for [(a), (d), (g)] 30 minutes, [(b), (e), (h)] 60 minutes, [(c), (f), (i)] 90 minutes. The characteristic of surface morphology are dot may contain high carbon in the form of nanocarbon.

Table 3. Peak position and intensity of chemical bond for Rice Husk as a function of temperature and time of heating were measured by using Fourier transform infra-red (FTIR).

Chemical bond	250°C Peak Position and Intensity (Time (minutes))	350°C Peak Position and Intensity (Time (minutes))	450°C Peak Position and Intensity (Time)
O-H stretching (Lignin, Hemicellulose, Cellulose)	3412 cm^{-1}, 27.21 (30), 38.26 (60), 48.67 (90)	3423 cm^{-1}, 40.68 (30), 48.19(60) 55.58(90)	3452cm^{-1} 30.64(30), 47.89(60), 64.25 (90)
C=O Stretching (Lignin)	1710cm^{-1}, 54.39(30), 66.11(60), 66.32(90)	1705 cm^{-1} 61.386(30), 58.85(60) 83.57(90)	1697cm^{-1} 0(30), 84.03(60) 95.21(90)
C=C Asymmetric Stretching (Lignin)	1612 cm^{-1}, 49.82(30), 65.11(60), 68.47(90)	1620 cm^{-1} 48.55(30), 65.43(60) 72.69(90)	1627 cm^{-1} 55.62(30), 70.57(60) 84.47(90)
C-O Stretching (Hemicellulose)	1383 cm^{-1}, 85.03(30), 87.05(60), 85.93(90)	1382 cm^{-1} 89.24(30), 86.93(60) 84.94(90)	1382cm^{-1} 89.41(30), 95.88(60) 92.62(90)
C-C Stretching (Hemicellulose)	1421cm^{-1} 66.63(30), no observed(60) no observed(90)	Peak no observed	1423cm^{-1} no observed(30), no observed(60) 97.47(90)
C-O Stretching C-OH Bending (Hemicellulose, Cellulose)	1095 cm^{-1} 2.56(30), 4.83(60) 11.59(90)	1087 cm^{-1} 14.26(30), 6.007(60) 9.654(90)	1095 cm^{-1} 0.50(30), 1.05(60) 5.93(90)
C-H Bending (Lignin)	792 cm^{-1} 55.81(30), 59.97(60) 70.15(90)	796cm^{-1} 51.05(30), 59.40(60) 61.42(90)	798 cm^{-1} 36.91(30), 41.81(60) 45.47(90)

Fourier Transform Infrared (FTIR) spectroscopy used to analyses the effect of temperature and time of heating to the surface functional groups of biochar. Besides the porosity, adsorption behavior is influenced by the chemical reactivity of the surface especially in the form of chemisorbed oxygen in various forms of functional groups. The temperature effect on biochar was successfully investigated using FTIR but not a significant change in intensity with time varied. Figure 3 shows the FTIR spectra as a function of temperature and time of heating in free oxygen environments. Detailed of chemical bonding, peak position, and intensity of Fig. 3 are shown in Table 3. Commonly, most FTIR gives features from organic functional groups for the examination of the organic components of biochar. The peak at 3452 cm^{-1} expected from organic O-H stretching with contribution from any water might retained in the sample or other mineral derived from hydroxyl group. The intensity of the hydroxyl peak which decrease for the temperature increment from 250°C to 450°C indicates that the loss of hydrogen and oxygen atoms due to the breaking bond from hydroxyl group. The aromatic group from lignin give rise to C=C asymmetric stretching at 1612 cm^{-1} indicated as a G band in Refs. [8], corresponds to the sp2-hybradization bonding of carbon atoms. C-H bending modes at 872 cm^{-1} decreases and produced CH_4 as a gas with increasing temperature from 250°C to 450°C .

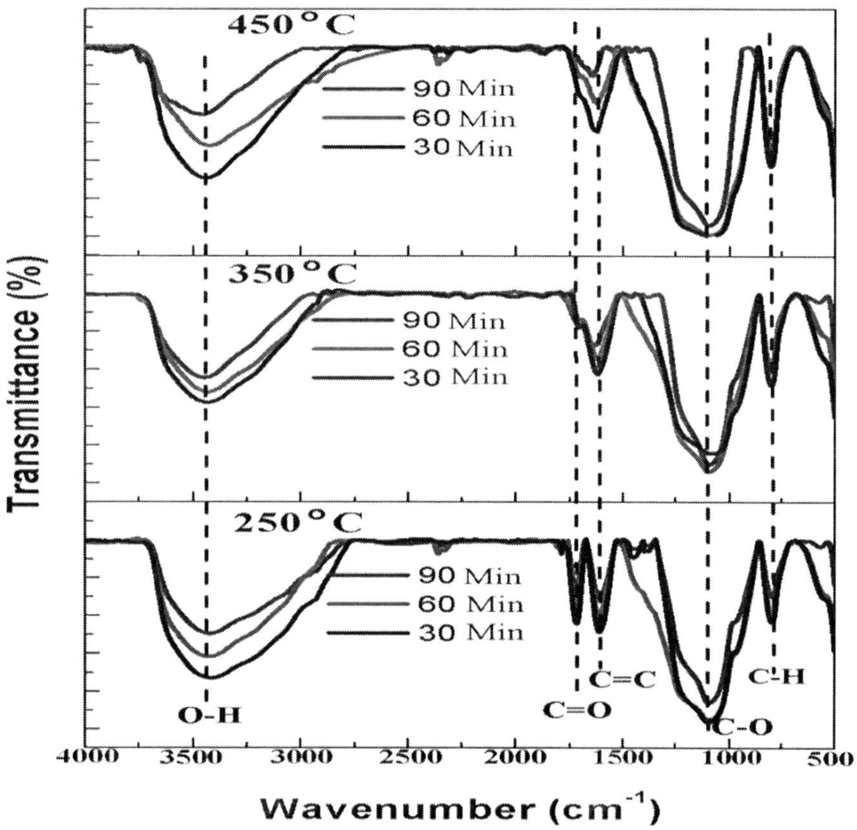

Figure 3. Fourier transform infra-red spectra (FTIR) spectra for biochar from Rice husk as a function of wave number for different temperature and time of heating (see Table 3 for the detail chemical bonding and intensity detected by FTIR).

As shown in Fig. 3 and Table 3, the vibration of C=O stretching of a cyclic and acid anhydride provide the high intensity at 1716 cm^{-1} for temperature 250°C. As the temperature increase the intensity decreases and disappear at the temperature of 450 ^{o}C due to thermal degradation with substantial loss of oxygen atoms and produced CO_2 gas. This is consistent with the result of hydrogen and oxygen content from EDS data in Figure 1. As the heating temperature increase to 350 ^{o}C, the breaking bond of C=O may be also increased.

The transmittance at 1087 cm^{-1} is due to the sp3-hybradization bonding of carbon atoms [8-10], and 1098 cm^{-1} is symmetric C-O stretching for cellulose, hemicellulose, and lignin. The transmission peak of 1421 cm^{-1} recommends the occurrence of aromatic with C-C stretching (ester and phenol). On the other hand, the peak observed at 792 cm^{-1} reveals alkynes with C-H bending is present. The FTIR was confirmed as all cellulose, hemicellulose and some lignin content in bamboo leaf biomass assigned peaks increase with increasing the heating temperature produced gas CO_2 and CH_4.

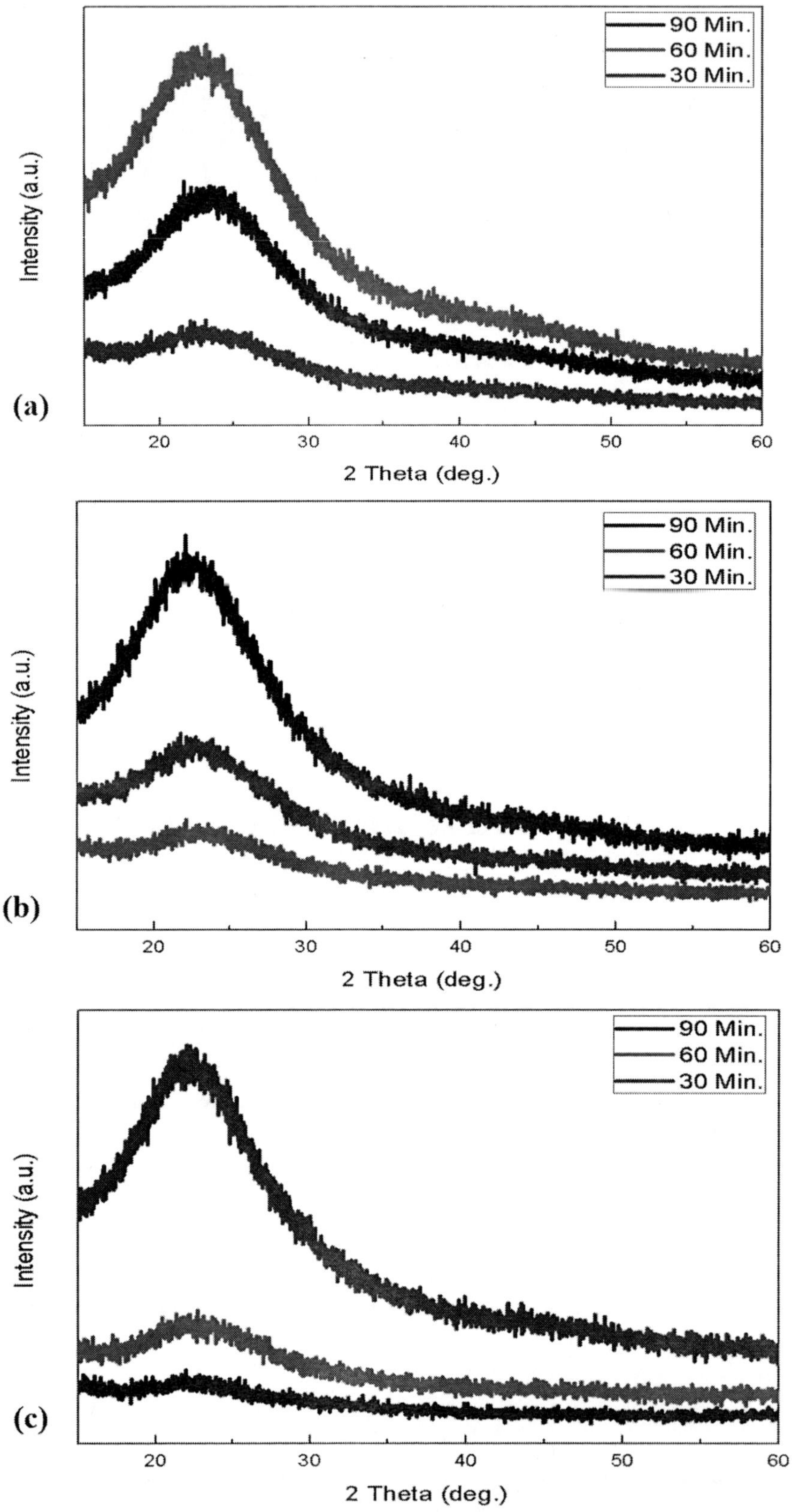

Figure 4. X-ray diffraction (XRD) patterns of the biochar from rice husk for heating time 30 and 90 minutes with temperature (a) 250 °C, (b) 350°C, and (c) 450°C

Figure 4 shows the XRD patterns of the biochar for heating time of 30 and 90 minutes with the temperature of 250 °C, 350°C, and 450°C. The rice husk samples were burnt in a furnace for the temperatures of 250^0C, 350^0C, 450^0C, continue with the XRD plot observation (Fig. 4). The XRD analysis performs for the selected samples to identify the formation differences of amorphous or crystalline silica for different combustion temperatures. A qualitative assessment of the crystallinity of the samples can be obtained from the intensity of the narrow reflections with the comparation to the broad band around 22^0(2theta) for Rice husk burned at 450^0C. The intense broad peak observed for the Rice Husk at 450^0C samples indicates the amorphous nature of silica (Fig. 4). The start of crystallization of all Rice Husk samples burnt below 450^0C shows the amorphous nature. They are typically disordered carbons with high carbon content as can be seen clearly by the broad (002) reflections between 20^0 and 30^0 [8,14,15] which is consistent with morphology surface in Fig. 2 from SEM image.

4. Conclusion

Chemical and physical properties of Rice Husk were analyzed by means, SEM-EDS, FTIR, and XRD. The EDS analysis shows that the ash composition for temperatures 350°C and 450°C increases as the heating time increase due to the release of CO_2 and CH_4 gasses from hemicelluloses, cellulose, and lignin decomposition. Carbon and nitrogen content determined by the EDS shows that the time of heating play an important role as a parameter. It shows that the carbon and nitrogen decrease when the heating time increase. While on the contrary, the oxygen increase when the heating time increase. The XRD shows broad (002) reflections between 20° and 30° and surface morphology from SEM indicated high carbon content. FTIR shows aromatic group corresponds to the sp2-hybradization bonding of carbon atoms C=C asymmetric stretching at 1612 cm^{-1}. These results of the chemical composition characteristic, calorific value, bonding formation and structural properties of the Rice husk shows that the prominent source of useful energy can eventually replace the fossil fuel.

Referances

[1] Werther J, Seeger M, Hartge U, Ogada T, and Siagi Z 2000 *J. Progress in Energy and Combustion Science* **26** 1–27
[2] McKendriya P 2002 *J. Bioresource Technology*. **1** 83- 37
[3] Van den borek R, Faaij A, van and Wijk A 1996 *J. Biomass and bio energy*.11 (4) 271-281
[4] Narendra Prasad G, Srinath S and Venkat R G 2010 *Proc. Ist International Conference on New Frontiers in Biofuels, DTU* 350-358
[5] Perm C W and Kuprianov VI 2004 *J. Bioresource Technology*. **92** 83–91
[6] McKendry P 2012 *J. Energy production from biomass (part2): conversion technology* **83** 47–54.
[7] Sjaak V Land Jaap K, 2008 *The Handbook of Biomass Combustion and Co-firing*
[8] Keiluweit M, Nico P S and Johnson M, Kleber M. 2010 *J. Environ. Sci. Technol* **44(4)** 1247-1253
[9] Zhongqing M, Jiewang Y, Chao Z, and Qisheng Z 2015 *J. BioResourrces* **10**2888-2902
[10] Nattaporn P, Robert J G, Wanpen W, Apnya D, and Tintuing D 2013 *IJMSCI* **3**
[11] Anand M, Kangueng T, Michael H, Yoshiyuki S 2017 *J. Agricultural Science* **8** 1014-1032
[12] Kyle C, Andrey M, Saran P S, Piter B and Andrew C 2013 *J. GCB Bioenergy* **5** 122-131
[13] Ajay K, Kalyani M, Devandra K, and Om P 2012 *IJETAE*. **2**
[14] Singh K, Das K C, and Worley J 2007 *Proc. 15th North American Waste to Energy Conference May* 21-23 (Miami, Florida US)
[15] Kumar N, Prakash P, Kalachelvi P and Sheeba N K 2016 *J. Energy Source* **38** 1428-1434.

Solar Pond Potential as A New Renewable Energy in South Sulawesi

Nur Fadliah Baso, Indar Chaerah Gunadin, Yusran

Postgraduate Student of Electrical Engineering Department, [2]Department of Electrical Engineering.
Jl. Perintis Kemerdekaan KM.10 Makassar, Sulawesi Selatan, 90245, Indonesia

E-mail : nfadliahb@gmail.com

Abstract. Renewable energy sources need to be developed to maintain the electric energy availability by utilizing oceanic energy, namely solar pond energy. This energy is highly influenced by several factors including salinity, air temperature and solar radiation. This study was focused on finding the potential of solar pond in South Sulawesi, a region with fairly high solar radiation and abundant salt water raw materials availability. The method used in this study was analyzing the values from the mathematic models of daily horizontal solar radiation, air temperature, wind speed, relative humidity and atmospheric pressure for the last 22 years which were finalized using MATLAB. The findings of this study will show the areas with good potentials to apply solar pond in South Sulawesi that can be utilized in various fields including power generator, industrial heating process, desalination and heating for biomass conversion.

1. Introduction

One of the potentials from oceanic energy that is still underdeveloped in Indonesia is the energy source from solar pond. Compared to others, the method of using solar thermal energy for power generation is more efficient [1]. Solar pond is a salt water mining able to store solar radiation and heat energy for a long time [2]. Solar ponds have been suggested to be simple and economical in terms of collecting and storing energy on a large scale [3]. Generally solar pond has an area of 2000 m^2 to 250.000 m^2 with depth of 4 m.

When the solar heat energy get into water, the water become hot and the hot water part will decrease its density due to expansion in the water making the hot water move upward to the surface and the cooler part moves to the bottom. Due the presence of contact between hot water and surface air, there is a calorie release by conduction and convection in the hot water, wasting the calorie in the water. In order for the heat in the water is not wasted, salts are added into the water. When the salted water is heated, the salt can be easily soluble in the water, so that the formerly light hot water can gain more weight because the salt solution make the water density heavier and the cooler water move upward to the surface because it is lighter. Figure 1 represents the layer modeling in solar pond.

The heat produced in LCZ which later can be used for power generation, industrial heating process, desalination and heating for biomass conversion. The utilization of LCZ heat energy is realized after the temperature on LCZ has reached 70 ^{0} C [4]. On the power generation, the growth of widely-opened transmission access has a significant side effect for modern utilities [5]. The planning development scheme to ancipate the distributed generation growth is a necessity for electrical network operator. Thus, the planning will regulate the capacity, location and type of transmission and distributed generation [6].

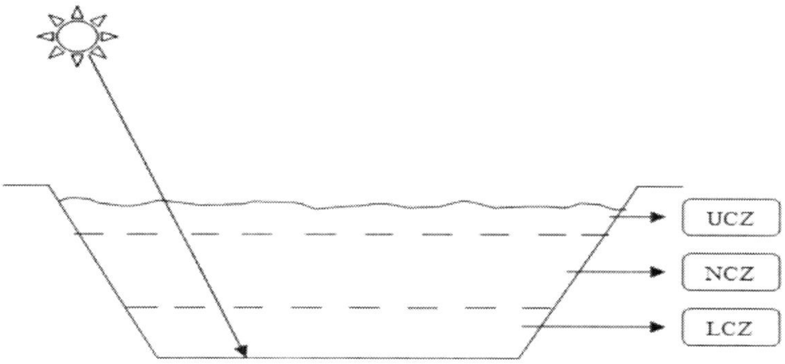

Figure 1. Solar Pond layer modeling

The solar pond layer modeling is divided into three parts:

a. Upper Convective Zone (UCZ) / Upper Zone has a depth of 0.3 – 0.5 m, receives direct solar radiation, has a well-distributed low salinity and temperature along its entire depth.
b. Non-Convective Zone (NCZ) / Gradient Zone has a depth of 1- 1.5 m, has varied salinity and temperature distribution as depth function. The higher the depth in NCZ the greater its salinity and temperature.
c. Lower Convective Zone (LCZ) / Storage Zone has a depth of 1.5 - 2 m, has constant salinity and temperature and well-distributed along its entire depth. This area functions as thermal storage of solar pond, where the LCZ temperature can be utilized based on its need.

2. Material and Method

2.1 Study Area

The selected study areas were located in South Sulawesi, namely Jeneponto at coordinates of 5°35'17.5"S 119°33'53.9"E and Pangkep at coordinates of 4°43'50.6"S 119°31'42.8"E because these two areas are salt producer and have thermal potential to explore. Salinity is one of the important parameters to catch heat in solar pond. The higher the salinity in solar pond the greater the heat produced [7,8]. Figure 2 shows the maps of Jeneponto and Pangkep in South Sulawesi.

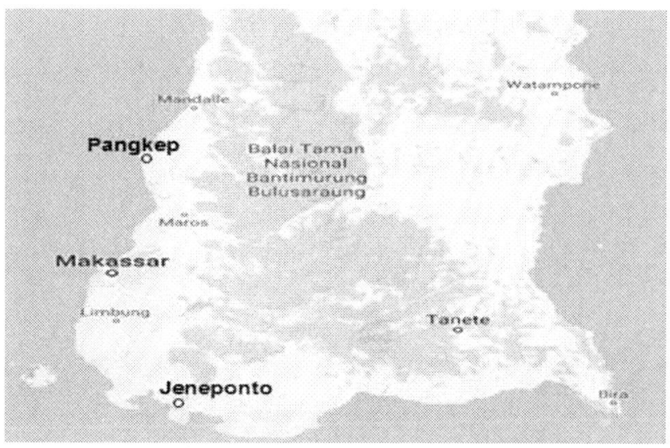

Figure 2. Map of Jeneponto and Pangkep, South Sulawesi

2.2 Data Source

Data were obtained from NASA website that records average data; daily solar radiation horizontal, air temperature, wind speed, relative humidity, and atmospheric pressure for the last 22 years [9,10]. The obtained data were then processed to obtain storage zone heat in solar pond.

Table 1. The Value Of Parameter In Jeneponto

Month	Daily Solar Radiation Horizontal ($kWh/m^2/day$)	Air Temperature (^{0} C)	Wind Speed (m/s)	Relative Humidity (%)	Atmospheric Pressure (kPa)
January	4.57	26.4	4.36	81.1	100
February	4.84	26.3	4.43	80.3	100
March	5.75	26.4	3.29	81.5	100
April	5.91	26.6	3.36	80.9	100
May	5.97	26.6	4.76	79.4	100
June	5.67	26.1	5.46	79.0	100
July	5.95	25.7	5.92	76.6	100
August	6.70	25.8	6.09	72.7	100
September	7.22	26.3	5.59	71.5	100
October	7.05	26.7	4.25	73.7	100
November	6.09	26.7	3.09	78.6	100
December	4.75	26.5	3.46	80.4	100
Annual	5.87	26.3	4.50	78.0	100

*Data Source = NASA 2017 [9]

Table 2. The Value Of Parameter In Pangkep

Month	Daily Solar Radiation Horizontal ($kWh/m^2/day$)	Air Temperature (^{0} C)	Wind Speed (m/s)	Relative Humidity (%)	Atmospheric Pressure (kPa)
January	4.68	25.8	3.62	81.5	99.4
February	4.93	25.9	3.65	79.7	99.4
March	5.54	26.0	2.70	81.2	99.4
April	5.80	26.0	2.57	81.9	99.4
May	5.73	25.9	3.85	80.9	99.4
June	5.53	25.5	4.42	80.1	99.5
July	5.74	25.2	4.83	76.7	99.5
August	6.48	25.8	5.01	69.8	99.5
September	6.83	26.6	4.56	67.0	99.5
October	6.64	26.8	3.33	70.9	99.4
November	5.63	26.2	2.46	78.7	99.4
December	4.46	25.9	2.95	80.7	99.4
Annual	5.67	25.9	3.66	77.4	99.4

*Data Source = NASA 2017 [10]

IOP Conf. Series: Journal of Physics: Conf. Series **979** (2018) 012039 doi:10.1088/1742-6596/979/1/012039

2.3 Model Desrcription

Heat losses is an important factor that affects solar pond performance. Heat losses can be convective loss, radiation loss, evaporation and side loss [4]. Figure 3 shows the heat balance in the Upper Zone.

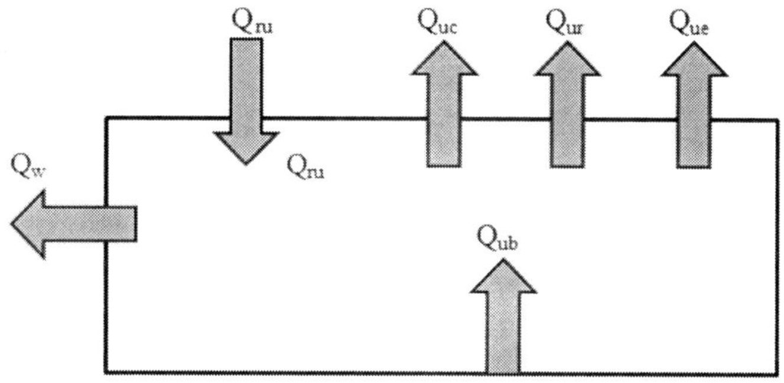

Figure 3. Heat Balance On Upper Zone

The steady state of model of Upper Zone can be written as :

$$\rho_u\, C_{pu} A_u\, X_u \frac{dT_u}{dt} = Q_{ru} + Q_{ub} - Q_{uc} - Q_{ur} - Q_{ue} - Q_w \quad (1)$$

The left hand side of Equation (1) represents the useful heat accumulated in the upper convective zone. For the right hand side of the equation, Q_w is the heat loss through walls of the pond. In this work $Q_w = 0$ (i.e. it is supposed that walls are well insulated) [4]. Equation (1) which represents energy conservation in the UCZcan therefore be rewritten as:

$$\rho_u\, C_{pu} A_u\, x_u \frac{dT_u}{dt} = A_u \left[Q_{ru} + \frac{[T_s - T_u]}{\frac{1}{h_1} + \frac{X_{NCZ}}{K_w} + \frac{1}{h_2}} - \{(5.7 + 3.8v)[T_u - T_a]\} - 4.708\, x\, 10^{-8}\{T_u^4 - [0.0552(T_a)^{1.5}]^4\} - [\lambda h_c (P_u - P_a)]/[(1.6 c_s P_{atm})] \right] \quad (2)$$

Heat losses occurred in Storage Zone is derived from heat transfer by conduction. Appropriate isolation can be used for side edge and lower part of Storage Zone to improve the performance of Storage Zone. Gradient Zone can also influence the performance by inhibiting the ascending heat loss from Storage Zone to Upper Zone [3]. Figure 4 shows the heat balance in the Storage Zone.

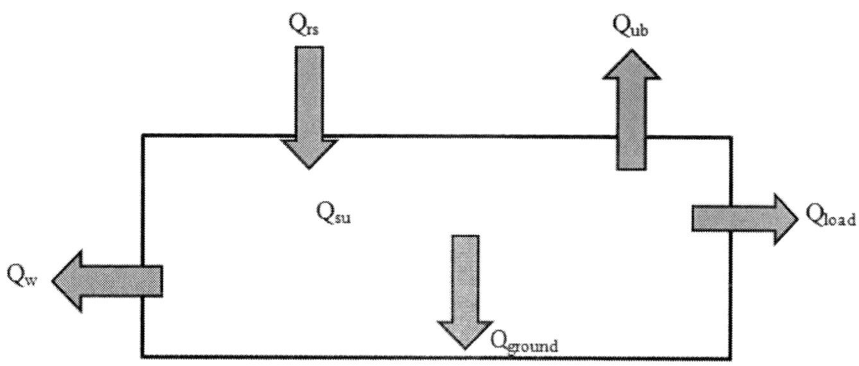

Figure 4. Heat Balance On Storage Zone

The steady state of model of Storage Zone can be written as :

$$\rho_l\, C_{pl} A_l\, X_l \frac{dT_s}{dt} = Q_{rs} - Q_{ub} - Q_{ground} - Q_{load} - Q_w \tag{3}$$

It is assumed to begin with that there is no load $Q_{load} = 0$. This corresponds to the initial warming periodof the pond. In addition, it is assumed that $Q_w = 0$. Equation (3) which represents energy conservation in the LCZ can therefore be rewritten as:

$$\rho_l\, C_{pl}\, A_l\, x_l \frac{dT_s}{dt} = A_l \left[Q_{rs} - \frac{[T_s - T_u]}{\frac{1}{h_1} + \frac{X_{NCZ}}{K_w} + \frac{1}{h_2}} \right] - \frac{A_b [T_s - T_g]}{\frac{1}{h_3} + \frac{X_g}{K_g} + \frac{1}{h_4}} \tag{4}$$

Further details on each parameter in Equation (2) and Equation (4) are provided by Jerome, Ahmed, Wongsakorn [3] and Assad, Hazim, Alasdir [4]

3. Results and Discussion

According to NASA data, Jeneponto and Pangkep have good radiation. Radiation has a direct impact on temperatures in storage zones in solar ponds subject to gradient thickness. The deeper the gradient layer, the less radiation that enters the pool. In the gradient layer with a depth of 1 m then 36% of the incoming radiation pools are available and at a depth of 2 m will be reduced by up to 30%. Increased radiation will increase the temperature as more heat will be absorbed as mentioned before depth of the pond is an important factor in that along with the evaporation losses which can be reduced by introducing a cover on the pond [3].

The comparison of irradiation data between Jeneponto and Pangkep can be seen from Table 3 as follow:

Table 3. Irradiation data between Jeneponto and Pangkep

Month	Jeneponto Radiation ($MJ/m^2/day$)	Pangkep Radiation ($MJ/m^2/day$)
January	16.452	16.848
February	17.424	17.748
March	20.700	19.944
April	21.276	20.880
May	21.492	20.628
June	20.412	19.908
July	21.420	20.664
August	24.120	23.328
September	25.992	24.588
Oktober	25.380	23.904
November	21.924	20.268
December	17.100	16.056
Annual	21.132	20.412

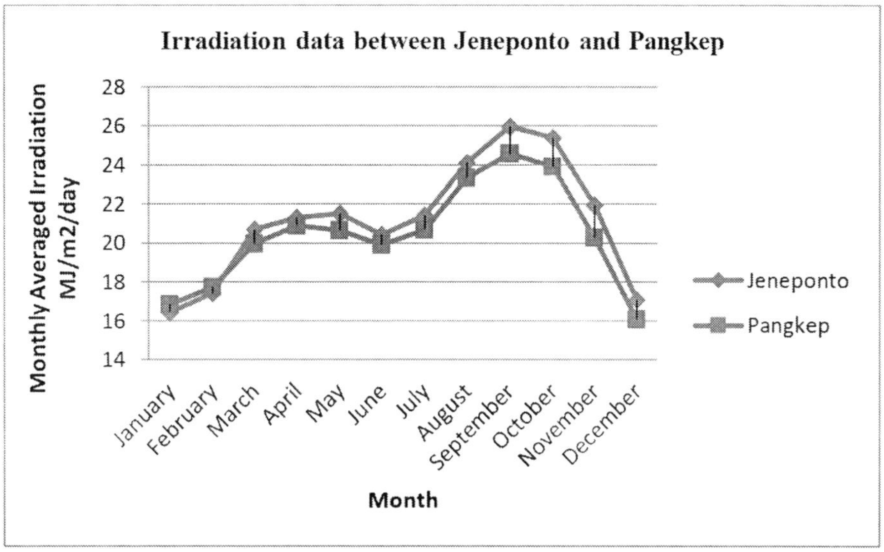

Figure 5. The Plot of Jeneponto and Pangkep Irradiation data

Figure 5 shows that Jeneponto has the highest radiation value in September (25.992 MJ/m2/day) and lowest radiation value in January (16.452 MJ/m2/day). Whereas, Pangkep has the highest radiation in September (24.588 MJ/m2/day) and the lowest radiation value in December (16.056 MJ/m2/day).

Equations (2) and (4) have been solved by using MATLAB. By this method Equations (2) and (4) can be solved depending on the initial values of the unknown temperatures *Tu* and *Ts*. These initial values vary with the location of the pond and the time of year when the pond starts working. The values of the constants which are used in the model are as follows ρ_u = 1000 kg/m^3, ρ_l = 1200 kg/m^3, c_{pu} = 4180 J/kg K, c_{pl} = 3300 J/kg K, $A_u = A_l = A_b$ = 1 m^2, h_1 = 56.58, h_2 = 48.279, h_3 = 78.12, h_4 = 185 (all values in W/m^2 K) and k_w = 0.596 W/m K, T_g = 23 ^{0} C. The value of x_g and k_g depends on the soil properties under the pond [4].

Figure 6 and 7 shows the temperature of storage zone for one year in Jeneponto and Pangkep areas.

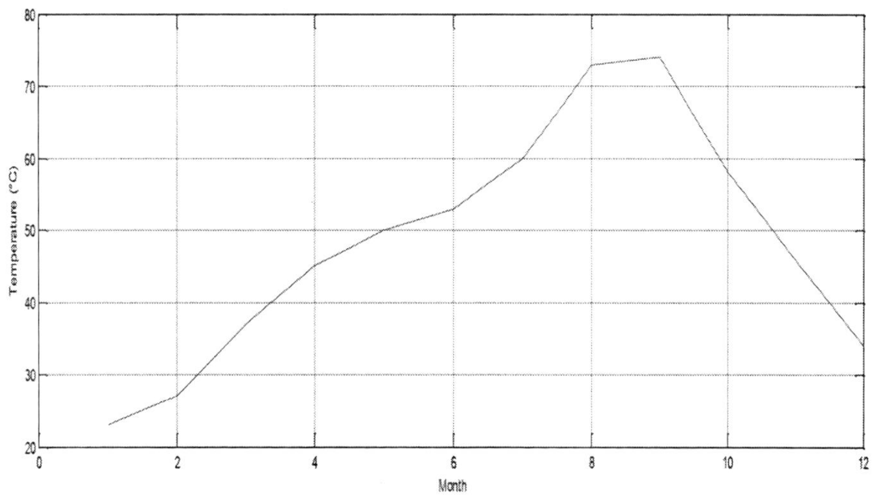

Figure 6. The Storage Zone Temperature during 12 month in Jeneponto

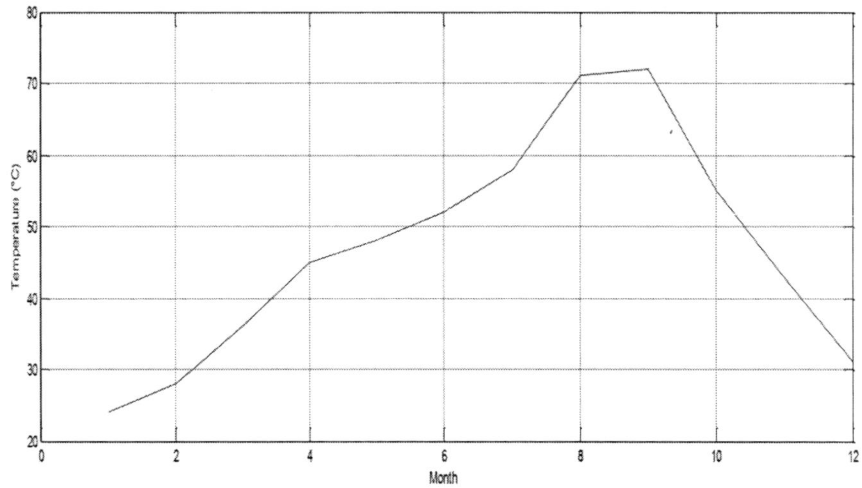

Figure 7. The Storage Zone Temperature during 12 month in Pangkep

From figure 6 and 7 it can be seen that the maximum temperatures in Storage Zone (LCZ) in Jeneponto and Pangkep were observed in August to September ranging from 71 ^{0} C to 74 ^{0} C. Whereas, the minimum temperatures were observed in January to February, ranging from 23 ^{0} C to 28 ^{0} C.

With the Storage Zone (LCZ) temperature above 70 ^{0} C, the solar pond has been able to be used as its intended purpose in various fields such as power generation, industrial heating process, desalination and heating for biomass conversion.

4. Conclusion

Solar Pond is a cheap and environment-friendly technology that can be applied in various fields including power generation, industrial heating process, desalination and heat for biomass conversion. This paper highlights two different locations in South Sulawesi, Jeneponto and Pangkep. From the study findings it can be concluded that the two areas have potential for solar pond assignment. Jeneponto has average maximum irradiance of 25.992 MJ/m^2/day and temperature at storage zone reached 74 ^{0} C, whereas Pangkep has average irradiance of 24.588 MJ/m^2/day and temperature of storage zone reached 72 ^{0} C.

References

[1] Aizaz A and Yousaf R 2013 *European Scientific Journal Edition* **9** 1857-7881
[2] Khaterin I, Bambang I W, Ridho H M, and Awan E S M 2013. *Studi Pemanfaatan Closed Cycle Salt Gradien Solar Pond (CCSGSP) Untuk Pembangkit Listrik di Pesisir Kabupaten Gresik.* (Institut Teknologi Sepuluh November)
[3] Egbe J G, Khan A H, and Wisatesajja W 2013 *IOSR-JMCE* **6** 22-32.
[4] Sayer A H, Al-Hussaini H, and Campbell A N. 2016. *J ScienceDirect Publisher. Solar Energy* **125** 207-218
[5] Gunadin I C, Abdillah, Soetrijanto M A, and Penangsang, O. 2012. *Steady State stability assesment using extreme learning machine based on modal analysis* (International Review Of Electrical Engineering)
[6] Yusran, Ashari M and Soeprijanto A 2013 *J. Theoretical and Applied Information Technology* 1992-8645
[7] Tundee S, Srihajong N and Charmongkolpradit S 2014 *J. ScienceDirect Publisher. Energy Procedia* **48** 453-463

[8] Dah M.M.O, Ouni M, Guizani A and Belghith A 2005 *J. ScienceDirect Publisher. Desalination* **183** 179-185
[9] NASA *Data NASA untuk Kabupaten Jeneponto* https://eosweb.larc.nasa.gov/cgi-bin/sse/grid.cgi (accesed on 14 Agustus 2017)
[10] NASA *Data NASA untuk Kabupaten Pangkep* https://eosweb.larc.nasa.gov/cgi-bin/sse/grid.cgi (accesed on 14 Agustus 2107)

IOP Conf. Series: Journal of Physics: Conf. Series **979** (2018) 012040 doi:10.1088/1742-6596/979/1/012040

The Habitat of Yellow Mouth Turban *Turbo Chrysostomus*, Linnaeus, 1758

E Soekendarsi

Department of Biology, Faculty of Mathematics and Natural Sciences, Hasanuddin University, Jl. Perintis Kemerdekaaan Km 10, Tamalanrea, Makassar, Indonesia

E-mail: eddy@science.unhas.ac.id

ABSTRACT. In general, yellow mouth turban snail *Turbo chrysostomus* L. 1758 was found in intertidal and coral reef area. This animal is active at night (nocturnal) and settles the coral reef-flats area to do its activity as substrate. In doing its activity, yellow mouth turban snail can be found in the depth of 50 cm until 4 m of tidal area. The adult yellow mouth turban snails are found in great number at intertidal area's border and at coastal area of coral reef-flats. Methodology that was used in this study is visual analysis (descriptive method), and divided into two parameters which were observed, i.e. abiotic and biotic. Abiotic components that were measured are; Oxygen (ppm), pH, Water Temperature (°C), Salinity (ppm), Ammonia (mg/L), Nitrate (mg/L), Nitrite (mg/L), and Calsium Carbonat (mg/L).Whereas, biotic components that were measured are; substrates, seaweeds, other organisms, and epilithon. The observation's result of yellow mouth turban snail's environmental condition showed: abiotic condition of the waters consists of oxygen 3-5 ppm, seawater pH 7-8, seawater temperature 23-26°C, and the salinity of 32-33 ppm. The Habitat of yellow mouth turban snail settled the reef-flats area that is overgrown covered by seaweed *Sargassum sp.* as the place to do its activity.

1. Introduction

Mollusks are one of the potential marine biological resources in the waters of Indonesia. Some animal species have a high economic value and can be used either meat or shell [1, 2]. Latama [3] found that the shells of 59 species of mollusks (95%) of gastropod in South Sulawesi have commercial value. Krisnamurthy [4] adds that more than 30 species of mollusks from the Indian shell used for souvenirs. Utilization of a larger mollusks found in Thailand, which the mollusks are about 225 species, they are dominated by gastropods (174 species) and traded on the market (found in the souvenir market and a meat market) [5, 6].

The Yellow Turban snail (*Turbo chrysostomus*) is one of the most common types of marine gastropods found in tropical climates. This snail is known by international name as yellow-mouth turban or gold-mouth turban. Common names that are known by Indonesian people are Siput Bulan (Ambon), and Siput Mata Lembu (Pelabuhan Ratu and Pangandaran, West Jawa) [7].

The yellow turban snails are found in coastal waters in Indo-Pacic region including coastal waters of the Indian Ocean (Kenya, Seychelles, Chagos, Andaman and Nicobar islands), Southeast Asia (Malaysia, Indonesia, Thailand and the Philippines) and the Fijian Islands in the South sand [8]. This type of snail is also found in the coastal waters of the Ryukyu Islands, Japan [9], and in the waters of Northern Melanesia to southern New Caledonia [10, 11].

As well as batulaga snail (*Turbo marmoratus*), *Turbo chrysostomus* is one of the marine biological resources with the potential to be developed in Indonesia in the future. Utilization for the sake of national consumption is still relatively low because it has not been entrenched. In Indonesia, Ujung Genteng snails live in certain coastal waters, the waters of southern coast of Java, while in the waters of northern coast of Java island has been no information about the discovery of this type of snail.

2. Materials and Methods

This research was conducted in the waters of Ujung Genteng, Gulf of Pelabuhan Ratu, Sukabumi, West Java. Ujung Genteng located in the South of Java Island overlooking the Indian Ocean with position on 7°17'08" - 7°21'50" south latitude and 106°23'40 "- 106°24'10" East Longitude. This study was conducted for one year, starting from January to December respectively. The tides in the waters of Ujung Genteng have the range from one to two meters that occurred during the rainy season, and one to two meters occurred at dry season. The environmental conditions influence the physiological activity of gastropod snail in nature in order to survive. Thus, it can inhibit or otherwise stimulate the reproduction activity. Some parameters of water conditions' snail in situ are measured using the method or apparatus as follows:

Table 1. The observed environmental aspects, the use of methods / tools and the place of observation

No.	The Observed Variables	The Use of Method / Tools	Place of Observation
Abiotic Components			
1	Oxygen (ppm)	Winkler/Titration	*In-situ*
2	pH	pH meter	*In-situ*
3	Water Temperature (°C)	Thermometer	*In-situ*
4	Salinity (ppm)	Refractometer	*In-situ*
5	Ammonia (mg/L)	Phenate/Spectrometer	*In-situ*/Lab.
6	Nitrate (mg/L)	Brusin Sulfat/Spectrometer	*In-situ*/Lab.
7	Nitrite (mg/L)	Sulfanilik/Spectrometer	*In-situ*/Lab.
8	Calcium Carbonate (mg/L)	Permanganometri/Titration	*In-situ*/Lab.
Biotic Components			
8	Substrate	Visual	*In-situ*
9	Algae	Visual	*In-situ*
10	Other Organisms	Visual	*In-situ*

Some parameters cannot be measured directly *in situ*. In fact, the measurements were done at The Laboratory of Oceanic Biology, Department of Fisheries Resource Management (MSP), Faculty of Marine Sciences and Fisheries (FIKP) IPB.

3. Results and Discussion

The results of water quality analysis (chemical water components) during the study in one year observation are presented in Table 2. Sampling for water quality analysis conducted between three to two days before the full moon is complete. The results of abiotic component showed that the levels of dissolved oxygen in the snail habitat of Turban snail is around 3 to 5 ppm, while the pH of sea water is ranged from 7 to 8. Similarly, the salinity of sea water is around 32 to 33 ppm. The content of ammonia in the waters of Ujung Genteng is recorded with the lowest value of 0.031 mg/L and the highest around 0.951 mg/L. Nitrate which is also one of the important components in the water body, the lowest is in 0.092 mg/L and the highest is in 0.676 mg/L. Nitrite content in seawater, the lowest is recorded in 0.001 mg/L and the highest is in 0.029 mg/L respectively. Deposits of calcium carbonate in which this component is indispensable to the process of growth, the lowest value is recorded in 145.88 mg/L and the highest is in 182.48 mg/L.

The results analysis of biological situation in the waters of Ujung Genteng, Sukabumi was presented in Table 2. The substrates which are preferable by snail at Ujung Genteng are fragments of rock, dead coral flats which are covered by some types of seaweed, and coarse sand that only a small fraction in the region reef flats that were carried away by the waves of the sea water. The monitoring results on the ground were taken during the inactive time of snail at the day in which the snail chose the protected or hiding area into rock to hide or rest.

At night or at sunset, the Yellow Mouth Turban shellfish came near the bank for feeding or doing other activities such as looking for a partner at the time to spawn. However, such substrates are generally covered by some types of seaweed. In addition, it is suspected that the condition of the substrate is to avoid currents or waves of sea water at high tide, so, it will help the snail to avoid the blows of sea currents or waves that can cause death or damage to the shell.

Table 2. The results of environmental conditions at sampling sites

No.	The Observed Variables	Result
Abiotic Components		
1	Oxygen (ppm)	3 – 5
2	pH	7 – 8
3	Water Temperature (°C)	23 – 26
4	Salinity (ppm)	32 – 33
5	Ammonia (mg/L)	0.031 – 0.951
6	Nitrate (mg/L)	0.092 – 0.656
7	Nitrite (mg/L)	0.001 – 0.029
8	Calsium Carbonate (mg/L)	145.88 - 182.48
Biotic Components		
9	Substrate	Rubble; Death Coral reef; Rough sands
10	Seaweeds	Green Algae: *Boergesenia; Bornetella; Caulerpa; Chaetomorpha; Codium; Halimeda; Neomeris; Udotea; Ulva* Brown Algae: *Dictyopteris; Hormophysa; Padina; Sargassum; Turbinaria* Red Algae: *Acanthophora; Amansia; Amphiroa; Corralina; Galaxaura; Gracilaria; Halymenia; Hypnea; Laurencia.*
11	Other Organism	Nematoda; Oligochaeta; Polichaeta; Diptera; Crustacea; Gastropoda
12	Epilithon	Tebellariaceae; Nitzschiaceae; Naviculaceae; Oscillatoria; Prasiolaceae; Ankisbodesmus; Chaetophora; Ulotrichaceae; Cymboloporidae; Schizomeridaceae; Nematoda; Nauplius

The Yellow Mouth Turban liked reef flats area that covered by seaweed *Sargassum sp*. In which as feeding ground, because the snail is the herbivorous animal with marked by the teeth scrape for the seaweed with the radulas. From the field of observation, are obtained nine species of green seaweed which is always found during the observation, but its abundance during the dry season in contrast to the rainy season. The green seaweed in the dry season is more abundant than in the rainy season, while the brown and red seaweed in the dry season are less than in the rainy season. All observations were made only in the area of coral reef in accordance with specimen collection which was only done in the area of coral reef-flat from the slope areas toward the beach that was grown by the three types of seaweed. At the slopes to the depths of the sea, there is no supporting information regarding the state of the vegetation. When colleting specimens of yellow tubinid snail done in the field, a few specimens were caught still eating the fresh of thalus seaweed *Sargassum sp*. [12, 13]. Being in the laboratory (in the aquarium), the yellow mouth turban snail grazing the talus of *Sargassum*, but prefers diatomaceous attached to the aquarium. Asano [14] and Soekendarsi *et al*. [15] have examined the content of the stomach which contains the Topshell snail of *Trochus niloticus*. Similarly, [13] and [16], examining the content of the stomach which contains the Topshell snail of *Trochus niloticus* and batulaga snail had the same result that the kind of food Topshell and also some kinds of yellow – silver snails were algae within namely Cyanophyta, Phaeophyta, Rhodophyta and Chlorophyta and some types of foraminifera, sand and detritus.

In the habitat's substrate of yellow mouth turban snail, it was acquired several other animal species that live together with the snail that is from the Nematoda Class; Oligochaeta; Polichaeta; Diptera; Crustaceans; and other gastropods, and other gastropod species here are Turbo argyrostoma inhabiting or overlapping the same area with yellow mouth turban snail. Allegedly, some kinds of crustacean are predators of yellow mouth turban snail, whereas the animals of the class Polichaeta and Diptera, thought to be one of the yellow mouth turban's food that are ingested when grazing for feeding time.

During the study, we have been analyzed 12 types epilithon which attached to the *Sargassum sp*. So, presumably some kind epilithon are ingested by snails when grazing the tallus of *Sargassum*. There are several types of epilithon which are the food that commonly eaten by several species of gastropods during the juvenile phase and phase veliger, such as: Nitzschia; Navicula; and Oscillatoria [17].

There are some environmental factors that may limit the activity of the Sub-Class Prosobranchia on habitat [6, 18, 19], namely:

- Oxygen content and temperature of the water, not a limiting factor that is essential for the activity of the sub-class snail Prosobranchia. These snails require dissolved oxygen in the water for at least 5 mg /L or more, and only a few types of these which still survive with dissolved oxygen content lower than 2 mg / L.
- pH, generally ranging from 6.7 - 9.0. However, snails are particularly fond of waters having a pH in the range of 7.0 - 8.4. When the pH is lower than this range, shell growth can be disrupted to damage the snail shell
- Calcium carbonate, alkalinity conditions will be good when the content of calcium carbonate in the water is in the range of 200 - 600 mg /L. While the calcium carbonate that is lower than 200 mg /L can cause the stress of snail.

Protho [20] in his book entitled 'Ambient Quality Criteria for Ammonia water', states that some species of mollusks that live in the sea-water have enough power for good environmental tolerance of the content of ammonia is 3.08 - 19.10 mg/L; pH: 7 - 8; water temperature is around 24 - 25^{0}C and salinity is around 31 - 33 ppm. The content of ammonia which are dissolved in seawater are directly related to salinity, temperature and pH, whereas the dissolved nitrite that can be tolerated by the snail is between 1.3 - 3.5 mg/L. As well as nitrate dissolved is between 0.37 to 2.30 mg/L. In addition, according to [20,19], if the content of ammonia waters is between 0.005 – 0.01 mg/L, on sea urchin *Lytechinus Pictus*, fertilization process will not occur.

Ujung Genteng waters is an area of coral reef with a fairly high tides and ocean currents are quite strong. The content of ammonia measured here is around 0.031-0.881 mg /L. (So it does not matter if the content of ammonia increased significantly) the increased content of ammonia in addition affected by salinity, pH and temperature, are also influenced by the detritus decay that occurs in the bottom waters at any time which can be washed away by the flow of seawater from the coast towards the sea. Thus, we can say that the process of ammonia formation will be different in an area where flooded by the open area (washing the substrate by a current).

Yi and Lee [21], which examines Topshell snail *Trochus niloticus* get the result that the temperature is one of the factors to sustain the life of Topshell, especially in oxygen consumption. 31^{0}C temperature is the optimum temperature for activity and growth. This is in accordance with the opinion of Heslinga [22], that the difference in temperature also affect the growth of Topshell, mainly related to the abundance and quality of natural food. Topshell snail tolerates to temperature 28 - 34^{0}C and changes in the salinity of 31 – 37 ppm.

As well as Yellow mouth Turban snail, Hatta [23] examined the existence of conch Topshell *Trochus niloticus* where the conch Topshell commonly found on the substrate rocks and sandy stone and the conch Topshell that found in the rocks have a weight higher than Topshell living in sandy rock areas. This is in accordance with the opinion of Schuster [18] and Islay [24] that the calcium carbonate is very influential on the growth of several species of gastropods. It was also known that the substrate of rocks which is overgrown covered by seaweed is the food of the family snails and slugs Turbinidae Trochidae and also other herbivores.

4. Conclusion

The habitat of Yellow Mouth Turban snail in their activities have dissolved oxygen 3 – 5 ppm, seawater pH 7 – 8, seas water temperatures 23 – 26^{0}C, and the salinity of 32 – 33 ppm, with its substrate in the form of reef flats covered overgrown with seaweed *Sargassum sp.* near the edge of the slope of the area directly opposite the sea.

Acknowledgments
We are very grateful to Dean Faculty of Mathematics and Sciences University of Hasanuddin for grand this research and Dr. Ari Setiarso (Insitute Technology of Bandung) for correcting the manuscript. We also thanks to all staffs in Laboratory of Oceanic Biology, Department of Fisheries Resource Management (MSP), Faculty of Marine Sciences and Fisheries (FIKP) Institute Agriculture of Bogor, Bogor.

References

[1] Abbot R T and Boss K J 1989 A Classification of the living mollusca *American Malacologist Van Nostrand Reinhold* (New York) 663pp
[2] Ghesquiere S 2007 Gastropoda: snails and slugs Palo Alto California (Blackwell Scientific Publications and Pacific Grove California) The Boxwood Press 251–329
[3] Latama G 1995 *Journal of Phuket Marine Biological Center Special Publication* **16** 31-33
[4] Krishnamurthy V, Soundararajan R 1999 *Journal of Phuket Marine Biological Center Special Publication* **19** 345-352
[5] Bussarawit S 1995 *Jurnal Phuket Marine Biological Center Spec. Publ* **15** 35-38
[6] Ainis A F, Vellanoweth R L, Lapeña Q G, Thornber C S 2014 *Journal of Archaeological Science* YJASC 4082
[7] Arifin Z 1994 Green snail (*Turbo marmoratus*) habitat and live cycle Bulletin of Lonawarta **I** 1 - 8
[8] Bowen B W, Rocha L A, Toonen R J, and Karl S A 2013 j.tree.2013.01.018
[9] Yamaguchi M 1993 Green snail. In: A Wright and L Hill. (eds). Nearshore marine resources of the South Pacific (Institute for Pacific Studies, Suva) 497-511
[10] FAO 1998 The living marine resources of the western central pacic **1** seaweeds, corals, bivalves and gastropods e*d*: K E Carpenter dan Volker H. Niem (Food and Agriculture Organization Of The United Nations: Roma)
[11] Lazzeri A M 2017 *Estuarine, Coastal and Shelf Science* **188** 12-17
[12] Niamaimandi N, Javadzadeh N, Kabgani N, Vaghei R G, Rajabzadeh I 2017 *Journal of Regional Studies in Marine Science* RSMA **237**
[13] Nifong R L 2017 *Journal of Aquatic Botany* **139** 57–64
[14] Asano N 1944 *Journal of Fisheries* **35**
[15] Soekendarsi E, Palinggi A, Santosa S 1998 *Journal of Phuket Marine Biological Center Special Publication* **18** 73 - 76.
[16] Yamaguchi M 1993 Biology of the greensanil (*Turbo marmoratus*) and its resource management. Workshop on pacic inshore fisheries resources working paper no. 11. South Pacic Commission (Noumea, New Caledonia) 9 p
[17] Dwiono S A P, Pradina, Makatipu P C 2003 Spawning and seed production of the green snail (*Turbo marmoratus* L.) in Indonesia. SPC Trochus Information Bulletin #7. http://www.spc.org.nc/coastfish/index.html
[18] Schuster 2003 Phylum Mollusca: Snails, Chitons, Bivalves, Squid, and Octopods. http://www.biology.eku.edu/SCHUSTER/bio20802/mollusca.htm
[19] Şahin S K, Zeybek M 2013 *Turkish Journal of Fisheries and Aquatic Sciences* 1303-2712-v161-19

IOP Conf. Series: Journal of Physics: Conf. Series **979** (2018) 012040 doi:10.1088/1742-6596/979/1/012040

[20] Protho M G 1989 Ambient water quality criteria for ammonia (salt water) (United States Environmental Protection Agency, Washington) 67 p
[21] Yi S K, CJ Lee C J 1997 Effect of temperature and salinity on the oxygen consumption and survival of hathcery-reared juvenile top shell *Trochus niloticus* (Mollucs:Gastropoda) *ACIAR Proceedings* **79** 69 – 75
[22] Heslinga G A 1981 *Trochus niloticus Journal of Malacologia* **20(2)** 349 – 357
[23] Hatta M 1992 *Possibility study of cockle shells cultivation lola Trochus niloticus that observed from the food and habitat and also other some biological aspects at the coastal waters of bojo village Barru regency* (Thesis Hasanuddin University) 66p
[24] Islay D. Marsden, Baharuddin N 2014 *Journal of Estuarine, Coastal and Shelf Science* YECSS 4461

Status of Suspended Particulate Matters Pollution at Traditional Markets in Makassar City

Sri Suryani, Fahrunnisa

Department of Physics, Hasanuddin University, Makassar, 90245 Indonesia

Email: tamalanreah3@gmail.com

Abstract. Research on the status of suspended particulate matters pollution in four traditional markets located in Makassar city has been done. The purpose of this research is to know the air quality in the traditional market areas, especially caused by suspended particulate matters. The background of this research is because traders who trade in traditional markets generally peddle their goods along dusty roads and suspended particulate matters in dust can be inhaled when the vehicle passes. These suspended particulate matters pollutant can cause lung diseases. The results showed that the level of suspended particulate matters pollution fluctuates every year depending on the local wind speed, humidity, and temperature. Research results also showed the values were over the standard value according to the governor of South Sulawesi regulation.

1. Introductions

Air is an important material in human life, humans cannot live without air. However, the air around us is not always in a clean state, whereas the human body needs clean air free of pollutants. Dirty air can be caused by the entry of pollutants from transportation, industry, or nature, such as smoke from forest fires, volcanic lava eruptions and etc.

Especially in urban areas, air pollution is mainly caused by industrial activity and transportation. The use of gasoline fuel in transportation, not only produces emissions in the form of gas, but also produces particulate matters that can be suspended in the air. Suspended particulate matters or particulate matters or more commonly referred to as dust in the air can be toxic and interfere respiratory and cardiovascular process. Suspended particulate matters can also be used as an air pollution indicator that shows the level of hazards to both the environment and health and safety [1, 2].

Suspended particulate matters present in the dust are of several sizes, from 0.001 μm to 150 μm. This particle is a gas emission having a size of 1 - 150 μm, bacteria with size 0.3 - 60 μm, wood burning (0.2 - 3), grain dusts (5 - 1000), human sneeze (10 - 100) viruses (0.005 - 0.3) [3, 4]. The size of suspended particulate matters that is often used as an air pollution indicator has a size of 1μm to 100 μm [4], while the health hazard has a size between (0.1 - 10) μm [1, 2, 5, 6]. Sedimentation of suspended particulate matters on the leaf surface can cause leaf photosynthesis capability decreased [7]. Furthermore, the air contaminated by these suspended particulate matters when exposed to the body, can cause various diseases, such as respiratory infections, coughing, irritation of the eyes and throat, cancer, and heart disease [1, 2, 8].

Traditional markets are markets that have no special buildings. Generally, traditional markets arose accidentally, because of the construction of new settlements and the location of modern markets far from these settlements. The market has very simple form of buildings, even in some parts there are no buildings at all (Figure 1). Merchants hawk their merchandise using only plastic-coated tables, even some traders put their merchandise on the sidewalk. This situation often makes road jams. Vehicles that go through the road become very close to the merchant, which causes the dust to fly. Dust from vehicles that pass near the place of trade, can cause traders at risk for disturbances in the respiratory system due to inhalation of dust particles.

Figure 1: Traditional market (source: CELEBESONLINE).

To maintain public health within the traditional market environment, local governments should periodically monitor air quality measurements. This is done as a consideration for decision-making whether it is time to relocate the trader or improve market conditions. One of the parameters in air quality is the suspended particulate matters. These parameters are important to be measured and known, because the effects of inhalation of these particles can be seen in a short time.

2. Materials and Methods

The measurement data, which are concentration data of total suspended particles (TSP), were taken directly by using the Air Pollution Collector tool and analysed by gravity meter in laboratory [9]. Samples were taken from five traditional markets: Makassar Central Market, Gowa Central Market, Sungguminasa Market, Daya Market and Maros Central Market (Figure 2).

Figure 2. Research location

The data of wind speed, humidity, and temperature are carried out at the same day, with measurements twice per day, during high activity to determine the climate conditions of the location of the measurement.

2.1 Results and Discussion

Air quality is strongly influenced by climatic conditions, especially the local climate. High temperatures make the air more dilute, because it corresponds to the physics principle of fluid which says the air viscosity decreases with increasing temperature [10, 11], so the particles are easily dispersed. In addition, wind velocity also affects the distribution of particles in the air, the greater the wind speed, the more distant the particle [12]. The humidity of the air, will make the particles react with moisture, so that the particle size becomes larger [1, 13], similarly, rain will remove the suspended particles in the air by make them falling to the earth.

The air temperature in the five study sites is approximately 30 °C (Table 1). The lowest air temperature is in the Central Market of Gowa which is an average of 30 °C, this market is located in the southern city of Makassar, while the highest temperature is measured in the Central Market of Makassar which is located in the centre of Makassar. The same situation also appears at wind speed. Gowa's Central Market lies close to the mountainous areas, as well as the Sungguminasa and Maros markets. Instead the Daya Market is located near the bus terminal in the direction of out of town, so the passing vehicles are quite crowded, many buildings, and less plants growing on the roadside, cause the air temperature to be large. Similarly, the Central Market of Makassar is located in the center of the city and surrounded by tall buildings, so the Central Market looks like it is in an enclosed space. That's why the air temperature in this place to be high and the wind rate is very low.

Table 1. Microclimate state at the study site.

	Temperature (°C)	Wind Speed (m/s)	Humidity (%)
Maros Central Market	30.5 - 32	1.8 – 1.9	55 - 75
Gowa Central Market	29 - 31	3.1 – 7	55 - 85
Sungguminasa Market	29 - 32	1.5 - 2.7	60 - 75
Daya Market	31 - 32.75	1.4 – 3.2	50 - 74
Makassar Central Market	32 - 36	0.4 - 0.6	60 - 80

Suspended particulate matters concentration in each traditional market fluctuates in every market, sometimes increasing and occasionally decreasing each year (Figure 3). This is due to fluctuations of traders who sell in the market. At certain times, as at the approaching feast, the number of merchants increases. Increasing numbers of traders will increase market activity, so traffic and pedestrian traffic friction will increase the concentration of suspended particulate matters in the air. This situation is worsened by road repair activities around the market location, as happened in 2006 in the Maros Central Market, and the Central Market of Makassar in 2010.

In 2009, several tens of meters from the traditional Daya Market began to be built modern market. These development activities lead to increased concentrations of airborne dust. This situation makes the dust concentration to reach more than the air quality standard for dust that is 230 μg/m^3 [14].

Although the concentration of suspended particulate matters in the air is largely determined by the existence of development activities or motor vehicle traffic, but human traffic alone can also increase the concentration of suspended particulate matters in the air. Traditional markets that merchants only use the table or put merchandise on the ground very easily exposed to suspended particulate matters. As it was explained in the background, the size of the dust is vary in shape and size. The particle size ranges from >2 mm to <63 μm with approximately one third of the dust being smaller than 500 μm [4]. Atmospheric dust has a diameter of 0.001 – 40 micron [3]. Measured suspended particulate matters at the site has a diameter of ≥ 50 μm, which is a particle with a size large enough. Particles with diameter size of this magnitude only takes 4 - 5 seconds in the air from a height of about 2 meters [15]. The presence of these particulate matters in the air are shorter when the humidity of the air at the site is high, because the these particulate matters can absorb water that will form a bigger size [1, 13].

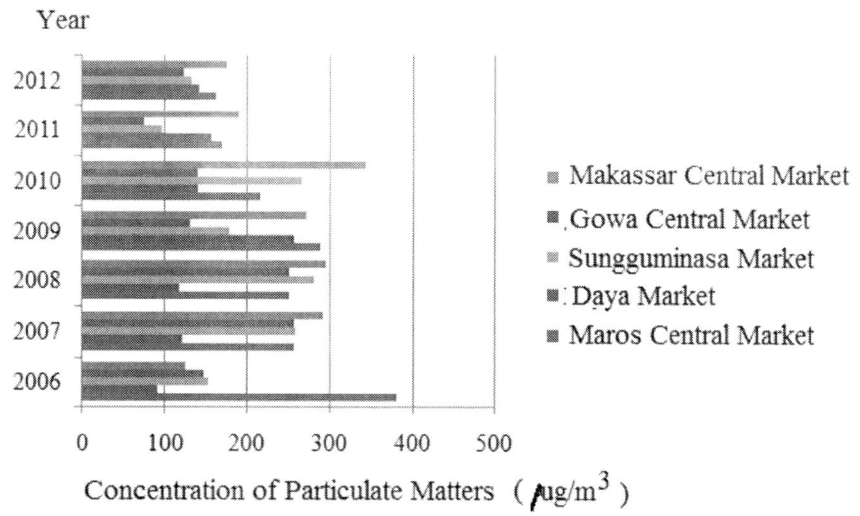

Figure 3: Fluctuations particle concentration in five traditional markets in Makassar in period of 2006 - 2012.

However, the particulate matters are not only derived from the land around where merchants are located but also come from motor vehicle emissions, so the particulate matters size reaches 2.5 μm. Particulate matters with this size can enter the alveoli that will cause respiratory problems [16]. This has been proved by Nurjazuli [17] who found a disturbance in respiratory function in street vendors in the city of Semarang.

This research, only measured the total concentration of suspended particulate matters that have a large enough particle size, while that is very influential on the respiratory system is a particle with a size of 2.5 μm. Therefore, this study cannot be used to infer the cause of respiratory system disorders in traders caused by exposure to suspended particulate matters. However, the results of this research indicate that the quality of suspended particulate matters contamination at the site has exceeded the permitted threshold, so the location is not good for trading. Therefore, local government should consider the relocation of the market or re-arrange the location of the market.

3. Conclusion.

Status of suspended particulate matters contamination in five traditional markets located in Makassar city, namely Daya Market, Maros Central Market. Sungguminasa Market, Gowa Central Market, and Makassar Central Market fluctuate every year. Sometimes it exceeds the threshold, sometimes below the threshold. Nevertheless, the concentration of dust particles often exceeds the threshold limit, which can disrupt the health of traders in those markets. Therefore, the local government should consider the relocation of the market or re-arrange the location of the market so as not to be directly on the side of the road.

References

[1] Bernstein J.A 2004 *J. Alergy Clin. Immunol.*, **114(5)** 1116-1123
[2] Kelly F.J 2015 *Environ.Geochem.Health.*, **37(4)** 631-649
[3] Engineering Toolbox *Particle Size*. https://www.engineeringtoolbox.com/particle-sizes-d_934.html
[4] Lanzerstorfer C 2017 *Journal of Environmental Science and Health, Part A Toxic/Hazardous Substances and Environmental Engineering* **52(8)** 770-777
[5] Volk H.E, Fred Lurnman, Bryan Penvold, Iva Hertz-Picciotto, Rob Mc.Connell 2013 *JAMA Psychiatry* **70(1)** 71-77
[6] Phalen R.F 2004 *Nonlinearity Bi.Toxicol.Med.* **2(4)** 259-292
[7] Rahul J., Manish Kumar Jain 2014 *Research Journal of Environmental Sciences.* **8** 356-372
[8] Azam A.G, Bamdad Riahi-Zanjani, Mahdi Balali-Mood 2016 *J.Res.Med.Sci.* **21** 65
[9] Center for Environmental Research Information Office of Research and Development U.S. Environmental Protection Agency 1999 *Compendium of Methods for the Determination of Inorganic Compounds in Ambient Air : Chapter 10-2 : Integrated sampling of suspended particulate matter (spm) in ambient air, overview*
[10] Dow R.B 1937 *Journal of Applied Physics* **8** 367.
[11] Montgomery R.B 1947 *Journal of Meteorology* **4** 193-196
[12] Kim K.H, Seung Bok Lee, Daekwang Woo, Gwi Nam Bae 2015 *Atmospheric Pollution Research* **6(6)** 1024-1034
[13] Cynthia Lin C.Y 2013 *Journal of Environmental* **4** 24-27
[14] Menteri Lingkungan Hidup. Peraturan menteri negara lingkungan hidup nomor 12 tahun 2010
[15] Aluko O., Kenneth E. Noll 2006 *Aerosol Science and Technology.* **40(7)** 503-513
[16] Lonnati G., Senem Ozgen, Giovanna Ripamonti, Stefano Cernuschi, Michele Giugliano 2011 *Journal of The Air & Waste Management Association* **61(11)** 1273-1280
[17] Nurjazuli., Onny Setiani 2010 *J.Kesehat.Masy.Indones.* **6(1)** 66-75

IOP Conf. Series: Journal of Physics: Conf. Series **979** (2018) 012042 doi:10.1088/1742-6596/979/1/012042

Distribution of Escherichia Coli as Soil Pollutant around Antang Landfills

Andi Artiningsih[1], Hazairin Zubair[2], A.M. Imran[3], Sri Widodo[4]

[1]Chemical Engineering, Faculty of Industrial Technology, University of Muslim Indonesia 90231, Indonesia
[2]Agricultural Sciences, Faculty of Agricultural Science, Hasanuddin University 90245, Indonesia
[3]Geological Engineering Department, Faculty of Engineering, Hasanuddin University 90245, Indonesia
[4]Mining Engineering Department, Faculty of Engineering, Hasanuddin University 90245, Indonesia

E-mail: srwd007@yahoo.com

Abstract. Tamangapa Antang Landfill locates around the residential area and faces an air and water pollution due to an open dumping system in its operation. The system arises a potential pollution in air, water and soil. Sampling was done surround the landfill in two parts, parallel and perpendicular to the ground water flow. This study shows the abundance of E. coli bacteria in soil around the Antang Landfills at depth of 10 to 20 cm ($93x10^5$ cfu/gr of soil) in the direction of groundwater flow. While in other locations the E. coli bacteria is not detected. The abundance of E. coli bacteria is a conjunction factor from landfill and human activities surround the area. The absence of E. coli bacteria in other location highly interpreted that the landfill is the major contributor of pollutant.

1. Introduction

Landfill for Makassar people is an important requirement to create environmental relieve in a big city. If the waste a day in Makassar city is not transferred to the Landfill, the city will be full of waste so it is very disturbing for the public convenience. The production of the waste in Makassar City is constantly and drastically increased [1]. Piles of the waste in the Landfill are the media for breeding of pathogenic and non-pathogenic microbes [2]. Tamangapa Antang Landfill is the center of all Landfills for all areas in Makassar City. Since the opening of Tamangapa Antang Landfill, it is estimated as many as 1.240.000 tons of inorganic and organic waste has been disposed [3]. In its operation, it is doing the Open Dumping way, so it is potentially contaminates the soil.

The location of Tamangapa Antang Landfill located in Bangkala Sub-District, Manggala District, Makassar City, South Sulawesi Province, has been used since 1995 until now with the area of 18.8 hectare. According to the planing, Tamangapa Landfill that was originally designed for the needs of 10 years, but in fact, until now the Landfill is still in use, it means that the age of the Landfill is 21 years, and cannot accommodate the volume of waste that is in Makassar City is reaching 800 tons or about 4,000 cubic per day. Based on the records of the Department of Hygiene and Environment, Makassar with an approximate total population of 1.3 million people, is producing about 3,800 m^3 or equal to 300 tons of public waste every day. Whereas the maximum capacity of Tamangapa Landfill is only about 2,800 m^3 for accommodate public waste every day. Additional landfill is needed for the disposal of 1000 m^3 of the over waste. About 87% of waste in Makassar City is organic waste and about 13% is inorganic waste, such as plastic and paper.

The impacts that can be caused by Tamangapa Antang Landfill are mainly on the shallow-groundwater quality. According to [1], the direction of groundwater flow carrying the leachate at Antang Landfill is in line with the slope of the rock layers, namely Northwest-Southeast. The pollution to shallow-groundwater is caused by leachate from the landfill, and can contaminate the wells of the nearby residents. This pollution has been felt by residents around the landfill, especially for those who use the groundwater as a source of drinking water. The leachate movement has spread to the Northwest-Southeast, about 300-450 meters from the Tamangapa Landfill following the flow of the groundwater [4].

E. coli as the pollutant of land around the landfill has been done, but in Indonesia especially in Makassar City, South Sulawesi Province is still lacking. The pollution in the landfill area is not only on water but also on the soil. One of the pollutants that can be detected on the soil is E. coli. Some previous studies only focused on leachate and air pollution. The purpose of this study is to determine how far the distribution of E. coli bacteria can contaminate the soil around the Antang Landfill. The study focuses on the distribution of E. coli bacteria as pollutant of the soil around TPA Tamangapa Antang Makassar, South Sulawesi Province, Indonesia, at Point 0 m for Line A, Line B, Line C and Line D with at the depth of 0 to 10 cm, 10 to 20 cm and 20 to 30 cm [5].

2. Research Methods

2.1 Research Location

The location of this study is administratively included in the Bangkala Sub-District, Manggala District, and Makassar City, South Sulawesi Province, Indonesia. Geographically, Tamangapa Antang Landfill is situated 119°29'10" to 119°29'40" East Longitude and 5°10'20" to 5°10'40" South Latitude. The location of the study area is shown in Figure 1.

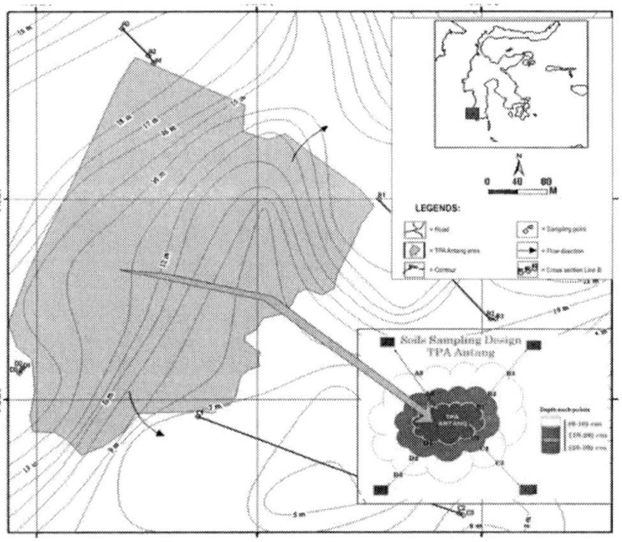

Figure 1. Map and location of sampling point Line A, line B, Line C and Line D at point 0 m at the depth of 0 to10 cm, 10 to 20 cm and 20 to 30 cm on land TPA Tamangapa Antang Makassar, South Sulawesi Province, Indonesia.

2.2. Sampling Point

The soil samples taken were the surface soil (Top Soil) around TPA Tamangapa Antang Makassar, at point 0 m at the depth of 0 to10 cm, 10 to 20 cm, and 20 to 30 cm [5], by dividing four Lines. Two Line are in the direction to the ground water flow, namely: Line A and Line C, and two other Lines are perpendicular to the direction of the ground water flow: Line B and Line D [4], (Figure 1). Twelve soil samples were collected in the study area and analyzed by use of biochemically.

Each sample was collected as many as 0.5 kg by using a stainless steel shovel. Then, the soil samples were put into sterile containers and labeled containing the location, date of taking, and the soil depth, then transported the laboratory, stored at 4°C for microbial analysis. When taking the samples, the soil samples are recorded about the soil type, the soil color, the soil conditions and the temperature [6, 7].

2.3. Microbiological analyses
In the laboratory, soil samples were mixed and the coarser particles removed, one gram of soil was suspended in 100 ml water and then serially diluted to five folds that were used for the microbial analysis. After solidification of media, they were incubated at 37°C for 24 h for bacteria, then colonies were counted and isolates were identified on the basis of cultural, microscopic, and biochemical characteristics with reference to Bergey's manual of systematic bacteriology for bacteria [8].

3. Result and Discussion

3.1. Result
To calculate the number of E. coli bacteria population is using Ec. Broat media with reference to Bergey's manual of systematic bacteriology for bacteria [8], (Figure 2). Mean total bacterial count was $20x10^5$ cfu/gr of soil, $93x10^5$ cfu/gr of soil and $23x10^5$ cfu/gr of soil on Line D. The result of calculating the population of E. Coli bacteria at point 0 m for Line A, Line B, Line C and Line D at the depth of 0 to10 cm, 10 to 20 cm and 20 to 30 cm [5], on land TPA Tamangapa Antang Makassar, South Sulawesi Province, Indonesia (Figure 2).

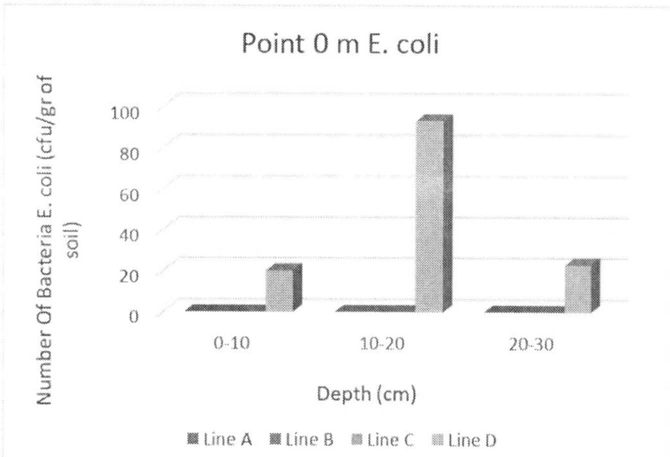

Figure 2. Number of E. coli bacteria (cfu/gr of soil) at point 0 m for Line A, Line B, Line C and Line D with depth from 0 to 10 cm, 10 to 20 cm and 20 to 30 cm on land TPA Tamangapa Antang Makassar, South Sulawesi Province, Indonesia.

Figure 2 show that the population of E. coli bacteria for Line A, Line B and Line C, at point 0 m at the depth of 0 to10 cm, 10 to 20 cm and 20 to 30 cm [5], the soil around TPA Tamangapa Antang Makassar, indicated that there is no bacterial growth. This is because Line A, Line B and Line C contain no soil organic compounds, no oxygen intake, and no nutrients, to support the survival of bacteria, so that the bacteria growth and reproduction cannot breed in the soil.

For Line D at the depth of 0 to10 cm is as many as 20 $x10^5$ cfu/gr of soil, at the depth of 10 to 20 cm increases as many as 93 $x10^5$ cfu/gr of soil, while at the depth of 20 to 30 cm begins to decrease to 23 $x10^5$ cfu/gr of soil. This is because at the depth of 0 to10 cm and 20 to 30 cm just contains little soil organic compounds, the intake of oxygen and nutrients are not sufficient to maintain the survival of the bacteria, so that the growth and reproduction of bacteria in the soil will be slow. Whereas, at the depth of 10 to 20 cm contains many soil organic compounds, get sufficient intake of oxygen and nutrients, and have a relative humidity, so that the growth and reproduction of bacteria in the soil will be very quick.

3.2. Discussion

From the results of the study, it can be concluded that the abundant population of E. coli, these results corroborate with findings [8], is on Line D, with the depth of 10 to 20 cm (93 x10^5 cfu/gr of soil), 20 to 30 cm (23 x10^5 cfu/gr of soil) and 0 to 10 cm (20 x10^5 cfu/gr of soil). This is because Line D is located close to the people environment, and is influenced by human activities. The soil around the landfill can be defiled bacteriologically through the permeation of public residual channels, such as septic tank. At the time of sampling in the rainy season, so that soil moisture and temperature conditions favorable for microbial activity to multiply rapidly [5].

Distribution of soil bacterial populations such as E. coli is influenced by several of environmental factors such as humidity, soil pH, temperature, light, soil moisture, and organic compounds [5]. The average soil pH of the landfill ranges from 4.82-6 (Acidic), with an average temperature ranges from (T = 25-36°C). At low pH, the availability of nutrients is so high that bacterial activity will increase. Furthermore the temperature also plays an important role [5], in controlling the metabolic reactions of all human beings like bacteria.

The layers of soil at the depth of 10 to 20 cm contain high soil organic compounds, sufficient humidity to provide an opportunity for microorganisms to decompose the complex organic residues into [5], the simpler ones, so that, the amount of the microbe is generally higher in the layer of soil under the surface, compared to the dry season, according to the studies of [5, 8]. The abundance of E. coli bacteria at the certain depth and the point around the landfill site indicates the effect of the humidity. Line D that is perpendicular to the direction of groundwater flow or leachate indicates that the development of E. coli is not affected by the direction of the leachate flow.

Bacteria E. coli can survive under different environmental conditions such as in soil environment, water bodies and manure [9, 10, 11], or the possibility of bacteria habitat displacement E. coli itself. Bacteria E. coli can survive on the surface soil layer (Top Soil), if in the environment there are plenty of adequate water intake [12].

Provided resource availability and key abiotic conditions are propitious, E. coli populations can survive and even grow in open environments. However, under fluctuating environmental conditions, such as those present in many soils and aquatic environments, growth may be differential and gross bacterial death may ensue if the death rate exceeds the growth rate. Both growth and death rates are determined by the environmental conditions at the local scale and by how the microorganism is able to cope with these local conditions by regulating its gene expression patterns [13].

The availability of carbon compounds is important because it affects bacterial growth E. coli and interconnected with the conditions of origin for the life of microorganism in the soil and in water bodies [14]. The combination of abiotic conditions such as availability of nutrients, soil pH, soil moisture and soil temperature and biotic conditions such as micro flora is a very good situation for bacterial populations E. coli to survive in both conditions [15, 16].

E. coli is a gram-negative, rod-shaped, 2.4 x (0.4-0.7) μm-sized, encapsulated, active and non-spore, and anaerobic facultative bacteria, has a peritrichous flagellum and can grow at optimum temperature of 37°C and maximum pH 9.0. E. coli is distinguished on its serologic characteristics based on (somatic) O, (Capsule) K and (Flagella) H [17].

E. coli bacteria when contaminate into the gastrointestinal tract in large quantities can endanger the health. Although E. coli is part of the normal microbe of the gastrointestinal tract, it is now proven that certain strains are capable of causing a temperate to severe gastroenteritis in humans and animals. E. coli bacteria can cause bloody diarrhea and invade the large intestine, and is present in feces that is filled with leukocytes and erythrocytes [18].

It has been traditionally assumed that E. coli shows natural declines in open environments. However, forming an environmental reservoir of a given size in different conditions, the bacterium can cause risks to human health. The factors that determine the rate of survival of particular E. coli strains, and thus the risks, are not (yet) easily predictable, and our capability to understand the effects of the secondary habitat (especially soil and water resources) on E. coli behavior will be paramount to our abilities to manage the organism from both environmental and public health perspectives [13].

4. Conclusion

The study results show that the population of E. coli bacteria in Antang Landfill soil had spread to the under layer of soil surface at the depth of 10 to 20 cm. The largest population of E. coli bacteria is found on Line D, with values of $93x10^5$ cfu/gr of soil.

Acknowledgments

We would like to thank the Head of Health Laboratory of Makassar, the Laboratory assistant, and the Head of soil Science Laboratory of Agriculture Faculty of Muslim University of Indonesia for all the facilities provided, for our togetherness, their support and assistance during our research. Our special thank goes to Rector of Muslim University of Indonesia and the dean of Faculty of Industrial Technology who have provided research fund and given support so that we can finished our research and prepared the manuscript. The authors also thank to editor and anonymous reviewers for their comments on the manuscript.

References

[1] Arifin M F, Imran A M, Ramli M and Hatta P M 2014 *Proc. Seminar Nasional Geofisika.* (Makassar)
[2] Nasution H I 2012 *J. Saintika* **12(2)** 165-169
[3] Peraturan daerah Kota makassar Nomor 6 tahun 2006. Tentang Rencana tata ruang Wilayah Kota Makassar 2005-2015
[4] Imran A M 2008. *J. Penelitian Enjiniring Fakultas Teknik Unhas* **12(2)** 187-410
[5] Das K, Nathh R and Azad P 2013 *Global Journal of Science Frontier Research Biological Science* **13(3)**
[6] Mc Grath D E, Mc Cormack R J 1999 *The significance of heavy metal and organik micropollutants in soil* ISBN 1 84170 27 4
[7] Mtunzi F M, Dikoi E D, Moja S J 2015 *Int. J. Of Environmental Monitoring and Analysis* **3(2)** 44-49
[8] Shah S A H, Ahmed K 2013 *Academia journal of Scientific Researc* **1(1)** 6-9
[9] Jiang X, Morgan J and Doyle M P 2002 *J. Appl Environ Microbiol* **68** 2605–2609
[10] Kudva I T, Blanch K, Hovde C J 1998 *J. Appl Environ Microbiol* **64** 3166–3174
[11] Vital M, Hammes F, Egli T 2008 *J. Environ Microbiol* **10** 2387–2396
[12] Mankin K R, Wang L, Hutchinson S L and Marchin G L 2007 *Transactions of the ASABE* **50** 1159–1165
[13] Van Elsas J D, Semenov A V, Costa R, Trevors J T 2011 *ISME J.* **5(2)** 173-18
[14] Franchini A G, Egli T 2006 *J. Microbiology* **152** 2111–2127
[15] Semenov A V, Franz E, van Overbeek L, Termorshuizen A J and van Bruggen A H C 2008 *J. Environ Microbiol.* **10** 1450–1459
[16] Van Elsas J D, Jansson J K, Trevors J T 2007 *Modern Soil Microbiology 2edn* (USA: CRC Press)
[17] Fatimawali, Kepel B, Yusuf I, Natsir R and Baharuddin F 2009 *J. Kedokteran Yarsi* **17(2)** 134-141
[18] Suharyono 2008 *Diare akut klinik dan laboratorik* (Jakarta: PT.Rineka Cipta)

IOP Conf. Series: Journal of Physics: Conf. Series **979** (2018) 012043 doi:10.1088/1742-6596/979/1/012043

The Effect of Deflector Angle in Savonius Water Turbine with Horizontal Axis on the Power Output of Water Flow in Pipe

Ari Prasetyo, Budi Kristiawan, Dominicus Danardono, Syamsul Hadi

Mechanical Engineering Department, SebelasMaret University.
Jl. Ir. Sutami 36 A, Surakarta, Central Java, Indonesia.
Jl. Mojo No 1 PO BOX 215 Surakarta, Central Java, Indonesia.

E-mail: syamsulhadi@ft.uns.ac.id

Abstract. Savonius turbine is one type of turbines with simple design and low manufacture. However, this turbine has a relatively low efficiency. This condition can be solved by installing fluid deflectors in the system's circuit. The deflector is used to direct the focus of the water flow, thus increasing the torque working moment. In this study, a single stage horizontal axis Savonius water turbine was installed on a 3 inch diameter pipeline. This experiment aims to obtain optimal deflector angle design on each water discharge level. The deflector performance is analyzed through power output, TSR, and power coefficient generated by the turbine. The deflector angles tested are without deflector, 20°, 30°, 40°, and 50° with a deflector ratio of 50%. The experimental results at $10.67x10^{-3}m^3/s$ discharge show that turbine equipped with 30° deflector has the most optimal performance of 18.04 Watt power output, TSR of 1.12 and power coefficient 0.127. While with the same discharge, turbine without deflector produces only 9.77 Watt power output, TSR of 0.93, and power coefficient of 0.09. Thus, it can be concluded that the deflector increases power output equal to 85%.

1. Introduction

Buildings drainage water has the potential for a picohydro scale power plant. However, its relatively small head and discharge conditions require a particular turbine design. The Savonius turbine is one of the drag type wind turbines which has a simple and inexpensive design to manufacture [1]. Based on this, Savonius turbine design is a solution to the problem. The water momentum which is greater than air is also a reason for the application of the Savonius turbine for water [2]. Drag turbines such as the Single Stage Savonius which is applied to water flow in a vertical pipe have produced the biggest power output [3]. Various methods have been undertaken to increase the output power provided by the Savonius turbine [4]. One method used to increase the turbine's power output is adjusting the fluid velocity to increase torque [5]. The fluid flowing within the pipe is directed by a steering mechanism to increase the turbine's efficiency [6].

The fluid flow direction arrangement has the principle to reduce the negative torque on the turbine convex surface and increase the positive torque on the turbine concave surface [7]. In this condition, the fluid flow will completely blade the concave blades, and the flow rate that will blade the convex blades of the turbine blade is inhibited by the steering. The angled design of the steering position is one of the important methods for obtaining fluid flow profile which will increase the torque. A simulated study has also been performed to obtain a profile image of the fluid direction that will blade the turbine blades so that the shape and direction of the fluid flow can be identified [8]. Furthermore, to obtain accurate data, simulated research and experiments on the design of deflector forms have been carried out over several generations [9]. In the last generation, 88.2 Watt power output result has been obtained, but this method resulted in a pressure drop of ±5m.

2. Research Method

The designs of the Savonius turbine have been studied in previous references to produce the best efficiency. Turbine design uses 5 blades. Moreover, the curve angle of the turbine blades applies an angle of 70° [10], as in the following Figure 1.

Figure 1. Turbine Design.

The Savonius turbine was manufactured by a 3D printing process in order to obtain the appropriate size of the designed model, see Figure 1. Turbine and deflector are installed by adjusting to the direction of turbine rotation so that it can give positive torque effect on the direction of turbine rotation. Data retrievals are rpm, electrical current, and electrical voltage. The rpm data is obtained from the tachometer reading, while the electrical current data and electrical voltage are obtained from the multimeter readings [11]. Figures 2 show the specimen of deflector variation for data retrieval.

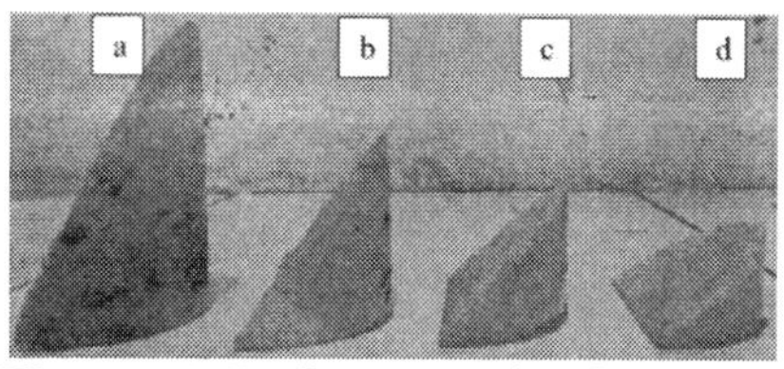

Figures 2. Deflector angle of (a) 20°, (b) 30, (c) 40°, and (d) 50°.

The study used test apparatus arranged with the working principle of moving the fluid from the bottom reservoir which is then pumped upward. The fluid flow from the upper tub is a simulation of rainwater and wastewater from a multi-story building. Water from the upper tub will flow to turn the turbine to produce electricity. The water will rotate the turbine that is connected directly to the generator. The amount of discharge is set by using the valve to get the correct data.

3. Results and Discussion

In this section will discuss and analyze the performance of each deflector of any variation of discharge. The data of turbine without deflector are used as references to compare the efficiency generated on each deflector. This study tested the angle of deflector 20°, 30°, 40°, and 50°. Then, each deflector is tested with four large variations of discharge. Each deflector will produce different input discharge. The amount of discharge produced by each deflector is shown in Figure 3 below.

IOP Conf. Series: Journal of Physics: Conf. Series **979** (2018) 012043 doi:10.1088/1742-6596/979/1/012043

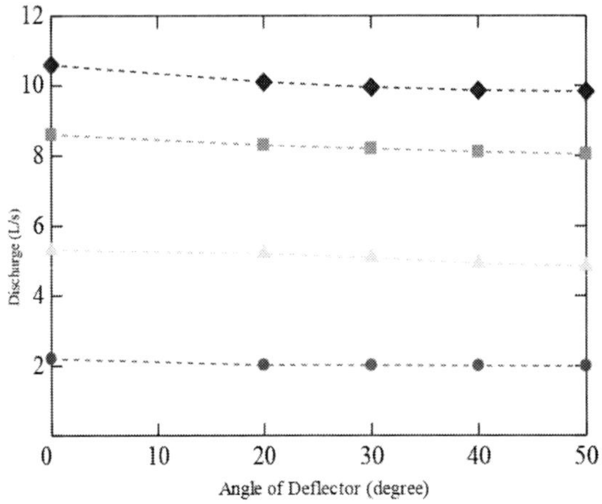

Figure 3. Relation between deflector angle and discharge.

Figure 3 shows a trend which illustrates that the discharge decreases as the deflector angle increases. This is due to the difference in the shape of the reflector resulting in a change in the direction of fluid flowThe turbine without deflector (0°) produces the greatest discharge which then decreases as the deflector angle reach 50°. This will affect the input power which will be generated at each variation of the deflector angle. The input power is the magnitude of the potential power of the fluid to generate electricity. Figure 4 shows the input power graph on each deflector.

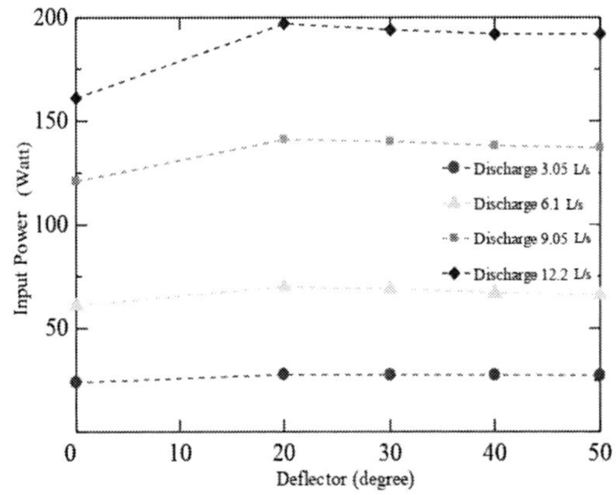

Figure 4. Relation between deflector angle and input power.

Figure 4 shows the amount of input power generated by each deflector. On each variation of the discharge, the 20° deflector always produces the highest input power. This is due to the slope plane of each deflector variation causes different fluid drag forces. In this condition, water experiences resistance due to the deflector. On the other hand, with the deflector, the water can be set to blade the concave turbine blades and block the water that will blade the convex turbine blades. This will increase the working torque on the turbine.

The deflector performance is analyzed based on the generated electrical power. The electric power observed is the multiplication of electrical current and electrical voltage. Figure 5 below shows the electrical power generated by the deflector on each variation of the discharge.

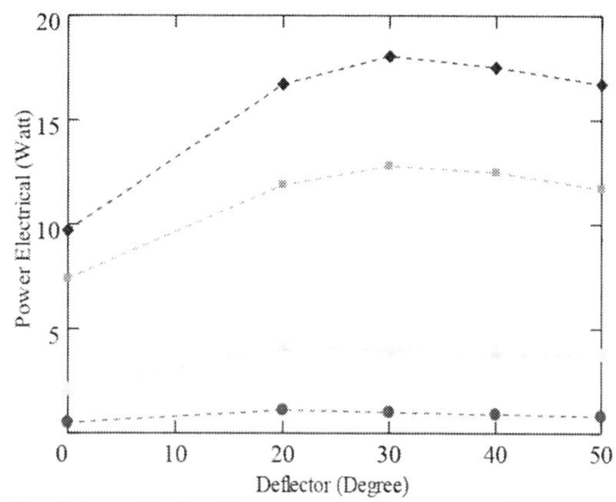

Figure 5. Graph of the relation between deflector angle and electrical power.

Figure 5 shows the trend of electrical power generated by the discharge variation for each deflector angle. Turbines without deflectors always produce the lowest power on any variation of discharge. This is caused by the fluid flow bladeing the concave and convex surface of the turbine blades, resulting in unbalanced turbine rotation. These conditions will affect the generated electrical power. At $3.05 \times 10^{-3} m^3/s$ and $6.10 \times 10^{-3} m^3/s$ discharge, the largest electric power is generated at 20° deflector angle with 1.18 Watt value and 4.13 Watt value. Then the electric power decreases as the deflector angle increases up to 50° of angle. This is due to the flow of water at the discharge which does not fill the pipe space so that it can be directed with a 20° deflector. While at $9.05 \times 10^{-3} m^3/s$ and $12.2 \times 10^{-3} m^3/s$ discharge, the largest power is generated by 30° deflector angle with a value of 12.65 Watts and a value of 18.04 Watts. This is because the 30° deflector can direct the water flow right at the turbine blades which produce the highest torque.

Power coefficient illustrates the actual power that the turbine can extract from the working fluid. While Tip Speed Ratio (TSR) is the value of comparison between blade's tip velocity and fluid velocity which is one of the important parameters to identify turbine performance.

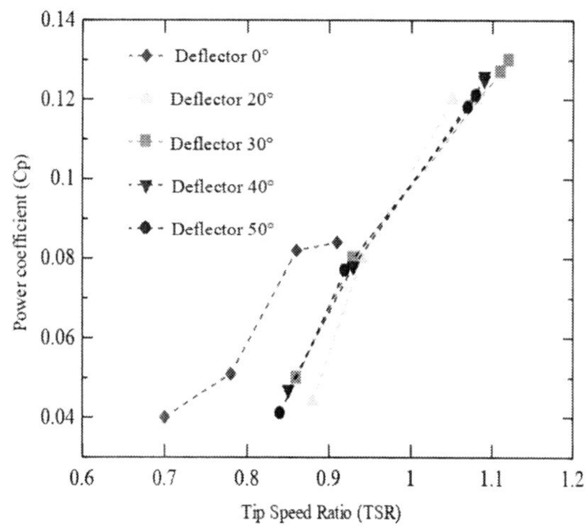

Figure 6. Graph of the relation between TSR and Cp.

Figure 6 shows the relationship between power coefficient and tip speed ratio in which the power coefficient increases as the increase of the tip speed ratio. The 20° deflector angle has the best performance on the $2.2 \times 10^{-3} m^3/s$ and $5.1 \times 10^{-3} m^3/s$ discharge, while the 30° deflector angle has the best performance on $8.5 \times 10^{-3} m^3/s$ and 10.8×10^{-3} m^3/s discharge. Savonius turbine with deflector has significant power increase when compared to turbine without the deflector. This proves that, with the deflector installation, the turbine performance increases [7]. This turbine will produce greater electrical power output than turbine without the equipped deflector.

4. Conclusion

We have investigated the effect of deflector Angle in Savonius Water Turbine with Horizontal Axis on the power output. Based on the experiment, the deflector installation decreases discharge until of 5.8%. Furthermore, the deflector angle produces different power at the various discharge. Turbine without deflector produces power output of 9.77 Watt only, but using installed deflector of 30° the power output increases up to 84.6% of 18.04 Watt with 10.75 $x10^{-3} m^3/s$ discharge and 2 meter head as the best performance of the turbine.

References

[1] Golecha K, Eldho T I, Prabu S V 2011 *Int. J. Applied Energy, Elsevier Inc.* **88** 3207-3217
[2] Biswas A, Sarma N K, Misra R D 2014 *Int. J. Energy Conversion and Management, Elsevier Inc.* **83** 88-98
[3] Norzanah R, Jauhari A S, Mustaamal A H, Husin F, Hasan M Y 2015 *J. Energy Procedia* **68** 274 – 281
[4] Saini A K R P 2016 *Int. J. Renewable and Sustainable Energy Reviews* **64** 289–310
[5] Yaakob, Ismail O M A and Ahmed Y M 2013 *Parametric Study for Savonius Vertical Axis Marine Current Turbine using CFD Simulation* **3** 200-205
[6] Rus L 2012 *Journal of Sustainable Energy* **3** 128-130
[7] Kailash G, Eldho T I, Prabu S V 2012 *Hindawi Publishing Corporation International Journal of Rotating Machinery* **10** 1155-1167
[8] El-Askary W, Nasef M H, Abdel-Hamid A A, Gad H E 2015 *J. Wind Eng. Ind. Aerodyn.* **139** 8-15.
[9] Chen J, Yang H X, Liu C P, Lau C H, Lo M 2013 *Int. J. Energy, Elsevier Inc.* **54** 184-193
[10] Raffiudin M, Faizal M and Lee Y H 2013 *Int. J. Ocean Engineering, Elsevier Inc.* **65** 32-38
[11] Hadi S, Apdila R J, Purwono A H, Budiana E P, Tjahjana D D D P 2017 *AIP Conference Proceedings*. 1788,030004

Effect of Blade Curvature Angle of Savonius Horizontal Axis Water Turbine to the Power Generation

Taufan Apha Sanditya, Ari Prasetyo, Budi Kristiawan, Syamsul Hadi

Mechanical Engineering Department, SebelasMaret University.
Jl. Ir. Sutami 36 A, Surakarta, Central Java, Indonesia.
Jl. Mojo No 1 PO BOX 215 Surakarta, Central Java, Indonesia.

E-mail: syamsulhadi@ft.uns.ac.id

Abstract. The water energy is one of potential alternative in creating power generation specifically for the picohydro energy. Savonius is a kind of wind turbine which now proposed to be operated utilizing the energy from low fluid flow. Researches about the utilization of Savonius turbine have been developed in the horizontal water pipelines and wave. The testing experimental on the Savonius Horizontal Axis Water Turbine (HAWT) by observing the effect of the blade curvature angle (ψ) of 110°, 120°, 130°, and 140° at the debit of 176.4 lpm, 345 lpm, 489.6 lpm, and 714 lpm in order to know the power output was already conducted. The optimal result in every debit variation was obtained in the blade curvature angle of 120°. In the maximum debit of 714 lpm with blade curvature angle of 120° the power output is 39.15 Watt with the coefficient power (Cp) of 0.23 and tip speed ratio (TSR) of 1.075.

1. Introduction

Indonesia needs enormous energy to meet its energy requirements for supporting high economic growth. Until now, fossil power plants still supply most of the electricity needs in Indonesia. Petroleum still ranks highest as energy source supplier at 51.66%, natural gas ranks second at 28.57%, and the rest is provided from oil energy of 15.34%, and renewable energy 4.43%. However, Indonesia has the greatest renewable energy potential (EBT), namely water energy (hydro) of 75,670 MW. Nevertheless, the utilization only reached 4,200 MW or about 5% of the existing potential [1]. Rainwater is one of a potency to create a pico-hydropower plant. Pico-hydro is a hydroelectric power plant with electric power generated no more than 5kW. Experiments for rainwater harvesting (RWH) using Savonius single stage turbine type have done. The result is that it can turn on LED lights of 0.3 Watt. The power plant with this RWH system depends heavily on the efficiency of blade rotation capability [2]. Research for the utilization of water energy also had been done on Vertical Axis Water Turbine (VAWT) with drag type applied to the flow in the pipe to generate electrical power as a monitor supply in controlling water quality in the tube. Maximum power achieved is 88.2 W with a water speed of 1.5 m / s and pressure drop less than 5 m [3].

Efforts to improve turbine efficiency continued with some research. Tunisian scientists conducted a study of the angle of blade curvature with numerical modeling using software SolidWorks to obtain the angle of optimal blade curvature. The study was carried out by observing the increasing effect of the blade curvature angle (ψ) against the depression zone and the acceleration zone [4]. Another study observed the curvature of the Savonius turbine blade with the angular variation of the blade angle (ψ) 60°, 65°, 70°, 75°, 80°, 85°, and 90° tested in the wave channel. The results show that the Savonius turbine with the angle of curvature of the 70° blade is the most optimal [5]. Experiments by modifying

IOP Conf. Series: Journal of Physics: Conf. Series **979** (2018) 012044 doi:10.1088/1742-6596/979/1/012044

the Savonius turbine without the shaft and with the end plate have also been performed, to increase the power coefficient and to obtain static torsional uniformity [6].

2. Research Method

The savonius turbine was designed by 5 number of blades [5] aspect ratio (*H/D*) of 1, and the end plate parameter (*Do/D*) of 1.1 [7]. The diameter of the turbine (D) is 82 mm while the thickness of the blade is 2 mm. The curvature angle is designed using the principle of arc angle in the circle as shown in Figure 1.

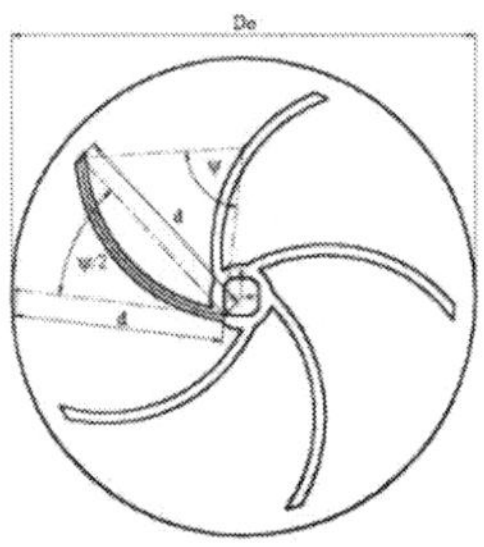

Figure 1. Schematic design of turbine blade curvature angle.

Savonius turbine made by a 3D printing method as the fastest prototyping techniques and most precision manufacturing process for complicated shape.

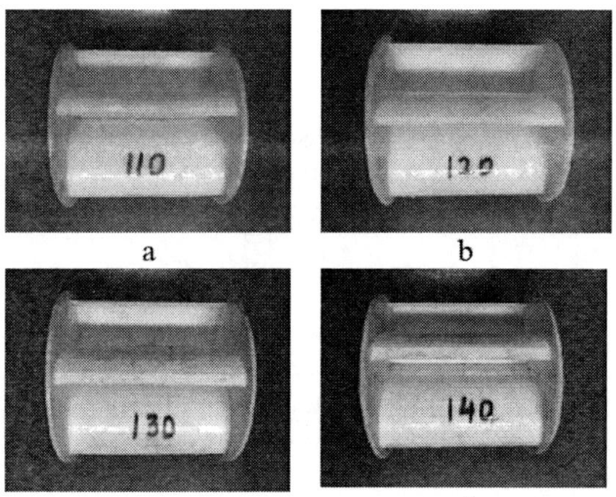

a b

c d

Figures 2. Savonius turbine (a) Blade arc angle 110 (b) Blade arc angle 120 (c) Blade arc angle 130 (d) Blade arc angle 140

3. Result and Discussion

This section will discuss the performance of turbine savonius. Turbine performance is evaluated from Cp, TSR, and output power based on the same input. The input power is the power generated by water due to altitude and discharge. Maximum head height is 2 m, and maximum flow rate (*Q*) is 714 lpm obtained from experimental data. Power input calculated by density air, gravity, head, flow rate. The power of the input power (*P*i) of each of the incoming fluid flow rate is shown in Figure 3 as follows.

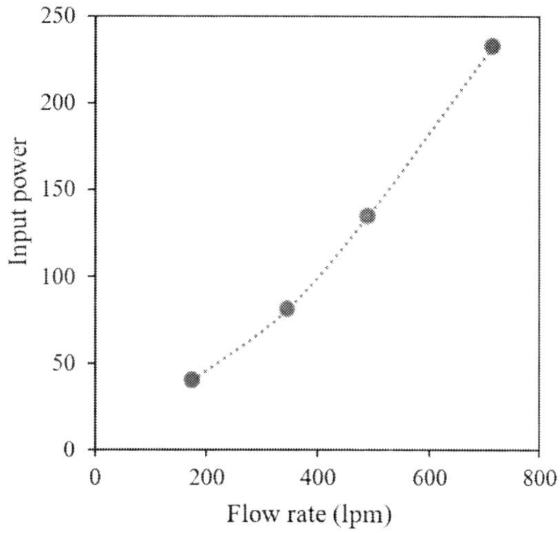

Figure 3. Graph of the ratio of the input power to the flow rate of water.

Figure 3 above illustrates the amount of input power of each flow rate. At the flow rate of 176.4 lpm, 345 lpm, 489.6 lpm, and 714 lpm, the input power was 40.23 Watt, 81.51 Watt, 135.6 Watt and 233.36 Watt, respectively. The smallest input power is obtained at a flow rate of 176.4 lpm, while the highest input power is achieved at a flow rate of 714 lpm. From the results obtained, it appears that the greater the amount of water entering the greater the input power by water.

Tip speed ratio is the tangential velocity at the tip of the blade against the actual speed of the fluid, while the power coefficient is defined as the ratio of the comparison between actual power generated by the rotor blade with the power acting on the fluid. The graph of tip speed ratio (TSR) effect on power coefficient (Cp) can be seen in Figure 4 below.

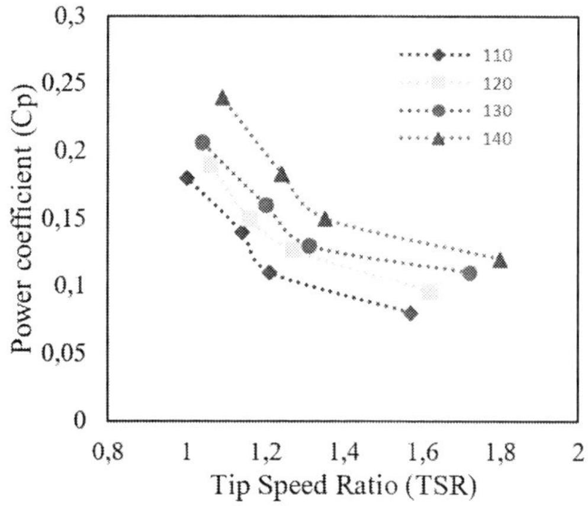

Figure 4. The relationship between TSR to Cp at the curvature angle of 110° to 140°.

Figure 4 describes that in all angle variations of the blade curvature decreases the value of *Cp* followed by the increase in TSR value. The decline in *Cp* value is due to the smaller flow rate change. Debit variation is very influential on the value of *Cp* because if the flow rate is low, then the value of Cp will be low while if the flow rate is high, then the value of *Cp* will be high. Figure 4 shows that maximum *Cp* obtained at a 120° angle of blade curvature is 0.23 with indigo TSR 1.07 whereas minimum Cp value is obtained at 110° of blade curvature angle that is 0,081 with a value of TSR 1.55. Optimal rotor performance can be seen from the amount of electric power output generated by the alternator. Figure 5 below will explain the magnitude graph of the output power produced by each angle of the blade curvature variation.

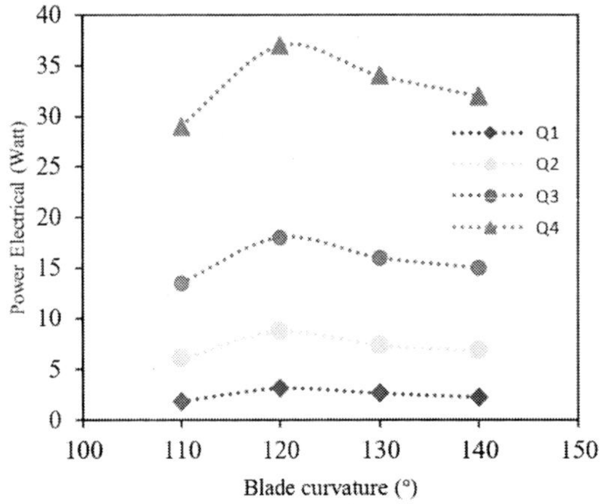

Figure 5. Graph of the relationship between the blade curvatures with power electrical.

Based on Figure 5 it can be seen that the angle of the blade curvature affects the electrical energy output produced by the Savonius turbine. The result of the trend between the angle of the blade curvature with the electric power output at each variation of the flow rate is the same as the magnitude of the electrical energy output at each angle of the blade curvature has increased according to the addition of the amount.

On Figure 5, the trend result shows that along with the increased angle of the blade curvature is not directly proportional to the increase in output power generated, where the angle of the blade curvature has an optimum point at the angle of the particular blade curvature. At a flow rate of 714 lpm, the electric power output value of the blade curvature angle of 110 ° is 29.9 Watt, increased at 120 ° blade curvature angle with an output power of 39.15 Watt, then decreased until blade curvature angle of 140 ° with an electrical energy output of 31.52 Watt. This result is in accordance to research conducted [6] with the angle curvature 110°, 124°, 135°, and 150° obtained that at the angle 124° gives the effect of the most power coefficient. Other studies corresponding to the data obtained were carried out by [8], where the angle variations of blade curvature that studied were 150°, 160 °, 170 °, 180 °, 190 °, and 200 °. That, the highest positive torque on the blade can be increased by reducing the ellipticity blade, whereas the lowest negative torque is to increase the elliptic blade, where turbines with a blade curvature angle of 160° generate a maximum coefficient of power (Cp).

4. Conclusion

Effect of the blade curvature angle in the Savonius Horizontal Axis Water Turbine (HAWT) were already investigated and discussed. The highest electric power output of 39.15 Watt was obtained at blade curvature angle of 120° at maximum discharge of 714 lpm the with power coefficient (Cp) and tip speed ratio (TSR) of 0.23 and 1.075 respectively.

References

[1] Yuningsih A and Masduki A 2011 *J. Ilmu dan Teknologi Kelautan Tropis* **3(1)** 13-2.
[2] Norzanah R, Jauhari A S, Mustaamal A H, Husin F, Hasan M Y 2015 *J. Energy Procedia* **68** 274 – 281
[3] Chen J, Yang H X, Liu C P, Lau C H, Lo M 2013 *Int. J. Energy, Elsevier Inc.* **54** 184-193
[4] Driss Z, Mlayeh O, Driss S, Driss D, Maaloul M, and Abid M S 2015 *J. Elsevier Inc. Energy.* **89** 708-729
[5] Ahmed M R, Faizal M and Lee Y H 2013 *J. Elsevier Inc. Ocean Engineering.* **65** 32-38

[6] Kamoji M A, Kedare S B and Prabhu S V 2009 *J. Elsevier Inc. Applied Energy.* **86** 1064-1073
[7] Saha U K and Roy S 2015 *J. Elsevier Inc. Applied Energy.* **137** 117-125
[8] Mao Z and Tian W 2015 *J. Advances in Mechanical Engineering.* **7(5)** 1-10
[9] Hadi S, Apdila R J, Purwono A H, Budiana E P, Tjahjana D D D P 2017 *AIP Conference Proceedings.* 1788,030004

IOP Conf. Series: Journal of Physics: Conf. Series **979** (2018) 012045 doi:10.1088/1742-6596/979/1/012045

Determination Hypocentre and Focal Mechanism Earthquake of Oct 31, 2016 in Bone, South Sulawesi

Muhammad Altin Massinai, Muhammad Fawzy Ismullah M

Geophysics Dept., Hasanuddin University

E-mail: altin@science.unhas.ac.id

Abstract. Indonesian Meteorology, Climatology and Geophysics Agency (BMKG) recorded an earthquake with M4.6 on at October 31, 2016 at Bone District, around 80 Km northeast form Makassar, South Sulawesi. The earthquake occurred 18:18:14 local time in 4.7^0S, 120^0E with depth 10 Km. Seismicity around location predicted caused by activity Walennae fault. We reprocessed earthquake data to determine precise hypocentre location and focal mechanism. The P- and S-wave arrival time got from BMKG used as input HYPOELLIPSE code to determine hypocentre. The results showed that the earthquake occurred 10:18:14.46 UTC in 4.638^0S, 119.966^0E with depth 24.76 Km. The hypocentre resolved 10 Km fix depth and had lower travel time residual than BMKG result. Focal mechanism determination used Azmtak code based on the first arrival polarity at earthquake waveform manually picked. The result showed a reverse mechanism with strike direction 38^0, dip 44^0, rake angle 134^0 on fault plane I and strike direction 164^0, dip 60^0, rake angle 56^0 on fault plane II. So, the earthquake which may be related to a reverse East Walennae Fault.

1. Introduction

Monday October 31, 2016, tectonic earthquake occurred at Bone, South Sulawesi. The analyzed result from BMKG (Indonesia Meteorology, Climatology and Geophysics Agency) showed that earthquake occurred on 18:18:14 LT (Central Indonesia Time) with M4.6 and epicentre located at 4.7^0S, 120^0E with 10 km focus depth, 80 km eastern Makassar. The earthquake impact felt in many areas like Makassar, Soppeng, Bone, Camming, Watanlamuru, Watampone, Mare', Soreang, Maros, Mandai, Campae and Rala on intensity scale II SIG-BMKG (III MMI). Citizen of this area reported run out to rescue [1].

The epicentre located at Walennae depression, which located between West Walennae Fault (WWF) and East Walennae Fault (EWF) as part of Walennae Fault. Walennae Fault is reverse fault [2]. WWF is a sinistral strike-slip fault and EWF as a reverse fault with a dextral component of slip. This area covered by Camba volcanic and sandstone Neogene, volcanic and Salokalupang fm (sedimentary and volcaniclastic rock) Paleogene and plutonic and metamorphic rock, volcanic rock, sedimentary rock [3].

The first step to study about seismicity and his relations to fault is hypocentre determination. Hypocentre required to study structure, tectonic regime and processes that trigger seismic activity [4]. Hypocentre accuracy depends on factors, one of them is small travel time residual. Many techniques had developed to determine hypocentre become precisely. One of them is Geiger's method [5]. The basic of this method are minimizing travel time residuals between observation data and calculated travel times by calculating using distant seismic networks by employing spherical earth travel time [6].

IOP Conf. Series: Journal of Physics: Conf. Series **979** (2018) 012045 doi:10.1088/1742-6596/979/1/012045

The second step are moment tensor analysis. Moment tensor described by focal mechanism, is method to know tectonic force regime which caused earthquake.

This research used seismogram data onto BMKG stations on Sulawesi. The aim at this research is to get more geological information on earthquake analysis: hypocentre determination and focal mechanism.

2. Data and Methods

We used HYPOELLIPSE [7, 8] code to determine hypocentre using arrival time data from BMKG. HYPOELLIPSE code locates a hypocentre based on Geiger's method [5] which minimizes travel time residuals between observation data and calculated travel times [6]. This code also allows to calculate using distant seismic networks by employing spherical earth travel time tables [9]. We used IASP91 velocity model was adopted from [10].

Focal mechanism of the 2016 Bone earthquake was determined using Azmtak [11] code employed first arrival polarity data. We determined first arrival polarity data from seismogram by manual picking. A total of six stations were used to calculate the focal mechanism.

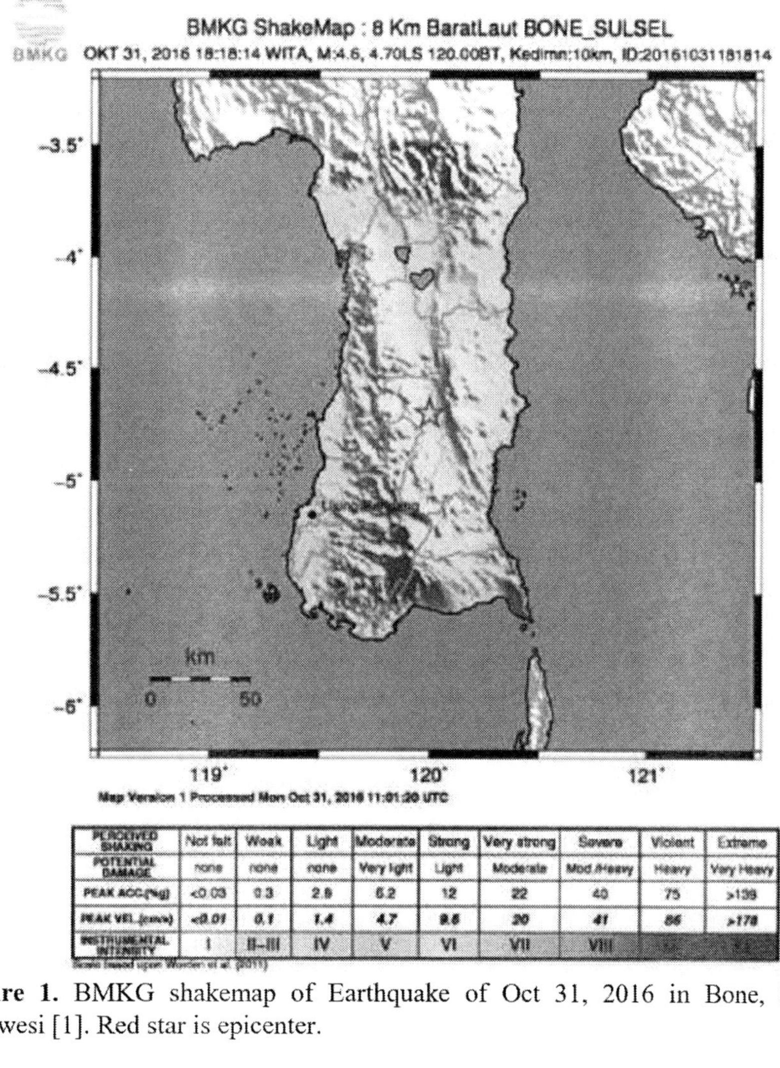

Figure 1. BMKG shakemap of Earthquake of Oct 31, 2016 in Bone, South Sulawesi [1]. Red star is epicenter.

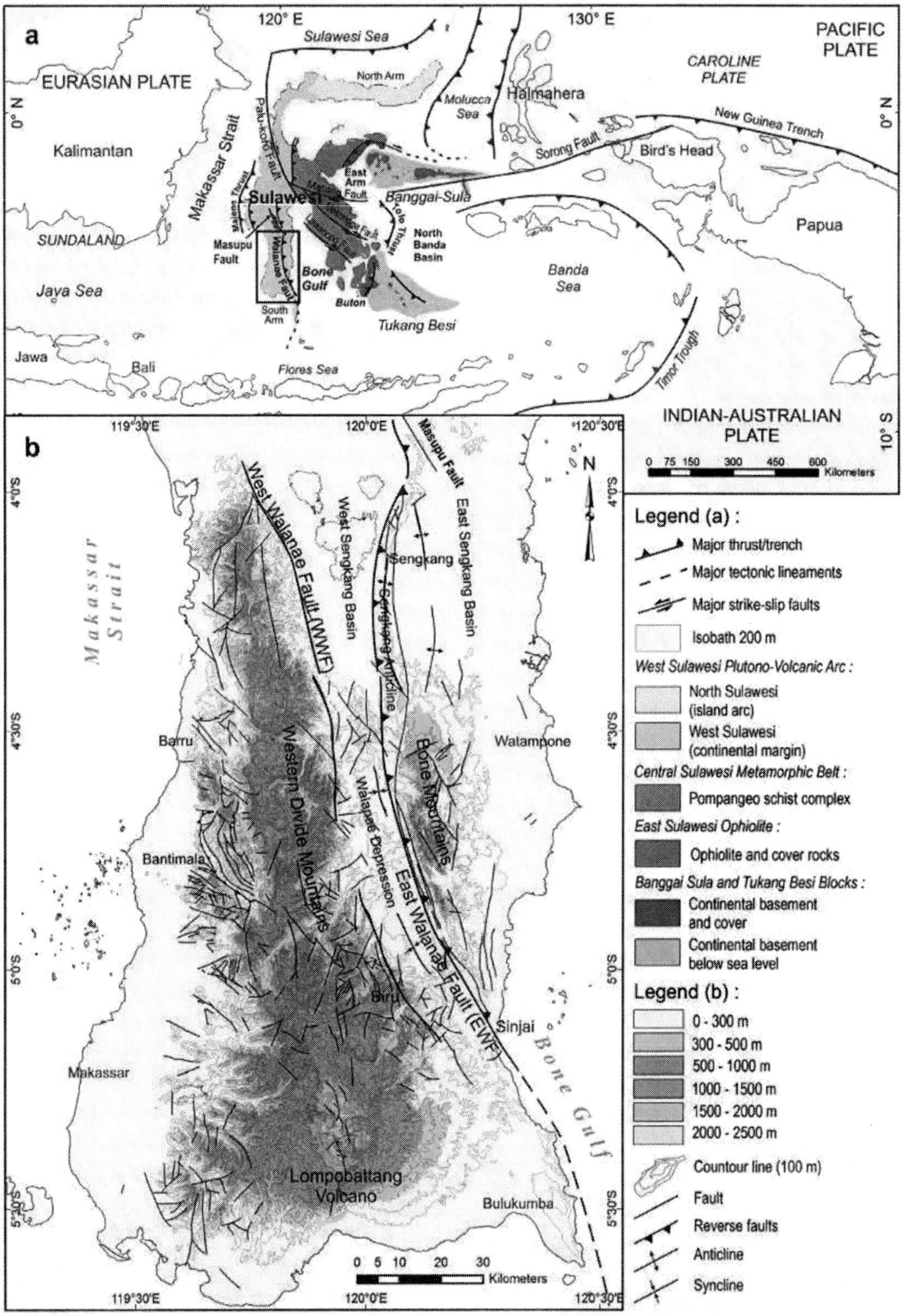

Figure 2. (a) Tectonic map showing the distribution of plates, continents and island arcs in the eastern part of Indonesia. The area of the rectangle corresponds to South Sulawesi shown in (b). (b) Structural and topographic map of South Sulawesi [2].

3. Results and Discussion

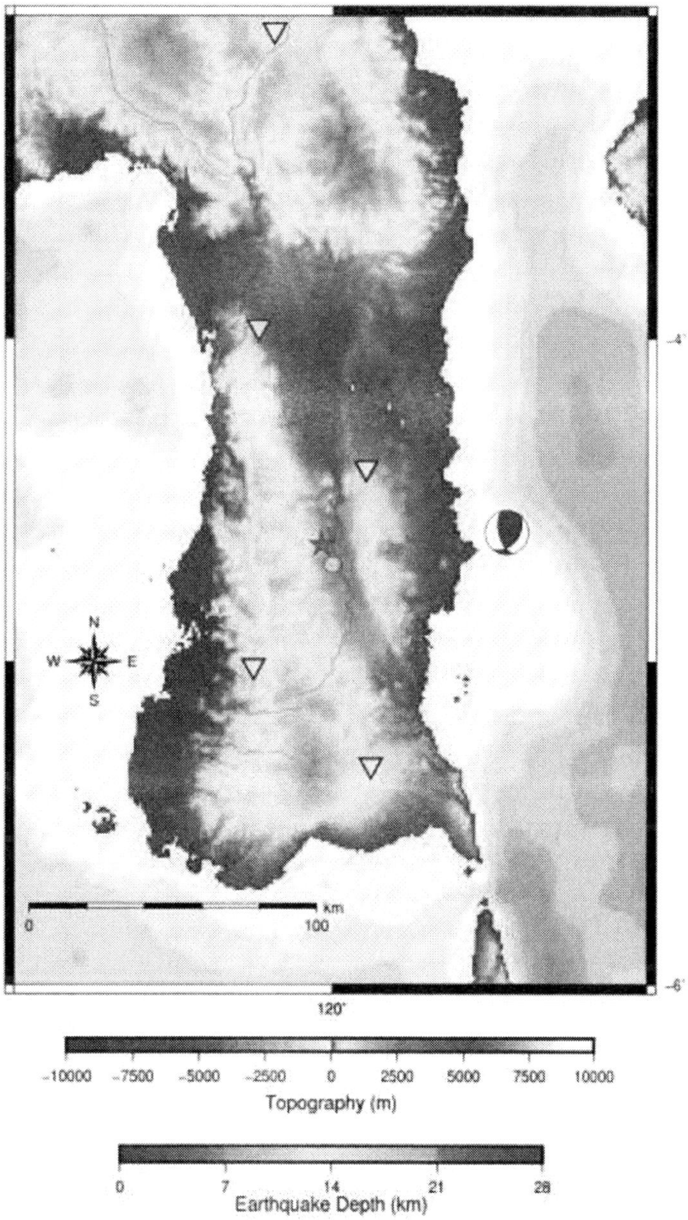

Figure 3. BMKG station (yellow triangle), Epicentre of BMKG (orange circle) and result of this research (blue star).

Hypocentre of the Bone earthquake based on our inversion is 4.638^0S, 119.9663^0E with 24.76 km focus depth (figure 1). Our location is located 7.85 km northwest of BMKG location. Furthermore, our inversion is able to resolve the fix-depth. Our location showed lower travel time residual, 0.1385 ms, compared to the BMKG result, 1.318 ms. It showed that our hypocentre is more accurate because able to resolve the fix-depth and have lower travel time residual.

Focal mechanism solution showed a reverse mechanisms with strike direction 38^0, dip 44^0 and rake angle 134^0 on fault plane I and strike direction 164^0, dip 60^0 and rake angle 56^0 on fault plane II. This result agree with [3] that said the EWF is reverse fault. We chose strike direction 164^0 based on fault feature of the topography map of South Sulawesi.

That is the last earthquake in several years later. This can give information about EWF activity nowadays. The tectonic activity still active until now in this area. EWF dimension can be assumed have activity on 24.76 km. It means activity of EWF is very shallow and can be affected on human activity in surface. It's possible if EWF activity can make other bigger earthquake in shallower depth. The direction of EWF signed tectonic force activity that created characteristic of EWF.

There was collision of East Sulawesi with the Banggai-Sula and Buton-Tukangbesi blocks, includes thrusting in the northern part of Sulawesi Island, and folding and uplifting in some parts of Bone Basin during the Pliocene. The convergence directions of the micro-continents have persisted in an approximately E-W direction at a high angle to the Walennae fault systems since the Late Miocene. This collision event could have affected the South Sulawesi and activated the EWF [3].

This result showed the activity of EWF caused this earthquake, although this location is in Walennae depression. Sandstone and other sedimentary rock made damage to this earthquake spread into surrounding EWF, but volcanic rock from Bone mountain and Walennae depression can be solved hold the shaking. So, the government and citizen must carefully planning city development around Walennae depression.

4. Conclusions

The Bone earthquake with magnitude 4.6 occurred on October 31, 2016 and located 4.638^0S, 119.9663^0E with 24.76 km depth. The mechanism of Bone earthquake is reverse with strike 38^0, dip 44^0 and rake angle 134^0 on fault plane I and strike direction 164^0, dip 60^0 and rake angle 56^0 on fault plane II. We suggest that this earthquake is may be related to EWF.

Acknowledgements

We gratefully acknowledge the BMKG for arrival time and waveform data. We thank to geophysics dept., Faculty of Mathematics and Natural Science Hasanuddin University (UNHAS) for supporting this research. Figures in this manuscript were created using Generic Mapping Tools [12].

References

[1] BMKG 2016 *Gempabumi Darat Guncang Makassar, Dipicu Sesar Aktif* www.bmkg.go.id

[2] Massinai M F I, Lantu, Aswad S and Massinai M A 2014 *AIP Conf. Proc.* **1658** 030013-1 – 030013-10

[3] Jaya A and Nishikawa O 2013 *Journal of Structural Geology* **55** 34-49

[4] Massinai M F I, Nugraha A D, Ramdhan M and Wandono 2017 *IOP Conf. Series: Earth and Enviromental Science* **62** 012056

[5] Lahr J C 1979 HYPOELLIPSE: A Computer Program for Determining Local Earthquake Hypocentral Parameters, Magnitude and First Motion Pattern *Open-le Report* 79-431 US Geological Survey Menlo Park California 310

[6] Lahr J C 1999 HYPOELLIPSE: A Computer Program for Determining Local Earthquake Hypocentral Parameters, Magnitude and First Motion Pattern (Y2K Compliant Version) *Open-File Report* 99-023 (US Geological Survey Golden Colorado paper and on-line editions) 112 pp

[7] Geiger L 1912 *Bull. St. Louis. Univ.* **8** 60-71

[8] Lahr J C and Snoke J A 2002 *Handbook of Earthquake and Engineering Seismology* Chapter 85 7

[9] Nugraha A D, Supendi P, Shiddiqi A H, Widiyantoro S 2016 *AIP Conf. Proc.* **1739** 020001-1 – 020001-4

[10] Kennett B L N and Engdahl E R 1991 *Geophys. J. Int.* **105** 429-465

[11] Suetsugu D 1997 Source Mechanism Practice *Lecture notes International Institute of Seismology and Earthquake Engineering, Tsukuba, Japan*

[12] Wessel P, Smith W H F, Scharroo R, Luis J and Wobbe F 2013 *EOS Trans. American Geophys. Union* **94** 45 409 – 410

Shallow Depth Study Using Gravity & Magnetics Data in Central Java - Yogyakarta

Muhammad Fawzy Ismullah M, Muhammad Altin Massinai, Maria

Geophysics Dept., Hasanuddin University

E-mail: altin@science.unhas.ac.id

Abstract. Gravity and magnetics measurements carried out in Karangsambung – Bayat – Wonosari track, Central Java – Yogyakarta region as much as 34 points for subsurface identification. Modeling and interpretation using both data at 3 sections. Section A lies on Karangsambung area and reach to 1900 m. Section A showed formation of 0.000001 – 0.0014 nT and 2.00 – 2.80 g/cm^3 like alluvium, basalt and tuff. Section B lies on Wates - Yogyakarta area and reach to 1700 m. Section B showed formation of (-0.01) – 0.02 nT and 2.40 – 3.00 g/cm^3 like andesite intrusive and Merapi volcano sediments. Section C lies on Bayat - Wonosari area and reach to 2000 m. Section C showed formation of 0.00016 – 0.0005 nT and 2.30 – 3.14 g/cm^3 like limestone, tuff and diorite intrusive. Based on modeling results from 2D structure inversion method can identify the formation of sediments from volcano activity on Karangsambung – Bayat – Wonosari track, Central Java – Yogyakarta region. The method of this study shows potential application for identify the formation of volcano activity from 2D structure.

1. Introduction

Gravity and magnetics measurements carried out along Karangsambung – Bayat – Wonosari. Gravity or/ and magnetics are two methods which usually used together to identify natural resource (like identify geothermal [1]), geo-engineering and environment (like detection of underground cavities [2]) or study general geology condition (like identify Makassar subsurface condition [3]). The gravity and magnetics methods are geophysical method which used potential field on subsurface to identify physical properties. Gravity surveying measures variations in the Earth's gravitational field caused by differences in the density of subsurface rocks [4]. Magnetics is generally more complex and variations in the magnetic field are more erratic and localized [5]. The advantages used gravity and magnetic together is to get precisely about subsurface object with depth and dimension.

Karangsambung – Bayat – Wonosari included Southern Mountain Zone in Eastern Java. The geology condition on this area is chaotic caused by volcano activity in the past. The aim at this study is to identify general geology subsurface condition on research area. The method of this study shows potential application to identify the formation of volcano activity from 2D structure.

2. Regional Geology

Karangsambung – Bayat – Wonosari lies on Central Java and Yogyakarta region, from geology condition included Eastern Java. The Eastern Java based on structure and stratigraphy [6] divided into four tectonostratigraphic zone (figure 1). There are, from south to north: (1) Southern Mountain Zone,

(2) Present-day volcanic arc, (3) Kendeng Zone and (4) Rembang Zone. The research area lies on Southern Mountain Zone.

This zone is volcano Eocene – Miocene arc which his sediments composed siliclastic, volcanoclastic, volcano and carbonates with dip to southern. Southern Mountain Zone based by Cretaceous bed rock like outcrop in Karangsambung and Bayat. Outcrop in Karangsambung covered accretion complex rock called Luk Ulo Melange Complex, form phyllite, blue schist, eclogite, ultramafic, ophiolite, basalt, calcilutite, chert planted at shale matrix. In Bayat, bed rocks outcrop covered phyllite, schist and marble. The oldest sediment rocks which sediment above angular unconformity covered conglomerate with bed rock and sandstone fragment like at Middle Eocene Nanggulan Formation and Wungkal-Gamping Formation. Above conglomerate and quartz sandstone, there are transgression sediments covered coal, sandstone and siltstone. Volcano material Southern Mountain Zone has to increase northern whereas bed rocks material proportion decreased. The sediment thickness until 100 m with limited outcrops in western, are in Karangsambung (Karangsambung Formation), Nanggulan (Nanggulan Formation) and Bayat (Wungkal-Gamping Formation).

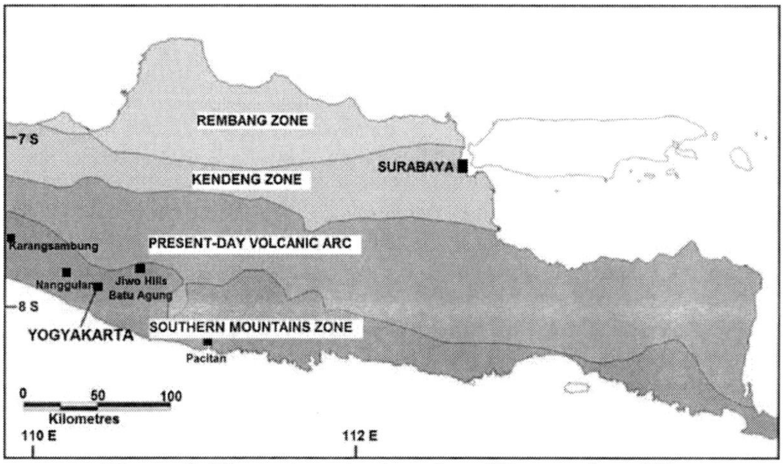

Figure 1. The tectonostratigraphy of Eastern Java [6].

Above that unconformity settled sediment series composed volcanoclastic sediment from Kaligesing Formation in Kulonprogo, Kebobutak in Bayat and Besole and Mandalika Formation in Pacitan, Oligo-Miocene age and covered all of Southern Mountain Zone. This volcanoclastic sediments sequent record development Southern Mountain Oligo-Miocene volcanic arc. The volcano activity covered wide area, explosive and assumed Plinian-type. This sediment composed andesite until rhyolite and this lithology composed thick volcano ash, tuff, pumice breccia, andesite breccia, lava dome and lava flow with thickness 250 until more than 2000 m. The upper boundary from this volcanoclastic signed by short volcano activity assumed super eruption produced Semilir Formation. [6].

When Oligo-Miocene volcanism activity decreased, even death, than erosion and the material re-sedimented as next sediment sequent. Besides that the next sediment sequent signed by wide carbonate shelf like in Wonosari (Wonosari Formation) and Pacitan (Punung and Campurdarat Formation). The thickness around 500 m and reefs developed in highlands limited by fault or in volcano former area. In top there are volcanic ash layer contains zircon, based on U-Pb time indicated 10 – 12 million year ago [6]. This age assumed related to re-appeared volcano activity on Late Miocene, in Sunda Arc position in recent [7].

3. Data and Methods

Gravity surveying measures variations in the Earth's gravitational field caused by differences in the density of subsurface rocks. Although known colloquially as the "gravity" method, it is in fact the variation on the acceleration due to gravity that is measured [4]. One hopes to locate local mass of greater or lesser density than the surrounding formations and learn something about them from the irregularities in the earth's field. It is not possible, however, to determine a unique source for an observed anomaly. Observations normally are made at the earth's surface, but underground surveys also are carried out occasionally [5].

Magnetic and gravity methods have much in common, but magnetics is generally more complex and variations in the magnetic field are more erratic and localized. That is partly due to the difference between the dipolar magnetic field and the monopolar gravity field, partly due to the variable direction of the magnetic field, whereas the gravity filed is always in the vertical direction, and partly due to the time-dependence of the magnetic field, whereas the gravity field is time-invariant (ignoring small tidal variations). Whereas a gravity map usually is dominated by regional effects, a magnetic map generally shows a multitude of local anomalies. Magnetic measurements are made more easily and cheaply than most geophysical measurements and corrections are practically unnecessary. Susceptibility is the fundamental rock parameter in magnetic prospecting. The degree to which a body is magnetized is determined by its magnetic susceptibility. The magnetic response from rock and minerals is determined by the amounts and susceptibilities of magnetic materials in them [5].

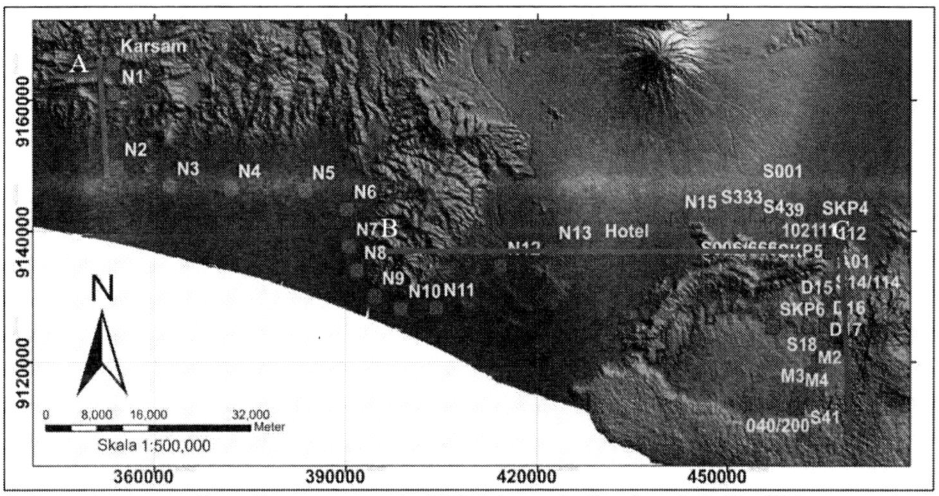

Figure 2. Location map. Coordinate system in UTM (meter). Section A,B,C will interpret in next part.

Gravity and magnetics measurements carried out in Karangsambung – Bayat – Wonosari track, Central Java – Yogyakarta region as much as 34 points on Tuesday, May 12th 2015. We began from LIPI campus (Karsam at figure. 2) at Karangsambung to Gunungkidul area (S41 at figure 2). We used Gravimeter Scintrex CG-5 to get gravity respond and Magnetometer GSM 19-T to get magnetic respond. We took data on the road along Karangsambung to Gunungkidul. The data covered gravity respond, magnetic respond, time and elevation. The data processing covered several steps, showed at figure 3.

We used several corrections on gravity and magnetic data. Just as the water in the oceans responds to the gravitational pull of the moon, and to a lesser extent of the sun, so too does the solid earth. Earth tides give the rise to a change in gravity. This can correct by Tidal Correction. Drift correction used because gravimeter reading change (drift) with time as a result of elastic creep in the springs, producing an apparent change in gravity at a given station. The drift can be determined simply by repeating measurements at same stations at different times of the day. Elevation correction is normally adequate to cope with slight topographic effect on the acceleration due to gravity. Latitude correction is negative with distance northwards in the northern hemisphere or with distance southwards in the southern hemisphere. This is to compensate for the increase in the gravity field from the equators the poles. The basis of free-air correction is that it makes allowance for the reduction in magnitude of gravity with height above the geoid, irrespective of the nature of the rock below. Whereas the free-air correction compensates for the reduction in that part of gravity due only to increased distance from the centre of mass, the Bouguer correction is used to account for the rock mass between the measuring station and sea level. Bouguer correction calculates the extra gravitational pull exerted by a rock slab of thickness h metres and mean density which results in measurements of gravity called Simple Bouguer Anomaly [4]. In magnetic processing, we used IGRF correction. This correction produced by International Geomagnetic Reference Field (IGRF), who recalculated a theoretical value for the field strength of the Earth's magnetic field for any location on earth. Survey data at any given location can be corrected by subtracting th theoretical field value, obtained from the IGRF from measured value [4].

Modeling used 2D structure inversion method. The first step is section determination. The chose section that have SBA and magnetics respond near measurement point to reduce interpolated value. Based on geology and geophysics observation track, chose three sections on Karangsambung, Wates – Yogyakarta and Bayat – Gunungkidul. The input data area SBA value, magnetics respond and true elevation. The output model is cross-section based susceptibility and density rock information.

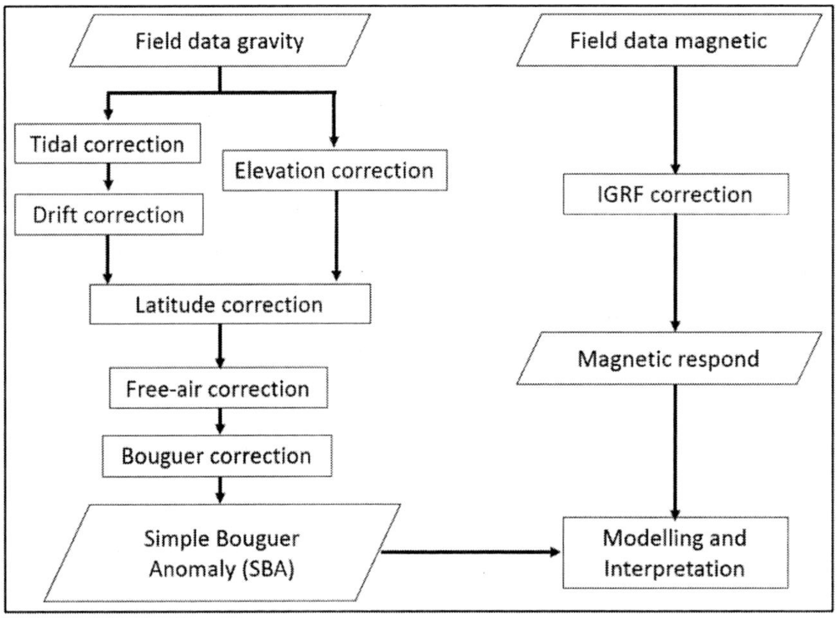

Figure 3. Workflow diagram.

4. Results and Discussion

Section A lies on Karangsambung area and reach to 1900 m. Section A showed formation of 0.000001 – 0.0014 nT and 2.00 – 2.80 g/cm^3 (figure 4). There are 10 models divided into eight units. Qa – Alluvium sediments composed of clay, silt, sand, pebble and gravel. Teok – Karangsambung Formation composed of claystone with limestone block, conglomerate, sandstone, limestone and basalt. Karangsambung Formation is the oldest formation which assumed as Java basement [7]. This formation had anticline structure. Tomt – Totogan Formation composed of breccia with claystone component, sandstone, limestone and basalt. There is thrust fault that is force interaction on anticline structure of Karangsambung Formation. Tmw – Waturanda Formation had coarse sandstone on lower, and upper become breccia with andesite component; basalt and sandstone base mass, tuff. Tmp – Penosogan Formation signed by interludes between lime-sandstone, claystone, tuff, marl and calcarenite, influenced by turbid flow. There was reverse fault influenced by tectonic effect on past. Tmpb – breccia Halang Formation had breccia with andesite, basalt and limestone, coarse tuff sandstone mass base, sandtone layer and basalt lava. Tmph – Halang Formation had interludes between sandstone, limestone, marl and tuff with breccia layered, influenced by turbid flow and sea water slide down [8]. Overall, section A composed alluvium, basalt and tuff.

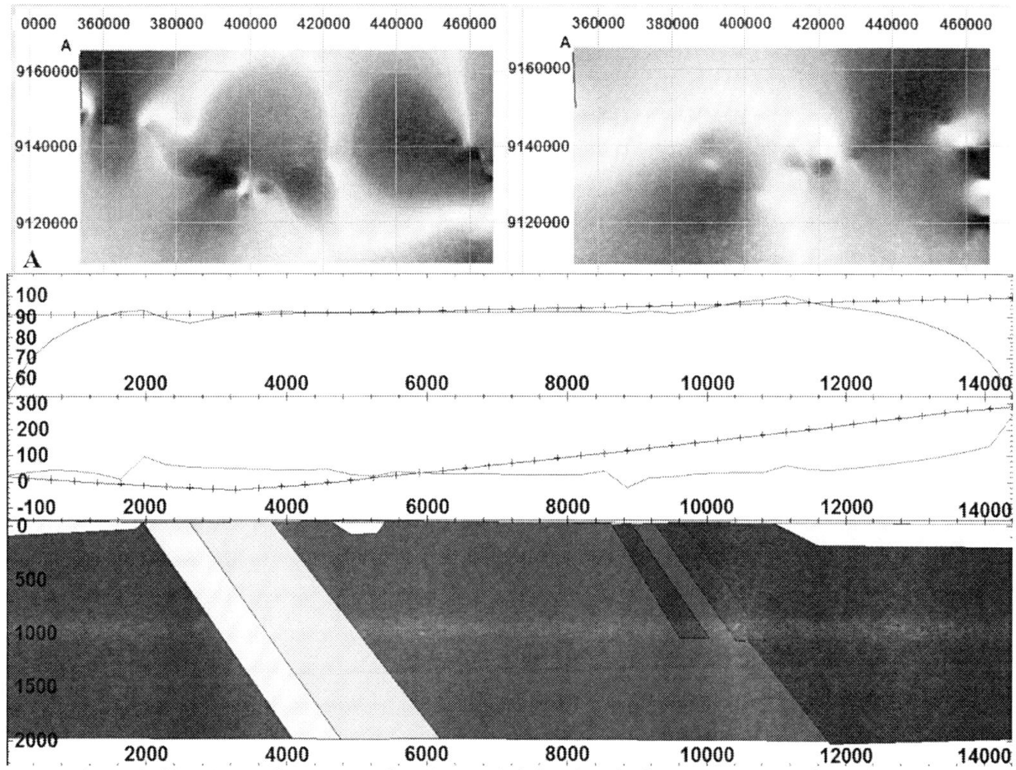

Figure 4. (Top left) Magnetics anomaly, (Top right) Simple bouguer anomaly. The colour refer to value of parameter, blue is low and red is high. (Bottom) Cross-section model from Section A. Blue line is simple bouguer anomaly (mGal) and red line is magnetic anomaly (nT).

Table 1. Susceptibility and density from cross-section A.

Label	Colour	Susceptibility (nT)	Density (g/cm^3)
Qa – Alluvium sediments		0.0013	2.00
Teok – Karangsambung Formation		0.00019	2.80
Tomt – Totogan Formation		0.00001	2.10
Tmw – Waturanda Formation		0.000015	2.71
Tmp – Penosogan Formation		0.000011	2.57
Tmpb – Breccia Halang Formation		0.00012	2.37
Tmph – Halang Formation		0.000001	2.80

Section B lies on Wates - Yogyakarta area and reach to 1700 m. Section B showed formation of (-0.01) – 0.02 nT and 2.40 – 3.00 g/cm^3 (figure 5). There are eight models divided into seven units. a – andesite located in here is outcrop from intrusion. Tmok – Kebobutak Formation had negative susceptibility value, influenced by mineral on rock composer this formation. Several minerals known had negative susceptibility value like quartz and anhydrite (-0.01) nT and calcite (-0.001 – (-0.01)) nT [5]. This formation composed of andesite breccia, tuff, lapilli tuff, agglomerate and andesite lava flow inset. Tmps – Sentolo Formation composed of limestone and clay-sandstone. Qml – young Merapi volcano sediment is tuff, ash, breccia, agglomerate, lava indivisible. Tms – Semilir Formation composed of interludes between breccia and tuff, pumice breccia, dacite tuff, andesite tuff and tuff marl [9]. KTm – metamorphic rock signed by schist, marble, metamorphic volcanic rock, metamorphic sediment rock and slate [10]. Tpdi – Pendul diorite is igneous rock signed by red color. Overall, section B covered volcano product like andesite intrusive and Merapi volcano sediments.

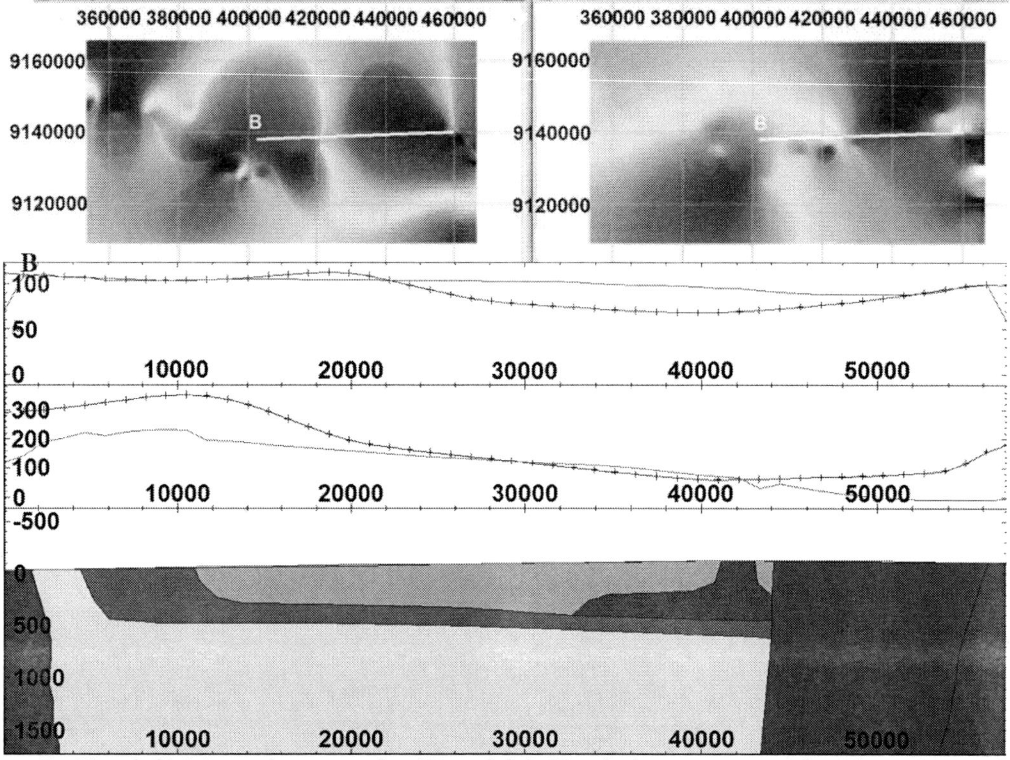

Figure 5. (Top left) Magnetics anomaly, (Top right) Simple bouguer anomaly. The colour refer to value of parameter, blue is low and red is high. (Bottom) Cross-section model from Section B. Blue line is simple bouguer anomaly (mGal) and red line is magnetic anomaly (nT).

Table 2. Susceptibility and density from cross-section B.

Label	Colour	Susceptibility (nT)	Density (g/cm^3)
a – Andesite intrusion		0.0013	3.00
Tmok – Kebobutak Formation		-0.01	2.60
Tmps – Sentolo Formation		0.0012	2.50
Qml – Young Merapi Volcano sediments		0.02	2.70
Tms – Semilir Formation		0.008	2.40
KTm – Metamorphic Rock		0.0023	2.43
Tpdi – Pentul Diorite		0.0013	3.00

Section C lies on Bayat - Wonosari area and reach to 2000 m. Section C showed formation of 0.00016 – 0.0005 nT and 2.30 – 3.14 g/cm^3 (figure 6). There are nine models divided into eight units. Tpdi – Pentul Diorite signed by red color. The outcrop of this formation is micro-diorite intrusion (figure 7a). KTm – metamorphic rock composed hill geomorphology on Gunungajah Village result from schist and phyillite weathering (figure 7b). Tomk – Kebobutak Formation, upper part: sandstone interlude, claystone and acid tuff thin layer, lower part: sandstone, siltstone, claystone, shale, tuff and agglomerate. Tms – Semilir Formation signed by tuff, dacite – pumice breccia, tuff sandstone and shale [10]. Geomorphology in Tancep Village, Gunungkidul regency showed tuff outcrop which part of Semilir Formation (figure 7c). Tmng–Nglanggran Formation covered volcano breccia, agglomerate, andesite – basalt lava and tuff. Tmss – Sambipitu Formation had sandstone and claystone. Tmwi – Wonosari Punung Formation had limestone, marble – tuff limestone, conglomerate limestone, tuff sandstone and siltstone. Tmo – Oyo Formation of tuff – marble, andesite – tuff and conglomerate – limestone [10]. Clastic limestone outcrop in Oyo River is the part of Oyo Formation (figure 7d). Overall, section C covered diorite intrusive, limestone and tuff.

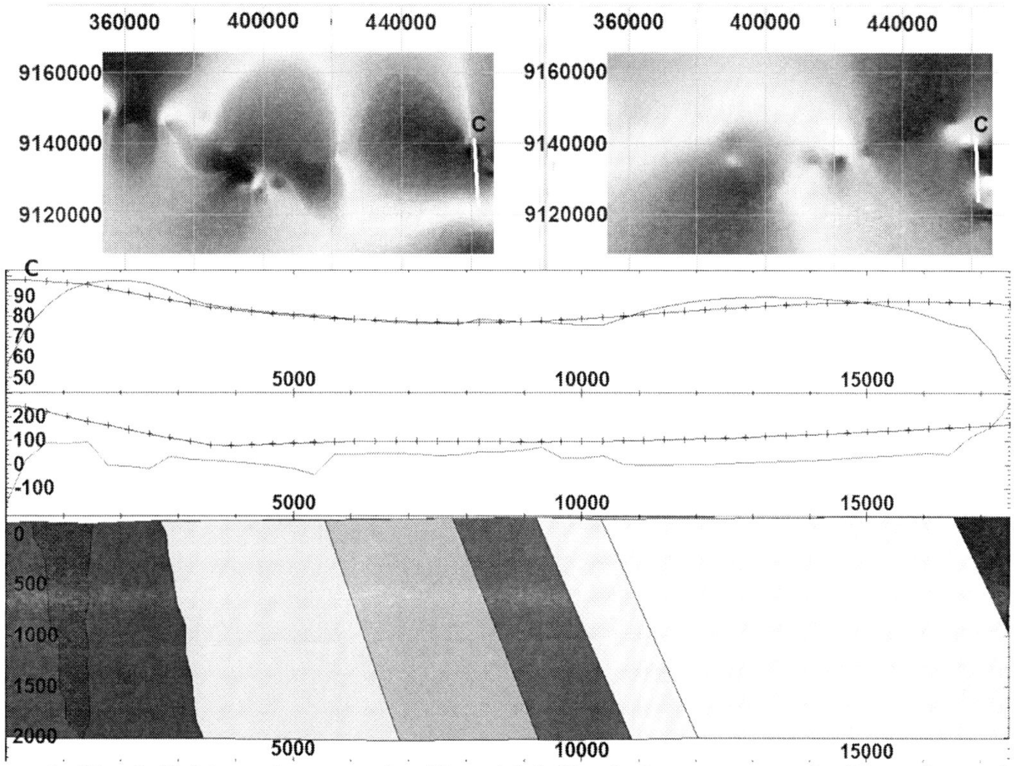

Figure 6. (Top left) Magnetics anomaly, (Top right) Simple bouguer anomaly. The colour refer to value of parameter, blue is low and red is high. (Bottom) Cross-section model from Section C. Blue line is simple bouguer anomaly (mGal) and red line is magnetic anomaly (nT).

Table 3. Susceptibility and density from cross-section C.

Label	Colour	Susceptibility (nT)	Density (g/cm^3)
Tpdi – Pentul Diorite		0.0005	3.00
KTm – Metamorphic Rock		0.00027	3.14
Tomk – Kebobutak Formation		0.0003	2.51
Tms – Semilir Formation		0.000019	2.42
Tmng – Nglanggran Formation		0.0004	2.47
Tmss – Sambipitu Formation		0.0003	2.77
Tmwi – Wonosari Punung Formation		0.00016	2.80
Tmo – Oyo Formation		0.00018	2.30

Section A – B – C represented research area, which showed several formations. The depth until 2,000 m, is shallow depth. The formation composed the volcano product and sediment on research area. The existence of basalt, andesite, dacite and pyroclastic, is tuff, sign the volcano product. The sedimentation processed on this area product sediment rock like alluvium, limestone, sandstone, siltstone, claystone and shale. Breccia and conglomerate on this area composed by volcano fragment, formed pumice breccia and agglomerate. High pressure and temperature from subsurface made metamorphic on this area, like schist and phyllite. This distribution signed high volcano activity in past, may Oligo-Miocene [7], and covered wide area, assumed Plinian-type [6].

Figure 7. (a) The outcrop of micro-diorite intrusion on Pentul (near D15 at figure 2). (b) The schist and phyllite weathering on Gununggajah Village (near D16 at figure 2). (c) The tuff outcrop on Tancep Village, Gunungkidul (near D17 at figure 2). (d) Clastic limestone in Oyo River (near M4 at figure 2).

5. Conclusion

Formation distribution of Karangsambung – Bayat – Wonosari showed similar type. This showed on the formation of sediments from volcano activity on Karangsambung – Bayat – Wonosari track, Central Java – Yogyakarta region. The volcano product is igneous rock, are basalt, andesite, dacite and pyroclastic, is tuff. This distribution signed high volcano activity in past and covered wide area. The gravity and magnetics method in this study shows potential application to identify the formation of volcano activity from 2D structure in Karangsambung – Bayat – Wonosari, Central Java – Yogyakarta region.

Acknowledgement

We gratefully acknowledge the Kemenristek-DIKTI for funding this research. We thank to Geophysical Engineering Dept., Bandung Institute of Technology (ITB) and Geophysics Dept., Faculty of Mathematics and Natural Science Hasanuddin University (UNHAS) for supporting this research.

References

[1] Arif N, Lantu, Aswad S, Maria 2014 *Proc. Seminar Nasional Geofisika Universitas Hasanuddin* p 94 – 99
[2] Mochales T, Casas A M, Pueyo E L, Pueyo O, Roman M T, Pocovi A, Soriano M A, Anson D 2008 *Environ Geol* **53** p 1067-1077
[3] Syahruddin M H 2015 *Proc. 2nd Makassar International Conference on Civil Engineering*
[4] Reynolds J M 1997 *An Introduction to Applied and Enviromental Geophysics* John Wiley & Sons Ltd.
[5] Telford W M, Geldart L P, Sheriff R R 1990 *Applied Geophysics* (Cambridge: Cambridge University Press)
[6] Smyth H, Hall R, Hamilton J and Kinny P 2005 *Proc. 30th Indonesian Petroleum Association Annual Convention and Exhibition* p 251
[7] Prasetyadi C 2007 *Dissertation: Evolusi Tektonik Paleogen Jawa Bagian Timur* (Bandung)
[8] Asikin S, Handoyo A, Busono H and Gafoer S 1992 *Peta Geologi Lembar Kebumen Jawa* (Bandung: Pusat Penelitian dan Pengembangan Geologi Bandung)

[9] Rahardjo W, Sukandarrumidi and Rosidi H M D 1995 *Peta Geologi Lembar Yogyakarta Jawa* (Bandung: Pusat Penelitian dan Pengembangan Geologi Bandung)
[10] Surono, Toha B and Sudarno I 1992 *Peta Geologi Lembar Surakarta – Giritontro Jawa* (Bandung: Pusat Penelitian dan Pengembangan Geologi Bandung)

IOP Conf. Series: Journal of Physics: Conf. Series **979** (2018) 012047 doi:10.1088/1742-6596/979/1/012047

Ore Reserve Estimation of Saprolite Nickel Using Inverse Distance Method in PIT Block 3A Banggai Area Central Sulawesi

Muhammad Khaidir Noor

Mining Engineering Department, Universitas Pejuang Republik Indonesia, Makassar, 90234

E-mail: ipungnoor70@gmail.com

Abstract. Reserve estimation is one of important work in evaluating a mining project. It is estimation of the quality and quantity of the presence of minerals have economic value. Reserve calculation method plays an important role in determining the efficiency in commercial exploration of a deposit. This study was intended to calculate ore reserves contained in the study area especially Pit Block 3A. Nickel ore reserve was estimated by using detailed exploration data, processing by using Surpac 6.2 by Inverse Distance Weight: Squared Power estimation method. Ore estimation result obtained from 30 drilling data was 76453.5 ton of Saprolite with density of 1.5 ton/m^3 and COG (Cut Off Grade) Ni $\geq$ 1.6 %, while overburden data was 112,570.8 tons with waste rock density of 1.2 ton/m^3 . Striping Ratio (SR) was 1.47 : 1 smaller than Stripping Ratio (SR) were set of 1.60 : 1.

1. Introduction

1.1 Background

Exploration activities and calculation of mineral deposits reserves constitute an important step in the mining operation plan, so that the quantity and quality of mineral deposits must be known by value of SR *(Stripping Ratio)* and COG *(Cut Of Grade)*, in order to result high degree of geological confidence and economy. The use of a computerized system in certain *Mine Software* can calculate the reserves and even design a mine with a faster and better approach to calculating mineral deposits.

Surpac 6.2 is one type of GIS Software (Geologic Information System), the CAD group (Computer Aided Design) which is a software from A Gemcom Software Company Australia. CAD (Computer Aided Design) is software that provides utility to input data, display visualization, and representation in 3 dimensions [1].

PT. Kumala Mining is a contractor of PT. Aneka Nusantara International which is a mining company with nickel ore commodities. This company located in Banggai, Central Sulawesi. The aims of this research are to calculate tonnage reserves of nickel ore deposits, tonnage of overburden, and stripping ratio in the Block PIT 3A. Estimating the reserves and stripping ratio in PIT Block 3A at PT. Kumala Mining Site Project Bunta given limit value of COG *(Cut Off Grade)*: >1.6, density of rock: 1.5 ton/m^3, waste: 1.2 ton/m^3, standard of SR *(Striping Ratio)* in company: 1.60 : 1.

2. Basic Theory

2.1 Model Block Method

Block model is one of the existing tool functions in Surpac 6.2, this method effectively used on ore mineral as it uses attributes of content, quality, density, and lithology as consideration of reserve estimation [2].

- *Geological Database (GDB)*
  Store drillhole survey, lithology and quality data, generate standardized summary and customizable reports, and display graphical drillhole data. GDB can display lithology, interval, geophysical and correlation as well as provide the function of compositing and washability,
- *Constraints*
  All block model functions can be done with constraints. Constraint is a logical combination of one or more spatial objects in the selected block. Objects that can be used in constraints are plane surface, DTM, solids (body solids), closed strings and attribute block values. Constraints can be stored in files for reuse and may be used as a component of other constrains.
- *Estimations*
  Once the block model is created and all the required attributes exist, it should then be done with some estimation methods. This can be done by estimating and assigning attribute values from sample data that have XYZ coordinates and the value of the main attribute estimation methods.

2.2 Pit Design

It is one of the applications in Surpac 6.2 for the design of mining pits. Mining pit design is done after getting areas that have a stripping ratio in accordance with the set. These areas are then formed into mining blocks by naming such as Block 01, Block 02, and so on. Each of these blocks is limited by a polygon with different extents [3].

Based on the analysis of the area using the software, the limit of the area of mining (pit limit) and the limits of mining elevation can be determined. Based on the data, it can be done the overall mining pit design and the estimated the reserve that has a stripping ratio that has been determined. In preliminary reserve estimation, it aims to estimate the amount of reserves that can be mined with the appropriate stripping ratio and obtain ore distribution data. The ore distribution is based on quality of laboratory analysis from explorotion drilling data [4].

Some stages in pit design using Surpac 6.2 namely: calculation of reserve blocks, field topography identification, making mine layout and design pit, data of bench geometry (high bench, wide bench and tilt bench), and data of road geometry (road width and slope of road) [5].

2.3 Stripping Ratio.

The stripping ratio is defined as the ratio of the of material cover (waste) to the ore material (ore). At ore mines, this ratio is usually expressed in tonnage of waste / tonnage of ore. In coal mines are often used waste/ton of coal [4].

3. Methodology

The research method is to collect accurate data at field is associated with reserve estimation. The research methods include; data collection, data processing, and analysis of data [5].

4. Results and Discussion

4.1 Reserve Calculation Method

To find out the nickel ore reserves contained in the research area of PIT Block 3A, calculated using non-conventional methods that use Surpac 6.2. Firstly, this research is done by making geological database and block model to calculate the area and volume of existing reserves as reference reserves estimates or to determine the value of nickel ore tonnage. COG *(Cut Off Grade)* in the calculation of reserves in this study determined by the company is 1.6%. The level or stage of exploration conducted at PIT Block 3A is an advanced stage. The collected data indicates that the level of confidence in reserves is higher.

In the use of Surpac 6.2 Software, the step to make a reserve estimation is to first create a geological database with core drilling data, rock description, and laboratory analysis which further creates and calculates the volume of block model and its function of *Constraint* with *Inverse Distance Weight Power 2* method with previous Geological Database created.

Geological database on the outline is the data that contains the geological information representatives of core drilling *(core drilling)* exploration, rock description and analysis of the sample in the preparation laboratory. With the geology database we can visualize 30 drilling point in Pit Block 3A, even making a cross-section and shape of nickel ore deposit.

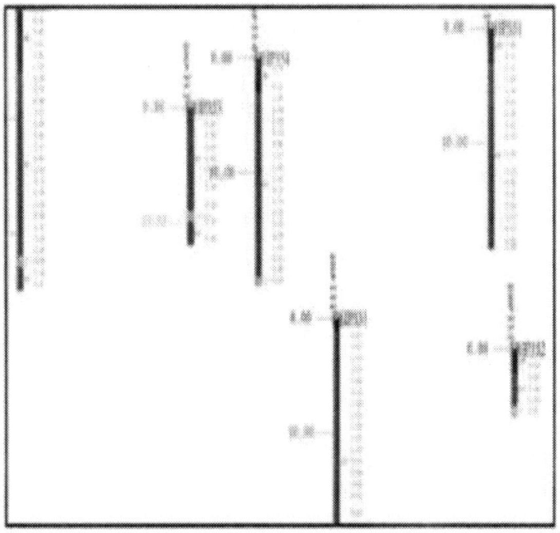

Figure 1. Geological database PIT Block 3A.

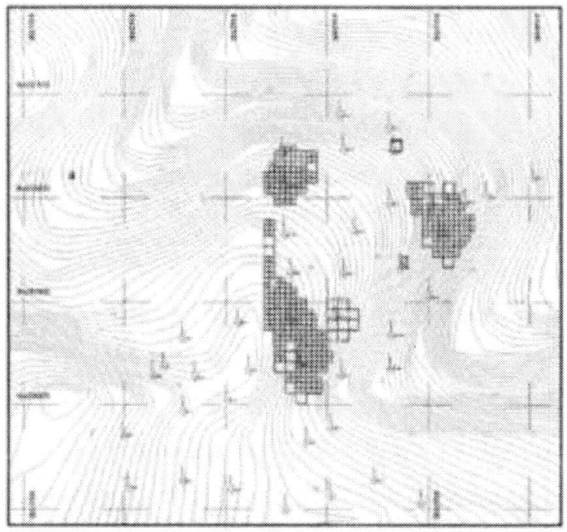

Figure 2. Shape of nickel ore distribution.

By a pre-prepared geological database, it can created block model. Due to the distribution of laterite nickel deposits located in Pit Block 3A is not uniform then sediment block model required *constraint* with modify *constraint select* with functions that can be modified according to the needs with grade and lithology parameter.

First *constraint* with lithologic attributes will form a block corresponding to lithology, with upper and lower boundary on the saprolite zone only, using the top ore and bottom ore boundary of saprolite zone and the outer boundary of the specified drilling point using the formula, so that the data is estimated only at lithology/geological database of the saprolite zone.

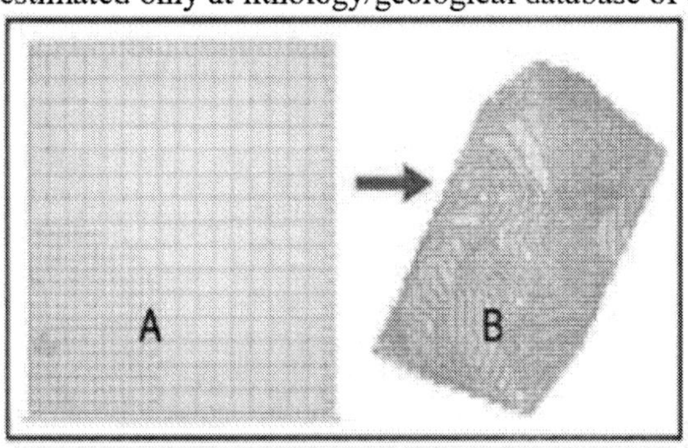

Figure 3. A Block without constraints.

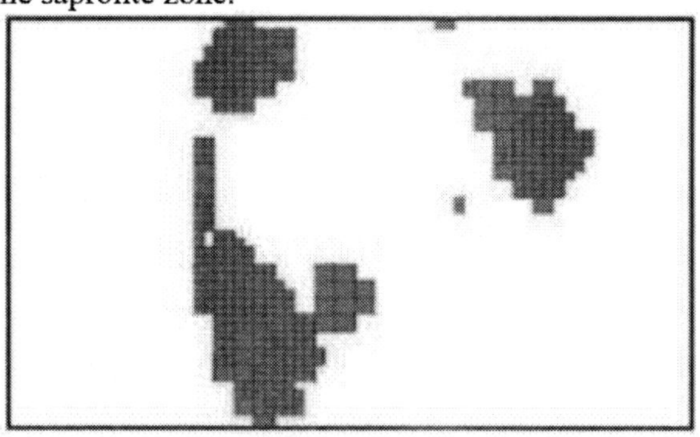

Figure 4. Block with grade and lithology *constraints.*

Both *constraint select* with this grade attribute will separate the appropriate blocks with the desired grade starting from the lowest grade of 1.6 which is still economical when mined to the highest grade available in accordance with COG *(cut off grade)*.

Next, marked with some color on the block representing a certain grade of ore located in the saprolite zone, then the block with color will separate the blocks according to the lowest grade available in the database that is 1.6 to the highest grade available. It is also intended to separate the visual distribution of the deposit profile according to the color.

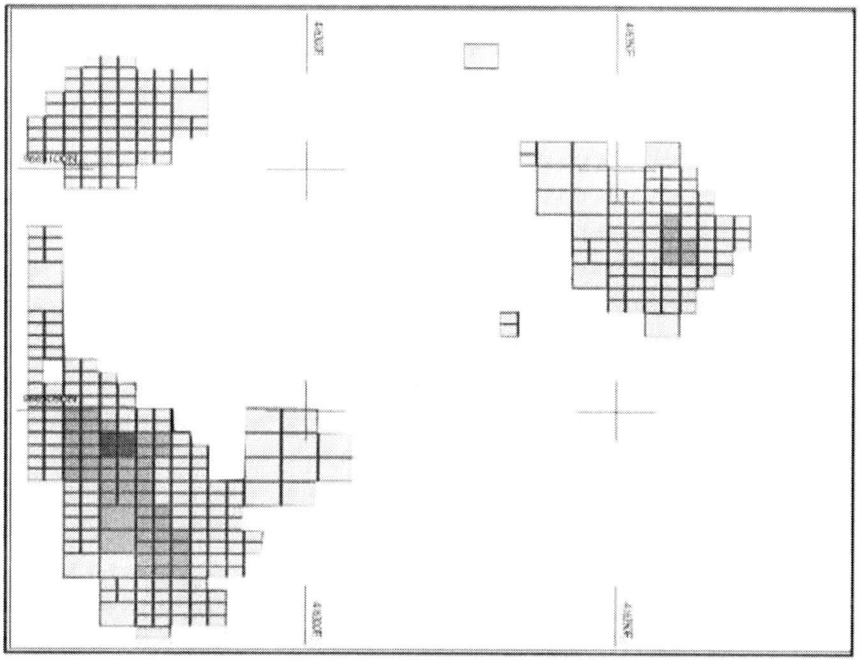

Figure 5. Block with grade constraints based on colours yellow >1.6, green >2.0 and dark green >2.5.

The volume and tonnage calculated by using *Inverse Distance Weight Power 2*, with density value of 1.5 tons/m3 and the lowest grade of COG *(cut off grade)* determined by the company is 1.6. Then the calculation of volume only on the saprolite zone and tonnage will only be calculated starting from the level of >1.6 up. Reserve tonnage is obtained by multiplying the volume of saprolite 50969 m^3 with density ore 1.5 ton/m^3. So that the ore tonnage is 76453.5 tons in accordance with the calculation using Surpac 6.2.

4.2 Design of Pit Mine

The construction of mining bench designs is based on safety factors with geotechnical recommendations by using open pit mining method. Based on the direction of ore distribution. Design of the geometry of the mining bench requires several important parameters, such parameters are: Standard Stripping Ratio (SR) 1.60: 1, height of bench 3 m, width of bench 1.5 m, and single slope 56°.

Construction bench width design that is up to 78.959 mdpl does not change the value of the geotechnical recommendation decisions, but at the elevation of 73.968 mdpl, width of the bench on the pit block design changes as far as 15,534 meters and at elevation of 63.908 md as far as 37.998 meters. This is caused by the uneven distribution of the ore at the elevation. The shape of the pit design on Pit Block 3A can be seen on figure 6.

Based on the factors described above and the consideration of the mining design and recommendations from geotechnical study that the design of Block 3A pit using surpac 6.2 applies the open pit mine system. So in getting the volume from the design of Block 3A pit by using Software Surpac 6.2 is 144778 m^3.

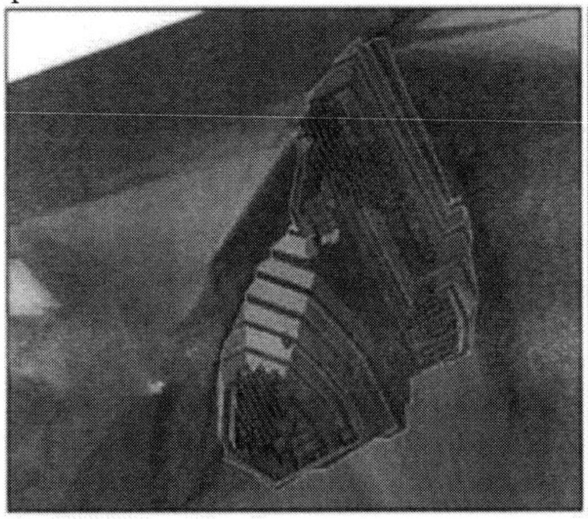

Figure 6. Design of PIT Block 3A in DTM.

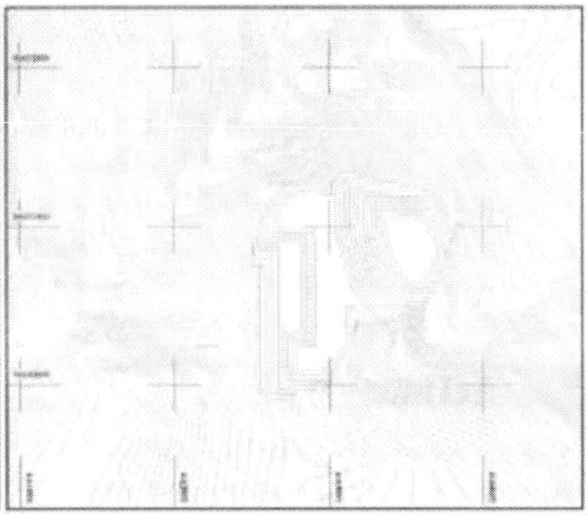

Figure 7. Of PIT Block 3A in String.

4.3 Waste Tonnage/Overburden (OB)

The calculation of overburden is also done based on the calculation of pit volume. Tonnage overburden is influenced by the extent of pit design results. The overburden tonnage from the pit mine can be calculated by reducing the pit volume, 144778 m^3 with the ore volume, which is 50969 m^3 and multiplying with the waste rock density of 1.2 m^3/ton. So in the tonnage can be over burden pit block 3A is 112570.8 tons.

4.4 Striping Ratio (SR) Pit Block 3A

The determination of the stripping ratio relates to the feasibility of ore deposits in PIT 3A to be mined profitably. The shape of ore and waste deposit illustration can be seen in figure 8. The value of stripping ratio is calculated by comparing the waste tonnage to ore tonnage. Based on the calculation of ore tonnage and waste tonnage, namely:

Tonnage Ore = 76453.5 tons
Tonase Waste = 112570.8 tons

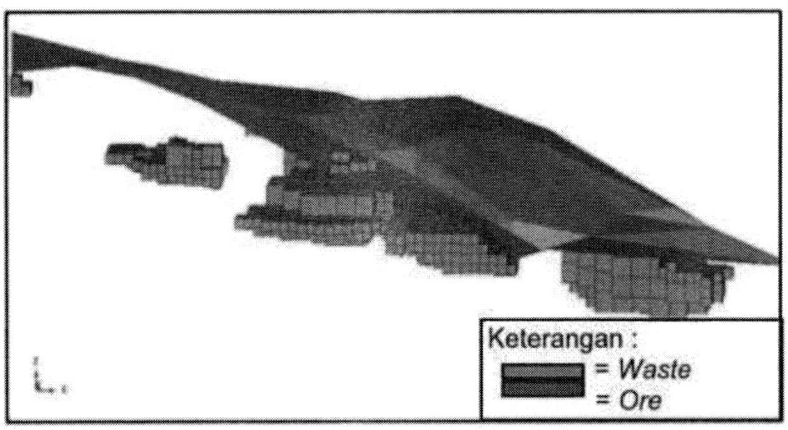

Figure 8. Modelling of Ore and Waste.

After the tonnage of waste and tonnage of ore calculated in research area obtained value of SR *(stripping ratio)*, that is 1,47:1. From the value of stripping ratio obtained (1.47:1), it is known that the value of stripping ratio is smaller than the stripping ratio set by the company, which is 1.60: 1. So that the reserves of nickel contained Pit Block 3A is feasible to proceed to phase of mining operations planning and include the category *of Inferred Ore Reserve* in accordance with the provisions of the Code-KCMI 2011.

5. Conclusion

From the results of the description and discussion in the previous chapters, it can be concluded as follows:

1. Based on the results of the analysis, the volume of ore reserves is 50969 m^3 and tonnage of saprolite reserves for PIT Block 3A PT. Kumala Mining Site Project Bunta with 30 drilling points is 76453.5 tons with density of ore 1.5 ton/m^3 calculated using Surpac 6.2 with COG *(cut off grade)* Ni $\geq$ 1.6%.
2. Based on pit design and some analysis, pit volume is 144778 m3, waste volume is 93809 m^3 and tonnage from overburden PIT Block 3A is 112570.8 tons with waste rock density 1.2 m^3/ton calculated using Surpac 6.2.
3. Stripping Ratio (SR) of the research area is 1.47: 1. The reserves of nickel deposits contained Blocks 3A are feasible to be mined, because the SR of the study area is smaller than the SR value set by PT. Kumala Mining is 1.60: 1.

References

[1] Group Surpac Minex 2006 *Solid Modelling In Surpac Vision* (Western Australia: A Gemcom Company)
[2] Group Surpac Minex 2006 *Geological Database* (Western Australia: A Gemcom Company)
[3] Komite Bersama PERHAPI (Perhimpunan Ahli Pertambangan Indonesia) & IAGI (Ikatan Ahli Geologi Indoensia) 2011 *Kode-KCMI 2011* (Indonesia: PERHAPI)
[4] Rauf A 1998 *Perhitungan Cadangan Endapan Mineral* (Yogyakarta: UPN)
[5] Usman N D 2004 *Perhitungan Cadangan Dan Geostatistik* (Bandung: Diklat Perencanaan Tambang Terbuka UNISBA)

Coal Layer Identification using Electrical Resistivity Imaging Method in Sinjai Area South Sulawesi

Andi Ilham Samanlangi

Mining Engineering Department, Universitas Pejuang Republik Indonesia, Makassar, 90234

E-mail: andhismining@gmail.com

Abstract. The purpose of this research is to image subsurface resistivity for coal identification in Panaikang Village, Sinjai, South Sulawesi.Resistivity measurements were conducted in 3 lines of length 400 meters and 300 meter using resistivity imaging, dipole-dipole configuration. Resistivity data was processed using Res2DInv software to image resistivity variation and interpret lithology. The research results shown that coal resistivity in Line is about 70–200 Ωm, Line 2 is about 70–90 Ωm, and Line 3 is about 70–200 Ωm with average thickness about 10 meters and distributed to the east of research area.

1. Introduction

1.1 Background

Coal is a mineral material composed of organic mascerals and little nonorganic elements. Coal was formed by plants that lived in freshwater environments, commonly grew in the tropics. Coal as one of the substitute for petroleum fuels, in the future will be expected to play a big role in Indonesia, as an alternative fuel.

Based on the Regional Geology of Sulawesi, coal in South Sulawesi can be found in several formations, such as the Mallawa Formation, Walanae Formation, Camba Formation and Toraja Formation, with different characteristics in each formation. Generally coal in South Sulawesi includes lignite, it was formed into Tertiary, Eocene and Neogen [1]. Coal outcrops in Sinjai distributed irregularly. Coal is a type of lignite deposited in ancient tidal environments. One part of the area that has an outcrop of coal is administratively located in Panaikang Village.

One of geophysical methods can be used to identify subsurface lithology is resistivity method. This methods consists of some acquisition techniques, such as sounding (VES) and imaging. Sounding (VES) measures resistivity variation vertically using Schlumberger configuration and imaging measures resistivity variation vertically and laterally using Wenner, Wenner-Schlumberger, and Dipole-dipole [2]. In this research, imaging acquisition technique was conducted using Dipole-dipole configuration. This configuration can image resistivity variation laterally well and moderate depth accuracy [3]. So it is better to identify coal distribution laterally in research area.

1.2 Geological Consideration

Based on geological map (figure 1), Panaikang area lies on Walanae Formation (Tmpw) and Alluvium, Swamp, and Coastal Deposits (Qac). Walanae Formation (Tmpw) aged late Miocene–Pliocene, this formation overlapped with the Camba Volcano Rock Formations. The Walanae Formation is composed of sandstone, conglomerates, and tuffs with limestone inserts, claystone, limestones, napal and lignite, medium to coarse grained sandstones. The thickness of this unit is estimated to be about 1.200 meters. Alluvium, Swamp, and Coastal Deposits (Qac) consists of clay, silt, sand, mud, and gravel along large rivers and beaches. Local coastal deposits contain seashells and coral limestones [1]. Coal outcrop in research area is lignite, part of lithology of Walanae Formation (figure 2). This outcrop reveals coal potency in research area and to be interesting to investigate the distribution of coal as pre-eliminary study.

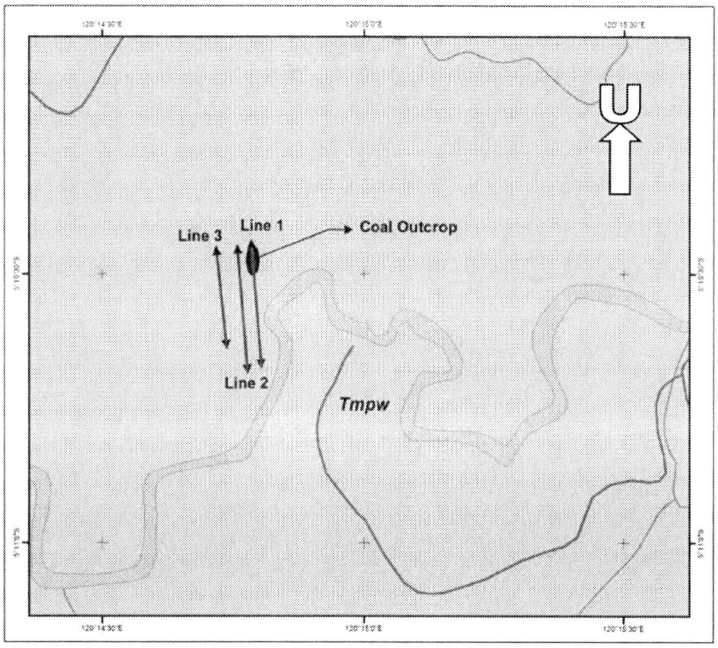

Figure 1. Geological Formation of Research Area [1].

Figure 2. Coal Outcrop in Research Area.

2. Methods

Panaikang Village, Sinjai Regency, South Sulawesi geographically located between 120° 14' 39.4" E – 120° 16' 59.2" E and 5° 9' 28.5" S – 5° 11' 26.4" S. Resistivity measurements were conducted using resistivity meter Naniura NR 300, four electrodes consist of two current and potential electrodes. These electrode were injected into the soil surface based on predetermined distance, then the electrical current was flowed through current electrodes into subsurface. The potential electrodes.will measure potential difference from the electrical current injected [4]. The apparent resistivity value is calculated using potential difference data, electrical current injected, and geometry factor of dipole-dipole configuration and the real resistivity is calculated by using inversion method applied by Res2DInv software [5]. The resistivity measurement in this research consist of three lines. They are Line 1 and

Line 2 have 400 meters in length, and Line 3 has 300 meters in length. The acquisition in Line 1 was conducted on coal outcrop to get resistivity variation on coal layer directly and this resistivity value became reference to interpret coal layer in other lines. Line 2 and Line 3 was conducted parallel with Line 1 to identify continuity and distribution direction of coal. The minimum distance of electrodes for each line is 5 meters.

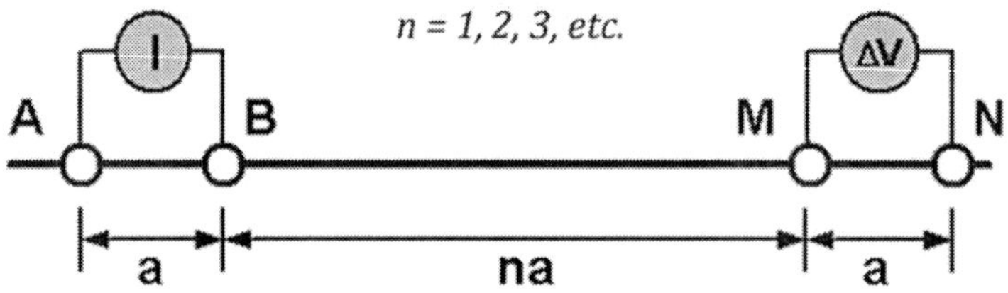

Figure 3. Schematic Diagram of Dipole-dipole Configuration [6].

The apparent resistivity is calculated by using the following equation where V (potential difference), I (electrical current) and k (geometry factor of dipole-dipole configuration) [7].

$$\rho = k\frac{V}{I} \quad (1)$$

$$k = \pi\, n\, (n + 1)(n + 2)a \quad (2)$$

Data inversion was processed by using Res2DInv software to image real resistivity of subsurface vertically and laterally.

3. Results and Discussion

Result of resistivity processing using Res2DInv software for Line 1 shown on the following figure 4.

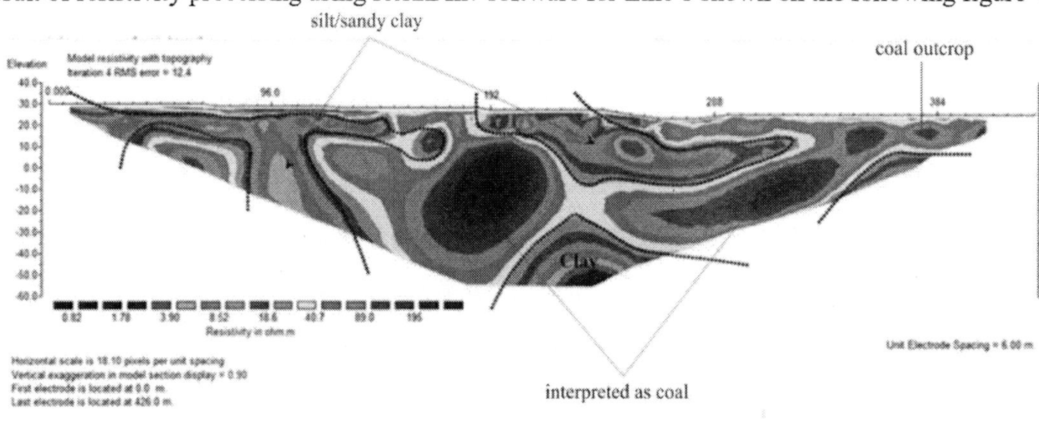

Figure 4. Distribution of Resistivity in Line 1.

Figure 4 shows distribution of resistivity in Line 1. The variation of resistivity starts from 0.82 Ωm to 200 Ωm with the depth about 80 meters from surface. Interpretation of resistivity imaging in Figure 4 revealed that distribution of coal resistivity is about 70–200 Ωm, this resistivity variation will become reference to interpret coal layer in other lines. The thickness of coal layer is about 10 meters and tends to distribute to the east of research area. Others lithology can be interpreted from the resistivity value are wet clay with resistivity value about 0.82–20 Ωm and silt or sandy clay with resistivity value about 20–70 Ωm.

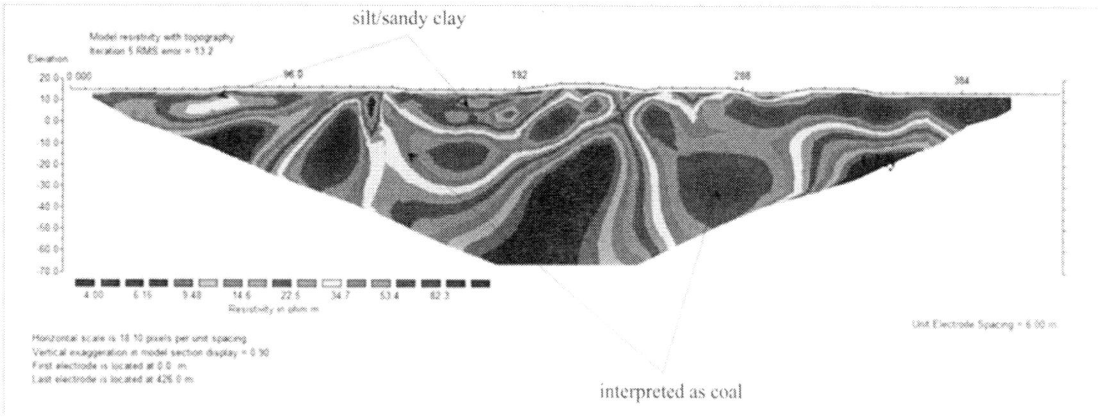

Figure 5. Distribution of Resistivity in Line 2.

Figure 5 shows distribution of resistivity in Line 1. The variation of resistivity starts from 4 Ωm to 90 Ωm with the depth about 80 meters from surface. Interpretation of resistivity imaging in Figure 5 revealed that distribution of coal resistivity is about 70–90 Ωm..The thickness of coal layer is about 10 meters and tends to distribute to the east of research area. Others lithology can be interpreted from the resistivity value are wet clay with resistivity value about 4–10 Ωm and silt or sandy clay with resistivity value about 10–70 Ωm. This line is near and parallel with Line 1 and reveals similarity of coal resistivity distribution.

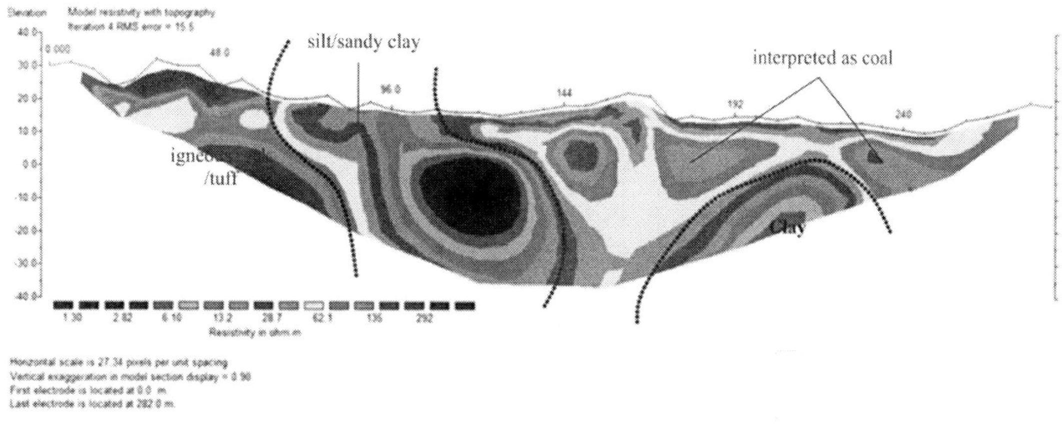

Figure 6. Distribution of Resistivity in Line 3.

Figure 6 shows distribution of resistivity in Line 1. The variation of resistivity starts from 1.3 Ωm to 300 Ωm with the depth about 60 meters from surface. Interpretation of resistivity imaging in Figure 6 revealed that distribution of coal resistivity is about 70–200 Ωm. The thickness of coal layer is about 10 meters and tends to distribute to the east of research area. Others lithology can be interpreted from the resistivity value are wet clay with resistivity value about 1.3–10 Ωm, silt or sandy clay with resistivity value about 10–70 Ωm, and tuff with resistivity value about >200 Ωm.

Compared with relevant research about coal resistivity value using resistivity imaging shown that coal resistivity ranging from 100 Ωm to 400 Ωm [8]. So, coal resistivity in this research is 70 Ωm to 200 Ωm adequately similar with other reference.

4. Conclusion

Resistivity investigation of coal resistivity distribution in Panaikang Village, Sinjai, and South Sulawesi had conducted successfully. Resistivity measurements were conducted in 3 lines with length 400 meters for Line 1 and Line 2 and 300 meter for Line 3 using resistivity imaging, dipole-dipole configuration. Resistivity data was processed using Res2DInv software to image resistivity distribution variation. The research results shown that coal resistivity in Line is about 70–200 Ωm, Line 2 is about 70–90 Ωm, and Line 3 is about 70–200 Ωm with average thickness about 10 meters and distributed to the east of research area. Based on this research, the coal in Panaikang Village is potential for mining.

Reference

[1] Sukamto R 1982 *Geological Map Sheet Ujung Pandang, Benteng and Sinjai, Center for Geological Research and Development, Directorate General of General Mining and Energy* (Bandung)
[2] Dahlin, T, Zhou B 2003 *Multiple-gradient array resistivity measurements for multichannel 2D imaging.* (Near Surface Geophysics, European Association of Geoscientists & Engineers)
[3] Loke M H 2004 *Tutorial: 2-D and 3-D Electrical Imaging Surveys*
[4] Wardah 2015 *Metode Geolistrik untuk Prediksi Arah Sebaran Kontaminan Limbah TPA Antang* (Makassar: Universitas Hasanuddin)
[5] Levieux 2014 *Electrical Tomography Survey in Soroako.* Geophysical. (Private reports for PT.VALE Indonesia Tbk)
[6] Santoso D 2002 *PengantarTeknik Geophysics* (Bandung: ITB)
[7] Mutmainnah 2014 *Determination of the depth of saprolite layer for Optimizing Drilling Region* Thesis (Makassar: University of Hasanuddin)
[8] Ngkoimani L, et al. 2017 *Journal of Physics: Conference Series* **846** 012015

Upwelling Dynamic Based on Satellite and INDESO Data in the Flores Sea

Reski Kurniawan, D. A. Suriamihardja, Alimuddin Hamzah Assegaf

Study Program in Geophysics, Faculty of Mathematics and Natural Sciences, Hasanuddin University

E-mail: reskur.r@gmail.com

Abstract. Upwelling phenomenon is crucial to be forecasted, mainly concerning the information of potential fishery areas. Utilization of calibrated model for recorded upwelling such as INDESO gives benefit for historical result up to the present time. The aim of this study is to estimate areas and seasons of upwelling occurrences in the Flores Sea using data assimilation of satellite and modeling result. This study uses sea surface temperature, chlorophyll-a data from level 3 of MODIS image and sea surface height from satellite Jason-2 monthly for three years (2014-2016) and INDESO model data for sea surface temperature, sea surface height, and chlorophyll-a daily for three years (2014-2016). The upwelling is indicated by declining of sea surface temperature, sea surface height and increasing of chlorophyll-a. Verification is conducted by comparing the model result with recorded MODIS satellite image. The result shows that the area of southern Makassar Strait having occurrences of upwelling phenomenon every year starting in June, extended to July and August. The strongest upwelling occurred in 2015 covering more or less the area of 23,000 km^2. The relation of monthly data of satellite has significantly correlated with daily data of INDESO model

1. Introduction

Oceanographic characteristics are determined by various oceanographic parameters, namely temperature, chlorophyll-a, and sea surface height. Temperature and chlorophyll-a are crucial resources for life of fish, and sea surface height is important to predict upwelling area [1]. Temperature is environmental parameter which is most often measured at sea because of its usefulness in studying physics, chemistry, and biology processes that happened at the sea. The distribution pattern of sea surface temperature can be used for identification other oceanographic parameters such as current, upwelling, and front [2]. Some of the parameters for describing oceanographic characteristic at the sea can be measured in two ways. The first way is direct measured oceanographic parameter at sea for example ship survey or by placing measuring instrument on research location (mooring, tide gauge, etc). The second way is using satellite technology [3].

Provision of oceanographic data through direct measurements are relatively involved large operating costs. Information data obtained from the satellite could estimate chlorophyll-a distribution, sea surface temperature, and the pattern of sea waves. Assimilation data from satellite and modeling data could interpret ocean phenomenon especially for the potential presence of schooling fishes which is associated with the upwelling phenomenon [4]. Some researchers have investigated for reviewing upwelling areas at the Flores Sea, predicted the areas of upwelling phenomena just using satellite data and have not yet compared with the modeling data.

An accurate monitoring and forecasting system for the ocean is certainly vital to managing Indonesia waters and its resources. To meet these challenging objectives the INDESO (Infrastructure Development of Space Oceanography) project was developed for the Government of Indonesia [5]. The Operational INDESO physical Ocean forecast system, at 1/12° horizontal resolution, provides 10 days of 3D ocean forecast and is updated weekly. The output of the INDESO model covers daily mean fields of atmospheric fluxes, physical parameters (temperature, salinity, currents, sea level), hourly surface

IOP Conf. Series: Journal of Physics: Conf. Series **979** (2018) 012049 doi:10.1088/1742-6596/979/1/012049

variables such as Sea Surface Temperature (SST), Sea Surface Height (SSH), and currents, as well as hourly values of all fields related to selected mooring sites and validation metrics [6].

The main objective of this study aims to obtain proper areas and significant months providing potential fishing grounds, based on SST, SSH, and chlorophyll-a maps using the assimilation of MODIS satellite images with INDESO model. Therefore, it is necessary to examine the correlation between satellite data and model data to construct reasonable judgment.

2. Methodology

The study area covers the Flores Sea as the whole depicted in figure 1 with southern latitude between 5°50'22.69" S and 8°30'32.43'' S, and with longitude 119°18'56.20'' E - 125°10'2.61'' E. Soft-wares used in this study for processing data are Ferret NOAA, SeaDAS 7.4, and Ocean Data View. Data to be analysed are (1) SST, chlorophyll-a from level 3 of MODIS image (obtained from www.oceancolor.gsfc.nasa.gov); (2) SSH from satellite Jason-2 (obtained from www.podaac-tools.jpl.nasa.gov) monthly for three years (2014-2016); and (3) INDESO model data for SST, SSH, and chlorophyll-a daily for three years (2014-2016) (obtained from www.indeso.web.id).

The ocean physical model named INDO12 used for the INDESO project is a regional configuration of the NEMO-OPA primitive equation OGCM (Madec et al., 1998; Madec, 2008). This model is forced at the surface using 3-hourly ECWMF atmospheric analysis and forecast fields and at the lateral open boundaries using Mercator Ocean global operational analysis and forecast fields. This regional ocean physical model provides physical fields for the past two weeks (forced by a global 2-weeks analysis) and ocean forecasts (forced by a global 10-day forecast) for the next 10 days. The regional INDESO physical ocean model is implemented with a horizontal resolution of 1/12° and 50 vertical layers with increased resolution near the surface. The horizontal grid is defined on a curvilinear ORCA grid but outputs are provided on a regular 1/12° grid. The regional INDESO physical ocean model has the following additional characteristics that is partial step representation of the bathymetry, explicit resolution of the tides, non-linear free surface to allow a good representation of tidal waves in coastal regions where their amplitude is large compared to the local depth (Levier et al., 2007), parameterization of internal tidal mixing as developed by Kock-Larrouy et al. (2007) for the Indonesian sea [6].

Processing data of SST and chlorophyll-a from Aqua-MODIS image is done by downloading the data in Hierarchical Data Format (HDF). The downloaded data is cropped according to the study area using software SeaDAS 7.4. The cropping result is then converted into * txt format in order to be able to export to Ocean Data View that serves as a data filter (quality control) to eliminate extreme high data and low extreme data that is not supposed to be the value of the parameters being searched. Quality control is used for SST in the ranges of $25^0C < SST \leq 32^0C$ and for chlorophyll-a in the ranges of $0\ mg/m^3 <$ chlorophyll-a $\leq 1.0\ mg/m^3$. The result of this data processing will be in a two-dimensional spatial distribution profile of SST and chlorophyll-a concentration in the * tif format.

Data processing for INDESO model data using Ferret application from NOAA, this application run under Ubuntu Linux operating system. The SST and chlorophyll-a data are processed to produce spatial profiles for two seasons to analyze upwelling phenomena, then SST and SSH data are created in order to be able to overlay through the obtained graphs in time series to examine the relationship between the two parameters to upwelling phenomena.

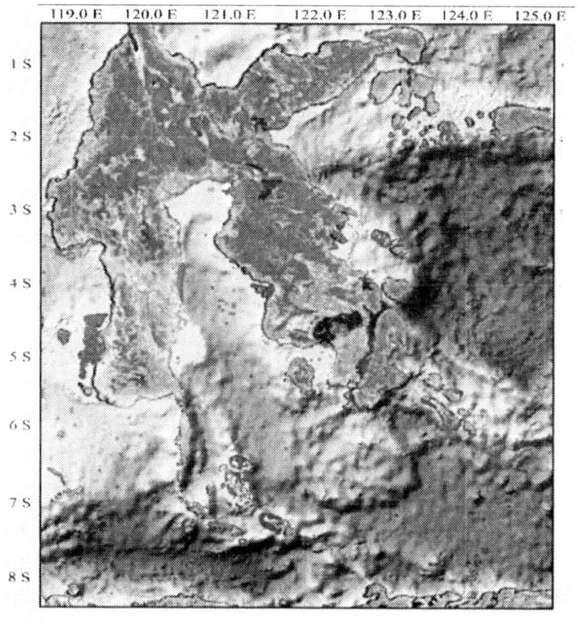

Figure 1. Map of the Study area.

3. Results

3.1. Verification Data

Figure 2 depicts the data from satellite and INDESO model that shows a similar pattern of SST and chlorophyll-a. Root Mean Square Error (RMSE) value between the two data origins is about 0.5 for SST and 0.57 for chlorophyll-a concentrations. Those results informed that the model has a good level of trust. The two data origins show strong correlation value, as for SST is about 0.91 and as for chlorophyll-a concentrations is about 0.94.

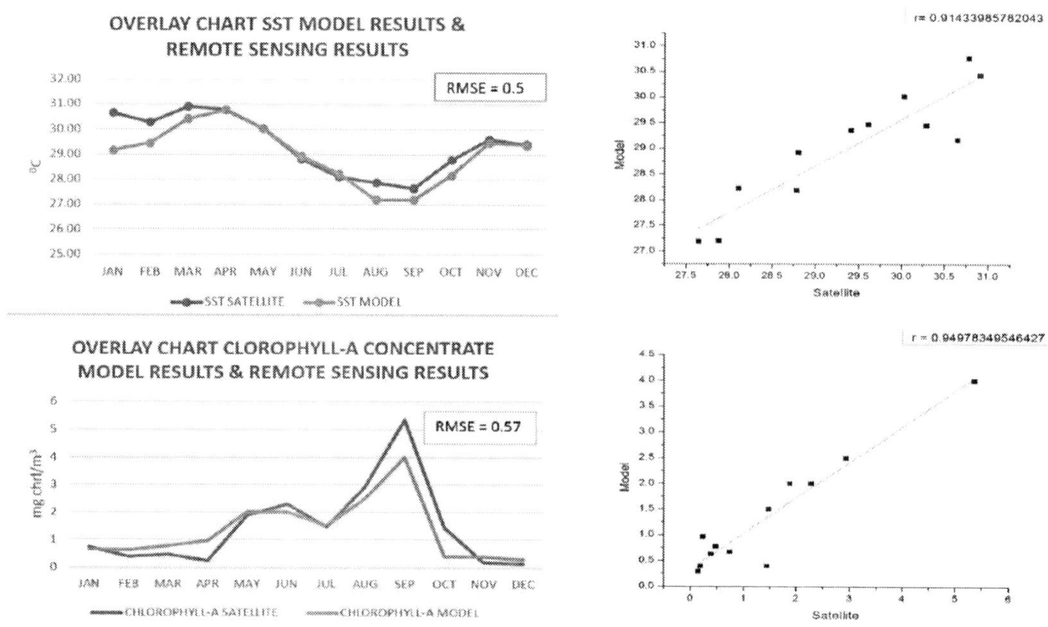

Figure 2. Chart comparison between model results and satellit.

IOP Conf. Series: Journal of Physics: Conf. Series **979** (2018) 012049 doi:10.1088/1742-6596/979/1/012049

3.2, Distribution Sea Surface Temperature (SST) and Sea Surface Height (SSH)

The result of monthly satellite image from level 3 and INDESO model at the Flores Sea showed homogeneous distribution pattern of SST. The more homogeneous distribution pattern of SST at the Flores Sea is found in the west monsoon (December-February). To be the relatively high temperature in the range of 28-31^{0}C. The east monsoon begins in June characterized by decreasing temperature at southern Makassar Strait as depicted in Figure 3 and figure 4.

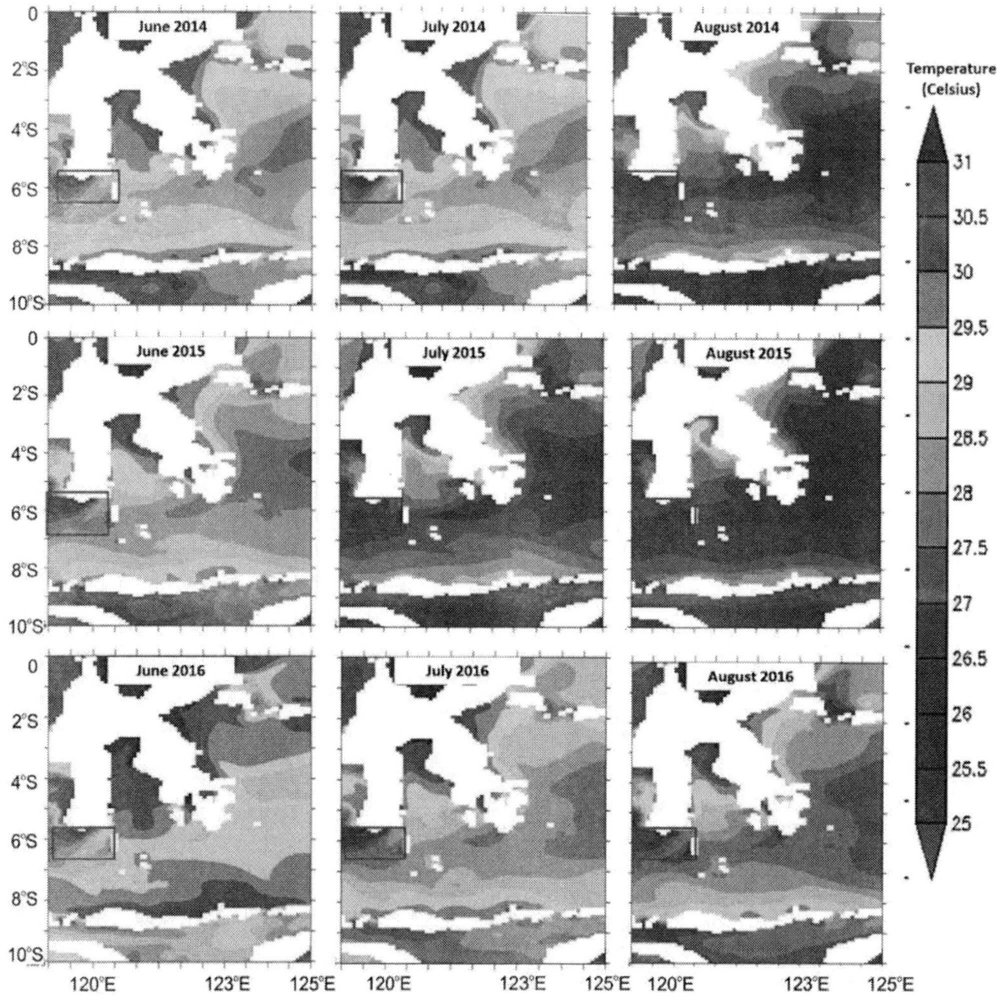

Figure 3. INDESO model results in distribution pattern of SST on east monsoon (June-August).

The more visible decreasing temperature in the area is found in July and August, it is supposed to indicate upwelling phenomena. The low value of SST indicates upwelling phenomenon reached its culmination in the first week on July with water mass temperature is about 23^{0}C-24 ^{0}C and continued until the fourth week on August, then weakened until disappear in the fifth week on September which is in agreement with the results of Bidawi (2010) [7].

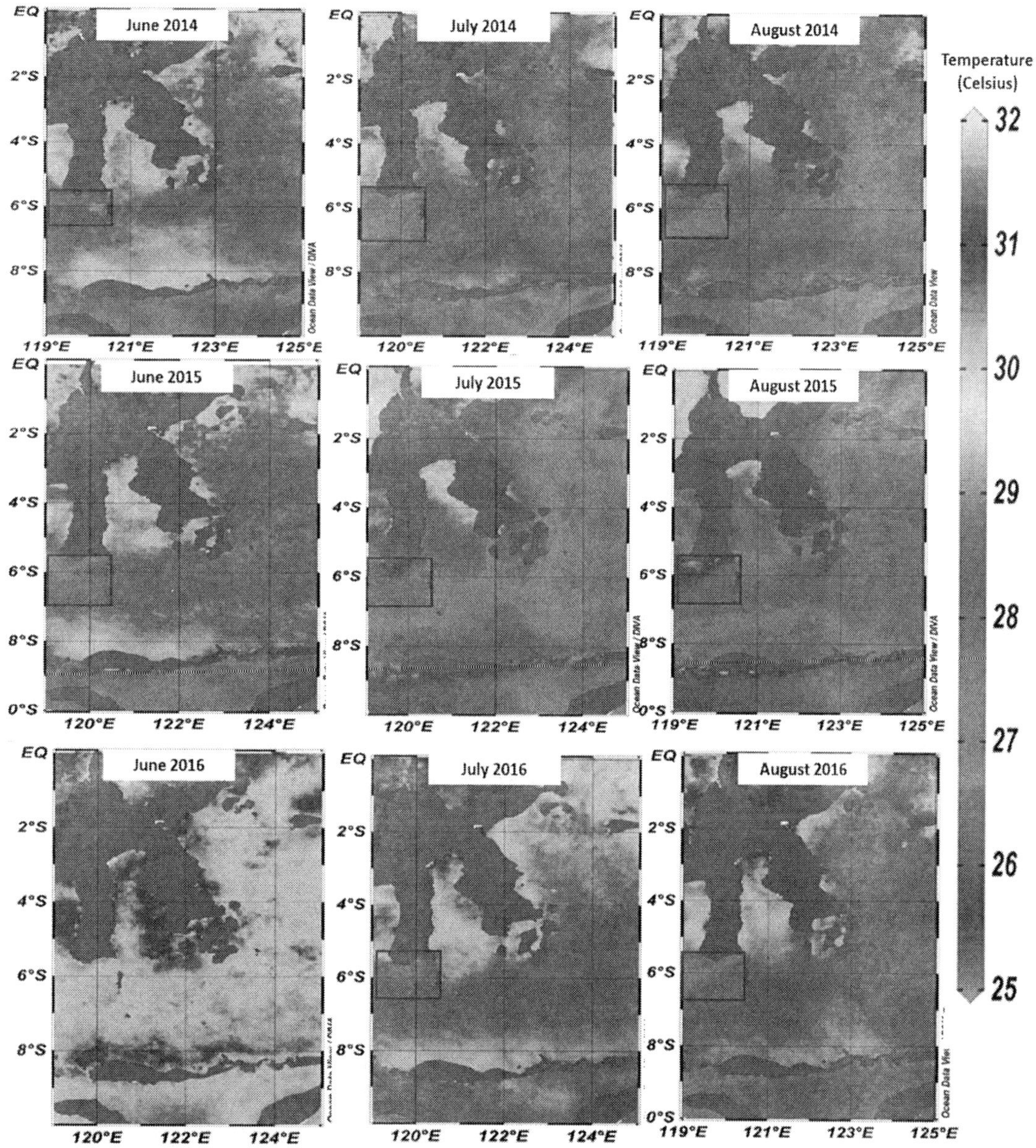

Figure 4. Satellite results in distribution pattern of SST on east monsoon (June-August).

Generally, the percentage rate of distribution SST for upwelling phenomena on 2014, 2015, and 2016 begins on July that happened on southern Makassar Strait and reached its culmination on August, which showed the phenomenon of expanding area of low SST value indicated by increasingly expanding of upwelling phenomena. Decreasing SST value is most visible in June and July which indicates the existence of upwelling phenomena. On July-August these phenomena are most visible with the distribution pattern of SST which distributed horizontally on southern Makassar Strait area, this pattern is in agreement with Inaku (2015) [8].

IOP Conf. Series: Journal of Physics: Conf. Series 979 (2018) 012049 doi:10.1088/1742-6596/979/1/012049

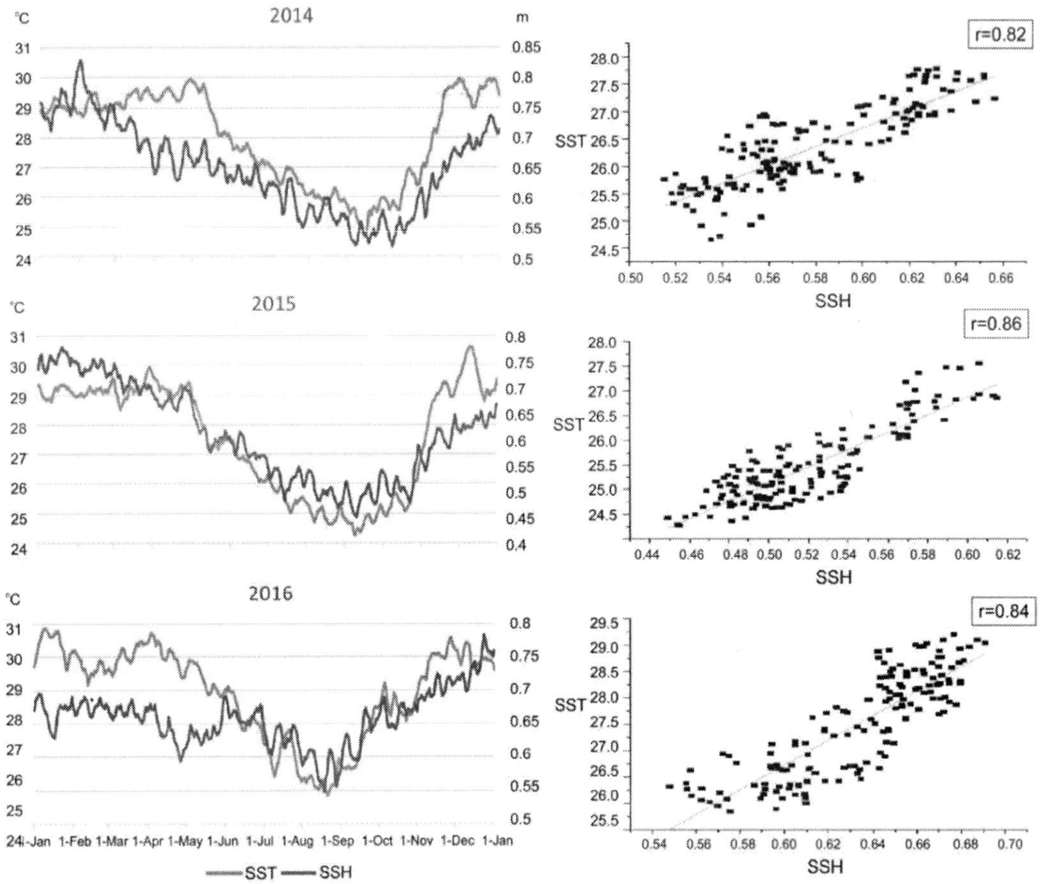

Figure 5. Overlay chart between SST & SSH (left), correlation chart between SST & SSH (right).

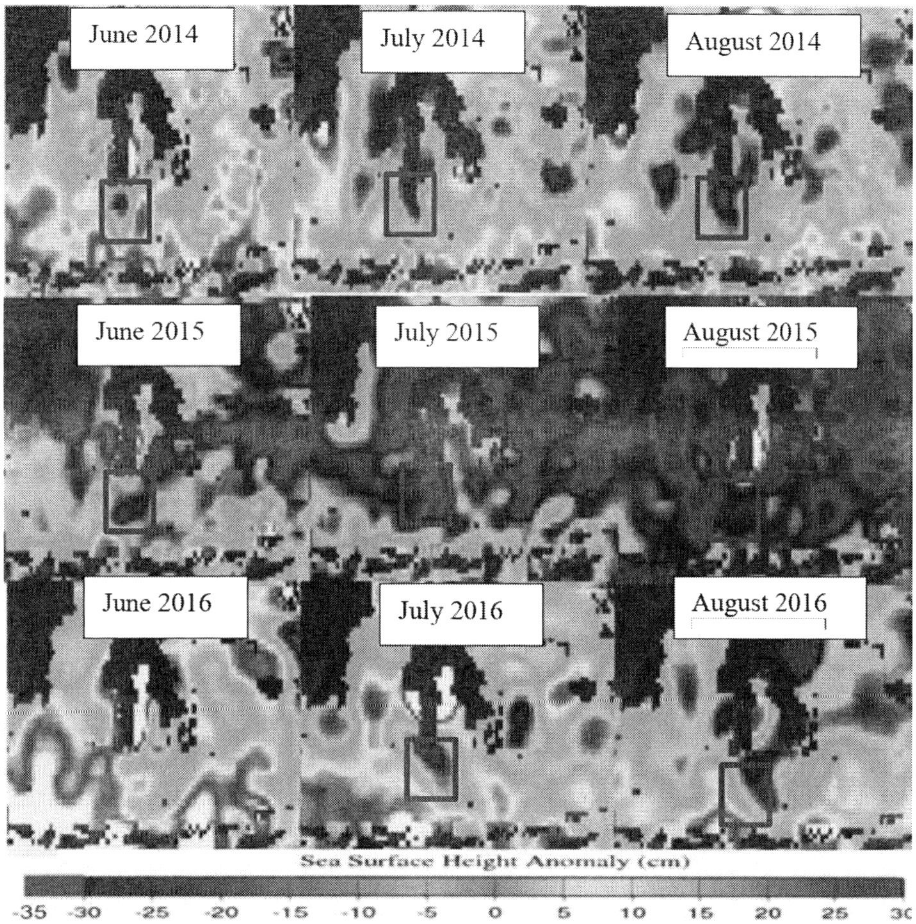

Figure 6. Jason-2 satellite results of sea surface height (June-August).

IOP Conf. Series: Journal of Physics: Conf. Series **979** (2018) 012049 doi:10.1088/1742-6596/979/1/012049

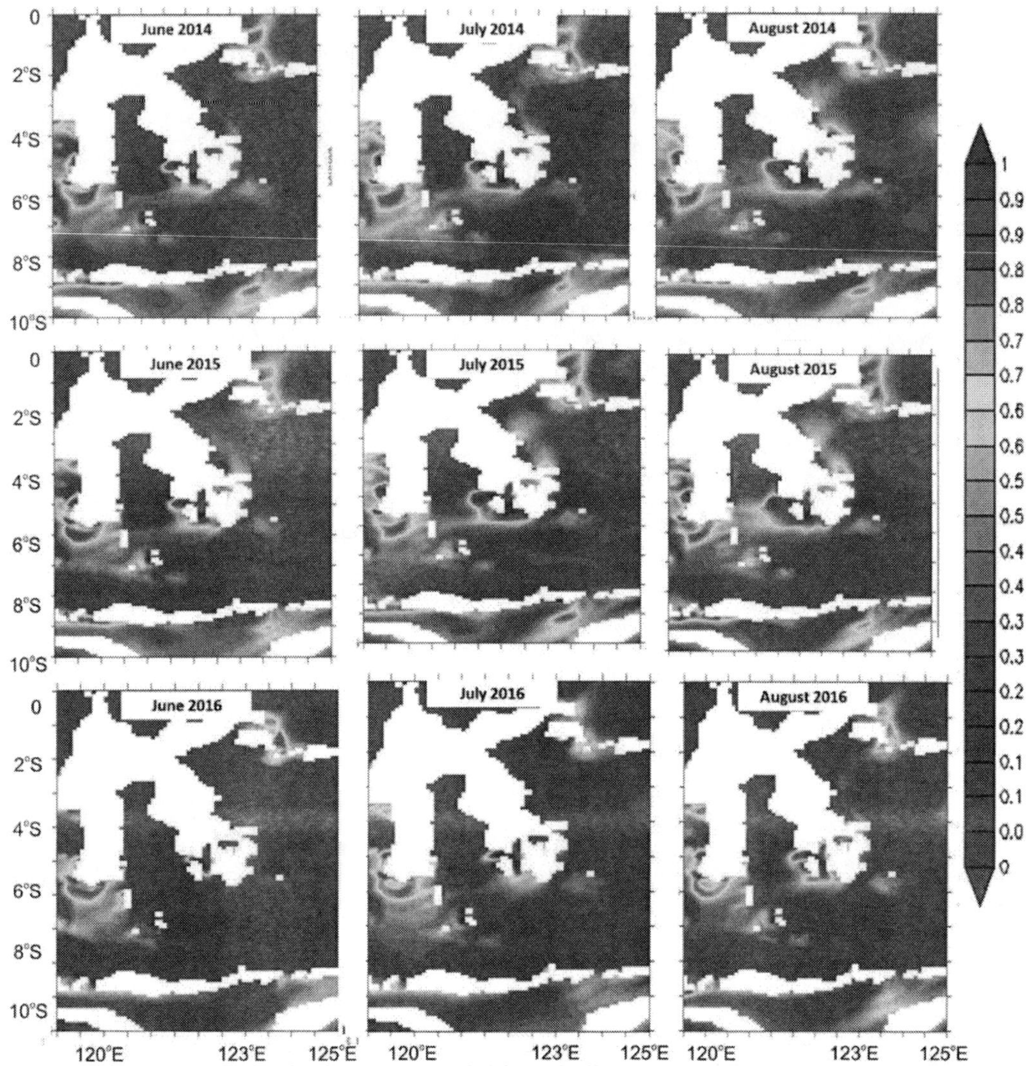

Figure 7. INDESO model distribution pattern of chlorophyll-a concentrations on east monsoon (June-August).

IOP Conf. Series: Journal of Physics: Conf. Series **979** (2018) 012049 doi:10.1088/1742-6596/979/1/012049

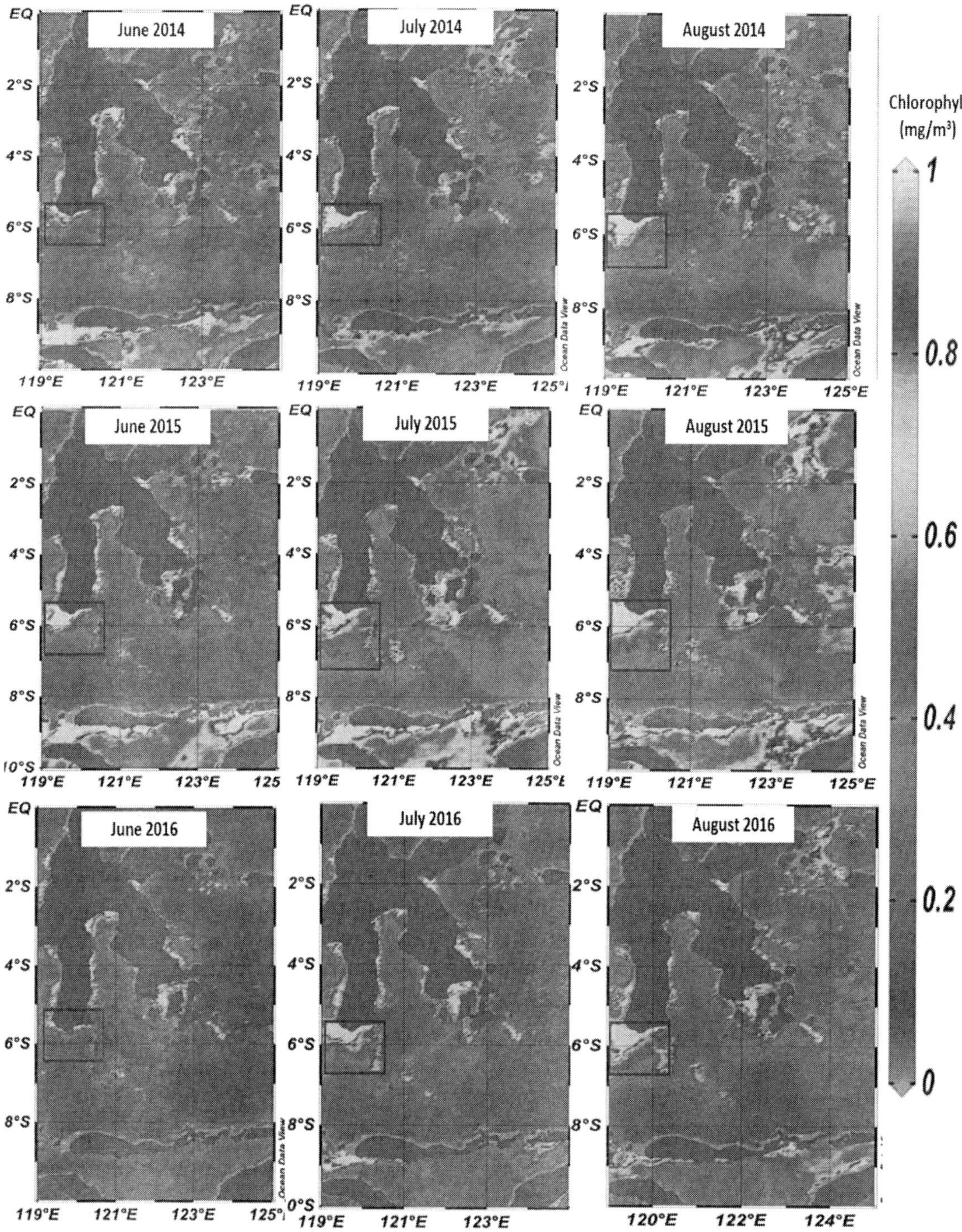

Figure 8. Satellite result distribution pattern of chlorophyll-a concentrations on east monsoon (June-August).

This decreasing temperature followed by low SSH value on southern Makassar Strait during east monsoon, this trend gives a way to overlay between SSH and SST which indicates the upwelling phenomena, this trend can be seen in figure 6. Coefficient correlation between the coordinate of SST and coordinate of SSH has strong value that is 0.8 every year. It means that the troughs and peaks of SST and SSH are relatively coincided.

SSH profile in the ocean is dynamic always changing depending on wind speed blowing over sea surface and strongly influenced by the melting ice in the polar region. The existence of upwelling phenomena also affects the profile of SSH, which causes the value of SSH to be low or even minus. Figure 6 shows areas in southern Makassar Strait has a dynamic SSH and low SSH value which indicates that area is experienced upwelling formation.

3.2. Distribution chlorophyll-a

Apparently, the level of chlorophyll-a concentrations looks different every season. During west monsoon, the level of chlorophyll-a concentrations on the Flores Sea reaches a low value. However, the level of chlorophyll-a concentrations relatively high encountered in coastal waters instead of offshore areas. This fact is reasonable due to the influence of nutrient input from the land caused by relatively high rainfalls levels in west monsoon.

The spatial distribution of chlorophyll-a concentrations during the east monsoon shows a relatively high increase in June and its culmination occurs in August. The high concentrations of chlorophyll-a in east monsoon previously preceded by low SST that indicates upwelling formation, this is in accordance with Wyrtki explanation (1961) [9] which stated that the upwelling in this area occurs in the east monsoon (June-August). At the beginning of the second transition season in September, the pattern of upwelling is still apparent, estimated in October as the end of the upwelling phenomena, this is seen from the appearance of chlorophyll-a concentrations that begin to decrease again this is shown in figure 7 and figure 8.

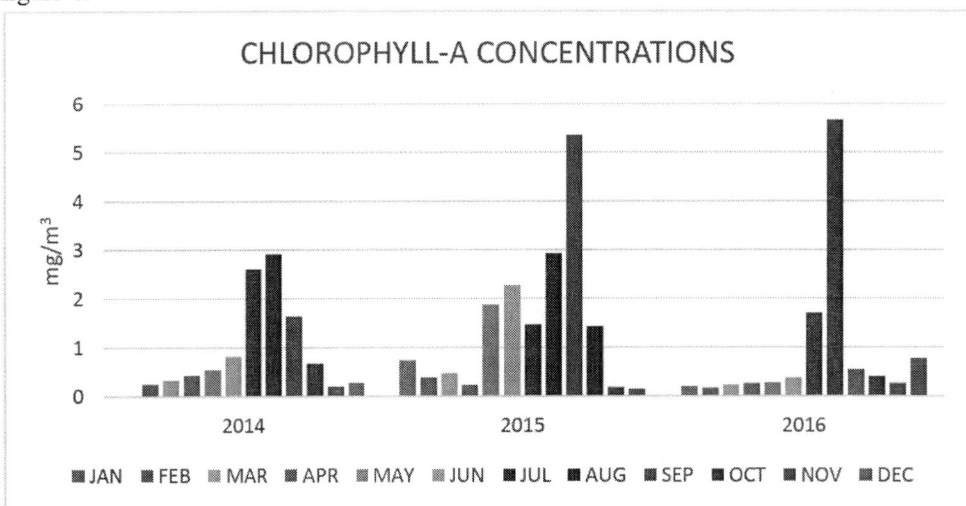

Figure 9. Chlorophyll-a concentrations distribution on southern Makassar Strait.

In July it was seen that the pattern of chlorophyll-a concentrations distribution is relatively high in the surrounding coastal waters especially the southern Makassar Strait, while in July until August distribution pattern began to look widespread to the southwest of Sulawesi Island towards the Flores Sea. Figure 8 depicts the level of chlorophyll-a concentrations in the southern Makassar Strait in 2014, 2015, and 2016 that shows the chlorophyll-a concentrations begins to increase and reaches its culmination in the east monsoon each year, the increased chlorophyll-a concentration is obviously given evidence to the widespread distribution pattern in southern Makassar Strait.

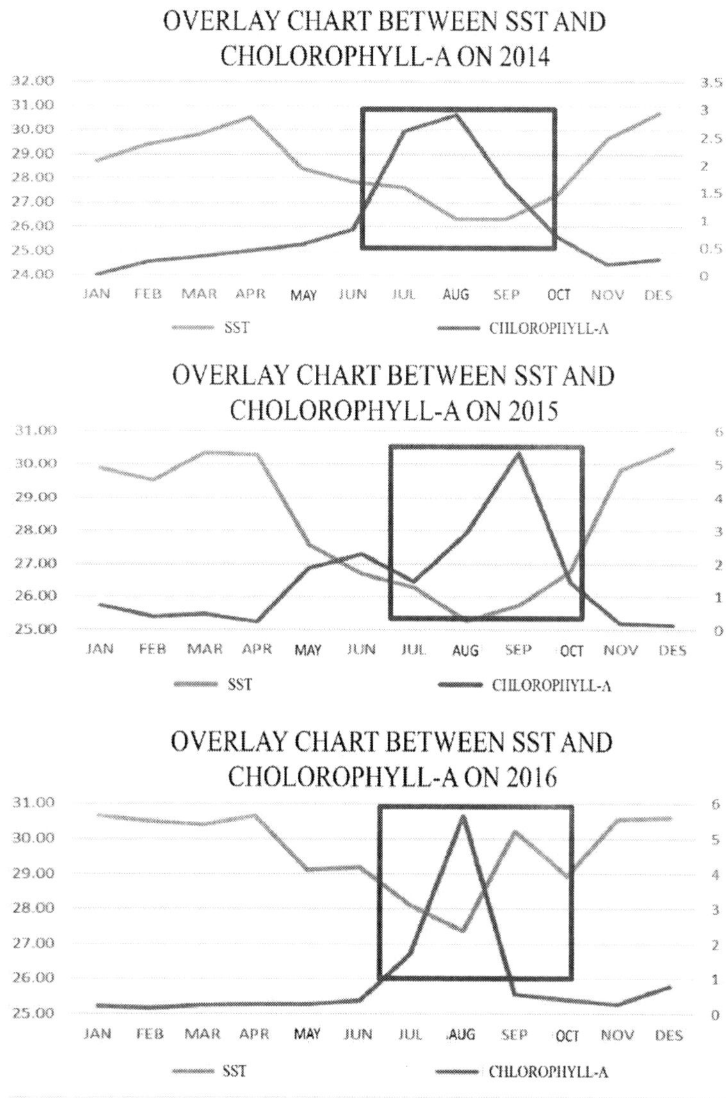

Figure 10. Overlay chart between SST and chlorophyll-a.

3.3. Upwelling Fluctuations

Distribution pattern obtained from the analysed results of SST, SSH, and chlorophyll-a in 2014, 2015, and 2016 indicates that the average SST decrease begins in May or when it enters the east monsoon, which is then followed by decreasing SSH and increasing chlorophyll-a concentrations on southern Makassar Strait which spread to the Flores Sea as depicted in.

The decline of SST in 2014 begins in the last week of May then reaches its culmination in the first week of September and ends in the third week of October, followed by increased chlorophyll-a concentration in August and September which then declines again in October. In 2015 the decrease of SST begins in the second week of May then climax in the second week of July and ends in the last week of October, followed by increased chlorophyll-a in August and September which then declines in October. The phenomena that occurred in 2016 were not much different from the previous two years, the decrease in SST starting in the last week of May then climax in the second week of August and ending in the second week of September, as well as the increase in chlorophyll-a concentrations that occurred in August and decreased at September. In the three years, 2015 was the peak of upwelling with the spreading area of the decreasing SST from the southern part of the Makassar Strait to the southwest Makassar Strait towards the Flores Sea. The time of upwelling phenomena are peaked in all years (2014-2016), but the pattern of upwelling spread clearly pointed towards the southwest of Sulawesi Island. The total upwelling area for 2014 is estimated to be about 15,000 km^2, then by 2015 to be about 23,000 km^2, and by 2016 to be about 6,000 km^2.

4. Conclusion

As mentioned beforehand, this study aims to obtain proper areas and significant months providing fishing ground which give benefit to fishermen. According to the maps of SST and Chlorophyll-a which have been shown in figure 3, 4, 7 and 8 can be concluded that the significant fishing grounds are in the area of offshore waters of Jeneponto and Takalar Regencies covering more or less 23,000 km^2. According to the data recorded in three years, the most probable season in providing fishing ground is the months of June, July and August which is mainly during east monsoon.

This study involved the two sources of data, the first is obtained from monthly satellite data, and the second is obtained from daily INDESO model. Based on the fact provided by figure 2, the two sources are considered to be significantly related due to the value of correlation is achieving 0.9.

Acknowledgments

We appreciate to Balai Penelitian dan Observasi Laut Jembrana Bali for letting us use INDESO data for three years (2014-2016), and especially for Ms. Desy Berlianty for her introducing and guidance in using the program of INDESO for whom we are indebted.

References

[1] Hakim M R 2011 *Karakteristik Oseanografi di Permukaan Perairan Utara Jawa, Selatan Lombok Hingga Sorong, Papua Barat Pada Musim Timur 2010 Skripsi* (Bandung: Institut Pertanian Bogor)

[2] Amri K, Asep P and Suprapto 2014 Karakteristik Oseanografi dan Kelimpahan Fitoplanton di Perairan Selat Sunda Pada Musim Timur *Jurnal BAWAL* **6** (1)

[3] Wilopo M D *2005 Karakter Fisik Oseanografi di Perairan Barat Sumatera dan Selatan Jawa-Sumbawa Dari Data Satelit Multi Sensor Skripsi* (Bandung: Institut Pertanian Bogor)

[4] Sukoraharjo S S, Djisman Manurung, Indra Jaya, Bonar P. Pasaribu and Jonson L G 2011 Menduga Penaikan Massa Air Dengan Menganalisis Pola Pergerakan Angin di Perairan Selat Makassar *Jurnal Kelautan Nasional* **6** (3)

[5] Nugroho D, Koch-Larrouy A, Gaspar P, Lyard F, Reffray G and Tranchant B 2016 Modelling Explicit Tides in the Indonesian Seas: An Important Process for Surface Seawater Properties *Marine Pollution Bulletin* https://doi.org/10.1016/j.marpolbul.2017.06.033

[6] INDESO 2013 A System Implemented by CLS For Balitbang KP all rights reserved

[7] Hasyim B, Sayidah S and Maryani H 2010 Kajian Dinamika Suhu Permukaan Laut Global Menggunakan Data Penginderaan Jauh *Majalah Sains dan Teknologi Dirgantara* **5** (4) 130-143

[8] Inaku D F 2015 Analisis Pola Sebaran dan Perkembangan Area Upwelling di Bagian Selatan Selat Makassar *Torani (Jurnal Ilmu Kelautan dan Perikanan)* **25** (2) 67-74

[9] Wyrtki K 1961 Physical Oceanography of the South East Asian Water *NAGA Report* **2** Scripps Inst.Oceanography La Jolla (California: The University of California)

Earthquake Macro-zonation Based Peak Ground Acceleration, Modified Mercalli Intensity, And Type of Rocks around Matano Fault

Muh Karnaen, D A Suriamihardja, A Maulana, A Jaya
Hasanuddin University, Makassar, Indonesia

Email: muh.karnaen@gmail.com

Abstract. This study aims to determine earthquake vulnerable zones. We conducted research on earthquake macro-zonation based on PGA, Modified Mercalli Intensity (MMI), and type of rocks around Matano Fault in the area of 1.60 S to 2.990 S and 120.50 E to 122.470 E. We have acquired Maximum PGA and Modified Mercalli Intensity (MMI) for each observation point on the ground from the four major earthquake events. The empirical model is used due to lack of acceleration data recorded. We tried some empirical methods, but the McGuire method is found to be acceptable for this area. The result gives the maximum variation of PGA which is ranged between 18.40 - 363.54 gals. While the variation of MMI using empirical Wald attenuation gives values ranging from 2.9 - 7.7 MMI. The most vulnerable zone is located around Sorowako city with PGA value of 326.55 gals and MMI value of 7.5 MMI. This area is between ultra-basic rock and metamorphic rock formation. The vulnerable zone is near largest earthquake 6.2 M on 15-02-2011.

1. Introduction

Macro-zonation and micro-zonation are efforts to evaluate and to map disaster prone area in the region. The zonation figures out how strong the effect of earth movement caused by earthquake [1]. Macro-zonation gives basic data and regulatory information on earthquake vulnerability in the level of region under consideration. Macro-zonation must be prepared in advance prior to evolve micro-zonation [2]. Calculations on earthquake shocks can be conducted using intensity, PGA, PGV, and spectral analysis. To measure the strength of shock, the PGA approach is more reliable than the scale of seismic intensity [3]. In this connection, intensity values are obtainable theoretically derived from PGA through empirical calculations [4].

High PGA values correspond to most vulnerable areas of earthquake hazard. The value of PGA is influenced by attenuation factors related to distance of hypocentre to observation area as well as local geological factors and site local effect [5]. Variations of attenuation and acceleration factors depend on the strength, location, and movement of the faults [6]. Shake-map is a zone of vulnerable zonation generated by the model to illustrate the level of shock hazard and be a means to give estimates and respond to potential hazards [7].

Intensity is a qualitative scale of observation of people and buildings due to earthquake shocks this is related to vulnerability in the zone [8]. One of the most important factors is the sediment in the area [9]. The PGA of rock is the most vulnerable to earthquake hazard, and it is related to rock formation

[10]. This study used PGA, intensity, and geological factors for macro-zonation earthquake investigation.

This macro-zonation study of the vulnerability of natural disasters caused by this earthquake will be applied to the Matano Fault that extends horizontally in the central part of Sulawesi Island. The PGA calculation model is a simple way of determining the vulnerable zone around the fault because it only uses the earthquake analysis parameter input. The empirical PGA formulas produced by [11, 12, 13] can be applied to this study because they have the same characteristics in the research zone. The PGA model to be generated in this study will be converted to the intensity to be combined with local geological data.

2. Theory

The lack of data from the acceleration instrument around the Matano Fault makes the empirical method a feasible study in obtaining PGA values. Some experts have provides some empirical formulas that fit the criteria in their research zone. The models listed below are some of the empirical equations that will apply to this research.

1. McGuire (1977)

$$E_v = \alpha 10^{bM}(R+25)^{-c} \tag{1}$$

where E_v = eventuality of acceleration (gal), alpha (α) is a constant of 472, b is 0.278, and c is 1.301. While M = Magnitude, and R = distance of hypocentre. The depth of the earthquake being input from 9 to 70 km. This empiric is more efficient for an earthquake with strength of 6.5 and hypocentre not more than 50 km.

2. Humbert and Viallet (2008)

$$\log(A) = aM + bR - \log(R) + c \tag{2}$$

The A is a PGA calculated value (cm/s^2), the alpha (a) is a constant of 0.31, b is -0.00091, and c is 1.57. While M = Magnitude, and R = distance of hypocentre. Focal depths of earthquakes are between 0 and 30 km. This empiric is more efficient for earthquakes with magnitudes between 4.0 – 7.8 M_S and distances of hypocentre less than 140 km.

3. Akkar and Bommer (2010)

$$\log y = b_1 + b_2 M + b_3 M^2 + (b_4 + b_5 M)\log\sqrt{R_{jb}^2 + b_6^2} + b_7 S_S + b_8 S_A + b_9 F_N + b_{10} F_R \tag{3}$$

This empirical has result y that shows PGA in cm/s^2, the constant b1 = 1.04159, b2 = 0.91333, b3 = 0.08140, b4 = 2.92728, b5 = 0.28120, b6 = 7.86638, b7 = 0.08753, b8 = 0.01527, b9 = ¡0.04189, b10 = 0.08015, σ1 = 0.2610 (intra-plate event) and σ2 = 0.0994 (inter-plate event). Most records from earthquakes with magnitudes between 4.0 – 7.6 M_S and the distance of hypocentre less than 200 km.

4. Wald (1999)

$$I_{MM} = 3.66\log\alpha - 1.66 \tag{4}$$

where I_{MM} is intensity of earthquake in Modified Mercalli Intensity (MMI). The range intensity is between 1 to 10 MMI. Log α is PGA logarithmic at the study area.

The geological structure will affect the value of the variables of seismic activity, the acceleration of the ground, the intensity of seismicity, and attenuation. These variables can be used as a method to describe the risk parameter of earthquake disaster vulnerability in a region. PGA is the suitable choice among the method for this research, besides that is the simple, quick access data from regional earthquake centre in BMKG agency. Perhaps this method is more accurate if complementary other methods.

IOP Conf. Series: Journal of Physics: Conf. Series **979** (2018) 012050 doi:10.1088/1742-6596/979/1/012050

3. Data and Processing

An empirical approach is needed due to the lack of accelerated ground data from accelerograph instruments for strong earthquakes. Using the empirical model obtained the calculation of the intensity and maximum ground acceleration for each point of coordinate observation. From the data of intensity and maximum ground acceleration for each coordinate point, we get the contour of intensity and maximum land acceleration.

The data of earthquakes used in this research is the earthquake BBMKG database Region IV Makassar in the period of year 2009-2017. These earthquake parameters are the position of the epicenter (latitude and longitude), depth, magnitude, and earthquake occurrence time. During 8 years of observations, we obtained data of 547 earthquakes which is depicted in figure1.

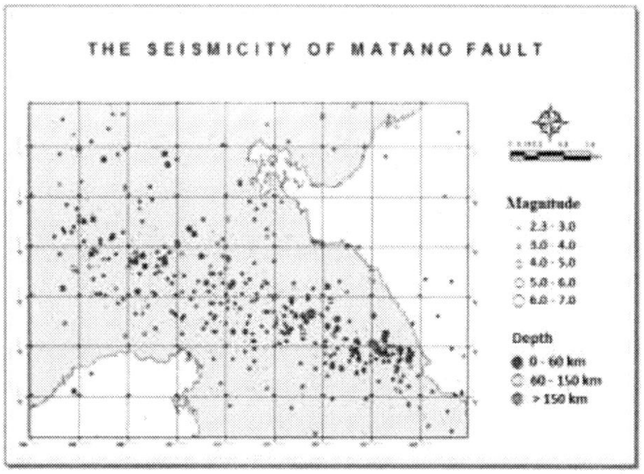

Figure 1. The Seismicity of Matano Fault.

Data processing includes:

1. We selected four significant earthquakes events as PGA variable input (table 1).

Table 1. Parameter of the four significant earthquakes.

No.	Date	Time	Location	Depth (km)	Magnitude (M_b)
1.	15/02/2011	13.33.55 GMT	2.47 S - 121.55 E	34	6.2
2.	03/12/2014	00:27:05 GMT	2.87 S - 122.39 E	10	5.9
3.	16/04/2012	02:17:50 GMT	2.63 S - 121.85 E	10	5.8
4.	24/05/2017	09:10:16 GMT	2.78 S - 122.17 E	10	5.7

2. Determining the observation point with grid interval 0.10 x 0.10 (longitude-latitude) at the research location.
3. Calculating the PGA value of by the equation (1), (2), and (3) at the observation point.
4. Mapping the results of the maximum PGA calculation at the latitude - longitude of the observation point then determining the contour and interpolation
5. Plot the intensity (MMI) value on each grid with the Wald attenuation equation.

4. Result and Discussion

4.1 Result

From the empirical calculation data the PGA of 4 earthquake events produce 68 grids of point observations, where in each grid there is the maximum value of PGA in Matano Fault as depict figure 2 below.

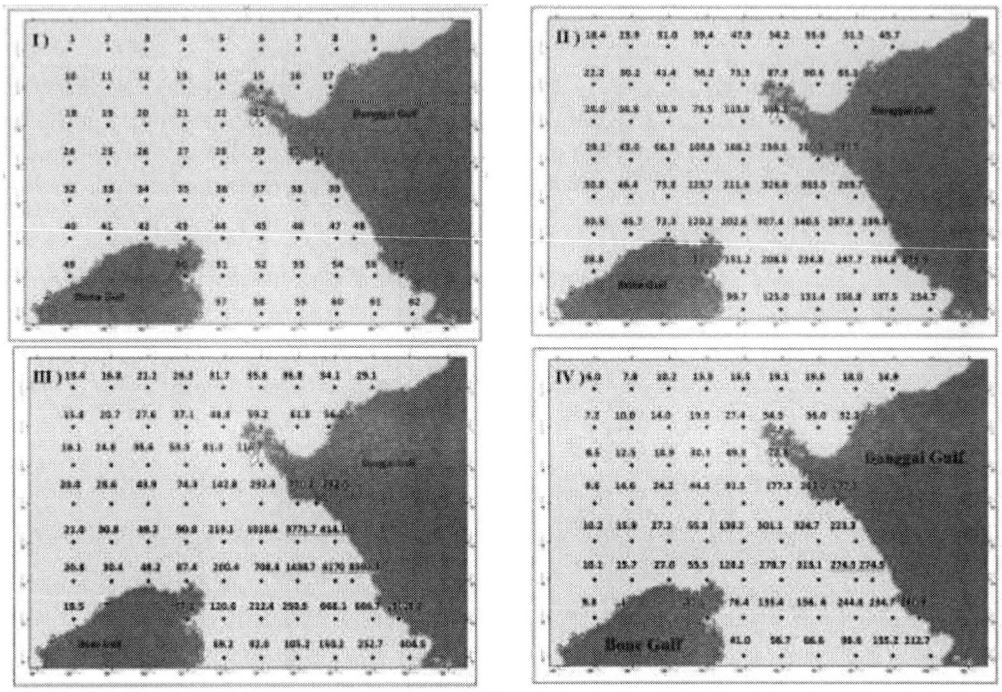

Figure 2. The number of observation point (I), the result of PGA at each point from McGuire empiric method (II), the result of PGA from Humbert and Viallet empiric method (III), and the result of PGA from Akkar and Bommer empiric method (IV).

From the results it is shown that the McGuire equation is more acceptable than other empiric equation, because the value of PGA is appropriate with the PGA region of Indonesia [14] and the deviation not too big. The PGA result of Humbert and Viallet empiric method has big deviation (between 13.4 and 15025.3) and the PGA result of Akkar and Bommer empiric method has very small value. The result from McGuire equation gives lowest PGA at the observation point of 1 (18.4 gal) and the highest PGA at the observation point of 38 (363.5 gal). This highest PGA is near the epicentre 18 km northeast of Sorowako on February 15, 2011 at 10:33:55 WITA depth of 34 km with magnitude of 6.2 M.

The McGuire's PGA distribution is as follows:

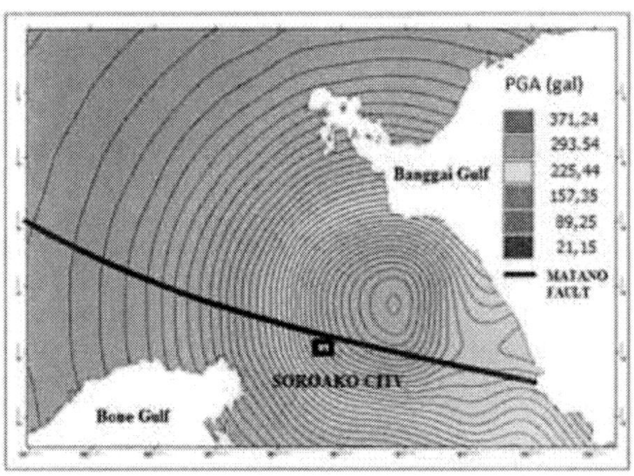

Figure 3. Macro-zonation Maximum PGA around Matano Fault.

IOP Conf. Series: Journal of Physics: Conf. Series **979** (2018) 012050 doi:10.1088/1742-6596/979/1/012050

Intensity is obtained from the PGA of [13] Wald equation intensity is between II to VII MMI. The PGA results and intensity can be compared with the geological map (figure4). It is shown that the intensity correlate to the direction of Matano Fault, which decrease inland from east to west.

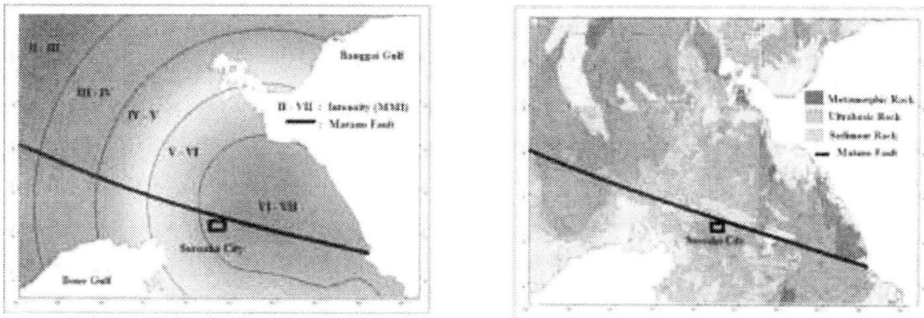

Figure 4. The zonation of Intensity (MMI) and geology map of Matano Fault [15].

Based on comparison between the intensity and local rock type (figure 3), it is shown that highest intensity is dominated by ultrabasic rocks, medium intensity with a composite combination between metamorphic and ultra-basic rocks and lowest intensity has composition of metamorphic rocks. It can be concluded that the vulnerable areas are dominated by ultrabasic rocks with high PGA and MMI.

4.2 Discussion
Matano fault is an active area of the earthquake (figure 1). Several large earthquakes at Matano Fault (table 1) caused the loss of property and caused some people to be injured. Soroako City is the area most affected by the earthquake because it is nearest to the epicentre of the largest earthquake in the area. To describe the level of disaster in Matano Fault required some PGA analysis using PGA interpolation by means of empiric equation.

Employed some empirical methods (equation 1, 2 and 3 in theory) can be used to conclude that McGuire is most suitable than the others [11, 12]. Obviously, the parameter data used by McGuire have similarities with the data parameter in this study. The similarity data is the focal depth of earthquakes at shallow layer and the distribution of earthquakes in the land area. The focal depth in Matano Fault dominated by shallow earthquakes and input data for PGA calculation are between 10-34 km depth of earthquakes.

The largest earthquake (6.2 M) in February, 15, 2011 about 19 km North East direction of Soroako City gives maximum PGA at least 363.5 gal. It causes Soroako City is the most vulnerable area. The city has a dense population and heterogeneous building structures. Vulnerability factors are influenced by residents and building structures [8]. The largest PGA is near the source of the earthquake [6]. The variations in PGA values are increasing to east direction of Soroako city where it's dominated by ultrabasic rock. By using Wald equation, maximum intensity can obtained so at least VI-VII MMI felt around eastern part. Therefore, eastern Soarako city is most vulnerable area.

The macro-zonation of earthquake in this study based on PGA, intensity, and type of rock has developed to describe the risk parameters of earthquake disaster vulnerability in Matano Fault. However, it is not satisfied to know exactly the condition of earthquake disaster in this area. It still needs other macro-zonation method and should be complemented by micro-zonation of earthquake vulnerability.

5. Conclusions
From the paper have presented, it can be deduced that PGA is simply method for macro-zonation study because only used magnitude and distance of earthquakes for input data. Empirical calculations are done because of lack of instruments. This study is preliminary research of macro-zonation then

combined with other methods. McGuire's PGA empirical method is more suitable to study earthquake macro-zonation in Matano Fault compared to other methods. It has appropriate criteria for earthquake data of this area and the deviation from the result is not big.

The value of PGA and intensity is decreasing in the direction of the Matano Fault pattern that stretches from east to west. Geological observations show that high PGA values are in the area dominated by Ultra-basic rock. in the middle of a matano fault there is a town of sorowako that is threatened by earthquakes. The city is the most vulnerable area because it has high PGA and densely populated settlements.

Reference

[1] Wang Z 2008 Understanding Seismic Hazard and Risk Assessments: An Example in the New Madrid Seismic Zone of the Central United States *Proceedings of the 8th U S National Conference on Earthquake Engineering April 18-22 2006* 416 (San Francisco California USA)
[2] Nath S K and Thingbaijam K K S 2009 Seismic Hazard Assesment – A Holistic Micro-zonation Approach *Natural Hazard and Earth System Science* **9** 1445-1449
[3] Mihalic S, Ostric M and Krkac M 2011 Seismic Micro-zonation: A review and Principle Practice *Geofizica* **28** 1
[4] Wald D J, Quitoriano V, Heaton T H and Kanamori H 1999 Relationships Between Peak Ground Acceleration, Peak Ground Velocity and Modified Mercalli Intensity in California *The Professional Journal of the Earthquake Engineering Research Institute* **15** 3 557-564
[5] Ganapaty G P 2010 First Level Seismic Micro-zonation Map of Chennai city – a GIS Approach *Natural Hazard and Earth System Science* **11** 549-559
[6] Ardeleanu L, Grecu B and Raileanu V 2012 Peak Ground Acceleration, Velocity, and Displacement from Moderate Magnitude Undercrustal Earthquakes of Vrancea Region *Romanian Report on Physics* **64** 2 555-570
[7] Atkinson G M and Kaka S I 2007 Relationships between Felt Intensity and Instrumental Ground Motion in the Central United States and California *Bulletin of the Seismological Society of America* **97** 2 497-510
[8] Edward M R, Robinson D, McAneney K J and Schneider J 2004 Vulnerability of Residential Structures In Australia *13th World Conference on Earthquake Engineering, B C Canada* 1-6 2985
[9] Langston C A 2003 Local Earthquake Wave Propagation Through Mississippi Embayment Sediments, Part I: Body-Wave Phases and Local Site Responses *Bulletin Seismological Society of America* **93** 2664-2684
[10] Schanabel P, Seed H B and Lysmer J 1973 Modification of Seismograph Records for Effects of Soil Conditions *Bulettin Seismology Society of America* **62** 1649- 1664
[11] Akkar S and Bommer J J 2010 Empirical Equations for The Prediction of PGA, PGV and Spectral Accelerations in Europe the Mediterranean Region and The Middle East *Seismological Research Letters* **81** (2) 195–206
[12] Humbert N and Viallet E 2008 An Evaluation of Epistemic and Random Uncertainties Included in Attenuation Relationship Parameters *Proceedings of Fourteenth World Conference on Earthquake Engineering* 07-0117
[13] McGuire R K 1997 Seismic Design Spectra and Mapping Procedures Using Hazard Analysis Based Directly on Oscillator Response *Earthquake Engineering Structural Dynamics* **5** 211–234
[14] Irsyam M, Sengara I W, Aldiamar F, Widiyantoro S, Triyoso W, Natawidjaja D H, Kertapati E, Meilano I, Suhardjono, Asrurifak M and Ridwan M 2010 *Ringkasan Hasil Studi Tim Revisi Peta Gempa Indonesia* (Bandung-Indonesia)
[15] Simanjutak T O, Rusmana E, Surono and Supandjono 1991 *Peta Geologi Lembar Malili Sulawesi Selatan* (Bandung: PPPG Bandung)

Response of Gravity, Magnetic, and Geoelectrical Resistivity Methods on Ngeni Southern Blitar Mineralization Zone

Sunaryo

Geophysical Engineering Program Study, Physics Department, Sciences Faculty, University of Brawijaya, Malang, Indonesia

E-mail: sunaryo@ub.ac.id

Abstract. The research with entitle response of gravity, magnetic, and geoelectrical resistivity methods on Ngeni Southern Blitar mineralization zone has been done. This study aims to find the response of several geophysical methods of gravity, magnetic, and geoelectrical resistivity in an integrated manner. Gravity data acquisition was acquired 224 data which covers the whole region of Blitar district by using Gravity Meter La Coste & Romberg Model "G", and magnetic data acquisition were acquired 195 data which covers the southern Blitar district only by using Proton Precession Magnetometer G-856. Meanwhile geoelectrical resistivity data only done in Ngeni village which is the location of phyropilite mining with the composition content of Fe, Si, Ca, S, Cu, and Mn by using ABEM Terrameter SAS 300C. Gravity data processing was performed to obtain the Bouguer anomaly value, which included unit conversion, tidal correction, drift correction, correction of tie point, base station correction, free air correction, and Bouguer correction. Magnetic data processing has been done by some corrections i.e daily, drift, and IGRF(International Geomagnetic Refference Field) to obtain the total magnetic anomaly. From gravity data processing has been obtained the simple Bouguer anomaly value in range from -10mGal until 115mGal. From this data processing has been obtained the total magnetic anomaly value in range from -650nT until 800nT. Meanwhile from geoelectrical resistivity 3.03Ωm until 11249.91 Ωm. There is a correlation between gravity anomaly, magnetic anomaly, and geoelectrical resistivity anomaly that are associated with deep anomaly, middle anomaly , and shallow anomaly.

1. Introduction

Physiographically, the southern Blitar region is part of the southern mountains that stretches from the Wonosari region of Central Java to the Blambangan peninsula in East Java. On the north of these mountains lined the volcanoes of quaternary in the Solo zone. While in the south is the Indian Ocean [1]. The southern Blitar area is bordered by Tulungagung district in the west, Kediri regency in the north, Malang regency in the east, and Indian Ocean in the south. In general these mountains are Tertiary (Oligocene, Miocene, to Pliocene) as well as Quarter, some formations are intruded by diorite, granodiorite, and dacite [2]. As a result of this intrusion, the surrounding rocks become hydrothermal alteration. As a result, in the area there are many mineralized zones that produce propylite, pyrite, or other ore minerals.

Metal minerals are generally deposit products as a result of the differentiation process and magma crystallization that occurs when the activity of the formation of igneous rocks will end. The beginning of this process begins with an intrusion process of high concentrations of hydrothermal solutions to older rocks. When the solution enters through the fractures and/or faults, an alteration processes to occur to the rock. Alteration processes to occur to high temperatures and pressures, followed by changes in mineral composition from rocks due to the enter of hydrothermal solutions. The process of alteration can be grouped into three kinds [3], namely: argilitisation, silisication, and propilitisation.

At the time of cool accompanied by the process of chemical and physical reaction, it will be deposited newly formed minerals. The minerals formed to depend on the type of hydrothermal solution, the type of rock being penetrated, and the level of cool process that occurs to the rock itself. Rock that vary the mineral content will produce several kinds of newly formed minerals. The end result of this process will accumulate in rock pores, fractures, faults, or rock layers. While the main characteristic of the alteration region is the formation of CO_2 which is the product of the alteration process. The presence and amount to CO_2 depend on the type of metal, magnesium and / or potassium of rocks that intruded. In this case the alteration process is followed by the formation of pyrite (FeS_2). This genesis is usually found around the body of the sulphide mineral. Other possible sulphide minerals may form to include chalcopyrite (CuS_2) and galena (PbS) [4].

Physically, the alteration zone has a higher density and susceptibility, but lower resistivity value compared to the host rocks in the surrounding. This occurs because the alteration zone which is the accumulation of metal minerals will have a low resistivity value or high conductivity value compared with nonmetallic minerals. The existence of these physical values contrasts that resulted in the application of integrated geophysical methods can be used to determine the distribution of underground mineral deposits [5]. Therefore in this paper will be presented a research result of the response method gravity, magnetic, and geoelectrical resistivity to mineralized zone located in Ngeni of south.Blitar area.

2. Methods

In this research, three geophysical methods are integrated, namely gravity, magnetic, and geoelectrical resistivity method.

2.1. *Gravity Method*

In principle, gravity geophysical surveys are based on Newton's law of gravity. Due to the reality that the mass of the earth is strongly influenced by other celestial bodies, especially the moon and sun, some correction in order to eliminate and reduce the error of measurement values due to them must be done in step of data processing and interpretation.

The study of gravity is based on Newton's law of gravity published by Newton in 1687 under the title Philosophiae Naturalis Principia Mathematica which states that the magnitude of the force of gravity of two masses is proportional to the multiplication of the second mass and inversely proportional to the square of the distance between the two center of mass.

$$\overrightarrow{F(r)} = G\frac{m.m_o}{r^2}\hat{r} \tag{1}$$

The massa of m and m_o experience a common force of gravity proportional to m, m_o, and r^{-2}. The direction of the unit vector $\hat{r}$ is from the source of gravity to the observation point, in this case lies in the test mass of m_o with:

m, m_o = mass
r = distance from m to m_o
G = Newton's gravitational constant (6.672 x 10^{-11} Nm^2/kg^2)

$$r = \left[(x-x')^2+(y-y')^2+(z-z')^2\right]^{1/2} \tag{2}$$

$\hat{r}$ = unit vector towards m_o [6]

If m_o is assumed to be a specimen of a certain mass, then dividing the gravitational force in equation (1) by m_o will result in the acceleration of gravity generated by the m at the point of the test object being, i.e.:

$$\overrightarrow{g(r)} = -\,G\,\frac{m}{r^2}\,\hat{r} \tag{3}$$

with $\hat{r}$ is the unit vector that direction from mass m to the observatios point P, in the Cartesian coordinate,

$$\hat{r} = \frac{1}{r}\left[(x-x')\hat{i} + (y-y')\hat{j} + (z-z')\hat{k}\right] \tag{4}$$

The minus sign of equation (3) is necessary because of the convention factor, i.e $\hat{r}$ leading from source to the observation point opposite to the gravitational acceleration field (g).

When the mass of the earth is M_e with the radius of the earth is R_e, spherical and non-rotating, so the magnitude of the gravitational acceleration on the surface of the earth is [6]:

$$\overrightarrow{g(r)} = -G\,\frac{M_e}{R_e^2}\,\hat{r} \tag{5}$$

The gravitational field is conservative when the attempt to move a particle of the field is independent of the passage, but depends only on the starting and ending points. The force of gravity is a vector with a direction along a line connecting the centerof two mass pieces. This force gives rise to a conservative field which can be derived from a scalar potential function with the following relations:

$$g(r) = \nabla\,U(r) \tag{6}$$

If the mass is continuously distributed with density $\rho(r_0)$ at volume V, then the gravitational potential at point P is:

$$U(r) = \int_{\infty}^{R} g.dr = -\int_V \frac{G\,dm}{|r-r_0|} = -G\int_V \frac{\rho(r_0)\,d^3r_0}{|r-r_0|} \tag{7}$$

with $|r-r_0| = \sqrt{r^2 + r_0^2 - 2rr_0\cos\gamma}$

which states the attempt to move a mass from an infinite point away with any passage to a point r distance from the center of mass m. The function of U is called the potential of gravity or Newton's potential and the acceleration of gravity g is the potential field. Some books establish the potential of gravity as an attempt by the test particles, so in equation (6) is written $g(r) = -\nabla U(r)$. Equation (6) follows the convention by Kellog [7] which states that the gravitational potential is the effort made by the gravitational field on the test particle and the negative of the particle's potential energy.

At the gravitational potential apply the principle of superposition, namely: the gravitational potential of a distribution of masses is the sum of the potential for each mass. The magnitude of the potential on the test particle is the sum of the vectors of the mass potential. The superposition principle can be applied to determine the gravitational potential for a continuous mass distribution. A continuous mass distribution m is a collection of very small and large masses, $dm = \rho(x,y,z)dv$, with $\rho(x,y,z)$ is the mass distribution density. By applying the principle of superposition gained:

$$U(r) = G\int_v \frac{dm}{r} = G\int_v \frac{\rho(Q)}{r}\,dv \tag{8}$$

with the integration of v, the actual volume of the mass. As usual P is an observation point, Q is the integration point and r is the distance between P and Q. The density ρ has units of kg.m^{-3} in the SI system and gr.cm^{-3} in the cgs system [6].

Implementation of this gravity geophysical survey must be applied some correction: drift correction, normal gravity, free air, Bouguer, terrain, and tide. Likewise, some filtering should be applied in the data processing step, either low pass filter or high pass filters depending on the value of the acquisition data obtained. At the interpretation stage is also very needed modeling that fits with local geological target conditions. The main equipment for this gravity geophysical survey is the use of Gravitymeter La Coste & Romberg.

2.2. Magnetic Method

In principle, magnetic geophysical surveys are based on theories about the main magnetic field of the earth, the earth's magnetic field elements, and the variations that influence the measurement of the magnetic field. Rocks, including mineral resources, have unique rocks and mineral susceptibility ranges that can be used as rock or mineral conversion values to their susceptibility values.

Research on the source of the main magnetic field of the earth is the source from the external and from internal of the earth carried out by Gauss in 1838 which concluded that the earth's main magnetic field measured on the surface is almost entirely caused by the source from internal of the earth, while the source from the external of the earth influence is very small [6].

Some classical theories state that the Earth's magnetic field arises as a result of the flow of electricity in the earth's crust. In rotation of the earth as a whole can be described as a giant dynamo. A magnetic field arises as a result of a combination of motion and electric current. In the core of the earth there is a self-exciting dynamo action called self-exciting dynamo actions, this theory proposed by Elsasser in 1950 [8].

The earth's magnetic field is a vector quantity having magnitude and direction, this magnitude can be decomposed into its components, namely: total component ($\vec{F}_o$),horizontal component ($\vec{H}$), horizontal component of X ($\vec{H}_X$), horizontal component of Y ($\vec{H}_Y$), vertical component ($\vec{H}_Z$), inclination angle (*I*), and declination angle (*D*) [9]. Contours that describe variations of geomagnetic field elements are called isomagnetic maps. The Isodynamic map illustrates the same field intensity contours, such as total intensity, vertical intensity, or horizontal intensity. The Isoclinal map shows the same inclination contours, and the Isogonic map shows the same declination.

The magnitude of this magnetic field varies with time varying from hundreds of milliseconds to hundreds of years. Earth's magnetic field variations in time are grouped into secular variations and daily variations (diurnal). Secular variation is a variation caused by the internal changes of the earth, where changes are very slow (order of tens to hundreds of years) to be able to affect the price of the magnetic field. Diurnal variation is a variation that is predominantly caused by the sun's interference. The nature of this daily change is random but periodic with an average period around the length of the day (24 hours). Another variation is the magnetic storm caused by solar activity, is random, valuable so that it can disrupt the observation data.

The total magnetic field anomaly can be determined from the total intensity of the measured magnetic field on the earth's surface less the effect of the terrain and the magnitude of the main magnetic field of the earth in the research area using data onto IGRF (International Geomagnetic Reference Field). If an object is in a magnetic field, a new magnetic field will develop in an object (induced) resulting in anomalous magnetic field, so that in the presence of rocks containing magnetic minerals, the Earth's normal magnetic field will be disturbed by anomalous magnetic fields as a result magnetization of rocks [10]. If an object (rock body) is uniform magnetized due to the induction of the earth's magnetic field, then the direction of the magnetic field is approximately equal to the direction of the earth's magnetic field for the concept of dipole earth and monopole [11].

The magnet will always consist of two poles. The poles are always paired in nature, always in the form of bipolar or dipole. The field intensity is proportional to the flux density of the line. At the magnetic poles have a greater intensity than the middle. In the earth's main magnetic field in the equator region the field intensity is about 30,000 gamma and 60,000 gamma in the polar region. The direction of the magnetic field at the edge of the dipole is parallel to the direction of the field at its center but opposite direction [11]. In reality there is no monopolized magnetic field, but only the dipole field at the far ends of each other [9].

Implementation of this magnetic geophysical survey must be applied some corrections: drift, diurnal, and International Geomagnetic Reference Field (IGRF). Likewise, some filtering should be applied to the data processing step, either low pass filters or high pass filter depending on the value of the acquisition data obtained. At the interpretation stage is also very needed modeling that fits with local geological target conditions. The equipment used in this magnetic survey is the Proton Precession Magnetometer (PPM).

2.3. Geoelectrical Resistivity Method

The geoelectrical resistivity method detects subsurface conditions based on the contrast of electrical properties of the earth's rocks. The application of this geoelectrical resistivity method is based on the electrical properties of the earth's resistance which assumes that if on earth surface is injected an electric current, then the earth as a conductor that has resistance will cause a certain potential difference [10].

The geoelectrical resistivity method is one of the active methods of geophysics, therefore the measurement of earth resistivity is done by injecting the current into the earth through two current electrodes (C1 and C2), and measuring the earth response to the injection current in the form of potential difference and resistance it is generated at two potential electrodes (P1 and P2). From the current value data (I), the potential difference (V), and resistance (R), we can calculate the apparent resistivity value (ρa) as follows :

$$\rho_a = k\frac{V}{I} \tag{9}$$

where k is the geometry factor which depends on the configuration of the four electrodes that used.

The calculated resistivity value is not the actual subsurface resistivity value, but is the apparent resistivity value which is the result of the contribution resistivity response from the earth as a whole, according to the range of equipotential and equipotential currents considered to be homogeneous which gives the same resistance value for the electrode arrangement the same one. Subsequently, apparent resistivity values must be processed to be true resistivity values for each layer of Vertical Electrical Sounding (VES). The equipment used in this geoelectrical resistivity survey is OYO Mc-Ohm.

3. Results and Discussion

Gravity data is distributed among Blitar district and is located between geographical coordinating (112.43952°E; 7.91870 °S) and (111.96268°E; 8.33120 °S) with a total of 224 points of measurement data. Data processing is done to get observation value (after conversion unit, tidal correction, drift correction, tie point correction, and base station correction). Free air correction is done to return the datum to topography as a result of normal gravity correction (latitude correction). In the end the Bouguer anomaly was obtained by involving all of the corrections above and Bouguer correction which considers the dominant local rock density. Contour map of Bouguer anomaly can be seen in figure 1.

The Bouguer anomaly map is showing the rocks that cause the mineralization process. Based on the Bouguer anomaly data, the potential of mining materials primarily for primary reserves can be localized. The red color contour is a mineralized zone, otherwise the blue color contour is not a zone of mineralization. Based on the contours of the Bouguer anomaly it shows that the Ngeni zone has a high Bouguer value of about 115mGal (low anomaly located in Kelud volcano area with a value of -10mGal). Based on the local geological map, the relatively high value of Bouguer anomaly is a representation of andesite-diorite-granodiorite and limestone.

The magnetic data is distributed over the district of south Blitar and is located between the geographical coordinates (112.02731°BT; 8.33114°LS) and (112.43811°BT; 8.12586°LS) with a total of 195 measuring points. The processing of the above data must be performed to obtain the value of the total magnetic field free of noise influence, which is the effect of daily variation, drift, and anomaly in IGRF (International Geomagnetic Reference Fields). The total anomaly contour of the total processing magnetic field can be seen in figure 2. The total anomaly contours of the processing magnetic field as shown in figure 2 is a description of the distribution of mined materials having magnetic contrast values. Based on the contour of the magnetic anomaly as shown in figure 2 indicates that the Ngeni zone is a mineralized zone having a magnetic value and magnetic dipole magnitude (a value range of about -650nT up to 800 nT). Based on the local geological map, the relatively high magnetic field anomaly value is also a representation of andesite-diorite-granodiorite and limestone rocks. The presence of magnetic dipole pairs tending to the southwest indicates that magnetic rock anomalies which associate with metals is forward to the southwest.

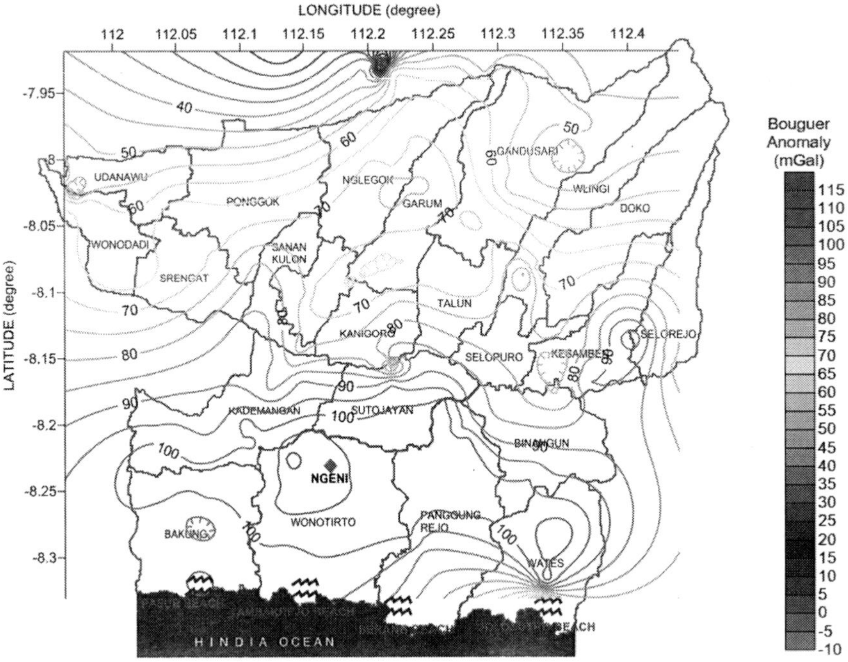

Figure 1. Contour anomaly of Bouguer on the administration map.

Based on the Bouguer anomaly of gravity data and the total magnetic field anomaly of magnetic data it appears that the deep anomaly that is thought to be the cause of the potential for minerals or mineralized zones in Blitar district is distributed among southern Blitar and as the center is the Ngeni zone.

To obtain a subsurface structure that describes a shallow anomaly, a near surface geophysics approach is called geoelectrical resistivity has applied. Topographic contours and distribution of geoelectrical resistivity points in Ngeni zone can be seen in figure 3.

Based on the results of geochemical tests, the rocks in the Ngeni zone contain Mn, Cu, S, Fe, Si, and Ca content with the content values of 1.88% (spectrophotometry method), 278.63 ppm (AAS method), 18.37% spectrophotometry method), 967.319ppm (AAS method), 36.53% (gravimetry method), and 8651.145 ppm (AAS method) respectively.

Based on the modeling results and the interpretation of the geoelectrical resistivity method showed that for the ngeni_wontir_01 point the lowest resistivity value is 3.2 Ωm and the highest resistivity value is 5042.37 Ωm. For the ngeni_wontir_02 point the lowest resistivity value is 3.03 Ωm and the highest resistivity value is 542.57 Ωm. As for the point ngeni_wontir_03 the lowest resistivity value is 22.8 Ωm and the highest resistivity value is 11249.91 Ωm. Overall based on all points the lowest resistivity value is 3.03 Ωm and the highest resistivity value is 11249.91 Ωm. Based on local geological conditions, it is generally accepted that rocks with metal mineral content have very small resistivity values of less than 1 Ωm. Resistivity values between 1 Ωm to 1000 Ωm is the value of water, alluvium, and clay. Resistivity values greater than 1000Ωm are generally values of lava, andesite, diorite, granodiorite and limestones. Figure 4 is resistivity cross section of the Ngeni zone.

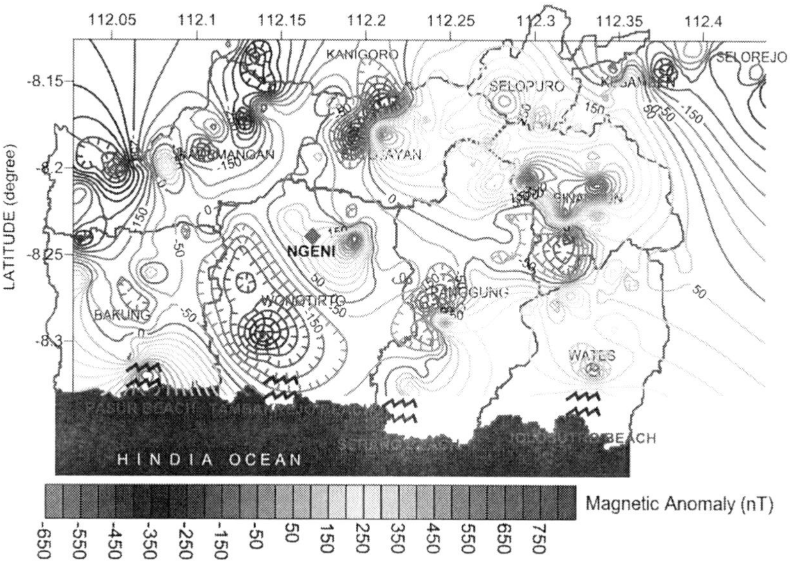

Figure 2. Contour anomaly of the total magnetic field on the administration map.

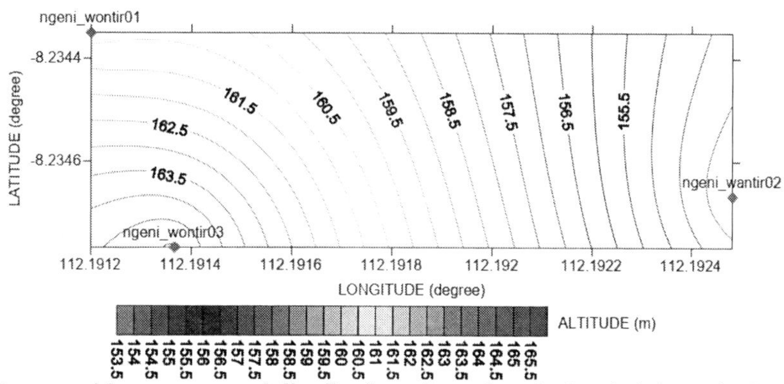

Figure 3. Topographic contours and distribution of geoelectrical resistivity point in Ngeni zone.

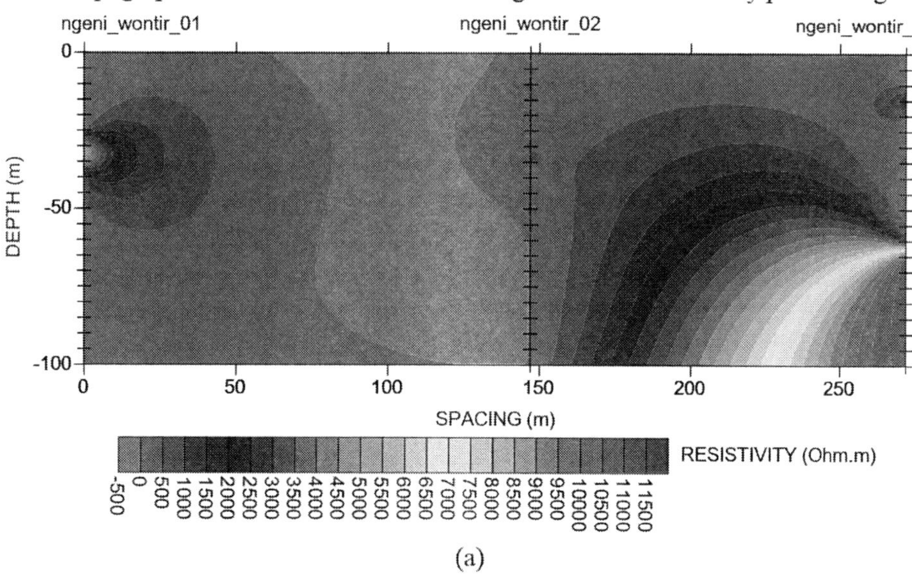

(a)

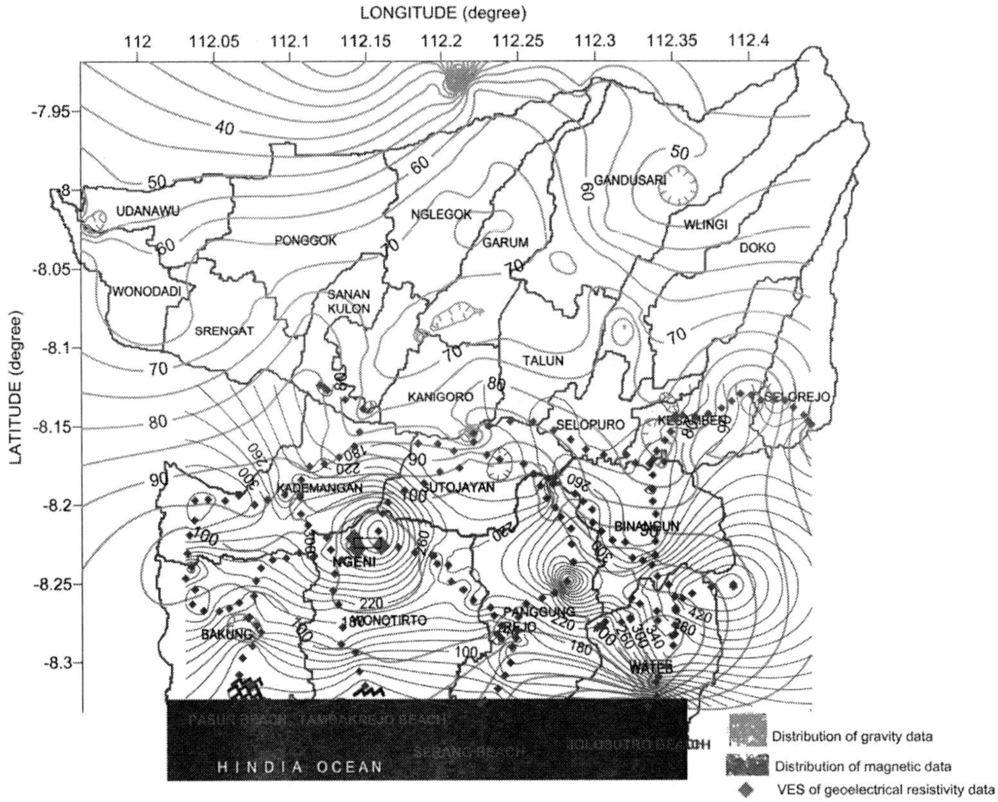

(b)

Figure 4. Resistivity cross section of the Ngeni zone.

4. Conclusion

Based on the research that has been done, it can be concluded the things as follows:

1) Based on the contours of the Bouguer anomaly shows that the Ngeni zone has a high Bouguer value of about 115mGal, which is a representation of andesite-diorite-granodiorite and limestone.
2) Based on the contours of magnetic anomaly indicates that the Ngeni zone is a mineralized zone that has a magnetic value value magnetic value and a large magnetic dipole (a value range of about -650nT up to 800 nT). It is a representation of andesite-diorite-granodiorite and limestone rocks. The presence of magnetic dipole pairs tending to the southwest indicates that magnetic rock anomalies associate with metals heading southwest.
3) Based on the results of modeling and interpretation of the geoelectrical resistivity method showed that overall the lowest resistivity value is 3.03Ωm and the highest resistivity value is 11249.91 Ωm. This suggests that rocks with metal mineral content have very small resistivity values of less than 1 Ωm. Resistivity values between 1Ωm to 1000 Ωm is the value of water, alluvium, and clay. Resistivity values greater than 1000Ωm are generally values of lava, andesite, diorite, granodiorite and limestone.
4) Based on the results of geochemical tests, rocks in the Ngeni zone contain Mn, Cu, S, Fe, Si, and Ca contents with content of 1.88%, 278.63ppm, 18.37%, 967.319ppm, 36.53 %, and 8651.145ppm respectively.
5) There is a correlation between anomalous gravity, magnetic, and geoelectrical resistivity that associated with an anomaly of deep, middle, and shallow respectively.

Acknowledgement
I would like to thank the ICOS 2017 committees and the proceeding editors for the technical and language editing of this manuscript.

References

[1] Bemmelen R W V 1949 A General Geology of Indonesia and Adjacent Archipelagos *The Geology of Indonesia* **1** (The Hague: Government Printing Office)
[2] Sjarifudin M Z and Hamidi S 1992 *Peta Geologi lembar Blitar, Jawa skala 1:100*.000 (Bandung: Pusat Penelitian dan Pengembangan Geologi)
[3] Gary M, McAfee R J and Wolf C L 1974 *Glossary of Geology* (Washington D C: American Geological Institute)
[4] Park C F J and MacDarmid R A 1976 *Ore deposits* (San Francisco: W. H. Freeman and Company)
[5] Sunaryo 2007 *Study Zona Mineralisasi di Malang selatan-Jawa Timur Menggunakan Metode Geofisika Terpadu Disertasi S-3*
[6] Blakely R J 1995 *Potential Theory in Gravity and Magnetic Applications* (Cambridge: University Press)
[7] Kellogg O D 1953 *Foundations of Potential Theory* (New York: Dover)
[8] Clark S P Jr 1971 *Structure of The Earth* Prentice-Hill Inc (New Jersey: Englewood Cliffs)
[9] Dobrin M B and Savit C H 1988 *Introduction to Geophysical Prospecting 4th ed* (New York: McGraw-Hill Book Company)
[10] Telford W M, Geldart L P and Sheriff R E 1976 *Applied Geophysics* (Cambridge: Cambridge University Press)
[11] Breiner S 1973 *Applications Manual for Portable Magnetometers* (California USA: GeoMetrics 395 Java Drive Sunnyvale)

Preliminary Result of Earthquake Source Parameters the Mw 3.4 at 23:22:47 IWST, August 21, 2004, Centre Java, Indonesia Based on MERAMEX Project

Y A Laksono[1], K S Brotopuspito[2], W Suryanto[2], Widodo[3], R A Wardah[2], I Rudianto[2]

[1]Department of Physics, Universitas Negeri Malang, Jl. Semarang 5 Malang 65145, Indonesia
[2]Department of Physics, Universitas Gadjah Mada, Sekip Utara, Bulaksumur, DIY 55281, Indonesia
[3]Department of Mathematics, Universitas Gadjah Mada, Sekip Utara, Bulaksumur, DIY 55281, Indonesia

E-mail: yoyok.adisetio.fmipa@um.ac.id

Abstract. In order to study the structure subsurface at Merapi Lawu anomaly (MLA) using forward modelling or full waveform inversion, it needs a good earthquake source parameters. The best result source parameter comes from seismogram with high signal to noise ratio (SNR). Beside that the source must be near the MLA location and the stations that used as parameters must be outside from MLA in order to avoid anomaly. At first the seismograms are processed by software SEISAN v10 using a few stations from MERAMEX project. After we found the hypocentre that match the criterion we fine-tuned the source parameters using more stations. Based on seismogram from 21 stations, it is obtained the source parameters as follows: the event is at August, 21 2004, on 23:22:47 Indonesia western standard time (IWST), epicentre coordinate -7.80°S, 101.34°E, hypocentre 47.3 km, dominant frequency f0 = 3.0 Hz, the earthquake magnitude Mw = 3.4.

1. Introduction

Merapi Lawu anomaly (MLA) is the location having an anomaly Vp/Vs = 1.9 found by [1, 2, 3, 4, 5] from seismograms from MERAMEX (MERapi AMphibious EXperiment) project. In order to study the structure of MLA using waveform forward modeling or full waveform inversion it needs high precision source parameters. The result from catalog searching at the time where MERAMEX project conducted, it is found only one event at near MLA which has hypocenter in 160 km depth located at -7.5440°S, 110.7046°E with Mb=3.6 on August 24, 2014 on 07:08:31 IWT from [6].

The magnitude of the event is weak so the quality of the seismogram must be checked which the good seismogram is if it has SNR value higher and equal to 5 dB [7].

The hypocenter of this event is far from MLA that could reduce the accuracy. The faraway location makes the wave propagation time longer, the spatial coordinates get bigger, and the waves will pass through various unknown subsurface structure. These combinations increase the accumulation of errors and lead to incorrect results. To improve the accuracy, the station location is chosen where the seismic wave travel from the source does not pass through the MLA.

For the weak event in which the magnitude < 4 the focal mechanism is difficult to calculate, which the error is higher and not reliable [8]. For weak event, the seismic source wave modelling is approached by the point source [9] so that focal mechanism on this research is not computed.

The next parameter needed is the source frequency where the frequency is obtained from the dominant frequency from the P-wave phase because the value of this frequency is the highest compared to the others phases.

Based on these arguments, this research looks for an event that occurs as close as possible to the MLA location, has high SNR value, and the station location chosen is out of the MLA region so that forward modeling or full waveform inversion produce accurate results.

2. Seismogram and Location of Investigation

Seismogram is taken from the MERAMEX project from GFZ German Research Centre for Geosciences. The seismogram are obtained from the 130 seismographic stations which were installed and operated from May to September 2004. The installed station were scattered on both onshore and offshore in the Central Java, and Yogyakarta region. For this research the station that are onshore are chosen.

MLA is located between (-7.01°E, 110.52°S) to (-7.67°E, 111.19°S) at Centre Java, Indonesia. figure 1 shows MLA location and the contour based on [1]. From figure 1 it can be seen that MLA is irregularly shaped. The black shape has the highest Vp/Vs = 1.9. The grey colour indicates decreased gradation of Vp/Vs <1.9.

3. Method

There are two steps to find the event; i.e. find the event roughly from a few stations, and fine-tuned the selected event that matched the criterion using many more stations. The purpose of the 1st step is finding the event location nearest of MLA quickly, and the 2nd step is fine-tuned the event location and source frequency. Many stations are used on the second steps with the hope the result will be more accurate.

SEISAN [10] is used to find the location of event. This software consists of various softwares to process seismograms for many purposes. To compute the hypocentre, SEISAN uses many alternatives ways using HYPOCENTER [11], HYPO71 [12], or HYPOINVERSE [13].

To calculate the SNR value the $SNR = 20 \log_{10} (A_s/A_n)$ formula is applied, where A_s is a root mean square (rms) signal value and A_n is rms noise. The signal to be used to compute SNR is P-wave phase as A_s and the background signal (the signal before the P-wave phase comes) which considered as noise as A_n.

Spectrum analyser function from SEISAN is used to find the dominant frequency of P-wave phase. The P-wave phase is choose because this wave phase has the highest frequency in seismogram. This frequency is very important in waveform forward modelling or full waveform inversion because the size of grid or mesh depends to this frequency. Not all the seismogram are used to calculate the dominant frequency. The choice of seismogram is based on the representative of the farthest, medium and the closest distance toward the epicentre.

4. Result and Discussion

The hypocentre at 1st step is searched using five stations from July to the end of September 2004 from the MERAMEX project seismogram data. There are more or less 500 events are found both onshore and offshore. One of the events occurred which fulfil the requirement is on 21 August 2004, at 23:22:49 IWT, Mw = 3.2, epicentre coordinate is (-7.668°S, 110.967°E), and hypocentre is 37.5 km. This result is obtained from five stations namely AI1, AI3, AK1, BK3, and CK6 as shown in figure 2. In this figure the triangle is the station, the red dot is epicentre, and the black ellipsoid is the epicentre error. The epicentre error is about 20 km.

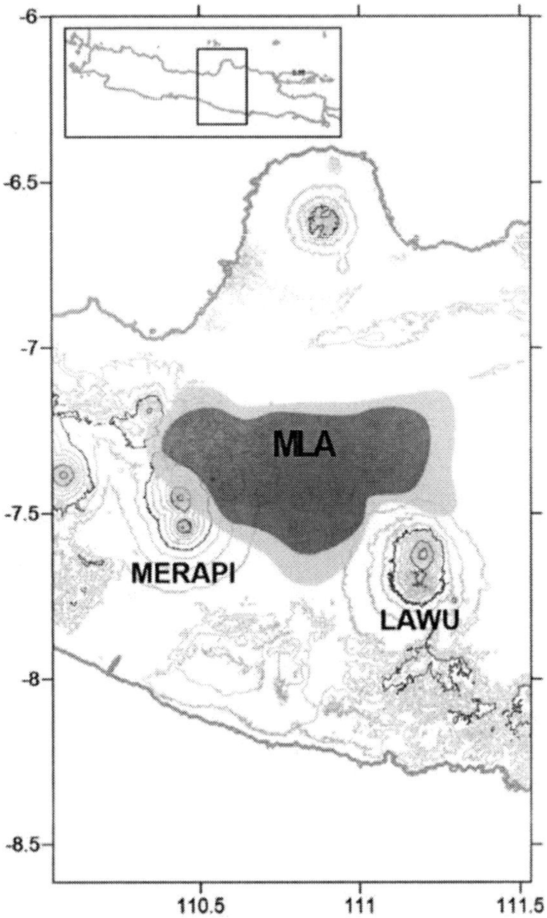

Figure 1. MLA location based from [1].

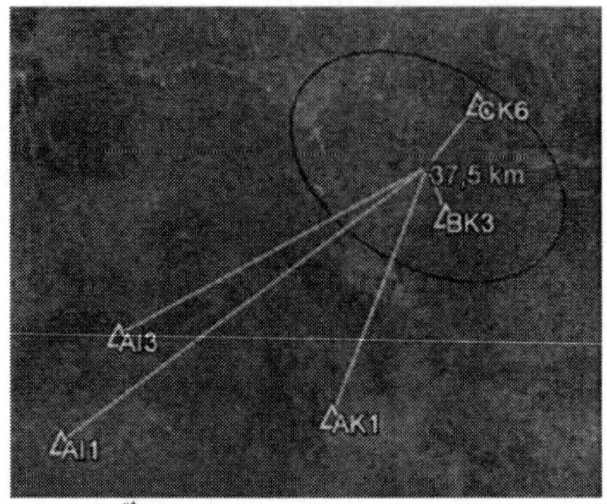

Figure 2. 1st epicentre from 5 stations, namely AI1, AI3, AK1, BK3, and CK6. The hypocentre at 37.5 km.

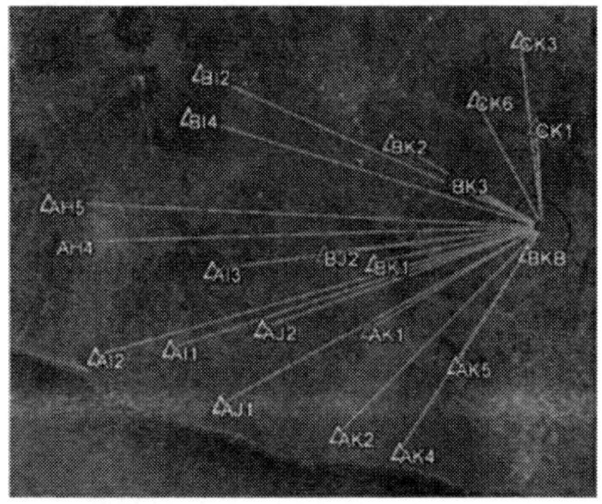

Figure 3. Fine-tuned epicentre from 21 stations. The epicentre is slighty toward south. The hypocenter is 47.3 km.

The SNR of the P-wave phases of all stations has a value of more than 20 dB which means it qualifies as a good signal. Table 1 shows the value of SNR for five stations.

Table 1. The value of SNR from five stations.

Station	SNR (dB)
AI1	35.0
AI3	29.7
AK1	45.8
BK3	25.3
CK6	24.0

21 stations is used to fine-tune this event namely BJ2, BK1, BKB, BK2, BK3, AK1, AK2, AK4, AK5, AJ1, AJ2, AI1, AI2, AI3, AH4, AH5, BI2, BI4, CK1, CK3 and CK6. The result of fine-tuned are follow. The origin time on 23:22:47 IWT, epicentre coordinate becomes (-7.808°S, 111.134°E), hypocentre is 47.3 km, dominant frequency f0=5.04 Hz, and Mw=3.4. Figure 3 shows the fine-tuned epicentre and stations location. The different colour at all of the stations in figure 3 shows the error which the green colour is the lowest error, yellow colour is moderate error, and the red colour is

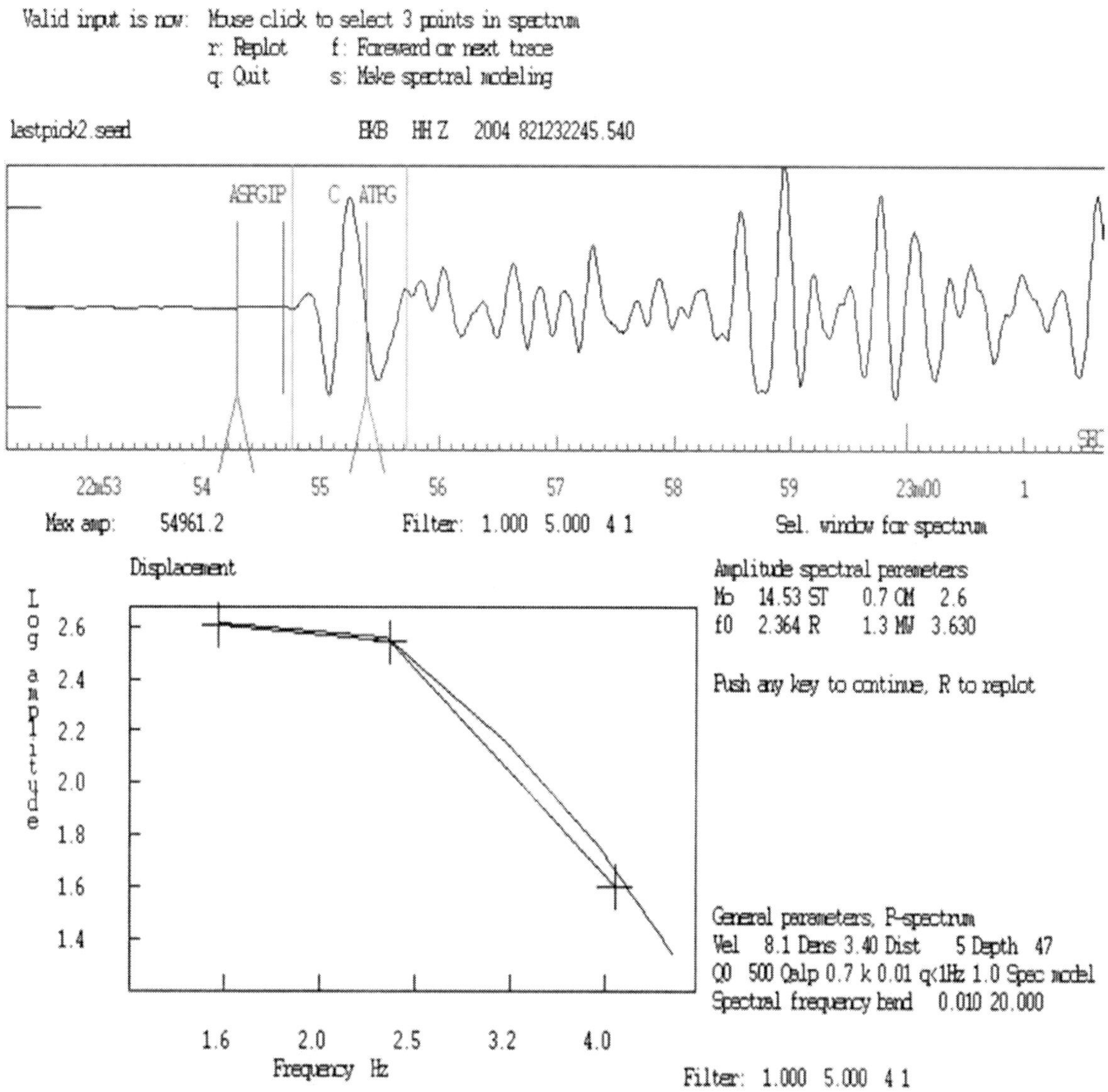
Valid input is now: Mouse click to select 3 points in spectrum
r: Replot f: Foreward or next trace
q: Quit s: Make spectral modeling
lastpick2.seed
BKB HH Z 2004 821232245.540
ASPGIP
C ATPG
22m53 54 55 56 57 58 59 23m00 1
Max amp: 54961.2
Filter: 1.000 5.000 4 1
Sel. window for spectrum
Displacement
Log amplitude
2.6 2.4 2.2 2.0 1.8 1.6 1.4
1.6 2.0 2.5 3.2 4.0
Frequency Hz
Amplitude spectral parameters
Mo 14.53 ST 0.7 OM 2.6
f0 2.364 R 1.3 MW 3.630
Push any key to continue, R to replot
General parameters, P-spectrum
Vel 8.1 Dens 3.40 Dist 5 Depth 47
Q0 500 Qalp 0.7 k 0.01 q<1Hz 1.0 Spec model
Spectral frequency band 0.010 20.000
Filter: 1.000 5.000 4 1

(a)

IOP Conf. Series: Journal of Physics: Conf. Series **979** (2018) 012052 doi:10.1088/1742-6596/979/1/012052

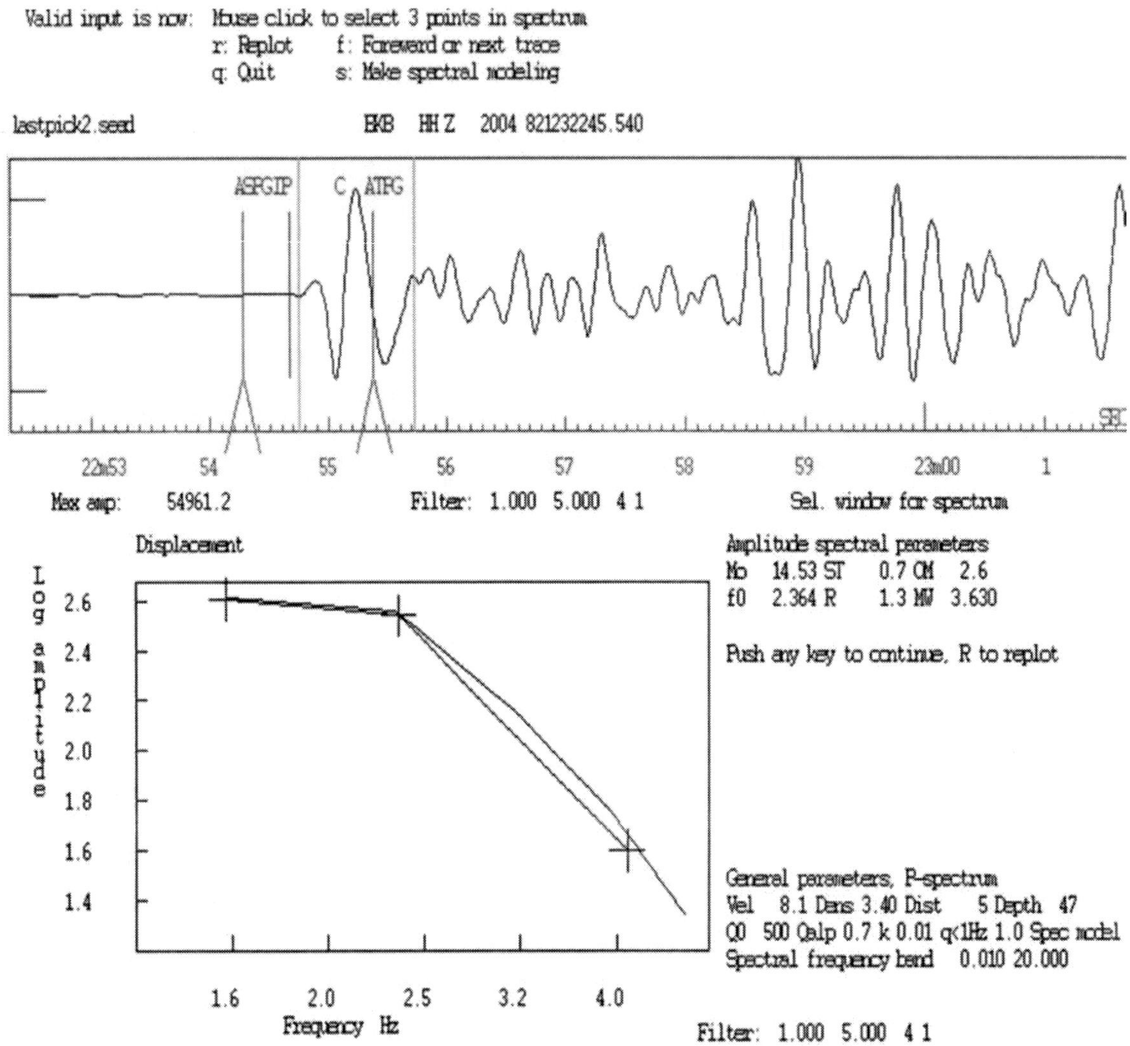

(b)

Figure 4. P-wave phase spectrum analyser for station (a) BKB, and (b)

the highest error. The epicentre error is about 12 km nearly halves from result the 1st step which is means the accuracy in the 2nd step is improve.

The dominant frequency of source is fine-tuned using spectrum analysis as shown in figure 4. In this figure we shows P-wave phase spectrum which give f0 = 2.364 Hz. By using the same processing method, we process another selected stations, i.e. BK3, AK1, and AH5 stations, where the final dominant frequency is 3.0 Hz.

5. Conclusion

We have found an event that can be used for forward modelling or full waveform inversion to investigate the structure of MLA. This event is located near the MLA region derived from seismograms located outside the MLA.

Future Work

The source parameters obtained will be validated by performing forward modelling of the stations that have been used to find the hypocenter.

Acknowledgment

Seismic data were obtained from the GEOFON data centre of the GFZ German Research Centre for Geosciences. We would like to thank you for being able to access MERAMEX project data.

Reference

[1] Koulakov I, et al 2007 P and S Velocity Structure of the Crust and the Upper Mantle Beneath Central Java from Local Tomography Inversion *J. Geophys. Res. Solid Earth* **112** B8 p. B08310

[2] Wagner D, et al 2007 Joint Inversion of Active and Passive Seismic Data in Central Java *Geophys. J. Int.* **170** 2 923–932

[3] Bohm M, Haberland C and Asch G 2013 Imaging Fluid-related Subduction Processes Beneath Central Java (Indonesia) Using Seismic Attenuation Tomography *Tectonophysics* **590** 175–188

[4] Luehr B G, et al 2013 Fluid Ascent and Magma Storage Beneath Gunung Merapi Revealed by Multi-scale Seismic Imaging *J. Volcanol. Geotherm. Res.* **261** 7–19

[5] Zulfakriza Z, et al 2014 Upper Crustal Structure of Central Java, Indonesia, from Transdimensional Seismic Ambient Noise Tomography *Geophys. J. Int.* **197** 1 630–635

[6] International Seismological Centre *On-line Bulletin* Thatcham, United Kingdom: Internatl. Seismol. Cent. 2014

[7] Reif C, Shearer P M and Astiz L 2002 Evaluating the Performance of Global Seismic Stations *Seismol. Res. Lett.* **73** 1 46–56

[8] Zahradník J, Janský J and Papatsimpa K 2001 Focal Mechanisms of Weak Earthquakes from Amplitude Spectra and Polarities *Pure Appl. Geophys.* **158** 4 647–665

[9] Lee S J, Chen H W, Liu Q, Komatitsch D, Huang B S, and Tromp J 2008 Three-Dimensional Simulations of Seismic-Wave Propagation in the Taipei Basin with Realistic Topography Based upon the Spectral-Element Method *Bull. Seismol. Soc. Am.* **98** 1 253–264

[10] Ottemöller L, Voss P and Havskov J 2016 Seisan Earthquake Analysis Software For Windows, Solaris, Linux and Macosx, 2016

[11] Lienert B R, Berg E and Frazer L N 1986 HYPOCENTER: An Earthquake Location Method using Centered, Scaled and Adaptively Damped Least Squares *Bull. Seismol. Soc. Am.* **76** 3 771–783

[12] Lee W H K and Lahr J C 1972 HYPO71: A Computer Program for Determining Hypocenter, Magnitude and First Motion Pattern of Local Earthquakes *USGS Publ. Wareh*

[13] Klein F W 2014 User's Guide to HYPOINVERSE-2000, A Fortran Program to Solve for Earthquake Locations and Magnitudes, Open File Report 02-171*USGS*

IOP Conf. Series: Journal of Physics: Conf. Series **979** (2018) 012053 doi:10.1088/1742-6596/979/1/012053

Microtremor Study of Site Effect for Disaster Mitigation and Geotechnical Purpose

Sabrianto Aswad, Muh. Altin Massinai , Syamsuddin

Geophysics Program Study, Hasanuddin University

E-mail: sabri@science.unhas.ac.id

Abstract. Makassar city have relatively lower earthquake vulnerability compared to other regions in Indonesia, however detailed mapping related to seismic wave amplification needs to be done in the interest of geotechnical, regional planning and disaster mitigation. It is generally known that the magnitude of the damage during the occurrence of earthquakes or tremor occur periodically is influenced by the dynamic characteristics of the building as a function of seismic wave amplification. The degree of seismic wave amplification depends on several factors, including the thickness of the sediment layer, the level of compaction and the geological age factor. The purpose of this research is to investigate seismic vulnerability in Makassar by using spectral comparison through microtremor measurement. There are several of the approaches that can be done and microtremor is the easiest and cheapest method to understand these dynamic characteristics without causing damage effects. Spectra comparison technique used was popular by Nakamura, which is comparison technique of horizontal component noise spectra and the vertical component in sediment areas (H/V spectra). Results from seismic vulnerability index (SVI) distribution maps show values ranging from 0, 14 – 158, 31. In general, the eastern part of the city of Makassar near from coastal areas is more vulnerable to damage especially earthquakes or periodic earth tremor with certain dominant frequency compared with the western part of Makassar City.

1. Introduction

Almost all big cities in Indonesia have high levels of earthquake vulnerability but are not supported by detailed mapping of earthquake risk in that area. One mapping method that nowadays gains popularity because cheaper and easy to apply is a method of using microtremor surveys. Although microtremor surveys have not been widely applied in Indonesia, in some countries such as Europe and Japan have conducted these surveys and contribute significantly for disaster mitigation, geotechnical purpose, and urban planning. The city of Makassar, even though data seismicity zone issues by Ministry of Public Works [1] is relatively lower than in other areas, lay on sediment are which gives site effect on the effects of damaged buildings due to the vibrations that occur on the surface. It is also found in several cities in the world that lay on sediment are, like in Turkey, Lebanon, Kroasi and Italy [2-6]. The subsurface topography and geological conditions of an area gave significantly on the extent of damage caused by earthquakes in the area. Several studies have shown that major damage is caused by the effects of the site [7, 8].

Spectra comparison technique used is Nakamura technique, that is comparison technique of horizontal component noise spectra and the vertical component in sediment areas (H / V spectra).

This technique can also be combined with other geophysical measurement methods and geotechnics to produce unique solutions [3, 4].

The use of HVSR method is quite suitable for Makassar City, considering the geology of Makassar city area located on the sediment of alluvium and marine sedimentary rocks, so it is the right area to do the research in determining the effect of the local site (site effect). Basemen (hard rock) from Makassar based on geological map consist of Camba formation rock rocks (breccia, lava, conglomerate, and tuffa) in Maros and volcanic eruptions, Lompobattang volcano (conglomerate, lava, breccia, sediment, lava, and tuff) especially volcanic lava rocks found in the area around Gowa. This technique is also suitable to use in Makassar since this area based on seismicity maps issued by USGS [9] is relatively small so it is difficult to determine the effect of local sediment (site effect) using earthquake data.

This method provides information about the characteristics of the sediment layer such as the dominant frequency and amplification factors, so that the potential damage to the city of Makassar can be known and very useful for disaster mitigation, especially due to the fault with the north-east of Makassar potential source of earthquakes which is very damaging effect on the city of Makassar.

2. Data and Methodology

2.1. Geological Setting

Makassar City's bedrock located in the area around Maros District well known as Camba formation volcanic rock consists of breccia, lava, conglomerate, and tuff. Makassar's landscape is a coastal plain found in the west and north of the main street of Makassar City-Maros District and directly adjacent to the river plains and floodplains.

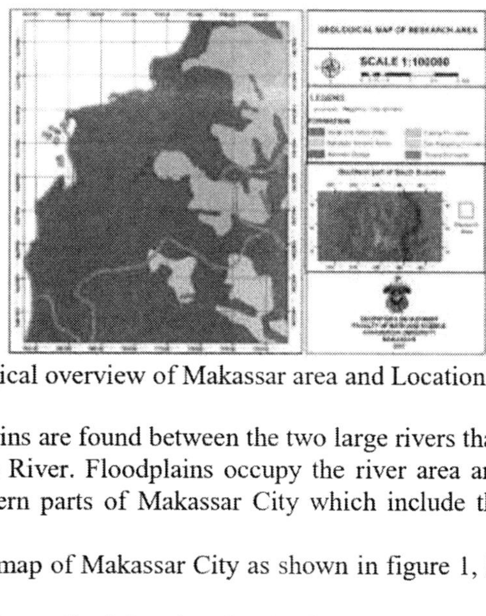

Figure 1. Geological overview of Makassar area and Location of sites measured.

River plains and floodplains are found between the two large rivers that divide the city of Makassar Jeneberang River and Tallo River. Floodplains occupy the river area and the highland found in the northern, eastern and southern parts of Makassar City which include the area of Biringkanaya and Panakukang.

Based on the geological map of Makassar City as shown in figure 1, it is generally composed of 3 rock units:

1. Alluvial sediment, these alluvial rock units dominate almost all areas of the city with an area of 11,693.83 ha. Its spread covers the area around the mainland to the coast.
2. Basal Rock, the distribution of basalt rock units is located in two districts of Tamalate sub-district with an area of 3.201 ha and in Biringkanaya District with an area of 25.027 ha.
3. Sediment comes from seas interbedded with Bawakaraeng volcanic rocks, and Camba Formation consisting of lava, breccia, tuff and conglomerate result from Gunungapi Batturappe-Cindako in the form of tuff and breccias, the distribution of tuff and breccias is located in Biringkanaya Subdistrict, Tamalate Sub-district, and Kecamatan Panakukang.

IOP Conf. Series: Journal of Physics: Conf. Series **979** (2018) 012053 doi:10.1088/1742-6596/979/1/012053

2.2. Microtremor

Okada [10] defined microtremor is the constant vibration of earth's surface. The source of microtremor comes from the active source and passive source. Active source created by human activity (vehicles or plant machinery vibration) and passive source rise from natural phenomena (ocean waves, wind, rain, and variations in atmospheric pressure). Vibration in term of microtremors refers to not including short duration events such as earthquakes and explosions [11].

Microtremor surveys can be performed in two ways depending on seismic activity in the area [12]. The first approach is well known as the classical method which is good for high seismic activity. This method needs two or more station locations that recorded simultaneously, one station should be put in an area of hard rock (basement) that doesn't indicate a strengthening of frequency due to a ground vibration and others should be put in soft rock (sediment) indicate the amplification of seismic wave. These two or more Station in two area will be compared to get response microtremor spectrum as a transfer function. This approach quite difficult to perform since we need an area with hard rock (basement) and also need sufficient earthquake that record in both area.

The second approach popular by Nakamura [13], he named it as H/V spectral ratio. The advantage of this method only needs one station, the procedure very simple the spectra ratio of horizontal and the vertical components recorded at the same site (H/V spectral ratio).This technique attractive since it gives ease of data collection and it can be applied to an area of low or even no seismicity. Nakamura assumes that the H/V ratio reflects the amplification of ground motion. With this method of measurement does not need to be done with the requirements of the hard rock.

Stasiun site can be seen in figure 2. A measurement performed using a set of short-period seismograph-303S portable type TDL (3 components), each of which consists of a digitizer, sensors, laptops, batteries, and GPS.

Ambient noise data (microtremor) obtained from 106 measurement points are scattered in the area Makassar City, Maros District and Gowa District (figure 2). Recording the data onto each point measurements was performed for approximately 30 minutes. An example of one station of data collection can be seen in figure 2.

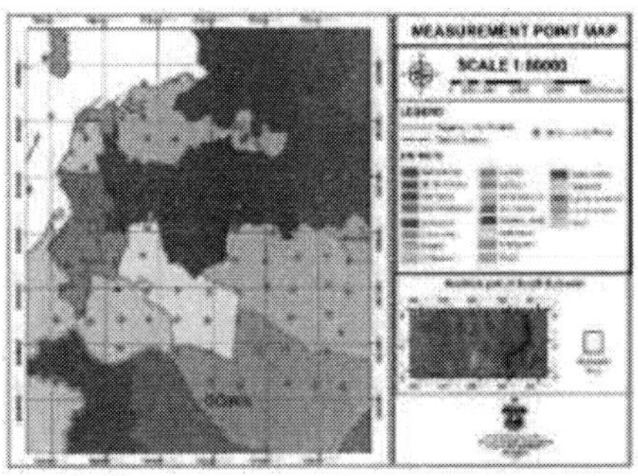

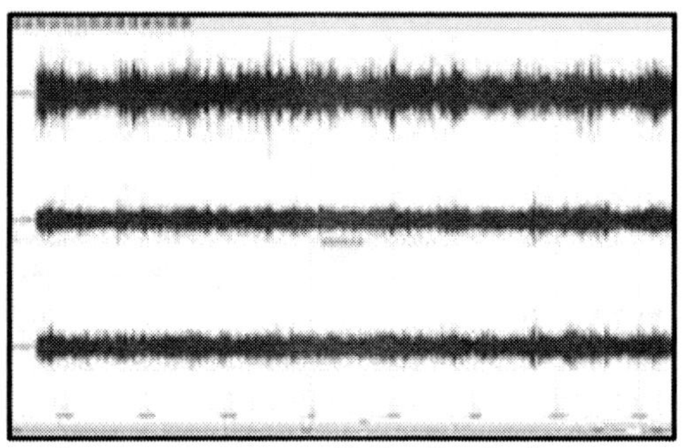

Figure 2. Map of research site and An example raw data measurement (Arka Station).

In processing data, the Geopsy software we used. This software contains information on recording time, the amount of data, and other supporting data. The result is a spectrum at each station will then be analyzed to obtain the HVSR peak value (A) and predominant frequency (f0).

At the time of processing using GEOPSY software, data is divided into several windows (figure 3). If the data is large enough sorting windows is automatically provided by the software. Windowing is the process to sort data onto the tremor signals and transient event (especially specific sources such as footsteps and vehicle passes). If signal quite difficult to sort automatically, event detect manually to prevent the transient event.

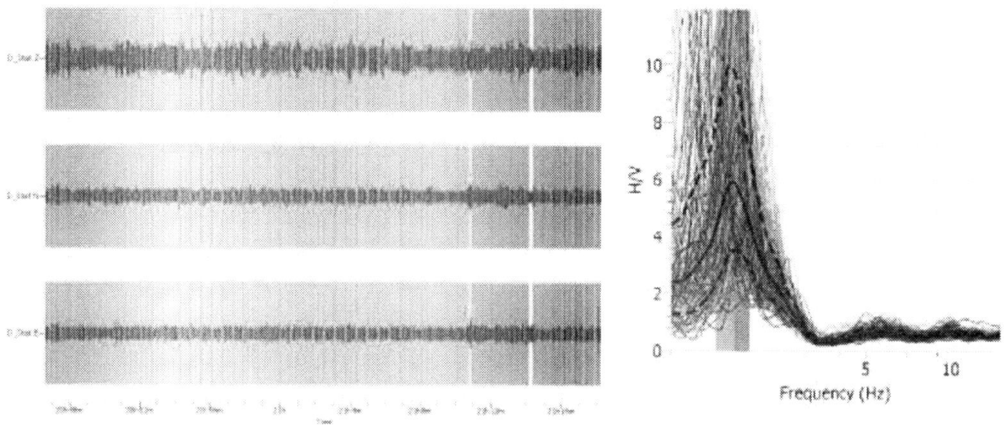

Figure 3. Windowing the raw data before doing FFT.

The next are fast Fourier transform (FFT), which involves the process of smoothing. The smoothing process is done by using Konno and Omachi algorithms [14] with a bandwidth b coefficient of 40. In this process, we also did cosine taper to minimize border effect or boundary effects due to the selection process window. The end results of the processing of data onto the form of spectral ratio H / V (figure 3). From this spectrum, we can determine the value of the dominant frequency (f_o) and peak spectral ratio H / V (A) at the measurement site microtremors. Based on the relationship $T = 1/f_0$ then we will get the value of a dominant period in the measurement site [15].

2.3. *HVSR Technique*

This technique widely used for micro-zoning and engineering purposes proposed by Nakamura [13]. The H/V spectral ratio [$T_{H/V}(\omega)$] was obtained by dividing the average spectra of the horizontal component of sediment site [$s_{NS}(\omega) + s_{EW}(\omega)$] by the spectrum of the vertical component [$S_v(\omega)$] of the sediment site :

$$T_{H/V(\omega)} = \frac{s_{NS}(\omega) + s_{EW}(\omega)/2}{S_V(\omega)} \tag{1}$$

Based on Nakamura Idea, amplitude effect of Source can be estimated by the ratio:

$$A_S = \frac{V_S}{V_B} \tag{2}$$

where amplitude spectrum of the vertical component of motion at the surface and amplitude spectrum of the vertical component of motion at the half-space defined as V_S and V_B. Nakamura also estimates site effect of interest in earthquake engineering (S_E), as a ratio

$$S_E = \frac{H_S}{H_B} \tag{3}$$

where H_S and H_B are Fourier amplitude spectra of the horizontal component of motion at the surface and the base of soli layer.

Using assumption:

$$\frac{H_B}{V_B} = 1 \tag{4}$$

We can calculate modified site affect spectral ratio (S_M) as

$$S_M = \frac{S_E}{A_S} = \frac{\frac{H_S}{V_S}}{\frac{H_B}{V_B}} = \frac{H_S}{V_S} \quad (5)$$

Assumption $H_B/V_B = 1$ was verified by Nakamura experimentally using microtremor measurement at depth in a borehole. A more detailed about this method can be found in Lermo dan Chaves-Garcia [16, 17].

The H/V spectral ratio [$T_{H/V}(\omega)$] was obtained by dividing the averaged spectra of the horizontal components of the sediment site $s_{NS}(\omega)$ and $s_{EW}(\omega)$ by the spectrum of the vertical component $S_V(\omega)$ of the sediment site:

$$T_{H/V}(\omega) = \frac{s_{NS}(\omega) + s_{EW}(\omega)/2}{S_V(\omega)} \quad (6)$$

3. Result and Discussion

3.1. Dominan Frequency (f_0) and Amplitudo Maximum (A)

The dominant frequency of the area of Makassar and its surroundings have an interval of 1.11 - 18.69 Herzt. Dominant frequency values for all measurement point were contours using the kriging interpolation method produced contour map distribution of dominant frequency (f_0) in the study area (figure 4). The high dominant frequency (over 10 Hz) is in the eastern part of Makassar City, which includes Manggala, Tamalanrea, western Panakkukang, eastern Somba Opu and Biringkana. These information matches with the geological information and Makassar's groundwater basin is relatively close to the base layer. Low frequencies (below 5 Hz) generally in the west, the area close to the coast, this area has relatively thick sediments.

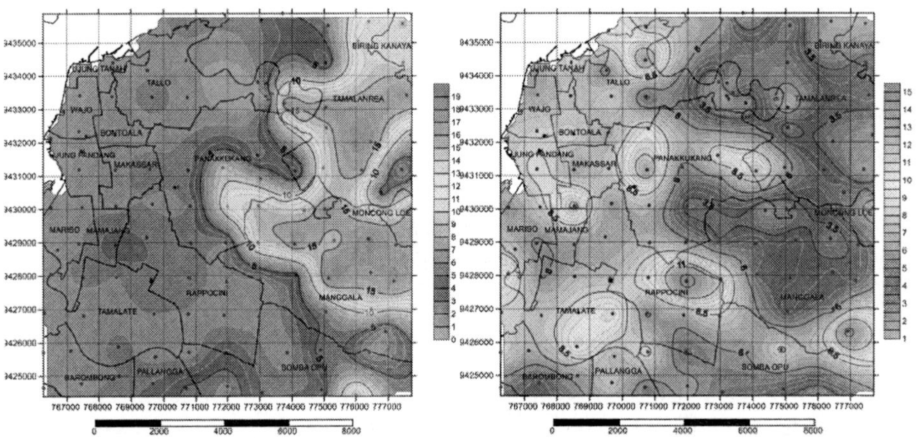

Figure 4. Contour map distribution of dominant frequency (f_0) and Maximum amplitude (A).

Analysis of the result of H / V spectrum analysis can be seen that the research area has Amplitude Maximum of HVSR at intervals value of 1,14 - 15,03. Nakamura [13] uses the peak value of HVSR as a function of the amplification factor in case of earthquakes or periodic vibrations happen at that location. The lowest value was found at Batu Duang station and the highest were located at Permata Hijau residence station. In general, the low maximum amplitude value of the dominant frequency was in the eastern part of Makassar, meanwhile, the distribution of maximum amplitude can be seen in figure 4. The western part of Makasar City have a larger amplitude value, This means that the western part of Makasar City have a surface vibration with the low dominant frequency with high intensity.

This information is very important for geotechnical planning to avoid the same building frequency as the dominant frequency because it will cause resonance effect on the building.

3.2. Seismic Vulnerability Index (SVI)

SVI values can be obtained from the dominant frequency (f_0) and Maximum amplitude (A), research areas has SVI value at interval value 0,14 - 158,31. The lowest value is at Batu Duang station and the highest are located at Permata Hijau and Singkabatu Residences stations. Generally, Makassar City and surrounding areas have SVI value below 60. SVI values in the western part of Makassar arebelow 10 while the western part is generally above 10. This means the western part is vulnerable to deformation, which means that this area is an unconsolidated sediment at the time of the earthquake.

Information on several destructive earthquake events, Nakamura [18] shows that areas that are often affected by major earthquake damage have an SVI value of between 20-100 whereas for non-destructive areas are completely in areas with SVI below 5. So this result shows the areas of the west of Makassar city is more vulnerable to damage than the eastern region.

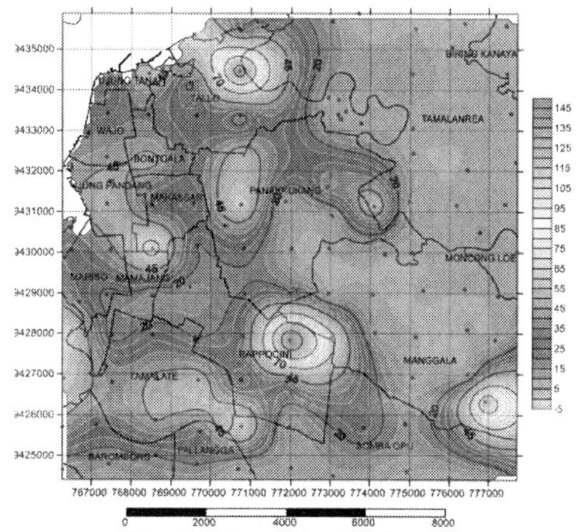

Figure 4. Contour map distribution of Seismic Vulnerability Indeks (SVI).

4. Conclusion

The result of the dominant frequency distribution map provides information on the western (near shore) areas of Makassar has a thicker sediment compared to the west. This information also provides information on the planning of buildings to create buildings that have a frequency that is not equal to the dominant frequency in order to avoid resonance from vibrations generated from periodic vibration. Maximum amplitudes distribution mapped shows the intensity of the dominant frequency that occurs larger to the western part of Makassar City. This indicates that the area has a lot of vibrations with the dominant frequency it has. The western part of Makassar have a larger Seismic Vulnerability Index than others. In other words, the western part of the city of Makassar are more vulnerable to damage than the eastern region.

References

[1] Ministry of Public Works 2010 *Earthquakes Zonation Maps*

[2] Akkaya I 2015 The Application of HVSR Microtremor Survey Method in Yüksekova (Hakkari) Region, Eastern Turkey *Journal of African Earth Sciences* **109** 87–95

[3] Maresca R and Berrino G 2016 Investigation of the Buried Structure of the Volturara Irpina Basin (Southern Italy) by Microtremor and Gravimetric Data *Journal of Applied Geophysics* **128** 96-109

[4] Salloum N, Chehade F H, Jongmans D, Voisin C, Cornou C, Massih D Y A and Mariscal A 2014 The Shear Wave Velocity Structure of the Heterogeneous Alluvial Plain of Beirut

(Lebanon): Combined Analysis of Geophysical and Geotechnical Data *Geophysical Journal International* **199** 894-913

[5] Stanko D, Markušic S, Strelec S and Gazdek M 2017 HVSR Analysis of Seismic Site Effects and Soil-structure Resonance in Varaždin City (North Croatia) *Soil Dynamic Earthquake Engeneering* **92** 666–677

[6] Tun M, Pekkan E, Ozel O and Guney Y 2016 An Investigation Into the Bedrock Depth in the Eskisehir Quaternary Basin (Turkey) Using the Microtremor Method *Geophysical Journal International* **207** 589-607

[7] Chávez García F J and Raptakis M 2017 Local Amplification and Subsoil Structure at A Difficult Site: Understanding Site Effects from Different Measurements *Soil Dynamic Earthquake Engeneering* **92** 334–344

[8] Mihaylov D, El-Naggar M H, Dineva S 2016 Separation of High and Low-Level Ambient Noise fo HVSR: Application in City Conditions for Greater Toronto Are *Bulletin of the Seismological Society of America* **106** 1-8

[9] USGS 2006 *Earthquake summary Map*

[10] Okada H 2004 *The Microtremor Survey Method* (United State of America: Society of Exploration Geophysicists)

[11] Seht M I and Wohlenberg J 1999 Microtremor Measurement Used to Map Thickness of Soft Sediments *Bulletin of the Seismological Society of America* **89** 250-259

[12] Mukhopadhyay S and Bormann P 2004 Low Cost Seismic Microzonation Using Microtremor Data: An Example from Delhi, India *Journal of Asian Earth Sciences* **24** 271–280

[13] Nakamura 1989 A Method for Dynamic Characteristic Estimatimation of Subsurface Using 51 Microtremor on the Ground Surface *Q.R. of R.T.I.* **30** 125-33

[14] Konno K and Ohmachi T 1998 Ground-Motion Characteristics Estimated from Spectral Ratio between Horizontal and Vertical Components of Microtremor *Bulletin of the Seismological Society of America* **88** 228-241

[15] Sesame 2004 Guidelines for the Implementation of the H/V Spectral Ratio Technique on Ambient Vibration Measurements *Processing and Interpretation, European Commission-Research General Directorate*

[16] Lermo J and Chavez-Garcia J 1993 Site Effect Evaluation using Spectral Ratios with Only One Station *Bull. Seism. Soc. Am.* **83** 1574-1594

[17] Lermo J and Chavez-Garcia J 1994 Are Microtremors Useful in Site Response Evaluation *Bull Seism. Soc. Am.* **84** 1350-1364

[18] Nakamura Y 1997 Seismic Vulnerability Indices for Ground and Structures Using Microtremor *Proceeding World Congress on Railway Research, Florence*

IOP Conf. Series: Journal of Physics: Conf. Series **979** (2018) 012054 doi:10.1088/1742-6596/979/1/012054

Earthquake Hazard Analysis Use Vs30 Data In Palu

Muhammad Rusydi[1], Rustan Efendi[1], Sandra[1], Rahmawati[2]

[1] Physics Department, Matematics and Natural Scienses Faculty, Tadulako University, Palu
[2] Geografi Department, Faculty of Teacher Training and Education, Tadulako University, Palu, Indonesia

E-mail : rusydiutd@gmail.com

Abstract. Palu City is an area passed by Palu-Koro fault and some small faults around it, causing the Palu of city often hit by earthquake. Therefore, this study is intended to mapped the earthquake hazard zones. Determination of this zone is one of aspect that can be used to reducing risk of earthquake disaster. This research was conducted by integrating Vs30 data from USGS with Vs30 from mikrotremor data. Vs30 data from microtremor used to correction Vs30 from USGS. This Results are then used to determine PeakGround Acceleration (PGA) can be used to calculate the impact of earthquake disaster. Results of the study shows that Palu City is in high danger class. Eight sub-districts in Palu City, there are 7 sub-districts that have high danger level, namely Palu Barat, PaluTimur, Palu Selatan, Palu Utara, Tatanga, Mantikulore and Tawaeli.

1. Background

One of the cities that often hit by the earthquake is Palu City. This is caused by an active fault that crosses this region. The fault is known as the Palu-Koro Fault (figure 1). Based on earthquake history since 1927 in this region, shows that there have been 9 (nine) times destructive earthquake events in Palu valley. The eighth earthquake caused the death toll not less than 250 people and property damage [1]. A study on the level of earthquake hazard that is interpreted in the form of zoning area of earthquake disaster hazard has been done. Determination of seismic hazard level is done by analyzing the response of geological characteristics through Determination of Vs30 and peak ground acceleration [2].

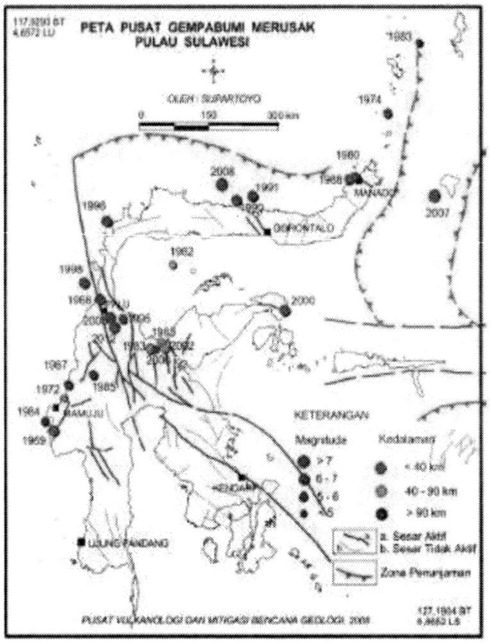

Figure 1. Points of earthquakes destructive and tectonic of Sulawesi Island [3].

Vs30 is one of the most commonly used parameters in predicting shear wave velocity in sedimentary layers at a depth of 30 meters[4]. This parameter is very significant in identifying the characteristics of rocks for earthquake resistant building design[5, 6]. One classification is often used in decrypt geological conditions that experience earthquake vibration effects, namely Site Class (SC) introduced National Earthquake Hazard Reduction Program (NEHRP) which is defined by the Building Seismic Safety Council [7, 8].

The natural period is estimated to be four times the speed of shear wave propagation in each soil layer, assuming that each part is represented by a soil layer with a constant shear wave velocity[8]. There are four classes in this classification, namely SC I, II, III and IV, which means, I, instead of rock, II hard soil, III medium soil and IV soft soil. The class determination based on shear wave (Vs30) can be seen in table 1:

Table 1. *Site class* NEHRP which define by BSSC (2000) [7].

Site Classes	Site Natural Period (s)	V_{s30}	NEHRP class
SC I : (Rock/Stiff Soil)	$T_G < 0.2s$	$V_{s30} > 600$ m/s	A + B
SC II : (Hard Soil)	$0.2s \leq T_G < 0.4s$	$300 < V_{s30} \leq 600$ m/s	C
SC III : (Medium Soil)	$0.4s \leq T_G < 0.6s$	$200 < V_{s30} \leq 300$ m/s	D
SC IV : (Soft soil)	$T_G \geq 0.6s$	$V_{s30} \leq 200$ m/s	E

One of methods that commonly used in determining shear wave velocity is the HVSR, this method uses microtremor observations that can display subsurface profile [9, 10]. The results obtained are the horizontal-to-vertical (HVSR) spectrum ratio that can estimate the shear wave velocity Vs. This value is then used to predict Vs30.

2. Geological Conditions

Palu is one of the cities on the island of Sulawesi which has various geological conditions, this is caused the island of Sulawesi is a collision of three plates, the Indian-Australian Plate, the Pacific Ocean Plate and the Eurasia Plate. As a result of the collision Sulawesi has many active faults, one of which is the active Palu-Koro fault located in Palu City[11]. In addition to its unique geological structure, Palu City is composed of alluvial deposits in the Valley, while in the northwest part it consists of granite fragments. In the western to north part it consists of granite and granodiorite rocks, to the south composed of schistphyllitic rocks and to the east consisting of mollasses[12]. Distribution of geological rocks of Palu City can be seen in figure 2.

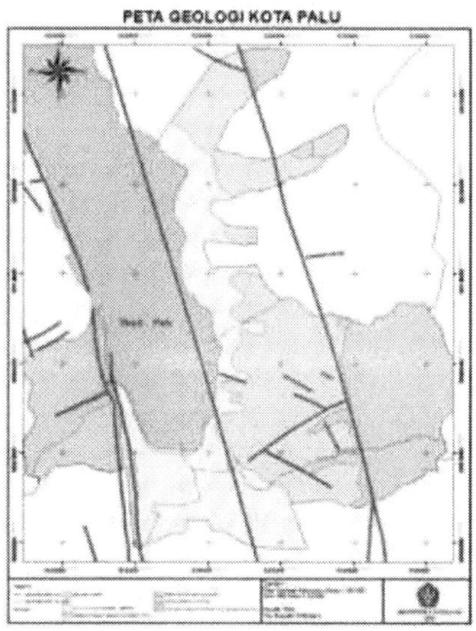

Figure 2. Geologycal conditions of Palu city.

3. Method

The research begins by analyzing the global Vs30 data. Further correction is done using microtremor field measurement results. This is done to improve unfamiliar data on global data. The result of this improvement yields a corrected Vs30 value and can be considered as the wave propagation velocity at the study site. Vs30 measurements in the field using a portable microtremor set of tools consisting of a Portable Digital Seismograph 3 component (2 horizontal components: EW-NS and 1 vertical components) short-period Taurus (Canada) brand with type of Feedback Short Period Seismometer type DS-4A sensor and equipped with digitizer (Data logger). Measurement of microtremor done at 6 point of Palu City. The flow of this study is more fully shown in figure 3.

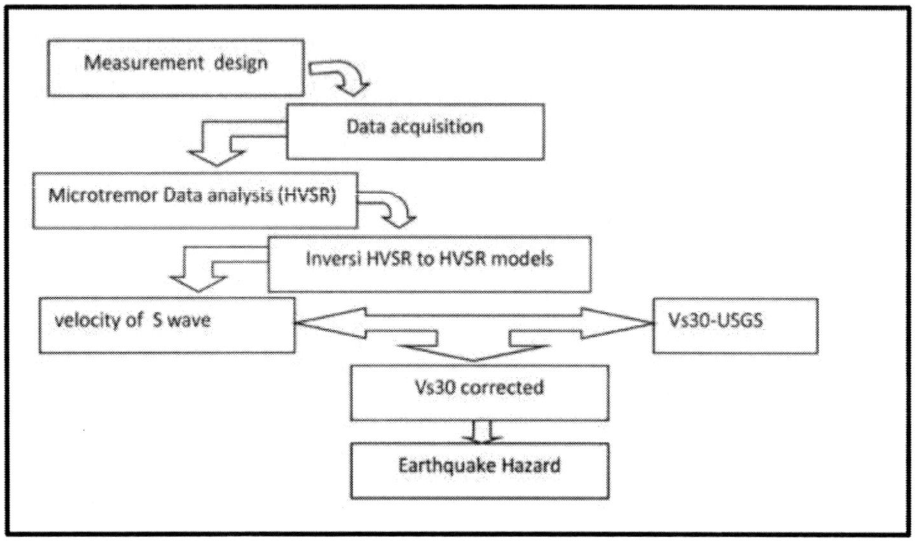

Figure 3. Flowchart Method.

Microtremor data processing is done by designing the concept of field acquisition, so that obtained a detailed description related to the acquisition will be done. The next stage is the retrieval of microtremor data at the point of measurement that has been designed previously. The next microtremor data is processed to obtain natural frequency and amplification values. Data processing is used to analyze the data HVSR (Horizontal to Vertical Spectral Ratio). Furthermore the HVSR curve of the HVSR analysis is converted to obtain a VS value which is then used to estimate the value of Vs30 [13].

Knowledge of the elastic properties of material near the surface and its effect on seismic wave enlargement is essential in the design of earthquake resistant buildings for disaster risk reduction. The most important factor that causes the amplification of the earthquake movement is the increase in amplitude in soft sediments [14]. The VS30 value can be determined from the formulation:

$$V_{s30} = \frac{30}{\sum_{i=1}^{N}\left(\frac{d_i}{v_i}\right)} \quad (1)$$

With, d_i is thickness of the sediment layer (m), Vi is the shear wave velocity (m / s) and N is the number of layers above 30 m depth. Vs30 value is used to determine rock classification based on vibration strength of earthquake due to local effect and used for purposes in earthquake resistant building design. Basically the Vs30 value alone can not represent the site effect because the shear wave velocity is only at a depth of 30 meters from the surface, where the bedrock depth and the impedance ratio between the soil and bedrock layers all contribute significantly to the site response. However, only layers of rock to a depth of 30 m only can determine the magnification of earthquake waves. This wave magnification effect on earthquake hazard assessment as recommendation material for earthquake resistant building construction

To produce an accurate picture of the distribution of Vs30 values, sampling should be done at each geological condition. Since some regions can not be measured because of unreachable location factors, to obtain the value, it is done by using Vs30 data from USGS then corrected with VS30 data on field measurements. Correction value of Vs30 is done with the following formula:

$$V_{s30}\text{ USGS} - V_{s30}\text{ Ukur} = \Delta V \quad (2)$$

$$\overline{\Delta V} = \frac{v_i}{n} \quad (3)$$

$$V_{s30}\text{ terkoreksi} = V_{s30}\text{ USGS} - \overline{\Delta V} \quad (4)$$

where the value ΔV is difference of shear wave velocity (m/s), $\overline{\Delta V}$ is average shear wave velocity (m/s) with v_i is the shear wave velocity at point i (m/s) and n is point of measurement at i. and then the Vs30 value is then converted to get the PGA value. The value of PGA is used to determine the area of earthquake hazard according to Perka BNPB No. 3 Year 2012. The PGA formulation according to Takahashi in 2000[15], as follows:

$$Log_{10}[y] = aM - bx - log_{10}(x + c10^{dM}) + e\,(h - h_c)\delta_h + S_k \quad (5)$$

y is the value PGA (cm/s^2), $a = 0.446$, $b = 0.00350$, $c = 0.012$, $d = 0.446$, $e = 0.00665$, $S = 0.941$, S_r = 0.751, $S_m = 1.003$, $Ss = 0.995$

4. Results and Discussion

The analysis Vs30 - USGS shows that the value of Vs30 distribution in Palu City is in the range of 268-760 m / s2. After corrected with Vs30 microtremor data shows a lower Vs30 value, with a range of values from 137 - 456 m / s2. Referring from the NEHRP site class, indicates that Palu City is located on SC II (hard ground) SC III (moderate ground) and SC IV (Soil Soil).Earthquakes that occur tend to have an enlarged amplitude in soft soil. While on hard soil amplitude decreases. This incident greatly affects the condition of the building above the ground, where the building located on soft soil will experience a greater degree of damage than the building located on hard soil[12].

4.1. Integration Vs30 Microtremor and Vs30 USGS

Palu City has a topographic shape of basin or valley so that most of the land in Palu Valley area is alluvial soil. This is in line with the Vs30 correction results indicating that almost all sub-districts in Palu City are located in soft soil, especially in the Palu Valley area. The area of Site Class Vs30 in each sub-district in Palu City can be seen in table 2.

Table 2. An area *Site Class* V_{s30} in Palu city.

No	Sub district	SCV$_{s30}$	Area (Ha)	No	Sub district	SCVs30	Area(Ha)
1		2	12265	5		2	0
	Mantikulore	3	4583		Palu Utara	3	918
		4	2393			4	2077
2		2	56	6		2	250
	Palu Barat	3	73		Tatanga	3	358
		4	610			4	918
3		2	0	7		2	1163
	Palu Selatan	3	0		Taweli	3	2119
		4	1969			4	2786
4		2	0	8		2	5884
	PaluTimur	3	0		Ulujadi	3	429
		4	601			4	55

IOP Conf. Series: Journal of Physics: Conf. Series **979** (2018) 012054 doi:10.1088/1742-6596/979/1/012054

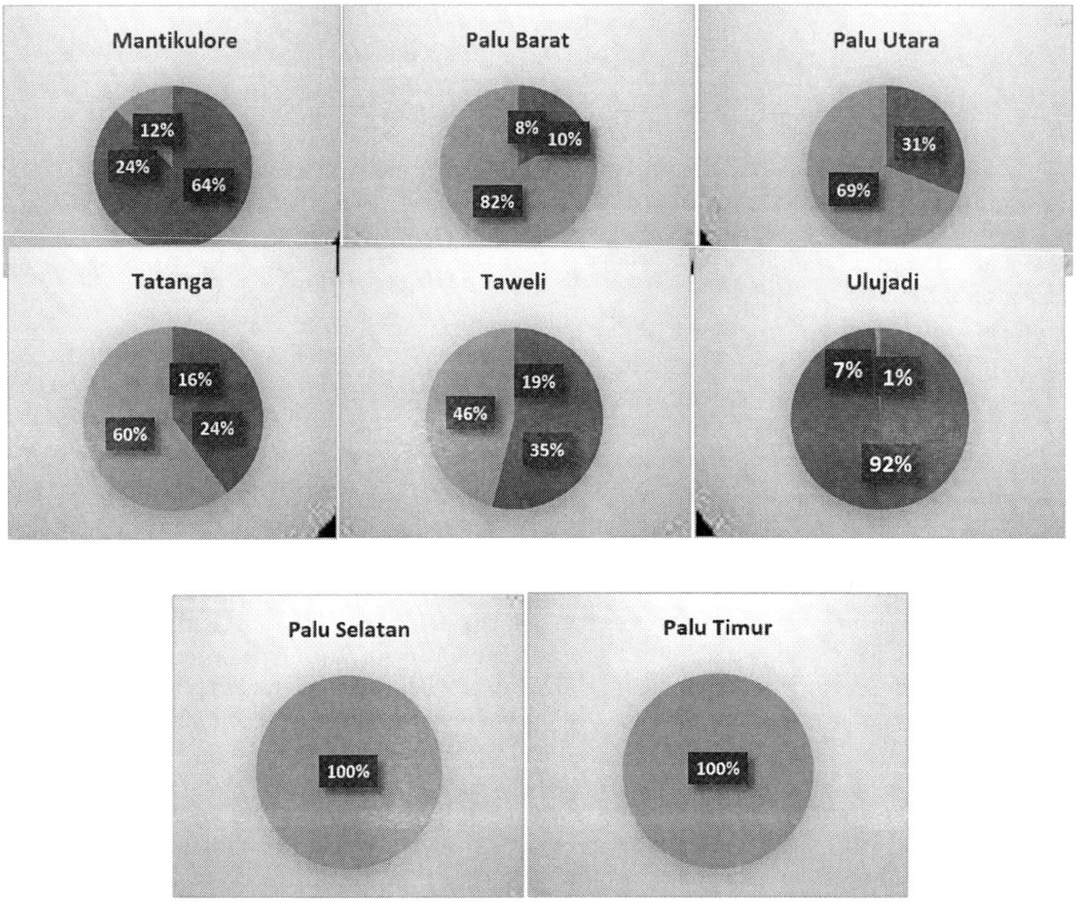

Keterangan : SC II SCIII SC IV

Figure 4. Percentage Chart *Site Class*V_{s30} in Palu City each sub district

Figure 4 shows a diagram illustrating the site class of each sub-district in Palu City. This diagram shows the percentage of site class of each sub-district in Palu City. District of South Palu and District of East Palu 100% of its territory is in SC IV. Furthermore, sub-districts with a significant percentage of SC IV are Palu Barat District 82% followed by Palu Utara sub-district 69%, then Tatanga District 60% and Kecamatan Tawaeli 46%. Kecamatan kecamatan which dominant region SC IV is subdistrict with rock dominated by soft soil. Ulujadi sub district has 92% area located on site class II, while Mantikulore District has SC big enough that is 64%. This picture shows that these two sub-districts are dominated by hard rocks such as granite and granodiorote rocks that belong to intrusive rocks. This condition resulted in shrinking earthquakes tend to shrink in the District Ulujadi and Matikulore Subdistrict. Distribution of Vs30 showing site class in each region in Palu City can be seen in figure 5:

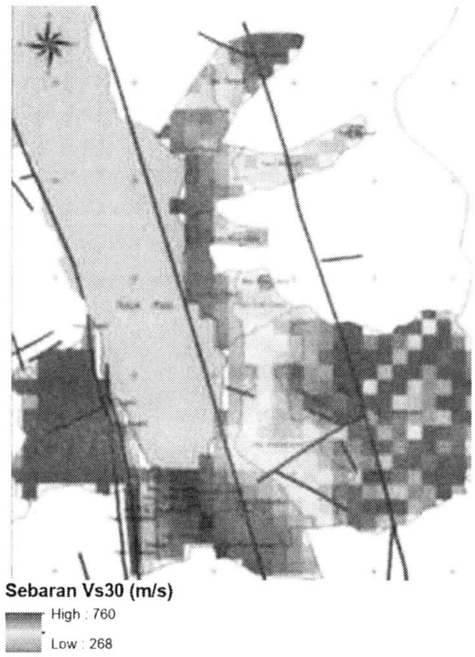

(a)

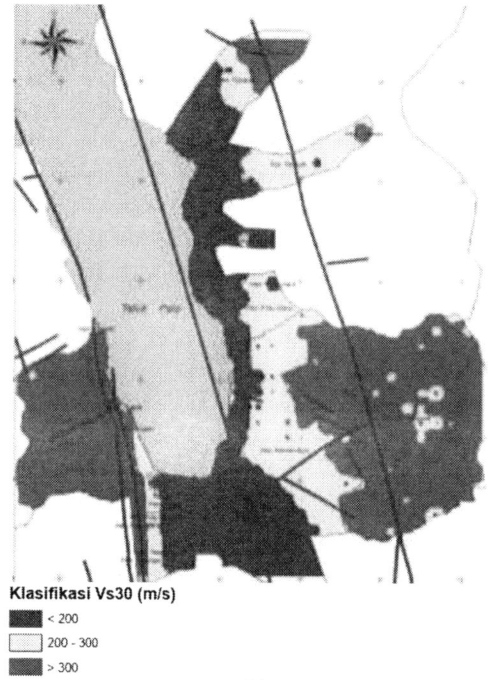

(b)

Figure 5. Vs30 from USGS (a) Vs30 Corection of microtremor (b).

Figure 5 shows the distribution of Vs30 both uncorrected (a), as well as those already in Correction (b). The results show that most of Palu City is hard land. This is inversely proportional to the results of research conducted by Nurahmi, where research conducted by using mictremor shows that in the lowland areas into the category of soft soil [16]. Since the Vs30 value of the USGS is considered not to reflect the real condition of the field, a Vs30 value correction is made using mictrotremor data taken at some point which is considered to represent the geological condition of the local area.This spatial data illustrates that the valley area, in Palu City has a low shear wave velocity, ie the average value is < 200 m/s. This data illustrates that the valley area, in Palu City is a location with soft soil class. While in the east and west side shows that the more toward the hill speed of shear wave propagation has increased 200-600 m / s, which correlates with the level of hardness of the rock toward the hills. Associated with the carrying capacity of rocks against the building in case of earthquake, then the lowland area or the valley area is the area with the most low its carrying capacity increases the carrying capacity to the hills.

Taking into account this condition, the area experiencing retrofitting in case of an earthquake is a lowland or valley area, while hilly areas in the west, as well as in the eastern part, will not experience significant reinforcement. This can be seen from the low shear wave velocity in the valley area whereas in the hilly region the wave velocity is quite high.

Unfortunately, the inhabitants of Palu City mostly live in the lowlands or Palu valleys, where this location is an area with a very low velocity of Vs30 shear wave propagation, which means the rocks in this location are soft rocks. The condition of soft rock locations like this carries the consequences of propagation of the waves will undergo high amflification so that the buildings on it are more shaken than the buildings located in hard rock [14]. This figure shows that the level of risk in the Valley or in the lowlands in Palu City is higher than in the hills from the threat of earthquake disaster.

4.2. Earthquake Hazard analysis

Vs30 data processing results are then processed using the Takahashi equation to obtain the Peak Ground Acceleration (PGA) value. The analysis was performed on earthquake occurrence on January 24, 2005 with magnitude 6.3 Mw and obtained 3 classes showing the increase of spectral peak value from SC II to SC IV (figure 6). The highest PGA value obtained is 73.8 cm/sec2 in SC IV indicating that this region is categorized as dangerous if it refers to BNPB PERKA No.2, (2012)[17]. The distribution of PGA values in Palu City shows results that are not much different from the distribution of Vs30. The highest PGA values are in the Valley area, and decrease toward the hill, either towards the West, or to the east. The higher the PGA value, the higher the dangers of earthquakes and can result in large losses for the buildings in the area.

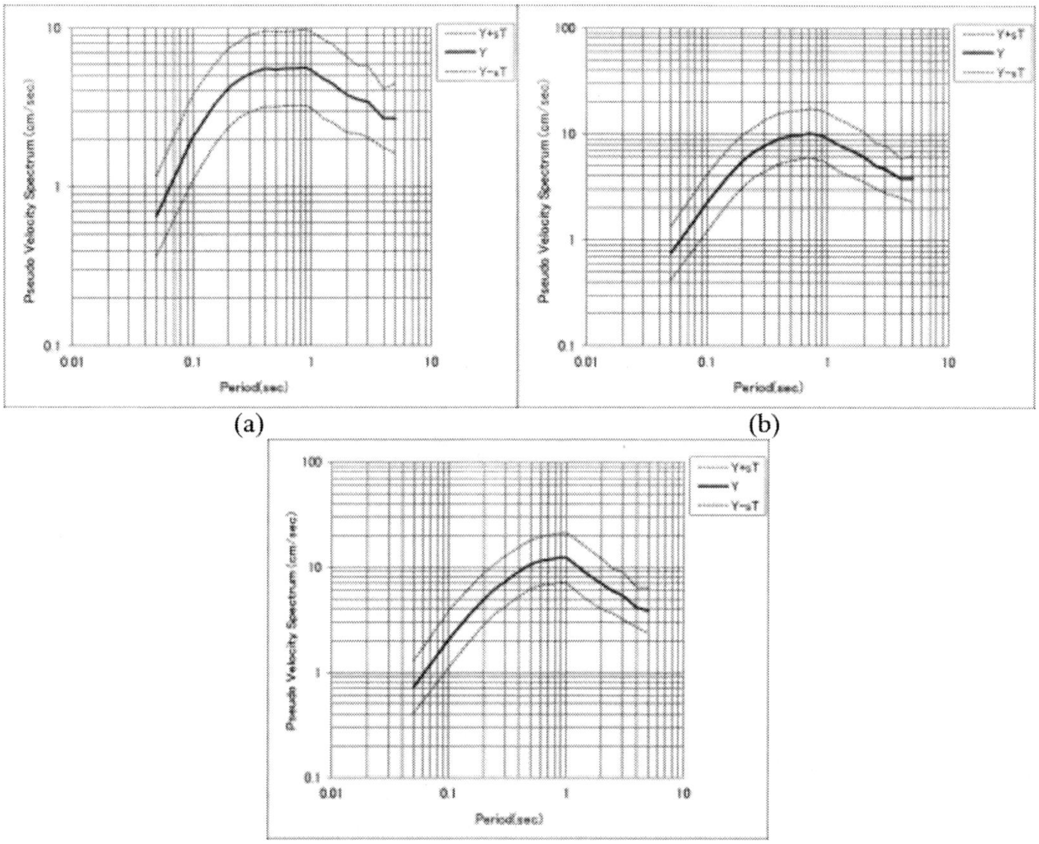

Figure 6. Graphic PGA by magnitude 6.3 Mw at SC II-*hard soil* (a), for SC III-*medium soil* (b), and SC IV-*soft soil* (c).

The picture above shows the relationship of pseudo speed with PGA value. The points depicted on the graph show the maximum value of PGA. The results obtained in three geological conditions show the value of pseudo speed on hard soil below 10 cm / s, whereas on medium soil 10 cm / s and in soft soil are at the apparent speed above 10 cm / s. Based on the hazard class published by BNPB by referring to the guideline published by National Geological Agency[17]. Palu City is divided into two classes of earthquake hazard indices: moderate hazard class with PGA 25-70 cm/sec2 and high hazard class with PGA value greater than 70 cm/sec2. In line with the division of the class, then the eight districts in Palu City can be divided into, seven districts are in areas with high hazard and one districts level of moderate hazard. The sub-districts located in the high hazard are Palu Barat, Palu Timur, Palu

Selatan, Palu Utara, Tatanga, Mantikulore and Taweli sub-districts, while sub-districts entering the dangers are districts Ulujadi (figure 7b).

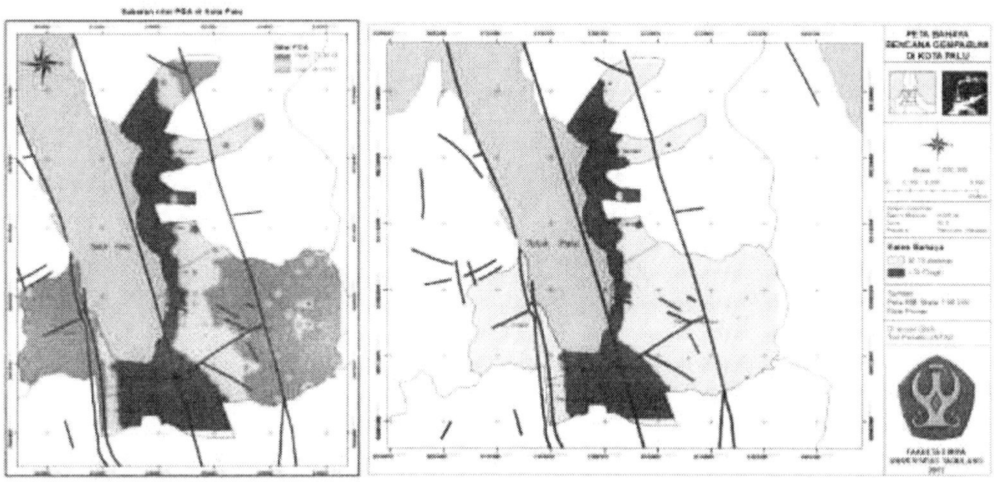

(a) (b)

Figure 7. Dstributon of PGA value in di Kota Palu (a) Map of Earthquake hazard in Kota Palu (b)

5. Conclution

The use of the USGS Vs30 can be used properly when corrected with Vs30 microtremor measurements in the field. The results obtained are lower, with a range of values from 137 - 456 m / s2 corresponding to actual field conditions, where Palu City is located on SC II (hard soil) SC III (medium soil) and SC IV (Soil Soil). Palu City has a topographic shape of basin or valley so that most of the land in Palu Valley area is alluvial soil. This is in line with the Vs30 results indicating that almost all sub-districts in Palu City are located in soft soil, especially in the Palu Valley area.

Palu city is one of the areas that have high threat of earthquake. From eight sub-districts in Palu City, there are 7 sub-districts that have high danger level, namely Palu Barat, Palu Timur, Palu Selatan, Palu Utara, Tatanga, Mantikulore and Taweli sub-districts while Ulujadi District is in moderate danger class.

Peack Ground Acceleration (PGA) or maximum ground motion acceleration is one of the parameters that become the reference in earthquake resistant building planning. Analysis of these parameters is important especially in areas where the threat of earthquakes is quite high. Particularly for Palu City which passes by active fault of Palu Koro, where most of its area is categorized as high hazard, the building should be built with earthquake resistant construction.

References

[1] Rusydi H, Rahmawati R E 2017 Vulnerability Zoning of Earthquake Disaster of Palu *International Journal Of Science and Applied Science* **1(2)** 137-143

[2] Zillur M R S S, Maksud K A S M 2015 Liquefaction Hazard Mapping by Liquefaction Potential Index for Dhaka City, Bangladesh *Engineering Geology* **188** 137–147

[3] Supartoyo 2014 Tectonic class of Palu Koro Fault, Central Sulawesi *Jurnal Lingkungan dan Bencana Geologi* **5** 111-128

[4] Chun-Hsiang K C T C, Che-Min L, Kuo-Liang W, Jyun-Yan H and Chang S C 2016 S-wave Velocity Structure and Site Effect Parameters Derived from Microtremor Arrays in the Western Plain of Taiwan. *Journal of Asian Earth Sciences* **128** 27–41

[5] Pelekis P C, G A A 2013 Seismic Microzonation of Chania, Crete(Greece) Based on SASW Measurement Sand non-linear Site Response Analyses *Soil Dynamics and Earthquake Engineering* **53** I45–159

[6] Mohamed E K, Shokry M M F, Hassoup A, Helal A M A 2016 Local Site Effect Evaluation of the Western Side of the Suez Canal Area by Applying H/V and MASW Techniques *Journal of African Earth Sciences* **123** 403-419

[7] (BSSC) B S S C 2000 The 2000 NEHRP Recommended Provisions for New Buildings and Other Structures *Part I (Provisions) and Part II (Commentary)* FEMA **368/369**

[8] Zhao J X, K I, Zhang J, Fukushima Y, Somerville P G, Asano A, Saiki T, Okada H and Takahashi T 2004 Site Classification For Strong-Motion Stations In Japan Using H/V Response Spectral Ratio *13th World Conference on Earthquake Engineering Vancouver* (Canada: B.C) p 1278

[9] Nakamura Y 2000 Clear Identification Of Fundamental Idea Of Nakamura's Technique And Its Applications *12th World Conference on Earthquake Engineering* **14** 2656 (New Zeland: Auckland)

[10] Stephenson W J, S H, Frankel A D, Asten M, Carver D L and Kim W Y 2009 Site Characterization for Urban Seismic Hazards in Lower Manhattan, New York City from Microtremor Array Analysis *Geophysical Research Letters* **36** L03301

[11] Suliyanti P D N, Masturyono, Supriyanto R, Rasmid, Handi S W, Pupung S 2015 Seismicity Analysis in Palu Koro Fault Zone, Central Sulawesi *Journal of Environment and Geological Hazards* **6** 253-264

[12] Thein P S, S P, Brotopuspito K S, Kiyono J, Wilopo W, Furukawa A, Setianto A 2014 Estimation of Seismic Ground Motion and Shaking Palu City, Central Sulawesi Province, Indonesia *International Journal of Geological and Environmental Engineering* **8** 308-319

[13] Abdelmoneam R, S G, Tucker-Kulesza S 2015 Multichannel Analysis of Surface-waves and Integration of Downhole Acoustic Televiewer Imaging, Ultrasonic Vs and Vp, and Vertical Seismic Profiling in an NEHRP-standard Classification, South of Concordia, Kansas, USA. *Journal of Applied Geophysics* **121** 149–161

[14] Uma M R, A B, Dodagoudar G R 2010 Seismic Site Classification and Site Period Mapping of Chennai City using Geophysical and Geotechnical Data. *Journal of Applied Geophysics* **72** 152–168

[15] Takahashi S K , Fukushima J, Zhao J X, Nakamura H and Somerville P G 2000 A Spectral Attenuation Model for Japan using Strong Motion Data Base. *In Proceedings of the Sixth International Conference on Seismic Zonation*

[16] Nurrahmi, R E, Sandra 2015 Analysis of Shear Wave Velocity to A Depth of 30 m (Vs30) inTalise Village using Refraction Mikrotremor Method *Gravitasi 2015* **14** 1 1-2

[17] BNPB, K 2012 *PERKA BNPB* **2**

Determination of Binding Energy For Cu and Cu_2o Based X-Ray Diffraction Spectrum

Heryanto, Bualkar Abdullah, Dahlang Tahir

Department of physics, Hasanuddin University, Makassar, 90245 Indonesia

E-mail: dtahir@fmipa.unhas.ac.id

Abstract. Binding Energy of Cu and Cu_2O was determined by using semi-emperical method based on X-ray diffraction (X-RD) spectra. Parameter Method 3 (PM3) single point method in hyperChem used to analysis Binding Energy of Cu and Cu_2O. Cu was annealing at temperature 600^0C, 800^0C for 30 minutes to get the oxide form. X-RD use to obtain lattice and distance between neighbor atoms formed in the samples as input parameter for empirical method by using Material Analysis Using Diffraction (MAUD). The Binding Energy for Cu is -87,69 kcal/mol at room temperature (RT). The binding energy for Cu_2O are -156.29 kcal/mol and -159.36 kcal/mol, at 600^0C and 800^0C respectively. The advantage of the method used in this study is not required the input parameter of ionic energy interaction, which is very difficult to determine based on X-RD data. This mean, the method in this study is effectively and efficiently for determining the binding energy of element metal and their oxides.

1. Introductions

Process of chemical bonds never break or form with spontaneously, but it require the role of energy. calculate the energy, it takes a study of the mechanical properties of the nanoscale system that now presents new theoretical and experimental challenges [1]. Relation of the crystal structure and the binding energy from the analyzed XRD data was reported by [2]. Processing computationally to define crystals that have lattice information and atomic spacing serve as one of the foundations in an attempt to estimate of binding energy a molecule. Negative value on the calculation of energy binding means the process of releasing energy, whereas if positive signifies the process experienced in the formation of molecules of the test material occurs the process of binding / receiving energy. The potential energy equations for molecules can be written [3].

$$Up = \left[\frac{Az^+z^-}{4\pi\varepsilon_0 r}e^2\right] + \left[\frac{B}{r^n}\right] \tag{1}$$

U_p is a potential energy (joule), A is constant's madelung, r distance between atoms (Å) and B is interaction of atomic repulsion as well as z^+ as cation and z^- as anion. The Coulomb (U_c) energy be obtained by the equation [4].

$$Uc = \frac{z^2e^2}{4\pi\varepsilon_0} \tag{2}$$

One of the factor determinants value of binding energy is influenced by distance between atoms in a molecule. The interaction between electrons determines the structure and properties of matter from molecules to solids [5]. Estimation of binding energy in this study does not require parameter input such as ionic interaction energy for simple models crystal. Calculation of binding energies metal with cubic structure using several mathematical approaches that facilitate [6]. Knowledge of binding energy of metal and their oxide is essential for search oxide material which suitable in the semiconductor and superconductor applications. In this work, we focus our attention for calculation of binding energy by using the size of lattice and the distance between atoms from the x-ray diffraction data. We used some applications free software MAUD, GaussView and HyperChem which facilitate the process of analysis and calculation in this study.

2. Material and Method

Cupper CAT 2703 powder samples were given thermal treatment with variation of 30^0C, 600^0C and 800^0C respectively in 30 minutes. the voltages input is 40kV and currents input is 30mA. Observation angle from are 20^0 to 80^0, appliance type of x-ray diffractions is Shimadzu X-RD 7000. MAUD applications are used to estimate lattice parameters in molecules. Estimation lattice parameters in MAUD applications using standards data [7, 8, 9, 10]. After adjustment of diffraction pattern with standard data has been completed in MAUD application, then determine distance betwen atoms using GausView5.0 applications.

GaussView5.0 applications reads format CIF data. Determination distance betwen atoms based geometri and crystal system. After modeling is completed by using GaussView applications, the data is stored in protein data base (pdb) formats. HyperChem applications read and process pdb format. Semi-Empirical Method Parameter Method 3 (PM3) [11] use in this paper as a basic technique for optimizing element parameters by utilizing available data in HyperChem applications [12].

Single point calculations process the static properties of molecules that include potential energy, electrostatic potential, molecular orbital energy and orbital coefficients. Single point calculations are used to calculate the magnitude of the bonding energies based on the geometry of molecular structures that have been modeled based on X-RD data.

2.1 Result and Discussion

Data of diffraction shows the composition and lattice parameters, by thermal variations for Cu and Cu_2O.

Table 1. Lattice parameters and percentage of Cu and Cu_2O.

Compound	Lattice Parameter	Composition (%)
Cu (RT)	3,58	99
Cu_ref [7]	2,55	none
Cu_ref [16]	2,53	none
Cu_2O (600^0C)	4,24	46,5
Cu_2O_ref [13]	4,25	none
Cu_2O (800^0C)	4,25	22
Cu_2O_ref [14]	4,26	none

a)

b)

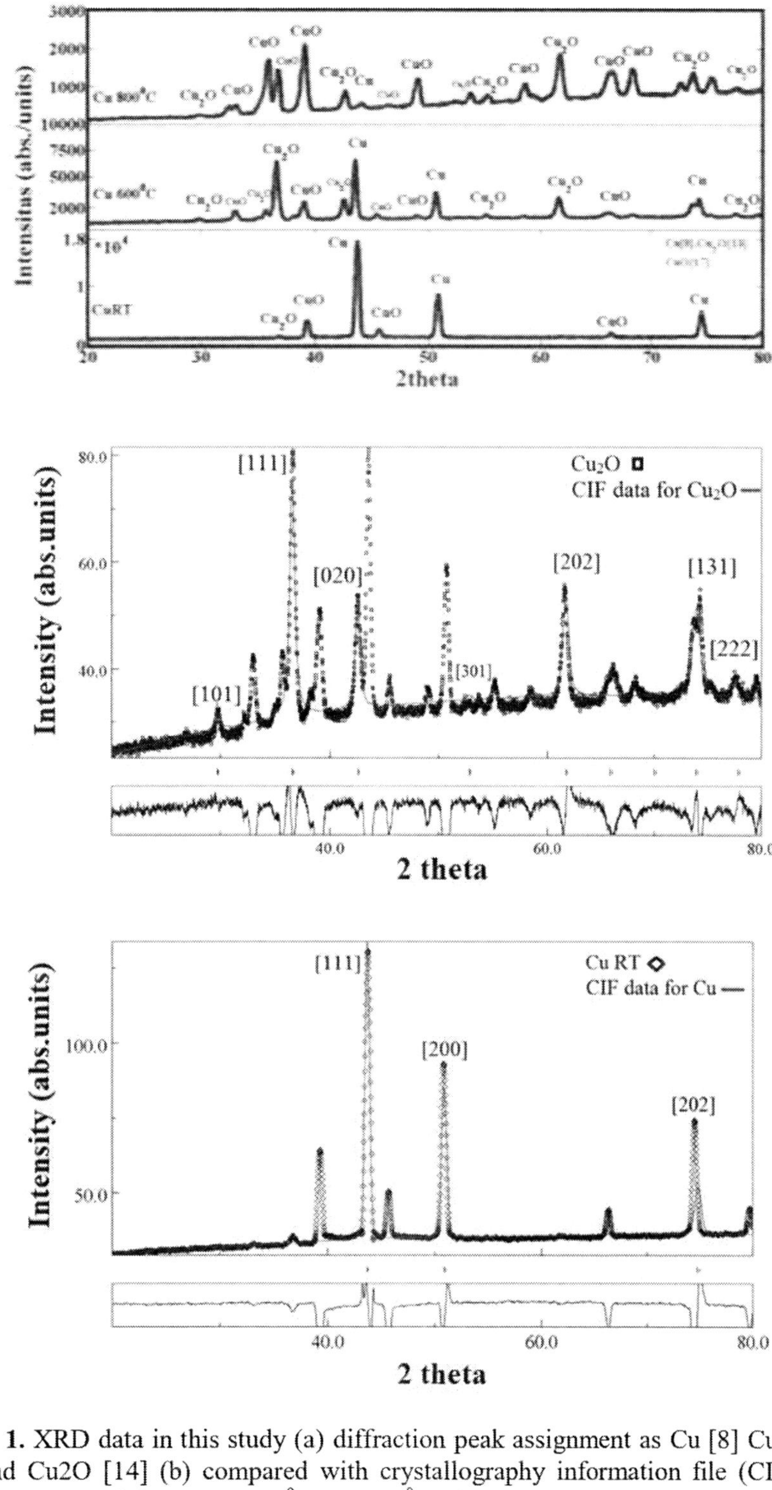

Figure 1. XRD data in this study (a) diffraction peak assignment as Cu [8] CuO [13] and Cu2O [14] (b) compared with crystallography information file (CIF) data for Cu at RT [7] Cu_2O at 600^{0}C and 800^{0}C [9].

Table 1 shows the lattice parameter and composition of Cu as the effect of annealing Cu powder from (RT) to 800^{0}C. We can see clearly from the table 1, Cu at room temperature is 99% and decrease to 46.5% and 22% for the temperature 600^{0}C and 800^{0}C, respectively. Lattice parameter shows slowly increase with increasing the annealing temperature. We applied MAUD for determining the lattice parameters by fitting the peak diffractions until good agreement. Figure 1 shows the composition amount of for Cu shows variations for different annealing temperature.

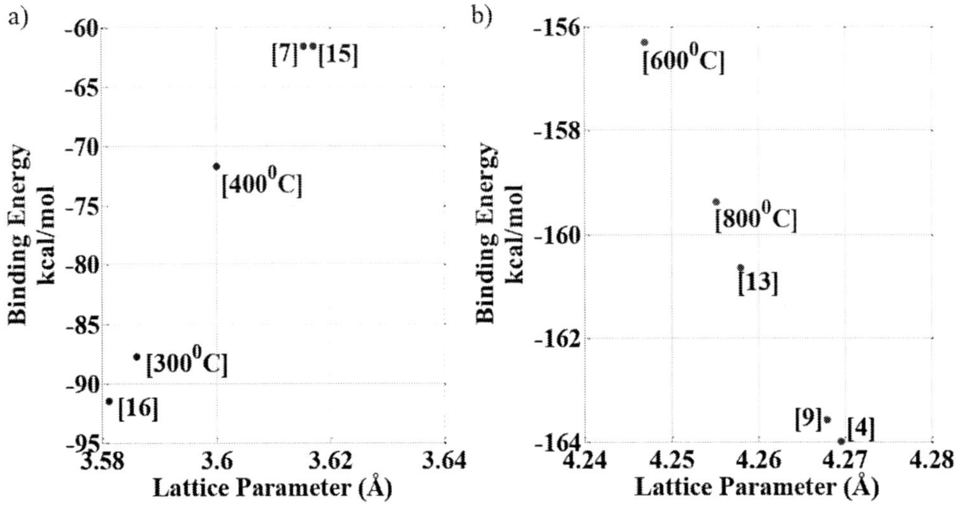

Figure 2. (a) The binding energy (kcal/mol) of Cu_2O variations in temperature with matching the database of [14], [15] dan [9]. (b) The binding energy (kcal/mol) of Cu variations in temperature with matching the database of [16], [17] dan [7].

Figure 2 Cu molecules gained energy (ΔE) from room temperature (RT) to 400^{0}C during in the thermal treatment process is about 15,92 kcal / mol and thermal treatments on Cu_2O molecule releases energy about 3,07 kcal / mol. From this results, we obtained for the distance between atomic of Cu elements at RT is 2,536Å. Almost Similarly with the distance between atoms reported [7] is 2,55Å and [17] is 2,532Å. Binding energy of Cu molecules at RT is -87 kcal / mol and fairly agreement reported [7, [17]. The lattice parameter in this study for Cu_2O at 600^{0}C and 800^{0}C are 4,24 Å and 4,25 Å, respectively, and very good agreemnet with [14] about 4,25 Å and also reported [15, 17] about 4,26 Å. Binding energy of Cu_2O molecules at 600^{0}C and 800^{0}C are -156,2 kcal / mol and -159,3 kcal/mol, respectively and good agreement with binding energy reported [9, 15]. Analysis of binding energy for Cu and Cu_2O by using free software MAUD, GaussView and HyperChem in this study shows useful method for applications in determining the binding energy of metal and their oxide.

3. Conclusions

We annealing Cu at 600^{0}C and 800^{0}C to form Cu_2O and succesfull analysis the lattice parameter and the binding energy. The lattice parameter for Cu at room temperature is 3,58 Å increase about 4,24 Å at 600^{0}C and 800^{0}C. Cu composition at room RT is 99%, Cu_2O is 46,5% and 22% for 600^{0}C and 800^{0}C, respectively. Binding energy for Cu is -87,6 kcal/mol. The lattice parameter for Cu_2O at 600^{0}C and 800^{0}C are 4,24 Å and 4,25 Å, respectively. Binding energy for Cu_2O are -156,2 kcal/mol and -159,3 kcal/mol at 600^{0}C and 800^{0}C.

IOP Conf. Series: Journal of Physics: Conf. Series **979** (2018) 012055 doi:10.1088/1742-6596/979/1/012055

References

[1] Autreto P A S, Legoas S B, Flores M Z S S, Galvao D S 2010 *America J. Chem. Phys.* **133** 124513
[2] Lee H U, Lee S C, Won J, Byung-Chul S, Soehoe C, Kim Y, Park S Y, Hee-Sik K, Young-Chul L, Lee J 2015 *Scientific reports* **5** 8691
[3] Johnson 2002 *Metals and chemical change* (Cambridge: Royal Society of Chemistry)
[4] Ismunandar 1996 *Kimia anorganik* (Indonesia: Portal Pendidikan Gratis Indonesia) p 34
[5] Aron J C, Mori-Sanchez P, Yang W 2008 *Sciencemag* **321**
[6] Shimada K 1974 *J. Physics* ***61*** 325
[7] Catherine F S, Dinah R P, Mohamed Z, Mark R J, Zoltán A G, Simon J C 2011 *J. American Chemical Society* **133** 2691-2705
[8] Suh I K, Ohta H, Waseda Y 1988 *J. Materials Science* **23** 757-760
[9] Kirfel A, Eichhorn K 1990 *Acta Crystallographica Section A* **46** 271-284
[10] Volanti, Diogo P, Orlandi, Marcelo O, Andrés, Juan, Longo, Elson 2010 *CrystEngComm.* **12** 1696
[11] James J P S 1988 Optimization of Parameters for Semiempirical Methods *J. Chomputational Chemistry* **10** 209-220
[12] James J P S 1990 Optimization of Parameters for Semiempirical Methods. I1l Extension of PM3 to Be, Mg, Zn, Ga, Ge, As, Se, Cd, In, Sn, Sb, Te, Hg, T1, Pb, and Bi *J. Computational Chemistry* **12** 320-341
[13] Niggli P 1922 *Zeitschrift fuer Kristallographie, Kristallgeometrie, Kristallphysik, Kristallchemie* **57** 253-299
[14] Yamaguti T 1938 *Proceedings of the Physico-Mathematical Society* **20** 230-241
[15] Hafner S S, Nagel S 1983 *Physics and Chemistry of Minerals* **9** 19-22
[16] Simon C, Oliver J R, Gareth R W 2006 *Chemical Communications*
[17] Fortes A D, Emmanuelle S, Lemée-Cailleau, Marie-Hélène, Pickard, Christopher J, Needs R J 2009 *J. of the American Chemical Society* **131** 13508-13515

Dye sensitized solar cell (DSSC) with natural dyes extracted from *Jatropha* leaves and purple *Chrysanthemum* flowers as sensitizer

Dahlang Tahir, Wilda Satriani, P L Gareso and B Abdullah

Department of Physics, Hasanuddin University, Makassar 90245 Indonesia

E-mail: dtahir@fmipa.unhas.ac.id

Abstract. DSSC (Dye-Sensitized Solar Cell) prototype has been investigated using Jatropha leaves and purple Chrysanthemum flowers as natural dyes. DSSC consists of working electrode and counter electrode. A working electrode composed of semiconductor nanoparticles TiO_2 that has been coated with dye molecules. Dye molecules serve as light photon catchers, while semiconductor nanoparticles TiO_2 function to absorb and forward photons into electrons. In the electrode counter given catalyst carbon, serves to accelerate the reaction kinetics of triiodide reduction process on transparent conductive oxide (TCO). DSSC using TiO_2 as a semiconductor material and natural dyes as sensitizer from Jatropha leaves and purple Chrysanthemum flowers are successful produced. The physical properties of the working electrode have been determined by using XRD and the chemical properties of the TiO_2 powder and dye powder using FTIR and dye solution using UV-Vis. The resulted fabrications are also examined its I-V characteristics. The best performance is generated by mixed dye 1.91×10^{-3} % compared than those DSSC for dye extracted from *Jatropha* leaves or purple *Chrysanthemum*. The characterization results show that the higher of the absorption wavelength of the DSSC efficiency is high.

1. Introduction

Dye-Sensitized Solar Cells (DSSC) was invented in 1991 and now become attractive attention for researcher in the world due to inexpensive devices for converting sunlight into electricity [1,2]. Wide band gap oxide semiconductor materials as photoanode for DSSC which is accepted electrons from the photo excited dye/sensitizer [3]. The absorption spectrum of the dye to the surface of titanium dioxide (TiO_2) is important parameters in determining the efficiency of the cell [4].

Previously, DSSC employs ruthenium poly-pyridyl complexes as the effective sensitizers. However, ruthenium is a heavy metal, which is undesirable from point of view of the environmental aspects and costly [5]. Another method is replacing it by using natural dyes pigments due to the extraction is easy, cost efficiency, environment friendly, non- toxicity, availability, and biodegradability [3]. Natural dyes are playing a key role in harvesting sunlight and transferring solar energy into electrical energy [6].

In recent years, several researches have been reported several part of plants as natural dyes. They used natural dyes extracted from the trees, flowers, fruits, and vegetables as sensitizers for fabrication of Dye Sensitized Solar Cell (DSSC) [7]. Peihui Luo et al. achieved conversion of visible light into electricity of dyes of C. indica L., S. splendens and S. nigrum L with the efficiency of the cells sensitized up to (0.29%), (0.26%), and (0.31%), respectively [5]. Afterward, Rajita Ramanarayanan et al. extracted dye from red amaranth leaves showed a better efficiency (0.53 %) [3].

In this paper, DSSCs were prepared with natural dyes extracted from Jatropha and Chrysanthemum (purple). These extracted dyes were characterized by X-Ray diffractometer to analyze the structure of a material, UV-Vis Spectrophotometer to observe on the absorption spectra and FT-IR spectral analysis to determine the functional group in the natural dye.

2. Experimental

2.1. Preparation of sensitizers using natural dyes/Natural Dyes preparation as Sensitizers of DSSC

Jatropha leaves and purple Chrysanthemum were cut about 2 cm x 2 cm, washed with distilled water and dried in an oven at 40 °C before being crushed into fine powder by using a mortar. Concisely, the powders are dissolved using ethanol, acid citrate, and distilled water 5: 1: 4 (10 grams of powder, 10 ml of ethanol, 2 ml of acid citrate, and 8 ml of distilled water) and stir uses a magnetic stirrer in temperature of 60 °C for 30 minutes. Then, the solution stored in the bottle for 24 hours at room temperature without light. The extract was then filtrated using filter paper (Whatman No. 42) to remove solid residues resulting dye solution was used as a sensitizer in the DSSC. Then, 7 g of TiO2 nano-powder is added in 15 ml of Ethanol. The paste was stirred it uses magnetic stirrer for 30 minutes. Powder Potassium Iodide (KI) of 0.8 grams mixed with 10 ml of polyethylene glycol (PEG). The solution was added with iodine (I2) and stored in the bottle. Furthermore, the scotch tape is applied on four sides of conducting ITO glass to control the thickness of the TiO_2 film. The scotch tape controlled the thickness and the area of the TiO_2 film. TiO_2 paste was immediately spread as evenly possible onto the 2 cm × 2 cm area of the conductive glass by using a dropper pipette and spin by using a spin coating at 1000 rpm for 60 seconds. The TiO_2 films have cooled for 15 minutes at room temperatures. After that, the TiO_2 was immersed into dye for 2 hours for "sensitized" process. The electrolyte solution was dripped into the film. Then, the scotch tape was removed from the film. For counter electrode, graphite is applies on the conducting surface of another ITO glass. The two electrodes were clipped together using two binder clips to prevent the electrolyte from leaking.

2.2. Characterization and measurements

In this research, the characterizations were used X-Ray Diffraction (XRD), Fourier Transform Infra-Red (FTIR), and Uv-Vis Spectrophotometer. a XRD measurement was used to determine the crystal structure and phase of TiO_2. The crystal size has been calculated using Scherrer's formula [9] :

$$D = \frac{k\lambda}{\beta \cos\theta} \tag{1}$$

where, D is the grain size, K is a constant taken to be 0.94, k (1.54 A°) is the wavelength of the X-ray radiation, β is the full width at half maximum and θ is the angle of diffrac- tion. FTIR is used to determine the functional groups of TiO_2 and dye and Uv-Vis Spectrophotometer is used to determine the dye absorbance spectrum. The performance of solar cells can be seen through current measurements and voltage variations. The voltage measurement circuit in the solar cell construction is carried out using an unhindered circuit and with resistance (100 kΩ, 200 kΩ, 300 kΩ, 390 kΩ, 470 kΩ, 680 kΩ, and 820 kΩ) and a digital multimeter. Measurements are performed outdoors using sunlight as a light source. As for the resulting currents is calculated using Ohm's law approach, namely:

$$I = \frac{V}{R} \tag{2}$$

The point on I-V curve yielding maximum current and voltage was called maximum power point (Max). The maximum power of solar cell was stated by [10]:

$$P_{max} = \frac{V_{max} I_{max}}{A} \tag{3}$$

Calculates the efficiency of the DSSC was calculating from the data given by the equation:

$$\eta = \frac{P_{max}}{i} \times 100\% \quad (4)$$

3. Result and discussion

3.1. XRD Characterization

Figure 1 (a) shows the XRD spectra of TiO_2. The figure shows the peaks of the TiO_2 characteristics that form the anatase phase at an angle of 25.9948 °; 37.6237 °; 38.4764 °; 39.2470 °; 48.7037 °; 54.5413 °; 55.7067 °; 63.3199 ° with the highest intensity at the angle 25,9948° which is clearly indicated phase of anatase as reported in Ref [11]. The maximum grain size is found to be 57.90 nm.The peaks of anatase are indicated by "A". Figure 1 (b) shows the XRD spectra of dye extracted from *Jatropha* leaves which have been deposition with TiO_2. Figure 1 (c) shows the XRD spectra of dye extracted from purple *Chrysanthemum* flowers which has been deposition with TiO_2.

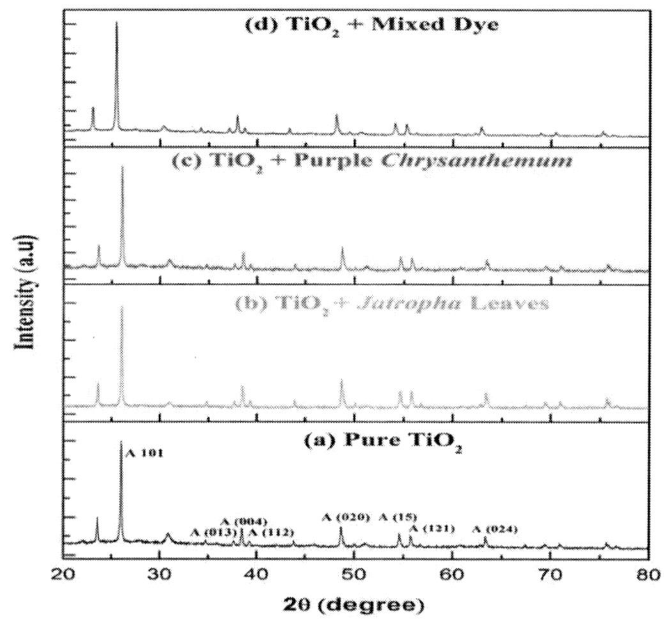

Figure 1. (a) Spectrum XRD (a) pure TiO_2, (b) TiO_2- *Jatropha* leaves, (c) TiO_2– purple *Chrysanthemum,* and (d) TiO_2–mixed dye.

Figure 1 (a) and (b) show that the absorption of dye on the TiO_2 molecule. The peak intensity of TiO_2 increased after the absorption of dye in TiO2 layer. As shown in table 1:

Table 1. Intensity from XRD spectra for all peaks of TiO2 and with dyes.

2θ	Intensity		
	TiO_2	TiO_2 - *Jatropha* leaves	TiO_2 - purple *Chrysanthemum*
25,99	1812 (011)	3710	1824
37,62	95 (013)	212	3185
38,48	338 (004)	759	293
39,25	81 (112)	220	1021
48,70	433 (020)	628	303
54,54	283 (015)	658	1004
55,71	262 (121)	642	882
63,32	239 (024)	591	854

3.2. FTIR Characterization

Figure 2 (a) shows the FTIR spectrum of TiO2 by using the optimal solvent (ethanol). The figure shows that spectrum range of 500 - 4000 cm^{-1}. The O-H group appears as the broadest peak at the wavelength of 3419,9 cm^{-1}. At the wave number 2926.01 cm^{-1} shows the C-H bond of the aromatic ring. The wave number 1645.28 cm^{-1} shows the C = C bond of the alkene compound and at the wave number 1514.12 cm^{-1} shows the C = C bond of the aromatic ring compound. In the TiO_2 spectrum, there is also an O-Ti-O bond at 500 cm^{-1}- 680 cm^{-1}.

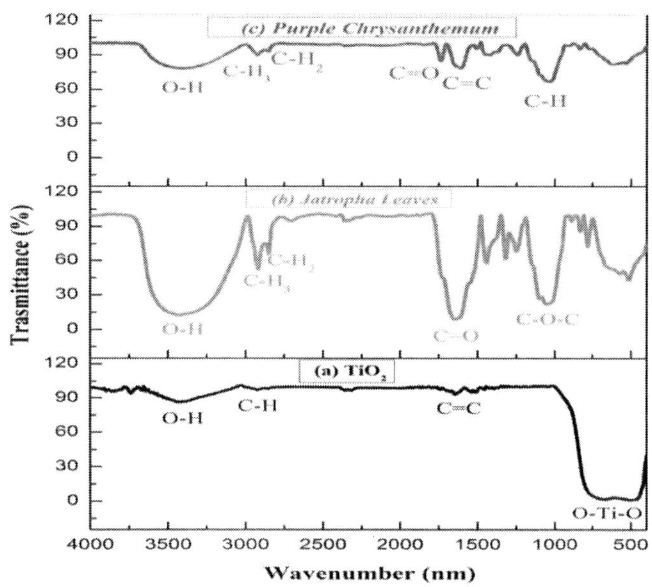

Figure 2. Spectrum FTIR (a) Pure TiO_2 (b) TiO_2 - *Jatropha* leaves (c) TiO_2 - purple *Chrysanthemum*

Figure 2 (b) shows that the FTIR spectrum of by Jatropha leaves using the optimal solvent (ethanol, citric acid, and aquades). The O-H group appears as the broadest peak at the wavelength of 3421,72 cm^{-1} which is also supported by the sharp peak at the wave number 1737.86 cm^{-1} for the bonding of C=O. The absorbance peak at 2920.23 cm^{-1} shown $C\text{-}H_3$ vibration and 2852.72 cm^{-1} indicates the presence of $C\text{-}H_3$ vibration. The absorption of C = C aromatic double bonds is shown at 1651.07 cm^{-1} and for the C-O-C bonding absorption at the wave number 779.24 cm^{-1}, which is indicated the existence of chlorophyll.

Figure 2 (c) shows that the FTIR spectrum of purple chrysanthemum by using the optimal solvent (ethanol, citric acid, and aquades). The O-H group appears as the broadest peak at the wave number of 3412.08 cm^{-1} which is also supported by the peak absorption at the wave number of 2924.09 $cm^{-1}$for the $C\text{-}H_3$, at the wave number 2854.65 shown $C\text{-}H_2$, and at the wave number 1737.86 cm^{-1} shown C=O double bond of alcohol. The absorption peak for C-N bond is shown at the wave number from 1441.61 cm^{-1} which is indicated that the existence of anthocyanin.

3.3. UV-Vis Characterization

Figure 3 (a) shows UV-Vis spectrum of dye extracted from, purple Chrysanthemum, and Jatropha leaves mixed with purple Chrysanthemum leaves used UV-Visible Spectrophotometer 300-800 nm. Figure 3 (a) shows the UV-Vis absorption spectra of dyes extracted from Jatropha leaves. Figure 3 (b) shows the UV-Vis absorption spectra of dyes extracted from Purple Chrysanthemum. Figure 3 (c) shows the UV-Vis absorption spectra of dyes extracted from Purple Chrysanthemum.

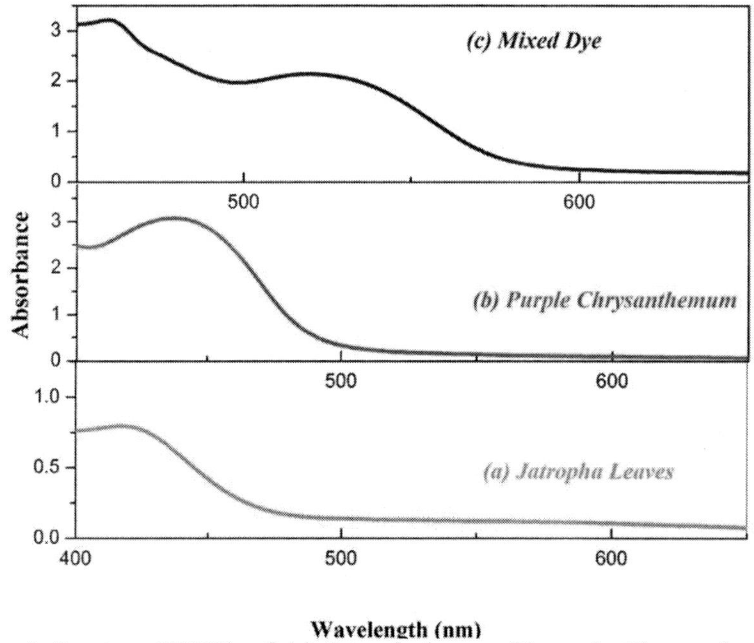

Figure 3. Spectrum UV-Vis of (a) *Jatropha* leaves, (b) purple *Chrysanthemum,* (c) *Jatropha* mixed with purple *Chrysanthemum*

3.4. Performance of DSSC

The DSSC prototype has been tested using a digital multimeter by placing a positive pole on the working electrode and the negative pole of the counter electrode to determine the resulting voltage. The resulting current is calculated using Ohm's law approach on equation (3). By varying, the value of resistance will be obtained the value of current and voltage generated from the solar cell is shown in table 2.

Table 2. The measurement results of DSSC

Dye	i (mW/cm^2)	V (mV)	I (mA)	P (mW/cm^2)	η (%)
Jatropha leaves	55.3	3.5	3.5×10^{-6}	3.06 x 10-6	5.53×10^{-6}
Purple *Chrysanthemum*	51.35	3.3	4.02 x 10-6	3.32 x 10-7	6.46×10^{-7}
Jatropha leaves + Purple *Chrysanthemum*	66.36	22.5	22.5 x 10-4	1.26 x 10-3	1.91×10^{-3}

After measured the current, then will be calculated the value P_m (Maximum power generated by voltage and current) for determine of the efficiency. Based on Table 3.2 can be seen that P_m produced of *Jatropha* leaves 3.06×10^{-6} mW / cm^{-2}, purple chrysanthemum 4.02 × 10-6 mW / cm^{-2}, and *Jatropha* leaves mixed with purple *Chrysanthemum* 3.32×10^{-7} mW / cm^{-2}.

DSSC conversion efficiency used equation (4) to determining efficiency by used dye extracted from Jatropha leaves is 5.53 × 10-6 %, purple Chrysanthemum is 6.46 × 10-7% and for mixed 1.91 x 10-3 %. From the analysis efficiency of DSSC based on the dye extracted from Jatropha leaves mixed with purple Chrysanthemum is higher than dye extracted from only Jatropha leaves or purple Chrysanthemum. DSSC is performed outdoors using sunlight as a light source. DSSC successfully convert solar energy into electrical energy based on current and voltage values generated.

The efficiency of DSSC is influenced by the absorbance of dye and semiconductor materials. Dye absorbance is directly proportional to the coefficient of absorption. The higher absorption coefficient is indicated the higher the electron-hole rate. In addition, the efficiency of DSSC is also influenced by the type of semiconductor, light source, conductivity, and phase and semiconductor structure.

4. Conclusion

Fabrication of DSSC used natural dyes Jatropha Leaves and Purple Crhysanthemum have been investigated. The dye gives the absorbance at wavelength 400-500 nm. It indicated that the dyes absorb visible light and have been found to be suitable for the use as sensitizer in solar cells. The best performance is generated by mixed dye 1.91×10^{-3} % compared than those DSSC for dye extracted from *Jatropha* leaves is 5.53×10^{-6} % or purple *Chrysanthemum* 6.46×10^{-7}%.

References

[1] Hao, S., Wu, J., Huang, Y., & Lin, J. (2006). *Natural dyes as photosensitizers for dye-sensitized solar cell,* Solar Energy, vol. **80**, no. 2, pp. 209–216
[2] El-Ghamri, H. S., Taya, S. A., El-Agez, T. M., Al-Kahlout, A. M., Al Dahoudi, N., & Abdel-Latif, M. S. (2015). Natural dyes as photosensitizers for dye-sensitized solar cells. *Journal of nano-and electronic physics*, vol. **7**, no. 3, pp. 3001-1
[3] Al-alwani, M. A. M., Mohamad, A. B., Kadhum, A. A. H., & Ludin, N. A. (2017). *Natural dye extracted from Pandannus amaryllifolius leaves as sensitizer in fabrication of dye-sensitized solar cells*, International Journal of Electrochemical Science, vol. **12**, pp. 747–761
[4] Torchani, A., Saadaoui, S., Gharbi, R., & Fathallah, M. (2015). Sensitized solar cells based on natural dyes. *Current Applied Physics*, vol **15**, no. 3, pp. 307-312
[5] Luo, P., Niu, H., Zheng, G., Bai, X., Zhang, M., & Wang, W. (2009). *From salmon pink to blue natural sensitizers for solar cells: Canna indica L., Salvia splendens, cowberry and Solanum nigrum L,* Spectrochimica Acta - Part A: Molecular and Biomolecular Spectroscopy, vol. **74**, no. 4, pp. 936–942
[6] Alhamed, M., Issa, A., & Doubal, A. (2012). *Studying of natural dyes properties as*

photo-sensitizer for dye sensitized solar cells (DSSC), Journal of Electron Devices, vol. **16**, pp. 1370–1383

[7] R. Syafinar, N. Gomesh, M. Irwanto, M. Fareq, Y. M. Irwan. (2015). *Chlorophyll pigments as nature based dye for dye sensitized solar cells(DSSC)*, Energy Procedia, vol. **79**, pp. 896- 902

[8] Abdel-Latif, M. S., Abuiriban, M. B., El-Agez, T. M., & Taya, S. A. (2015). *Dye-sensitized solar cells using dyes extracted from flowers, leaves, parks, and roots of three trees,* International Journal of Renewable Energy Research, vol. **5**, no. 1, pp. 294–298

[9] Gokilamani, N., et al. 2013. *Dye-sensitized solar cells with natural dyes extracted from rose petals.* Journal of materials science. Materials in electronics, vol. **24**, no. 9, pp. 3394

[10] Subodro, Rohmat, et al. 2017.*Dye-Sensitized Solar Cells (DSSCs) reengineering using TiO2 with natural dye (anthocyanin)*. AIP Conference Proceedings, vol. **1788**, no. 1, pp 030104

[11] Hanaor, D. A., & Sorrell, C. C. (2011). Review of the anatase to rutile phase transformation. *Journal of Materials science*, vol **46**, no. 4, pp. 855-874

Modifying Of Particle Boards From Rice Husk and Pinus Merkusii Sawdust And Using Soybean Waste Waters Based Adhesive.

Indah Raya[1], Nurfika Ramdani[1], Abd. Karim[1] & Musrizal Muin[2]

[1]Chemistry Department, Faculty of Mathematics and Natural Sciences, Hasanuddin University, Makassar, South Sulawesi, Indonesia 90245
[2]Faculty of Forestry, Hasanuddin University, Makassar, South Sulawesi, Indonesia 90245

E-mail: indahraya05@gmail.com

Abstract. Research of modifying particle board has been prepared by mixing of pinus merkusii sawdust and rice husk, where used of adhesive base on Boiled Soybean Water (BSW) has been done. The research utilize the rise husk and sawdust pines mixed, and used of a renewable and environmental adhesive to replace the toxic and carcinogenic one. The testing of adhesive included are; colour, pH, solid contain, gelatination time, density and viscosity. Result showed yellowish colour, 10 of pH, 44.70 % of solid contained, 56.29 minutes of gelatination time, and 1.1656 g/cm^3 and 182.4387 cP of viscocity respectively. While the particle boards testing include are density, moisture, immersion, thickness, Modulus of Rapture (MOR) and Modulus of Elasticity (MOE). The particle board best ratio it was 1:4 (Rice husk: pinus merkusii sawdust). Result of each parameter are, 0.7735 g/cm^3 of density, 5.79 % of moisture, however the immersion for 2 hours is 26.90 % and immersion for 24 hours is 39.77 %, 101.1592 kg/cm^2 of MOR and 18,248.3063 kg/cm^{2}of MOE.
The summary, the adhesive based on SNI 06-4567-1998 and the particle board based on SNI 03-2105-1996

1. Introduction

Indonesia is well-known as one of the forest areas in the world, but lowest supplier of furniture compare with Italia and China. The furniture productions in Indonesia has not optimal to supply necessary of world markets yet. Whereas, Italia and china has little raw material, but they can be supply necessary of world markets [1].

Based on the forestry of statistics in 2011, the progress of export products of forest in last five years, especially particle board, be through a fluctuating production. It is caused by many factors, dominated by particle board low quality which produced by woods industries in Indonesia [2].

In Indonesia the particle board generally has produced with using of formaldehid based woods adhesives, like Urea-Formaldehid (UF) and Phenol-Formaldehid (PF). Whereas, they are carcinogenic for human [3]. The formaldehid emission which carcinogenic, it is require to develop an environmental friendly and renewable adhesive [4].

Therefore, many scientists looking for an non carsinogenic adhesive. An example, Liu 2014 who has modifying soy protein become good and waterproof wood adhesives [4].

The others alternative, where using waste liquids from industries of tofu and tempe that contains percentage of protein is high sufficiently. It has not utilized by people, so that can be contaminate environment in the industrial area. Because industries of tempe and tempe just the little industry, so that it has not water treatments [5]. In this research were used Boiled Soybean Water (BSW), to make efficient in making process of particle board which according of Indonesia National Standards (SNI).

On the other hand, waste of rice husk and sawdust has not been utilized optimally by people. So that, economic value has not been increased yet. Based on the Statistic Centre Institution (BPS), production of rice plant in 2010 reaches 64.90 million tons rice husk. Process of rice plant milling produced rice husk between 20-30 %, husk between 8-12 % and rice between 50-63.5 % [6]. Therefore, we can see the most waste is rice husk in each year. So that, this research utilized it as one of the raw material in particle board.

Similarly rice husk, sawdust has taken from sawmill waste, was *pinus merkusii,* in PT Nelly Jaya Pratama, Tana Toraja. According of survey which has done in the PT Nelly Jaya Pratama, veneer s has produced as 72 % from raw material and sawdust as 28 % from raw material. It was not utilized by that industry although people in the industrial area [7]. Therefore, this research utilized it as raw material also. Besides base material from soybean which environmental and used waste rice husk and sawdust as raw materials for increasing economic value it.

2. Materials and Methods

2.1 Materials

The materials used in this research include: soybeans, rice husk, *pinus merkusii* sawdust, aquadest, sodium hydorxide 40 %, glycerin, gelatine, SiO_2 solution, acetic acid 5 %, pine latex, tissue roll and detergent.

2.2 Apparatus

The apparatus used are, analytical balance model NO AP 110, pycnometer 25 mL, stopwatch, viscometer ostwald canon 400, bulb, pumpkin spray, hot plate, grinder, oven, pH universal, FT-IR model SHIMADZU 820 1PC, UTM (Universal Testing Machine), SEM-EDX Tescan Vega3SB and glass tools which are generally used in the laboratory.

2.3 Procedures

2.3.1 Preparation of Adhesive Based Boiled Soybean Water

Soybean weighed as 40 g, added aquadest to 300 mL, and then boiled for 30 minutes and filtered. Filtrate that soybean taken as 60 mL and boiled. And then, gelatine was added as 20 g and homogenized to soluble. Glycerin was added as 25 mL while homogenized to coagulate. Acetic acid 5 % was added as 12 mL. Latex of *Pinus merkusii* was added as 33 g while homogenized. Sodium hydroxide 40 % was added dropwise as 8.5 mL while homogenized to coagulate. SiO_2 solution was added as 4.5 mL to homogen. Each addition of reagent analyzed interact happen using FT-IR model SHIMADZU 820 1PC.

2.3.2 Test of Adhesive Based Boiled Soybean Water

The test of quality the adhesive based Boiled Soybean Water according of SNI 06-4567-1998. Parameter of testing are appearing colour, pH, solid contain, gelatination time, density and viscosity.

2.3.3 Preparation of Particle board Raw Materials

Rice husk immersed for 24 hours and then boiled for 2 hours. Rice husk dried at room temperature and grinded using grinder. Moisture contain determined for working-out raw material that used in manufacture of particle board.

Sawdust dried at room temperature. Moisture contain determined for working-out raw material that used in manufacture of particle board.

2.3.4 Determined Ratio of Rice Husk and Sawdust as Particle board materials

Respectively rice husk and sawdust weighed with ratio 0:1; 1:4; 1:2; 1:1; 2:1; 4:1 and 1:0 with total both of them as 491.07 g based on oven dry weight. Respectively of ratio was homogenized, and then mixed with adhesive has been made and homogenized. It put into block sized 25 x 25 x 1 cm. And then, it made the process of pressing, conditioning and cutting the sample test.

2.3.5 Test of Physical and Mechanical Properties

Test of quality the particle board according of SNI 03-2105-1996. Parameter of testing are density, moisture contain, thickness, immersion, Modulus of Rapture (MOR) and Modulus of Elasticity (MOE).

2.3.6 Data Analysis

Data was obtained, be treated using descriptive method, qualitative and quantitative. Manufacture of adhesive and particle board were analyzed interact that happen using FT-IR model SHIMADZU 820 1PC. Morphology of the best ratio particle board was analyzed using SEM-EDX Tescan Vega3SB.

2.4 Scheme of Particle Board preparations

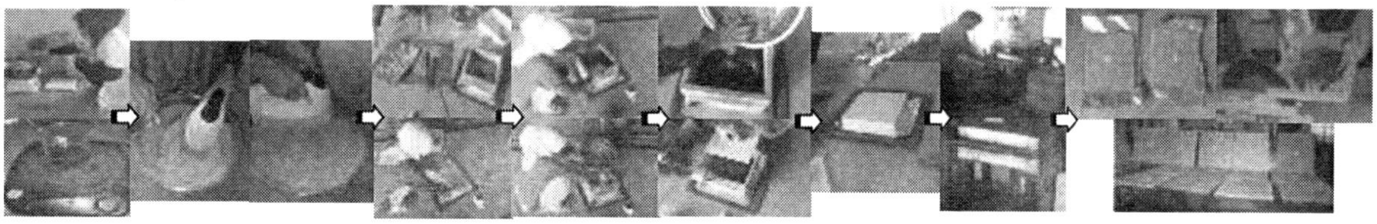

Figure 1. Scheme of Particle Board Processing.

3. Results and Discussions

3.1 Test of Adhesive Based Boiled Soybean Water

Test of adhesive based Boiled Soybean Water has been done, tested the quality according of the adhesive Urea-Formaldehid Adhesive SNI 06-4567-1998. The result were appearing colour is yellowish, pH of 10, solid contain of 44.70 %, gelatination time of 56 minutes 29.71 seconds, density of 1.1656 g/cm^3 and viscosity of 182.4387 cP.

That result has statisfied SNI 06-4567-1998 about adhesive Urea-Formaldehid (UF) standard. So that can be said that adhesion of adhesive based Boiled Soybean Water same with adhesion of Urea-Formaldehid adhesive.

3.2 Test of Physical and Mechanical Properties

Test of physical and mechanical properties with various ratio between ricehusk and sawdust are 0:1; 1:2; 1:4; 1:1; 2:1; 4:1 dan 1:0 based on SNI 03-2105-1996.

Histogram on figure 2 was shown that density of particle board from all ratio have statisfied SNI 03-2105-1996. The highest density value seen on 1:2 ratio of 0.7804 g/cm^3. Whereas the lowest density value seen on 1:0 ratio of 0.6943 g/cm^3.

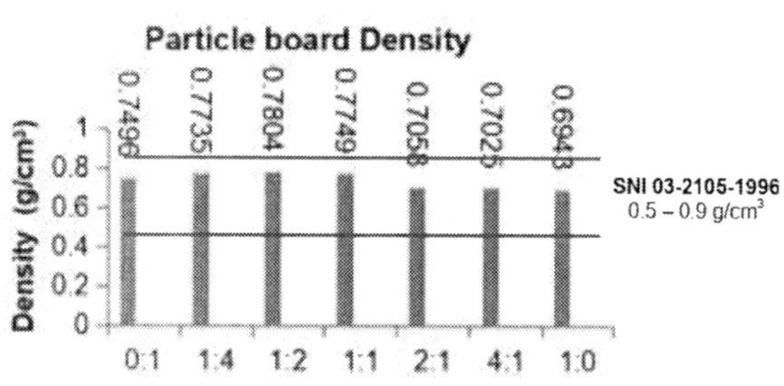

Figure 2. Density of Particle board.

Histogram was shown on figure 3 that particle board with 0:1 ratio has the lowest moisture contain of 5.49 %. Whereas the particle board with 4:1 has the highest moisture contain of 7.45 %. Although from all the ratio have moisture contain statisfied SNI 03-2105-1996. Based on figure 2 has shown that moisture contain of particle board affected by density from particle board, eventhough that effet has not significant.

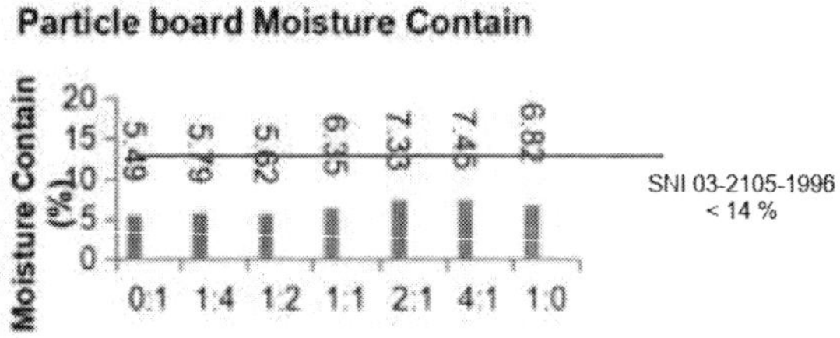

Figure 3. Particle board Moisture Contain.

The water immersed value it is shown on figure 4, it was founded that the 1:0 particle board ratios has the highest immersion, is 47.12 % for 2 hours immersed, and 83.00 % for 24 hours immersed. Whereas the particle board with 0:1 ratio has the lowest immersion of 26.52 % for 2 hours and 38.64 % for 24 hours immersed, it was caused by density of particle board which high fairly for 0:1 ratios. The higher density of the volume between particle became getting narrow, so that immersion of water more lower than the particle board that lower density [8]. Eventhough the 1:2 ratio has the highest density but effected there was rice husk then the immersion became higher. The rice husk has given a pore on particle board that resulting. So that, it affected on seeped of water into the particle board. The pore on particle board affected by particle size of variety raw materials.

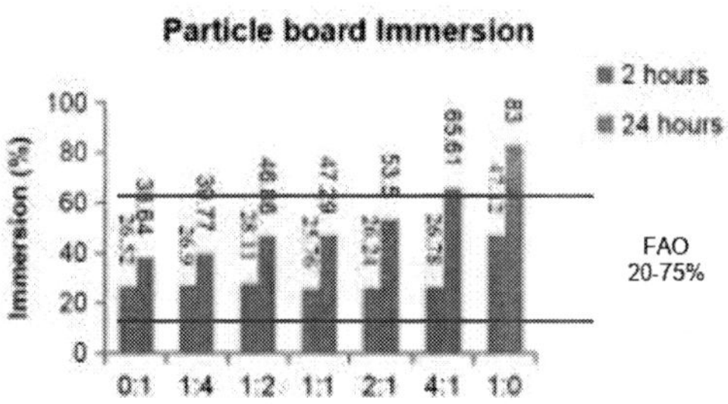

Figure 4. The Immersion of Particle board.

Histogram was shown on figure 5 that value of thickness was varying. Particle board thickness statisfied SNI 03-2105-1996 were particle board with 0:1, 1:4, 1:2, 1:1, and 2:1 (2 hours) ratio. If that ratio was observed, then ricehusk increasing, thickness will be increasing. Thickness and immersion of particle board were directly proportional. Therefore, the same factor affected.

Histogram was shown on figure 6 that sawdust increasing in specific ratio can be increased modulus of rapture (MOR). MOR was affected by many factors are contain and kind of adhesive that used, power tie adhesion, amount of adhesive was added, and particle size [8].

Histogram on figure 7 was shown that the highest of particle board MOE 1:4 ratio with 18248,3063 kg/cm^2. Whereas, the lowest of particle board MOE 4:1 ratio with 5315,6268 kg/cm^2. Ratio of rice husk and sawdust gave effect to the MOE. The 1:4 ratio was the highest of particle board MOE. It was caused rice husk can be increasing the elasticity of particle board. The rice husk with adhesive act as the bridge. It resulting in the highest of MOE.

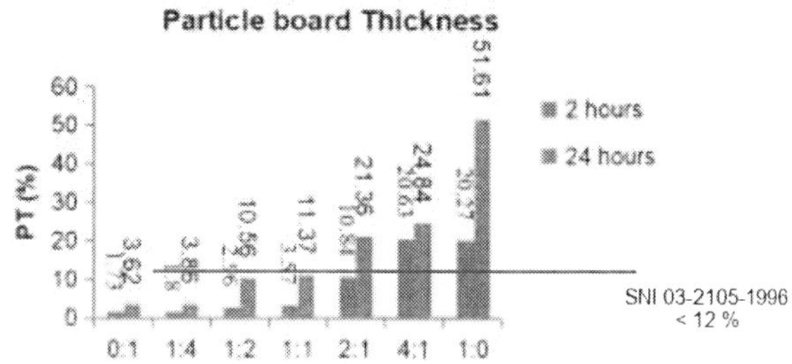

Figure 5. Thickness of Particle board.

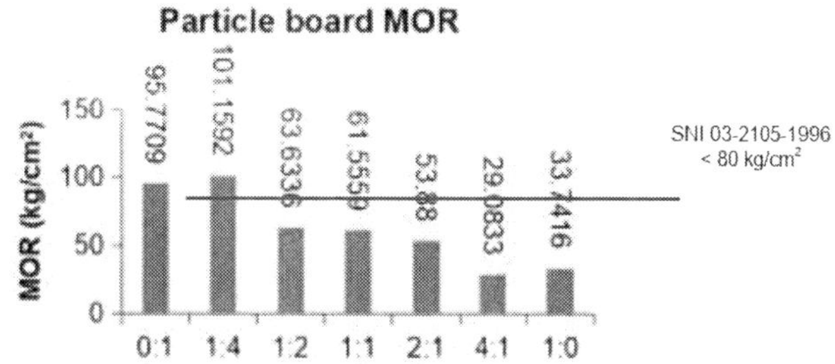

Figure 6. MOR of Particle board.

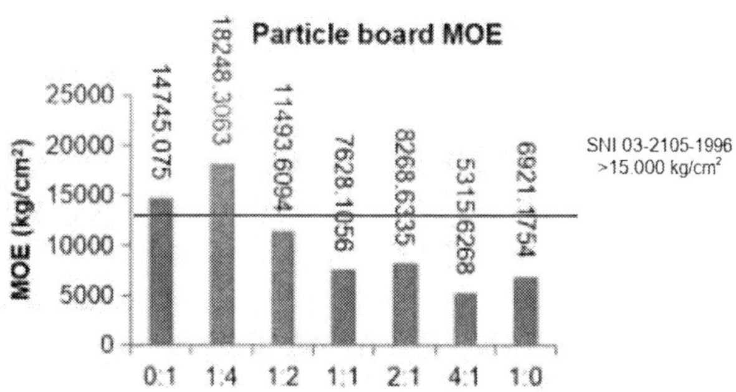

Figure 7. MOE of Particle board.

From the various parameters were tested, the particle board ratio that has the best quality was 1:4 ratio. Therefore, can be known with mixed sawdust and ricehusk of specific ratio resulting particle board quality was the best.

3.3 Analys of Adhesive and Particle boards Interactions by Using Infra-Red Spectroscopy

3.3.1 Interact Adhesive Analysis

- *Boiled Soybean Water (BSW)*

The Infra Red spectrum showed on figure 8 indicated that broadband on 3383.14 cm^{-1} area is O-H group. The area of 3732.6 cm^{-1} with weak intensity indicated as N-H group, where supported by double band in 2356.94 cm^{-1} and 2326.15 cm^{-1} shows as $-NH_3^+$. The area of 2929.87 cm^{-1} with medium intensity is C-H aliphatic. The area of 1597.06 cm^{-1} with strong intensity is C=O from carboxylic functional groups. The area of 1055.06 cm^{-1} is C-O that was supported by strong intensity band in 923.90 cm^{-1} as C-O-H deformation.

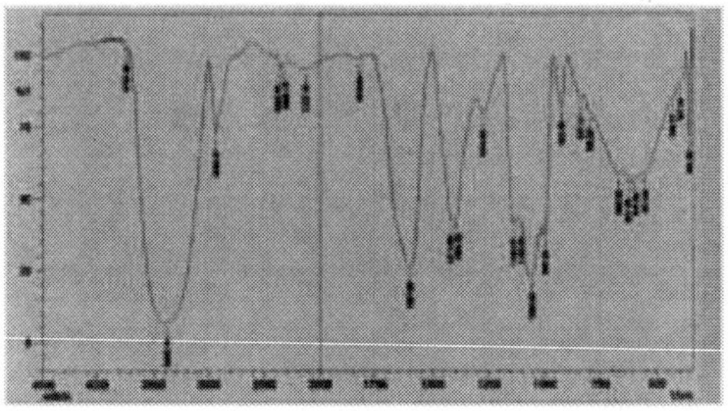

Figure 8. Boiled Soybean Water (BSW) FT-IR spectrum.

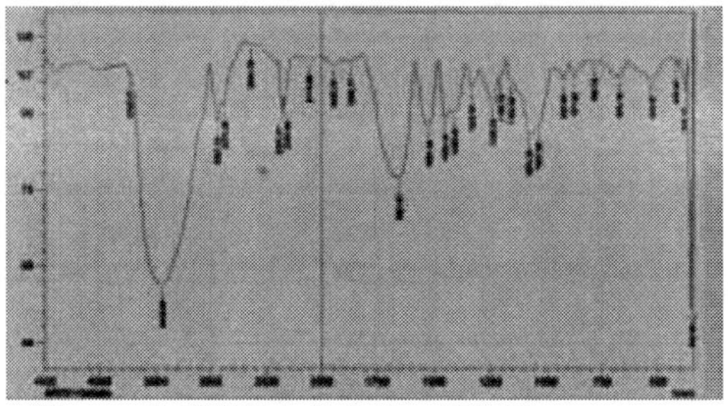

Figure 9. BSW + Gelatine FT-IR spectrum.

The Infra Red spectrum on figure 9 indicated that broadband on 3442.92 cm^{-1} is O-H group area. While the area of 2927.94 cm^{-1} is C-H aliphatic. There was N-H group with weak intensity on 3726.47 cm^{-1} area. Whereas the area of 1651.07 cm^{-1} is overlap between C=O and C=O amide band I. The area of 1074.35 cm^{-1} is C-O absorbs that supported by sharp band intensity on 920.05 cm^{-1} area as C-OH deformation.

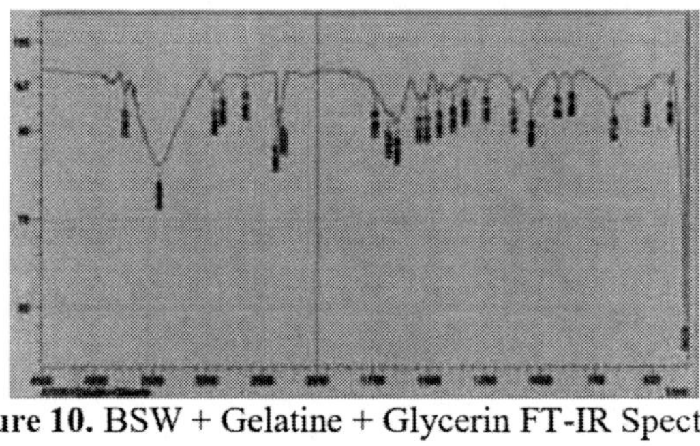

Figure 10. BSW + Gelatine + Glycerin FT-IR Spectrum.

IR spectrum on figure 10 indicated that decrease of intensity from band absorbs after glycerin was added. The area of 3442.94 cm^{-1} with broadband is O-H. The area of 2926.01 cm^{-1} is C-H aliphatic. The area of 3739.97 cm^{-1} is N-H with weak intensity that was supported by double band in 2380.87 cm^{-1} and 2333.87 cm^{-1} area is $-NH_3^+$. The area of 1645.26 cm^{-1} is stretch overlap between C=O and C=O amide band I. The area of 1678,07 cm^{-1} is C=O amide band II. The area of 1039,63 cm^{-1} is C-O absorbs that was supported by 921.97 cm^{-1} area is C-OH deformation.

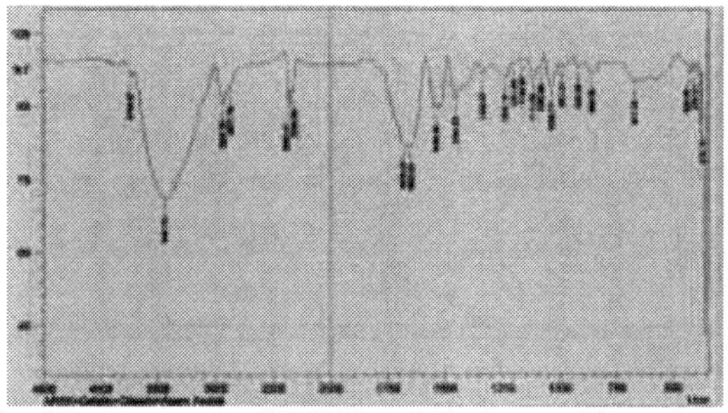

Figure 11. The IR spectrum of BSW + Gelatine + Glycerin + Acetic Acid.

The FT-IR spectrum showed on figure 11 indicated that there is -OH group on 3423.65 cm^{-1} area. There is N-H with weak intensity on 3739.97 cm^{-1} area that was supported by double band on 2360.67 cm^{-1} and 2335.60 cm^{-1} is $-NH_3^+$. The area of 2927.94 cm^{-1} is C-H aliphatic. The area of 1653.00 cm^{-1} is stretch overlap between C=O and C=O amide band I. The area of 1678.07 cm^{-1} is CThe =O amide band II. The area of 1035.77 cm^{-1} is C-O absorbs that was supported by 920.05 cm^{-1} area is C-O H.

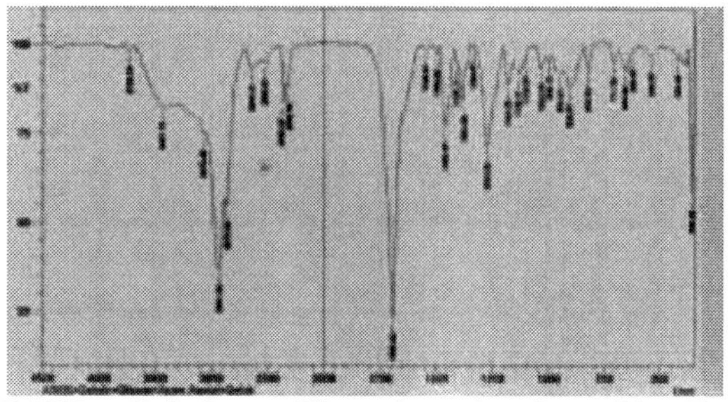

Figure 12. The FT-IR of BSW + Gelatine + Glycerin + Acetic Acid + *Pinus merkusii* Latex mixed.

Spectrum on figure 12 indicated there was spectrum change clearly after it was added *Pinus merkusii* latex. The spectrum showed intensity decrease on 3730.33 cm^{-1} area is N-H. It was supported by double band with weak intensity is $-NH_3^+$ on 2360.67 cm^{-1} and 2335.60 cm^{-1} area. It happened on O-H that intensity decrease on 3437.15 cm^{-1} area and weak intensity on 950.91 cm^{-1} area is C-OH deformation. Whereas strong band on 2931,60 cm^{-1} area that is C-H aliphatic. It is supported by CH_3 deformation on 1384.89 cm^{-1} area. The area of 1695.43 cm^{-1} is strong intensity C=O but it isn't overlap with C=O amide band I. Whereas the area of 1463.97 cm^{-1} with strong intensity is C=C.

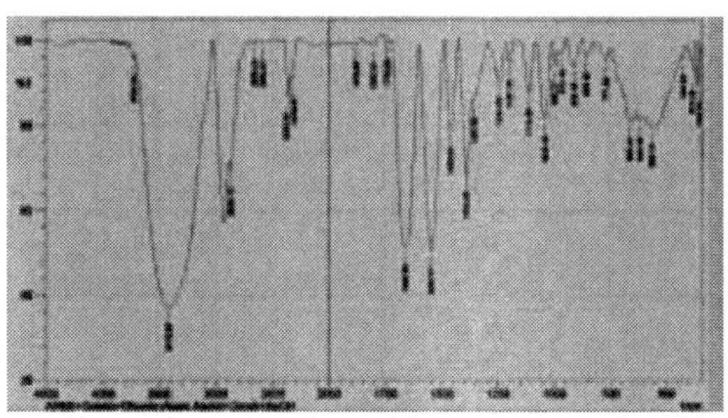

Figure 13. BSW + Gelatine + Glycerin + Acetic Acid + *Pinus merkusii* Latex + NaOH FT-IR Result.

The Spectrum on figure 13 indicated that N-H on 3724.54 cm^{-1} area was supported by double band on 2358.94 cm^{-1} and 2333.87 cm^{-1} areas is $-NH_3^+$. The area of 3415.93 cm^{-1} with broad band is O-H. The area of 2927.94 cm^{-1} is C-H aliphatic. The area of 1662.64 cm^{-1} is C=O. The area of 1546.91 cm^{-1} is C=O amide band I. The area of 1043.49 cm^{-1} is C-O absorbs that was supported by C-OH on 665,44 cm^{-1} area.

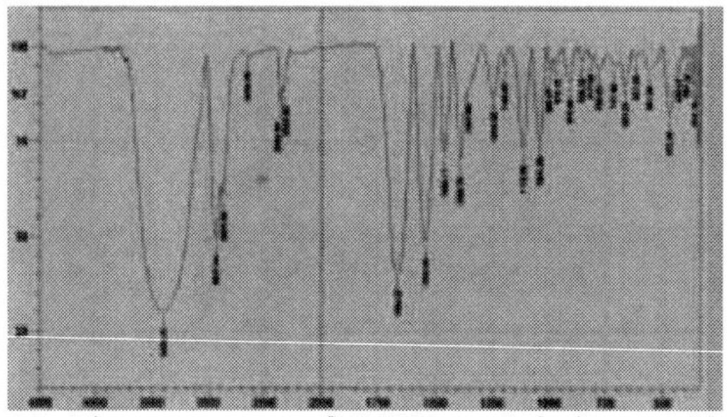

Figure 14. The spectrum of BSW + Gelatine + Glycerin + Acetic Acid + *Pinus merkusii* Latex + NaOH + SiO_2 mixed.

The Spectrum on figure 14 indicated that O-H and Si-OH on 3402.43 cm^{-1} area. There was double band on 2362.60 cm^{-1} and 2335.80 cm^{-1} areas is $-NH_3^+$. The area of 2927.94 cm^{-1} is C-H aliphatic. The area of 1664.57 cm^{-1} is C=O. The area of 1543.05 cm^{-1} is C=O amide band I. The area of 1041.56 cm^{-1} is Si-O.

3.3.2 Particle Board Interact Analysis

Figure 15 indicated that interact between raw material was added adhesive and became particle board with hot-pressing. Based on spectrum on figure 15 can be seen changing that happened before and after mixed. It explained that there was interact of adhesive and raw material. Figure 15a was C-O absorbs, whereas figure 15b was Si-O. The increase of intensity of Si-O shown when the raw material and adhesive was mixed obviously (figure 15c). It was indicated that Si on adhesive reacted with raw material of particle boards.

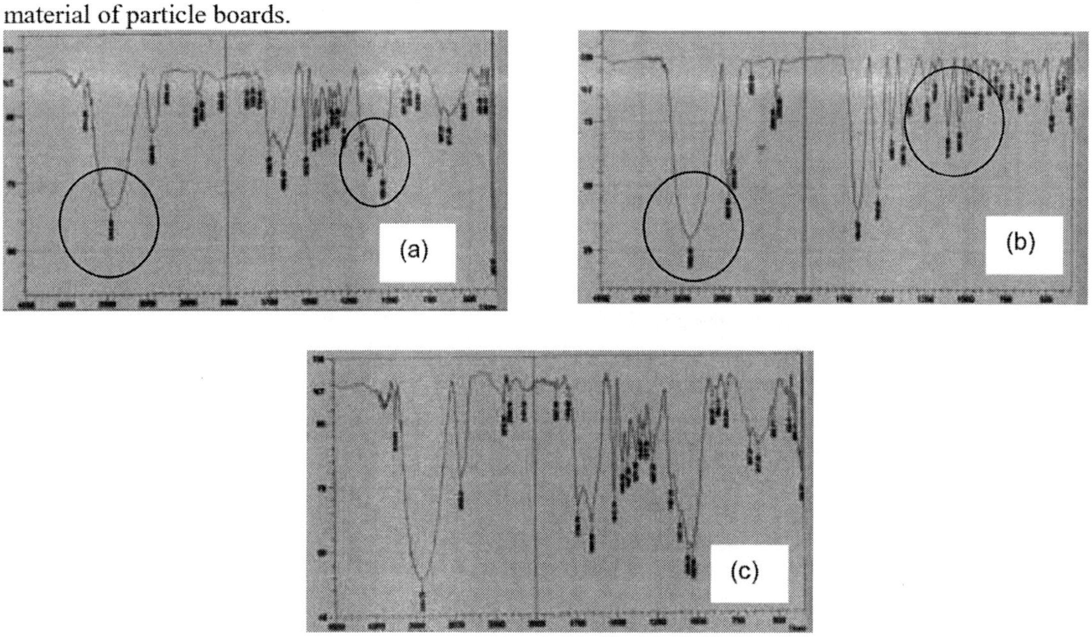

Figure 15. (a) The 1:4 ratio of Rice husk:Sawdust FT-IR Results (b) BSW Adhesives and (c) Particle board FT-IR spectrum results.

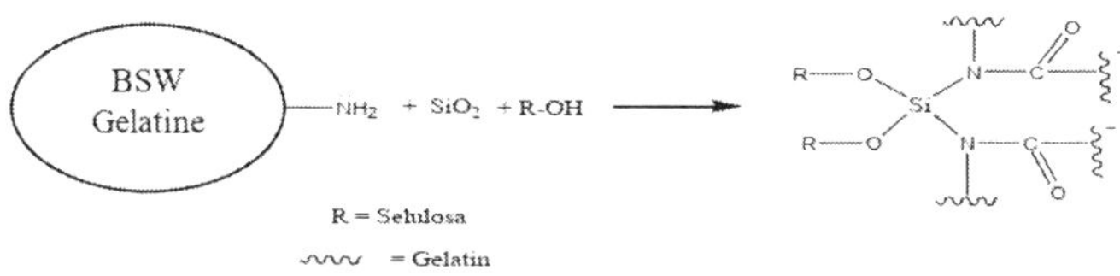

Figure 16. Reaction Hypotheses.

Based on FT-IR spectrums, suggest that the reaction hypotheses between adhesive and raw material as figure 16, while $-NH_2$ and SiO_2 in adhesive present could interact with -OH group from raw material (selulose), where the presence of SiO_2 as bridge between protein and selulose form particle board was stronger (figure 16).

3.4 Particle board Morphology Characterization Using SEM-EDX
SEM using on sample to analysis the sample surface. Emmision beam to sample will be reflected and diffracted. The presence of electron that was diffracted can be observed that diffraction pattern. SEM that connected by EDX (Energy Dispersive X-Ray). EDX has resulted from X-Ray that shooted on position which we will known the contain.

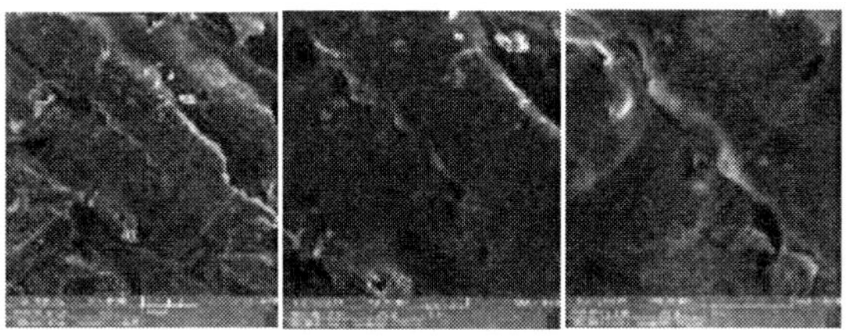

(a) (b) (c)

Figure 17. (a) Magnification 500x SEM Result (b) Magnification 1000x SEM Result (c) Magnification 2000x SEM Result.

The result of particle board SEM on figure 17 shown that still has pore that affected the immersion and thickness. It has explained about the effect of pore on particle board. But, it was not too big, so that the immersion and thickness were low.

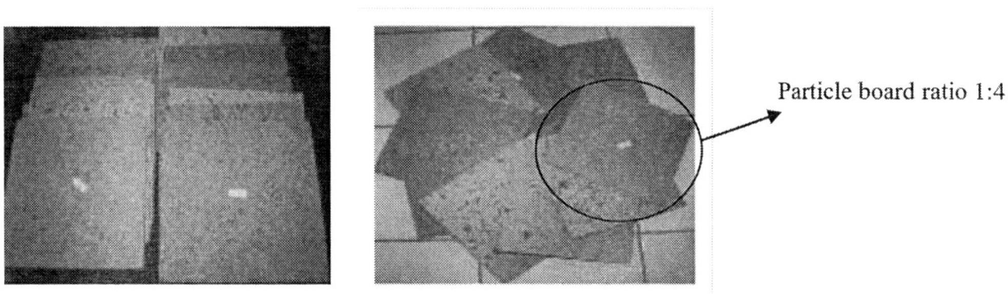

Figure 18. Particle Board Prototipe.

4 Conclussion

The best composition of adhesive was occupy based on SNI 06-4567-1998 and particle board from ricehusk and sawdust composite occupy based on SNI 03-2105-1996. The adhesive composition were BSW 60 mL, gelatine 20 g, glycerin 25 mL, acetic acid 12 mL, latex pine 33 g, NaOH 8.5 mL, and

SiO_2 4.5 mL. Quality of adhesive according of SNI 06-4567-1998 were appearing yellowish colour, pH of 10, solid contain of 44.70%, gelatination time of 56 minutes 29.71 seconds, density of 1.1656 g/cm^3, and viscosity of 182.4387 cP.

The best comparison of ricehusk and sawdust composite according of SNI 03-2105-1996 was 1:4. The result of particle board testing were density of 0.7735 g/cm^3, moisture contain of 5.79%, immersion for 2 hours of 26.90% and for 24 hours of 39.77%, thickness for 2 hours of 1.80% and for 24 hours of 3.86%, MOR of 101.1592 kg/cm^2 and MOE of 18,248.3063 kg/cm^2.

Conflict of interest

The author(s) declare(s) that there is no conflict of interests regarding the publication of this paper.

Acknowledgements

Author thanks to the R and D of South Sulawesi, Indonesian Government has fund this rsresearch. Haslindah, analyst of Inorganic Laboratory and Kartini, analyst of Integrated Chemistry Laboratory. And all of the participants who helped in this research.lou

References

[1] Tambunan T 2006 *Development and Export Competitiveness Indonesian Wood Furniture* p 5 (Jakarta: Industry Study Center dan UKM Economic Faculty Trisakti University, Jakarta)

[2] Indonesia Forestry Statistical 2011 Ministry of Forestry, Jakarta

[3] Li K 2007 Commercial Application of New Formaldehyde-free Wood Adhesive from Renewable Materials *U.S.Patent* 7 252 735 (Oregon: Department of Wood Science and Engineering Oregon State University)

[4] Liu Y 2006 Formaldehyde-Free Wood Adhesives from Soybean Protein and Lignin: Development and Characterization *Unpublished Dissertation* (Oregon: Department of Wood Science and Engineering, Oregon State University)

[5] Said I N and Wahjono D H 1999 *Wastewater Treatment Technology Tahu-Tempe by Anaerobic and Aerobic Biofilter Process,* Paper presented at a seminar of Water Management Technology Group and Liquid Waste in Jakarta, Directorate of Environmental Technology, Agency for the Assessment and Application of Technology, Jakarta

[6] Wahyuningsih S N 2011 Effect of Soaking and Geometry Particles on the Quality of Rice Husk Particle Board *Unpublished Essay* (Bogor: Department of Forest Products, Faculty of Forestry, Bogor Agricultural University)

[7] Parura P D 2012 Modified of Particle Board from Sawdust Pine (Pinus merkusii) Using Adhesive Soybean Meal-Based and Pine Latex with $CaCO_3$ as Hardener *Unpublished Essay* (Makassar: Chemistry Department Science Faculty Hasanuddin University)

[8] Setiawan B 2008 Rice Husk Particle Board Quality *Unpublished essay* (Bogor: Department of Forest Products, Faculty of Forestry Bogor Agricultural University)

Adsorption of Pb^{2+} on Thiol-functionalized Mesoporous Silica, SH-MCM-48

P Taba[1,3], R D P Mustafa[2], L M Ramang[1], A H Kasim[1]

[1] Department of Chemistry, Hasanuddin University, Jl. Perintis Kemerdekaan, Km 10, Tamalanrea, Makassar 90245, Indonesia
[2] Undergraduate student in Department of Chemistry Hasanuddin University, Jl. Perintis Kemerdekaan, Km 10, Tamalanrea, Makassar 90245, Indonesia

E-mail: paulinataba@unhas.ac.id

Abstract. Modification of mesoporous silica, MCM-48, by using 3-mercaptopropyltrimethoxysilane has been successfully conducted. MCM-48 and SH-MCM-48 were characterized using XRD and FTIR. SH-MCM-48 was used as an adsorbent of Pb^{2+} ions from solution. A number of Pb^{2+} ions adsorbed were studied as the function of time, pH, and concentration. The concentration of the ions after adsorption was determined by an Atomic Absorption Spectrophotometer. The removal of the adsorbed ions from the SH-MCM-48 was also studied using several desorbing agents. The result showed that the optimum time was 20 minutes and optimum pH was 4. The adsorption of Pb(II) ion followed the pseudo-second-order with the rate constant of 0,2632 $g{\cdot}mg^{-1}{\cdot}min^{-1}$. Adsorption of Pb(II) ion fitted the Langmuir isotherm with the adsorption capacity of 0,1088 mmol/g. The best desorbing agent to remove the adsorbed ion from SH-MCM-48 was 0.3 M HCl solution with the desorption percentage of 58.6%.

1. Introduction

The fast growth of industry can result in the positive impact into the community. However, it can also create negative impact to the environment due to its waste. One industry that can cause a problem to the water environment is electroplating where its waste contains heavy metals. Heavy metals are toxic and carcinogenic to a human being. Lead is an example of the metals that are *non-biodegradable* in the environment [1]. The metal can cause the toxic effect to people, especially at the concentration is above the threshold value. Therefore, their existence in the waters can be dangerous. Handling of heavy metal waste has been done to overcome pollution and the risk of poisoning for living things [2]. One method that widely used is the adsorption method [3, 4] which is one of the simplest and quite effective processes in removing heavy metals from liquid waste [5, 6, 7].

The adsorption method is generally based on the interaction of metal ions with functional groups present on the surface of the adsorbent through the formation of the complex and usually occurs on the surface of the solid containing the functional groups such as -OH, -NH, -SH and –OOH [8]. Several adsorbents of porous material have been developed to adsorb heavy metals. According to the classification of IUPAC, the porous material is divided into three kinds based on the diameter of the pore, i.e. micropore, material having pore diameter of <2 nm, mesopore with the pore diameter of between 2-50 nm and macropore with the pore diameter of > 50 nm [9]. The development of mesoporous material is now the concern of many researchers, especially since the discovery of M41S family, materials from silicates or aluminosilicates, by researchers from Mobil Corporation United States in 1992. [10]. This mesoporous silica consists of MCM-41 which have hexagonal structures, MCM-48 which has a cubic structure, and MCM-50 with a lamellar (flat) and unstable structure [11, 12]. MCM-48 have been used as an adsorbent for organic compound because MCM-48 has large surface area and regular pores [13]. The material has an advantage compared to MCM-41 and MCM-50. It consists of two mesoporous channel networks that form a three-dimensional channel system that allows the particles to enter easily into the material because the pores are not limited in one direction.

The MCM-48 surface has silanol and siloxane groups that have little role in the metal ion adsorption process so that modification by adding an active functional group will increase its effectiveness and selectivity [14, 15]. Several modifications have been made to increase the adsorption capacity of MCM-48. Wang and Li [16] have modified MCM-48 by adding a thiol group (-SH) on its surface to adsorb Hg^{2+} ions. Modified MCM-48 with thiol (-SH) groups to adsorb Ag^+ and Cd^{2+} ions [14]. The adsorption capacity for Ag^+ and Cd^{2+} ions was 2.73 mmol g^{-1} and 0.14 mmol g^{-1}, respectively.

Based on the principle of Hard Soft Acid Base (HSAB), modified MCM-48 with thiol (-SH) groups of the soft base can bind hard acidic of metal ions [17]. But it is possible that the material with the thiol group (-SH) can interact with borderline acidic metal ions such as Pb (II) ions due to other factors, such as surface area large and regular pores that can make it be able to interact with the borderline acid [13].

The adsorption process is usually followed by a desorption process to determine whether adsorbate can be recovered. Desorption can be carried out by contacting the desorbing agent with adsorbents containing the metal ion. The desorbing agent may be acidic, alkaline, or neutral [15].

Based on the above description, the study modified MCM-48 with thiol groups (-SH) of the 3-MPTMS compound. MCM-48-SH is then used to adsorb Pb^{2+} ions. Adsorption of metal ions is strongly influenced by pH, contact time, and solution concentration [14]. Therefore, the effect of these parameters was studied on the adsorption of Pb^{2+} ions to obtain the optimum conditions. To determine the possibility of reuse of the adsorbent, a desorption study of Pb^{2+} ions from MCM-48-SH was also performed.

2. Experimental

2.1. Materials

The materials used in this study were cetyl trimethyl ammonium bromide (CTAB) obtained from Sigma Aldrich, Ludox HS40 consisting of SiO_2 (35% w/w), Na_2O (0.4% w/w) and H_2O (60.1% w/w) obtained from Du Pont, Triton X-100 purchased from Sigma Aldrich, 3-mercaptopropyltrimethoxysilane, 3-MPTMS, (Sigma Aldrich), $Pb(NO_3)_2$, HCl, HNO_3, Na_2-EDTA, NaOH, CH_3COOH, ethanol, toluene (all chemicals were in pro analysis grades), double distilled water, Whatman 42 filter paper, and universal pH papers.

2.2. Equipment

Items of equipment used in this research were oven (SPNI 505FD), Magruch x-stirrer 15 (velp Scientifica), magnetic stirrer (cole-palmer), hotplate stirrer ceramag midi (Ika - Basic 1), NO_2 AP210 analytical balance, vacuum pumps (ABM 4EK F6 3CX-4), pH meters, Atomic Absorption Spectrophotometer (AAS) 205 VGP, Fourier Transform Infra Red (FTIR) (Shimadzu 820 IPC), and X-Ray Diffraction (Paralytical X'-Pert3).

2.3. *Procedure*

2.3.1. Preparation of Mesoporous Silica (MCM-48). Mesoporous silica, MCM-48, was prepared using previous procedure [11, 18]. The surfactant was removed from the white product using the HCl-ethanol mixture [18] without the addition of sodium chloride and calcination at high temperatures. The washed procedure was repeated two more times and the precipitate was filtered, washed with doubly distilled water and dried at 110 °C. The material was characterized using XRD, and FTIR instruments to ensure that the as-synthesized material has been MCM-48 and the template has been removed, respectively

2.3.2. Modifying MCM-48 with 3-MPTMS. MCM-48 was modified with a thiol group followed the procedures of Wang and Li [16]. The characterization of mesoporous silica materials produced was conducted by XRD and FTIR.

2.3.3. Influence of the contact time on adsorption of Pb^{2+} ion. The modified material, MCM-48-SH, of 0.1 gram was put into an Erlenmeyer containing 50 mL of 10 ppm Pb^{2+} ion solution. The mixture was stirred with a magnetic stirrer for 5 min and then filtered. The filtrate absorbance was measured using AAS. The treatments were repeated with varying stirring times of 10, 20, 40, 60 and 90 minutes respectively. The kinetic of adsorption was studied by a pseudo-first-order equation of Lagergren and a pseudo-order two of Hossain. The equations are given in equations (1) for the former and equation (2) for the later.

$$\log(q_e - q_t) = \log q_e - \left(\frac{k_1}{2{,}303}\right) \tag{1}$$

$$\frac{1}{q_t} = \frac{1}{k_2 q_e^2} + \frac{1}{q_e} t \tag{2}$$

where q_e and q_t are the amounts of ion adsorbed at equilibrium and at t, respectively, k_1 is the pseudo-first order rate constant and k_2 is the pseudo-second-order rate constant [19,20].

2.3.4. Influence of pH on adsorption of Pb^{2+} ion. The modified material, MCM-48-SH, of 0.1 g was inserted into an Erlenmeyer glass containing 50 mL of 10 ppm Pb^{2+} ion solution at various pH (2, 3, 4, and 5) the mixture was then stirred during the optimum time. The mixture was filtered and the filtrate absorbance was measured using AAS. Blank experiments (without adsorbent) were also conducted.

2.3.5. Influence of initial concentration of Pb^{2+} on adsorption. The modified material, MCM-48-SH, of 0.1 grams were put into 5 different Erlenmeyer containing 50 mL of Pb^{2+} ion solution with concentrations of 30, 60, 90, 180 and 270 ppm, separately. The solution was then stirred at the optimum pH during the optimum time. The mixture was filtered and the filtrate absorbance was measured using AAS.

2.3.6. Desorption studies. The modified material, MCM-48-SH, that has been in contact with 200 mL of 500 ppm Pb^{2+} ion solution was washed with double distilled water to take out the unbound ions on the surface. After that, 0.1 g of the dried material was put into 4 different Erlenmeyer containing 50 mL of doubly distilled water, 0.3 M HCl, 0.3 M HNO_3, and 0.3 M Na_2-EDTA. The solution was then stirred during the optimum time. The mixture was filtered and the filtrate absorbance was measured using AAS.

3. Results and Discussion

3.1. Characterization of MCM-48 Materials

The X-ray difragtogram of MCM-48 before and after washing as well after modification with 3-MPTMS is given in figure 1. The as-synthesized MCM-48 material has 8 peaks (figure 1a) which are characteristic of well-ordered material known as MCM-48 as has been previously reported [11,21]. After washing with HCl-ethanol, the intensities of the peaks increase because of the removal of

IOP Conf. Series: Journal of Physics: Conf. Series **979** (2018) 012058 doi:10.1088/1742-6596/979/1/012058

surfactant so that there is no scattering. The peaks shift to the lower 2Θ. The same result was reported by the mobile corporation caused by the condensation of silanol groups and constriction of pores [21]. After modification with 3-MPTMS, the diffraction pattern of the modified MCM-48 is similar to that of the unmodified one (figure 1c). This indicates that the structural order of the material remain maintained after the modification.

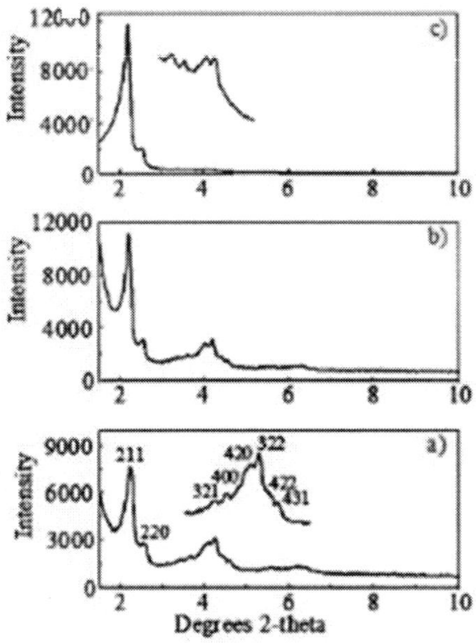

Figure 1. X-ray diffractograms of a) as-synthesized MCM-48, b) washed MCM-48, and c) MCM-48-SH.

FTIR spectra of the used materials are given in figure 2. It is clear that the as-synthesized MCM-48 (figure 1a) contains –OH groups indicated by the presence of a peak at 3442 cm^{-1} for O-H stretching vibration supported by a peak at 1652 cm^{-1} for –O-H bending vibration. Bands at 2852 and 2924 cm^{-1} are associated with C-H stretching vibrations of $-CH_2$ symmetric and $-CH_2$ antisymmetric and the one at 3012 cm^{-1} is for terminal $-CH_3$ asymmetric. The C-H bending vibration is shown at 1481 cm^{-1}. After washing with HCl-ethanol, the peaks of C-H stretching and bending vibrations have nearly been disappeared showing that the surfactants used as the structure-directing agents have almost been removed. The intense peaks at 1226 and 1084 cm^{-1} and the weak bands at 960 and 792 cm^{-1} associated with the vibrations of Si-O stretching in the lattice of silicate. The intense peaks shift to 1228 and 1085 cm^{-1} after washing with HCl-ethanol. This indicates that there is a constriction in the lattice during the removal of surfactants as has been reported before [15, 22, 23].

Figure 2c presents the FTIR spectrum of MCM-48 after modification with 3-MPTMS. It is clear that after modification the FTIR with 3-MPTMS, spectrum changes. Several new peaks appear. Peaks at 2852 and 2931 cm^{-1} are attributed to C-H stretching vibrations from the modifying agent, indicating that 3-MPTMS is grafted on the surface of MCM-48. The presence of methylene groups is supported by small peaks below 1500 cm^{-1}. The S-H vibration is shown at a wavenumber of 2575 cm^{-1}. The peaks at 800 – 900 cm^{-1} are due to the –Si-O (silanol) groups, demonstrating that not all silanol groups of the surface are used during the functionalization of MCM-48. The peak at 690 $cm-^{1}$ associates with the Si-C vibration. After adsorption of Pb^{2+} ion, the band at 2575 cm^{-1} moves to 2563 cm^{-1} with lower intensity. This indicates that there is an interaction between the -SH group with Pb^{2+} ion.

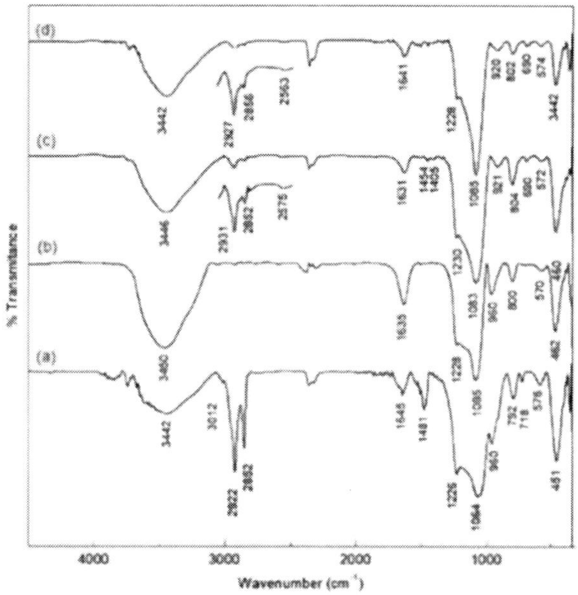

Figure 2. FTIR spectra of a) as-synthesized MCM-48 b) washed MCM-48 c) MCM-48-SH, and d) MCM-48-SH + Pb^{2+} ion.

3.2. *Effect of contact time on adsorption and kinetic study*

The influence of the contact time on the removal of Pb^{2+} by MCM-48-SH is given in figure 3. Adsorption of Pb^{2+} ion on MCM-48-SH enhances with increasing the contact time until achieving the optimum at 20 min. At that time, the equilibrium adsorption occurs, the active site of the adsorbent has been saturated to Pb^{2+} ion. The optimum time for the adsorption of the ion was lower than that of Ag^{+} and Cd^{2} using the same modified MCM-48 as has been reported previously [14].

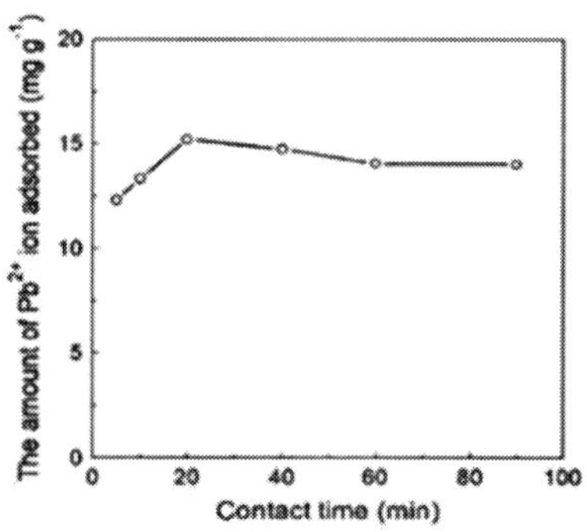

Figure 3. The effect of the contact time on adsorption of Pb^{2+} ion.

Data of the effect of the contact time on the amount of Pb^{2+} adsorbed were used to study the adsorption kinetics that can be seen in figure 4. Both graphs show that the adsorption follows the pseudo-second-order equation because the R^2 value obtained using the pseudo-first-order equation is 0.373, whereas that using the pseudo-second-order equation is 0.999. Calculation from the regression equation resulted in the pseudo-second-rate constant (k_2) of 0.26 g mg^{-1} min^{-1} and the amount of Pb^{2+} adsorbed is 14.06 mg g^{-1} which was closer to the experimental value (15.20 mg g^{-1}).

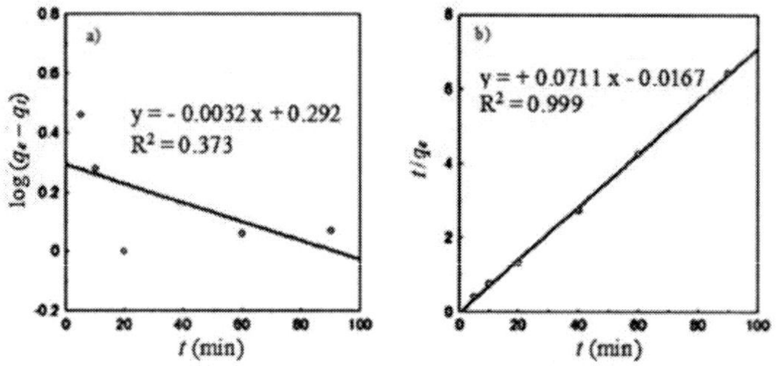

Figure 4. Kinetic study of Pb^{2+} adsorbed using a) pseudo-first-order and b) pseudo-second-order equations.

3.3. Effect of pH on the adsorption

The initial pH of solution affected the amount of ion adsorbed as can be seen in figure 5. The amount

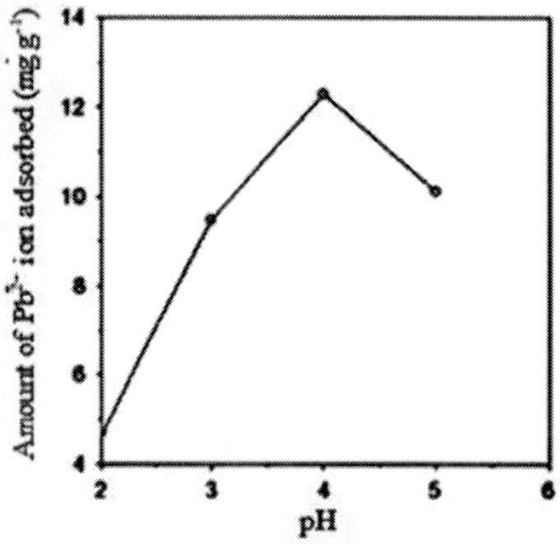

Figure 5. The effect of the initial pH of solution on the adsorption of Pb^{2+} ion.

of ion adsorbed at a pH of 2 was 4.69 mg g^{-1}. By increasing the solution pH, the amount of ion adsorbed increases until the solution pH of 4 (12.28 mg g^{-1}). The amount of ion adsorbed at a pH of 2 is low because –SH and –OH groups on the surface of adsorbent is protonated at low pH. Therefore, there is competition between H^+ and Pb^{2+} ions leading to the low ability of MCM-48-SH to interact with Pb^{2+} ions. At the pH of 4, the number of H^+ ions to be competitive with Pb^{2+} ions decrease. As a result, the ability of adsorbent to interact with Pb^{2+} ions increases. At higher pH, some of Pb^{2+} ions precipitate causing the concentration of Pb^{2+} ion in the solution decreases. Therefore, the amount of ion adsorbed decreases. Thus, the highest amount adsorbed is at the pH of 4 which is the optimum pH used for determining the adsorption capacity.

3.4. Effect of concentration of Pb^{2+} ion

The amount ion of Pb^{2+} adsorbed on MCM-48-SH as a function of the equilibrium concentration of the ion is presented in figure 6.

IOP Conf. Series: Journal of Physics: Conf. Series **979** (2018) 012058 doi:10.1088/1742-6596/979/1/012058

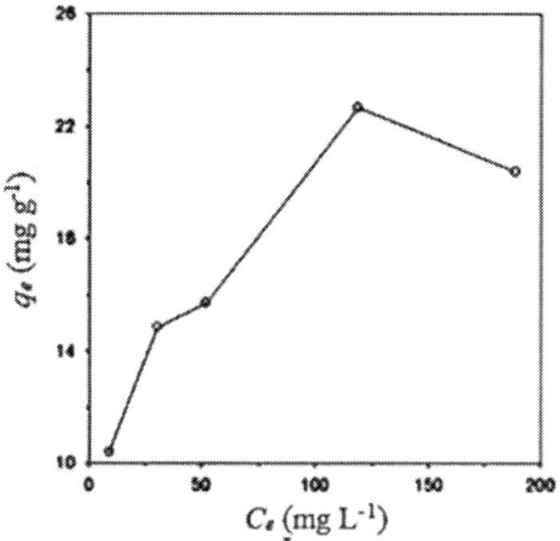

Figure 6. The influence of the equilibrium concentration on the removal of Pb^{2+} ion.

It is obvious that the higher the equilibrium concentration is, the higher the amount of the ion adsorbed is until achieving the maximum at the equilibrium concentration of 118.7 mg L^{-1} and the amount adsorbed of 22.67 mg g^{-1}. To obtain the adsorption capacity, the adsorption isotherms of Langmuir and Freundlich were used. The linear mathematic expression of Langmuir and Freundlich models are given in equations (3) and (4) [24, 25],

$$\frac{C_e}{q_e} = \frac{1}{Q_o b} + \frac{C_e}{Q_o} \quad (3)$$

$$\mathrm{l}q_e = log\,k + \frac{1}{n}\ log\ C_e \quad (4)$$

where C_e, q_e, Q_o, b, k, and n are the equilibrium concentration of the ion (mg L^{-1}), the amount of the ion adsorbed (mg g^{-1}), the adsorption capacity of Langmuir (mg g^{-1}), the Langmuir coefficient (L mg^{-1}), the Freundlich adsorption constant related to the adsorption capacity of adsorbent (mg g^{-1}), and the Freundlich adsorption intensity, respectively. The application of data as obtained to both adsorption isotherms can be seen in figure 7.

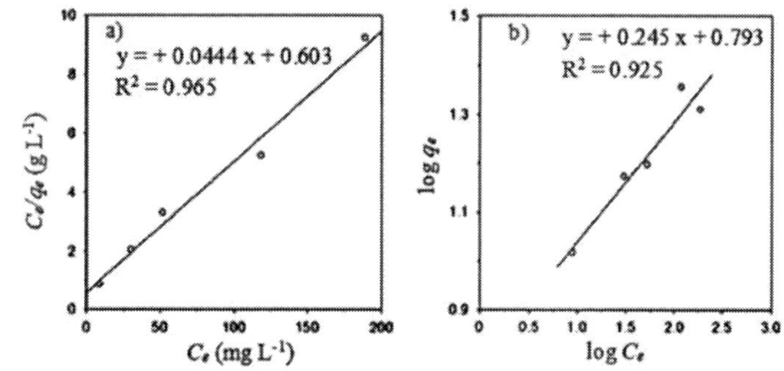

Figure 7. Adsorption isotherms of a) Langmuir and b) Freundlich.

The constants of Langmuir and Freundlich for the adsorption of Pb^{2+} ion on MCM-48-SH are given in table 1. The adsorption of Pb^{2+} ion better fits the model of Langmuir compared to the model of Freundlich shown by the higher of R^2 produced by the former model than that by the later one indicating the occurrence of monolayer adsorption. The same result has been performed when the same functionalized MCM-48 was used to adsorb Ag^+ ion. The adsorption of Pb^{2+} ion is less than that of Ag^+ ion because –SH is known as a soft base that can interact better with Ag^+ (a soft acid) than Pb^{2+} (a borderline one).

Table 1. Langmuir and Freundlich constant for the adsorption of Pb^{2+} ion on MCM-48-SH.

Ion	Langmuir model				Freundlich model			
	Q_o		b	R^2	k		n	R^2
	mg g^{-1}	mmol g^{-1}	(L mg^{-1})		mg g^{-1}	mmol g^{-1}	(g L^{-1})	
Pb^{2+}	25.52	0.11	0.06	0.98	6.21	0.03	4.08	0.92
Ag^{+}	294.12	2.73	9.06	1.00	49.91	0.46	2.35	0.93[a]

[a] Taken from Ref. [10]

3.5. Desorption of Pb^{2+} ion from MCM-48-SH

The process of desorption was studied by contacting the adsorbent with the desorbing agents used. The amount of Pb^{2+} ion desorbed from the used adsorbent is given in figure 8.

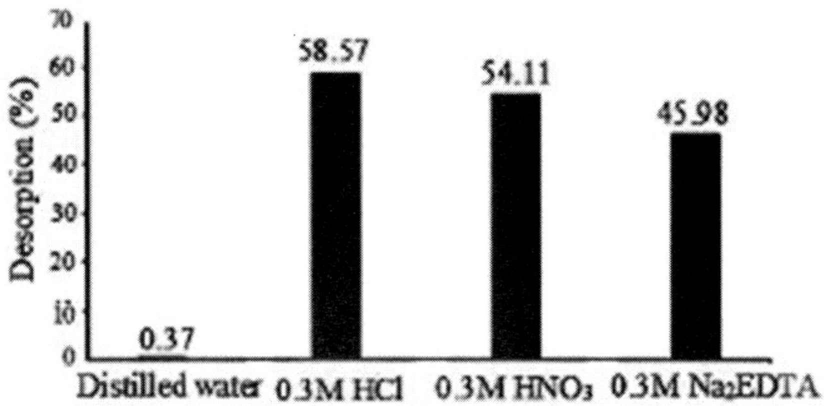

Figure 8. Desorption of Pb^{2+} ion from used MCM-48-SH.

It is clear that the amount of Pb^{2+} ion desorbed from the used MCM-48-SH using distilled water is very small. The amount of ion desorbed decreases using desorbing agents in the order of 0.3 M HCl > 0.3 M HNO_3 > 0.3M Na_2EDTA > distilled water. The fact indicated that the ion strongly interacted with the adsorbent meaning that the adsorption is mainly caused by chemical adsorption. This is supported by FTIR data as given in figure 2d which shows the interaction between the –SH group and the ion. The use of acid in removing the ion from the surface because they can protonate the surface. Therefore, the ion was removed from the surface of the adsorbent. The same trend was reported previously [15] in the removal of Cu^{2+} and Mn^{2+} from used MCM-48-NH_2.

4. Conclusion

Functionalized MCM-48 has been successfully conducted using 3-MPTMS. The material was able to be used as an adsorbent of Pb^{2+} ion with the adsorption capacity of 0.11 mmol g^{-1} at the optimum contact time of 20 min and the optimum pH of 4. The interaction between the adsorbent and the ion occurred between the –SH group of adsorbent and the Pb^{2+} ion.

Acknowledgments

Authors would like to acknowledge Prof. Russell F. Howe and Bayu Adi Samodro for the assistance in producing the XRD patterns of the materials as well as to colleagues and technical staff for their support.

References

[1] Sahu M K, Mandal S, Dash S S, Badhai P and Patel R K 2013 *J. Env. Chem. Eng.* **1** 1315-24
[2] Blais J F, Dufresne B and Mercier G 2000 *Rev. Sci. Eau.* **12** (4) 687-711
[3] Azmiyawati C, Nuryono, Narsito 2012 Adsorption of Mg(II) and Ca(II) on disulfonato-silica hybrid *Indo. J. Chem.* **12** (3) 223-228
[4] He R, Wang Z, Tan L, Zhong Y, Li W, Xing D, Wei C and Tang Y 2017 *Micropor. Mesopor. Mater.* doi: 10.1016/j.micromeso.2017.08.007
[5] Oliviera L C A, Petkowicz D I, Smaniotto A and Pergher S B C 2004 *Water Res.* **38** 3699-704
[6] Jamil T S, Ibrahim H S, El-Maksoud A I H and El-Wakeel S T 2010 *Desalination* **258** 34-40
[7] Motsi T, Rowson N A and Simmons M J H 2011 *Int. J. Miner. Process.* **101** 42-9
[8] Stum W and Morgan J J 1996 *Aquatic Chemistry* (New York: John Wiley and Sons)
[9] Alothman A Z 2012 *Material* **5** 2874-2902
[10] Kresge C T, Leonowicz M E, Roth W J, Vartuli J C and Beck J S 1992 *Nat.* **359** 710-71
[11] Ryoo R, Joo S H and Jun S 1999 *J. Phys. Chem. B* **103**(37) 7743-46
[12] Mokaya R 2001 *Micropor.Mesopor.Mater* **44-45** 119-27
[13] Beckmann M P, Betsholtz C, Heldin C H, Westermark B, Dimarco E, Difiore P P, Robbins K C, and Aaronson S A 2000 *Science* **77** 34-8
[14] Taba P, Hala Y, Pongsendana M and Ruru E 2013 *Proc. 2nd Int. Conf. Indonesian Chem. Soc.* 196-201
[15] Taba P, Budi P, Sari A Y P 2017 *IOP Conf. Ser.: Mater. Sci. Eng.* 188 012015
[16] Wang S and Li J 2006 *Chinese Chem. Lett.* **17** (2) 221-224
[17] Pearson R G 1963 *J. Am. Soc.* **85** 3533-39
[18] Taba P 2008 *Makara Sains* **12** (2) 120-5
[19] Beck S J, Vartli C J, Roth J W, Leonowicz E M, Kresge T C, Schimitt D K, Chu W T C, Olson H D, Sheppard W E, McMulley B S, Higgins B J and Schlenker L J 1992 *J. Am. Chem. Soc.* **114** 10834-43
[20] Deng S and Ting Y P 2005 *Environ. Sci. Technol.* **39** 8490-6
[21] Hossain K Z, Monreal C M, and Sayari A 2008 *Colloids Surf. B: Biointerfaces* **62** 42-50
[22] Chen F, Huang L and Li Q 1997 *Chem. Mater.* **9** 2685-266
[23] Yanicke M, Kumar D, Stucky G D, Chmelka B F 1994 *Zeolites and Related Microporous Materials: State of the Art, Stud. Surf. Sci. Catal.* **84** 243 (Amsterdam Elsevier Science)
[24] Temoçin Z and Yigitoglu M 2010 *Water Air Soil Polutt* **201** 463-72
[25] Namasivayam C, Radhika R and Suba S 2001 *Water Manag* **21** 381-7

Selectivity of β-Sitosterol Imprinted Polymers as Adsorbent

St Fauziah[1], Nunuk Hariani Soekamto[1], Paulina Taba[1], Muh Bachri Amran[2]

[1]Department of Chemistry, Faculty of Mathematics and Natural Science, Hasanuddin University, Jl. Perintis Kemerdekaan KM. 10, Makassar, Indonesia, 90245
[2]Department of Chemistry, Faculty of Mathematics and Natural Science, Institute of Technology Bandung, Jl. Ganesa No. 10, Bandung, Indonesia, 40132

E-mail: stfauziah@unhas.ac.id

Abstract. Molecularly Imprinted Polymers (MIPs) are smart materials that have been used as adsorbents in separation processes of compounds because they have a memorial effect to a certain compound. In this research, MIP synthesized was used as adsorbent for β-sitosterol. The objective of the research was to know the selectivity of MIP in adsorbing β-sitosterol. The concentrations of β-sitosterol after adsorption and desorption were analyzed by a UV-Vis spectrophotometer and the selectivity test was analyzed by HPLC. Result showed that the MIP had high adsorption ability (q_e). The recovery of β-sitosterol from MIP for the adsorption-desorption process was 68.48%. The MIP was very selective to β-sitosterol compared to cholesterol because it can adsorb β-sitosterol as many as 100%, whereas the adsorption of cholesterol was only 30.27 %.

1. Introduction

The β-sitosterol compound is one of the sterol compounds which are included in the phytosterol group [1]. The presence of sitosterol as the major compound in natural material samples often disrupts the process of isolation and purification of certain compounds [2]. This becomes one of the obstacles in the process of separation and purification of the desired compounds especially for compounds having similar structures. Therefore, one attempt to overcome this problem is to utilize the highly developed synthetic polymers as an adsorbent to assist the selective separation and purification of compounds. The material is known as molecularly imprinted polymers (MIP) [3]. The β-sitosterol compound is used as a template molecule in the MIP synthesis, in order to produce MIP that can be used as a selective adsorbent to β-sitosterol in natural material samples.

MIP is generally made through non-covalent interactions because it is simple and flexible [4]. Non-covalent interactions between monomers and templates can form the active sides of binding to MIP [5]. In addition, the release of template molecules from MIP is easier to occur due to the use of organic solvents [6]. The polymer will have a mold with the appropriate shape and size of the loose molecular template [7] and has functional groups corresponding to the target molecule [8], so that it is selective towards its target molecule.

Selective and highly adsorptive MIPs are influenced by the functional monomers and cross-linkers used. Appropriate functional monomers and cross-linking strongly determine the effectiveness of MIPs interacting with target molecules [9] and polymer complex stability formed during the polymerization process [10]. In this study, the functional monomer used was methacrylic acid (MAA).

This monomer is widely used in MIP synthesis because it ability interact with the template through hydrogen bonds and ionic bonds [11]. Cross-linkers used in this study were ethyleneglycol dimethacrylate (EGDMA). It is highly reactive and can produce polymers with high stiffness so that it is more stable [4].

2. Methods

2.1. Materials

In the research were used a functional monomer of 99% methacrylic acid, MAA, (Sigma Aldrich), template molecule of 97% β-sitosterol (Sigma Aldrich), etileneglycol dimethacrylate (Sigma Aldrich) as *cross linker*, toluene as porogen solvent, 2,2 '-azobisisobutyronitrile (Sigma Aldrich) as initiator, double distilled water, methanol (HPLC grade), tetrahydrofuran (THF), and acetic acid. Items of equipment used were glassware, shaker, digital balance, water bath, oven. Instruments used were High Pressure Liquid Chromatography (HPLC) Merck Agilent 1260 infinity (column Cronus RP E18 C, dimension 12.5 cm x 0.4 cm), Fourier Transform Infra Red (FTIR) : IR Prestige-21 (Shimadzu), Scanning Electron Micrographs (SEM) : Tescan Vega 3SB and Energy Dispersive X-ray Spectroscopy (EDS) 6510 (LA), Thermo Gravimetric Analysis (TGA) NETZSCH STA 449F1, Spectrophotometer Ultra Violet-visible (UV-Vis) Agilent 845 (Shimadzu).

2.2. Synthesis of MIP_MAA-co-EGDMA

A total 2 mL (23.58 mmol) of MAA and 50 mg (0.12 mmol) of β-sitosterol were dissolved in 5 mL (47.05 mmol) toluene into a round flask and the mixtures were left for some time. Then 3 mL (15.91 mmol) of EGDMA crosslinker and 1.5 mL of AIBN as an initiator were added to the mixture. The solution was sonicated and flowed with nitrogen gas.

Synthesis of MIP_MAA-co-EGDMA was conducted through a polymerization process for 24 hours in a water bath at a temperature of 55°C. The synthesis product was dried and then washed with tetrahydrofuran (THF), methanol/acetic acid (9:1), and double distilled water. The polymer has been washed and dried back at 50°C. Nonimprinting polymer (NIP) was made without β-sitosterol using the same synthesis procedure of MIP synthesis [12].

2.3. The Adsorption Capability Test

The adsorption ability of MIP to β-sitosterol was tested by adsorption process. The amount of β-sitosterol adsorbed by MIP was known by making various concentrations of β-sitosterol standard solutions of 2, 4, 6, 8 and 10 ppm. Five vials, each containing 20 mg of MIP, were prepared and 3 mL of β-sitosterol solution with different concentrations was added to each vial. The mixture was stirred using a shaker for 60 minutes as the optimum time [12] at room temperature, filtered and β-sitosterol concentration in the filtrate was analyzed by a UV-Vis spectrophotometer at a wavelength of 202 nm.

2.4. MIP Selectivity Test.

A minicolumn was prepared, filled with 50 mg MIP and then saturated with methanol before it was used on the selectivity test. A total of 0.5 mL of 10 ppm β-sitosterol standard solution was mixed with 0.5 mL of 10 ppm cholesterol standard solution, followed by the addition of 2 mL of a carrier carrier solvent consisting of methanol and water mixture. The mixture flowed to a minicolumn containing 50 mg of MIP. Solutions passing through the columns were collected in several vials as much as 1 mL, and 100% methanol was passed through the column. The methanol solvent that has passed through the column was also collected (1 mL each) in the vial and then the eluate was analyzed with HPLC. Results of the analysis with HPLC were calculated to determine the selectivity of MIP to β-sitosterol.

2.5. Adsorption-Desorption Test on MIP

The ability of MIP to adsorb and desorb β-sitosterol was tested using a minicolumn and a UV spectrophotometer. The minicolumn containing 50 mg of MIP was saturated with methanol. Furthermore, a standard solution of 10 ppm β-sitosterol (0.5 mL) was introduced into 2 ml of the carrier solvent (the best solvent obtained from the selectivity test) containing methanol and water mixture. The adsorption process of β-sitosterol was carried out by flowing the mixture on the minicolumn containing 50 mg MIP. The mixture was passed through the columns and 1 mL of eluate was collected. If the carrier solvent containing β-sitosterol was discharged then 100% methanol was applied to desorb β-sitosterol which has been adsorbed by MIP. The eluate (about 1 mL each) passing through the columns was also collected and analyzed by a UV spectrophotometer.

2.6. MIP application as adsorbent by minicolumn chromatography
The wood powder (50 g) was dissolved in 100 mL of n-hexane. The extraction process was carried out at temperature of room for 24 hours until obtained extract. The extract was filtered and concentrated using a rotary evaporator. The jackfruit extract (10 mg) was dissolved in 10 mL of methanol and filtered. The jackfruit extract solution (1 mL) was inserted slowly into a chromatographic column which contains 100 mg of MIP. The column containing MIP was conditioned first by passing methanol several times. The β-sitosterol compound in the jackfruit extract solution is expected to be retained on MIP in the column and unexpected components will be eluted and analyzed by HPLC. Furthermore, MIP that adsorbs β-sitosterol was washed with 1 mL of the best solvent used in the selectivity test to remove components that were not retained by MIP. The solvents that have passed through MIP in columns were collected and analyzed by HPLC. The final step was extracting the retained β-sitosterol in MIP using methanol 100%, the solvent out of column was collected as much as 1 mL and analyzed by HPLC.

3. Results and Discussion
NIP and MIP were synthesized using the bulk polymerization method through free radical polymerization reactions were white solids.

3.1. Characterization of MIP
3.1.1. Analysis FTIR. The FTIR spectra of MAA, NIP and MIP materials before and after extraction of the template can be seen in figure 1.

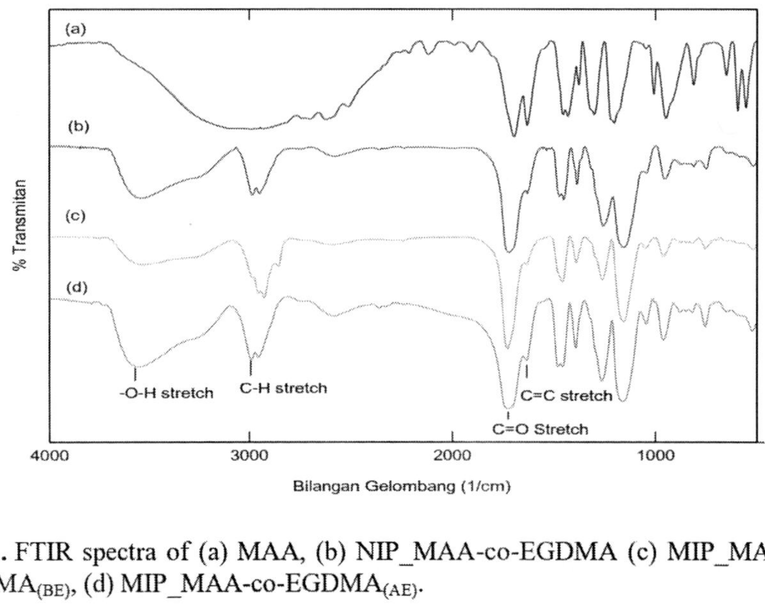

Figure 1. FTIR spectra of (a) MAA, (b) NIP_MAA-co-EGDMA (c) MIP_MAA-co-EGDMA$_{(BE)}$, (d) MIP_MAA-co-EGDMA$_{(AE)}$.

Figure 1 shows that the absorption peak of -OH stretching, C=O stretching -C=C of MAA shifts after the MAA molecule interacts with other MAA molecules, template molecules, and crosslinkers in the formation of polymers. There is also a considerable shift of absorption peaks of -OH in NIP_MAA-co-EGDMA and MIP_MAA-co-EGDMA$_{(BE)}$ compared to that in MAA. This is because the OH functional group in the MAA molecule interacts with the functional group in the MAA itself or with OH groups of β-sitosterol that forms the hydrogen bond. Similarly, the functional groups of -C=O in MAA and OH in β-sitosterol may interact and form hydrogen bonds, this causes the wavenumber of C=O also shift after the formation of NIP_MAA-co-EGDMA and MIP_MAA-co-EGDMA$_{(BE)}$. The shift of wavenumbers for the absorption peaks of -C=C in NIP_MAA-co-EGDMA and MIP_MAA-co-EGDMA$_{(BE)}$ is very small, but there is a considerable decrease in their intensity compared to that of MAA where the -C=C- functional group have a strong and sharp absorption intensity. This suggests that the -C=C interaction of monomers occurred at the polymer formation.

The absorption peak of the -OH stretching function group on the MIP_MAA-co-EGDMA$_{(BE)}$ shifts compared to that in MIP_MAA-co-EGDMA$_{(AE)}$, this indicates that the hydrogen bond termination has occurred between -OH functional group of β-sitosterol with the functional group contained in the MIP_MAA-co-EGDMA$_{(AE)}$ template.

3.1.2. Analysis SEM. The surface morphological characteristic of NIP_MAA-co-EGDMA, MIP_MAA-co-EGDMA$_{(BE)}$ and MIP_MAA-co-EGDMA$_{(AE)}$ using SEM is shown in figure 2.

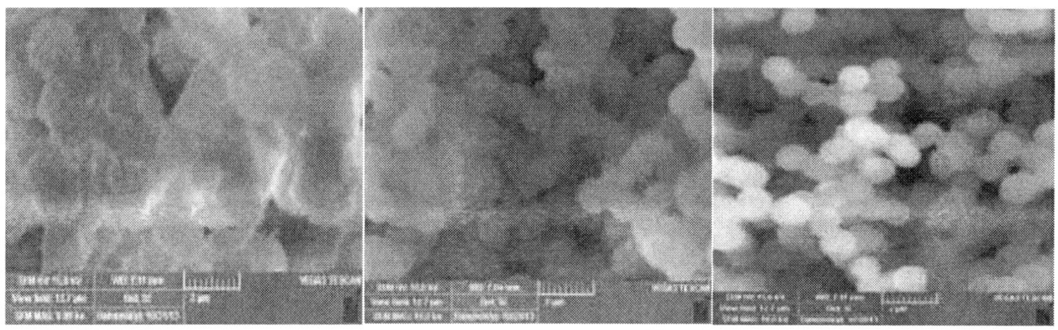

(a) (b) (c)

Figure 2. The SEM images of (a) NIP_MAA-co-EGDMA (b) MIP_MAA-co-EGDMA $_{(BE)}$ and (c) MIP_MAA-co-EGDMA$_{(AE)}$.

NIP_MAA-co-EGDMA is composed of irregularly shaped particles and they are very close each other indicating that there is no pore in the material. MIP_MAA-co-EGDMA$_{(BE)}$ is composed of spherical granules, but the particles are still close each other. By comparing with the surface morphology of MIP_MAA-co-EGDMA$_{(AE)}$, it is clear that the particles of the former are closer each other than that of the later. This indicates that before extraction of β-sitosterol, the surface is more tightly than that of MIP_MAA-co-EGDMA$_{(AE)}$ where the surface looks more porous. A more porous surface of MIP_MAA-co-EGDMA$_{(AE)}$ indicates that β-sitosterol has been detached from MIP. The SEM analysis results show that the surface of MIP_MAA-co-EGDMA$_{(AE)}$ is composed of highly regulated granules and resembles balls of similar size. All materials (NIP and MIP) looks rigid.

3.1.3. Analysis EDS. The EDS analysis was also conducted to determine the composition of the main constituent elements of each polymer. The EDS data in table 1 shows the percentage of C and O atoms in NIP and MIP materials.

Table 1. The EDS data of NIP, MIP_MAA-co-EGDMA$_{(BE)}$, MIP_MAA-co-EGDMA$_{(AE)}$.

Element	Mass %			Atomic %		
	NIP	MIP$_{(BE)}$	MIP$_{(AE)}$	NIP	MIP$_{(BE)}$	MIP$_{(AE)}$
C	82.17	79.56	77.79	85.99	83.83	82.35
O	17.83	20.44	22.21	14.01	16.17	17.65

Table 1 shows that after the extraction of β-sitosterol from MIP, the percentage of C atom decreases, while the percentage of O atom increases. This occurs because the number of C atoms in β-sitosterol is much higher than that of O atoms so that when β-sitosterol was still in the polymer, the C character increased, whereas when β-sitosterol was released from the polymer, the O character increased. A decrease in the percentage of C atom by 1.48% proves that β-sitosterol has been released from the polymer. This data is supported by the FTIR spectrum in which the absorption of the -OH functional group in the MIP_MAA-co-EGDMA$_{(AE)}$ spectrum undergoes a shift to a larger wavenumber compared to the one in the MIP_MAA-co-EGDMA$_{(BE)}$ spectrum. These results indicate that the hydrogen bonding between the functional groups of polymer and β-sitosterol broke after the removal of β-sitosterol from the MIP molecule.

3.1.4 Analysis TGA. Thermo Gravimetric analysis (TGA) was also conducted in this study to determine the thermal stability and different levels of decomposition of NIP and MIP at high temperatures. The thermogram of NIP_MAA-co-EGDMA, MIP_MAA_CO_EGDMA$_{(BE)}$ and MIP_MAA-co-EGDMA$_{(AE)}$ can be seen in figure 3.

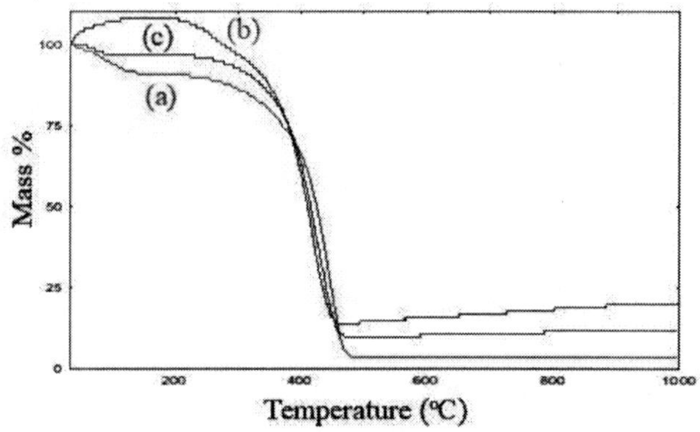

Figure 3. TGA thermograms of (a) NIP_MAA-co-EGDMA, (b) MIP_MAA-co-EGDMA$_{(BE)}$ and (c) MIP_MAA-co-EGDMA$_{(AE)}$.

The TGA thermograms of NIP_MAA-co-EGDMA and MIP_MAA-co-EGDMA$_{(AE)}$ show similar patterns because β-sitosterol in MIP_MAA-co-EGDMA$_{(AE)}$ has been removed from the polymer, while the thermogram of MIP_MAA-co-EGDMA$_{(BE)}$ is different that of the two materials because it still contains β-sitosterol.

3.1.5 *Analysis DTG.* The results of the DTG analysis (Derived Thermogravimetric Analysis) from MIP_MAA-co-EGDMA$_{(AE)}$ can be seen in figure 4.

IOP Conf. Series: Journal of Physics: Conf. Series **979** (2018) 012059 doi:10.1088/1742-6596/979/1/012059

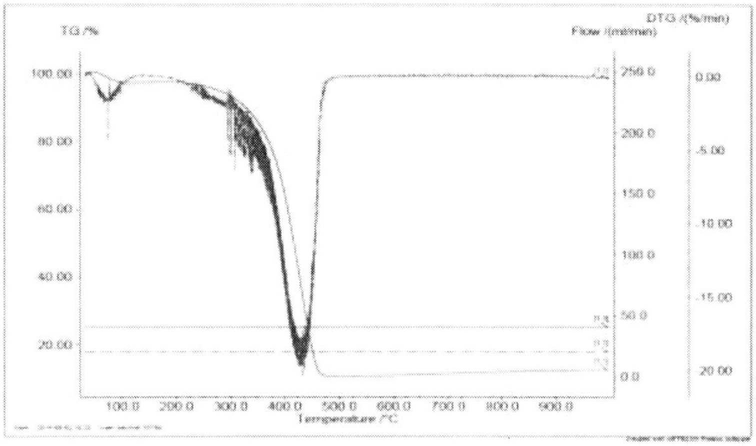

Figure 4. DTG thermogram of MIP_MAA-co-EGDMA$_{(AE)}$.

The DTG thermogram of MIP_MAA-co-EGDMA$_{(AE)}$ shows that there are several mass losses during heating of the material. The mass loss starts at a low temperature of 46.91 °C to 225.91 °C. The mass loss at this initial temperatures is suspected to be due to water loss at the temperature range. The great mass loss occurs at temperatures of 380.91 °C to 470.91 °C in which at this range, MIP is suspected to be degraded to CO and CO_2. At higher temperatures, bonds between monomers broken and only carbon remained. The high final temperature degradation indicates that MIP_MAA-co-EGDMA is stable.

3.2. The adsorption ability test

The amount of β-sitosterol adsorbed MIP_MAA-co-EGDMA and NIP_MAA_co-EGDMA was 0.72 and 0.06 mg/g, respectively. This shows that the adsorption ability of the former is better than that of the later. The same trend was also reported in the previous study [12]. The adsorption ability shown by the adsorption capacity values obtained from calculations using Langmuir and Freundlich isothermal *i.e.* 2.03 and 1.05 mg/g, correspondingly.

3.3. Selectivity test

The selectivity test was performed using the best carrier solvent composition. The solvent was mixed with β-sitosterol and cholesterol to carry both materials through a minicolumn containing MIP. The MIP can adsorb more β-sitosterol than cholesterol. The amount of β-sitosterol that can be adsorbed by MIP depends on the polarity properties of the solvent composition and the type of MIP. The polarity difference of the solvent was made from the composition ratio of methanol(M):water(W). The amount of β-sitosterol and cholesterol adsorbed was analyzed with HPLC to determine the percentage of effectiveness. The effectiveness percentage of β-sitosterol and cholesterol adsorption by MIP can be seen in table 2.

Tabel 2. The adsorption effectiveness of β-sitosterol dan cholesterol with several carrier solvent compositions.

MIP type	Carrier Solvent Composition	Adsorpstion Effectiveness Cholesterol (%)	Adsorpstion Effectiveness β-sitosterol (%)	Difference Adsorpstion Effectiveness (%)
MIP_MAA-co-EGDMA	M:W (90:10)	100	100	0
	M:W (80:20)	30.27	100	69.73

MIP_MAA-co-EGDMA can adsorb 100% cholesterol and β-sitosterol with the methanol:water composition of 90:10 (% v/v). When the polarity of solvent increases polarity such as in methanol ratio: water of 80: 20 (% v/v), MIP can adsorb 100 % β-sitosterol while cholesterol adsorbed is only 30.27 %. This because cholesterol is more polar than, whereas MIP_MAA-co-EGDMA is nonpolar. Based on these data, the carrier solvent composition used for MIP to be selective against β-sitosterol is the methanol:water ratio of 80:20.

The data of the effectiveness percentage proves that MIP_MAA-co-EGDMA has better selectivity against β-sitosterol than cholesterol. The effectiveness of a polymer is strongly influenced by the stability of the polymer. The mold is unchanged, therefore, the functional group in the polymer will only bind to the molecule whose functional group corresponds to the mold in the polymer. The MIP selectivity also relies greatly on the monomers and crosslinkers used [10]. Hashim *et al.* [13] reported that polymers made from MAA and EGDMA had high selectivity to stigmasterol used as their mold molecules, as in this study using MAA and EGDMA has a high selectivity for β-sitosterol. However, compared to MIP_TFMAA-co-EGDMA [14], MIP_MAA-co-EGDMA is more selective because the polymer adsorbed more cholesterol.

3.4. Adsorption-Desorption Test of MIP
The adsorption-desorption test of MIP was conducted in order to know the possibility of MIP to be reused. The recovery percentage of β-sitosterol from the used MIP_MAA-co-EGDMA in the second adsorption-desorption process reached 68.48%. Fauziah *et al.* [14] also tested the adsorption-desorption ability of MIP_TFMAA-co-EGDMA against β-sitosterol. The results obtained showed that the recovery percentage of β-sitosterol from MIP_MAA-co-EGDMA were better than that from MIP_TFMAA-co-EGDMA (35.56 %) for the second adsorption-desorption process.

3.5. MIP application
The MIP produced in this study was utilized as an adsorbent in a solid phase extraction (SPE) method in order to get information on the ability of the material to separate β-sitosterol from natural product samples and simultaneously can be used to facilitate the separation of other compounds from the β-sitosterol which is often the dominant compound in natural product samples. Table 3 shows the peak areas data of β-sitosterol obtained from chromatograms. The peak areas are used as data which are proportional to the amount of β-sitosterol present in the filtrate that has passed the SPE adsorbent. The greater the value of the area obtained, the greater the amount of β-sitosterol presents in the filtrate. The SPE method used for MIP applications as adsorbents was carried out through several important stages, *i.e.* retention, washing, and elution.

Table 3. Peak areas of β-sitosterol obtained from chromatograms at several stages in the SPE method.

Types	Stage on SPE method	Solvent	Peak area
Jackfruit		Methanol	775.6778
MIP_MAA-co-EGDMA	Retention	Jackfruit extract solution	28.7730
	Washing	methanol:water (8:2)	344.3110
	Elution	Methanol 100%	247.5000

4. Conclusion
Characterization of MIP_MAA-co-EGDMA using SEM, TGA and DTG showed that MIP_MAA-co-EGDMA was composed of highly regulated granules and resembled balls of similar size. The

structural geometry of polymers was more rigid and the polymers were stable to temperature. The functional groups that play a role in MAA synthesis and characterization were OH, C=O and C=C.

MIP_MAA-co-EGDMA had high adsorption ability (q_e) and the recovery percentage of β-sitosterol from MIP were 68.48 %. The MIP was very selective to β-sitosterol compared to cholesterol with the adsorption effectiveness to β-sitosterol of 100% and that to cholesterol of 30.27 %. The adsorption ability of MIP_MAA-co-EGDMA (0.72 mg/g) was better than that of NIP_MAA-co-EGDMA (0.06 mg/g).

Acknowledgement
Thank you to the Directorate of Higher Education of Indonesia and Hasanuddin University for the funding support and facilities provided in the research.

References

[1] Zhang Z, Tan W, Hu Y and Li G 2011 Simultaneous Determination Trace Sterol in Complicated Biological Samples by Gas Chromatography-mass Spectrometry Coupled with Extraction Using β-sitosterol Magnetic Molecularly Imprinted Polymers Beads *J Chromatogr A.* **1218** 4275-4283
[2] Chuang Y H, Ju Y H and Widjaja A 2006 Separation of Campesterol and β-Sitosterol from a Sterol Mixture *Separ Sci Technol* **4** 3027-3038
[3] Yan H, Wang R, Han Y and Liu S 2012 Screening Recognition, and Quantitation of Salbumol Residues in Ham Sausages by Molecularly Imprinted Solid Phase Extraction Coupled with High Performance Liquid Chromatography-UV Detection *J Chromatogr B* **900** 18-23
[4] Walsh R 2010 *Development and characterization of molecularly imprinted suspension polymers.* Thesis. Pharmaceutical and Molecular Biotechnology Research Centre Waterford Institute of Technology
[5] Wulff G, Vesper R, Grobe E and Sarhan A 1977 Enzyme-analogue Built Polymers on the Synthesis of Polymers Containing Chiral Cavities and Their Use for the Resolution of Racemates *Makromol Chem* **178** 2799-2816
[6] Hongyuan Y and Kyung H R 2006 Characteristic and Synthetic Approach of Molecularly Imprinted Polymer *Int J Mol Sci* **7** 155-178
[7] Sellergren B 2001 Molecularly Imprinted Polymers: Man-made Mimics of Antibodies and Their Applications in Analytical Chemistry *Techniques and instrumentation in analytical chemistry* **23** (Amsterdam Elsevier: Sellergren B) pp 305–322
[8] Gema P G, Pilar F H and Durand J S A 2006 A Molecularly Study of Molecularly Imprinted Polymers Using the Scanning Electron Microscope *Anal Chim Acta* **557** 179-183
[9] Kim H and Spivak D A 2003 New Insight Into Modeling Non-covalently Imprinted Polymers *J Am Chem Soc* **125** 11269-11275
[10] Beltran A, Borrull F, Cormack P A G and Marce R M 2010 Molecularly-imprinted Polymers: Useful Sorbents for Selective Extractions *Trac-Trends Anal Chem* **29** (11) 1363-1375
[11] Mayes A G and Whitcombe M J 2005 Synthetic Strategies for the Generation of Molecularly Imprinted Organic Polymers *Adv Drug Delive Rev* **57** 1742-1778
[12] Soekamto N H, Fauziah S T, Taba P and Amran M B 2017 Adsorption of β-sitosterol on Molecularly Imprinted Polymer *IOP Conf Series Mater Sci Eng* 188 012048
[13] Hashim S N N S, Boysen R I, Schwarz L J, Danylec B and Hearn M T W 2014 A Comparison of Covalent and Non Covalent Imprinting Strategies for the Synthesis of Stigmasterol Imprinted Polymers *J of Chromatography A* **1359** 35-43
[14] Fauziah S, Hariani N S, Amran M B and Taba P 2016 Capacity and Selectivity Study of MIP_TFMAA-co-EGDMA as Adsorbent *Indonesia Chimica Acta* **9** (2) 1-7

IOP Conf. Series: Journal of Physics: Conf. Series **979** (2018) 012060 doi:10.1088/1742-6596/979/1/012060

Optical and Structural Characterization of ZnO/TiO_2 Bilayer Thin Films Grown by Sol-Gel Spin Coating

P.L. Gareso, Musfitasari, and Eko Juarlin

Department of Physics, Faculty of Mathematics and Natural Science, Hasanuddin University, Jl. Perintis Kemerdekaan KM. 10 Tamalanrea-Makassar 90245, Indonesia.

E-mail: pgareso@fmipa.unhas.ac.id

Abstract. Structural and optical properties of ZnO/TiO_2 bilayers thin films have been investigated using x-ray diffraction (X-RD), scanning electron microscopy (SEM), and optical transmittance UV-Vis measurements. ZnO thin films were prepared by dissolving zinc acetate dehydrated into a solvent of ethanol and then added triethanolamin. In the case of TiO_2 layers, tetraisoproxide was dissolved into ethanol and then added an acetate acid. The layer of ZnO was deposited first followed by TiO_2 layer on a glass substrate using a spin coating technique. The ZnO/TiO_2 bilayers were annealed at various temperatures from 300°C until 600°C for 60 minutes. The X-ray diffraction results show that there was an enhancement of the x-ray spectra as annealed temperature increased to 600°C in comparison to the samples that were annealed at 300°C. Based on the optical measurement of UV-Vis, the band gap energy of ZnO/TiO_2 bilayer is around 3.2 eV at temperature of 300°C. This value is similar to the band gap energy of ZnO. SEM results show that there is no cluster in the surface of ZnO/TiO_2 bilayer.

1. Introduction

The optical properties of semiconductor can be enhanced by coupling of two semiconductors in the terms of hetero-structures than a single semiconductor. The two semiconductors that are combined in this study ZnO and TiO_2. Zinc oxide thin film is one of the II-VI compound semiconductors that has been received considerable attention for many researchers due to its application in thin film sensors [1], light emitting diodes [2], spintronic devices [3], and nanolasers [4, 5]. Also, ZnO thin films used as a solar cell window since it has high optical transmittance in the visible region. ZnO has a wide band gap of 3.3 eV, large exitonic binding energy of 60 meV, and high carrier mobility at room temperature.

Titanium dioxide (TiO_2) is important semiconductor having a huge application in photo catalysis, solar cell and photonic devices [6]. Moreover, TiO_2 can be applied in optical filters, antireflection coatings, sensors and catalysts [7]. This wide range of the application of TiO_2 is due to its unique electronic and structural properties. TiO_2 can be classified into three crystalline phases: anatase, rutile, and brookite. Rutile is the most stable phase and it is usually obtained after annealing at temperature above 500°C [8]. Also, TiO_2 is in the visible light region and its band gap is 3.0 eV for rutile and 3.2 eV for anatase crystalline phase.

In this present work, we investigate the optical and structural characterization of ZnO/TiO_2 bilayers thin films grown by sol-gel processing. For these purposes, the samples of ZnO/TiO_2 were deposited using spin coating method and the samples were annealed at various temperatures from

400°C to 600°C for 60 minutes. The films were characterized by optical transmittance UV-Vis, and X-ray diffraction (X-RD) measurements, scanning electron microscopy (SEM).

2. Experimental Set-Up

ZnO/TiO_2 bilayers were prepared using-sol-gel spin coating method with the speed rate of 300 rpm for 15 second. The first layer was ZnO thin films followed by TiO_2 layers. For ZnO layers, Zinc acetate dehydrate ($Zn(CH_3COO)_2.2H_2O$) was dissolved with ethanol of 1.5M, and diethanolamine (DEA) as starting material, solvent, and stabilizer, respectively. For preparing TiO_2 precursor, 1.5 ml titanium tetraisiproxide was dissolved into 10 ml ethanol and then added 0.1 ml acetate acid. These solvent were stirred on the magnetic hotplate at 80°C during 5 hours with the speed rate of 1100 rpm.

The glass substrate was cleaned with HCL to remove the unwanted particles prior to the ZnO layers were coated on it. The first layer of ZnO solution was deposited on the glass substrate and then it was heated at 300°C for 10 minutes. After that, TiO_2 dissolution was coated on the top of ZnO and then it was heated at 300°C for 10 minutes. Finally, the ZnO/TiO_2 bilayers were annealed at various temperatures from 300°C to 600°C for 60 minutes.

X-ray diffraction measurement were carried to investigate the structural characterization using a single scan diffractometer with Cu Kα (λ = 1.5406 Å) radiation and scanning range of 2θ between 20° and 50°. During the measurement, the current and the voltage of X-RD were maintained at 30 mA and 40 kV, respectively, and the scan speed was 2°/min. Optical transmittance measurements were performed to characterize the optical properties using a single beam UV-Vis spectrophotometer with a wavelength range of 250 – 800 nm.

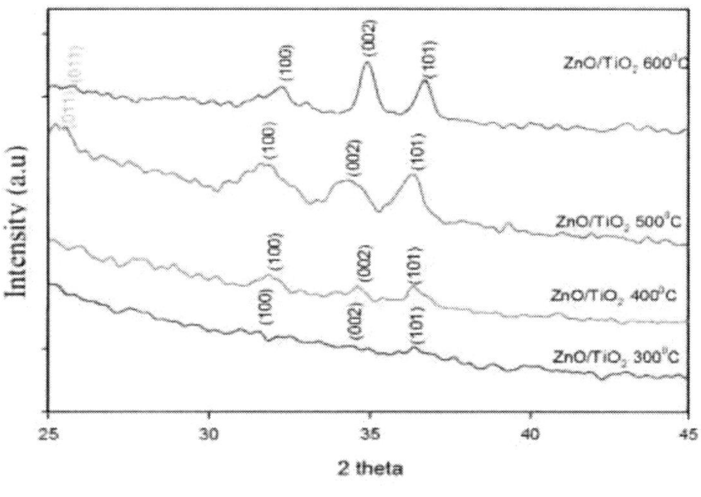

Figure 1. X-ray diffraction patterns of ZnO/TiO_2 bilayer thin films after annealing at various temperatures from 300°C to 600°C during 60 minutes.

3. Results and Discussion

3.1. Structural characterization

The structural characterization of ZnO/TiO_2 bilayer thin films were taken place used the X-ray diffraction measurements. Figure 1 shows the X-ray diffraction pattern of bilayer ZnO/TiO_2 after annealing at various temperatures from 300°C to 600°C during 60 minutes. As shown in this figure that, after annealing the samples at 300°C, the X-ray diffraction peak were observed in the very small intensity. Annealing the samples at 400°C shows the increase in the peak intensity. As the annealing temperature was further increased, the peak intensity of X-ray diffraction pattern becomes higher and narrower in comparison to the samples that were annealed at 400°C. These results indicate that there is an improvement in the crystalline of the films after annealing. In addition to this, the X-ray diffraction patterns of ZnO/TiO_2 bilayer after annealing at 400°C (see figure 1) appears at diffraction angles of 31.2°, 34.8°, and 36.6° which corresponds to the plane of (100), (002), and (101) respectively. These peaks were shifted to higher diffraction angle of 32.4°, 35.1°, and 36.9°. Also, the plane of (100), (002), and (101) are corresponded to the ZnO layers. Meanwhile, the X-ray diffraction peak which are related to the TiO_2 layers were not clearly observed after annealing at 300°C and 400°C, but this peak emerged when the samples were annealed at 600°C although its intensity too small.

The lattice parameters of ZnO/TiO_2 layers are determined with *Crystallography Open Database* (COD) that are confirmed from the X-ray diffraction patterns of ZnO/TiO_2 bilayer. Based on the database, the lattice parameter of both ZnO and TiO_2 layers after annealing at various temperatures from 300°C to 600°C during 60 minutes are tabulated in table 1. As can be seen in table 1, the lattice parameters of TiO_2 layers are higher than that of ZnO layers. Also, the lattice parameters of ZnO and TiO_2 layers are higher after annealing at 600°C than that annealing at 300°C. The increase of the lattice parameters of ZnO/TiO_2 bilayer thin films after annealing the samples at 600°C is probably due to the contribution from X-ray diffraction peak appears in the TiO_2 layers after annealing at 600°C. Annealing the samples below 600°C results the x-ray diffraction pattern was not clearly observed.

Table1. The lattice parameters of ZnO/TiO_2 at various temperatures from 300°C to 600°C during 60 minutes based on *COD* data.

Annealing Temperatures	**ZnO layers**		**TiO_2 layers**	
	a(Å)	**c(Å)**	**a(Å)**	**c(Å)**
300°C	3.205	5.122	3.797	9.579
400°C	3.219	5.149	3.797	9.579
500°C	3.219	5.149	3.797	9.579
600°C	3.219	5.149	3.804	9.614

From the XRD data measurements, the grain size (D) of the ZnO/TiO_2 bilayer thin films can be estimated using the Debye-Scherrer formula,

$$D = \frac{K\lambda}{\beta \cos\theta} \tag{1}$$

where k is a constant, λ is the x-ray wavelength (Cu Kα, λ = 1.5406 Å), β is the full width at half-maximum (FWHM), and θ is the Bragg angle, respectively. Based on figure 1, the plane of (002) and (101) are preferred orientation for ZnO layers, while the plane of (011) is belong to TiO_2 layers; therefore we used these planes to calculate the grain size (D) of the ZnO/TiO_2 bilayer thin films. Base on the Debye-Scherrer formula, the grain size and FWHM are tabulated in table 2.

As shown in table 2, the grain size of the ZnO/TiO_2 decreased as the annealing temperatures increased. However, when the annealing temperature was further increased up to 600°C, the grain size of the ZnO/TiO_2 increased significantly. The grain size is reflected to the X-ray diffraction patterns of the samples. It can be seen in figure 1, the x-ray diffraction peak from TiO_2 was not clearly observed after annealing the samples at 300°C to 500°C. This TiO_2 peak would contribute the grain size of the ZnO/TiO_2 bilayer, although there was an improvement in the ZnO diffraction peak at these temperatures range. After annealing the samples at 600°C, the TiO_2 diffraction peak was observed. Also in table 2, the grain size of the samples annealed at low temperature is higher than annealed the samples at high temperature, although the X-ray diffraction peak did not show the ZnO or TiO_2 layers, these indicates that at the lower annealing temperatures the samples were amorphous. Thereby, we must rely on the samples annealed at 500°C and 600°C. Increase the grain size of ZnO/TiO_2 bilayers that is followed by increase of the intensity indicates that there is an improvement in the crystalline of the films after annealing at 600°C. Previous studies were also observed the improvement of crystalline of the films after annealing in the ZnO samples [9].

Table 2. Parameters crystal of ZnO/TiO_2 bilayers includes crystal planes, FWHM, diffraction angle, and the grain size of the crystal.

Annealing temperatures (°C)	Crystal planes (hkl)	FWHM	Diffraction angle (2Θ)	Grain Size (D) (nm)
300	(002)	0.082	34.60	202.47
	(101)	0.120	36.50	139.53
400	(002)	0.083	34.77	201.83
	(101)	0.120	36.66	139.50
	(011)	0.127	25.71	128.04
500	(002)	0.179	34.93	93.32
	(101)	0.198	36.74	84.78
	(011)	0.120	25.33	164.35
600	(002)	0.120	35.11	154.04
	(101)	0.099	36.93	139.72

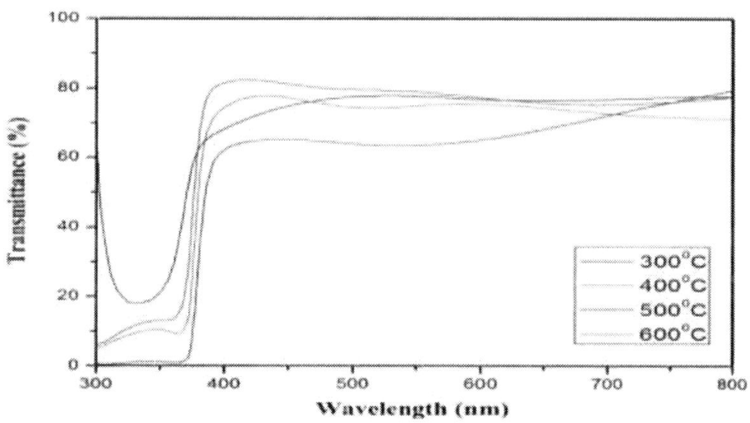

Figure 2. The transmittance spectra of ZnO/TiO_2 bilayers after annealing at different temperatures from 300°C to 600°C for 60 minutes.

3.2. Optical characterization

Figure 3 displays the optical transmittance of ZnO/TiO_2 bilayer thin films after annealing at different temperatures from 300°C to 600°C during 60 minutes. The transmittance value of the samples after annealing at 300°C is around 40 - 50%. After annealing at 500°C, it is around 55-60% in the transmittance value. This value slightly increased to 65-70% when the annealing temperature was further increased to 600°C. The increase of the transmittance value after annealing indicates the improvement of the crystalline of the films which is also confirmed in the X-ray measurements. In addition to this, there was a sharp absorptions edge was observed that located at 340-360 nm.

The optical band gap of ZnO/TiO_2 bilayer thin films are determined by extrapolation of the linier relationship between $(\alpha h\upsilon)^2$ and hυ according to the equation [10] $(\alpha h\upsilon)^2 = A(h\upsilon - E_g)$, where α is the absorption coefficient, *hυ* is the photon energy, E_g is the optical band gap and *A* is a constant. The optical band gap of ZnO/TiO_2 bilayer thin films are obtained using this equation by plotting $(\alpha h\upsilon)^2$ in regard to photon energy (hυ). Figure 3 depicts the plot of $(\alpha h\upsilon)^2$ versus photo energy (hυ) for ZnO/TiO_2 bilayer at various annealing temperatures from 300°C to 600°C for 60 minutes. The optical band gaps of ZnO/TiO_2 bilayers thin films were determined by extrapolating the linier portion of $(\alpha h\upsilon)^2$ in regard to (hυ) abscissa. After annealing the samples at 300°C, the optical band gap of

ZnO/TiO_2 is 3.24 eV. This optical band gap slight decreased to 3.17 eV and 3.14 eV after annealing the samples at 400°C and 500°C, respectively. Narrowing the optical band gap as annealing temperatures increased indicates that there is a reduction of the amorphous phase in the annealed samples at these temperatures range [11].

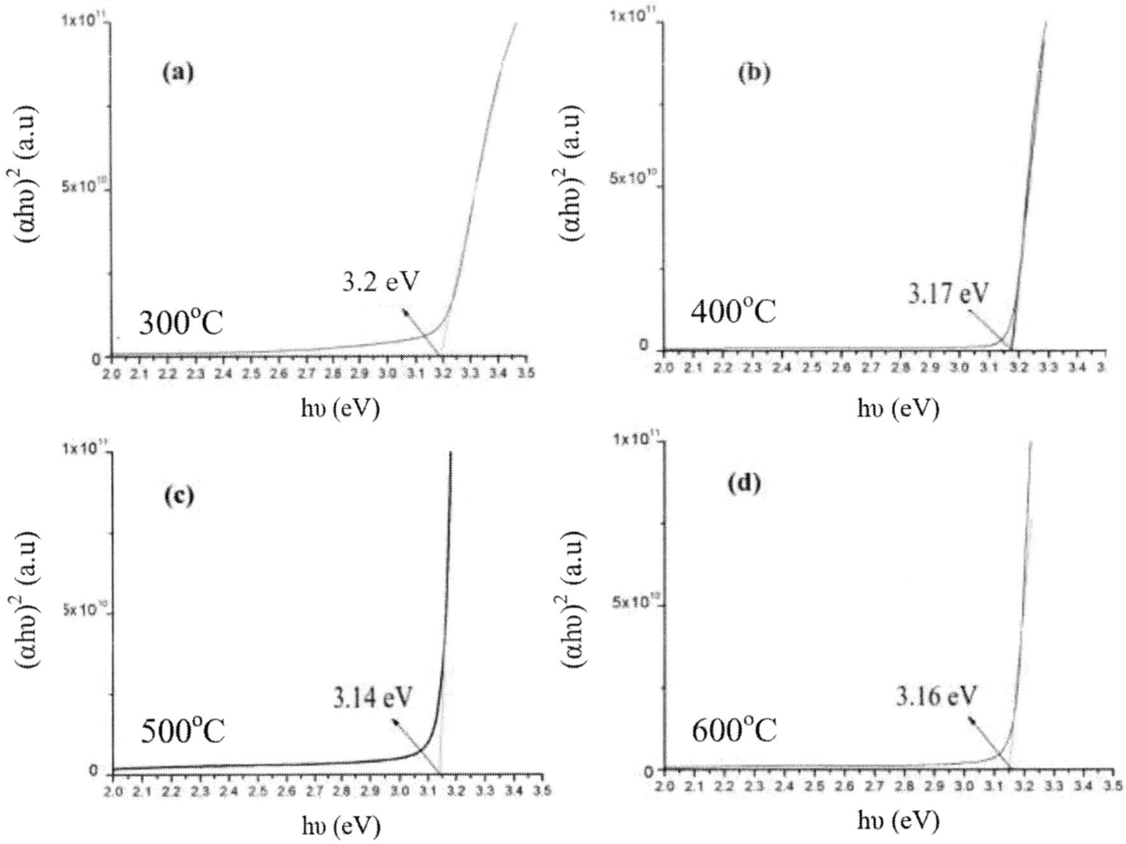

Figure 3. Plot of $(\alpha h\upsilon)^2$ vs photon energy ($h\upsilon$) of ZnO/TiO_2 bilayer thin films after annealing (a) 300°C, (b) 400°C, (c) 500°C, and (d) 600°C.

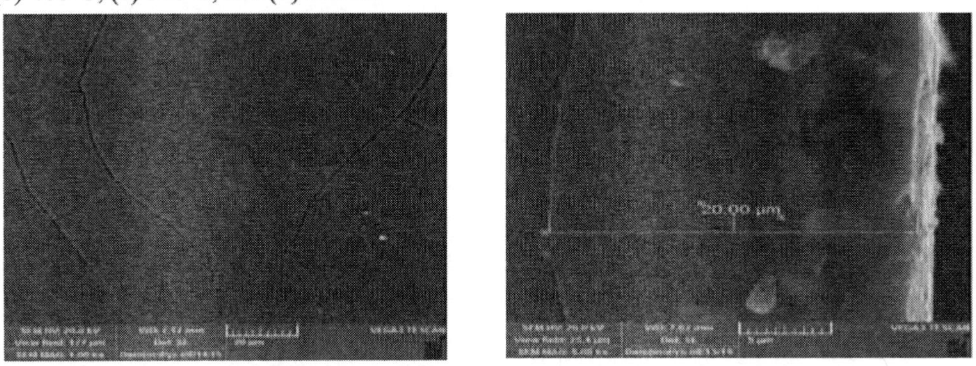

(a) (b)

Figure 4. (a) SEM image of ZnO/TiO_2 bilayer annealing at 600°C (b) Cross-sectional of SEM images of ZnO/TiO_2 bilayer.

The surface of ZnO/TiO_2 bilayer can be observed using scanning electron microscopy (SEM). Figure 4(a) shows the micrographs SEM of the morphology surface of ZnO/TiO_2 bilayer. It was clearly observed that the surface is smooth without any cluster forming at the surface. However, there is a crack on the surface that is due to the high temperature annealing to the substrate. Figure 4(b) displays a cross-section of SEM images of ZnO/TiO_2 bilayer to measure the thickness of the films. From the figure 4(b) the thickness of the film was 20 μm.

4. Conclusion

We have studied the structural and optical characterization of ZnO/TiO_2 bilayer thin films after annealing at various temperatures from 300°C to 600°C for 60 minutes. The ZnO/TiO_2 thin films were coated on the glass substrate using sol-gel method. Based on the X-ray diffraction measurements, the grain size of the ZnO/TiO_2 decreases as annealing temperature increased, but it decreases after high annealing temperatures. The transmittance value of ZnO/TiO_2 increases with increasing the temperature that indicates that the improvement of the crystalline of the films. SEM results show that there is a crack on the surface due to high annealing temperature.

References

[1] Shishiyanu S T, Shishiyanu T S and Lupen O I 2005 *Sensors and Actuators B:Chemical* **101** 379
[2] Saito N, Haneda N, Sekiguchi T, Ohashi N, Sakaguchi I and Koumoto K 2002 *Advanced Materials* **14** 418
[3] Meron T and Markovich G 2005 *J. Phys. Chem B* **109** 20232
[4] Huang M H, Mao S, Feick H, Yan H, Yang P 2001 *Science* **292** 1897
[5] Dhere N G 2006 *Solar Energy Materials and Solar Cells* **90** 2181
[6] Matthew N R, Morales E R, Cortes-Jacome M A and Antonio J A T 2009 *Solar Energy* **83** 1499
[7] Reddy B M, Ganesh I, Khan A, Mol J 2004 *Catal A:Chemical* **223** 295
[8] Abozovic N D, Mirenghi L, Jamkovic I A, Bibic N, Sojic D V, Abramovic B F and Cmor M L 2009 *Nanoscale. Res. Lett* **4** 518
[9] Wang M, Kim E J, Chung J S, Shin E W, Hahn S H, Lee K E and Park C 2006 *Phys.Stat.Sol. (a)* **203** 2418
[10] Serpone N, Lawless D and Khairutdinov R 1995 *J. Phys. Chem* **99** 16646
[11] Hong R, Huang J, He H, Fan Z and Shao J 2005 *Appl. Surf. Sci* **79** 3767

IOP Conf. Series: Journal of Physics: Conf. Series **979** (2018) 012061 doi:10.1088/1742-6596/979/1/012061

Innovated Conceptual Design of Loading Unloading Tool for Livestock at the Port

Achmad Mustakim, Firmanto Hadi

Department of Marine Transportation Engineering, Institut Teknologi Sepuluh Nopember (ITS)

E-mail: mustakim@seatrans.its.ac.id

Abstract. The condition of loading and unloading process of livestock in a number of Indonesian ports doesn't meet the principle of animal welfare, which makes cattle lose weight and injury when unloaded. Livestock loading and unloading is done by throwing cattle into the sea one by one, tying cattle hung with a sling strap and push the cattle to the berth directly. This process is against PP. 82 year 2000 on Article 47 and 55 about animal welfare. Innovation of loading and unloading tools design offered are loading and unloading design with *garbarata*. In the design of loading and unloading tools with *garbarata*, apply the concept of semi-horizontal hydraulic ladder that connects the ship and truck directly. This livestock unloading equipment design innovation is a combination of fire extinguisher truck design and bridge equipped with weightlifting equipment. In 10 years of planning *garbarata*, requires a total cost of **IDR 321,142,921** ; gets benefits **IDR 923,352,333**; and BCR (Benefit-Cost Ratio) Value worth **2.88**. BCR value >1 means the tool is feasible applied. The designs of this loading and unloading tools are estimated up to 1 hour faster than existing way. It can also minimize risks such as injury and also weight reduction livestock agencies significantly.

1. Introduction

1.1. Background

Beef is one source of animal protein with an average consumption level of 5.1 kg / capita / year [1]. The demand for meat in Indonesia is mostly (± 65%) still fulfilled from domestic production and the rest is obtained from import. However, the condition of loading and loading process of livestock in a number of Indonesian ports does not meet the principle of animal welfare, which causes cattle to lose weight and also injury when unloaded. Transportation or transportation used by inter-island cattle entrepreneurs or cities in general can lead to stress, so that it can affect decreased appetite and in the end can lose weight cattle. It is therefore necessary to have careful handling of inter-island, city, and local transportation, so that not only the road factors affect it but the condition of the vehicles used, the density of livestock, the nation or the type of cow, the climate / weather at the time of transportation, and the availability of food on the long journey [2].

The way to unloaded is done by throwing cattle into the sea one by one then TKBM (*Tenaga Kerja Bongkar Muat*/ loading and unloading labors) ready to herd cows from the sea to the edge of the beach to be taken to the quarantine. Or by tying a cow hung with a rope to be transferred to the truck. This process is contrary to PP. 82 year 2000 that is on Article 47 and Article 55 about animal welfare [3]. The absence of SOPs (Standard Operating Procedures) also worsened the treatment of livestock that caused cattle to lose weight up to 20-30% or even die during the trip until it often reap protests from other countries.

A study by Aletha Yuliana contained in the journal Indonesia Medicus Veterinus 2016, reveals that the means of transportation and the way of loading and unloading cattle from NTT to the Animal Market Beringkit Bali cause cattle stress and injury. Transportation modes used are ships, large trucks, medium trucks and pick-ups. In the Animal Market Beringkit Bali consists of male and female calves both young and adult. Of the 6,881 incoming cows, there were about 400 incidents of cattle due to trauma related to transportation. Lame cattle have low prices, because cows cannot be sold out of the region or island and are not productive if they are to be used as breeders, breeding or fattening. Breeders will experience losses ranging from one million rupiah to three million rupiah per tail due to the incident [4].

Along with technological advances, Indonesia should immediately clean up to improve the quality of domestic livestock. In addition to nurseries and livestock care facilities, livestock loading and unloading facilities should be built to facilitate the loading and unloading activities of animal welfare, user friendly and integrated. Therefore, a loading and unloading tool is required by applying the concept of semi-horizontal hydraulic ladder connecting the ship and truck directly. This livestock unloading tool design innovation is a combination of fire extinguisher truck design and bridge equipped with weightlifting equipment. With this tool, it is expected to reduce the risk of cow accidents and increase the speed of loading and unloading that is safe for animals.

1.2. Problem formulation

Based on the background that has been described previously, then the problems studied are:

1. How is the process of loading and unloading cattle in the Indonesian port today?
2. What factors affecting the loading and unloading activities of cattle do not meet the animal welfare principles in Indonesian ports?
3. How to design an innovative conceptual design of integrated loading and unloading tools between ships and trucks to minimize animal unwelfare treatment at Indonesian ports?

1.3. Objective

Based on the description of the above problem formulation, the purpose of this study are:

1. Knowing the existing condition of loading and unloading cattle process at the port of Indonesia
2. Being able to identify the factors that affect the loading and unloading activities of cattle does not meet the animal welfare principles in Indonesian ports
3. Designing a conceptual design of integrated loading and unloading tools between ships and trucks at Indonesian ports to minimize the number of incidents of cows that are not treated to meet the principle of animal welfare.

2. Literature Review

2.1. Animal Welfare Principal for Livestock

There are 5 (five) principles of freedom in the rules of animal welfare that need to be considered as already adopted and contained in Government Regulation No. 95 of 2012 are [5]:

1. Free from thirst and starvation
2. Free from the discomfort
3. Free from pain
4. Freely express normal behavior
5. Free from fear and depression

2.2. Port Planning

In a study conducted by Arrazi Rustam (2016), reveal that to reduce the existence of treatment that does not meet the principle of animal welfare as stated in Government Regulation No. 95 of 2012, the

IOP Conf. Series: Journal of Physics: Conf. Series **979** (2018) 012061 doi:10.1088/1742-6596/979/1/012061

port planning is adjusted with the concept of animal welfare. The development of the port is done with the construction of floating dock HDPE. HDPE pier is a pier that can be a floating cube or pipe (cylinder) which is the latest innovation to replace the three pontoons above because it is more durable and not damage the environment / Green Technology [6]. The floating dock HDPE has advantages over the concrete dock in general, such as:

1. Environmental friendly because it is made of HMW HDPE (High Molecular Weight- High Density Polyethylene) materials so it can be recycled and resistant to chemicals
2. Safer because it can adjust eith the tides of sea water
3. Easy to use
4. Quite cheap with range square meter reaches 1.5-2.5 million
5. Anti-corrosion
6. Complete accessories that are very supportive to change the function of the floating cube/poontoon
7. Minimize the risk of loss of livestock due to loading and unloading that is not animal welfare resulting in maintenance cost
8. Minimize the cost of livestock depreciation

2.3. Ship Route Planning

In a study conducted by Silvia Dewi Kumalasari (2016) that to reduce the existence of treatment that does not meet the principle of animal welfare, the ship's planning is done [7]. The operating pattern for optimum selected mode of cattle vessel is obtained by the optimization result of port to port and multiport operation pattern, with details as follows:

1. For the port to port operation pattern is obtained route 2 that is Larantuka-Jakarta with the number of fleet needed in a year that is as much as 1 fleet and route 4 namely Waingpau-Jakrta with the number of fleet required in a year that is 4 fleet
2. For the multiport operation pattern, the route 16 is Tenau Kupang-Ende-Jakarta with the number of fleet required for one year of 43 fleets and 26 routes is Maumere-Labuan Bajo-Jakarta with the number of fleet required for one year is 4 fleets.
3. The total fleet required in ayear of 52 fleets of cattle ships

In terms of the fastest time, by ignoring the administrative time, that is for the port of origin Tenau Kupang using general cargo ship with total delivery time of 6.8 days, whereas with the port of origin Waingapu obtained the fastest delivery time with results using the same mode that is for 5.5 days. While in terms of meat production, the least experienced shrinkage with the port of origin of Tenau Kupang is using the mode of cattle ship with the weight of cows produced per tail of 378 kg, and 106 meat production produced by 137 kg per cow, while with the port of origin Waingapu yielded cow weight per tail of 381 kg and meat production of 138 kg.

3. Research Methods

In doing this research begins with the observation and analysis of existing conditions then continues on the primary and secondary surveys. For more details can be seen in figure 1.

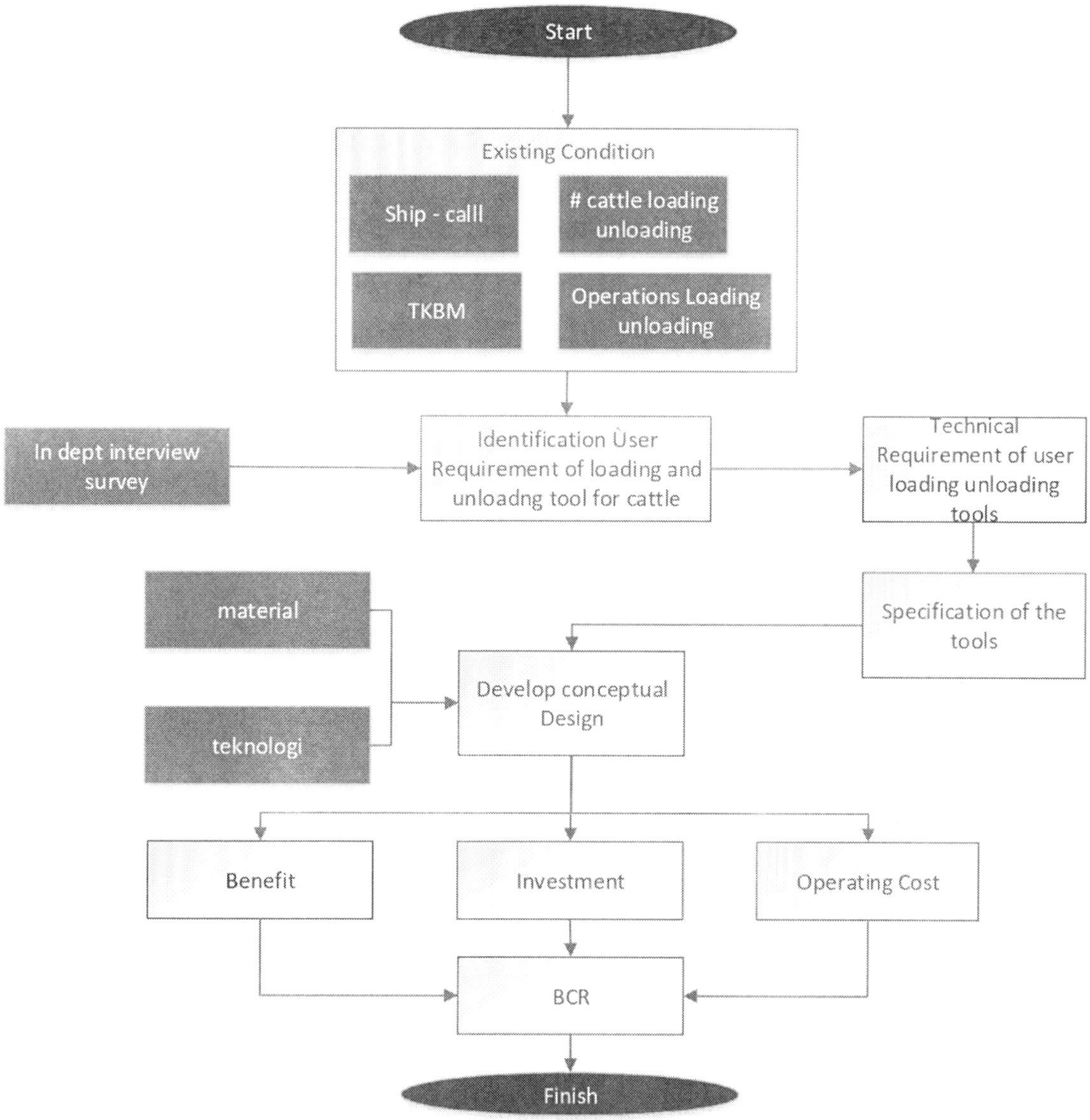

Figure 1. Research Methodology.

4. Result and Analysis

4.1. Existing conditions analysis

From the results of the primary survey that has been done to the Port of Tanjung Perak Surabaya, then obtained data on the dimensions of ships, containers, trucks and cattle. In the method of transporting / loading and unloading activities using containers at the Port of Rakas Kalimas, Surabaya with KM PASENDANA 8 which can carry up to 500 cows. This ship is dimensionless with length (Lpp) 65m, width 10m, height 7m and full 5m ship. While the container with a length of 380cm, 156cm wide and 205cm high. Box truck with length 555cm, width 240cm and height 165cm. Average size of cow is 200cm long, 60cm wide and 110cm high.

Figure 2. Visualisation of ship, truck, and container.

Figure 2 shows that loading and unloading of cattle in the port. Red line represent visualisation of ship, blue line represent visualisation of truck and green line represent visualisation of container for cows. Cattle is carried out through the transport of containers with a capacity of 7 cows to trucks with a capacity of 21 cows per truck. Delivery of cattle to the Port of Rakas Kalimas, Surabaya comes from several regions, among others Kupang, Bima, Waingapu, and Atapupu through PT. EMKL Kamadjaja Logistic. Loading and unloading activities which were then transported by trucks transported 21 cows per truck. Therefore, 1 cattle truck can be filled by 3 containers of each container containing 6-7 cows. The speed of loading and unloading per container is 4 minutes which consumes 0.05 liter per box.

Figure 3. Visualisation container for cows.

Figure 4. Unloaded cattle in the truck.

There are many kind of cow that loading and unloading in Kalimas port. The cows are Bali cow, Ongole cow, Madura cow and Kupang cow. The size of the cows are different each kind of cow. The siza of the cows can be seen on table 1 below:

Table 1. Size of Cows.

Size (cm)	Cows Kind			
	Bali	Ongole	Madura	Kupang
Heigth	115	122	110	105
Length	120	132	126	187
breast perimeter	36	39	37	60
Age (year)	4-8	4-8	4-8	4-8

4.2. User Requirement Analysis
After conducting the primary survey, Voice of Customer obtained from potential users is as follows:

1. Fast loading and unloading
2. Unharmed cattle during loading and unloading process
3. Inexpensive B / M tools in operation
4. Easy in tool storage
5. Easy operation of the tool
6. Low manufacturing cost
7. In accordance with the regulation

Furthermore, after obtaining the needs of prospective users as written design conceptualized design of loading and unloading tools Livestock as much as possible to accommodate the needs of potential users.

4.3. Design of Loading and Unloading Tools
Innovation of the planned loading tool is a loading and unloading system wherein a building will be constructed like a bridge equipped with a weight-weighting tool and also a stockman beside a bridge to drive in or out of livestock to and from the vessel. The function of digital weight scales placed in the bridge / garbarata is to weigh the weight of the cow. With this scale we can know the weight of cows when loaded at the port of origin until the destination port whether experiencing significant weight loss or not so that if the weight loss, the solution to overcome this can be quickly done by the animal quarantine hall of Port of Tanjung Perak. In addition to livestock transportation, livestock loading and unloading facilities at the port are also indispensable for improving livestock distribution processes both to Surabaya and throughout Indonesia. This supporting infrastructure is only a garbarata, but its application in real life strongly supports the process of loading and unloading livestock to conform with government autonomy to support the self-sufficiency of livestock. The design of this bridge or garbarata can be seen in figure 5.

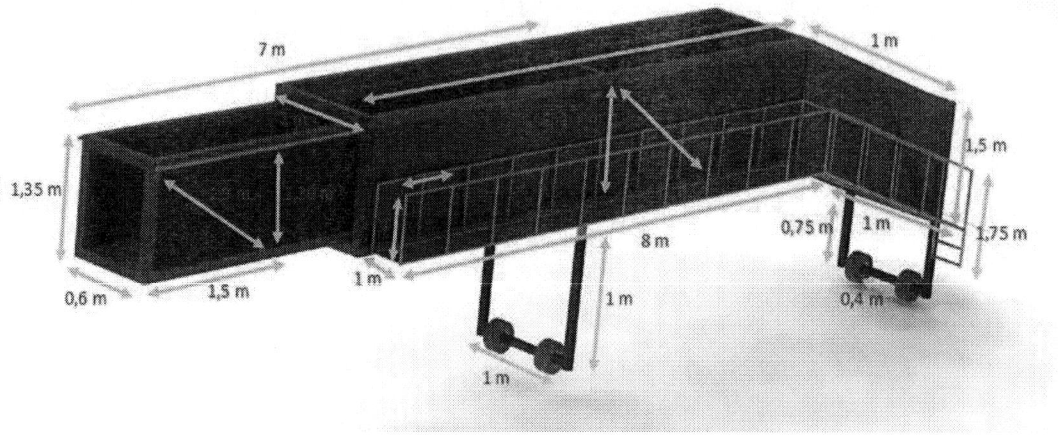

Figure 5. Conceptual design of loading and unloading tools.

Innovation of loading and unloading tool design that is offered is 2 (two) that is loading and unloading design with garbarata and loading and unloading design with container. In the design of loading tool unloading garbarata apply the concept of semi-horizontal hydraulic ladder that connects the ship and truck directly.

4.4. Benefit and Cost Ratio (BCR)

BCR is such of analysis which compare the benefit cost with investment and operation cost. The economic age of this tool is planned for 10 years. The calculation of benefit and cost with 10 years planning. The operations cost increase 10% per year. The calculation of the benefit and the investmest costs are described below:

Table 2. Investment Cost.

Investment	
Item	Cost (IDR)
Design	5,000,000
Raw Material	17,760,568
Manufacture	67,900,000
Mobilize	1,000,000
Total	**91,660,568**

Table 3. Operations Cost.

Operations Cost	
Item	Cost (IDR)
Labor	27,200,000
Gas	310,000
Maintenance	5,000,000
Total	**32,510,000**

From the cost calculation above the Present Value of cost **is IDR 321,142,921**

The design of this loading and unloading tool will make the loading and unloading of the cow progressing smoothly and also efficiently because the time required for loading and unloading is faster. If the loading and unloading time is usually spent up to 2 - 3 hours, then the development of the port and also the procurement of special loading and unloading equipment will make the loading and unloading time faster up to 1 hour and also minimize risks such as injury and also weight reduction livestock agencies significantly. Then it can be calculated the benefit of this tool with assumption: Decreasing price of the cow because of the injury is 20% from cow price which is IDR 10.000.000 per cow. From the interview 1% cow has serious injury. The price of the cow increase 10% per year. With those assumption the benefit is **IDR 923,352,333.**

So the Benefit cost ratio (BCR) of this tool is **2.88**. The BCR value is more than 1 so it can be concluded that this tool is feasible to apply.

5. Conclusion

There were some conclusion that we can get throughout the result and analysis that has done before, such as:

a. The innovation of this loading and unloading tool is a garbarata which in its application strongly supports the process of loading and unloading livestock to conform with government regulations to support self-sufficiency of livestock in Indonesia. Innovation of the planned loading tool is a loading and unloading system wherein a building will be constructed like a bridge equipped with a weight-weighting tool and also a stockman beside a bridge to drive in or out of livestock to and from the vessel. The function of digital weight scales placed in the bridge / garbarata is to weigh the weight of the cow. With this scale we can know the weight of cows when loaded at the port of origin until the destination port whether experiencing significant weight loss or not so that if the weight loss, the solution to overcome this can be quickly done by the animal quarantine hall of Port of Tanjung Perak.

b. Loading and unloading of cattle carried out through the transport of containers with a capacity of 7 cows to trucks with a capacity of 21 cows per truck. Delivery of cows to the Port of Rakas Kalimas, Surabaya comes from several regions, among others Kupang, Bima, Waingapu, and Atapupu through PT. EMKL Kamadjaja Logistic. Loading and unloading activities which were then transported by trucks transported 21 cows per truck. Therefore, 1 cattle truck can be filled by 3 containers of each container containing 6-7 cows. The speed of loading and unloading per container is 4 minutes which consumes 0.05 liter per box.

c. The factors considered in making the design of loading and unloading equipment are as follows:
   1. Fast loading and unloading
   2. Livestock that is not unjured during loading and unloading process
   3. Inexpensive B/M tools in operation
   4. Easy in tool storage
   5. Easy operation of the tool
   6. Low manufacturing cost
   7. In accordance with the regulation

d. The design innovation in 10 years of garbarata planning requires a total cost of **IDR 321,142,921** While the benefits obtained in the 10 years of planning of **IDR 923,352,333**. The value of BCR (Benefit Cost Ratio) worth **2.88** which means the tool is feasible applied.

Acknowledgments
We are pleased to acknowledge funding support from the BNPB Dana Lokal ITS for this paper.

References

[1] Animal Husbandry Service 2015 *Beef Consumption in Annual Report Book Annual*
[2] Ginting N 2006 *Personal Communications About Depreciation Body Weight On Beef Cattle Due to Freight* (Jakarta: Spreading Self-Helpers)
[3] Government of the Republic Indonesia Government Regulation no. 82 year 2000 Article 47 and Article 55 on Animal Welfare
[4] Yuliana A, et al. 2016 Animal Welfare Implementation Appraisal on Cow Slaughtering Process at Slaughterhouse of Animal Mambal of Badung Regency *Indonesia Medicus Veterinus* p-ISSN: 2301-7848: eISSN: 2477-6637
[5] Government of the Republic of Indonesia Government Regulation No. 95 of 2012 on Welfare of Veterinary Society and Animal Welfare
[6] Rustam A 2016 *Conceptual Design of Livestock Special Port Development: Case Study of Kalbut Port Situbondo Thesis* (Surabaya: Institute of Technology Sepuluh Nopember)
[7] Kumalasari S D 2016 *Comparative Analysis of Inter-Cow Distribution Modes: Case Study of Nusa Tenggara Timur – Jakarta Thesis* (Surabaya: Institute of Technology Sepuluh Nopember)

Piper Ornatum and *Piper Betle* as Organic Dyes for TiO_2 and SnO_2 Dye Sensitized Solar Cells

Azwar Hayat[1], A. Erwin E. Putra[1], Novriany Amaliyah[1], Shuzi Hayase[2] and Shyam. S. Pandey[2]

[1]Mechanical Engineering Department, Faculty of Engineering, Hasanuddin University, Poros Malino Km. 6 Gowa, South Sulawesi, Indonesia.
[2]LSSE, Kyushu Institute of Technology, Japan, 2-4 Hibikino, Wakamatsu-ku, Kitakyushu-shi, Fukuoka, Japan.

E-mail: azwar.hayat@unhas.ac.id

Abstract. Dye sensitized solar cell (DSSC) mimics the principle of natural photosynthesis are now currently investigated due to low manufacturing cost as compared to silicon based solar cells. In this report, we utilized Piper ornatum (PO) and Piper betle (PB) as sensitizer to fabricate low cost DSSCs. We compared the photovoltaic performance of both sensitizers with Titanium dioxide (TiO2) and Tin dioxide (SnO2) semiconductors. The results show that PO and PB dyes have higher Short circuit current (Jsc) when applied in SnO2 compared to standard TiO2 photo-anode film even though the Open circuit voltage (Voc) was hampered on SnO2 device. In conclusion, from the result, higher electron injections can be achieved by choosing appropriate semiconductors with band gap that match with dyes energy level as one of strategy for further low cost solar cell.

1. Introduction

Solar energy is one of the greenest types of energy resources. It is free and abundant that makes this energy form one of candidate to replace fossil fuel. Silicon based solar cells have dominated the installed solar cell today. However, Silicon based solar cell required huge energy resources to purify the silicon and high production cost in order to achieve high efficiency cell. Low cost solar cells, such as Dye sensitized solar cells (DSSCs) are gaining more and more attention owing to their high efficiency and relatively lower cost of production [1]. A DSSC consist of two sandwiched electrodes, working and counter electrode, made from transparent conducting glass which mainly made of Fluorine doped tin oxide (FTO). Working electrode is made of thin layer of semiconducting material such as titanium dioxide (TiO_2), tin dioxide (SnO_2) Zinc oxide (ZnO) etc, coated with dyes as photo anode on top of FTO [2-9]. Counter electrode is made of FTO with thin layer of platinum or carbon [10-13]. Electrolyte is placed between both electrodes to connect and maintain electrons transfer. One of the main components to enhance the Photo-conversion efficiency (PCE) is dye sensitizer. The most widely used in DSSCs research area are Ruthenium based sensitizer. In the recent past Ruthenium based as well as metal free dyes have been reported having nearly quantitative photon harvesting mainly in the visible region of solar spectrum [14-17]. High quality ruthenium dyes are produced in sophisticated ways that increase the production cost of the DSSC. Most of sensitized dyes need anchoring groups in order to attach in nanoporous surface. Although attachment dyes on the nanoporous oxide semiconductor via a suitable anchoring group is common practice in DSSCs but introduction of such anchoring groups in dye structure at suitable position is synthetically cumbersome. In order to attach dyes on nanoporous oxide semiconductor surface, it could be achieved by several types of bonding, for example hydrophobic or electrostatic interactions and co-valent bonding [18]. Natural dye has widely used in textiles industry and recently in DSSC as sensitizer due to easier to prepare and low in cost [19-23]. In this works, the extracts of Piper ornatum and Piper betle as dye sensitizer in DSSC. The absorption spectra of PO and PB dyes solution in ethanol was measured with UV-VIS Spectrophotometer. The DSSC was made using two types of semiconductors which are TiO_2 and SnO_2. The current-voltage (I-V) characteristics ware measured along with Incident photon-to-current Efficiency (IPCE) to determine the photo-conversion efficiency and internal quantum efficiency.

2. Experimental

2.1. Dye preparation

The natural dyes used in this work extracted from Piper Ornatum (PO) and Piper Betle (PB) leafs. PO and PB fresh leafs was cleaned with distilled water and dried in shaded place for more than 3 weeks. After dried, the dried leaf ware crushed with Food grinder FTC-Z100 with output fineness 20-180 mesh. Powdered PO and PB ware diluted in ethanol. 1 gr of powdered leafs was immersed in 10 ml of ethanol and stored for 48 hours in dark place. After that, dye solution was filtered from remained powder using vacuum pump. Dye solution was kept in dark container to protect it from direct light exposure.

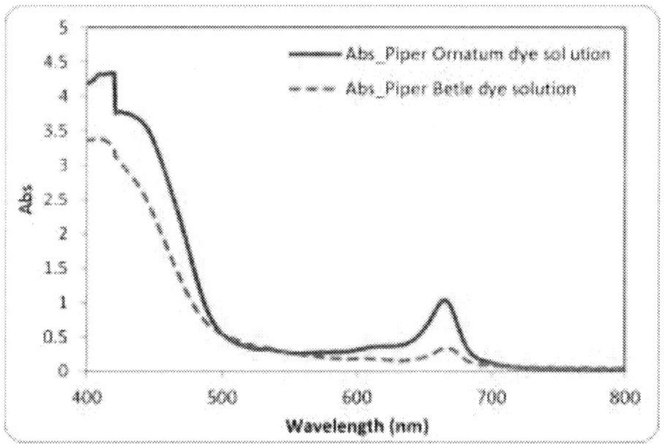

Figure 1. Absorption spectra of Natural dyes: Piper Ornatum in solution (red solid lines) and Piper Betle in solution (green dashed lines).

2.2. Preparation of dye sensitized solar cell

DSSCs ware fabricated with standard sandwich-like structure. Working electrode was prepared using Fluorine doped Tin Oxide (FTO) glass coated with 7 μM thick TiO_2 (30NRD) and SnO_2 followed by backing at 450°C for 30 minutes. Baked working electrode immersed in dye solution for 20 hours (figure 2). The cell area was 0.25 cm^2. The thickness of the spacer used to create a gap between working electrode and counter electrode was 25 μm (Solaronix Spacer). Counter electrode using Platinum (Pt) sputtered on FTO glass and electrolyte containing 500 mM LiI, 50 mM I_2, 600 mM Ethylmethylimidazolium dicyanoimide (MeEtIm-DCA), and 580 mM *tert*-butylpyridine (TBP) in Acetonitrile was used to fabricate the DSSCs. Natural dye solution in ethanol was used for sensitizer on the TiO_2 and SnO_2 surface. Photovoltaic characterization the DSSCs were made using a solar simulator (Bunko-Keiki Co. Ltd., Model Solar Simulator CEP-2000SRR) and black metal mask used to avoid the effect of optical reflection on photovoltaic measurement.

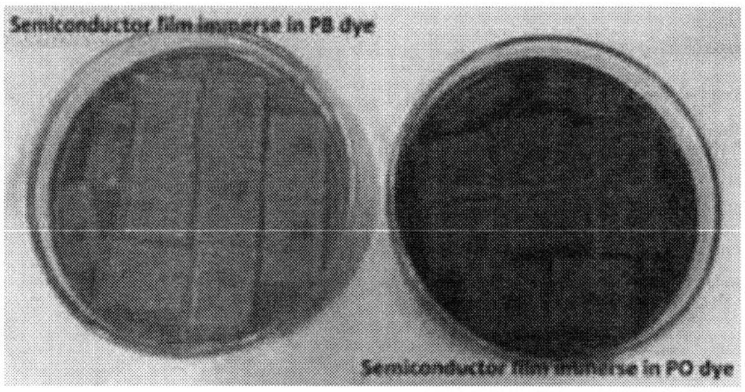

Figure 2. Immersion of semiconductor films on PO and PB dye.

3. Results and Discussions

Two kinds of Natural dyes (PO and PB) have been utilized as sensitizers in the present work. Two kind of semiconducting materials (TiO_2 and SnO_2) ware utilized to accommodate energy level of the dyes. Both PO and PB dyes used in the present work exhibits intense light absorption having absorption maximum at 667 nm and 671 nm in ethanol solution for the PB and PO dyes respectively. Light absorption spectra of dyes were characterized using UV-VIS spectrophotometer (Jusco V-670) to confirm the dye absorption (figure 1). Due to lower solubility of the natural dyes, it has been found that 48 hours are required to immerse in ethanol solution before proper usage in DSSC fabrication.

Table 1. I-V Characteristic of DSSCs with PO and PB dyes.

	PO-TiO_2	PO-SnO_2	PB-TiO_2	PB-$Sn0_2$
Efficiency [%]	0.032	0.022	0.034	0.014
FF	0.553	0.357	0.598	0.325
Voc[V]	0.463	0.238	0.455	0.202
Jsc[mA/cm^2]	0.124	0.261	0.125	0.207

Figure 3 exhibits the current-voltage characteristics of the DSSCs fabricated using these two dyes as sensitizer. It can be clearly seen that PO and PB exhibits relatively same PCE on TiO_2 as compared to the SnO_2 fabricated under similar experimental conditions. This phenomenon occurred maybe due to deeper LUMO (Low Unoccupied Molecular Orbital) of these dyes that suppress the electron injection to semiconductor conduction band. The main factor controlling the efficiency is the short circuit density (Jsc) since open circuit voltage (Voc) and fill factors (FF) are nearly the same for PO and PB for each semiconductor. Table 1 shows result of PO-TiO_2 have Jsc of 0.124 mA/cm^2, Voc 0.463 V, FF 0.553 and PCE 0.032% and PB-TiO_2 have Jsc of 0.125 mA/cm^2, Voc 0.455 V, FF 0.598 and PCE 0.034% for device with TiO_2 semiconductor. The results were comparatively higher that dye from same genus reported before utilizing Piper Nigrum which was exhibit only 0.009% of PCE [24].

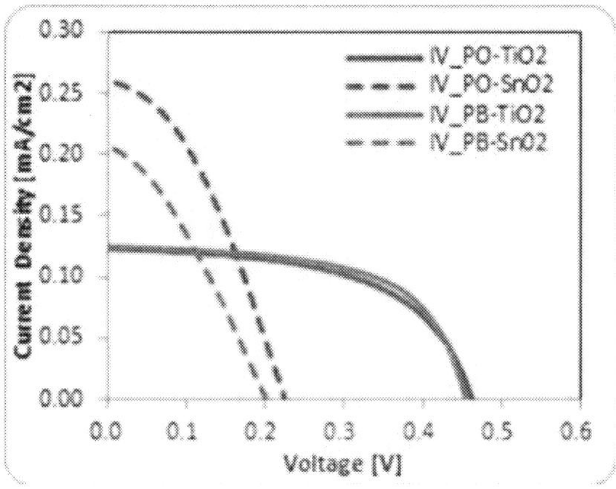

Figure 3. I-V Characteristic of DSSCs.

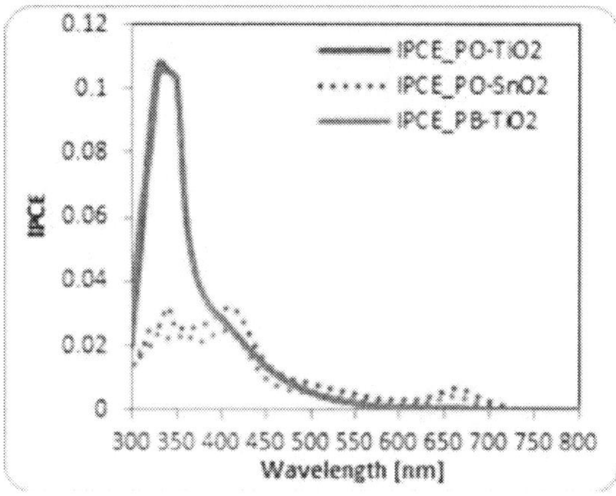

Figure 4. IPCE Spectra of DSSCs.

Device with SnO_2 semiconductor, PO-SnO_2 have Jsc of 0.261 mA/cm^2, Voc 0.238 V, FF 0.357 and PCE 0.022% and PB-SnO_2 have Jsc of 0.207 mA/cm^2, Voc 0.4202 V, FF 0.325 and PCE 0.014%. The results are interpreted in terms for facile dye regeneration in SnO_2 film was higher as compared to TiO_2 film due to its deeper conduction band. IPCE graph shown that photon to current density on SnO_2 film was higher. Higher electron injection was confirmed from peaks shown in 600 to 700 nm wavelength that agreed with absorption peak of the dyes shown in figure 4. Cells fabricated with SnO_2 film have lower Voc due to high traps on SnO_2 semiconductor lead to higher recombination that affect the Voc and FF.

4. Conclusions

DSSCs using Natural dyes Piper Ornatum and Piper Betle were fabricated for with two type semi conducting material Titanium dioxide and Tin dioxide. Fabricated cells produced relatively same PCE on TiO_2 film. Deeper conduction band of SnO2 compared to TiO2 lead to higher dye regeneration in SnO_2 film. IPCE graph confirmed that photon to current density on SnO_2 film was higher shown from peaks shown in 600 to 700 nm wavelength. Traps on SnO_2 semiconductor lead to higher recombination that affect the Voc and FF make both parameters lower compared on TiO_2 film. In conclusion, from the result, higher electron injections can be achieved by choosing appropriate semiconductors with band gap that match with dyes energy level as one of strategy for further low cost solar cell.

Acknowledgment

The work was part of Grand "PUPT 2017" from Indonesia Directorate General of Higher Education.

References

[1] Brian O and Grätzel M 1991 A Low-cost, High-efficiency Solar Cell Based on Dye Sensitized Colloidal TiO_2 Films *Nature* **353** 6346 737-740

[2] Chandiran A K, et al 2014 Analysis of Electron Transfer Properties of ZnO and TiO_2 Photoanodes for Dye-sensitized Solar Cells *ACS nano* **8** 3 2261-2268

[3] Ahn S H, et al 2013 One-Dimensional Hierarchical Nanostructures of TiO_2 Nanosheets on SnO_2 Nanotubes for High Efficiency Solid-State Dye-Sensitized Solar Cells *Advanced Materials* **25** 35 4893-4897

[4] Niu H, et al 2014 Dye-sensitized Solar Cells Employing a Multifunctionalized Hierarchical SnO_2 Nanoflower Structure Passivated by TiO_2 Nanogranulum *The Journal of Physical Chemistry C* **118** 7 3504-3513

[5] Thapa A, et al 2014 TiO_2 Coated Urchin-like SnO_2 Microspheres for Efficient Dye-sensitized Solar Cells *Nano Research* **7** 8 1154-1163

[6] Lv H, Xiaowei Y and Can C 2017 Enhancing the Photovoltaic Performance of Dye-sensitized Solar Cells by Modifying TiO_2 Photoanodes with Layered Structure g-C3N4 *AIP Conference Proceedings* **1820** 1

[7] Dongre J K 2017 A Comparative Study on the CDS Sensitized and Dye-sensitized ZnO Photoelectrodes Based on Solar Cell Performance *IJAR* **3** 3 504-507

[8] Abdalla J T, et al 2017 $TiCl_4$ Surface-treated SnO_2 Photoanodes for Self-powered UV Photodetectors and Dye-sensitized Solar Cells *Materials Technology* **32** 7 443-450

[9] Samsuri S A M, et al 2017 Influence of ZnO Growth Temperature on the Performance of Dye-sensitized Solar Cell Utilizing TiO_2-ZnO Composite Film Photoanode *Ionics* 1-12

[10] Ahmad I, et al 2014 Carbon Nanomaterial Based Counter Electrodes for Dye-sensitized Solar Cells *Solar Energy* **102** 152-161

[11] Zheng M, et al 2015 Flowerlike Molybdenum Sulfide/Multi-walled Carbon Nanotube Hybrid as Pt-free Counter Electrode Used in Dye-sensitized Solar Cells *Electrochimica Acta* **173** 252-259

[12] Ma J, et al 2014 3 D Single-Walled Carbon Nanotube/Graphene Aerogels as Pt-Free Transparent Counter Electrodes for High Efficiency Dye-Sensitized Solar Cells *ChemSusChem* **7** 12 3304-3311

[13] Arbab A A, et al 2017 A Complete Carbon Counter Electrode for High Performance Quasi Solid State Dye Sensitized Solar Cell *Journal of Power Sources* **343** 412-423

[14] Parisi A, et al 2017 Anomalous Performance Enhancement Effects in Ruthenium-based Dye Sensitized Solar Cells *Clean Electrical Power (ICCEP) 2017 6th International Conference on* IEEE

[15] Parisi A, et al 2017 Electro-optical Characterization of Ruthenium-based Dye Sensitized Solar Cells: A study of light soaking, ageing and temperature effects *Optik-International Journal for Light and Electron Optics* **135** 227-237

[16] Nagarajan B, et al 2017 Novel Ethynyl-pyrene Substituted Phenothiazine Based Metal Free Organic Dyes in DSSC with 12% Conversion Efficiency *Journal of Materials Chemistry A* **5** 21 10289-10300

[17] Chiu K Y, et al 2017 A New Series of Azobenzene-bridged Metal-free Organic Dyes and Application on DSSC *Dyes and Pigments* **146** 512-519

[18] Kalyanasundaram K and Grätzel M 1998 Applications of Functionalized Transition Metal Complexes in Photonic and Optoelectronic Devices *Coordination chemistry reviews* **177** 1 347-414

[19] Lim A, et al 2016 Co-dominant Effect of Selected Natural Dye Sensitizers in DSSC Performance *Spectrochimica Acta Part A: Molecular and Biomolecular Spectroscopy* **167** 26-31

[20] Jinchu I, et al 2016 Photovoltaic Parameters of DSSCs Using Natural Dyes with TiO_2 Nanopowder and Nanofiber as Photoanodes: A Comparative Study *Electrical, Electronics, and Optimization Techniques (ICEEOT) International Conference on* IEEE

[21] Teoli F, et al 2016 Role of pH and Pigment Concentration for Natural Dye-sensitized Solar Cells Treated with Anthocyanin Extracts of Common Fruits *Journal of Photochemistry and Photobiology A: Chemistry* **316** 24-30

[22] Sathyajothi S, Jayavel R and Dhanemozhi C A 2017 The Fabrication of Natural Dye Sensitized Solar Cell (DSSC) Based On TiO_2 Using Henna And Beetroot Dye Extracts *Materials Today: Proceedings* **4** 2 668-676

[23] Yoon S J, et al 2016 Deprotonated Curcumin as A Simple and Quick Available Natural Dye for Dye Sensitized Solar Cells *Energy Sources, Part A: Recovery, Utilization, and Environmental Effects* **38** 2 183-189

[24] Abdel-Latif M S, El-Agez T M, Taya S A, El-Ghamri H S and Batniji A Y 2015 Dyes Extracted

from Biota orientalis, Piper nigrum, and Glycyrrhiza glabra as Photosensitizers for Dye-Sensitized Solar Cells *International Journal of Renewable Energy Research (IJRER)* **5**(4) 1034-1040

IOP Conf. Series: Journal of Physics: Conf. Series **979** (2018) 012063 doi:10.1088/1742-6596/979/1/012063

The Isolation of Nanofibre Cellulose from Oil Palm Empty Fruit Bunch Via Steam Explosion and Hydrolysis with HCl 10%

S Gea, Z Zulfahmi, D Yunus, A Andriayani and Y A Hutapea

Department of Chemistry, Faculty of Mathematics and Natural Scinces, University of Sumatera Utara, Jl. Bioteknologi No. 1 Kampus Padang Bulan Medan 20155, Indonesia

E-mail: s.gea@usu.ac.id

Abstract. Cellulose nanofibrils were obtained from oil palm empty fruit bunch using steam explosion and hydrolized with 10% solution of HCl. Steam explosion coupled with acid hydrolysis pretreatment on the oil palm empty fruit bunch was very effective in the depolymerization and defibrillation process of the fibre to produce fibers in nanodimension. Structural analysis of steam exploded fibers was determined by Fourier Transform Infrared (FT-IR) spectroscopy. Thermal stability of cellulose measured using image analysis software image J. Characterization of the fibers by TEM and SEM displayed that fiber diameter decreases with mechanical-chemical treatment and final nanofibril size was 20-30 nm. FT-IR and TGA data confirmed the removal of hemicellulose and lignin during the chemical treatment process.

1. Introduction

Along with the production of palm oil, OPEFB is produced as a by-product of solids. OPEFB is a fibrous mass left after the separation of fruit due to sterilization (steam treatment at 294 kPa for 1 hour) [1]. It is estimated that for every ton of palm oil produced, yielding 1.1 tons of OPEFB approximately. In year 2011, the world production was 14.5 million tons (dry basis), and half of it was produced by Indonesia [2]. The thingthat can be done to overcome the production of OPEFB obtained from palm oil is through its traditional management. OPEFB is commonly used as a source of heat and driving the incinerator palm oil mills. However, this may cause environmental pollution [3]. In addition, there are other things that can be performed to overcome the production of OPEFB, such as; it can be used as materials for the development of science and further utilization that can increase the value of OPEFB. Given the development of nanotechnology, empty palm fruit bunches can be used as an ingredient to isolate the cellulosenanofiber.

Cellulose is a major component within the cell wall of all plants, and is a linear polymer of a poly-(1→4)-D-glucose unitwith a syndiotactic configuration [4]. The cellulose chain is organized into cellulosic microfiberssurrounded by a non-cellulose matrix within the cell wall of a plant [5]. Cellulosenanofiber is a very thin fiber; for example, the thickness of the nanofiber contained in the wood cell is 3-4 nm. Their characteristic structures allow them to be used in new applications such as reinforcing nanomaterialson plastics [6] drug delivery systems [7] biosensors, and packaging [8]. Steam explosion is an effective method for separating nanofiber from biomass [9]. High-pressure hot steam followed by rapid decompression is called a steam explosion. The steam explosive process involves saturation of a dry material with steam at high pressure and temperature, followed by a sudden release of pressure, during which the evaporation of water gives the thermo mechanical force causing the material to break. The steam explosive has been shown to be effective for a variety of ligno-cellulosic biomass, including hardwood, softwood, herbaceous residue and wheat straw [10]. In this case, biomass is heated for a short time with saturated vapor at high pressure, followed by sudden decompression. High-pressure steam provides a means to co-fund biomass, penetrate and modify cell wall structures, ensuring effective hydrolysis especially on hemicelluloses that makes cellulose easier to obtain. Decompression abruptly leads to an adiabatic expansion that makes the material is mechanically disrupted from modified fibers in plant cell walls [11, 12, 13].

The advantages of steam explosion are giving the significantly lower environmental impacts, lower energy consumption, lower capital investment, and less harmful chemical processes[9]. Bibin et al. (2010) have isolated cellulosic nanofibers from pineapple fibers by steam explosion methods[14]. Cherian et al. (2008) explained that the steam explosion method was an effective method for separating the nanofiber from biomass. The method of steam explosion with an alkaline medium resultedthe occurrence of hydrolysis of hemicellulose present in the fibers and produces a sugar which can then be washed with water, leaving a residue of cellulose with lignin[9]. Kaushik et al. (2011) reported that the method of steam explosion with alkaline medium and followed by a treatment with hydrochloric acid was an effective method for depolymerization and fiber defibrillation process to obtain the cellulosic nanofiber[15]. Based on research conducted by Bibin et al. (2010), cellulosic nanofibers have a broad uses in biomedical and biotechnology applications, such as tissue engineering, drug delivery systems, wound closure and medical implants [14].

Based on the above description, the researcher was interested to study the isolation of cellulose nanofiber by using steam explosion method from empty palm bunches and hydrolyzed with 10% HCl and then characterized by SEM and TEM morphology test, FT-IR functional group analysis and TGA thermal endurance test.

2. Research method

2.1. Equipments

In this study, the equipments used were a set of autoclave tools, a set of glassware, homogenizer, ultrasonic bath, analytical balance, thermometer, hotplate, magnetic stirrer, oven, universal pH, a set of TEM and SEM, a set of FT-IR spectroscopy and TGA machine.

2.2. Materials

The materials used in this study were including, Oil Palm Bunches, NaOH, $CH_3COOH_{(c)}$, $NAOCl_{(c)}$, $H2O2_{(c)}$, $HCl_{(c)}$ and distilled water.

2.3. Procedure

2.3.1. Preparation of empty palm bunches. Empty palm bunches were torned apart and dried then chopped into small pieces.

2.3.2. The treatment of OPEFB fiber with steam explosion. OPEFB fibers were immersed with 2% NaOH overnight and then NaOH was disposed after immersion process. Then, itwas added with 12% NaOH (with the ratio 1:10) and placed into an autoclave with temperature of 130°C and pressure of 168 kPa. This wasperformed to reduce the amount of lignin contained in the OPEFB fiber. The fibers were removed from the autoclave and washed with distilled water for several times until they were free from alkaline materials.

2.3.3. Bleaching process of fiber. The fibers that have been vaporized and liberated from the alkali were washed with a mixture of 17.5% NaOH solution and 7.4% acetic acid (1: 1 ratio) and with a 1: 3 mixture of NaOCl solution. Bleaching process was performed using $H_2O_2$8%. Then, the fibers were washed with distilled water and dried in the oven.

2.3.4. The acid hydrolysis of fiber. HCl 10% was added into the fiber that has been dried. Then, it was putted in ultrasonicator for 3 hours. After that, it was taken from ultrasonicator and washed until the fiber reached a neutral pH. Then, it was suspended in water and stirred using a homogenizer for 15 minutes until the fiber was agglomerated.

2.3.5. Morphological analysis. The morphological characterization of the resultant cellulosic nanofiber was determined using a TEM and SEM analysis equipments. In this case, the TEM equipment used was the JEM-1400 model. Images weretaken with the voltage acceleration of 120 kV. Nanofiber cellulose was a dilute liquid and dropped over a carbon coating. As for SEM analysis, the equipment used was ZEIZZ SEM. The analyzed samples were coated with platinum and analyzed using sputtering technique.

2.3.6. FTIR Spectroscopy analysis. This analysis was performed to investigate the functional group of fibers before the acid hydrolysis of cellulosic nanofibers. The FTIR spectra were determined using an FTIR ABB 3000 spectrometer and using a triglycerin sulfate detector. Meanwhile, the FTIR spectra obtained were determined in the 4000-450 cm-1 absorption area with a resolution of 8 cm-1 and using 32 scans for each sample.

2.3.7. Thermal analysis. The curve of the thermal analysis for fibers prior to the acid hydrolysis of cellulosic nanofiber was determined using a thermogravimetric analyzer (TGA) type SDT Q600 V20.9 Build 20. Samples were placed on a platinum cup sample. Nitrogen gas was used to the sample and heated at a rate of 10°C/ min from room temperature up to 650°C.

3. Results and Discussions

3.1. The results of cellulose nanofiber from OPEFB

Empty palm fruit bunches was soaked in clean water, dried and cutted into small pieces have been used as an ingredient in the manufacture of cellulosic nanofibers. The OPEFB fibers used were 50 g and the yield of cellulose nanofiber was 14.54 g. Table 1 shows the data obtained from the results of cellulose nanofiber with a sample of OPEFB. The result of cellulose nanofiber from OPEFB can be seen in figure 1.

Table 1. The weight of cellulose nanofiber obtained from the isolation of OPEFB.

Weight of OPEFB (g)	Weight of Cellulose Nanofiber (g)	*Yield* (%)
50	14,54	29,08

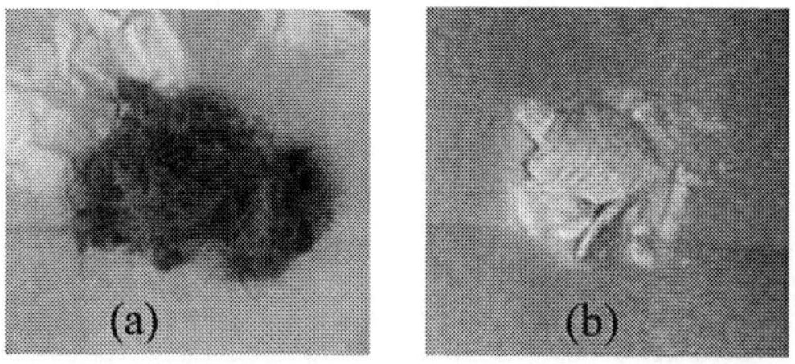

Figure 1. (a) OPEFB fibers (b) cellulosic nanofibers.

3.2. Cellulose nanofiber manufacture
The manufacture of cellulose nanofiber was done by inserting a soaked fiber with 12% NaOH into the autoclave device, then bleached with acetate buffer and NaOCl and using H_2O_2 as well. This was done to remove the remaining lignin and hemicellulose parts. Then, it was followed by hydrolysis treatment with 10% HCl. It was aimed to increase the degree of crystallinity and change fibers into nanoscale fibers. The acid-treated fibers were then suspended with water and stirred using a homogenizer device. This treatment was aimed to break down the pile (Agglomeration). The resulting cellulose nanofiber was analyzed using TEM (Transmission Electron Microscopy). The TEM result in figure 2 showing that the resulting fiber had a size of 20-30 nm and can be classified as nanofiber because its size was below 100 nm.

3.3. Transmission Electron Microscopy (TEM) analysis
Figure 2 shows the result of TEM analysis of cellulose nanofiber which has treated through chemical and mechanical treatment. Mechanical treatment caused the defibrillation of cellulose nanofiber in the cell wall and the TEM images showedthe separation of nanofiber from fiber bundles in micro scale size. In addition, the accumulating fibers may also be observed via Transmission Electron Microscopy (TEM). The average diameter of the nanofiber can be calculated from the electron micrograph by using digital image analysis software. Most of the particles were found in the area (range) diameter from 10-70 nm.

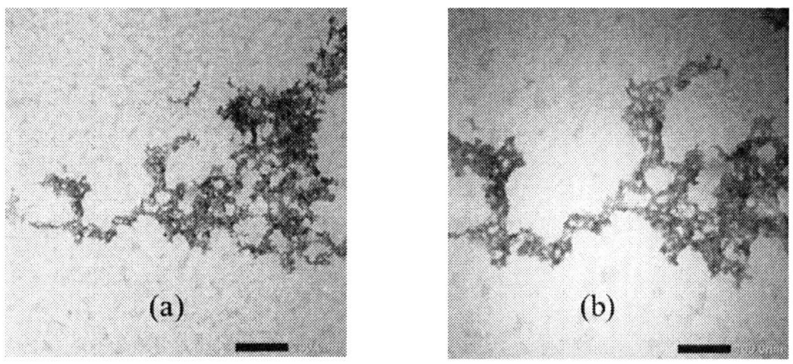

Figure 2. The results of TEM analysis with magnification of (a) 20000x and (b) 50000x.

Table 2 and figure 2 show the distribution of diameter dimensions of nanometric fibers after passing the final treatment. A total of 20% fiber had a diameter size of 20-30 nm. While 17.89 and 16.84% were fiber with diameter of 40-50 nm and 10-20 nm respectively. The rest were fibers with diameter size of 30-40 nm, 50-60 nm, and 60-70 nm.

Table 2. Distribution of OPEFB fiber diameter after chemical and mechanical treatment.

No	Diameter (nm)	Frequency (percentage)
1	10-20	16,84
2	20-30	20
3	30-40	13.68
4	40-50	17,89
5	50-60	11,57
6	60-70	9,48
7	>70	10,53

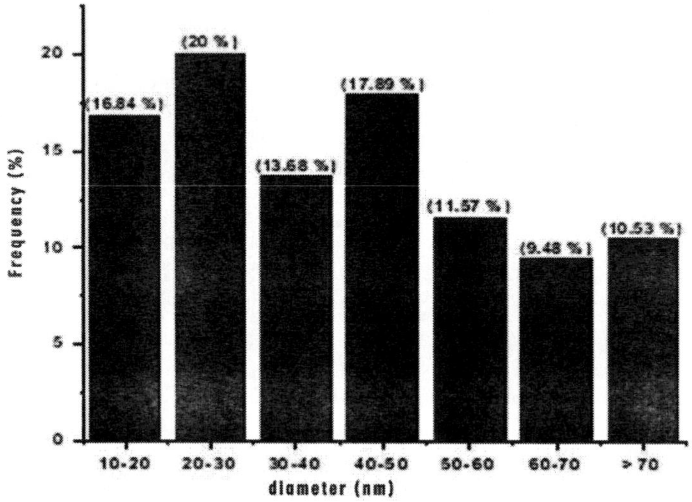

Figure 3. The chart of distribution of OPEFB fiber diameter after chemical and mechanical treatment.

3.4. Scanning Electron Microscopy (SEM)

The steam explosion method with an alkaline medium produced a structural change on the fiber surface. The results on the SEM image of the OPEFB after treated with the steam explosion process can be observed to see the structural changes. SEM micrograph was shown in figure 3.4. In this case it was clear that the fibers became separated individually because of the loss of hemicelluloses, lignin, and pectin after the chemical treatment, and they were adhesives around the bundles of fibers. In addition, the figure explained that the average diameter size of the fibers was about 10-20 μm, which was lower than the average size of the bundles of fibers prior to the chemical treatment. The reduced particle size due to the breakdown of hemicelluloses and lignin are also clearly seen in FT-IR data.

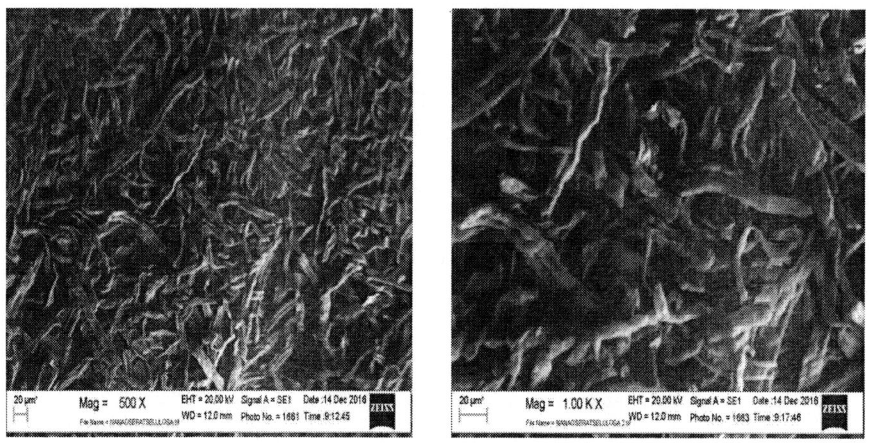

(a) (b)

Figure 4. The SEM results of cellulose nanofiber with (a) 500 and (b) 100,000 magnification.

3.5. Fourier Transform Infrared Spectroscopy analysis

The following discussion refers to the results of the study. The fibers, both before hydrolysis and cellulose nanofiber were analyzed with FT-IR to see the changes in the constituents or chemical constituents of the fiber before and after a chemical treatment (hydrolysis) and mechanics. Figure 3.5 shows the FT-IR spectrum of fibers prior to the treatment of hydrolysis and cellulose nanofibers. The peak in the region of 3348 cm^{-1} corresponds to the O-H stretching that is, due to the vibration of the hydrogen attached to the hydroxyl group. The peak at 2916 cm^{-1} shows vibrational stretching of the lignin polysaccharide (Cellulose and hemicelluloses) in the form of saturated aliphatic C-H. The tendency of hydrophilic fiber before the hydrolysis and cellulose nanofiber is reflected on the occurence of broad band around the 3700-3100 cm^{-1}, which shows the O-H group exists in their main component. The Peak area of 1715 cm^{-1} on fiber before the hydrolysis was caused by the presence of acetyl groups from hemicelluloses and uronat ester linkage or ester of the carboxylic group on ferulic acid and p-coumaric acid from lignin or hemicelluloses.

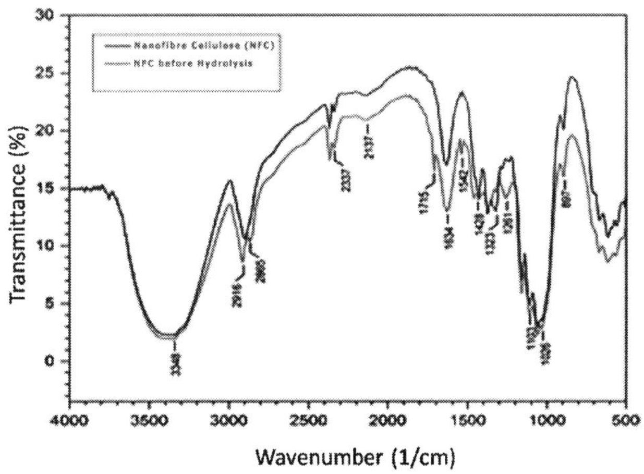

Figure 5. The comparison of FTIR result of OPEFB before and after the hydrolysis process.

It can also be seen in figure 5 that this peak is virtually absent in the cellulosenanofiber spectrum, indicating the breaking of the ester bond. At the peak of the 1634 cm^{-1} region there is a possibility of bending due to the absorption of water and the contribution of the carboxylate group [15]. The C=C aromatic stretching of the aromatic ring of lignin gave two peaks at 1542 cm^{-1} and 1428 cm^{-1}. It can be observed in the fiber before the hydrolysis treatment [16]. Then, cellulosenanofibers peak of 1542 cm^{-1} has almost disappeared and the intensity at peak of 1428 $cm^{-1}$was significantly decreased. This corresponded to the loss of the lignin portion of the fiber. The peak area of 1323 cm^{-1} represents an asymmetrical C-H shape change. In the cellulose nanofiber, the peak intensity in the area of 1261 cm^{-1} has diminished in its sharpness, indicating the loss of hemicelluloses. The area between 1103-1026 cm^{-1}represents the C-O stretch and band changes in cellulose, lignin and hemicelluloses residues. The band on the 897 cm^{-1} region indicates the type of structure of cellulose (caused by the β-glycoside bond of the glucose ring on cellulose) [15].

3.6. Thermogravimetric Analysis (TGA)

The result of TGA for Oil Palm Empty Fruit Bunch (OPEFB) with different stages is shown in figure 6. This result clearly shows that the thermal stability of OPEFB fiber increases after chemical and mechanical treatment. The degradation temperature in OPEFB fibers increased after chemical and mechanical treatment. This is caused by the amount of non-cellulosic materials that have been lost and the high structural degrees can be well maintained. In this case, it explains the relationship between the structure and thermal degradation in cellulose, where a large crystallinity structure requires a higher degradation temperature.

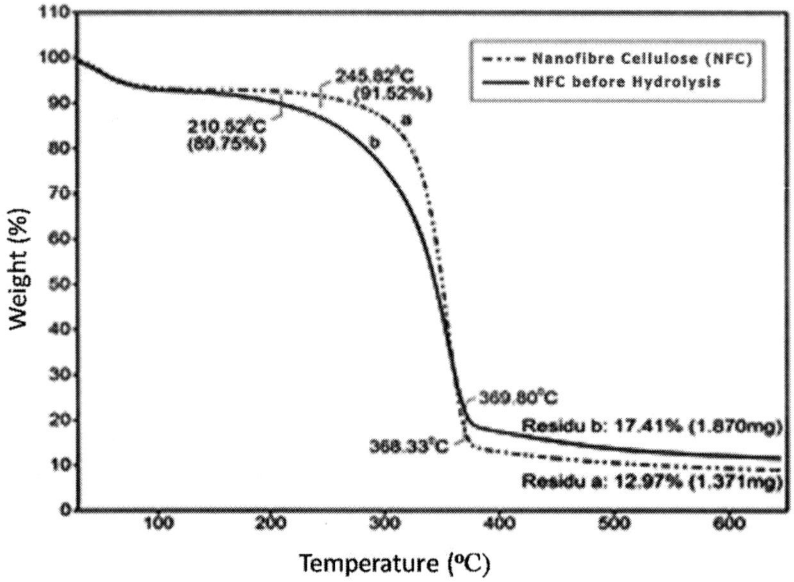

Figure 6. The TGA curve of: (a) cellulosic nanofiber (b) fiber prior to hydrolysis treatment.

In addition, there is a difference in the amount of residue produced from the residual fibers after heating up to 600°C against the untreated fibers and nanofibers that have undergone the chemical and mechanical treatment. The maximum residue produced on fiber before hydrolysis treatment was 17.41% (1.870 mg). While the cellulose nanofiber produced as much as 12.97% (1.371 mg) residue. This happened due to the high temperatures that cause the occurrence of thermal decomposition and reduced residual mass of the fibers produced after chemical-mechanical treatment. This occurs due to the missing parts of hemicelluloses and lignin in fiber and high degree of crystallinity in cellulose. These results also relate to the results of FT-IR measurements.

4. Conclusion

The process of steam explosion can break up the bundles of fibers, and with the alkaline medium, it will remove some of the lignin in the OPEFB fiber. The hydrolysis treatment using 10% HCl can increase the degree of crystallinity in the fiber and forming a fiber with nanoscale diameter size. The use of autoclave with a temperature of 130°C and pressure of 168.9 kPa was a proper treatment, because high temperature and pressure can affect the fiber of OPEFB, namely the defibrillation of the fiber. SEM analysis results showed the change of structure in fiber. In addition, TEM analysis showed a fiber diameter of 20-30 nm. Moreover, the FT-IR data proved that the release of lignin and hemicelluloses in the fiber and resulting in thermal stability in cellulose nanofiber to increase compared to the untreated fiber.

Acknowledgement

Authors would like to express gratitude to Ministry of Research, Technology, and Higher Education of Indonesia for funding this research via the funding scheme of "Penelitan Unggulan Universitas 2017".

References

[1] Shinoj S, Visvanathan R, Panigrahi S and Kochubabu M 2011 *Industrial Crops and Products* **33** 1 7–22
[2] Purwandari F A, et al. 2013 *Bioresour. Technol.* **128** 461–6
[3] Cotana F, et al. 2014 *Energy Procedia* **45** 52–60
[4] Dufresne A, Cavaille J Y and Vignon M R 1997 *J. Appl. Polym. Sci.* **64** 6 1185–94

[5] Frey-Wyssling A 1954 *Science* **119** 80-2
[6] Yano H, et al. 2005 *Adv. Matterials* **17** 2 153-5
[7] Roman M, Dong S, Hirani A and Lee Y W 2010 *Polysaccharide Materials: Performance by Design* 81–91
[8] De Azeredo H M C 2009 *Food Res. Int.* **42** 9 1240–53
[9] Cherian B M, Pothan L A, Nguyen-Chung T, Mennig G, Kottaisamy M and Thomas S 2008 *J. Agric. Food Chem.* **56** 14 5617–27
[10] Fernandes M C et al. 2015 *Bioresour. Technol.* **186** 309–15
[11] Hendriks A T W M and Zeeman G 2009 *Bioresour. Technol.* **100**(1) 10–18
[12] Mosier N et al. 2005 *Bioresour. Technol.* **96** 6 673–86
[13] Ramos L P 2003 *Quimica Nova* **26** 6 863–71
[14] Cherian B M, Leão A L, De Souza S F, Thomas S, Pothan L A and Kottaisamy M 2010 *Carbohydr. Polym.* **81** 3 720–5
[15] Kaushik A and Singh M 2011 *Carbohydr. Res.* **346** 1 76–85
[16] Alemdar A and Sain M 2008 *Bioresour. Technol.* **99** 6 1664–71

The Effect of Synthesis Temperature on Physical and Magnetic Properties of Manganese Ferrite ($MnFe_2O_4$) based on Natural Iron Sand

E A Setiadi[1], Rahmat[2], S Simbolon[1], M Yunus[1], C Kurniawan[1], A P Tetuko[1], S Zelviani[2], Rahmaniah[2] and P Sebayang[1]

[1] Research Center for Physics, Indonesian Institute of Sciences (LIPI), Komplek Puspiptek, Serpong, Tangerang Selatan, Banten 15314, Indonesia.
[2] Department of Physics, Alauddin State Islamic University of Makassar, Jalan Sultan Alauddin No. 63, Makasar, Indonesia.

E-mail: eko.arief.setiadi@lipi.go.id

Abstract. The magnetic material of Manganese Ferrite ($MnFe_2O_4$) based on natural Iron sand have been successfully prepared by using co-precipitation method. The natural Iron sand, $MnCl_2.4H_2O$, HCl and NH_4OH were used as raw materials to synthesize the $MnFe_2O_4$. The synthesis was carried out at various temperatures of 70, 100 and 130°C, respectively. The physical and the magnetic properties of the samples were analysed by using X-Ray Diffraction (XRD), Scanning Electron Microscopy (SEM) and Vibrating Sample Magnetometer (VSM). The diffraction pattern indicates that the $MnFe_2O_4$ is the dominant phase. The crystallite size tends to increase as the synthesis temperature increased while the saturation magnetization of $MnFe_2O_4$ tends to decrease.

1. Introduction

Ferrite is soft magnetic material that have spinel stucture. The formula of ferrite is M-Fe_2O_4 where M is Cu, Zn, Ni, Co, Fe, Mn, and Mg. Ferrite has high permeability and thermal resistance and low coercivity [1]. Soft ferrite can be applied in many applications, such as the immobilization and the separation of proteins or enzymes, and drug administration or purification of Deoxyribonucleic Acid (DNA) [2]. Ferrite also can be used in drugs delivery, hyperthermia treatment, magnetic resonant imaging, cell membranes manipulation, biosensors, bio-labelling and metal ions adsorption [3,4,5]. In their research, Lee *et al.* [6] reported that the magnetic suseptibility of $MnFe_2O_4$ (manganese ferrite, 5 μ_b) was higher than others substances, such as $CoFe_2O_4$ and $NiFe_2O_4$. This can cause the magnetic moment of manganese ferrite correspond to the Neel's coupling scheme [7]. Beside that, the resistivity of $MnFe_2O_4$ is much lower than $CoFe_2O_4$ and $NiFe_2O_4$ [8]. The $MnFe_2O_4$ also have higher biocompatibility than Fe_3O_4, γ-Fe_2O_3, $CoFe_2O_4$, and $NiFe_2O_4$ to be applied as *magnetic resonance imaging* (MRI) [9].

The study of manganese ferrite continues to grow in recent decades. Many methods were developed to synthesize $MnFe_2O_4$ that resulted in various grain size, e.g. 4 nm at 320°C to 154.1 nm at temperature 420°C [10,11]. Generally, co-precipitation is often used for synthesizing $MnFe_2O_4$ because the method is simple, and only requires a low temperature reaction [12]. Co-precipitation method is a chemical process that simultaneously precipitate the normally soluble component with a macro-component from the solution. This is often used to separate the solution from the impurities.

In this research, $MnFe_2O_4$ nanoparticles will be synthesized from natural iron sand by using co-precipitation method. The synthesis of $MnFe_2O_4$ was conducted using temperature variations of 70, 100, 130 °C. In order to find the characteristic of $MnFe_2O_4$ nanoparticles, the sample will be analysed by using XRD, SEM and VSM to determine their physical properties, morphology and magnetic properties.

2. Experimental Method

The raw materials used in the synthesis of $MnFe_2O_4$ were natural iron sand from Jeneberang river (Gowa district, Sulawesi Selatan) and $MnCl_2.4H_2O$. The synthesis process was done using co-precipitation method by mixing 8 g of iron sand, 1.017 g of $MnCl_2.4H_2O$ and 25.5 mL of HCl (32%) in the 24.5 mL of aquades until the homogeneous solution obtained. Then, the solution was added dropwise into 100 mL of 2 M NH_4OH, followed by stirring the mixture of solution at 250 rpm for 2 hours. The synthesis temperatures were varied at 70, 100 and 130°C which was referred to T70, T100, and T130 sample, respectively.

After the precipitation of each sample was formed, the precipitation was separated from the solution. Then, the precipitate sample was washed by using aquades to purify the residue salt. Then, $MnFe_2O_4$ powder sample was dried in the oven at temperature of 70°C. $MnFe_2O_4$ sample was characterized by using X-ray Diffraction (XRD – Rigaku SmartLab) at a wavelength ($\lambda_{CuK\alpha}$) of 1.5418 Å, Scanning Electron Microscopy (SEM - Hitachi SU 3500), Vibrating Sample Magnetometer (VSM - Electromagnetic VSM250) to obtain the physical and magnetic properties.

3. Results and Discussion

The X-ray diffraction pattern of $MnFe_2O_4$ sample is shown in figure 1. The crystallography database of $MnFe_2O_4$ show that the peaks of each three samples is related to $MnFe_2O_4$ phase (cubic spinel crystal structure). Based on the analysis result using XRD, the lattice parameter value, crystallite diameter and dislocation density of sample can be calculated using equations 1-3 [13] and the calculated parameters are shown in table 1.

$$a^2 = \frac{\lambda^2}{4\sin^2\theta}(h^2 + k^2 + l^2) \quad (1)$$

$$D = \frac{0,89\lambda}{B\cos\theta} \quad (2)$$

$$\delta = \frac{1}{D^2} \quad (3)$$

where a is lattice parameter, D is crystallite diameter, δ is dislocation density, θ is diffraction angle, hkl is Miller index, K is Scherrer constant (0.89), λ is wavelength of the x-ray and B is full width at half maximum (FWHM).

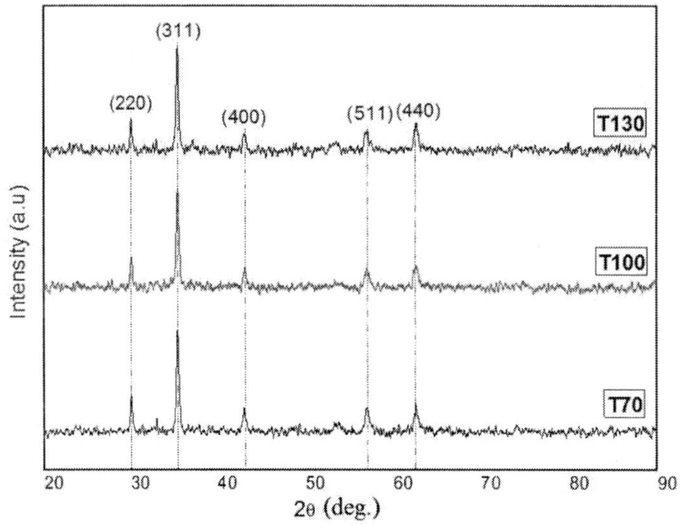

Figure 1. XRD pattern of $MnFe_2O_4$ (synthesis temperature variations of 70, 100 and 130°C).

Table 1. Lattice parameter, Crystallite diameter and dislocation density of $MnFe_2O_4$.

Sample	Lattice parameter α (nm)	Crystallite diameter D (nm)	Dislocation density δ (10^{15}line/m^2)
T70	0.84	26.37	1.44
T100	0.84	27.06	1.37
T130	0.84	29.80	1.13

Based on the analysis result, the three samples have an identical lattice parameter (0.84 nm). This lattice parameter value has a good agreement compared to the database value of $MnFe_2O_4$ (0.85 nm). This indicates that the sample of $MnFe_2O_4$ is well formed. The peaks of the $MnFe_2O_4$ suggest that higher synthesis temperature can increase the diffraction peak of $MnFe_2O_4$, thus larger crystallite diameter and unit cell volume could be obtained. This occurs due to the higher synthesis temperature that increases the crystal growth, thus larger diameter crystallite was formed and reduced the dislocation density [13,14].

The magnetic properties analysis are shown in figure 2 and table 2. The results suggested that the magnetisation saturation of $MnFe_2O_4$ in three different samples: T70, T100 and T130 are 38.75, 37.45, and 33.22 emu/g, respectively. The results shows that the magnetization saturation is inversely proportional to the grain size of the particle where the smallest grain size have a magnetic moment of $MnFe_2O_4$ that tends to be more unstable. The instability of the magnetic moment on the particles with smaller grain size caused by the anistropic energy possessed by the particles is much smaller than that of large particles. As a result when an external magnetic field is given, a magnetic moment with smaller grain size will be more reactive in response to the given external field, thus the magnetic moment is easily to be switched [14]. Based on the characterization using VSM, various coercivity values have been obtained: 233.86, 172.41, and 221.71 Oe for the T70, T100, and T130 samples, respectively.

The T70 sample shows the highest value compared to the other samples (T100 and T130), this indicates that the absence of the conformity with the grain size of $MnFe_2O_4$. The magnitude of the external magnetic field required by the magnetized particles to return to zero is directly proportional to the size of the resulting particles that also affect the coercivity value. However, in this research, the T70 sample has the highest coercivity value caused by the agglomeration effect.

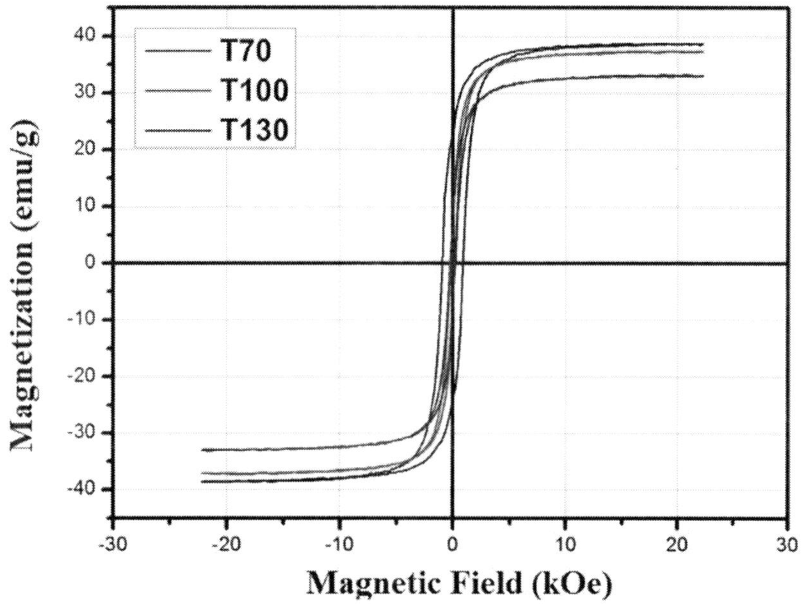

Figure 2. Hysteresis curve of $MnFe_2O_4$ (synthesis temperature variations of 70, 100 and 130°C).

Table 2. Magnetic properties of $MnFe_2O_4$.

Sample	Coercivity Hc (Oe)	Remanence Mr (Oe)	Magnetization saturation Ms (emu/g)
T70	233,86	7,70	38,75
T100	172,41	6,59	37,45
T130	221,71	6,90	33,22

The morphology analysis of the T70 sample using SEM-EDX spectrum is shown in figure 3. The results suggested that the particles tend to be round in agglomerate condition. This agglomeration occurs as a result of the interaction forces between the surfaces of the magnetic particles. This can be the cause of higher coercivity value in T70 sample compared to other samples (T100 and T130).

Based on the diameter measurement using SEM-EDX, the particle size of 26.45 nm was obtained. This is consistent with the calculation of the crystallite diameter using the scherrer equation (26.37 nm). Based on the EDX spectrum showed that the T70 sample consists of three elements: Mn, Fe, and O that have weight percentages of 6.97, 52.81 and 40.22 wt%, respectively. This corresponds to a sample molecular formula composed of the three elements.

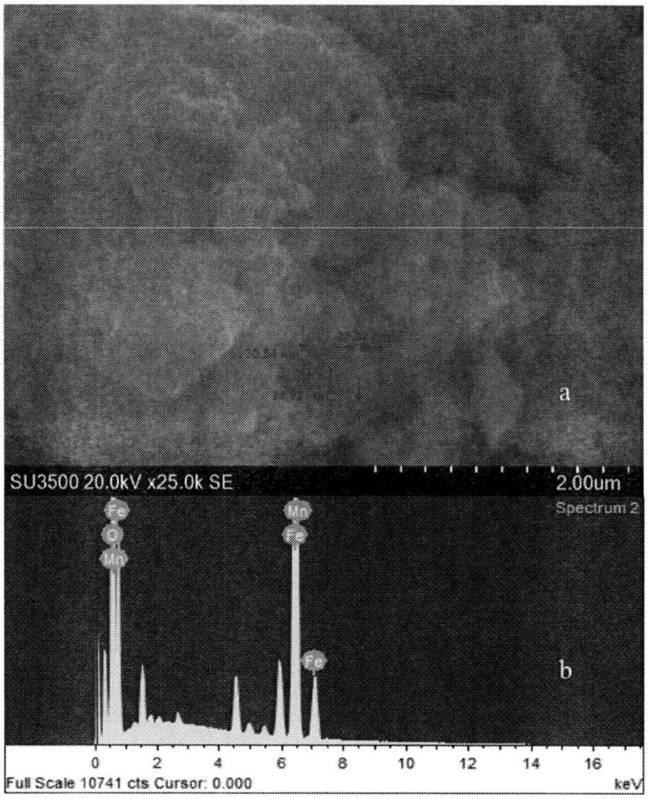

Figure 3. SEM imaging of $MnFe_2O_4$ sample T70: (a) morphology (b) EDX spectrum.

Table 3. Element of $MnFe_2O_4$ for sample T70.

Element	Weight (%)
Fe	52.81
O	40.22
Mn	6.97
Total	100

4. Conclusion

$MnFe_2O_4$ based natural iron sand from the Jeneberang river, Sulawesi Selatan has been successfully synthesized using co-precipitation method with various temperatures: 70, 100 and 130°C. XRD pattern suggested that the sample have a dominant phase of $MnFe_2O_4$. The physical properties of the sample indicated that the crystallite diameter increase while the dislocation density decrease. The magnetic properties of the sample suggested that lower synthesis temperature have higher magnetization saturation. The morphological analysis indicated that the $MnFe_2O_4$ powder samples tend to be round and agglomerate.

Acknowlegment

The authors acknowledge The Research Center for Physics, Indonesian Institute of Sciences (LIPI) for the financial support and the characterization facility.

References

[1] Maity D, Agrawal D C 2007 *Journal of Magnetism and Magnetics* **308** 46-55

[2] Saoud F 2010 *Superparamagnetic Nanoparticles for Synthesis and Purification of Polymers Prepared Via Controlled Living Radical Polymerization (CLRP)* (Stellenbosch: The University of Stellenbosch)
[3] Feng B, Hong R Y, Wang L S, Guo L, Li H Z Ding J Zheng Y and Wei D G 2008 *Colloids Surfaces A Physicochem. Eng. Asp.* **328** 52
[4] Anbarasu M, Anandan M, Chinnasamy E, Gopinath V and Balamurugan K 2015 *Spectrochim. ACTA Part A Mol. Biomol. Spectrosc.* **135** 536
[5] Larraza I, López-gónzalez M, Corrales T and Marcelo G 2012 *J. Colloid Interface Sci.* **38** 24
[6] Lee J H, Huh Y M, Jun Y W, Seo J W, Jang J T, Song H T, Kim S J, Cho E J, Yoon H G, Suh J S, and Cheon J 2006 *Nature Me* **13** 95
[7] Takadate K, Yamamoto Y, Makino A, Yamaguchi T and Sasada I 1998 J. *Appl. Phys.* **83** 6854.
[8] Lotger F K 1964 *J. Phys. Chem. Solids* **25** 345
[9] Tomsdorf U I, Bigall N C, Kaul M G, Bruns O T, Nikolic M S, Mollwitz B, Sperling R A, Reimer R, Hohenberg H, Parak W J, Förster S, Beisiegel U, Adam G, and Wellder H 2007 *NanoLett* **7** 2422
[10] Nikumbh A K, Nagawade A V, Tadke V B and Bakare P P 2001 *J. Mater. Sci.* **36** 653
[11] Ahmed M A, Okasha N, El-Dek S I 2008 *Preperation and Characterization of nanometricmn ferrite via Different Methode, Nanotechnology* (Egypt: Cairo University)
[12] Setiadi E A, Sebayang P, Ginting M, Sari A Y, Kurniawan C, Saragih C S and Simamora P 2016 *Journal of Physics: Conference Series* **776** 012020
[13] Shah S Z H, Khan U, Riaz S and Naseem S 2015 *Proc. Materials Today: Proceedings 2* 5214
[14] Setiadi E A, Amriani F and Sebayang P 2017 *Proc. AIP Conf.* **1904**, 020039-1–020039-7

Edge Antimagic Total Labeling on Two Copies of Path

Nurdin, A M Abrar, A R M Bhayangkara, Muliani, A U Samsir and M R An Nahdi

Department of Mathematics, Faculty of Mathematics and Natural Sciences, Hasanuddin University, Indonesia

E-mail: ahmad.muchlas312@gmail.com

Abstract. A graph G = (V(G), E(G)) denotes the vertex set and the edge set, respectively. A (p,q)-graph G is a graph such that $|V(G)| = p$ and $|E(G)| = q$. Graph of order p and size q is called (a,d)-edge-anti magic total if there exists a bijection f: V(G) ∪ E(G)→$\{1,2,..., p + q\}$ such that the edge weights $w(u,v) = f(u) + f(uv) + f(v)$ form an arithmetic sequence $\{a, a + d, a + 2d,...,a + (q - 1)d\}$ with the first term a and common difference d. Two copies of path is disjoint union of two path graph with same order ($P_n \cup P_n$) denoted by $2P_n$. In this paper we construct the (a,d)-edge-anti magic total labeling in two copies of path for some differences d.

1. Introduction

In 1967, the concept of graph labeling was introduced by Rosa [1]. Graph labeling is an assignment of integers to the edges or vertices or both subject to certain conditions. Labeled graphs serve as useful models in broad range of applications such as circuit design, communication network addressing, X-ray crystallography, radar, astronomy, data base management and coding theory. Graph labeling have found application in partitioning complete graphs into isomorphic subgraphs, in combinatorial and algebraic structures such as cyclic difference sets, characterizing neo-fields, generalized and near-complete mappings [2,3]. In most applications labels are positive (or nonnegative) integers, though in general real numbers could be used [4]. Over the past three decades various labeling of graphs such as cordial labeling, prime labeling, binary labeling, magic labeling, anti-magic labeling, bi-magic labeling, mean labeling, arithmetic labeling, graceful labeling, harmonious labeling etc., have been investigated in the literature [5].

Simanjuntak, et al introduced (a,d)-edge-antimagic total labeling. They proved that C_{2n} has $(2n+2, 1)$- and $(3n+2,1)$-edge-antimagic total labeling; C_{2n} has $(4n+2,2)$- and $(4n+3,2)$-edge-antimagic total labeling; C_{2n+1} has $(3n+4,3)$- and $(3n+5,3)$-edge-antimagic total labelings; P_{2n+1} has $(3n+4,2)$-$(3n+4,3)$-, $(2n+4,4)$-, $(5n+4,2)$-, $(3n+5,2)$- and $(2n+6,4)$-edge-antimagic total labelings; P_{2n} has $(6n,1)$- and $(6n+2,2)$-edge-antimagic total labelling [6].

A labeling of graph G is any mapping that sends some set of graph elements to a set of non-negative integers. If the domain is the vertex set or the edge set, the labelings are called vertex labelings or edge labelings, respectively. Moreover, if the domain is $V(G) \cup E(G)$ then the labeling is called a total labeling [7].

Let f be a vertex labeling of a graph G. Then we define the edge-weight of $uv \in E(G)$ to be $w(uv) = f(u) + f(v)$. If f is a total labeling then the edge-weight of uv is $w(uv) = f(u) + f(uv) + f(v)$.

An (a,d)-edge-anti magic total labeling of a (p,q)-graph G is a bijective function $f: V(G) \cup E(G) \rightarrow \{1,\ldots, p + q\}$ with the property that the edge-weights $w(uv) = f(u) + f(uv) + f(v)$, $uv \in E(G)$, form an arithmetic progression $\{a, a+d, a+2d, \ldots, a+(q-1)d\}$, where $a > 0$ and $d \geq 0$ are two fixed integers. If such a labeling exists then G is said to be an (a,d)-edge-antimagic total graph[7].

Two copies of path is disjoint union of two path graph with same order $(P_n \cup P_n)$ denoted by $2P_n$. Though (a,d)-edge anti magic total labeling have been studied for different kinds of graphs, (a,d)-edge anti magic total labeling for two copies of path have not been investigated. In this paper we construct the (a,d)-edge anti magic total labeling on two copies of path for some differences d.

2. Main Result

Let $G = G(V,E)$ be a finite, simple and undirected graph. By a labeling we mean one-to-one mapping that carries a set of graph elements onto a set of numbers, called labels (usually the set of integers). In this paper we deal labeling with domain as the set of all vertices and egdes [1].

Lemma 1. *Let G be a finite,simple,and undirected graph with p vertices and q edges. If G has (a,d)-edge-antimagic total labeling then*

$$d \leq \frac{3p+3q-9}{q-1}$$

Proof.
Assume that G has (a,d)-edge-antimagic total labeling $f : V(G) \cup E(G) \rightarrow \{1,2,\ldots,p+q\}$ and the set of edge-weights $w(u,v) = f(u) + f(uv) + f(v)$ form an arithmetic sequence $\{a, a+d, a+2d, \ldots, a+(q-1)d\}$. The minimal edge-weight is 6. Consequently $a \geq 6$. On the other hand, the maximum edge-weight is $p+q-2+p+q-1+p+q=3p+3q-3$. Thus, $a+(q-1)d \leq 3p+3q-3$.

Using the last inequality and fact that $a \geq 6$, we have

$$d \leq \frac{3p+3q-9}{q-1}$$

From now, every edge $v_{i,j}, v_{i+1,j} \in E(2P_n)$ will be denoted by $e_{i,j}$. Let $2P_n$ be a graph with $p = 2n$ and $q = 2(n-1)$. According to the inequality above, the following observation is hold.

Observation 1. *If path $2P_2$ has an (a,d)-EAT labelling, then $d \leq 9$.*

Theorem 1. *For $2P_2$, has (10,1)-EAT, (9,3)-EAT, (8,5)-EAT, (7,7)-EAT, and (6,9)-EAT labelings.*

Proof. Let $2P_2$ be a graph with four vertices. Let $\{v_{1,1}, v_{2,1}, v_{2,1}, v_{2,2}\}$ be a set of vertices of $2P_2$. Label the vertices and the edges of $2P_2$ as follow :

$$f_1(v_{i,j}) = \begin{cases} 1 & \text{if } i,j = 1,1 \\ 2 & \text{if } i,j = 1,2 \\ 5 & \text{if } i,j = 2,1 \\ 6 & \text{if } i,j = 2,2 \end{cases},$$

$$f_1(e_{i,j}) = \begin{cases} 4 & \text{if } i,j = 1,1 \\ 3 & \text{if } i,j = 1,2 \end{cases}.$$

We can see that f_1 is a bijection and the edge weights are :

$$wt(e_{1,1}) = f_1(v_{1,1}) + f_1(v_{2,1}) + f_1(e_{1,1}) = 10,$$

$$wt(e_{1,2}) = f_1(v_{1,2}) + f_1(v_{2,2}) + f_1(e_{1,2}) = 11.$$

The edge weights form an arithmetic sequence with difference 1. Thus, f_1 is a $(10,1)$-EAT labeling for $2P_2$.

Define a new labeling f_2 for the vertices and the edges of $2P_2$ as follow :

$$f_2(v_{i,j})=\begin{cases}1 & if\ i,j=1,1\\ 2 & if\ i,j=1,2\\ 3 & if\ i,j=2,1\\ 4 & if\ i,j=2,2\end{cases}$$

$$f_2(e_{i,j})=\begin{cases}5 & if\ i,j=1,1\\ 6 & if\ i,j=1,2.\end{cases}$$

We can see that f_2 is a bijection and the edge weights are :

$$wt(e_{1,1})=f_2(v_{1,1})+f_2(v_{1,2})+f_2(e_{1,1})=9,$$

$$wt(e_{1,2})=f_2(v_{1,2})+f_2(v_{2,2})+f_2(e_{1,2})=12.$$

The edge weights form an arithmetic sequence with difference 3. Thus, f_2 is a $(9,3)$-EAT labeling for $2P_2$.

Define a new labeling f_3 for the vertices and the edges of $2P_2$ as follow :

$$f_3(v_{i,j})=\begin{cases}1 & if\ i,j=1,1\\ 2 & if\ i,j=1,2\\ 4 & if\ i,j=2,1\\ 6 & if\ i,j=2,2\end{cases},$$

$$f_3(e_{i,j})=\begin{cases}3 & if\ i,j=1,1\\ 5 & if\ i,j=1,2\end{cases}$$

We can see that f_3 is a bijection and the edge weights are :

$$wt(e_{1,1})=f_3(v_{1,1})+f_3(v_{2,1})+f_3(e_{1,1})=8,$$

$$wt(e_{1,2})=f_3(v_{1,2})+f_3(v_{2,2})+f_3(e_{1,2})=13.$$

The edge weights form an arithmetic sequence with difference 5. Thus, f_3 is a $(8,5)$-EAT labeling for $2P_2$.

Define a new labeling f_4 for the vertices and the edges of $2P_2$ as follow:

$$f_4(v_{i,j})=\begin{cases}1 & if\ i,j=1,1\\ 3 & if\ i,j=1,2\\ 4 & if\ i,j=2,1\\ 6 & if\ i,j=2,2\end{cases},$$

$$f_4(e_{i,j})=\begin{cases}2 & if\ i,j=1,1\\ 5 & if\ i,j=1,2\end{cases}.$$

We can see that f_4 is a bijection and the edge weights are :

$$wt(e_{1,1})=f_4(v_{1,1})+f_4(v_{2,1})+f_4(e_{1,1})=7,$$

$$wt(e_{1,2})=f_4(v_{1,2})+f_4(v_{2,2})+f_4(e_{1,2})=14.$$

The edge weights form an arithmetic sequence with difference 7. Thus, f_4 is a $(7,7)$-EAT labeling for $2P_2$.

Define a new labeling f_5 for the vertices and the edges of $2P_2$ as follow :

$$f_5(v_{i,j})=\begin{cases}1 & if\ i,j=1,1\\ 4 & if\ i,j=1,2\\ 3 & if\ i,j=2,1\\ 6 & if\ i,j=2,2\end{cases},$$

$$f_5(e_{i,j})=\begin{cases}2 & if\ i,j=1,1\\ 5 & if\ i,j=1,2\end{cases}.$$

We can see that f_5 is a bijection and the edge weights are :

$$wt(e_{1,1})=f_5(v_{1,1})+f_5(v_{2,1})+f_5(e_{1,1})=6,$$

$$wt(e_{1,2})=f_5(v_{1,2})+f_5(v_{2,2})+f_5(e_{1,2})=15.$$

The edge weights form an arithmetic progression with difference 9. Thus, f_5 is a $(6,9)$-EAT labeling for $2P_2$.

Observation 2. *If path $2P_3$ has an (a,d)-EAT labeling.Then $d \leq 7$.*

Observation 3. *If path $2P_n$,$n \geq 4$, has an (a,d)-EAT labeling.Then $d \leq 6$.*

Theorem 2. *For every $2P_n$,$n \geq 3$ has $(4n+2,1)$-EAT and $(6n+2,1)$ -EAT labelings.*

Proof. Let us define the vertices and edges labelings of $2P_n$ as follow:

$$f_6(v_{i,j})=\begin{cases}2i-1, & j=1, i=1,\dots,n\\ 2i, & j=2, i=1,\dots,n\end{cases}$$

$$f_6(e_{i,j})=\begin{cases}4n-2i, & j=1, i=1,\dots,n-1\\ 4n-2i-1 & j=2, i=1,\dots,n-1.\end{cases}$$

It is easy to verify that f_6 is bijective. For all edge $e_{i,j}$, $i = 1,\dots,n-1$, $j = 1,2$ of $2P_n$, the edge weights are :

$$\begin{aligned}wt(e_{i,1}) &= f_6(v_{i,1})+f_6(e_{i,1})+f_6(v_{i+1,1})\\ &= (2i-1)+(4n-2i)+(2(i+1)-1)\\ &= 4n+2i,\\ wt(e_{i,2}) &= f_6(v_{i,2})+f_6(e_{i,2})+f_6(v_{i+1,2})\\ &= 2i+(4n-2i-1)+(2i+2)\\ &= 4n+2i+1.\end{aligned}$$

We can see that $e_{1,1}$ has the smallest weight $4n + 2$ and set of all edge weights form a consecutive sequence $4n + 2, 4n + 3, \ldots, 6n-1$ with $a = 4n+2$ and difference $d=1$. Thus, f_6 is $(4n+2,1)$-EAT labeling of $2P_n$, $n \geq 3$.

Define new labeling f_7 as follow :

$$f_7(v_{i,j}) = \begin{cases} 2i+2n-1, & j=1, \ i=1,\ldots,n \\ 2i+2n, & j=2, \ i=1,\ldots,n, \end{cases}$$

$$f_7(e_{i,j}) = \begin{cases} 2n-2i, & j=1, i=1,\ldots,n-1 \\ 2n-2i-1, & j=2, i=1,\ldots,n-1. \end{cases}$$

It is easy to verify that f_7 is bijective. For all edge $e_{i,j}$, $i = 1,\ldots,n-1$, $j = 1,2$ of $2P_n$, the edge weights are :

$$\begin{aligned} wt(e_{i,1}) &= f_7(v_{i,1}) + f_7(e_{i,1}) + f_7(v_{i+1,1}) \\ &= (2i+2n-1) + (2n-2i) + (2n+2i+1) \\ &= 6n+2i, \\ wt(e_{i,2}) &= f_7(v_{i,2}) + f_7(e_{i,2}) + f_7(v_{i+1,2}) \\ &= (2i+2n) + (2n-2i-1) + (2n+2i+2) \\ &= 6n+2i+1. \end{aligned}$$

We can see that $e_{1,1}$ has the smallest weight $6n+2$ and set of all edge weights form a consecutive sequence $6n+2, 6n+3, \ldots, 8n-1$ with $a = 6n+2$ and difference $d = 1$. Thus, f_7 is $(6n+2,1)$-EAT labeling of $2P_n$, $n \geq 3$. ■

Theorem 3. *For every $2P_n$, $n \geq 3$ has a $(2n+5,3)$-EAT labeling*

Proof. Let us define the vertices and edges labelings of $2P_n$ as following way :

$$f_8(v_{i,j}) = \begin{cases} 2i-1, & j=1, i=1,\ldots,n \\ 2i, & j=2, i=1,\ldots,n \end{cases},$$

$$f_8(e_{i,j}) = \begin{cases} 2n+2i-1, & j=1, i=1,\ldots,n-1 \\ 2n+2i, & j=2, i=1,\ldots,n-1 \end{cases}.$$

It is easy to verify that f_8 is bijective. For all edge $e_{i,j}$, $i = 1,\ldots,n-1$, $j=1,2$ of $2P_n$, the edge weights are :

$$\begin{aligned} wt(e_{i,1}) &= f_8(v_{i,1}) + f_8(e_{i,1}) + f_8(v_{i+1,1}) \\ &= (2i-1) + (2n+2i-1) + (2i+1) \\ &= 2n+6i-1, \\ wt(e_{i,2}) &= f_8(v_{i,2}) + f_8(e_{i,2}) + f_8(v_{i+1,2}) \\ &= 2i + (2n+2i) + (2i+2) \\ &= 2n + 6i + 2. \end{aligned}$$

We can see that $e_{1,1}$ has the smallest weight $2n+5$ and set of all edge weights form a consecutive sequence $2n+5, 2n+8, \ldots, 8n-4$ with $a=2n+5$ and difference $d=3$. Thus, f_8 is $(2n+5,3)$-EAT labeling of $2P_n$, $n \geq 3$.

Define new labeling of $2P_n$ as follow:

$$f_9\left(v_{i,j}\right)=\begin{cases}2i+2n-1, & j=1,\ i=1,\dots,n\\ 2n+2i, & j=2,\ i=1,\dots,n\end{cases}$$

$$f_9\left(e_{i,j}\right)=\begin{cases}2i-1, & j=1,\ i=1,\dots,n-1\\ 2i, & j=2,\ i=1,\dots,n-1\end{cases}.$$

It is easy to verify that f_9 is bijective. For all edge $e_{i,j}$, $i=1,\dots,n-1, j=1,2$ of $2P_n$, the edge weights are :

$$\begin{aligned} wt\left(e_{i,1}\right) &= f_9\left(v_{i,1}\right)+f_9\left(e_{i,1}\right)+f_9\left(v_{i+1,1}\right)\\ &=\left(2n+2i-1\right)+\left(2i-1\right)+(2n+2i+1)\\ &=4n+6i-1,\\ wt\left(e_{i,2}\right) &= f_9\left(v_{i,2}\right)+f_9\left(e_{i,2}\right)+f_9\left(v_{i+1,2}\right)\\ &=(2n+2i)+2i+(2n+2i+2)\\ &=4n+6i+2.\end{aligned}$$

We can see that $e_{1,1}$ has the smallest weight $4n+5$ and set of all edge weights form a consecutive sequence $4n+5, 4n+8, \dots, 10n-4$ with $a=4n+5$ and difference $d=3$. Thus, f_9 is $(4n+5,3)$-EAT labeling of $2P_n$, $n\geq 3$.

3. Conclusion

By the research we have done obtained edge antimagic total labeling on two copies of path. we obtain the edge antimagic total labeling for $n\geq 3$ with common different are $d=1$ and $d=3$,and common different are $d=1$, $d=3$, $d=5$, $d=7$, and $d=9$ for $n>2$. This result can be an important reference for further research, especially in graph theory research and generally for other applied science researchers in order to produce useful applications in various areas of society.

Acknowledgment

This research was supported by grant of Program Kreativitas Mahasiswa (PKM) 2017 KEMENRISTEKDIKTI.

References

[1] Ngurah A 2003 On (a, b)-edge-antimagic total labeling of odd cycle *J. Indones, Math. Soc* **9** 9-12
[2] Bloom G S and Hsu D F 1982 On graceful digraphs and a problem in network addressing *Congr. Numer* **35** 91–103
[3] Bloom G S and Hsu D F 1985 *On graceful directed graphs that are computational models of some algebraic systems* (New York: Wiley)
[4] Babujee J B and Babitha S 2012 New Construction on Edge-Antimagic Labeling *Applied Mathematical Sciences* **6** 44 2149-2157
[5] Gallian J A 2010 A Dynamic Survey of graph labeling *The Electronic Journal of Combinatories* **17(D)** S6
[6] Simanjuntak R, Bertault F and Miller M 2000 *Proc. Eleventh Australia Workshop Combin. Algor* 179-189 (Australia: Hunrer Valley)
[7] Dafik, Miller M, Ryan j and Bača M 2009 On super (a,d)-edge-antimagic total labeling of disconnecter graphs *Discrete Mathematics* **309** 4909-4915

A Kinetic Modelling of Enzyme Inhibitions in the Central Metabolism of Yeast Cells

Kasbawati, A Kalondeng, N Aris, N Erawaty and M I Azis

Department of Mathematics, Hasanuddin University, Jl. P. Kemerdekaan Km. 10, Makassar, Indonesia

E-mail: kasbawati@unhas.ac.id

Abstract. Metabolic regulation plays an important role in the metabolic engineering of a cellular process. It is conducted to improve the productivity of a microbial process by identifying the important regulatory nodes of a metabolic pathway such as fermentation pathway. Regulation of enzymes involved in a particular pathway can be held to improve the productivity of the system. In the central metabolism of yeast cell, some enzymes are known as regulating enzymes that can be inhibited to increase the production of ethanol. In this research we study the kinetic modelling of the enzymes in the central pathway of yeast metabolism by taking into consideration the enzyme inhibition effects to the ethanol production. The existence of positive steady state solution and the stability of the system are also analysed to study the property and dynamical behaviour of the system. One stable steady state of the system is produced if some conditions are fulfilled. The conditions concern to the restriction of the maximum reactions of the enzymes in the pyruvate and acetaldehyde branch points. There exists a certain time of fermentation reaction at which a maximum and a minimum ethanol productions are attained after regulating the system. Optimal ethanol concentration is also produced for a certain initial concentration of inhibitor.

1. Introduction

Mostly genetic engineering used metabolic engineering as a powerful tool for optimization and introduction of a new cellular process [1-5]. Since it has been studied in a multidisciplinary approach, success researches keep accumulating [5-10]. Biochemical information which can be accessed freely was underlying the strategies for experimental design to find the manipulation targets, and today, increasing of yield and productivity were being the main objectives of metabolic engineering [11]. Many researchers gave some reviews about theoretical and practical aspects of metabolic engineering [12-17]. Information about metabolic networks or enzymes would be underlying the essential prerequisite of a successful metabolic regulation [14]. Internal regulation of a metabolic pathway is one of the best practical ways to improve the productivity of the cell. It refers to the metabolic regulation level by regulating the metabolic fluxes in a metabolic network.

In order to coordinate the metabolic fluxes, activity of the key enzymes can be engineered such that cellular activities involving multiple enzymes can be improved. Introducing a molecule of inhibitor is one of the engineering choices for internal regulation since it acts to reduce the velocity of an enzymatic reaction. It exists naturally or can be introduced experimentally by using synthetic molecules. For instance, various kinds of inhibitor have been studied for designing and optimization performance of microbial yeast cell factories [7, 18-21]. It becomes an interesting study since different inhibitor may has different mechanism. Furthermore the mechanism can be approximated mathematically by using kinetic modeling such that theoretical aspect of this regulation can be captured.

In this paper, modeling the metabolic process of a yeast cell is studied as a premise to capture an optimal fermentation design such that an optimal productivity of yeast cell can be achieved. It is an interesting metabolic system since it consists of a multiple branched metabolic pathway that interact biochemically and a multiple reaction that is catalyzed by different enzymes [19]. Our study is focused on the main metabolic pathways, at the pyruvate and acetaldehyde branch points, where the overflow metabolism was occurred. Since enzymes became a key point in the metabolic network, regulation can be taken place through controlling the enzyme activity. Based on the study results of Kasbawati et al. [22], the key regulating enzymes reside at the pyruvate and acetaldehyde branch points. These enzymes have a significant effect in reducing and increasing the flux of ethanol. Here, we carry out kinetic modeling of ethanol production by taking into consideration effects of inhibitor molecules to the regulating enzymes in the culture medium. We assume that the inhibitor is introduced experimentally to regulate the metabolic fluxes of yeast cell such that flux of ethanol can be maximized.

This paper is organized as follows: in Section 2 we formulate the kinetic model of fermentation system by taking into consideration the inhibition of regulating enzyme at the pyruvate and acetaldehyde branch points. In Section 3, analytical results are presented to study the steady state condition of the system. Numerical simulation and some discussions are presented in Section 4, and conclusions are presented in the last section.

2. Kinetic Formulation

In this paper we propose a mathematical model describing the chemical conversion of metabolites in the central metabolism of yeast cell. We assume that the continuous fermentation process occurred at the ideal conditions. Carbon metabolism on the yeast metabolic pathway is started with the uptake of glucose that be converted becomes pyruvate via glycolysis pathway (r_1). We assume the glycolytic pathway as a lumped reaction [19]. Chemical conversion of glucose into pyruvate as the last product of glycolytic pathway is modeled using Michaelis-Menten [23] that is

$$r_1 = v_1 \frac{s_1}{s_1 + k_1} \tag{1}$$

where s_1 is the concentration of glucose, v_1 is the maximum uptake reaction and k_1 is the affinity constant for the glucose uptake reaction. In the yeast Saccharomyces cerevisiae metabolic pathway, pyruvate as the end-product of glycolysis is a branch-point between fermentative and respiratory sugar metabolism (see figure 1). The key enzymes for regulation purposes are located at the pyruvate branch-point. Here, pyruvate (s_2) is converted into three intermediate metabolites, i.e. oxaloacetate (s_3) and acetyl-coa (s_4) which lead to the key part of aerobic respiration of the yeast (TCA cycle), and acetaldehyde (s_5) which leads to the fermentative metabolism of sugar by yeast cells. These chemical conversion processes involve enzymes to catalyze the reactions and the kinetics follow the Michaelis-Menten kinetic equation, i.e.

$$r_i = v_i \frac{s_2}{s_2 + k_i}, i = 2, \ldots, 4 \tag{2}$$

where v_i is the maximum velocity of reaction and k_i is the affinity constant for each reaction. At the fermentative metabolism, acetaldehyde is converted to become acetate (s_6) and ethanol (s_7) where the conversion reactions also follow the Michaelis-Menten kinetic for reversible and irreversible mechanisms, respectively, i.e.

$$r_5 = \frac{v_{5f}k_{5b}s_5 - v_{5b}k_{5f}s_7}{k_{5f}k_{5b} + k_{5b}s_5 + k_{5f}s_7} \tag{3}$$

$$r_6 = v_6 \frac{s_5}{s_5 + k_6}, \tag{4}$$

where v_i is the maximum velocity of the every reaction and k_i is the affinity constant for each reaction (reversible or irreversible reaction). Under some conditions, enzyme acetyl-CoA synthetase then converts the acetate to become acetyl-coa that is leading to TCA cycle for respiration process. The conversion process fullfills

$$r_7 = v_7 \frac{s_6}{s_6 + k_7}. \tag{5}$$

where v_i is the maximum velocity of enzyme acetyl-CoA synthetase and k_i is the affinity constant of the enzyme.

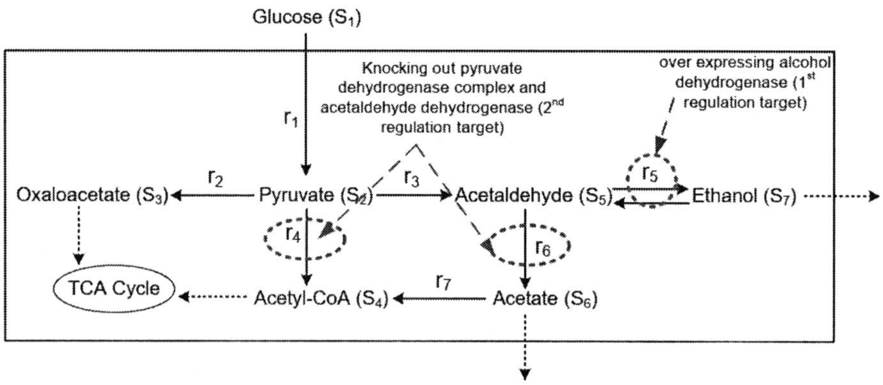

Figure 1. Central metabolism of *S. cerevisiae*. Reaction r_1 is a lumped process of glucose conversion to become pyruvate and reaction r_i, i=2, ..., 7 is a chemical conversion catalyzed by: (2) pyruvate carboxylase (3) pyruvate decarboxylase, (4) pyruvate dehydrogenase complex, (5) alcohol dehydrogenase, (6) acetaldehyde dehydrogenase, (7) acetyl-CoA synthetase. Reactions with cycle present enzymes for regulation target based on the sensitivity anaylisis results taken from [22].

In the previous study Kasbawati et al. [22] have showed that a fraction change of r_6 causes ethanol flux to decrease extremely. This reaction is catalyzed by acetaldehyde dehydrogenase and resides in the acetaldehyde branch point. Reducing the maximal rate of this reaction may lead the ethanol production to increases significantly. In fact, regulating r_6 leads also to regulate acetaldehyde branch point which involves r_4 and r_5. It is due to the stability condition for this branch, i.e. its maximum inflow should be less than its maximum outflow (see [22] for detail about the sensitivity results). Both reactions are respectively catalyzed by pyruvate decarboxylase and alcohol dehydrogenase which give positive control for ethanol flux. Therefore we have that enzymes in the acetaldehyde branch point are the key enzymes in the central metabolism of the yeast cell. An experimental observation showed that aspartate is an inhibitor for pyruvate dehydrogenase complex with inhibition constant, $k_i \approx$ 1.9mM [19]. Some researchers also showed that furfural is another inhibitor for pyruvate dehydrogenase complex (r_4), alcohol dehydrogenase (r_5), and acetaldehyde dehydrogenase (r_6) with different inhibition pattern. It inhibits enzymes at the dehydrogenase classes. Activity of pyruvate dehydrogenase complex (PDH) and alcohol dehydrogenase (AlDH) decreased by more than 90% at a concentration of less than 2 mM furfural whereas the acetaldehyde dehydrogenase (ADH) activity decreased by less than 20% at the same concentration [20]. Mathematically, competitive inhibition model can be formulated to describe the inhibition effects of furfural to the ADH and AlDH activities whereas noncompetitive inhibition model can be formulated to describe the inhibition effects of furfural to the PDH activity. By taking into consideration the furfural inhibition effects to the ADH, AlDH, and PDH, we then have the kinetic equations for these enzymes as follow

$$r_4 = v_4 \frac{s_2}{(s_2 + k_4) f(s_8)},\ r_5 = \frac{v_{5f} k_{5b} s_5 - v_{5b} k_{5f} s_7}{k_{5f} k_{5b} f(s_8) + k_{5b} s_5 + k_{5f} s_7},\ \text{and}\ r_6 = v_6 \frac{s_5}{s_5 + k_6 f(s_8)}$$

where $f(s_8) = 1 + s_8/(1+K)$ is an inhibition function that depends on the concentration of inhibitor s_8 with inhibition constant K. We assume that the concentration of inhibitor s_8 decreases with respect to the time t such that we have

$$\frac{ds_8(t)}{dt} = -\alpha s_8, \tag{6}$$

with the rate constant α. We assume that the inhibitor is only introduced to the system at the initial of reaction. There is no continuously inhibitor supply. If we assume that the initial concentration of inhibitor s_8 is $s_8(0) = p$ then from Equation (6) we get $s_8(t) = p\exp(-\delta t)$. For a long time of reaction, the inhibition function becomes $\lim_{t\to\infty} f(s_8) = \lim_{t\to\infty}\left(1 + \frac{pe^{-\delta t}}{1+K}\right) = 1$. Therefore by using the mass action rate law for the central metabolism of the yeast cell, we have the following nonlinear differential equation system,

$$\begin{aligned}
\dot{s}_1(t) &= \alpha G - v_1 \frac{s_1}{s_1 + k_1} = f_1,\\
\dot{s}_2(t) f_2 &= v_1 \frac{s_1}{s_1 + k_1} - \left(v_2 \frac{s_2}{s_2 + k_2} + v_3 \frac{s_2}{s_2 + k_3} + v_4 \frac{s_2}{(s_2 + k_4)\left(1 + \frac{s_8}{[1+K]}\right)} \right) = f_2,\\
\dot{s}_3(t) &= v_2 \frac{s_2}{s_2 + k_2} - \alpha s_3 = f_3,\\
\dot{s}_4(t) &= \left(v_4 \frac{s_2}{(s_2 + k_4)\left(1 + \frac{s_8}{[1+K]}\right)} + v_7 \frac{s_6}{s_6 + k_7} \right) - \alpha s_4 = f_4,\\
\dot{s}_5(t) &= v_3 \frac{s_2}{s_2 + k_3} - \left[\frac{v_{5f} k_{5b} s_5 - v_{5b} k_{5f} s_7}{k_{5f} k_{5b}\left(1 + \frac{s_8}{[1+K]}\right) + k_{5b} s_5 + k_{5f} s_7} + v_6 \frac{s_5}{s_5 + k_6\left(1 + \frac{s_8}{[1+K]}\right)} \right] = f_5,\\
\dot{s}_6(t) &= v_6 \frac{s_5}{s_5 + k_6\left(1 + \frac{s_8}{[1+K]}\right)} - v_7 \frac{s_6}{s_6 + k_7} - \alpha s_6 = f_6,\\
\dot{s}_7(t) &= \frac{v_{5f} k_{5b} s_5 - v_{5b} k_{5f} s_7}{k_{5f} k_{5b}\left(1 + \frac{s_8}{[1+K]}\right) + k_{5b} s_5 + k_{5f} s_7} - \alpha s_7 = f_7,\\
\dot{s}_8(t) &= -\delta s_8 = f_8,
\end{aligned} \tag{7}$$

with initial concentration for each metabolite: $s_1(0) = s_{1_0}, s_2(0) = s_{2_0}, s_3(0) = s_{3_0}, s_4(0) = s_{4_0}, s_5(0) = s_{5_0}, s_6(0) = s_{6_0}$, and $s_7(0) = s_{7_0}$. Since we model a continues fermentation system, the term αG in the model denotes the continues inflow of glucose supply and αs_i denotes the continuous outflow of products, $i = 3,4,6,7$, where α is a dilution rate that represents the volumetric flow rate of glucose supplied to the culture medium divided by the volume of the culture, i.e. $0 < \alpha < 1$ [24].

3. Mathematical Analysis

In this section we analyze the stability of the positive steady state solution of system (7) defined in the region $R_+^8 = \{(s_1, s_2, s_3, s_4, s_5, s_6, s_7, s_8) \in R^8 : s_i \geq 0, i = 1,...,8\}$. At the equilibrium condition we have $f_i = 0,\ i = 1,...,7$. Therefore from $f_1 = 0$ we have

$$s_1^* = \frac{\alpha G k_1}{v_1 - \alpha G}. \tag{8}$$

If $v_1 > \alpha G$ then we have $s_1^* > 0$ defined in R_+^8. Since $s_8^* = 0$ then for $f_2 = 0$ we get cubic polynomial in s_2 as follow,

$$a_1 (s_2)^3 + a_2 (s_2)^2 + a_3 (s_2) + a_4 = 0 \tag{9}$$

with $a_1 = (v_2 + v_3 + v_4) - \alpha G,\ a_2 = (v_3 + v_4 - \alpha G)k_2 + (v_2 + v_4 - \alpha G)k_3 + (v_2 + v_3 - \alpha G)k_4,$
$a_3 = (v_2 - \alpha G)k_3 k_4 + (v_3 - \alpha G)k_2 k_4 + (v_4 - \alpha G)k_2 k_3,\ a_4 = -\alpha G k_2 k_3 k_4.$
If $v_i > \alpha G, i = 1,...,4$ then $a_i > 0, i = 1,2,3$. Since a_4 is negative then there exists one sign change of the coefficients of polynomial (9). Therefore based on Descartes' rule of sign, polynomial (9) has only one positive root. Furthermore consider $f_3 = 0$ and $f_4 = 0$. From these equations we get $s_3^* = \frac{v_2 s_2^*}{(s_2^* + k_2)\alpha}$ and $s_4^* = \frac{1}{\alpha}\left[\frac{v_3 s_2^*}{(s_2^* + k_3)} + \frac{v_7 s_6^*}{(s_6^* + k_7)}\right]$, which depend on s_2^* and s_6^*. Meanwhile from $f_7 = 0$ we have $s_5^* = \frac{k_{5f} s_7^* (v_{5b} + \alpha k_{5b} + \alpha s_7^*)}{k_{5b}(v_{5f} - \alpha s_7^*)}$ which depends on s_7^* that will be positive if it fulfils the condition $s_7^* < \frac{v_{5f}}{\alpha}$.
By substituting s_5^* to the equation $f_5 = 0$ we get a cubic polynomial in s_7 as follows

$$b_1 (s_7)^3 + b_2 (s_7)^2 + b_3 (s_7) + b_4 = 0 \tag{10}$$

with $b_1 = -\alpha^2 k_{5f},\ b_2 = k_{5b}\alpha^2(k_6 - k_{5f}) - k_{5f}\alpha\left(v_6 + v_{5b} - \frac{v_3 s_2^*}{s_2^* + k_3}\right), b_3 = -\left(v_6 - \frac{v_3 s_2^*}{s_2^* + k_3}\right)(v_{5b}k_{5f} + \alpha k_{5b} k_{5f}) - \left(\frac{\alpha k_{5b} k_6 v_3 s_2^*}{s_2^* + k_3} + v_{5f}\alpha k_{5b} k_6\right)$ and $b_4 = \frac{v_3 s_2^*}{s_2^* + k_3} k_6 k_{5b} v_{5f}$. Since $b_1 < 0$ and $b_4 > 0$ then polynomial (14) will produce at least one positive root defined in R_+^8. This positive root fulfils the condition $0 < s_7^* < \frac{v_{5f}}{\alpha}$. Last equation is $f_6 = 0$. This equation generates a quadratic polynomial

$$c_1 (s_6)^2 + c_2 (s_6) + c_3 = 0 \tag{11}$$

with $c_1 = \alpha,\ c_2 = \alpha k_7 + v_7 - \left(\frac{v_6 s_5^*}{s_5^* + k_6}\right),\ c_3 = -\left(\frac{v_6 s_5^* k_7}{s_5^* + k_6}\right)$. Since $c_1 > 0$ and $c_3 < 0$ then polynomial (15) will produce one positive root defined in R_+^8. Therefore in conclusion, system (11) has only one positive steady state solution defined in R_+^8 if it fulfils the conditions $v_i > \alpha G, i = 1,...,4$ and $0 < s_7^* < \frac{v_{5f}}{\alpha}$. We define the steady state solution as follows,

$$E = (s_1^*, \cdots, s_8^*) = \left(\frac{\alpha G k_1}{v_1 - \alpha G}, s_2^*, \frac{v_2 s_2^*}{(s_2^* + k_2)\alpha}, \frac{1}{\alpha}\left[\frac{v_3 s_2^*}{(s_2^* + k_3)} + \frac{v_7 s_6^*}{(s_6^* + k_7)}\right], \frac{k_{5f} s_7^* (v_{5b} + \alpha k_{5b} + \alpha s_7^*)}{k_{5b}(v_{5f} - \alpha s_7^*)}, s_6^*, s_7^*, 0\right)$$

where s_6^* and s_7^* are the solutions of polynomial (10) and (11). By linearizing system (7) arround the equilibrium point E, it gives a linearized system $\dot{\mathbf{s}}(t) = J\mathbf{s}(t)$ where $\mathbf{s}(t) = (s_1(t),...,s_8(t))$ and $J = \begin{bmatrix} J_1 & 0 \\ J_2 & J_3 \end{bmatrix}$ is a Jacobian matrix with

$$J_1 = \begin{bmatrix} -\frac{(v_1-\alpha G)^2}{v_1k_1} & 0 & 0 \\ \frac{(v_1-\alpha G)^2}{v_1k_1} & -\sum_{i=2}^{4}\frac{v_ik_i}{(s_2^*+k_i)^2} & 0 \\ 0 & \frac{v_2k_2}{(s_2^*+k_2)^2} & -\alpha \end{bmatrix}, J_2 = \begin{bmatrix} 0 & \frac{v_3k_3}{(s_2^*+k_3)^2} & 0 \\ 0 & \frac{v_4k_4}{(s_2^*+k_4)^2} & 0 \\ 0 & 0 & 0 \\ 0 & 0 & 0 \\ 0 & 0 & 0 \end{bmatrix}, J_3 = \begin{bmatrix} -\alpha & 0 & \frac{v_7k_7}{(s_6^*+k_7)^2} & 0 & 0 \\ 0 & -\left(\frac{v_6k_6}{(s_5^*+k_6)^2}+\Delta_1\right) & 0 & \Delta_1 & 0 \\ 0 & \frac{v_6k_6}{(s_5^*+k_6)^2} & -\frac{v_7k_7}{(s_6^*+k_7)^2}-\alpha & 0 & 0 \\ 0 & \Delta_2 & 0 & -(\Delta_2+\alpha) & 0 \\ 0 & 0 & 0 & 0 & -\delta \end{bmatrix}$$

$\Delta_1 = \frac{k_{5f}k_{5b}(k_{5f}v_{5b}+s_5^*(v_{5f}+v_{5b}))}{(k_{5f}k_{5b}+k_{5b}s_5^*+k_{5f}s_7^*)^2}$, $\Delta_2 = \frac{k_{5f}k_{5b}(k_{5b}v_{5f}+s_7^*(v_{5f}+v_{5b}))}{(k_{5f}k_{5b}+k_{5b}s_5^*+k_{5f}s_7^*)^2}$. Jacobian matrix J generates a characteristic equation, $\det(\lambda I_3 - J_1) = 0$ or $\det(\lambda I_5 - J_3) = 0$. From $\det(\lambda I_3 - J_1) = 0$ we get

$$\lambda_1 = -\frac{(v_1-\alpha G)^2}{v_1k_1},\ \lambda_2 = -\sum_{i=2}^{4}\frac{v_ik_i}{(s_2^*+k_i)^2},\ \lambda_3 = -\alpha,\ \text{and}\ \left(\lambda+\frac{(v_1-\alpha G)^2}{v_1k_1}\right)\left(\lambda+\sum_{i=2}^{4}\frac{v_ik_i}{(s_2^*+k_i)^2}\right)(\lambda+\alpha) = 0.$$

While from $\det(\lambda I_5 - J_3) = 0$ we get $\lambda_4 = -\left[\frac{v_7k_7}{(s_6^*+k_7)^2}+\alpha\right]$, $\lambda_5 = -\alpha$, $\lambda_6 = -\delta$, and λ_7 and λ_8 which are the roots of polynomial $\lambda^2 + d_1\lambda + d_2 = 0$ with $d_1 = \left(\alpha+\Delta_1+\Delta_2+\frac{v_6k_6}{(s_5^*+k_6)^2}\right)$, $d_2 = \frac{\Delta_1 v_6k_6}{(s_5^*+k_6)^2}+\frac{\alpha v_6k_6}{(s_5^*+k_6)^2}+\alpha\Delta_2$.

Since $d_1>0$ and $d_2>0$, then we get negative eigenvalues such that the equilibrium point E is locally asymptotically stable.

4. Numerical Results

In this section we present numerical simulation of the model using kinetic parameter values given in table 1. This simulation is generated to capture the dynamical behavior of the solutions of the model and to analyze the inhibition effect of furfural with respect to the ethanol production. In this simulation we assume that the initial concentrations of variables are $s_1(0)$=83 g/l, $s_8(0)$=10 g/l, and $s_i(0)$=0 for all i=2,...7. Figure 2 shows the comparison of the model with and without inhibition effects. We can observe that the existence of furfural in fermentation system has significantly affected the production of acetyl-coa, acetate and ethanol. As shown in figure 2, regulation of the enzymes at the pyruvate and acetaldehyde branch points decreases the concentration of acetyl-coa and acetate. In contrast, regulation of the fermentation system increases ethanol production as expected. Therefore this inhibitor has reduced the production of acetyl-coa as well as acetate. These simulation results are in agreement with the fact that furfural can inhibit the activity of pyruvate dehydrogenase complex in r_4, alcohol dehydrogenase in r_5, and acetaldehyde dehydrogenase in r_6. For ethanol observation, although both solutions converge to the same equilibrium solution, their transient behaviors are different (see figure 2). After regulating the fermentation system, ethanol concentration increases. However there exists a certain time at which the ethanol production decreases and goes to zero. It means that there exists a certain time at which the ethanol production reaches a maximum production and a minimum production (see figure 3).

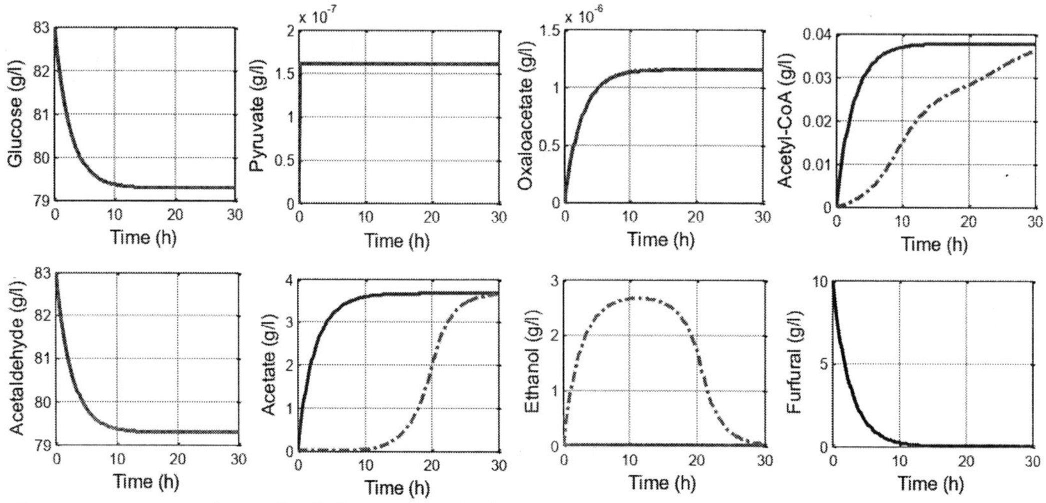

Figure 2. Comparison of solutions with (dashed line) and without (solid line) inhibition effects.

Table 1. Parameter value of parameter of the model.

Parameter	Value	Ref.	Parameter	Value	Ref.
α	0.38 1/h	[25]	vf	2.82 g/lh	[25]
v1	1.43 g/lh	[25]	kf	0.034 g/l	[25]
k1	0.94 g/l	[25]	vb	0.0125 g/lh	[25]
v2	648 g/lh	[26]	kb	0.057 g/l	[25]
k2	237.5 g/l	[26]	v6	4.8 g/lh	[25]
v3	5.81 g/lh	[25]	k6	2.64x10-4	[25]
k3	5x10-7 g/l	[25]	v7	0.0104 g/lh	[25]
v4	0.501 g/lh	[25]	k7	0.0102 g/l	[25]
k4	2x10-5 g/l	[25]	K	0.0019	[19]

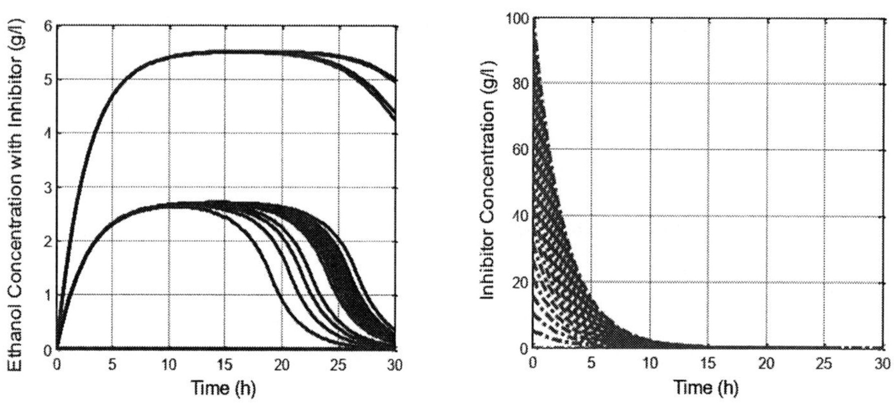

Figure 3. Simulations of ethanol concentration (left) and inhibitor concentration (right) for different initial concentration of inhibitor.

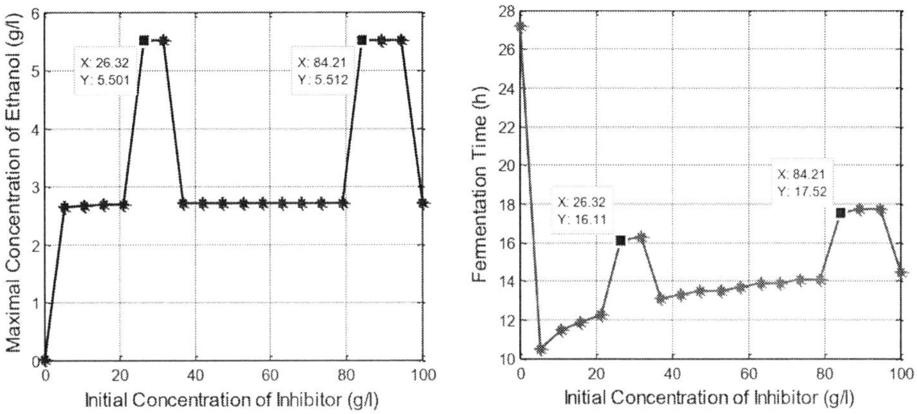

Figure 4. Graphs of maximal ethanol concentration (left) and fermentation time (right) with respect to the initial concentration of inhibitor. The fermentation time indicates the time span of fermentation required to produce a maximal ethanol concentration.

Furthermore, in figure 4, we present the simulation results for different initial concentration of inhibitor. We can observe that different initial concentration of inhibitor will produce different maximal concentration of ethanol (see the left picture in figure 4). Some initial concentrations of inhibitor produce almost similar maximal concentration of ethanol. While some other initial concentrations of inhibitor produce maximal concentration of ethanol which is higher than the previous one. For instance, by using 26 g/l initial concentration of inhibitor and by running the fermentation about 16 hours, it produces ethanol concentration about 5.5 g/l. Similar result is attained when we set the initial concentration of inhibitor becomes 84 g/l and run the fermentation reaction about 17.5 hours. Although the two schemes of fermentation system generate a similar result, the first scheme is more efficient than the second one since it only needs lower initial inhibitor concentration and shorter fermentation time to produce maximum ethanol concentration. Therefore experimentally, initial concentration of inhibitor and length of fermentation time should be set appropriately such that a maximum concentration of ethanol can be attained efficiently.

5. Conclusion
In this paper we formulated a mathematical model of the central metabolism of yeast cell by taking into consideration the inhibitions of some regulation enzymes at the pyruvate and acetaldehyde branch points. The inhibition mechanisms followed the competitive and noncompetitive inhibition mechanisms. Analytically the system produced one stable steady state if it fulfilled some existences and stability conditions that restricted the maximum reaction of enzymes at the pyruvate and acetaldehyde branch points. Numerically we found that inhibitions of the regulating enzymes affected the production of ethanol significantly. There existed a certain fermentation time at which ethanol production attained a minimum and a maximum concentration after regulating the system. There also existed a certain initial concentration of inhibitor that produced a maximum ethanol concentration in a short time of fermentation reaction.

Acknowledgment
The authors would like to thank DIKTI via Institute for Research and Community Services Hasanuddin University that supported this research through Penelitian Dasar Unggulan Perguruan Tinggi (PDUPT) research grant with grant number: 005/ADD/SP2H/LT/DRPM/VIII/2017.

References
[1] Ostergaard S, Olsson L and Nielsen J 2000 Metabolic Engineering of *Saccharomyces cerevisiae*

Microbiol Mol Biol Rev **64** 1 34–50
[2] Nevoigt E 2008 Progress in metabolic engineering of Saccharomyces cerevisiae *Microbiol Mol Biol Rev* **72** 3 379–412
[3] Copeland W B, Bartley B, Chandran D, Galdzicki M, Kim K H, Sleight S C, Maranas C D and Sauro H M 2012 Computational tools for metabolic engineering *Metab Eng* **14** 3 270–280
[4] Fong S S 2014 Computational approaches to metabolic engineering utilizing systems biology and synthetic biology *Comput Struct Biotechnol J* **11** 18 28–34
[5] Tan Z, Khakbaz P, Chen Y, Lombardo J, Yoon J M, Shanks J V, Klauda J B and Jarboe L R 2017 Engineering Escherichia coli membrane phospholipid head distribution improves tolerance and production of biorenewable *Metab Eng* **44** 1–12
[6] Hoefnagel M H N, Starrenburg M J C, Martens D E, Hugenholtz J, Kleerebezem M, Swam I I V, Bongers R, Westerhoff H V and Snoep J L 2002 Metabolic engineering of lactic acid bacteria the combined approach: kinetic modelling metabolic control and experimental analysis *Microbiology* **148** (Pt 4):1003-13
[7] Kern A, Tilley E, Hunter I S, Legisa M and Glieder A 2007 Engineering primary metabolic pathways of industrial microorganisms *J Biotechnol* **129** 6–29
[8] Hou J, Lages N F, Oldiges M and Vemuri G N 2009 Metabolic impact of redox cofactor perturbations in *Saccharomyces cerevisiae Metab Eng* **11** 4–5 253–261
[9] Hou J, Scalcinati G, Oldiges M and Vemuri G N 2010 Metabolic impact of increased NADH availability in saccharomyces cerevisiae *Appl Environ Microbiol* **76** 3 851–859
[10] Sun J, Jeffryes J G, Henry C S, Bruner S D and Hanson A D 2017 Metabolite damage and repair in metabolic engineering design *Metab Eng*
[11] Kern A, Tilley E, Hunter I S, Legiša M and Glieder A 2007 Engineering primary metabolic pathways of industrial micro-organisms *J Biotechnol* **129** 1 6–29
[12] Cameron D C and Chaplen F W 1997 Developments in metabolic engineering *Curr Opin Biotechnol* **8** 175–180
[13] Cameron D C and Tong I T 1993 Metabolic engineering: techniques for analysis of targets for genetic manipulations *An overview Appl Biochem Biotechnol* **38** 105–140
[14] Nielsen J and Olsson L 2002 An expanded role for microbial physiology in metabolic engineering and functional genomics:moving towards systems biology *FEMS Yeast Res* **2** 175–181
[15] Stephanopoulos G 1999 Metabolic engineering *Curr Opin Biotechnol* **5** 196–200
[16] Stephanopoulos G 1999 Metabolic fluxes and metabolic engineering *Metab Eng* **1** 1–11
[17] Stephanopoulos G and Vallino J J 1991 Network rigidity and metabolic engineering in metabolite overproduction *Science (80-)* **252** 1675–1681
[18] Gathercole P S, Thiel P G and Hofmeyr J H 1986 Inhibition of pyruvate dehydrogenase complex by moniliformin *Biochem J* **233** 719–723
[19] Pronk J T J J T, Steensma H Y H H Y Y and Van D J P 1996 Pyruvate metabolism in Saccharomyces cerevisiae *Yeast* **12** 1607–1633
[20] Modig T, Lidén G and Taherzadeh M J 2002 Inhibition effects of furfural on alcohol dehydrogenase aldehyde dehydrogenase and pyruvate dehydrogenase *Biochem J* **363** 769–776
[21] Fisher A K, Freedman B G, Bevan D R and Senger R S 2014 A review of metabolic and enzymatic engineering strategies for designing and optimizing performance of microbial cell factories *Comput Struct Biotechnol J* **11** 18 91–99
[22] Kasbawati, Gunawan A Y, Hertadi R and Sidarto K A 2015 Metabolic regulation and maximal reaction optimization in the central metabolism of a yeast cell in *AIP Conference Proceedings* **1651**
[23] Benthin S, Nielsen J and Villadsen J 1992 Anomeric Specificity of Glucose Uptake and Implications
[24] Kasbawati, Gunawan A Y and Sidarto K A 2017 Washout and non-washout solutions of a

system describing microbial fermentation process under the influence of growth inhibitions and maximal concentration of yeast cells *Math Biosci* **289**

[25] Lei F, Rotboll M, Jorgensen S 2001 A biochemically structured model for *Saccaromyces cerevisiae J Biotechnol* **88** 205–221

[26] Schomburg I, Chang A, Placzek S, Sohngen C, Rother M, Lang M, Munaretto C, Ulas S, Stelzer M, Grote A, Scheer M and Schomburg D 2013 BRENDA in 2013: integrated reactions kinetic data enzyme function data improved disease classification: new options and contents in BRENDA *Nucleic Acids Res* **41** D764–D772

IOP Conf. Series: Journal of Physics: Conf. Series **979** (2018) 012067 doi:10.1088/1742-6596/979/1/012067

On Super Edge-magic Total Labeling of Modified Watermill Graph

Nurdin, T S Ungko, J Gormantara, A Abdullah, S Aulyah and Nikita

Department of Mathematics, Hasanuddin University, Makassar, 90245, Indonesia

E-mail: triungko@gmail.com

Abstract. An *edge-magic total labeling* on a graph G is one-to-one map from $V(G) \cup E(G)$ onto the set of integers $1,2,\dots,v+e$, where $v = |V(G)|$ and $e = |E(G)|$, with the property that, given any edge uv, $f(u) + f(\{u,v\}) + f(v) = k$ for every $u, v \in V(G)$, and k is called *magic valuation*. An *edge-magic total labeling* f is called *super edge-magic total* if $f(V(G)) = \{1,2,\dots,|V(G)|\}$ and $f(E(G)) = \{|V(G)|+1, |V(G)|+2,\dots,|V(G)+E(G)|\}$. In this paper we investigate *edge-magic total labeling* of a new graph called modified *Watermill graph*. Furthermore, the *magic valuation* of the modified *Watermill graph* $WM(n)$ is $k = \frac{1}{2}(21n+3)$, for n odd, $n \geq 3$.

1. Introduction

All graph in this paper are finite, simple, and have no loops and multiple edge. A general reference of graph theory can be seen in [1].

Labeling is one of topic in the graph theory. Labeling graph is a map from graph elements to numbers [2], in this paper we discuss about edge total magic labeling which domain is the set of all vertices and edges that map to the natural numbers.

For graph G with vertex-set $V(G)$ and edge-set $E(G)$ an *edge-magic total labeling* is a bijection $\lambda: V(G) \cup E(G)$ to the set integers $1,2,\dots,|V(G) \cup E(G)|$ with the property that, for each edge $\{u,v\}$,

$$f(u) + f(\{u,v\}) + f(v) = k$$

For a fixed integers k. Call $f(u) + f(\{u,v\}) + f(v)$ the *edge sum* of $\{u,v\}$, and k is the magic valuation sum of graph G. An edge-magic total labeling f is called super edge-magic total if $f(V(G)) = \{1,2,\dots,|V(G)|\}$ and $f(E(G)) = \{|V(G)|+1, |V(G)|+2,\dots,|V(G)+E(G)|\}$. A graph is called edge-magic total or super edge-magic total if it admits any edge-magic total labeling or super edge-magic total labeling, respectively [3].

Various authors have introduce labeling that generalized the idea of magic square. Labeling was introduced by Sadlàčk, Sedlack [4] defined a graph to be magic if it had an edge-labeling, with range the real numbers, such that sum of the labels around any vertices equalled constant independent of the choice of vertices.

The notion of edge-magic total graph was introduced and studied by Kotzig and Rosa [5] with a different name as graphs with magic valuation. In 1996 Ringel and Llado [6] redefine this type of labeling called the labeling edge magic. After that Wallis et al (2000) [7], found the concept to distinguish by another *magic labeling*. Recently Enomoto et al [8] defines the *super edge-magic total labeling*.

There are several research studies has been discussed about the super edge-magic total labeling. In [8] proved that every cycle C_n, $n \geq 3$ are edge-magic. then in [8] proved that cycle , $n \geq 3$ are Super Edge Labeling if only if n is odd. In [9] show that all paths P_n and all n-suns are edge-magic total. Wijaya and Baskoro [10] studied edge magic total labeling for a product of two graphs. They showed the product $P_m \times C_n$ admits an edge-magic total labeling for n odd, $n \geq 3$. For n even, we only know that $P_2 \times C_n$ is not edge-magic total.

In [11] Ngurah, Baskoro Tomescu gave methods for construction new super edge-magic total graphs from old ones by adding some new pendent edges. They also prove that $K_{1,m} \cup P_n{}^m$ is super edge-magic total. Wallis proves that a cycle with one pendent edge is edge-magic total.

Ngurah and Baskoro [3] studies about magic total labeling of generalized Petersen graph. They showed that if $n \geq 3$, then the generalized Petersen Graph $P(n, 1)$ has an Super edge-magic total labeling with the magic valuation $k = \frac{1}{2}(11n + 3)$. In this paper, we shall discuss super edge-magic total labelling of modified watermill graph.

2. Result & Discussion

This section explains the research outcome on super edge-magic total labeling of modified watermill graph.

2.1 Sun Graph

Before discussing about the watermill graph, first we have to know about definition of the sun graph. The **sun** graph $(C_n \odot \overline{K_1})$ is a graph that constructed from cycle graph C_n where every vertices on that cycle graph is added a vertex with degree 1 such that every vertices on the sun graph have degree 3, except on the endvertices that have degree 1. The sun graph is the product of corona between two graphs, cycle graph with n vertices $(C_n, n \geq 3)$ and the complement of complete graph with one vertex $(\overline{K_1})$. Sun graph is denoted by $C_n \odot \overline{K_1}$ with n is the number of vertices on cycle graph. Example for $C_6 \odot \overline{K_1}$ is shown by figure 1.

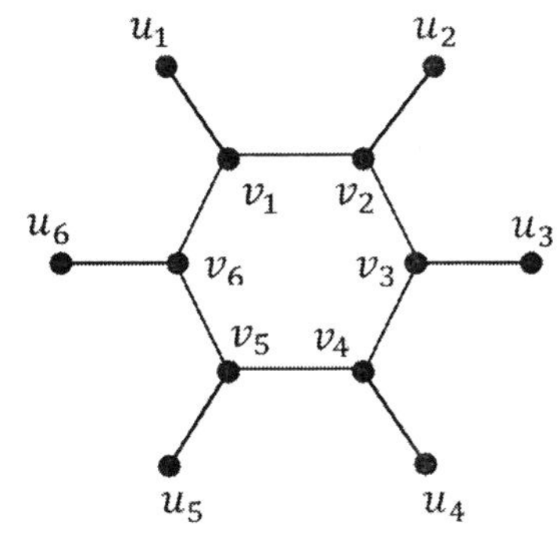

Figure 1. Sun Graph $C_6 \odot \overline{K_1}$.

2.2 Watermill Graph

The Watermill graph is denoted by $WM(n)$ with set of vertices $V(WM(n))$ and set of edges $E(WM(n))$. Watermill graph is defined as follows :

$$WM(n) = (V(WM(n)), E(WM(n)))$$

with,

$$V(WM(n)) = \{u_1, u_2, \dots, u_{2n}, v_1, v_2, \dots, v_{2n}\}$$

and

$$E(WM(n)) = \{u_i u_{i+1} \cup v_i v_{i+1} | i = 1,2,\dots,n-1\} \cup \{u_i u_{i+n} \cup v_i v_{i+n} | = 1,2,\dots,n\}$$
$$\cup \{u_1 u_n \cup v_1 v_n\} \cup \{u_i v_i | 1 \le i \le n\} \cup \{u_{i+n} v_{i+n} | 1 \le i \le n\}$$

Watermill graph is a graph formed by two copy sun graphs $(C_n \odot \overline{K_1})$, where that graphs are made parallel with the parallel vertices is adjacent each other. For example, Watermill graph with $n = 5$ $(WM(5))$.

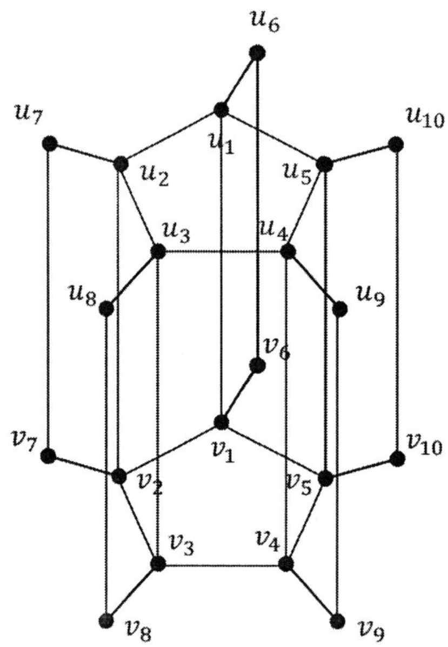

Figure 2. Watermill Graph $WM(5)$.

In this paper we will discuss about super edge-magic total labeling on modified Watermill graph. Modification for this graph is omit some edges on Watermill graph, such that set of vertices and set of edges for this graph is defined as follows,

$$V(WM(n)) = \{u_1, u_2, \dots, u_{2n}, v_1, v_2, \dots, v_{2n}\}$$

$$E(WM(n)) = \{u_i u_{i+1} \cup v_i v_{i+1} | i = 1,2,\dots,n-1\} \cup \{u_i u_{i+n} \cup v_i v_{i+n} | i = 1,2,\dots,n\}$$
$$\cup \{u_1 u_n \cup v_1 v_n\} \cup \{u_i v_i | 1 \le i \le n\} \cup \{u_{i+n} v_{i+n} | 1 \le i \le n\}$$
$$-\left\{u_{2(n-i)} v_{2(n-i)} \middle| i = 0, \dots, \tfrac{1}{2}(n-3)\right\}$$

For example modified Watermill graph with $n = 5$ is given by figure 3.

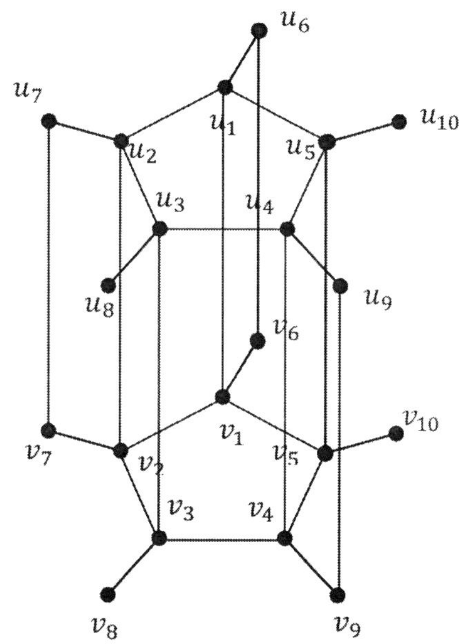

Figure 3. Modified Watermill Graph $WM(5)$.

Theorem

If n odd, $n \geq 3$, then the modified watermill graph $WM(n)$ has an edge-magic total labeling with the magic valuation $k = \frac{1}{2}(21n + 3)$.

Proof Label the vertices and edges of $WM(n)$ in the following way

$$f(u_i) = \begin{cases} i, & i \leq n, \quad i \equiv 1 \pmod 2 \\ i + n, & i < n, \quad i \equiv 0 \pmod 2 \\ i, & i > n, \quad i \equiv 0 \pmod 2 \\ i - n, & i > n, \quad i \equiv 1 \pmod 2 \end{cases}$$

$$f(v_i) = \begin{cases} \dfrac{6n - i + 1}{2}, & i \leq n, \quad i \equiv 1 \pmod 2 \\ \dfrac{5n - i + 1}{2}, & i < n, \quad i \equiv 0 \pmod 2 \\ \dfrac{8n - i}{2}, & i > n, \quad i \equiv 0 \pmod 2 \\ \dfrac{9n - i}{2}, & i > n, \quad i \equiv 1 \pmod 2 \\ 4n, & i = 2n \end{cases}$$

$$f(v_i v_{i+1}) = 5n + i + 1, \qquad i < n$$

$$f(v_1 v_n) = 5n + 1$$

IOP Conf. Series: Journal of Physics: Conf. Series **979** (2018) 012067 doi:10.1088/1742-6596/979/1/012067

$$f(v_i v_{i+n}) = \begin{cases} 4n+i+1, & i<n \\ 4n+1, & i=n \end{cases}$$

$$f(u_i v_i) = \begin{cases} \frac{15n-i+2}{2}, & i \le n, \quad i \equiv 1 \pmod 2 \\ \frac{14n-i+2}{2}, & i < n, \quad i \equiv 0 \pmod 2 \\ \frac{14n-i+3}{2}, & i > n, \quad i \equiv 1 \pmod 2 \\ \frac{13n-i+3}{2}, & i = n+1 \end{cases}$$

$$f(u_i u_{i+1}) = \frac{19n-4i+1}{2}, \quad i<n$$

$$f(u_1 u_n) = \frac{19n+1}{2}$$

$$f(u_i u_{i+n}) = \frac{19n-4i+3}{2}, \quad i \le n$$

To prove the function above is bijective, we see that

$$f(u_i) = \begin{cases} i, & 1,3,\dots,n \\ i+n, & n+2, n+4, \dots, 2n-1 \\ i, & n+1, n+3, \dots, 2n \\ i-n, & 2,4,\dots,n-1 \end{cases}$$

$$f_1 = u_i = \{1,2,3,\dots,2n\}$$

$$f(v_i) = \begin{cases} \frac{6n-i+1}{2}, & \frac{5n+1}{2}, \frac{5n+3}{2}, \dots, 3n-2, 3n-1, 3n \\ \frac{5n-i+1}{2}, & 2n+1, 2n+2, \dots, \frac{5n-3}{2}, \frac{5n-1}{2} \\ \frac{8n-i}{2}, & 3n+1, 3n+2, \dots, \frac{7n-3}{2}, \frac{7n-1}{2} \\ \frac{9n-i}{2}, & \frac{7n+1}{2}, \frac{7n+2}{2}, \dots, 4n-2, 4n-1 \\ 4n, & 4n \end{cases}$$

$$f(v_i v_{i+1}) = 5n+i+1, \quad 5n+2, 5n+3, \dots, 6n$$

$$f(v_1 v_n) = 5n+1, \quad 5n+1$$

$$f(v_i v_{i+n}) = \begin{cases} 4n+i+1, & 4n+2, 4n+3, \dots, 5n \\ 4n+1, & 4n+1 \end{cases}$$

$$f(u_iv_i)=\begin{cases}\frac{15n-i+2}{2}, & 7n+1,7n+2,\dots,\frac{15n-3}{2},\frac{15n-1}{2},\frac{15n+1}{2}\\ \frac{14n-i+2}{2}, & \frac{13n+3}{2},\frac{13n+5}{2},\dots,7n-2,7n-1,7n\\ \frac{14n-i+3}{2}, & 6n+2,6n+3,\dots,\frac{13n-3}{2},\frac{13n-1}{2},\frac{13n+1}{2}\\ \frac{13n-i+3}{2}, & 6n+1\end{cases}$$

$$f(u_iu_{i+1})=\frac{19n-4i+1}{2}, \quad \frac{15n+1}{2},\frac{15n+5}{2},\dots,\frac{19n-11}{2},\frac{19n-7}{2},\frac{19n-3}{2}$$

$$f(u_1u_n)=\frac{19n+1}{2}, \quad \frac{19n+1}{2}$$

$$f(u_iu_{i+n})=\frac{19n-4i+3}{2}, \quad \frac{15n+3}{2},\frac{15n+7}{2},\dots,\frac{19n-5}{2},\frac{19n-1}{2}$$

Clearly, every generated labels by the function are different each other.

Define $f: f(u_i)\cup f(v_i)\cup f(v_iv_{i+1})\cup f(v_1v_n)\cup f(v_iv_{i+n})\cup f(u_iv_i)\cup f(u_iu_{i+1})$

$$\cup f(u_1u_n)\cup f(u_iu_{i+n})=V(WM(n))\cup E(WM(n))=\left\{1,\dots,\frac{19n+1}{2}\right\}$$

since,

$$|V(WM(n))\cup E(WM(n))|=\frac{19n+1}{2}$$

then, f is bijective.

Clearly, for every i :

$$\begin{aligned} f(u_i)+f(u_iu_{i+1})+f(u_{i+1}) &= f(u_i)+f(u_iu_n)+f(u_n)\\ &= f(u_i)+f(u_iu_{i+n})+f(u_{i+n})\\ &= f(v_i)+f(v_iv_{i+1})+f(v_{i+1})\\ &= f(v_i)+f(v_iv_n)+f(v_n)\\ &= f(v_i)+f(v_iv_{i+n})+f(v_{i+n})\\ &= f(u_i)+f(u_iv_i)+f(v_i)\\ &= \frac{1}{2}(21n+3). \end{aligned}$$

Here is example for super edge-magic total labeling of modified graph $WM(5)$. For the graph see figure 3.

Let G is modified graph $WM(5)$.

Graph G will be labeled super edge-magic total labeling based on contructed function that we found before.

IOP Conf. Series: Journal of Physics: Conf. Series **979** (2018) 012067 doi:10.1088/1742-6596/979/1/012067

Labels for $V(G)$:

$f(u_1) = 1$	$f(v_1) = 15$
$f(u_2) = 7$	$f(v_2) = 12$
$f(u_3) = 3$	$f(v_3) = 14$
$f(u_4) = 9$	$f(v_4) = 11$
$f(u_5) = 5$	$f(v_5) = 13$
$f(u_6) = 6$	$f(v_6) = 17$
$f(u_7) = 2$	$f(v_7) = 19$
$f(u_8) = 8$	$f(v_8) = 16$
$f(u_9) = 4$	$f(v_9) = 18$
$f(u_{10}) = 10$	$f(v_{10}) = 2$

Label for $E(G)$:

$f(v_1v_2) = 27$		$f(u_1u_2) = 46$
$f(v_2v_3) = 28$	$f(u_1v_1) = 38$	$f(u_2u_3) = 44$
$f(v_3v_4) = 29$	$f(u_2v_2) = 35$	$f(u_3u_4) = 42$
$f(v_4v_5) = 30$	$f(u_3v_3) = 37$	$f(u_4u_5) = 40$
$f(v_1v_5) = 26$	$f(u_4v_4) = 34$	$f(u_1u_5) = 48$
	$f(u_5v_5) = 36$	
$f(v_1v_6) = 22$	$f(u_6v_6) = 31$	$f(u_1u_6) = 47$
$f(v_2v_7) = 23$	$f(u_7v_7) = 33$	$f(u_2u_7) = 45$
$f(v_3v_8) = 24$	$f(u_9v_9) = 32$	$f(u_3u_8) = 43$
$f(v_4v_9) = 25$		$f(u_4u_9) = 41$
$f(v_5v_{10}) = 21$		$f(u_5u_{10}) = 39$

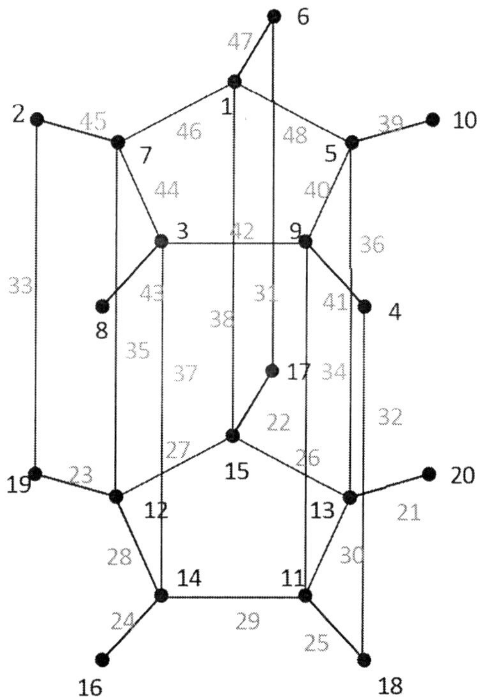

So we obtained graph G which has been labeled as follows:

Figure 4. Labeled Modified Watermill Graph $WM(5)$.

For every *edge sum* from graph G as follows:

$f(u_1)+f(u_1u_2)+f(u_2)=54$

$f(u_2)+f(u_2u_3)+f(u_3)=54$

$f(u_3)+f(u_3u_4)+f(u_4)=54$

$f(u_4)+f(u_4u_5)+f(u_5)=54$

$f(u_1)+f(u_1u_5)+f(u_5)=54$

$f(u_1)+f(u_1u_6)+f(u_6)=54$

$f(u_2)+f(u_2u_7)+f(u_7)=54$

$f(u_3)+f(u_3u_8)+f(u_8)=54$

$f(u_4)+f(u_4u_9)+f(u_9)=54$

$f(u_5)+f(u_5u_{10})+f(u_{10})=54$

$f(v_1)+f(v_1v_2)+f(v_2)=54$

$f(v_2)+f(v_2v_3)+f(v_3)=54$

$f(v_3)+f(v_3v_4)+f(v_4)=54$

$f(v_4)+f(v_4v_5)+f(v_5)=54$

$f(v_1)+f(v_1v_5)+f(v_5)=54$

$f(v_1)+f(v_1v_6)+f(v_6)=54$

$f(v_2)+f(v_2v_7)+f(v_7)=54$

$f(v_3)+f(v_3v_8)+f(v_8)=54$

$f(v_4)+f(v_4v_9)+f(v_9)=54$

$f(v_5)+f(v_5v_{10})+f(v_{10})=5$

$$f(u_1)+f(u_1v_1)+f(v_1)=54$$
$$f(u_2)+f(u_2v_2)+f(v_2)=54$$
$$f(u_3)+f(u_3v_3)+f(v_3)=54$$
$$f(u_4)+f(u_4v_4)+f(v_4)=54$$
$$f(u_5)+f(u_5v_5)+f(v_5)=54$$
$$f(u_6)+f(u_6v_6)+f(v_6)=54$$
$$f(u_7)+f(u_7v_7)+f(v_7)=54$$
$$f(u_9)+f(u_9v_9)+f(v_9)=54$$

As we can see, every sum of labels on an edge and its two endpoints have the same value, which is 54. So the magic value for this graph is $k = 54$. This corresponds to the theorem that already proved for magic value,

$$k = \frac{21n+3}{2}$$
$$k = \frac{21(5)+3}{2}$$
$$k = 54$$

3. Conclusion
According to result and discussion we found the *magic valuation* of the modified *Watermill graph* $WM(n)$ is $k = \frac{1}{2}(21n+3)$, for n odd, $n \geq 3$.

Acknowledgments
This research was supported by grant of Program Kreativitas Mahasiswa (PKM) 2017 KEMENRISTEKDIKTI.

References
[1] Harstfield N and Ringel G 1990 *Pearls in Graph Theory, Academic Press*, (New York/London: Boston/San Diego)
[2] McDougall J, Slamin M M and Wallis W D 2002 Vertex-magic total labelings *Utilitas Math* **61** 3-21
[3] Ngurah A A G and Baskoro E T 2003 On Magic and Antimagic Total Labelling of Generalized Petersen Graph *Utilitas Math* **63**
[4] Sedlacek J 1964 Problem 27 *Theory of graphs and its applications (Smolenice, 1963)* (Prague: House Czechoslovak Acad. Sci) 163-164
[5] Kotzig A and Rosa A 1970 Magic valuation of finite graphs *Canad. Math. Bull* **13** 451-461
[6] Ringel G and Llada A S 1996 Another tree conjecture *Bull ICA* **18** 83-85
[7] Wallis W D, Baskoro E T, Miller M and Slamin *Edge-Magic Total Labeling, Submitted*
[8] Enomoto H, Llado A S, Nakamigawa T and Ringel G Super edge-magic graphs *SUT J. Math* **2** 105-109
[9] Godbold R D and Slatter P J 1998 All cycle are edge-magic *Bull ICA* **22** 93-97
[10] Wijaya K and Baskoro E T 2000 *Proc. Seminar MIPA* (Bandung: ITB Bandung)
[11] Ngurah A A G, Baskoro E T and Tomescu I 2011 Magic graphs with pendant edges *Ars Combin* **99** 149-160

IOP Conf. Series: Journal of Physics: Conf. Series **979** (2018) 012068 doi:10.1088/1742-6596/979/1/012068

Some Properties of Fixed Point for Contraction Mappings in Quasi αb-metric Space

Budi Nurwahyu[1], **Asriadi Nasrun**[2] **and Naimah Aris**[3]

[1,3] Dept. of Mathematics, Fac. of Math. And Nat. Sci, Hasanuddin University
[2] Graduate of Mathematics, Fac. of Math. And Nat. Sci, Hasanuddin University

E-mail: budinurwahyu@unhas.ac.id

Abstract
We propose fixed point theorems for some contraction mappings in quasi αb-metric space. Especially, the sufficient conditions to obtain an existence and uniqueness of fixed point on certain contraction mappings. The quasi αb-metric space is extension of quasi b-metric space.

1. Introduction

We know that b-metric space was expressed by Bakhtin in 1989 [1], then Czewick in 1993 applied it for his study on contraction mappings in the b-metric space [2]. However, many authors used the b-metric space for existence and uniqueness of fixed point on several contraction mappings [3, 4, 5] and developed the b-metric space to become quasi b-metric space. Most of authors used the quasi b-metric space as a dislocated quasi b-metric space, and used it on fixed point theorems for several contraction mappings [6, 7, 8, 9]. While, quasi αb-metric space was developed from quasi b-metric space, with modifying triangular inequality condition of quasi b-metric. This work is motivated by the results on fixed point for contraction mappings in b-metric space, especially in creating of sufficient condition for existence and uniqueness of fixed point in the quasi αb-metric space [10].

For showing some properties of quasi αb-metric space for fixed point on contraction mapping, it is needed some notions as follows.

2. Preliminaries

Definition 2.1. [1, 2] *Let X be a nonempty set and $b \geq 1$.*

Let $d : X \times X \to [0, \infty)$ be and for all $x, y, z \in X$ the following conditions are satisfied:

(1) $d(x,y) = d(y,x) = 0$ *if and only if* $x = y$;
(2) $d(x,y) = d(y,x)$;
(3) $d(x,y) \leq b\big(d(x,z) + d(z,y)\big)$.

Then d is called as b-metric on X and (X,d) is called as b-metric space. And (X,d) is called as quasi b-metric space if only (1), (3) hold.

Definition 2.2.[10] *Let X be a nonempty set,* $0 \leq \alpha < 1$ *and* $b \geq 1$.

Let $d : X \times X \to [0, \infty)$ be a mapping which satisfying the conditions:

(1) $d(x,y) = d(y,x) = 0$ *if and only if* $x = y$;
(2) $d(x,y) \leq \alpha d(y,x) + \frac{1}{2} b\big(d(x,z) + d(z,y)\big)$ *for all* $x, y, z \in X$ (1)

Then d is called as a *quasi αb-metric* on X and (X,d) is called as a *quasi αb-metric space*.

Example 2.4.[10] *Let* $X = \{0,1,2\}$. *Defined* $d: X \times X \to R^+$ *as follows* $d(0,0) = d(1,1) = d(2,2) = d(0,2) = d(2,1) = 0$, $d(1,0) = 4$, $d(2,0) = 1$, $d(0,1) = 2$, *and* $d(1,2) = 3$. *It shows that d is quasi αb-metric with* $\alpha = \frac{1}{2}$ *and* $b = 4$, *because* $2 = d(0,1) \leq \frac{1}{2}d(1,0) + 2(d(0,2) + d(2,1))$, *but* $2 = d(0,1) > c(d(0,2) + d(2,1))$ *for every* $c \geq 1$, *so d is not quasi b-metric.*

It shows that quasi αb-metric space is not necessarily a quasi b-metric.

Example 2.5. [10] *Let* $X = R$ *and defined* $d: X \times X \to R^+$ *as* $d(x,y) = \begin{cases} 2x^2 + y^2, & x \neq y \\ 0, & x = y \end{cases}$
For the first condition of quasi αb-metric is obvious from above definition of function d(x,y), and for the second condition, we can show that
For $x \neq y$, *and every* $z \in X$, *we have*

$$d(x,y) = 2x^2 + y^2 \leq \frac{5}{2}x^2 + 2y^2 + 3z^2 = \frac{1}{2}(2y^2 + x^2) + ((2x^2 + z^2) + (2z^2 + y^2))$$
$$= \frac{1}{2}d(y,x) + \frac{2}{2}(d(x,z) + d(z,y))$$

So we obtain $d(x,y) = \frac{1}{2}d(y,x) + \frac{2}{2}(d(x,z) + d(z,y))$
Hence d is a quasi αb-metric with $\alpha = \frac{1}{2}$ *and* $b = 2$.

Definition 2.6.[10] *Let* (X,d) *be a quasi αb-metric space, a sequence* $\{x_n\}$ *in* (X,d) *is called convergent to* $x \in X$ *if* $\lim_{n\to\infty} d(x_n,x) = \lim_{n\to\infty} d(x,x_n) = 0$, *we write* $\lim_{n\to\infty} x_n = x$.

Definition 2.7.[10] *Let* $\{x_n\}$ *be a sequence in quasi αb-metric space* (X,d), $\{x_n\}$ *is called Cauchy sequence if* $\lim_{n,m\to\infty} d(x_n,x_m) = \lim_{n\to\infty} d(x_m,x_n) = 0$

Definition 2.8. Let (X,d) *be a quasi αb-metric space,* (X,d) *is called complete if every Cauchy sequence in X is convergent in X.*

Definition 2.9. Let X be a nonempty set and let T be a self mapping on X, $x \in X$ *is called a fixed point of T, if* $Tx = x$. *And for every* $x \in X$ *be defined* $TT^{n-1}x = T^n x$ *with* $T^0 x = x$

For proving next theorems in main results, we need sufficient conditions that a sequence to be a Cauchy sequence in quasi αb-metric space as follows

Theorem 2.10. [10] *Let (X,d) be a quasi αb-metric space with* $0 \leq \alpha < 1$ *and* $b \geq 1$, *and* $\{x_n\}$ *be a sequence in X satisfying the conditions as follows:*

1. $d(x_n,x_{n+1}) \leq ad(x_{n-1},x_n)$, *where* $0 < a < 1$;
2. $d(x_{n+1},x_n) \leq cd(x_n,x_{n-1})$, *where* $0 < c < 1$;
3. $ba + \alpha^2 < 1$ *and* $bc + \alpha^2 < 1$.

Then $\{x_n\}$ *is a Cauchy sequence in X*

3. Main Results

Some properties of existence and uniqueness of fixed point in quasi αb-metric space are shown on some theorems for certain contraction mappings as follows.

Theorem 3.1. Let (X, d) be a quasi αb-metric space with $0 \leq \alpha < 1$ and $b \geq 1$. Then for every $x, y, z \in X$ holds

1) $d(x,y) \leq \frac{b}{2(1-\alpha^2)}[(d(x,z) + d(z,y)) + \alpha(d(y,z) + d(z,x))]$
2) $d(x,y) + d(y,x) \leq \frac{b}{2(1-\alpha)}(d(x,z) + d(z,y) + d(y,z) + d(z,x))$

Proof. For every $x, y, z \in X$, and from (1) we have

$$d(x,y) \leq \alpha\, d(y,x) + \frac{1}{2}b\left(d(x,z) + d(z,y)\right)$$
$$\leq \alpha\left(\alpha\, d(x,y) + \frac{1}{2}b\left(d(y,z) + d(z,x)\right)\right) + \frac{1}{2}b\left(d(x,z) + d(z,y)\right)$$
$$= \alpha^2\, d(x,y) + \frac{1}{2}\alpha b\left(d(y,z) + d(z,x)\right) + \frac{1}{2}b\left(d(x,z) + d(z,y)\right)$$

Thus,

$$d(x,y) \leq \frac{b}{2(1-\alpha^2)}\left[\left(d(x,z) + d(z,y)\right) + \alpha\left(d(y,z) + d(z,x)\right)\right] \quad (2)$$

Furthermore, from (1) we have

$$d(y,x) \leq \frac{b}{2(1-\alpha^2)}\left[\left(d(y,z) + d(x,y)\right) + \alpha\left(d(x,z) + d(y,x)\right)\right] \quad (3)$$

Therefore from (2) and (3) we obtain

$$d(x,y) + d(y,x) \leq \frac{b}{2(1-\alpha^2)}\left[(1+\alpha)\left(d(x,z) + d(z,y)\right) + (1+\alpha)\left(d(y,z) + d(z,x)\right)\right]$$
$$= \frac{(1+\alpha)b}{2(1-\alpha^2)}\left[\left(d(x,z) + d(z,y)\right) + \left(d(y,z) + d(z,x)\right)\right]$$
$$= \frac{b}{2(1-\alpha)}\left[\left(d(x,z) + d(z,y)\right) + \left(d(y,z) + d(z,x)\right)\right]$$

It implies that part 2) is true □

Theorem 3.2. Let (X, d) be a complete quasi αb-metric space with $0 \leq \alpha < 1$ and $b \geq 1$. $T: X \to X$ be a continuous mapping which satisfying the conditions

$$d(Tx, Ty) \leq \frac{pq\, d(x,Tx)\, d(Ty,y)}{p\, d(x,Tx) + q\, d(Ty,y) + r} \quad (4)$$

For every $x, y \in X$, $0 < p < 1$, $0 < q < 1$, $r > 0$, $bp + \alpha^2 < 1$ and $bq + \alpha^2 < 1$. Then there exists a unique fixed point of T in X.

Proof. First, we will show that T has a fixed point in X. Taken $x_0 \in X$ and made a sequence $\{x_n\}$ where $x_1 = Tx_0$ and $x_{n+1} = Tx_n$ for $n = 0,1,2, \ldots$. Then using (4) we get

$$d(x_n, x_{n+1}) = d(Tx_{n-1}, Tx_n) \leq \frac{pq\, d(x_{n-1}, x_n)\, d(x_{n+1}, x_n)}{p\, d(x_{n-1}, x_n) + q\, d(x_{n+1}, x_n) + r}$$
$$< \frac{pq\, d(x_{n-1}, x_n)\, d(x_{n+1}, x_n)}{q\, d(x_{n+1}, x_n)} = pd(x_{n-1}, x_n)$$

So we have

$$d(x_n, x_{n+1}) < p\, d(x_{n-1}, x_n) \quad (5)$$

With similarly way and using (5) we get

$$d(x_{n+1},x_n) = d(Tx_n,Tx_{n-1}) \le \frac{pq\, d(x_n,x_{n+1})\, d(x_n,x_{n-1})}{p\, d(x_n,x_{n+1}) + q\, d(x_n,x_{n-1}) + r}$$
$$< \frac{pq\, d(x_n,x_{n+1})\, d(x_n,x_{n-1})}{p\, d(x_n,x_{n+1})}$$

So we have

$$d(x_n,x_{n+1}) < q\, d(x_n,x_{n-1}) \tag{6}$$

So from (5) and (6) we obtain that

$d(x_{n+1},x_n) < p\, d(x_{n-1},x_n)$,
$d(x_n,x_{n+1}) < q\, d(x_n,x_{n-1})$

Since $bp + \alpha^2 < 1$, $bq + \alpha^2 < 1$, and according to *Theorem 2.10*, then we conclude that $\{x_n\}$ is a Cauchy sequence in *X*. Since *X* complete then there exists $x^* \in X$ such that $\lim_{n\to\infty} x_n = x^*$.

Since *T* continuous on *X*, then

$Tx^* = T \lim_{n\to\infty} x_n = \lim_{n\to\infty} Tx_n = \lim_{n\to\infty} x_{n+1} = x^*$.

Hence, x^* is the fixed point of *T*.

To show the uniqueness of fixed point of *T*, we suppose that there exists $y^* \in X$ such that $Ty^* = y^*$. Then from (4), we have

$$d(x^*,y^*) = d(Tx^*,Ty^*) \le \frac{pq\, d(x^*,Tx^*)\, d(Ty^*,y^*)}{p\, d(x^*,Tx^*) + q\, d(Ty^*,y^*) + r}$$
$$= \frac{pq\, d(x^*,x^*)\, d(y^*,y^*)}{p\, d(x^*,x^*) + q\, d(y^*,y^*) + r}$$
$$= 0$$

This implies $d(x^*,y^*) = 0$. And using similarly way we have $d(y^*,x^*) = 0$. So we obtain

$x^* = y^*$

Thus *T* has a unique fixed point in X □

Theorem 3.3. Let (X, d) *be a complete quasi* αb *– metric space with* $0 \le \alpha < 1$ *dan* $b \ge 1$. *Let* $T: X \to X$ *be a continuous function satisfies the conditions*

$$d(Tx, Ty) \le min\{pd(x, Tx), q\, d(Ty, y)\} \tag{7}$$

For every $x, y \in X$, $0 < p < 1$, $0 < q < 1$, $bp + \alpha^2 < 1$, $bq + \alpha^2 < 1$.
Then T has a unique fixed point in X.

Proof. Taken $x_0 \in X$, $x_1 = Tx_0$ and $x_{n+1} = Tx_n$ for $n = 0,1,2,\ldots$.
So $\{x_n\}$ is a sequence in *X*, then using (7) we obtain

$d(x_n,x_{n+1}) = d(Tx_{n-1},Tx_n) \le \min\{p\, d(x_{n-1},x_n), q\, d(x_{n+1},x_n)\} \le p\, d(x_{n-1},x_n)$

So we have

$$d(x_n,x_{n+1}) \le p\, d(x_{n-1},x_n) \tag{8}$$

Once again, we use condition (7) then we obtain

$d(x_{n+1},x_n) = d(Tx_n,Tx_{n-1}) \le \min\{p\, d(x_n,x_{n+1}), q\, d(x_n,x_{n-1})\} \le q\, d(x_n,x_{n-1})$

So we have

$$d(x_{n+1},x_n) \le q\, d(x_n,x_{n-1}) \tag{9}$$

From (8) and (9) , since $p + \alpha^2 < 1$, $bq + \alpha^2 < 1$ and using *Theorem 2.10*, thus we have $\{x_n\}$ is a Cauchy sequence in *X*. Since *X* complete, thus we have $x^* \in X$ such that $\lim_{n\to\infty} x_n = x^*$.

And since *T* continuous in *X*, then we obtain

$Tx^* = T \lim_{n\to\infty} x_n = \lim_{n\to\infty} Tx_n = \lim_{n\to\infty} x_{n+1} = x^*$.

Hence x^* is the fixed point of T.
To show x^* as an unique fixed point of T, we suppose there exists $y^* \in X$ such that $Ty^* = y^*$.
From condition (7) we can obtain

$$\begin{aligned} d(x^*,y^*) = d(Tx^*,Ty^*) &\leq \min\{p\, d(x^*,Tx^*), q\, d(Ty^*,y^*)\} \\ &= \min\{p\, d(x^*,x^*), q\, d(y^*,y^*)\} \\ &= 0 \end{aligned}$$

This implies $d(x^*,y^*) = 0$. Similarly we have also $d(y^*,x^*) = 0$. This implies that $x^* = y^*$.
Thus T has a unique fixed point in X □

Theorem 3.4. Let (X,d) be a complete quasi αb-metric space with $0 \leq \alpha < 1$ and $b \geq 1$. Let $T: X \to X$ be a continuous function satisfies conditions as follows

$$d(Tx,Ty) \leq k\, d(x,y) + d(Tx,y)d(x,Ty)d(x,Tx) \tag{10}$$

For every $x,y \in X$, where $0 < k < 1$ and $bk + \alpha^2 < 1$.
Then there exists an element in X as a unique fixed point of T.

Proof. Let $x_0 \in X$ and $x_1 = Tx_0$ and $x_{n+1} = Tx_n$ for $n = 0,1,2, \dots$, then $\{x_n\}$ is a sequence in X.
From condition (10) we obtain

$$\begin{aligned} d(x_n,x_{n+1}) &= d(Tx_{n-1},Tx_n) \\ &\leq k\, d(x_{n-1},x_n) + d(Tx_{n-1},x_n)\, d(x_{n-1},Tx_n)\, d(x_{n-1},Tx_{n-1}) \\ &= k\, d(x_{n-1},x_n) + d(x_n,x_n)\, d(x_{n-1},x_{n+1})\, d(x_{n-1},x_n) \\ &= k\, d(x_{n-1},x_n) \end{aligned}$$

So, we have

$$d(x_n,x_{n+1}) \leq k\, d(x_{n-1},x_n) \tag{11}$$

And then once again we use condition (10), so we get

$$\begin{aligned} d(x_{n+1},x_n) &= d(Tx_n,Tx_{n-1}) \\ &\leq k\, d(x_n,x_{n-1}) + d(Tx_n,x_{n-1})\, d(x_n,Tx_{n-1})\, d(x_n,Tx_n) \\ &= k\, d(x_n,x_{n-1}) + d(x_{n+1},x_{n-1})\, d(x_n,x_n)\, d(x_n,Tx_{n+1}) \\ &= k\, d(x_n,x_{n-1}) \end{aligned}$$

So, we have

$$d(x_{n+1},x_n) \leq k\, d(x_n,x_{n-1}) \tag{12}$$

And from using (11) and (12) we obtain
$d(x_n,x_{n+1}) \leq k\, d(x_{n-1},x_n)$ and $d(x_{n+1},x_n) \leq k\, d(x_n,x_{n-1})$
Since $0 < k < 1$, $bk + \alpha^2 < 1$ and using *Theorem 2.10*, this implies that $\{x_n\}$ is a Cauchy sequence in X.
Since X is a complete quasi αb-metric space, then there exists $x^* \in X$ such that $\lim_{n\to\infty} x_n = x^*$.
Since T continuous on X then

$$Tx^* = T \lim_{n\to\infty} x_n = \lim_{n\to\infty} Tx_n = \lim_{n\to\infty} x_{n+1} = x^*.$$

To show the uniqueness of fixed point of T, it can be shown as follows
We Suppose that there exists $y^* \in X$ such that $Ty^* = y^*$. From condition (10) we obtain

$$\begin{aligned} d(x^*,y^*) = d(Tx^*,Ty^*) &\leq k\, d(x^*,y^*) + d(Tx^*,y^*)\, d(x^*,Ty^*)\, d(x^*,Tx^*) \\ &= k\, d(x^*,y^*) + d(x^*,y^*)\, d(x^*,y^*)\, d(x^*,x^*) \end{aligned}$$

Then we have

$$d(x^*,y^*) \leq k\, d(x^*,y^*) \tag{13}$$

Since $0 < k < 1$ and from (13) then it is possible if $d(x^*,y^*) = 0$.
Similarly, using (10) we can have

$$d(y^*,x^*) \leq k\, d(y^*,x^*) \tag{14}$$

And since $0 < k < 1$ and from (14) then is possible if $d(y^*,x^*) = 0$.

So we have $(x^*, y^*) = d(y^*, x^*) = 0$, it concludes that $x^* = y^*$. Thus T has a unique fixed point in X.□

Acknowledgements
Thank full to our colleagues in department of mathematics, Faculty of Mathematics and Natural Sciences that has supported the study of extension of metric space, and Hasanuddin University through the research program BMIS 2017 that has funded this research.

References
[1] Bakhtin I A 1989 *Funct. Anal.Unianowsk Gos. Ped. Inst.* **30** 26-37
[2] Czerwik S 1993 *Acta Math inf. Univ.Ostravinsis* **1** 5-11
[3] Kamran T, Samreen M and Ain U L Q 2017 *Mathematics* **5** 19 1-7
[4] Mishra P K, Sachdeva S and Banerjee S K 2014 *Turkish J. Anal. and Number Theory* **2** 1 19-22
[5] Demmaa M, Saadatib R, Vetro P 2016 *Iranian J. Math. Sci. and Inf.* **11** 1 123-136
[6] Wu H and Wu D 2016 *J. Math.Research* **8** 4 68-73
[7] Rahman M U 2017 *Appl. Math. Inf. Sci. Lett.* **5** 1 7-11
[8] Klin-eam C and Suanoom C 2015 *J. Fixed Point Theory and Appl.* **2015** 74 1-12
[9] Dekic D D, Došenovi T′, Huang H and Radenovi S 2016 *J. Fixed Point Theory and App.* **2016** 74 2-10
[10] Nurwahyu B 2017 *Far East J. Math. Sci.* **101** 8 1813-32

Stability and Optimal Harvesting of Modified Leslie-Gower Predator-Prey Model

S Toaha and M I Azis

Department of Mathematics, Hasanuddin University, Tamalanrea, 90245, Makassar, Indonesia

E-mail: syamsuddint@yahoo.com

Abstract. This paper studies a modified of dynamics of Leslie-Gower predator-prey population model. The model is stated as a system of first order differential equations. The model consists of one predator and one prey. The Holling type II as a predation function is considered in this model. The predator and prey populations are assumed to be beneficial and then the two populations are harvested with constant efforts. Existence and stability of the interior equilibrium point are analysed. Linearization method is used to get the linearized model and the eigenvalue is used to justify the stability of the interior equilibrium point. From the analyses, we show that under a certain condition the interior equilibrium point exists and is locally asymptotically stable. For the model with constant efforts of harvesting, cost function, revenue function, and profit function are considered. The stable interior equilibrium point is then related to the maximum profit problem as well as net present value of revenues problem. We show that there exists a certain value of the efforts that maximizes the profit function and net present value of revenues while the interior equilibrium point remains stable. This means that the populations can live in coexistence for a long time and also maximize the benefit even though the populations are harvested with constant efforts.

1. Introduction

Mathematical modeling associated with the dynamics of predator and prey populations has become a great research theme in the field of mathematical ecology and in the field of fisheries management. There are some fenomena in the fisheries includes population as a predator and the other population as a prey. The dynamics of the predator and prey populations does not include only two populations, but some fenomena show the involvement of more than two populations in the environment. Since the population, for example, fish gives benefit then the population is considered as a stock. The population is then managed to give more benefits and the population remains sustainable. There are some prey-predator models with harvesting. The predator-prey fishery model with selective harvesting for prey, see [1-4], selective harvesting for predator, see [5- 8], and the two populations are harvested, see [9].

The dynamics of predator and prey populations via Lotka-Volterra model have been extensively considered by many authors. In Lotka-Volterra model, the predator just depends on the size of prey population. This model does not consider the situation when the size of prey decreases then the predator will seek other prey. Leslie modeled the effect of this fenomena. Leslie-Gower predator-prey model is another approach to the dynamics of predator-prey. This model is structurally different from the classical Lotka-Volterra model. Some researchers have considered and modified the Leslie-Gower model with various aspects, see [10-12]. Most of predator-prey fisheries models with harvesting are related to the economic problems including maximum profit and total discounted net revenue, see [1, 6, 9].

Based on the works above, we present a modified Leslie-Gower predator-prey model with constant efforts of harvesting for the two populations. We analyse the stability of the interior equilibrium point and not only to determine the critical value of the efforts that maximize the profit but also to maintain the stability of the equilibrium point. The optimal harvesting policy of present value of revenue is solved using Pontryagin's maximal principle.

2. Leslie-Gower predator-prey model

We consider a predator-prey fishery in an environment which includes two populations, predator and prey. If we let $x(t)$ denotes the size of prey and let $y(t)$ be the size of predator at time t, then the growth rate of predator-prey populations based on the Leslie-Gower model is given by following system

$$\begin{aligned} \frac{dx}{dt} &= (r - bx)x - p(x)y \\ \frac{dy}{dt} &= \left(s - \frac{ay}{x}\right)y. \end{aligned} \tag{1}$$

The function $p(x)$ in model (1) is a predation functional response which measures the level of predator consumption to the prey. Firstly, Leslie considered the function $p(x)$ is proportional to the size of prey. The meaning of all parameters in model (1) can be seen, for example, in [6, 11]. By considering the Holling type II functional response of the predator to the prey, the model is then modified and becomes

$$\begin{aligned} \frac{dx}{dt} &= \left(r - bx - \frac{a_1 y}{k_1 + x}\right)x \\ \frac{dy}{dt} &= \left(s - \frac{a_2 y}{k_2 + x}\right)y. \end{aligned} \tag{2}$$

We consider that predator and prey populations are economically valuable, thus these two populations are harvested with constant efforts. Model (2) is then extended and becomes

$$\begin{aligned} \frac{dx}{dt} &= \left(r - bx - \frac{a_1 y}{k_1 + x}\right)x - q_1 E_1 x \\ \frac{dy}{dt} &= \left(s - \frac{a_2 y}{k_2 + x}\right)y - q_2 E_2 y. \end{aligned} \tag{3}$$

In the model (3), parameters q_1 and q_2 state the cathability coefficient for the prey and the predator populations respectively. Parameters E_1 and E_2 state the harvesting efforts satisfying the conditions $0 \le E_i \le E_{i\max}$ for $i = 1, 2$ and some values of $E_{i\max}$. For the analysis, we let $r_1 = r - q_1 E_1$ and $s_1 = s - q_2 E_2$. From the model (3), we have five non negative equilibrium points, namely $T_0 = (0, 0)$, $T_1 = \left(0, \frac{s_1 k_2}{a_2}\right)$, $T_2 = \left(\frac{r_1}{b}, 0\right)$, $T_3 = (x_3, y_3)$, and $T_4 = (x_4, y_4)$, where

$$x_3 = \frac{-A_2 + \left(A_2^2 - 4A_1A_3\right)^{1/2}}{2A_1},\quad y_3 = \frac{s_1(x_3 + k_2)}{a_2},\quad x_4 = \frac{-A_2 - \left(A_2^2 - 4A_1A_3\right)^{1/2}}{2A_1},\quad y_4 = \frac{s_1(x_4 + k_2)}{a_2},$$

$A_1 = ba_2 > 0$, $A_2 = s_1a_1 - r_1a_2 + k_1ba_2$, and $A_3 = s_1k_2a_1 - r_1k_1a_2$. In order for the equilibrium points T_3 and T_4 be exist in the first quadrant we assume that $A_2^2 - 4A_1A_3 \geq 0$.

Case 1. $A_3 = s_1k_2a_1 - r_1k_1a_2 < 0$, the equilibrium point T_3 becomes an interior equilibrium point while the equilibrium point T_4 will not be considered because $x_4 < 0$.

Case 2. $A_2 = s_1a_1 - r_1a_2 + k_1ba_2 < 0$ and $0 < A_3 < \frac{A_2^2}{4A_1}$, the equilibrium points T_3 and T_4 become interior equilibrium point. In this case, the equilibrium point T_3 may be stable or not stable while the equilibrium point T_4 is an unstable saddle point.

For analysis we use the Jacobian matrix to get the linear model and by evaluating the Jacobian matrix at the equilibrium point $T_3 = (x_3, y_3)$, we have

$$J_T = \begin{pmatrix} r_1 - 2bx_3 + \frac{a_1y_3x_3}{(k_1+x_3)^2} - \frac{a_1y_3}{(k_1+x_3)} & -\frac{a_1x_3}{(k_1+x_3)} \\ \frac{a_2y_3^{\,2}}{(k_2+x_3)^2} & s_1 - \frac{2a_2y_3}{(k_2+x_3)^2} \end{pmatrix}. \quad (4)$$

The polynomial characteristic (4) associated with the Jacobian matrix J_T is given by $f(\lambda) = \det(\lambda I - J_T)$, i.e $f(\lambda) = \lambda^2 - (d_1 + d_4)\lambda + (d_1d_4 - d_2d_3)$, where $d_1 = r_1 - 2bx_3 + \frac{a_1y_3x_3}{(k_1+x_3)^2} - \frac{a_1y_3}{(k_1+x_3)}$, $d_2 = -\frac{a_1x_3}{(k_1+x_3)} < 0$, $d_3 = \frac{a_2y_3^{\,2}}{(k_2+x_3)^2} > 0$, and $d_4 = s_1 - \frac{2a_2y_3}{(k_2+x_3)^2}$. Referring to the Routh-Hurwitz criterion [13], the interior equilibrium point T_3 is locally asymptotically stable when the conditions $d_1 + d_4 < 0$ and $d_1d_4 - d_2d_3 > 0$ are satisfied.

Example 1. For model (2), we set the parameter values as $r = 12$, $b = 0.001$, $s = 9$, $a_1 = 2.5$, $a_2 = 2.1$, $k_1 = 5$, and $k_2 = 7$ in appropriate units. Then we have equilibrium points $T_0 = (0, 0)$, $T_1 = (0, 30)$, $T_2 = (1200, 0)$, $T_3 = (1268.89, 5468.11)$, and $T_4 = (11.82, 80.66)$. For the equilibrium point T_3, we get eigenvalues from the Jacobian matrix as $0.2100 \pm 3.3503i$ which means the equilibrium point T_3 is unstable spiral. For the equilibrium point T_4, we get the eigenvalues from the associated Jacobian matrix as 0.5414 and -3.1284 which means that the equilibrium point T_4 is unstable saddle point, see figure 1. It is easy to check from the phase portrait that the positive x axis and positive y axis are stable manifold while the equilibrium point T_4 is unstable saddle point and equilibrium point T_3 is unstable spiral. In this situation, there exist a stable limit cycle in the first quadrant.

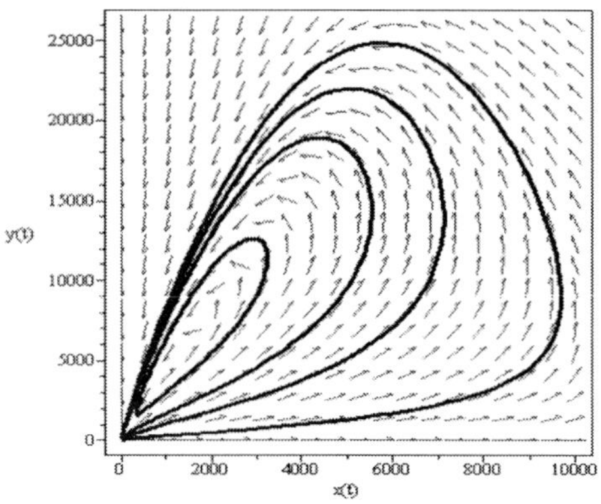

Figure 1. Plot of trajectories around the interior equilibrium points.

Example 2. For model (2), we set the parameter values as $r=12$, $b=0.001$, $s=8$, $a_1=1$, $a_2=3$, $k_1=10$, and $k_2=5$ in appropriate units. The equilibrium points of the model are $T_0=(0,0)$, $T_1=(0,13.33)$, $T_2=(12000,0)$, $T_3=(9334.76,24906.02)$, and $T_4=(-11.42,-17.13)$. In this case, the equilibrium point T_4 does not exist in the first quadrant. The equilibrium point T_3 is asymptotically stable with the eigenvalues $-7.3362\pm 4.5683i$, see figure 2.

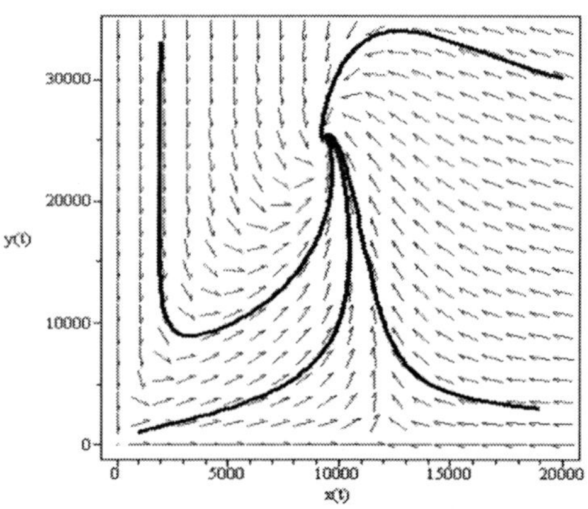

Figure 2. Plot of trajectories around the equilibrium point T_3

3. Bionomic equilibrium

The biological equilibrium is found by solving the system $\frac{dx}{dt}=0$ and $\frac{dy}{dt}=0$. The economic equilibrium is obtained when the proceeds from the sale of harvested biomass equals to the costs used in harvesting activities. We assume that $TR=pY(E)$, where $Y(E)=EqN$ is the yield of harvesting, p is the price of the biomass, E is harvesting effort, q is the cathability coefficient, and N is biomass. We also assume that total cost of harvesting is proportional the harvesting effort, $TC=cE$. The profit function is defined as $\pi=TR-TC$. The profit function of harvesting for the modified Leslie-Gower predator-prey model is $\pi(E_1,E_2)=(p_1q_1x)E_1+(p_2q_2y)E_2-(c_1E_1+c_2E_2)$. The bionomic equilibrium (x^*,y^*,E_1^*,E_2^*), see [14], is found by solving simultaneously the equations $\left(r-bx-\frac{a_1y}{k_1+x}\right)x-q_1E_1x=0$, $\left(s-\frac{a_2y}{k_2+x}\right)y-q_2E_2y=0$, and $\pi(E_1,E_2)=0$.

We then relate the stable interior equilibrium point T_3 to the maximum profit problem. The equilibrium point T_3 is an interior equilibrium if the conditions $0 \le E_i \le E_{i\max}$ for $i = 1, 2$ and $A_3 = s_1 k_2 a_1 - r_1 k_1 a_2 < 0$, that is

$$q_1 k_1 a_2 E_1 - q_2 k_2 a_1 E_2 < r k_1 a_2 - s k_2 a_1, \tag{5}$$

are satisfied. The profit function at the equilibrium point T_3 is written in the form

$$\pi(E_1, E_2) = (p_1 q_1 x) E_1 + (p_2 q_2 y) E_1 - (c_1 E_1 + c_2 E_2).$$

The problem is determining a pair of efforts (E_1, E_2) which satisfies condition (5) and $0 \le E_i \le E_{i\max}$ for $i = 1, 2$ that maximizes the profit function $\pi(E_1, E_2)$. We also need to maintain the interior equilibrium point T_3 is always asymptotically stable.

Example 3. For the problem of maximum profit function, we set the parameter values as $r = 12$, $b = 0.001$, $s = 8$, $a_1 = 1$, $a_2 = 3$, $k_1 = 12$, $k_2 = 5$, $q_1 = 1$, and $q_2 = 1$ in appropriate units. Take $p_1 = 10$, $p_2 = 15$, $c_1 = 5$, and $c_2 = 5$ in appropriate units. Then we have the equilibrium point $T_3 = (x_3, y_3)$, where

$$x_3 = 4661.6667 + 166.6667 E_2 - 500 E_1 + 166.6667\left[(-27.970 - E_2 + 3E_1)^2 + 3.840 - 0.360 E_1 + 0.060 E_2\right]^{1/2},$$

$$y_3 = (2.6667 + 0.3333 E_2)\left\{4666.6667 + 166.6667 E_2 - 500 E_1 + 166.6667\left[(-27.970 - E_2 + 3E_1)^2 + 3.840 - 0.360 E_1 + 0.060 E_2\right]^{1/2}\right\}.$$

In order for equilibrium point T_3 becomes an interior point, the harvesting efforts E_1 and E_2 must satisfy the conditions $A_3 < 0$ where $A_3 = 30E_1 - 5E_2 - 320$ and $0 \le E_i(t) \le E_{i\max}$ for $i = 1, 2$. We put $E_{1\max} = 5$ and $E_{2\max} = 5$. In the other words, the equilibrium point T_3 becomes an interior point when $(E_1, E_2) \in D_1$, where $D_1 = \{(E_1, E_2) : 0 \le E_1 \le 5, 0 \le E_2 \le 5, 30E_1 - 5E_2 < 320\}$. The profit function associated with the equilibrium point T_3 is given by

$$\pi(E_1, E_2) = (p_1 q_1 x_3) E_1 + (p_2 q_2 y_3) E_2 - (c_1 E_1 + c_2 E_2).$$

After substituting the values of x_3 and y_3, and then simplifying we get

$$\begin{aligned}\pi(E_1, E_2) = &\Big\{46611.6667 + 1666.6667 E_2 + 5000 E_1 \\ &+ 1666.6667\left[(-27.970 - E_2 + 3E_1)^2 + 3.840 - 0.360 E_1 + 0.060 E_2\right]^{1/2}\Big\} E_1 \\ &+ \Big\{(40 - 5E_2)\Big[4666.6667 + 166.6667 E_2 - 500 E_1 \\ &+ 1666.6667\left((-27.970 - E_2 + 3E_1)^2 + 3.840 - 0.360 E_1 + 0.060 E_2\right)^{1/2}\Big] - 5\Big\} E_2 .\end{aligned}$$

We have a stationary point $(E_1^*, E_2^*) = (1.41693, 4.33906)$ in D_1. The only critical point $(E_1^*, E_2^*) = (1.41693, 4.33906)$ satisfies the conditions and also maximizes the profit with the value of $\pi(E_1^*, E_2^*) = 876733.5362$. By applying the value of harvesting efforts

$(E_1^*, E_2^*) = (1.41693, 4.33906)$ we get the equilibrium point $T_3 = (9361.3614, 11449.1077)$. The polynomial characteristic of the Jacobian matrix at the equilibrium point T_3 is given by $f(\lambda) = \lambda^2 + 11.80804\lambda + 34.33137$ which has the eigenvalues -6.62936 and -5.17869. Under this situation, with the values of efforts at the level of $E_1^* = 1.41693$ and $E_2^* = 4.33906$, then the predator and the prey populations will sustain for a long period of time and also maximize the profit function.

4. Optimal harvesting policy

Our objective in this problem is to maximize the net present value of revenues which is given by

$$J = \int_0^{\infty} e^{-\delta t}\left[(p_1 q_1 x - c_1)E_1(t) + (p_2 q_2 y - c_2)E_2(t)\right] dt. \tag{6}$$

The symbol δ states the instantaneous rate of discount. We need to maximize J subject to the constraint equation (3) by using Pontryagin's maximum principle [15]. The control variables $E_1(t)$ and $E_2(t)$ are subject to the constraints $0 \le E_i(t) \le E_{i\max}$, for $i = 1, 2$.

The Hamiltonian equation is

$$H = e^{-\delta t}\left[(p_1 q_1 x - c_1)E_1 + (p_2 q_2 y - c_2)E_2\right] + \lambda_1\left(rx - bx^2 - \frac{a_1 xy}{k_1 + x} - q_1 E_1 x\right) + \lambda_2\left(sy - \frac{a_2 y^2}{k_2 + x} - q_2 E_2 y\right). \tag{7}$$

We set $\frac{\partial H}{\partial E_1} = 0$ and $\frac{\partial H}{\partial E_2} = 0$ as the necessary conditions for the control variables E_1 and E_2 to be optimal. From the Hamiltonian function (7), we have $\frac{\partial H}{\partial E_1} = e^{-\delta t}(p_1 q_1 x - c_1) - \lambda_1 q_1 x = 0$ and $\frac{\partial H}{\partial E_2} = e^{-\delta t}(p_2 q_2 y - c_2) - \lambda_2 q_2 y = 0$. Then we get $\lambda_1 = \frac{e^{-\delta t}(p_1 q_1 x - c_1)}{q_1 x}$ and $\lambda_2 = \frac{e^{-\delta t}(p_2 q_2 y - c_2)}{q_2 y}$.

From the Hamiltonian equation we also have

$$\frac{\partial H}{\partial x} = e^{-\delta t} p_1 q_1 E_1 + \lambda_1\left[r - 2bx + \frac{a_1 xy}{(k_1 + x)^2} - \frac{a_1 y}{(k_1 + x)} - q_1 E_1\right] + \frac{\lambda_2 a_2 y^2}{(k_2 + x)^2} \text{ and}$$

$$\frac{\partial H}{\partial y} = e^{-\delta t} p_2 q_2 E_2 - \frac{\lambda_1 a_1 x}{(k_1 + x)} + \lambda_2\left[s - \frac{2a_2 y}{(k_2 + x)} - q_2 E_2\right].$$

From the Pontryagin's maximum principle $\dot{\lambda}_1 = -\frac{\partial H}{\partial x}$, $\dot{\lambda}_2 = -\frac{\partial H}{\partial y}$, and we get

$$\frac{\delta e^{\delta t}(-p_1 q_1 x + c_1)}{q_1 x} + e^{-\delta t} p_1 q_1 E_1 + \lambda_1\left[r - 2bx + \frac{a_1 xy}{(k_1 + x)^2} - \frac{a_1 y}{(k_1 + x)} - q_1 E_1\right] + \frac{\lambda_2 a_2 y^2}{(k_2 + x)^2} = 0, \tag{8}$$

$$\frac{\delta e^{-\delta t}(-p_2 q_2 y + c_2)}{q_2 y} + e^{-\delta t} p_2 q_2 E_2 - \frac{\lambda_1 a_1 x}{(k_1 + x)} + \lambda_2\left[s - \frac{2a_2 y}{(k_2 + x)} - q_2 E_2\right] = 0. \tag{9}$$

By substituting $\lambda_1 = \frac{e^{-\delta t}(p_1 q_1 x - c_1)}{q_1 x}$ and $\lambda_2 = \frac{e^{-\delta t}(p_2 q_2 y - c_2)}{q_2 y}$ into the equations (8) and (9) we get $E_1 = E_1(x, y)$ and $E_2 = E_2(x, y)$. Again, after substituting $x = x_3 = \frac{q_3 E_3 + k}{a_5}$ and

$y = y_3 = \frac{s_2 + \left(s_2{}^2 + 4s_1a_1x_3\right)^{1/2}}{2s_1}$ into the equations $E_1 = E_1(x,y)$ and $E_2 = E_2(x,y)$ we get the value of control variables E_1 and E_2. The values of E_1, E_2, x_3, and y_3 maximize the present value of revenues J.

Example 4. For the problem of maximizing present value of the net revenues, we set the parameter values as $r = 12$, $b = 0.001$, $s = 8$, $a_1 = 1$, $a_2 = 3$, $k_1 = 10$, $k_2 = 5$, $q_1 = 1$, and $q_2 = 0.9$ in appropriate units. Take $p_1 = 15$, $p_2 = 20$, $c_1 = 10$, $c_2 = 15$, and $\delta = 0.005$ in appropriate units. Then we have the equilibrium point $T_3 = (x_3, y_3)$, where

$$x_3 = 4661.6667 + 150.0000E_2 - 500E_1 + 166.6667\left[(-27.970 - 0.9E_2 + 3E_1)^2 + 3.840 - 0.360E_1 + 0.0540E_2\right]^{1/2},$$

$$y_3 = (2.6667 + 0.3000E_2)\left\{4666.6667 + 150.0000E_2 - 500E_1 + 166.6667\left[(-27.970 - 0.9E_2 + 3E_1)^2 + 3.840 - 0.360E_1 + 0.0540E_2\right]^{1/2}\right\}.$$

The adjoint variables are $\lambda_1 = \frac{5e^{-0.005t}(3x_3 - 2)}{x_3}$ and $\lambda_2 = \frac{3.3333e^{-0.005t}(6y_3 - 5)}{y_3}$. We get the efforts $E_1 = 2.63403$ and $E_2 = 8.80361$, $T_3 = (9340.39754, 239.09227)$. The eigenvalues relates to the equilibrium are -9.31464 and -0.07696. The adjoint variables are $\lambda_1 = 14.99893e^{-0.005t}$ and $\lambda_2 = 19.93029e^{-0.005t}$. Therefore we get the maximum value of present value of the net revenues $J = \int_0^{\infty} 4.0677297 \cdot 10^5\ e^{-0.005t} dt = 8.135459 \cdot 10^7$.

5. Conclusions

The modified Leslie-Gower predator-prey population model with constant efforts of harvesting may has two interior equilibrium points, namely $T_3 = (x_3, y_3)$ and $T_4 = (x_4, y_4)$. When the equilibrium point T_3 exists in the first quadrant, it may be stable or unstable. When the equilibrium T_4 exists in the first quadrant, it is always unstable saddle point. When the equilibrium points T_3 and T_4 exist in the first quadrant where the equilibrium point T_3 is unstable spiral, there exists a stable limit cycle in the first quadrant.

With the restriction of harvesting efforts, $0 \le E_i \le E_{i\max}$, there exists a certain condition such that the interior equilibrium point T_3 remains stable and also gives maximum profit. The predator and the prey populations can live in coexistence, although the two populations are harvested with constant efforts. By using Pontryagin's maximum principle, we found that there exists a certain value of harvesting efforts, E_1 and E_2, which is associated with the stable equilibrium point T_3 that maximize the net present value of revenues.

Acknowledgments
This research is supported by Ministry of Research and Higher Education of Republic Indonesia via LP2M Hasanuddin University grant competition of Professorship Research Scheme year 2017. This financial support is highly appreciated.

References

[1] Hou H F, Jiang H M and Meng X Y 2012 A dynamic model for fishery resource with reserve area and taxation *Journal of Applied Mathematics* http://dx.doi.org/10.1155/2012/794719

[2] Chakraborty K, Chakraborty M and Kar T K 2011 Bifurcation and control of a bioeconomic model of a prey-predator system with time delay *Nonlinear Analysis: Hybrid Systems* **5** 613-625

[3] Gupta R P, Banerjee M and Chandra P 2012 Bifurcation analysis and control of Leslie–Gower predator–prey model with Michaelis–Menten type prey-harvesting *Journal Differential Equations and Dynamical Systems* **22** 3 339-366

[4] Kar T K 2010 A Dynamic reaction model of a prey-predator system with stage-structure for predator *Modern Applied Science* **4** 5 183-195

[5] Laksmi G M V, Gunasekaran M and Vijaya S 2014 Bifurcation analysis of predator model with harvested predator *International Journal of Engineering Research and Development* **10** 6 42-51

[6] Chakraborty K, Chakraborty M and Kar T K 2011 Optimal control of harvest and bifurcation of a prey-predator model with stage structure *Applied Mathematics and Computation* **217** 21 8778-8792

[7] Qu Y and Wei J 2010 Bifurcation analysis in a predator-prey system with stage structure and harvesting *Journal of Franklin Institute* **347** 7 1096-1113

[8] Chakraborty K, Jana S and Kar T K 2012 Global dynamics and bifurcation in a stage structured prey-predator fishery model with harvesting *Applied Mathematics and Computation* **218** 5 9271-9290

[9] Toaha S, Kusuma J, Khaeruddin and Mawardi 2014 Stability analysis and optimal harvesting policy of prey-predator model with stage structure for predator *Applied Mathematical Sciences* **8** 159 7923-7934

[10] Ji C, Jiang D and Shi N 2009 Analysis of a predator-prey model with modified Leslie-Gower and Holling-type II schemes with stochastic perturbation *Journal of Mathematical Analysis and Applications* **359** 482-498

[11] Hou H F, Wang X and Chavez C C 2011 Dynamics of a stage-structural Leslie-Gower predator-prey model *Mathematical Problems in Engineering* doi:10.1155/2011/149341

[12] Yue Q 2016 Dynamics of a modified Leslié-Gower predator-prey model with Holling-type II schemes and a prey refuge *Springerplus*. 5: 461. doi:10.1186/s40064-016-2087-7

[13] Keshet L E 2005 *Mathematical models in biology* (Philadelphia: SIAM)

[14] Clark C W 1985 *Bioeconomic modelling and fisheries management* (New York: John Wiley & Sons)

[15] Grass D, Caulkins J P, Feichtinger G, Tragler G and Behrens D A 2003 *Optimal control of nonlinear processes*. (Berlin Heidelberg: Springer-Verlag)

IOP Conf. Series: Journal of Physics: Conf. Series **979** (2018) 012070 doi:10.1088/1742-6596/979/1/012070

Irregular Assignment of Series Parallel Networks

Nurdin

Department of Mathematics, Faculty of Mathematics and Natural Sciences, Hasanuddin Hasanuddin

E-mail: nurdin1701@unhas.ac.id

Abstract. In this paper, we determined that irregularity strength and the total vertex irregularity strength of the parallel networks, $sp(m, r, 2)$ for the natural number $m, r \geq 3$, is $mr + 1$ *for* $m \geq 4, r \geq 3$ and $\left\lceil\frac{2mr+2}{3}\right\rceil$, *for* $m, r \geq 3$, respectively.

1. Introduction

In 1736, a Swiss mathematician, Leonhard Euler was introduced graph theory. Research on graph theory continues to grow along with the development of information technology. One of the most developed is a graphs labelling. The study object of graph labelling are vertex set, edge set, and number (generally positive integers). Graph labelling introduced by Sadlack, then Stewart, [1], Kotzig, and Rosa [2].

Formally, *graph labelling* is a function that pairing graph elements to a set of numbers (usually integer number). If the domain of this labelling is vertex set, then this labelling called *vertex labelling*. If the domain of this labelling is edge set, then this labelling called *edge labelling*. If the domain of this labelling is a joint of vertex and edge sets, then called *total labelling* [3].

An edge labelling of a graph can be done in many ways. One of them is irregular labelling. The irregular labelling on a graph was introduced by Chartrand. A edge labelling $f: E \rightarrow \{1, 2, \cdots, s\}$ is called a *irregular s-labelling* of G if every two distinct vertices x and y in V satisfy $wt(x) \neq wt(y)$, where $wt(x) = \sum_{xz \in E(G)} f(xz)$. The minimum s for which a graph G has an irregular s-labelling, denoted by (G), is called the *irregularity labelling* of G [4].

In 2007, Bača [5] introduced a vertex and edge irregular total labellings. A total labelling $f: V \cup E \rightarrow \{1, 2, \cdots, k\}$ is called a *vertex irregular total k-labelling* of G if every two distinct vertices x and y in G satisfy $wt(x) \neq wt(y)$, where $wt(x) = f(x) + \sum_{xz \in E(G)} f(xz)$. The minimum k for which a graph G has a vertex irregular total k-labelling, denoted by $tvs(G)$, is called the *total vertex irregularity strength* of G [5].

Some experts have determined the value of irregularity labelling of some graphs. Ahmad et al, have determined irregular labelling of helm and sun graphs [6]. Besides that, Ahmad et.al have determined the total vertex and the total edge irregularity strength of Halin graph [7]. In 2012, Anolcher and Palmer determined the irregular labelling of circulant graphs [8]. Chartrand et al. [4], has determined irregular labelling of regular-d graph and posed the following.

Theorem 1. *Let G is a regular-d graph, then the lower bound for irregularity strength of G, for* $d \geq 2$ *is* $s(G) \geq \left\lceil\frac{n+d-1}{d}\right\rceil$.

The total vertex irregularity strength of a graph has been done by some experts. Rajasingh have determined the total vertex irregularity strength of triangular related graph [9]. Nurdin et al. determined the total vertex irregularity strength of tree graphs and posed the following [10]. In [11], Nurdin et. al have determined the total irregularity strength of network constructed by cycle. Recently, Nurdin have determined the total vertex irregularity strength of butterfly network [12].

Theorem 2. *Let G be a connected graph having n_i vertices of degree $i(i = \delta, \delta + 1, \delta + 2, \dots, \Delta)$ where δ and Δ are minimum and maximum degree of G, respectively. Then*

$$tvs(G) \geq maks\left\{\left\lceil\frac{\delta + n_\delta}{\delta + 1}\right\rceil, \left\lceil\frac{\delta + n_\delta + n_{\delta+1}}{\delta + 2}\right\rceil, \dots, \left\lceil\frac{\delta + \sum_{i=\delta}^{\Delta} n_i}{\Delta + 1}\right\rceil\right\}.$$

In 2015, Rajasingh have determined the total edge irregularity strength of series parallel graphs [9]. It has not yet determined the irregular labelling of series parallel graph and the total vertex irregularity strength of series parallel graph. This study aims to determine the greatest lower bound and the smallest upper bound to obtain the irregular labelling of series parallel graphs and the total vertex irregularity strength of series parallel graphs are exact.

2. Series Parallel Graph

Rajasingh [9, 13] gave the definition of series parallel graph as follows.

Definition 1. *The series parallel graph of G is a chain graph in which each block is generalized theta graph. The series parallel is denoted by $sp(m, r, l)$ for $m, r \geq 3$ and $l \geq 2$.*

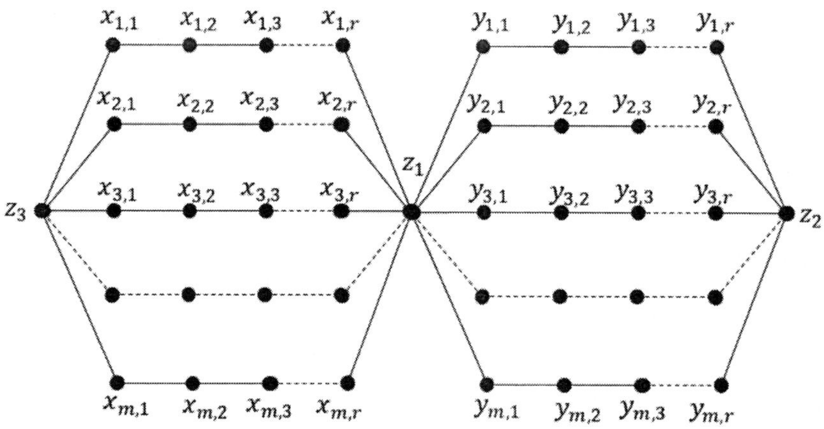

Figure 1. Series Parallel Graph $\mathrm{sp(m, r, 2)}$.

Figure 1 show the series parallel graph for $\mathrm{l} = 2$. Defined vertex set of series parallel graph $\mathrm{V} = \{x_{i,j}, y_{i,j} \mid j = 1,2, \dots, r\} \cup \{z_1, z_2, z_3\}$ and edge set of is

$$\mathrm{E} = \{z_3x_{i,1}, z_1x_{i,r}, z_1y_{i,1}, z_2y_{i,r}\} \cup \{x_{i,j}x_{i,j+1}, y_{i,j}y_{i,j+1} \mid j = 1,2, \dots, r-1\}, \text{ for } i = 1,2, \dots, m.$$

3. Irregular Labelling of Series Parallel Graph

In determining of the irregular labelling of series parallel graph, begin with determining the lower and upper bound of irregularity strength of the graph. The lower bound of irregularity strength of a series parallel graph $sp(m, r, 2)$ for $\mathrm{m} = 4,5,6$ and $\mathrm{r} = 3$ is shown in table 1 which is analyzed by using series parallel properties and based on Theorem 1.

Table 1. Lower Bound of Irregularity Strength sp(m, 3,2).

m	δ	n_δ	$\left\lceil \frac{n_\delta + \delta - 1}{\delta} \right\rceil \leq s(sp(m,3,2))$
4	2	24	13
5	2	30	16
6	2	36	19

The upper bound is analyzed by edge labelling on the series parallel sp(m, r, 2) for m = 4,5,6 and r = 3 by maintaining pattern labelling as follow.

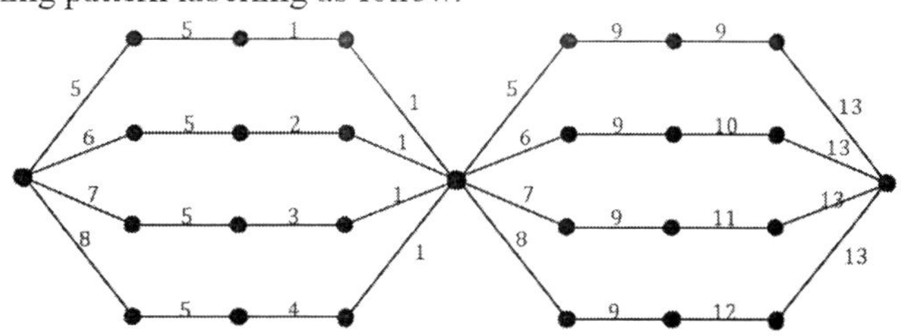

Figure 2. Irregular Labelling-13 of sp(4,3,2).

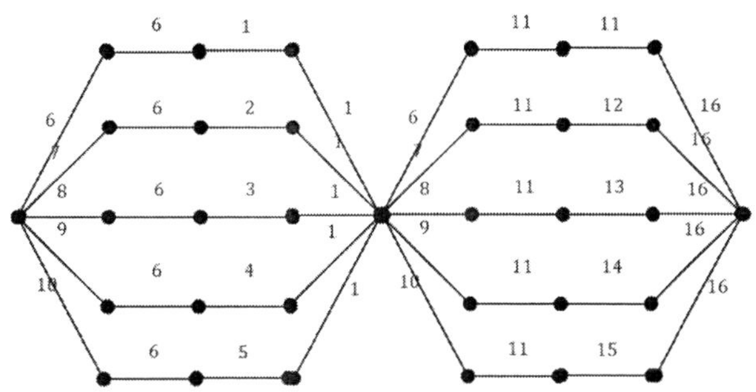

Figure 3. Irregular Labelling-16 of sp(5,3,2).

Based on table 1, figure 2 and figure 3, in general for any m ≥ 4, r ≥ 3, irregular labelling of series parallel (sp(m, r, 2)) can be estimated as

Table 2. Irregular Labelling $sp(m, r, 2)$.

m \ r	**3**	**4**	**5**	⋯	**r**
4	13	17	21	⋯	$(4.r) + 1$
5	16	21	26	⋯	$(5.r) + 1$
6	19	25	31	⋯	$(6.r) + 1$
⋮	⋮	⋮	⋮	⋱	⋮
M	$(m.3) + 1$	$(m.4) + 1$	$(m.5) + 1$	⋯	$mr + 1$

Based on table 2 it is assumed that irregular labelling of series parallel graph is $(sp(m,r,2)) = mr + 1$. These results are written on the theorem as follows.

Theorem 3. *For $m \geq 4$ and $r \geq 3$, then irregular labelling of series parallel $sp(m,r,2)$ is*

$$s(sp(m,r,2)) = mr + 1.$$

Proof
To prove that $s(sp(m,r,2)) \geq mr + 1$, use Theorem 1. Since the minimum degree of $sp\,(m,r,2)$ is $\delta = 2$, the number of vertex of degree 2 is $n_2 = 2mr$. Then based on Theorem 1 obtained

$$s(sp(m,r,2)) \geq \left\lceil \frac{n_\delta + \delta - 1}{\delta} \right\rceil = \left\lceil \frac{2mr + 2 - 1}{2} \right\rceil = mr + 1.$$

Next, we will determine the upper bound of $s(sp\,(m,r,2))$ is $mr + 1$. To prove it, we will construction an irregular labelling on the series parallel $sp(m,r,2)$ as follows. Let $s = mr + 1$.

Case I. For r is odd
For $i = 1,2,\dots,m$, defined λ as follows:

$$\lambda(z_3x_{i,1}) = s - \left(\frac{m}{2}(r+1) - i + 1\right)$$

$$\lambda(x_{i,j}x_{i,j+1}) = \begin{cases} s - \left(\frac{m}{2}(r+j)\right), & j = 1,3,\dots,r-2 \\ s - \left(\frac{m}{2}(r+j+1) - i + 1\right), & j = 2,4,\dots,r-1 \end{cases}$$

$$\lambda(z_1x_{i,r}) = 1$$

$$\lambda(z_1y_{i,1}) = s - \left(\frac{m}{2}(r+1) - i + 1\right)$$

$$\lambda(y_{i,j}y_{i,j+1}) = \begin{cases} s - \left(\frac{m}{2}(r-j)\right), & j = 1,3,\dots,r-2 \\ s - \left(\frac{m}{2}(r-j+1) - i + 1\right), & j = 2,4,\dots,r-1 \end{cases}$$

$$\lambda(z_2y_{i,r}) = s.$$

Case II. For r is even
For $i = 1,2,\dots,m$, defined λ as follows:

$$\lambda(z_3x_{i,1}) = s - \left(\frac{mr}{2}\right)$$

$$\lambda(x_{i,j}x_{i,j+1}) = \begin{cases} s - \left(\frac{m}{2}(r+j+1) - i + 1\right), & j = 1,3,\dots,r-1 \\ s - \left(\frac{m}{2}(r+j)\right), & j = 2,4,\dots,r-2 \end{cases}$$

$$\lambda(z_1x_{i,r}) = 1$$

$$\lambda(z_1y_{i,1}) = s - \left(\frac{mr}{2}\right)$$

$$\lambda(y_{i,j}y_{i,j+1}) = \begin{cases} s - \left(\frac{m}{2}(r-j+1) - i + 1\right), & j = 1,3,\dots,r-1 \\ s - \left(\frac{m}{2}(r-j)\right), & j = 2,4,\dots,r-2 \end{cases}$$

$$\lambda(z_2y_{i,r}) = s.$$

Based on the labelling function λ, we obtained the weight of all vertices of graph $sp(m,r,2)$ for $i = 1,2,\dots,m$ and $j = 1,2,\dots,r-1$ as follows.

Case I. For r is odd

$$wt(x_{i,r}) = \lambda(x_{i,r-1}x_{i,r}) + \lambda(z_1x_{i,r}) = s - mr + i.$$

$$wt(x_{i,j}) = \lambda(x_{i,j-1}x_{i,j}) + \lambda(x_{i,j}x_{i,j+1}) = 2s - 2m(r+j) - \frac{m}{2} + i - 1.$$

$$wt(x_{i,1}) = \lambda(z_3x_{i,1}) + \lambda(x_{i,1}x_{i,2}) = 2s - mr - m + i - 1.$$

$$wt(y_{i,1}) = \lambda(z_1y_{i,1}) + \lambda(y_{i,1}y_{i,2}) = 2s - mr + i - 1.$$

$wt(y_{i,j}) = \lambda(y_{i,j-1}y_{i,j}) + \lambda(y_{i,j}y_{i,j+1}) = 2s - m(r-j) - \frac{m}{2} + i - 1.$
$wt(y_{i,r}) = \lambda(y_{1,r-1}y_{1,r}) + \lambda(z_2y_{1,r}) = 2s - m + i - 1.$
$wt(z_3) = \sum_{i=1}^{m}\lambda(z_3x_{i,1}) = \sum_{i=1}^{m}\left(s - \left(\frac{m}{2}(r+1) - i + 1\right)\right).$
$wt(z_1) = \sum_{i=1}^{m}\lambda(z_1x_{i,r}) + \sum_{i=1}^{m}\lambda(z_1y_{i,1}) = m + \sum_{i=1}^{m}\left(s - \left(\frac{m}{2}(r+1) - i + 1\right)\right).$
$wt(z_2) = \sum_{i=1}^{m}\lambda(z_2y_{i,r}) = ms.$

Case II. For r is even
$wt(x_{i,r}) = \lambda(x_{i,r-1}x_{i,r}) + \lambda(z_1x_{i,r}) = s - mr + i.$
$wt(x_{i,j}) = \lambda(x_{i,j-1}x_{i,j}) + \lambda(x_{i,j}x_{i,j+1}) = 2s - 2m(r+j) - \frac{m}{2} + i - 1.$
$wt(x_{i,1}) = \lambda(z_3x_{i,1}) + \lambda(x_{i,1}x_{i,2}) = 2s - mr - m + i - 1.$
$wt(y_{i,1}) = \lambda(z_1y_{i,1}) + \lambda(y_{i,1}y_{i,2}) = 2s - mr + i - 1.$
$wt(y_{i,j}) = \lambda(y_{i,j-1}y_{i,j}) + \lambda(y_{i,j}y_{i,j+1}) = 2s - m(r-j) - \frac{m}{2} + i - 1.$
$wt(y_{i,r}) = \lambda(y_{1,r-1}y_{1,r}) + \lambda(z_2y_{1,r}) = 2s - m + i - 1.$
$wt(z_3) = \sum_{i=1}^{m}\lambda(z_3x_{i,1}) = \sum_{i=1}^{m}\left(s - \left(\frac{mr}{2}\right)\right).$
$wt(z_1) = \sum_{i=1}^{m}\lambda(z_1x_{i,r}) + \sum_{i=1}^{m}\lambda(z_1y_{i,1}) = m + \sum_{i=1}^{m}\left(s - \left(\frac{mr}{2}\right)\right).$
$wt(z_2) = \sum_{i=1}^{m}\lambda(z_2y_{i,r}) = ms.$

Based on the definition of the weight of the vertices, obtained

$$wt(x_{1,r}) < wt(x_{2,r}) < \cdots < wt(x_{m,r}) < wt(x_{1,r-1}) < wt(x_{1,r-1}) < \cdots < wt(x_{m,r-1})$$
$$< wt(x_{1,r-2}) < \cdots < wt(x_{1,1}) < wt(x_{2,1}) < \cdots < wt(x_{m,1}) < wt(y_{1,1}) < wt(y_{2,1}) < \cdots$$
$$< wt(y_{m,1}) < wt(y_{1,2}) < wt(y_{2,2}) < \cdots < wt(y_{m,2}) < wt(y_{1,3}) < \cdots < wt(y_{1,r})$$
$$< wt(y_{2,r}) < \cdots < wt(y_{m,r}) < wt(z_3) < wt(z_1) < wt(z_2).$$

So it can be concluded that the weight of each vertex on $sp(m,r,2)$ is different. Then the λ constructed is an irregular labelling on $sp(m,r,2)$. Thus λ is an irregular labelling$-s$ with $s = mr + 1$ for $m \geq 4, r \geq 3$. That is $s(sp(m,r,2)) \leq mr + 1$.

4. Total Vertex Irregularity Strength of Series Parallel

In determining the total vertex irregularity strength of series parallel graph, begin from determine the lower and upper bound of the total vertex irregularity strength. The lower bound of series parallel for $m = 3,4,5$ and $r = 3$ is shown in table 2 which is analyzed by using series parallel properties and based on Theorem 2.

Table 3. Lower Bound of Total Labelling $sp(m,3,2)$

m	δ	n_δ	$\left\lceil\frac{\delta+n_\delta}{\delta+1}\right\rceil \leq tvs(sp(m,3,2))$
3	2	18	7
4	2	24	9
5	2	30	11

The upper bound is analyzed by total labelling on the series parallel $\mathrm{sp}(\mathrm{m}, \mathrm{r}, 2)$ for $\mathrm{m} = 3,4,5$ and $\mathrm{r} = 3$ by maintaining pattern labelling as follows.

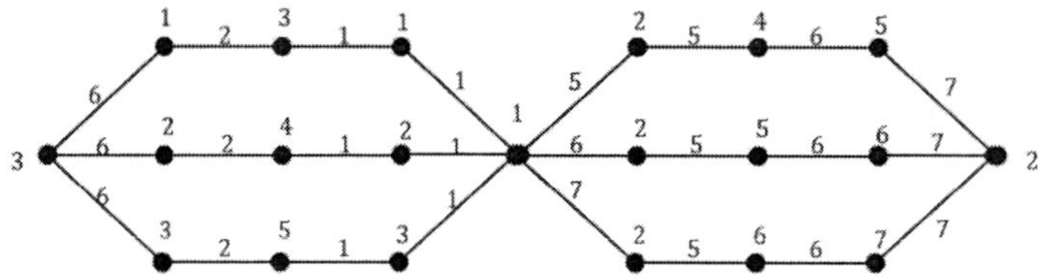

Figure 4. Irregular total vertex 7 – labelling sp(3,3,2).

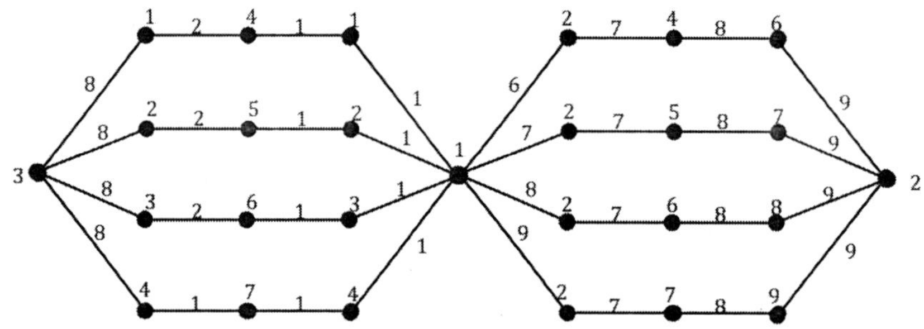

Figure 5. Irregular Total Vertex 9 – labelling sp(4,3,2).

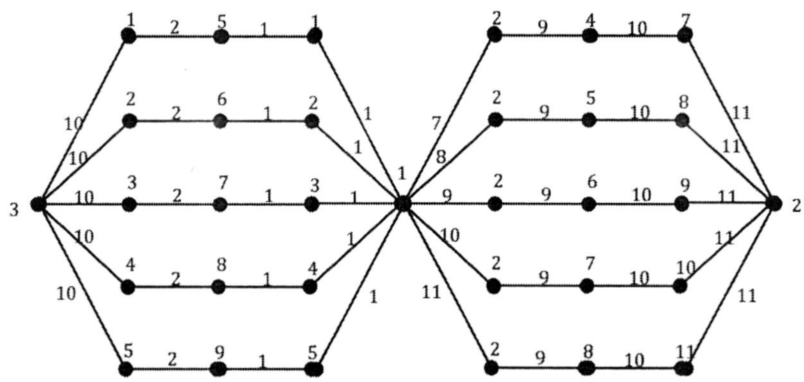

Figure 6. Irregular Total Vertex 11 – labelling sp(5,3,2).

Based on table 3, figure 3 and figure 4, in general for any $\mathrm{m}, \mathrm{r} \geq 3$, the total vertex irregularity strength of series parallel $(\mathrm{sp}(\mathrm{m}, \mathrm{r}, 2))$ can be estimated like in table 4.

Based on table 3 it is assumed that total vertex irregularity strength of series parallel graph is $(\mathrm{sp}(\mathrm{m}, \mathrm{r}, 2)) = \left\lceil \frac{2\mathrm{mr}+2}{3} \right\rceil$. These results are written on the theorem as follows.

Theorem 4. *For $m, r \geq 3$, then total vertex irregularity strength of series parallel $sp(m, r, 2)$ is*

$$tvs(sp(m, r, 2)) = \left\lceil \frac{2mr + 2}{3} \right\rceil.$$

Proof.
To prove that $tvs(\mathrm{sp}(\mathrm{m}, \mathrm{r}, 2)) \geq \left\lceil \frac{2\mathrm{mr}+2}{3} \right\rceil$, use Theorem 2. The minimum degree of $\mathrm{sp}\ (\mathrm{m}, \mathrm{r}, 2)$ is $\delta = 2$, the number of vertex of degree 2 is $\mathrm{n}_2 = 2\mathrm{mr}$. Then based on Theorem 2 we obtained

$$\mathrm{tvs}(\mathrm{sp}(\mathrm{m},\mathrm{r},2)) \geq \mathrm{maks}\left\{\left\lceil\frac{2\mathrm{mr}+2}{3}\right\rceil, \left\lceil\frac{2\mathrm{mr}+\mathrm{m}+4}{\mathrm{m}}\right\rceil, \left\lceil\frac{2\mathrm{mr}+3\mathrm{m}+3}{2\mathrm{m}+1}\right\rceil\right\} = \left\lceil\frac{2\mathrm{mr}+2}{3}\right\rceil.$$

Table 4. Total Vertex Irregularity Strength $\mathrm{sp}(\mathrm{m},\mathrm{r},2)$.

	3	**4**	**5**	⋯	**r**
3	7	9	11	⋯	$\left\lceil\frac{(2.3.r)+2}{3}\right\rceil$
4	9	12	14	⋯	$\left\lceil\frac{(2.3.r)+2}{3}\right\rceil$
5	11	14	18	⋯	$\left\lceil\frac{(2.3.r)+2}{3}\right\rceil$
⋮	⋮	⋮	⋮	⋱	⋮
m	$\left\lceil\frac{(2.m.3)+2}{3}\right\rceil$	$\left\lceil\frac{(2.m.4)+2}{3}\right\rceil$	$\left\lceil\frac{(2.m.5)+2}{3}\right\rceil$	⋯	$\left\lceil\frac{(2.m.r)+2}{3}\right\rceil$

Next determine the upper bound of $\mathrm{tvs}(\mathrm{sp}\,(\mathrm{m},\mathrm{r},2)) \leq \left\lceil\frac{2\mathrm{mr}+2}{3}\right\rceil$. To prove it we will construct a vertex irregular total labelling on the series parallel $\mathrm{sp}(\mathrm{m},\mathrm{r},2)$. Let $\mathrm{t} = \left\lceil\frac{2\mathrm{mr}+2}{3}\right\rceil$.

Case I. For m = 0 (mod 3)

$$f(y_{i,j}) = t - \left(\frac{m}{3}(r-j+4) - i - 1\right)$$

$$f(y_{i,r}) = t - (m-i)$$

$$f(z_i) = i, \quad i = 1,2,3$$

$$f(z_3x_{i,1}) = t - (m-2)$$

$$f(x_{i,j}x_{i,j+1}) = t - \left(\frac{m}{3}(r+j+1)\right)$$

$$f(z_1x_{i,r}) = 1$$

$$f(z_1y_{i,1}) = t - (m-i)$$

$$f(y_{i,j}y_{i,j+1}) = t - \left(\frac{m}{3}(r-j-1) + 1\right)$$

$$f(z_2y_{i,r}) = t$$

$$f(x_{i,1}) = t - \left(\frac{m}{3}(2r-2) - i + 3\right)$$

$$f(x_{i,j}) = t - \left(\frac{m}{3}(r+j-1) - i + 1\right)$$

$$f(x_{i,r}) = t - \left(\frac{2mr}{3} - i + 1\right)$$

$$f(y_{i,1}) = t - \left(\frac{m}{3}(2r-1)\right)$$

IOP Conf. Series: Journal of Physics: Conf. Series **979** (2018) 012070 doi:10.1088/1742-6596/979/1/012070

Case II. For m = 1 (mod 3)

$$f(x_{i,1}) = \begin{cases} t - \left(\frac{m}{3}(2r-2) - \frac{r}{3} - i + \frac{14}{3}\right), & r = 0 \bmod 3 \\ t - \left(\frac{m}{3}(2r-2) - \frac{r}{3} - i + \frac{16}{3}\right), & r = 1 \bmod 3 \\ t - \left(\frac{m}{3}(2r-2) - \frac{r}{3} - i + \frac{12}{3}\right), & r = 2 \bmod 3 \end{cases}$$

$$f(x_{i,j}) = \begin{cases} t - \left(\frac{m}{3}(r+j-1) - \frac{2(j-r)}{3} - i + \frac{4}{3}\right), & r = 0 \bmod 3 \\ t - \left(\frac{m}{3}(r+j-1) - \frac{2(j-r)}{3} - i + \frac{5}{3}\right), & r = 1 \bmod 3 \\ t - \left(\frac{m}{3}(r+j-1) - \frac{2(j-r)}{3} - i + \frac{3}{3}\right), & r = 2 \bmod 3 \end{cases}$$

$$f(x_{i,r}) = \begin{cases} t - \left(\frac{2mr}{3} - i + 1\right), & r = 0 \bmod 3 \\ t - \left(\frac{2mr}{3} - i + \frac{4}{3}\right), & r = 1 \bmod 3 \\ t - \left(\frac{2mr}{3} - i + \frac{2}{3}\right), & r = 2 \bmod 3 \end{cases}$$

$$f(y_{i,1}) = \begin{cases} t - \left(\frac{m}{3}(2r-1) + \frac{1}{3}\right), & r = 0 \bmod 3 \\ t - \left(\frac{m}{3}(2r-1) + \frac{2}{3}\right), & r = 1 \bmod 3 \\ t - \left(\frac{m}{3}(2r-1) + 1\right), & r = 2 \bmod 3 \end{cases}$$

$$f(y_{i,j}) = \begin{cases} t - \left(\frac{m}{3}(r-j+4) - \frac{2(r-j)}{3} - i - \frac{4}{3}\right), & r = 0 \bmod 3 \\ t - \left(\frac{m}{3}(r-j+4) - \frac{2(r-j)}{3} - i - \frac{1}{3}\right) & r = 1 \bmod 3 \\ t - \left(\frac{m}{3}(r-j+4) - \frac{2(r-j)}{3} - i - \frac{1}{3}\right) & r = 2 \bmod 3 \end{cases}$$

$$f(y_{i,r}) = \begin{cases} t - (m-i), & r = 0 \bmod 3 \\ t - (m-i+1), & r = 1 \bmod 3 \\ t - (m-i), & r = 2 \bmod 3 \end{cases}$$

$$f(z_i) = i, \quad i = 1,2,3$$

$$f(z_3 x_{i,1}) = t - (m-3)$$

$$f(x_{i,j} x_{i,j+1}) = \begin{cases} t - \left(\frac{m}{3}(r+j+1) + \frac{r-j-1}{3}\right), & r = 0 \bmod 3 \\ t - \left(\frac{m}{3}(r+j+1) + \frac{r-j}{3}\right), & r = 1 \bmod 3 \\ t - \left(\frac{m}{3}(r+j+1) + \frac{r-j-2}{3}\right), & r = 2 \bmod 3 \end{cases}$$

$$f(z_1 x_{i,r}) = 1$$

$$f(z_1y_{i,1}) = \begin{cases} t-\left(m+\dfrac{r-3}{3}-i\right), & r=0 \bmod 3 \\ t-\left(m+\dfrac{r-1}{3}-i\right), & r=1 \bmod 3 \\ t-\left(m+\dfrac{r-5}{3}-i\right), & r=2 \bmod 3 \end{cases}$$

$$f(y_{i,j}y_{i,j+1}) = \begin{cases} t-\left(\dfrac{m}{3}(r-j-1)+\dfrac{j-r+4}{3}\right), & r=0 \bmod 3 \\ t-\left(\dfrac{m}{3}(r-j-1)+\dfrac{j-r+4}{3}\right), & r=1 \bmod 3 \\ t-\left(\dfrac{m}{3}(r-j-1)+\dfrac{j-r+1}{3}\right), & r=2 \bmod 3 \end{cases}$$

$$f(z_2y_{i,r}) = t$$

Case III. For m = 2 (mod 3)

$$f(x_{i,1}) = \begin{cases} t-\left(\dfrac{m}{3}(2r-2)-\dfrac{2r}{3}-i+\dfrac{19}{3}\right), & r=0 \bmod 3 \\ t-\left(\dfrac{m}{3}(2r-2)-\dfrac{2r}{3}-i+\dfrac{17}{3}\right), & r=1 \bmod 3 \\ t-\left(\dfrac{m}{3}(2r-2)-\dfrac{2r}{3}-i+\dfrac{21}{3}\right), & r=2 \bmod 3 \end{cases}$$

$$f(x_{i,j}) = \begin{cases} t-\left(\dfrac{m}{3}(r+j-1)-\dfrac{4(j-r)}{3}-i+\dfrac{5}{3}\right), & r=0 \bmod 3 \\ t-\left(\dfrac{m}{3}(r+j-1)-\dfrac{4(j-r)}{3}-i+\dfrac{4}{3}\right), & r=1 \bmod 3 \\ t-\left(\dfrac{m}{3}(r+j-1)-\dfrac{4(j-r)}{3}-i+\dfrac{6}{3}\right), & r=2 \bmod 3 \end{cases}$$

$$f(x_{i,r}) = \begin{cases} t-\left(\dfrac{2mr}{3}-i+1\right), & r=0 \bmod 3 \\ t-\left(\dfrac{2mr}{3}-i+\dfrac{2}{3}\right), & r=1 \bmod 3 \\ t-\left(\dfrac{2mr}{3}-i+\dfrac{4}{3}\right), & r=2 \bmod 3 \end{cases}$$

$$f(y_{i,1}) = \begin{cases} t-\left(\dfrac{m}{3}(2r-1)+\dfrac{2}{3}\right), & r=0 \bmod 3 \\ t-\left(\dfrac{m}{3}(2r-1)+\dfrac{4}{3}\right), & r=1 \bmod 3 \\ t-\left(\dfrac{m}{3}(2r-1)+1\right), & r=2 \bmod 3 \end{cases}$$

$$f(y_{i,j}) = \begin{cases} t-\left(\dfrac{m}{3}(r-j+4)-\dfrac{4(r-j)}{3}-i-\dfrac{5}{3}\right), & r=0 \bmod 3 \\ t-\left(\dfrac{m}{3}(r-j+4)-\dfrac{4(r-j)}{3}-i-\dfrac{2}{3}\right) & r=1 \bmod 3 \\ t-\left(\dfrac{m}{3}(r-j+4)-\dfrac{4(r-j)}{3}-i-\dfrac{2}{3}\right) & r=2 \bmod 3 \end{cases}$$

$$f(y_{i,r}) = \begin{cases} t-(m-i), & r=0 \bmod 3 \\ t-(m-i), & r=1 \bmod 3 \\ t-(m-i+1), & r=2 \bmod 3 \end{cases}$$

$$f(z_i) = i, \quad i = 1,2,3$$

$f(z_3x_{i,1}) = t - (m-4)$

$$f(x_{i,j}x_{i,j+1}) = \begin{cases} t - \left(\frac{m}{3}(r+j+1) + \frac{2(r-j)}{3} - \frac{2}{3}\right), & r = 0 \bmod 3 \\ t - \left(\frac{m}{3}(r+j+1) + \frac{2(r-j)}{3} - \frac{3}{3}\right), & r = 1 \bmod 3 \\ t - \left(\frac{m}{3}(r+j+1) + \frac{2(r-j)}{3} - \frac{1}{3}\right), & r = 2 \bmod 3 \end{cases}$$

$f(z_1x_{i,r}) = 1$

$$f(z_1y_{i,1}) = \begin{cases} t - \left(m + \frac{2(r-3)}{3} - i\right), & r = 0 \bmod 3 \\ t - \left(m + \frac{2(r-4)}{3} - i\right), & r = 1 \bmod 3 \\ t - \left(m + \frac{2(r-2)}{3} - i\right), & r = 2 \bmod 3 \end{cases}$$

$$f(y_{i,j}y_{i,j+1}) = \begin{cases} t - \left(\frac{m}{3}(r-j-1) + \frac{2(j-r)}{3} + \frac{5}{3}\right), & r = 0 \bmod 3 \\ t - \left(\frac{m}{3}(r-j-1) + \frac{2(j-r)}{3} + \frac{2}{3}\right), & r = 1 \bmod 3 \\ t - \left(\frac{m}{3}(r-j-1) + \frac{2(j-r)}{3} + \frac{5}{3}\right), & r = 2 \bmod 3 \end{cases}$$

$f(z_2y_{i,r}) = t$

Based on the labelling function f, then we obtained that the weight of all vertices of a graph $sp(m, r, 2)$ for $i = 1,2, \dots, m$ and $j = 1,2, \dots, r-1$ as follows.

Case I For m = 0 (mod 3)

$wt(x_{i,r}) = f(x_{i,r}) + f(z_1x_{i,r}) + f(x_{i,r-1}x_{i,r}) = 2t - \frac{4mr}{3} + i$

$wt(x_{i,j}) = f(x_{i,j}) + f(x_{i,j-1}x_{i,j}) + f(x_{i,j}x_{i,j+1}) = 3t - m(r+j) + i - 1$

$wt(x_{i,1}) = f(x_{i,1}) + f(z_3x_{i,1}) + f(x_{i,1}x_{i,2}) = 3t - mr - m + i - 1$

$wt(y_{i,1}) = f(y_{i,1}) + f(z_1y_{i,1}) + f(y_{i,1}y_{i,2}) = 3t - mr + i - 1$

$wt(y_{i,j}) = f(y_{i,j}) + f(y_{i,j-1}y_{i,j}) + f(y_{i,j}y_{i,j+1}) = 3t - m(r-j+1) + i - 1$

$wt(y_{i,r}) = f(y_{i,r}) + f(z_2y_{i,r}) + f(y_{i,r-1}y_{i,r}) = 3t - m + i - 1$

$wt(z_3) = \sum_{i=1}^{m} f(z_3x_{i,1}) + f(z_3) = m(t - (m-2)) + 3$

$wt(z_1) = \sum_{i=1}^{m} f(z_1x_{i,r}) + f(z_1) + \sum_{i=1}^{m} f(z_1y_{i,1}) = m + 1 + \sum_{i=1}^{m} t - (m-i)$

$wt(z_2) = \left(\sum_{i=1}^{m} f(z_2y_{i,r})\right) + f(z_2) = mt + 2$

Case II For m = 1 (mod 3). There are three subcases as follows.

Subcase 1.For r = 0 (mod 3)

$wt(x_{i,r}) = f(x_{i,r}) + f(z_1x_{i,r}) + f(x_{i,r-1}x_{i,r}) = 2t - \frac{4mr}{3} + i$

$wt(x_{i,j}) = f(x_{i,j}) + f(x_{i,j-1}x_{i,j}) + f(x_{i,j}x_{i,j+1}) = 3t - m(r+j) + i - 1$

$wt(x_{i,1}) = f(x_{i,1}) + f(z_3x_{i,1}) + f(x_{i,1}x_{i,2}) = 3t - mr - m + i - 1$

$wt(y_{i,1}) = f(y_{i,1}) + f(z_1 y_{i,1}) + f(y_{i,1} y_{i,2}) = 3t - mr + i - 1$

$wt(y_{i,j}) = f(y_{i,j}) + f(y_{i,j-1} y_{i,j}) + f(y_{i,j} y_{i,j+1}) = 3t - m(r - j + 1) + i - 1$

$wt(y_{i,r}) = f(y_{i,r}) + f(z_2 y_{i,r}) + f(y_{i,r-1} y_{i,r}) = 3t - m + i - 1$

$wt(z_3) = \sum_{i=1}^{m} f(z_3 x_{i,1}) + f(z_3) = m(t - (m - 3)) + 3$

$wt(z_1) = \sum_{i=1}^{m} f(z_1 x_{i,r}) + f(z_1) + \sum_{i=1}^{m} f(z_1 y_{i,1}) = m + 1 + \sum_{i=1}^{m} t - (m + \frac{r-3}{3} - i)$

$wt(z_2) = (\sum_{i=1}^{m} f(z_2 y_{i,r})) + f(z_2) = mt + 2$

Subcase 2. For r = 1 (mod 3)

$wt(x_{i,r}) = f(x_{i,r}) + f(z_1 x_{i,r}) + f(x_{i,r-1} x_{i,r}) = 2t - \frac{4mr}{3} + i - \frac{2}{3}$

$wt(x_{i,j}) = f(x_{i,j}) + f(x_{i,j-1} x_{i,j}) + f(x_{i,j} x_{i,j+1}) = 3t - m(r + j) + i - 2$

$wt(x_{i,1}) = f(x_{i,1}) + f(z_3 x_{i,1}) + f(x_{i,1} x_{i,2}) = 3t - mr - m + i - 2$

$wt(y_{i,1}) = f(y_{i,1}) + f(z_1 y_{i,1}) + f(y_{i,1} y_{i,2}) = 3t - mr + i - 2$

$wt(y_{i,j}) = f(y_{i,j}) + f(y_{i,j-1} y_{i,j}) + f(y_{i,j} y_{i,j+1}) = 3t - m(r - j + 1) + i - 2$

$wt(y_{i,r}) = f(y_{i,r}) + f(z_2 y_{i,r}) + f(y_{i,r-1} y_{i,r}) = 3t - m + i - 2$

$wt(z_3) = \sum_{i=1}^{m} f(z_3 x_{i,1}) + f(z_3) = m(t - (m - 3)) + 3$

$wt(z_1) = \sum_{i=1}^{m} f(z_1 x_{i,r}) + f(z_1) + \sum_{i=1}^{m} f(z_1 y_{i,1}) = m + 1 + \sum_{i=1}^{m} t - (m + \frac{r-1}{3} - i)$

$wt(z_2) = (\sum_{i=1}^{m} f(z_2 y_{i,r})) + f(z_2) = mt + 2.$

Subcase 3. For r = 2 (mod 3)

$wt(x_{i,r}) = f(x_{i,r}) + f(z_1 x_{i,r}) + f(x_{i,r-1} x_{i,r}) = 2t - \frac{4mr}{3} + i + \frac{2}{3}$

$wt(x_{i,j}) = f(x_{i,j}) + f(x_{i,j-1} x_{i,j}) + f(x_{i,j} x_{i,j+1}) = 3t - m(r + j) + i$

$wt(x_{i,1}) = f(x_{i,1}) + f(z_3 x_{i,1}) + f(x_{i,1} x_{i,2}) = 3t - mr - m + i$

$wt(y_{i,1}) = f(y_{i,1}) + f(z_1 y_{i,1}) + f(y_{i,1} y_{i,2}) = 3t - mr + i$

$wt(y_{i,j}) = f(y_{i,j}) + f(y_{i,j-1} y_{i,j}) + f(y_{i,j} y_{i,j+1}) = 3t - m(r - j + 1) + i$

$wt(y_{i,r}) = f(y_{i,r}) + f(z_2 y_{i,r}) + f(y_{i,r-1} y_{i,r}) = 3t - m + i$

$wt(z_3) = \sum_{i=1}^{m} f(z_3 x_{i,1}) + f(z_3) = m(t - (m - 3)) + 3$

$wt(z_1) = \sum_{i=1}^{m} f(z_1 x_{i,r}) + f(z_1) + \sum_{i=1}^{m} f(z_1 y_{i,1}) = m + 1 + \sum_{i=1}^{m} t - (m + \frac{r-5}{3} - i)$

$wt(z_2) = (\sum_{i=1}^{m} f(z_2 y_{i,r})) + f(z_2) = mt + 2$

Case III For m = 2 (mod 3). In this case, there are three cases as follows.

Subcase 1. For r = 0 (mod 3)

$wt(x_{i,r}) = f(x_{i,r}) + f(z_1 x_{i,r}) + f(x_{i,r-1} x_{i,r}) = 2t - \frac{4mr}{3} + i$

$wt(x_{i,j}) = f(x_{i,j}) + f(x_{i,j-1} x_{i,j}) + f(x_{i,j} x_{i,j+1}) = 3t - m(r + j) + i - 1$

$$wt(x_{i,1}) = f(x_{i,1}) + f(z_3x_{i,1}) + f(x_{i,1}x_{i,2}) = 3t - mr - m + i - 1$$

$$wt(y_{i,1}) = f(y_{i,1}) + f(z_1y_{i,1}) + f(y_{i,1}y_{i,2}) = 3t - mr + i - 1$$

$$wt(y_{i,j}) = f(y_{i,j}) + f(y_{i,j-1}y_{i,j}) + f(y_{i,j}y_{i,j+1}) = 3t - m(r - j + 1) + i - 1$$

$$wt(y_{i,r}) = f(y_{i,r}) + f(z_2y_{i,r}) + f(y_{i,r-1}y_{i,r}) = 3t - m + i - 1$$

$$wt(z_3) = \sum_{i=1}^{m} f(z_3x_{i,1}) + f(z_3) = m(t - (m - 4)) + 3$$

$$wt(z_1) = \sum_{i=1}^{m} f(z_1x_{i,r}) + f(z_1) + \sum_{i=1}^{m} f(z_1y_{i,1}) = m + 1 + \sum_{i=1}^{m} t - (m + \frac{2(r-3)}{3} - i)$$

$$wt(z_2) = \left(\sum_{i=1}^{m} f(z_2y_{i,r})\right) + f(z_2) = mt + 2.$$

Subcase 2. For r = 1 (mod 3)

$$wt(x_{i,r}) = f(x_{i,r}) + f(z_1x_{i,r}) + f(x_{i,r-1}x_{i,r}) = 2t - \frac{4mr}{3} + i + \frac{2}{3}$$

$$wt(x_{i,j}) = f(x_{i,j}) + f(x_{i,j-1}x_{i,j}) + f(x_{i,j}x_{i,j+1}) = 3t - m(r + j) + i$$

$$wt(x_{i,1}) = f(x_{i,1}) + f(z_3x_{i,1}) + f(x_{i,1}x_{i,2}) = 3t - mr - m + i$$

$$wt(y_{i,1}) = f(y_{i,1}) + f(z_1y_{i,1}) + f(y_{i,1}y_{i,2}) = 3t - mr + i$$

$$wt(y_{i,j}) = f(y_{i,j}) + f(y_{i,j-1}y_{i,j}) + f(y_{i,j}y_{i,j+1}) = 3t - m(r - j + 1) + i$$

$$wt(y_{i,r}) = f(y_{i,r}) + f(z_2y_{i,r}) + f(y_{i,r-1}y_{i,r}) = 3t - m + i$$

$$wt(z_3) = \sum_{i=1}^{m} f(z_3x_{i,1}) + f(z_3) = m(t - (m - 4)) + 3$$

$$wt(z_1) = \sum_{i=1}^{m} f(z_1x_{i,r}) + f(z_1) + \sum_{i=1}^{m} f(z_1y_{i,1}) = m + 1 + \sum_{i=1}^{m} t - (m + \frac{2(r-4)}{3} - i)$$

$$wt(z_2) = \left(\sum_{i=1}^{m} f(z_2y_{i,r})\right) + f(z_2) = mt + 2.$$

Subcase 3. For r = 2 (mod 3)

$$wt(x_{i,r}) = f(x_{i,r}) + f(z_1x_{i,r}) + f(x_{i,r-1}x_{i,r}) = 2t - \frac{4mr}{3} + i - \frac{2}{3}$$

$$wt(x_{i,j}) = f(x_{i,j}) + f(x_{i,j-1}x_{i,j}) + f(x_{i,j}x_{i,j+1}) = 3t - m(r + j) + i - 2$$

$$wt(x_{i,1}) = f(x_{i,1}) + f(z_3x_{i,1}) + f(x_{i,1}x_{i,2}) = 3t - mr - m + i - 2$$

$$wt(y_{i,1}) = f(y_{i,1}) + f(z_1y_{i,1}) + f(y_{i,1}y_{i,2}) = 3t - mr + i - 2$$

$$wt(y_{i,j}) = f(y_{i,j}) + f(y_{i,j-1}y_{i,j}) + f(y_{i,j}y_{i,j+1}) = 3t - m(r - j + 1) + i - 2$$

$$wt(y_{i,r}) = f(y_{i,r}) + f(z_2y_{i,r}) + f(y_{i,r-1}y_{i,r}) = 3t - m + i - 2$$

$$wt(z_3) = \sum_{i=1}^{m} f(z_3x_{i,1}) + f(z_3) = m(t - (m - 4)) + 3$$

$$wt(z_1) = \sum_{i=1}^{m} f(z_1x_{i,r}) + f(z_1) + \sum_{i=1}^{m} f(z_1y_{i,1}) = m + 1 + \sum_{i=1}^{m} t - (m + \frac{2(r-2)}{3} - i)$$

$$wt(z_2) = \left(\sum_{i=1}^{m} f(z_2y_{i,r})\right) + f(z_2) = mt + 2.$$

So it can be concluded that the weight of each vertex on $sp(m, r, 2)$ is different. Thus f is a total irregular t −labelling with $t = \left\lceil \frac{2mr+2}{3} \right\rceil$ for $m, r \geq 3$. That is, $tvs(sp(m, r, 2)) \leq \left\lceil \frac{2mr+2}{3} \right\rceil$.

5. Conclusion

There are two theorems in this paper as conclusions, i.e. the irregularity strength of the parallel networks is $\mathbf{mr+1}$ for the natural numbers $\mathbf{m, r \geq 3}$, and the total vertex irregularity strength of the parallel networks is $\left\lceil \frac{2mr+2}{3} \right\rceil$, for the natural numbers $\mathbf{m, r \geq 3}$, and the total vertex irregularity strength of the parallel networks, $\mathbf{sp(m, r, 2)}$ for the natural number $\mathbf{m, r \geq 3}$, is $\mathbf{mr+1}$ for $\mathbf{m \geq 4, r \geq 3}$ and $\left\lceil \frac{2mr+2}{3} \right\rceil$, for $\mathbf{m, r \geq 3}$, respectively.

Acknowledgement
The research for this article was supported by PPI-WCU Hasanuddin University Grant Number: 14613/UN4.1/TU.21/2017.

References

[1] Stewart B M 1966 Magic graphs *Canadian Journal of Mathematics* **18** 1031-1059
[2] Kotzig A and Rosa A 1970 Magic valuations of finite graphs *Canadian Mathematical Bulleting* **13** 451–323
[3] Wallis W D 2001 *Magic graphs* (New Work: Birkhäuser Boston)
[4] Chartrand G, Jacobson M S, Lehel J, Oellermann O R, Ruiz S and Saba F 1988 Irregular Networks *Congressus Numerantium* **64** 197-210
[5] Bača M, Jendrol, Miller M and Ryan J 2007 On Irregular total labellings *Discrete Mathematics* **307** 1378-1388
[6] Ahmad A, Arshad M and Izaricova G 2015 Irregularity labellings of helm and sun graphs *International Journal of Graphs and Combinatorics* **12** 161-168
[7] Ahmad A, Nurdin and Baskoro E T 2015 On total irregularity strength of generalized Halin graph **122** 319 – 332
[8] Anolcher M and Palmer C 2012 Irregular labellings of circulant graphs *Discrete Mathematics* **312** 3416-3466
[9] Rajasingh I and Arockiamary S T 2015 Total edge irregularity strength of series parallel graphs *International Journal of Pure and Applied Mathematics* **99** 11-21
[10] Nurdin, Baskoro E T, Salman A N M and Gaos N N 2010 On the total vertex irregularity strength of trees *Discrete Mathematics* **310** 3043-3048
[11] Nurdin, Massalesse J and Yulandary B B 2016 The total edge irregular labelling of network constructed by some copies of cycle on three vertices corona a vertex *Indian Journal of Science and Technology* **9** 28 1 – 3
[12] Nurdin 2017 Total irregular labelling of butterfly network on level two *AIP Conference Proceedings* 020067-1 - 020067-3
[13] Rajasingh I, Rajan B and Annama V 2012 On total vertex irregularity strength of triangle related graphs *Annals of Pure and Applied Mathematics* **2** 108-116

Prediction of Tidal Elevations and Barotropic Currents in the Gulf of Bone

Rika Purnamasari, Agustinus Ribal and Jeffry Kusuma

Department of Mathematics, Faculty of Mathematics and Natural Sciences, Hasanuddin University, Makassar, Indonesia, 90245

E-mail: agus.ribal@gmail.com

Abstract. Tidal elevation and barotropic current predictions in the gulf of Bone have been carried out in this work based on a two-dimensional, depth-integrated Advanced Circulation (ADCIRC-2DDI) model for 2017. Eight tidal constituents which were obtained from FES2012 have been imposed along the open boundary conditions. However, even using these very high-resolution tidal constituents, the discrepancy between the model and the data from tide gauge is still very high. In order to overcome such issues, Green's function approach has been applied which reduced the root-mean-square error (RMSE) significantly. Two different starting times are used for predictions, namely from 2015 and 2016. After improving the open boundary conditions, RMSE between observation and model decreased significantly. In fact, RMSEs for 2015 and 2016 decreased 75.30% and 88.65%, respectively. Furthermore, the prediction for tidal elevations as well as tidal current, which is barotropic current, is carried out. This prediction was compared with the prediction conducted by Geospatial Information Agency (GIA) of Indonesia and we found that our prediction is much better than one carried out by GIA. Finally, since there is no tidal current observation available in this area, we assume that, when tidal elevations have been fixed, then the tidal current will approach the actual current velocity.

1. Introduction

Tidal information is very important for those who work around coastal waters. In practical applications, the knowledge of the times, heights, and extent of the ebb and flood of tidal waters is of importance. This includes navigation for professional workers such as intra-coastal waterways; construction of bridges, docks, breakwaters, and deep water channels within bays and harbours and commercial fishing. This is also important leisure activities, for example recreational boating, surfing, and swimming [1]. Moreover, prediction of tide is also important for military in which they can use the information for operational planning and execution in marine environments. In addition, knowing the tidal elevation, at high tide, it can be used as beach landing, while at low tide at the same location may strand amphibious vehicles on muddy tidal flats. In addition, high tide can be used to generate electricity as well as wave height as renewable energy [2].

Similarly, prediction of tidal current is also important as it can affect the movement under the sea-floor and the most important is for tidal current energy if they are very strong as shown in [3]. Therefore, in this present work, both tidal elevation and current will be predicted particularly for 2017. However, we will only consider barotropic current.

In order to achieve this goal, **AD**vanced **CIRC**ulation (ADCIRC) model will be employed but it will be limited to two-dimensional depth integrated (ADCIRC – 2DDI) [4-6]. This model has been widely used and verified for ocean circulation and particularly to study tidal elevation and current, see for example in [1,7-10] and the references therein. As pointed out by some researchers, such as Zang et al. [11] , open boundary conditions have significant impact for the regional tidal model. In fact, model output strongly depends on the tidal open boundary conditions. In order to overcome this problem, a Green's function approach will be used. This method was developed by [12,13] and has been proven that the method is very powerful to optimize open boundary conditions for a regional tidal model [3,14,15].

This work will be presented in the following way. Short description of the location and model setup will be presented in section 2. Furthermore, model validation will be presented in section 3. In section 4, we will describe and show the method of optimizing open boundary conditions. Predictions of tidal elevation and barotropic current will be presented in section 5 and followed by summary and conclusion in section 6.

2. Location and methods

2.1. Location of the Gulf of Bone

Gulf of Bone is located between south Sulawesi and southeast Sulawesi provinces in Sulawesi Island which is one of the largest islands in Indonesia as shown in Figure 1. The area of this gulf is about 28,632 km^2 which are between longitudes 120.277^0 E and 121.688^0 E and latitudes 5.417^0 S and 2.610^0 S.

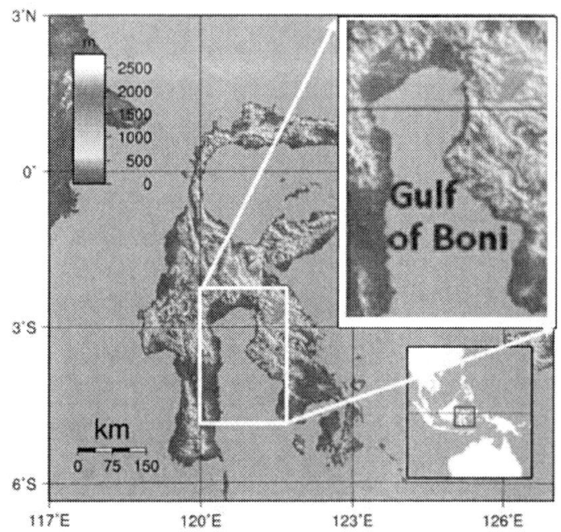

Figure 1. Location of the study (https://fr.wikipedia.org/wiki/Golfe_de_Bone, https://upload.wikimedia.org/wikipedia/commons/9/93/Buton_Topography.png).

2.2. ADCIRC model

A two-dimensional, depth-integrated version of Advanced Circulation model (ADCIRC – 2DDI) uses a depth-integrated equation of mass and momentum conservation which subject to Boussinesq, incompressible and the approximation of hydrostatic pressure. Even though this equation is based on shallow water equation, ADCIRC does not solve the primitive equation. Instead, it is based on the generalized wave continuity equation of the shallow water equation [5]. A complete set of equation can be found in [7,16,17] and the numerical implementations are documented in [4,5] and [18].

2.3. Model setup

The bathymetry data used in this research were obtained from the General Bathymetric Chart of the Oceans (GEBCO) in which its resolution is about 30 arc-seconds (~ one kilometre). Moreover, the bathymetry is freely available for entirely globe (https://www.bodc.ac.uk/data/hosted_data_systems/gebco_gridded_bathymetry_data/). For our study location, the bathymetry is presented in Figure 2b and the finite element mesh is shown in Figure 2a. The number of elements and nodes of the domain are 30,695 and 15,722, respectively. Moreover, the maximum grid is 8,495 *m* while the minimum grid is 397 *m*. Usually the minimum grid's spacing is around the coastline and the maximum grid's spacing is on the open boundary.

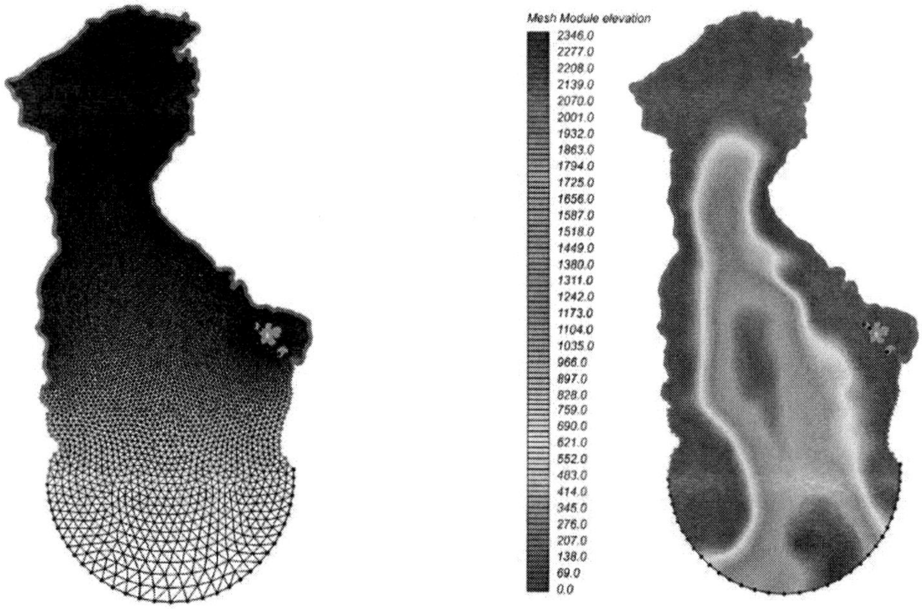

(*a*) (*b*)

Figure 2. Features of Gulf of Bone (a) Mesh and (b) Bathymetry.

Eight major tidal constituents are imposed along open boundary conditions as it has been argued by some researchers that these eight constituents will be enough for tidal elevation and current simulation. From all eight major tidal constituents, four of them namely K_1, O_1, P_1 and Q_1 are diurnal tides while other four constituents which are M_2, N_2, K_2, S_2 are semi diurnal tides. Moreover, the phases and amplitudes of these tidal constituents will be obtained from FES2012 (*Finite Element Solutions* 2012) tidal database in which the resolution is $1/16^0$. This database was produced by Noveltis, Legos and CLS Space Oceanography Division and distributed by Aviso, with support from Cnes (http://www.aviso. altimetry.fr/).

3. Model validation

In order to verify the accuracy of the model, observation is needed to validate the model output. Fortunately, tide gauge is available in this location, even though the data are not free. In fact, the data have to be purchased from Geospatial Information Agency in Indonesia. This is the national surveying and mapping agency of Indonesia. The tide gauge is located near Palopo city which is the capital city of Luwu regency. The longitude and the latitude of the observation are 120.2098^0 *E* and 2.9834^0 *S*, respectively. Tidal component of the model and observation have been extracted using a harmonic analysis tool so-called *T_TIDE* [19]. The comparison between tidal elevations obtained from the model and observation for 2015 and 2016 are presented in Figure 3 and Figure 4, respectively. The root mean square errors (RMSE) for each case are 11.58 *cm* and 11.46 *cm*, respectively. Moreover, correlation coefficient for case 2015 is 0.9765 while the correlation coefficient for 2016 is 0.9768. It should be noted that we have run two cases for prediction purposes. At the end, it is expected that two predictions with different starting time should not be different significantly. As a matter of fact, the model has been run for 200 days for each case but for simple presentation, only first 800 hours are presented.

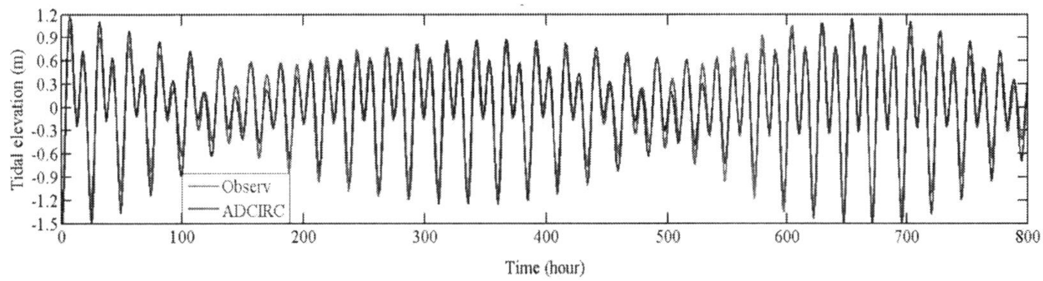

Figure 3. Tidal elevation comparisons between tide gauge data and ADCIRC model in 2015.

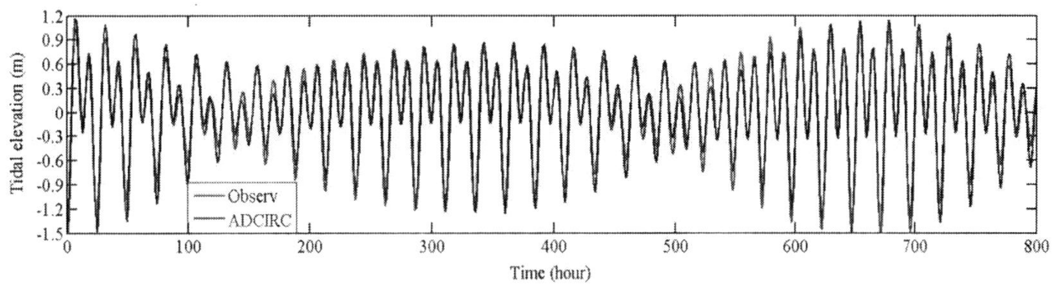

Figure 4. Tidal elevation comparisons between tide gauge data and ADCIRC model in 2016.

4. Improving open boundary condition

Based on the previous results, we have been using high resolution tidal database but the discrepancy between model and observation is still large. In fact, as aforementioned, RMSEs for the cases 2015 and 2016 are 11.58 *cm* and 11.46 *cm*, respectively. Therefore, in this section, the amplitudes and phases of tidal constituents will be improved using Green's functions approach. Full description of this method can be found in [12,13] and the implementation method can be found in [3,14,15]. Basically, in order to apply this method, one has to have an observation inside the domain of the simulation. This observation will be used to tune the tidal amplitude and phase of the tidal constituents imposed along open boundary conditions. Moreover, one needs to run perturbation for each parameter. Therefore, since we employ eight tidal constituents, 16 perturbations have to be run. Finally, the change of amplitude and phase of each constituent will be determined by Green's function.

Furthermore, the corrected values of amplitude and phase for each constituent will be re-imposed along open boundary conditions. The comparison between tidal elevation and observation after applying Green's function for the years 2015 and 2016 are presented in the Figure 5 and Figure 6, respectively. Root-mean square error (RMSE) for the case of 2015 decreased dramatically from 11.58 *cm* to 2.86 *cm* which is about 75.30%. Similarly, Root-mean square error (RMSE) for the case of 2016 decreased significantly from 11.46 *cm* to 1.30 *cm* which is about 88.65%. The correlation coefficients for these two cases are 0.9988 and 0.9997, respectively.

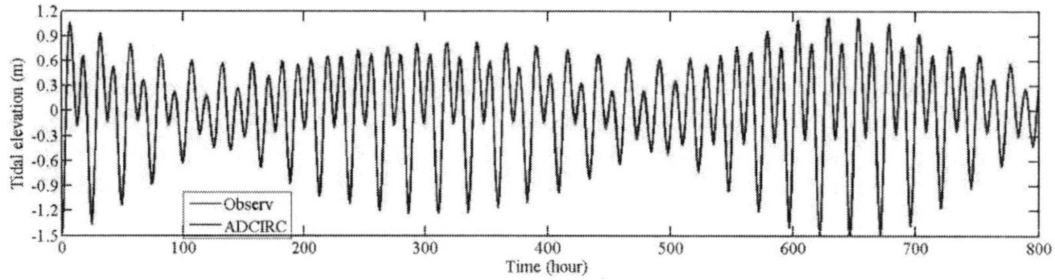

Figure 5. Tidal elevation between observation and ADCIRC in 2015 after Green's function.

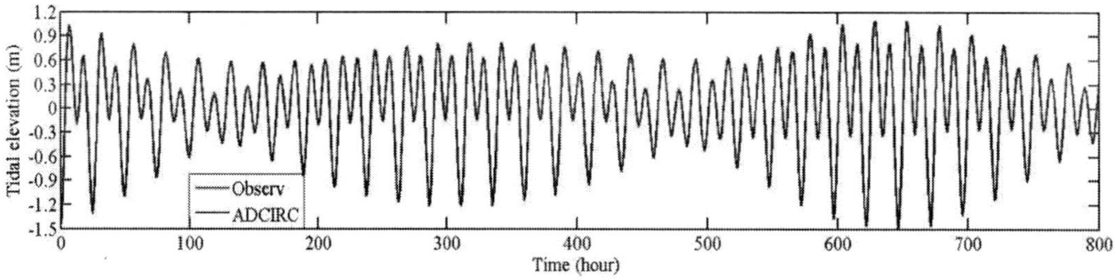

Figure 6. Tidal elevation between observation and ADCIRC in 2016 after Green's function.

5. Prediction of tidal elevation and currents

Prediction of tidal elevation including validation and currents will be presented in this section. First of all, the validation of tidal elevation will be shown for two different cases. Then, the predictions of tidal elevation for 2017 will be presented. Finally, tidal current approximation will also be shown.

5.1. Prediction of tidal elevation and its validation

After applying Green's function approach, the model, then, be run for 11 months which is from February to December 2015. In this case we suppose that the prediction was for three months. Furthermore, the model's output is validated based on the observation data. This comparison is presented in Figure 7. As seen from the figure, it is difficult to differentiate between tidal elevation from model and observation. In fact, RMSE is quite small that is 2.47 *cm*.

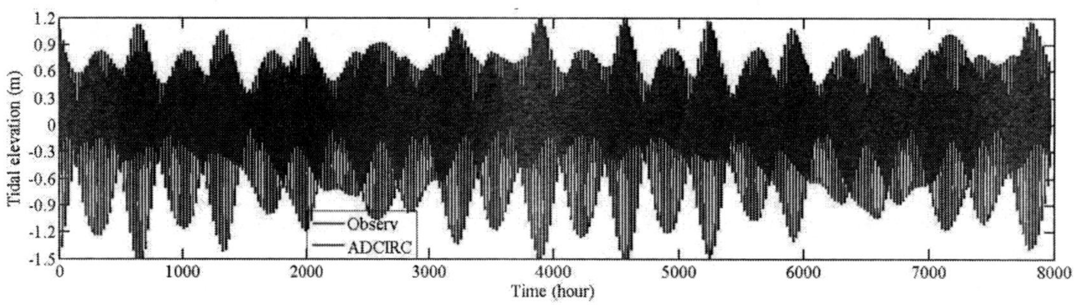

Figure 7. Validation of the tidal elevation's prediction in 2015.

Similar case has been carried out for the case of 2016 and the comparison with respect to observation is shown in Figure 8. The RMSE for this case is also very small, that is 1.84 *cm* which indicates that model's output and observation are in an excellent agreement.

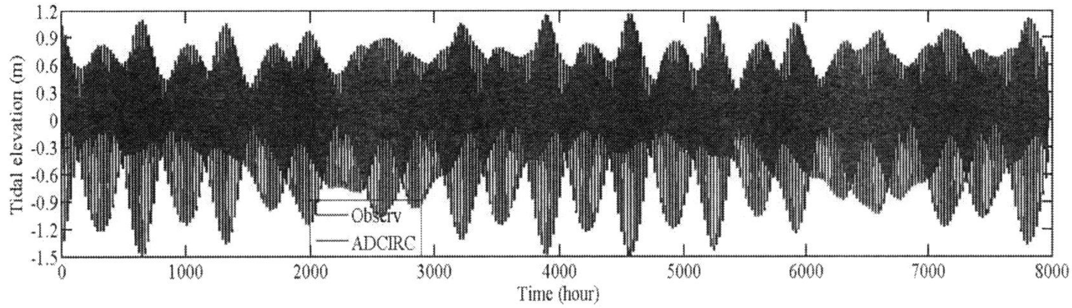

Figure 8. Validation of the tidal elevation's prediction in 2016.

5.2. Predicted tidal elevation in 2017

As shown in the previous sub-section, the results for the cases in 2015 and 2016 are in a perfect agreement with respect to the observation. Using the improved tidal constituents that have been obtained from Green's function approach for cases 2015 and 2016, in this part, the prediction for 2017 will be presented. Therefore, in order to use data from 2015, the model has to be run for 1065 days. The comparison between tidal prediction obtained from Geospatial Information Agency (GIA) of Indonesia and the model for 2017 which is based on 2015 data assimilation is shown in Figure 9. Although the figure shows a small different between GIA's prediction and our prediction, we have shown that our prediction has much higher accuracy the GIA's prediction. In fact, using the data in 2015 and 2016, we have shown that the RMSEs between observation and GIA's prediction are 14.21 *cm* and 16.02 *cm* respectively (figures not shown).

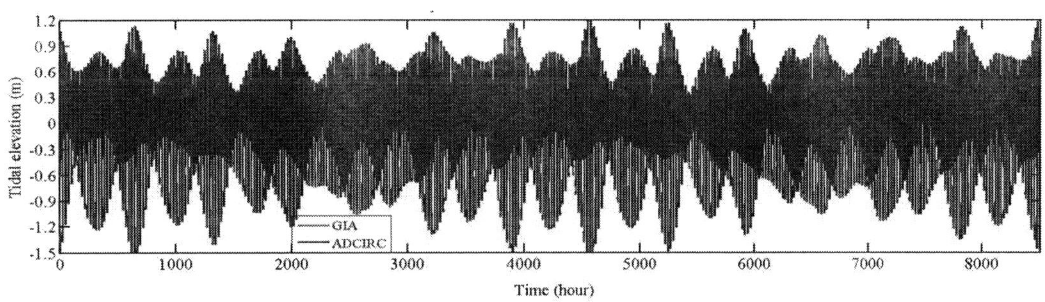

Figure 9. Tidal prediction in 2017 based on GIA and the 2015 data assimilation.

Similarly, using the improved tidal constituents for 2016, the model has been run from February 2016 until December 2017. Tidal prediction based on Geospatial Information Agency (GIA) and model which is based on 2016 data assimilation is presented in Figure 10. Figure 11 reveals the comparison prediction for 2017 using two different starting time, namely based on 2015 and 2016 data assimilations. It is expected that both prediction should not have a very significant different. To this end, RMSE and correlation coefficient of these two predictions were calculated and we found that RMSE is very small, that is 3.52 *cm*. Similarly, correlation coefficient is very close to 1 as expected, which is 0.9979.

5.3. Barotropic currents

Since there is no observation for tidal current available in the study location, we could not compare our results with respect to observation. However, since tidal elevation has been validated which is in an excellent agreement with tide gauge data, it is a plausibility to assume that tidal current from the model is very close to the actual current. From our simulation, tidal current is quite weak around the location which is less than 50 *cm*. One selected location has been chosen in which longitude and latitude are 120.7788^0 *E* and 2.9384^0 *S*, respectively. Predicted meridional and zonal velocities for 2017 are presented in Figure 12 and Figure 13, respectively. Moreover, the absolute tidal current at the location is less than 25 *cm*. The figure is not shown for simplicity.

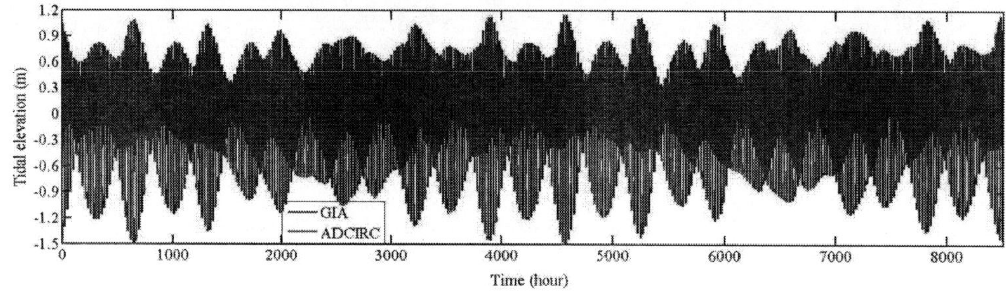

Figure 10. Tidal prediction in 2017 based on GIA and the 2016 data assimilation.

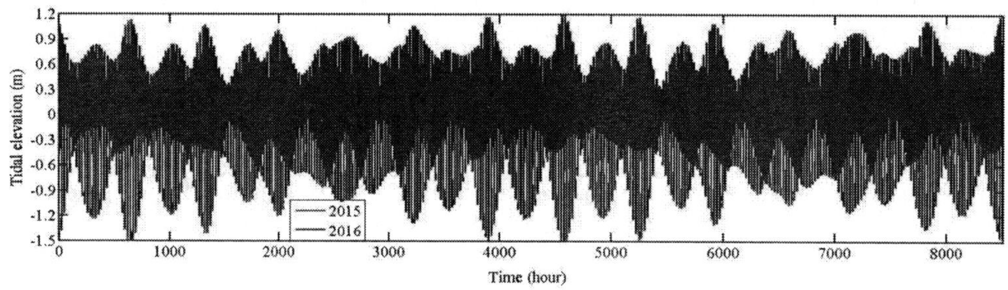

Figure 11. Comparison prediction for 2017 based on 2015 and 2016 data assimilation.

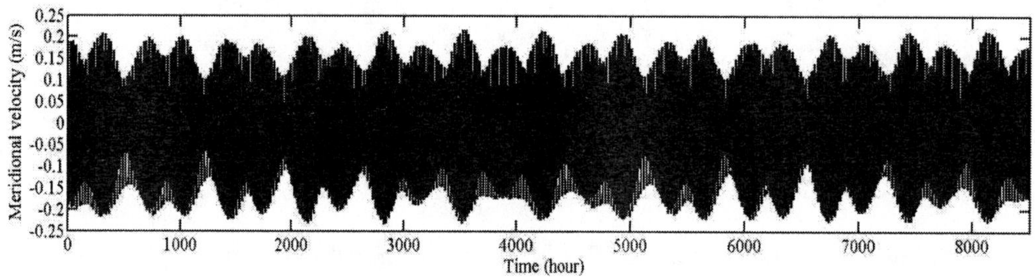

Figure 12. Predicted meridional current velocity in the selected location in 2017.

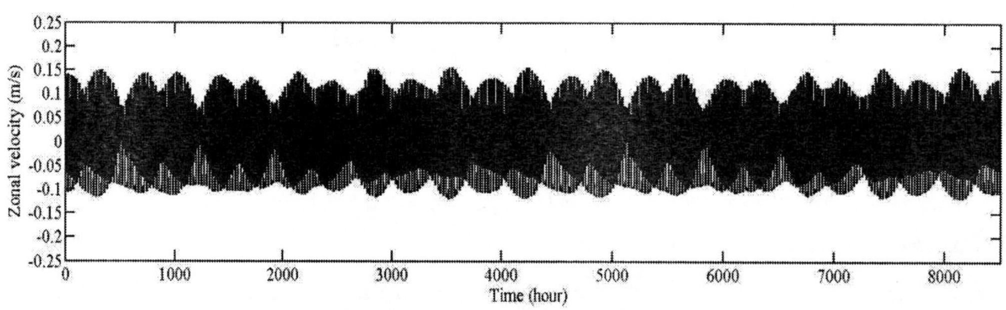

Figure 13. Predicted zonal current velocity in the selected location in 2017.

6. Summary and conclusions

Tidal elevation and baroptropic current for 2017 has been predicted in this work. Two different starting times namely 2015 and 2016 have been used and it is found that both prediction does not have a significant different. In fact, RMSE is only about 3.52 *cm*. We also found that our prediction is better compared to the prediction of Geospatial Information Agency.

References

[1] Blain C, Preller R and Rivera A 2002 Tidal prediction using the advanced circulation model (ADCIRC) and a relocatable PC-based system *Oceanography* **15** 77

[2] Ribal A, Zieger S 2016 *Proc. 3rd AUN/SEED-NET Regional Conference on Energy Engineering and the 7th International Conference on Thermofluids* (Yogyakarta: AIP Conference Proceedings) 060001

[3] Ribal A, Amir A K, Toaha S, Kusuma J and Khaeruddin 2017 Tidal Current Energy Resource Assessment Around Buton Island Southeast Sulawesi Indonesia *International Journal of Renewable Energy Research* **7** 857

[4] Luettich R A, Westerink J J and Scheffner N W 1992 ADCIRC: an Advanced Three-dimensional Circulation Model for Shelves Coasts and Estuaries *Report 1: Theory and Methodology of ADCIRC-2DD1 and ADCIRC-3DL US Army Engineer Waterways Experiment Station*

[5] Westerink J J, Blain C A, Luettich R A and Scheffner N W 1994 ADCIRC: An Advanced Three-dimensional Circulation Model for Shelves Coasts and Estuaries: *Report 2: Users Manual for ADCIRC-2DDI US Army Engineer Waterways Experiment Station*

[6] Westerink J J, Luettich R A and Scheffner N W 1993 ADCIRC: An Advanced Three-Dimensional Circulation Model for Shelves Coasts and Estuaries *Report 3: Development of a Tidal Constituent data base for the Western North Atlantic and Gulf of Mexico US Army Engineers aterways Experiment Station Vicksburg*

[7] Bacopoulos P 2005 *Master of Science* (Florida: University of Central Florida)

[8] Westerink J J 1993 Tidal prediction in the Gulf of Mexico/Galveston Bay using model ADCIRC-2DDI Contractors *Report to the US Army Engineer Waterways Experiment Station Vicksburg MS*

[9] Luettich R A, Birkhahn R H and Westerink J J 1991 Application of ADCIRC-2DDI to Masonboro Inlet North Carolina: A brief numerical modeling study Contractors *Report to the US Army Engineer Waterways Experiment Station Vicksburg*

[10] Westerink J J and Luettich R A 1991 Tide and storm surge predictions in the Gulf of Mexico using model ADCIRC-2D Contract *Report prepared for the US Army Engineer Waterways Experiment Station Vicksburg MS*

[11] Zhang A, Wei E and Parker B B 2003 Optimal estimation of tidal open boundary conditions using predicted tides and adjoint data assimilation technique *Continental Shelf Research* **23** 1055

[12] Menemenlis D and Wunsch C 1997 Linearization of an Oceanic General Circulation Model for Data Assimilation and Climate Studies *Journal of Atmospheric and Oceanic Technology* **14** 1420

[13] Menemenlis D, Fukumori I and Lee T 2005 Using Green's functions to calibrate an ocean general circulation model *Monthly weather review* **133**

[14] Moon J H, Hirose N and Morimoto A 2012 Green's function approach for calibrating tides in a circulation model for the East Asian marginal seas *J Oceanogr* **68** 345

[15] Ribal A, Waseda T, Wada R and Kiyomatsu K 2017 Optimal control of open boundary conditions for a regional tidal model *in preperation*

[16] Kolar R L, Gray W G, Westerink J J and Luettich R A 1994 Shallow water modeling in spherical coordinates: equation formulation numerical implementation and application *Journal of Hydraulic Research* **32** 3

[17] Kolar R L, Westerink J J, Cantekin M E and Blain CA 1994 Aspects of nonlinear simulations using shallow-water models based on the wave continuity equation *Computers & Fluids* **23** 523

[18] Luettich R and Westerink J 2004 Formulation and Numerical Implementation of the 2D/3D ADCIRC Finite Element Model Version 44XX *http://wwwuncedu/ims/adcirc/adcirc_theory_2004_12_08pdf*

[19] Pawlowicz R, Beardsley B and Lentz S 2002 Classical tidal harmonic analysis including error estimates in MATLAB using T_TIDE *Computers & Geosciences* **28** 929

IOP Conf. Series: Journal of Physics: Conf. Series **979** (2018) 012072 doi:10.1088/1742-6596/979/1/012072

A boundary element method with analytical integration for deformation of inhomogeneous elastic materials

Moh. Ivan Azis, Syamsuddin Toaha, Mawardi Bahri and Nirwan Ilyas
Department of Mathematics, Hasanuddin University, Makassar, INDONESIA
E-mail: mohivanazis@yahoo.co.id

Abstract. This paper is concerned with obtaining solutions to the equation governing static deformations of inhomogeneous elastic materials. The material parameters are assumed to vary continuously with the spatial variables. A boundary element method (BEM) with analytical integration is used to find the solutions. The results show that the BEM is feasible to be used to find the solutions of the problems and with analytical integration the BEM gives very accurate solutions in a shorter computation time.

1. Introduction

A various classes of problems have been solved numerically using BEM. The classes of problems include heat conduction problems, infiltration problems, deformation problems of elastic materials, pollutant transport problems and many others. The use of BEM has also been applied to homogeneous as well as to inhomogeneous materials, and not only to isotropic but also to anisotropic materials. For example, [1] used the BEM for solving infiltration problems of isotropic homogeneous media, [2] derived a fundamental solution, [3] considered diffusion-convection problems for anisotropic homogeneous materials, [4] solved elasticity problems for isotropic inhomogeneous materials and [5] studied elasticity problems, [6] worked on elliptic boundary value problems, [7] solved heat conduction problems, [8] worked on elasticity problems and [9] considered transient heat conduction problems for anisotropic inhomogeneous materials.

Apparently in the implementation of the BEM the partial differential equation which governs the system has to be converted to a boundary integral equation. And eventually after a discretisation of the boundary into a number of segments, integration over each segment has to be calculated either numerically or analytically if possible. To some extent analytical integration has of course advantages over numerical integration in the aspect of accuracy and computation time.

In recent years some progress has been made toward solving numerically the deformation problem of inhomogeneous elastic materials using the BEM. The work done by [4], for example, deals with the case of isotropic materials, whereas in [5, 8] considered the case for anisotropic materials. However, in all of the papers the authors used numerical integration for finding the BEM solutions. The current study is concerned with finding the solutions of the problems using the BEM with analytical integration.

2. The boundary value problem

Referred to a Cartesian frame $Ox_1x_2x_3$ the equilibrium equations governing small deformations of an inhomogeneous anisotropic elastic material occupying a region Ω in R^3 with boundary $\partial\Omega$ which consists of a finite number of piecewise smooth closed curves may be written in the form

$$\frac{\partial}{\partial x_j}\left[c_{ijkl}(\mathbf{x})\frac{\partial u_k}{\partial x_l}\right]=0 \tag{1}$$

where $i,j,k,l = 1,2,3$, $\mathbf{x} = (x_1,x_2,x_3)$, u_k denotes the displacement, $c_{ijkl}(\mathbf{x})$ the elastic parameters and the repeated summation convention (summing from 1 to 3) is used for repeated Latin suffices. Troughout the paper it is assumed that u_k is independent of x_3 and takes the form

$$u_k = A_k f(x_1+\tau x_2)$$

where A_k, τ are constants and f is a twice differentiable function. The stress displacement relations are given by

$$\sigma_{ij} = c_{ijkl}\frac{\partial u_k}{\partial x_l}$$

and the traction vector P_i on the boundary $\partial\Omega$ is defined as

$$P_i \equiv \sigma_{ij}n_j = c_{ijkl}\frac{\partial u_k}{\partial x_l}n_j \tag{2}$$

where $\mathbf{n} = (n_1,n_2)$ denotes the outward pointing normal to the boundary $\partial\Omega$.

For all points in Ω the coefficients $c_{ijkl}(\mathbf{x})$ are required to satisfy the usual symmetry conditions and also sufficient conditions for the strain energy density to be positive. This requirement ensures that the system of partial differential equations is elliptic throughout Ω.

A solution to (1) is sought which is valid in the region Ω and satisfies the boundary conditions that on $\partial\Omega_1$ the displacement u_k is specified and on $\partial\Omega_2$ the traction P_i is specified, where $\partial\Omega = \partial\Omega_1 \cup \partial\Omega_2$.

3. Reduction to a linear constant coefficients equation

The coefficients in (1) are now required to take the form

$$c_{ijkl}(\mathbf{x}) = c^{(0)}_{ijkl}g(\mathbf{x}) \tag{3}$$

where the $c^{(0)}_{ijkl}$ are constants. Equation (1) thus may be written

$$c^{(0)}_{ijkl}\frac{\partial}{\partial x_j}\left[g(\mathbf{x})\frac{\partial u_k}{\partial x_l}\right]=0 \tag{4}$$

Consider the transformation

$$u_k = g^{-1/2}\psi_k \tag{5}$$

Use of (5) in (4) provides the equation

$$c^{(0)}_{ijkl}g^{1/2}\frac{\partial^2\psi_k}{\partial x_j\partial x_l} - c^{(0)}_{ijkl}\psi_k\frac{\partial^2 g^{1/2}}{\partial x_j\partial x_l}=0 \tag{6}$$

Thus if

$$c^{(0)}_{ijkl}\frac{\partial^2\psi_k}{\partial x_j\partial x_l} = 0 \tag{7}$$

$$c^{(0)}_{ijkl}\frac{\partial^2 g^{1/2}}{\partial x_j\partial x_l} = 0 \tag{8}$$

then (6) will be satisfied. Thus when g satisfies the system (8) the transformation given by (5) transforms the linear system with variable coefficients (4) to the linear system with constant coefficients (7).

Equation (7) comprises a system of three constant coefficients linear elliptic partial differential equations in the three dependent variables u_k for $k = 1, 2, 3$. A solution to this system may be readily obtained in terms of arbitrary analytic functions.

As a result of the symmetry property $c_{ijkl} = c_{klij}$ equation (8) consists of a system of six constant coefficients partial differential equations in the one dependent variable $g^{1/2}$. In general the solution to this system consists of a linear function of the three independent variables x_1, x_2, x_3. Thus in the general case g may be written in the form

$$g(\mathbf{x}) = (\alpha_i x_i)^2 \tag{9}$$

where the α_i are constants which may be used to fit the elastic parameters $c_{ijkl} = c^{(0)}_{ijkl} g(\mathbf{x})$ to given numerical data.

Although in the general case it is necessary for g to be restricted to the form given by (9) it is appropriate to note at this stage that for particular classes of deformations the tranformation from (4) to (7) may be achieved for a more general class of functions $g(\mathbf{x})$. Some of these cases are now considered.

Now substitution of (3) and (5) into (2) yields

$$P_i = -P^{[g]}_{ik}\psi_k + P^{[\psi]}_i g^{1/2} \tag{10}$$

where

$$P^{[g]}_{ik} = c^{(0)}_{ijkl}\frac{\partial g^{1/2}}{\partial x_l}n_j \tag{11}$$

$$P^{[\psi]}_{i} = c^{(0)}_{ijkl}\frac{\partial \psi_k}{\partial x_l}n_j \tag{12}$$

The boundary integral equation for the solution of (7) may be written in the form

$$\eta\psi_m(\mathbf{x}_0) = -\int_{\partial\Omega}\left[\Phi_{im}P^{[\psi]}_i - \Gamma_{im}\psi_i\right]ds \tag{13}$$

Use of (5) and (10) in (13) yields

$$\eta g^{1/2}u_m(\mathbf{x}_0) = -\int_{\partial\Omega}\left[(g^{-1/2}\Phi_{im})P_i - (g^{1/2}\Gamma_{km} - P^{[g]}_{ki}\Phi_{im})u_k\right]ds$$

This equation provides a boundary integral equation for determining u_m and P_m at all points of Ω.

Also Φ_{im} and Γ_{im} are given by

$$\Phi_{im} = \frac{1}{2\pi}\Re\left\{\sum_{\alpha=1}^{3} A_{i\alpha}N_{\alpha k}\log(z_\alpha - c_\alpha)\right\}d_{km} \qquad \Gamma_{im} = \frac{1}{2\pi}\Re\left\{\sum_{\alpha=1}^{3} L_{ij\alpha}N_{\alpha k}(z_\alpha - c_\alpha)^{-1}\right\}n_j d_{km} \tag{14}$$

where $\Re$ denotes the real part of a complex number, $z_\alpha = x_1 + \tau_\alpha x_2$, and $c_\alpha = a + \tau_\alpha b$, where τ_α are the three roots with positive imaginary part of the polynomial in τ

$$|c^{(0)}_{i1k1} + c^{(0)}_{i2k1}\tau + c^{(0)}_{i1k2}\tau + c^{(0)}_{i2k2}\tau^2| = 0$$

The $A_{i\alpha}$ occuring in (14) are the solutions of the system

$$\left(c^{(0)}_{i1k1} + c^{(0)}_{i2k1}\tau_\alpha + c^{(0)}_{i1k2}\tau_\alpha + c^{(0)}_{i2k2}\tau^2_\alpha\right) A_{k\alpha} = 0$$

Also the $N_{\alpha k}$, $L_{ij\alpha}$ and d_{km} are defined by

$$\delta_{ik} = \sum_\alpha A_{i\alpha}N_{\alpha k} \quad L_{ij\alpha} = (c^{(0)}_{ijk1} + \tau_\alpha c^{(0)}_{ijk2})A_{k\alpha} \quad \delta_{im} = -\frac{1}{2}i\sum_{\alpha=1}^{3}\left\{L_{i2\alpha}N_{\alpha k} - \overline{L}_{i2\alpha}\overline{N}_{\alpha k}\right\} d_{km}$$

where bar denotes the complex conjugate.

4. A further perturbation method

In this section a procedure is obtained when the coefficients $c_{ijkl}(\mathbf{x})$ is perturbed about its expression presented in (3) while retaining the condition (8). Specifically the coefficient $c_{ijkl}(\mathbf{x})$ is supposed to have the form

$$c_{ijkl}(\mathbf{x}) = c^{(0)}_{ijkl}g(\mathbf{x}) + \epsilon c^{(1)}_{ijkl}(\mathbf{x}) \tag{15}$$

where the $c^{(0)}_{ijkl}$ are constants, ϵ is a small parameter, $c^{(1)}_{ijkl}$ is a differentiable function and $g(\mathbf{x})$ satisfies (8).

Substitution of (15) into (1) and use of the transformation (5) give

$$c^{(0)}_{ijkl}\frac{\partial}{\partial x_j}\left[g\frac{\partial}{\partial x_l}(g^{-1/2}\psi_k)\right] = -\epsilon\frac{\partial}{\partial x_j}\left[c^{(1)}_{ijkl}\frac{\partial}{\partial x_l}(g^{-1/2}\psi_k)\right] \tag{16}$$

Use the analysis used to derive (7) from (4) for the left hand side and a simplification for the right hand side of (16) yields

$$c^{(0)}_{ijkl}\frac{\partial^2\psi_k}{\partial x_j\partial x_l} = -\epsilon\left[\frac{\partial A_{ijk}}{\partial x_j}\psi_k + (A_{ijk} + \frac{\partial B_{ijkl}}{\partial x_l})\frac{\partial\psi_k}{\partial x_j} + B_{ijkl}\frac{\partial^2\psi_k}{\partial x_j\partial x_l}\right] \tag{17}$$

where

$$A_{ijk}(\mathbf{x}) = c^{(1)}_{ijkl}\frac{\partial g^{-1/2}}{\partial x_l} \quad , \quad B_{ijkl}(\mathbf{x}) = c^{(1)}_{ijkl}g^{-1/2}$$

A solution to equation (17) is sought in the form

$$\psi_k(\mathbf{x}) = \sum_{r=0}^{\infty}\epsilon^r\psi^{(r)}_k(\mathbf{x}) \tag{18}$$

Substitution of (18) into (17) and equating coefficients of powers of ϵ yields

$$c^{(0)}_{ijkl}\frac{\partial^2\psi^{(0)}_k}{\partial x_j\partial x_l} = 0 \tag{19}$$

$$c^{(0)}_{ijkl}\frac{\partial^2\psi^{(r)}_k}{\partial x_j\partial x_l} = -\left[\frac{\partial A_{ijk}}{\partial x_j}\psi^{(r-1)}_k + (A_{ijk} + \frac{\partial B_{ijkl}}{\partial x_l})\frac{\partial\psi^{(r-1)}_k}{\partial x_j} + B_{ijkl}\frac{\partial^2\psi^{(r-1)}_k}{\partial x_j\partial x_l}\right] \tag{20}$$

for $r = 1, 2, \ldots$.

IOP Conf. Series: Journal of Physics: Conf. Series **979** (2018) 012072 doi:10.1088/1742-6596/979/1/012072

The integral equations for (19) and (20) are respectively

$$\eta\psi_m^{(0)}(\mathbf{x}_0) = -\int_{\partial\Omega}\left[\Phi_{im}P_i^{[\psi^{(0)}]} - \Gamma_{im}\psi_i^{(0)}\right]ds \tag{21}$$

where $P_i^{[\psi^{(0)}]} = c_{ijkl}^{(0)}(\partial\psi_k^{(0)}/\partial x_l)n_j$ and

$$\eta\psi_m^{(r)}(\mathbf{x}_0) = -\int_{\partial\Omega}\left[\Phi_{im}P_i^{[\psi^{(r)}]} - \Gamma_{im}\psi_i^{(r)}\right]ds + \int_{\Omega} h_i^{(r)}\Phi_{im}dS \tag{22}$$

where $P_i^{[\psi^{(r)}]} = c_{ijkl}^{(0)}(\partial\psi_k^{(r)}/\partial x_l)n_j$ and $h_i^{(r)}$ is the right hand side of (20) for $r = 1, 2, \ldots$.

The corresponding value of P may be written as

$$P_i = -P_{ik}^{[g]}\psi_k^{(0)} + P_i^{[\psi^{(0)}]}g^{1/2} + \sum_{r=1}^{\infty}\epsilon^r\left[-P_{ik}^{[g]}\psi_k^{(r)} + P_i^{[\psi^{(r)}]}g^{1/2} + G_i^{(r)}\right]$$

where $P_{ik}^{[g]}$ and $P_i^{[\psi]}$ are given by (11) and (12), and

$$G_i^{(r)} = c_{ijkl}^{(1)}\frac{\partial(g^{-1/2}\psi_k^{(r-1)})}{\partial x_l}n_j \tag{23}$$

To satisfy the boundary conditions in Section 2 it is required that $\psi_k^{(0)} = g^{1/2}u_k$ where u_k takes on its specified value on $\partial\Omega_1$. Also it is required that on $\partial\Omega_2$ $P_i^{[\psi^{(0)}]} = g^{-1/2}(P_i + P_{ik}^{[g]}\psi_k^{(0)})$ where P_i takes on its specified value on $\partial\Omega_2$. It then follows from (12) and (18) that $\psi_i^{(r)} = 0$ on $\partial\Omega_1$ and $P_i^{[\psi^{(r)}]} = g^{-1/2}(P_{ik}^{[g]}\psi_k^{(r)} - G_i^{(r)})$ on $\partial\Omega_2$ for $r = 1, 2, \ldots$.

The integral equations (21) and (22) may thus be written in the form

$$\begin{aligned}\eta\psi_m^{(0)}(\mathbf{x}_0) &= -\int_{\partial\Omega_1}\left[\Phi_{im}P_i^{[\psi^{(0)}]} - (\Gamma_{im}g^{1/2})u_i\right]ds \\ &\quad -\int_{\partial\Omega_2}\left[(g^{-1/2}\Phi_{im})P_i - (\Gamma_{km} - g^{-1/2}P_{ki}^{[g]}\Phi_{im})\psi_k^{(0)}\right]ds \end{aligned}\tag{24}$$

$$\begin{aligned}\eta\psi_m^{(r)}(\mathbf{x}_0) &= -\int_{\partial\Omega_1}\Phi_{im}P_i^{[\psi^{(r)}]}ds \\ &\quad -\int_{\partial\Omega_2}\left[-(g^{-1/2}\Phi_{im})G_i^{(r)} - (\Gamma_{km} - g^{-1/2}P_{ki}^{[g]}\Phi_{im})\psi_k^{(r)}\right]ds \\ &\quad +\int_{\Omega} h_i^{(r)}\Phi_{im}dS \quad \text{for } r = 1, 2, \ldots\end{aligned}\tag{25}$$

The boundary integral equation (24) serves to determine the unknown $\psi_i^{(0)}(\mathbf{x})$ for $\mathbf{x} \in \partial\Omega_2$ and the unknown $P_i^{[\psi^{(0)}]}(\mathbf{x})$ for $\mathbf{x} \in \partial\Omega_1$. Once these boundary values are known equation (24) serves to determine $\psi_i^{(0)}$ and its derivatives for all points in the domain Ω.

Similarly the integral equation (25) serves to determine the unknown $\psi_i^{(r)}(\mathbf{x})$ for $\mathbf{x} \in \partial\Omega_2$ and the unknown $P_i^{[\psi^{(r)}]}(\mathbf{x})$ for $\mathbf{x} \in \partial\Omega_1$ for $r = 1, 2, \ldots$. Once these boundary values are known equation (25) serves to determine $\psi_i^{(r)}$ and its derivatives for all points in the domain Ω. At each stage in using (25) to determine $\psi_i^{(r)}$ and the $G_i^{(r)}$ occuring in (25) may be obtained from (23) which may be evaluated from the previously determined derivatives of $\psi_i^{(r-1)}$.

Having determined the values of $\psi_i^{(r)}$ for all r equations (18) and (5) then provide successively, the values of $\psi_i(\mathbf{x})$ and $u_i(\mathbf{x})$.

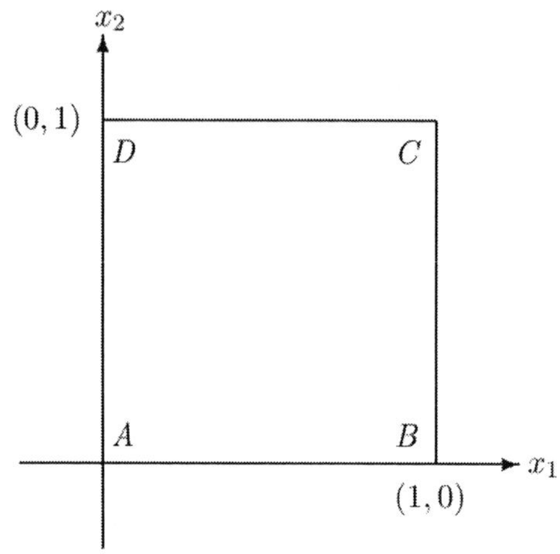

Figure 1. The geometry for the numerical examples

5. Numerical results

To show the usefulness of the formulas and the validity of the techniques derived in the previous sections some particular plane strain problems, for which the displacements $u_1 = u_1(x_1, x_2)$, $u_2 = u_2(x_1, x_2)$ and $u_3 = 0$, of inhomogeneous anisotropic materials will be considered.

The constant BEM is used to find numerical solutions to the problems. The domain Ω is taken to be a unit square (see Figure 1). Each side of the boundary $\partial\Omega$ is divided into N segments of equal length and nodal points are taken to be the mid points of each segment.

To save the computation time and to increase the accuracy, the evaluation of the line integrals of Φ_{im}, Γ_{im} and its derivatives along a segment $[\mathbf{q}_{k-1}, \mathbf{q}_k]$ joining the point $\mathbf{q}_{k-1}$ to the point $\mathbf{q}_k$ and excluding the singular points of Φ_{im} and Γ_{im} is done analytically by utilising the following results,

$$\int_{\mathbf{q}_{k-1}}^{\mathbf{q}_k} \log(z_\alpha - c_\alpha) ds(\mathbf{x}) = R_k \left[-1 + \log(A_\alpha + B_\alpha) + \frac{A_\alpha}{B_\alpha} \log(1 + \frac{B_\alpha}{A_\alpha}) \right],$$

$$\int_{\mathbf{q}_{k-1}}^{\mathbf{q}_k} (z_\alpha - c_\alpha)^{-1} ds(\mathbf{x}) = \frac{R_k}{B_\alpha} \log(1 + \frac{B_\alpha}{A_\alpha})$$

$$\int_{\mathbf{q}_{k-1}}^{\mathbf{q}_k} \frac{\partial}{\partial a} \left[\log(z_\alpha - c_\alpha)\right] ds(\mathbf{x}) = \frac{-R_k}{B_\alpha} \log(1 + \frac{B_\alpha}{A_\alpha})$$

$$\int_{\mathbf{q}_{k-1}}^{\mathbf{q}_k} \frac{\partial}{\partial b} \left[\log(z_\alpha - c_\alpha)\right] ds(\mathbf{x}) = \frac{-R_k \tau_\alpha}{B_\alpha} \log(1 + \frac{B_\alpha}{A_\alpha})$$

$$\int_{\mathbf{q}_{k-1}}^{\mathbf{q}_k} \frac{\partial}{\partial a} \left[(z_\alpha - c_\alpha)^{-1}\right] ds(\mathbf{x}) = R_k \left[A_\alpha (A_\alpha + B_\alpha)\right]^{-1}$$

$$\int_{\mathbf{q}_{k-1}}^{\mathbf{q}_k} \frac{\partial}{\partial b} \left[(z_\alpha - c_\alpha)^{-1}\right] ds(\mathbf{x}) = R_k \tau_\alpha \left[A_\alpha (A_\alpha + B_\alpha)\right]^{-1}$$

where

$$\mathbf{q}_k = (x_k, y_k), \quad x = x_1, \quad y = x_2 \qquad R_k = |\mathbf{q}_k - \mathbf{q}_{k-1}|$$
$$A_\alpha = x_{k-1} - a + \tau_\alpha (y_{k-1} - b) \qquad B_\alpha = x_k - x_{k-1} + \tau_\alpha (y_k - y_{k-1})$$

The last four formulas are used for the calculation of the stress distribution in the domain Ω. Along a segment $[\mathbf{q}_{k-1}, \mathbf{q}_k]$ which includes a singular point (ie. the nodal or mid point of a segment) the evaluation of the integral of Φ_{im} is done by splitting the segment into two parts on either side of the singular point and summing the values of integrals along both sub segments. This then gives the following formula

$$\int_{\mathbf{q}_{k-1}}^{\mathbf{q}_k} \log(z_\alpha - c_\alpha) ds(\mathbf{x}) = R_k(-1 + \log\frac{1}{2} + \log B_\alpha)$$

Meanwhile the integral of Γ_{im} is zero due to the fact that in this case the vectors $\mathbf{x} - \mathbf{x}_0$ and $\mathbf{n}$ are perpendicular and therefore the normal derivative of Φ_{im} (and so Γ_{im}) is zero.

For the methods requiring domain integrals the domain Ω is discritised into N^2 equal sub-squares and the following analytical results are used for the evaluation of area integrals over a particular sub square $\{(x_1, x_2) : s \leq x_1 \leq t, v \leq x_2 \leq w\}$

$$\begin{aligned}
&\int_s^t \int_v^w \log\left[x_1 - a + \tau_\alpha(x_2 - b)\right] dx_2\, dx_1 = \\
&\quad q_1\left[-(v-b)(t-a) - \frac{1}{2\tau_\alpha}(t-a)^2 - \frac{1}{2}\tau_\alpha(v-b)^2\right] + \\
&\quad q_2\left[(v-b)(s-a) + \frac{1}{2\tau_\alpha}(s-a)^2 + \frac{1}{2}\tau_\alpha(v-b)^2\right] + \\
&\quad q_3\left[-(w-b)(s-a) - \frac{1}{2\tau_\alpha}(s-a)^2 - \frac{1}{2}\tau_\alpha(w-b)^2\right] + \\
&\quad q_4\left[(w-b)(t-a) + \frac{1}{2\tau_\alpha}(t-a)^2 + \frac{1}{2}\tau_\alpha(w-b)^2\right] - \\
&\quad \frac{3}{2}(t-s)(w-v)
\end{aligned}$$

$$\begin{aligned}
&\int_s^t \int_v^w x_1 \log\left[x_1 - a + \tau_\alpha(x_2 - b)\right] dx_2\, dx_1 = \\
&\quad \frac{1}{12\tau_\alpha}\Big[q_1\{2q_5^3 - 6aq_5^2 + 6(a^2 - t^2)q_5 - 2a^3 - 4t^3 + 6at^2\} + \\
&\qquad q_2\{-2q_5^3 + 6aq_5^2 - 6(a^2 - s^2)q_5 + 2a^3 + 4s^3 - 6as^2\} + \\
&\qquad q_3\{2q_6^3 - 6aq_6^2 + 6(a^2 - s^2)q_6 - 2a^3 - 4s^3 + 6as^2\} + \\
&\qquad q_4\{-2q_6^3 + 6aq_6^2 - 6(a^2 - t^2)q_6 + 2a^3 + 4t^3 - 6at^2\} + \\
&\qquad \tau_\alpha q_7\{-4b + 2(v + w)\} + q_7\{-4a - 7(s + t)\}\Big]
\end{aligned}$$

$$\begin{aligned}
&\int_s^t \int_v^w x_2 \log\left[x_1 - a + \tau_\alpha(x_2 - b)\right] dx_2\, dx_1 = \\
&\quad \frac{-1}{12\tau_\alpha^2}\Big[q_1\{2\tau_\alpha^3 q_8 + 6\tau_\alpha^2(v^2 - b^2)(t-a) + 6\tau_\alpha b(t-a)^2 - 2(t-a)^3\} + \\
&\qquad q_2\{-2\tau_\alpha^3 q_8 - 6\tau_\alpha^2(v^2 - b^2)(s-a) - 6\tau_\alpha b(s-a)^2 + 2(s-a)^3\} + \\
&\qquad q_3\{2\tau_\alpha^3 q_9 + 6\tau_\alpha^2(w^2 - b^2)(s-a) + 6\tau_\alpha b(s-a)^2 - 2(s-a)^3\} + \\
&\qquad q_4\{-2\tau_\alpha^3 q_9 - 6\tau_\alpha^2(w^2 - b^2)(t-a) - 6\tau_\alpha b(t-a)^2 + 2(t-a)^3\} + \\
&\qquad \tau_\alpha q_7\{4b + 7(v + w)\} + q_7\{4a - 2(s + t)\}\Big]
\end{aligned}$$

where

$$\begin{aligned}
&q_1 = \log\left[t - a + \tau_\alpha(v - b)\right] \quad q_2 = \log\left[s - a + \tau_\alpha(v - b)\right] \quad q_3 = \log\left[s - a + \tau_\alpha(w - b)\right] \\
&q_4 = \log\left[t - a + \tau_\alpha(w - b)\right] \quad q_5 = \tau_\alpha(v - b) \quad q_6 = \tau_\alpha(w - b) \\
&q_7 = \tau_\alpha(t - s)(w - v) \quad q_8 = b^3 + 2v^3 - 3v^2 b \quad q_9 = b^3 + 2w^3 - 3w^2 b
\end{aligned}$$

Table 1. Numerical and analytical results

Position	BEM 12 segments		BEM 40 segments		Analytical	
(x_1, x_2)	$u_1/\overline{u}$	$u_2/\overline{u}$	$u_1/\overline{u}$	$u_2/\overline{u}$	$u_1/\overline{u}$	$u_2/\overline{u}$
(0.1,0.5)	0.0542	0.1477	0.0493	0.1486	0.0495	0.1485
(0.3,0.5)	0.1469	0.2450	0.1456	0.2428	0.1456	0.2427
(0.5,0.5)	0.2383	0.3333	0.2381	0.3333	0.2381	0.3333
(0.7,0.5)	0.3261	0.4181	0.3271	0.4204	0.3271	0.4205
(0.9,0.5)	0.4086	0.5047	0.4130	0.5043	0.4128	0.5046

5.1. A test problem for Section 4
Consider the isotropic case for which the elastic moduli take the form (15) with $A/\overline{c} = \lambda^{(0)}+2\mu^{(0)}$, $B/\overline{c} = \lambda^{(0)}$, $D/\overline{c} = \lambda^{(0)} + 2\mu^{(0)}$ and $F/\overline{c} = \mu^{(0)}$,

$$\epsilon = 0.1 \qquad g(\mathbf{x}) = (1 + 0.1x_1/l)^2 \qquad \lambda^{(0)} = \mu^{(0)} = 0.5$$
$$\lambda^{(1)} = (1 + 0.1x_1/l)^2 \qquad \mu^{(1)} = 0$$

and the boundary conditions

$$\begin{array}{lll} u_1/\overline{u} = 5 - 5/(1 + 0.1x_1/l) & u_2/\overline{u} = 5 - 4.9/(1 + 0.1x_1/l) & \text{on AB} \\ u_1/\overline{u} = 0.4545 & u_2/\overline{u} = 0.5454 & \text{on BC} \\ u_1/\overline{u} = 5 - 5/(1 + 0.1x_1/l) & u_2/\overline{u} = 5 - 4.9/(1 + 0.1x_1/l) & \text{on CD} \\ u_1/\overline{u} = 0 & u_2/\overline{u} = 0.1 & \text{on AD} \end{array}$$

This problem admits the analytical solution $u_1/\overline{u} = 5-5/(1+0.1x_1/l)$ and $u_2/\overline{u} = 5-4.9/(1+0.1x_1/l)$.

An approximate solution is taken to be the sum of the first two terms of the series solution (18). The so called regular method is used to avoid a singular point by locating the source points outside the domain at the position of a distance as long as the length of the boundary segment measured from the mid point of each boundary segment. It should be noted that $\eta = 0$ if $\mathbf{x}_0 \notin \Omega$. The area integral over a domain cell Ω_l is approximated with

$$\int_{\Omega_l} h_i^{(r)} \Phi_{im} dS \approx h_i^{(r)}(\mathbf{x}) \int_{\Omega_l} \Phi_{im} dS$$

where $\mathbf{x}$ is the centre point of Ω_l.

Table 1 compares the analytical and BEM results for some interior points. It is observed that the accuracy improves as the number of segments increases. It also improves as we move further from the boundary points to the centre point of the domain.

6. Summary
Some BEMs for the solution of certain classes of boundary value problems of elasticity for anisotropic inhomogeneous media has been derived. The methods are generally easy to implement to obtain numerical values for particular problems. They can be applied to a wide class of important practical problems for inhomogeneous anisotropic materials. The numerical results obtained using the methods indicate that they can provide accurate numerical solutions.

Acknowledgments
This work was financially supported by the LP2M (Lembaga Penelitian dan Pengabdian pada Masyarakat) of Hasanuddin University Makassar and The Higher Education Ministry of Indonesia.

References

[1] Azis M I, Clements D L and Lobo M 2003 A boundary element method for steady infiltration from periodic channels *ANZIAM J.* **44(E)** C61
[2] Azis M I 2017 Fundamental solutions to two types of 2D boundary value problems of anisotropic materials *Far East Journal of Mathematical Sciences* **101** 11 2405
[3] Haddade A, Salam N, Khaeruddin and Azis M I 2017 A Boundary Element Method for 2D Diffusion-Convection Problems in Anisotropic Media *Far East Journal of Mathematical Sciences* **102** 8
[4] Clements D L and Azis M I 2000 A Note on a Boundary Element Method for the Numerical Solution of Boundary Value Problems in Isotropic Inhomogeneous Elasticity *Journal of the Chinese Institute of Engineers* **23** 3 261
[5] Azis M I and Clements D L 2001 A Boundary Element Method for Anisotropic Inhomogeneous Elasticity *International Journal of Solids and Structures* **38** 5747
[6] Azis M I, Clements D L and Budhi W S 2003 A boundary element method for the numerical solution of a class of elliptic boundary value problems for anisotropic inhomogeneous media *ANZIAM J.* **44(E)** C79
[7] Azis M I and Clements D L 2008 Nonlinear transient heat conduction problems for a class of inhomogeneous anisotropic materials by BEM *Engineering Analysis with Boundary Elements* **32** 1054
[8] Azis M I and Clements D L 2014 On some problems concerning deformations of functionally graded anisotropic elastic materials *Far East Journal of Mathematical Sciences* **87** 2 173
[9] Azis M I and Clements D L 2014 A Boundary Element Method for Transient Heat Conduction Problem of Nonhomogeneous Anisotropic Materials *Far East Journal of Mathematical Sciences* **89** 1 51

IOP Conf. Series: Journal of Physics: Conf. Series **979** (2018) 012073 doi:10.1088/1742-6596/979/1/012073

The Ramsey Numbers for Star Versus Wheel of Even Order

Hasmawati, Nur Rohmah Oktaviani P and Loeky Haryanto

Department of Mathematics Faculty of Mathematics, Hasanuddin University

E-mail : hasmaba97@gmail.com

Abstract. The Ramsey Numbers of $R(G,H)$ is the smallest integer k such that for any graph F of order k, either F contains G or the complement of F (or $\bar{F}$) contains H. Let S_n denoted star of order n and W_m a wheel of order $m+1$. In this paper, we show that $R(S_{20}, W_{10}) = 43$, and if k is even, $R(S_{2k+1}, W_{2k}) = 5k - 1$ for $k \geq 6$.

1. Introduction

All graphs considered in this paper are finite simple graphs. For two graphs G and H, the Ramsey numbers $R(G,H)$ is the smallest positive integer n such that for any graph F of order n, either F contains G as a subgraph or the complement of F (or $\bar{F}$) contains H as a subgraph.

Let G be a graph with $G = (V(G), E(G))$, $V(G) = \{u_1, u_2, \dots, u_n\}$ and $S \subseteq V(G)$. Sometimes we write $G_n := u_1, u_2, \dots, u_n$. Subgraph induced by S denoted $G[S]$ and subgraph with vertex set $V(G) - S$ denoted by $G - S$. Let H is subgraf of G. The neighborhood $N_H(v)$ of a vertex v is the set of vertices in H adjacent to v and $N_H[v] = N_H(v) \cup \{v\}$. If $H = G$, we only write $N_H(v)$ with $N(v)$. The minimum degree, maximum degree, independence number, and connectivity of G are denoted by $\delta(G), \Delta(G), \alpha(G)$, and $\kappa(G)$, respectively. For two graph H and G. If graph H is a subgraph of G, we denoted $G \supseteq H$. Chavatal and Harary [1] established a useful lower bound for finding the exact Ramsey numbers $R(G,H)$, namely $R(G,H) \geq (\chi(G) - 1)(C(H) - 1) + 1$, where $\chi(G)$ is the chromatic number of G and $C(H)$ is the number of vertices of the largest component of H.

A cycle and a path of order n are denoted by C_n and P_n, respectively. The cycle C_n, sometimes we write $C_n := u_1, u_2, \dots, u_n$. Graph G that the vertex set can be divided into several subsets $V_1, V_2, \dots, V_k$, such that every edge of G connecting the vertex on V_i to the vertex on $V_j$with $i \neq j$, called multipartite graph denoted by $B_{n_1,n_2,\dots,n_k}$with $|V_i| = n_i, i = 1,2,\dots,k$. If $k = 2$, the graph is called bipartite and can be denoted by $B(V_1, V_2)$ besides B_{n_1,n_2}. A multipartite graph $B_{n_1,n_2,\dots,n_k}$ is a complete multipartite graph if every vertex of V_i is adjacent to every vertex of V_j and this graph is denoted by $K_{n_1,n_2,\dots,n_k}$. A star S_n is the graph of n vertices with one vertex of degree $n-1$, called the $center$ and $n-1$ vertices of degree 1. A wheel W_m is the graph on $m+1$ vertices obtained from a cycle C_m on m vertices by adding one vertex x, called the hub of the wheel. For the Ramsey number of stars with wheels, several results have obtined. Korolova, A. have shown in [2], that $R(S_n, W_m) = 3n - 2$ for $n = m, m+1, m+2$. She has also shown that $R(S_n, W_m) \geq 2n + 1$ for m is even. One year later, Hasmawati et al. [3], determined the Ramsey number $R(S_n, W_m)$ for even $m, m \geq 3$ and odd $n, n \geq 3$. In [4], Chen at al. show that $R(S_n, W_6) = 2n + 1$ for odd $n, n \geq 3$.

Theorem 1. (Hasmawati et al.[3]). If n is even and $n \geq 4$, then

$$R(S_{n+1},W_m) = \begin{cases} 3n-1, \text{for } m = 2n-2\,, or\ \ m = n \\ 3n-3, \text{for } m = 2n-6\ or\ m = 2n-4. \end{cases}$$

They also determined the Ramsey number $R(S_n,W_m)$ for odd m, $m \geq 3$, as follows.

Theorem 2. (Hasmawati et al.[3]).
If $n \geq 2$, then $R(S_{n+1},W_m) = 3n+1$ for even m, $3 \leq m \leq 2n+1$.

In [5], Zhang et al. established $R(S_{n+1},W_8)$ for n, $6 \leq n \leq 11$, as follows.

Theorem 3. (Zhang et al. [5]). $R(S_{n+1},W_8) = \begin{cases} 2n+3, \text{for } even\ n, \\ 2n+4, \text{for } odd\ n. \end{cases}$

Recently, Binlong Li and Ingo Schiermeyer [5] obtained the lower bound and the exact value on $R(S_n,W_m)$ for $6 \leq n \leq m \leq 2n-4$.

Theorem 4. (Li and Schiermeyer [6]). If $6 \leq m \leq 2n-2$ and m is even, then

$$R(S_{n+1},W_m) \geq \begin{cases} 2n+\dfrac{m}{2}-1, for\ both\ n\ and\ \dfrac{m}{2}\ are\ even, \\ 2n+\dfrac{m}{2}, \quad otherwise. \end{cases}$$

Theorem 5. (Li and Schiermeyer [6]). If $n+2 \leq m \leq 2n-2$ and m is even, then

$$R(S_{n+1},W_m) = \begin{cases} 2n+\frac{m}{2}-1, for\ n\ and\ \frac{m}{2}\ are\ even, \\ 2n+\frac{m}{2}, \quad otherwise. \end{cases}$$

In this paper, we consider the value $R(S_n,W_m)$ for $10 \leq m \leq n$ for even m. Firstly, we give the exact value of (S_{20},W_{10}), and secondly we give the exact value of $R(S_{2k+1},W_{2k})$ for even $k, k \geq 6$.

Theorem 6. $R(S_{20},W_{10}) = 43$.

Theorem 7. If k is even, then $R(S_{2k+1},W_{2k}) = 5k-1$ for $k \geq 6$.

2. Some Lemmas

In order to prove of the main result, we need the following lemmas.

Lemma 1. Graph G is a bipartite graph if and only if every cycle on G has even length.
Lemma 2 (Bondy, [7]). Let G be a graph of order n. If $\delta(G) \geq \frac{n}{2}$, then either G is pancyclic or n is even and $G = K_{\frac{n}{2},\frac{n}{2}}$.
Lemma 3 (Brandt, [8]). Every non-bipartite graph G of order n with $\delta(G) \geq \frac{n+2}{3}$ is weakly pancyclic with $g(G) = 3$ or 4.
Lemma 4 (Dirac, [9]). Let G be a 2-connected graph of order $n \geq 3$ with $\delta(G) = \delta$. Then $c(G) \geq \min\{2\delta, n\}$.

3. Proofs of the main results

3.1. Proof of Theorem 6

Let F be a graph with $|F| = 43$. Suppose F contains no S_{20}. So for every vertex $v \in V(F)$, $d_F(v) \leq 18$. Next, we will show that $\bar{F}$ contains a W_{10}. Because $|F| = 43$, , there are at least one vertex $v \in V(F)$, with $d_F(v) = 2l$, $9 \geq l \geq 0$. Choose any vertex $u \in F$ and write $A = V(F)\backslash N[u]$, $T = F[A]$. Since $d_T(u) = 2l$, and $9 \geq l \geq 0$, then $|T| = |\bar{T}| \geq 24$, and $\delta_{\bar{T}}(v) \geq 5$ for every vertex $v \in V(F)$. Let $\bar{T}$ be bipartite. For $0 \leq r \leq 7$, we can write $\bar{T} = (V_1, V_2) = B_{5+r,24-(5+r)}$. For $r = 0, \bar{T} = K_{5,19}$ (complete bipartite) contains C_{10}. The vertex u together with C_{10} forms a wheel W_{10} in $\bar{F}$. For $1 \leq r \leq 7$, we can order the vertices in $\bar{T}$ such that $V_1 = \{v_1, v_2, \dots, v_{5+r}\}$, and $V_2 = \{u_1, u_2, \dots, u_{24-(5+r)}\}$ with $u_i v_{i+j} \notin E(\bar{T})$ for $i = 1,2, \dots, 24-(5+r)$, $j = 0,1, \dots, r-1$, $(i+j)$ mod $(5+r)$, the other vertices are adjacent. Then, now we have a cycle

$$C_{10} := u_i, v_{i+r+1}, u_{i+1}, v_{i+r+2}, u_{i+2}, v_i, u_{i+4}, v_{i+1}, u_{i+3}, v_{i+r+4}, u_i.$$

where u_r means $u_{r \bmod 24-(5+r)}$ and v_s means $v_{s \bmod (5+r)}$. Thus there is W_{10} in $\bar{F}$ with hub u.

Next, let $\bar{T}$ be non bipartite. Now we have the following cases.

Case I. Let $d_F(u) = 2l$, with $0 \leq l \leq 6$. Since $v \in V(T)$, $d_T(v) \leq 18$ and $|T| = |\bar{T}| \geq 30$, then $\delta_{\bar{T}}(v) \geq 11 \geq \frac{|\bar{T}|+2}{2}$. Next, there are three subcases.

Subcase 1.1. Let $\kappa(\bar{T}) = 0$, which implies $\bar{T}$ is not connected. Suppose $\bar{T}$ consists of two components, write $\bar{T} = \bar{T}_1 \cup \bar{T}_2$. Because $\delta(\bar{T}) \geq 11$, then $\delta(\bar{T}_1) = \delta(\bar{T}_2) \geq 11$. Next, because $|\bar{T}| = 30$, there are $i \in \{1,2\}$, $12 \leq |\bar{T}_i| \leq 15$, and so $\delta(\bar{T}_i) = 11 \geq \frac{|\bar{T}_i|}{2}$. By Lemma 2, $\bar{T}_i$ is pancyclic and contains all of the cycle with length of 3 to $|\bar{T}_i|$. Therefore, $\bar{T}_i$ contains a C_{10} and form W_{10} on $\bar{F}$ with hub u.

Subcase 1.2. Let $\kappa(\bar{T}) = 1$. Let T_i be components of $T\backslash\{v\}$ with cut vertex v. For some i, $12 \leq |V(T_i) \cup \{v\}| \leq 15$, subgraph induced by $V(T_i) \cup \{v\}$ is pancyclic. So it contains C_{10}. Hence, graph $\bar{F}$ contains a W_{10} with hub u.

Subcase 1.3. Let $\kappa(\bar{T}) \geq 2$, which means $\bar{T}$ is a 2-connected graph. Note that $\bar{T}$ is nonbipartite. By Lemma 3, $c(\bar{T}) \geq \min\{2(\delta), |\bar{T}|\} = \min\{2(11), 30\} = 22$, where $c(\bar{T})$ is the longest cycle on $\bar{T}$. Because $\delta(\bar{T}) \geq \frac{|\bar{T}|+2}{3}$, by Lemma 2, $\bar{T}$ is weakly pancyclic and contains all cycles C_l where $g(\bar{T}) \leq l \leq 22 \leq c(\bar{T})$ with $g(\bar{T})$ equal 3 or 4. Hence, $\bar{T}$ contains C_{10} and the graph $C_{10} + u$ forms W_{10} on $\bar{F}$.

Case II. Let $d_F(u) = 2l$ with $7 \leq l \leq 9$. Write $B = V(F)\backslash N[u]$ and $T = F[B]$. Since $v \in V(T)$, $d_T(v) \leq 18$ and $19 \geq |N[u]| \geq 15$, then $29 \geq |T| = |\bar{T}| \geq 24$ and $10 \geq \delta_{\bar{T}}(v) \geq 5$, respectively. Since $\bar{T}$ is non-bipartite, there are three subcases as follows.

Subcase 2.1. Let $\kappa(\bar{T}) = 0$, which means $\bar{T}$ is a disconnected graph. Let G_i, $i = 1,2, \dots r$, be the components of $\bar{T}$ or $\bar{T} = \bigcup_{i=1}^{r} \bar{G}_i$. Because $\delta(\bar{T}) \geq 5$ and $|\bar{T}| \geq 24$, then $2 \leq r \leq 4$. For $r = 4$, there at least one G_i with $|G_i| = 6$. This means every vertex in G_i is adjacent to every vertex in G_j with $i \neq j$. Hence, G_i is dependent on $N[u]$ in F. Hence, every vertex in $N[u]$ is adjacent to every vertex in G_i. Therefore, subgraph induced by $V(G_i) \cup N[u]$ contain W_{10} in $\bar{F}$, with any vertex in G_i being the hub of the wheel W_{10}. For $r \leq 3$ and if there are at least one G_i with $|G_i| \geq 10$ and contains no cut vertex, then by Lemma 2, G_i is pancyclic and so G_i contains a C_{10}, such that graph $\bar{F}$ contains a W_{10} which the hub u. Otherwise, we let there is G_i without cut vertex with $6 \leq |G_i| \leq 8$. For this case, every vertex in G_i is adjacent to all vertices in G_j at T for all $j \neq i$, such that every vertex in G_i is adjacent to at most two vertices in $N[u]$ of F. Hence in $\bar{F}$, every vertex in G_i is adjacent to at least $|N[u]| - 2$

vertices of $N[u]$. This means, for every two vertices v_i, v_j in G_i, $|N(v_i) \cap N(v_j)|_{N[u]} \geq |N[u]| - 4$ of $\bar{F}$. So, obviously $\bar{F}[V(G_i) \cup N[u]]$ contains W_{10} with any vertex in G_i as its hub. Next, $|G_1| = 9$ and $|G_2| = 15$. Since $\delta_{\bar{T}}(v) \geq 5$, there are at least two vertices v_i, $v_j \in G_1$, such that $N(v_i) \cap N(v_j) = \{u_j\}$. Thus, subgraph $\bar{F}[(G_1) \cup \{u\}]$ contains W_{10} of $\bar{F}$ with any vertex in G_1 as its hub.

Subcase 2.2. Let $\kappa(\bar{T}) = 1$. Let T_i be components of $T\backslash\{v\}$ with v as cut vertex where $i = 1,2..,s$. If there are $i \in \{1,2,3\}$ such that $6 \leq |V(T_i) \cup \{v\}| \leq 8$, then every vertex in T_i is adjacent to all vertices in $T_j, j \in \{1,2,3\}, j \neq i$ of F. Hence in F, every vertex in T_i is adjacent to at most three vertices of $N[u]$. Thus in $\bar{F}$, every vertex in T_i is adjacent to at least $|N[u]| - 3$ vertices in $N[u]$. Therefore, $\bar{F}[V(T_i) \cup N[u] \cup \{v\}]$ contains W_{10} with any vertex in T_i as its hub. Otherwise, $|V(T_i) \cup \{v\}| \geq 9$ for every i. If there are i, $10 \leq |V(T_i) \cup \{v\}| \leq 13$, then $\bar{F}[V(T_i) \cup \{v\}]$ contains C_{10} and forms W_{10} on $\bar{F}$ with hub u. Not that if $|V(T_i) \cup \{v\}| = 10$, then $|V(T_j) \cup \{v\}| = 14$. Next, let $|V(T_1) \cup \{v\}| = 9$ and $|V(T_2) \cup \{v\}| = 16$. If every vertex $v_i \in V(T_2) \cup \{v\}$, satisfies $d(v_i) \geq 6$ and contains no cut vertex, then $\bar{F}[V(T_2) \cup N[u] \cup \{v\}]$ is weakly pancyclic with $g(G) = 3$ or 4. By Lemma 3 and 4, $\bar{F}[V(T_2) \cup N[u] \cup \{v\}]$ contains C_{10}. This is C_{10} together vertex u forms W_{10} of $\bar{F}$. Furthermore, if there is vertex $v_0 \in V(T_2)$ with $d(v_0) = 5$. Write $N_{T_2}(v_0) = \{v_1, v_2, ..., v_5\}$. Since $d_{\bar{F}}(v_i) \geq 24$ and $d_{\bar{T}}(v_i) \leq 15$, vertex v_i can be adjacent to at least 9 vertices in $N(u)$ of $\bar{F}$. Let $N_{N(u)}(v_i) = \{u_1, u_2, u_3, u_4 u_5, u_6, u_7, u_8, u_9\}$ and $N_{N(u)}(v_j) = \{u_{10}, u_{11}, u_{12}, u_{13}, u_{14}, u_{15}, u_{16}, u_{17}, u_{18}\}$.

Then for every $v_r, r \neq i, j$, $|N_{N(u)}(v_r) \cap N_{N(u)}(v_i)| \geq 5$ or $|N_{N(u)}(v_r) \cap N_{N(u)}(v_j)| \geq 5$. Without loss of generality, we can write $N_{N(u)}(v_i) \cap N_{N(u)}(v_j) = \{u_1, u_2, u_3, u_4 u_5\}$ for every i and j. Vertex set $\{v_1, v_2, ..., v_5\} \cup \{u_1, u_2, u_3, u_4 u_5\}$ forms C_{10} and C_{10} together vertex v_0 forms W_{10} of $\bar{F}$. Then, we have a wheel W_{10} of $\bar{F}$ with the hub v_0.

Subcase 2.3. Let $\kappa(\bar{T}) \geq 2$, i.e. $\bar{T}$ is a 2-connected graph. Let for every vertex $u \in V(\bar{T})$, $d(u) \geq 9$, then $\bar{T}$ is weakly pancyclic and so $\bar{T}$ contains C_{10} (since $\kappa(\bar{T}) \geq 2$, $\bar{T}$ is nonbipartite, Lemma 3 and 4). Cycle C_{10} with vertex u form a W_{10} in $\bar{F}$. Now, suppose there is vertex $v \in V(\bar{T})$, with $d(v) \leq 8$. Since, $d_{\bar{F}}(v_i) \geq 24$ and $d_{\bar{T}}(v) \leq 8 = \{v_1, v_2, ..., v_8\}$, vertex v_i can be adjacent to at least 16 vertices in $N(u)$ of $\bar{F}$. Let $N_{N(u)}(v_i) = \{u, u_1, u_2, u_3, ..., u_{15}\}$. If v_i is not adjacent to the other vertices v_j in $\bar{T}$, then $d_{\bar{T}}(v_i) \geq 9$ or $d_{\bar{T}}(v_i) \leq 8$. Let $d_{\bar{T}}(v_i) \geq 9$. Then $\bar{T}\backslash\{v_1, v_2, ..., v_s\}$ is weakly pancyclic. For $s \leq 5$, $|\bar{T}\backslash\{v_1, v_2, ..., v_s\}| \geq 10$ and so, $\bar{T}\backslash\{v_1, v_2, ..., v_s\}$ contains a C_{10}. The cycle C_{10} together with vertex u form W_{10} with the hub u. Next, let $d_{\bar{T}}(v_i) \leq 8$ for some $i = 1,2, ...,5$. For this case, vertex v_i can be adjacent to at least 16 vertices of $N(u)$ in $\bar{F}$, e.g. vertex v_0. Write $N_{N(u)}(v_i) = \{u, u_{16}, u_{17}, u_{18}, u_1 ..., u_{12}\}$. Consequently, the induced subgraph $\bar{F}[N_{N(u)}(v_0) \cap N_{N(u)}(v_i) \cup \{v_1, v_2, ..., v_s\}]$ contains C_{10}. Therefore, graph $\bar{F}$ contains a W_{10} which the hub v_0.

Now, for all cases, graph $\bar{F}$ always contains W_{10}. Hence, we have $R(S_{20}, W_{10}) \leq 43$. By Li and Schiermeyer's lower bound, $R(S_{20}, W_{10}) \geq 2(20) + \frac{10}{2} - 2 = 43$. Hence, we have $R(S_{20}, W_{10}) = 43$.

3.2. Proof of Theorem 7

By Li and Schiermeyer's lower bound we have $R(S_{2k+1}, W_{2k}) \geq 2(2k) + \frac{2k}{2} - 1 = 5k - 1$. Next, we only need to show that $R(S_{2k+1}, W_{2k}) \leq 5k - 1$, for even k, $k \geq 6$. Let F' be a graph with $|F'| = 5k - 1$, for even k, $k \geq 6$. Suppose graph F' contains no S_{2k+1}. We will show that $\bar{F}'$ contains W_{2k} for even $k, k \geq 6$. Note that $F' \not\supseteq S_{2k+1}$, such that every vertex $v \in V(F')$, $d_{F'}(v) \leq 2k - 1$. Because $d_T(v) \leq 2k - 1$ (odd) and $|F'| = 5k - 1$ (odd), then there are $r \in F'$ with $\delta_F(r) = 2l, 0 \leq l \leq k - 1$. Write $A = V(F')\backslash N[r]$ and $T = F'[A]$ and so $|T| = |\bar{T}| \geq 5k - 1 - (2k - 2) - 1 = 3k$. Because $d_T(v) \leq 2k - 1$, every $v \in \bar{T}$ satisfies $\delta_{\bar{T}}(v) \geq 3k - (2k - 1) - 1 = k$. Now, we have the following three cases.

Case I. Let $\bar{T}$ be a bipartite graph. Since for every vertex in $\bar{T}$, $d_T(v) \geq k$ and $|\bar{T}| \geq 3k$, then we can write $\bar{T} = (V,U) = B_{k+r,\ 3k-(k+r)}$ for $0 \leq r \leq \frac{k}{2}$. For $r = 0, \bar{T}=K_{k,2k}$ (complete bipartite) that contains C_{2k}. The vertex r together C_{2k} form W_{2k} in $\bar{F}$. For $1 \leq r \leq \frac{k}{2}$, we can order the vertices in $\bar{T}$ such that $V = \{v_1, v_2, \dots, v_{k+r}\}$, $U = \{u_1, u_2, \dots, u_{2k-r}\}$ with $u_i v_{i+j} \notin E(\bar{T})$ for $i = 1,2,\dots,2k-r$, $j = 0,1,\dots,r-1$, $(i+j) \bmod (k+r)$, and u_i is adjacent to the other vertices in $\bar{T}$.
Then, we have a $C_{2k} := u_i, v_{i+r+1}, u_{i+1}, v_{i+r+2}, u_{i+2}, v_{i+r+3}, \dots, v_{i+r+s}, u_{i+j}, v_{i+r}, u_i$, $s = k - i$ and $j = k - i$. In this case, $i + r + s \bmod (k+r)$ and $i + j \bmod (2k - r)$. Hence we have W_{2k} of $\bar{F}'$ with the hub r.

Case II. Let $\bar{T}$ be a nonbipartite graph.
Subcase 2.1. For $\kappa(\bar{T}) = 0$, $\bar{T}$ is not connected. Because for every vertex in $\bar{T}$, $\delta_{\bar{T}}(v) \geq k$ and $|\bar{T}| \geq 3k$, $\bar{T}$ consists two components. Write $\bar{T} = \bar{T}_1 \cup \bar{T}_2$ with $k+1 \leq |\bar{T}_i| \leq 2k-1$. Let $|\bar{T}_1| = k+1+j, j = 0,1,\dots,\frac{k-2}{2}$. Note that every vertex in T_1 is adjacent to every vertex in T_2 and $d_T(v) \leq 2k-1$ for every $v \in V(T)$. Then, in F', every vertex in T_i is adjacent to at most $\frac{k-2}{2}$ vertices in $N[r]$. This is means, in $\bar{F}'$, every vertex in $\bar{T}_i$ is adjacent to at least $2k - \left(\frac{k-2}{2}\right) = \frac{3k+2}{2}$ vertices in $N[r]$. Thus, every two vertices $u,v \in V(\bar{T}_i)$, $|N_{N(r)}(u) \cap N_{N(r)}(v)| \geq k+2$. Therefore, $\bar{F}'[V(\bar{T}_i) \cup N[r]]$ contains W_{2k} with any vertex in $\bar{T}_1$ as the hub.

Subcase 2.2. For $\kappa(\bar{T}) = 1$, graph $\bar{T}$ have one cut vertex. Let T_i be components of $T \backslash \{u\}$ with cut vertex u. For some i, $k+1 \leq |V(T_i) \cup \{u\}| \leq 2k-1$. Subgraph T' induced by $V(T_i) \cup \{u\}$ is isomorphic with $\bar{T}_i$ described in Subcase 2.1. It means that $\bar{F}'[[V(T') \cup N[r]]$ contains W_{2k} with any vertex in T_i as the hub.

Subcase 2.3. Now, let $\kappa(\bar{T}) \geq 2$. it means that $\bar{T}$ is a 2-connected graph. If for every vertex $u \in V(\bar{T})$, $d(u) \geq k + \frac{k}{2}$, then $\bar{T}$ is weakly pancyclic and so, $\bar{T}$ contains C_{2k} (since $\kappa(\bar{T}) \geq 2$, $\bar{T}$ is nonbipartite, and by Lemma 3 and 4). Cycle C_{2k} with vertex u form a W_{2k} in $\bar{F}'$. Next, if there is a vertex $v \in V(\bar{T})$, with $k \leq d(v) \leq \frac{3k-2}{2}$. Since v_i, $d_{\bar{F}'}(v_i) \geq 3k-1$, vertex v_i is adjacent to at least $\frac{3k}{2}$ vertices in $N(r)$. Let $N_{N(r)}(v_0) = \{u, u_1, u_2, u_3, \dots, u_{2k-1}\}$ and $N_{\bar{T}}(v_0) = \{v_j : j = 1,2,\dots,k\}$. If v_i is not adjacent to any other vertices v_j in $N(v_0)$, then $d_{\bar{T}\backslash N(v_0)}(v_i) \geq k + \frac{k}{2}$, or $d_{\bar{T}\backslash N(v_0)}(v_i) \leq \frac{3k-2}{2}$. Let $d_{\bar{T}\backslash N(v_0)}(v_i) \geq k + \frac{k}{2}$. Then $\bar{T}\backslash\{v_0\}$ is weakly pancyclic. Thus, $\bar{T}\backslash\{v_0\}$ contains a C_{2k}. The cycle C_{2k} together with vertex r form W_{2k} of $\bar{F}'$ with the hub r. Next, if $d_{\bar{T}}(v_i) \leq \frac{3k-2}{2}$ for some $i = 1,2,\dots,k-1$, vertex v_i can be adjacent to at least $\frac{3k}{2}$ vertices in $N(r)$ of $\bar{F}'$ e.g. to vertex v_0. Write $N_{N(r)}(v_i) = \left\{u, u_{\frac{k-2}{2}}, u_{\frac{k}{2}}, u_{\frac{k+2}{2}}, u_{\frac{k+4}{2}} \dots, u_{2k-1}\right\}$ So the induced subgraph $\bar{F}[N(v_0) \cap N_{N(r)}(v_i)]$ contains a C_{2k}. Hence, graph $\bar{F}'$ contains a W_{2k} which the hub v_0.

Therefore, we have $R(S_{2k+1}, W_{2k}) \leq 5k-1$ for every even k, $k \geq 6$. The proof is complete.

References
[1] Chavatal V and Harary *F 1972* Generalized Ramsey Theory for graphs, III, small off-diagonal numbers *J. Math.* **41** 335-345
[2] Korolova A 2005 Ramsey Numbers of Stars Versus Wheels of Similar Size *Discrete Mathematics* **292** 107-117

[3] Hasmawati 2007 *Ramsey Numbers For the Joint Graf Stars* Unpublished dissertation (Bandung: ITB Graduate Program)

[4] Chen Y J, Zhang Y Q and Zhang K M 2004 The Ramsey numbers of star versus wheels *European J. Combin* **25** 7 1067-1075

[5] Zhang Y, Chen Y and Zhang K 2008 The Ramsey Numbers For Stars of Even Orders Versus A Wheel of Order Nine *European Journal of Combinatorics* **29** 1744-1754

[6] Li, Binlong and Schiermeyer I 2016 On Star-Wheel Ramsey Numbers *Graphs and Combinatorics* **32** 733-739

[7] Bondy J and Murty U S 1976 *Graph Theory With Applications* (USA: Elsevier Science Publishing Co. Inc)

[8] Brand S, FAU Dree R. J and Goddard W 1998 Weakly pancyclic graph *J. Graph Theory* **27** 141-176

[9] Dirac G A 1952 Some theorems on abstract graphs. *Proc. Lond. Math. Soc.* 2 69-81

Flattening Property and the Existence of Global Attractors in Banach Space

Naimah Aris, Sitti Maharani, Massalesse Jusmawati and Budi Nurwahyu

Mathematics Department, Hasanuddin University, Jalan Perintis Kemerdekaan KM. 10 Tamalanrea, Makassar 90245, Indonesia.

E-mail: aris.naimah@unhas.ac.id

Abstract. This paper analyses the existence of global attractor in infinite dimensional system using flattening property. The earlier stage we show the existence of the global attractor in complete metric space by using concept of the ω-limit compact concept with measure of non-compactness methods. Then we show that the ω-limit compact concept is equivalent with the flattening property in Banach space. If we can prove there exist an absorbing set in the system and the flattening property holds, then the global attractor exist in the system.

1. Introduction

Today many of the physical problems in infinite dimensional system can be represented in mathematical models studied in a strongly continuous semigroup, C^0 semigroups, i.e. the operator family $\{S_t\}$ that maps a metric space M to itself and satisfies the properties [1, 2]:

1). $S_{(0)}$ is the identity map on M,

2). $S_{(t+s)} = S_t \circ S_s$ for all $t, s \geq 0$,

3). the function

$$S: [0, \infty) \times M \to M$$
$$(t, x) \mapsto S_t(x) \in M$$

continous at each point $(t, x) \in [0, \infty) \times M$.

The long time dynamics of an infinite dimensional system can be described by the global attractor. Global attractors is the set of connected, compact, and invariant set which attracts all bounded set [2, 3]. Normally, the existence of the global attractor can be proved by showing that:

1). the absorbing set exist in the system, and

2). the semigroup of the system is compact uniform.

Unfortunately, it is very difficult to show that the semigrup is uniformly compact in many problems. Thus in [3], it is given conditions which can be easily proven by using the measure of non-compactness (MNC) called ω-limit compact. The existence of the global attractor for a C^0 semigroup can be proved if and only if:

1). the absorbing set exist in the system, and

2). the semigroup is ω-limit compact.

The use of the MNC concept as well, is used by [3] to show the ω-limit compact of a C^0 semigroup in a convex Banach space X. It is shown to be equivalent with the condition known as flattening property. In [4], flattening property is defined as:

Let $S_t: X \to X$ be a semigroup, and B be any bounded set of X. Then for any $\mathcal{E} > 0$, there exist $t(B) > 0$ and a finite dimensional subspace X_1 of X, such that $(P)\left(\bigcup_{t \geq t_B} S_t(B)\right)$ is bounded in X and

$$(I - P)\left(\bigcup_{t \geq t_B} S_t(B)\right) \subset B(0, \varepsilon) \quad \textit{for } t \geq t(B),$$

where $P: X \to X_1$ is a bounded projection, and $I: X \to X$.

Furthermore, from the paper [2] the authors provide that the existence of a global attractors can be proved by showing that

1). the absorbing set exist in the system, and
2). the flattening property holds.

From that explanation, we are interested to re-examine the concept in detail to show the existence of a global attractor by using the measure of non-compactness property. The assessment was carried out in this paper.

2. Measure of Non-compactness (MNC)

Some concepts of measure of non-compactness and its basic property will be given in this section, (see [4]).

Definition 2.1 Let M be a complete metric space and B be a bounded subset of M. The measure of non-compactness $\gamma(B)$ of B is defined by

$$\gamma(B) = \inf\{\delta > 0 | B \text{ admits a finite cover by sets of diameter } \leq \delta\}.$$

Lemma 2.1 *Let M be a complete metric space and B be a bounded subset of M. A mapping $\gamma: B \to [0, \infty)$ is called measure of noncompactness of M if it satisfies the following property i.e. for every $Q, Q_1, Q_2 \in B$*

1). $\gamma(Q) = 0$ *if and only if* $\bar{Q}$ *compact*
2). $\gamma(Q) = \gamma(\bar{Q})$
3). $\gamma(Q_1 \cup Q_2) = max\{\gamma(Q_1), \gamma(Q_2)\}$
4). $\gamma(Q_1 + Q_2) \leq \gamma(Q_1) + \gamma(Q_2)$
5). $\gamma(Q_1) \leq \gamma(Q_2)$, *whenever* $Q_1 \subset Q_2$.

Lemma 2.2 *Let X be a Banach space and γ be the measure of non-compactness, $B(0, \varepsilon)$ is a open ball of radius ε then $\gamma(B(0, \varepsilon)) = 2\varepsilon$.*

Lemma 2.3 *Let M be a complete metric space and γ be the measure of non-compactness. Assume that (F_n) is a sequence of bounded and closed subset of M, and satisfying*

1). $F_n \neq \emptyset$
2). $F_{n+1} \subset F_n$, $n = 1,2, \ldots$
3). $\gamma(F_n) \to 0$, *as* $n \to \infty$

Then $F = \bigcap_{n=1}^{\infty}(F_n)$ is nonempty compact set.

Definition 2.2 Let X be a convex Banach space. For any bounded set B of X, $S_t: X \to X$ and for any $\mathcal{E} > 0$, there exists $t(B) > 0$ and a finite dimensional subspace X_1 of X, such that $(P)\left(\bigcup_{t \geq t_B} S_t(B)\right)$ is bounded in X and

$$(I-P)\left(\bigcup_{t\geq t_B} S_t(B)\right) \subset B(0,\varepsilon) \quad \text{for } t \geq t(B),$$

where $P: X \to X_1$ is a bounded projection, and $I: X \to X$.

3. The Concept of Global Attractor

In this section, we will provide some definitions from global attractors in dynamic systems related to our paper, ([1, 4, 5])

Definition 3.1 Let $\{S_t\}_{t\geq 0}$ be a C^0 semigroup in a complete metric space M. A subset B_0 of M is called an absorbing set in M if for any bounded subset B of M, there exists some $t_1 \geq 0$ such that $S_t B \subset B_0$ for all $t \geq t_1$.

Definition 3.2 For any subset W_1, W_2 in complete metric space M, the subset W_1 is called absorbs W_2 under $\{S_t\}$ if there exists a number $t_0 \geq 0$ such that $S_t W_2 \subset W_1$, for all $t \geq t_0$.

Notice that W_1 attracts W_2 if and only if each open neighborhood $\mathcal{N}_{W_1}$ of W_1 in M absorbs W_2.

Definition 3.3 Let $\{S_t\}_{t\geq 0}$ be a C^0 semigroup in a complete metric space M, the semigroup is called ω-limit compact if for every bounded subset B of M and for any $\varepsilon > 0$, there exists a $t_0 > 0$ such that

$$\gamma\left(\bigcup_{t\geq t_0} S_t(B)\right) \leq \varepsilon.$$

Definition 3.4 Let M be a metric space. A subset $\mathcal{A} \subset M$ is called global attractor if $\mathcal{A}$ is attractor compact that attracts all bounded sets of M.

Lemma 3.1 *Let $\bigcup_{t\geq t_B} S_t(B)$ is bounded, then $\overline{\bigcup_{t\geq t_B} S_t(B)}$ is bounded.*
Let $\{S_t\}_{t\geq 0}$ be a C^0 semigroup in a metric space M. If S_t is ω-limit compact, then $\mathcal{A} = \omega(B)$ is compact set.

Lemma 3.2 *Let $\{S_t\}_{t\geq 0}$ be a C^0 semigroup in a metric space M. Then $\mathcal{A} = \omega(B)$ is an invariant positive, i.e. $S_t\mathcal{A} \subset \mathcal{A}$.*

Lemma 3.3 *Let $\{S_t\}_{t\geq 0}$ be a C^0 semigroup in a metric space M, and $\mathcal{A} = \omega(B)$ is compact set. Then $\mathcal{A}$ is an invariant negative, i.e. $S_t\mathcal{A} \supset \mathcal{A}$.*

Lemma 3.4 *Let $\{S_t\}_{t\geq 0}$ be a C^0 semigroup in a metric space M. If there exists a bounded absorbing subset B of M, then $\mathcal{A} = \omega(B)$ attracts B of M.*

4. Main Result

In this section, firstly we recall some basic lemmas in [3, 5], secondly we show the existence of global attractor by concept ω-limit compact with measure of noncompactness methods in a metric space complete. Then we show that concept ω-limit compact equivalent with the flattening property in Banach space.

Theorem 4.1 Let $\{S_t\}_{t\geq 0}$ *be a C^0 semigroup in a complete metric space M. Assume that*

1). *S_t is ω −limit compact,*
2). *there exists a bounded absorbing subset B of M.*

Then the ω −limit set of B, $\mathcal{A} = \omega(B)$ is compact attractor which attracts all bounded subsets of M.

Proof:
In proving that $\mathcal{A} = \omega(B)$ is compact attractor, firstly we show that $\mathcal{A}$ is compact and $\mathcal{A}$ attractor. From Lemma 3.2, we prove that $\mathcal{A}$ is compact. To show that $\mathcal{A}$ attractor, as well as showing $\mathcal{A}$ invariant set and $\mathcal{A}$ attracts B of M. $\mathcal{A}$ invariant set, it is shown that $\mathcal{A}$ is invariant positive and $\mathcal{A}$ is invariant negative. From Lemma 3.3 and Lemma 3.4, it can be concluded that $\mathcal{A}$ is invariant.
Furthermore, we will show that $\mathcal{A}$ attract B of M.
Since there exists a bounded subset B of M. From Definition 3.2, we get that $\mathcal{A}$ attracts B of M. Besides that, we can show that $\mathcal{A}$ attracts B of M by using contradictions like Lemma 3.4. Thus, $\mathcal{A} = \omega(B)$ is a compact attractor that attracts all bounded subsets in M.

Theorem 4.2 *Let $\{S_t\}_{t\geq 0}$ be a C^0 semigroup in a complete metric space M. Then S_t has a global attractor $\mathcal{A}$ in M if and only if*
1). *$\{S_t\}_{t\geq 0}$ is ω −limit compact, and*
2). *there is a bounded absorbing set $B \subset M$.*

Proof:
($\Leftarrow$) From Theorem 4.1 we have proved that $\mathcal{A}$ is a compact attractor. Referring to Definition 3.4, $\mathcal{A}$ is a global attractor.
($\Rightarrow$) Now we will prove the opposite. Since $\mathcal{A}$ is a global attractor, then $\mathcal{A}$ attractor and $\mathcal{A}$ compact (see Definition 3.4). So that the neighborhood $\mathcal{N}_\varepsilon(\mathcal{A})$ is absorbing set. According to the Definition 3.4 $\mathcal{N}_\varepsilon(\mathcal{A})$ absorbs all bounded sets B of M. Next we will prove that the ω −limit compact.
Since B is bounded set, then $\bigcup_{t\geq t_\varepsilon B} S_t(B)$ is bounded for $t_\varepsilon(B) \geq 0$, and for any $t \geq t_\varepsilon(B)$.
Since $\mathcal{N}_\varepsilon(\mathcal{A})$ is the absorbing set then

$$\bigcup_{t\geq t_\varepsilon B} S_t(B) \subset \mathcal{N}_\varepsilon(\mathcal{A}).$$

For $\varepsilon > 0$ take $\frac{\varepsilon}{4} > 0$, such that

$$\bigcup_{t\geq t_\varepsilon B} S_t(B) \subset \mathcal{N}_{\frac{\varepsilon}{4}}(\mathcal{A}).$$

From Lemma 2.2, we have

$$\gamma\left(\bigcup_{t\geq t_\varepsilon B} S_t(B)\right) \leq \gamma\left(\mathcal{N}_{\frac{\varepsilon}{4}}(\mathcal{A})\right).$$

To show that the semigroup is compact, then it is the same as we show that

$$\gamma\left(\bigcup_{t\geq t_\varepsilon B} S_t(B)\right) \leq \gamma\left(\mathcal{N}_{\frac{\varepsilon}{4}}(\mathcal{A})\right) \leq \varepsilon.$$

Since $\mathcal{A}$ is compact, there exists a finite number of elements $x_1, x_2, \dots, x_n \in M$ such that

$$\mathcal{A} \subset \bigcup_{i=1}^{n} \mathcal{N}\left(x_i, \frac{\varepsilon}{4}\right).$$

On the other hand, for every $y_j \in \mathcal{N}_{\varepsilon_1}(\mathcal{A}), j = 1,2,\dots,k$ then $y_j \in \bigcup_{i=1}^{n} \mathcal{N}(x_i, \varepsilon_2)$, $i = 1,2,\dots,n$ with $y_j, x_i \in M$, and $\varepsilon_2 \geq \varepsilon_1$. So that

$$\mathcal{N}_{\frac{\varepsilon}{4}}(\mathcal{A}) \subset \bigcup_{i=1}^{n} \mathcal{N}\left(x_i, \frac{\varepsilon}{2}\right).$$

Which implies that

$$\gamma\left(\bigcup_{t\geq t_\varepsilon B} S_t(B)\right) \leq \gamma\left(\mathcal{N}_{\frac{\varepsilon}{4}}(\mathcal{A})\right) \leq \varepsilon.$$

Therefore, $\gamma\left(\bigcup_{t\geq t_\varepsilon B} S_t(B)\right) \leq \varepsilon$ has a semigroup which is ω −limit compact.

Theorem 4.3 *Let X be a Banach space and* $\{S_t\}_{t\geq 0}$ *be a* C^0 *semigroup in X. Then* $\{S_t\}_{t\geq 0}$ *is* ω *−limit compact if and only if flattening property holds true.*

Proof. ($\Leftarrow$) From Definition 3.3, we know that to show $\{S_t\}_{t\geq 0}$ is ω −limit compact equal to indicating that for any bounded absorbing sets B in X, and for every $\varepsilon > 0$ there exists $t_B > 0$ such that

$$\gamma\left(\bigcup_{t\geq t_B} S_t(B)\right) \leq \varepsilon.$$

Note that

$$\gamma\left(\bigcup_{t\geq t_B} S_t(B)\right) = \gamma\left((P + I - P)\left(\bigcup_{t\geq t_B} S_t(B)\right)\right)$$

$$= \gamma\left(P\left(\bigcup_{t\geq t_B} S_t(B)\right) + (I - P)\left(\bigcup_{t\geq t_B} S_t(B)\right)\right). \qquad (1)$$

From Definition 2.2, $(P)\left(\bigcup_{t\geq t_B} S_t(B)\right)$ is bounded, and we get $(P)\left(\overline{\bigcup_{t\geq t_B} S_t(B)}\right)$ bounded (see Lemma 3.1).

From closure definition [6], we have

$$\overline{\bigcup_{t\geq t_B} S_t(B)} = \bigcup_{t\geq t_B} S_t(B) \cup \left(\bigcup_{t\geq t_B} S_t(B)\right)'$$

In other word,

$$\left(\bigcup_{t\geq t_B} S_t(B)\right)' \subset \overline{\bigcup_{t\geq t_B} S_t(B)}.$$

From the theorem of the close set on [6], it is found that $\overline{\bigcup_{t\geq t_B} S_t(B)}$ closed and $(p)\left(\overline{\bigcup_{t\geq t_B} S_t(B)}\right)$ also closed. So we can conclude that $(p)\left(\overline{\bigcup_{t\geq t_B} S_t(B)}\right)$ is compact.

From Lemma 2.2 (1), we have

$$\gamma\left(p\left(\overline{\bigcup_{t\geq t_B} S_t(B)}\right)\right) = 0.$$

From the conclusion of Lemma 2.2, Definition 2.2, and (1), we have

$$\gamma\left(\bigcup_{t\geq t_B} S_t(B)\right) = \gamma\left(P\left(\bigcup_{t\geq t_B} S_t(B)\right) + (I - P)\left(\bigcup_{t\geq t_B} S_t(B)\right)\right)$$

$$\leq \gamma\left(P\left(\bigcup_{t\geq t_B} S_t(B)\right)\right)+\gamma\left((I-P)\left(\bigcup_{t\geq t_B} S_t(B)\right)\right)$$
$$\leq 0+\gamma\big(B(0,\varepsilon)\big)=2\varepsilon.$$

Therefore, S_t is ω −limit compact.

($\Longrightarrow$) Since S_t is ω −limit compact, then for any bounded absorbing sets B in X, and for every $\varepsilon > 0$ there exists $t_B > 0$ such that

$$\gamma\left(\bigcup_{t\geq t_\varepsilon B} S_t(B)\right)\leq \frac{\varepsilon}{2}.$$

Let $A_1, A_2, \ldots, A_n$ be subset of A with diameter less than $\frac{\varepsilon}{2}$, such that

$$\bigcup_{t\geq t_\varepsilon B} S_t(B)\subset \bigcup_{i=1}^{n} A_i.$$

Let $x_i \in A_i$, and diameter less than ε then

$$\bigcup_{t\geq t_\varepsilon B} S_t(B)\subset \bigcup_{i=1}^{n} \mathcal{N}(x_i,\varepsilon).$$

Since X is uniformly convex Banach space, and let $X_1 = \text{span}\{x_1, \ldots, x_n\}$, there exists a projection $P: X \to X_1$ such that $\text{dist}(x, Px) = \|(I-P)x\| = \text{dist}(x, X_1)$, for any $x \in X$. Hence,

$$(I-P)\left(\bigcup_{t\geq t_\varepsilon B} S_t(B)\right)\subset \mathcal{N}(0,\varepsilon). \quad (2)$$

Since B is bounded, then $\bigcup_{t\geq t_\varepsilon B} S_t(B)$ is bounded in X (see theorem of close set on [6]). Hence the following equation is bound in X.

$$(P)\left(\bigcup_{t\geq t_\varepsilon B} S_t(B)\right). \quad (3)$$

From (2) and (3), Definition 2.2 is met. Namely flattening property holds true. ■

Theorem 4.4 *Let $\{S_t\}_{t\geq 0}$ be a C^0 semigroup in a Banach space X. If the following conditions*

1). *There exist a bounded absorbing set $B \subset X$, and*

2). *$\{S_t\}_{t\geq 0}$ satisfies the flattening property*

hold true, then there is a global attractor for $\{S_t\}_{t\geq 0}$ in X.

Proof:

Using Theorem 4.3, it has been shown that $\{S_t\}_{t\geq 0}$ meets the flattening property. In addition, referring to Theorem 4.2 we can conclude that there is a bounded absorbing set $B \subset X$. Therefore the conditions in Theorem 4.4 have been met. The existence of a global attractor for $\{S_t\}_{t\geq 0}$ in X is guaranteed to be true.

References

[1] Robinson R C 2004 *An Introduction to Dynamical Systems: Continuous and Discrete* (America: Person Education, Inc)

[2] Xu L and Shi Y 2013 Notes on the Global Attractors for Semigroup *International Journal of Modern Nonlinear Theory and Application* 219-222

[3] Ma Q F, Wang S and Zhong C 2002 Necessary and Sufficient Conditions for Existence of Global Attractors for Semigroups and Applications *Indiana University Mathematics Journal* **51** 1541-

1559.
[4] Deimling K 1985 *Nonlinear Functional Analysis* (Berlin: Springer-Verlag)
[5] Cholewa J W and Dlotko T 2000 *Global Attractors in Abstract Parabolic Problem.* (Cambridge: Cambridge University Press)
[6] Kreyszig E 1989 *Introductory Functional Analysis with Aplications* (Ohio: John Wiley & Sons, Inc)

On some examples of pollutant transport problems solved numerically using the boundary element method

Moh. Ivan Azis[1], **Kasbawati**[1], **Amiruddin Haddade**[2] **and Sri Astuti Thamrin**[1]

[1]Department of Mathematics, Hasanuddin University, Makassar, INDONESIA
[2]Department of Physics, Hasanuddin University, Makassar, INDONESIA

E-mail: mohivanazis@yahoo.co.id

Abstract. A boundary element method (BEM) is obtained for solving a boundary value problem of homogeneous anisotropic media governed by diffusion-convection equation. The application of the BEM is shown for two particular pollutant transport problems of Tello river and Unhas lake in Makassar Indonesia. For the two particular problems a variety of the coefficients of diffusion and the velocity components are taken. The results show that the solutions vary as the parameters change. And this suggests that one has to be careful in measuring or determining the values of the parameters.

1. Introduction

The boundary element method (BEM) has been used extensively for solving numerically a wide range of problems. This includes deformation problems of elastic materials, heat conduction problems, infiltration problems, pollutant transport problems and many others. And the use of the BEM has also been applied not only to homogeneous but also to inhomogeneous materials, and not only to isotropic but also to anisotropic materials. For example, [1] used the BEM for solving infiltration problems of isotropic homogeneous media, [2] derived a fundamental solution, [3] considered diffusion-convection problems for anisotropic homogeneous materials, [4] solved elasticity problems for isotropic inhomogeneous materials and [5] studied elasticity problems, [6] worked on elliptic boundary value problems, [7] solved heat conduction problems, [8] worked on elasticity problems and [9] considered transient heat conduction problems for anisotropic inhomogeneous materials.

In this paper, some more examples of pollutant transport problems will be solved using the BEM, in addition to those considered by [3]. In the following sections the analysis used in [3] for deriving a boundary integral equation will be shown again.

2. The boundary value problems

The dimensionless governing equation of pollutant transport problems for anisotropic media with constant coefficients may be written as

$$\lambda_{ij}\frac{\partial^2 p(\mathbf{x})}{\partial x_i \partial x_j} - v_i\frac{\partial p(\mathbf{x})}{\partial x_i} = 0 \qquad (1)$$

Figure 1. Circumstances of Tello River (left) and Unhas Lake (right).

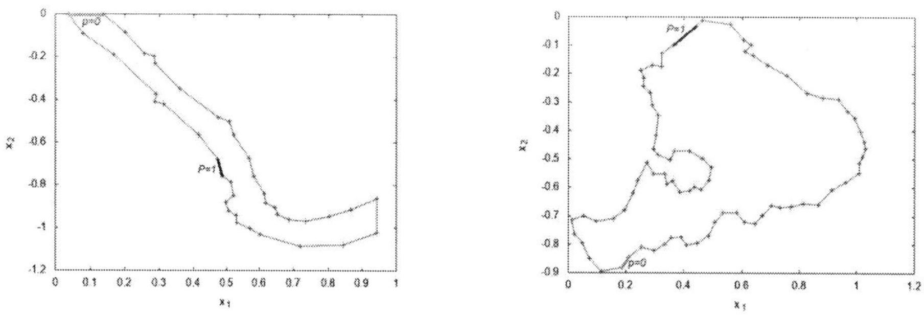

Figure 2. Discretization of the boundary Γ of Tello River (left) and Unhas Lake (right)

where $\lambda_{ij} = \kappa_{ij}/k$, $p = C/C'$, $x_i = X_i/l$, $v_i = (l/k)V_i$, κ_{ij} is the components of dispersion/diffusion coefficients (L^2T^{-1}), C is dissolved concentration of the pollutant (ML^{-3}), X_i is the component of the point coordinates $\mathbf{X}$ (L), V_i is the component of the seepage velocity (LT^{-1}), k is a reference dispersion coefficient, C' is a reference concentration of pollutant and l is a reference length. See [10] for the governing equation of pollutant transport problems.

Referred to the Cartesian frame Ox_1x_2 we will consider the boundary value problems governed by (1) where $\mathbf{x} = (x_1, x_2)$, λ_{ij} and v_i are constant coefficients, $i, j = 1, 2$. The coefficient $[\lambda_{ij}]$ is a real positive definite symmetrical matrix. Also, in (1) the summation convention for repeated indices apply, so that (1) can be written explicitly as

$$\lambda_{11}\frac{\partial^2 p}{\partial x_1^2} + 2\lambda_{12}\frac{\partial^2 p}{\partial x_1 \partial x_2} + \lambda_{22}\frac{\partial^2 p}{\partial x_2^2} - v_1\frac{\partial p}{\partial x_1} - v_2\frac{\partial p}{\partial x_2} = 0$$

A solution p to (1) is sought which is valid in a region Ω in R^2 with boundary Γ which consists of a finite number of piecewise smooth curves. On Γ_1 the dependent variable $p(\mathbf{x})$ is specified, and $P = \lambda_{ij}(\partial p/\partial x_i)n_j = l/(kC')\kappa_{ij}(\partial C/\partial X_i)n_j$ is specified on Γ_2 where $\Gamma = \Gamma_1 \cup \Gamma_2$ and $\mathbf{n} = (n_1, n_2)$ denotes the outward pointing normal to Γ. The method of solution will be to obtain a boundary integral equation from which numerical values of the dependent variable p and P may be obtained for all points in Ω.

3. The boundary integral equation

Multiplying both sides of (1) by function p^* and then integrating it over the domain Ω yields

$$\int_\Omega \left(\lambda_{ij}\frac{\partial^2 p}{\partial x_i \partial x_j} - v_i\frac{\partial p}{\partial x_i}\right) p^* \, d\Omega = 0 \qquad (2)$$

IOP Conf. Series: Journal of Physics: Conf. Series **979** (2018) 012075 doi:10.1088/1742-6596/979/1/012075

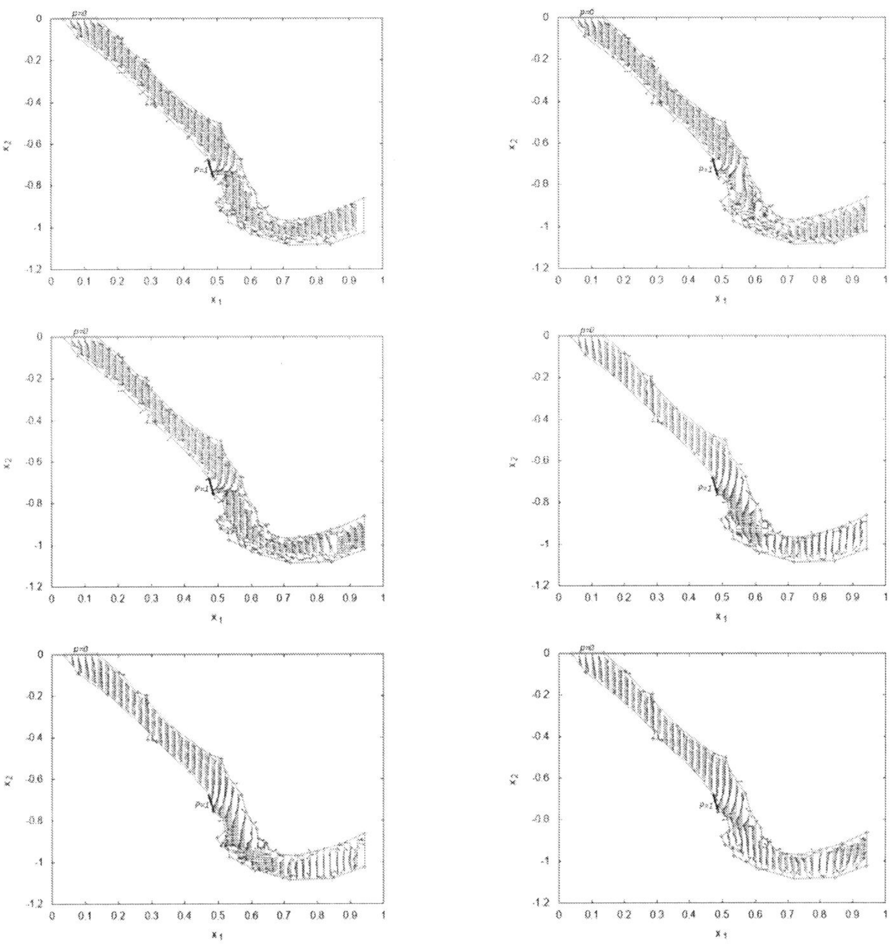

Figure 3. Vectors of $(\partial p/\partial x_1, \partial p/\partial x_2)$ for the Tello River. **Upper, left:** $\lambda_{11} = 1, \lambda_{12} = 1, \lambda_{22} = 2, v_1 = 0, v_2 = 1$. **Upper, right:** $\lambda_{11} = 1, \lambda_{12} = 1, \lambda_{22} = 2, v_1 = 0, v_2 = -1$. **Middle, left:** $\lambda_{11} = 1, \lambda_{12} = 1, \lambda_{22} = 2, v_1 = 1, v_2 = 1.5$. **Middle, right:** $\lambda_{11} = 1, \lambda_{12} = 0, \lambda_{22} = 1, v_1 = 0, v_2 = 1$. **Lower, left:** $\lambda_{11} = 1, \lambda_{12} = 0, \lambda_{22} = 1, v_1 = 0, v_2 = -1$. **Lower, right:** $\lambda_{11} = 1, \lambda_{12} = 0, \lambda_{22} = 1, v_1 = 1, v_2 = 1.5$.

Using Gauss divergence theorem in (2) we obtain

$$\int_\Gamma \left(\lambda_{ij} \frac{\partial p}{\partial x_i} n_j - p v_i n_i \right) p^* \, d\Gamma - \int_\Omega \left(\lambda_{ij} \frac{\partial p}{\partial x_i} \frac{\partial p^*}{\partial x_j} - p \, v_i \, \frac{\partial p^*}{\partial x_i} \right) d\Omega = 0 \tag{3}$$

Use of Gauss divergence theorem once again for the first integrand function $\lambda_{ij} \frac{\partial p}{\partial x_i} \frac{\partial p^*}{\partial x_j}$ in the domain integral in (3) results in

$$\int_\Gamma \left(\lambda_{ij} \frac{\partial p}{\partial x_i} n_j - p v_i n_i \right) p^* \, d\Gamma - \int_\Gamma p \lambda_{ij} \frac{\partial p^*}{\partial x_i} n_j d\Gamma$$
$$+ \int_\Omega \left(p \lambda_{ij} \frac{\partial^2 p^*}{\partial x_i \partial x_j} + p \, v_i \, \frac{\partial p^*}{\partial x_i} \right) d\Omega = 0 \tag{4}$$

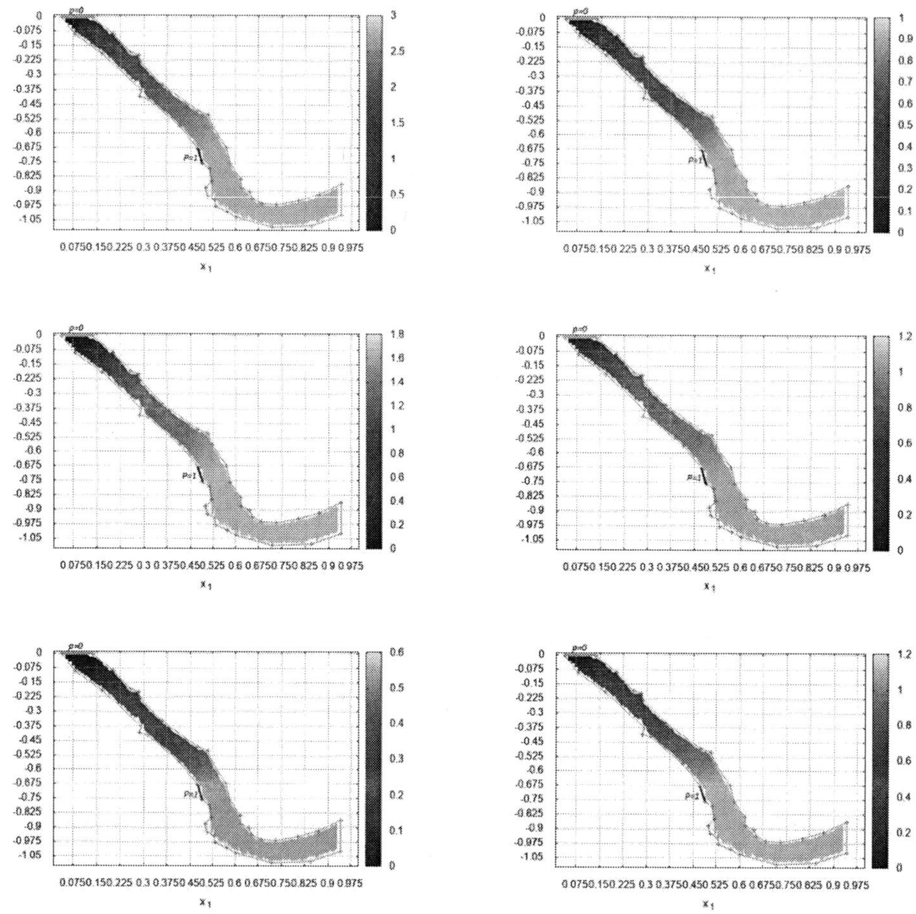

Figure 4. Values of p for the Tello River. **Upper, left:** $\lambda_{11} = 1, \lambda_{12} = 1, \lambda_{22} = 2, v_1 = 0, v_2 = 1$. **Upper, right:** $\lambda_{11} = 1, \lambda_{12} = 1, \lambda_{22} = 2, v_1 = 0, v_2 = -1$. **Middle, left:** $\lambda_{11} = 1, \lambda_{12} = 1, \lambda_{22} = 2, v_1 = 1, v_2 = 1.5$. **Middle, right:** $\lambda_{11} = 1, \lambda_{12} = 0, \lambda_{22} = 1, v_1 = 0, v_2 = 1$. **Lower, left:** $\lambda_{11} = 1, \lambda_{12} = 0, \lambda_{22} = 1, v_1 = 0, v_2 = -1$. **Lower, right:** $\lambda_{11} = 1, \lambda_{12} = 0, \lambda_{22} = 1, v_1 = 1, v_2 = 1.5$.

or

$$\int_\Omega \left(\lambda_{ij} \frac{\partial^2 p^*}{\partial x_i \partial x_j} + v_i \frac{\partial p^*}{\partial x_i} \right) p \, d\Omega = - \int_\Gamma [Pp^* - (P_v p^* + P^*)p] \, d\Gamma \tag{5}$$

where

$$P_v(\mathbf{x}) = v_i \, n_i(\mathbf{x}) \qquad \text{and} \qquad P^*(\mathbf{x}, \boldsymbol{\xi}) = \lambda_{ij} \frac{\partial p^*(\mathbf{x}, \boldsymbol{\xi})}{\partial x_i} n_j(\mathbf{x})$$

If the function p^* is taken such that

$$\lambda_{ij} \frac{\partial^2 p^*}{\partial x_i \partial x_j} + v_i \frac{\partial p^*}{\partial x_i} = -\delta(\mathbf{x} - \boldsymbol{\xi}) \tag{6}$$

where δ is the Dirac delta function and $\boldsymbol{\xi} = (\xi_1, \xi_2)$, then (5) may be written as

$$\int_\Omega p(\mathbf{x}) \, \delta(\mathbf{x} - \boldsymbol{\xi}) \, d\Omega = \int_\Gamma \{P(\mathbf{x}) \, p^*(\mathbf{x}, \boldsymbol{\xi}) - [P_v(\mathbf{x}) \, p^*(\mathbf{x}, \boldsymbol{\xi}) + P^*(\mathbf{x}, \boldsymbol{\xi})] \, p(\mathbf{x})\} \, d\Gamma \tag{7}$$

IOP Conf. Series: Journal of Physics: Conf. Series **979** (2018) 012075 doi:10.1088/1742-6596/979/1/012075

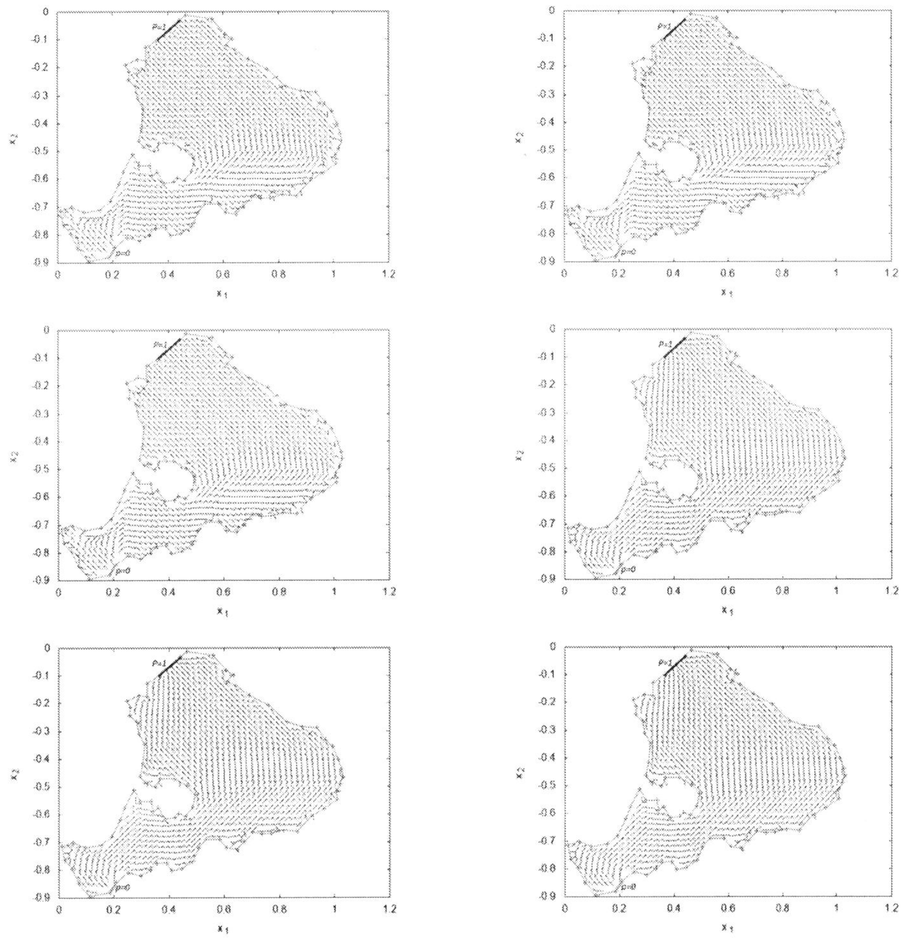

Figure 5. Vectors of $(\partial p/\partial x_1, \partial p/\partial x_2)$ for the Unhas Lake. **Upper, left:** $\lambda_{11} = 1, \lambda_{12} = 1, \lambda_{22} = 2, v_1 = 0, v_2 = 1$. **Upper, right:** $\lambda_{11} = 1, \lambda_{12} = 1, \lambda_{22} = 2, v_1 = 0, v_2 = -1$. **Middle, left:** $\lambda_{11} = 1, \lambda_{12} = 1, \lambda_{22} = 2, v_1 = 1, v_2 = 1.5$. **Middle, right:** $\lambda_{11} = 1, \lambda_{12} = 0, \lambda_{22} = 1, v_1 = 0, v_2 = 1$. **Lower, left:** $\lambda_{11} = 1, \lambda_{12} = 0, \lambda_{22} = 1, v_1 = 0, v_2 = -1$. **Lower, right:** $\lambda_{11} = 1, \lambda_{12} = 0, \lambda_{22} = 1, v_1 = 1, v_2 = 1.5$.

As one of the Dirac delta function's properties, the following equation holds

$$\int_\Omega p(\mathbf{x})\, \delta(\mathbf{x} - \boldsymbol{\xi})\, d\Omega = \eta(\boldsymbol{\xi})\, p(\boldsymbol{\xi}) \tag{8}$$

with $\eta = \frac{1}{2}$ if $\boldsymbol{\xi}$ lies on the boundary Γ, $\eta = 1$ if $\boldsymbol{\xi}$ is inside of the domain Ω, $\eta = 0$ if $\boldsymbol{\xi}$ is outside of the domain Ω. By substituting (8) into (7) we obtain a boundary integral equation

$$\eta(\boldsymbol{\xi})\, p(\boldsymbol{\xi}) = \int_\Gamma [P(\mathbf{x})\, p^*(\mathbf{x}, \boldsymbol{\xi}) - (P_v(\mathbf{x})\, p^*(\mathbf{x}, \boldsymbol{\xi}) + P^*(\mathbf{x}, \boldsymbol{\xi}))\, p(\mathbf{x})]\, d\Gamma \tag{9}$$

The so called fundamental solution p^* which satisfies (6) may be written (see Azis [2])

$$p^*(\mathbf{x}, \boldsymbol{\xi}) = \frac{K}{2\pi} \exp\left(-\frac{\dot{\mathbf{v}}.\dot{\mathbf{R}}}{2D}\right) K_0\left(\frac{\dot{v}\dot{R}}{2D}\right) \tag{10}$$

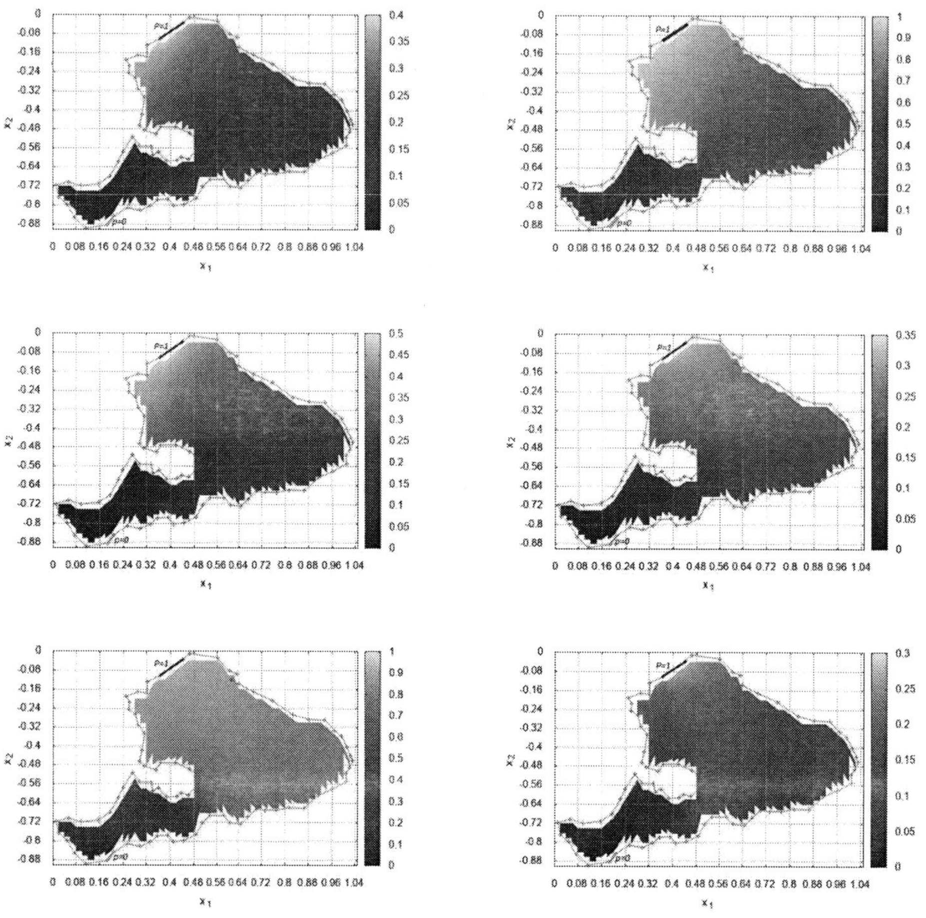

Figure 6. Values of p for the Unhas Lake. **Upper, left:** $\lambda_{11} = 1, \lambda_{12} = 1, \lambda_{22} = 2, v_1 = 0, v_2 = 1$. **Upper, right:** $\lambda_{11} = 1, \lambda_{12} = 1, \lambda_{22} = 2, v_1 = 0, v_2 = -1$. **Middle, left:** $\lambda_{11} = 1, \lambda_{12} = 1, \lambda_{22} = 2, v_1 = 1, v_2 = 1.5$. **Middle, right:** $\lambda_{11} = 1, \lambda_{12} = 0, \lambda_{22} = 1, v_1 = 0, v_2 = 1$. **Lower, left:** $\lambda_{11} = 1, \lambda_{12} = 0, \lambda_{22} = 1, v_1 = 0, v_2 = -1$. **Lower right:** $\lambda_{11} = 1, \lambda_{12} = 0, \lambda_{22} = 1, v_1 = 1, v_2 = 1.5$.

where

$$D = [\lambda_{11} + \lambda_{12}(\rho + \bar{\rho}) + \lambda_{22}\rho\bar{\rho}]/2 \qquad K = \ddot{\rho}/D \qquad \dot{\mathbf{R}} = \dot{\mathbf{x}} - \dot{\boldsymbol{\xi}} \qquad \dot{\mathbf{x}} = (x_1 + \dot{\rho}x_2, \ddot{\rho}x_2)$$

$$\dot{\boldsymbol{\xi}} = (\xi_1 + \dot{\rho}\xi_2, \ddot{\rho}\xi_2) \qquad \dot{\mathbf{v}} = (v_1 + \dot{\rho}v_2, \ddot{\rho}v_2)$$

$$\dot{R} = \sqrt{(x_1 + \dot{\rho}x_2 - \xi_1 - \dot{\rho}\xi_2)^2 + (\ddot{\rho}x_2 - \ddot{\rho}\xi_2)^2} \qquad \dot{v} = \sqrt{(v_1 + \dot{\rho}v_2)^2 + (\ddot{\rho}v_2)^2}$$

where$\dot{\rho}$ and $\ddot{\rho}$ are respectively the real and the positive imaginary parts of the complex root ρ of the quadratic equation

$$\lambda_{11} + 2\lambda_{12}\rho + \lambda_{22}\rho^2 = 0$$

and the bar notation ($^-$) denotes complex conjugate and K_0 is the modified Bessel function.

Now equation (9) may be used to calculate the solutions p and P for all points $\mathbf{x}$ on the boundary Γ. Once the boundary solutions are found then we can calculate solutions p and its derivatives $\partial p/\partial x_1$, $\partial p/\partial x_2$ in the domain Ω.

4. Numerical results

An example of problem with analytical solution had been shown in Haddade et. al [3]. The example showed the feasibility and the accuracy of the BEM in solving diffusion-convection problems. In this section two more problems will be solved numerically by using the boundary integral equation (9). The two problems will consider simulation adopting the circumstances of the Tello River around which an electricity generating industry (in the map written in Indonesian language as PLN - Sektor Tello), assumed as a source of pollutant, stands and Unhas Lake (in the map written as Danau Unhas) (see figure 1) around which a source of pollutant is assumed to be existing. Both sites mentioned are located in Makassar city of Indonesia.

The standard BEM is used to find the numerical solutions of p and the derivatives $\partial p/\partial x_1$ and $\partial p/\partial x_2$ in the domain Ω. The boundary Γ is divided into a number of sides and each side is divided into a number of elements of equal length. See figure 2. The constant element is used for which it is assumed that the values of p and P are constant along an element. The integrals over each element are evaluated numerically using Bode's quadrature with error of order $O(h^{11})$.

One side of the boundary Γ is where the pollutant flux is located, that is $P = 1$. Another side is where the boundary condition is set to be $p = 0$. All other sides are assumed to be parts of the boundary of no flux that is $P = 0$. See figure 2.

For each problem we consider the cases of isotropic ($\lambda_{11} = \lambda_{22} = 1, \lambda_{12} = \lambda_{21} = 0$) and anisotropic ($\lambda_{11} = 1, \lambda_{22} = 2, \lambda_{12} = \lambda_{21} = 1$) media. And for each case we take a variation of the values of the velocity (v_1, v_2), that is $(v_1, v_2) = (0, 1)$, $(v_1, v_2) = (0, -1)$ and $(v_1, v_2) = (1, 1.5)$. The results are shown in figures 3 – 6. The results show that the change of the parameters λ_{ij} and v_i to some extent effects the values of p and its derivatives. Therefore as a recommendation, for implementation it is very important to measure carefully the values the parameters, and also of course the values of the boundary data.

Acknowledgment

This work was financially supported by the LP2M (Lembaga Penelitian dan Pengabdian pada Masyarakat) of Hasanuddin University Makassar and The Higher Education Ministry of Indonesia.

References

[1] Azis M I, Clements D L and Lobo M 2003 A boundary element method for steady infiltration from periodic channels *ANZIAM J.* **44(E)** C61
[2] Azis M I 2017 Fundamental solutions to two types of 2D boundary value problems of anisotropic materials *Far East Journal of Mathematical Sciences* **101** 11 2405
[3] Haddade A, Salam N, Khaeruddin and Azis M I 2017 A Boundary Element Method for 2D Diffusion-Convection Problems in Anisotropic Media *Far East Journal of Mathematical Sciences* **102** 8
[4] Clements D L and Azis M I 2000 A Note on a Boundary Element Method for the Numerical Solution of Boundary Value Problems in Isotropic Inhomogeneous Elasticity *Journal of the Chinese Institute of Engineers* **23** 3 261
[5] Azis M I and Clements D L 2001 A Boundary Element Method for Anisotropic Inhomogeneous Elasticity *International Journal of Solids and Structures* **38** 5747
[6] Azis M I, Clements D L and Budhi W S 2003 A boundary element method for the numerical solution of a class of elliptic boundary value problems for anisotropic inhomogeneous media *ANZIAM J.* **44(E)** C79
[7] Azis M I and Clements D L 2008 Nonlinear transient heat conduction problems for a class of inhomogeneous anisotropic materials by BEM *Engineering Analysis with Boundary Elements* **32** 1054
[8] Azis M I and Clements D L 2014 On some problems concerning deformations of functionally graded anisotropic elastic materials *Far East Journal of Mathematical Sciences* **87** 2 173
[9] Azis M I and Clements D L 2014 A Boundary Element Method for Transient Heat Conduction Problem of Nonhomogeneous Anisotropic Materials *Far East Journal of Mathematical Sciences* **89** 1 51
[10] Meenal M and Eldho T I 2012 Two-dimensional contaminant transport modeling using mesh free point collocation method (PCM) *Engineering Analysis with Boundary Elements* **36** 551

IOP Conf. Series: Journal of Physics: Conf. Series **979** (2018) 012076 doi:10.1088/1742-6596/979/1/012076

Source to Skin Distance (SSD) Characteristics from Varian CX Linear Accelerator

Wira Bahari Nurdin, Aji Purnomo and Syamsir Dewang

Physics Department, Hasanuddin University, Makassar, Indonesia 90245

E-mail: wbnurdin@unhas.ac.id

Abstract. This study aims to describe the characteristics of the source to skin distance (SSD) of Varian CX linear accelerator (LINAC) using the X-ray beam of 6 MV and 10 MV. The variation of the source to the SSD are 90, 100 and 110 cms; the depth of the water phantom used are 5, 10, 15, 20, and 25 cms, respectively. The depth of the water phantom was created for analysis of percentage depth dose (PDD) and profile dose. It can be concluded from the tests that from the measured SSD, SSD of 110 cm with the depth water phantom of 20-25 cm for energy beam of 6 MV and at all levels of depth for 10 MV energy corresponding tolerance limits to be used in clinical radiotherapy. For the SSD 90 and 100, the values beam symmetry and flatness obtained slightly beyond the limits of tolerance.

1. Introduction

Source to skin distance (SSD) is almost used as an important reference to examine the depth of an x-ray in radiotherapist treatment [1]. The aim of this study is to characterize and to measure the dosimetric output of 6 and 10 MV Varian Clinac CX linear accelerator using the variation of SSD. Varian Clinac CX linear accelerator is a very popular linac in Indonesian hospitals. The measurement of output was conducted using devices with ion chambers. Linac used a multi-high energy beam of electrons and photons, the electron with the energy range of 4-22 MeV for the purpose of radiotherapy and the energy range of 6-18 MV for photons. Since the nature of ionizing radiation can damage tissue, then the radiation dose given to the tumor cells must be distributed uniformly or homogeneously in accordance with the rules ICRU that the maximum dose in the range of 95% -107% [2, 3]. Therapy planning and delivery of the appropriate dose will determine the success of treatment. The maximum dose values for photons do not come on the surface of the skin but is at a certain depth called the maximum depth dose (d_{max}). The value of d_{max} depends on the energy of the primary photon. The area between the surface and the maximum depth is known as the build-up. Linac uses high-frequency electromagnetic waves to accelerate the electrons into a high energy through a linear and conductive tube which contains an array of cells form cavities made of copper [4].We performed in this study a complete description of the characteristics of the SSD of Varian CX linear accelerator (LINAC) using the X-ray beam of 6 MV and 10 MV.

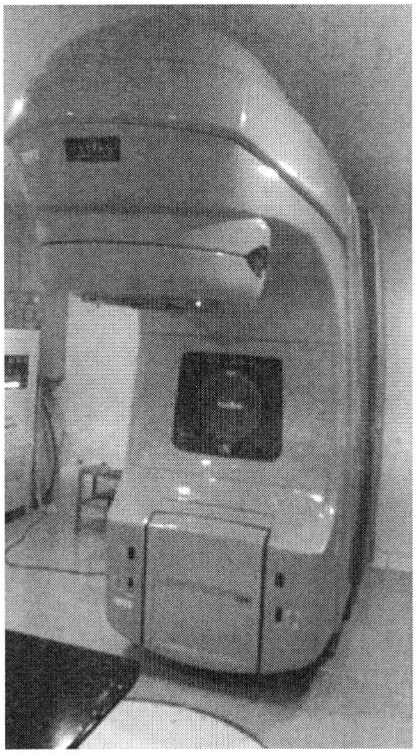

Figure 1. Varian Clinac CX linear accelerator.

2. Material and Methods

The measurements were conducted using Varian Clinac CX (Varian Oncology, Palo Alto, Ca) at Hasanuddin University Hospital. The procedures of the dosimetric measurements follow a few steps. A phantom containing distilled water was set up and water phantom placed and positioned parallel to the laser position. Position each side of the water phantom was positioned at the same level. The ionization chamber was put in a water phantom and the ionization chamber was connected with an Omni-Pro software on a computer via an optical cable. The irradiation field area was set 10 x10 cm^2. Later, the SSD was set at a distance of 90 cm and the energy 6 MV. The depth of the ionization chamber in a water phantom was positioned with a depth of 5 cm from the surface of the water. The beam focus would set on and PDD and dose profile would be analyzed using the Omni-Pro software. The procedure would be repeated by adjusting the SSD, energy, and depth.

The variations of SSD was 90, 100, and 110 cm. the depth of the water phantom was 5, 10, 15, 20, and 25 cms. The depth of the water phantom was created for analysis of percentage depth dose (PDD) and profile dose. Radiation dose profile characteristics of the radiation beam and provide important information for planning radiotherapy. Dose profile also declared a curve showing the form of face-rays on the horizontal axis that is perpendicular to the direction of x-rays. That curve interprets the distribution of radiation dose relatively on a particular radiation field. Percentage depth dose is the absorbed dose is given on the main depth as a percentage of absorbed dose at the depth of the pointer on the main axis of the area. Dose distribution at the point in the main axis of the files inside phantom usually normalized to D_{max} as 100% at a depth of maximum dose, D_{max} corresponding to the reference depth. Value percentage depth dose can be defined as a result, in the form of a percentage, the dose absorbed at a certain depth where is a depth d. d_0 is the reference dose along the beam axis. The percentage depth dose (PDD) is defined as:

$$PDD = \frac{D_d}{D_{d0}} \times 100\% \qquad (1)$$

where D_d and D_{d0} are the doses at a depth d and d_0 respectively. In use, the peak dose absorbed at a central axis referred to as the maximum doses given, D_{max}, is:

$$D_{max} = \frac{D_d}{P_{dd}} \times 100\% \quad (2)$$

percentage depth dose for X-rays decreases with increasing depth as the maximum dose is achieved.

3. Results and Discussion

PDD and dose profile measurement is done by using energy X-ray beam 6 MV and 10 MV with a variety of SSD and depth, while the width of the irradiation field is constant (10x10 cm^2). All of the measurement method using a water phantom and the ionization detector chamber. Based on figures (3-4) obtained the results of the comparison PDD can be observed in table 1. The value of symmetry and flatness of the dose profile can be obtained in the figures (2-3).

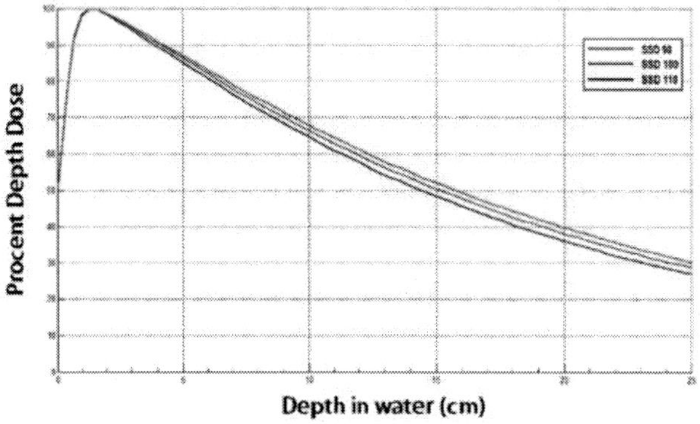

Figure 2. Percent Depth Dose (PDD) of X-ray beam 6 MV for SSD 90, 100, and 110 cms at a depth of 5, 10, 15, 20 and 25 cms, respectively.

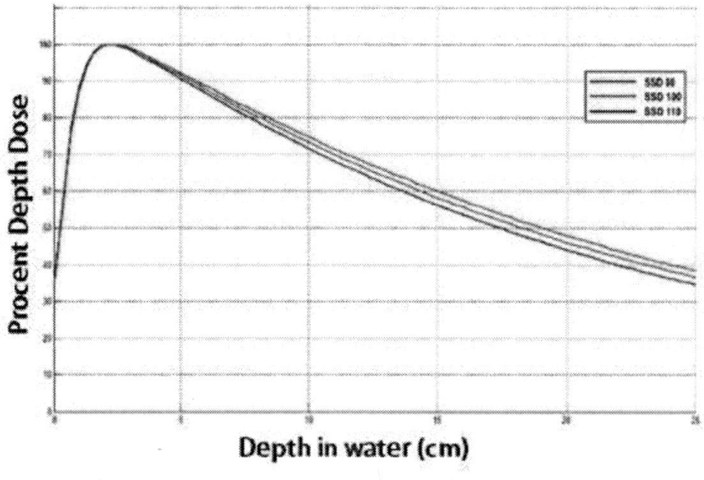

Figure 3. Percent Depth Dose (PDD) of X-ray beam 10 MV for SSD 90, 100, and 110 cms at a depth of 5, 10, 15, 20 and 25 cms, respectively.

Table 1. Comparison of Percent Depth Dose in the SSD 90, 100 and 110, respectively to the energies of 6 MV X-rays in varying degrees of depth.

PPD X-Ray 6 Mev			
SSD (cm)	Depth	Relative Dose (%)	Reference Dose (%)
90	1,28	100	1,38
100	1,28	100	1,58
110	1,58	100	1,78

D_0 is an area starting from the center point (medium surface) up to the point where it can be absorbed dose maximum. From all SSD it can be seen, that the dose absorbed in the area will increase until maximum. The higher energy it was given, the longer distance would be required to reach it. Then, after passing through the maximum point, the absorbed dose would decrease with increasing depth.

Beam symmetry depicted in figure 5 and figure 6 showed dose conformity profile area between left and right. While the flatness indicates the maximum dose value and the value of the minimum dose at every level depth. Beam symmetry and flatness value obtained from these figures is determined by the value of SSD. The greater the energy, SSD, and depth are used, the smaller the value of symmetry and flatness obtained. The asymmetry was determined as less than 2% over 80% over all of the field sizes. Using PC Program ion chamber, the maximum asymmetry measured more than 75% of any field for a 40x40 cm field size.

From all the SSD for 6 MV and 10 MV energy beams, the value of symmetry in the SSD 90 in general, is the largest depth, then SSD 100, and the smallest is SSD 110. None of this SSD reached the value of symmetry with the ideal value 0%. Only SSD 110 was still within tolerable limits (S <2%), which is at a depth of 20 and 25 cm for 6 MV beam of energy, whereas for the 10 MV beam of energy to meet the standards of tolerance for all levels of depth used. This indicates that there is an imbalance between the regional distribution of doses left and right areas, this can be seen in figures 4 and figure 5.

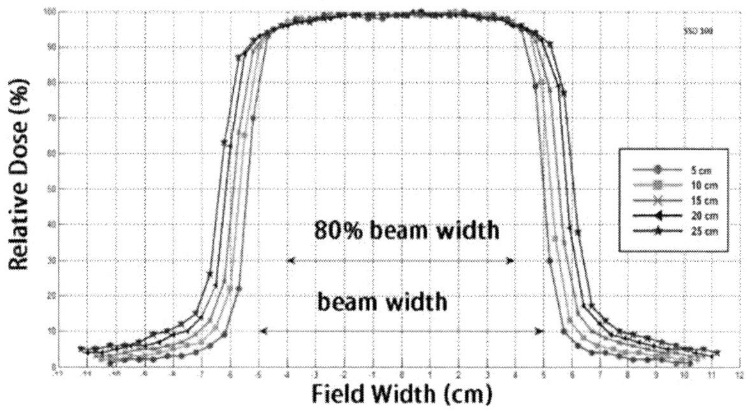

Figure 4. Graph of beam profile for 6 MV beam for SSD 90 cm at a depth of 5 to 25 cm.

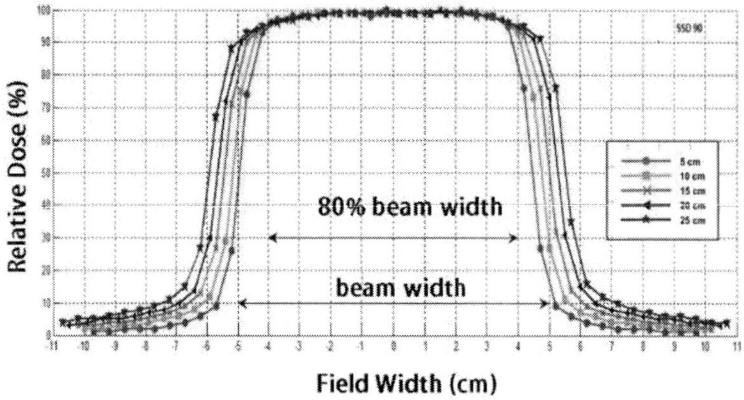

Figure 5. Graph of beam profile for 10 MV dose for SSD 90 cm at a depth of 5 to 25 cm.

Similarly, for the beam symmetry, none of the SSD reached the ideal value of flatness. For SSD 90 with X-ray energy beam of 6 MV and 10 MV, the average value of flatness is high enough. Only at a depth of 25 cm for X-ray energy 6 MV and at a depth of 20 to 25 cm for X-ray energy 10 MV that remain within tolerable limits (less than 3%). This is due to the distance between the energy source and the surface of the medium (SSD) which is too close. It results in flatness value that approaches the ideal value as it was away from the depth of the surface. For the SSD 100 and 110 for both energy level has a flatness values which are within tolerance at every level of depth.

4. Conclusion

We have measured the dosimetric parameter of the 6-MV X-Ray beams from Varian Clinac CX linear accelerator and compared with other linear accelerators. The SSD characteristics of Varian Clinac CX are similar and allowed to be used in regular clinical radiotherapy. From the three measured SSD, only SSD 110 with the depth of 20 - 25 cm for energy beam of 6 MV and at all levels of depth for 10 MV energy corresponding tolerance limits, which is feasible for use. For the SSD 90 and 100, the values beam symmetry and flatness values obtained slightly beyond the limits of tolerance.

References

[1] Delaney G, Jacob S, Featherstone C and Barton N 2005 Cancer 104 1129.
[2] Davidson, Scott E, Cui J, Kry S F, Deasy J O, Ibbott G S, Vicic M, White R A and Followill D S 2016 Medical physics **43** 8 4842
[3] Fontenla D P, Napoli J J and Chui C S 1992 Medical Physics 19 343
[4] Feng, Yuangming et al 2005 Radiation Oncology *Int. Jour* **61** 1507

Evaluation of Energy Dose and Output Power Optimum of Diode's Laser of 450 nm and 650 nm in Photoantimicrobial Mechanisms Against Inhibition of C. Albicans Biofilm Cells

S Dewi-Astuty[1,4], Suhariningsih[2], S Dyah-Astuti[2], A Baktir[3]

[1] Student of Science and Technology Faculty Doctoral Program, Airlangga University, Surabaya
[2]Biophysics Laboratory, Physics Department of Airlangga University, Surabaya
[3] Chemistry Department of Airlangga University, Surabaya
[4] Physics Department, Hasanuddin University, Makassar

E-mail: tuty_ilyas@yahoo.com

Abstract. Photoantimicrobial as a pathogenic microbial inhibitory therapy system such as C. albicans in biofilms forms has been studied in vitro. Mechanisms of inhibiting called inactivating used the absorb principles of a dye agents such as chlorophyll against the photon energy of diode laser which any number of ROS product depend on energy doses of a laser, time of irradiation, concentration and time of incubation the dyes agent. The inactivation profile of C. albicans biofilm cells was observed based on cell viability reduction after photoantimicrobial treatment with or without oxygenation by *XTT assay* test. Results show that the inhibiting significantly with the time incubation of the dye agents and the oxygen degree inside the sample. The inhibition for oxygenation biofilm's group 10% lower than without oxygenation biofilm's group at the maximum of reduction of cell viability occurred in the 3-hour incubation group. Optimum of inactivation are 89.6% (without oxygenation) and 94.8% (with oxygenation) after irradiation with 450 nm laser (power output 128.73 at energy dose 86.09 J/cm^2), While, by 650 nm laser (power output 164.53 mW at energy dose 92.52 J/cm^2) irradiation treatment obtained optimum of inactivation are 89.5% (without oxygenation) and 92.3% (with oxygenation).

1. Introduction

The photodynamic therapy known as PDT began to be studied in 1900 by Oscar Raab which saw the effects of light on dyestuffs in medicine, followed by Lipson and Schwartz in 1960 which found a neoplasm lesions emit fluorescence light during surgery after *hematoporphyrin* injection [1]. Initially, PDT therapy was studied to kill cancer cells (malignant cells) but has evolved and proven effective against infectious diseases caused by pathogenic microbes known as *antimicrobial photodynamic therapy* (aPDT) which popular kind photoinactivation [2].

In the photonic, aPDT mechanisms are therapy that involves interaction between light and biomolecule and throught photophysics process and photochemistry process like absorbtion of photon energy, electronic transition, energy or electron transfer against other molecule, sustainable of chemical reaction and produce the radical compounds. The radical compounds produced in the aPDT mechanism can be superoxide anions, hydrogen peroxide, and hydroxyl radicals or oxygen singlet (called *reactive oxygen species* (ROS)) that property very reactive and toxic, that believed can be killing the pathogen microbial cell [3,4,5].

The principle of the aPDT mechanism is different from that of specific antibiotic therapy because inhibiting process on the aPDT therapy is inactivating cells division system and break DNA chain the cells that active by ROS. There are three pathways of photoinactivation to microbes are 1) light is absorbed by a sensitizer outside the cell and forms a ROS that will be lysing the cell membrane; 2) the light who penetrates the cell wall and is absorbed by the chromophore inside cell and forms another ROS; 3) the light which directs inside cell breaks the DNA chain [6].

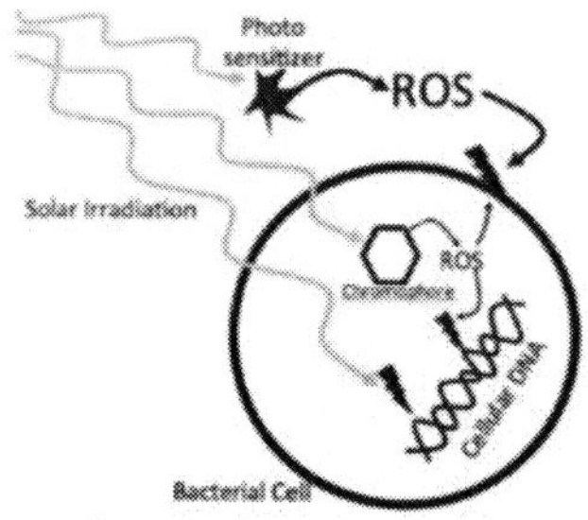

Figure 1. Three pathways for bacterial photoinactivation [6].

A component in a target molecule that acts as a photon light absorber called a chromophore, generally in the form of a molecule called photosensitizer (PS). PS can be from target cell self (endogenous PS) or can be from outer of target cell (exogenous PS) that added on target cell before irradiating. The success of the aPDT mechanisms depends on the suitability between the wavelength of light and the optimum uptake of chromophores to light and the presence of oxygen molecules around the cell target. Types of PS that have been shown to be effective as sensitizers in photoinactivation [7-11] include methylene blue, toluidine blue, phthalocyanine, protoporphyrin IX and deuteroporphyrin IX. In this study used chlorophyll as PS because it has the characteristics absorbtion optimum at two wavelength area (413 nm and 650 nm) [12] so that it can be combined with different source of light. Chlorophyll molecule is PS of the third generation applied on PDT because of their properties as a pure chemical compound, maximum absorbance below 650 nm, a high-value yield to produce ROS as well as long lifetime, no toxic in dark, stable and easily isolate [13,14]. Excess chlorophyll as a sensitizer is also used in photoinactivation of pathogenic microbes because many plants in Indonesia have active substances as antimicrobials.

The increasing problem of pathogenic microbial cell resistance to antibiotic drug types and the antibiotic drugs treatment's inconsistently by patient causing aPDT mechanisms as a promising alternative therapy. aPDT studies have been widely developed to inactivate pathogenic microbes especially oral infections caused by C. albicans microbes. The pathogenicity of C. albicans is caused by microbial characteristics such as dimorphism and biofilm formation. Characteristics of biofilm C. albicans is the main object will be investigated through the application of aPDT mechanism.

The light source will be used on aPDT must is known the output power characteristics and exposure area. This characteristic is important for determining the irradiation dosimetry which will be adjusted with the absorption characteristics of the photosensitizers will be used. In this research, used two types of the laser diode with a specification of wavelength 450 nm and 650 nm, for determining optimum spectrum that effective to inhibit the growth of C. albicans biofilm cells. The irradiating parameter of the light source to be determined as the characteristic of the light source is the stability of the power against change of time and distance, the intensity and exposure area, the energy dose of irradiation. The analysis conducted in this study is the wavelength spectrum and the energy dose of irradiation that optimum is effectively reduced the viability of C. albicans cells through optical density measurement by XTT assay method.

2. Experiment

Papaya leaf chlorophyll extract was isolated by maceration with a mixture of methanol solvent, petroleum ether and diethyl ether which was further fractionated with n-hexane and acetone [12]. The maximum absorbance characteristics of chlorophyll extracts obtained at two wavelengths were 405 nm (Soret band) and 650 nm (Q-band) and the MIC values obtained were 0.5 mg.mL^{-1}, 3.75 mg.mL^{-1} and 5.0 mg.mL^{-1}.

The laser diode fabrication specifications used with a power output of 100 mW, subsequently characterized in advance to determine the proper research design. The diode laser characterization test includes optimum output power to wavelength (monochromator JASCO CT-10), exposure time, change of radiation distance (PM100D Thorlabs Optical Power Meters), light intensity, exposure area and radiation dose, whereas PS used is a chlorophyll molecule isolated from papaya leaf and had been tested for characteristics with fluorescence spectroscopy and UV- Vis Shimadzu.

The optimum characteristic results are adjusted for the duration of irradiation set to determine the radiation dose as the treatment variation. The prescribed radiation time is 60 s, 120 s, 180 s, 240 s, and 300 s, while the duration of PS incubation before aPDT is applied is 1 hour and 3 hours.

Biofilm C. albicans used as research object was grown for 6 days in a 96-well microplate container. Inocula C. albicans 0.5 McF in sterile PBS pH 7.4 was planted in each well as 100 μL then incubated for 90 min in a 37^{0}C incubator shaker, 120 rpm. The well was washed with sterile PBS pH 7.4 twice and added BHI-B 8% glucose into each well as 200 μL. The biofilm suspension was incubated in a 37^{0}C incubator shaker, 120 rpm for 6 days. At harvest, microplate contents are poured over a container containing tissue to remove C. albicans planktonic cells that do not form biofilms. 200 μL chlorophyll extracts of various concentrations were injected on the biofilm of C. albicans in a predetermined well (the biofilm group using the sensitizer as the treatment).

As a comparison of the aPDT treatment, in the biofilm of C. albicans before the chlorophyll extract was first extracted oxygen gas for 30 seconds with flowrate variation of 1 L.min^{-1} (80 psi), 2 L.min^{-1} (70 psi), 3 L.min^{-1} (60 psi) and 4 L.min^{-1} (50 psi) at pressure adjusted to the regulator scale listed for each oxygen gas stream (pressure decreases as oxygen exits from the tube). The oxygenation mechanisms performed to overcome the limitations of the oxygen zone in the biofilm. Irradiated biofilms (microplates with laser treatment groups only and laser-sensitizer combinations) are placed under pre-prepared laser devices, and light exposure during the time set for each appropriate well. Each well-treated microplate-96 wells treated for all treatment groups were washed with sterile PBS pH 7.4 twice subsequently stained 40 μL XTT (1 mg.mL^{-1}) + 2 μL menadione (10 mg.mL^{-1}) + 158 μL sterile PBS pH 7.4 and incubated for 2 hours. 100 μL result of staining transfer to a new microplate for readability of viability cells with ELISA READER.

3. Result and discussion

The diode laser characteristic from both of laser (450 nm and 650 nm) that result in the maximum of power respectively is 35,790 mW at 445 nm wavelength (red laser) and 45,499 mW at 650 nm wavelength (blue laser). This output power is the true power value of both lasers based on the determination of each wavelength using a monochromator.

The power stability test against to the exposure time is measured to see power stability for one hour operating. It is important to determine the length of exposure that still produces stable power.

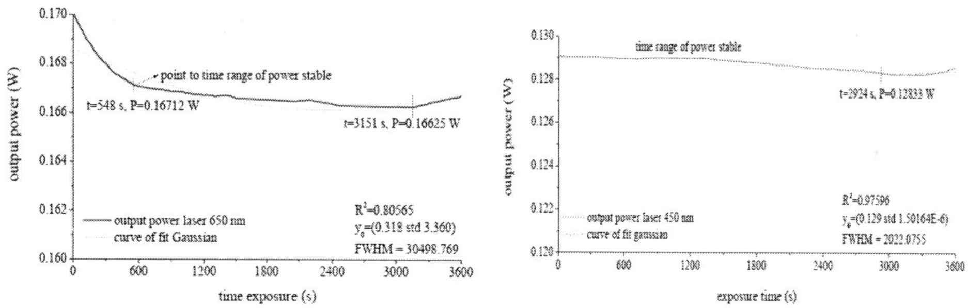

(a) (b)

Figure 2. Characteristic of power stable against time exposure of laser 650 nm (red) and 450 nm (blue).

Figure 2 shows that blue laser relative more stable than the red laser. The blue laser stable starting from the time start and increasing after 2924 s exposure time with a stable power of 0.12833 W, while the red laser has a stable area after exposure time 548 s with a stable power of 0.16712 W.

These research did not use a light filter adjusted to the maximum absorbance of the sensitizer so that the possibility of photosensitization for all range wavelengths absorbance of papaya chlorophyll allowed it to occur. The light photon energy coming out of each laser is $3x10^{-19}$ J (laser 650 nm) and $4.47x10^{-19}$ J (450 nm laser).

The characteristic of a stable power to change in distance is determined by adjusting the distance between samples with light from a distance of 1 cm to 5 cm. The measurements of the two lasers obtained different distances of light samples at maximum power, where the measurements for the red laser, optimum power occurred at a distance of 3 cm (power $0.16453\pm4.22\times10^{-6}$ W) whereas for blue laser, optimum power occurred at a distance of 1 cm (power $0.12873\pm1.17\times10^{-5}$ W). Theoretically, the further the distance between the light source and the sample, the strength will remain stable but the intensity changes as the exposure area also changes. When compared to a distance of 5 cm (power $0.15839\pm5.2\times10^{-6}$ W for the red laser) and a distance of 2 cm (power $0.12790\pm2.1\times10^{-5}$ W for the blue laser) can be said laser power remains stable.

The result of extensive measurement laser beam, the intensity of the laser light, exposure time and dose of radiation energy as a variable dosimetry in aPDT for any distance corresponding to the power stability test against more distance shown in table 1. The result of measuring the area of irradiation for a red laser at a distance of 3 cm is 0.453 cm^2 while for blue laser 0.357 cm^2. Intensity at 3 cm distance of is 0.363 W.cm^{-2} (red laser) and 0.360 W.cm^{-2} (blue laser). Value of intensity to be reference for determining the energy dose yang sampai di permukaan biofilm.

Table 1. Dosimetry of aPDT C. albicans biofilm mediated *carica papaya* L. chlorophyll extract with energy dose difference for both laser .

No	time irradiance (s)	energy of dose (J.cm^{-2})	
		650nm	450nm
1	60 (L1)	28.70	30.84
2	120 (L2)	57.39	61.68
3	180 (L3)	86.09	92.52
4	240 (L4)	114.78	123.36
5	300 (L5)	143.48	154.20

Results aPDT treatment for all treatment groups: control group, sensitizer group only, only the laser groups and the combination group laser-sensitizer in three variations concentration of sensitizer, time of sensitizer incubation and energy dose irradiation.

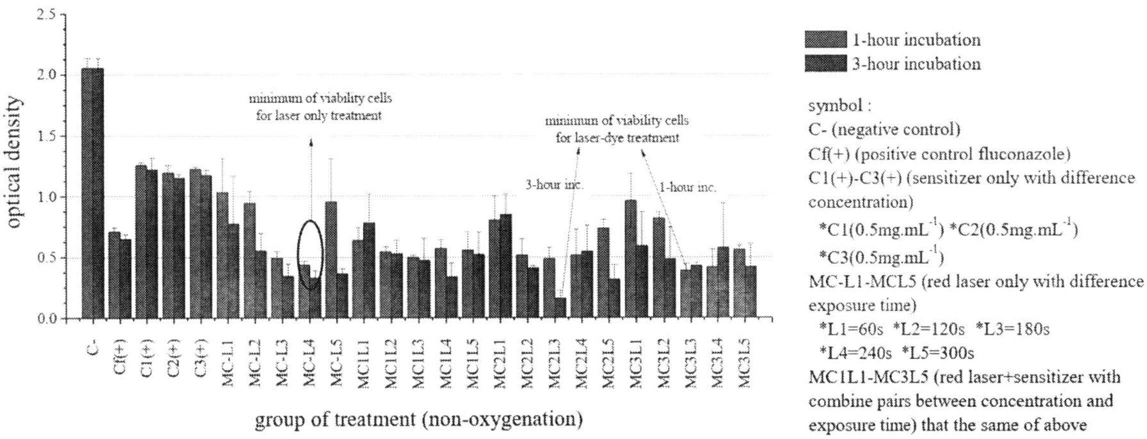

Figure 3. Viability C. albicans biofilms cell for 650 nm laser treatment.

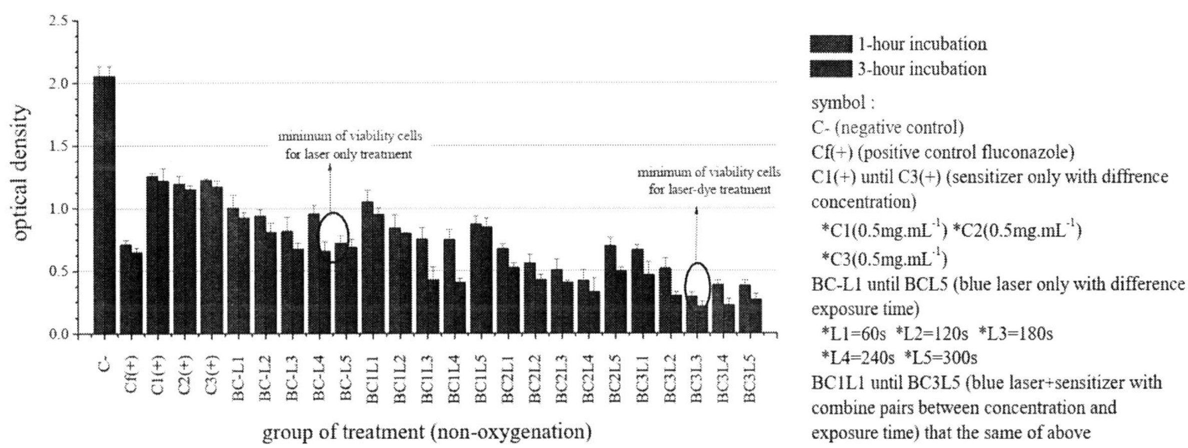

Figure 4. Viability C. albicans biofilms cell for 450 nm laser treatment.

Figures 3 and 4 show the results of optical density reading of aPDT treatment results on C. albicans biofilm cells without oxygenation respectively for 650 nm lasers (figure 3) and for 450 nm lasers (figure 4) at significance ($p>0.05$). In general, the laser only groups are more effective than the sensitizer only groups and the chlorophyll mediation laser groups is most effective than another group (groups of sensitizers only or laser only). The effect of inactivation of the red laser can reduce viability until 92% than the blue laser. The effect of time of incubation also showed significant results where 3-hours of incubation more effective than time 1-hour of incubation. The concentration effect of chlorophyll does not show a uniform pattern, which theoretically, higher concentration correlating with the greater absorption of light photons thus leads to more opportunities for the formation of radical compounds that will affect to be success of inactivation. Figure 3 shows the maximum reduction occurring in the MC2L3 group (1-hour incubation) and in the MC3L3 group (3-h incubation). Figure 4 shows the maximum reduction occurring in the same group is the BC3L3 group for different the time of incubation.

The optimum inactivation with the red laser irradiating occurs at doses of 86.09 $J.cm^{-2}$ for both time incubation of sensitizer with percent inactivation obtained respectively 81.2% (1-hour incubation) and 92.3% (3-hours incubation). Both inactivation optimum value is obtained from the treatment with the sensitizer concentration of each 3.75 $mg.mL^{-1}$ and 5.0 $mg.mL^{-1}$. The result of optimum inactivation with the blue laser irradiating obtained 85.6% (1-hour incubation) and 89.6% (3-hour incubation) at an energy dose of 92.52 $J.cm^{-2}$.

For the additional treatment with oxygenation mechanisms, see figure 5 below. The condition of treatment which applied on with oxygenation groups only the group with optimum inactivation i.e concentration of PS is C3=5.0 $mg.mL^{-1}$ at 3-hour incubation.

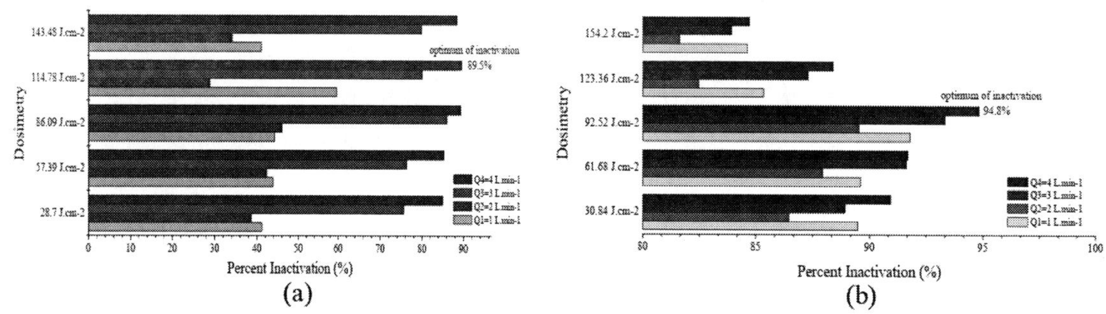

Figure 5. Inactivation of 650 nm laser based on dosimetry (energy dose) with oxygenation mechanisms on C. albicans biofilm (a) red laser treatment; (b) blue laser treatment.

The highest of percent inactivation for red laser (650 nm) induction is for flow rate Q4 with the energy of dose 114.78 $J.cm^{-2}$ is 89.5% and the lowest value for flow rate Q2 with the same energy of dose is 28.8%. The highest of percent inactivation for red laser (450 nm) induction is for flow rate Q4 with the energy of dose 92.52 $J.cm^{-2}$ is 94.8% and the lowest value for flow rate Q2 with the same energy of dose 154.20 $J.cm^{-2}$ is 81.6%.

Before used, laser (light source) must be performance tested to determine the dosimetry to be applied in aPDT. In dosimetry of light, total energy be expressed in units of Joule that depend of exposure time and power of light. Intensity are fluence rate is expressed in W/area. The fluence rate should not more than 0.200 W/cm2 (for microlens output light) and 0.400 W/cm2 (for cylindrical diffuser) because more than the value specified can be damaging cell normal that caused by the thermal effect. The laser that used in this study still include safe.

The time incubation of sensitizer has a different effect against C albicans biofilms cell because of antifungal properties of a sensitizer. The longer the incubation of the sensitizer before the treatment, the more chlorophyll substances that diffuse into the biofilm layer. In general, increasing the dose of irradiation energy inconsistent with increasing percentage of inactivation. It is assumed that when treatment and staining can be affected by human errors such as inhomogeneous exposure at the sample and the long exposure can decrease laser output power so that the energy of dose be different with the calculation result. Another difference between theory and practice is the time of staining which using three components of the reagent wherein one component of menadione (in a very small volume scale of 2 μL) is not expected to be sucked by a micropipette or while injection can exceed the volume so that sample has more or few dense coloration.

Figures 5(a) and 5(b) represent inactivation profiles of C. albicans biofilm cells with additional treatment being oxygen streaming in order to increase oxygen levels in biofilm inner layers. With the pressure and flow of certain flow of oxygen can diffuse to a certain depth and even have the opportunity to reach the bottom of the biofilm in the well. The inactivation results showing percent inactivation significant with the increase of flow and pressure, where for the Q1 and Q2 although at low volume but flowed at high pressure, while for the Q3 and Q4 even at low pressure but the volume of oxygen relatively higher.

4. Conclusion

The percent value of inactivation from the aPDT treatment results from the second treatment of different wavelengths and power shows significantly different results. In the no oxygenated sample group, the inactivation percent for laser treatment was 650 nm higher than for the 450 nm laser treatment. In the sample group with an oxygenation mechanism prior to the aPDT treatment, the inactivation effect of the laser 450 nm was more effective than the inactivation effect of the 650 nm laser. For 450 nm laser irradiation with photon energy is 4.47×10^{-19} J and power output 0.12873 W, optimum inactivation is 89.6% (without oxygenation) and 94.8% (with oxygenation) at energy dose 86.09 $J.cm^{-2}$. At an irradiation of 650 nm laser with photon energy 3×10^{-19} J and power output 0.16453 W, optimum inactivation is 89.5% (without oxygenation) and 92.3% (with oxygenation) at energy dose 92.52 $J.cm^{-2}$.

Acknowledgments

Thanks to all those who have assisted in the completion of this research, so that all the data in this paper is eligible to be published. Especially the party who has sponsored this research through doctoral dissertation grant.

References

[1] Sibata C H, Colussi V C, Oleinick N L and Kinsella T J 2000 Photodynamic therapy: a new concept in medical treatment *Brazilian Journal of Medical and Biological Research* **33** 869-880
[2] Prasad P N 2003 *Introduction to Biophotonic* (John Wiley and Sons Inc)
[3] Danis T G S and Hamblin M R 2011 An Introduction to Photoantimicrobials: Photodynamic Therapy as a Novel Method of Microbial Pathogen Eradication Science against microbial pathogens: communicating current research and technological advances *A Mendez-Vilas edition* 675-683
[4] Plaetzer K, Krammer B, Berlanda J, Berr F and Kiesslich T 2009 Photophysics and photochemistry of photodynamic therapy: Fundamental aspects *Lasers in Medical Science* **24** 2 259-268
[5] Apsari R, Maghfiroh I, Setyawati H, Winarno, Arifiant D and Zaidan 2017 A The dosage optimization of He-Ne Laser energy as a candidate for photodynamic therapy of cancer cells with exogenous photosensitizer variations *Journal of Physics: Conference Series*
[6] Grimaldi M, Maraccini P, Boehm A and Rucker S The Role of Exogenous Sensitizers in the Photoinactivation of Pathogenic Bacteria unpublished
[7] Calzavara-Pinton P G, Rossi M T and Ortel B 2011 Photodynamic Antifungal Chemotherapy Photodynamic Inactivation of Microbial Pathogens *Medical and Environmental Applications Chapter's book Editors Michael R Hamblin and Giulio Jori European Society for Photobiology Published by the Royal Society of Chemistry*
[8] Brandis A S, Salomon Y and Scherz A 2006 Chlorophyll Sensitizer in Photodynamic Therapy Chlorophyll and bacteriochlorophylls chapter's book of Biochemistry *Biophysics Function and applications B Grimm R J Forra W Rudiger and H Scheer Springer printed in Netherlands* 461-483
[9] Kolarova H, Nevrelova P, Tomankova K, Kolar P, Bajgar R and Mosinger J 2008 Production of reactive oxygen species after photodynamic therapy by porphyrin sensitizers *Gen Physiol Biophys* **27** 101–105
[10] Kochevar I E and Redmond R W 2000 Photosensitized production of singlet oxygen *Methods Enzymol* **319** 20–28

[11] Cieplik F, Spath A, Regensgurger J, Gollmer A, Tabenski L, Hiller K A, Baumler W, Maisch T and Schmalz G 2013 Photodynamic Biofilm Inactivation by SAPYR – An Exclusive Singlet Oxygen Photosensitizer *Free Radical Biology and Medicine* **65** 477-487

[12] Astuty S D and Baktir 2017 A The effectiveness of laser diode inducting to carica papaya L chlorophyll extract to be ROS generating in the photodynamic inactivation mechanisms for C albicans biofilms *International Conference on Physical Instrumentation and Advanced Materials IOP Publishing Conf series : Journal of Physics : Conf Series* 853/012026

[13] Budiyanto A W, Notosudarmo and Limantara L 2008 Effect of Acidification on Photodegradation of Chlorophyll-a *Jurnal Matematika dan Sains* September **13** 3 66-75

[14] Astuti S D 2011 Potential of Light Emitting Diode (LED) Blue For Photoinactivation of Staphylococcus Aureus Bacteria with Endogenous Porphyrins *Dissertation MIPA of Doctoral Programs Airlangga University Surabaya*

IOP Conf. Series: Journal of Physics: Conf. Series **979** (2018) 012078 doi:10.1088/1742-6596/979/1/012078

Study of Image Quality, Radiation Dose and Low Contrast Resolution from MSCT Head by Using Low Tube Voltage

Jumriah[1,2]**, Syamsir Dewang**[2]**, Bualkar Abdullah**[2] **and Dahlang Tahir**[2]

[1]Health Facility Hall of Makassar, Makassar 90245 Indonesia

[2]Department of Physics, Hasanuddin University, Makassar 90245 Indonesia

E-mail: dtahir@fmipa.unhas.ac.id

Abstrak. Effect of low tube voltage (80 kV) on image quality, radiation dose, and low contrast (LCD) detection capability on CT Scan head was studied by using a solid phantom contains four modules with scanning CT at 80 and 120 kV tube voltages, with tube current timing at 200, 250 and 300 mAs. The difference between image noise, noise to ratio contrast and LCD values obtained with 80 and 120 kV at 200, 250 and 300 mAs were then compared. Substantially the image noise increases with the use of low tube voltage. However, with the same dose, the LCD values are between 120 kV at 300 mAs and 80 kV at 200, 250 and 300 mAs (P> 0.05). The relative dose given at 80 kV against 200, 250 and 300 mAs is equal to 68%. By reducing the tube voltage from 120 kV to 80 kV at head CT, the radiation dose can be reduced by 32% without CNR and LCD degradation.

1. Introduction

One of the X-ray modalities often used in radiodiagnostics is CT Scan. In the development of the CT Scan has progressed quite rapidly starting from the first generation that only has one detector and uses pencil files, until now have used Multi-Slice Detector (MSCT) [1]. CT has undergone evolution leading to an increase in spatial resolution and temporal resolution. Spatial resolution alone is one of the parameters of image quality. The image quality produced by CT Scan becomes very important to note because the image will be used to diagnose a disease. If the resulting image is not good then it will have an impact on diagnostic errors that can result in biological effects on the patient. The estimated risk of dying from cancer for those undergoing CT is 12.5 / 10,000 population for each CT scan through the abdomen. Therefore, concern about the reduction of irradiation at CT doses is rare [2]. Although the decrease in tube currents is the most common way to reduce the radiation dose, it will reduce the factors that influence the diagnosis. Some studies have shown that scanning with low tube stress is possible to reduce the dose without reducing image quality. Thus, the purpose of this study was to determine the effect of low tube voltage usage on 80 kV on image noise (SD, standard deviation from CT Number), CNR and radiation dose from Multi-Slice Detector (MSCT) head.

CT Scan is a fairly complex medical imaging system so there is a risk of calibration error, failure of generating system function and X-ray detection. Therefore, CT Scan aircraft require QC (quality control) program to ensure image quality while maintaining the dose is still below the allowable limit [3].

2. Materials and Methods

2.1. Phantom Description
Phantom ACR accreditation CT (Gammex 464) is a solid phantom that contains four modules and is made of materials equivalent to water. Each module with a thickness of 4 cm and a diameter of 20 cm. There is a transverse and perpendicular line of white to reflect the laser marker on the phantom so that there are no slope of the axial, coronal and sagittal axes. There is also a "HEAD" and "FOOT" sign in the phantom to help position the front and back.[2]

2.2. CT Scanning Phantom
This study uses axial head protocol. Scanning is done on a standard 120kV tube voltage and a low tube voltage of 80 kV with a tube current arrangement of 200, 250 and 300 mAs.

2.3. Radiation Dose Measurement
This study used CTDIvol to estimate the radiation dose. CTDIvol was obtained on standard tube voltage and low tube voltages at various mAs (tube currents).

2.4. CNR measurements
In phantom ACR section module, 2 is used to assess low-contrast resolution. This module consists of a series of cylinders of different diameters. There are four cylinders of diameter: 2 mm, 3 mm, 4 mm, 5 mm and 6 mm. The space between each cylinder is equal to the diameter of the cylinder and a cylinder 25 mm in diameter. CT number measurements were performed on a 25 mm diameter cylinder and mid between a 25 mm and 6 mm diameter cylinder. [4,5]

The low-resolution contrast (CNR) is calculated by the formula: CNR = A-B / SD, In this case, A records the ROI signal inside the 25 mm cylinder (target), B records the ROI signal outside the 25 mm cylinder (Background),and SD isa Standard Deviation (SD) from ROI outside cylinder 25 mm.

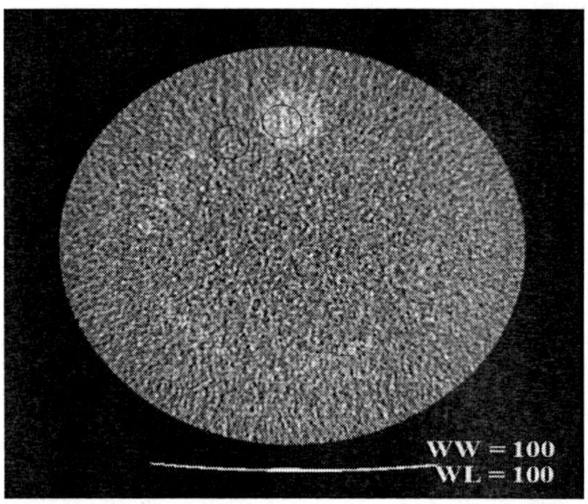

Figure 1. Low Contrast Resolution.

2.5. Statistical analysis

We used t-student test to evaluate the image quality (CNR) and radiation dose (CTDIvol) at a standard 120 kV tube voltage and a low tube voltage of 80 kV with a tube current arrangement of 200, 250 and 300 mAs.the relationship between CNR and CTDIvol was investigated using linear regression analysis and Pearson correlation coefficient (r). All statistical analyzes were performed with commercially available software packages [2].

Table 1. CTDIvol value on various kVp and mAs.

Tube current-time product (mAs)	CTDIvol (mGy)	
	120 kVp	80 kVp
200	30.69	9.93
250	38.36	12.37
300	45.86	14.81

Table 2. The CT numbers, image noise, and CNR obtained at each set of acquisition condition.

Tube current-time product (mAs)	CT Number		P Value	Image noise		P Value	CNR		P Value
	120 kVp	80 kVp		120 kVp	80 kVp		120 kVp	80 kVp	
200	-4.0	-5.7	< 0.001	3.8	9.0	< 0.001	1.64	0.76	0.043
250	-5.4	-5.7	< 0.001	3.7	7.2	< 0.001	1.54	0.79	0.036
300	-4.1	-6.8	< 0.001	3.6	6.4	< 0.001	1.94	0.93	0.049

3. Results and Discussion

3.1. Radiation Dose

CTDIvol is an estimate of the radiation dose obtained from each of the irradiation conditions. The results obtained are shown in table 1. In the same tube time arrangement with different tube voltages, CTDIvol obtained at 80 kV is approximately 68% of 120 kV, thereby reducing the received radiation by 32%.

3.2. Image Quality Results

The noise and CNR values in the CT scan image of each of the irradiation conditions can be seen in table 2. At the same irradiation of tubular currents and different tube voltages, the noise values from the lowest to the highest can be seen in figure 2. The noise value obtained at 120 kV with 300 mAs as the standard tuning condition for CT scan is higher than the noise value obtained at 80 kV with 200-300 mAs, so its significance value (P <0.001) (table 2). There was a strong correlation between CNR and CTDIvol with Pearson correlation coefficient r = 0.875 (P <0.001) for irradiation at 80 kV tube voltage, while r = 0,510 (P <0.001) on 120 kV tube tension. CTDIvol on the use of 80 kV tube voltage produces a lower CNR compared to CNR at 120 kV (figure 3). Using a two-tailed Student t-test, the CNR obtained at 80 kV with 200-250 mAs was significantly lower than 120 kV with 300 mAs ($P < 0.05$) (table 2). However, there was no statistically significant difference between CNR obtained at 120 kV with 300 mAs and CNR obtained with 80 kV at 200 mAs and 250 mAs ($P > 0.05$) (table 2).

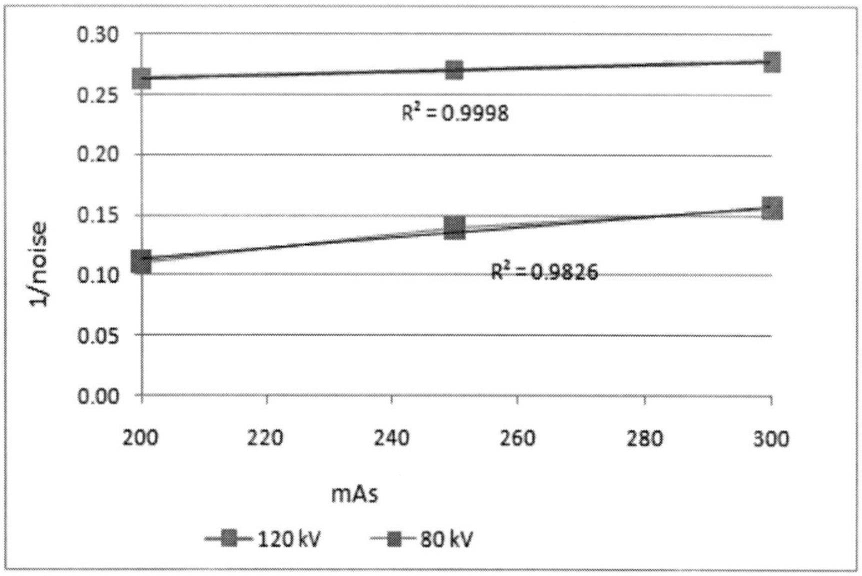

Figure 2. The relationship of noise value with mAs.

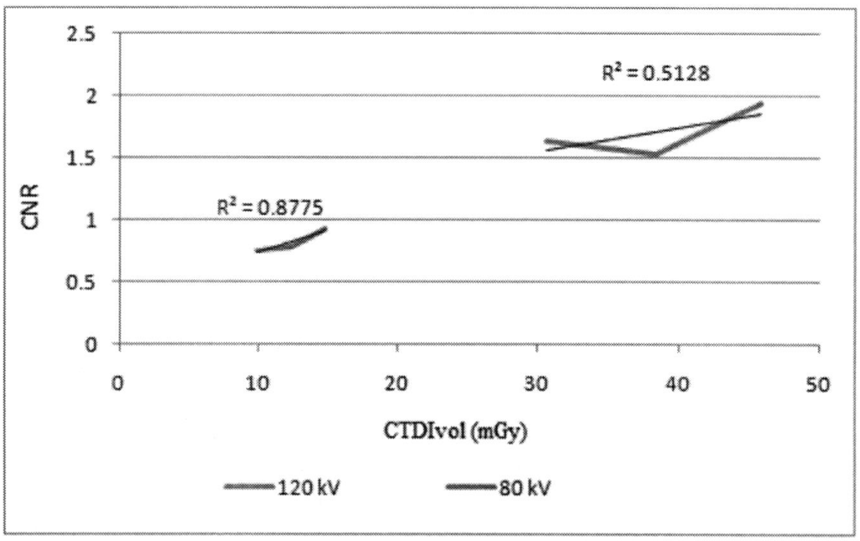

Figure 3. Relation of CNR value to CTDIvol (mGy).

3.3. CT Scan is one of the medical methods that are now widely used in the medical world
CT Scan or commonly referred to as Computed Tomography is a method of medical depiction using tomography, which relies on X-rays in producing images from the inside of an object, which in this case is human. Current CT scans are often used to confirm the diagnosis of patients suspected of having cancer disorders. Besides, CT scan can cause the risk of death from cancer, especially for patients who have CT scan more than 1 times. The management of patient doses is of major concern at the head MSCT examination.

In previous studies conducted on abdominal CT with theoutcome the CNR value decreases when the CT acquisition is performed at an 80 kV tube voltage and an identical tube current setting, CNR increases substantially when identical CTDIvol is used. Compared to CNR obtained at 120 kVand 300 mAs, there was no statistically significant difference at 80 kV and 550mAs, 600mAs, and 650mAs (P >0.05). [6]
In this study, we used the CT dose index (CTDIvol) for the radiation dose present in the monitor expressed in mGy. CTDIvol as the intensity of the radiation dose will facilitate the comparison of the accuracy of the radiation dose at different tube voltages. For example, the radiation dose between the tube voltage is 80 kV and 120 kV.Our results show that to reduce radiation exposure, substantially by lowering tube voltage from 120 kV to 80 kV. However, it has its limitations. Because radiation exposure is the average dose of a homogeneous cylindrical phantom. Therefore, the patient dose should be estimated with CTDIvol[2]. In our study, our findings show that there is a direct correlation between CNR and CTDIvol. The higher the CTDIvol, the higher the CNR value. CNR obtained at 120 kV and 300 mAs, there was no statistically significant difference at 80 kV and 200 mAs and 250 mAs (P> 0.05). This indicates that to reduce the radiation dose received by the patient does not always rest on the use of a 120 kV tube voltage, with the use of a tube voltage below 120 kV the patient's received radiation dose can be reduced to 32%.

3.4. Limitations of the study
This study uses phantom as a substitute for the human body, so it does not take into account the variability of body composition, therefore, whether these results are appropriate for clinical use. Marin et al. [7] showed that techniques with low tube stress at 80 kV could be applied to significantly reduce the radiation dose received by the patient. Possible to do with head CT scan.

5. Conclusions
In this phantom CT study, we have shown that although the image noise increases at low tube voltage, it is possible to reduce radiation doses by up to 32% without CNR decrease by reducing tube voltage from 120 kV to 80 kV and increasing tube current of more than 250 mAs. CT Scan examination with low tube voltage technique has a huge advantage for patients, especially patients who undergo CT scan more than once so that negative effects of radiation can be avoided.

References
[1] Castellano I and Geleijns J 2007 *Computed Tomography Physicsfor Medical Imaging Applications* (UK and Netherlands: Springer)
[2] Yusanti W, Budi W S and Adi K 2014 *Penentuan Quality Control (QC) ResolusiSpasialPada Citra CT Scan DenganMetode Line Spread Function (LSF) Menggunakan Phantom AAPM CT*
[3] AAPM 2002 *Quality Control in Diagnostic Radiology, AAPM Report NO. 74* (Medison: Medical Physics Publishing)

[4] ACR 2017 *American College of Radiology CT Accreditation Program Testing Instructions* Revised: 1- 06 – 2017

[5] Gulliksrud K, Martin A C T and Stokke C 2014 How to measure CT image quality:Variations in CT-numbers, uniformity and low contrast resolution for a CT quality assurance phantom, Committee Task Grup No. 30 *Medical Physics Jurnal* **30** 1-6 1120 – 1797

[6] Tang K, Wang L, Li R, Lin J, Zheng X and Cao G 2012 Effect of Low Tube Voltage on Image Quality, Radiation Dose, and Low-Contrast Detectability at Abdominal Multidetector CT: Phantom Study *Journal of Biomedicine and Biotechnology*

[7] Marin D, Nelson R C and Samei E 2009 Hypervascular liver tumors: low tube voltage, high tube current multi-detector CT during thelate hepatic arterial phase for detection—initial clinical experience *Radiology* **251** 3 771–779

IOP Conf. Series: Journal of Physics: Conf. Series **979** (2018) 012079 doi:10.1088/1742-6596/979/1/012079

Improve Image Quality of Transversal Relaxation Time PROPELLER and FLAIR on Magnetic Resonance Imaging

N Rauf, D Y Alam, M Jamaluddin and B A Samad

Department of Physics, FMIPA, University of Hasanuddin, Makassar, Indonesia

E-mail: n-rauf@fmipa.unhas.ac.id

Abstract. The Magnetic Resonance Imaging (MRI) is a medical imaging technique that uses the interaction between the magnetic field and the nuclear spins. MRI can be used to show disparity of pathology by transversal relaxation time (T2) weighted images. Some techniques for producing T2-weighted images are Periodically Rotated Overlapping Parallel Lines with Enhanced Reconstruction (PROPELLER) and Fluid Attenuated Inversion Recovery (FLAIR). A comparison of T2 PROPELLER and T2 FLAIR parameters in MRI image has been conducted. And improve Image Quality the image by using RadiAnt DICOM Viewer and ENVI software with method of image segmentation and Region of Interest (ROI). Brain images were randomly selected. The result of research showed that Time Repetition (TR) and Time Echo (TE) values in all types of images were not influenced by age. T2 FLAIR images had longer TR value (9000 ms), meanwhile T2 PROPELLER images had longer TE value (100.75 – 102.1 ms). Furthermore, areas with low and medium signal intensity appeared clearer by using T2 PROPELLER images (average coefficients of variation for low and medium signal intensity were 0.0431 and 0.0705, respectively). As for areas with high signal intensity appeared clearer by using T2 FLAIR images (average coefficient of variation was 0.0637).

1. Introduction

The Magnetic Resonance Imaging (MRI) is one of the most accurate diagnostic tools in the medical world [1]. The ability of MRI to be able to provide anatomical imaging of soft tissue with high resolution makes it a very important tool for imaging at the molecular and cell level [2]. The basic principle of MRI is based on nuclear magnetic resonance (NMR) along with the proton spin-spin relaxation process [1]. Relaxation process is divided into two parts, namely relaxation T1 and T2. Some techniques for generating T2 weighted images are Periodically Rotated Overlapping Parallel Lines with Enhanced Reconstruction (PROPELLER) and Fluid Attenuated Inversion Recovery (FLAIR).

Studies on PROPELLER and FLAIR have been widely practiced. Forbes et al. examined the measurements of the patient's head movements and their effect on image quality. The examination was done in the form of axial T2-W PROPELLER (P-CR corrected and uncorrected P-UNCR) and conventional MRI (CONV) in five normal volunteers and thirty-five randomized patients. The results obtained are the PROPELLER MR sequence (P-CR and P-UNCR) offering better image quality than the CONV sequences for all the cases studied [3]. Rathi and Palani also examined the weighting images of T1, T2, and FLAIR of twenty patients with glioma through distribution values using

MATLAB software. Both of these studies concluded that the image with the FLAIR technique provides more information for pathological disorders than with T1 weighted images as well as ordinary T2 weighted images [4].

Based on the above studies, we conducted a study by comparing the parameters of transverse relaxation time of PROPELLER and the relaxation time of FLAIR in MRI image. This study focuses on brain images with variables Time Repetition (TR) and Time Echo (TE). Image comparison is clarified by using the ENVI software via image segmentation method by Roy and Bandyopadhyay [5] and Region of Intererest (ROI).

1.1. Periodically Rotated Overlapping Parallel Lines with Enhanced Reconstruction (PROPELLER)
One significant problem in brain Magnetic Resonance Imaging is head movement. There is no data to measure movement, complexity, or influence on image quality. PROPELLER (Periodically Rotated Overlapping Parallel Lines with Enhanced Reconstruction) offers a new way of measuring and balancing head movements. The basic idea of PROPELLER is to try k-space in rotation using radial trellised strips.

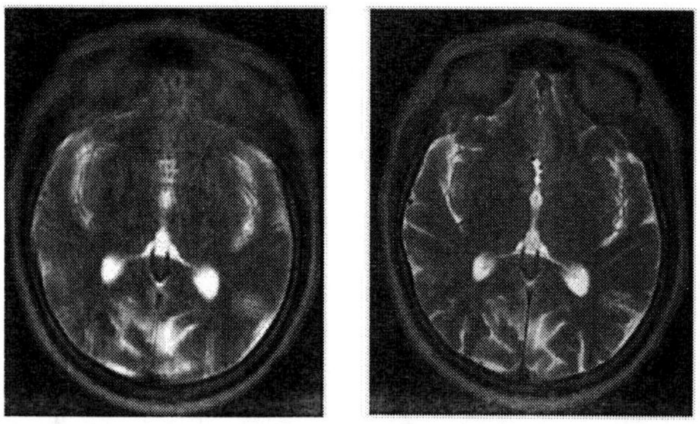

Figure 1. Patient movement artifacts (left) and motion artifacts substantially reduced with PROPELLER (right).

1.2 Fluid Attenuated Inversion Recovery (FLAIR)
Although the imaging time is very long (usually 15 ± 20 min), the T2-FLAIR technique is proven to repeatedly show a variety of lesions, including cortical, periventricular, and meninges disease that are difficult to see in conventional imagery. In addition, the intensity of edema in the FLAIR image differs significantly from the intensity of Cerebra Spinal Fluid (CSF), unlike in T2 weighted images. CSF produces weaker signals than gray matter, tumors, or edema [6]. FLAIR is also substantially more sensitive to demyelinating diseases, such as multiple sclerosis.

1.3 Region Of Interest (ROI)
ROI can be drawn on top of MRI imagery by using any combination of polygons, dots, or vectors. In the ENVI software, the user can specify multiple ROI and describe it to either the Image, Scroll, or Zoom window. Additionally, users can develop an ROI to an adjacent pixel that is at a certain pixel value limit. When determining ROI, users assign pixels from imagery to enter or exit ROI [7].

2. Methodology

This research was conducted in Radiology Installation room, General Hospital in Makassar from October to December 2016. The tools used are laptop with Windows 7 operating system, RadiAnt DICOM Viewer software version 3.0.2.12209 (32-bit), and ENVI software version 4.5. The materials used are axial images of transversal relaxation Periodically Rotated Overlapping Parallel Lines with Enhanced Reconstruction (T2 PROPELLER) and axial images of transversal relaxation time Fluid Attenuated Inversion Recovery (T2 FLAIR). The first step in this research is randomly collecting patient data in the form of brain image with various parameter types in the form of DICOM extension (.dcm). The data are then grouped by age category. Once grouped, the selected image from the patient data is the axial image of T2 PROPELLER and the axial image of T2 FLAIR in the form of .jpg extension using RadiAnt DICOM Viewer software version 3.0.2.12209 (32-bit). Then, the Time Repetition and Time Echo values of the images are analyzed.

The next step is image processing with ENVI version 4.5. Image processing steps include axial imagery T2 PROPELLER and axial imagery T2 FLAIR, image conversion into gray-scale, high pass filter usage on gray-scale image, gray-scale image fusing and filtered image through histogram matching, and use of median filter in composite image. Then, the image is colored by using Region of Interest (ROI).

3. Result and discussion

The axial image of T2 PROPELLER and the axial image of T2 FLAIR are grouped. Then Time Repetition (TR) and Time Echo (TE) selected from the image, see table 1.

Table 1. Time Repetition (TR) and Time Echo (TE) of the patient.

Patient	T2 PROPELLER		T2 FLAIR	
(year old)	TR (ms)	TE (ms)	TR (ms)	TE (ms)
A (10)	4100.12	100.75	9000	91.26
B (20)	4592.62	102.10	9000	90.72
C (21)	4115.78	101.14	9000	91.70
D (34)	4147.40	101.90	9000	91.34
E (34)	4115.30	102.10	9000	91.08
F (36)	4588.33	102.00	9000	92.04
G (52)	4155.22	102.10	9000	91.96
H (56)	4155.22	102.10	9000	90.55
I (63)	4151.20	102.00	9000	91.70
J (65)	4811.40	102.10	9000	91.78
K (71)	4588.16	102.00	9000	92.92

The TR value of the axial image of T2 FLAIR is the same for all age groups. However, the value of each patient's TE varies 90.55 – 92.92 msec. This shows that there is no age effect on the TR and TE values of patients, but is influenced by the sequence type and the use of radio frequency inversion pulses 180° or reversal of gradient polarity at TE/2 [8].

The TR value of T2 PROPELLER and T2 FLAIR of each patient is long (greater than 1500 ms) [9]. The long TR makes the difference of T1 do not affect to the signal. The axial image of T2 FLAIR has a TR value longer than the axial image of T2 PROPELLER, thus allowing radio frequency excitation energy to be extended for longer by protons via spin relaxation. Thus, the axial image of T2 FLAIR has the smallest T1 weighting and the signal intensity difference between the tissues is not significant.

Furthermore, the TE values in the axial images of T2 PROPELLER and T2 FLAIR of each patient are also relatively long (more than 80 milliseconds) [9]. The long TE is required to produce a good T2 weighting image. The axial image of T2 PROPELLER has a TE longer than the axial image of T2 FLAIR, thus allowing the proton to dephasing or to have longer T2 relaxation and generate more signals from the tissue having a long T2 value. Thus, the axial image of T2 PROPELLER has the largest T2 weighting. However, the longer TE values will cause the total signal intensity to decrease and the image look rough.

3.1 Image Segmentation and Region of Interest (ROI)

The axial image of T2 PROPELLER and the axial image T2 FLAIR patient is processed by using segmentation method, which includes image conversion into gray-scale, high pass filter usage, image integration through histogram matching, and median filter usage. Then, the image is colored by using ROI.

Table 2. Comparison of PROPELLER and FLAIR images of patient A.

Image Processing	T2 PROPELLER	T2 FLAIR
Original Image	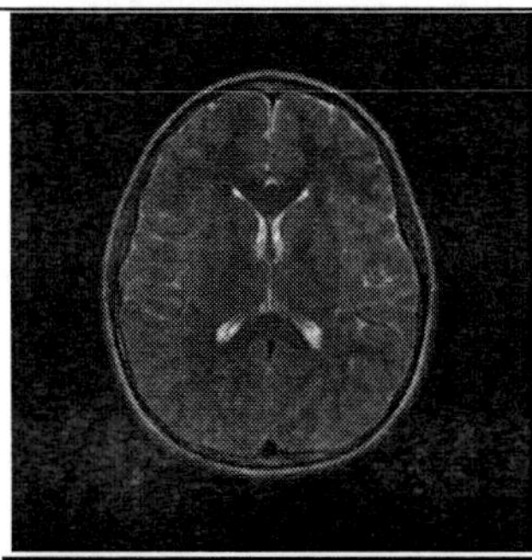	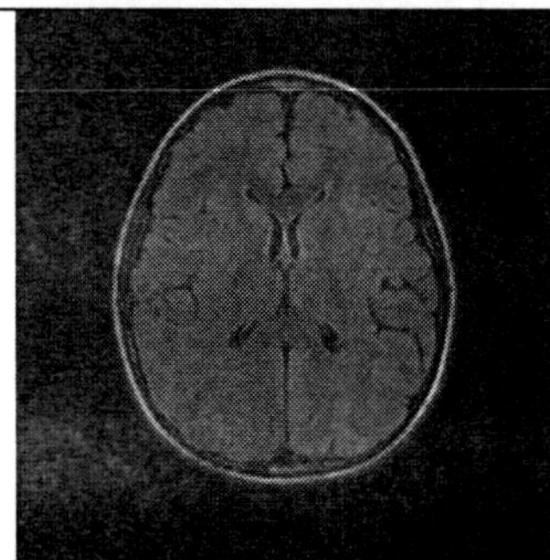
Grayscale Image	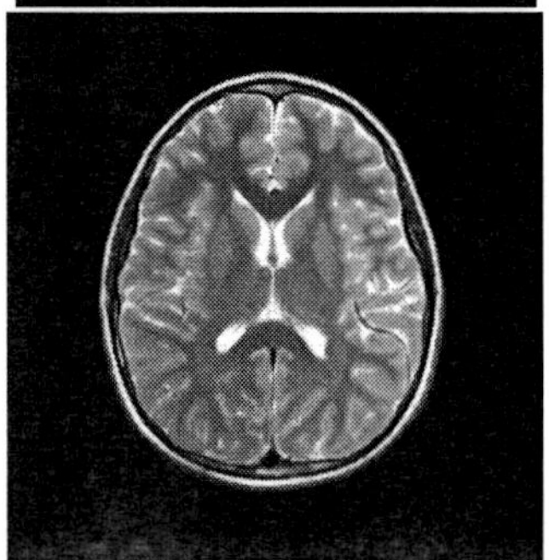	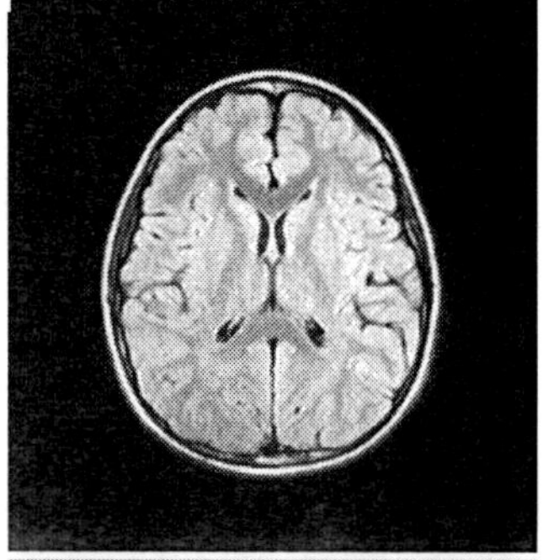
High Pass Filter	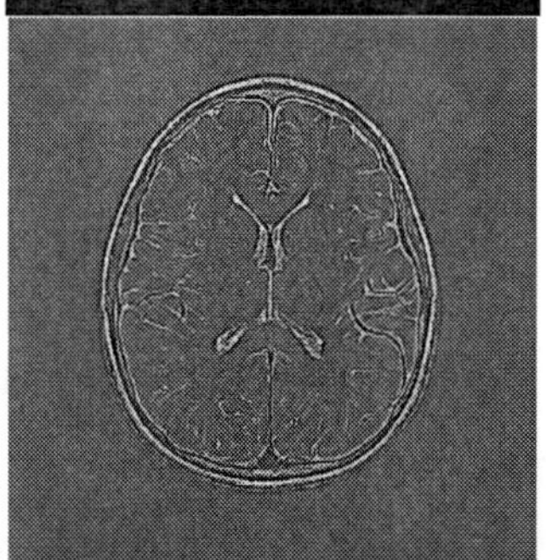	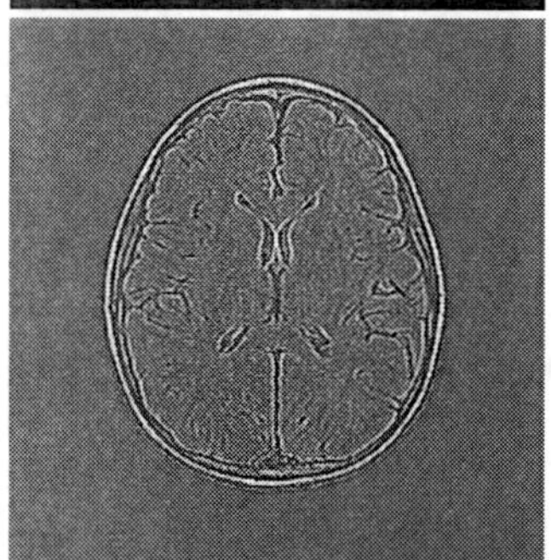
Histogram Matching	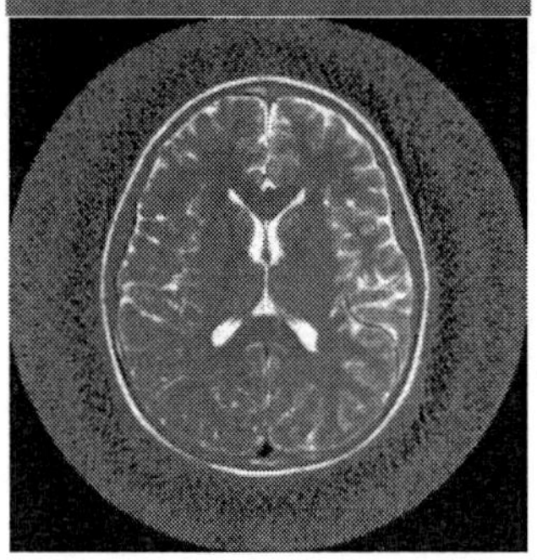	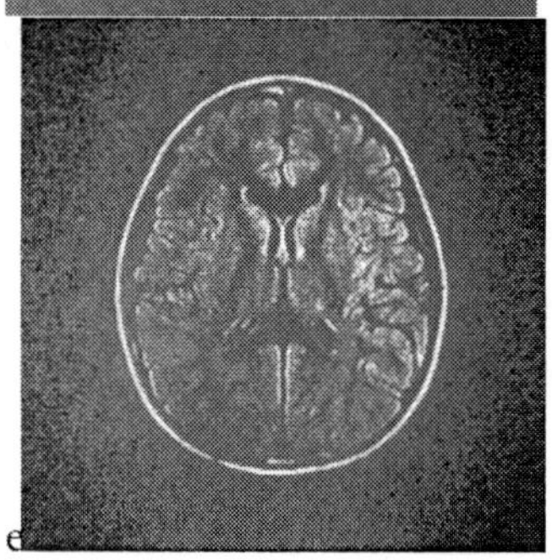

Median Filter

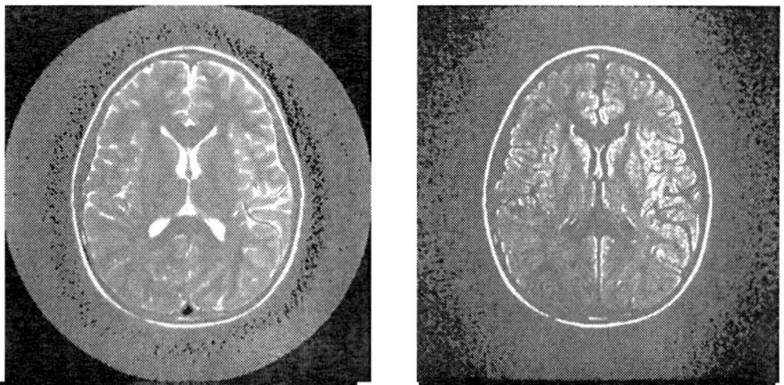

ROI

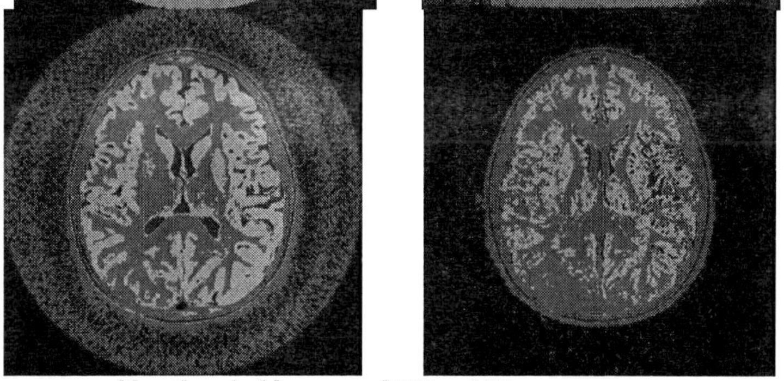

The red color indicates an area with a threshold range of 133 – 152
The green color indicates an area with a threshold range of 153 - 203
The blue color indicates an area with a threshold range of 204 - 255

Table 2 shows the brain image of one patient, namely patient A. The original image has dimensions of 512 x 512 pixels with three bands (Red, Green, and Blue). When the image is converted into gray scale, the bands are used only one (blue). Then, the gray scale image is filtered with a high pass filter to sharpen the image. The Kernel size selected on the filter is 3x3, and the image does not have significant blurring. Once filtered, the image is combined with gray scale image through a matching histogram. To reduce noise, the image is filtered again using the median filter. The Kernel size is 3x3. Then the image is colored using ROI. The selected colors are red, green, and blue.

In the axial image of T2 PROPELLER, the red color is the part (white matter and fat) that has a short T2 value, green color is gray matter, and blue color has a long T2 value or a lot of water content (CSF and high edema tissue) [8].

Whereas on the axial image of T2 FLAIR, the red color is CSF and the tissues, which have value T1 and T2 relative the same [10]. Weighting FLAIR with long TI (in this case 2200 ms for all patients) reduced the intensity level of relative signals of CSF and the tissues, resulting in better assessment of the anatomy around. The blue color is a tissue that has a short T2 value [8]. Each patient image has its own threshold value range for staining. The value range data is shown in tables 3 and 4.

Table 3. Threshold Value Range of Axial Image T2 PROPELLER.

Patient	Red				Green				Blue			
	Min	Max	Mean	Stdev	Min	Max	Mean	Stdev	Min	Max	Mean	Stdev
A (10)	133	152	141.2	5.8	153	203	166.3	119	204	255	241.2	16.7
B (20)	130	160	146.6	7.4	161	208	173.7	11.3	209	255	243.0	15.1
C (21)	130	153	141.9	5.6	154	206	165.8	11.6	207	255	243.3	15.4
D (34)	141	164	150.3	6.3	165	210	178.2	11.8	211	255	243.4	14.4

E (34)	135	154	142.9	5.0	155	209	169.1	14.3	210	255	242.8	14.8
F (36)	137	150	142.7	3.7	151	211	166.6	14.6	212	255	243.6	14.5
G (52)	130	163	149.7	7.8	164	204	175.9	10.4	205	255	239.9	17.3
H (56)	130	157	144.2	6.3	158	219	173.8	16.0	220	255	247.2	10.8
I (63)	134	163	149.1	7.1	164	204	175.9	10.4	205	255	241.8	16.9
J (65)	135	167	148.2	7.5	168	205	181.5	10.8	206	255	240.5	16.7
K (71)	135	163	151.3	6.9	164	204	176.1	10.7	205	255	240.4	17.6

Table 4 .Threshold Value Range of Axial Image T2 FLAIR.

Patient	Red				Green				Blue			
	Min	Max	Mean	Stdev	Min	Max	Mean	Stdev	Min	Max	Mean	Stdev
A (10)	133	152	1418	5.5	153	203	167.8	13.5	204	255	240.2	17.1
B (20)	130	160	141.7	8.5	161	208	177.0	13.3	209	255	242.3	15.1
C (21)	130	153	140.5	6.6	154	206	169.8	13.9	207	255	243.0	15.1
D (34)	141	164	151.7	6.0	165	210	179.0	12.4	211	255	243.1	14.5
E (34)	135	154	144.0	5.7	155	209	170.4	14.3	210	255	242.6	14.8
F (36)	137	150	144.2	3.9	151	211	166.8	15.3	212	255	242.1	14.5
G (52)	130	163	142.4	9.0	164	204	179.1	11.3	205	255	240.5	16.9
H (56)	130	157	144.0	8.0	158	219	174.6	15.8	220	255	245.6	11.4
I (63)	134	163	145.4	7.9	164	204	178.8	11.6	205	255	237.1	17.4
J (65)	135	167	147.6	8.3	168	205	181.8	10.6	206	255	241.5	16.4
K (71)	135	163	145.4	7.4	164	204	177.8	11.3	205	255	240.9	16.5

The coefficient of variation is used to compare the data in table 3 and 4. The coefficient of variation (CV) is calculated from mean (μ) and standard deviation (σ) of data. The average variation coefficient (CV) for T2 PROPELLER and T2 FLAIR is shown in table 5.

Table 5. The average variation coefficient image of T2 PROPELLER and T2 FLAIR.

Color	T2 PROPELLER			T2 FLAIR		
	$\bar{\mu}$	$\bar{\sigma}$	$\overline{CV}$	$\bar{\mu}$	$\bar{\sigma}$	$\overline{CV}$
Red	146.2	6.3	0.0431	144.4	7.0	0.0485
Green	173.0	12.2	0.0705	174.8	13.0	0.0744
Blue	242.5	15.5	0.0639	241.7	15.4	0.0637

In table 5, the axial image of T2 PROPELLER has the smaller average variation coefficient value for red and green, while the axial image T2 FLAIR has the smaller average variation coefficient value for blue. If a data set has a small coefficient of variation, the data points are collected around the mean or in other words the data is uniform [11].

It means the axial image of T2 PROPELLER shows areas with low and medium intensity signals (red and green) better than axial image of T2 FLAIR, since axial image T2 PROPELLER has the larger T2 weighted value that makes a tissue with a short T1 appear dark due to the decay of signals. In addition, the axial image of T2 PROPELLER can overcome significant problems in brain magnetic resonance imaging, in this case the head motion artifact, through a series of concentric blades that rotate through k-space centers.

The axial image of T2 FLAIR indicates areas with high signal intensity (blue color) is better than axial image of T2 PROPELLER, because FLAIR weighting has the longer Time Repetition (TR) that makes the difference in T1 do not affect to signal, so the tissues produce similar signals. In addition, the TI factor on the axial image of T2 FLAIR can reduce the signal intensity level of CSF and the tissue having relatively equal T1 and T2. This causes tissues with shorter T2 values to appear brighter, such as edema. The image T2 FLAIR is more sensitive in showing pathological abnormalities.

4. Conclusion

The conclusion of this study is no age influence on the selection of Time Repetition (TR) and Time Echo (TE) values for all types of images. In T2 FLAIR image, the TR value is longer than T2 PROPELLER image, so the T1 weighting value is the smallest and the signal intensity difference between the tissues is not significant. The T2 PROPELLER image has a TE value longer than the T2 FLAIR image, so the weighted value of T2 is greatest and generates more signals from the tissue with long T2. The T2 PROPELLER image is better used in areas with low and medium signal intensity, whereas areas with high signal intensity preferably use T2 FLAIR image.

References

[1] Brown M A, Semelka R C 2003 *MRI: Basic Principles and Applications Third Edition* (Hoboken, New Jersey: John Wiley and Sons, Inc.)

[2] Na H B, Song I C, Hyeon T 2009 Inorganic Nanoparticles for MRI Contrast Agents *Adv. Mater.* **21** 2133–48

[3] Forbes K P, Pipe J G, Bird R and Heiserman J E 2001 PROPELLER MRI: Clinical Testing of a Novel Technique for Quantification and Compensation of Head Motion. *Proc. Intl. Soc. Mag. Reson. Med* **9**

[4] Rathi V P, Palani S 2012 Brain Tumor MRI Image Classification with Feature Selection and Extraction using Linear Discriminant Analysis *International Journal of Information Sciences and Techniques (IJIST)* **2** 4 131–146

[5] Roy S, Bandyopadhyay S K 2012 Detection and Quantification of Brain Tumor from MRI of Brain and it's Symmetric Analysis *Journal of Information and Communication Technology (JICT)* **2** 6 477– 483

[6] Dvořák P, Bartušek K, Kropatsch W G and Smékal Z 2015 Automated Multi-Contrast Brain Pathological Area Extraction from 2D MR Images *Journal of Applied Research and Technology (JART)* **13** 58–69

[7] Anonim. Region of Interest (ROI) Tool. http://www.harrisgeospatial.com/docs/RegionOfInterestTool.html. Access on 25 September 2016.

[8] Bushberg J T, Seibert J A, Leidholdt E M and Boone J M 2012 *The Essential Physics of Medical Imaging: Third Edition* (Philadelphia: Lippincott Williams & Wilkins)

[9] Schild HH 1990 *MRI: Made Easy* (Berlin: Schering AG)

[10] Damanik A M, Azam M and Nur M 2005 Pengaruh Parameter Teknis TR, TE, dan TI dalam Pembobotan T1, T2, dan FLAIR Pencitraan Magnetic Resonance Imaging (MRI) *Berkala Fisika.* **8** 1 15–20

[11] Al-Saleh M F, Yousif A E 2009 Properties of the Standard Deviation that are Rarely Mentioned in Classrooms *Austrian Journal of Statistics.* **38** 3 193–202

IOP Conf. Series: Journal of Physics: Conf. Series **979** (2018) 012080 doi:10.1088/1742-6596/979/1/012080

Study of Image Quality From CT Scanner Multi-Detector by using Americans College of Radiology (ACR) Phantom

Mulyadin[1,2], Syamsir Dewang[2], Bualkar Abdullah[2], Dahlang Tahir[2]

[1]Health Facility Hall of Makassar, Makassar 90245 Indonesia
[2]Department of Physics, Hasanuddin University, Makassar 90245 Indonesia

E-mail: dtahir@fmipa.unhas.ac.id

Abstract. In this study, the image quality of CT scan using phantom American College of Radiology (ACR) was determined. Scanning multidetector CT is used to know the image quality parameters by using a solid phantom containing four modules and primarily from materials that are equivalent to water. Each module is 4 cm in diameter and 20 cm in diameter. There is white alignment marks painted white to reflect the alignment laser and there are also "HEAD", "FOOT", and "TOP" marks on the phantom to help align. This test obtains CT images of each module according to the routine inspection protocol of the head. Acceptance of image quality obtained for determination: CT Number Accuracy (CTN), CT Number Uniformity and Noise, Linearity CT Number, Slice Technique, Low Contrast Resolution and High Contrast Resolution represent image quality parameters. In testing CT Number Accuracy (CTN), CT Uniform number and Noise are in the range of tolerable values allowed. In the test, Linearity CT Number obtained correlation value above 0.99 is the relationship between electron density and CT Number. In a low contrast resolution test, the smallest contrast groups are visible. In contrast, the high resolution is seen up to 7 lp/cm. The quality of GE CT Scan is very high, as all the image quality tests obtained are within the tolerance brackets of values permitted by the Nuclear Power Control Agency (BAPETEN). Image quality test is a way to get very important information about the accuracy of snoring result by using phantom ACR.

1. Introduction

The CT Scan (Computed Tomography Scanning) imaging technology was first introduced in 1972 by Sir Godfrey New Bold Hounsfield and Allan Cormack. The first CT scan was produced by EMI Mark 1, with image resolution using an 80 x 80-pixel matrix, each pair of slices taking 4.5 minutes per scan. Due to the length of the examination, it is unable to perform cardiac and respiratory examinations. This first-generation CT scan is only capable of performing a head examination (brain) only [1].

CT Scan is a fairly complex medical imaging system so there is a risk of calibration error, failure of generating system function and X-ray detection as well as the risk for the patient. Therefore, CT Scan aircraft require QC (quality control) program to ensure image quality while maintaining the dose is still below the allowable limit [2,3].

Quality assurance program on the use of CT Scan aircraft using phantom ACR (gammex). This program has the highest radiation safety standards and has basic details, enabling optimal achievement of results [4].

2. Materials

2.1. CT Scan. The CT used as an object is a CT Scan dual slice (General Electric, GE)

2.2. Phantom ACR. The ACR CT accreditation phantom (Gammex 464) is a solid phantom containing four modules and is constructed primarily from a water-equivalent material. Each module is 4 cm in depth and 20 cm in diameter. There are external alignment markings scribed and painted white (to reflect alignment lights) on EACH module to allow centering of the phantom in the axial (z-axis, cranial/caudal), coronal (y-axis, anterior/posterior), and sagittal (x-axis, left/right) directions. There are also "HEAD", "FOOT" and "TOP" markings on the phantom to assist with positioning [5].

Figure 1. Phantom ACR.

There are 4 phantom ACR CT modules as follows :

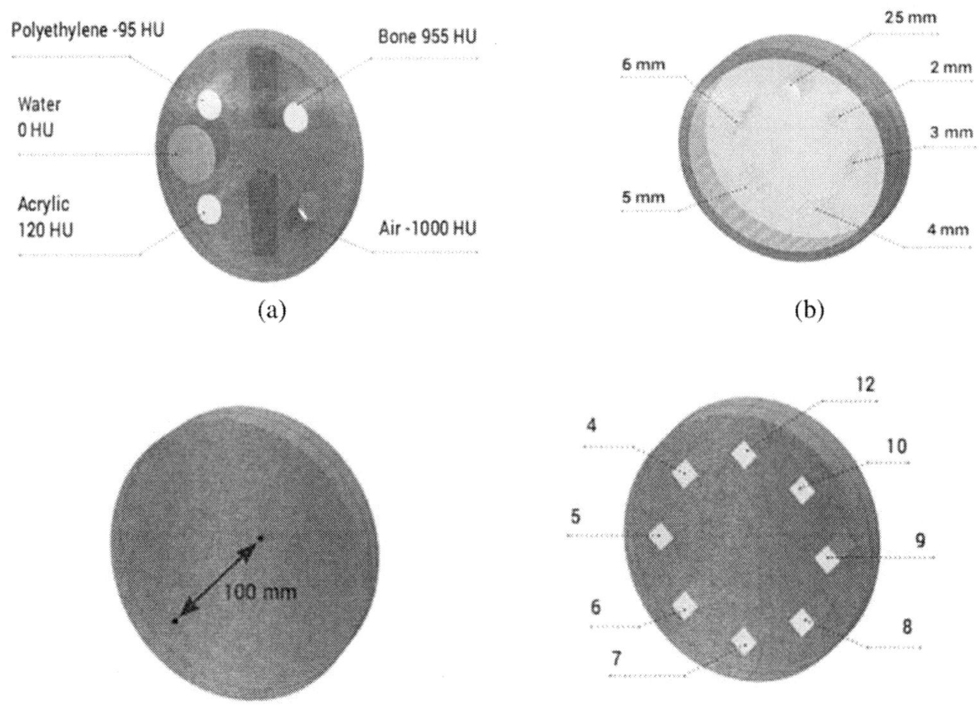

Figure 2. Phantom ACR of module.

a). Module 1 is used to assess positioning and alignment, CT number accuracy, and slice thickness, b). Module 2 is used to assess low contrast resolution, c). Module 3 consists of a uniform, tissue-equivalent material to assess CT number uniformity, d). Module 4 is used to assess high contrast (spatial) resolution.

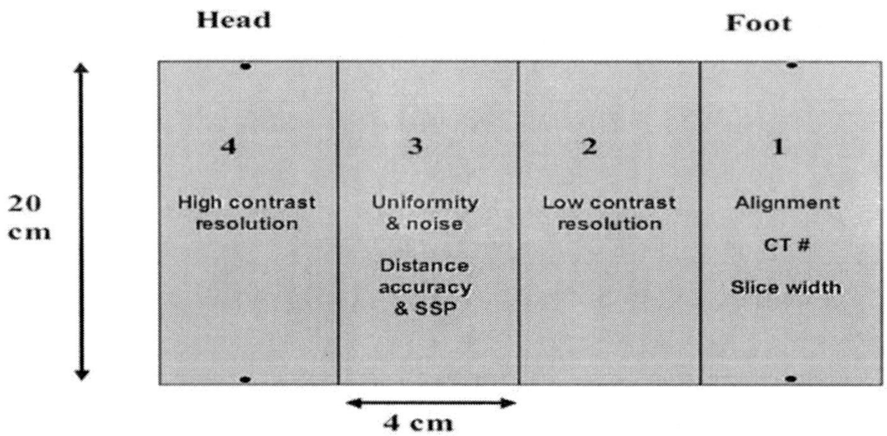

Figure 3. "HEAD" and "FOOT" for phantom.
1). Module 1 is used to assess positioning and alignment, CT number accuracy, and slice thickness, 2). Module 2 is used to assess low contrast resolution, 3). Module 3 consists of a uniform, tissue-equivalent material to assess CT number uniformity, 4). Module 4 is used to assess high contrast (spatial) resolution.

3. Method

We observed seven images of the Phantom ACR for the head protocol. We analyzed the image in May 2017 at the Testing laboratory, BPFK Makassar, Indonesia.

3.1. Scanning Procedures. The parameters used in all of the following tests, derived from the Accredited Phantom Instructions of ACR Instructions.

3.1.1. Compatibility of scanners and phantom. The phantom is aligned with the coronal, sagittal axis and CT Scan plane on the axial axis then centered on the center of the table. Make sure that Phantom will not move [5].

3.1.2. Scanning phantom gammex 464. This study uses axial head protocol.

3.2. Determination of linearity of CT number and slice thickness. To calibrate the CT number, the Phantom is placed at the center of the rotation of the gantry, for "Head" to be placed at the front while the "Foot" is the back than the center point above the module 1. The scanning parameters used are FOV = 22.0 cm, slice thickness = 5 mm and scanning time = 1.5 s. on the monitor, the image is set at WW = 100 HU, WL = 0 HU, ROI = 200.01 mm2. ROI was placed on air, polyethylene, water, acrylic, and bone. Each value of the CT number was recorded. Then made a curve correlation between the electron density with CT number. As for the thickness of the slice setting tested are 3, 5, and 7 mm. to determine the thickness of the slice in the image that is counting the number of bar patterns that appear then divided into two.

IOP Conf. Series: Journal of Physics: Conf. Series **979** (2018) 012080 doi:10.1088/1742-6596/979/1/012080

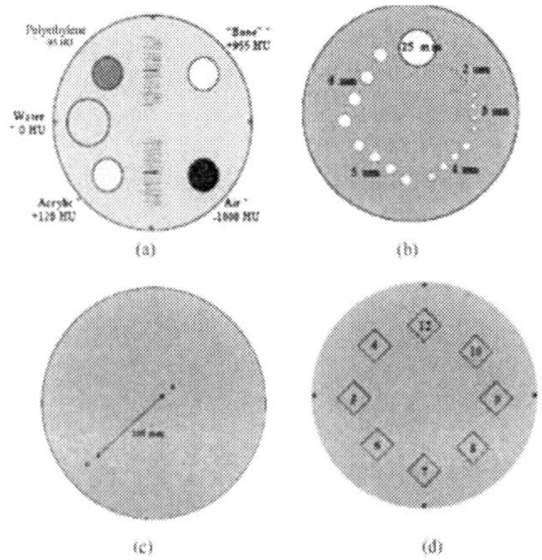

Figure 4. ACR phantom of image a) linearity position of CT number and slice thickness b) low contrast resolution c) accuracy and uniformity CT number d) high contrast resolution.

3.3. Determination of low-contrast resolution. Phantom is placed at the center of the rotation of the gantry, for the "Head" is placed in front while the "Foot" back then center point above module 2. on the monitor, the image is set at WW = 114 HU, WL = 114 HU, ROI = 105.57 mm2. Then place the ROI on a large cylinder image (25 mm) and place also the ROI between the 2 cylinders ie the big and small cylinders. Record the value of CT number and noise into the worksheet. To calculate the low-contrast resolution using the formula of the contrast ratio value to noise. CNR = I A - B I / SD [6].

3.4. Accuracy, Uniformity CT number and Uniformity Noise

The phantom is placed in the center of the rotation of the gantry, for the "Head" is placed on the front while the "Foot" is the back then the center point above module 3. on the monitor, the image is set at WW = 100 HU, WL = 0 HU, ROI = 208,81 mm2. to get the CT number, place each ROI: center, edge 1 (clock 12), edge 2 (clock 3), edge 3 (clock 6) and edge 4 (clock 9). Record the value of CT number and noise into the worksheet.

3.5. High contrast resolution. The phantom is placed at the center of the rotation of the gantry, for the "Head" is placed on the front while the "Foot" is the back then the center point above module 4. on the monitor, the image is set at WW = 1785 HU, WL = 606 HU. to determine high-contrast resolution is to observe clear and separate bars with each other. And noted in the worksheet.

3.6. Data Analysis. The obtained image quality parameters were compared with the tolerance values set by the Nuclear Power Control Agency (BAPETEN). Analysis of image quality parameters using the phantom ACR guide for scanned protocols.

4. Results and Discussion

Here's the result of parameter determination Image quality is obtained by using axial brain protocol.

4.1. Module 1

4.1.1. Linearity CT number. In this test obtained the value of CT numbers: air, polyethylene, water, acrylic, and bone are strongly correlated with the electron density value of each material. The correlation value that is equal to = 0.996 as shown in figure 6.

IOP Conf. Series: Journal of Physics: Conf. Series **979** (2018) 012080 doi:10.1088/1742-6596/979/1/012080

4.1.2. Determination of slice thickness. Table 1 shows that all the slice thicknesses measured according to the slice thickness. This indicates that all slice thicknesses tested are still within the tolerance limit.

4.2. Module 2: Measurement of low-contrast resolution on axial head ACR

After evaluated using CNR equation, it turns out the value obtained is = 1.75. This can be seen in the figure 7.a.

4.3. Module 3: Determination of Accuracy, Uniformity of CT number and Noise Uniformity

This is shown in figure 7.c.d. the accuracy of CT numbers at center area = -2.26 HU, while the number of CT and noise numbers occupies values of 1.7 and 0.23. Tolerance value for the accuracy of CT number ± 4, uniformity of CT number and noise 2. Thus, all parameters tested are still within the tolerance range.

4.4. Module 4: High contrast resolution in Axial brain ACR

In figure 7.b. shows the pattern bar is 7 cm/lp which shows that the resulting image modality is still very good. The tolerance value set is to be 6 lp/cm.

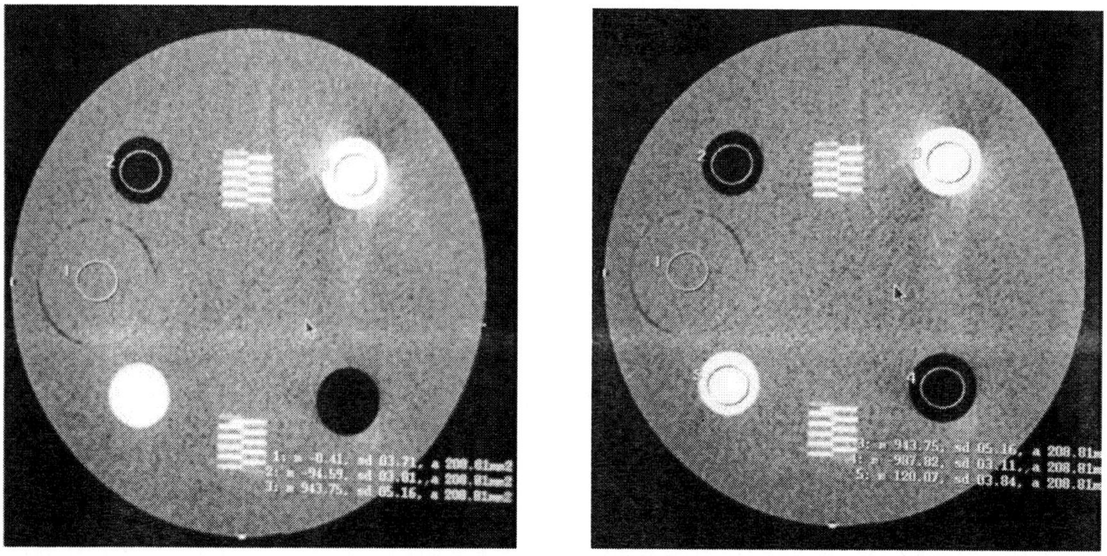

Figure 5. Module 1, linearity image of CT number and slice thickness.

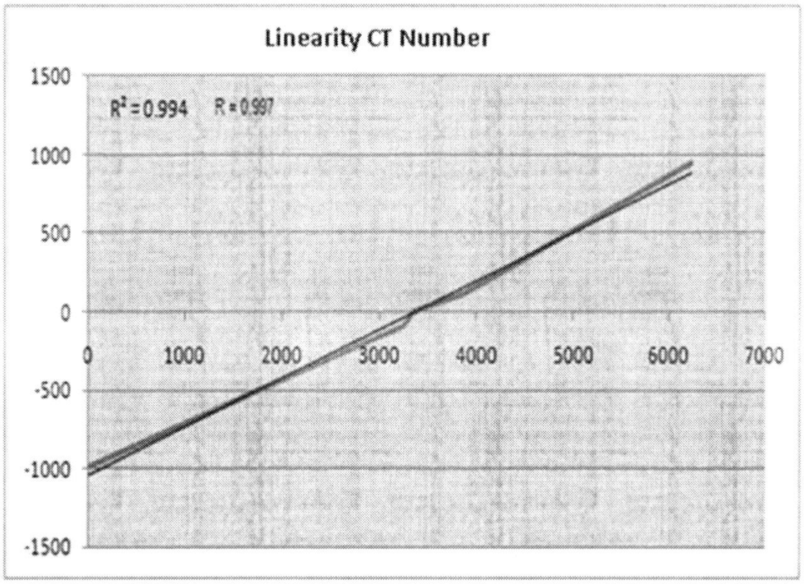

Figure 6. Graph of the relationship between electron density and CT number.

4.5. Discussion. Image quality generated using phantom ACR is acceptable for all test parameters. CT linearity test is accepted due to the CT number is proportional to the electron density in each material ie air, polyethylene, water, acrylic, and bone with a very strong correlation value of 0.997. the allowed tolerance is 0.99. the materials are selected, as they are very close to the tissues of the human body. Slice thickness test is also accepted because for 3 mm thickness the number of wires visible at the top and bottom is divided into two equal to the thickness of the slice. Similarly for the thickness of 5 and 7 mm slices. The allowable tolerance value is (± 0.5 mm). The thickness of the slice has a direct effect on the resulting spatial resolution of the image. Spatial resolution is the ability to reveal objects/organs with high contrast levels. The thinner the image spatial image is getting better, and vice versa. But the thinner the slice, the bigger the noise [7]. The low contrast test is acceptable because the value generated after through the calculation = 1.75, while the tolerable value is permitted 1. Test Accuracy, Uniform CT number, and Noise Uniformity are acceptable. The accuracy of CT number is very influential in determining the density of the organs so as to give a big influence on the diagnosis of a disease. This test is aimed to know the average value of CT in a water medium, uniformity of CT mean value and noise value. In module 4, a high contrast resolution test is accepted, since the Tolerance value of the head protocol for this test is 6 lp/cm and 7 lp/cm visible.

Table.1 The result of measurement of slice thickness.

Thickness set (mm)	Thickness sett (mm)	Slice (mm)	Tolerance Value	Results
3	3	0.0		Pass
5	5	0.0	0.5 mm	Pass
7	7	0.0		Pass

IOP Conf. Series: Journal of Physics: Conf. Series **979** (2018) 012080 doi:10.1088/1742-6596/979/1/012080

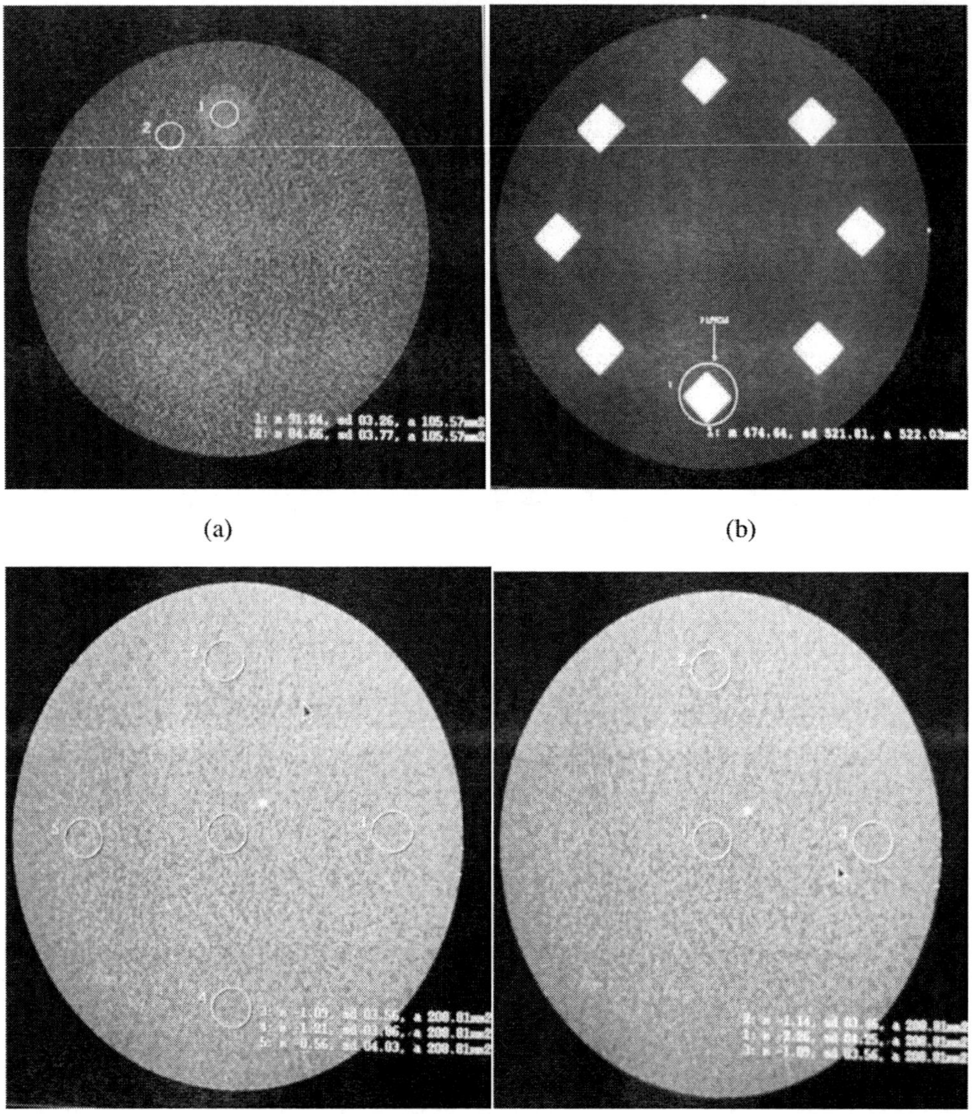

(a) (b)

(c) (d)

Figure 7. Citra Phantom ACR a. Low contrast, b. High contrast, c.d. Result of determination of CT number.

6. Conclusions

The quality control of the equipment is very important for the quality of the results, as is the case with the X-ray CT scans. It has obtained satisfactory results on the performance of the equipment so that radiological results can be accounted for. All test results on quality control using ACR Phantom are acceptable.

References

[1] Bushberg and Jerold T 2001 *The Essential Physics of Medical Imaging* ed third Lippincott William and Wilkins Philadelphia (USA)
[2] AAPM 2002 *Quality Control in Diagnostic Radiology* AAPM Report NO. 74 (Medison : Medical Physics Publishing USA)
[3] Weirna Y, Wahyu S B dan Kusworo A 2014 Penentuan Quality Control (QC) ResolusiSpasialPada Citra CT Scan Dengan Metode Line Spread Function (LSF) Menggunakan Phantom AAPM CT
[4] Mansour Z, Mokhtar A, Sarhan A, Ahmed M T, El-Diasty T 2016 Quality control of image using American Collage of Radiology (ACR) phantom *Journal of Radiology and Nuclear Medicine*
[5] ACR 2017 American College of Radiology CT Accreditation Program Testing Instructions Revised : 1- 06 – 2017
[6] Kristine G, Anne C T M and Caroline S 2014 How to measure CT image quality: Variations in CT-numbers, uniformity and low contrast resolution for a CT quality assurance phantom *Medical Physics Jurnal* **30**(1-6) 1120 – 1797
[7] Bushberg J T, Seibert J A, Leidholdt E M and Boone J M 2002 *The Essential Physics of Medical Imaging* Lippicott W and Wilkins (Philadelphia)

Deviation Value for Conventional X-ray in Hospitals in South Sulawesi Province from 2014 to 2016

Ilham Bachtiar, Bualkar Abdullah, Dahlan Tahir

Postgraduate Student of Physic Department Hasanuddin University, [2]Department of Physic of Hasanuddin University.
Jl. Perintis Kemerdekaan KM. 10 Makassar, Sulawesi Selatan, 90245, Indonesia

E-mail: liseilham@gmail.com

Abstract. This paper describes the conventional X-ray machine parameters tested in the region of South Sulawesi from 2014 to 2016. The objective of this research is to know deviation of every parameter of conventional X-ray machine. The testing parameters were analyzed by using quantitative methods with participatory observational approach. Data collection was performed by testing the output of conventional X-ray plane using non-invasive x-ray multimeter. The test parameters include tube voltage (kV) accuracy, radiation output linearity, reproducibility and radiation beam value (HVL) quality. The results of the analysis show four conventional X-ray test parameters have varying deviation spans, where the tube voltage (kV) accuracy has an average value of 4.12%, the average radiation output linearity is 4.47% of the average reproducibility of 0.62% and the averaged of the radiation beam (HVL) is 3.00 mm.

1. Introduction

Radiological services related to radiation protection aim to obtain a timely diagnosis and the lowest possible treatment for patients [1]. The achievement of these objectives is determined by two determinants of radiology equipment quality factors and their support systems and human resource factors. Human resource factors include installation procedures, operational procedures, maintenance procedures and improvements and procedures Quality Assurance (QA) and Quality Control (QC) [2], [3]. The procedure of carrying QA / QC therein is the implementation of the test or the so-called X-ray suitability test [4].

Quality Assurance (QA) is a quality assurance of all management programs organized to achieve excellent radiology health services by means of data collection and systematic evaluation. Quality Control (QC) is the quality control of a quality assurance program that focuses its program activities on the techniques necessary for monitoring, maintaining and maintaining technical elements of a radiographic equipment system that may affect the quality of a radiograph. One form of quality control program is an X-ray plane matching test [2]. The X-ray suitability test is a test to ensure the X-ray is in a reliable condition, both for diagnostic and interventional radiology activities and to comply with legislation [4].

Head of BAPETEN Rules No. 9 of 2011 about Compliance Testing X-ray Diagnostic and Interventional Radiology states that the purpose of X-ray plane suitability test is to realize reliable and safe operation of X-ray machine for patients, workers and the public. One X-ray plane tested for conformity is a conventional X-ray machine. Conventional X-ray is a permanently fixed X-ray machine in a room used for routine general inspection [5]. This type of this plane is most widely owned by both government-owned and private hospitals and clinics doctors and clinical laboratories, especially in the province of South Sulawesi. The plane is declared reliable if all test parameters are within the specified tolerable range. The tolerance limit is established by both national and international nuclear inspectors. In Perka BAPETEN number 9 of 2011, the test parameters for conventional X-ray plane are tube voltage accuracy (kV), irradiation time accuracy (s), radiation output linearity, kV reproducibility, time, and radiation output and radiation beam quality value (HVL). The maximum deviation value (maximum deviation) of tube voltage accuracy (kV) and irradiation time accuracy (s) is not allowed from 10%, the linearity coefficient of the radiation output and the reproducibility variation coefficient shall not exceed 10%, while the radiation beam quality value (HVL) is more than or equal to 2.1 mm HVL [4, 6, 7]

The purpose of this study is to analyze several factors that affect the conventional X-ray plane quality assurance as well as the distribution of data of deviation test results of X-ray plane in South Sulawesi province. These factors include tube voltage (kV) accuracy, radiation output linearity, radiation output reproducibility (mGy) and radiation beam quality (HVL) value [8].

2. Research Methodology

Conventional X-ray testing was performed using a non-invasive x-ray multimeter [9]. This tool can measure tube voltage (kV), radiation output linearity, tube voltage reproducibility and radiation output along with radiation beam quality values. The measurement accuracy is the proximity level of the quantity measurement produced by the actual value. Tubular voltage accuracy measurements and time accuracy are performed for each X-ray plane. The value of deviation (Rx) tube voltage accuracy is obtained by the equation:

$$R_x = \frac{X_m - X_n}{X_n} \times 100\% \quad (1)$$

Where: Xm is the measured value of the tube voltage (kV) and Xn is the value of tube voltage regulation (kV) on a conventional X-ray plane [1].

The linearity test of the radiation output is performed to know the radiation output at some radiation conditions and the linearity of the current generated by the X-ray tube at a certain distance. The value of irradiance linearity of radiation output is coefficient of linearity (CL). This CL value is obtained by the equation:

$$CL = \left|\frac{X_{max} - X_{min}}{X_{max} - X_{min}}\right| \times 100\% \quad (2)$$

Xmax and Xmin are the calculations of the distribution of radiation dose measurements (mGy) and the milliampere second (mAs) setting values on the largest and smallest conventional X-ray plane [8].

The reproducibility distortion value is the coefficient of variety (CV) of the tube voltage (kV), and the radiation output (mGy) is obtained by the equation:

$$CV = \frac{SD}{Z_{av}} \times 100\% \quad (3)$$

Where SD is the standard deviation estimate of a series of voltage measurements (kVp), time (s) and dose (mGy), Zav is the mean value of the measured parameters (kVp), time (s) and dose (mGy) [8].

The value of the quality of the radiation beam (HVL) in a conventional X-ray plane can be obtained in two ways: direct and indirect measurement [10]. This measurement uses direct measurement as an estimate of HVL values in millimeters (mm).

3. Study Case

The study case in this study was to analyze the value of deviation of conventional X-ray test results conducted on 24 hospitals located in South Sulawesi Province from 2014 to 2016 with the number of conventional X-ray plane tested by 27 units of equipment.

The test parameters use four parameters: kV accuracy, radiation output linearity, reproducibility of radiation output and HVL value [10, 11], each of which will be tested for deviation value and compared to the allowable tolerance limit ie kV accuracy and radiation output linearity equal to 10%, reproducibility of radiation output of 5% and HVL value must be greater than or equal to 2.1 mmAl [4].

4. Results and Discussion

The study used four parameters namely tube voltage accuracy (kV), radiation output linearity, reproducibility and radiation beam quality value (HVL) to be evaluated using a quantitative method approach where the results will be described below:

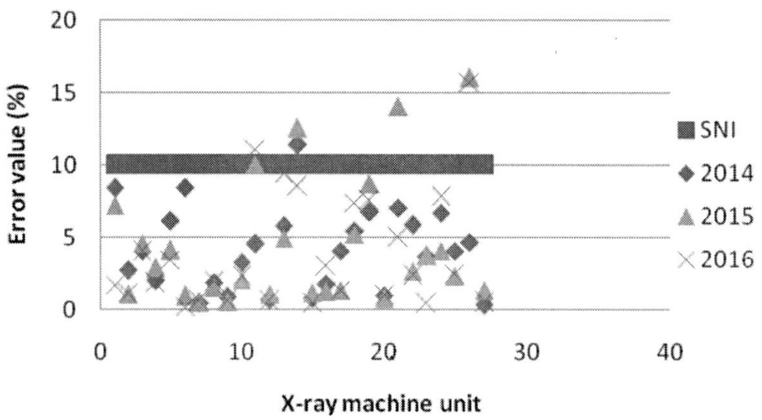

Figure 1. Data distribution of kV accuracy deviation value.

From figure 1 it can be seen that kV accuracy deviation values from 2014 to 2016 are still within the permissible tolerable range of 10%. The red line indicates the allowable tolerance threshold. The results of the highest deviation occurred in 2015 is 11% with the highest deviation value is at 16.06% while the lowest occurred in 2016 with the value deviation 7% with the highest deviation value is at the value of 15.76%. For the year 2014 the value of deviation is at 4% with deviation value of reading result of highest tool is at value 11,4%. Thaha's research results kilo-voltage accuracy percentage was ranged from 1.5 to 3.5 % and time accuracy percentage was ranged from 0.5 to 4.1 % respectively. This study concluded that as the kilo-voltage increases byone, the dose increases by 28%[1].

From figure 1 it can be seen that overall from 2014 to 2016 the conventional x-ray device under test is still within the permitted tolerance limit of 10% from Indonesia National Standard (SNI), where in 2014 the distribution of 96% of the tools is eligible for tolerance, by 2015 93% and 2016 as many as 89% of the categorized tools are eligible.

Table 1. Deviation Values of kV accuracy from 2014 to 2016.

Year	Accuracy kV		
	Mean (%)	SD	SNI (%)
2014	4.20	2.86	10
2015	4.35	4.36	10
2016	3.81	3.97	10

For analysis purposes, the participation factor used to determine the bus and generator weakly damped and most affected by instability. From the analysis found that the generator 2, 6, 7, 12 and 13 represents the weakest bus with eigenvalue levels approaching critical value. There are 111 state and 31 pairs resulting complex to be initialized and observed.

From table 1 kV accuracy it can be seen that the highest kV accuracy deviation value is in 2015 of 4.35% and the lowest deviation value of kV accuracy is in 2016 with 3.81%.

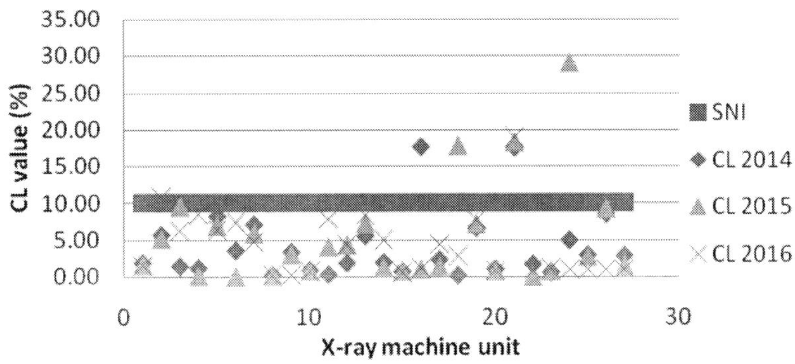

Figure 2. Data distribution of irradiance value of radiation output irregularities.

From the figure above it can be seen that the irregularity value of the radiation output from 2014 to 2016 is largely within the allowable tolerance limit of 10%. The red line indicates the allowable tolerance threshold. The results of the highest deviation occurred in 2015 is 11% with the highest deviation value is at 28.99% while the lowest occurred in 2016 with a value of deviation of 4% with the highest deviation value is at the value of 18.95%. For the year 2014 value deviation is at 7% with deviation value of the highest reading result of the instrument is at value 17.58%.

From the figures it can be seen that overall from 2014 to 2016 the conventional x-ray device under test is still within the permitted tolerance limit of 10% from Indonesia National Standard (SNI), where in 2014 the percentage of 93% of equipment is eligible for tolerance, by 2015 by 89% and in 2016 96% of the tools categorized as eligible are permitted.

Table 2. The value of linearity irregularities from 2014 to 2016

Year	Radiation Output Linearity		
	Mean (%)	SD	SNI (%)
2014	4.04	4.61	10
2015	5.18	6.83	10
2016	4.19	4.34	10

IOP Conf. Series: Journal of Physics: Conf. Series **979** (2018) 012081 doi:10.1088/1742-6596/979/1/012081

From table 2 Linearity of Radiation Output can be seen that the average deviation value of Radiation Output Linearity is highest in 2015 of 5.18% and the lowest average deviation value Linearity Radiation output is in 2016 with value of 4.47%.

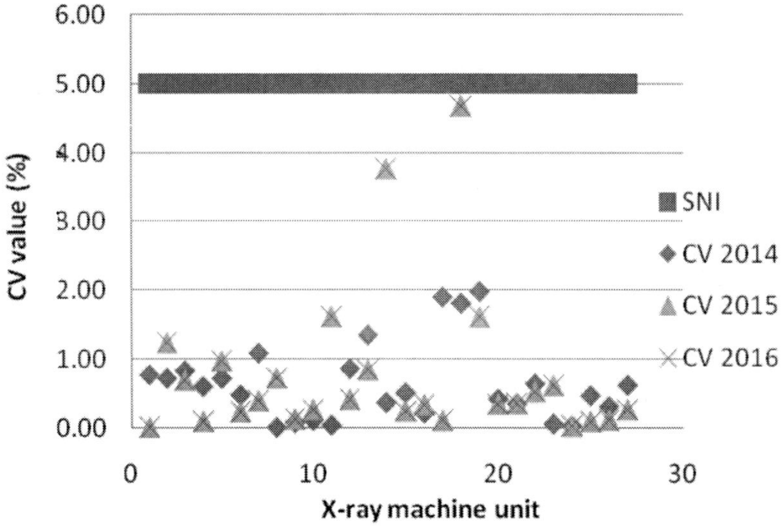

Figure 3. Distribution of the value data radiation output (mGy).

From figure 3 shows the reproducibility value of the radiation output tested from 2014 to 2016 is within the tolerable range of 5% from Indonesia National Standard (SNI).

Table 3. Deviation value Reproducibility of radiation output (mGy) from 2014 to 2016

Year	Reproducibility (mGy)		
	Mean (%)	SD	SNI (%)
2014	0.64	0.56	5
2015	0.76	1.10	5
2016	0.45	0.68	5

Table 3 shows the highest reproducibility rate of radiation output in 2015 of 0.76%, while the lowest was in 2016 and 2014 at 0.45% and 0.64%, respectively.

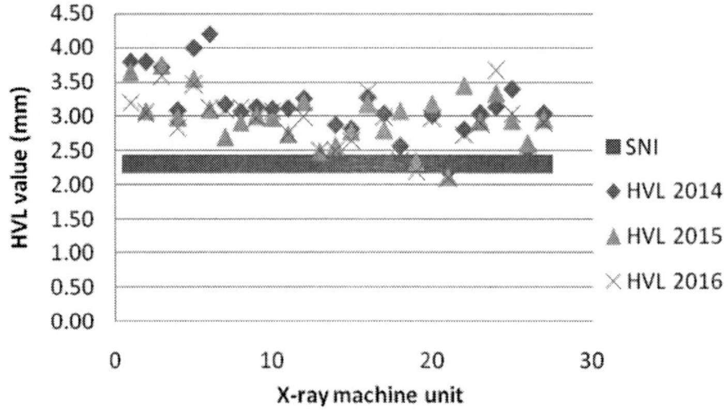

Figure 4. Data distribution of deviation value of HVL radiation. beam quality value.

Figure 4 shows the results of radiation beam quality (HVL) quality measurement from 2014 to 2016. From the chart it can be seen that HVL values that are below the allowable tolerance limit of 2.1 mm from Indonesia National Standard (SNI) are in 2016 and 2014. The deviation rate in 2016 is 7% and 2014 by 4%.

From figure 4 we can also see that HVL values are largely within the allowable tolerance range of above or equal to 2.1 mm from 2014 to 2016. In 2014 96%, 2015 by 100% and by 2016 by 93% of HVL values entering at the allowed tolerance limit.

Table 4. Deviation value of HVL from 2014 to 2016.

Year	Radiation Beam Quality (HVL)		
	Mean (%)	SD	SNI (%)
2014	3.11	0.49	5
2015	2.98	0.38	5
2016	2.91	0.40	5

Table 4 shows the quality of the radiation beam (HVL) from 2014 to 2016 which in 2014 showed the highest average HVL value of 3.11 HVL at 2.91%.

5. Conclusions

From the results of data analysis deviation values of four test parameters ie kV accuracy, radiation output linearity, reproducibility of radiation output and HVL values largely meet the allowable tolerance limits. By 2015 the largest deviation value for kV accuracy, linearity and reproducibility is given, for the HVL value which is out of the maximum allowable tolerance limit in 2014. While for the average deviation, the value for kV accuracy is 4.12%, linearity of 4.47%, reproducibility of 0.62% and HVL of 3.00 mm. For further research, it can use more complete data and the latest method by using computation process or Artificial Intelligence such as Firefly Algorithm (FA) and others.

References

[1] Taha M T 2011 Study the Quality Assurance of Conventional X-ray Machines Using Non Invasive KV meter *Int. J. Sci. Res.* **4** 27–30

[2] Al-Kinani and Yasin M 2013 Study the quality uality assurance of conventional onventional x-ray machines ines at medical city in Baghdad *Merit Res. Journals* **1** 194–200

[3] H. A. Ismail, O. A. Ali, M. A. Omer, M. E. Garelnabi, and N. S. Mustafa 2013 Evaluation of Diagnostic Radiology Department in Term of Quality Control (QC) of X-Ray Units at Khartoum State Hospitals *Int. J. Sci. Res.* **4** 1875–1878

[4] Perka BAPETEN 2011 *Peraturan Kepala Badan Pengawas Tenaga Nuklir Nomor 9 Tahun 2011: Uji Kesesuaian Pesawat Sinar-X Radiologi Diagnostik dan Intervensional* (Indonesia) pp 2–3

[5] E. Babikir, H. A. Hasan, A. Abdelrazig, M. A. Alkhorayef, E. Manssor, and A. Sulieman 2015 Radiation dose levels for conventional chestand abdominal X-ray procedures in elected hospitals in Sudan *Radiat. Prot. Dosimetry* **165** 102–106

[6] A. M. Costa, R. S. Badin, M. S. Leite, and L. V. E. Caldas 2008 Constancy check of beam quality in conventional diagnostic X-ray equipment *Appl. Radiat. Isot.* **66** 1403–1405

[7] P. M. C. de Oliveira, P. do Carmo Santana, M. A. de Sousa Lacerda, and T. A. da Silva 2016 Radiation levels and image quality in patients undergoing chest X-ray examinations *Radiat. Phys. Chem.* 0–1

[8] C. T. Equipment 2006 *Radiation Safety Act 1975 Diagnostic X-Ray Equipment Compliance Testing* (Australia)

IOP Conf. Series: Journal of Physics: Conf. Series **979** (2018) 012081 doi:10.1088/1742-6596/979/1/012081

[9] IAEA 2014 *Dosimetry in Diagnostic Radiology: an Internacional Code of Practice* (Austria) pp 1–5
[10] Sulieman A, et al. 2017 Assessment of medical radiation exposure to patients and ambient doses in several diagnostic radiology departments *Radiat. Phys. Chem.* 1–5
[11] Gholamhosseinian-Najjar H, Bahreyni-Toosi M T, Zare M H, Sadeghi H R and Sadoughi H R 2014 Quality control status of radiology centers of hospitals associated with mashhad university of medical sciences *Iran. J. Med. Phys.* **10–11** 182–187

IOP Conf. Series: Journal of Physics: Conf. Series **979** (2018) 012082 doi:10.1088/1742-6596/979/1/012082

Thermodynamic and Mechanic Consideration on the stability of Anti-symmetric Schaefer's equation

D.A. Suriamihardja, Amiruddin and Saaduddin

Study Program of Geophysics, Faculty of Mathematics and Natural Sciences, Hasanuddin University, Jl. P. Kemerdekaan Km. 10 Tamalanrea, Makassar, South Sulawesi, Indonesia 90245

E-mail: dahmaduh@gmail.com

Abstract. Schaefer's equation relates an interaction between population of fishes and the number of units of fishing effort. The population growth of fishes is reduced by the number of units of fishing effort, while the population growth of units of fishing effort depends on the existence of fishes. This paper aims to examine the stability of an anti-symmetric Schaefer's equation through thermodynamic and mechanic procedure using a formula of entropy production near equilibrium which is recognized as Onsager's relation. The results confirm that entropic approach (thermodynamics) and dissipative approach (mechanics) are usable to be applied as Lyapunov's procedure in examining the stability of systems of differential equations.

1. Introduction

Schaefer's equation. Originally interaction between a school's population of fish (N_1) and the number of units of fishing effort (N_2) is described by [1] as follows:

$$\left.\begin{aligned} \frac{dN_1}{dt} &= kN_1(L-N_1) - k_1N_1N_2 \\ \frac{dN_2}{dt} &= k_2N_2(N_1-b) \end{aligned}\right\} \qquad (1)$$

Where L stands for an environmental carrying capacity to support the population growth of fish. Multiplying k and L form a coefficient $(r = kL)$ of population growth of fish. In the second term of upper equation (1), $k_1N_1N_2$ represents the rate of catching and consequently reducing the fish population. In the first term lower equation (1), $k_2N_1N_2$ represents the rate of population growth of fishery. The value of b should be far less than the value of L in order to recover sooner from disturbances by fishing efforts to maintain system's sustainability of fishes and fisheries. Schaefer's equation in the form of anti-symmetric appearance required the condition that:

$$q = k_1 = k_2 \qquad (2)$$

By dividing the upper by the lower equation in (1) using condition (2), we have:

$$\frac{dN_1}{dN_2} = \frac{kN_1(L-N_1)-qN_1N_2}{qN_2(N_1-b)} = \frac{N_1[k(L-N_1)-qN_2]}{qN_2(N_1-b)} \qquad (3)$$

Observing the value of (dN_1/dN_2) through equation (3), Schaefer [1] stated that:

$$\left.\begin{array}{l} N_1 > b \text{ and } qN_2 > k(L-N_1) \text{ then } (dN_1/dN_2) < 0 \\ N_1 > b \text{ and } qN_2 < k(L-N_1) \text{ then } (dN_1/dN_2) > 0 \\ N_1 < b \text{ and } qN_2 > k(L-N_1) \text{ then } (dN_1/dN_2) > 0 \\ N_1 < b \text{ and } qN_2 < k(L-N_1) \text{ then } (dN_1/dN_2) < 0 \end{array}\right\} \quad (4)$$

Schaefer found that N_1 and N_2 appear as functions of time in some damped oscillatory functions, where the former fluctuates in the vicinity of $N_1{}^* = b$, and the latter fluctuates in the vicinity of $N_2{}^* = (k/q)(L-b)$. Both curves are approaching the crossing point between $N_1{}^* = Q_1 = b$ and $N_2{}^* = Q_2 = (k/q)(L-b)$ as a point of stable equilibrium.

Objective and method. The equation (1) has been confirmed mathematically by Schaefer that its behaviour tends toward a stable equilibrium point. This paper tries to search its stability through thermodynamic and mechanic consideration which is the main question to be discussed. The aim of this study is to examine the applicability of thermodynamic consideration on the stability of an anti-symmetric Schaefer's equation through examining the production of entropy which was applied by [2] based on the work of [3], and the examination will be compromised in transformed formulation to Lyapunov function.

2. Thermodynamic open systems

Bertalanffy [4] that a living system should be treated as open thermodynamic systems. The upper equation in (1) is a general transport equation in which the first term has a role to import materials rich in exergy that can avoid the increase of entropy which cannot be averted in closed systems. It gives that the population of N_1 will increase with time due to this factor, where the coefficient $(r = kL)$ is its birth coefficient and has a role to import materials rich in exergy and to receive negative entropy from the environment. This importation of a negative entropy is not unlimited, but there is limitation of it by the value of L which stands for the carrying capacity of the environment. While the second term of (1) is reducing part to N_1 as units of fishing efforts, but acts as its growing factor to the population of fishery, N_2. Both populations maintain themselves in steady state exchange between the system and the environment, and are approaching a point of stable equilibrium.

Prigogine [5] ensured that there is a thermodynamic function S called entropy which increases monotonically until it reaches its maximum at the state of thermodynamic equilibrium, in which its formulation can be expressed as:

$$\frac{dS}{dt} \geq 0 \quad (5)$$

Extending this thought into open system that required increasing entropy, first we think that internal system representing irreversible interaction system between N_1 and N_2 should has monotonically increasing entropy as:

$$\frac{dS_{int}}{dt} \geq 0 \quad (6)$$

This increasing entropy should be counterbalanced by negative entropy which is imported from the environment with the amount of:

$$\frac{dS_{ext}}{dt} < 0 \quad (7)$$

The sum of these two entropy-productions will be:

$$\frac{dS_{ext}}{dt} + \frac{dS_{int}}{dt} = \frac{dS}{dt} \quad (8)$$

Equation (8) could be negative, positive or zero depending on continual existence, continual disappearance or steady state of the system being considered.

In this relation, Wolfgang *et.al* [6] explained equation (8) further in a process of metabolism. According to Yourgrau, during the period of growth, dS_{ext}/dt exceeds dS_{int}/dt in absolute value, so that dS/dt becomes negative. When the steady state is established, $dS/dt = 0$, a positive dS_{int}/dt is exactly counterbalanced by a negative value of dS_{ext}/dt. Based on these two processes, we able to think that when a positive value of dS_{int}/dt is greater than a negative value of dS_{ext}/dt in absolute values, it makes dS/dt becomes positive and means that the carrying capacity L of the environment has been exceeded, and the system goes into the period of decay into disappearance.

3. Production of entropy.

Chakrabarti [2] divided the rate of entropy-production dS_{int}/dt into two parts: the first is the rate of entropy production per individual species-*j*; and the second $\tilde{P}$ is the rate of ecological entropy-production which is connected with the change of population densities and is the evolutionary part, playing crucial role in the study of stability and evolution of the ecosystem.

$$P = \frac{dS_{int}}{dt} = \sum_j N_j\, P_j^0 + \frac{d\tilde{S}_{int}}{dt} = \sum_j N_j\, P_j^0 + \tilde{P} \tag{9}$$

Based on Prigogine and Onsager as pioneer thinkers, Chakrabarti [2] proposed an expression for entropy-production as macroscopic thermodynamics of irreversible processes in local equilibrium. The expression derived from Schaefer's anti-symmetric equation as follows:

$$\tilde{P} = \frac{d\tilde{S}_{int}}{dt} = \sum_j J_j \chi_j < 0 \tag{10}$$

In equation (10), J_j is irreversible thermodynamic fluxes, and χ_j is irreversible thermodynamic forces and those terms are related into the form of Onsager's phenomenological expression:

$$J_i = \sum_j \ell_{ij} \chi_j \tag{11}$$

Wolfgang *et.al* [6] mentioned that equation (11) may be referred as *thermodynamic equation of motion*, but it is more customary to designate them as *phenomenological equations* of thermodynamics. Correspondingly, the constants ℓ_{ij} are known as *phenomenological coefficients*. Prigogine ensured that the formula of (11) can only be established in some neighbourhood of equilibrium. This neighbourhood defines the region of local equilibrium.

Chakrabarti [2] interpreted local equilibrium based on Le Chatelier-Braun principle. One of many various statement of the principle in the list explained by [7] encountered in a statement as follows:

'Whenever stress is placed on any system in a state of equilibrium, the system will always react in a direction which will tend to counteract the applied stress'

Considered a transition from the stationary state at $N_1{}^* = Q_1$ and $N_2{}^* = Q_2$ to a state N_1 and N_2 which leads to an assumption that the deviation $\delta N = N_1 - Q_1$ is to be small [2]. This assumption makes the entropy-production reduces to the form:

$$\delta^2 \tilde{S}_{int} = \sum \frac{(\delta N_i)^2}{Q_i} \tag{12}$$

The value of (12) is positive definite and can be applied to study local stability of the stationary state. So the criteria of stability of the stationary state (Q_1, Q_2) are:

$$\sum \frac{(\delta N_i)^2}{Q_i} > 0 \tag{13a}$$

$$\frac{d}{dt}\left(\sum \frac{(\delta N_i)^2}{Q_i}\right) < 0 \tag{13b}$$

The second condition is necessary in order that the deviation from the stationary state cannot grow in time so that the system will remain within the domain of local equilibrium around (Q_1, Q_2) which clearly apparent if a phase plane.

4. Transformed formulation of anti-symmetric Schaefer's equation

Considering minor modification to equation (1) stated above, and by using (2) we will have the anti-symmetric Schaefer's equation in the form as written below:

$$\left.\begin{aligned} \frac{Q_1}{N_1}\frac{d}{dt}\left(\frac{N_1}{Q_1}\right) &= kL - kQ_1\left(\frac{N_1}{Q_1}\right) - qQ_2\left(\frac{N_2}{Q_2}\right) \\ \frac{Q_2}{N_2}\frac{d}{dt}\left(\frac{N_2}{Q_2}\right) &= qQ_1\left(\frac{N_1}{Q_1}\right) - qQ_1 \end{aligned}\right\} \tag{14}$$

Each variables in equation (14) will be transformed using the following relations:

$$\left.\begin{aligned} (N_1/Q_1) = e^{x_1};\ (N_2/Q_2) = e^{x_2} \\ x_1 = \ln\left(\frac{N_1}{Q_1}\right);\ x_2 = \ln\left(\frac{N_2}{Q_2}\right) \end{aligned}\right\} \tag{15}$$

Using (15), equation (14) can be written as:

$$\left.\begin{aligned} \dot{x}_1 &= -kQ_1(e^{x_1} - 1) - qQ_2(e^{x_2} - 1) \\ \dot{x}_2 &= qQ_1(e^{x_1} - 1) \end{aligned}\right\} \tag{16}$$

In the case of no limitation for the value of carrying capacity or if the value of $L \to \infty$, the equation (16) will take the form as the famous Lotka-Voltera equation:

$$\left.\begin{aligned} \dot{x}_1 &= -qQ_2(e^{x_2} - 1) \\ \dot{x}_2 &= qQ_1(e^{x_1} - 1) \end{aligned}\right\} \tag{17}$$

By dividing the first by the second of equation (17), we have:

$$\frac{dx_1}{dx_2} = \frac{-qQ_2(e^{x_2}-1)}{qQ_1(e^{x_1}-1)} \quad \text{or} \quad qQ_1(e^{x_1} - 1)dx_1 + qQ_2(e^{x_2} - 1)dx_2 = 0 \tag{18}$$

Integrating equation (18) in the transformed variable: from 0 to x_i or in the primitive variable from Q_i to N_i, then we will have a constant of integration, later it will be called as total energy E for the equation (17) as a constant of integration:

$$E = Q_1 \int_0^{x_1}(e^{x_1} - 1)dx_1 + Q_2 \int_0^{x_2}(e^{x_2} - 1)dx_2$$

$$E = Q_1(e^{x_1} - x_1 - 1) + Q_2(e^{x_2} - x_2 - 1) \tag{19}$$

The Lotka-Voltera equation (17) will give curves of oscillatory function without damping with the total energy E (19) which is time-independent.

Equation (16) can be rewritten using (11), where thermodynamical forces are written as:

$$\left.\begin{aligned} \dot{x}_1 &= -k\chi_1 - q\chi_2 \\ \dot{x}_2 &= q\chi_1 \end{aligned}\right\} \quad \text{or} \quad \left.\begin{aligned} \dot{x}_1 &= \ell_{11}\chi_1 + \ell_{21}\chi_2 \\ \dot{x}_2 &= \ell_{12}\chi_1 + \ell_{22}\chi_2 \end{aligned}\right\} \tag{20}$$

Four coefficients in (20) $\ell_{11} = -k$; $\ell_{21} = -q$; $\ell_{12} = q$; and $\ell_{22} = 0$ are called as Onsager's *phenomenological coefficients*, where $\chi_i = Q_i(e^{x_i} - 1)$ is used in formulating (20).

Entropy's production will be calculated through Onsager's formulation as a product between thermodynamic fluxes and thermodynamic forces. For the non-equilibrium thermodynamic model of the system, defines the thermodynamic flux (or flow) as the intrinsic rate of population growth as follows[2]:

$$J_i = \dot{x}_i \; ; \; i = 1{,}2 \tag{21}$$

Using the corresponding thermodynamical forces which have been defined previously in (20), the rate of entropy's production as formulated previously in (10) will be:

$$\tilde{P} = \chi_1 J_1 + \chi_2 J_2 \tag{22}$$

Using (21) and (22), the rate of entropy production is derived as follows:

$$\tilde{P} = \chi_1[-k\chi_1 - q\chi_2] + \chi_2[q\chi_1] \tag{23a}$$

$$\tilde{P} = -k{\chi_1}^2 - q\chi_1\chi_2 + q\chi_1\chi_2 \tag{23b}$$

Finally, we have that the produced entropy is decreasing with time:

$$\tilde{P} = -k{Q_1}^2(e^{x_1} - 1)^2 \leq 0 \tag{24a}$$

Equation (24) informs us that the ecological entropy-production is decreasing towards the value of $x_1 \approx 0$ or in the primitive variable towards the value of $N_1 \approx b$. According to the limit of growth, population of N_1 could be extended actually towards $N_1 \approx L$, where $L > b$. But beyond the value of b, the disturbed population of fishes by harvesting becomes threatening the sustainability of the fishes and fishery's system. Expressing (24) in the primitive variables will be:

$$\tilde{P} = -k(N_1 - Q_1)^2 = -k(N_1 - b)^2 \leq 0 \tag{24b}$$

Again, equation (24) confirms that the anti-symmetric Schaefer's equation is thermodynamically stable which confirms the Le Chatelier-Braun principle. The value of ecological entropy production, $\tilde{P}$, tends to be zero at equilibrium point $N_1 \approx b$ as depicted in figure 1.

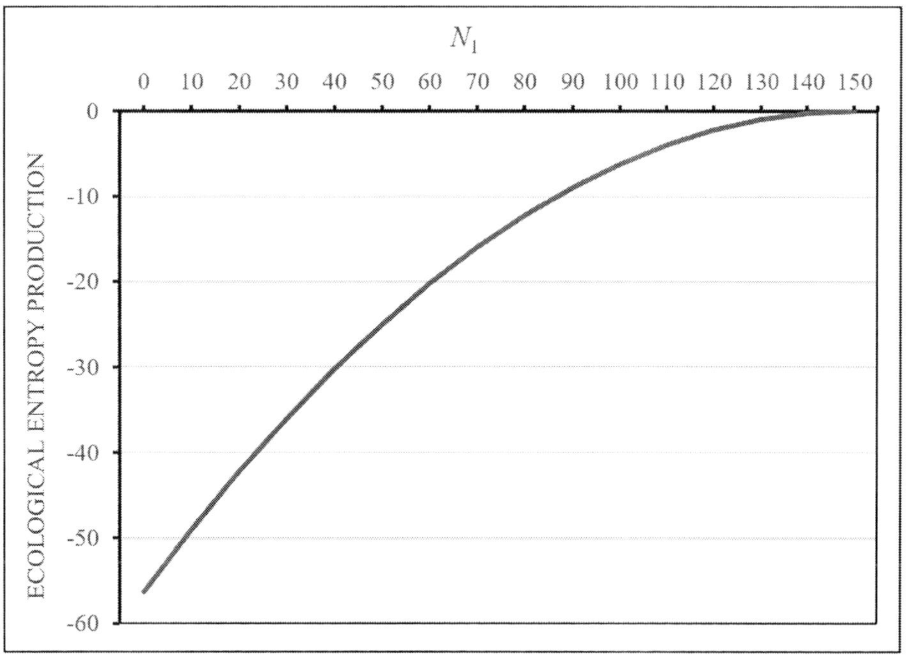

Figure 1. The curve of ecological entropy production tends to be zero at $b = 150$.

5. Lyapunov's procedure as mechanical energy approach

Equation (20) is manipulated to be linearized equations by using approximation of $(e^x - 1) \approx x$, we have:

$$\left.\begin{aligned} \dot{x}_1 &= -kQ_1x_1 - qQ_2x_2 \\ \dot{x}_2 &= qQ_1x_1 \end{aligned}\right\} \tag{25}$$

Differentiate (25) to time, it becomes:

$$\left.\begin{aligned} \ddot{x}_1 + kQ_1\dot{x}_1 + q^2Q_1Q_2x_1 = 0 \\ \ddot{x}_2 + kQ_1\dot{x}_2 + q^2Q_1Q_2x_2 = 0 \end{aligned}\right\} \tag{26}$$

Each component x_1 and x_2 will behave as oscillatory function of time. The second terms in equation (26) contribute to be a damped term to the system of oscillation. Its characteristic equation will give the roots for the solution to (26), that is:

$$\lambda_{1,2} = -\frac{1}{2}kQ_1 \pm \frac{1}{2}kQ_1\sqrt{1 - \frac{4q^2}{k^2}\frac{Q_2}{Q_1}};\ Q_1 = b;\ Q_2 = (k/q)(L - b) \tag{27a}$$

$$\lambda_{1,2} = -\frac{1}{2}kb \pm \frac{1}{2}kb\sqrt{1 - \frac{4q}{k}\left(\frac{L}{b} - 1\right)} \tag{27b}$$

To maintain both equations in (26) are expressing fluctuating behaviour and decreasing by factor of $\mathrm{Exp}[-kbt/2]$ with time, the value under the root should be negative, hence it specifies the value of b as follows:

$$1 - \frac{4q}{k}\left(\frac{L}{b} - 1\right) < 0 \quad \text{or} \quad \left(1 + \frac{k}{4q}\right)b < L \tag{27c}$$

Damped oscillatory function for primitive variables N_1 and N_2 are plotted in a phase space as depicted in figure 2.

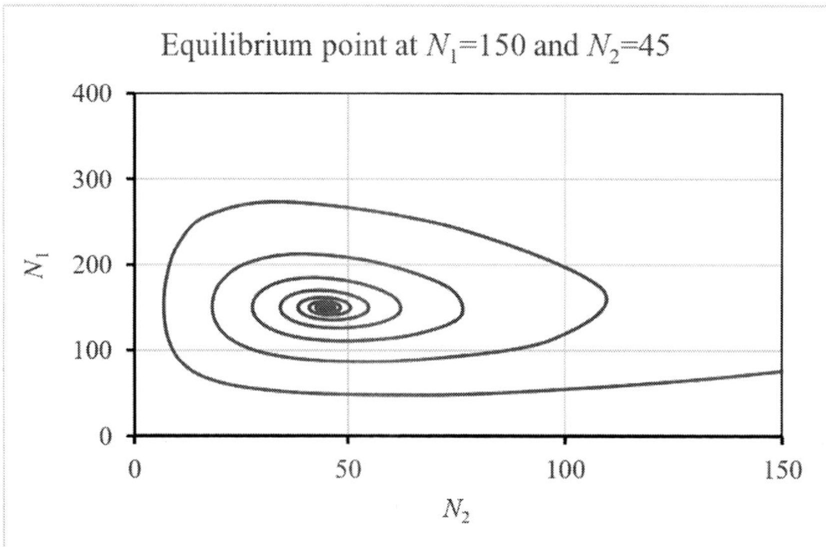

Figure 2. Numerical values of $L = 600$; $b = 150$; $q = 0.025$; and $k = 0.0025$ are used to plot a phase plane of N_1 and N_2. The fluctuation tends to the equilibrium point at $N_1 = 150$ and $N_2 = \frac{k}{q}(L - b) = 45$.

Following [8] and [9] the Lagrangians for equation (26) can be formulated based on the characteristics equations of (27a, b and c) as follows:

$$\mathfrak{L} = e^{kbt}\left\{\frac{1}{2}(\dot{x}_1^2 + \dot{x}_2^2) - \frac{1}{2}\omega^2(x_1^2 + x_2^2)\right\} ;\ \omega^2 = q^2 Q_1 Q_2 = kq\left(\frac{L}{b} - 1\right)$$

$$\mathfrak{L} = e^{kbt}\mathbb{L} ;\ \mathbb{L} = \left\{\frac{1}{2}(\dot{x}_1^2 + \dot{x}_2^2) - \frac{1}{2}\omega^2(x_1^2 + x_2^2)\right\} \quad (28)$$

The related mechanical energy E or Hamiltonian $\mathbb{H}$ to Lagrangian $\mathbb{L}$ in (28) is as follows:

$$\mathbb{H} = \left\{\frac{1}{2}(\dot{x}_1^2 + \dot{x}_2^2) + \frac{1}{2}\omega^2(x_1^2 + x_2^2)\right\} \quad (29)$$

Euler-Lagrange equations to produce (26) are as follows:

$$\frac{d}{dt}\left(\frac{\partial \mathfrak{L}}{\partial \dot{x}_i}\right) - \frac{\partial \mathfrak{L}}{\partial x_i} = 0 \quad \text{or} \quad \frac{d}{dt}\left(\frac{\partial \mathbb{L}}{\partial \dot{x}_i}\right) - \frac{\partial \mathbb{L}}{\partial x_i} = -\frac{\partial \mathbb{F}}{\partial \dot{x}_i} \quad (30)$$

Physical term $\mathbb{F}$ stands for friction as dissipative terms to the system under consideration which is formulated by

$$\mathbb{F} = \frac{1}{2}kQ_1(\dot{x}_1^2 + \dot{x}_2^2) \quad (31)$$

Lyapunov's procedure uses a formulation of dissipative energy function given by [10] using (29), (30) and (31) as follows:

$$\frac{dE}{dt} = -2\mathbb{F} = -kQ_1(\dot{x}_1^2 + \dot{x}_2^2) < 0 \quad (32)$$

Since kQ_1 is a positive constant, $\dot{E}$ is a negative quantity, so that E is decreasing function with time. As long as there are values for thermodynamic fluxes, E can never become negative, but attains its stable rest point at disappearance of $J_1 = \dot{x}_1$ and $J_2 = \dot{x}_2$ or at the crossing between the line of $N_1 = b$ and $N_2 = (k/q)(L - b)$. According to equation (31), when $L \approx b$ there will be no growth for the fishery population (N_2), but for the fishes (N_1) rich their maximum population without catching by fishing efforts. Equation (32) confirm that the anti-symmetric Schaefer's equation is also mechanically stable and supports biological stability.

6. Conclusions

The main question is concerning the applicability of thermodynamic and mechanic consideration on the stability of an anti-symmetric Schaefer's equation .The results are as follows:

a) To keep profitable investment in fishing needs to monitor and to control the population of fishes. It means that N_1 should not exceed the value of b which should be less than the value of limited resources as environmental carrying capacity, L, and the population of N_2 should not exceed the value of $(k/q)(L - b)$. The crossing point $(N_1 = b,\ N_2 = (k/q)(L - b))$ is a position having thermodynamical stability and supports economic stability. This approach as has been shown by equation (24) which is also can be used as Lyapunov's procedure due to decreasing value with time: $\tilde{P} = \frac{d}{dt}\tilde{S}_{in} = -kQ_1{}^2(e^{x_1} - 1)^2 \le 0$.

b) Decreasing value of total mechanical energy is confirmed by the existence of dissipative function expressed by equation (30) that depends on thermodynamical fluxes $J_1 = \dot{x}_1$ and $J_2 = \dot{x}_2$. Using Lyapunov's procedure, equation (31) confirms the anti-symmetric Schaefer's equation is mechanically stable towards the crossing point $(N_1 = b,\ N_2 = (k/q)(L - b))$ and supports biological stability.

c) Entropic approach (thermodynamics) and dissipative approach (mechanics) are usable to be applied as Lyapunov's procedure in examining the stability of systems of differential equations.

d) Both equations (24b) and (27c) are confirming the usable area of Le Chatelier-Braun principle in interaction between population of fishes and fisheries. The former supports its fluctuating behaviour near equilibrium, and the latter supports damped oscillatory functions toward the point of equilibrium.

References

[1] Schaefer M B 1954 *Inter-American Tropical Tuna Commision* **1**(2) 27-51
[2] Chakrabarti C G and Ghosh K 2009 *Ecological Modelling* **220** 1950-1956
[3] Onsager L 1931 *Physical Review* **37** 405-426
[4] Bertalanffy V L 1950 *Science* **111** 23-29
[5] Prigogine I 1978 *Science* **201** 777-785
[6] Wolfgang Y, Alwyn v d M and Gough R 1966 *Treatise on Irreversible and Statistical Thermophysics: An Introduction to Nonclassical Thermodynamics* (Toronto: General Publishing Company)
[7] Gold J and Gold V 1985 *Education in Chemistry* 22 82-85
[8] Smith C E 2010 *J. Physics: Conference Series* **237** 012021 (IOP Publishing)
[9] Pritula G M, Petrenko E V and Usatenko O V 2017 *Physics Letters A* **382** 548-553
[10] Landau L L D 1960 *Mechanics* vol I of Course of Theoretical Physics. 9 vols. (London: Pergamon Press)

IOP Conf. Series: Journal of Physics: Conf. Series **979** (2018) 012083 doi:10.1088/1742-6596/979/1/012083

Design of Oil Viscosity Sensor Based on Plastic Optical Fiber

Muhammad Yunus, A. Arifin

Department of physics, Hasanuddin University, Makassar, 90245 Indonesia

E-mail: yunusmuhammad0808@gmail.com, arifinpide@gmail.com

Abstract. A research of plastic optical fiber based sensors have been studied for measurement of oil viscosity. This sensor was made with straight configuration, U configuration, and gamma configuration have two types, there are optical fiber sensor with cladding and without cladding. Viscosity sensor was made, dipped into an oil sample with a concentration of viscosity percentage about 270 mPa.s – 350 mPa.s. The light from the LED propagated into the optical fiber, then it was received by the photodetector converted to output power. When plastic optical fiber dipped into an oil sample, viscosity of oil affect increase of refractive index on optical fiber resulting in a bigger loss of power so the light intensity will be smaller, consequences the measured output power will be smaller. Sensitivity and resolution viscosity sensor without cladding peel showed the best result rather than viscosity sensor with cladding peel. The best result in the measurement showed in gamma configuration with 3 cm length of cladding peel and the diameter of bending 0,25 cm is the range 103,090 nWatt, sensitivity 1,289 nWatt/mPa.s, and resolution 0,776 mPa.s. This method is effectively and efficiently used as an oil viscosity sensor with high sensitivity and resolution.

1. Introductions

The rapid development of science and technology in the field of electronics and instrumentation has made possible measurements that work precisely and practically. Optical fiber is developed towards a sensor system that can be used for the measurement of various types of physical variables [1]. Scientists have developed many optical fiber based sensors as versatile sensors with advantages such as not using electrical signals, low noise due to no interference with electromagnetic waves, high sensitivity and measurement accuracy, smaller size and lighter, and can be connected with communication systems data via interface device [2]. Application of optical fiber as a sensor one of them can be used to measure the level of oil viscosity. Optical fiber has many advantages such as speed, resistance to rusting, and relatively affordable price [3].

The testing on plastic optical fiber as a sensor has been made in various applications such as detecting UV radiation [4], temperature measurement [5]. In the application the measurement of the solution concentration level [6], water turbidity [7], in load measurements [8, 9], to the application of viscosity sensors on engine oil [10]. So far, the testing of optical fiber as the oil viscosity sensor has been done by using the method of peeling and bending on the fiber optic sensor. This method is used to measure the refractive index of oil on the effect of temperature on oil viscosity. The disadvantage of this method is to use only one type of oil and one type of sensor model, having a more complex measurement process and a more difficult manufacturing process so that the resulting data is not very varied in the determination of sensor characteristics [11].

Based on the results of previous researchers above, it will be developed about the determination of the characteristics of plastic optical fiber sensor to the variation of the level of oil viscosity. The novelty of this study is using gamma configuration, it is expected to produce better sensitivity and sensor resolution. The advantage of this oil viscosity sensor is to measure the level of oil viscosity more accurately with a simple measurement system, easy fabrication, low cost, and can be connected with other measurement system devices.

2. Experimental SET-UP

In the design study for the oil viscosity sensor uses a circuit of power supply ± 12 volt and 5 volt as a source of electrical voltage to power the light source. The light source used is the IF-E91A infrared LED that transmits light at a wavelength of 950 nm. While the light receiver uses photodetector type S120C with wavelength 400-1100 nm and power range 50 nW-50 mW at Optical Power Meter (OPM). Both ends of optical fiber are connected with LED and photodetector. The optical fiber used is made of polymethyl metacrylate (PMMA) polymethyl metacrylate (PMMA) plated fiberglass optical fiber with coat, cladding, and core diameters of 2,2 mm, 1 mm, and 0,98 mm, respectively. The refractive index of the core and cladding of the plastic optical fiber are respectively n_{core} = 1,492 and $n_{cladding}$ = 1,402 with numerical on slot value NA = 0,5. The oil samples of the Mesran SAE 20W-40 and 20W-50 with various mixtures yielding 10 samples and having viscosity variation from 270 mPa.s to 350 mPa.s. The viscosity of oil is measured using a viscometer.

The schematic circuit of oil viscosity sensor based on plastic optical fiber shown in figure 1. The sensor is made with 3 configurations is the straight, U and gamma configurations shown in figure 2. Each configuration is made by varying the length of peeling and the diameter of bending with cladding and without cladding.

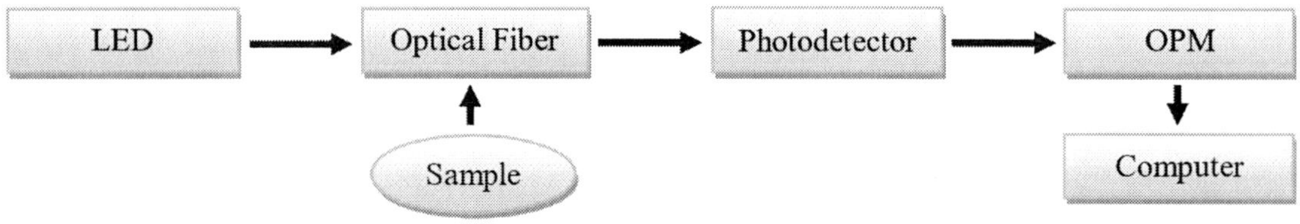

Figure 1. Schematic circuit of oil viscosity sensor based on optical fiber plastic.

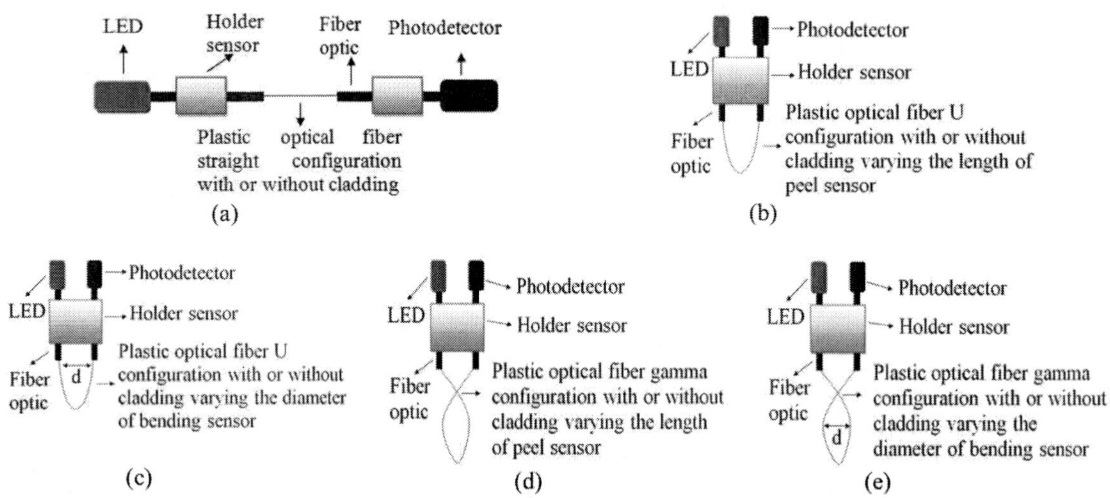

Figure 2. (a) The straight configuration, (b) the U configuration of the length of peeling variation, (c) the U configuration of the diameter of bending variation, (d) the gamma configuration of the length of peeling variation, (e) the gamma configuration of the diameter of bending variation.

The working principle of oil viscosity sensor based on optical fiber based on figure 1 is that LED light is transmitted through plastic optical fiber and received by photodetector, then forwarded to the amplifier and will be read on OPM and computer. Changes of oil viscosity levels causes a change in the refractive index of oil around the sensor. The higher of the oil viscosity causes the higher of refractive index. This will cause the loss of power be increases and the sensor output power be decreases.

3. Result and Discussion

The results obtained in this study, optical fiber are made as oil viscosity sensors in straight, U, and gamma configurations with cladding and without cladding. Each sensor configuration of the length of peel is made at 1 cm, 2 cm, and 3 cm. The sensor is dipped into an oil sample at 10 different viscosity levels. The measured results on OPM are the sensor output power to oil viscosity for straight configuration in the length of peel variation as shown in figure 3 below:

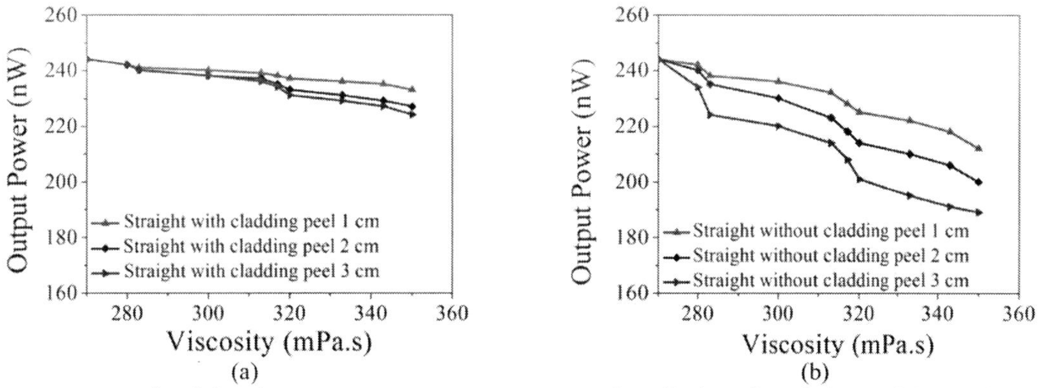

Figure 3. Graph of the changes output power sensor on the oil viscosity of the straight configuration (a) with cladding, (b) without cladding.

Figure 3 shows the response changes the output power of the straight configuration sensor. Increased levels of oil viscosity results in decreased output power. The graph illustrates the characteristics of the oil viscosity sensor based on plastic optical fiber. Characteristics of sensor can be analyzed using the calculation of output voltage range values, sensitivity, and resolution. The value of the output voltage range for the sensor is determined using the following equation [8,12]:

$$\Delta = V_{max} - V_{min} \tag{1}$$

With V_{max} as the output voltage at maximum viscosity and V_{min} as the output voltage at minimum viscosity.

Sensitivity of the sensor is a measure to find out how much sensor sensitivity to the quantity of a measured solution. The sensitivity of the oil viscosity sensor can be measured using the following equation [8,12]:

$$S = \frac{V_{max} - V_{min}}{K_{max} - K_{min}} \tag{2}$$

Next calculate the sensor resolution to know the value of the smallest quantity that can be measured by the viscosity sensor. The sensor resolution of the oil viscosity can be measured using the following equation [8,12]:

$$R = \frac{N}{S} \tag{3}$$

With N as the smallest scale value of the Optical Power Meter used is 0,001 μWatt and S is the sensitivity value of the oil viscosity sensor.

Determination of oil viscosity sensor characteristics for each configuration can be known by using the above equation. Characteristic of oil viscosity sensors for straight configuration can be seen in table 1 below:

Table 1. Characteristics of oil viscosity sensor on straight configuration in the length of peel variation.

Characteristics of Sensor	With Cladding			Without Cladding		
	1 cm	2 cm	3 cm	1 cm	2 cm	3 cm
Output Power Range (nW)	11,000	17,000	20,000	32,000	44,000	55,000
Sensitivity (nW/mPa.s)	0,138	0,212	0,250	0,400	0,550	0,688
Resolution (mPa.s)	7,273	4,706	4,000	2,500	1,818	1,455

Table 1 shows the calculation of range, sensitivity, and resolution of variations in length of the peel sensor with cladding and without cladding. Can be seen that the range, sensitivity, and resolution of the sensor the better with the length of peel be increase. The best results were obtained on the lenght of peel at 3 cm without cladding, sensitivity 0,688 nW/mPa.s and resolution 1,455 mPa.s.

The oil viscosity sensor based on plastic optical fiber on U configuration of the length of peel variation and the diameter of bending using the same sample with oil viscosity sensor in straight configuration. The same way of testing is by dipping the sensor into the sample at different viscosities. The result of the testing of the oil viscosity sensor based on plastic optical fiber for U configuration is shown in figure 4 below:

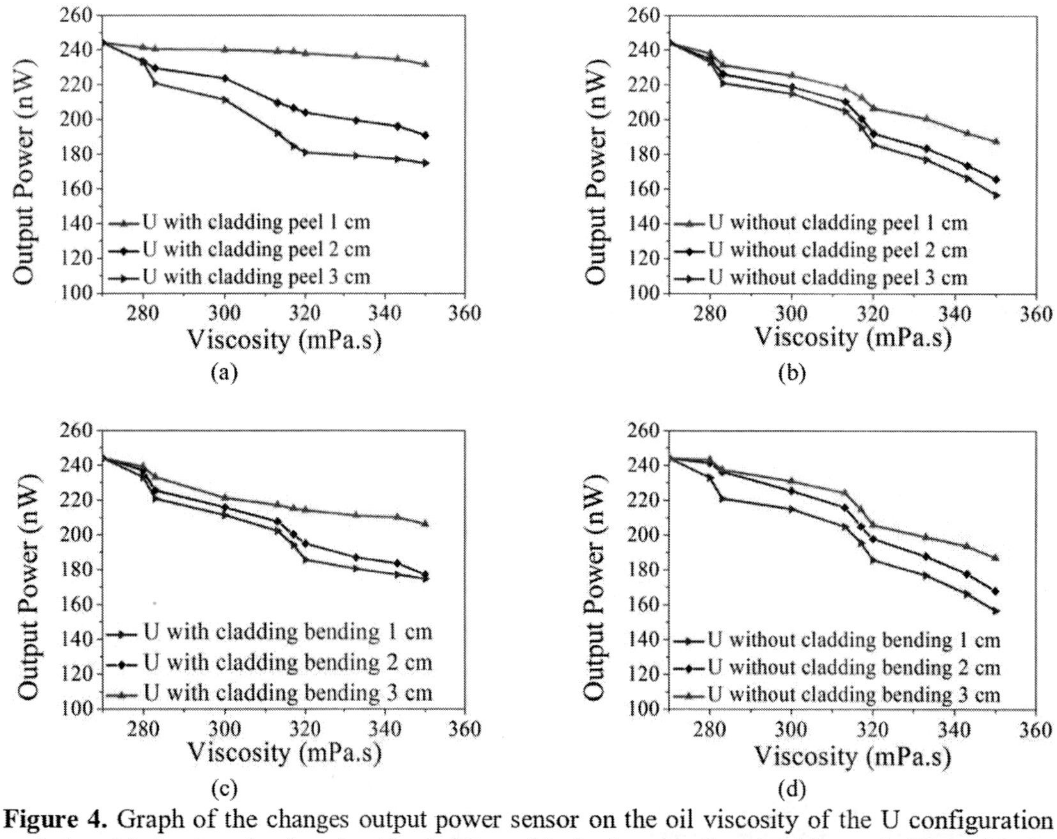

Figure 4. Graph of the changes output power sensor on the oil viscosity of the U configuration variation (a) the length of peel with cladding, (b) the length of peel without cladding, (c) the diameter of bending with cladding, (d) the diameter of bending without cladding.

Figure 4 shows the response changes output power sensor of U configuration. Increasing of the oil viscosity level causes result in decreasing output power. The characteristics of oil viscosity sensor on U configuration can be seen in the following table 2 and table 3:

Table 2. Characteristics of oil viscosity sensor on U configuration for the length of peel variation.

Characteristics of Sensor	With Cladding			Without Cladding		
	1 cm	2 cm	3 cm	1 cm	2 cm	3 cm
Output Power Range (nW)	12,600	53,100	69,400	56,216	78,130	87,214
Sensitivity (nW/mPa.s)	0,158	0,664	0,868	0,703	0,977	1,090
Resolution (mPa.s)	6,349	1,506	1,153	1,423	1,024	0,917

Table 3. Characteristics of oil viscosity sensor on U configuration for the diameter of bending variation.

Characteristics of Sensor	With Cladding			Without Cladding		
	1 cm	2 cm	3 cm	1 cm	2 cm	3 cm
Output power Range (nW)	69,400	66,900	38,000	87,214	75,881	56,820
Sensitivity (nW/mPa.s)	0,868	0,836	0,475	1,090	0,948	0,710
Resolution (mPa.s)	1,153	1,196	2,105	0,917	1,054	1,408

In table 2 shows the values of range, sensitivity, and resolution of sensors affected by the length of peel variation with cladding and without cladding. The longer of the peel sensor has results of the sensor characteristics will be better. The best sensor characteristics were obtained on a 3 cm peel. While table 3 on at 3 cm constant length of peel indicates that the diameter of bending affects the sensor characteristics. The smaller the diameter of bending sensor causes the higher the range and sensitivity of the sensor and the smaller the resolution sensor. The best measurements of the U configuration are shown in 3 cm length of peel and 3 cm in diameter of bending without cladding, with a sensitivity value of 1,090 nW/mPa.s and a resolution of 0,917 mPa.s.

Testing of oil viscosity sensor based on plastic optical fiber on gamma configuration with cladding and without cladding using the length of peel and the diameter of bending variation sensor. The test results of the oil viscosity sensor based on plastic optical fiber in gamma configuration are shown in figure 5 below:

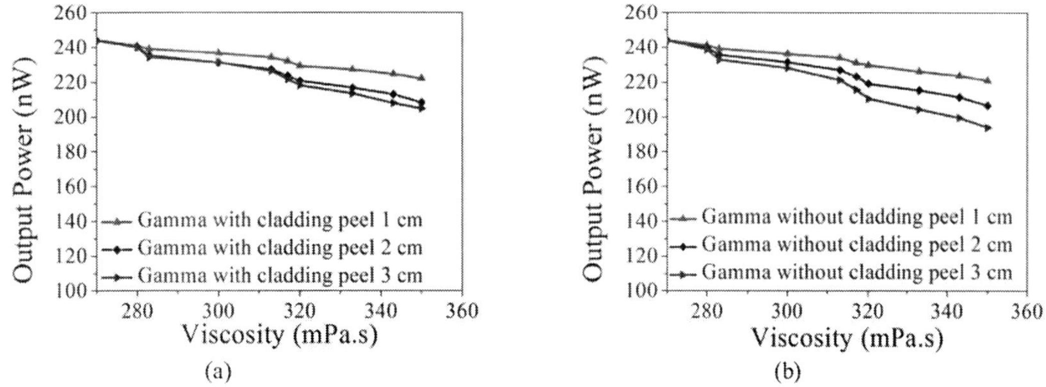

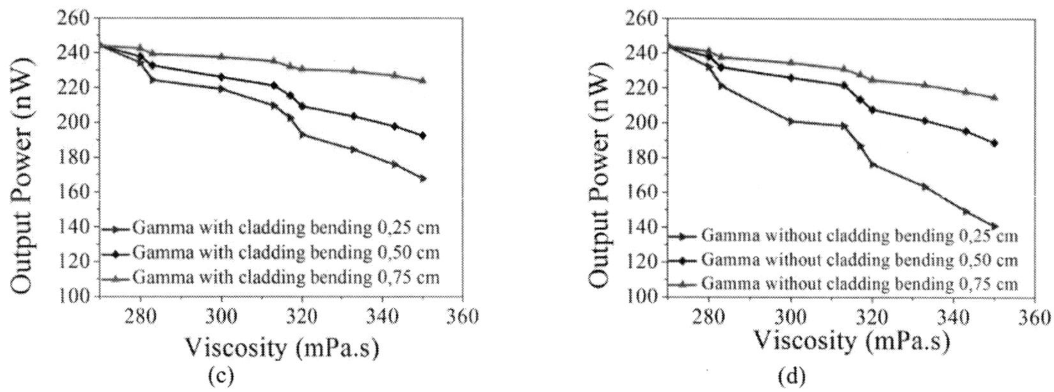

Figure 5. Graph of the changes output power sensor on the oil viscosity on the gamma configuration variation (a) the length of peel with cladding, (b) the length of peel without cladding, (c) the diameter of bending with cladding (d) the diameter of bending without cladding.

In figure 5 for the measurement on gamma configuration in the length of peel and diameter of bending variations with cladding and without cladding indicates that the sensor without cladding has the highest change of output power response compared to the sensor with cladding. Characteristics of the gamma configuration oil viscosity sensors are shown in the following table 4 and table 5:

Table 4. Characteristics of oil viscosity sensor on gamma configuration for the length of peel variation.

Characteristics of Sensor	With Cladding			Without Cladding		
	1 cm	2 cm	3 cm	1 cm	2 cm	3 cm
Output power Range (nW)	22,000	35,700	39,200	23,046	37,406	49,987
Sensitivity (nW/mPa.s)	0,275	0,446	0,490	0,288	0,468	0,625
Resolution (mPa.s)	3,636	2,241	2,041	3,471	2,139	1,600

Table 5. Characteristics of oil viscosity sensor on gamma configuration for the diameter of bending variation.

Characteristics of Sensor	With Cladding			Without Cladding		
	0,25 cm	0,50 cm	0,75 cm	0,25 cm	0,50 cm	0,75 cm
Output power Range (nW)	76,400	51,600	20,400	103,090	55,206	29,201
Sensitivity (nW/mPa.s)	0,955	0,645	0,255	1,289	0,690	0,365
Resolution (mPa.s)	1,047	1,550	3,922	0,776	1,449	2,739

The results of the measurements in table 4 show the range values, sensitivity and resolution of the sensors affected by in the length of peel variation with cladding and without cladding. While in table 5 for variations in the diameter of bending sensor shows that the smaller of the diameter of bending sensor better the characteristics of sensor. The best results were obtained on sensors without cladding at the length of peel 3 cm with the diameter of bending 0,25 cm is sensitivity value of 1,289 nW/mPa.s and resolution of 0,776 mPa.s.

Testing of oil viscosity sensors in straight, U, and gamma configurations obtained response changes values of varying the power output sensor. The best of output power sensor change response of the various configurations is shown in figure 6 below:

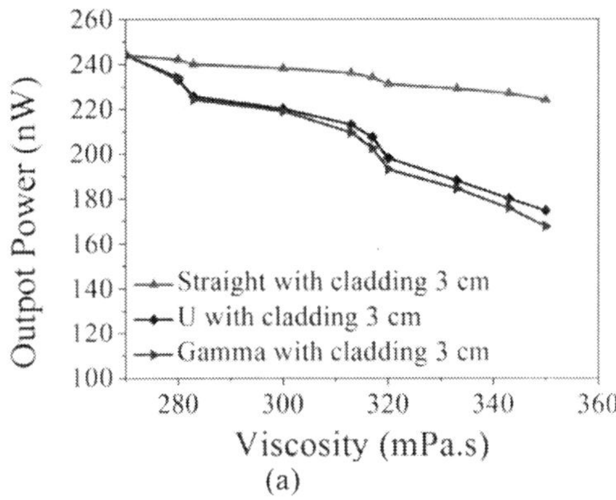

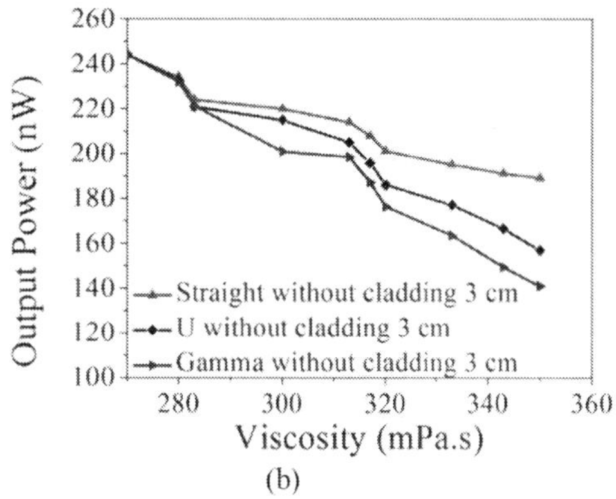

Figure 6. Graph of the changes output power sensor on oil viscosity with various configurations (a) with cladding, (b) without cladding.

In figure 6 shows the result of various configurations sensor at a constant length of peel 3 cm. It can be seen that the sensor without cladding has the highest output power changes response compared to the sensor with cladding. The characteristics of the oil viscosity sensor of various configurations are shown in table 6 below:

Table 6. Characteristics of oil viscosity sensor with various configurations.

Characteristics of Sensor	With Cladding			Without Cladding		
	Straight	U	Gamma	Straight	U	Gamma
Output power Range (nW)	20,000	69,400	76,400	55,000	87,214	103,090
Sensitivity (nW/mPa.s)	0,250	0,868	0,955	0,688	1,090	1,289
Resolution (mPa.s)	4,000	1,153	1,047	1,455	0,917	0,776

From figure 6 and table 6 are a composite result of straight, U, and gamma configuration sensor of the length of peel 3 cm with cladding and without cladding. The graph illustrates the characteristics of the sensitivity sensor. The best characteristic was obtained in the gamma configuration sensor without cladding of length of peel 3 cm and diameter of bending 0,25 cm sensor has results the range 103,090 nWatt, sensitivity 1,289 nWatt/mPa.s and resolution 0,776 mPa.s.

The results of this study has consistent with result of previous studies on mixing the concentration of glycerin solution to water using on U configuration based on plastic optical fiber sensor for measuring the transmission of sensor output power using OPM. Increasing the glycerin solution to the water solution results in a refractive index of the solution increasing so that the sensor output power decreases [6]. Likewise, on the measurement of water turbidity using a straight configuration based on plastic fiber optic sensor. The increased concentration of water turbidity causes the refractive index of the solution to increase so that the sensor output voltage decreases [7]. The use of plastic optical fiber sensors for straight configuration on increasing the concentration of sugar and salt solution to water causes the refractive index of the solution to increase and the lost power value is greater [13]. Neither the suitability is the smaller the diameter of bending sensor causes the sensitivity and resolution of the sensor the better [8]. So in this research has experimented the measurement of sensor output power to the level of oil viscosity based on plastic optical fiber sensor using straight configuration and U configuration with a novelty method using gamma configuration. The results show that the best characteristics of sensor are obtained on gamma configuration on plastic optical fiber sensors compared with straight and U configurations. The advantage of this method is that it has a simpler measurement and easy fabrication.

4. Conclusion

The results showed that the change of sensor output power is inversely proportional to the level of oil viscosity. The higher of oil viscosity levels causes the output power decreases. The longer of peel sensor and the smaller the diameter of bending sensor causes result sensitivity and resolution sensor the better. The result of the measurement of oil viscosity sensor based on plastic optical power shows that the gamma configuration sensor without cladding length of peel 3 cm and diameter of bending 0,25 cm has the best result output power range, sensitivity and resolution is range 103,090 nWatt, sensitivity 1,289 nWatt/mPa.s, and resolution 0,776 mPa.s compared with straight and U configurations.

Acknowledgements
The Research was supported by: BPOPTN-PPKPS, Hasanuddin University-2017.

References

[1] Herwig K 1964 Coupling and Conversion Coefficients for Optical Modes in Quasi-Optics *Microwave Research Institute Symposia Series* **14** 333-347

[2] Fidanboylu K, Efendioglu H S 2009 Fiber Optic Sensors and Their Applications *J. 5th International Advanced Technologies Symposium (IATS)* 1-6

[3] A Arifin, A M Hatta, M S Muntini, A Rubiyanto 2014 Bent Of Plastic Optical Fiber With Structural Imperfection For Displacement Sensor *J. Indian Journal Of Pure & Applied Physics* **52** 520-524

[4] Ana V J, Bajic, Jovan S B, Dragan Z S, Milos P S, Miodrag J, Milos B Z 2012 Simple and Low-Cost Fiber-Optic Sensors for Detection of UV Radiation *J. Telfor* **4** 2

[5] Mustapha R, Francis G, Patrick M 2012 Flexible Optical Waveguide Bent Loss Atenuation Effects Analysis and Modeling Aplication to an Intrinsic Optical Fiber Temperature Sensor *J. Optics and Photonics* **2** 1-7

[6] B G Lumanta, R T Candidato Jr, R L Reserva 2015 Characterization of POF for Liquid Level and Concentration Sensing Applications *IOP Conference Series: Materials Science and Engineering* **79** 012018

[7] Pedro A, Jon A, Joseba Z 2009 Design of a Sensor Based on Plastic Optical Fibre (POF) to Measure Fluid Flow and Turbidity *J. Sensors* **9** 790-3800

[8] A Arifin, Yusran, Miftahuddin, Bualkar A, Dahlang T 2017 Comparison of Sensitivity and Resolution Load Sensor at Various Configuration Polymer Optical Fiber *The 6th International Conference on Theoretical and Applied Physics (The 6th ICTAP)* **1801** 1-6

[9] D Tosi, M Olivero, A Vallan, G Perrone 2010 Weigh In Motion Through Fibre Bragg Grating Optical Sensor *Electronic Letters* **46** 17

[10] A Agoston, C Otsch, B Jakoby 2005 Viscosity sensors for engine oil condition monitoring-Application and interpretation of results *Sensor and actuator A* **121** 327–332

[11] S Bordoloi and S Laskar 2016 A Study on Effect of Temperature on Lubricating Oil Using Bare and Bent Optical Fibre Sensor: A Theoretical and Experimental Approach *Journal of Computer Sciences and Engineering* **4** 7

[12] A Arifin, A M Hatta, Sekartedjo, M S Muntini, A Rubianto 2015 Long-Range Displacement Sensor Based on SMS Fiber Structure and OTDR *Photonic Sensors* **5** 166-171

[13] C Venkatasubbaiah, V Sreeharibabu, A J Kumar, S Chandralingam 2013 Study of Characterization of Intensity Modulated Fiber Optic Sensor *International Journal of Innovative Research in Science, Engineering and Technology* **2** 2319-8753

Study on steam pressure characteristics in various types of nozzles

Firman, Muhammad Anshar

Mechanical Engineering Department, Politeknik Negeri Ujung Pandang, Makassar, Indonesia 90245

E-mail: firman@poliupg.ac.id

Abstract. Steam Jet Refrigeration (SJR) is one of the most widely applied technologies in the industry. The SJR system was utilizes residual steam from the steam generator and then flowed through the nozzle to a tank that was containing liquid. The nozzle converts the pressure energy into kinetic energy. Thus, it can evaporate the liquid briefly and release it to the condenser. The chilled water, was produced from the condenser, can be used to cool the product through a heat transfer process. This research aims to study the characteristics of vapor pressure in different types of nozzles using a simulation. The Simulation was performed using ANSYS FLUENT software for nozzle types such as convergent, convrgent-parallel, and convergent-divergent. The results of this study was presented the visualization of pressure in nozzles and was been validated with experiment data.

1. Introduction

Steam injector refrigeration system can replace vapor compression of refrigeration system. In general, the refrigerant thermodynamic characteristics of the two systems are the same. The fact indicates that only water that has been used so far and available resources to drive injectors. Some research has been done on the use of fluorocarbon in vapor compression systems. These systems use water as a refrigerant in vapor compression cycle and every compression obtained from compression jet principles, known as refrigeration water-steam or steam-injector refrigeration (SJR) system.

SJR system utilizes remaining steam of boilers that is then channeled through nozzles into a tank (flash tanks) containing water. The nozzles convert pressure energy into kinetic energy so that it can quickly evaporate water. Cold water resulting from condensers (chilled water) is used to cool products through heat transfer. Generally, evaporating 1% of water in the tank can reduce 6°C of water temperature [1].

SJR system can be operated at the boiler temperature of 120°C up to 140°C and evaporator temperature of 5°C up to 15°C [2]. The temperature of evaporators depends on the pressure of evaporators. Low pressure or vacuum condition of evaporators is highly dependent on the design of injectors. According to [3], one of the factors that affect vacuum pressure of evaporators is the geometry between water surface and the suction side of injectors. It is therefore required that an injector be designed properly for the applications of refrigeration steam jet systems. The design of optimum angle of steam injector inclination is 2° for convergent type and 3° for divergent type with a throat length of 137 mm [4]. However, geometry factor causes the efficiency or performance (COP) of steam jet system refrigeration is still very low.

A quick evaporation of 1 kg of water results in a decline in temperatures of 5, 7°C [1]. If such a condition persists, the temperature of the water will get lower and lower. This should be allowed to take place to achieve a desired water temperature. The shortage of SJR refrigeration is associated with the fact that water freezes at the temperature of 0°C; if water freezes, the circulation in the system will not occur. Accordingly, the SJR refrigeration applications at low temperatures is very limited.

The performance of SJR system relies heavily on the efficiency of nozzles. Meanwhile, the efficiency of nozzles are influenced by geometry and fluid pressure in the injector. Sahni [5] reported that a the COP of steam jet refrigeration system is influenced by the geometry of nozzles and pressure drop.

According to Mitchley [3], one of the influential factors in the design of injector is geometry between water surface and the suction side of injector. On the output side the parameter that affects the efficiency of injector is pressure drop [5]. Other factors to consider are critical pressure and shock waves on the injector [6]. The ratio between intake pressure and critical pressure also affects the efficiency of injector and depends on geometry [7]. Vadalia [8] states that the optimization of nozzle geometry (throoat diameter) is very influential towards the performance of steam jet injector. Petel [9] conducted research on the geometry optimization of injector in order to obtain the best efficiency by computational fluid dynamics (CFD) analysis and by the reduction of the pressure drop through the geometry injector.

2. Materials and methods

In this experiment there are 4 types of nozzle analyzed theoretically by simulations of convergent, parallel-convergent, 1° inclination-angle convergent-divergent, and 3° inclination-angle convergent-divergent nozzles. However, only the simulation of 1° inclination-angle convergent-divergent nozzles which were validated by the data of experiments.

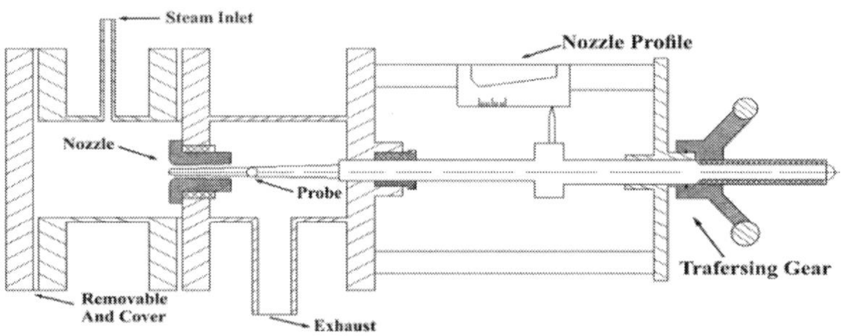

Figure 1. Experiment scheme.

The experiments were begun by placing a 1° inclination-angle convergent-divergent on certain positions, as shown at figure 1, then followed by setting the intake pressure of nozzle as much as 5.4 bars, which was maintained constant. Furthermore, the profile arm of nozzles was arranged to read the pressure at each point along a range of 5 mm. The exit pressure obtained from the nozzles in the experiment was 2 bars. The data of experiment were used as the boundary condition in the simulation as shown in figure 2.

To view the condition of fluid inside the nozzle visually with the simulations, ANSYS 16.0 with an system analysis feature FLUENT was used. The simulation steps were, firstly, drawing CAD nozzle by using solid works. Secondly, the CAD nozzle was imported to the ANSYS by using the geometric features in the ANSYS. Thirdly, the geometry of nozzle was imported to the mesh. Fourthly, the mesh was imported to the setup then to set solver based on pressure and steady state of problems in time types; then the energy equation was activated and the inlet and outlet pressures in the boundary conditions were set up. Finally, iterative solver and mixed initialization were selected and the program was run.

Figure 2. Boundary Condition.

3. Results and discussion

The use of ANSYS 16.0 CFD with the feature of simulation system FLUENT in ANSYS 16.0 was applied in to order to simulate the flows of fluid in the form of nozzle geometry in the experiment. In the experiment the inlet and outlet sides of the nozzles were the parameter derived from the results of laboratory testing, which were 5.5 bars and 2 bars on the inlet side and outlet side respectively. From figure 12, it can be seen that pressure graph on each wall of the nozzle with pressure changes at each position.

To see the characteristics of each varied nozzle, various step models were validated in accordance with the data of experiment. The contour parameters shown in figure 3, 5, 7 and 9 were found after inputting the inlet and outlet pressures on the 1° inclination-angle convergent-divergent, 3° inclination-angle convergent-divergent, convergent, and divergent-parallel nozzle models by using the method of fluent simulation. To compare the results of laboratory experiments, the results of simulations using ASNYS 16.0 can be seen below.

Figure 3. Contour of 1° inclination angle convergent-divergent Nozzle.

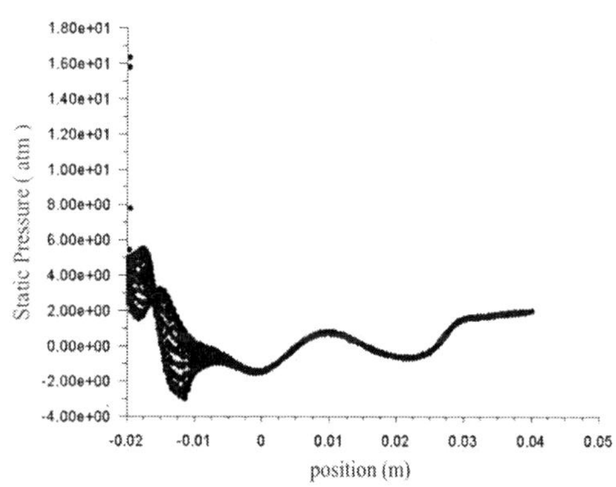

Figure 4. Pressure Graph of 1° inclination angle convergent-divergent Nozzle.

On the 1° inclination-angle convergent-divergent the pressure will decrease at when its position was after 30 mm as shown in figure 4. As shown in figure 3, the pressures began to go down after shocks took place. This is in line with the results of the study of Pansari [10] that the decrease moved from the shock location headed the exit side.

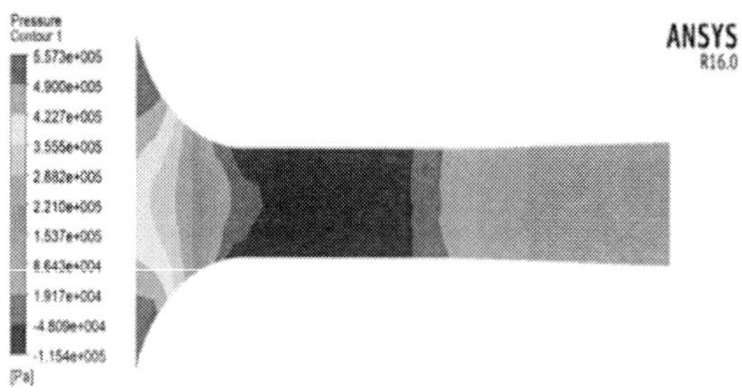

Figure 5. Contour Pressure of 3° inclination angle convergent-divergent Nozzle.

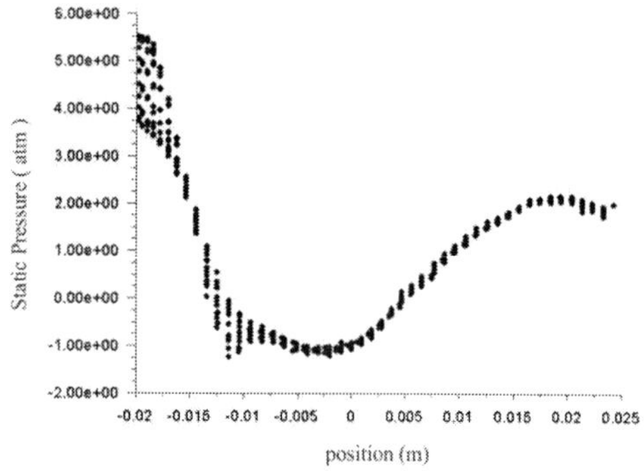

Figure 6. Pressure Graph of 3° inclination angle convergent-divergent Nozzle.

On the contrary, the 3° inclination-angle convergent-divergent underwent pressure decline when its position reached 20 mm; the differences were also due to the angle of inclination. The influence of inclination angle is accorded to the research results of Surya [11] claiming that inclination angle affected the performance of nozzles.

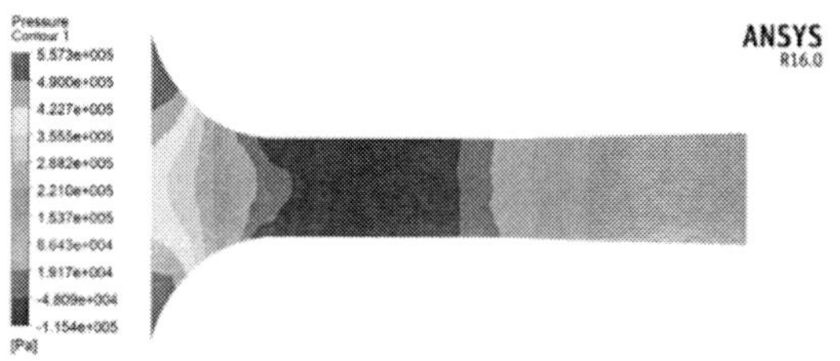

Figure 7. Contour pressure of convergent nozzle.

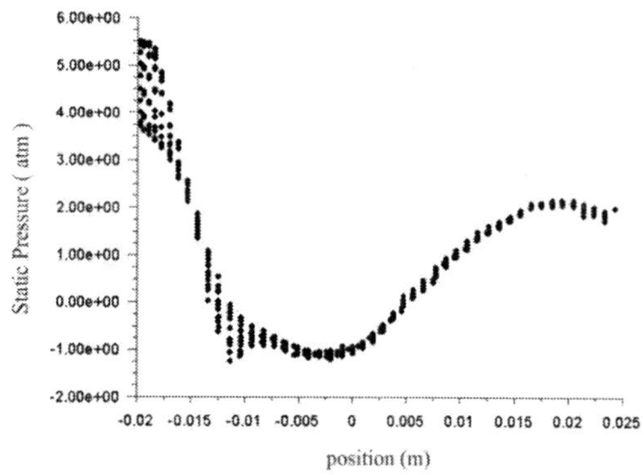

Figure 8. Pressure graph of convergent nozzle.

The characteristics of pressure on convergent nozzles were almost the same with those of convergent-parallel nozzle where the decrease of pressure occurred when they reached the same position. The decrease of pressure happened because the flow of fluid in the nozzles of both types underwent a sudden change of pressure on the exit sides of such nozzles. These conditions were in line with the research results of Rao [12] that analyzed flows in convergent nozzles using the CFD for the ratio of different nozzles.

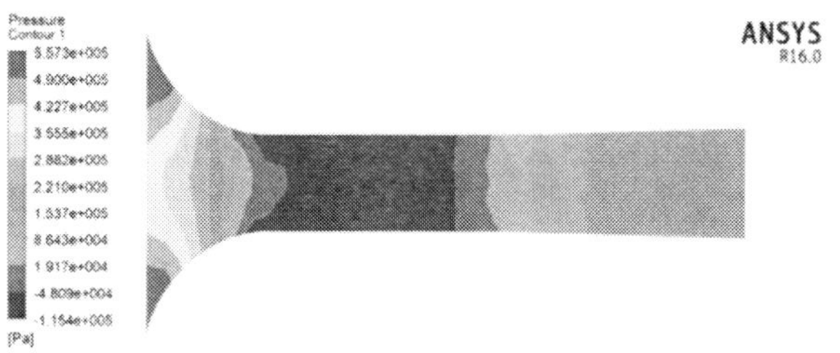

Figure 9. Contour pressure of convergent-parallel nozzle.

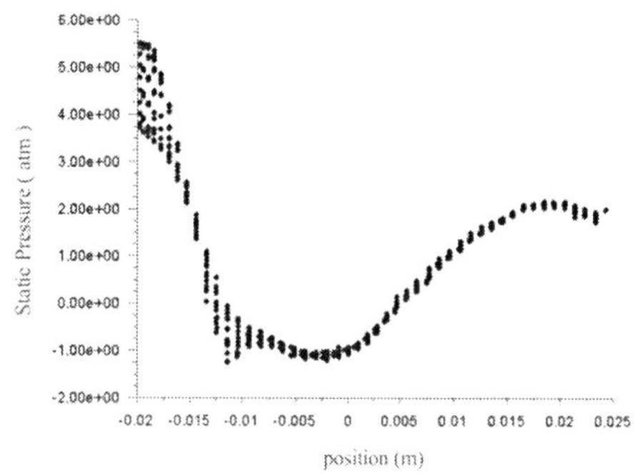

Figure 10. Pressure graph of convergent-parallel nozzle.

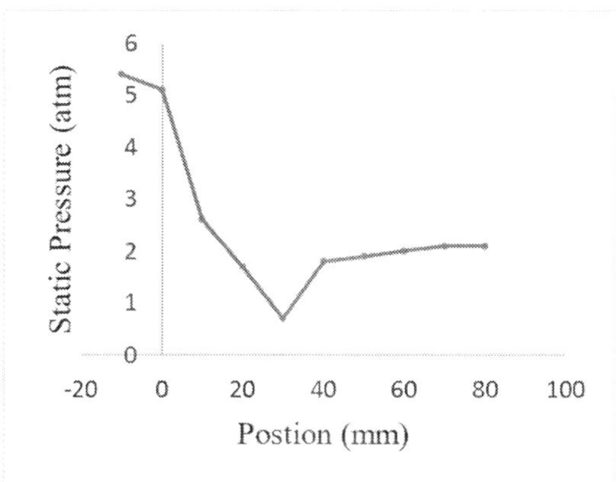

Figure 11. Graph of experiment results.

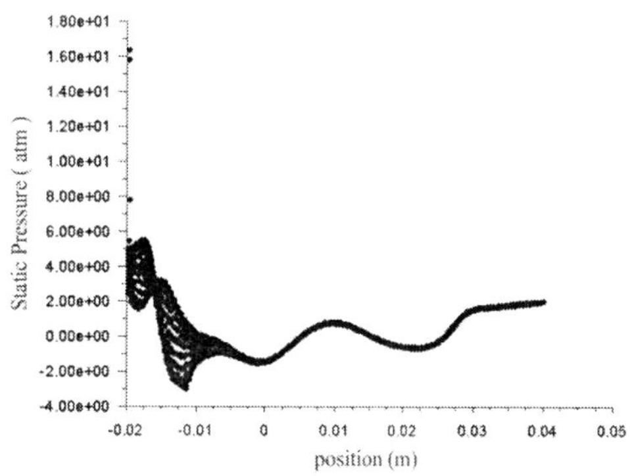

Figure 12. Graph of simulation results.

The graph in figure 11, the results of the experiment, demonstrated the suitability of characteristics with those of the simulation in figure 12. The contours of pressure as in figure 3 up to figure 12 show the variation of pressure along the walls of nozzles; the pressures increased at inlet sides but decreased at outlet sides. The analysis results were in line with the view of Satyanarayana [13] that CFD analysis on convergent-divergent nozzle with rectangular, square, and round surfaces showed the result of the computation was very suitable with the experimental results. Sudhakar [14] has also made such a modeling with theoretical analysis by using CFD with parameters of pressure, temperature, and flow speed that resulted in a suitability of flow characteristics in convergent-divergent nozzles. Similarly, the suitability was demonstrated by the modeling of fluid flow in convergent-divergent nozzles assuming that the Quasi-one of dimensional isentropis flows using CFD [15].

4. Conclusion

Based on the results of experiments by simulations, it was found that various pressures occurred along the nozzles. The characteristic analysis on pressures in every type of nozzle showed different results; and this was subject to the difference of geometry. It was concluded that analysis using FLUENT indicated conformity with the results of the experiments in this study.

Acknowledgement
Author would like thanks Ministry of Research, Technology, and Higher Education for the financial support for Applied Product Research Fiscal Year 2017.

References

[1] Shet U S P, T Sundararajan, J M Malkerjuna 2014 *Refrigeration Cycles* (Mandroa: Indian Institute of Technology)

[2] Pianthong K, W Seehanam, M Behnia, T Sriveerakul, S Aphonratana 2007 Investigation and Improvement of Ejector Refrigeration system Using Computational Fluid Dynamics Technique *Energy conversion &Manajement* **48** 2556-2564

[3] Mitchley S R 1998 Vacum Boiling of Water in a Steam Jet Refrigeration System *Dissertation* (Johannesburg: Faculty of Engineering University of The Witwatersand)

[4] Saengmanee C, Pianthong K 2010 Design of a Steam ejector by co-operating the ESDU design method and CFD Simulation *The First TSME International Conference on Mechanical Engineering* 20-22 October Ubon Ratchathani

[5] Sahni R 2015 Ejector Expansion Refrigeration Systems Reaserch Inventy *International Journal of Engineering and science* **5** (2) 25-29

[6] Elbel S, Hanjak P 2008 Ejector Refrigeration: An Overview of Historical and Present Development with an Emphasis on Air- Conditioning Applications *International refrigeration and Air Conditioning Conference at Purdue* July 14-17 (2350) 1-9

[7] Chunnannond K, S Aphornratana 2004 Ejectors: Application in refrigeration technology *Renewable and Sustainable Energy, Reviews* **8** 129-155

[8] Vadalia D R 2017 Performance Optimization of Steam Jet Ejector Using CFD A Review *International Research Journal of Engineering and technology (IRJET)* **4** (2)

[9] Petel A R and Khunt J 2013 Performance Optimization of Steam Jet Ejector using CFD *International Journal Innovative Research in Science and Technology* **2** (1)

[10] Pansari K and Jilani S A K 2013 *Numerical Investigation of the Performance of Convergent Divergent Nozzle* **3** (3) 2001-2006

[11] Surya S D, Vesu T A, Raghavan K S, Chavali M 2017 CFD Simulation of Ejector in Steam Jet Refrigeration *Journal of Applied Mechanical Engineering* **6** 263

[12] Rao G R, Ramakanth U S, Lakshman A 2013 Flow Analysis in a Convergent-Divergent Nozzle Using CFD *International Journal of Research in Mechanical Engineering* **1**

[13] Satyanarayana G, Varun C, Naidu S S 2013 CFD Analysis of Convergent-Divergent Nozzle. *Acta Technica Corviniensis- Bulletin of Engineering, Tome VI (Year 2013) Vascicule* **3**

[14] Sudakhar B V V N, Sekhar B P C, Sekhar B P, Mohan P N, Ahmad M D S 2016 Modelling and simulation of Convergent-Divergent Nozzle Using Computational Fluid Dynamics *International Research Journal of Engineering and Technology (IJRET)* **3** (8)

[15] Lakshmi K S, Vakatesh K 2016 Modelling and Simulation of Supersonic Nozzle using Computational Fluid Dynamics *Technology Management and Research* **3** (9)

Characteristic Analysis Light Intensity Sensor Based On Plastic Optical Fiber At Various Configuration

A. Arifin, Lusiana, Muhammad Yunus, Syamsir Dewang

[1]Department of physics, Hasanuddin University, Makassar, 90245 Indonesia

E-mail: arifinpide@gmail.com

Abstract. This research discusses the light intensity sensor based on plastic optical fiber. This light intensity sensor is made of plastic optical fiber consisting of two types, namely which is cladding and without cladding. Plastic optical fiber used multi-mode step-index type made of polymethyl metacrylate (PMMA). The infrared LED emits light into the optical fiber of the plastic and is subsequently received by the phototransistor to be converted to an electric voltage. The sensor configuration is made with three models: straight configuration, U configuration and gamma configuration with cladding and without cladding. The measured light source uses a 30 Watt high power LED with a light intensity of 0 to 10 Klux. The measured light intensity will affect the propagation of light inside the optical fiber sensor. The greater the intensity of the measured light, the greater the output voltage that is read on the computer. The results showed that the best optical fiber sensor characteristics were obtained in U configuration. Sensors with U-configuration without cladding had the best sensitivity and resolution values of 0.0307 volts/Klux and 0.0326 Klux. The advantages of this measuring light intensity based on the plastic optical fiber instrument are simple, easy-to-make operational systems, low cost, high sensitivity and resolution.

1. Introductions

The rapid development of technology in the present era causes a very significant need for a sensor system that has reliability in detecting a parameter. Various sensor systems have been created by scientists. Up to the optical fiber technology which is a new breakthrough as a multi-function sensor system in various fields. Starting from the medical field, civil, communication, etc. As a sensor system, optical fibers can also be applied in various measuring areas for physical and chemical parameters, such as temperature sensors, pH sensors, electric current sensors and hydrogen detection sensors [1, 2].

The advantages fiber optic sensors over conventional electronic sensors that provide solutions as sensors in small size, simple, lightweight, robust, easy to integrate in various forms because they are flexible, low noise due to resistance to electromagnetic interference, high sensitivity and measurement accuracy and can be interconnected with data communication system through interface device [3-6].

Many research has been done using fiber optics as a sensor system in an attempt to develop it. As an example of sensor shift studies based on imperfection methods on optical fiber structures [3], the development of optical fiber-based sensors applied for measuring human respiration rates [7], the manufacture of load sensors using microbending methods on optical fibers with a certain distance [8], the use of plastic optical fibers as a medium for detection of load changes [9] and many other studies using fiber optics.

Fiber optic sensors have been developed to detect the effect of changes in the intensity of light on a particular light source. One example is a straightforward fiber optic configuration and U configuration for UV detection using Thorlabs SP1 spectrometer [10]. In addition, optical fiber sensors are also used to detect UV by utilizing a particular material that is bonded to a fiber optic configuration structure using a highly complex system as it passes through several processes to obtain materials that can be combined with optical fiber structures [11].

In this paper will be discussed about the light intensity sensor based on plastic optical fiber. Optical fiber is used as a medium for detecting changes in the light intensity of the measured light source used. Sensors are made with 3 configurations of straight configuration, U configuration and gamma configuration. Sensors are made to provide solutions for sensors in small sizes, simple systems, as well as sensitivity and resolution of good sensor measurements.

2. Experimental SET-UP

The plastic optical fiber is made of polymethyl metacrylate (PMMA) type of step index with coat, cladding, and core each other diameters of 2.2 mm, 1 mm, and 0.98 mm respectively used as transmission medium and detector of intensity change from source measurable light used. The core refractive index and the plastic optical fiber cladding are respectively 1.492 and 1.402 with NA numerical slot values of 0.5. The measured light source used is a white high power LED with a power of 30 Watt.

The light intensity sensor scheme of the plastic optical fiber is shown in figure 1. The sensor is made with 3 configurations of straight configuration, U configuration and the gamma configuration shown in figure 2. The U configuration sensor is made with some variations is the length with cladding and without cladding. Optical fiber is peeled at the curve area along 1 cm, 2 cm and 3 cm. In addition, there are also variations in diameter of bending 0.5 cm, 1.0 cm and 1.5 cm.

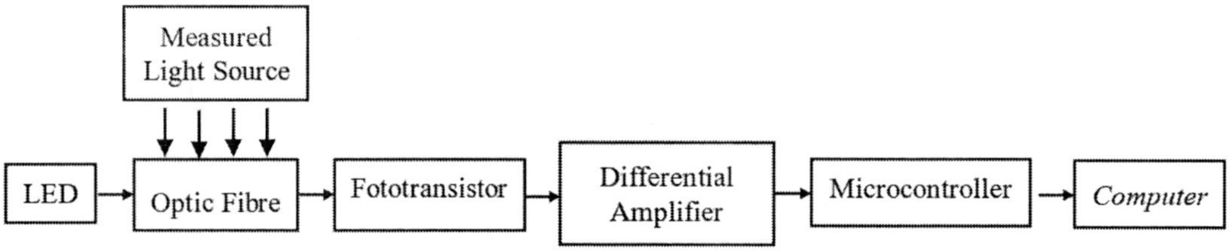

Figure 1. Scheme of light intensity sensor based on plastic optical fiber.

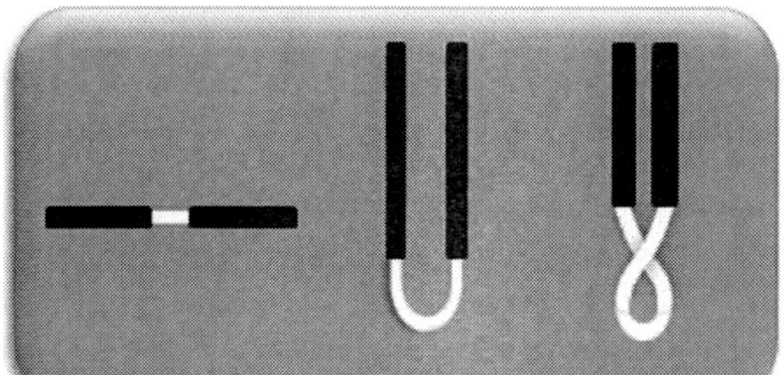

Figure 2. Sensor on straight, U, and gamma configurations.

Appropriate scheme in figure 1. The measured light source is placed perpendicular to the optical fiber sensor. The light signals propagating inside the core of the optic fiber will be passed to the phototransistor to be converted into electrical signals. To enlarge the electrical signal is used a differential amplifier circuit. Then the electrical signal in the form of analog signal is converted into digital signal using microcontroller to be readable on computer.

When the measured light source illuminates the optical fiber sensor, it will cause some light beam to enter into the fiber-optic core. The light entering the fiber-optic core will cause the intensity of light that was originally constant to increase. This increase in light intensity is proportional to the change in sensor output voltage. When measurements are made, the light intensity of the measured light source illuminating the optical fiber sensor is adjusted from the intensity of 0 Klux to 10 Klux with a change of 0.5 Klux

3. Result and Discussion

The optical fiber sensor is made to measure the light intensity of a light source illuminating the optical fiber portion which is peeled by the measurement result of a voltage. Some beams of light from the measured light source illuminate the sensor into the fiber-optic core and cause the intensity of the original constant to increase. This change in light intensity is proportional to the sensor output voltage.

The result of the change in the output voltage of the plastic optical fiber sensor to the intensity change of the measured light source illuminating the sensor is shown in figure 3 below:

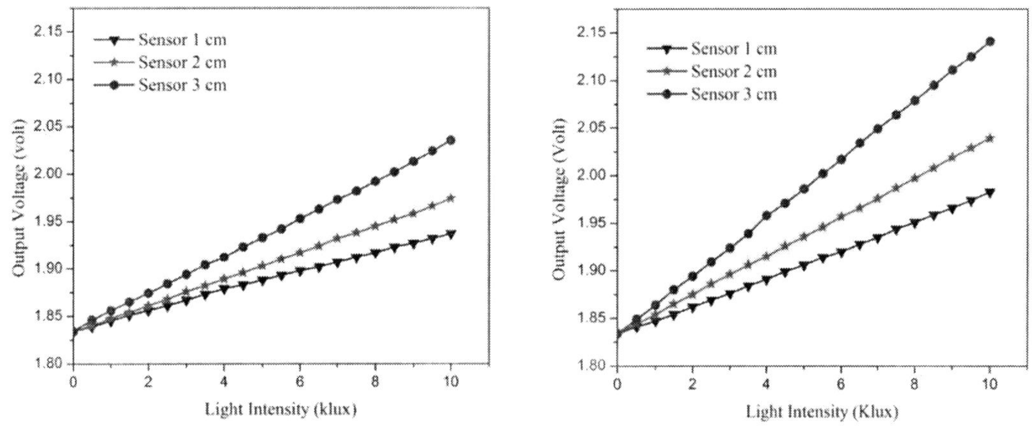

Figure 3. Graph of the output voltage of the sensor on the change of light intensity with variation of length of peel (a) with cladding and (b) without cladding.

Measurements using a U-configuration sensor with variations of peel shows the results where figure 3 shows that the higher the intensity of the measured light source the higher the sensor output voltage. The longer of the length of peel the greater the voltage change the sensor output. The characteristics of the light intensity sensor U configuration in the length of the peel variations are shown in table 1 below:

Table 1. Characteristics of the light intensity sensor U configuration in the length of peel variation.

Type of Sensor	Length of peel	Range (Volt)	Sensitivity (Volt/Klux)	Resolution (Klux)
With Cladding	1 cm	0.1030	0.0103	0.0971
	2 cm	0.1400	0.0140	0.0714
	3 cm	0.2010	0.0201	0.0497
Without Cladding	1 cm	0.1490	0.0149	0.0671
	2 cm	0.2050	0.0205	0.0488
	3 cm	0.3070	0.0307	0.0326

In table 1 above shows the range values, sensitivity and resolution of the sensors influenced by the length of peels of the fiber optic structure. The longer of the sensor peels the better the sensitivity and the resolution results. The best results were obtained at length of 3 cm peel sensor without cladding with sensitivity value 0.0307 Volt/Klux and resolution of 0.0326 Klux.

The results of the light intensity sensor based plastic optical fiber testing for U configuration are shown in figure 4 below:

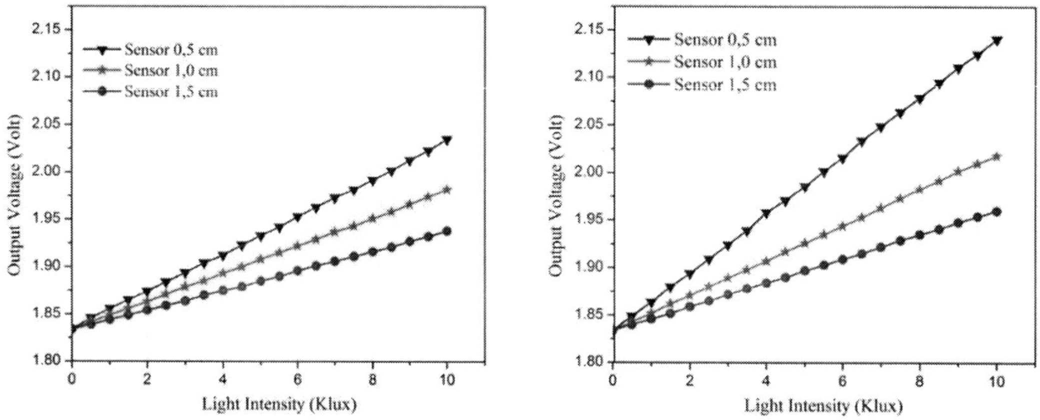

(a) (b)

Figure 4. Graph of the output voltage of the sensor on the change of light intensity with the diameter of bending variation (a) with cladding and (b) without cladding.

Figure 4 shows the response of changes in sensor output voltage is higher when the intensity of light from white LEDs is greater. In the graph can be seen that the smaller diameter curve sensor U configuration, the greater the voltage change sensor output. The characteristics of the light intensity sensor U configuration are shown in table 2 below:

Table 2. Characteristics of light intensity sensor based on plastic optical fiber U configuration with variation of indentation diameter.

Type of Sensor	Diameter of bending	Range (Volt)	Sensitivity (Volt/Klux)	Resolution (Klux)
With Cladding	0.5 cm	0.2010	0.0201	0.0497
	1.0 cm	0.1480	0.0148	0.0676
	1.5 cm	0.1040	0.0104	0.0961
Without Cladding	0.5 cm	0.3070	0.0307	0.0326
	1.0 cm	0.1840	0.0184	0.0543
	1.5 cm	0.1260	0.0126	0.0794

Table 2 above shows the range values, sensitivity and resolution of the sensors influenced by the diameter of bending sensor of the fiber optic structure. The smaller the diameter of bending sensor the better the sensitivity and the resulting resolution. The best results were obtained at 0.5 cm the diameter of bending without cladding sensor with a sensitivity value of 0.0307 Volt/Klux and a resolution of 0.0326 Klux.

Some configurations are made to produce sensors with the best sensitivity and resolution of the straight, U and gamma configurations. Each configurations using a sensor without cladding with a length of 3 cm peel. The response of the output voltage change to the change of light intensity for the straight configuration sensor, gamma configuration and U configuration is show in figure 7 below:

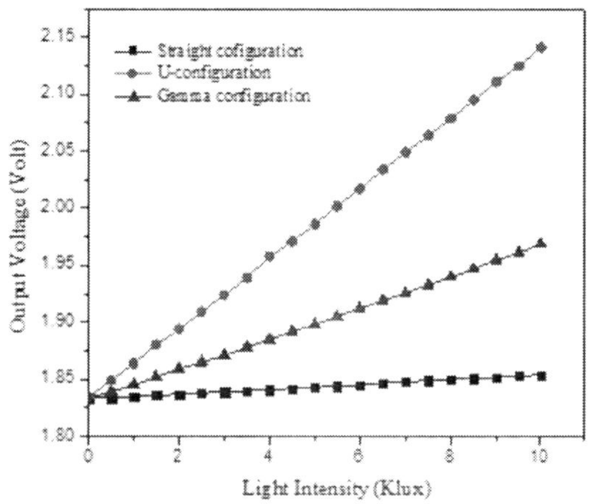

Figure 5. Graph of the output voltage of the sensor in straight configuration, gamma configuration and U configuration on the change of light intensity.

Figure 5 above is the result of comparison of straight configuration, U configuration of 0.5 cm diameter of bending, and gamma configuration of 1 cm in diameter of bending. Each configuration has a length of 3 cm each without cladding. The graph shows the greater the intensity of white LED light then the greater the sensor output voltage. Based on these graph, the U configuration sensor has the highest sensor output voltage change response compared to other configuration sensors.

Characteristics of light intensity sensor based on plastic optical fiber with straight configuration, U configuration and gamma configuration are shown in table 3 below:

Table 3. Characteristics of light intensity sensor of various configurations.

Configurations	Range (Volt)	Sensitivity (Volt/Klux)	Resolution (Klux)
Straight	0.020	0.002	0.500
U	0.307	0.031	0.033
Gamma	0.135	0.014	0.074

Table 3 shows that sensors with U configuration have the best sensitivity and resolution value compared to the straight configuration and gamma configuration. This is causes the U configuration has the smallest of the diameter of bending compared to other configurations. The smaller the diameter of the sensor indentation the better the sensitivity and resolution. The best results were obtained with sensitivity values of 0.031 Volt/Klux and resolution of 0.033 Klux.

The results of this study has consistent with result of previous studies on the comparison of sensitivity and resolution of load sensors based on plastic optical fiber using loop configuration. The addition of the load resulted in the smaller diameter of bending the sensor. The results show the smaller the diameter of bending the sensor the better the sensitivity and resolution [9].

4. Conclusion

The change in the sensor output voltage is proportional to the measured light source intensity change. The greater the intensity of the white LED light (measurable external light source) that illuminates the optical fiber the larger the sensor output voltage. The longer of the peel and the smaller the diameter of bending, the sensitivity and the sensor resolution will get better. The result of measurement using light intensity sensor based on plastic optical fiber shows that U configuration sensor without cladding 3 cm length of peel on diameter of bending 0.5 cm has the best sensitivity of 0.031 volts/Klux and 0.033 Klux resolution.

Acknowledgements
The Research was supported by: BPOPTN-PPKPS, Hasanuddin University-2017.

References

[1] Aranda A R M, Pensado M A B, Ceron E E A, Gómez L L C, Beltran G U, Rodriguez J A, García J C, Mondragón J J S, Pérez V I R 2013 Fiber Optic Pressure Sensor of 0–0.36 psi by Multimode Interference Technique *Journal of Applied Research and Technology* **11** 695-701

[2] Lúcia B, Nélia A, João L P, Rogério N 2012 Optical Sensors Based on Plastic Fibers *J. Sensor* **12** 12184-12207

[3] Arifin A, Hatta A M, Muntini M S, Rubiyanto A 2014 Bent of Plastic Optical Fiber with Structural Imperfection for Displacement Sensor *Indian Journal of Pure & Applied Physics (IJPAP), NISCAIR Publication* **52** 520-524

[4] Arifin A, Hatta A M, Sekartedjo, Muntini M S, Rubiyanto A 2015 Long Range Displacement Sensor Based on SMS Fiber Structure and OTD *Photonic Sensors* **5** 166-171

[5] Yingtian H, Yizhang W, Xiaoping W 2015 Novel Method Of Turbidity Compensation for Chemical Oxygen Demand Measurement by Using UV-Vis Spectrometry *Sensor and Actuators B* 393-398

[6] Dziuda L, Lewandowski J, Skibniewski F, Nowicki G 2012 Fiber-Optic Sensor For Respiration and Hear Rate Monitoring in The MRI Environment *Procedia Engineering* **59** 1291-1294

[7] Marek K, Michel S, René M R, Luciano F B, Gian L B, Lukas J S 2014 An Optical Fiber-Based Sensor For Respitory Monitoring *Laboratory for Protection and Physiology* **13** 13088-13101

[8] Červeňová J, Iglarčík M 2013 Weight Measurements Using Microbending Optical Fibre Sensor and OTDR *Proceedings of the 9th International Conference* 243-246

[9] Arifin A, Yusran, Miftahuddin, Abdullah B, Tahir D 2017 Comparison of Sensitivity and Resolution Load Sensor at Various Configuration Polymer Optical Fiber *The 6th International Conference on Theoritical an Applied Physics* 1-6

[10] Ana V J, Bajic, Jovan S B, Dragan Z S, Milos P S, Miodrag J, Milos B Z 2012 Simple and Low Cost Fiber-Optic Sensors for Detection of UV Radiation *J. Telfor* **4** 2

[11] Ana V J, Dragan Z S, Jovan S B, Bojan M D, Zoran M, Miloš P S, Miloš B Z 2013 An End-Type Fiber-Optic UV Sensor Covered with Mixture of Two UV Sensitive Materials *Engineering Materials* **543** 265-268

Probing critical behavior of 2D Ising ferromagnet with diluted bonds using Wang-Landau algorithm

N. A. Ridha, M. F. Mustamin and T. Surungan

Department of Physics, Hasanuddin University, Makassar, South Sulawesi 90245, Indonesia

E-mail: tasrief@unhas.ac.id

Abstract. Randomness is an important subject in the study of phase transition as defect and impurity may present in any real material. The pre-existing ordered phase of a pure system can be affected or even ruined by the presence of randomness. Here we study ferromagnetic Ising model on a square lattice with a presence of randomness in the form of bond dilution. The pure system of this model is known to experience second order phase transition, separating between the high temperature paramagnetic and low-temperature ferromagnetic phase. We used Wang-Landau algorithm of Monte Carlo method to obtain the density of states from which we extract the ensemble average of energy and the specific heat. We observed the signature of phase transition indicated by the diverging peak of the specific heat as system sizes increase. These peaks shift to the lower temperature side as the dilution increases. The lower temperature ordered phase preserves up to certain concentration of dilution and is totally ruined when the bonds no longer percolates.

1. Introduction

Phase transitions are very common natural phenomena. Although they may appear in various realizations, they can be grouped into two main categories, namely the first and the second order phase transitions (SOPT) [1]. A first order transition is characterized by discontinuous changes in free energy at critical temperature, firmly exemplified by transition of the PVT systems such as water. In this type of transition, we can always find latent heat in the course of transition as there is co-existence phases. The SOPT which is also called a continues phase transition is characterized by the continuous change of the free energy during the transition. The famous example of these is the spontaneous magnetization of ferromagnetic material, which change from disorder spin configuration at higher temperature to ordered phase below Curie (T_c) temperature, commonly referred to as critical point, and the phenomena nearby this point is called critical phenomena. Spontaneous magnetization is a typical type critical phenomena of systems consisting of many microscopic interacting components. The interesting aspect of studying critical phenomena is that the behavior of system (the critical properties) usually indicates universality, in the sense that totally different systems may exhibit similar behavior, which is characterized by similar set of exponents called critical exponents.

In the study of critical phenomena, the exponent values or essentially the critical properties of the system, may change due to variation of a number of macroscopic variables such as spatial dimension, symmetry of spin and the type of coupling interactions. The critical properties of system can also be affected by the presence of randomness. In fact randomness is an important subject to probe as defect or impurity is always present in any real material. There exists an important theoretical conjecture called Harris criterion [2] which predict the role played by a randomness in affecting the critical properties of system. According to this criterion, the critical behavior of disordered system can only be affected by the presence of randomness if the specific heat exponent α of the corresponding pure system is positive. For $\alpha < 0$, randomness is irrelavant, in other words, it will not affect the critical properties of the system. The presence of randomness is called marginal if $\alpha = 0$, meaning that the Harris criterion is inconclusive.

The two dimensional (2D) Ising model [3] has long been the object for benchmarking study in many fields. The present study is probing whether certain type of randomness is relevant or not for the existing phase transition. This model is well known to experience SOPT with positive exponent α, hence some type of randomness should be relevant, according to Harris criterion. A type of randomness by adding extra bonds, i.e., rewiring regular lattices with antiferromagnetic coupling have been studied for probing the so-called non-canonical spin glasses and the effect on the existing Neél phase [4, 5, 6, 7]. In the case of bond dilution, There has been several numerical studies on 2d Ising model, for example the old report by [8, 9]. However, the data are not very acurate due to the limitation of computer power in the previous decades. The effect of diluted bonds for system experiencing Kosterlitz-Thouless transition was also studied [10].

In this work, we study diluted bonds of 2D Ising model by using Wang-Landau (WL) algorithm [11] of Monte Carlo method. This algorithm have been widely used in studying phase transition of various magnetic models [12, 13]. Typical characteristics of this method is to obtain the system density of states (DOS) for the caculation of ensemble average of physical quantity. Recently, the DOS obtained from WL algorithm also used to determine the order of phase transition, whether first or second order by analysing the dinference of DOS between neighboring energy states [14, 15].

In general, to analyse the critical behavior of the system, one calculates the order parameter, which is the magnetization for the case of ferromagnetic system. The correlation ratio, i.e., the ratio of correlation functions of order parameter for two different distances [16] is another interesting quantity for probing critical behavior. This compelling physical quantity had been applied in several previous studies [10, 13, 17]. However, other quantities such as the energy and the specific heat can also be used to detect the existence of phase transition. In particular, for system experiencing SOPT. the diverging specific heat can be used to estimate the critical temperature. The paper is organized as follows: Section II will discribe the model the method used, Section III presents and discusses the the results and Section IV is devouted for the summary and concluding remarks.

2. Model and Method

Ising model is the simplest magnetic model yet mostly studied in probing phase transition [3]. The model consists all important concepts related to phase transition, for example, the broken symmetry, the existence of critical exponents of various physical quantities, order parameters, etc. The model is represented with the following Hamiltonian

$$H = -\sum_{<ij>} J_{ij} s_i \, s_j \tag{1}$$

where spins s_i occupy lattice sites; here we consider two dimensional (2D) lattice (square). As the underlining symmetry for the Ising spin is Z_2, the value of s_i can be either +1 or -1, respectively associated with up and down spins. J_{ij} is the exchange coupling between spin on site i and j. We consider ferromagnetic system, therefore with this notation J_{ij} is positive. We apply the nearest neighbor interaction and the summation are performed to all pairs of nearest neighboring spins. However, as we probe diluted problem, some bonds are depleted, i.e., $J_{ij} = 0$.

In this study, Monte Carlo method of Wang-Landau (WL) algorithm [11] is used. It is an extended ensemble sampling algorithm where the random walk is performed in the whole energy

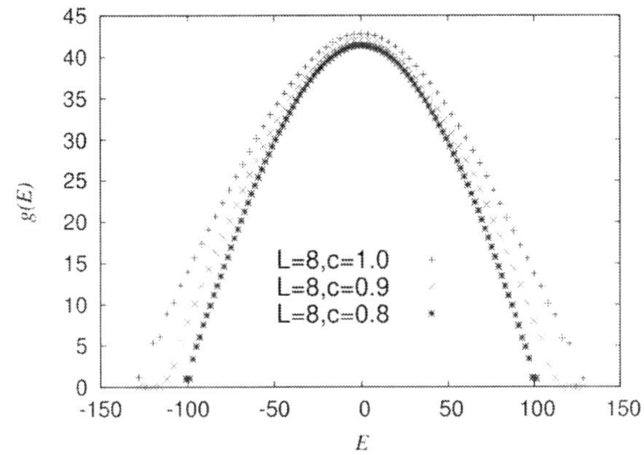

Figure 1. (Color online) Density of states for system size L=8, with various bond concentrations.

space, which is essential to obtain the energy density of states (DOS). However, because of a wide range of energy space, two common problems are frequently encountered by the random walker: (i) the random walker tends to be blocked by the edge of the already visited energy states [12], (ii) it takes time from one end of the area to reach the other end, which is a common feature of random walks. The main idea of WL is to get away from the problems and to accelerate the dynamics of the random walk so that the DOS $g(E)$ can be efficiently and precisely evaluated. This is done through the introduction of the transition probability defined as follows

$$P(E \to E') = \min\left[1, \frac{g(E)}{g(E')}\right] \tag{2}$$

where $g(E)$ is the DOS for state E. Since this $g(E)$ is not known at the beginning of the simulation, it is presumed to be $g(E) = 1$ for the whole states. The DOS is then evaluated through iterative random walk and its exact value is obtained when the histogram $h(E)$ of energy achieves flatness condition *i.e.* the histogram for all possible E is not less than some value of the average histogram, say, 0.80. The DOS and $h(E)$ are updated every time E is visited through the relation

$$\ln g(E_i) \to \ln g(E_i) + f_i \tag{3}$$
$$h(E_i) \to h(E_i) + 1 \tag{4}$$

where f_i is the modification parameter which is gradually reduced to zero when the DOS reaches the convergence.

Thermal average of physical quantity Q is obtained through relation

$$\langle Q \rangle_\beta = \frac{\int dE\ g(E)\ Q(E)\ \exp(-\beta E)}{\int dE\ g(E)\ \exp(-\beta E)} \tag{5}$$

where $\beta = 1/kT$, k and T respectively being the Boltzmann constant and temperature. The detail steps of the algorithm are as follows

(i) Set (a) initial spin configuration and initial iteration factor $f_i = 1$; (b) assume that the initial values of the DOS and histogram as $\ln g(E_i) = 1$ and $h(E_i) = 0, \forall_i$.

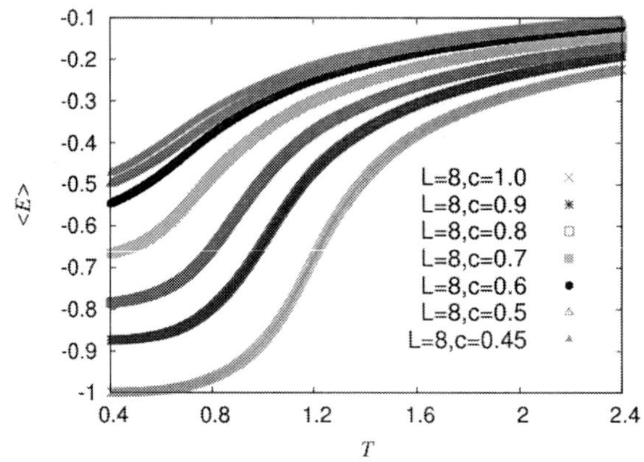

Figure 2. (Color online) Temperature dependence of energy of system size $L = 8$ for several bond concentartions. It is to be noticed that we used normalized scale, therefore the minimum energy $E_{min} = 1$, instead of 2.

(ii) Update the configuration based on the transition probability (2).

(iii) Update the DOS and the histogram as (3) and go to step 2 until the histogram reaches the flatness condition.

(iv) Refine the iteration factor $f_i = f_i/2$, rescale the DOS by using condition $\sum_{E_i} g(E_i) = 2^N$ (for the case of the Ising model); and reset the histogram $h(E_i) = 0$ before going back to step 2.

(v) Do steps 1 to 4 until the DOS achieves convergence, then perform measurement of physical quantity Q_i with a definite number of MCS's.

For the application of the WL algorithm to the large system size, one performs window break, where the whole range of the energy is cut into several reasonable widths. The result from each window is later joined to obtain the resultant.

3. Results and discussion

We studied the ferromagnetic Ising model on square lattice with diluted bonds using Wang-Landau algorithm. Several system sizes, namely $L = 8, 10, 12$ and 16 were simulated in order to probe how the size dependent of physical quantities. This is essential for extracting the thermodynamics behavior from the data of finite system sizes. As we are investigating the role played by randomness in the form of bond dilution, we took several number of bond concentrations, namely p=1.0, 0.9, 0,8, 0.7, 0,6, 0.5 and 0,45; where p=1.00 corresponds to pure case, i.e., all bonds exist. With p=0.90, there are 10% bonds deleted. The density of state (DOS) for the pure and diluted case, i.e., for P=1.0, 0.9 and 0.8 for system size L=8, are shown in figure 1.

The ensemble (thermal) average of any physical quantity $\langle Q \rangle$ is calculated according to standard relation in statistical mechanics written as follows

$$\langle Q \rangle_\beta = \frac{\int Q(E)g(E)e^{-\beta E}dE}{\int g(E)e^{-\beta E}dE}. \tag{6}$$

Using the density of state we can extract ensemble average energy $\langle E \rangle$ and its n-th order moment $\langle E^n \rangle$. Using $\langle E \rangle$ and the second order moment $\langle E^2 \rangle$, we can obtain the specific heat defined as

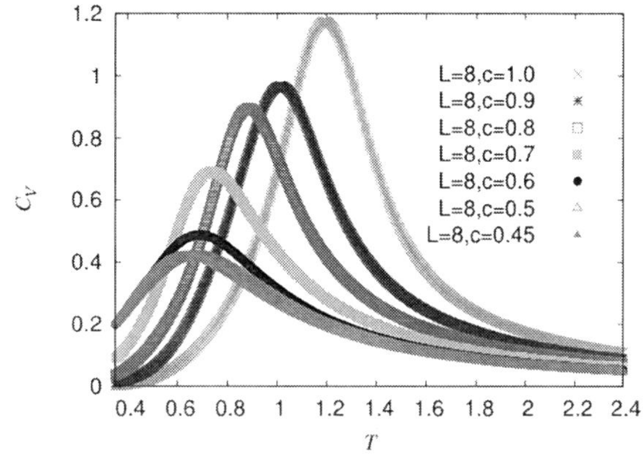

Figure 3. (Color online) Temperature dependence of specific heat of system size $L = 8$ for several bond concentartions. The peak of the specific heat shifted to the left as bond concentration c decreases. The Error bar is in the order smaller than the symbol size.

follows

$$C_v(T) = \frac{1}{Nk_BT^2}(\langle E^2\rangle - \langle E\rangle^2) \tag{7}$$

where E is the energy in unit of J. All temperatures are expressed in unit of J/k_B where k_B is the Boltzmann constant.

The temperature dependence of energy and specific heat for various system sizes as well as for different bond concentrations are plotted in figure 2 and figure 3, respectively. As shown, there is a systematic shift of the energy as well as the peak of specific heat to the lower temperature as concentration dcereases. For pure size for example, $c = 1.0$, the critical temperature, which is asscociated with the abscissa of peak as $L \to \infty$ is roughly estimated around T=2.48(3), after the energy being rescaled to the original value. For system with $c = 0.90$, the estimate of T_c is 2.04(2). The complete estimate of our critical temperatures is tabulated in table 1. Our rough estimates are in good agreement with the previous reports although always smaller due to large correction to finite size scaling as we probe quite small sizes. However, this already indicates the elegance of the WL algorithm. We have yet to report our estimate on critical exponents, such as the ν and η for the correlation length and magnetizaition. Our precise estimate and more conprehensive report will be published elsewhere.

It is to be noticed that the typical characteritics of physical quantity extracted from the WL calculation in principle can be continuously temperature dependence, whic is different from that of obtained from such microcanonical ensemble methods as the Metropolis algorithm. Therefore, Using WL algorithm, we can estimate critical temperatures for each concentration more precisely, compared to all previous results, such as that of reported in [8].

4. Summary and concluding remarks

We have investigated the role played by randomness in affecting the existing phase transition by probing the 2D ferromagnetic Ising model with bond dilution. The macrocanonical ensemble WL algorithm is used. Various concentrations of bond dilution are simulated. W calculated the ensemble average of energy and the specific heat which are extracted from the DOS directly obtained from the simulation using the WL algorithm. Then, we estimate the critical

Table 1. The estimate of critical temperature T_c for each value of c. The number in bracket is the error bar of the last digit.

Bond concentration (c)	T_c
1.0	2.48(3)
0.9	2.04(2)
0.8	1.80(2)
0.7	1.44(2)
0.6	1.36(2)
0.5	1.30(2)
0.45	1.24(2)

temperature for each bond concentration. We start from the pure case (where no bond depleted) down to lower concentration of bonds close to percolation ratio. We observe the shift of critical temperature to the lower side as the bond concentration c decreases.

Our estimate of critical temperatures are tabulated in table 1. These results are consistent with most previous work. We observe SOPT all the way down until it disappear at certain concetration. The concertration bonds c=0.45 is close to the critical value where the order phase disappear. This value is related to percolation ratio. For better estimate of our results, we perform a comprehensive study, which is in progress, and calculate the order parameter and consider larger system sizes. The results will be reported elsewhere.

Acknowledgments

The authors wish to thank Bansawang BJ, D. Tahir and A. W. Sandvik for valuable discussions. The extensive computation was performed using parallel computer facility at the Department of Physics, Hasanuddin University and that belongs to GRID-LIPI of the Indonesian Institute of Science. The present work is supported by a research grant of scheme: Priority Research for Tertiary Education, Indonesian Ministry of Research, Technology and Tertiary Education, Fiscal Year 2017.

References

[1] Ehrenfest P and Ehrenfest T 1907 Phys. Z **8** 311
[2] Harris A B 1974 J. Phys. C **7** 1671
[3] Ising E 1925 Z. Physik **31** 253
[4] Surungan T, Zen F P and Williams A G 2015 JPCS 640 012005
[5] Surungan T, Bansawang B J, Tahir D 2016 *Proc. AIP Conf.* 1719 030006
[6] Surungan T 2016 JPCS 759 012012
[7] Surungan T, Bansawang B J and Yusuf M (*accepted*)
[8] Zobin D 1978 Phys. Rev. B 18, 2387
[9] McGurn A R and Thorpe M F 1978 J. Phys. C 11 3667
[10] Surungan T and Okabe Y 2005 Phys. Rev. B **71**, 188428
[11] Wang F and Landau D P 2001 Phys. Rev. Lett. 86 2050 (2001); Phys. Rev. E 64 056101
[12] Yamaguchi C and Kawashima N 2002 Phys. Rev. E **65** 056710
[13] Surungan T, Okabe Y and Tomita Y 2004 J. Phys. A: Math. Gen. 37 4219
[14] Komura Y and Okabe Y 2012 Phys. Rev. E 85 010102(R)
[15] Surungan T, Komura Y and Okabe Y 2014 *AIP Conference Proceedings* 1617 79
[16] Tomita Y and Okabe Y 2001 Phys. Rev. Lett. **86** 572
[17] Surungan T, Kawashima N, and Okabe Y 2008 Phys. Rev. Rev. B **77** 214401

The Application of Extended Cox Proportional Hazard Method for Estimating Survival Time of Breast Cancer

Hartina Husain [1], Sri Astuti Thamrin[1], Sulaiha Tahir [1], Ahmad Mukhlisin [1], Mirna Apriani M [2]

[1.] Statistics, Faculty of Mathematics and Natural Sciences, Hasanuddin University
[2]Department of Epidemiology, Faculty of Public Health, Hasanuddin University

E-mail: hartina.husainn@gmail.com

Abstract. Breast cancer is one type of cancer that is the leading cause of death worldwide. This study aims to model the factors that affect the survival time and rate of cure of breast cancer patients. The extended cox model, which is a modification of the proportional hazard cox model in which the proportional hazard assumptions are not met, is used in this study. The maximum likelihood estimation approach is used to estimate the parameters of the model. This method is then applied to medical record data of breast cancer patient in 2011-2016, which is taken from Hasanuddin University Education Hospital. The results obtained indicate that the factors that affect the survival time of breast cancer patients are malignancy and leukocyte levels.

1. Introduction
Preventing the death rate from illness is the main goal of treatment. Prevention is done by considering the external and internal factors that affect the pattern of human life. Non-infectious diseases are still the main concern of one cancer. Cancer is the leading cause of death and disability worldwide, affecting more than 14 million people each year [1]. One of cancer types that is breast cancer. According to the [2] each year there are 1.7 million cases of breast cancer and 552,000 died. According to data Globocan (IARC) [3] breast cancer has a percentage of 43.3% with a mortality rate of 12.9%. Unlike lung cancer with a percentage of 34.3% with a 30.0% death rate. Breast cancer (Carsinoma mammae) occurs due to abnormal growth of breast cells (Sutjipto, 2006). In South Sulawesi, breast cancer cases are ranked first among many cancer diseases suffered by women. Based on data from medical record of Dr. Wahidin Sudirohusodo Makassar, the numbers of patients treated from 2010 to 2013 are 132, 360 and 573 cases of breast cancer, respectively.

Some of the factors that are suspected to be the trigger of breast cancer are affected by five behavioral and dietary risks, namely high growth mass index, less consumption of fruits and vegetables, less physical activity, cigarette use, and excessive alcohol consumption. These five factors have a rate of causing 30% cancer death rate. This unhealthy behavior that causes cancer risk becomes high. The existence of these factors indicates that there is a correlation between breast cancer disease and survival of patients [4]. Therefore, it is needed to see the relationship between survival time and the factors that its influence.

The survival analysis is a series of useful statistical procedures to analyze the timing of events. Explains that the survival analysis is a statistical analysis that model the length of time until an event occurs [5]. The survival analysis is used with emphasis on individual survival time or duration of individual survival time until an event occurs (time to event data). Such events may be the development of a disease, the response to treatment, the recurrence of the disease, death or even other occurrences determined by the researcher. In the hazard ratio model, the resulting value is no longer time-dependent so that the ratio of the two objects remains at all times proportional hazard. The proportional cox regression model emphasizes the fulfillment of the proportional hazard assumption which means that the ratio between individual hazard functions of one and other individual hazard functions is constant. However, in some cases, the covariate hazard ratio is not over time or not fixed. Therefore, proportional hazard assumptions are not met. To see the effect of a covariate that does not meet the proportional hazard assumption, the covariate must be converted into a time-consuming covariate, called a folded proportional hazard. This model is referred to as an extension of the proportional cox model containing the time-bound or multiplication variable of the independent variable with the time function. Some of the time functions that can be used in the extendend model are g (t) = 0, g (t) = t and g (t) = log (t) . This model aims to determine how much influence of variables on the survival time of a breast cancer patient.

Research related to non-proportional hazard has been done by Feriana and Dwi [6]. Modeled lung cancer data using cox stratification model [6]. Furthermore, Arini and Isna [7] has done research on extended cox model-to-model patient data of kidney transplanted. This study found that patients with catheters were more rapidly infectious than those with percutaneous instances. According to the study, Iskandar and Bayu [8] examined the cox proportional hazard model on the Joint occurrence in the case of traffic accidents in the United States obtained results that age and use of seat belts increase the risk of accidents. In addition, Pahlevi et al. [9] has conducted research on cox stratified regression model on unspecified Hemorrhagic Stroke. In addition, Vitriana et al. [10] examined the procedure for establishing the extended Cox model on joint events and its application in the case of individuals quitting work. Provided by Cox extended model with significant variables, namely age variable, marital status, age bound time and marital status bound time. Based on these reviews, this paper is interested to model the breast cancer patients' survival time based on the factors influenced using the cox extended proportional hazard.

This paper outlines in section 2 a materials and method such as survival analysis, proportional hazard model, estimated cox proportional hazard model and data collect. Section 3, we illustrate the performance of our models with a breast cancer data set from the Unhas Education Hospital database. Then, section 4 concludes the paper with a summary and some possible areas of extension.

2. Materials and Methods

2.1 Study Area

The total population of Makassar City based on data from Statistics Center Bureau called Badan Pusat Statistik (BPS) in 2015 was recorded at 1,449,401 inhabitants. Based on sex ratio, female population was more than male population with sex ratio of 97.84. Which means that every 100 souls of female population there are 97 male population. South Sulawesi case of breast cancer is the first rank of cancer which suffered by many women. The number of patients treated during 2010 found 132 cases of breast cancer, in 2011 found 360 cases of breast cancer, in 2012 there was an increase to 573 cases of breast cancer and in 2013 as many as 592 cases of breast cancer.

2.2. Data Collection

Data of this research is medical record of patient of Breast Cancer at RSUP Universitas Hasanuddin Makassar City with number of patient counted 101 people. The covariates of this study were age, medication, occupation, marital status, malignancy, histology, stage, leukite, hemoglobin, and sensor indicators with 6.94% censored data and 93.06% uncensored data. Interval age of breast cancer patients is ranged in age 21-75 years with the average patient age is 43.77 years.

2.3 Survival Analysis

Survival analysis is a statistical technique used to analyze data. This technique aims to determine the results of variables that affect an event beginning to the end of the event, such as time (day, week, month or year). Forms of initial events for example the initial infected patients and the end events such as death or recovery of patients [11]. Let X denote the time of failure or death time, then T can be viewed as a nonnegative random variable. Survival function is the chance that an individual can survive till time (experience after time t) (Thamrin, 2013a). The function of Survival is defined as $S(t) = P(T>t) = \int_x^\infty f(t)dt$ and $f(t) = -\frac{dS(t)}{dt}$, where $F(x) = P(X \le x)$ is a cumulative distribution function.

According to Miller (1998), the life function S (t) is a function that does not rise or monotone down with properties:

1. $S(t) = 1$ for $t = 0$, it means individual opportunities to live at the moment $t = 0$ is 1
2. $S(t) = 0$ for $t = \infty$, it means individual opportunities to live at the moment $t = \infty$ is 0

The other functions related to the function of survival is the hazard function. The hazard rate function is defined as the conditional failure rate i.e. the limit of the probability of an individual failing to persist in very short intervals of time t to $t + \Delta t$, if the individual has survived until time t [4] This function is defined as $h(t) = \lim_{\Delta t \to 0} \frac{P[t \le X < t+\Delta t | T \ge t]}{\Delta t}$ where the density function is probability

$$f(t) = h(t).S(t) \quad (1)$$

The proportional hazard model depends on the values $x_1, x_2, \ldots, x_p$ form p explanatory variables, $X_1, X_2, \ldots, X_p$. The values of the explanatory variables in the proportional hazard model are expressed in vector form $\boldsymbol{x}$, so that $\boldsymbol{x} = (x_1, x_2, \ldots, x_p)'$. Suppose $h_0(t)$ is a hazard function for an individual that all vector explanatory variables **x** has a value of zero, then the function $h_0(t)$ called baseline hazard function. The general proportional hazards model is as follows

$$h_i(t) = \exp(\beta_1 x_1 + \beta_2 x_2 + \cdots + \beta_p x_p)\, h_0(t) \quad (2)$$

can also be expressed in terms of log equations is

$$\log \frac{h_i(t)}{h_0(t)} = (\beta_1 xi_1 + \beta_2 xi_2 + \cdots + \beta_p xi_p), \quad (3)$$

Cox proportional model is used to determine the effect of variables on survival of cancer patients. This model makes it possible to isolate variables that have little effect on survival. In addition, the model allows to estimate the risk or danger of death for an individual based on the prognostic variable.

The Weibull distribution is one of the continuous distributions in probability theory and statistics. Weibull distribution can be used if in a data survival hazard function increases or decreases monotonically with an increase in survival time. Cox proportional hazard model uses Weibull distribution:

$$h(t, X) = \lambda p t^{p-1}, \quad (4)$$

where $\lambda = \exp\left[\sum_{i=1}^{p} \beta_i X_i\right]$ and

$$h_o(t) = p t^{p-1}, \quad (5)$$

with the shape of Weibull hazard is a parametric model with form parameters (λ), scale parameters (p) and covariate (β_i) not known.

2.4 Extended Cox

In time dependent covariate is, $X_b(t) = X_b.g_b(t)$, where X_b is a covariate that has no time dependency. However, because it does not meet the assumption Proportional Hazard, X_b must be interacted with the time function. If all p covariate does not meet the proportional hazard assumption, then the number of p covariates must be interacted with time. The time function for covariate-b is defined as $g_b(t)$, obtained by the form of extended cox as follows:

$$h(t,X(t)) = h_o(t)\exp\left[\sum_{b=1}^{p}\beta_b X_b + \sum_{b=1}^{p}\delta_b X_b g_b(t)\right] \quad (6)$$

where available p_1 covariate that meets the proportional hazard assumptions and exists p_2 which does not meet the proportional hazard assumption with $p_1 + p_2$, then obtained the model as follows:

$$h(t,X(t)) = h_o(t)\exp\left[\sum_{a=1}^{p_1}\beta_a X_a + \sum_{b=p_1+1}^{p_2}\beta_b X_b + \sum_{b=p_1+1}^{p_2}\delta_b X_b g_b(t)\right] \quad (7)$$

with time function $g_b(t)$ which can be used is: $g_b(t) = 0, g_b(t) = t, g_b(t) = \ln t$

3. Results

The formation of Cox proportional hazard model was conducted to determine the relationship between survival time and the variables suspected to influence survival time. Estimation of Cox proportional hazard model parameters was done using Breslow method approach.

Table 1. Parameter Estimation Results Cox Proportional Hazard Model of Breast Cancer.

Variable	Coefficient	exp (β_j)	SE
Age	0,018915	1,019095	0,009109
Treatment	0,153104	1,165447	0,113618
Work	0,397254	1,487734	0,251960
Marital status	0,827081	2,286634	0,350766
Degree of malignancy	0,053600	1,055062	0,163627
Histology	0,096296	1,101081	0,144184
Stadium	0,122131	1,129902	0,116122
Haemoglobin levels	0,052927	1,054352	0,207227
Leukocyte levels	0,146134	1,157351	0,247730

Based on Table 1 we get the Cox proportional hazard model, that is:

$$h_i(t_i) = h_0(t)\exp(0{,}018915X_1 + 0{,}153104X_2 + 0{,}397254X_3 + 0{,}827081X_4 + 0{,}0536X_5 + 0{,}096293X_6 + 0{,}122131X_7 + 0{,}052927X_8 + 0{,}146134X_9)$$

Furthermore, based on the model formed will be tested partial likelihood ratio log obtained, ln L(0) = -353,4687 and ln L($\hat{\beta}$) = -335,8804 so that the value can be obtained G, that is 35,1766. Earned value G = 35,1766 $\geq \chi^2_{(0,05;9)}$ with value $\chi^2_{(0,05;9)}$ = 16,9190, then H_0 rejected and concluded that there is at least one influential variable in the model established.

All covariates affecting breast cancer are tested for parameters. Partial test results parameters with Wald test using Breslow method approach as follows:

Table 2. Partial test results parameters.

Variable	Coefficient	SE	Z^2_{hitung}
Age	0,020912	0,008823	5,6169
Treatment	0,1376	0,1115	1,52276
Work	0,3263	0,2173	2,256
Marital status	0,544	0,319	2,90703
Degree of malignancy	-0,08573	0,15004	0,32604
Histology	0,1959	0,1273	2,37468
Stadium	0,1578	0,1015	2,41802
Haemoglobin levels	-0,1168	0,1922	0,36966
Leukocyte levels	0,3217	0,2126	2,28917

Based on Table 2, it is found that the overall variables affect the survival time because of the value $Z^2_{hitung} \geq \chi^2_{(0,05;1)}$ with $\chi^2_{(0,05;1)}$ = 0,004. So obtained model:

$$h_i(t_i) = h_0(t)\exp(0{,}020912X_1 + 0{,}1376X_2 + 0{,}3263X_3 + 0{,}544X_4 \pm 0{,}08573X_5 + 0{,}1959X_6 + 0{,}1578X_7 \pm 0{,}1168X_8 + 0{,}3217X_9) \quad (8)$$

3.1 Testing Proportional Hazard Assumption

Testing proportional hazard assumption in this study using Schoenfeld residual as follows:

Table 3. Correlation, χ^2_{hit}, and p-value variable is free

Variable	Correlation	χ^2_{hit}	p-value
Age	0,00742	0,00395	0,95
Treatment	-0,0657	0,528	0,468
Work	0,00522	0,0028	0,958
Marital status	0,106	1,06	0,302
Degree of malignancy	**-0,197**	4,45	**0,035**
Histology	0,137	1,57	0,211
Stadium	-0,161	2,94	0,0863
Haemoglobin levels	-0,0504	0,252	0,616
Leukocyte levels	**0,284**	6,44	**0,0111**

Table 3 shows that age, medication, occupation, marriage, histology, stage, and Hemoglobin levels variables meet assumptions Proportional Hazard with p-value > α (0,05). Meanwhile, the variables Degree of malignancy and Leukocyte levels have p-value respectively, that is 0,035 and 0,0111 smaller than α (0,05) so it does not fulfill the Proportional Hazard assumption. Based on the test using Schoenfeld residual obtained the conclusion that the variable degree of malignancy and Leukocytes did not meet the assumption of proportional hazard so this study was continued to the extended proportional Hazard cox model.

IOP Conf. Series: Journal of Physics: Conf. Series **979** (2018) 012087 doi:10.1088/1742-6596/979/1/012087

3.2 Cox Extended Proportional Hazard (PH) model

Based on the assumption of proportional hazard test, it is found that the variable that does not meet the assumption, that is the variable of malignancy degree and the leukocyte level does not meet the proportional hazard assumption so that new model is needed to overcome the variable that does not meet the cox proportional hazard that is Cox Extended. Therefore, added the time function $(t) = t$ on variable Degree of malignancy and Leukocyte levels, that is as follows:

$$h_i(t_i) = h_0(t)\exp(\beta_1 X_1 + \beta_2 X_2 + \beta_3 X_3 + \beta_4 X_4 + \beta_5 X_5 g(t) + \beta_6 X_6 + \beta_6 + \beta_8 X_8 + \beta_9 X_9 g(t)) \quad (9)$$

Then, this model estimated parameter, with $g(t) = t$ using the Breslow method approach

Table 4. Estimation of Model parameters *Cox Extended* with $g(t) = t$ using the Breslow method approach

Variable	Coefficient	$\exp(\beta_j)$	SE
Age	0,01649	1,01663	0,00929
Treatment	0,15967	1,17313	0,11561
Work	0,44844	1,56587	0,25622
Marital status	0,88150	2,41451	0,35139
Degree of malignancy	0,06355	1,06561	0,17211
Degree of malignancy g(t)	**0,27402**	**1,31524**	**0,53929**
Histology	0,05199	1,05337	0,14669
Stadium	0,14433	1,15526	0,11566
Haemoglobin levels	0,06206	1,06402	0,20781
Leukocyte levels	-0,06370	0,93829	0,27107
Leukocyte levels g(t)	**1,52563**	**4,59806**	**0,69343**

Based on Table 4 we get the Cox extended model equation with time function $g(t) = t$ as follows:

$$h_i(t_i) = h_0(t)\exp\,[0{,}01649X_1 + 0{,}15967X_2 + 0{,}44844X_3 + 0{,}88150X_4 + 0{,}06355X_5 + 0{,}27402X_5 g(t) + 0{,}05199X_6 + 0{,}14433X_7 + 0{,}06206X_8 - 0{,}06370X_9 + 1{,}52563X_9 g(t)].$$

From the model obtained is done partial likelihood logging, so obtained ln $L(\hat{\beta}) = -353{,}4687$ and ln $L(\hat{\delta}) = -333{,}3845$, so that the value can be obtained G, that is 40,1684. Earned value $G = 40{,}1684 \geq \chi^2_{(0,05;11)}$ with value $\chi^2_{(0,05;11)} = 4{,}575$, then H_0 rejected and concluded that there is at least one influential variable in the model established. Then, then this model is tested partially with Wald Test, as follows

Table 5. Parameter Test Results Partially with Wald Test

Variable	**Coefficient**	**SE**	Z^2_{hitung}
Degree of malignancy g(t)	-0,1248	0,8827	0,082944
Leukocyte levels g(t)	1,4206	4,1397	6,431296

Based on Table 5 it is found that the malignant degree variable is bound by time and affects the survival time with Z^2_{hit} = 0,082944 > $\chi^2_{(0,05;1)}$ with $\chi^2_{(0,05;1)} = 0{,}004$ other than that variable Leukocyte levels Time bound and effect on survival time with Z^2_{hit} = 6,431296 > $\chi^2_{(0,05;1)}$ with $\chi^2_{(0,05;1)} = 0{,}004$.

4. Conclusion

In this paper, we modeled survival time for breast cancer patients using extended cox proportional hazard. Based on the results of the data processing interpretation of the single unit, the individual has the risk of death is equal to $\exp(0{,}06355 + 0{,}27402) = 0{,}27757$, So each addition of one degree will decrease the risk of dying in breast cancer patients by 0,2. Every individual has a leukocyte level < 4500 /mm^3 has a risk of unity is as big $\exp(0{,}06370 + 1{,}5463) = 0{,}33757$. Therefore, individuals who have leukocyte levels > 4500 /mm^3 has a lower risk of death rate of 0,3.

Acknowledgments

The authors acknowledge to the Directorate of Research and Community Service, Directorate General Strengthening Research and Development Ministry of Research, Technology and Higher Education with Student Creativity Program (PKM) with the grant number 11153/UN4.1.3/PM.05/2017.

Appendix

```
cox<-
coxph(Surv(Dirawat,Status)~Usia+Pengobatan+Pekerjaan+Pernikahan+Keganasan+Histologi+Stadium+
Hb+wbc, method="breslow",data=DataFIX)
summary(cox)
Call:
coxph(formula = Surv(Dirawat, Status) ~ Usia + Pengobatan + Pekerjaan +
  Pernikahan + Keganasan + Histologi + Stadium + Hb + wbc,
  data = DataFIX1, method = "breslow")
```

For full ODC files contact the author directly.

References

[1] World Health Organisazation. 2013a Cancer Control: A Global Snaptshot in 2015 http://www.who.int/cancer/cancer-snapshot-2015/en/ Accesed on 9 November 2016

[2] World Health Organisazation 2013b Breast Cancer Awareness Month: increased awareness, equitable access to early diagnosis and timely, effective, and affordable treatment needed globally http://www.who.int/gho/publications/world_health_statistics/2013/en/ Accesed on 9 November 2016

[3] Kementrian Kesehatan Republik Indonesia 2015 *Infodatin: Pusat Data dan Informasi Kementrian Kesehatan RI.* http://www.depkes.go.id/resources/ download/pusdatin/infodatin/infodatin-kanker.pdf Accesed on 12 Oktober 2016

[4] Basu S and Tiwari R C 2010 Breast cancer survival, competing risk and mixture cure model: A Bayesian analysis *Journal of the Royal Statistical Society* **173**(2) 307–329

[5] Kleinbaum D C and Klein M 2005 *Survival Analysis* (New York: Springer)

[6] Feriana and Dwi A 2011 *Model Cox Stratifikasi* (Jakarta: Universitas Indonesia)

[7] Arini and Isna N 2011 *Extended Cox Model untuk Time-Independent Covariate yang Tidak Memenuhi Asumsi Proportional Hazard pada Model Cox Proportional Hazard* (Jakata: Universitas Indonesia)

[8] Iskandar and Bayu M 2015 *Model Cox Proportional hazard Pada Kejadian Bersama* (Jojakarta: Universitas Negeri Yogyakarta)

[9] Pahlevi, Mohammad R, Mustafid, Triastuti W 2016 Model Regresi Cox Stratified pada Data Ketahanan *Junal Gaussian,* **5**(3) 455-464

[10] Vitriana, Anita N, Rosita K 2016 Model Cox Extended dengan g(t) = t untuk Mengatasi Nonproportional Hazard pada Kejadian Bersama Jogjakarta

[11] Kleinbaum D C and Klein M 2011 Survival Analysis (New York: Springer)

The Application Law of Large Numbers That Predicts The Amount of Actual Loss in Insurance of Life

Georgina Maria Tinungki

Department of Mathematics Faculty of Mathematics and Natural Science Hasanuddin University Makassar 90245, Indonesia

E-mail : ina_matematika@yahoo.co.id

Abstract. The law of large numbers is a statistical concept that calculates the average number of events or risks in a sample or population to predict something. The larger the population is calculated, the more accurate predictions. In the field of insurance, the Law of Large Numbers is used to predict the risk of loss or claims of some participants so that the premium can be calculated appropriately. For example there is an average that of every 100 insurance participants, there is one participant who filed an accident claim, then the premium of 100 participants should be able to provide Sum Assured to at least 1 accident claim. The larger the insurance participant is calculated, the more precise the prediction of the calendar and the calculation of the premium. Life insurance, as a tool for risk spread, can only work if a life insurance company is able to bear the same risk in large numbers. Here apply what is called the law of large number. The law of large numbers states that if the amount of exposure to losses increases, then the predicted loss will be closer to the actual loss. The use of the law of large numbers allows the number of losses to be predicted better.

1. Introduction

The law of large numbers describes the stability of large random variables. In the law of large numbers it is stated that if given a random sample of a distribution with its mean and its variance limited, then the average sample would be close to the population average [1]. The insurance business has a close relationship with statistical theory, especially the probability theory of managing the risks it assures [2].

The importance of mobile insurance to take over the risks that may occur due to destruction, theft or train accidents or when accidents cause direct damage to third parties. The principle brought forth from the theory of probability is named after a laws of large number [3]. The method undertaken by the insurance business in applicants of the laws of large numbers is increasing the number of risks themselves. If there are random variables so the random variable must be stable for a long time. That is, in a long time the value of observation will always be close to the value of expectation [4].

Insurance companies have principles in setting premiums. In calculating the possibility of certain disasters, such as accidents occurring with a home fire event, thereby increasing this value in return for being gained in such an event [5]. This amount is the amount that the company will give in paying the average for each person they cover. They then set interest rates at a rate that includes these "costs" in addition to their benefits. The policyholder acquires peace in mind because the insurance company can provide effective results in minimizing the risk of potential loss in a disaster [5].

Insurance companies will provide regular payments in exchange for massive benefits in the absence of substantial claims.

The Law of Large Numbers is a very powerful tool that is expected to say the definite things about the real-world results of unexpected un expected events. This useful tool represents only one example of how the Great Numbers Law can be used to handle randomness. The Law of Large Numbers applies to specific results and their probabilities, but what about the overall range of possible outcomes and their associated possibilities? Just as the frequency of a particular event will tend toward its probability in the long run, a series of possible outcomes each tends to match its own probability. Studying the probability distribution of results and probabilities will give us more powerful tools to predict long-term average behavior [6].

2. Life insurance

Life insurance is a business. But it is only a business for those companies who are able to maintain their financial strength while paying out claims. While insurance helps you manage risk by protecting against things that would significantly impact your financial future if they occurred, the law of large numbers helps insurance companies by making predictable, with reasonable accuracy, the claims it will pay from year to year [7].

In the insurance business, the risks faced by each individual are transferred to the life insurance company, which agrees to indemnify the amount specified in the policy contract. To compensate for this loss, the insurer sets the premium to be paid by the insured individual. In setting a premium, there are a few things to note: possible losses, the value of any loss, administrative costs, required to run a business, such as collecting premiums from each member, measuring losses, paying claim, threshold errors that may arise when predicting losses, other factors such as financial, health, and social factors. Mistakes in measuring these factors can cause losses for life insurance companies, such as setting a smaller premium than they should [8]. Life insurance, as a tool to spread risk, can only work if a life insurance company is able to bear the same risk in large numbers. Here apply what is called the law of large numbers [9].

3. The Law of Large Numbers

The law of large numbers is the principle of statistics and probability theory which states that the more the number of samples used from an event, the monitoring results may be closer to the average population. Simply put in the world of insurance is: the more people who join the insurance, then the likelihood of losses will be close to the expected loss [4].

The law of large numbers if linked in insurance means that the insurer can determine the mortality rate (mortality rate) and morbidity level (the level of illness, injury and the occurrence of health failure) of the insured person. The mortality rate and morbidity are used in the calculation of the insurance company's premium rate to various types insurance products issued [10].

The purpose of this first requirement is for the insurance company to predict the potential for losses based on "the law of large number". Loss data can be complicated over time, and losses can be accurately predicted. So the cost of losses that can be distributed evenly distributed to all units of the insured in accordance with the class guaranteed by the insurance.

When given a random sample taken from an identical and free random variable. While the mean and its variance are, then the average of the sample will be close to the average of the population [11].

Let, $Y_1, Y_2, \ldots\ldots, Y_n$ is a random variable array (for n $\geq$ 1) defined in the sample space Ω the same one. Let Y Other random variables are defined in the same sample space. We will discuss what is meant by Y_n "toward (or convergent) to " Y.

We say Y_n converges to Y with probability one if $P\left[\lim_{n\to\infty} Y_n = Y\right] = 1$. (this also called almost sure convergence, or almost everywhere, or strong convergence).

Definition 1 :

We say Y_n converges to Y with probability one if $P\left[\lim_{n\to\infty} Y_n = Y\right] = 1$. (this also called almost sure convergence, or almost everywhere, or strong convergence.

IOP Conf. Series: Journal of Physics: Conf. Series **979** (2018) 012088 doi:10.1088/1742-6596/979/1/012088

Definition 2 :

We say Y_n converges to Y in probability ($Y_n \xrightarrow{p} Y$) if for every $\varepsilon > 0$

$$\lim_{n\to\infty} P[|Y_n - Y| > \varepsilon] = 0$$

(This is also called stochastic convergence, or convergence in measure, or weak convergence).

Definition 3 :

We say Y_n converges to Y in mean square if

$$\lim_{n\to\infty} E(Y_n - Y)^2 = 0$$

(we also say $Y_n \to Y$ in L_2 or $Y_n \xrightarrow{2} Y$)

Definition 4:

We say Y_n converges to Y in distribution ($Y_n \xrightarrow{d} Z$) if

$$\lim_{n\to\infty} F_{Y_n}(t) = F_Y(t)$$

at each point t where $F_Y(.)$ is continuous. In such a case , we also say that $F_n(t) = F_{Y_n}(t)$ converges weakly to $F(t) = F_Y(t)$ and write $F_n \to F$

4. Proof of Weak Large Number Law with Chebyshev's Inequality

The most important type of convergence in statistics is convergence in probability and distribution, these two concepts of convergence are called "weak" convergence types, and are generally easier to prove than "strong" convergence.

The weak large number law explain that mean sample converges in probability to expectation value $\lim_{n\to\infty} P(|\overline{X}_n - \mu| < \varepsilon) = 1$

It is shown to every $\varepsilon > 0$,

It is mentioned weak law because the converges in probability is indeed weak in convergence random variable.

The proof of weak large number law by using Chebyshev's inequality

Known: $$\overline{X}_n = \frac{1}{n}(X_1 + X_2 + \ldots + X_n)$$

Will be Proofed : $$\overline{X}_n \xrightarrow{P} \mu \quad for \quad n \to \infty$$

$$\text{var}(\overline{X}_n) = \frac{n\sigma^2}{n^2} = \frac{\sigma^2}{n}$$

$$E(\overline{X}_n) = \mu$$

$$P(|\overline{X}_n - \mu| \geq \varepsilon) \leq \frac{\sigma^2}{n\varepsilon^2}$$

or :

$$P(|\overline{X}_n - \mu| < \varepsilon) \geq 1 - \frac{\sigma^2}{n\varepsilon^2}$$

As long as n to ∞, according to definition the converges to probability can be:

$$\overline{X}_n \xrightarrow{P} \mu \quad for \quad n \to \infty$$

5. The Strong Law of Large Number

The strong law of large numbers explain that mean sample certainly will convergence to expectation value.

$$\overline{X}_n \xrightarrow{a.s} \mu \quad for \quad n \to \infty$$

It is explained to every ε>0,

$$\lim_{n\to\infty} P\left(\overline{X}_n = \mu\right) = 1$$

The proof of this law is more complex than the weak law. This law makes true the interpretation that based on the intuition from expectation value of random variable as *"Long-term average when sampling repeatedly"*. It is mentioned the Strong Law because the convergence almost sure is the strong convergence on random variable.

Definition 5 :

The sequence random sample Y_n is called converges in probability to Y, written $Y_n \xrightarrow{p} Y$

If :

(i) $\lim_{n\to\infty} P_r[|Y_n - Y| < \varepsilon] = 1$

(ii) $\lim_{n\to\infty} P_r[|X_n - X| \geq \varepsilon] = 0$

Theorem 1 :

If $Y_n \xrightarrow{p} c$, then for any function g(y) that continue on c,

$$g(y) \xrightarrow{p} g(c)$$

Proof :

Because g(y) is continue on c, then

$\forall\varepsilon > 0\ \exists\delta > 0 \ni |y - c| < \delta$ where $|g(y) - g(c)| < \varepsilon$

Because $|Y_n - c| < \delta \subset |g(Y_n) - g(c)| < \varepsilon$

Then $P(|g(Y_n) - g(c)| < \varepsilon) \geq P(|Y_n - c| < \delta)$

Because $Y_n \xrightarrow{p} c$, this case means for $\delta > 0$

$$P(|Y_n - c| < \delta) = 1$$

$$P(|g(Y_n) - g(c)| < \varepsilon) \geq P(|Y_n - c| < \delta) = 1$$

Because $P(|g(Y_n) - g(c)| < \varepsilon)$ impossible more than 1

Then $P(|g(Y_n) - g(c)| < \varepsilon) = 1$

$\therefore g(Y_n) \xrightarrow{p} g(c)$

Theorem 2 :

If X_n and Y_n is sequence of random variable where $X_n \xrightarrow{p} c$ and $Y_n \xrightarrow{p} d$, then :

1. $aX_n + bY_n \xrightarrow{p} ac + bd$
2. $X_nY_n \xrightarrow{p} cd$
3. $\frac{X_n}{c} \xrightarrow{p} 1\, p(X_n \geq 0)\ \forall n$, for $c \neq 0$
4. $\frac{1}{X_n} \xrightarrow{p} \frac{1}{c}$, if $p(X_n \neq 0) = 1,\ \forall n,\ c \neq 0$
5. $\sqrt{X_n} \xrightarrow{p} \sqrt{c}$, if

Proof theorem 2.

3. If
$X_n \xrightarrow{p} c \neq 0$ X_n is sequence of random variable

Conclusion $\frac{X_n}{c} \xrightarrow{p} 1$

Proof :
Known : $\lim_{n\to\infty} Pr(|X_n - c| \geq \varepsilon) = 0$
Will shown :
$\lim_{n\to\infty} Pr\left(\left|\frac{X_n}{c} - 1\right| \geq \varepsilon\right) = 0$

Solution :
Take $\varepsilon > 0$ any
$\lim_{n\to\infty}\left(\left|\frac{X_n}{c} - 1\right| \geq \varepsilon\right) = \lim_{n\to\infty}(|X_n - c| \geq \varepsilon|c|)$
Example : $\varepsilon = \varepsilon|c|$
Applying $\lim_{n\to\infty} Pr(|X_n - c| \geq \varepsilon') = 0$
$\therefore \lim_{n\to\infty} Pr(|X_n - c| \geq \varepsilon) = \lim_{n\to\infty} Pr(|X_n - c| \geq \varepsilon|c|) = 0$

4. Known : X_n is random variable
$p(X_n \neq 0) = 1,\ \forall n$
$X_n \xrightarrow{p} c$ then $\forall \varepsilon \geq 0 \ni \lim_{n\to\infty} Pr(|X_n - c| < \varepsilon) = 1$
Will proof : - $\frac{1}{X_n} \xrightarrow{p} \frac{1}{c}, c \neq 0$
- $\lim_{n\to\infty} Pr(|X_n - c| < \varepsilon) = 1$
Take $\varepsilon > 0$ any

$$|X_n - c| < \varepsilon \quad = \left|(X_n - c)\frac{X_n c}{X_n c}\right| < \varepsilon$$
$$= \left|\left(\frac{1}{c} - \frac{1}{X_n}\right)X_n c\right| < \varepsilon$$
$$= \left|\left(\frac{1}{c} - \frac{1}{X_n}\right)\right| |X_n c| < \varepsilon$$
$$Pr(|X_n - c| < \varepsilon) = Pr\left(\left|\frac{1}{c} - \frac{1}{X_n}\right| |X_n c| < \varepsilon\right)$$
$$= Pr\left(\left|\frac{1}{X_n} - \frac{1}{c}\right| |X_n c| < \varepsilon\right)$$
$$= Pr\left(\left|\frac{1}{X_n} - \frac{1}{c}\right| < \frac{\varepsilon}{|X_n c|}\right)$$

Because
$$Pr\left(\left|\frac{1}{X_n} - \frac{1}{c}\right| < \frac{\varepsilon}{|X_n c|}\right) \leq Pr\left(\left|\frac{1}{X_n} - \frac{1}{c}\right| < \frac{\varepsilon}{|c|}\right) \leq 1$$

Example

$\varepsilon' = \frac{\varepsilon}{|c|}$, *untuk* $c \neq 0$

$\lim_{n\to\infty} Pr(|X_n - c| < \varepsilon) \leq \lim_{n\to\infty} Pr\left(\left|\frac{1}{X_n} - \frac{1}{c}\right| < \varepsilon'\right) = 1 \leq 1$

$1 \leq \lim_{n\to\infty} Pr\left(\left|\frac{1}{X_n} - \frac{1}{c}\right| < \varepsilon'\right) \leq 1$

$\therefore \lim_{n\to\infty} Pr\left(\left|\frac{1}{X_n} - \frac{1}{c}\right| < \varepsilon'\right) = 1$

$\therefore \frac{1}{X_n} \xrightarrow{p} \frac{1}{c}$

5. Known : X_n is random variable
   $p[X_n \geq 0] = 1, \forall n$
   $X_n \xrightarrow{p} c$ means $\lim_{n\to\infty} Pr(|X_n - c| < \varepsilon) = 1$
   Is asked $\sqrt{X_n} \xrightarrow{p} \sqrt{c}$ =.......?
   The meaning will proof $\lim_{n\to\infty} Pr\left(|\sqrt{X_n} - \sqrt{c}| < \varepsilon\right) = 1$
   Take $\varepsilon > 0$ any

$$Pr(|X_n - c| < \varepsilon) = Pr\left(|(\sqrt{X_n} - \sqrt{c})(\sqrt{X_n} + \sqrt{c})| < \varepsilon\right)$$
$$= Pr\left(|\sqrt{X_n} - \sqrt{c}| < \frac{\varepsilon}{\sqrt{X_n} + \sqrt{c}}\right)$$
$$\geq Pr\left(|\sqrt{X_n} - \sqrt{c}| < \frac{\varepsilon}{\sqrt{c}}\right)$$
$$\geq 1$$

   For example $\varepsilon' = \frac{\varepsilon}{\sqrt{c}}$ for $c > 0$
   Applying :

$$1 = \lim_{n\to\infty} Pr(|X_n - c| < \varepsilon) \leq \lim_{n\to\infty} Pr\left(|\sqrt{X_n} - \sqrt{c}| < \varepsilon'\right) \leq 1$$

   Because $1 \leq \lim_{n\to\infty} Pr\left(|\sqrt{X_n} - \sqrt{c}| < \varepsilon\right) \leq 1$
   So, $\lim_{n\to\infty} Pr\left(|\sqrt{X_n} - \sqrt{c}| < \varepsilon\right) = 1$
   $\therefore \sqrt{X_n} \xrightarrow{p} \sqrt{c}$

6. Application of Law of Large Numbers in Life insurance

The gross premium collected by the insurance industry in 2014 reached the amount of Rp247.29 trillion, increased by 28.0% compared to the previous year figure of Rp193.06 trillion. Within the last five years, the average of gross premium annual growth was around 18.5%. If that gross premium is compared to Indonesian population in 2014, that is 252 million people, this will be resulted in insurance density of Rp981,297. It means, in average, each Indonesian person spent Rp981,297 for insurance premium.

In Life Insurance to know the amount of risk, widely used mathematical / statistical formulas, namely the theory called probability theory. The degree of risk depends on the size of the deviation between the expected and the actual event.

$$Degree\ of\ Risk = \frac{Area\ of\ Uncertainty}{Exposures}\ x\ 100\%$$

The prescribed premium is based on the mortality table which is a table describing the number of possible deaths by age in which the table is based on a large group of people. In order for an insurance company to experience the exact mortality table, the submissions or risks received must be dispersed. This means that the risk should not be concentrated on a particular loss even more that is catatrophic (explosive risk), Meanwhile, the contribution of insurance industry to Gross Domestic Product (GDP), as measured by ratio of gross premium to GDP (penetration), increased by 0.22% from 2.13% in 2013 to 2.35% in 2014. Table 2 shows the ratio of gross premium relative to Indonesian GDP from 2010 to 2014 [12].

IOP Conf. Series: Journal of Physics: Conf. Series **979** (2018) 012088 doi:10.1088/1742-6596/979/1/012088

Table 1. Gross Premiums (In trillion Rupiah).

Year	Gross Premiums		Gross *Domestic Product*		Ratio a / b (%)
	Total	Growth (%)	Total	Growth (%)	
2010	125,12	17,5	6.422,90	14,4	1,95
2011	153,13	22,4	7.427,10	15,6	2,06
2012	175,89	14,9	8.241,90	11,0	2,13
2013	193,06	9,8	9.084,00	10,2	2,13
2014	247,29	28,0	10.542,70	16,1	2,35

Source : BPS, 2017

Based on table 1 above shows that the gross premium collected by the insurance industry in 2014 reached the amount of Rp247.29 trillion, increased by 28.0% compared to the previous year figure of Rp193.06 trillion. Within the last five years, the average of gross premium annual growth was around 18.5%. If that gross premium is compared to Indonesian population in 2014, that is 252 million people, this will be resulted in insurance density of Rp981,297. It means, in average, each Indonesian person spent Rp 981,297 for insurance premium [12].

The highest increase of gross premium in 2014 was reached by agencies administering social insurance (69.4%), and followed by non life and reinsurance (27.8%), life insurance companies (-0.3%), companies administering mandatory Insurance (-36.9%). The biggest contribution to gross premium of insurance industry in 2014 was gross premium of life insurance companies (45.6%), followed by those of agencies administering social insurance (28.1%), gross premium of non life and reinsurance companies (22.1%); and premium collected [12].

7. The Application in insurance :

The probability of a 20-year-old and another 40 years old, both going to live another 20 years is 0.6. Of the 50,000 people who lived at the age of 20, 3000 of them died before the age of 25 years. So, the chances that a 25-year-old will now die before reaching the age of 60.

For example, lx = the exact number of people aged x

l_{20} = 50.000 people

l_{25} = 50.000 – 3000 = 47.000 people

$$X_i = \begin{cases} 0{,}6, & \text{live another 20 years} \\ 0, & \text{others} \end{cases}$$

To calculate the probability that a 25-year-old would die before the age of 60, it would take (1) l25 and (2) l25 - l60 (25 people die before the age of 60). Or the opportunity you want to look for is

$$\frac{(l_{25} - l_{60})}{(l_{25}} = ...?$$

Suppose too $_nP_x$ = probability of people aged x years

Is known, $_{20}P_{20}$ .... $_{20}P_{40}$=0,6, ie the probabilities of 20-year-olds will live 20 more years and 40-year-olds will live another 20 years, because both events are independent then multiplied. However, it can be said also: Opportunities of 20-year-olds will live 20 more years) to 40 years) are $_{20}P_{20}$ and the person will reach the age of 60 if he reaches the age of 40 years is $_{40}P_{20}$

Thus, the probability of a person aged 20 years reaching the age of 60 years is:

$$_{40}P_{20} = {}_{20}P_{20} \cdot {}_{20}P_{40} = 0{,}6$$

Obtained,

$$l_{60} = l_{20} \cdot {}_{40}P_{20} = (50.000)(0{,}6) = 30.000 \text{ people}$$

So the probability 25-year-old would die before reaching the age of 60 is (47.000 – 30.000) / 47.000 = 17/47.

Based on the above discussion, it is seen that to convergence while n → ∞ because even though the real n may be small or medium magnitude. However it is important to know whether the workings used are good enough for a very large n (n → ∞). Intuitively if a procedure is unlikely to be good although with so many observations it may also not be good with only a few observations.

8. Conclusion

Based on the above discussion can be concluded that :

(a) If random sampling of size n of an arbitrary population with a mean of μ and finite variance, the greater the sample n, the closer to the population, according to the law of large numbers definition, or in accordance with the equation : $\overline{X}_n \xrightarrow{P} \mu \quad for \quad n \to \infty$

(b) Law of large numbers is important because it "guarantees" a stable long-term outcome for the average of some random events.The greater the sample n, the closer to the population, according to the law of large numbers definition, the more people who join the insurance, then the likelihood of losses will be close to the expected loss. Simply put in the world of insurance is: the more people who join the insurance, then the likelihood of losses will be close to the expected loss.

(c) To calculate the probability that a 25-year-old would die before the age of 60, it would take (1) l_{25} and (2) $l_{25} - l_{60}$ (25 people die before the age of 60).

References

[1] Bain L J and Max E 1992 *Introduction To Probability and Matehematical Statistics. second edition* (California: Duxbury Press. An Imprint of Wadsworth Publishing Company Belmont)

[2] Billigsley Patrick 1998. *Probability and Measure second edition* (New York.: John W and Sons)

[3] Doherty, Neil A 1991 The Design of Optimal Insurance Contracts when Liability Rules are Unstable. *The Journal of Risk and Insurance* **58** 227–246

[4] Dudewicz E J and Mishra S N 1988. *Modern Mathematical Statistics.* (John W & Sons, Ltd. *Inc*)

[5] Cummins, David J. 1974 Insurer's Risk: A Restatement. *Journal of Risk and Insurance* **41** 147–157

[6] Yeneng S 2006 The exact law of large numbers via Fubini extension and characterization of insurable risks *Journal of Economic Theory* **126** (2006)001 pp 31-69

[7] Bickelhaupt and David L 1983 *General Insurance eleventh edition* (Homewood, Ill.: Richard D. Irwin, Inc)

[8] Gschlößl and Czado 2007 Spatial modelling of claim frequency and claim size in non-life insurance Scand. (Actuar) pp 202-225

[9] Shi et al 2015 Ivantsova Dependent frequency-severity modeling of insurance claims Insurance Math. *Insurance Math. Econom.*, **64**(2015) pp 417-428

[10] Beard R E, Teivo P, Erkki P 1984 *Risk Theory third edition* (New York: Chapman and Hall, Ltd)
[11] Casella G and Roger L B 1990 *Statistical Inference* (California: Wadsworth. Inc. Belmont) 94002
[12] BPS 2014 *Statistik Perasuransian Industri Keuangan Non Bank Otoritas Jasa Keuangan*

Financial Distress Prediction using Linear Discriminant Analysis and Support Vector Machine

Noviyanti Santoso, Wahyu Wibowo

Department of Business Statistics, Institut Teknologi Sepuluh Nopember
Kampus ITS Sukolilo-Surabaya 60111, Indonesia

E-mail: noviyanti_s@statistika.its.ac.id

Abstract. A financial difficulty is the early stages before the bankruptcy. Bankruptcies caused by the financial distress can be seen from the financial statements of the company. The ability to predict financial distress became an important research topic because it can provide early warning for the company. In addition, predicting financial distress is also beneficial for investors and creditors. This research will be made the prediction model of financial distress at industrial companies in Indonesia by comparing the performance of Linear Discriminant Analysis (LDA) and Support Vector Machine (SVM) combined with variable selection technique. The result of this research is prediction model based on hybrid Stepwise-SVM obtains better balance among fitting ability, generalization ability and model stability than the other models.

1. Introduction

Economic problems in a country indirectly have an impact on the financial condition of the company. At the time of the exchange rate of the currency weakening, inflation, and soaring fuel prices will affect the performance of the company. It takes effort to monitor, evaluate, and control corporate finance in order to avoid financial distress. The condition of the financial distress that is not immediately corrected can lead to bankruptcy of the company that caused the economic problems to the company's management, investors, creditors, suppliers, and employees. Based on this reason, the financial distress prediction became an important topic of discussion in financial affairs.

A diverse range of research both in the field of accounting or finance is working to develop a model prediction of financial distress. Nowadays, the researches of prediction of financial distress are also increasingly interested in the field of statistics and machine learning. Firstly, the prediction of bankruptcy starts by [1] that makes the model using univariate analysis of financial ratios and concluded that the company's financial ratios can be used to distinguish between bankrupt and non-bankrupt companies. Then [2] identified the companies went bankrupt with combined 22 financial ratios using Multiple Discriminant Analysis (MDA). As many as five significant variables resulted in such research, i.e. working capital/total assets, retained earnings/total assets, EBIT/total assets, market value of equity/total liabilities, and sales/total assets. Ohlson [3] using a logistic regression to find out the possibilities of a company facing financial distress.

In the development of data mining in the era of 1990 was emerging variety of classification methods, such as a decision tree (DT), case-based reasoning (CBR), artificial neural networks (ANNs), evolutionary genetic algorithm (GA) and a new technique of support vector machine (SVM). The technique was applied to predict financial distress of companies [4]. Shin and Lee using SVM to predict the bankruptcy of several companies in South Korea and concluded that SVM show highest prediction accuracy than logistic regression, MDA, and NN [5]. [6] and [7] also use SVM for predicting financial trouble in companies in China and reported that SVM outperformed other classifiers. Nisa [8] using SVM and LDA for predicting financial trouble in manufacturing company in Indonesia and obtained results that SVM give accuracy appropriately.

According to [9] variable that will be used as input will affect the accuracy. Therefore, the selection of variable financial ratios is an important step in pre-processing. Variable selection technique that is commonly used is the Principal Component Analysis (PCA), Factor Analysis, t-test, stepwise regression and correlation matrix [10]. Considering what mentioned above, the purpose of this study is to apply stepwise for selecting significant variables to use in prediction financial distress of industrial companies listed in the Indonesia Stock Exchange using LDA and SVM.

2. Materials and Methods

Classifications are statistical methods used for constructing models of predictive with the aim of separate and classify new data. All the classification methods use set features or parameters be characteristic of objects. There are two kinds of the classification methods that supervised learning and unsupervised learning. Supervised learning is an approach to where there are trained data, and there are variable targeted so the purpose of this approach has is made a data to the existing, while unsupervised learning has no data trainer, that data from, we classify this data then becomes 2 part or 3 parts and so on [11]. In this study the classification methods used are LDA and SVM where included in supervised learning.

2.1. Linear Discriminant Analysis

The discriminant analysis is a method used to distinguish two or more groups based on some independent variables. In this method, the independent variable can be either metric or non-metric but different in terms of the nature of the dependent variable. In the linear discriminant analysis, the dependent variable is the categorical / non-metric variable. If the number of categories is two, discriminant analysis is similar to binary logistic regression analysis [12]. The linear discriminant function takes form of a linear combination of coefficients of variables and their respective variables in the study as equation 1.

$$Z = \alpha + \beta_1 X_1 + \beta_2 X_2 + \ldots\ldots + \beta_i X_i \tag{1}$$

where, Z= discriminant score, α = a constant term, β_i = the discriminant coefficient or weight of the variable, X_i = predictor or independent variable, i = number of predictor variables; i = 1,2,3,....k.
Function of discriminant worthy to form if there are the differences in value the average between groups. Hence prior to the formation of discriminant function we need to done testing distinction vector the average of the groups has. Discriminant analyses having the assumption that must be fulfilled of them are normal multivariate distribution, covariance matrices to all groups is homogeneous, and every group has the average different [13]. But R.A. Fisher proposed another method for linear discriminant analysis that did not presuppose any known distribution of the training data, namely Fisher's discriminant. The idea is finding projection of $\boldsymbol{a}^T\boldsymbol{x}$ like best separating that result ratio between group sums of square within group sum of square maximal. An object will classifying into group l if $\boldsymbol{a}^T\overline{\boldsymbol{x}}_j$ is near to $\boldsymbol{a}^T\boldsymbol{x}$. $x \rightarrow \Pi_l$, where $l = \arg\min_i |a^T(x - \bar{x}_i)|$.

2.2. Support Vector Machine

Support vector machine, a more recent learning algorithm that has been developed from statistical learning theory [14, 15], has a very strong mathematical foundation and has been shown to exhibit excellent performance in time series forecasting [16] and in classification [17]. The SVM technique has been effectively used to estimate parameters of multivariate function, non-linear regression problem, enhances generalization capability and uncommon representation of the solution [18]. SVM is a promising technique that follows the principle of structural risk minimization wherein the problem of over fitting by balancing the model complexity [19]. The basic of SVR is that non linearity the original dataset x_i is mapped into a high-dimensional feature space. Consider a training set $G =$

(x_i, y_i) where $i = 1,2,\ldots,\lambda$ with $x = \{x_1, x_2, x_3\} \ldots \subseteq \Re^N$ as input variable vector and $y = \{y_t, \ldots, y_\lambda\} \ldots \subseteq \Re$ are output variable. The hyperplane function is as equation 2:

$$f(x) = w \times \varphi(x) + b \tag{2}$$

where $\varphi(x)$ denotes the high-dimensional feature space, which is non-linearity mapped from input space x, w and b are coefficients that estimated by minimizing the regularized risk function as follows:

$$min\frac{1}{2}\|w\|^2 + C\frac{1}{\lambda}\sum_{i=1}^{\lambda} L_e(y_i, f(x_i)) \tag{3}$$

$$L_\varepsilon(y_i, f(x_i)) = \begin{cases} |y_i - f(x_i)| - \varepsilon & |y_i - f(x_i)| \geq 0 \\ 0 & otherwise \end{cases} \tag{4}$$

Factor $\|w\|^2$ is called regularization, minimizing the $\|w\|^2$ will make the flatness function, which can control the function capacity. The parameter ε represent the distance between actual values and values calculated from regression function that measured with ε-insentive loss function. Using ε-insentive loss function must be minimizing norm from w in order to get appropriate generalization for regression function. C denotes a cost function measuring empirical risk. The constant C > 0 determine trade off between the flat of function and limit of deviation ε. Then the constrained form can be formulated as follows:

$$min\frac{1}{2}\|w\|^2 + C\frac{1}{\lambda}\sum_{i=1}^{\lambda}(\xi_i, \xi_i^*) \tag{5}$$

The two slack variables ζ_i and ζ_i* can be introduced to represent the distance from actual values to the corresponding boundary values of ε-tube. Finally, the constrained optimization problem is solved using the following form:

$$\max R(\alpha_i, \alpha_i^*) = \sum_{j=1}^{\lambda}(\alpha_i - \alpha_i^*)y_i - \epsilon\sum_{j=1}^{\lambda}(\alpha_i - \alpha_i^*) - \frac{1}{2}\sum_{i=1}^{\lambda}\sum_{j=1}^{\lambda}(\alpha_i - \alpha_i^*)(\alpha_j - \alpha_j^*)K(x_i, x_j) \tag{6}$$

Here α_i and α_i^* are Lagrange multipliers. $K(x_i, x_j)$ is kernel function. In this study, a linear is used as kernel function in SVM which denote as $K(x_i, x_j) = {x_i}^T x_j$. Hence, non linear regression function is formulated as equation 7:

$$f(x) = \sum_{i=1}^{\lambda}(\alpha_i - \alpha_i^*)K(x_i, x_j) + b \tag{7}$$

2.3. Dataset

The dataset that used in this research are financial report of industrial company that listed on the Indonesia stock exchange. Companies that become object observations in this research are company that have a complete report financial consist of the report loser profit, the balance, equity changes report, and reports cash flow. A financial report will be used in this research was the annual financial report that has been audited in the period 2008-2014 by the number of company at any period is 31. The input variables that used refer to the previous research, adopting 20 financial ratio indexes shown in table 1 as follows:

Table 1. Financial ratio indexes.

Variable	Definition	Variable	Definition
X_1	EBIT/Total Asset	X_{11}	Net Profit Ratio
X_2	Sales/Total Asset	X_{12}	EBIT/Sales
X_3	Sales/Fixed Asset	X_{13}	ROI
X_4	Earning/DEBT	X_{14}	Working Capital/Long Term DEBT
X_5	Current Ratio	X_{15}	DEBT to Equity
X_6	Working Capital/Total Asset	X_{16}	Book Equity/Total Capital
X_7	ROE	X_{17}	Market Value Equity/Total Capital
X_8	Retained Earning/ Total Asset	X_{18}	Market Value Equity/Total Liabilitas
X_9	Gross Profit Ratio	X_{19}	PER
X_{10}	Operating Profit Ratio	X_{20}	PBV

The company financial distress prediction as output variables will be 0 and 1. There are 0 for financially healthy company and 1 for financial crisis company. The criteria used to determine the condition financial distress of company there are two, namely interest coverage ratio and profit negative. By due to count interest coverage ratio required value interest burden interest expense but an enterprise not always displayed in the report a finance company so by negative profit of company for a year or at year *t* we can determine bankruptcy.

2.4. Methodology

In order to obtain a model and prediction, the data were parted into two parts, which were training set and testing set. The data used for modelling is training set that applied in three scheme dynamic periods (*k*), there are k = 0, 1, and 2. Dynamic period with k = 0, means when input variables are in period *t* then output variable is in period *t* too. Additionally, k = 1 means when input variables are in period *t* then output variable is in period *t-1* and so on. Afterward, stepwise was applied as selection variables technique to see the effectiveness. So, we have combined the scheme and approach using LDA and SVM to build model of prediction. Meanwhile, the data used for prediction is testing set that consist of one last period. The last step was to compare the prediction accuracy using area under Receiver Operating Characteristic (ROC) curve that appropriate for unbalance data [20]. That curve commonly known as AUC calculated using this following equation:

$$AUC = \frac{1 + TP_{rate} - FP_{rate}}{2} \quad (8)$$

Table 2. Confusion Matrix

Actual	Prediction	
	Positive = class 0	Negative = class 1
Positive = class 0	*True Positive (TP)*	*False Negatif (FN)*
Negative = class 1	*False Positive (FP)*	*True Negatif (TN)*

Where TP_{rate} and FP_{rate} can be obtained by use equation 9 and equation 10 as follow:

$$TP_{rate} = Sensitivitas = recall = \frac{TP}{TP + FN} \quad (9)$$

$$FP_{rate} = \frac{FP}{FP + TN} \quad (10)$$

IOP Conf. Series: Journal of Physics: Conf. Series **979** (2018) 012089 doi:10.1088/1742-6596/979/1/012089

3. Results and Discussion

This study use 186 data obtained from financial report of industrial company in Indonesia. There are 3 points that become focus discussion. Firstly, we would find the greatest dynamic period of data. Secondly, found the effectiveness of selection variables technique and got some significant variables in prediction model and the last was find out the appropriate prediction model for financial distress. Based on Table 3 and Table 4, the highest AUC both training and testing set occur when dynamic period is $k = 0$, but in fact is ineffective for company, because to predict the financial distress in the end of year, they need financial report in the same year. Whereas, the evaluation of company's performance occurs on December even early next year. But, comparing the score of AUC between k = 1 and k = 2, the first is more appropriate. It means, the company have to prepare the report of achievement on previous year to predict the financial condition in one period. So, the early warning financial can be useful for company's financial.

Table 3. AUC score without selection variables.

Dynamic Periods (k)	Methods	Training	Testing
0	LDA	0.89	0.69
	SVM	1	0.93
1	LDA	0.71	0.62
	SVM	0.64	0.63
2	LDA	0.70	0.54
	SVM	0.61	0.50

The aim variable selection was to accomplish the important variables that affect the prediction model. In some research, selected variables can increase the accuracy of prediction. There are 5 financial ratio indexes that preferred using stepwise i.e. EBIT/Total Asset (X_1), Retained Earning (X_8), Operating Profit Ratio (X_{10}), ROI (X_{13}), and PER (X_{19}). Only one variable that relevant with the results of Herlina [21] and Wulandari [22] who have obtained PER, current ratio, ROE, leverage ratio, and EPS as significant variables in financial distress prediction. Herlina used analytic hierarchy process (AHP) as variable selection methods, while Wulandari used LP-SVM. Different variable selection method will obtain various results. Nevertheless, table 4 shown that there is enhancement score of AUC both LDA and SVM methods. It proves that variables selection is worthwhile to boost accuracy.

Table 4. AUC score with selection variables.

Dynamic Periods (k)	Methods	Training	Testing
0	LDA	0.90	0.75
	SVM	1	0.97
1	LDA	0.74	0.68
	SVM	0.71	0.70
2	LDA	0.60	0.48
	SVM	0.69	0.50

In general, financial condition of Indonesia listed companies can be better predicted using the proposed SVM model since the AUC score is 0.70 and prediction accuracy of SVM is as high as 74%. The comparison accuracy was shown completely in table 5. However, the proposed SVM model still include in "poor classification" category. Probably, it happens as a consequence the linear kernel that use in the SVM model.

Table 5. Comparison the accuracy.

Methods	Training		Testing	
	AUC	Accuracy	AUC	Accuracy
LDA	0.74	79%	0.68	70%
SVM	0.71	78%	0.70	74%

4. Conclusion

The LDA and SVM models are successful to predict financial distress. Their predictions are able to get a nearly result. The SVM model with dynamic period $k = 1$ and hybrid with stepwise as selection variables method becomes the best model among these four combination models with maximum AUC and accuracy. It shows that the variables selection could significantly increase the prediction accuracy of both LDA and SVM models. We consider that this method could be the promising method to improve the prediction performance, especially in bankruptcy cases. However, in this study the realization accuracy of SVM models still "poor classification" category. Perhaps it caused by kernel function that used in the SVM models is linear. We consider for the further research to apply polynomial and radial basis function to improve the accuracy. Moreover, it requires optimizing the parameters of SVM using genetic algorithm.

Acknowledgments

The author would like to thanks to Indonesia Stock Exchange for provide data set and this work was supported by Grant for Beginner Research 2017 ITS Surabaya.

References

[1] Beaver W 1966 Financial ratios as Predictors of Failure *Journal of Accounting Research* **4** 71-111
[2] Altman E I 1968 Financial Ratios Discriminant Analysis and The Prediction of Corporate Bankruptcy *Journal of Finance* **23** 1-22
[3] Ohlson J A 1980 Financial Ratios and the Probabilistic Prediction of Bankruptcy *Journal of Accounting Research* **18** 109-131
[4] Chen M Y 2011 Bankruptcy Prediction in Firms with Statistical and Intelligent Techniques and a Comparison of Evolutionary Approaches *Computers and Mathematics with Applications* **62** 4514-4524
[5] Shin K S, Lee T S and Kim H J 2005 An Application of Support Vector Machine in Bankruptcy Prediction Model *Expert Systems with Applications* **28** 127-135
[6] Hui X F and Sun J 2006 An Application of Support Vector Machine to Companies Financial Distress Prediction *Lecture Notes Artificial Intelligent* **3885** 274-282
[7] Ding Y, Song X and Zeng Y 2008 Forecasting Financial Condition of Chinese Listed Companies based on Support Vector Machine *Expert Systems with Applications* **34** 3081-3089
[8] Nisa U Z 2013 Model Prediksi Finansial Distress Pada Perusahaan Manufaktur Go Public di Indonesia *Thesis Program Magister Bidang Optimasi Sistem Industri Jurusan Teknik Industri Institut Teknologi Sepuluh Nopember*
[9] Lin F, Liang D and Chen E 2011 Financial Ratio Selection for Business Crisis Prediction *Expert Systems with Applications* **38** 15094-15102
[10] Tsai C F 2009 Feature Selection in Bankruptcy Prediction *Knowledge Based Systems* **22** 120-127

[11] Sthya R and Abraham A 2013 Comparison of Supervised and Unsupervised Learning Algorithms for Pattern Classification *International Journal of Advanced Research in Artificial Intelligence* **2**(2) 34-38

[12] Uddin N, Meah M S and Hossain R 2013 Discriminant Analysis as an Aid to Human Resource Selection and Human Resource Turnover Minimization Decisions *International Journal of Business and Management* **8** (17) 153-169

[13] Johnson R and Wichern D 2007 *Applied Multivariate Statistical Analysis* (New Jersey: Prentice Hall Inc.)

[14] Vapnik V N 1995 *The Nature of Statistical Learning Theory* (New York: Springer)

[15] Vapnik V, Golowich S and Smola A 1997 Support Vector Method for Function Approximation, Regression Estimation and Signal Processing In: Mozer M, Jordan M and Petsche T Eds *Advance in Neural Information Processing System* (Cambridge : MIT Press) 281-287

[16] Muller K R, Smola A J, Ratsch G, Scholkopf B, Kohlmorgen J and Vapnik V 1997 Prediction Time Series with Support Vector Machines Lecture Notes in Computer Science **1327** 999-1004

[17] Yu L, Wang S and Cao J 2009 A Modified Least Squares Support Vector Machine Classifier with Application to Credit Risk Analysis *International Journal of Information Technology & Decision Making* **8**(4) 697-710

[18] Vapnik V N 1999 An overview of statistical learning theory IEEE Transactions on Neural Networks / *Publication of the IEEE Neural Networks Council* **10**(5) 988–99

[19] Scholkopf B and Smola A 2002 *Learning with Kernels* (Cambridge: The MIT Press)

[20] Wang S and Yao X 2013 Using Class Imbalance Learning for Software Defect Prediction *IEEE Transactions on Reliability* 434-443

[21] Herlina, Santosa B and Ciptomulyono U 2014 Pemodelan Prediksi Financial Distress Menggunakan Metode Support Vector Machine *Prosiding Seminar Nasional Manajemen Teknologi XXI* (ITS)

[22] Wulandari L M C, Daru A and Andrian D 2016 Aplikasi Data Mining dengan Metode Support Vector Machine (SVM) untuk Prediksi Financial Distress pada Industri Jasa Go Public yang Terdaftar di Bursa Efek Indonesia *Prosiding Seminar Nasional Inovasi dan Aplikasi Teknologi di Industri* (ITS)

IOP Conf. Series: Journal of Physics: Conf. Series **979** (2018) 012090 doi:10.1088/1742-6596/979/1/012090

Interevent Time Distribution of Renewal Point Process, Case Study: Extreme Rainfall in South Sulawesi

Nurtiti Sunusi

Mathematics Department, Faculty of Mathematics and Natural Sciences, Hasanuddin University

Jl. Perintis Kemerdekaan Km 10 Tamalanrea, Makassar, South Sulawesi, Indonesia 90245

E-mail: nurtitisunusi@unhas.ac.id

Abstract.

The study of time distribution of occurrences of extreme rain phenomena plays a very important role in the analysis and weather forecast in an area. The timing of extreme rainfall is difficult to predict because its occurrence is random. This paper aims to determine the inter event time distribution of extreme rain events and minimum waiting time until the occurrence of next extreme event through a point process approach. The phenomenon of extreme rain events over a given period of time is following a renewal process in which the time for events is a random variable τ. The distribution of random variable τ is assumed to be a Pareto, Log Normal, and Gamma. To estimate model parameters, a moment method is used. Consider R_t as the time of the last extreme rain event at one location is the time difference since the last extreme rainfall event. if there are no extreme rain events up to t_0, there will be an opportunity for extreme rainfall events at $(t_0, t_0 + \delta t_0)$. Furthermore from the three models reviewed, the minimum waiting time until the next extreme rainfall will be determined. The result shows that Log Nrmal model is better than Pareto and Gamma model for predicting the next extreme rainfall in South Sulawesi while the Pareto model can not be used.

1. Introduction

Many phenomena in nature are random and complex, such as extreme rainfall events which can affect the occurrence of extreme weather. This is very influential on the early determination of the growing season of various agricultural products. Knowledge of the accuracy of the time distribution of rain events is a fundamental issue in meteorological hydrological studies such as the use of the surface runoff in the design of the hydraulic structures. This requires the right approximation to be able to explain the phenomenon. One stochastic model that can explain the random nature phenomena both in space and time is known by the name of the process. In the point process, each event is considered a point in a particular location [1]. Many ways of assessing point process of modeling random phenomena, for example: point process as homogeneous Poisson process, non homogeneous Poisson process, renewal process, and others.

Some definitions of the occurrence of extreme rainfall are described as events characterized by deviations from great intensity from the average condition [2]. The Indian Meteorological Department says that rain with intensity over 64.5 mm per day is referred to as heavy rain

[3]. Three different methods have commonly been used to identify extreme rainfall events [4]. The first method is based on the actual rainfall amounts. A second way to define extreme precipitation events is to use specific thresholds such as the 90th and 99th percentiles of precipitation days to define heavy and very heavy events, respectively. A third way of defining extreme precipitation events is to calculate return periods of the event based on the annual maximum 24-h precipitation series [5]. This research used the first method to identify extreme rainfall events.

Several approaches were proposed to determine the temporal distribution of rainfall. In 1975, Pilgrim and Cordery proposed temporal rainfall distribution by selection of intensity rainfalls with different time bases in Sidney of Australia [6]. Some researchers have proposed theoretical methods using statistical approach, that are [6], [4], and [7]. Several previous studies on the prediction of the timing of recurrence on some events have been done by [8],[9], [10], and [11]. Although various distribution models have been used in modeling the timing of recurrence of an event, it has not found the most appropriate distribution to state the time of occurrence of events at a site. This paper aims to determine the inter event time distribution of extreme rain events and minimum waiting time until the occurrence of next extreme event through a renewal point process approach. In this study, the process point is approximated by a renewal process with inter events time are independent. For that matter, this paper will examine the forecast of the time of occurrence of extreme rainfall for two extreme rain events follows Gamma, Log Normal, and Pareto distributions.

2. Point Process

One dimensional point process can be expressed as a model that useful for random timing sequences when an event occurs. A process of spatial point is a useful model for a random pattern of dots in the d dimensional space, where $d \geq 2$. As an example, if a map of the location of extreme rain events at any given time, then this map is a random pattern of dots in two dimensions.

There are several differences between the one-dimensional point process and the dimension point process higher. In point process one-dimensional, it has found natural properties which is not found on the point process of higher dimension. One-dimensional point process can expressed mathematically in a variety of different ways, including: arrival times $t_i < t_{i+1} < \cdots$ with t_i is the occurrence time of event-i. Another way that can be used to define a point process is through time between events $\tau_i = t_{i+1} - t_i$. The length of time interval between subsequent events are known as inter event times.

One-dimensional point process can expressed mathematically in a variety of different ways, including: arrival times $t_i < t_{i+1} < \cdots$ with t_i is the occurrence time of event i. Another way that can be used to define a point process is through time between two events $\tau_i = t_{i+1} - t_i$. The lengths of the time intervals between subsequent events are known as inter event times. For some special models (Poisson and renewal process), $\tau_1, \tau_2, \cdots$ are mutually exclusive.

2.1. Renewal Process

An important type of temporal point process is renewal process. Let $t_1, t_2, \cdots, t_n$ is the series of occurrences events. Consider $\{\tau_i\}$ in interval $(0, \infty)$ where $\tau_i = t_{i+1} - t_i$ is time difference between two consecutive events and τ_i as a stochastic process that has a positive value. Let $\{\tau_i\}$ is Independent and identically distributed random variables, then τ_i is a renewal process. If its marginal is exponentially distributed, then $\{\tau_i\}$ is Poisson stationer. A renewal process in line is simple point process so that inter event time $\{\tau_1, \tau_2, \cdots, \tau_n\}$ are independent random variables. Temporal renewal process with probability density function f, conditional intensity is defined by $\lambda(t) = v(t - \tilde{t})$ where $\tilde{t}$ is the last time before time t, $S(t)$ is survival function that associated

Table 1. Renewal Point Process Model.

No	Model	pdf	cdf
1	Eksponential	$f(\tau) = \theta e^{-\theta\tau}$	$F(\tau) = 1 - e^{-\theta\tau}$
2	Weibull	$f(\tau) = (\alpha\beta)\tau^{\beta-1}e^{-\alpha\tau^{\beta}}$, $\alpha > 0,\ \beta > 0$	$F(\tau) = 1 - e^{-\alpha\tau^{\beta}}$
3	Gamma	$f(\tau) = \frac{\beta^{\alpha}}{\Gamma(\alpha)}\tau^{\alpha-1}e^{-\beta\tau}$, $\tau > 0$	$F(\tau) = \Gamma(r, c\tau)/\Gamma(r)$ $\Gamma(\alpha, x) = \int_0^x e^{-u}.u^{k-1}du$
4	Log normal	$f(\tau) = \frac{1}{\sigma\tau\sqrt{2\pi}}\exp\left\{\frac{-(\ln\tau-\bar{Y})^2}{2\sigma^2}\right\}$, $m > 0$	$F(\tau) = \Phi\left\{\frac{(\ln\tau-m)}{\sigma}\right\}$; $\Phi = \left(\frac{1}{\sqrt{2\pi}}\right)\int_{-\infty}^{x} e^{\frac{-u^2}{2}}du$
5	Pareto	$f(\tau) = \frac{\alpha.x_m^{\alpha}}{\tau^{\alpha+1}}$, $\alpha > 0\ \ x_m > 0$	$F(\tau) = 1 - \left(\frac{x_m}{x}\right)^{\alpha}$
6	Rayleigh	$f(\tau) = \frac{\tau}{\alpha^2}e^{\left(-\frac{\tau^2}{2\alpha^2}\right)}$, $\tau > 0,\ \alpha > 0$	$F(\tau) = 1 - e^{-\frac{-\tau^2}{2\alpha^2}}$

by survival function f. Survival function is defined by $S(t) = \frac{f(t)}{1-F(t)}$, where $F(t) = \int_0^t f(u)du$ is a cumulative distribution function of f.

The renewal model is often used for estimate the long-term time dependent probability. In the renewal model, it is assumed that the time between successive extreme rain events follows a certain probability distribution. Some distribution opportunities have used as an renewal model in reliability issues and failure time are Pareto, Weibull, exponential, Gamma, log normal, and Rayleigh [8].

If $f(\tau)$ denotes the probability density function (pdf), $F(\tau)$ as the cumulative distribution function (cdf), and τ as the inter event time, then renewal models for the extreme rainfall occurrence are described in Table 1.

Notes:
τ : inter event time extreme rainfall.
pdf : probability density function of renewal models.
cdf : cumulative distribution function.

3. Results and Discussion

3.1. Time Recurrence of Extreme Rainfall

Time recurrence of extreme rainfall defined as the time interval until the next rain extreme event [7]. In this section we will model the timing of the occurrence of extreme rainfall with the inter event time follows Log Normal, Pareto, and Gamma distribution.

3.1.1. Log Normal Model Let **T** as random variable of the inter event time of extreme rainfall is log normal distributed with parameters σ^2 and $\bar{Y}$. The probability density function (pdf) is defined by [12]:

$$f(\tau) = \frac{1}{\tau\sigma\sqrt{2\pi}}e^{-\frac{(\ln\tau-\bar{Y})^2}{2\sigma^2}}. \qquad (1)$$

Cumulative distribution of log normal random variable is:

$$F(\tau) = \int_0^s \frac{1}{u\sigma\sqrt{2\pi}} e^{-(\ln u - \overline{Y})^2/2\sigma^2} du. \tag{2}$$

Cumulative distribution of equation (2) have to be evaluated by numerical method. Conditional probability density function of extreme rainfall as follows:

$$f(\tau|\mathbf{T} \geq t_0) = \frac{f(\tau)}{1 - F(t_0)} \tag{3}$$

so that

$$f(\tau|\mathbf{T} \geq t_0) = \frac{\frac{1}{\tau\sigma\sqrt{2\pi}} e^{-(\ln\tau - \overline{Y})^2/2\sigma^2}}{1 - \int_0^\tau \frac{1}{u\sigma\sqrt{2\pi}} e^{-(\ln u - \overline{Y})^2/2\sigma^2} du}. \tag{4}$$

Cumulative distribution of $F(\tau)$ for log normal distribution is constan D. Thus, conditional probability of log normal distribution for (4) is defined as:

$$f(\tau|\mathbf{T} \geq t_0) = Q\frac{1}{\tau\sigma\sqrt{2\pi}} e^{-(\ln\tau - \overline{Y})^2/2\sigma^2} \tag{5}$$

where $Q = \frac{1}{1-D}$.

To estimate recurrence time for $\hat{\tau}$ until the occurrence of the next extreme rainfall, then the maximum of conditional probability density function of log normal $f(\tau|\mathbf{T} \geq t_0)$ will be determined. Next, we have solution as follows:

$$\frac{d}{d\tau} f(\tau|\mathbf{T} \geq t_0) = Q\frac{d}{d\tau}\left\{\frac{1}{\tau\sigma\sqrt{2\pi}} e^{-(\ln\tau - \overline{Y})^2/2\sigma^2}\right\} = 0. \tag{6}$$

Based on equation (6) we have

$$\left[\left(\ln\tau - \overline{Y}\right) + \sigma^2\right] e^{-(\ln\tau - \overline{Y})^2/2\sigma^2} = 0. \tag{7}$$

and

$$\left(\ln\tau - \overline{Y}\right) + \sigma^2 = 0. \tag{8}$$

Thus, the solution is:

$$\ln\tau = \overline{Y} - \sigma^2. \tag{9}$$

or,

$$\hat{\tau} = e^{\overline{Y} - \sigma^2}. \tag{10}$$

Equation (10) can be used to estimate waiting time $\hat{\tau}$ until the next extreme rainfall. By equation (10), it can be knowing that the forecasting waiting time $\hat{\tau}$ until the next extreme rainfall does not depend on elapsed time since the last extreme rainfall. To estimate waiting time $\hat{\tau}$ until the next extreme rainfall using log normal model, then model parameters should be estimated. Furthermore, let mean and variance of the log normal model is given by

$$\mu = e^{\overline{Y} + \frac{1}{2}\sigma^2} \quad \text{dan} \quad \nu = e^{2\overline{Y} + \sigma^2}\left(e^{\sigma^2} - 1\right). \tag{11}$$

Based on equation (11) we have

$$\mu^2 = \left\{e^{\overline{Y} + \frac{1}{2}\sigma^2}\right\}^2 = e^{2\overline{Y} + \sigma^2} \tag{12}$$

and

$$\sigma^2 = \ln\left\{\frac{\nu}{\mu^2} + 1\right\}. \tag{13}$$

so

$$\overline{Y} = \ln\mu - \frac{1}{2}\sigma^2. \tag{14}$$

Thus, we have:

$$\overline{Y} = \ln\mu - \frac{1}{2}\ln\left\{\frac{\nu}{\mu^2} + 1\right\}. \tag{15}$$

3.1.2. Pareto Model Let $\mathbf{T}$ is random variable for inter event time of extreme rainfall that modeled by Pareto distribution with pdf as[12]:

$$f(\tau) = \frac{\alpha x_m^\alpha}{\tau^{\alpha+1}}, \quad \alpha > 0 \quad x_m > 0 \tag{16}$$

Mean and variance sample are $E[\mathbf{T}] = \mu_\tau = \alpha, x_m$, for $\alpha < 1$ and variance $Var[\mathbf{T}] = \frac{\alpha x_m^2}{(\alpha-1)^2(\alpha-2)}$ respectively. Cumulative distribution for Pareto random variable is given by:

$$F(\tau) = 1 - \left(\frac{x_m}{x}\right)^\alpha \tag{17}$$

Based on equation (4), we have conditional pdf as:

$$f(\tau|\mathbf{T} \geq t_0) = \frac{\alpha x_m^\alpha \tau^{-(\alpha+1)}}{1 - \int_0^\tau \alpha x_m^\alpha \tau^{-(\alpha+1)} d\tau}. \tag{18}$$

Since cumulative distribution function for random variable of Pareto distribution in equation (17) is constant C. Thus, pdf of extreme rainfall occurrence can be written as

$$f(\tau|\mathbf{T} \geq t_0) = W.\alpha x_m^\alpha \tau^{-(\alpha+1)}, \tag{19}$$

with $W = \frac{1}{1-C}$. Furthermore, we determined waiting time for $\hat{\tau}$ until the occurrence of the next extreme rainfall by maximizing the conditional pdf for the extreme rainfall occurrence as follows

$$\frac{\partial f(\tau|\mathbf{T} \geq t_0)}{\partial \tau} = \frac{-\alpha x_m^\alpha (\alpha+1)}{\tau^{\alpha+2}}. \tag{20}$$

so,

$$-\alpha x_m^\alpha (\alpha+1)\tau^{-\alpha-2} = 0. \tag{21}$$

Since $\alpha > 0$ and $x_m > 0$, then $\tau^{(-\alpha-2)} = 0$, so we have $\hat{\tau} = 0$. Thus it can be said that the Pareto model can not be used to forecast time recurrence of extreme rainfall.

3.1.3. Gamma Model Gamma is a distribution that plays an important role in statistics. Let random variable $\mathbf{T}$ is the inter event time of extreme rainfall, the pdf of Gamma distribution is [12]:

$$f(\tau) = \frac{\beta^\alpha}{\Gamma(\alpha)} \tau^{\alpha-1} e^{-\beta\tau}, \qquad \tau > 0. \tag{22}$$

This distribution depends on shape parameter α and *scale* parameter β, where $\alpha > 0$ and $\beta > 0$. The mean and variance for this distribution as follows

$$E[\mathbf{T}] = \frac{\alpha}{\beta} \quad \text{dan} \quad Var[\mathbf{T}] = \frac{\alpha}{\beta^2}. \tag{23}$$

Cumulative distribution for random variable of Gamma is

$$F(\tau) = \int_0^\tau \frac{\beta^\alpha}{\Gamma(\alpha)} u^{\alpha-1} e^{-\beta u} du. \tag{24}$$

Substitute equation (22) and equation (24) to equation (3), we found conditional probability density function of Gamma for extreme rainfall occurrence is

$$f(\tau|\mathbf{T} \geq t_0) = \frac{\frac{\beta^\alpha}{\Gamma(\alpha)} \tau^{\alpha-1} e^{-\beta\tau}}{1 - \int_0^\tau \frac{\beta^\alpha}{\Gamma(\alpha)} u^{\alpha-1} e^{-\beta u} du}. \tag{25}$$

Cumulative distribution function for random variable which Gamma distributed in equation (25) is constant C, then pdf of extreme rainfall occurrence can be written by:

Table 2. Renewal Model for Extreme Rainfall Occurrence.

No	Model	Mean and Variance	Parameters
1	Log Normal	62.85 and 8557.608	$\sigma^2 = 1.154$ and $\bar{Y} = 3.564$
2	Gamma	62.85 and 8557.608	$\alpha = 0.459$ and $\beta = 0.007$
5	Pareto	62.85 and 8557.608	$\alpha > 0$ and $x_m > 0$

$$f(\tau|\mathbf{T} \geq t_0) = W\frac{\beta^\alpha}{\Gamma(\alpha)}\tau^{\alpha-1}e^{-\beta\tau}, \tag{26}$$

with $W = \frac{1}{1-C}$. Furthermore, we determined recurrence time for $\hat{\tau}$ until the occurrence of the next extreme rainfall by maximizing the conditional pdf for the extreme rainfall occurrence as follows

$$\frac{\partial f(\tau)}{\partial \tau} = \frac{\beta^\alpha}{\Gamma(\alpha)}\tau^{\alpha-1}e^{-\beta\tau}\left\{\frac{(\alpha-1}{\tau} - \beta\right\} = 0. \tag{27}$$

Using equation (27) we found that $f(\tau|\mathbf{T} \geq t_0)$ value reaches maximum at

$$\hat{\tau} = \frac{\alpha-1}{\beta}. \tag{28}$$

If $\alpha < 1$, then maximum value occurs outside the definition range and for this case the conditional probability density for occurrence of extreme rainfall is a decreasing monotonous function to zero for t_0.

3.2. Case Study

The data used in this case study is the historical extreme daily rainfall data onto the South Sulawesi region period 1 January 2000 to 17 February 2017. We have 124 extreme rainfall data. There are 70% for training data and 30% for validation data. From the sample used it is obtained that the mean sample are 62,85 and the sample variance is 8557,608. Summary of parameter estimation of the three models is described in the following Table 2:

Based on the parameter values of each model given in the Table 2, we know that using log Normal model and equation $\hat{\tau} = e^{\bar{Y}-\sigma^2}$, we have the approximate of waiting time $\hat{\tau}$ until the next extreme rainfall is $\hat{\tau} = 11.14$ days. It is therefore possible to estimate the time of occurrence of the next extreme rainfall as follows: It is known that the time of the last extreme rainfall occurrence is February 17, 2017. Since $\hat{\tau} = 11.14$, then the next extreme rainfall prediction occurs before March 1, 2017.

For Gamma model and equation $\hat{\tau} = \frac{\alpha-1}{\beta}$, it is found that $\alpha = 0.459$. Because of $\alpha < 1$ then the maximum occurs outside the definition range. Especially for this case, conditional probability density of extreme rainfall is a monotone decreasing function to zero for $t_0 \to \infty$.

In Pareto model, $-\alpha.x_m^2(\alpha+1)\tau^{-\alpha-2} = 0$. Since $\alpha > 0$ and $x_m > 0$, then the equation is true if $\tau^{-\alpha-2} = 0$. So that $\frac{\tau^{-\alpha}}{\tau^2} = 0$, or $\tau^{-\alpha} = 0$. Thus, $\hat{\tau} \to \infty$. In other words, the inter events time that Pareto distributed can not be used to forecast the recurrence time of extreme rain.

3.3. Acknowledgments

The authors are grateful to all those who had provided suggestions for the completion of this paper. The authors would like to thank the ministry of technology research and higher education Indonesia for financial funding PUPT research grant 005/SP2H/LT/DRPM/IV/2017 through decentralization research.

4. Conclusion

1. The minimum waiting time until the next extreme rainfall is not depend on elapsed time since the last extreme rainfall.

2. The Log Normal model is better than Gamma model for predicting the next extreme rainfall in South Sulawesi. The minimum waiting time until next extreme rainfall is approximately 11 days.

3. Gamma model can not be used for predicting the next extreme rainfall in South Sulawesi.

References

[1] Daley D J and Vere-Jones D 2003 *An Introduction to the Theory of Point Processes* Springer Vol(I) ed 2
[2] Meehl G A, Karl T, Easterling D R, Changnon, et.al 2000 An Introduction to Trends in Exreme Weather and Climate Events: Observations, Socioeconomic Impacts, Terrestrial Ecological Impacts, and Model Projections *Bulletin of AMS*
[3] Massachussets Institute of Technology 2017 *Climate Change: Extreme Rainfall will Vary Between Regions: Intensification of Extreme Rainfall Varies from Region, Study Shows.* ScienceDaily
[4] Chu P S, Zhao X, Ruan Y and Grubbs M 2009 Extreme Rainfall Events in the Hawaiian Islands *J. App. Meteorology and Climatology* **48** 3
[5] 9. Kunkel K E, Andsager K and Easterling D R 1999 Long-term trends in extreme precipitation events over the conterminous United States and Canada *J. Climate* **12** 2515-2527.
[6] Pilgrim D H and Cordery I 1975 Rainfall Temporal Patterns for Design Floods *Journal of Hydraulics*, **101** 1 81-95
[7] Sunusi N, Herdiani E T and Nirwan 2017 Modeling of Extreme Rainfall Recurrence Time by Using Point Process Models *J. Environ. Sci. Technol.*, **10** 6 320-324
[8] Ellsworth W L 1955 *Characteristic Earthquake Forecasts: Implications of Central California Seismicity: in Urban Disaster Mitigation: the Role of Science and Technology* Elsevier Science Ltd
[9] Ogata Y 1998 Estimating the Hazard of Rupture Using Uncertain Occurrence Times of Paleoearthqua-kes*Journal of Geophysical Research*, **104** 17 995
[10] Sunusi N, Darwis S, Triyoso W and Mangku I W 2010 Study of Earthquake Forecast through Hazard Rate Analysis *International Journal of Applied Mathematics and Statistics* **17** J10 96-103
[11] Sunusi N, Herdiani E T, Giarno and Nawawi F 2015 Exploratory Analysis of Rainfall Occurrence in South Sulawesi Region Using Spatial Point Process *Journal of Mathematics and Statistics*, **11** 4 113-118
[12] lugman S A, Panjer H H and Wilmot G E 2002 Loss Models: From Data to Decisions Wiley

Preliminary Study of Perception and Consumer Behaviour Towards Energy Saving for Household Appliances: A Case of Makassar

Yusri Syam Akil[1], Saiful Mangngenre[2], Sri Mawar[1], Kifayah Amar[3]

[1]Department of Electrical Engineering, Hasanuddin University, Gowa, Indonesia
[2]Department of Industrial Engineering, Hasanuddin University, Gowa, Indonesia
[3]Department of Industrial Engineering, UIN Sunan Kalijaga, Yogyakarta, Indonesia

E-mail: yusakil@unhas.ac.id

Abstract. Electricity load has tendency to increase over the time. Therefore, efforts to maintain a balance between electricity supply and demand such as increasing energy saving related to the use of home electricity appliances are urgently needed. In general, one of the household appliances which consumes relatively high electricity energy is refrigerator. The purpose of this study is to analyze residential consumers perceptions and their behaviours about electricity energy saving in relation to the usage of household appliances in Makassar, Indonesia particularly for refrigerator. Moreover, typical relationship between perceptions and consumers behaviours is also analyzed by composed two regression models, namely model for usage behaviour ($UREF_m$ model) and model for habitual behaviour ($HREF_m$ model) by using general perception, specific perception, and external factors as explanation variables. To collect data, a questionnaire was designed for survey which involved 40 respondents as a preliminary study and then statistical tests including regression analysis were applied to analyze usable data. The target of respondent was an owner of a house in Makassar with installed power capacity at least 900 VA. Reliability test shown that all items in the developed questionnaire can be used for main survey as obtained Cronbach's alpha values were above 0.6. Evaluation for consumers perceptions on energy saving in relation to demographic aspect using mean and Standard Deviation values indicated some significant differences. Other results regarding regression analysis shown that both composed models were well validated and had quite good fitness degree with adjusted R-squared values around 49.31% for $UREF_m$ model and 80.90% for $HREF_m$ model. Among considered variables, specific perception, and external factors were found have significant influence to the usage and habitual behaviours of consumers as confirmed by their p-values in each model below 0.05. Findings of this research can be used as a reference in developing programs for residential consumers such as electricity energy conservation program.

1. Introduction

As in other countries, residential area is one of the major electricity load sectors in Makassar, Indonesia which contributes significantly to total electricity load demand. Considering electricity load has tendency to increase over the time, and depleting fossil fuel to produce electricity energy including environmental issue and fuel price make penetration of more renewable energy sources into power system and or other efforts to maintain balancing between electricity production and load demand in any level in power systems are needed. They are important to overcome energy crisis and continuing usage of electricity energy for a long time. At consumer level, increasing usage of energy saving electricity devices or managing the using of appliances at home is a possible option. In general, although energy saving in one home is relatively small, but in accumulation for large residential area can give significance reduction to the total electricity load at certain place such as in Makassar. Regarding condition of residential electricity demand in Makassar, number of residential consumers for each group in terms of installed power capacity (900 VA; 1,300 VA; 2,200 VA; 3,500 VA – 5,500 VA; and 6000 VA above) has been increased by time and same tendency occurred for electricity consumption (sold electricity energy in kWh). Based on the data from State Electricity Company in Makassar (PT. PLN (Persero)), among the groups, majority of consumers has installed power capacity 900 VA in their homes which proportional to the level of needed electricity consumption. Number of consumers particularly for group 900 VA in Year 2012 is 251,370 and increased 319,330 in Year 2016. Meanwhile electricity consumption for the group is 383,863,181 kWh in Year 2012 and become 562,927,231 kWh in Year 2016.

In residential area, electricity load derives from utilization of many household appliances such as refrigerator, air-conditioner, washing machine, rice cooker, and lighting. However, refrigerator is known as one of appliances which consumes relatively big electricity energy. As refrigerator is a main electricity device at home which operate continuously and used by majority of the houses, therefore, it is meaningful to analyze and manage the use of refrigerator as an effort to reduce monthly electricity consumption. Basically, increasing of electricity usage efficiency is not only help to keep balancing between electricity supply and demand in one place, but also to mitigate climate change in the fastest, cheapest, and safest way [1]. Some previous studies regarding household refrigerator can be found in [2-5]. In this research, we have interesting to analyze consumers perceptions and their behaviours related to efficiency of electricity usage for household appliances in Makassar, Indonesia. To address the research objective, a questionnaire is employed to gather data from respondents through a pilot survey. Next, statistical approach is applied to analyze usable data. As consumers behaviours in residential sector can affect pattern and big of electricity consumption, therefore, findings of this study are expected to provide a contribution for better energy usage pattern at consumers side. Further, it also can be used as a basis to developing suitable programs for residential consumers in Makassar such as program for electricity energy conservation. Previous studies on electricity demand analysis in Makassar can be found in [6,7].

2. Materials and Methods

2.1. Questionnaire Design and Dataset Collection

As an initial step, a questionnaire was designed to collect data from respondent who has energy saving refrigerator at home. Information from previous studies on refrigerator or other related studies were used as a basis to develop the questionnaire [4,8]. Specifically, the questionnaire was designed to measure perceptions of consumers including their typical behaviours towards energy saving through the usage of refrigerator at their homes. Study area was Makassar city and target of respondent was house owner with installed power capacity at least 900 VA. Purposive sampling technique was used to collect sample data for 40 respondents, and then usable data were analyzed using statistical approach incuded such as Cronbach's alpha value test, mean value, Standard Deviation, and regression analysis as described further in each related section. The period of the data collection was in August 2017.

In the questionnaire, systematic questions were composed to gather required information which consists of four main parts. Part 1 regarding general information, Part 2 regarding general perception towards energy saving and external factors, Part 3 regarding specific perception on attribute of energy saving refrigerator, and Part 4 regarding usage and habitual behaviours of the consumers.

For example in Part 1, questions included age of respondent, house size, household income per-month, etcetera. For general perception (Part 2), composed questions such as "Implementation energy saving lifestyle reduces monthly electricity bill at home", meanwhile the questions for external factors (social factor, media, and weather factor) such as "The usage of electrical devices is influenced by social factor like neighbors, relatives, and friends". For questions regarding specific perceptions on attribute of energy saving refrigerator (in terms of technology, aesthetics, and cost attribute) include "Energy saving refrigerator relatively produce low noise and the compressor is more durable". To asses items in the questionnaire, we used a seven point Likert scale which range from 1 to 7. Here, 1 shown "strongly disagree" and 7 for "strongly agree". Next to identify typical relationship between perceptions and consumers behaviours (usage and habit) regarding refrigerator, two regression models were composed as in [8]. Both of the models can be formulated as follows:

$$UREF_m = \alpha_0 + \alpha_1 GP + \alpha_2 SP + \alpha_3 EF + u_t \quad (1)$$

$$HREF_m = \beta_0 + \beta_1 GP + \beta_2 SP + \beta_3 EF + u_t \quad (2)$$

In the Equations (1) and (2), $UREF_m$ and $HREF_m$ are usage and habitual behaviour about energy saving refrigerator, respectively. As explanation variables in the models, GP is general perception, SP is specific perception, and EF is external factors. α_0 and β_0 are intercept values, and other α and β values are regression coefficients. u_t is residual term of models.

2.2. Reliability and Validity Tests

For reliability testing, Cronbach's alpha was used to measure internal consistency between items under the same dimensions and the Cronbach's alpha value of this pilot survey can be seen in table 1. All dimensions shown the Cronbach's alpha values are above 0.6. It means that all the questionnaire items can be used for main survey. For validity, we used expert validity as the questionnaire items were designed mainly based on the prior study on energy-efficient lighting [8] and it was modified to fulfill Indonesian lifestyles of household appliances usage in particular for refrigerator. All the items were reviewed and appropriate improvement were made by experts to validate them.

Table 1. Reliability test results.

No	Dimension	Cronbach's alpha value
1	General perception	0.891
2	Specific perception	0.909
3	External factors	0.890
4	Usage behaviour	0.910
5	Habitual behaviour	0.919

3. Results

3.1. Demographic Aspect of Respondents

Demographic of 40 respondents in this study is shown in table 2. The information includes gender, education, age, house size, household income per-month, installed power capacity at home, and lenght of using energy saving refrigerator. It can be seen from the table 2 that majority of respondents were female. Of the fourty respondents, 55% were hold bachelor degree, 25% were graduated from high school, 17.5% were hold other degree, and 2.5% were hold master and doctoral degree. Majority of respondents have house with size was 60 m^2 and power installed capacity at home was 1,300 VA. About 48% of respondents were using energy saving refrigerator for 1 – 3 years.

Table 2. Demographic of respondents.

No	Demographic aspect		Percentage	Total
1	Gender	Female	52.5 %	100 %
		Male	47.5 %	
2	Education	High School	25.0 %	100 %
		Bachelor Degree	55.0 %	
		Master and Doctoral Degree	2.5 %	
		Others	17.5 %	
3	Age	20 – 30 Years	20.0 %	100 %
		31 – 40 Years	42.5 %	
		41 – 50 Years	30.0 %	
		50 Years above	7.5 %	
4	House size	45 m^2	22.5%	100 %
		60 m^2	50.0 %	
		120 m^2	27.5 %	
		Others	0.0 %	
5	Household income	< 2,500,000 IDR	27.5 %	100 %
		> 2,500,000 – 5,000,000 IDR	55.0 %	
		> 5,000,000 – 7,500,000 IDR	12.5 %	
		> 7,500,000 IDR	5.0 %	
6	Power installed capacity at home	900 VA	32.5 %	100 %
		1,300 VA	47.5 %	
		2,200 VA	17.5 %	
		3,500 VA – 5,500 VA	2.5 %	
		6,000 VA above	0.0 %	
7	Length of using energy saving refrigerator	< 1 Year	37.5 %	100 %
		1 – 3 Years	47.5 %	
		> 3 Years	15.0 %	

3.2. Perception of Residential Consumers on Energy Saving Refrigerator

Consumers perceptions with regards to general perception, specific perception, and external factors particularly on energy saving refrigerator were analyzed. As results, gender and length of using energy saving refrigerator have significant difference on external factors as indicated by obtained p-values which were less then 0.05, namely 0.036 and 0.018, respectively. It is found that male respondents more perceived external factors as a variable which determine the usage of energy saving refrigerator at home (as shown by Mean value is 5.745 and Standard Deviation (SD) is 1.069) than female respondents (Mean value = 5.139 and SD = 1.113). Beside that, respondents which used refrigerator less then 1 year more positively perceive the influence of external factors on the usage of energy saving refrigerator than two other categories with Mean value is 5.977 and SD is 0.963.

3.3. Factors Influence on Usage and Habitual Behaviours of Energy Saving Refrigerator

Table 3 and table 4 show regression results for $UREF_m$ and $HREF_m$ models which well validated. From results, typical relationship between consumers perceptions and usage or habitual behaviours about energy saving refrigerator can be identified. The fitness degree of composed models were relatively good as adjusted R-squared (adj. R^2) around 49.31% for usage behaviour model and 80.90%

for habitual behaviour model. Next, at least one of involved variables in related models affect usage and habitual behaviours as indicated by their Prob. F-Statistic values are 0. The autocorrelation was not existed in the models. It is shown by Durbin-Watson test (D-W) around 2 for each composed model.

Table 3. Regression results for $UREF_m$ model.

Variable	$UREF_m$ model (Model for usage)			
	Coefficient	Prob. (p-value)	Standar error	t-statistic
α_0	2.2072	0.0007	0.5973	3.6953
GP	-0.4229	0.1081[a]	0.2566	-1.6479
SP	0.8176	0.0112	0.3057	2.6744
EF	0.2114	0.0369	0.0975	2.1667
R^2	0.5321			
Adj. R^2	0.4931			
S.E Reg.	0.5620			
Prob. (F-Statistic)	0.0000			
D-W	2.0709			

[a] is not significant.
S.E. Reg. is standard error regression.

Table 4. Regression results for $HREF_m$ model.

Variable	$HREF_m$ model (Model for habit)			
	Coefficient	Prob. (p-value)	Standar error	t-statistic
β_0	0.5345	0.0766[a,b]	0.2931	1.8233
GP	-0.1712	0.2560[a]	0.1483	-1.1543
SP	0.8564	0.0000	0.1656	5.1703
EF	0.2097	0.0014	0.0607	3.4512
R^2	0.8237			
Adj. R^2	0.8090			
S.E Reg.	0.3522			
Prob. (F-Statistic)	0.0000			
D-W	1.8038			

[a] is not significant.
[b]significance at 10% significance level.

By using 5% significance level, SP and EF variabels have significance, meanwhile GP variable is not significant as p-value for each model is above 0.05. This means specific perception (SP) and external factors (EF) are two important variables or factors which affect significantly usage and habitual behaviours of consumers regarding energy saving refrigerator in Makassar. Among the significance variables, specific perception has higher influence than external factors as indicated by regression coefficient value of SP variable for each model is bigger than regression coefficient of EF variable. Elimination of non-significance explanation variabel (GP variable) in both models gives almost the same regression results. Next, compared between the two models, adj. R^2 value for $HREF_m$ model (80.90%) is higher than $UREF_m$ model (49.31%). The difference of the values shown if considered variables affect usage and habitual behaviours differently. The variables can explain habitual behaviour better than usage behaviour of consumers.

4. Conclusion

This paper studies perceptions and consumers behaviours about electricity energy saving in relation to the usage of refrigerator at home in Makassar. It can be concluded that developed questionnaire to collect data in this study were reliable and can be used for main survey and further analysis. Both composed regression models (usage and habit models) to investigate relationship between consumers perceptions including external factors and their behaviours were well validated and had quite good fitness degree. Considered variables (general perception, specific perception, and external factors) affected the two models differently. They can explain habitual behaviour better than usage behaviour of consumers, and only specific perception and external factors had significance. Therefore, specific perception on attribute of refrigerator and external factors (social factor, media, and weather factor) were two important variables which affect highly usage and habitual behaviours of consumers in Makassar regarding the use of refrigerator towards energy saving. As consumers behaviours can affect pattern and level of electricity consumption, it is hoped findings of this research contribute to a better pattern of energy use from the consumer side. The information gained can be used as a basis in developing suitable programs for residential consumers such as electrical energy conservation programs. Future research will involve more respondents and analyze other household appliances to get more information with regards to typical behaviours of consumers and potential of energy saving at household level in Makassar.

Acknowledgements
This work was supported by Institute for Research and Community Services (LPPM) – Hasanuddin University (RUNAS 2017 grant scheme – UNHAS).

References

[1] Sorrell S 2015 Reducing energy demand: A review of issues, challenges and approaches *Renewable and Sustainable Energy Reviews* **47** 74-82
[2] Floarea A and Sgârciu V 2016 Smart refrigerator: A next generation refrigerator connected to the IoT *ECAI 2016 – Int. Conf.– 8th Edition, Electronics, Computers and Artificial Intelligence* (Ploiesti, Romania) 19-24
[3] Guo J, Luo J, Guo Y, Pan X, Fang X and Wu X 2015 Noise test and control for household refrigerator compressor *Proc. of the 2015 IEEE Int. Conf. on Information and Automation* (Lijiang, China) 112 – 5
[4] Khan Md I H, Afroz H M M, Rohoman Md A, Faruk M and Salim M 2013 Effect of different operating variables on energy consumption of household refrigerator *International Journal of Energy Engineering* **3** 144-150
[5] Lakshmanan V, Marinelli M, Kosek A M, Sossan F and Nørgård P 2013 Domestic refrigerators temperature prediction strategy for the evaluation of the expected power consumption *4th IEEE PES Innovative Smart Grid Technologies Europe* (Copenhagen) 1 – 5
[6] Akil Y S, Syafaruddin, Waris T and Lateko A A H 2014 The influence of meteorological parameters under tropical condition on electricity demand characteristic: Indonesia case study *Proc. of the 1st Int. Conf. on Information Technology, Computer, and Electrical Engineering* (Semarang, Indonesia) 381-5
[7] Akil Y S and Mitani Y 2017 Seasonal short-term electricity demand forecasting under tropical condition using fuzzy approach model *Journal of Telecommunication, Electronic and Computer Engineering Special Issue* **9** 77-82
[8] Park N and Lee E 2013 Energy-efficient lighting: Consumers' perceptions and behaviors *International Journal of Marketing Studies* **5** 26-35

IOP Conf. Series: Journal of Physics: Conf. Series **979** (2018) 012092 doi:10.1088/1742-6596/979/1/012092

On The Impact of Climate Change to Agricultural Productivity in East Java

Heri Kuswanto, Mutiah Salamah, Sri Mumpuni Retnaningsih, Dedy Dwi Prastyo

Department of Statistics, Institut Teknologi Sepuluh Nopember (ITS)
Kampus ITS Sukolilo Surabaya 60111 Indonesia

E-mail: heri_k@statistika.its.ac.id

Abstract. Many researches showed that climate change has significant impact on agricultural sector, which threats the food security especially in developing countries. It has been observed also that the climate change increases the intensity of extreme events. This research investigated the impact climate to the agricultural productivity in East Java, as one of the main rice producers in Indonesia. Standard regression as well as panel regression models have been performed in order to find the best model which is able to describe the climate change impact. The analysis found that the fixed effect model of panel regression outperforms the others showing that climate change had negatively impacted the rice productivity in East Java. The effect in Malang and Pasuruan were almost the same, while the impact in Sumenep was the least one compared to other districts.

1. Introduction

Agriculture is one of the important sectors to support Indonesian economic which significantly depends on climate. Numerous studies have been conducted to predict the future climate condition. Since IPCC report in 1990, maximum effort has been carried out to understand the clmate change impact on agricultural system. As the results, better changes have been gained to support studies on the impact of climate change such as better data collection, development of new methodology or model, as well as observation about actual change on climate and its impact. All of them are very useful to support the technological development and policy to mitigate and adapt the climate change.

Latest research on climate change impact on agricultural sector concludes that the climate change has negatively impacted the food production, as well as lead to a public perception that climate change causes food degradation globally. However, the impact is different every country with also different vulnerability level. Some countries such as US and European countries are able to adapt against climate change by change in a farmer level such as potential product, crop manajement, as well as anticipation in the form of investment. Meanwhile in developing countries, there are several reasons why farmers and agricultural supporting institutions are less adaptation to climate change.

Indonesia is one of the countries that is highly vulnerable to climate change with refer to agricultural and food sector. Increasing intensity of exteme climate events over the year has caused significant losses in agricultural sector. Many studies that have been conducted in Indonesia were limited on qualitative study or description about the fact. Nevertheless, due to the fact that climate change is a dynamic event with various impact over the time, it is necessary to carry out a study which is able to generate information about the impact of climate change to Indonesian agriculture over time. In statistical language, it is necessary to develop model which is able to explain the dynamic effect of climate change. With this model, mitigation strategy will be able to be formulated in order to reduce the negative impact which threats the food security.

Several studies have been conducted e.g. [1] which studied the impact of climate change on agricultural sector in Semarang. This study used cross sectional data modeled with log linear regression and hence, the dynamic impact cannot be identified. Rasmikayati [2] investigated the climate change impact to farmer behavior. Santoso [3] used simple time series model to forecast agricultural production in Maluku as the result of climate change, however the study quantified only the food production data without clear effect of climate change. Widayat *et.al* [4] studied the climate change impact on the coffee production in Aceh using Structural Equation Modeling (SEM). This paper differs from the aforemetioned studies in term of the model as well as the obtained information from the dynamic model using panel data analysis in order to investigate the effect of the climate change to agriculture in Indonesia especially East Java.

Researches in the developed countries have considered the dynamic effect. Mendelsohn [5] studied the potential impact of climat echange inclusing the temperature and precipitation variation to food production. Another research tried to simulate climate change effect to agriculture with several inputs such as [6, 7]. Several latest researches have used statistical and econometric approach on time series, cross section and panel data (see [8, 9] among others). Another strength of panel approach is the location-specific fixed effects information to absorbse time invariant missing variable. Furthermore, panel model is able to identify shock happening over the years, which is different from climate permanent shift [10]. This research focuses on East Java Province as one of the provinces in Indonesia with significant contribution of agricultural producton in Indonesia.

2. Research Methodology

This research analyzes two kinds of data as follow:

- Data to characterize the climate change

  The variable used as the basis of describing climate change is daily rainfall intensity. In the model, the daily rainfall data is aggregared into monthly data. The data spans from 2006 to 2016 which can be downloaded from the BMKG website at www.dataonline.bmkg.go.id.
- Agricultural Data

  This is the data of rice production, rice productivity and yield. The data is mostly collected from Statistics Office and Minisry of Agriclture East Java Province. The rice productivity and yield are used for descriptive statistic, while the rice production is used for panel data modelling.

The steps of the analysis can be described as follows. We firstly perform descriptive statistic about the variable i.e rainfall as well as rice yield and productivity during the periods of observation in four districts. Furthermore, we investigated the best model by performing regression models i.e. OLS regression and panel regression models. The panel regression models include fixed and random effect model by which the model can be written as follow:

$$Y_{it} = \beta_1 X_{it} + \alpha_i + u_{it} + \varepsilon_{it}$$

where α_i is the unknown intercept for each entity (in this case district), Y_{it} is dependent variable for entity i at time t, X_{it} represents one independent variable with hcoefficient β, while u_{it} and ε_{it} are between and within entity error respectively. The fixed effect model control for, or partial out, the effects of time-invariant variables with time-invariant effects so that

$$Y_{it} = \beta_1 X_{it} + \alpha_i + u_{it}$$

while in a random effects model, the unobserved variables are assumed to be uncorrelated with all the observed variables

$$Y_{it} = \beta X_{it} + \alpha + u_{it} + \varepsilon_{it}$$

The best model is the one with highest R-square.

3. Results and Discussion

3.1. Descriptive Statistic

Many researches concluded that agricultural production (in this case is rice) in a region depends on the weather and climate in that region. Therefore, information about climate variability is important in order to formulate the strategy to reduce th eimpact of climate change. Figure 1 depicts the cumulative monthly rainfall observed in several stations in East Java.

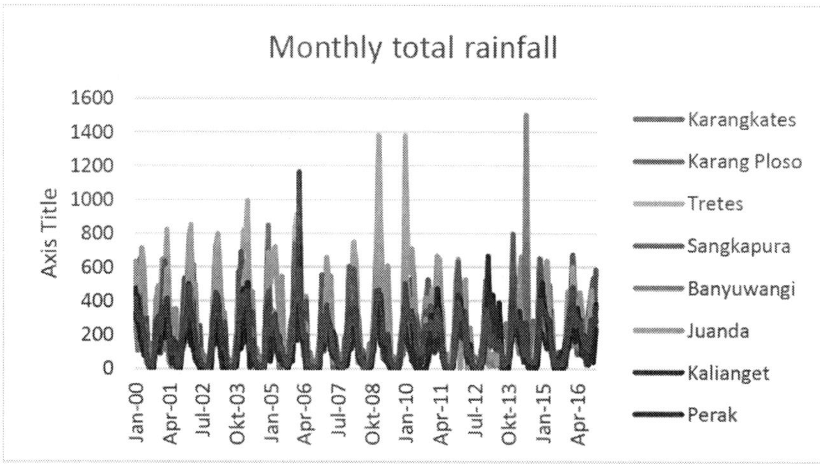

Figure 1. Monthly total rainfall in East Java.

Figure 1 clearly shows the seasonal pattern of the monthly rainfall. There was a consistency or similarity in the pattern among those 8 stations. October, November and December were the months with high rainfall intensity. The Figure covers only observations from 8 stations with most completed and reliable dataset. Morever, the distribution of the stations has represented the East Java province well.

3.2. Descriptive statistics of rice producttion, yield and productivity

This research focuses on rice as the main source of food in Indonesia especially in East Java. Figure 2 summarizes total production of rice in five highest rice producer districts during 2005-2014 (in Ton).

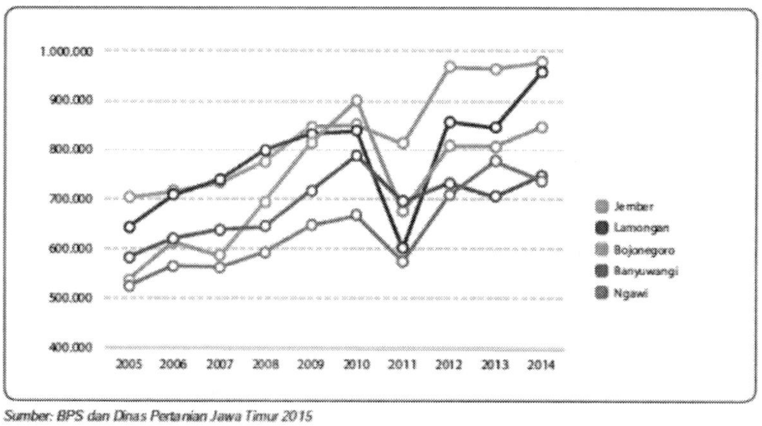

Figure 2. Largest total rice production in East Java (source BPS and Agricultural Office East Java, 2015).

As we can see from the figure that Jember, Lamongan, Bojonegoro, Banyuwangi and Ngawi have been main rice producers in Indonesia. There have been significant increase in the rice production.

However, in 2011 there was a significant decrease caused by a specific factor. The information about rice production is equipped with figure about total yield as performed in figure 3 below.

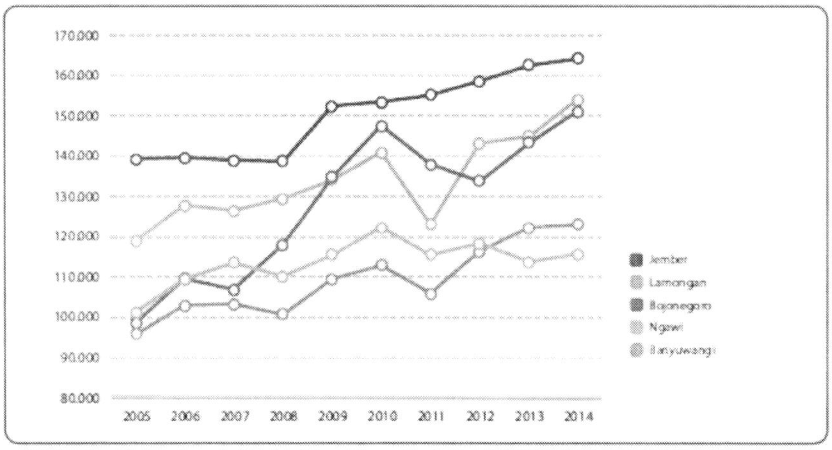

Figure 3. Largest yield in East Java (source BPS and Agricultural Office East Java, 2015).

In line with the total rice production, the yields increased also over the year. There was significant decrease also in 2011. In general the yield of rice at East Java in 2008 to 2014 can be seen on figure 4, which gives information that there were a regular seasonality pattern where the yield on March is always the highest compared to others.

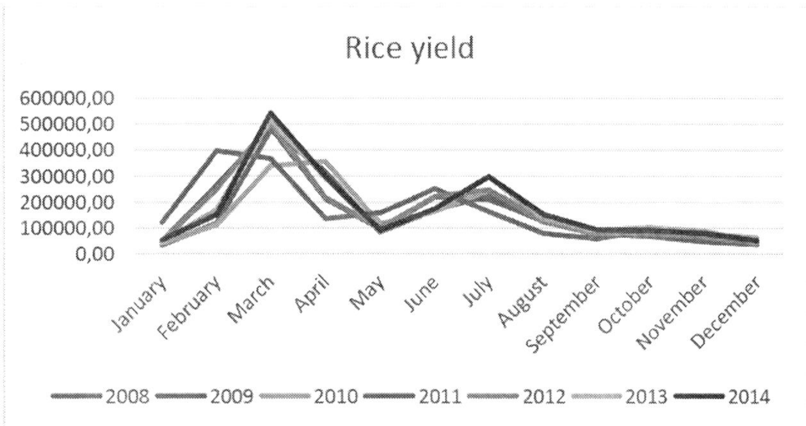

Figure 4. Total of rice yield in East java.

3.3. Panel Analysis

This subsection discusses the result of analyzing the dataset using panel regression. There are many advantages of panel regression such as the information about specific time effect, fixed effect, etc. Figure 5 presented the rainfall pattern observed from four meteorological stations, representing the climate condition in the corresponding districst. We see that the quarterly rainfall did not show significant increase or decrease over the year. Moreover, the rainfall intensity in Banyuwangi was higher than in other three districts. We can see also an indication of seasonality, although the regular pattern was somehow does not appear. Different pattern was observed for the rice productivity in those four districts. We can clearly see that the productivity increased over the years. In line with the rainfall intensity, the rice productivity in Banyuwangi was higher than the others.

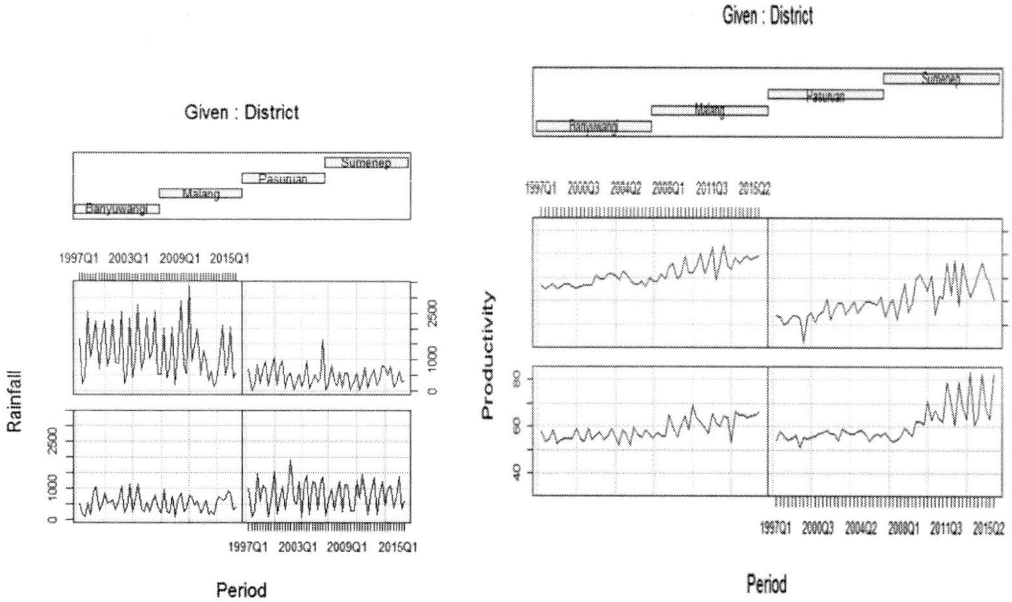

Figure 5. Heterogeneity plot of rainfall (leaf) and productivity (right) across districts.

Another important information is about the heterogeneity of the productivity and rainfall across districts as shown in figure 5. The productivity of heterogeneity gives information about how the productivity differs among the districts. As can be seen from the left panel of figure 5, the heterogeneity of rice productivity was increasing, which also means that the productivity distincted significantly. Meanwhile, the rainfall heterogeneity seemed to be constant indicating that the climate change impact is almost the same across the district. Figure 6 showed that the variability of rainfall intensity in in Pasuruan was higher than the others during the period of 1997 to 2015, while Sumenep had the least rainfalll with low heterogeneity too. From the heterogeneity plots, we see that the pattern of the rice productivity showed linear increasing pattern especially during the past five years. However, the yield was constant over the time.

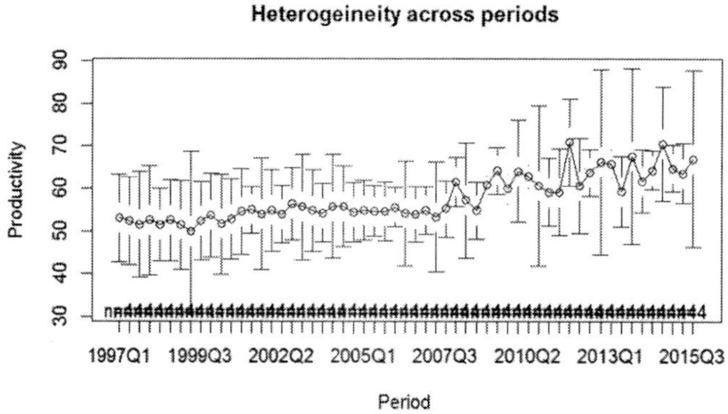

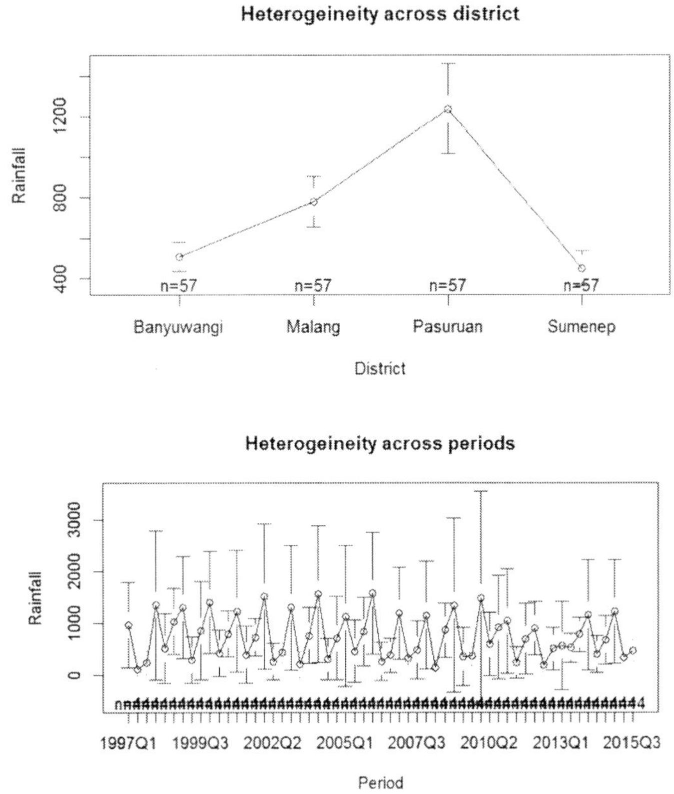

Figure 6. Productivity and rainfall heterogeneity plots across periods.

Table 1 below summarizes the regreesion result. The analysis is begin with OLS regression without neglecting the time or location specific effect. The regression shows that rainfall significantly influenced the rice productivity in East Java positively. However, the coefficient determination of the model is very low i.e. 3%. It indicated that the variability of the data cannot be covered by the model well. Furthermore, the fix effect on the districts can be seen form the following table.

Table 1. Summary of regression results

Model								
OLS			fixed Effect			Random Effect		
Variable	coefficient	p-value	Variable	coefficient	p-value	Variable	coefficient	p-value
Intercept	55.94	<2e-16 ***				Intercept	57.9270	<2e-16 ***
Rainfall	0.00232	0.00586	Rainfall	-0.0004613	0.57	Rainfall	-0.00033	0.6782
			Banyuwangi	58.914	<2e-16 ***			
			Malang	61.051	<2e-16 ***			
			Pasuruan	61.895	<2e-16 ***			
			Sumenep	50.221	<2e-16 ***			
R-square	0.003312		R-square	0.9883		R-square	0.00076316	

The panel regression indicates that the coefficient of the rainfall variable is negative, which can be seen as the negative effect of climate change to rice productivity. Moreover, the p-value greater than 0.05 indicating that rainfall does not significantly influence the rice productivity. Meanwhile, the rice productivity in each district significantly greater than zero where Pasuruan and Malang have higher level than Banyuwangi and Sumenep. Surprisingly the coefficient determination of the model is very high i.e. 98.83%.

In order to know which model is actually better to describe the climate change effect to rice productivity, we can performs F-test with the null hypothesis of OLS model better than fixed effect model. As the p-value is significantly lower than 0.05, thus we conclude that the fixed effect model is better than the OLS model. It suggests to analyze the specific location effect individually. Note that this paper did not focus on time specific effect as many researches agreed that the climate change impact influenced many sectors over the time. Moreover, the fixed effect model itself assumes that there is no autocorrelation among the time implying also there is no autocorrelation in the error term.

Another effect that we need to investigate within panel regression model is random effect. From the table, the coefficients can be interpreted as the within-entity and between-entity effects. In this case, it represents the average effect of rainfall over the rice production when rainfall changes across time and between districts by one unit. Furthermore, it against shows that the rainfall does not significantly influence the rice production as the P-value is greater than 0.05.

In order to know the effect, Hausmaan test has been performed with the null hypothesis that random effect is better than fixed effect. As the P-value is greater than 0.05, it means that the random effect is better than fixed effect. However, the P-value is only sightly different from 0.05 which suggest also that both models compete each other. From the result above, the panel regession model can be written as

$$\text{Rice Productivity}_{it} = -0.0004613 * \text{Rainfall}_{it} + \alpha_i$$

where α_i is the specific district effect with the values listed in table 1. The model suggests that rainfall happening at a specific time period will tend to decrease the rice productivity at that time. However, the decreasing level was very low. Furthermore, the model indicated that extreme or heavy rainfall will lead to higher decrease of rice productivity, which may lead to a puso.

4. Conclusion

This paper examined panel data analysis to investigate the impact of climate change to rice production in East Java. Using the dataset observed in four districts, it showed that the climate change indicated by the rainfall intensity does not significantly influence the rice productivity. The model indicated that the individual district effect is highly significant with different levels among the districts. Furthermore, the pool model indicated that rainfall intensity gave significant impact to rice productivity. However, the model has very low R-squares, which means also that the model has very low prediction ability. In order to improve the model, further study needs to be carried out by involving more data from other districts. Moreover, considering time lag influence may give better result.

References

[1] Hidayati I N and Suryanto 2015 Pengaruh perubahan iklim terhadap produksi Pertanian dan strategi adaptasi pada Lahan rawan kekeringan *Jurnal Ekonomi dan Studi Pembangunan* **16**(1) 42-52

[2] Rasmikayati E and Djuwendah E 2015 Dampak perubahan iklim terhadap perilaku dan pendapatan petani *(The Impact of Climate Change to Farmers' Behavior and Revenue) J. Manusia dan Lingkungan* **22**(3)

[3] Santoso A B 2016 Pengaruh Perubahan Iklim terhadap Produksi Tanaman Pangan di Provinsi Maluku *Perubahan iklim dan produksi tanaman pangan*

[4] Widayat H P, Anhar A and Baihaqi A 2015 Dampak perubahan iklim terhadap produksi, kualitas hasil dan pendapatan petani kopi arabika di aceh tengah *Agrisep* **16**(2)

[5] Mendelsohn R and Dinar A 2003 Climate Water and Agriculture *Land Economics* **79**(3) 328-341

[6] Adams, Richard, Ronald A. Fleming, Ching-Chang C, Bruce A. McCarl and Cynthia R 1995 A reassessment of the economic effects of global climate change on U.S. agriculture *Climatic Change* **30(**2) 147–167

[7] Kaufmann R K, Seth E S 1997 A Biophysical Model of Corn Yield: Integrating climatic and Social Determinants *American Journal of Agricultural Economics* **79**(1)

[8] Schlenker, W and David B. Lobell 2010 Robust negative impacts of climate change on African agriculture *Environmental Research Letters* **5**(1)

[9] Lobell DB, Schlenker WS, Costa-Roberts J 2011 Climate tren ds and global crop p roduction since 1980 *Science* **333** 616-620

[10] Fischer G, Shah M, Tubiello F N, van Velhuizen H 2005 Socio-economic and climate change impacts on agriculture: an integrated assessment,1990–2080 *Phil. Trans. R. Soc. B* **360** 2067–2083

Estimation of Covariance Matrix on Bi-Response Longitudinal Data Analysis with Penalized Spline Regression

A Islamiyati[1], Fatmawati[2] and N Chamidah[1]

[1]Department of Mathematics, Faculty of Mathematics and Natural Sciences, Hasanuddin University, Indonesia
[2]Department of Mathematics, Faculty of Sciences and Technology, Airlangga University, Indonesia

E-mail: annaislamiyati_701@gmail.com

Abstract. The correlation assumption of the longitudinal data with bi-response occurs on the measurement between the subjects of observation and the response. It causes the auto-correlation of error, and this can be overcome by using a covariance matrix. In this article, we estimate the covariance matrix based on the penalized spline regression model. Penalized spline involves knot points and smoothing parameters simultaneously in controlling the smoothness of the curve. Based on our simulation study, the estimated regression model of the weighted penalized spline with covariance matrix gives a smaller error value compared to the error of the model without covariance matrix.

1. Introduction

The covariance matrix is usually used as a weighting in the model estimation in case of violation of assumptions such as the correlation between errors. Error correlation can occur if the data contains more than one correlated response. The covariance matrix can be assumed to be known [1] and estimated with the smoothing spline estimator [2]. However, if the covariance matrix is unknown, then it is necessary to be estimated. The covariance matrix can be estimated from the data and it may have an impact on the estimation results of the smoothing spline estimator [3]. The estimation of the covariance matrix in cross sectional data to overcome the correlation between the responses has been estimated by using the maximum likelihood method in spline regression [4]. Besides that, it also has been estimated by using the expected value of the error based on the local polynomial estimators [5], and the kernel estimators [6].

Our study analyses longitudinal data which is a combination of cross sectional and time series data. The analysis of cross sectional data that should be performed on longitudinal data will provide a different interpretation of a problem [7]. The longitudinal data consists of several subjects that are measured over time based on measurement time. Each subject is assumed to be independent, but the measurements in the subject are mutually dependent. The approach used for addressing the correlations that occur in longitudinal data can be done through principal component analysis [8], varying coefficient models [9] and mixed-effect penalized spline models [10].

The longitudinal data can consist of two responses that are assumed to be correlated called bi-response [11]. The correlation between the responses was overcome by using the covariance matrix and it is estimated through the smoothing spline estimator [12]. In this article, we use the covariance matrix as a weighted in the penalized spline criteria to solve correlations that occur in the data, both correlations between responses, or the correlation between observed data bi-response longitudinal data. The advantage of a penalized spline is the ability of the estimator to produce a smooth and fitted curve, and also the pattern change of curve regression can be seen visually. This is due to knot involvement and smoothing parameters simultaneously in controlling the smoothness of the curve. The penalized spline estimators consist goodness of fit and penalty function. The function of goodness of fit is truncated polynomials that have an optimal rate of convergence [13]. Knot points in the goodness of fit selected based on fixed methods [14]. We use the penalty function in the penalized spline criterion based on quadratic polynomials form of spline regression coefficient [15]. The covariance matrix is estimated based on the error of the nonparametric bi-response regression model based on the penalized spline estimator un-weighted. This is why the use of covariance matrix can produce to accurate estimation results. The advantages of using the weighted of the covariance matrix are shown through a longitudinal data simulation study on a correlation value of 0.9. The results show that the use of the covariance matrix in the penalized spline criteria in the case of longitudinal bi-response data provides a smaller error value.

2. Bi-response longitudinal data

Bi-response longitudinal data are longitudinal data involving two correlated responses. Longitudinal data are measured data from subjects, where each subject is measured repeatedly in a time interval. Longitudinal data are assumed to be independent among subjects, however, on the inter-observations in the same subjects, the data are mutually dependent [16]. Futhermore, if t_{ij} is the observation at the jth time of the ith subject, $y_{1.ij}$ is the 1st response variable at a time t_{ij} and $y_{2.ij}$ is the 2nd response variable at a time t_{ij}, then the longitudinal data bi-response was given by $(t_{ij}, y_{1.ij}, y_{2.ij})$, $j = 1,2,\ldots,m_i$, $i = 1,2,\ldots,n$, where m_i is the number of repeated measurement of the ith individual. The relationship between t_{ij}, $y_{1.ij}$ and $y_{2.ij}$ can be expressed in the form of the regression model as follows.

$$y_{r.ij} = f_r(t_{ij}) + \varepsilon_{r.ij}, r = 1,2 \tag{1}$$

The longitudinal data structure of bi-response is shown in table 1.

Table 1. The structure of the longitudinal data bi-response to the predictor variable t_{ij}

r	i	j	t_{ij}	$y_{r.ij}$	r	i	j	t_{ij}	$y_{r.ij}$
1	1	1	t_{11}	$y_{1.11}$	2	1	1	t_{11}	$y_{2.11}$
		2	t_{12}	$y_{1.12}$			2	t_{12}	$y_{2.12}$
		$\vdots$	$\vdots$	$\vdots$			$\vdots$	$\vdots$	$\vdots$
		m_1	t_{1m_1}	$y_{1.1m_1}$			m_1	t_{1m_1}	$y_{2.1m_1}$
	$\vdots$	$\vdots$	$\vdots$	$\vdots$		$\vdots$	$\vdots$	$\vdots$	$\vdots$
	n	1	t_{n1}	$y_{1.n1}$		n	1	t_{n1}	$y_{2.n1}$
		2	t_{n2}	$y_{1.n2}$			2	t_{n2}	$y_{2.n2}$
		$\vdots$	$\vdots$	$\vdots$			$\vdots$	$\vdots$	$\vdots$
		m_n	t_{nm_n}	$y_{1.nm_n}$			m_n	t_{nm_n}	$y_{2.nm_n}$

Table 1 shows the structure of the longitudinal data containing two responses (r = 1,2). Each response consists of n subjects ($i = 1,2,\ldots,n$) which each subject is measured several times as much as m ($j = 1,2,\ldots,m_i$).

3. Penalized spline regression

The penalized spline is one of the estimators used in nonparametric regression in the estimation of the nonparametric regression function. Penalized spline estimator involves knot points and smoothing parameters simultaneously in controlling the smoothness of the curve. Penalty function in the penalized spline estimator takes the quadratic form of the truncated regression coefficient as recommended on the cross sectional data [17].

If the function $f_r\left(t_{ij}\right)$ in equation (1) is assumed unknown and estimated through a penalized spline estimator, then it is expressed as follows.

$$f_r\left(t_{ij}\right)=\sum_{u=0}^{q}\beta_{r.u}\left(t_{ij}\right)^u+\sum_{v=1}^{d}\beta_{r.(q+v)}\left(t_{ij}-K_v\right)_+^q . \quad (2)$$

The equation (2) can be made in matrix form, as in (3).

$$\vec{f}\left(\vec{t}\right)=\mathbf{X}\vec{\beta}, \quad (3)$$

where $\vec{f}(\vec{t})=\left[\vec{f}_1(\vec{t}),\vec{f}_2(\vec{t})\right]^T$ are an unknown form of regression function in the first and the second response.

Furthermore, the penalized spline regression model in equation (1) can be expressed to (4) based on the function at (3).

$$\vec{y}=\mathbf{X}\vec{\beta}+\vec{\varepsilon}, \quad (4)$$

where $\vec{y}$ is a response vector containing the first and the second response. $\mathbf{X}$ is a predictor matrix in the first and the second response. $\underset{\sim}{\varepsilon}$ is an error vector in the first and the second response with $E\left(\vec{\varepsilon}\right)=\vec{0}$ and $\text{Var}\left(\vec{\varepsilon}\right)=\mathbf{\Omega}$.

In the estimation of the covariance matrix, we need estimates of the nonparametric bi-response regression model un-weighted. The estimation criteria is called penalized least square (PLS) which can be expressed in the form of vector as in (5).

$$\text{PLS}=\left(\vec{y}-\mathbf{X}\vec{\beta}\right)^T\left(\vec{y}-\mathbf{X}\vec{\beta}\right)+\lambda\vec{\beta}^T\mathbf{D}\vec{\beta}. \quad (5)$$

4. Estimation of covariance matrix

The assumption for error random $\varepsilon_{r.ij}$ is related to the variance of the error.

$$E\left(\varepsilon_{r.ij}^2\right)=\sigma_{r.ij}^2, E\left(\varepsilon_{r.ij},\varepsilon_{s.i^*j}\right)=\begin{cases}\sigma_{r.ij}\sigma_{s.ij} & ; \; i=i^* \\ 0 & ; \; i\neq i^*\end{cases} \quad (6)$$

Equation (6) occurs at $r\neq s=1,2$. The model of bi-response nonparametric regression on longitudinal data is assumed to have a correlation between responses and to have a correlation between measurement data on the same subject. This led to the estimation of bi-response nonparametric regression model on longitudinal data using a weighted covariance matrix from $\vec{\varepsilon}$ is $\mathbf{\Omega}$.

$$\mathbf{\Omega}=\text{Var}\left(\vec{\varepsilon}\right)=E\left[\vec{\varepsilon}-E\left(\vec{\varepsilon}\right)\right]\left[\vec{\varepsilon}-E\left(\vec{\varepsilon}\right)\right]^T=E\left(\vec{\varepsilon}\vec{\varepsilon}^T\right), \quad (7)$$

where $\vec{\varepsilon}=\left(\vec{\varepsilon}_1,\vec{\varepsilon}_2\right)^T,\vec{\varepsilon}_1=\left(\vec{\varepsilon}_{1.1},\vec{\varepsilon}_{1.2},\ldots,\vec{\varepsilon}_{1.n}\right)^T,\ \vec{\varepsilon}_2=\left(\vec{\varepsilon}_{2.1},\vec{\varepsilon}_{2.2},\ldots,\vec{\varepsilon}_{2.n}\right)^T$.

Based on the assumption in (6), we can make a form the matrix of covariance as follows.

$$\mathbf{\Omega}=\begin{bmatrix}\mathbf{\Sigma}_{11.t} & \mathbf{\Sigma}_{12.t}\\ \mathbf{\Sigma}_{21.t} & \mathbf{\Sigma}_{22.t}\end{bmatrix}. \quad (8)$$

Equation (8) can also be expressed as in (9).

$$\boldsymbol{\Omega}=\begin{bmatrix} \boldsymbol{\Sigma}_{11.11} & \mathbf{0} & \cdots & \mathbf{0} & \boldsymbol{\Sigma}_{12.11} & \mathbf{0} & \cdots & \mathbf{0} \\ \mathbf{0} & \boldsymbol{\Sigma}_{11.22} & \cdots & \mathbf{0} & \mathbf{0} & \boldsymbol{\Sigma}_{12.22} & \cdots & \mathbf{0} \\ \vdots & \vdots & \ddots & \vdots & \vdots & \vdots & \ddots & \vdots \\ \mathbf{0} & \mathbf{0} & \cdots & \boldsymbol{\Sigma}_{11.nn} & \mathbf{0} & \mathbf{0} & \cdots & \boldsymbol{\Sigma}_{12.nn} \\ \boldsymbol{\Sigma}_{21.11} & \mathbf{0} & \cdots & \mathbf{0} & \boldsymbol{\Sigma}_{22.11} & \mathbf{0} & \cdots & \mathbf{0} \\ \mathbf{0} & \boldsymbol{\Sigma}_{21.22} & \cdots & \mathbf{0} & \mathbf{0} & \boldsymbol{\Sigma}_{22.22} & \cdots & \mathbf{0} \\ \vdots & \vdots & \ddots & \vdots & \vdots & \vdots & \ddots & \vdots \\ \mathbf{0} & \mathbf{0} & \cdots & \boldsymbol{\Sigma}_{21.nn} & \mathbf{0} & \mathbf{0} & \cdots & \boldsymbol{\Sigma}_{22.nn} \end{bmatrix} \quad (9)$$

The matrix $\boldsymbol{\Sigma}_{rr.i}$ is the matrix of the variance in the same response, $r=1,2$ with the same subject, $i=1,2,\ldots,n$ that is $\boldsymbol{\Sigma}_{rr.i}=\operatorname{diag}\left(\boldsymbol{\Sigma}_{rr.11},\boldsymbol{\Sigma}_{rr.22},\ldots,\boldsymbol{\Sigma}_{rr.nn}\right)$. The matrix $\boldsymbol{\Sigma}_{rs.ii}$ is the matrix of covariance in the different response, $r\neq s=1,2$ with the same subject, $i=1,2,\ldots,n$ that is $\boldsymbol{\Sigma}_{rs.i}=\operatorname{diag}\left(\boldsymbol{\Sigma}_{rs.11},\boldsymbol{\Sigma}_{rs.22},\ldots,\boldsymbol{\Sigma}_{rs.nn}\right)$. The matrix of covariance $\boldsymbol{\Omega}$ in equation (9) is used in estimating bi-response nonparametric regression model in longitudinal data with a penalized spline estimator. The matrix of covariance is used to overcome the correlation between responses. The estimation covariance matrix is related to the estimation of error variance in the measurement data. The covariance matrix at (9) is assumed to be unknown, so it estimated as set forth in Theorem 1.

Theorem 1
If the nonparametric regression model with two responses based on the un-weighted penalized spline in each sample $t_{r.ij}$ is

$$\vec{y}_{r.i}=\mathbf{X}_{r.t_{ij}}\vec{\beta}_r^{(0)}+\vec{\varepsilon}_{r.i}^{(0)}, r=1,2; i=1,2,\ldots,n,$$

then we get

1. $\hat{\sigma}^2_{r.ij}=\dfrac{\vec{y}_{r.i}^T\mathbf{P}_{rr}\vec{y}_{r.i}-\vec{y}_{r.i}^T\mathbf{A}_r^T\mathbf{P}_{rr}\mathbf{A}_r\vec{y}_{r.i}}{\operatorname{tr}\left(\mathbf{P}_{rr}\right)}, \mathbf{P}_{rr}=\left[\mathbf{I}-\mathbf{A}_r\right]^T\left[\mathbf{I}-\mathbf{A}_r\right]$
2. $\hat{\sigma}_{rs.ij}=\dfrac{\vec{y}_{r.i}^T\mathbf{P}_{rs}\vec{y}_{s.i^*}-\vec{y}_{r.i}^T\mathbf{A}_r^T\mathbf{P}_{rs}\mathbf{A}_s\vec{y}_{s.i^*}}{\operatorname{tr}\left(\mathbf{P}_{rs}\right)}$, $\mathbf{P}_{rs}=\left[\mathbf{I}-\mathbf{A}_r\right]^T\left[\mathbf{I}-\mathbf{A}_s\right]$

Proof
The coefficient of regression $\vec{\beta}_r^{(0)}$ is a vector of the regression coefficient on the un-weighted penalized spline. The covariance matrix is $\boldsymbol{\Omega}$ in the criteria of PLS. The PLS criterion in (5) can be described in the matrix form.

$$\mathrm{PLS}=\vec{y}_{r.i}^T\vec{y}_{r.i}-2\vec{\beta}_r^{(0)T}\mathbf{X}_{r.t_{ij}}^T\vec{y}_{r.i}+\vec{\beta}_r^{(0)T}\mathbf{X}_{r.t_{ij}}\mathbf{X}_{r.t_{ij}}\vec{\beta}_r^{(0)}+\vec{\lambda}_r^T\vec{\beta}_r^{(0)T}\mathbf{D}_r\vec{\beta}_r^{(0)} \quad (10)$$

If equation (10) is differentiable to $\vec{\beta}_r^{(0)}$, then obtained

$$\hat{\vec{\beta}}^{(0)}=\left(\mathbf{X}_{r.t_{ij}}^T\mathbf{X}_{r.t_{ij}}^T+\vec{\lambda}_r^T\mathbf{D}_r\right)^{-1}\mathbf{X}_{r.t_{ij}}^T\vec{y}_{r.i}.$$

Furthermore, the estimation of nonparametric bi-response regression function in longitudinal data through un-weighted penalized spline estimator can be expressed as follows.

$$\hat{\vec{f}}_r\left(t_{ij}\right)=\mathbf{X}_{r.t_{ij}}\hat{\vec{\beta}}^{(0)}=\mathbf{X}_{r.t_{ij}}\left(\mathbf{X}_{r.t_{ij}}^T\mathbf{X}_{r.t_{ij}}^T+\vec{\lambda}_r^T\mathbf{D}_r\right)^{-1}\mathbf{X}_{r.t_{ij}}^T\vec{y}_{r.i}. \quad (11)$$

If the matrix of the smoothing parameter is $\mathbf{A}_r=\mathbf{X}_{r.t_{ij}}\left(\mathbf{X}_{r.t_{ij}}^T\mathbf{X}_{r.t_{ij}}^T+\vec{\lambda}_r^T\mathbf{D}_r\right)^{-1}\mathbf{X}_{r.t_{ij}}^T$, then the equation (11) can be written into

$$\hat{\vec{f}}_r\left(t_{ij}\right)=\mathbf{A}_r\vec{y}_{r.i}.$$

Next, $E\left(\varepsilon_{r.ij}^{2(0)}\right)$ is obtained on the assumption of $\varepsilon_{r.ij}$ on (7).

$$E\left(\varepsilon_{r.ij}^{(0)}\varepsilon_{r.ij}^{(0)}\right)=E\left\{\vec{y}_{r.i}^{T}\left[\mathbf{I}-\mathbf{A}_{r}\right]^{T}\left[\mathbf{I}-\mathbf{A}_{r}\right]\vec{y}_{r.i}\right\}=E\left\{\vec{y}_{r.i}^{T}\mathbf{P}_{rr}\vec{y}_{r.i}\right\}, \quad (12)$$

where $\mathbf{P}_{rr}=\left[\mathbf{I}-\mathbf{A}_{r}\right]^{T}\left[\mathbf{I}-\mathbf{A}_{r}\right]$.

Equation (12) described by using the properties of the matrix and the result it is as follows.

$$\begin{aligned}E\left(\vec{y}_{r.i}^{T}\mathbf{P}_{rr}\vec{y}_{r.i}\right)&=\text{tr}\left(\mathbf{P}_{rr}\sigma_{r.ij}^{2}\mathbf{I}\right)+\left(\mathbf{A}_{r}\vec{y}_{r.i}\right)^{T}\mathbf{P}_{rr}\left(\mathbf{A}_{r}\vec{y}_{r.i}\right)\\&=\sigma_{r.ij}^{2}\text{tr}\left(\mathbf{P}_{rr}\right)+\vec{y}_{r.i}^{T}\mathbf{A}_{r}^{T}\left[\mathbf{I}-\mathbf{A}_{r}\right]^{T}\left[\mathbf{I}-\mathbf{A}_{r}\right]\mathbf{A}_{r}\vec{y}_{r.i}\\&=\sigma_{r.ij}^{2}\text{tr}\left(\mathbf{P}_{rr}\right)+\vec{y}_{r.i}^{T}\mathbf{A}_{r}^{T}\mathbf{P}_{rr}\mathbf{A}_{r}\vec{y}_{r.i}\end{aligned} \quad (13)$$

Based on (13), the estimated variance of the error is as follows.

$$\hat{\sigma}_{r.ij}^{2}=\frac{\vec{y}_{r.i}^{T}\mathbf{P}_{rr}\vec{y}_{r.i}-\vec{y}_{r.i}^{T}\mathbf{A}_{r}^{T}\mathbf{P}_{rr}\mathbf{A}_{r}\vec{y}_{r.i}}{\text{tr}\left(\mathbf{P}_{rr}\right)},\ \mathbf{P}_{rr}=\left[\mathbf{I}-\mathbf{A}_{r}\right]^{T}\left[\mathbf{I}-\mathbf{A}_{r}\right].\blacksquare$$

The covariance assumption of error is expressed by $E\left(\varepsilon_{r.ij}^{(0)},\varepsilon_{s.i^{*}j}^{(0)}\right)$, so $Cov\left(\varepsilon_{r.ij}^{(0)}\varepsilon_{s.i^{*}j}^{(0)}\right)=\sigma_{r.ij}\sigma_{s.ij}$ to $r=s=1,2$ and $i=i^{*}$, as for $Cov\left(\varepsilon_{r.ij}^{(0)}\varepsilon_{s.i^{*}j}^{(0)}\right)=0$ to $i\neq i^{*}$. Next, $E\left(\varepsilon_{r.ij}^{(0)},\varepsilon_{s.i^{*}j}^{(0)}\right)$ is described as follows.

$$E\left(\varepsilon_{r.ij}^{(0)},\varepsilon_{s.i^{*}j}^{(0)}\right)=E\left\{\vec{y}_{r.i}^{T}\left[\mathbf{I}-\mathbf{A}_{r}\right]^{T}\left[\mathbf{I}-\mathbf{A}_{s}\right]\vec{y}_{s.i^{*}}\right\}=E\left\{\vec{y}_{r.i}^{T}\mathbf{P}_{rs}\vec{y}_{s.i^{*}}\right\}, \quad (14)$$

where $\mathbf{P}_{rs}=\left[\mathbf{I}-\mathbf{A}_{r}\right]^{T}\left[\mathbf{I}-\mathbf{A}_{s}\right]$.

Furthermore, the equation (14) is described and the result is obtained at (15).

$$\begin{aligned}E\left(\vec{y}_{r.i}^{T}\mathbf{P}_{rs}\vec{y}_{s.i^{*}}\right)&=\text{tr}\left(\mathbf{P}_{rs}\sigma_{r.ij}\sigma_{s.i^{*}j}\mathbf{I}\right)+\left(\mathbf{A}_{r}\vec{y}_{r.i}\right)^{T}\mathbf{P}_{rs}\left(\mathbf{A}_{s}\vec{y}_{s.i^{*}}\right)\\&=\sigma_{r.ij}\sigma_{s.i^{*}j}\text{tr}\left(\mathbf{P}_{rs}\right)+\vec{y}_{r.i}^{T}\mathbf{A}_{r}^{T}\left[\mathbf{I}-\mathbf{A}_{r}\right]^{T}\left[\mathbf{I}-\mathbf{A}_{s}\right]\mathbf{A}_{s}\vec{y}_{s.i^{*}}\\&=\sigma_{r.ij}\sigma_{s.i^{*}j}\text{tr}\left(\mathbf{P}_{rs}\right)+\vec{y}_{r.i}^{T}\mathbf{A}_{r}^{T}\mathbf{P}_{rs}\mathbf{A}_{s}\vec{y}_{s.i^{*}}\end{aligned} \quad (15)$$

Let $\sigma_{r.ij}\sigma_{s.i^{*}j}=\sigma_{rs.ij}$, then the estimated variance of the error is as follows.

$$\hat{\sigma}_{rs.ij}=\frac{\vec{y}_{r.i}^{T}\mathbf{P}_{rs}\vec{y}_{s.i^{*}}-\vec{y}_{r.i}^{T}\mathbf{A}_{r}^{T}\mathbf{P}_{rs}\mathbf{A}_{s}\vec{y}_{s.i^{*}}}{\text{tr}\left(\mathbf{P}_{rs}\right)},\ \mathbf{P}_{rs}=\left[\mathbf{I}-\mathbf{A}_{r}\right]^{T}\left[\mathbf{I}-\mathbf{A}_{s}\right].\blacksquare$$

As a result of the theorem, estimation of covariance matrices $\mathbf{\Omega}$ on (9) in the nonparametric bi-response regression model on the longitudinal data based on the penalized spline estimator, it is as follows.

$$\hat{\mathbf{\Omega}}=\begin{bmatrix}\hat{\mathbf{\Sigma}}_{11.11}&\mathbf{0}&\cdots&\mathbf{0}&\hat{\mathbf{\Sigma}}_{12.11}&\mathbf{0}&\cdots&\mathbf{0}\\\mathbf{0}&\hat{\mathbf{\Sigma}}_{11.22}&\cdots&\mathbf{0}&\mathbf{0}&\hat{\mathbf{\Sigma}}_{12.22}&\cdots&\mathbf{0}\\\vdots&\vdots&\ddots&\vdots&\vdots&\vdots&\ddots&\vdots\\\mathbf{0}&\mathbf{0}&\cdots&\hat{\mathbf{\Sigma}}_{11.nn}&\mathbf{0}&\mathbf{0}&\cdots&\hat{\mathbf{\Sigma}}_{12.nn}\\\hat{\mathbf{\Sigma}}_{21.11}&\mathbf{0}&\cdots&\mathbf{0}&\hat{\mathbf{\Sigma}}_{22.11}&\mathbf{0}&\cdots&\mathbf{0}\\\mathbf{0}&\hat{\mathbf{\Sigma}}_{21.22}&\cdots&\mathbf{0}&\mathbf{0}&\hat{\mathbf{\Sigma}}_{22.22}&\cdots&\mathbf{0}\\\vdots&\vdots&\ddots&\vdots&\vdots&\vdots&\ddots&\vdots\\\mathbf{0}&\mathbf{0}&\cdots&\hat{\mathbf{\Sigma}}_{21.nn}&\mathbf{0}&\mathbf{0}&\cdots&\hat{\mathbf{\Sigma}}_{22.nn}\end{bmatrix} \quad (16)$$

with $\hat{\mathbf{\Sigma}}_{rr.ii}=\text{diag}\left(\hat{\sigma}_{r.i}^{2},\hat{\sigma}_{r.i}^{2},\ldots,\hat{\sigma}_{r.i}^{2}\right)$ and $\hat{\mathbf{\Sigma}}_{rs.ii}=\text{diag}\left(\hat{\sigma}_{r.i}\hat{\sigma}_{s.i},\hat{\sigma}_{r.i}\hat{\sigma}_{s.i},\ldots,\hat{\sigma}_{r.i}\hat{\sigma}_{s.i}\right)$.

5. Simulation Study

The ability of covariance matrix as a weighting in the nonparametric bi-response regression model of the longitudinal data is shown through a simulation study. We use a covariance matrix corresponding to equation (19). Simulated studies were performed on the number of subjects 5, 10 and 30, while each subject was measured from 3 to 5 times the measurement. The correlation between the first and the second response is 0.9 and $\varepsilon_r \sim N(0,1)$. The model of each response to the number of knot points of 2 and the lambda is 0.5 is shown in equation (17).

$$\begin{aligned} y_{1ij} &= f_1(t_{ij}) + \varepsilon_{1ij},\ \varepsilon_{1ij} \sim N(0,1), \\ y_{2ij} &= f_2(t_{ij}) + \varepsilon_{2ij},\ \varepsilon_{2ij} \sim N(0,1), \end{aligned} \qquad (17)$$

where $f_1 = 3.5 + 2t_{ij} - t_{ij}^2 + 2(t_{ij} - 2)_+^2 - 2(t_{ij} - 4)_+^2$ and $f_2 = 6.5 - t_{ij} + 2t_{ij}^2 + 1.5(t_{ij} - 2)_+^2 + 1.5(t_{ij} - 4)_+^2$

We compare the error value of the simulated data with the error value after weight through the covariance matrix. It is shown through Box Plot in figure 1.

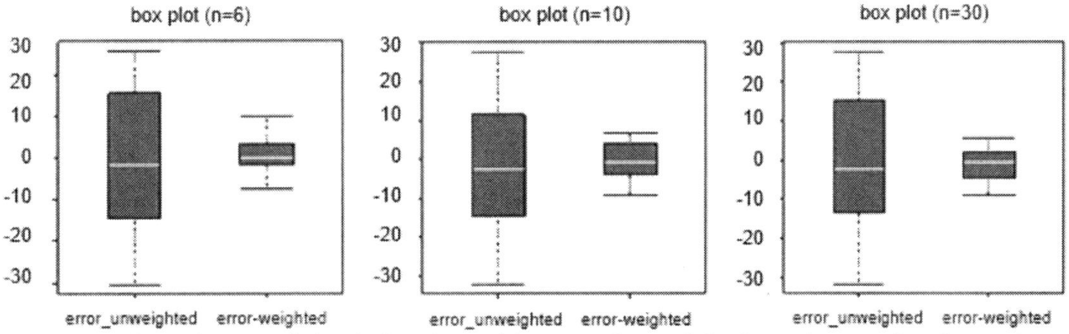

Figure 1. Plots of error of simulation data with error after weight through a matrix of covariance based on penalized spline estimators.

Figure 1 consists of three box plots for each number of subjects 5, 10 and 30. Each box plot shows an error from the model of simulated data without weighted (left) and error after weight with the covariance matrix (right). The results of the analysis show that the error of the simulated data without weighted is greater than the estimated error of the nonparametric bi-response regression model with the covariance matrix. The average error of the model of the covariance matrix rate close to zero with smaller error range. This result shows the superiority of the covariance matrix of error in overcoming the correlation that occurs in the longitudinal data of bi-response through the penalized spline estimator. The simultaneous process between the covariance matrix, knots and smoothing parameters are capable of produce estimation errors of smaller nonparametric bi-response regression models.

6. Conclusion

In this study, we described the covariance matrix estimation of the error in the bi-response longitudinal case through a penalized spline estimator. In addition, through simulation studies, we also demonstrated the ability of the covariance matrix as weight in correlated bi-response data. The result of the covariance matrix estimation through the penalized spline estimator is outlined in Theorem 1, and then it used in the simulation study. Our results have shown that the use of the covariance matrix as weight as set forth in Theorem 1 is capable of providing a smaller error value. The error value is shown through the box plot on several simulations of a number of different subjects.

Acknowledgments

Many thanks to the Directorate General of Higher Education, the Ministry of Research, Technology and Higher Education of the Republic of Indonesia passed through BPPDN grant and the University of Hasanuddin for financial support.

References

[1] Fessler J A 1991 Nonparametric fixed-interval smoothing with vector splines *IEEE Trans. Signal Processing* **39** 852-59

[2] Wahba G 1992 *Multivariate functional and operator estimation, based on smoothing splines and reproducing kernels Nonlinear Modelling and Forecasting SFI Studies in the Science of Complexity* ed by M. Casdagli and S. Eubank Vol XII (Addison Wesley Reading) pp 95-112

[3] Wang Y 1998 Smoothing spline models with correlated random errors *J. Amer. Statist. Assoc.* **93** 341-48

[4] Soo Y W and Bates D M 1996 Multi-response Spline Regression *Computational Statistics & Dana Analysis* **22** 619-31

[5] Chamidah N, Budiantara I N, Sunaryo S and Ismaini Z 2012 Designing of Child Growth Chart Based on Multi Response Local Polynomial Modeling *Journal of Mathematics and Statistics* **8** 342-347

[6] Chamidah N and Saifudin T 2013 Estimation of Children Growth Based on Kernel Smoothing in Multi response Nonparametric Regression *Applied Mathematical Sciences* **7** 1839-47

[7] Diggle P J, Liang Y K and Zeger Z L 2006 *Analysis of Longitudinal Data second edition* (New York: Oxford)

[8] Yao and Lee 2006 Penalized Spline Models for Functional Principal Component Analysis *Journal Royal Statistical Society Series B* **68** 3 – 25

[9] Liang H and Xiao Y 2006 Penalized for Longitudinal Data with an Application in AIDS Studies *Journal of Modern Applied Statistical Methods* **73** 13 – 22

[10] Chen H and Wang Y 2011 A Penalized Spline Approach to Functional Mixed Effects Model Analysis *Biometrics* **67** 861-70

[11] Fernandes A A R, Budiantara IN, Otok B W and Suhartono 2014 Spline Estimators for Bi-Responses Nonparametric Regression Model for Longitudinal Data *Applied Mathematical Sciences* **8** 5653 – 65

[12] Wang Y, Guo W and Brown M B 2000 Spline Smoothing for Bivariate Data with Application to Association Between Hormones *Statistica Sinica* **10** 377-397

[13] Claeskens G, Kribovokova T and Opsomer J D 2009 Asymptotic Properties of Penalized Spline Estimators *Biometrik* **96** 529-544

[14] Montoya L E, Ulloa N and Miller V 2014 A Simulation Study Comparing Knot Selection Methods with Equally Spaced Knot in a Penalized Regression Spline *International Journal of Statistic and Probability* **3** 96-110

[15] Ruppert D and Carrol R J 2000 Spatially-Adaptive Penalties for Spline Fitting *Australian and New Zealand Journal of Statistics* **42** 205-223

[16] Wu H and Zhang J T 2002 Local Polynomial Mixed Effects Models for Longitudinal Data *Journal of American Statistical Association* **97** 883-897

[17] Ruppert D 2002 Selecting the Number of Knot for Penalized Spline *Journal of Computational and Graphical Statistics* **11** 735-757

Risk evaluation on leading companies in property and real estate subsector at IDX: A Value-at-Risk with ARMAX-GARCHX approach and duration test

Dedy Dwi Prastyo[1], **Dwi Handayani**[1], **Soo-Fen Fam**[2], **Santi Puteri Rahayu**[1], **Suhartono**[1], and **Ni Luh Putu Satyaning Pradnya Paramita**[1]

[1] Department of Statistics, Institut Teknologi Sepuluh Nopember, Kampus ITS – Sukolilo, Surabaya 60111, Indonesia
[2] Department of Technopreneurship, Universiti Teknikal Malaysia Melaka, 76100 Durian Tunggal, Melaka, Malaysia

E-mail: dedy-dp@statistika.its.ac.id

Abstract. Risk assessment and evaluation becomes essential for financial institution to measure the potential risk of their counterparties. In middle of 2016 until first quarter of 2017, there is national program from Indonesian government so-called Tax Amnesty. One subsector that has potential to receive positive impact from the Tax Amnesty program is property and real estate. This work evaluates the risk of top five companies in term of capital share listed in Indonesia stock exchange (IDX). To do this, the Value-at-Risk (VaR) with ARMAX-GARCHX approach is employed. The ARMAX-GARCHX simultaneously models the adaptive mean and variance of stock return of each company considering exogenous variables, i.e. IDR/USD exchange rate and Jakarta Composite Index (JCI). The risk is evaluated in scheme of time moving window. The risk evaluation using 5% quantile with window size 500 transaction days perform better result compare to other scenarios. In addition, duration test is used to test the dependency between shortfalls. It informs that series of shortfall are independent.

1. Introduction

One of the macroeconomic indicators that can be used to determine the economic conditions of a country is the conditions of the capital market in that country. The capital market has two main functions in carrying out its role: (i) as a means for companies to obtain funds from investors, and (ii) as a means of investment for society. The rate of return on investment is associated with risk. There are two kind of return: (i) expected return, i.e. the return anticipated by investors in the future, and (ii) actual return, i.e. the return that has been obtained. Risk is the possibility of realization of actual return is lower than the expected minimum return. Risk and return have a positive relationship.

In the year 2016, the Indonesian government is being vigorous in carrying out the national development particularly in infrastructure which is also simultaneously deciding the rules concerning tax amnesty start from first of July. The tax amnesty was predicted have an effect on the property sector which was indicated by the increase of stock prices of companies in property sector at the closing of last transaction day before the tax amnesty start (28/06/2016) by 0.68%. Under the property sector there is property and real estate subsector. In IDX, the companies in these subsectors with highest market capitalization in 2015 include PT. Ciputra Development (CTRA), PT. Bumi Serpong

Damai (BSDE), PT. Lippo Karawaci (LPKR), PT. Summarecon Agung (SMRA), and PT. Pakuwon Jati (PWON). To invest in these five companies, it is required good risk management policies. Such risk analysis can be done through estimating stock risk level using Value-at-Risk (VaR) [1, 2]. Another subsector that possibly receives positive impact from tax amnesty program is construction and building as was analysed by Prastyo, *et. al* [3].

The VaR method encapsulates minimum losses at a certain level of confidence [2]. The stock prices are influenced by macroeconomic factors such as IDR/USD exchange rates and JCI, as exogenous variables in addition to historical return model with non-homogeneous variance in residual. Therefore, in this research the VaR estimation is done by ARMAX-GARCHX approach. The mean part in VaR is approximated by the ARMAX model, while the variance part is approached by using GARCHX model in order to capture the volatility. This study aims to analyse the risk of return on the shares of BSDE, PWON, LPKR, SMRA, and CTRA using VaR with ARMAX-GARCHX approach. Moreover, the VaR calculation is based on the moving window concept. The existence of this research is expected to provide information in the investment policy to the investors who will invest in companies listed in IDX belong to property and real estate subsector.

2. Method

The statistical methods employed in this research are written briefly in the following subsections.

2.1. Capital Asset Pricing Model (CAPM)

Stock return (R_t) is the profit or loss received by stock holder computed as follows:

$$R_t = \frac{P_t - P_{t-1}}{P_{t-1}}, \tag{1}$$

where P_t is closing price of stock at time t. The return of an individual stock as in (1) is compared with return of the market to measure sensitivity of that stock using CAPM.

The CAPM is used to describe the effect of the market on the return of an asset. In this study, the return of market is represented by the return of JCI or *Indeks Harga Saham Gabungan* (IHSG). Equation (2) is used to calculate the returns of a risky asset:

$$\bar{r}_a = r_f + \beta_a(\bar{r}_m - r_f) \tag{2}$$

with r_f is return from risk free investment, β_a so-called *Beta* is the risk measure of corresponding investment, and $\bar{r}_m$ is return from the market (JCI).

2.2. ARIMAX and GARCHX Model

The combination of Autoregressive (AR) and Moving Average (MA) processes is Autoregressive Moving Average (ARMA) process [4]. ARIMAX is a development of ARIMA model by adding exogenous variable to the model. In this study, the stages of the development of procedures for the formulation of ARIMAX model consist of the method forming the model with the input metric scale. This research employs ARMAX because return value is time series that stationary in mean (around zero). The general ARMAX model is presented in (3) as follows:

$$R_t = \beta X_t + \phi_1 R_{t-1} + \cdots + \phi_p R_{t-p} + a_t + \theta_1 a_{t-1} + \cdots + \theta_q a_{t-q}, \tag{3}$$

where X_t is an exogenous variable at time t with its corresponding coefficient β. More general, the X_t can be replaced by X_{t-1} or X_{t-p} depends on time lag giving reasonable interpretation or more accurate forecasting. Another form of ARIMAX model is applied by [5] that employ ARIMA model for residual of time series regression. The multivariate version of this approach (VARI-X) was proposed by [6].

The Generalized Autoregressive Conditional Heterocedasticity (GARCH) introduced by [7] is commonly employed for data with high volatility in finance. The ARCH model is built due to the influence of residuals in the past against conditional variance today. Today's conditional variance can be influenced not only by the squared residual at the past, but also can be affected by the residual's variance at past. Therefore, GARCH can handle the deficiency of ARCH model. The GARCH (r, s) model is formulated in (4) as follows [7]:

$$\sigma_t^2 = \varphi_0 \sum_{j=1}^{r} \varphi_j a_{t-j}^2 + \sum_{j=1}^{s} \lambda_j \sigma_{t-j}^2, \tag{4}$$

with $\lambda_j \geq 0$, $j=1,2,...,s$. The GARCH with Exogenous Variables (GARCHX) is formulated in (5) following [8, 9]. The GARCH(r, s)-X(0_1, 0_2, …, 0_m) has following form:

$$\sigma_t^2 = \varphi_0 + \sum_{j=1}^{r} \varphi_j a_{t-j}^2 + \sum_{j=1}^{s} \lambda_j \sigma_{t-j}^2 + \sum_{l=1}^{m} \gamma_l x_{lt}^2 , \quad (5)$$

where $\gamma_l \geq 0$. The part X(0_1, 0_2, …, 0_m) can be replaced by X(p_1, p_2, ...,p_m) if the significant lag is p_1 for X_1, lag p_2 for X_2, and p_m for X_m.

2.3. Value at Risk and Back testing

The calculation of the VaR value at time t using normal distribution approach is formulated in (6) as follows:

$$\widehat{VaR_\tau}(t) = \hat{\mu}_t + F^{-1}(\tau)\, \hat{\sigma}_t , \quad (6)$$

where $F^{-1}(\tau)$ is the inverse cumulative distribution function of standard normal distribution for quantile τ. In this study, VaR model is calculated by estimating $\hat{\mu}_t$ using ARMAX model, while $\hat{\sigma}_t$ is estimated using GARCHX model.

Back testing is one method that can be used to perform the validity or accuracy of a VaR model built over the realization of the market. The risk function for back testing is shown in the (7):

$$I_{\tau,t} = \begin{cases} 1, & r_t < -VaR_{\tau,t} \\ 0, & r_t \geq -VaR_{\tau,t} \end{cases} \quad (7)$$

The average of indicator function $I_{\tau,t}$ is compared with the risk preference with probability τ.

3. Data and Variable

The stock price data was downloaded from *finance.yahoo.com* for BSDE, PWON, LPKR, SMRA, CTRA, and IHSG. The JCI series was also sourced from the same website. The USD/USD exchange rate data was obtained from the *bi.go.id* website. The period of the data spans from January 1, 2010 until September 30, 2016.

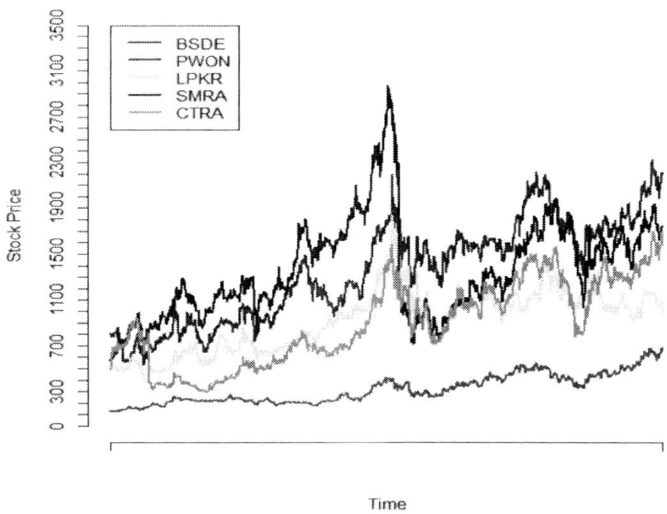

Figure 1. Time series plot of stock close price.

In figure 1, it is known that the stock price movement of the five companies has the same pattern. PWON's stock price is lower than that of other companies. It also showed that the highest stock price increase occurred in early 2013. However, in mid-2013 and 2015 there was a decline in stock prices. This is due to the policies of Bank Indonesia (BI), as a central bank of Indonesia, about loan to value (LTV) II in September 2013 and LTV III in June 2015 for property sector loans. In addition, there was an increase of interest rates in June 2013 by 7.5%. This BI policy resulted in a decline in demand for housing on credit which affected the decline in property stock prices. The descriptive statistic of returns of these five firms is summarized in table 1.

Table 1 Descriptive statistic of return

Firm	Mean	Variance	Coef. of Var.	Min	Max	Skewness	Kurtosis
BSDE	0.0010	0.00079	2763.9	-0.218	0.161	0.01	5.09
PWON	0.0014	0.00072	**1947.2**	-0.125	0.172	0.27	2.91
LPKR	0.0008	0.00061	3302.9	-0.177	0.135	-0.06	4.33
SMRA	0.0012	0.00097	2558.7	-0.478	0.127	-2.15	35.68
CTRA	0.0014	0.00117	2521.6	**-0.518**	**0.240**	-1.66	35.11

The PWON and CTRA have higher average stock returns. When it is viewed from the smallest coefficient of variance, it appears that the PWON has a more stable stock return. One company with the highest and lowest share return value is CTRA. This means that the maximum loss and profitability of CTRA is higher than other companies. Based on the value of skewness and kurtosis, it is known that the stock return data of the five companies do not follow the normal distribution because the value of skewness is not equal to zero and kurtosis value is not equal to 3.

Table 2. Descriptive statistics of daily return

Firm	Statistic	Monday	Tuesday	Wednesday	Thursday	Friday
BSDE	Mean	-0.0044	0.0022	0.0048	0.0017	0.0008
	Variance	0.0009	0.0008	0.0008	0.0009	0.0007
PWON	Mean	-0.0017	0.0003	0.0036	0.0019	0.0030
	Variance	0.0008	0.0007	0.0007	0.0008	0.0007
SMRA	Mean	-0.0013	0.0006	0.0017	0.0006	0.0022
	Variance	0.0008	0.0008	0.0008	0.0017	0.0007
LPKR	Mean	-0.0025	0.0006	0.0046	0.0018	0.0016
	Variance	0.0007	0.0006	0.0006	0.0007	0.0005
CTRA	Mean	-0.0029	0.0018	0.0049	0.0024	0.0007
	Variance	0.0010	0.0020	0.0011	0.0010	0.0008

The characteristics of daily stock price of the five companies can be seen in table 2. There are same characteristics on Monday which (in average) have negative returns. This means the returns tend to lose on Monday. On that day, stock return fluctuate highly indicated by their variances are high enough compared to other days. This phenomenon is called Monday effect. Moreover, when it is viewed from the monthly stock return, PWON and LPKR have the same median per month at about zero. The month with highest median return for BSDE and SMRA is in October at 0.00339 and 0.00602, respectively, while for CTRA is in January about at 0.00349. Hence, at that month the stocks tend to give profit.

Characteristics for exogenous variables of IDR/USD and JCI can be seen in figure 2. The pattern of movement of the IDR/USD exchange rate from day to day tends to increase, but it decreased in mid-2011 which reached the lowest value of Rp8,502.00. But since then, the IDR/USD exchange rate has always increased. The drastic increase occurred in mid-2013 to Rp12,331.00 and the highest IDR/ SD rate occurred in mid-2015 reaching Rp14,802.00. This phenomenon also coincided with the increase in BI rate to 7.5% which also affected the weakening of JCI at that time, as shown in figure 2(b). It appears that when the IDR weakened in mid-2013 and 2015, the rate of JCI precisely decreases. The decline of JCI is affecting the declining value of shares from several sectors such as property sector. Therefore, the movement of the IDR/USD exchange rate and JCI is able to influence the increase and decrease the stock price of a company.

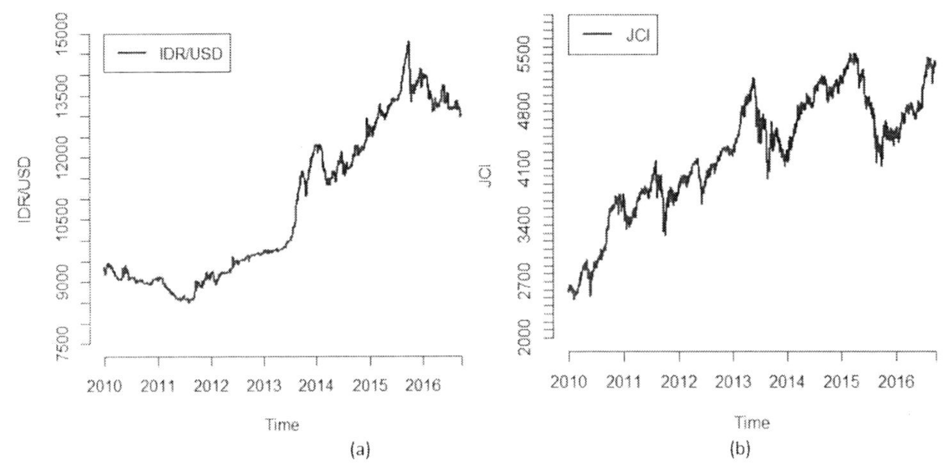

Figure 2. Time series plot of (a) IDR/USD exchange rate and (b) JCI.

4. Empirical Results

The CAPM aims to prove the theory about the influence of JCI as market price against the fifth stock of the company. The window used is 250 day transaction to estimate beta parameters of CAP as in figure 3. In the CAPM analysis, the variable used as risk free rate is the BI interest rate. In figure 3, it is known that CTRA has beta coefficient is greater than one in each window, which means that the sensitivity level of the stock is higher than the JCI. The BSDE, PWON, LPKR, and SMRA have beta coefficient values between zero and two in each window. This means that at the condition $0 < \beta < 1$, the stock sensitivity level is lower than JCI and at the condition $\beta > 1$ the sensitivity level of the stock is higher than JCI. The significance test of beta coefficient is summarized in table 3.

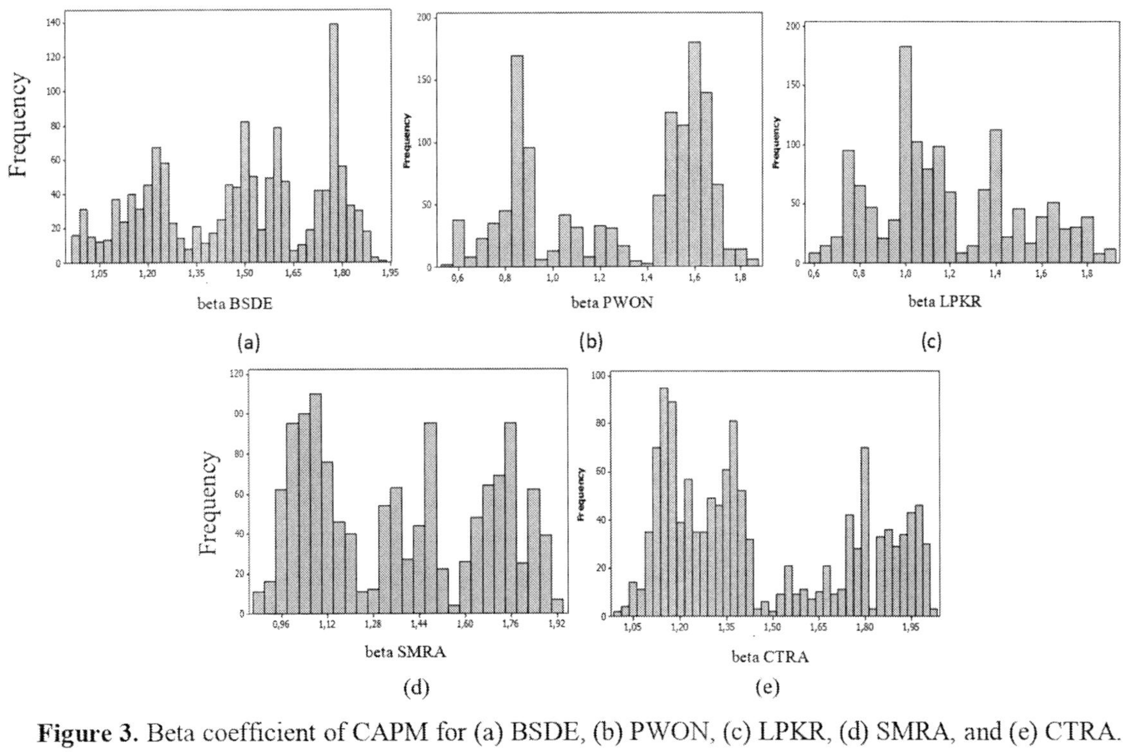

Figure 3. Beta coefficient of CAPM for (a) BSDE, (b) PWON, (c) LPKR, (d) SMRA, and (e) CTRA.

Table 3. Significance Test for Beta coefficient in CPAM.

Firm	Window	β	P-value	Decision
BSDE		1.5139731	1.57E-31	Significant
PWON		1.5112445	1.30E-21	Significant
LPKR	1	0.9413177	8.34E-12	Significant
SMRA		1.8372709	1.02E-27	Significant
CTRA		1.4159665	1.50E-15	Significant
⋮	⋮	⋮	⋮	⋮
BSDE		1.1964281	1.33E-14	Significant
PWON		0.8299523	8.48E-18	Significant
LPKR	1323	0.7454134	8.00E-08	Significant
SMRA		0.9746724	5.88E-15	Significant
CTRA		1.325943	0.0001859	Significant

4.1. ARMAX-GARCHX Modelling

As known from table 3, at α=5% the *p*-value of β for the five companies in each window has a value less than α such that it can be concluded that the JCI return significantly influence the return of each companies. Thus, the results of this CAPM analysis support the theory that JCI affects individual stocks. Estimation of ARMAX-GARCHX model with exogenous variables JCI (X_1) and the IDR/USD exchange rate (X_2) is summarized in table 4 for BSDE.

Table 4. Parameter estimation of GARCHX model for return of BSDE.

Model	Param.	Estimate	*t*-value	*p*-value
	ω	0.000002	957.49	≈ 0
GARCH(0,1) - X(- ,0)	λ_1	0.996586	6699.19	≈ 0
	γ_2	0.000169	4.19	2,80E-05

In table 4, it is known that model are significant because each parameter has |*t*-value| > |*t*-table| amounted to 1.96, such that the model for BSDE is GARCH(0,1)-X(- ,0) with significant exogenous variable is IDR/USD exchange rate. The model is formulated as follows.

$$\hat{\sigma}_t^2 = 0.000002 + 0.996586\,\hat{\sigma}_{t-1}^2 + 0.000169\,X_{2,t}^2 \quad (8)$$

Table 5. Parameter estimation of GARCHX model for return of PWON.

Model	Param.	Estimate	*t*-value	*p*-value
	ω	0.000002	9.87	≈ 0
GARCH(0,1) - X(- ,0)	λ_1	0.995397	39434.01	≈ 0
	γ_1	0.00458	414.13	≈ 0

Based on table 5, each parameter has |*t*-value| > |*t*-table = 1.96| which means the model parameters are significant. The model that is suitable for PWON is GARCH(1,0)-X(- ,0) with significant exogenous variable is IDR/USD exchange rate as formulated:

$$\hat{\sigma}_t^2 = 0.000002 + 0.995397\,\hat{\sigma}_{t-1}^2 + 0.00458\,X_{2,t}^2 \quad (9)$$

Table 6. Parameter estimation of ARMAX-GARCHX model for return of LPKR.

Model	Param.	Estimate	*t*-value	*p*-value
ARMA(1,0)-X(0,0)	ϕ_1	0.095401	3.70	0.000213
	β_1	1.016844	21.58	≈ 0
	β_2	-0.36511	-3.31	0.000945
& GARCH(1,1)-X(0, -)	ω	0.000002	2.22	0.026569
	φ_1	0,019499	308.19	≈ 0
	λ_1	0.976336	788.97	≈ 0
	γ_1	0.000553	2.05	0.040666

Table 6 list the parameters estimates of model for LPKR. The model is ARMA(1,0)-X(0,0) and GARCH(1,1)-X(1, -), where significant exogenous variable in ARMAX model are JCI and IDR/USD exchange rate while in the one in GARCHX model is JCI. The LPKR stock return model is as follows.

$$\hat{R}_t = 0.095401\, R_{t-1} + 1.01684\, X_{1,t} - 0.3651\, X_{2,t}$$
$$\hat{\sigma}_t^2 = 0.000002 + 0.01949\, \hat{a}_{t-1} + 0.976336\, \hat{\sigma}_{t-1}^2 + 0.000553\, X_{1,t}^2 \quad (10)$$

Table 7. Parameter estimation of GARCHX model for return of SMRA.

Model	Param.	Estimate	*t*-value	*p*-value
	ω	0.000009	756.3	≈ 0
GARCH(0,1)-X(- ,0)	λ_1	0.988899	2874.4	≈ 0
	γ_2	0.007913	128.5	≈ 0

In table 7, the SMRA stock return model is GARCH(0,1)-X(- ,0) where the IDR/USD exchange rate as its exogenous variable:

$$\hat{\sigma}_t^2 = 0.000009 + 0.988899\, \hat{\sigma}_{t-1}^2 + 0.007913\, X_{2,t}^2 \quad (11)$$

In CTRA stock return modelling, the result of GARCHX model parameter estimation using window length 250 transaction days, the 3 experiments is expected to represent the modelling in other window. The CTRA return model is listed in table 8.

Table 8. Parameter estimation of GARCHX model for return of CTRA in three windows.

Interval	Model	Par.	Estimate	*t*-value	*p*-value
		ω	0.000011	12.60	≈ 0
1 until 250	GARCH(0,1)-X(- ,0)	λ_1	1	6693.56	≈ 0
		γ_1	0.077516	1273.85	≈ 0
		ω	0.000009	39.99	≈ 0
2 until 251	GARCH(0,1)-X(- ,0)	λ_1	1	2525.63	≈ 0
		γ_1	0.067849	439.39	≈ 0
		ω	0.000009	57.01	≈ 0
3 until 252	GARCH(0,1)-X(- ,0)	λ_1	1	2661.24	≈ 0
		γ_1	0.092593	706.70	≈ 0

After re-modelling the window by 3 times it was obtained the model with the parameters that is always significant in the 3 experiments is GARCH(0,1)-X(- ,0) where the exogenous variable is the IDR/USD exchange rate:

Window 1 : $\hat{\sigma}_t^2 = 0.000011 + 1\,\hat{\sigma}_{t-1}^2 + 0.0775\,X_{2,t}^2$

Window 2 : $\hat{\sigma}_t^2 = 0.000009 + 1\,\hat{\sigma}_{t-1}^2 + 0.0678\,X_{2,t}^2$

Window 3 : $\hat{\sigma}_t^2 = 0.000009 + 1\,\hat{\sigma}_{t-1}^2 + 0.0926\,X_{2,t}^2$

4.2. VaR Estimation

The risk estimation of the five companies is done using ARMAX-GARCHX approach, where the mean part in VaR model is approached with ARMAX model while the variance part is approximated by GARCHX model. The VaR calculations are performed on each window with interval length 250, 375, and 500 days transactions and 5% quantile are used.

Based on table 9, it is found that at the 95% confidence level the biggest maximum loss value experienced by an investor when investing IDR 1 Billion in each company: IDR 43,580,000.00 (BSDE), IDR 46,480,000.00 (PWON), IDR 32,410,000.00 (LPKR), IDR 52,100,000.00 (SMRA), and IDR 51,600,000.00 (CTRA). It is also known that companies that provide the highest and lowest loss rates in window size 250 are CTRA and LPKR, while companies that provide the highest and lowest loss rates on windows size 375 and 500 are SMRA and LPKR, respectively. This VaR estimate applies when the economy is in a normal or non-crisis state. The accuracy of the VaR (quantile 5%) estimation of each window using back testing method is reported in table 10.

Table 9. Estimation of Risk (Loss) and Profit using VaR.

Firm	Statistic	Loss			Profit		
		Window length			Window length		
		250	375	500	250	375	500
BSDE	Mean	-0.0433	-0.0432	-0.0436	0.0433	0.0432	0.0436
	Variance	0.0002	0.0001	0.0001	0.0002	0.0001	0.0001
PWON	Mean	-0.0430	-0.0447	-0.0465	0.0430	0.0447	0.0465
	Variance	0.0002	0.0001	0.0001	0.0002	0.0001	0.0001
LPKR	Mean	-0.0324	-0.0318	-0.0320	0.0326	0.0319	0.0322
	Variance	0.0003	0.0002	0.0002	0.0002	0.0002	0.0002
SMRA	Mean	-0.0488	-0.0508	-0.0521	0.0488	0.0508	0.0521
	Variance	0.0003	0.0002	0.0001	0.0003	0.0002	0.0001
CTRA	Mean	-0.0505	-0.0507	-0.0516	0.0505	0.0507	0.0516
	Variance	0.0002	0.0002	0.0001	0.0002	0.0002	0.0001

Table 10. Back testing of VaR.

Firm	Window	Risk		Profit		| Difference | (%)	
		Loss	Expected Shortfall (%)	Loss	Expected Shortfall (%)	Risk	Profit
	250	63	4.62	79	5.80	0.38	0.80
BSDE	375	61	4.92	71	5.73	0.08	0.73
	500	49	4.40	53	4.76	0.60	0.24

	250	73	5.35	107	7.84	0.35	2.84
PWON	375	68	5.48	81	6.54	0.48	1.54
	500	54	4.85	68	4.65	0.15	0.35
	250	119	8.72	151	11.1	3.72	6.10
LPKR	375	109	8.80	134	10.82	3.80	5.82
	500	92	8.25	111	9.96	3.25	4.96
	250	69	5.06	77	5.64	0.06	0.64
SMRA	375	63	5.08	75	6.05	0.08	1.05
	500	51	4.58	62	5.57	0.42	0.57
	250	62	4.54	95	6.97	0.46	1.97
CTRA	375	53	4.28	76	6.13	0.72	1.13
	500	49	4.40	59	5.30	0.60	0.30

It is known that the results of risk and profit estimation are almost the same; this is due to symmetry VaR condition. The comparison of VaR estimates for BSDE, PWON, SMRA, and CTRA did not differ significantly. Risk and profit estimates in LPKR have a considerable difference from the 5% quantile, indicating that the VaR estimate in this firm tends to be under-estimate. Unlike the case of BSDE, PWON, SMRA, and CTRA which the difference is not too far away from the 5% reference value, which means that the VaR estimates of other companies provides fairly accurate results. It is also known that the calculation of risk and profit by using the window size 500 tends to produce a smaller loss and expected shortfall than using windows size 250 and 375. Hence, the VaR calculation by using observation within windows size 500 transaction days gives a more accurate VaR estimate.

In addition, the duration test is employed to test the dependency of shortfall series. The empirical test shows that there is no dependency between shortfalls. For further research, it is recommended to employ other nonlinear GARCH methods for VaR calculation. Conditional VaR (CVaR) method can be used in subsequent research to accommodate stock dependencies between companies. While other back testing methods are also suggested such as Kupic test, Quantile regression, and duration based model. Similar risk analysis as done by [10] about default prediction will be useful for financial analysis.

5. Conclusion

Companies with the highest and lowest share price averages are SMRA and PWON. The highest and lowest return value is owned by CTRA, which means that the maximum loss and profitability of the CTRA is higher than the others. The five stock returns tend to be negative on Monday which indicates Monday effect phenomenon. The BSDE and SMRA tend to experience profits in October, CTRA in January, while PWON and LPKR are stable every month. The movement pattern of the IDR/USD exchange rate and JCI is always increasing every year. In the CAPM analysis, it is proven that the JCI has a significant effect on the five stocks. It is also note that companies providing highest and lowest loss when estimated using window size 250 transaction days are CTRA and LPKR, while the highest and lowest loss estimated on window size 375 and 500 are SMRA and LPKR, respectively. The estimates of VaR score of LPKR tend to be under-estimate. Moreover, the VaR estimation using 500 transaction days in each window gives more accurate estimates of VaR. Moreover, the VaR estimate using the GARCHX approach yields a more accurate prediction than using the ARMAX-GARCHX. This is because the ARMAX model is redundant. Further research is suggested to use the GARCHX method by assuming the mean model is zero.

Acknowledgement

This reserach is supported by DRPM-RISTEKDIKTI through the Fundamental Research scheme (PDUPT under ITS) with the first author as the principal investigator.

References

[1] Chan N H and Wong H Y 2015 *Simulation Technique in Financial Risk Management, 2nd Edition* (New Jersey: John Wiley and Sons)

[2] Tsay R S 2010 *Analysis of Financial Times Series, 3rd Edition* (New Jersey: John Willey and Sons)

[3] Prastyo D D, Fam S F, Sudjati I L, Setiawan, Suhartono and Paramita N L P S P 2017 Value-at-Risk Modeling on Stock Return with Exogenous Variables Using ARMAX-GARCHX Approach, *The 2nd Int. Conf. on Statistics, Mathematics, Teaching, and Research* (Makassar)

[4] Wei W 2006 *Time Series Analysis, Univariate and Multivariate Method* (New York: Pearson Education, Inc.)

[5] Suhartono, Lee M H and Prastyo D D 2015 Two levels ARIMAX and regression models for forecasting time series data with calendar variation effects *AIP Conference Proceedings* **1691** 050026.

[6] Apriliadara M, Suhartono, Prastyo D D 2016 VARI-X Model for Currency Inflow and Outflow with Eid Fitr Effect in Indonesia, *AIP Conference Proceedings* **1746** 020041.

[7] Bollerslev T 1986 Generalized Autoregressive Conditional Heterocedasticity *Journal of Econometrics* **31** 3

[8] Apergis N and Rezitis A 2011 Food Price Volatility and Macroeconomic Factors: Evidence from GARCH and GARCH-X Estimates *Journal of Agricultural and Applied Economics* **43** 1

[9] Han H and Kristensen D 2014 Asymptotic Theory for the QMLE in GARCH-X Models with Stationary and Nonstationary Covariates *Journal of Business and Economic Statistics* **32** 3

[10] Haerdle H and Prastyo D D 2014 Embedded predictor selection for Default Risk Calculation: A Southeast Asian Industry Study *Handbook of Asian Finance vol. 1, Financial Market and Sovereign Wealth Fund* ed D L K Chuen and G N Gregorious (San Diego: Academic Press) chapter 7 131-148

Spatial Random Effects Survival Models to Assess Geographical Inequalities in Dengue Fever Using Bayesian Approach: a Case Study

Sri Astuti Thamrin and Irfan Taufik

Statistics, Hasanuddin University, Makassar, Indonesia, 90245

E-mail: tuti@unhas.ac.id

Abstract. *Dengue* haemorrhagic fever (DHF) is an infectious disease caused by dengue virus. The increasing number of people with DHF disease correlates with the neighbourhood, for example sub-districts, and the characteristics of the sub-districts are formed from individuals who are domiciled in the sub-districts. Data containing individuals and sub-districts is a hierarchical data structure, called multilevel analysis. Frequently encountered response variable of the data is the time until an event occurs. Multilevel and spatial models are being increasingly used to obtain substantive information on area-level inequalities in DHF survival. Using a case study approach, we report on the implications of using multilevel with spatial survival models to study geographical inequalities in all cause survival.

1. Introduction

Dengue hemorrhagic fever (DHF) is an infectious disease caused by dengue viruses, which are spread by Aedes aegepty beats. In a very short period of time, this disease can lead to the death of the patient. Globally, the number of DF cases has increased significantly every year [1]. To day, DHF still becomes a health problem in Indonesia with the increasing number of patient and wider spread of occurrences. Indonesia occupies the highest position in the case of dengue disease in Southeast Asia with 10,000 cases in 2011[2].

The increasing number of DHF patients relates to the characteristics of the surrounding of their homes such as a ward of sub district. This means that individuals are influenced by their home location and conversely, their location is formed by the individual living in it. Data containing individuals and location is hierarchal structured. Some researches have carried out analysis to such data without taking into consideration the level difference by using linier regression. This method has resulted in some problems such as the dissatisfaction in the analysis results due to the neglecting of other levels. Therefore, [3, 4] proposed a multilevel regression analysis to overcome the problem of hierarchical structured data.

The multilevel model is a regression analysis model taking into account the hierarchy of the data. In this model, the level in a hierarchical structure is defined as a level. The level 1 is the lowest level for example an individual and the level 2 is the higher one such as a location (sub district). A multi level analysis is trying to model the influence of predictor variables measured at the existing levels on response variables measured at the lowest level (level 1). The relationship between these variables with survival times is called the survival analysis.

A survival analysis is about data gathered from the records of a time, achieved by an object where this object has failed to survive. The survival data with hierarchical structured, the survival time of observed object in a group such as the same sub district tends to have a similar characteristic or correlate each other. This might happen because of the presence of an unobserved covariate. One way to solve this problem is to include this covariate in the model as frailty or random effects. In a multilevel analysis, there are two frailty types, including the level 1 and the level 2. However, when making sub districts into a level 2 types, as this research did so, a spatial random effect needs to be taken into account.

Researches on multilevel survival analysis have been previously carried out. Modelled recurrent infection of the urinary tract by using a residual maximum likelihood (REML) approach [5]. Also used multilevel survival analysis applied on recurrent infections of chronic granulomatous disease (CGD) data by implementing a Bayesian approach [6]. Next, [7] applied the same analysis on recurrent infections of CGD data employing a hierarchical likelihood approach and [8] used a spatial correlation as frailty in survival analysis applied on Minnesota's death birth data, but not involving frailty at multilevel analysis.

This paper discusses the implication of using multilevel with spatial survival models to study geographical inequalities in all cause DHF survival by using a Bayesian approach. In this approach, the population parameter is considered as a variable with a prior distribution. Before taking samples from a particular population, it is sometimes obtained some information with regard to parameter going to be estimated. This information is then combined with those of the sample used to estimate the population parameter. Given this, a Bayesian method is more flexibly applied due to the additional information to conclude the population's characteristics.

The paper is organised as follows. We explained introduction in the first section, followed by material and methods. In this second section, we describe the study area, the data collection, the multilevel model, multilevel survival model, multilevel spatial survival model, and Bayesian computation. In the next section, we present the result. The conclusions and the future work are presented further in the last section.

2. Material and Methods

2.1. *Study Area*

Makassar is a city found in South Sulawesi, Indonesia. It is located -5.14 latitude and 119.42 longitudes and is situated at elevation 1-25 meters above sea level. Makassar has a population of 1,747,562, making it the biggest city in South Sulawesi. The climate in Makassar City is tropical. There is significant rainfall in most months of the year. The short dry season has little effect on the overall climate.

The number of sub-districts in the Makassar city is 14, covering 143 villages. Among sub-districs, there are seven sub-districts bordering the coastal districts namely Tamalate, Mariso, Wajo, Ujung Tanah, Tallo, Tamalanrea and Biringkanaya. Makassar city residents based on the results from Socio-economic Survey National (SUSENAS) in 2013 approximately 1,408,072 people, made up of 696, 086 men and 711, 986 women [9].

2.2. *Data Collection*

We obtained data from the Dr. Wahidin Sudirohusodo hospital in Makassar City. The dataset comprises survival time's measurements, sex, age, hematocrit, thrombocyte, leukocyte, hemoglobin, grade and location of patients with DF. A total of 270 DF patients with positive DF, which the ones admitted to hospital, were collected from period January to December 2016 in the Makassar city. We assume that each DF patient in the data set that is not linked with a death must have been alive at the end of December 2016.

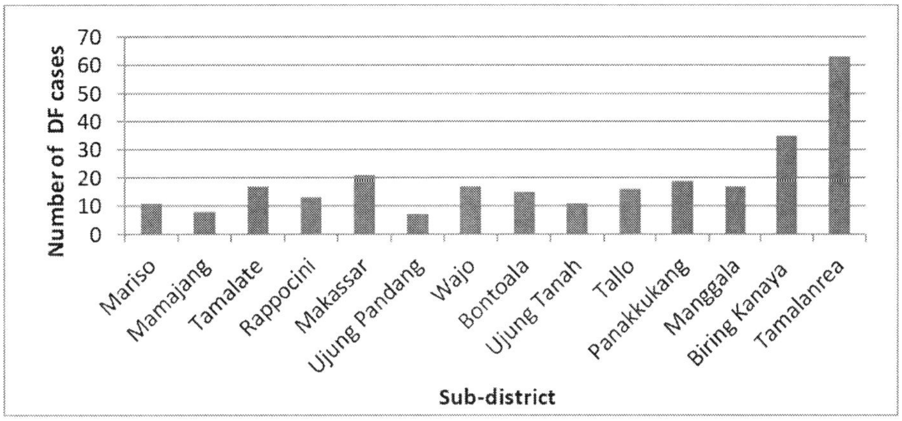

Figure 1. Number of DF cases for each sub-district in Makassar, Indonesia from January to December 2016.

The number of days they hospitalized until they recovered or were allowed to go home is treated as the response t_{ij} in our models, while the remaining survivors were treated as 'censored', or in other words, alive at the end of the study period. There are 8,89% censored patients and 91,1% uncensored patients, including111 (41.11%) men and 159 (46.4%) women with a mean age of 18.6 year. The minimum and maximum age of DF patients were 1 and 68 years, respectively and the average length of stay of DF patients at the hospital was 4 days. The annual dengue incidence rate (number of cases/10,000 people/year) for the January – December 2016 period for the Makassar City based on the sub-district can be seen in figure 1. The major epidemics occurred in Tamalanrea sub-district, whereas considerably fewer dengue cases were reported in Mamajang sub-district.

The response variable used in this study is the number of days they hospitalized and the predictor variables are sex, age, hematocrit, thrombocyte, leukocyte, hemoglobin, grade, and the population size. The data for the same time period for population size was obtained from the Indonesian Bureau of Statistics [9].

2.3. Multilevel Model

The multilevel model concerns populations with hierarchical structures, where samples from the population can be described as a multistage sample [3]. This model is useful for predicting relationships between variables observed at different levels in the multilevel data structure. The simplest multilevel model is a two level model where level 1 is the individual data and level 2 is the group data [10]. Generally, the regression model includes more than one predictor variable at the lowest level as well as at a higher level. The multilevel regression model for level 1 on each level 2 unit [4] is:

$$Y_{ij} = \beta_{0j} + \beta_1 X_{ij} + e_{ij}, \tag{1}$$

where Y_{ij} is the respond variable for level 1, β_{0j} is an intercept for level 1, β_1 is the regression coefficient for level 1, X_{ij} is the predictor variable for level 1, e_{ij} is error for level 1. Index i represents the unit of level 1 nested on level 2 and index j is the unit of level 2. The equation (1) at the level 2, that is intercept at level 1 β_{0j}, becomes the response variable at level 2 as follows:

$$\beta_{0j} = \gamma_{00} + \gamma_{01} Z_j + u_{0j}, \tag{2}$$

where γ_{00} is the intercept on level 2, γ_{01} is the regression coefficient at the level 2, Z_j is the predictor variable at the level 2, u_{0j} is the error at the level 2. By substitution (2) to (1), obtained multilevel regression model as follows:

$$Y_{ij} = \gamma_{00} + \gamma_{01} Z_j + \beta_1 X_{ij} + u_{0j} + e_{ij} \tag{3}$$

2.3.1. Multilevel Survival

Generally, the data type has a hierarchical structure such as survival data. If the response variable is the time for the occurrence of a particular occurrence, the hierarchy of such data may appear. Kim and Dey [6] use a hazard rate with the addition of frailty or random effects of individual effects and group effects in modeling multilevel survival defined as follows:

$$h(t, \boldsymbol{X}, U_{ij}) = h_0(t)\exp(\beta_{0j} + \beta_i X_{ij} + U_{ij}), \quad (4)$$
$$U_{ij} = u_j + e_{ij}$$

where t is the observational time, X is the covariate variable, β is the covariate parameter, $h_0(t)$ is the baseline hazard, u_j is the random effect group, e_{ij} the individual random effect, $i = 1,2,\dots,n$, $j = 1,2,\dots,J$, n are the numbers of individual and group, respectively.

2.3.2. Multilevel Spatial Survival

Frequently, the time data up to the occurrence of an event are grouped in strata or groups such as geographic areas. Hazard rate with the addition of frailty or spatial random effects [8] can be written as follows:

$$h(t, \boldsymbol{X}, w_j) = h_0(t)\exp(\beta_{0j} + \beta_i X_{ij} + w_j), \quad (5)$$

where w_j is the spatial random effect, $i = 1,2,\dots,n$, $j = 1,2,\dots,J$, *n is* the number of individuals, J is the number of locations. By combining a hazard model with the (4) and (5) equation we can obtain hazard model for the spatial survival multilevel as follows:

$$h(t, \boldsymbol{X}, w_j, u_j, e_{ij}) = h_0(t)\exp(\beta_{0j} + \beta_i X_{ij} + U_{ij}), \quad (6)$$
$$U_{ij} = w_j + u_j + e_{ij}$$

the function of likelihood full for the model on the (6) equation above is:

$$L(t_{ij}, X_{ij}, U_{ij}) = \prod_{i=1}^{n}\prod_{j=1}^{J}\left(\alpha t_{ij}{}^{\alpha-1}\exp(\beta_{0j}+\beta_i X_{ij} + U_{ij})\right)^{\delta_{ij}}\exp\{-\exp(\beta X_{ij} + U_{ij})t_{ij}{}^{\alpha}\}, \quad (7)$$

where δ_i is a death indicator (0 if alive, 1 if dead).

2.4. Bayesian Approach

From the perspective of an estimation theory, there are two approaches namely the classic statistical approach and the Bayesian statistical approach. The former one is mainly dependent on an inferential process on sample data from a population. While the latter one apart from making use of such sample data, also takes into consideration a prior distribution called prior statistical inference with Bayesian statistical approach. The classic statistical approach views the parameter θ as a parameter with a fixed value, while the Bayesian statistical approach regards the parameter θ as a random variable with distribution so called prior distribution from which we can determine a posterior distribution to then acquire Bayesian estimator as the average or modus of posterior distribution [10].

The establishment of a prior from multilevel model parameter is done in order (two stage prior) for a two stage hierarchical model. Making the stage-1 prior is the establishment of a prior for the parameter β at the lowest level (level 1). The next stage is to determine the stage-2 prior, which is a prior for the parameter γ at the higher level (level 2). This prior is also called hyperprior. A prior selected for the covariate β parameter is an uninformative prior with $\beta \sim N(0,100)$ and $\gamma \sim N(0,100)$.

Suppose, we implement the lattice frailty model (6). We know assume that we observe the possibly right-censored data with δ_i. The joint posterior distribution of this is given by:

$$p(\beta, W, U \mid t, x, \delta) \propto L(\beta, W, U \mid t, x, \delta)\, p(W|\lambda)p(\beta)p(\gamma) \quad (8)$$

where the first term in the right-hand side is the likelihood, the second is the joint distribution of the random frailties, and the remaining terms are prior distributions. The prior and posterior distributions for each of parameters can be seen in Table 1.

In general, the computation method for a Bayesian inference uses Markov Chain Monte Carlo (MCMC). In MCMC, there are some algorithm types including Gibbs Sampling which is used to withdraw samples form a complex distribution with high dimensions [11]. The main concept of Gibbs

Sampling is how to find the form of conditioned univariate distributions containing all random variables but only one variable is going to be taken its values.

Table 1. Prior Distribution and hyperprior for each of parameters

Parameter	Prior distribution form
$\beta \sim N(0,100)$	$f(\beta)=\frac{1}{10\sqrt{2\pi}}\exp\left(-\frac{\beta^2}{200}\right)$
$\alpha \sim \Gamma(1,1)$	$f(\alpha)=\frac{1}{\Gamma(1)}\exp(-\alpha)$
$e \sim \Gamma(1,1)$	$f(e)=\frac{1}{\Gamma(1)}\exp(-e)$
$\gamma \sim N(0,100)$	$f(\gamma)=\frac{1}{10\sqrt{2\pi}}\exp\left(-\frac{\gamma^2}{200}\right)$
$w_j \vert \lambda \sim N(\bar{w}_j, 1/(\lambda r_j))$	$f(w_j)=\frac{1}{\sqrt{2\pi/(\lambda_j)}}\exp\left(-\frac{(w_j-\bar{w}_j)^2}{2/(\lambda r_j)}\right)$
$\lambda \sim \Gamma(1,1)$	$f(u)=\frac{1}{\Gamma(1)}\exp(-\lambda)$
$u \sim \Gamma(1,1)$	$f(u)=\frac{1}{\Gamma(1)}\exp(-u)$

We choose the normal prior for β, $\rho \sim Gamma\left(\alpha, \frac{1}{\alpha}\right)$ and $\lambda \sim Gamma(a,b)$. Diffuse priors are represented by large positive values for α, a and b. The joint posterior density of (β, W, ρ) in (4) is analytically intractable because the integration of the joint posterior density is not easy to perform. Hence, the Gibbs sampler is used to update the parameter in the model. The complexity of the likelihood in (8) excludes closed-form full conditionals; the parameters may be updated conveniently using Metropolis-Hastings algorithm [12]. In this study, MCMC computations were implemented using the OpenBUGS system [13].

3. Results

The first stage is examining the Weibull distribution on the data using the Mann test, which is special for one distribution. This test shows that the DF survival time data is a weibull distribution. In this paper, spatial correlation was involved as a frailty term at the level 2 and using Moran'I with the queen contiguity neighborhood tested it. The Moran'I test shows that there is the autocorrelation spatial among location. Figure 2 illustrated queen contiguity neighborhood. Next step is to estimate the posterior parameters of the proportional hazard models for spatial survival multilevel use Bayesian approach via Gibbs Sampling algorithm. The posterior parameters that were estimated namely $\beta, \alpha, \sigma_{e_{ij}}, \gamma, w, \sigma_{u_j}, \lambda$.

We ran initially parallel MCMC for each of the models. Then, we monitored them using measurements of sample autocorrelations within the chains, cross-correlations between the parameters, and plots of the sample traces. For the spatial models, however, convergence was much slower. Here our diagnostic tools suggested discarding the first 10 000 iterations from each chain as pre-convergence burn-in. Then, we retained the remaining 50 000 iterations yielded a final sample of posterior analysis. We also note that the lattice model took shorter to run. This can be attributed to the computations involving matrix inversions and determinant evaluations of dimension 14 × 14 within each iteration of the sampler (there are 14 counties in Makassar). The CAR model, on the other hand, avoids this problem since it directly models the weight (inverse dispersion) matrix. We run our models in OpenBUGS, while R with map tool package was used for mapping the results.

IOP Conf. Series: Journal of Physics: Conf. Series **979** (2018) 012095 doi:10.1088/1742-6596/979/1/012095

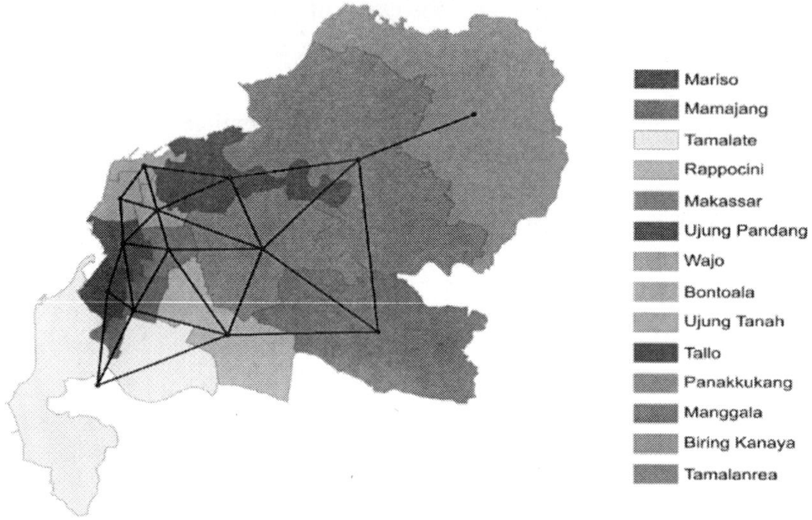

Figure 2. Map of Makassar City with queen contiguity.

Table 2. Posterior summaries for the spatial survival multilevel model.

Model	Parameter	Mean	95% Credible Interval (CI)
Level 1	β_{age}	-0.019	$(-0.033, -0.006)$
	β_{sex}	0.139	$(-0.132, 0.409)$
	$\sigma^2_{e_{ij}}$	2.866	$(0.8817, 6.422)$
	α	2.081	$(1.888, 2.274)$
Level 2	γ_{00}	-2.969	$(-4.257, -1.637)$
	$\gamma_{population\ density}$	7.372×10^{-7}	$(-9.842 \times 10^{-6}, 1.093 \times 10^{-5})$
	$\sigma^2_{u_j}$	2.878	$(0.845, 6.354)$

Table 2 provides 2.5, 50, and 97.5 posterior percentiles for the main effects in our hazard proportional multilevel models. In all two models, all of the predictors are significant at the 0.05 levels, except sex and density population. In the CAR model, the age covariate increases the posterior median hazard rate by a factor of $e^{-0.019} = 0.981$.

The mapping summaries of our results can be seen in Figure 3. We use the Maptool R package to map resulting fitted hazards rates. Figure 3 describes the DF disease mapping in Makassar City. This disease mapping presents the estimated result of the spatial effect for each sub-district to survival of DF patients in Makassar City. From Figure 3, we can see that the lowest spatial effect to survival of DF patients occur in Makassar sub-district (-0.101) and the highest spatial effect to survival of DF patients occur in Biringkanaya sub-district (0.099). The fitted model indicates excess mortality in the north, which is accentuated and extended to a generally increasing pattern from south to north by the survival CAR model.

We now consider several additional plots for our Weibull CAR frailty model with covariates to the observed survival times. Figure 4 illustrates the estimated survival probability curve for a selected set of covariate based on a sample in the DF dataset. The dashed curve is similar, except that it is under the CAR frailty model; for this model, 95% equal-tail point wise credible intervals are also shown as dotted curves. The estimated survival falls quite rapidly towards 0.10 when the survival time of this DF patient is about 7 days but then decays very slowly.

4. Conclusion

In this paper, we already presented the frailty modelling for spatially correlated and multilevel survival data. The Bayesian method coupled with MCMC computational technique was used as an approach to multilevel spatial survival models with CAR frailty. The case study involved DF survival period January to December 2016 at Dr. Wahidin Sudirohusodo hospital in Makassar, with covariates given by sex, age, and population density. Here, we proposed 'lattice' models with CAR.

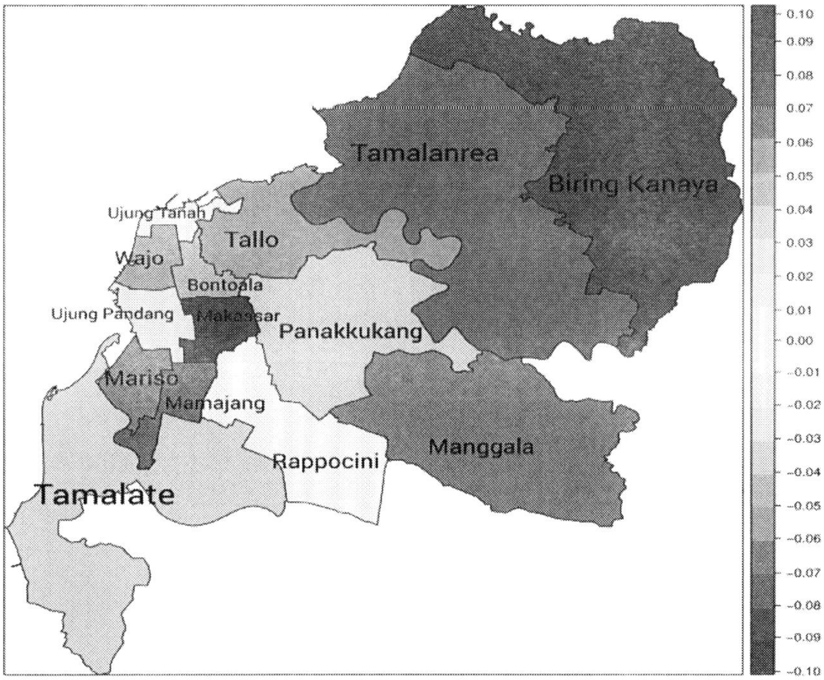

Figure 3. Posterior mean frailties with covariate, multilevel spatial survival, Makassar DF data.

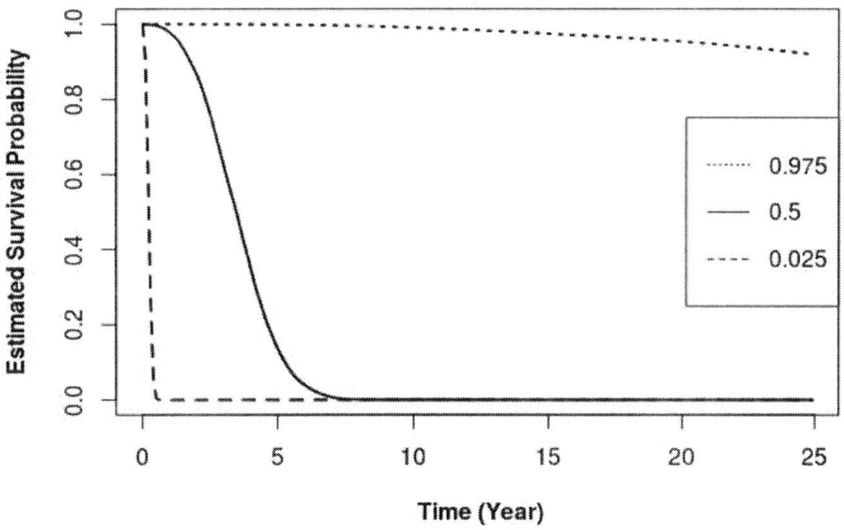

Figure 4. Kaplan-Meier curve for multilevel spatial survival model for DF dataset in Biringkanaya sub-district, Makassar, Indonesia.

Overall, this result shows that only the age covariate describe significantly the DF patients' survival in Makassar City. The lowest and the highest spatial effect to DF patients' survival occur in Makassar and Biringkanaya sub-districts, respectively. This trend, combined with the clear emergence of the Makassar urban area, strongly suggests the need for fitting covariates in our model, most of which vary spatially.

Acknowledgements
The authors acknowledge the Directorate of Research and Community Service, Directorate General Strengthening Research and Development Ministry of Research, Technology and Higher Education Subject Contract Number: 005/ ADD/ SP2H/ LT/ DPRM/ VIII/ 2017, for funding this research.

Appendix

```
model{
        for(i in 1:Nsubj) {
                obs.t[i] ~ dweib(alpha, mu[i])I(t.cen[i],)
                log(mu[i]) <- v[Id[i]] + beta1*Usia[i] + beta2*JenisKelamin[i] + e[Id[i]]}
        for(j in 1:14){ v[j] ~ dnorm(my[j], sigmau)
        my[j] <- gamma0 + gamma1*KepPend[j] + W[j]
        e[j] ~ dnorm(0.0, sigmae)}
        for (i in 1:nsum) {weights[i] <- 1}
        W[1:regions] ~ car.normal(adj[], weights[], num[], tau)
        W.mean <- mean(W[])
        gamma0 ~ dnorm(0.0,0.001)
        gamma1 ~ dnorm(0.0,0.001)
        beta1 ~ dnorm(0.0, 0.001)
        beta2 ~ dnorm(0.0, 0.001)
        alpha~ dgamma(1, 1)
        tau ~ dgamma(1, 1)
        sigmae ~ dgamma(1, 1)
        sigmau ~ dgamma(1, 1)
}
```

For full ODC files contact the author directly.

References

[1] World Health Organisation 2012 *Global Strategy for Dengue Prevention and Control (2012-2020)* WHO Library Cataloguing in Publication Data
[2] National Geographic Indonesia *Berita Kesehatan* [Online; accessed on 15 May 2016]
[3] Hox J J 2002 *Multilevel Analysis: Techiques and Applications* (New Jersey: Lawrence Erlbaum Asscociated Inc)
[4] Hox J J and Maas C J M 2005 *Multilevel Analysis:Encyclopedia of Social Measurement* **2**(1) 785-793
[5] Wang K, Yau K K, Lee A H and McLachlan G J 2007 Multilevel Survival Modeling of Recurrent Urinary Tract Infections *Computer Methods and Programs in Biomedicine* **87**(1) 225-229
[6] Kim S and Dey D K 2008 Modeling Multilevel Survival Data Using Fraity Models *Communications in Statistics-Theory and Methods* **37**(1) 1734-1741
[7] Mustafafi K D 2015 *Analisis Data Multilevel Survival dengan menggunakan Metode Hierarchical Likelihood* (Universitas Gadjah Mada)
[8] Banerjee S, Wall M M, and Carlin B P 2003 Frailty Modeling for Spatially Correlated Survival Data, with application to infant mortality in Minnesota *Biostatistics* **4**(1) 123-142
[9] Biro Pusat Statistik Kota Makassar 2014 *Indeks Pembangunan Manusia Kota Makassar 2014.*

BPS Catalog https://makassarkota.bps.go.id/websiteb/pdf_publikasi/Indeks-Pembangunan_-Manusia-Kota-Makassa-2014.pdf [Online; accessed on 31 October 2016]
[10] Thamrin S A, Amran, Kresna A K, Rahmi S, and Ansariadi 2017 Bayesian inference for spatial parametric proportional hazards model using spatsurv R. *AIP Conference Proceedings* **1827** 020015-1–020015-16
[11] Gilks W R dan Wild P 1992 Adaptive Rejection Sampling for Gibbs Sampling *Applied Statistics* **41**(2) 337-348
[12] Carlin, B P and Louis T A 2000 *Bayes and Empirical Bayes Methods for Data Analysis*ed Boca R (FL: Chapman and Hall/CRC Press)
[13] Lunn D, Spiegelhalter D, Thomas A and Best N 2009 The BUGS project: Evolution, critique and future directions (with discussion) *Statistics in Medicine* **28** 3049-3082

Forecasting Inflow and Outflow of Money Currency in East Java Using a Hybrid Exponential Smoothing and Calendar Variation Model

Ana Susanti[1], Suhartono[1], Hario Jati Setyadi[2], Medi Taruk[2], Haviluddin[3], Putut Pamilih Widagdo[3]

[1]Statistic Department, Faculty of Mathematics and Natural Science, Sepuluh Nopember Institute of Technology. Indonesia.
[2]Informatics Department, Faculty of Computer Science and Information Technology, Mulawarman University. Indonesia.
[3]Computer Science Department, Faculty of Computer Science and Information Technology, Mulawarman University. Indonesia.

E-mail: meditaruk@gmail.com

Abstract. Money currency availability in Bank Indonesia can be examined by inflow and outflow of money currency. The objective of this research is to forecast the inflow and outflow of money currency in each Representative Office (RO) of BI in East Java by using a hybrid exponential smoothing based on state space approach and calendar variation model. Hybrid model is expected to generate more accurate forecast. There are two studies that will be discussed in this research. The first studies about hybrid model using simulation data that contain pattern of trends, seasonal and calendar variation. The second studies about the application of a hybrid model for forecasting the inflow and outflow of money currency in each RO of BI in East Java. The first of results indicate that exponential smoothing model can not capture the pattern calendar variation. It results RMSE values 10 times standard deviation of error. The second of results indicate that hybrid model can capture the pattern of trends, seasonal and calendar variation. It results RMSE values approaching the standard deviation of error. In the applied study, the hybrid model give more accurate forecast for five variables : the inflow of money currency in Surabaya, Malang, Jember and outflow of money currency in Surabaya and Kediri. Otherwise, the time series regression model yields better for three variables : outflow of money currency in Malang, Jember and inflow of money currency in Kediri.

1. Introduction

Money currency is one of cash payment which has important functions to support economic transactions. The circulation of money currency in the bank and community are regulated by Bank Indonesia regulation No.14 /7/ PBI on June 27, 2012 [1]. Bank Indonesia is an independent state agency and central bank of Republic Indonesia, which has one single purpose to achieve and keep stability of rupiah [2]. Outflow of money currency in East Java on July 2015 is 13.456 trillion, while inflow of money currency in East Java reaches 10.719 trillion [3]. Therefore, it is important to research the inflow and outflow of money currency RO of BI in East Java. The objective of this research is to forecast the inflow and outflow of money currency in each RO of BI in East Java by using a hybrid exponential smoothing based on state space approach and calendar variation model.

Based on previous research showed that inflow and outflow of money currency contains patterns not only seasonal and trends but also calendar variation. If only use exponential smoothing models, it will give bad result. Therefore, this research will combine or hybrid exponential smoothing model with state space approach and calendar variation model. It hopes the hybrid model is a practical forecasting model and can capture three patterns to generate small forecasting error value. Previous research about hybrid modeling was made by Zhang [4] which showed that comparation method between ARIMA, NN, and hybrid ARIMA-NN the most accurate forecast is a hybrid ARIMA-NN model.

Previous research about netflow of money currency is made by Karomah [5] and Wulansari [6] using ARIMAX which showed that netflow of money currency is affected by Eid-holidays, seasonal effect and consumer price index. Hanim [7] and Masun [8] research about forecasting inflow and outflow of money currency using ARIMAX and regression time series. Their research showed that inflow and outflow of money currency is influenced by Eid-holidays and annual seasonal effects. Previous research about ARIMAX and regression two level model for forecasting sales data on retail company is made by Suhartono, Lee and Prastyo [9] which showed that ARIMAX and regression two level model can capture the effect of calendar variations and produce more accurate forecast. Research about exponential smoothing with state space approach was made by Koehler [10] and Nurhariyadi [11] which showed that exponential smoothing model with handling outlier is better than exponential smoothing model (basic model).

Until now, Bank Indonesia does not have a standard method to predict inflow and outflow of money currency in each RO of BI. This research are expected to be one of the policy for Bank Indonesia to forecast inflow and outflow of money currency in each RO of BI.

2. Literature Review

2.1. Time Series Regression Model

Time series regression model for data that contain trend pattern is following the equation [12].

$$Y_t = \beta_0 + \beta_1 t + a_t \quad , \quad t = 1,2,3,...,n \tag{1}$$

with a_t is a residual component that satisfy white noise assumption. While data which has seasonal pattern $S_{1,t}, S_{2,t}, ..., S_{s,t}$. It can be written as follows

$$Y_t = \beta_1 S_{1,t} + \beta_2 S_{2,t} + \cdots + \beta_s S_{s,t} + a_t \tag{2}$$

with $S_{1,t}, S_{2,t}, ..., S_{s,t}$ is a dummy variable to explain seasonal pattern. Linear regression model for data with calendar variation pattern follow this equation.

$$Y_t = \beta_1 V_{1,t} + \beta_2 V_{2,t} + \cdots + \beta_p V_{p,t} + a_t \tag{3}$$

with $V_{p,t}$ = dummy variable for calendar variation effects into-p.

2.2. Single Exponential Smoothing (N,N Method)

General equation single exponential smoothing can be written:

$$\hat{Y}_{t+1} = \hat{Y}_t + \alpha(Y_t - \hat{Y}_t) \tag{4}$$

Forecasting equation single exponential smoothing can be written:

$$\hat{Y}_{t+h|t} = \hat{Y}_{t+1} \ , \ h = 2,3,... \tag{5}$$

with state space model is

$$\ell_t = \hat{Y}_{t+1}$$

$$\widehat{Y}_{t+h|t} = \ell_t$$

$$\ell_t = \alpha Y_t + (1-\alpha)\ell_{t-1}. \tag{6}$$

So, single exponential smoothing model follow this equation [13].

$$\widehat{Y}_{t+1} = \ell_t = \widehat{Y}_t + \alpha(Y_t - \widehat{Y}_t) \tag{7}$$

$$\widehat{Y}_{t+1} = \alpha Y_t + (1-\alpha)\widehat{Y}_t. \tag{8}$$

2.3. *Addittive Seasonality Model : ETS (A,A,A)*

State space model for ETS (A,A,A) can be written as:

$$\mu_t = \ell_{t-1} + b_{t-1} + s_{t-m}$$
$$\ell_t = \ell_{t-1} + b_{t-1} + \alpha a_t$$
$$b_t = b_{t-1} + \beta a_t$$
$$s_t = s_{t-m} + \gamma a_t. \tag{9}$$

The forecast function for ETS (A,A,A) follow this equation [13].

$$\widehat{Y}_{t+h|t} = \ell_t + hb_t + s_{t-m+h_m^+}. \tag{10}$$

2.4. *Multiplicative Seasonality Model : ETS (M,A,M)*

State space model for ETS (M,A,M) can be written as:

$$\mu_t = (\ell_{t-1} + b_{t-1})s_{t-m}$$
$$\ell_t = (\ell_{t-1} + b_{t-1})(1 + \alpha a_t)$$
$$b_t = b_{t-1} + \beta(\ell_{t-1} + b_{t-1})a_t$$
$$s_t = s_{t-m}(1 + \gamma a_t). \tag{11}$$

The forecast function for ETS (M,A,M) follow this equation [13].

$$\widehat{Y}_{t+h|t} = (\ell_t + hb_t)s_{t-m+h_m^+}. \tag{12}$$

2.5. *Hybrid Exponential Smoothing with State Space Approach and Calendar Variation Model*

Hybrid exponential smoothing and calendar variation model can be written in two models.

2.5.1. *The First Model*

Level 1. The initial regression model

$$Y_t^1 = \delta t + \beta_1 M_{1,t} + \beta_2 M_{2,t} + \ldots + \beta_S M_{S,t}$$
$$+ \sum_g \alpha_g D_{g,t} + \sum_g \gamma_g D_{g,t+1} + u_t \tag{13}$$

Level 2. The regression model with calendar variation effect.

$$\widehat{\alpha}^*_t = a + bg$$
$$\widehat{\gamma}^*_g = c + dg \tag{14}$$

2.5.2. *The Second Model.*

$$\ell_t = \alpha y_t + (1-\alpha)\ell_{t-1}$$
$$\widehat{y}_t^* = \ell_t \tag{15}$$

2.5.3. The First Forecast

$$\hat{Y}_t = \hat{Y}_t^1 + \hat{\alpha}^*_g + \hat{\gamma}^*_g + \hat{y}_t^* \tag{16}$$

2.5.4. The Second Forecast

$$\hat{Y}_t = \hat{Y}_t^1 + \hat{y}_t^* \tag{17}$$

2.6. Selection The Best Model

In out-sample approach, the best model will be selected based on the smallest forecasting error value. This research use criteria RMSE (Root Mean Square Error). RMSE is defined as follows [14].

$$RMSE = \sqrt{\frac{1}{L}\sum_{l=1}^{L}(Y_{n+l} - \hat{Y}_n(l))^2} \tag{18}$$

with

Y_{n+l} : actual value for out-sample data into-l or data at time into-(n+l), $l = 1,2,...,L$

$\hat{Y}_n(l)$: forecast results of out-sample data into-l.

3. Research Methodology

3.1. Source of Data

This research use secondary data that consists of inflow and outflow of money currency in East Java from January 2003 into December 2014. It was obtained from Bank Indonesia. The period January 2003-December 2013 is used as in-sample data while January 2014-December 2014 is used as out-sample data. This research also use simulation study. Simulation data is obtained from the following equation:

$$Y_{t,a} = T_t + S_{t,a} + CV_t + \varepsilon_t \tag{19}$$

$$Y_{t,m} = (T_t \times S_{t,m}) + CV_t + \varepsilon_t \tag{20}$$

a. Generate data $\varepsilon \sim N(0, \sigma_\varepsilon^2)$ with n=120. This research use σ_ε=3.

b. Generate data T_t follow this equation

$$T_t = 10 + 2t. \tag{21}$$

c. Generate data $S_{t,a}$ follow this equation

$$S_{t,a} = 20 + 10\sin\frac{2\pi t}{12}. \tag{22}$$

d. Generate data $S_{t,m}$ that calculated by addition 12 for each year.

e. Determine calendar variation effect CV_t follow this equation

- effect of Eid-holidays

$$\alpha_g = 5 + 10g \tag{23}$$

- effect before Eid-holidays

$$\gamma_g = 100 - 5g. \tag{24}$$

f. Obtain simulation data with additive pattern that calculated by the sum of value ε_t, T_t, $S_{t,a}$ and CV_t so get the equation (19).

Obtain simulation data with multiplicative pattern that calculated by the sum of value ε_t, $(T_t \times S_{t,m})$ dan CV_t so get the equation (20).

3.2. Research Variables

The variables are used in this research involve:

Table 1. Research Variables

Research Variables	KPw			
	Surabaya	Malang	Kediri	Jember
Inflow	$Y_{1,1t}$	$Y_{1,2t}$	$Y_{1,3t}$	$Y_{1,4t}$
Outflow	$Y_{2,1t}$	$Y_{2,2t}$	$Y_{2,3t}$	$Y_{2,4t}$

While dummy variables are used in this research involve:

a. Dummy of trend following

$L_{1,t}$: The increase in second period, January 2007- December 2010
$L_{2,t}$: The increase in third period, January 2011-December 2014
$tL_{1,t}$: Trend and increase in second period
$tL_{2,t}$: Trend and increase in third period.

b. Dummy of seasonal (month) following

$M_{1,t}, M_{2,t}, \ldots, M_{12,t}$: January, February,...,December.

c. Dummy of calendar variation effect based on Eid-holidays with effect number of days before Eid-holidays with a period one month after (T+1), one month before (T-1) and the time Eid-holidays is occured (T). after (T+1), one month before (T-1) and the time Eid-holidays is occured (T).

$$D_{g,t} = \begin{cases} 1, & T = t \\ 0, & T \neq t \end{cases}$$

which g indicates the number of days before Eid-holidays that occured on (g + 1) and T indicates the time Eid-holidays is occured. This research use two level time series regression model The first level is time series regression equation that produced by initial model. Second level is made by using the coefficient parameter equation that produced by first level. Second level will be used to predict or forecast calendar variation effect on every possibility for many different previous day [8].

i. Level 1. The initial regression model.

$$Y_t = \delta t + \beta_1 M_{1,t} + \beta_2 M_{2,t} + \ldots + \beta_S M_{S,t} + \sum_g \alpha_g D_{g,t} + \sum_g \gamma_g D_{g,t+1} + u_t \quad (25)$$

ii. Level 2. The regression model with calendar variation effect.

$$\hat{\alpha}^*_g = a + bg \quad (26)$$

$$\hat{\gamma}^*_g = c + dg \quad (27)$$

with $\hat{\alpha}^*_g$ and $\hat{\gamma}^*_g$ is coefficient of parameter estimation results in the first level, and a, b, c, d is constants. Model at second level is performed if the number of days before Eid-holidays has not been found at the first level model. Forecasting of results can be calculated by the sum of forecast value at the regression model level 1 and level 2.

3.3. Step Analysis

Step study will be done in this research is as follows:

3.3.1. Simulation Study

a. Dividing the simulation data into two parts, in-sample and out-sample. Then apply exponential smoothing model for in-sample data.
b. Evaluation weakness of exponential smoothing model.
c. Combining exponential smoothing and calendar variation model. The step is as follows:
   - Calendar variation model for simulation data using time series regression. This research use two level regression model.
   - The first model is the sum of forecast value on level 1 and level 2.
   - Error that obtained by first model will analyze using exponential smoothing with state space approach model.
   - Combining predictions of both models. It is calculated by the sum of forecast value at the first model and second model.
   - Evaluate the hybrid model.

3.3.2. Applied Study

a. Modeling inflow and outflow of money currency in RO of BI Surabaya, in RO of BI Malang, in RO of BI Jember, and in RO of BI Kediri using hybrid model and time series regression model.
b. Comparing the accuracy of both models to determine the best model.
c. Forecast inflow and outflow of money currency in East Java by using the best model.

4. Analysis And Results

4.1. Exponential Smoothing Model on Simulation Data Trend and Seasonal Pattern

The combined graph trend and additive seasonal can be seen in figure 1.

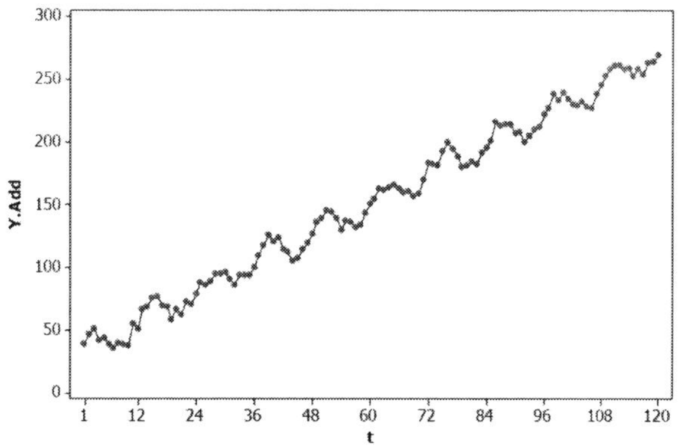

Figure 1. Time Series Plot Trend dan Additive Seasona.

Data with trend and additive seasonal patterns produce ETS (A,A,A). State space equation for ETS (A,A,A) can be written as:

$$y_t = \ell_{t-1} + b_{t-1} + s_{t-m} + a_t$$

$$\ell_t = \ell_{t-1} + b_{t-1} + 0.0001a_t$$

$$b_t = b_{t-1} + 0.0001a_t$$

$$s_t = s_{t-m} + 0.0001a_t$$

In-sample RMSE value for state space ETS (A,A,A) is 3.42. While out-sample RMSE value for state space ETS (A,A,A) is 3.18. Trend and multiplicative seasonal pattern can be seen in figure 2.

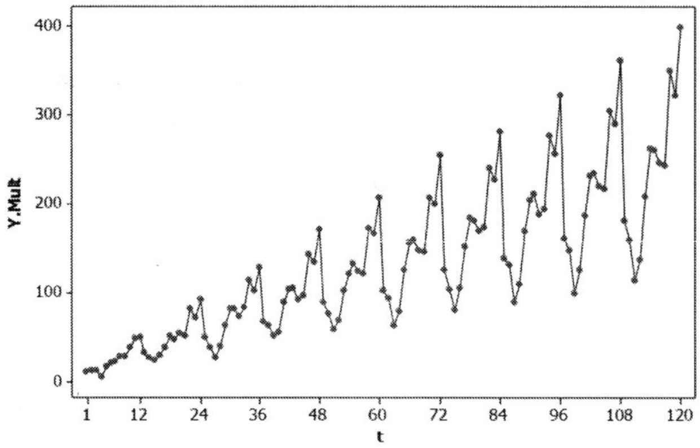

Figure 2. Time Series Plot Trend and Multiplicative Seasonal.

Data with trend and multiplicative seasonal pattern produce ETS (M,A,M). State space equation for ETS (M,A,M) can be written as:

$$y_t = (\ell_{t-1} + b_{t-1})s_{t-m} + a_t$$
$$\ell_t = (\ell_{t-1} + b_{t-1})(1 + 0.5565a_t)$$
$$b_t = b_{t-1} + 0.0238(\ell_{t-1} + b_{t-1})a_t$$
$$s_t = s_{t-m}(1 + 0.0001a_t)$$

State space model for ETS (M,A,M) produces in-sample RMSE is 4.88. While out-sample RMSE is 3.86. Based on the RMSE value can be obtained information that ETS (M,A,M) and ETS (A,A,A) method is an appropriate model to handle data with trend and seasonal pattern because RMSE value close to the standard deviation of error on simulation data. This is exponential smoothing model for data with trend, seasonal and calendar variation patterns.

4.2. Exponential Smoothing Model on Simulation Data Trend, Seasonal and Calendar Variation Pattern

The combined graph trend, additive seasonal and calendar variation can be seen in figure 3.

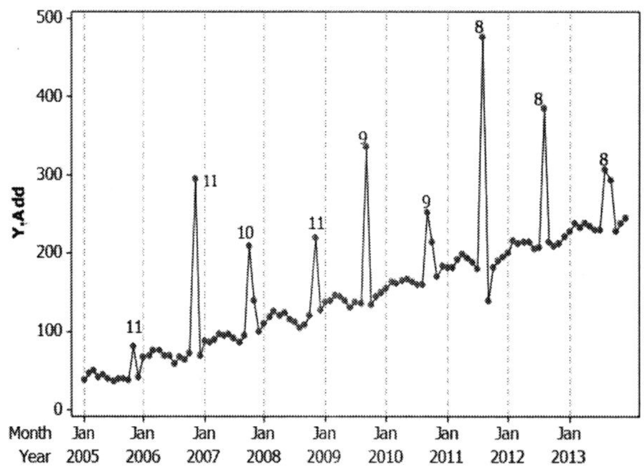

Figure 3. Time Series Plot Trend, Seasonal and Calendar Variation (Additive)

While the combined graph trend, multiplicative seasonal and calendar variation can be seen in figure 4.

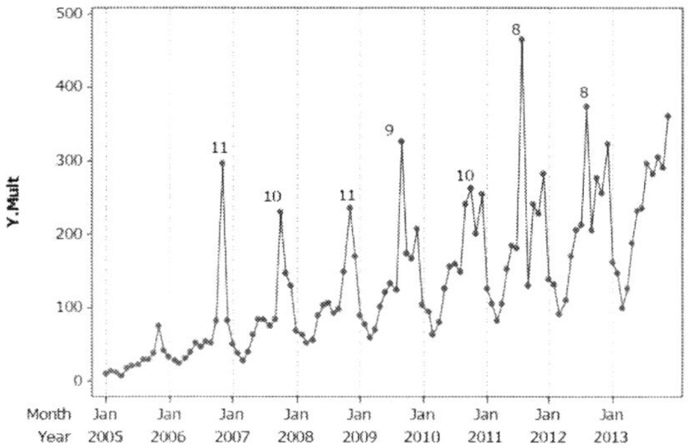

Figure 4. Time Series Plot Trend, Seasonal and Calendar Variation (Multiplicative)

When simulation data with trend, additive seasonal and calendar variation pattern were modeled by exponential smoothing method will produce ETS (A,A,A). But the state space model of ETS (A,A,A) produces RMSE value 42.98. Similarly, simulation data with trend, multiplicative seasonal and calendar variation pattern were modeled by exponential smoothing method will produce ETS (M,A,M) with RMSE value 56.82. It was the highest RMSE value.

This condition occurs because the winter's models was designed to handle data with trends and seasonal pattern.In this case, winter's models was not capture calendar variation pattern so RMSE value that produced was not approach standard deviation of the simulated data. Therefore, this research will use a hybrid exponential smoothing with state space approach and calendar variation model to handle this case.

4.3. Hybrid Model on Simulation Data Trend, Seasonal and Calendar Variation Pattern

Since there is constraint about calendar variation effect, hybrid model will be written into two equations model. The first model is regression equations with dummy variable calendar variation effect. First model consists of two levels as in equation (20) and (21). While the second model is exponential smoothing equation of residual that produced by the first model. There are the results time series regression (TSR) of the first model for every possibility:

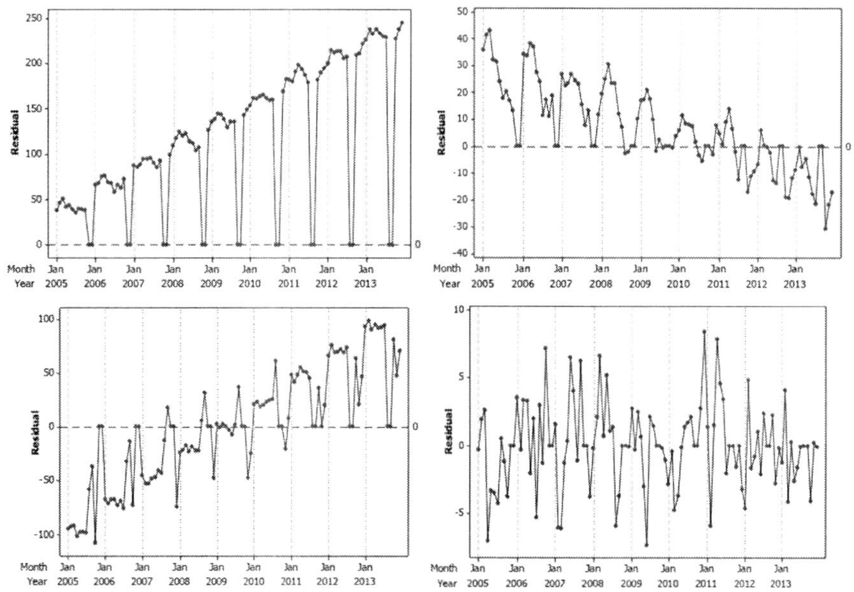

Figure 5. Time Series Plot Elimination CV_t (a), $T_t + CV_t$ (b), $S_t + CV_t$ (c), $T_t + S_t + CV_t$ (d) (Additive Pattern)

There were a summary of the RMSE value that generated by time series regression model to eliminate calendar variation, trend and seasonal additive effects.

Table 2 RMSE Value TSR Model on Simulation Data (Additive)

Time Series Regression Model	RMSE
elimination CV_t effect	34.44
elimination $T_t + CV_t$ effect	5.93
elimination $S_t + CV_t$ effect	22.42
elimination $T_t + S_t + CV_t$ effect	**3.15**

Based on the analysis that has been done, it can be concluded that TSR model for additive data with eliminate calendar variation, trend, and seasonal is appropriate model because it generated RMSE value 3.15. It is close to the standard deviation of error on simulation data. Furthermore, residual that produced by the first model will be modeled by using exponential smoothing model and obtained ETS (A,N,N) model. The results of hybrid time series regression model and ETS (A,N,N) produced in-sample RMSE values 3.15 while for out-sample RMSE value was 3.09. TSR model for multiplicative pattern was as follows.

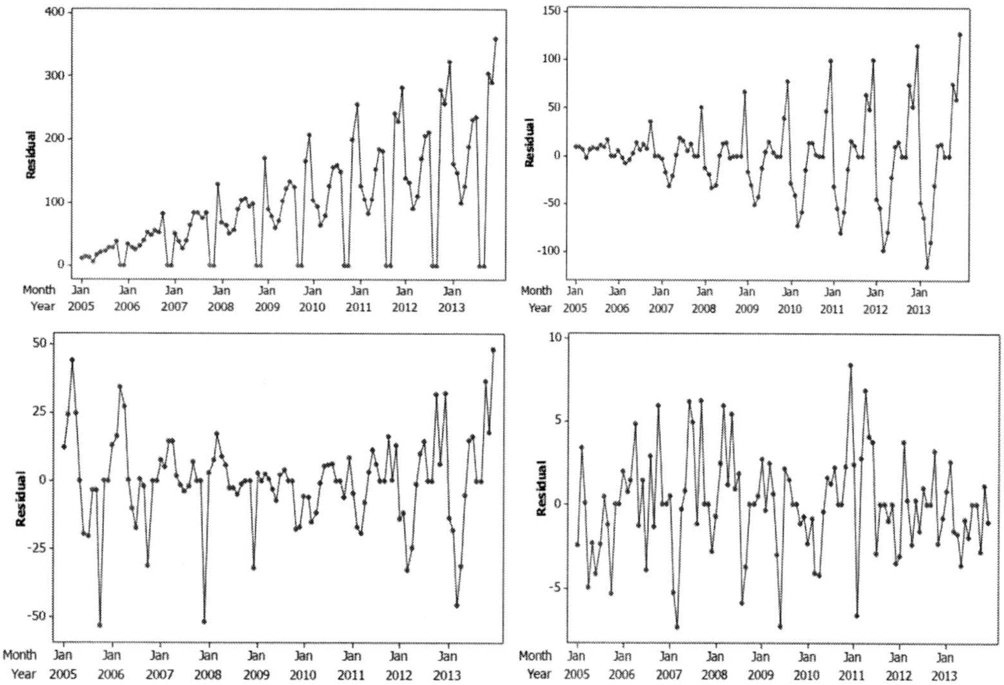

Figure 6. Time Series Plot Elimination CV_t (a), $T_t + CV_t$ (b), $T_t + S_t + CV_t$ (c), $T_t + S_t + CV_t + T_t\, S_t$ (d) (Multiplicative Pattern).

There were a summary of the RMSE value that generated by time series regression model to eliminate calendar variation, trend and seasonal multiplicative effects.

Table 3 RMSE Value TSR Model on Simulation Data (Multiplicative)

Time Series Regression Model	**RMSE**
elimination CV_t effect	40.09
elimination $T_t + CV_t$ effect	22.98
elimination $T_t + S_t + CV_t$ effect	16.97
elimination $T_t + S_t + CV_t + T_t\, S_t$ effect	**3.01**

Based on the analysis that has been done, it can be concluded that TSR model for multiplicative data with eliminate calendar variation, trend, seasonal and multiplication between trend and seasonal is appropriate model because it generated RMSE value 3.01. It is close to the standard deviation of error on simulation data. Furthermore, residual that produced by the first model will be modeled by using exponential smoothing model and obtained ETS (A,N,N) model. The results of hybrid time series regression model and ETS (A,N,N) produced in-sample RMSE values 3.01 while for out-sample RMSE value was 3.96.

4.4. Modeling Inflow dan Outflow of Money Currency by Using Hybrid Model

A hybrid model for inflow and outflow of money currency in East Java produced ETS (A,N,N) when analyzed the second model. Time series plot out-sample data and prediction value for inflow of money currency in East Java can be seen in figure 7.

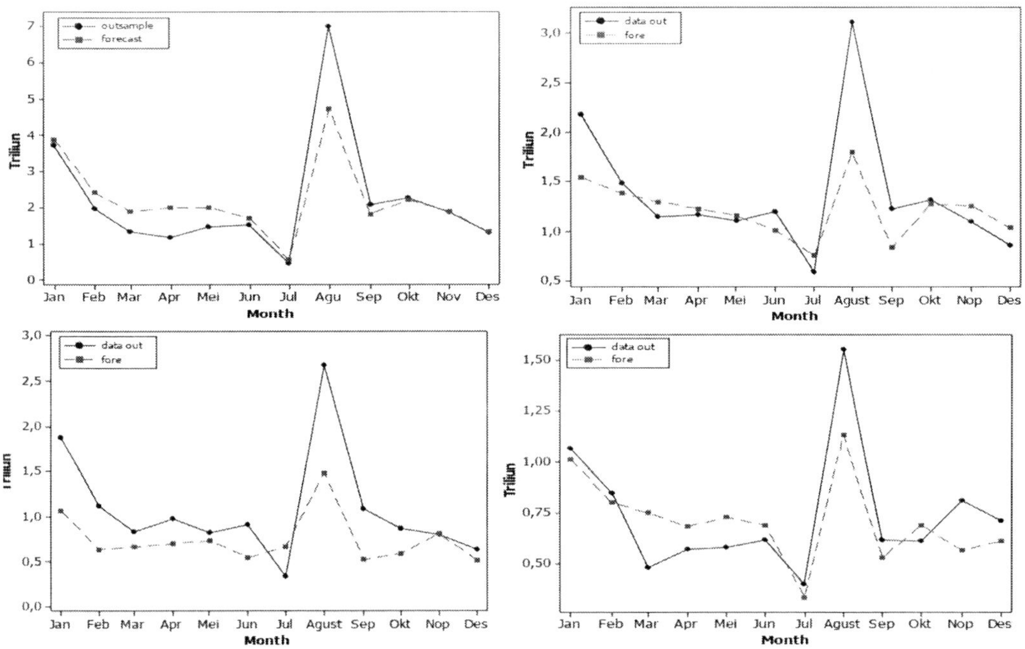

Figure 7. Actual data vs Forecasting inflow 2014 RO of BI Surabaya (a) Malang (b) Kediri (c) Jember (d).

4.5. Comparation between Hybrid and Time Series Regression Model

The model has been obtained will be compared using the criteria of goodness models. It can be seen by RMSE value as follows.

Table 4. Comparation RMSE Value.

RO of BI		Method	RMSE *in-sample*	RMSE *out-sample*
Surabaya	*Inflow*	Hybrid	0.189	**0.787**
		Time Series Regression	0.249	1.359
	Outflow	Hybrid	0.204	**1.022**
		Time Series Regression	0.346	1.189
Malang	*Inflow*	Hybrid	0.069	**0.451**
		Time Series Regression	0.082	0.705
	Outflow	Hybrid	0.056	0.529
		Time Series Regression	0.077	**0.498**
Kediri	*Inflow*	Hybrid	0.064	0.504
		Time Series Regression	0.085	**0.491**
	Outflow	Hybrid	0.092	**0.433**
		Time Series Regression	0.153	0.642
Jember	*Inflow*	Hybrid	0.028	**0.177**
		Time Series Regression	0.065	0.284
	Outflow	Hybrid	0.045	0.308
		Time Series Regression	0.063	**0.274**

Table 4 showed that hybrid model can provide better forecasting results than the time series regression model. Hybrid model for 5 variables while time series regression model for 3 variables. There were time series plot of forecasting inflow and outflow of money currency in 2015 for each RO of BI East Java.

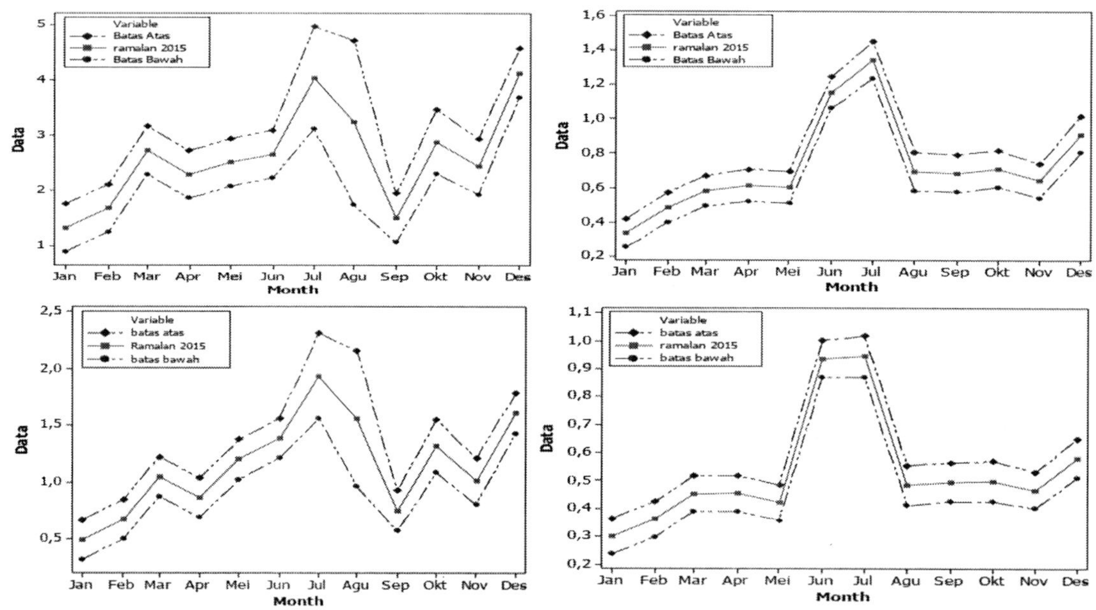

Figure 8. Forecasting Inflow and Outflow of Money Currency 2015 (a) RO of BI Surabaya (b) RO of BI Malang (c) RO of BI Kediri (d) RO of BI Jember

IOP Conf. Series: Journal of Physics: Conf. Series **979** (2018) 012096 doi:10.1088/1742-6596/979/1/012096

There were the results of adaptive RMSE value Inflow and Outflow of moner curency in East Java by using the best model.

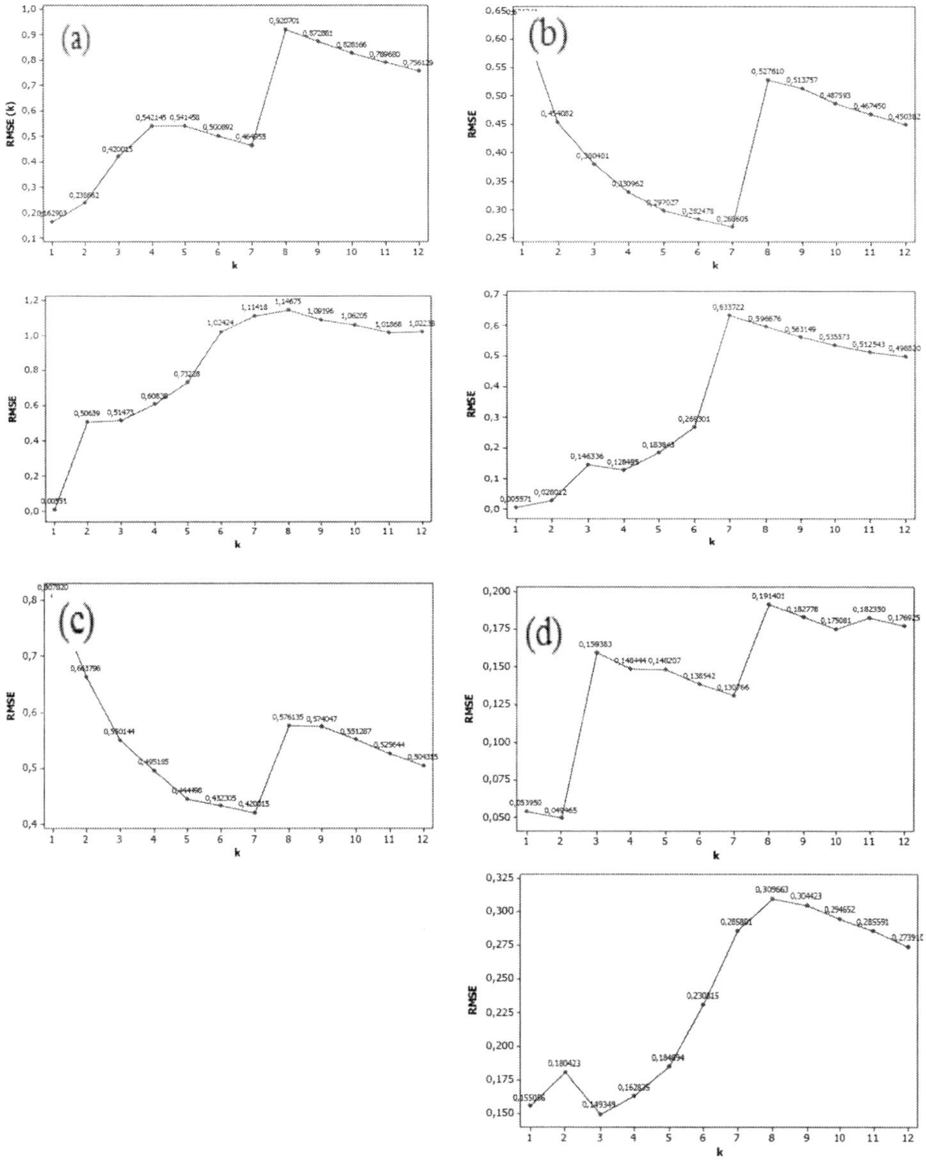

Figure 9. Adaptive RMSE Inflow and Outflow of Money Currency (a) RO of BI Surabaya (b) RO of BI Malang (c) RO of BI Kediri (d) RO of BI Jember

Based on figure 9 it can be seen that out-sample RMSE value increases when forecasted 6 or 7 periods. The best model in the case inflow of money currency uses to forecast 7 periods. While in the case outflow of money currency, the best model uses to forecast 6 periods.

5. Conclusion

Based on the analysis and results, it can be concluded that hybrid model can provide better forecasting results than time series regression model on simulation study and applied study. Hybrid model for inflow of money currency in RO of BI Surabaya, Malang, Jember with 3 variables while hybrid model for outflow of money currency in RO of BI Surabaya and Kediri with 2 variables.

In general, a hybrid model for out-sample data produce forecast not optimal because out-sample data consists of surge inflow and outflow of money currency which higher than previous years. To other research, it is suggested by using non-linear model on the second level. So, it can capture the effect of calendar variation better

References

[1] Bank Indonesia 2015a *Peraturan Bank Indonesia Nomor 14/7/PBI/2012 Tentang Pengelolaan Uang Rupiah* Diakses pada 20 September 2015 (Jakarta: Bank Indonesia)

[2] Bank Indonesia 2015b. *Peran Bank Indonesia*. Diakses pada 15 September 2015 dari http://www.bi.go.id/statistik/sistem-pembayaran/di-indonesia/peranbi

[3] Bank Indonesia 2015c. *Indikator Pengedaran Uang*. Diakses pada tanggal 15 September 2015 yang bersumber dari http://www.bi.go.id/id/publikasi/sistempembayaran/indikatorpengedaranuang/Contents/Default.aspx

[4] Zhang G P 2003 Time Series Forecasting Using a Hybrid ARIMA and Neural Network Model *International Journal of Forecasting, Neorocomputing* 153-175

[5] Karomah A 2014 Peramalan Netflow Uang Kartal dengan Model Variasi Kalender dan Model Autoregressive Distributed Lag (ARDL) (Surabaya: Tugas Akhir S1 Statistika Institut Teknologi Sepuluh Nopember)

[6] Wulansari R E 2014 Peramalan Netflow Uang Kartal dengan Metode ARIMAX dan Radial Basis Function Network (Surabaya: Tugas Akhir S1 Statistika Institut Teknologi Sepuluh Nopember)

[7] Hanim Y M 2015 Penerapan Regresi Time Series dan ARIMAX untuk Peramalan Inflow dan Outflow Uang Kartal di Jawa Timur,DKI Jakarta dan Nasional (Surabaya: Tugas Akhir S1 Statistika Institut Teknologi Sepuluh Nopember)

[8] Masun N H 2015 Peramalan Inflow dan Outflow Bulanan Uang Kartal di Kantor Perwakilan Bank Indonesia Wilayah IV Menggunakan Regresi Time Series dan ARIMAX (Surabaya: Tugas Akhir S1 Statistika Institut Teknologi Sepuluh Nopember)

[9] Suhartono, Lee M H and Prastyo D D 2015 Two Levels ARIMAX and Regression Models for Forecasting Time Series Data with Calendar Variation Effect *Procedings of the 2nd Innovation and Analytics Conference & Exhibition*

[10] Koehler A B, Synder R D, Ord J K and Beaumont A 2012 A Study of Outlier in the Exponential Smoothing Approach to Forecasting *International Journal of Forecasting* **28** 477-484

[11] Nurhariyadi 2015 Exponential Smoothing dengan Pendekatan State Space untuk Peramalan Data Inflasi (Surabaya: Tesis S2 Statistika Institut Teknologi Sepuluh Nopember)

[12] Lee M H, Suhartono and Hamzah N A 2010 Calendar Variation Model Based on Time Series Regression for Sales Forecasts: The Ramadhan Effect *Procedings of the Regional Conference on Statistical Sciences* 30-41

[13] Hyndman R J, Koehler A B, Ord J K and Snyder R D 2008 *Forecasting with Exponential Smoothing: The State Space Approach* (Berlin Heidelberg: Springer-Verlag)

[14] De Gooijer J G and Hynmand R J 2006 25 Years of Time Series Forecasting *International Journal of Forecasting* **22**(3) 443-473

Modified Exponential Weighted Moving Average (EWMA) Control Chart on Autocorrelation Data

Erna Tri Herdiani, Geysa Fandrilla, Nurtiti Sunusi

Study Program Statistics, Faculty Mathematics and Natural Sciences, Hasanuddin University

E-mail: herdiani.erna@gmail.com

Abstract. In general, observations of the statistical process control are assumed to be mutually independence. However, this assumption is often violated in practice. Consequently, statistical process controls were developed for interrelated processes, including Shewhart, Cumulative Sum (CUSUM), and exponentially weighted moving average (EWMA) control charts in the data that were autocorrelation. One researcher stated that this chart is not suitable if the same control limits are used in the case of independent variables. For this reason, it is necessary to apply the time series model in building the control chart. A classical control chart for independent variables is usually applied to residual processes. This procedure is permitted provided that residuals are independent. In 1978, Shewhart modification for the autoregressive process was introduced by using the distance between the sample mean and the target value compared to the standard deviation of the autocorrelation process. In this paper we will examine the mean of EWMA for autocorrelation process derived from Montgomery and Patel. Performance to be investigated was investigated by examining Average Run Length (ARL) based on the Markov Chain Method.

1. Introduction

In general, observations of the statistical process control are assumed to be mutually independence. However, this assumption is often violated in practically. Consequently, statistical process control observations developed for autocorrelation processes include Shewhart control charts, Cumulative Sum (CUSUM), and exponentially weighted moving average (EWMA)[1, 2]. State that this chart is not suitable if the same control limits are used in the case of independent variables [3]. For this reason, it is necessary to apply the time series model in building the control chart.

A classical control chart for independent variables is usually applied to residual processes. This procedure is permitted provided that residuals are independent. In 1978, [4] and Stamboulis introduce shewhart modifications to the autoregressive process. The distance between the sample mean and the target value is compared to the standard deviation of the autocorrelation process.

In the year of 1991, [5] has discussed the autocorrelation process for the Cumulative Sum (CUSUM) control chart, [6]) has examined the mean of EWMA for autocorrelation process,and Patel (2011) perfectly by searching for variance, control chart and ARL from EWMA with autocorrelation process. The patel formulation hereinafter referred to as patel modification, whose contents combine Shewart and EWMA control charts that are useful for detecting small variables of industrial processes that have data with a single-order autocorrelation process. Statistics proposed by adding values in EWMA statistics. The result shows that Patel modification is more sensitive than EWMA in autocorrelation process, where the formulation of EWMA used is in [2]. In this paper, we will propose

IOP Conf. Series: Journal of Physics: Conf. Series **979** (2018) 012097 doi:10.1088/1742-6596/979/1/012097

another modification of EWMA from Patel modification, which is then referred to as the Proposed modification. The result will be compared to modification of patel and Montgomery.

2. EWMA Control Chart

EWMA control chart was first introduced by Roberts in 1959. EWMA control chart is optimal for the process with the mean in the period t associated with the mean in the period (t–1) [2]. EWMA is also one of the Shewart control charts that are widely applied especially in case of time series [2]. EWMA control chart is a graphic alternative to the Shewart controller in detecting small shifts in the mean process. Specifically EWMA is used on individual observations. It is assumed observation of the process on the variable X distributed normal with mean μ and variance σ^2. Defined EWMA controls chart accordingly [2] to be:

$$Z_t = \lambda X_t + (1-\lambda)Z_{t-1} \quad (1)$$

With Z_t = average from historical data, λ = weighted parameters that are worth between 0 and 1, X_t = value of observation to- t, t = 1, 2, . ., Z_{t-1} = target. Upper control limit (UCL) and lower control limit (LCL) EWMA control chart as follows.

$$UCL = \mu_0 + L\sigma\sqrt{\left(\frac{\lambda}{2-\lambda}\right)(1-(1-\lambda)^{2t})}$$

$$CL = \mu_0$$

$$LCL = \mu_0 - L\sigma\sqrt{\left(\frac{\lambda}{2-\lambda}\right)(1-(1-\lambda)^{2t})}$$

With: μ_0 = average, σ = standard deviation, L = width of the control limit, λ = weighting parameters, t = 1, 2, . .

3. Modification EWMA Patel Control Chart

Modification EWMA control chart developed by Patel in the year 2011. The properties of the control chart are easy to apply and effective for detecting shifts of all sizes as per specifications. Patel combines Shewart and EWMA control charts that are useful for detecting small changes from industrial processes that have data with a single-order autocorrelation process. Statistics proposed by adding values in EWMA statistics. EWMA modification control chart according to Patel (2011) is:

$$Z_t = \lambda X_t + (1-\lambda)Z_{t-1} + (X_t - X_{t-1}) \quad (2)$$

With Z_t = average from historical data, λ = weighted parameters that are worth between 0 and 1, X_t = value of observation to-t, t = 1, 2, . ., X_{t-1} = value of previous observations, Z_{t-1} = target. Upper control limit (UCL) and lower control limit (LCL) of modification EWMA Patel is:

$$UCL = \mu_0 + L\sigma\sqrt{\frac{\lambda}{2-\lambda} + \frac{2\lambda(1-\lambda)}{2-\lambda}}$$

$$CL = \mu_0$$

$$UCL = \mu_0 - L\sigma\sqrt{\frac{\lambda}{2-\lambda} + \frac{2\lambda(1-\lambda)}{2-\lambda}}$$

With μ_0 = average, σ = standard deviation, L = width of control limit, λ = weighted parameter.

4. Modification EWMA Proposed Control Chart

Its Known autocorrelation process order 1 or AR (1) is as follows

$$\bar{x}_t = \phi x_{t-1} + \varepsilon_t \tag{3}$$

Modified EWMA patel is an ewma control chart with error following AR Process (1) which has value $\phi = 1$, the result can be seen in equation (2). Meanwhile, the proposed ewma modification is to add the process AR (1) to the data in the EWMA process. The advantages of EWMA modification proposed, the value of ϕ involved wider is $-1 \le \phi \le 1$, while modification EWMA patel only in the state $\phi = 1$. The results of the proposed EWMA modification are as follows:

$$Z_t = \lambda \bar{x}_t + (1-\lambda)Z_{t-1}$$

$$= \lambda(\phi x_{t-1} + \varepsilon_t) + (1-\lambda)Z_{t-1}$$

$$= \lambda\phi x_{t-1} + \lambda\varepsilon_t + \lambda(1-\lambda)\phi x_{t-1} + \lambda(1-\lambda)\varepsilon_{t-2}$$

$$+\lambda(1-\lambda)^2\phi x_{t-2} + \lambda(1-\lambda)^2\varepsilon_{t-2} + (1-\lambda)^3 Z_{t-3}$$

If set again for Z_{t-k}, $k = 4,5,6, \dots t$, so that Z_t can be expressed by

$$Z_t = \lambda\phi x_t + \lambda\phi(1-\lambda)x_{t-1} + \lambda\phi(1-\lambda)^2 x_{t-2} + \cdots$$

$$+\lambda\phi(1-\lambda)^{t-2}x_{t(t-2)} + \lambda\phi(1-\lambda)^{t-1}x_{t(t-1)} + (1-\lambda)^t Z_{t-t}$$

$$= \lambda\phi x_t + \lambda\phi(1-\lambda)x_{t-1} + \lambda\phi(1-\lambda)^2 x_{t-2} + \cdots$$

$$+\lambda\phi(1-\lambda)^{t-2}x_2 + (1-\lambda)^{t-1}x_1 + (1-\lambda)^t Z_0$$

Simply to $t = 1,2,3, \dots Z_t$ can be expressed by

$$Z_t = \lambda\phi \textstyle\sum_{k=0}^{t-1}(1-\lambda)^k\, x_{t-k} + \lambda(1-\lambda)^k\varepsilon_{t-k} + (1-\lambda)^t Z_0$$

In this chapter, we will determine the mean and variance of Z_t. Each subgroup, the average subgroup value is actually expected to be a value Z_0 or mathematically means for each subgroup $i, i = 1,2, \dots, (t-2), (t-1), t$ for $t = 1,2,3, \dots$ then $E[x_i] = \mu_0$. Because, $E[x_1] = E[x_2] = \cdots = E[x_{t-1}] = E[x_t] = \mu_0$ and Z_0 is value of expectation. So, $E[Z_0] = \mu_0$ part $\sum_{k=0}^{t-1}(1-\lambda)^k$ to be geometric series with initial value $a = 1$, comparison of the second value with initial value to be $r = (1-\lambda)$, and number of values $n = t$. So obtained,

$$\textstyle\sum_{k=0}^{t-1}(1-\lambda)^k = \frac{a(1-r^n)}{1-r} = \frac{1[1-(1-\lambda)^t]}{[1-(1-\lambda)]} = \frac{[1-(1-\lambda)^t]}{\lambda}$$

Therefore,

$$E(Z_t) = E\left(\lambda\sum_{k=0}^{t-1}(1-\lambda)^k\, x_{t-k} + (1-\lambda)^t Z_0\right) = \mu_0$$

So the mean value of EWMA to autocorrelation process is μ_0. After getting the centralization measureZ_t, next will be searched for variability value Z_t, namely variance. If for each x_t, $t = 1,2,3, \ldots$, independence and has a standard deviation $\sigma_{\bar{x}}$. Then variance of Z_t is

$$\sigma_{Z_t}^2 = Var(Z_t)$$

$$= Var\left(\lambda\phi \sum_{k=0}^{t-1} (1-\lambda)^k x_{t-k}\right) + Var((1-\lambda)^t Z_0)$$

$$+2cov\left(\lambda\phi \textstyle\sum_{k=0}^{t-1}(1-\lambda)^k x_{t-k}, (1-\lambda)^t Z_0\right)$$

Because, $Var(c) = c$ for c is constant then $Var((1-\lambda)^t Z_0) = 0$ and for each $\bar{x}_t$ independence, then

$$cov\left(\lambda\phi \sum_{k=0}^{t-1}(1-\lambda)^k x_{t-k}, (1-\lambda)^t Z_0\right) = 0$$

Furthermore,

$$\sigma_{Z_t}^2 = Var\left(\lambda\phi \sum_{k=0}^{t-1}(1-\lambda)^k x_{t-k}\right)$$

$$= \lambda^2\phi \sum_{k=0}^{t-1}(1-\lambda)^{2k} \sigma_{\bar{x}_t}^2$$

Part $\sum_{k=0}^{t-1}(1-\lambda)^{2k}$ is a geometric series with initial value $a = 1$, comparison of second value with initial value is $r = (1-\lambda)^2$, and number of value is $n = t$. So obtained

$$\textstyle\sum_{k=0}^{t-1}(1-\lambda)^{2k} = \frac{a(1-r^n)}{1-r} = \frac{1-(1-\lambda)^{2t}}{2\lambda-\lambda^2}$$

So that,

$$\sigma_{Z_t}^2 = \lambda^2\phi \textstyle\sum_{k=0}^{t-1}(1-\lambda)^{2k} \sigma_{\bar{x}_t}^2 = \lambda^2\phi\sigma_{\bar{x}_t}^2 \frac{1-(1-\lambda)^{2t}}{2\lambda-\lambda^2}$$

Let $\sigma_{\bar{x}_t}^2 = \frac{\sigma^2}{n}$ then $\sigma_{Z_t}^2 = \frac{\lambda\phi\sigma^2}{n}\left(\frac{1-(1-\lambda)^{2t}}{2-\lambda}\right)$.

So the value of the variance of EWMA is as follows:

$$\sigma_{Z_t}^2 = \frac{\lambda\phi\sigma^2}{n}\left(\frac{1-(1-\lambda)^{2t}}{2-\lambda}\right)$$

Lower Control Limit (LCL) and *Upper Control Limit* (UCL) to Modification EWMA control chart with autocorrelation data is

$$UCL = \mu_0 - 3\sigma \sqrt{\frac{\lambda\phi}{n}\left(\frac{1-(1-\lambda)^{2t}}{2-\lambda}\right)}$$

$$CL = \mu_0$$

$$LCL = \mu_0 + 3\sigma \sqrt{\frac{\lambda\phi}{n}\left(\frac{1-(1-\lambda)^{2t}}{2-\lambda}\right)}$$

Modification EWMA control chart on autocorrelation data, hereinafter referred to as Modification EWMA Proposed Control Chart. In the next section, we will determine the value of the average run length (ARL) to see which control charts are more sensitive in detecting the mean changes in the data.

5. Calculation of ARL Value

Calculation of ARL value will use the markov chain method as contained in the paper that has been written by [7] and [8]. The sequence is as follows:

1. It is known that the mean value of the data is value a = 0, the standard deviation value of the data is b = 1, n denotes the number of observations with n = 1, p addresses the selected sub interval of 5, and λ represents the weight by taking λ = 0.02. Based on the information obtained δ value as follows:

$$\delta = \frac{UCL - LCL}{2p} = \frac{5.4414 - 4.9266}{2(5)} = \frac{0.5148}{10} = 0.0515$$

2. The matrix transition will be determined by determining the matrix Q. The matrix elements Q for rows and columns are marked according to the five sub-intervals selected respectively -2, -1, 0, 1 and 2. Thus the matrix Q will contain elements as in below:

$$Q = \begin{pmatrix} Q_{-2,-2} & Q_{-2,-1} & Q_{-2,0} & 0 & 0 \\ 0 & Q_{-1-1} & Q_{-1,0} & Q_{-1,1} & 0 \\ 0 & 0 & Q_{0,0} & Q_{0,1} & Q_{0,2} \\ 0 & 0 & 0 & Q_{1,1} & Q_{1,2} \\ 0 & 0 & 0 & 0 & Q_{2,2} \end{pmatrix}$$

$$Q = \begin{pmatrix} 0.008 & 0.989 & 0.005 & 0 & 0 \\ 0 & 0.006 & 0.990 & 0.008 & 0 \\ 0 & 0 & 0.005 & 0.989 & 0.010 \\ 0 & 0 & 0 & 0.003 & 0.989 \\ 0 & 0 & 0 & 0 & 0.002 \end{pmatrix}$$

While the elements of $Q_{-2,-2}$ obtained by means

$$Q_{-2,-2} = F_N\left(\left(\frac{H_{-2} + \delta - (1-\lambda)H_{-2}}{\lambda} - a\right)\frac{\sqrt{n}}{b}\right) -$$

$$F_N\left(\left(\frac{H_{-2}-\delta-(1-\lambda)H_{-2}}{\lambda}-a\right)\frac{\sqrt{n}}{b}\right)$$

$$Q_{-2,-2}=F_N\left(\left(\frac{4.9781+0.0515-(1-0.02)4.9781}{0.02}-0\right)\frac{\sqrt{1}}{1}\right)-$$
$$F_N\left(\left(\frac{4.9781-0.0515-(1-0.02)4.9781}{0.02}-0\right)\frac{\sqrt{1}}{1}\right)$$

$$Q_{-2,-2}=F_N(7.5531)-F_N(2.4031)$$

$$Q_{-2,-2}=1-0.992=0.008$$

3. From the matrix results Q, ARL values can be obtained as follows

$$ARL=\ (Q^t*\mathbf{1})^t*((I-Q)^{-1}*\mathbf{1}\,)$$

Next ARL from each statistic can be seen in Table 4.

Table 1. ARL Value of EWMA Control Chart

λ value	Montgomery	Modification Patel	Modification Proposed
0.05	6.4268	2.8410e+03	1.4519
0.10	3.8301	59.7436	1.3433
0.20	2.1333	46.7014	1.2129
0.50	0.2981	7.2852	0.7586
0.70	0.0781	3.4342	0.4505

Table 1 describe ARL values obtained for modification EWMA proposed control charts is the smallest of the ARL values obtained for the EWMA Montgomery control chart and Modification EWMA Patel control chart, when the λ value is less than 0.5. While for λ is greater or equal with 0,5 then EWMA Montgomery is better used, because its ARL value is the smallest of modification EWMA Patel and modification EWMA proposed.

6. Conclusion

Modification EWMA Proposed Control Chart formed by considering the mean and standard deviation values on autocorrelation and standardized normal data. Mean data is μ_0 and deviation standard is $\sqrt{\frac{\lambda\phi}{n}\left(\frac{1-(1-\lambda)^{2t}}{2-\lambda}\right)}$. Based on this, we obtained a limit of control for the autocorrelation data as follows.

$$UCL=\mu_0+3\sigma\sqrt{\frac{\lambda\phi}{n}\left(\frac{1-(1-\lambda)^{2t}}{2-\lambda}\right)}$$

$$CL=\mu_0$$

$$LCL = \mu_0 - 3\sigma \sqrt{\frac{\lambda\phi}{n}\left(\frac{1-(1-\lambda)^{2t}}{2-\lambda}\right)}$$

ARL value comparison for$\lambda = 0.05$, $\lambda = 0.1$ and $\lambda = 0.2$ then the proposed Modification EWMA is more sensitive than Montgomery or Patel method. While $\lambda = 0.5$ and $\lambda = 0.7$ then Montgomery method is more sensitive than Patel and the proposed.

References

[1] Box G E P, Jenkins G M and MacGregor J F 1974 Some recent advances in forecasting and control *II J Roy Statist Soc Ser* C **23** 158-179
[2] Montgomery D C 2009 *Introduction to Statistical Quality Control Sixth edition* (New York: Wiley)
[3] Alwan L C and Roberts H V 1988 Time series modeling for statistical process control *J. Business and Economic Statist* **6** 87-95
[4] Vasilopoulos A V and Stamboulis A P 1978 Modifcation of control chart limits in the presence of data correlation *J. Quality Technology* **10** 20-30
[5] Harris T J and Ross W H 1991 Statistical process control procedures for correlated observations *Canadian J. Chemical Engineering* **69** 48-57
[6] Lu C W 1999 Control Chart for Monitoring the Mean and Variance of Autocorrelated Processes *Journal of Quality Technology* 259-274
[7] Yang S F, Tsai W C, Huang T M, Yang C C and Cheng S 2011 Monitoring process mean with a new EWMA control chart *Produção* vol 21 no 2 217-222
[8] Chananet C, Sukparungsee S, and Areepong Y 2014 The ARL of EWMA Chart for Monitoring ZINB Model Using Markov Chain Approach *International Journal of Applied Physics and Mathematic* **4**

Institute of Physics
Dirac House, Temple Back
Bristol BS1 6BE UK

ISSN: 1742-6588
ISBN 978-1-5108-5989-0
9 781510 859890